Taken from
Grant's ATLAS OF ANATOMY

GRANT'S Method of Anatomy

GRANT'S METHOD

by regions
descriptive
and deductive

OF ANATOMY

NINTH EDITION

JOHN V. BASMAJIAN, M.D.

Professor of Anatomy, Physical Medicine, and Psychiatry; Director, Regional Rehabilitation Research & Training Center; Emory University, Atlanta, Georgia

The Williams & Wilkins Company • Baltimore 1975

The Williams & Wilkins Company
428 E. Preston Street
Baltimore, Md. 21202 U.S.A.

Made in the United States of America

First Edition, October, 1937
Reprinted August, 1938
Second Edition, March 1940
Reprinted February, 1941
Reprinted November, 1943
Reprinted February, 1944
Third Edition October, 1944
Reprinted September, 1945
Reprinted April, 1946
Reprinted November, 1946
Reprinted November, 1947
Fourth Edition, August, 1948
Reprinted September, 1949
Reprinted February, 1951
Fifth Edition, July, 1952
Reprinted November, 1953
Sixth Edition, December, 1958
Reprinted April, 1962
Reprinted November, 1963
Reprinted August, 1964
Seventh Edition, May, 1965
Reprinted June, 1966
Eighth Edition, 1971
Reprinted December, 1972
Reprinted June, 1974
Ninth Edition, June, 1975

Library of Congress Cataloging in Publication Data

Grant, John Charles Boileau, 1886–1973.
Grant's Method of anatomy.

Bibliography: p. 622
Includes index.
1. Anatomy, Human. 2. Anatomy, Surgical and topographical. I. Basmajian, John V., 1921–
II. Title. III. Title: Method of anatomy. [DNLM: 1. Anatomy. QS4 G762m]
QM23.2.G7 1975 611 74-28409
ISBN 0-683-00372-0

Cover Illustration: Artist's rendering of "The Thinker" by Auguste Rodin. Drawn from the statue at the Baltimore Museum of Art.

Composed and printed at the
Waverly Press, Inc.
Mt. Royal & Guilford Aves.
Baltimore, Md. 21202 U.S.A

preface to the ninth edition

Any successful method devised to aid learning and remembering is like an oasis in a desert for students. This is especially true in this era of self-education in which laboratory exercises, lectures, and formal teaching are all being denounced. Not surprisingly, after 38 years this book continues to grow steadily in use among students of human anatomy. Its reliance on logic, analysis, and deduction—as opposed to dry memory-work—makes it unique among gross anatomy textbooks. The principles first established by Professor Grant have continued to guide me in shaping this edition to the needs of the late 1970's. As the curricula of medical and dental schools continue to shift, no text-book can stand still. Woe to any author who will not accommodate to the needs of the students—needs that are dictated by omnipotent Curriculum Committees.

The size of the eighth edition was greatly reduced from previous editions. In this edition further pruning has proved possible and profitable. Once more the test for emphasizing, downgrading, or eliminating every item was its significance. I have renewed my efforts to make the relative levels of importance of the material apparent by a number of devices, both typographic and literary. I feel that too many textbooks exhibit a cowardly fear of revealing what is important and what is not in their even-handed, bland treatment; this book is free of that taint at least. As in the earlier edition, small-type sections may be read once and skipped over by students thereafter. But a student would be foolish to gloss over sections that cover a topic thoroughly, for these are the parts that are important. Perhaps needlessly, I have often stated that a piece of information is either important or unimportant—this, at the risk of offending some professors!

As before, both the order of presentation of Sections and whether a student dissects a cadaver personally or not are immaterial in the use of this book. Any class should be able to start with any section of laboratory work or laboratory manual and encounter no difficulty resulting from the sequence of Sections here. An old complaint that some structures were previously described in a disjointed fashion because of the regional approach has been largely corrected. Nevertheless, the regional approach is the only one of real use to clinicians, and so it is emphasized. Students are well advised to read or skim over the Section of "General Considerations" from time to time; a clear knowledge of systematic anatomy will make study of regional anatomy easier.

Another area in which this book has been unique is its large numbers of simple drawings. In this edition a number of new ones have been added and a large number have been revised, corrected, or had colors added. I am grateful to Mr. G. B. Hogan for his careful artwork. The figures now exceed a thousand in number, a fact that is obscured by the decimal-numbering we have used to introduce new figures without requiring expensive re-numbering of all—expensive, that is, to the student.

By its nature, this book is student-oriented rather than being encyclopedic. Reference works also have their role in education, as do atlases, dissecting manuals, and handbooks; the eager student is wise to invest in those books to supplement his resources for learning. Yet it is his textbook and his teachers that will ease his burden by clearly emphasizing the relevant ideas and facts, by illuminating their interrelationships, and by offering logical and deductive methods for understanding and memorizing a great deal of information. The foundations laid by Professor Grant in earlier editions are so true and solid that the changing times and fashions have not changed what is fundamental here.

Finally, it is my sad duty to record that Professor Grant, the creator of this *Method*, died in 1973 at the age of 87. In the closing years of his fruitful life he concentrated his attention on his masterful "Atlas." Meanwhile he entrusted the last several editions of this textbook to me. As before, my work has been a labor of love, shared, I know by the publishers, represented by my ever-helpful friend and editor, Sara Finnegan. To the memory of Professor J. C. B. Grant this book is proudly dedicated.

J. V. Basmajian

Emory University, Atlanta
Spring, 1975

preface to the first edition

The study of human anatomy may be attempted in either of two ways. One consists in collecting facts and memorizing them. This demands a memory which is wax to receive impressions and marble to retain them. Even so endowed a student will not master the infinite complexities of the subject. The other way consists in correlating facts, that is studying them in their mutual relationships. This leads inevitably to the apprehending to the underlying principles involved, and the *raison d'être* of such relationships. The student will thus learn to reason anatomically and will find the acquisition of new and related facts an easier task. It is the purpose of this book to lead the student to approach the subject from this viewpoint, and it involves certain departures from tradition.

The human body is here considered by regions. In most regions some features predominates. It may be a muscle, a vessel, a nerve, a bony landmark, or other palpable structure, or it may be a viscus. The regions are for the most part built up around the dominant or central feature.

The markings, lines and ridges, depressions and excrescences on a bone tell a story as do the scars and irregularities of the earth's surface. Because they are in the main to be interpreted by reference to the soft parts that surround and find attachment to them, the bones are not described together under the heading "osteology" as though they were things apart. The shafts of the bones are considered with the surrounding soft parts; the ends with the joints into which they enter. The bones of the foot are primarily considered as a single mechanism—so are those of the hand and of the skull. The correct orientation of certain bones is given in cases where without this information the actions of certain muscles (e.g., Gluteus Medius, Teres Major) could not be understood.

It is not the mere presence of a ligament or its name that is of interest, but the functions it serves. These depend commonly on the direction of the fibers of the ligament; occasionally on their precise attachments. Many fibrous bands bearing individual names are really members of a community. A challenge thrown at one must be taken up by all. They act in unison, and therefore they are considered together as a unit.

In the consideration of viscera the subject is elucidated by reference to comparative anatomy and to embryology. These are cognate sciences which throw light about the existing structure of man. The positions of the viscera are referred to selected vertebral levels, the vertebral volume being an ever present and ever ready measuring rod.

Illustrations to be of value must be simple, accurate, and convey a definite idea. It is for these reasons that they consist entirely of line drawings. Their simplicity encourages the student to reproduce them; and though diagrammatic in nature they are based on measurements and observations of a great deal of carefully dissected material. Their accuracy, therefore, in those details they are intended to illuminate, has been the object of very considerable work.

The book is meant to be a working instrument designed to make Anatomy rational, interesting, and of direct application to the problems of medicine and surgery. The bare, dry and unrelated facts of Anatomy tend rapidly to disappear into forgetfulness. That is largely because its guiding principles are not grasped so as to capture the imagination. Once they are grasped it will be found that details and relationships will remain within certain and easy recall.

J. C. Boileau Grant

University of Toronto
September, 1937

contents

Preface to the Ninth Edition v
Preface to the First Edition vii
Introduction and Descriptive Terms xi

SECTION ONE

General Considerations

1 Locomotor Systems 3
2 Cardiovascular and Nervous Systems 30
3 Digestive and Respiratory Systems 52
4 Urogenital System and the Skin 60

SECTION TWO

Upper Limb

5 Pectoral Region and Axilla 75
6 Scapular and Deltoid Regions 89
7 Cutaneous Nerves and Superficial Veins 103
8 Arm 109
9 Flexor Region of Forearm 116
10 Hand 125
11 Extensor Regions of Forearm and Hand 142
12 Joints of Upper Limb 152

SECTION THREE

Abdomen

13 Anterior Abdominal Wall and Scrotum 175
14 Abdominopelvic Cavity 190
15 Stomach, Liver, and Related Structures 212
16 Mesenteric Vessels, Duodenum, and Pancreas 223
17 Three Paired Glands 231
18 Posterior Abdominal Structures 238

SECTION FOUR

Perineum and Pelvis

19 Perineum 257

20 Male Pelvis 268
21 Female Pelvis 290
22 Pelvic Autonomic Nerves and Lymphatics 302

SECTION FIVE

Lower Limb

23 Femur and Front of Thigh 309
24 Hip Bone and Gluteal Region 320
25 Back and Medial Region of Thigh 327
26 Leg and Dorsum of Foot 336
27 Bones and Sole of Foot 350
28 Joints of Lower Limb 364

SECTION SIX

Thorax

29 Walls of Thorax 391
30 Pleurae 400
31 Lungs 411
32 Heart and Pericardium 420
33 Superior and Posterior Mediastina 437

SECTION SEVEN

Head and Neck

34 Front of Skull, Face, and Scalp 451
35 Posterior Triangle of Neck 463
36 Back 469
37 Interior of Cranium 479
38 Orbital Cavity and Contents 493
39 Anterior Triangle of Neck 505
40 Root of Neck 516
41 Side of Skull, Parotid, Temporal, and Infratemporal Regions 521
42 Cervical Vertebrae, Prevertebral Region, and Exterior of Base of Skull 533
43 Great Vessels and Nerves of Neck: Review and Summary 542
44 Pharynx and Palate 552
45 Mouth, Tongue, and Teeth 566
46 Nose and Related Areas 575
47 Larynx 584
48 Ear 590
49 Lymphatics of Head and Neck 599
50 Bones of Skull: For Reference 602
References 622
Index 629

introduction and descriptive terms

There are few words with a longer history than the word *Anatomy*. If we write anatome, we use the name that Aristotle gave to the Science of Anatomy 2300 years ago. He made the first approach to accurate knowledge of the subject, although it was derived from dissections of the lower animals only. The word means cutting up—the method by which the study of the structure of living things is made possible.

The boundaries of the subject have widened. Through the use of the microscope and with the aid of stains the field of Anatomy has come to include microscopical anatomy or *histology* and the study of development before birth or *embryology*. The study of the anatomy of other animals, *comparative anatomy*, has been pursued exhaustively partly in an endeavor to explain the changes in form, *morphology*, of different animals including man. *Physical Anthropology*, or the branch of the study of mankind that deals chiefly with the external features and the measurements of different races and groups of people, and with the study of prehistoric remains commands interest of the anatomist. The hereditary, nutritional, chemical and other factors controlling and modifying the growth of the embryo, of the child, and of animals are within his legitimate field; so also is the growth of tissues in test-tubes, *tissue culture*. Feeding and other experiments on animals play leading parts in many investigations.

Individuals differ in outward form and features; for example, how varied are fingerprints and the arrangement of the veins visible through the skin; individuals differ also in their internal makeup. Textbooks, for the most part, describe average conditions where weights and measures are concerned, and the commonest conditions where arrangements and patterns are concerned. Owing to the variety of these the commonest may have less than a 50 per cent incidence; therefore, it may not be truly representative. As data on variations accumulate, the subject of *Statistical Anatomy* emerges. Some variations are so rare as to be abnormalities or *anomalies*. Among the different races of mankind there are percentage differences in the form and arrangement of structures, just as there are among the different races of the apes and other animals. But relatively little is known as yet of *Racial Anatomy*, which is a branch of physical anthropology.

The human body is generally dissected by regions, *Regional Anatomy*, and described by systems, *Systematic Anatomy*. The regions of the body comprise (1) the head and neck, (2) the trunk, and (3) the limbs. These can be divided and subdivided indefinitely. The trunk is divisible into thorax, abdomen, and pelvis. The systems of the body comprise the skeleton (the study of which is osteology), the joints (arthrology), the muscles (myology), the nervous system (neurology, which includes the brain, spinal cord, organs of special sense, the nerves, and the autonomic nervous sytem), the cardio-vascular system which includes the heart, blood vessels, and lymph vessels. The viscera of the body (exclusive of the heart and parts of the nervous system) comprise four tubular systems—the digestive, respiratory, urinary, and genital—and the ductless or endocrine glands. All these are wrapped up in the skin and subcutaneous tissue.

Anatomy considered with special reference to its medical and surgical bearing is called *Applied Anatomy*. Anatomy can be studied profitably, although to a limited extent, by means of cross-sections, *Cross-Section Anatomy*. In the living subject a great deal can be learned by inspection and palpation of surface parts. This and the relating of deeper parts to the skin surface, *Surface Anatomy*, are a necessary part of a medical education. And, *Radiographic Anatomy* relies on the X-ray to reveal much that cannot be investigated by other means.

Regarding nomenclature, it may be said briefly that over 30,000 anatomical terms were in use in the various textbooks of Anatomy and in the journals when, in the year 1895, the German Anatomical Society, meeting in Basle, approved a list of about 5000 terms known as the Basle Nomina Anatomica (B.N.A.). Six terse rules, set down by the Commission for its own guidance, are worth recording here. They are as follows: (1) Each part shall have only one name. (2) Each term shall be in Latin. (3) Each term shall be as short and simple as possible. (4) The terms shall be merely memory signs. (They need lay no claim to description or to speculative interpretation.) (5) Related terms shall, as far as possible, be similar, e.g., femoral nerve, femoral artery, and femoral vein. (6) Adjectives, in general, shall be arranged as opposites, e.g., major and minor, superior and inferior.

In the year 1933, The Anatomical Society of Great Britain and Ireland, meeting in Birmingham, adopted a revision of the B.N.A., known as the B.R.; in 1935, the German Anatomical Society, meeting in Jena, likewise adopted a revision, known as the J.N.A. or I.N.A. Despite their many excellent points, these found only local and restricted acceptance. In 1955, the Sixth International Congress of Anatomists, meeting in Paris, gave approval to a somewhat conservative revision of the B.N.A. which was submitted to it, and which contained many B.R. and I.N.A. terms. Subsequently, minor revisions and corrections were made at the Seventh and Eighth Congresses held in New York (1960), Wiesbaden, Germany (1965), and Moscow (1970). It is hoped and believed that this Nomina Anatomica will come to be used exclusively and universally and that it will become the truly international vocabulary and a boon to all concerned.

DESCRIPTIVE TERMS

In describing the relationship of one structure to another it is obviously necessary to avoid ambiguity and misunderstanding.

For descriptive purposes the human body is regarded as standing erect, the eyes looking forward to the horizon, the arms by the sides, and the palms of the hands and the toes directed forward; this is the **Anatomical Position.** The cadaver may be placed on the table lying on its back, on its side, or on its face, but for descriptive purposes it is assumed to be standing erect in the anatomical position. The palm of the hand is understood to be the anterior surface of the hand (*fig. 1*).

The body is divided into two halves, a right and a left, by the *median* or *midsagittal plane*. The anterior and posterior borders of this plane reach the skin

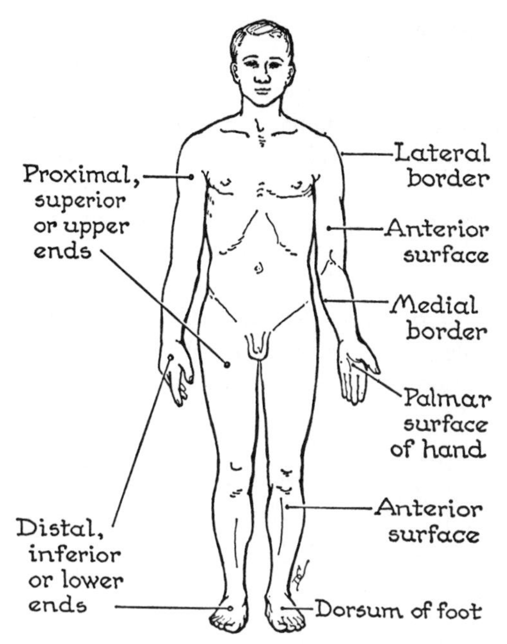

Fig. 1. The subject in the Anatomical Position—except for the right forearm, which is pronated.

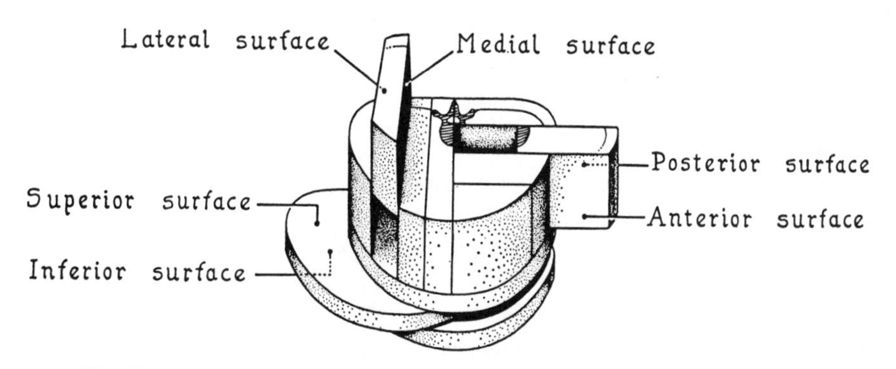

Fig. 2. Three pairs of surfaces involving six essential descriptive terms. They are related to the three fundamental planes in the body.

surface at the front and back of the body at the *median line* or *midline*.

Terms of Relationship. Three pairs of relative terms suffice to express the relationship of any given structure to another (*fig. 2*). They are:

1. Anterior or in front = nearer the front surface of the body.
   Posterior or behind = nearer the back surface of the body.
2. Superior or above = nearer the crown of the head.
   Inferior or below = nearer the soles of the feet.
3. Medial = nearer the median plane of the body.
   Lateral = farther from the median plane of the body.

The foregoing terms are applicable to all regions and all parts of the body—always provided that the body is, or is assumed to be, in the anatomical position.

Terms of Comparison. When it is desired to compare the relationship of some structure in man with the same structure in, for example, a dog, it is necessary to use a different set of terms, terms related not to space but to parts of the body, such as the head, tail, belly, and back. For example, in man standing erect the heart lies above the diaphragm; in the dog standing on all fours it lies in front of the diaphragm; but in both instances its position relative to other parts of the body is the same; so, *speaking comparatively*, one would say that both in man and in the dog the heart is on the head, cranial, or cephalic side of the diaphragm (*fig. 3*).

Hence, the terms *ventral* and *dorsal*, *cranial* and *caudal*, as well as *medial* and *lateral* are applicable to the trunk or torso (thorax, abdomen, and pelvis) irrespective of the position assumed by the body. Moreover, it is desirable to em-

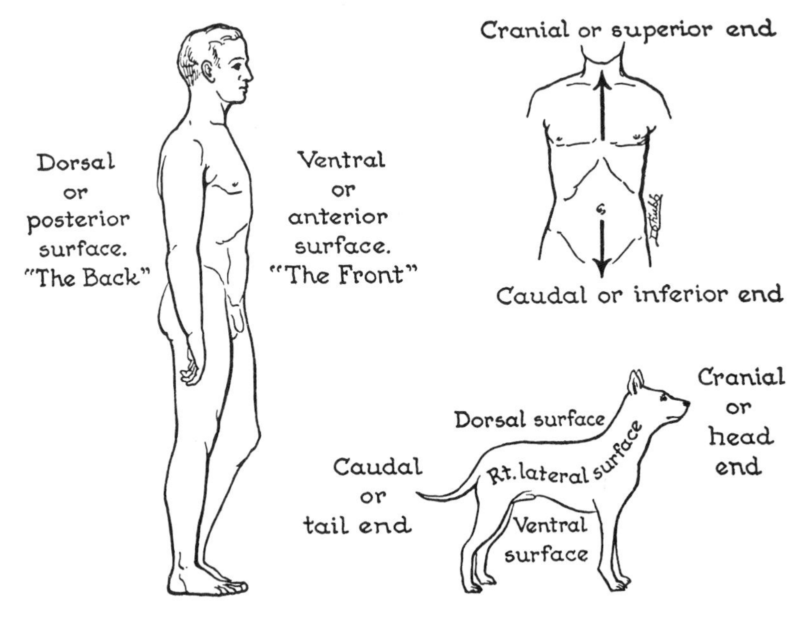

Fig. 3. Three pairs of terms necessary to comparative anatomy and of more general application than those given in *figure 2*.

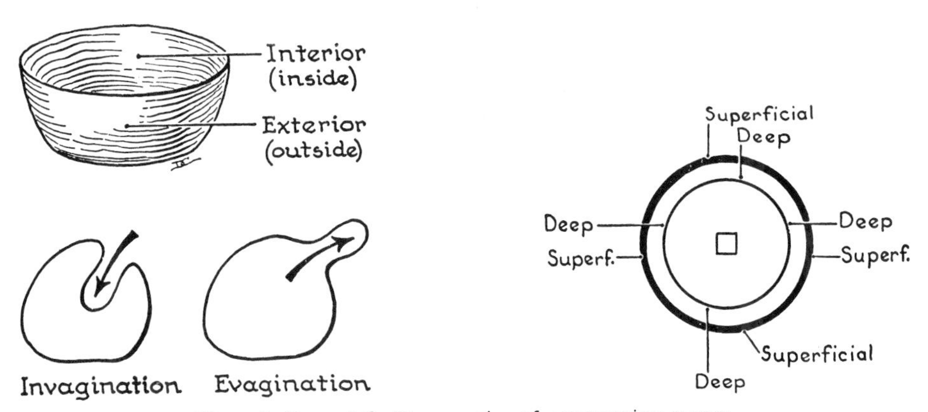

Figs. 4, 5, and 6. Three pairs of contrasting terms.

ploy these terms in embryology and comparative embryology, and it is quite correct to employ them in human anatomy—for no misunderstanding can arise from their use as synonyms for anterior, posterior, superior, and inferior.

In the limbs, terms are coupled with reference to (1) the proximity to the trunk, *proximal* = near the trunk and is synonymous with superior; *distal* = farther from the trunk and is synonymous with inferior; (2) the morphological borders, *preaxial* = the lateral or radial border (i.e., thumb side) of the upper limb and the medial or tibial border (i.e., big toe side) of the lower limb; *post-axial* = the medial or ulnar border of the upper limb and the lateral or fibular border of the lower limb; and (3) the functional surface, *flexor* and *extensor*, the flexor surface being anterior in the upper limb and posterior in the lower limb.

The anterior surface of the hand is generally called the *palmar* (or volar) surface, and the inferior surface of the foot the *plantar* surface. The opposite surfaces are called the *dorsum* of the hand and foot.

Other Terms. *Inside, interior,* or *internal* and *outside, exterior,* or *external,* are reserved (1) for bony cavities, such as the pelvic, thoracic, cranial, and orbital, and (2) for hollow organs, such as the heart, mouth, bladder, and intestine (*fig. 4*).

An *invagination* and an *evagination* (L. vagina = a sheath or scabbard) are inward and outward bulgings of the wall of a cavity (*fig. 5*).

Superficial and *deep* denote nearness to and remoteness from the skin surface irrespective of whether at the front, side, or back. These two may be applied to organs, such as the liver and lung (*fig. 6*).

On, over, and *under* are terms to beware of. They should be used in a general sense and without specific regard to the anatomical position. Carefully avoid using them loosely in place of "superior to" and "inferior to," for such misuse is the cause of much misunderstanding.

Ipsilateral refers to the same side of the body, e.g., the right arm and the right leg. *Contralateral* refers to opposite sides of the body.

Planes. (1) A *sagittal plane* is any vertical antero-posterior plane parallel to and including the median plane. (2 A *coronal* or *frontal plane* is any vertical side-to-side plane at right angles to the sagittal plane. (3) A *transverse plane* is any plane at right angles to 1 and 2, i.e., at right angles to the long axis of the body or limb. In the case of an organ or other structure a *transverse* or *cross section* is a section at right angles to the long axis of that organ or structure. (4) An *oblique plane* may lie at any other angle. (See *figs. 7–9.*)

Attachments of Muscles. Muscles are attached at both ends. The proximal attachment of a limb muscle is its *origin*; the distal end is its *insertion*. No function is implied by these terms.

Vessels. Arteries are likened to trees with *branches*; veins are likened to rivers with *tributaries*.

Movements at Joints. To *flex* is to bend or to make an angle.

To extend is to stretch out or to straighten. Movements of flexion and extension take place at the elbow joint.

To abduct is to draw away laterally from the median plane of the body.

To adduct is the opposite movement in the same plane (L. ab = from; ad = to; duco = I lead). Movements of abduction and adduction, as well as of flexion and extension, take place at the wrist joint.

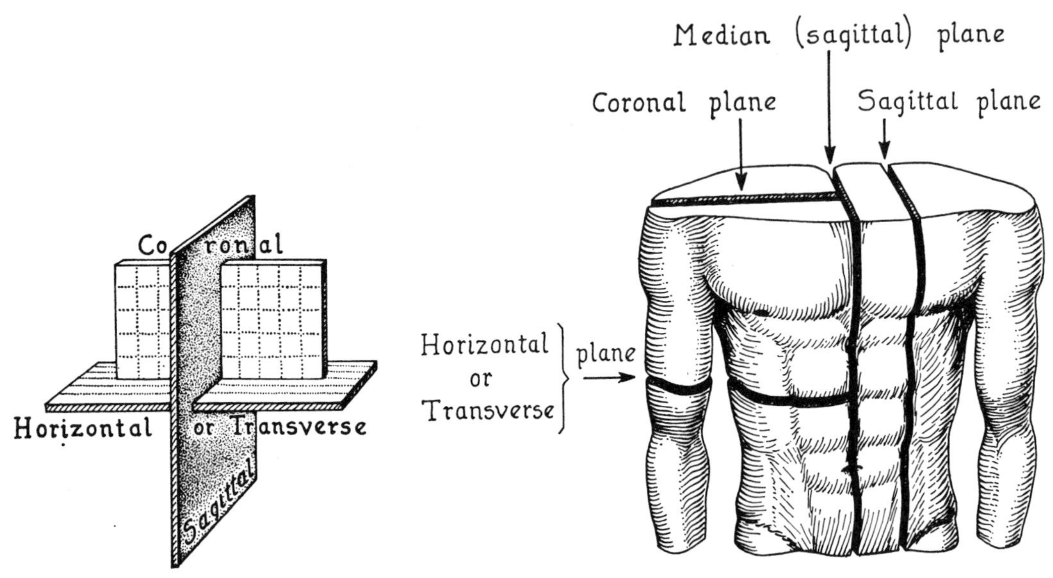

Fig. 7. The three fundamental planes.

Fig. 8. Fundamental planes in the body.

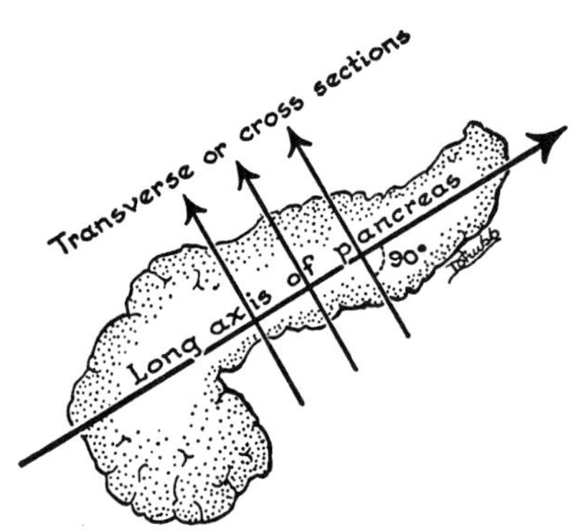

Fig. 9. A cross section of an organ or part is a section made at right angles to its long axis.

The middle finger is regarded as lying in the *axial line of the hand*; and the 2nd toe as lying in the *axial line of the foot*. Abduction and adduction of the fingers and toes are movements from and toward these axial lines, although, as discussed elsewhere, movements of the thumb are named differently.

To circumduct (L. circum = around) is to perform the movements of flexion, abduction, extension, and adduction in sequence, thereby describing a cone, as can be done at the shoulder, hip, wrist, and metacarpo-phalangeal joints.

To rotate is to turn or revolve on a long axis, as the arm at the shoulder joint, the femur at the hip joint, the radius on the ulna, and certain vertebrae on each other.

To pronate was originally to bend or flex the body forward as in obeisance in prayer, that is to face downward or prone. Applied to the forearm, *to pronate* means to turn it so that the palm of the hand faces downwards on a table, which is the equivalent of facing backward when it hangs by the side.

To supinate is to rotate the forearm laterally so that the dorsum of the hand rests on the table or faces backward when the limb hangs by the side. Supine = lying on the back.

To protract (L. pro = forward; traho = I pull) is to move forward.

To retract is to move backwards. Protraction and retraction are terms applied to the movements of the lower jaw and shoulder girdle.

SECTION ONE

GENERAL CONSIDERATIONS

1 Locomotor Systems 3

2 Cardiovascular and Nervous Systems 30

3 Digestive and Respiratory Systems 52

4 Urogenital System and the Skin 60

1 LOCOMOTOR SYSTEMS

BONE

Physical Properties; Functions; Structure; Classification.

Markings on Dried Bones: Terms; A Living Bone; The Parts of a Young Bone; Epiphyses; Nutrient Artery and Canal; Ossification; Bone Marrow; Vessels and Nerves; Historical.

CARTILAGE

Hyaline Cartilage; Fibrocartilage; Elastic Cartilage.

VERTEBRAL COLUMN

PARTS OF A TYPICAL VERTEBRA AND THEIR FUNCTIONS—*Body; Vertebral Arch; Processes; Ossification.*

ARTICULATED VERTEBRAL COLUMN—*Intervertebral Discs; Bodies of the Vertebrae; Curvatures; Varying Stature; Line of Gravity; Transverse Processes; Pedicles; Laminae; Vertebral Foramina and Vertebral Canal; Articular Processes; Spinous Process.*

ARTICULATIONS OR JOINTS

SKULL TYPE—*Suture and Synchondrosis; Synostosis.*

VERTEBRAL TYPE—*Symphysis; Symphysis between Two Vertebral Bodies; Syndesmosis; Vertebral Syndesmoses.*

LIMB TYPE—*Synovial Joint; articular capsule, joint cavity; synovial membrane and folds, synovia; fat-pads, articular discs, ligaments, articular cartilage, lubrication, labra, nerves, and vessels.*

Classification of Synovial Joints

MUSCLES

SKELETAL MUSCLES—*Fibers; Parts; Insertions; Synovial Bursa; Synovial Sheath.*

Internal Structure; Contraction; Investigation; Electromyography; Muscle Action

Blood Supply; Nerves; Nomenclature; Variations.

ACCESSORY MUSCLES.

BONE

In this chapter our main concern is with bone as a tissue and its general characteristics; but the vertebrae are described in detail because an understanding of them is fundamental to several regions. The student should not proceed with a study of regional anatomy until this chapter and the succeeding ones are read and understood.

A bone of a living man is itself a living thing. It has blood vessels, lymph vessels, and nerves. It grows. It is subject to disease. When fractured it heals itself; and if the fracture is so improperly set that the parts have lost their previous alignment, its internal structure undergoes remodeling in order that it may continue to withstand strains and stresses as it did before. Unnecessary bone is resorbed. For example, following the extraction of a tooth, the walls of the socket, thus rendered empty, disappear; also, the bones of a paralyzed limb atrophy (become thinner and weaker) from disuse. Conversely, when bones have increased weight to support, they hypertrophy (become thicker and stronger).

Bones have an *organic framework* of fibrous tissue and cells, among which *inorganic salts*—notably, phosphate of calcium—are deposited in a characteristic fashion. The fibrous tissue gives the bones resilience and toughness; the salts give them hardness and rigidity and make them opaque to X-rays. One-third is organic; two-thirds are inorganic.

Properties of Bone

Physical Properties. By submerging a bone in a mineral acid the salts are removed, but the organic material remains and still displays in detail the shape of the untreated bone. Such a specimen is flexible. For example, a decalcified fibula can be tied in a knot (*fig. 10*); when the knot is untied, the fibula springs back into shape.

The organic material of a bone, long buried near the surface of the earth, is removed by bacterial action (i.e., decomposition), and only the salts remain. The same result can be achieved more speedily by burning with fire. A bone so treated, being more brittle than porcelain, will crumble and fracture unless handled with care.

Bones that have lain buried in a limestone cave become petrified (i.e., calcium carbonate replaces the organic material); so, they endure; so do those that are mineralized through lying in soils containing, e.g., iron, lead, or zinc.

Moisture being necessary to bacterial action, bones that have remained thoroughly dry (mummified) retain their toughness.

Function of Bones. In addition to being (1) the rigid supporting framework of the body, bones serve as (2) levels for muscles; (3) they afford protection to certain viscera (e.g., brain and spinal cord, heart and lungs, liver and bladder); (4) they contain marrow, which is the factory for blood cells; and (5) they are the storehouses of calcium and phosphorus.

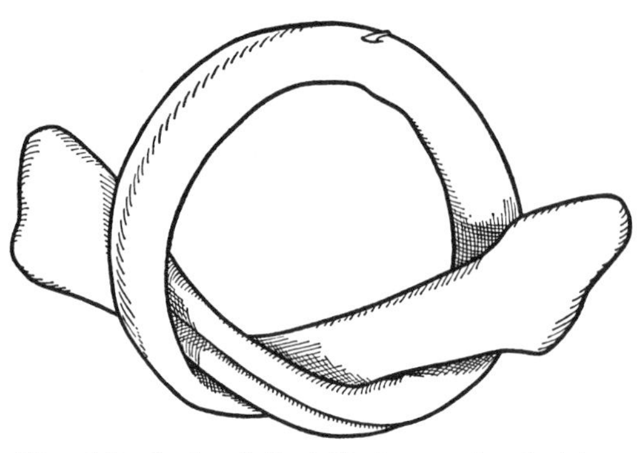

Fig. 10. A decalcified fibula can be tied in a knot.

Structure of a dried bone seen on section is shown in *figure 11*. Macroscopically, there are two forms of bony tissue (1) *spongy* or cancellous, and (2) *compact* or dense.

All bones have a complete outer casing of compact bone; the interior is filled with spongy or cancellous bone except when replaced by a medullary cavity or an air sinus (see below). In a long bone, such as the humerus, the compact bone is thickest near the middle of the shaft and it becomes progressively thinner as the bone expands toward its articular ends, these being covered with a mere shell of compact bone. Conversely, spongy bone fills the expanded ends and extends for a variable distance along the shaft but leaves a tubular space, the *medullary cavity*. The *lamellae* or plates of the spongework are arranged in lines of pressure and of tension, and in an X-ray photograph the pressure lines are seen to pass across joints from bone to bone (*fig. 496* of the hip joint).

Classification of Bones

The bones of the body may be classified variously.

Developmentally. According to whether they developed (1) in cartilage, or (2) in membrane.

Regionally.

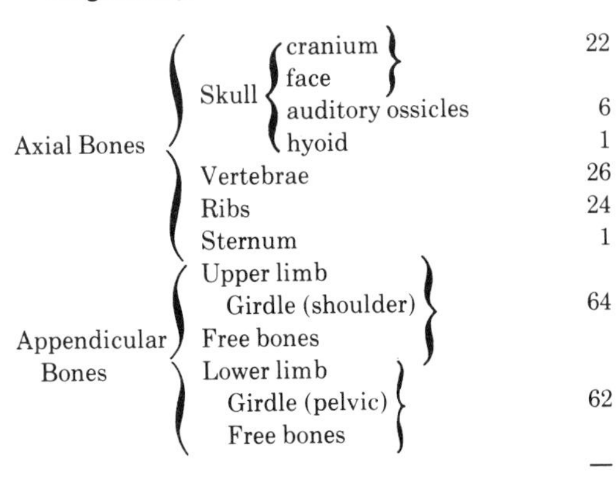

Axial Bones	Skull	cranium, face	22
		auditory ossicles	6
		hyoid	1
	Vertebrae		26
	Ribs		24
	Sternum		1
Appendicular Bones	Upper limb	Girdle (shoulder), Free bones	64
	Lower limb	Girdle (pelvic), Free bones	62
			—
			206

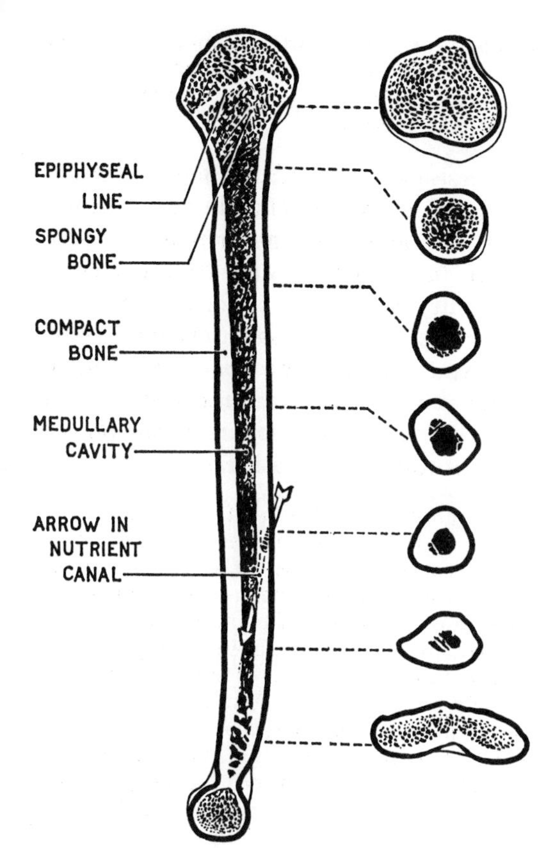

Fig. 11. The structure of a dried bone as shown by longitudinal and transverse sections of a humerus.

This number is not exact. It varies with age and with the individual, being larger in youth while the various parts of compound bones (e.g., frontal, sacrum) are still discrete and when accessory or supernumerary bones are present, and being smaller when two bones have fused (e.g., fusion of lunate and triquetrum, or of two vertebrae) and when a bone is suppressed or congenitally absent (e.g., absent phalanx or vertebra).

According to Shape.

1. Long } peculiar to the limbs.
2. Short }
3. Flat } peculiar to the axial skeleton and
4. Irregular } the girdles.
5. Pneumatic—peculiar to the skull.
6. Sesamoid—in certain tendons.

1. Long Bones are tubular. They are confined to the limbs, where they serve as levers for muscles.

A long bone has a body or shaft and two ends. The *ends*, being articular, are smooth, covered with cartilage, either convex or concave, and enlarged.

The *shaft* is hollow (medullary cavity) as a straw is hollow, thus obtaining most strength with least expenditure of material and with least weight. It, typically, has three borders which separate three surfaces, so on cross-section it is triangular rather than circular (*fig. 12*).

Long bones develop (are preformed) in cartilage. The shaft of every long bone begins to ossify (primary center) about the 2nd to 3rd month of intra-uterine life. One or both ends begin to ossify (secondary centers) soon after birth.

Exceptions. Every long bone does not conform to all the foregoing specifications. For example, the *clavicle* and the *ribs* have no medullary cavity, but they fulfil the functions of long bones. The *vertebrae* are classified as irregular bones, but their bodies possess most of the features of a long bone.

2. Short Bones are cubical or modified cubes. They are confined to the carpus and tarsus. They have six surfaces of which four (or less) are articular, leaving two (or more) free for the attachment of ligaments and for the entry of blood vessels. They develop in cartilage, and they begin to ossify soon after birth.

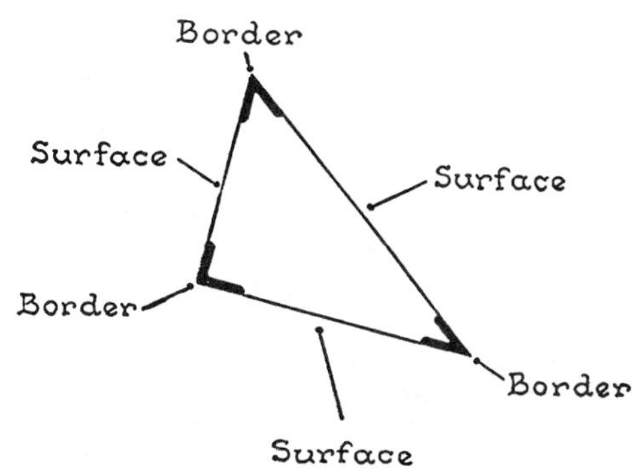

Fig. 12. The three borders are unbendable, like angle-iron.

Of the short bones, three (calcaneus, talus, and cuboid) start ossifying before birth; so do the epiphyses of three long bones (knee end of femur and of tibia and commonly the shoulder end of humerus).

3. *Flat Bones* resemble sandwiches. They consist of two layers or plates of compact bone with spongy bone and marrow spread between them. Many of the skull bones, the sternum, scapulae, and parts of other bones, are of the flat type. Most flat bones help to form the walls of rounded cavities and therefore are curved. At birth a flat bone consists of a single plate. In the flat bones of the skull the spongy bone, here called *diploe*, and its contained marrow appears some years later and splits the plate into two.

4. *Irregular Bones* have any irregular or mixed shape. All skull bones, not of the flat type, are irregular (e.g., sphenoid, maxilla); so are the vertebrae and the hip bones. They are composed of spongy bone and marrow within a compact covering.

5. *Pneumatic Bones*. Evaginations of the mucous lining of the nasal cavities and of the middle ear and mastoid antrum invade the diploe of certain flat and irregular bones of the skull thereby producing *air cells* or *air sinuses*. This pneumatic method of construction may be economical in bony material, but it invites "colds in the head" and other infections of the nose to extend to these sinuses.

6. *Sesamoid Bones* are nodules of bone that develop in certain tendons where they rub on convex bony surfaces ("Sesamoid" of Arabic origin = like a seed). The free surface of the nodule is covered with articular cartilage; the rest is buried in the tendon; it possesses no periosteum.

The largest, the *patella* or *knee-cap*, occurs in the Quadriceps Femoris tendon.

A pair of sesamoid bones lie below the head of the 1st metatarsal in the tendons of Flexor Hallucis Brevis. Two occur at the head of the first metacarpal, in the tendons of Adductor Pollicis and Flexor Pollicis Brevis. Others commonly (but not always) occur in the lateral head of Gastrocnemius, Peroneus Longus (at side of cuboid), Tibialis Posterior (behind navicular tuberosity), and at heads of metacarpals 2 and 5 (in palmar plates).

Accessory Bones. Certain bones normally ossify from several centers, and it sometimes happens that one or more of these centers fails to unite with the main mass of the bone; again, an abnormal or extra center of ossification may make its appearance and the resulting bone may remain discrete. In either case. the result is an accessory bone, and in an X-ray photograph it may simulate a fracture.

Examples: (1) The frontal bone may persist in right and left halves, i.e., persisting metopic suture. (2) The upper (interparietal) part of the occipital squama may remain discrete, and is typical of West Coast Indians. (3) The zygomatic bone may be in upper and lower parts notably in Asiatics. (4) Sutural bones, the size of a finger nail, may occur in the sutures of the skull. (5) The acromial epiphysis may remain a separate bone. (6) The 5th lumbar vertebra is commonly in two pieces (*fig. 321*). (7) The patella may be bipartite (*fig. 518*). (8) Supernumerary carpals and tarsals occur.

Markings on a Dried Bone

The surface of a dried bone is smooth, in fact almost polished, over areas covered with cartilage and where tendons play in grooves (e.g., head of humerus; upper and under surfaces of sustentaculum tali). Near the ends of a long bone there are large vascular foramina for veins and arteries, and piercing the shaft obliquely is the nutrient canal, for the nutrient vessels, which may be 5 cm long.

Markings occur wherever fibrous tissue is attached—no matter whether it be a ligament, tendon, aponeurosis, fascia, or intermuscular septum. Fibrous tissue markings are, however, not present at birth nor in the young. They appear about puberty and become progressively better marked. The fleshy fibers of a muscle make no mark on a bone.

Terms. Markings take the form of (1) elevations, (2) facets, and (3) depressions.

Elevations, in order of prominence: a linear elevation is a *line*, *ridge*, or *crest;* a rounded elevation is a *tubercle*, *tuberosity*, *malleolus*, or *trochanter;* a sharp elevation is a *spine* or *styloid process*.

Small, smooth, flat areas are called *facets* (*cf.* the facet of a diamond).

A *depression* is a *pit* or *fovea*, if small; a *fossa*, if large; a *groove* or *sulcus*, if it has length. A *notch* or *incisura*, when bridged by a ligament or by bone, is a *foramen* (i.e., a perforation or hole), and a foramen that has length is a *canal* or *meatus*. A canal has an *orifice* [*os* or *ostium*] at each end.

The portion of a notch, foramen, or orifice of a canal over which an emerging vessel or nerve rolls is rounded, but elsewhere it is sharp. Therefore, even on a dried bone the direction taken by the emerging occupant is evident (*cf.* lesser sciatic notch, anterior sacral foramina, infra-orbital canal).

Areas covered with articular cartilage are called *articular facets*, if approximately flat. Certain rounded articular areas are called *heads;* others are called *condyles* (= knuckles). A *trochlea* is a pulley.

A Living Bone or a Dissecting Room Specimen before Maceration

The articular parts are covered with *hyaline* (articular) *cartilage*.

Periosteum envelops all parts not covered with cartilage and not giving attachment to ligaments and tendons. It consists of two layers: (1) an outer, fibrous membrane and (2) an inner, vascular one lined with bone-forming cells, the *osteoblasts*. The periosteum is easily scraped off with the handle of the scalpel, leaving, however, many osteoblasts adhering to the bone. *Fibrocartilage* lines the grooves where tendons exert pressure. Some elevations seen in the macerated bone are but shadows of what they were before maceration, because in life they had fibrocartilaginous extensions, now shed (e.g., dorsal radial tubercle).

The Parts of a Young Bone (*fig. 13*)

At birth both ends of a long bone are cartilaginous, *cartilaginous epiphyses*. The part of the bone between the cartilaginous ends is the *diaphysis* (Gk. dia = in between, across). It comprises a casing of compact bone which encloses a medullary cavity at its middle and spongy bone at each end, and all is filled with red marrow. The diaphysis is clothed in *periosteum* (Gk. peri = around; osteon = bone).

Epiphyses (Gk. epi = upon, physis = growth). (1) During the 1st and 2nd years one (or both) of the cartilaginous ends begins to ossify subjacent to the site of articulation, constituting a *pressure epiphysis* (e.g., head of humerus, condyles of femur). (2) Later, generally about puberty, independent ossific centers appear in the cartilage at the sites of attachment of certain tendons, constituting *traction epiphyses* (e.g., tubercles of humerus, trochanters of femur). (3) A third type of epiphysis is the

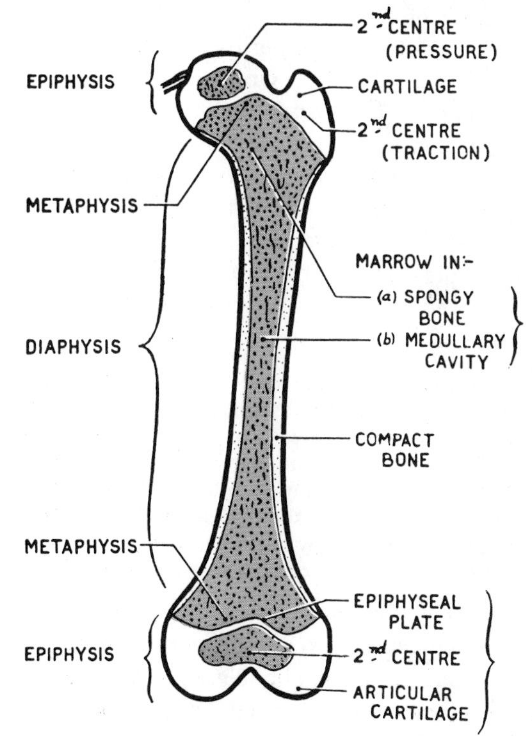

Fig. 13. The parts of a young bone as shown by a longitudinal section of a femur.

atavistic epiphysis. Atavistic epiphyses phylogenetically were independent bones now grafted on to other bones (e.g., coracoid process of scapula).

The epiphyseal cartilage between an epiphysis and a diaphysis is also known as an *epiphyseal "plate."* The region of the diaphysis adjacent to the plate, the *metaphysis* (Gk. meta = beyond), is the site where growth in length takes place.

Where there are two epiphyseal ends, the end that has the more work to do is the first to start work (ossifying) and the last to stop (to fuse with the diaphysis). When fusion (synostosis) takes place, growth in length ceases.

All long bones—including the metacarpals, metatarsals, phalanges, and ribs—have a pressure epiphysis at one end or the other, whereas five paired bones *always* have pressure epiphyses at both ends (viz., humerus, radius, femur, tibia, and fibula).

Nutrient Artery and Canal. The blood supply of living bones comes from many small vessels in the periosteum and from a large *nutrient artery* which enters the shaft through a *nutrient foramen* and is a constant feature. The nutrient canal (which early ran transversely) increasingly occupies an oblique position directed away from the epiphyseal end (*fig. 11*). Where there is an epiphysis at both ends, the canal is directed away from the more actively growing end.

It is roughly estimated that the shoulder end of the humerus and the wrist ends of the ulna and radius grow 3 to 4 times as much as their elbow ends; the knee end of the femur between 2 and 4 times as much as the hip end; and the knee end of the tibia slightly more than the ankle end (Digby and Phemister). Hence, in these bones the nutrient canals are directed "to the elbow I go, from the knee I flee" (Hughes).

Ossification. Except for certain bones of the skull and the clavicle, all the bones of the body pass through a cartilaginous stage. About the 8th intra-uterine week ossification of the long bones begins. There are two types of ossification: (1) intracartilaginous or enchondral and (2) periosteal or intramembranous.

In the shaft of a long bone both types take place concurrently, as described in text books of Histology (see Ham).

After birth, at the center of one or both cartilaginous ends, the process of enchondral ossification begins, as shown in *figure 13*, and a bony epiphysis takes form. Ossification progresses in the epiphysis until only two sheets of cartilage remain: (1) the *articular cartilage* which covers the end of the bone and persists throughout life, and (2) a residual plate, the *epiphyseal cartilage*, placed between the diaphysis and the bony epiphysis (forming a synchondrosis). Ultimately, when the bone has attained its adult length, the plate also ossifies—the site commonly being marked by an *epiphyseal line* (*fig. 11*). In technical terms, a synchondrosis has been converted into a synostosis.

Short bones (i.e., carpal and tarsal) ossify enchondrally like epiphyses.

The bones of the skull, except those of the base, do not pass through a cartilaginous stage but ossify directly from membrane.

Sexual Difference. Ossification starts earlier in females than in males and it is completed earlier—even by as much as 2 to 3 years.

Bone Marrow. Blood cells, made by bone marrow, have but a short life, the red cells living about 120 days, and the birth rate necessarily keeps pace with the death rate.

At birth, cancellous (spongy) bone, which at this age is limited in quantity, and the medullary cavities of the long bones are filled with red (blood-forming) marrow. By the 7th year, the amount of spongy bone has increased and the red marrow has extended into it, but at the same time has receded from the medullary cavities only to be replaced there by yellow (fatty) marrow. About the 18th year, red marrow is almost entirely replaced by yellow in the limb bones; thereafter, it is confined to the axial skeleton—skull, vertebrae, ribs, sternum, hip bones, and upper ends of femur and humerus (A. Piney and M. M. Wintrobe).

In certain conditions (e.g., some types of anemia) where the death rate of the red cells is high, the yellow marrow reverts to red in an endeavor to support the birth rate.

Vessels and Nerves. *Arteries* supply long bones thus: (1) *periosteal twigs* enter the shaft at many points, run in the small, longitudinal (Haversian) canals, and supply the outer part of the compact bone of the shaft (*fig. 13.1*); (2) twigs from *articular arteries*, which anastomose around the joint usually between the bone and the reflexion of the synovial membrane, supply the epiphyses, the metaphyseal region, and the capsule; (3) the *nutrient artery* (medullary a.), on entering the medullary cavity, divides into a proximal and a distal branch, each of which supplies the inner part of the compact bone, the marrow, and the metaphyseal region. It is the main artery of the shaft (Trueta and Cavadias). The blood flow through the cortex runs in a centrifugal and not in a centripetal direction (Brookes *el al.*).

The anastomoses between the branches of the nutrient and periosteal arteries seem to be feeble. Though many of the metaphyseal branches of the nutrient artery are end arteries, some of them anastomose with the metaphyseal branches of the articular arteries. Indeed, when the shaft of a long bone is fractured, one or other branch of the nutrient artery is necessarily torn across. It then falls to the anastomoses effected with the articular arteries to replenish the torn nutrient artery with blood.

Veins. There are periosteal veins and nutrient veins, but the chief veins, enriched with young blood cells, are said to escape by the large foramina near the ends of the bone.

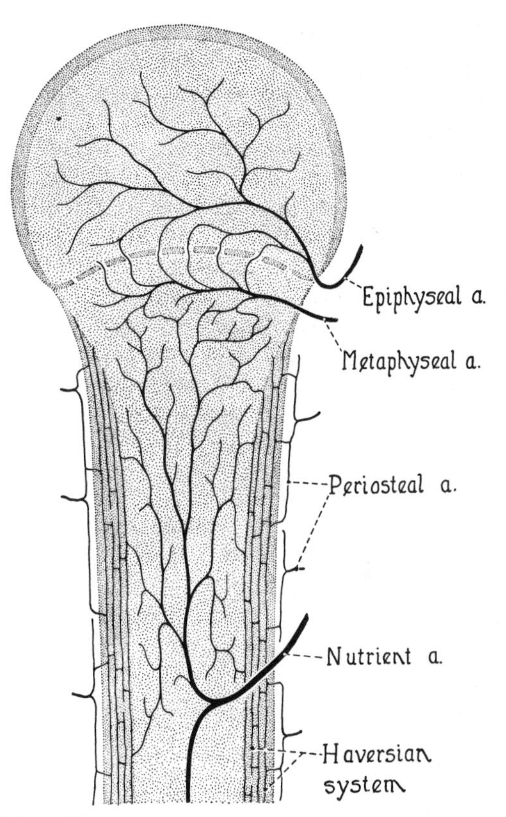

Fig. 13.1. Blood supply to a long bone (schematic).

Lymph Vessels exist in the periosteum and in the perivascular lymph spaces in Haversian canals.

Nerves. Sensory nerves are plentiful in the periosteum, and nerves (? to blood vessels) accompany the nutrient artery.

Fascinating Historical Background

One day in 1736, John Belchier, surgeon on the staff of Guy's Hospital, London, was dining with a friend. A joint of pork was served, and it was commented that the bones were red. The host, who was a calico-printer, explained that he utilized bran soaked in madder from his dye vats to feed his pigs, and to this fact he attributed the color. Belchier communicated this information to the Royal Society, and it was printed in its transactions.

Duhamel, a French squire, read Belchier's paper and, being curious, fed madder to some of his fowls and pigs; and with the same red result. He then conducted a number of experiments on pigs and found that if the animals were killed while the feeding of madder was in progress, the bones appeared red, and that if the feeding of madder had ceased for a period, the bones appeared white. On laying open the bones, he found that though they were white outside they were red inside. By alternatively feeding food with madder and without, he produced bones with alternating red and white rings or layers, so he concluded that *bones increase in girth like trees and that the periosteum is responsible for laying down the rings.* He encircled growing bones with rings of silver wire and in time he found the wire inside the medullary cavity, because the cavity too had enlarged.

In the shaft of a growing bone Duhamel bored holes at measured distances apart, and in them inserted silver stylets to keep them open. After a period he killed the animal and found that, although the length of the bone had increased, the holes remained the same distance apart, so he concluded that *growth in length takes place at the ends of the long bones.*

John Hunter sought further explanation. Knowing that the lower jaw has no epiphyses and that the milk teeth of a child fill the body of the jaw right back to the ramus, he wondered how space was found for the three additional teeth, the three permanent molars. He surmised that the *growth of bone entails two processes—one of deposition* (addition), *the other of absorption* (subtraction). Only thus could he account for the growth of the jaw, the formation and progressive enlargement of medullary cavities, and for changes in the neck of the femur. About 1764, John Hunter—employing madder-feeding experiments on pigs and using controls—put his theory to the test and proved it to be correct (*fig. 14*). (Consult "Menders of the Maimed" by Sir Arthur Keith.)

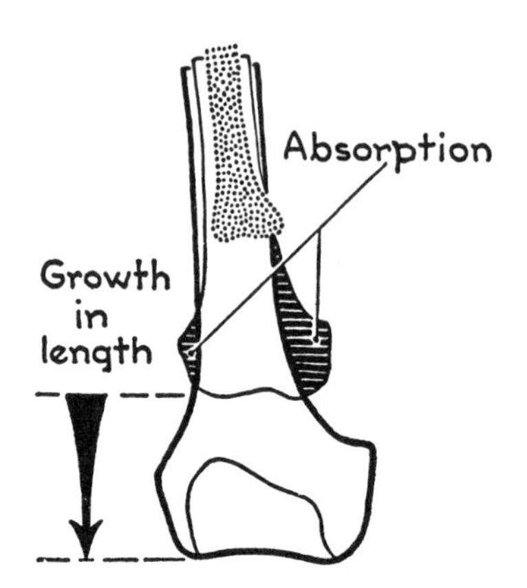

Fig. 14. Remodelling of bone. As a long bone grows, sites once occupied by the expanded ends become parts of the more slender shaft.

Skull. The bones of the base being mainly cartilaginous bones, grow as such, the chief epiphyseal plate being between the basi-occipital and the body of the sphenoid.

The bones of the vault of the skull are membranous bones. Regarding their growth there are two conflicting views: (1) In one view, the essential mode of growth is by deposition of bone on the exterior and resorption from the interior, modeling taking place as for long bones (J. C. Brash). (2) The other view is that growth is mainly sutural in all parts of the skull, with depositions and resorptions taking place in various areas both inside and out (J. P. Weinmann and H. Sicher, and L. W. Mednick and S. L. Washburn).

In support of the latter view, if the coronal suture closes prematurely and the metopic and sagittal remain open, the skull becomes unduly broad and high and short (R. K. Rau); whereas, if the metopic and sagittal sutures close prematurely, the result is a long narrow skull (B. H. Dawson and D. A. N. Hoyte).

Shadows of Past Events. The growing skeleton is sensitive to relatively slight and transient illnesses and to periods of malnutrition. When a child is ill or starved, his epiphyseal plates, ceasing to proliferate, become heavily calcified; and when growth is resumed, this line of arrested growth appears as a veritable scar (*fig.* 15). When a particulate radio-opaque substance, such as thorotrast, is injected into the bloodstream, it is taken up and retained by certain cells in the bone marrow. Employing this technique experimentally in young animals, Mortensen and Guest have shown that the part of the marrow infiltrated remains constant in length and the actively growing ends gradually recede from it.

CARTILAGE

Hyaline Cartilage; Fibrocartilage; Elastic Cartilage

Cartilage or gristle is a connective tissue in which a solid ground substance (more resilient than bone) forms the matrix. It has no blood vessels, lymph vessels, or nerves; so, it is insensitive. There are three types of cartilage: (1) hyaline, (2) fibro-, and (3) elastic.

Hyaline Cartilage (Gk. (h)ualos = a transparent stone) is white and resilient. It is potentially bone; in fact, all the bones, except certain skull bones and the clavicle, were preformed in hyaline cartilage.

Hyaline cartilage persists in the adult only at the articular ends of bones as articular cartilage, at the sternal ends of the ribs as costal cartilage, and as the cartilages of the nose, larynx, trachea, and bronchi. The thyroid, cricoid, and 1st costal cartilages commonly begin to calcify about the 40th year.

Fibrocartilage has the same structure as fibrous tissue with the addition of cartilage ground substance (*fig. 53*). Fibrocartilage bears the same resemblance to fibrous tissue as a starched collar bears to a soft collar. Wherever fibrous tissue is subjected to great pressure, it is replaced by fibrocartilage, which is tough, strong, and resilient.

Examples. It occurs in intervertebral discs, articular discs (e.g., semilunar cartilages of the knee), glenoid and acetabular labra, and the surface layers of tendons and ligaments that are pressed on by bone. It lines certain bony grooves in which tendons play, and it caps certain bony prominences.

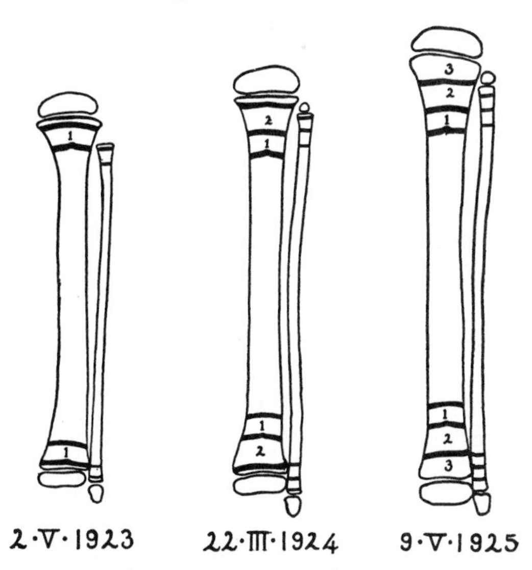

Fig. 15. Outlines of three radiograms of the leg bones of a young girl taken over a period of 2 years. Observe that the three lines of arrested growth, denoting three successive illnesses, remain equidistant. (After H. A. Harris).

Elastic Cartilage. Here cartilage cells are numerous and the solid ground work is pervaded by yellow elastic fibers making it more pliable. It is found only in the external ear, auditory tube, and small cartilages guarding the entrance to the larynx.

VERTEBRAL COLUMN

PARTS OF A TYPICAL VERTEBRA AND THEIR FUNCTIONS—*Body; Vertebral Arch; Processes; Ossification.*

ARTICULATED VERTEBRAL COLUMN—*Intervertebral Discs; Bodies of the Vertebrae; Curvatures; Varying Stature; Line of Gravity; Transverse Processes; Pedicles; Laminae; Vertebral Foramina and Vertebral Canal; Articular Processes; Spinous Processes.*

Parts of a Typical Vertebra and Their Functions

The vertebral column in made up of 33 vertebrae, arranged as follows: 7 cervical, 12 thoracic, 5 lumbar, 5 sacral, and 4 coccygeal. The sacral and the coccygeal vertebrae unite to form composite bones, called the os sacrum and coccyx. The 5 sacral vertebrae have completely fused to form a single mass by the 23rd year; a gap, however, often persists between the 1st and 2nd sacral bodies until the 32nd year (McKern and Stewart). The last three pieces of the coccyx fuse together in middle life and these in turn fuse with the first piece still later. There are, therefore, 24 presacral or true vertebrae, of which 12 bear ribs and 12 do not.

Not only have the bones of each region features characteristic of their particular region but every bone in each region has one or more distinguishing features of its own.

A vertebra is composed of the following parts (*fig. 16*):

1. A weight-bearing part—the *body*.
2. A part that protects the spinal cord—*the vertebral arch*.
3. Three levers on which muscles pull—the *spinous process* and the right and left *transverse processes*.
4. Four projections which restrict movements—two superior and two inferior *articular processes*.

The Body of a vertebra (*fig. 17*) resembles the long bones of the limbs in that it is weight-supporting, constricted about its "waist," and enlarged at its two ends, which are articular (although rather flat). Also, it has a primary center of ossification for the "diaphysis" which appears early, and secondary centers for the upper and lower epiphyses. It is, indeed, a long bone in miniature.

Large vascular foramina are found on the dorsal and lateral aspects.

The Vertebral Arch protects the spinal medulla (spinal cord) from injury. Immediately behind its attachment to its body, each half of the arch is crossed, both above and below, by a spinal nerve. Accordingly, this part of the arch, called the *pedicle* is grooved above and below, but especially below, to allow ample space for the passage of a nerve. The grooves are the *superior* and *inferior vertebral notches*.

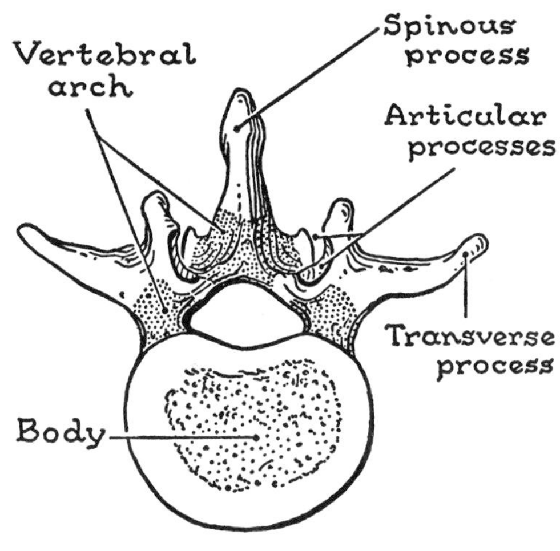

Fig. 16. The parts of a vertebra, from above.

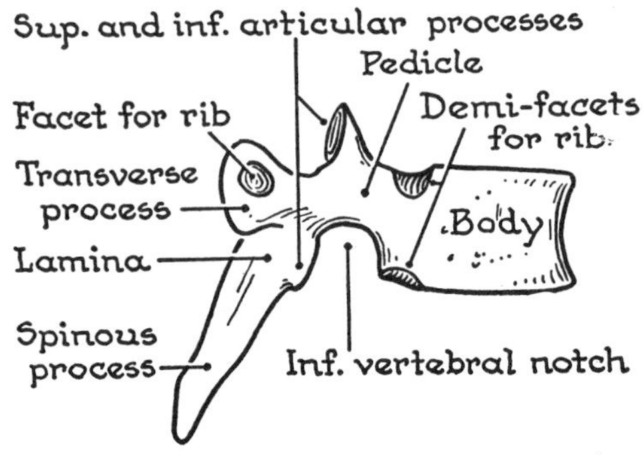

Fig. 17. A typical vertebra (side view).

The posterior band-like portions of each arch, the *right* and *left laminae*, meet behind in the median plane. A vertebral arch and the posterior aspect of a body enclose a space, the *vertebral foramen*, in which the spinal cord and its membranes are lodged.

Transverse and Spinous Processes. The movement of the body on another is effected in part through the actions of muscles on the lever-like transverse and spinous processes, which project like the spokes of a capstan. The transverse processes project laterally on each side from the junction of a pedicle and a lamina; the spinous process or spine projects backward in the median plane from the site of union of a right and a left lamina.

Articular Processes arise near the junction of pedicle and lamina. The superior processes spring rather from pedicles and face in a backward direction (backward and upward in the cervical region; backward and laterally in the thoracic; backward and medially in the lumbar), whereas inferior articular processes spring from laminae and face in contrary directions. It is evident that in all regions the contact established between upper and lower articular processes prevents forward displacement of an upper vertebra on a lower.

The upper and lower surfaces of the bodies are the real articular surfaces of the vertebrae. The articular processes (except those of the atlas and axis) do not transmit weight. Their presence interferes with the unrestricted mobility the bodies might otherwise enjoy and decrees in what direction movements between two adjacent vertebrae shall be allowed. There are, however, circumstances in which they bear weight, e.g., on rising from the stooping position.

Collectively, the vertebral foramina constitute the *vertebral canal.*

Collectively, two adjacent vertebral notches constitute an *intervertebral foramen.* Entering into the composition of an intervertebral foramen are: above and below, pedicles; in front, an intervertebral disc and parts of the two bodies it unites; and behind, two articular processes and the capsule uniting them (*fig. 18*).

Ossification. At birth a vertebra is in three parts—a *centrum* and the right and left sides of a *neural arch,* united to each other by hyaline cartilage (*fig. 19*). The site of union of a centrum and a neural arch is a *neuro-central synchondrosis.*

Synostosis of the two halves of the arch takes place posteriorly during the 1st year, and of the arch and centrum between the 3rd and 6th years.

Epiphyses. Pressure and traction epiphyses appear about puberty and fuse not later than the 24th year. In most mammals the pressure epiphyses take the form of plates, but in man they are rings (*fig. 27*). Scale-like traction epiphyses appear on the tips of the spinous and transverse processes.

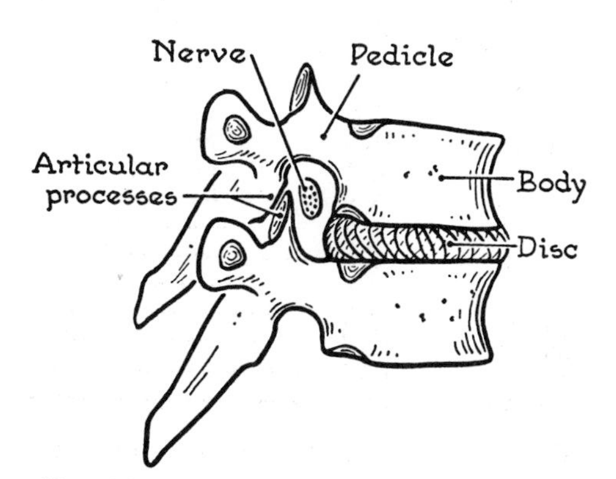

Fig. 18. Composition of an intervertebral foramen.

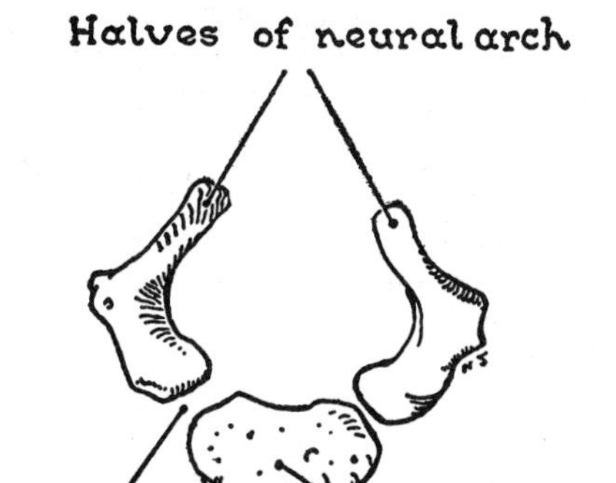

Fig. 19. Bony parts of a vertebra at birth.

The advanced student requiring details should consult the following: Cervical vertebrae (Chap. 42), thoracic vertebrae (Chap. 29), lumbar vertebrae (p. 244), sacrum and coccyx (p. 271).

Articulated Vertebral Column

The bodies of the vertebrae contribute three-fourths to the total length of the presacral portion of the articulated column; the intervertebral discs contribute one-fourth (*fig. 20*).

Intervertebral Discs. Movement between two vertebrae is most free where the disc is thickest (vertical height greatest), namely, in the cervical and lumbar regions, where the vertebral column is convex forward. Conversely, movement is least where the disc is thinnest—in the midthoracic region (T. 2–6) (*fig. 21*).

Further, in the cervical and lumbar regions each disc is thicker ventrally than dorsally whereas in the thoracic region the converse is the case; hence, in each region the disc contributes to the curvature of the column.

Bodies of the Vertebrae. As might be

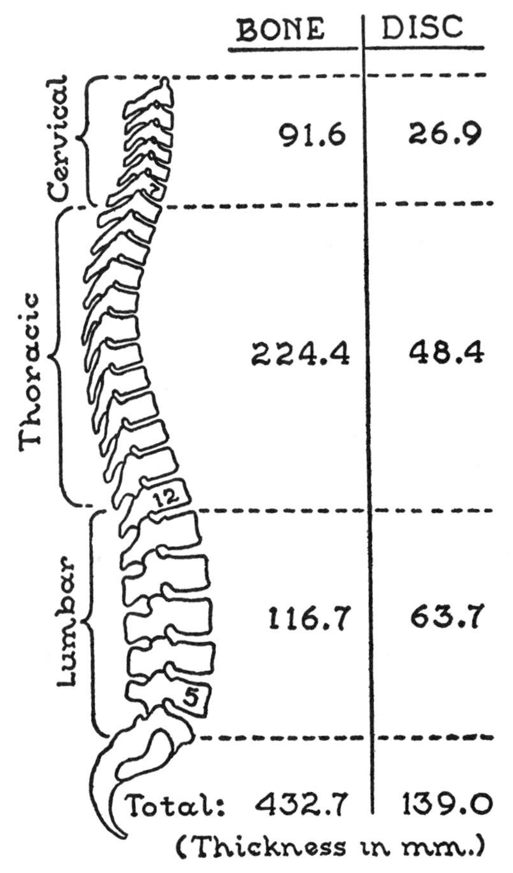

Fig. 20. Proportions of bone and disc in the presacral parts of the vertebral column (with the use of Todd's data).

expected, the weight-bearing surfaces increase progressively from above downward to the first piece of the sacrum. From there to the tip of the coccyx they diminish rapidly because the weight is transferred from the first three pieces of the sacrum laterally to the pelvis.

The 1st and 2nd cervical vertebrae—the atlas and axis—are highly specialized and their bodies are modified. The support of the skull is shifted to a pair of concave facets on the atlas.

The upper and lower surfaces of the body of a *cervical* vertebra are oblong; those of the *thoracic* are heart-shaped with long diameter anteroposterior and the *lumbar*, kidney-shaped with long diameter transverse.

Curvatures. In prenatal life the vertebral column is uniformly curved so as to be concave ventrally (*fig. 22*). In the thoracic and sacro-coccygeal regions these concavities persist. The cervical curvature (convexity) appears when the infant learns to hold its head erect and to direct its visual axes forward, about the 3rd month. The lumbar curvature (convexity) appears when the child acquires the art of walking erect, about the 18th month. The thoracic and sacral curvatures, therefore, are *primary curvatures;* the cervical and lumbar are *secondary or compensatory*.

In the cervical region the bodies are of equal depth in front and behind; so, the cervical curvature is due solely to disc. In the lumbar region only the 5th and 4th bodies are always deeper in front.

Varying Stature. One may be shorter in the evening than in the morning because with fatigue (1) the curvatures of the spine may increase, (2) the turgor of the pulp of the intervertebral discs may be reduced; and (3) the height of the arches of the feet may be lessened.

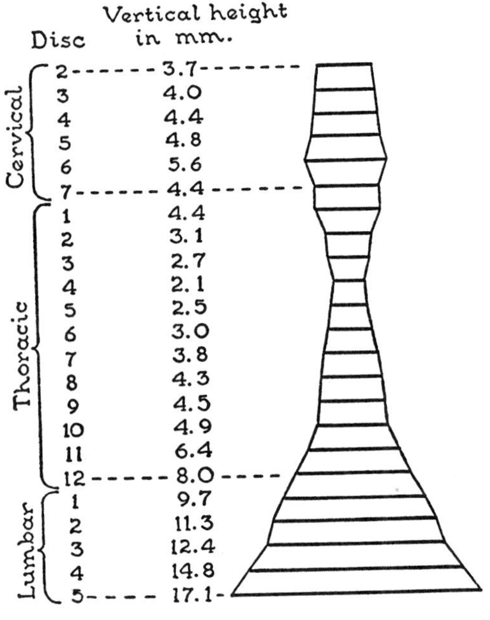

Fig. 21. Graph of the vertical heights of the intervertebral discs (with the use of Todd's data).

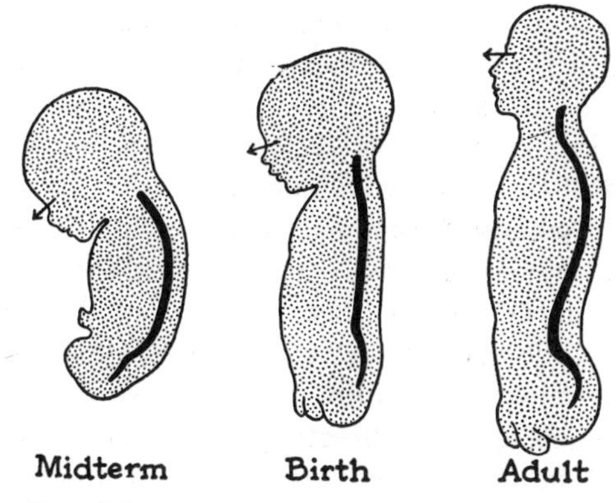

Fig. 22. Development of the curvatures of the spine; the thoracic and sacral curvatures are primary, and the cervical and lumbar are secondary.

On the other hand, the stature increases when one lies down.

The Line of Gravity passes through the body of the axis, just in front of the sacrum, behind the centers of the hip joints, and in front of the knee and ankle joints (*fig. 23*).

Transverse Processes arise between upper and lower articular processes at the junctions of pedicles with laminae, and project laterally. In the *thoracic* region they act not only as levers for muscles but also as fulcra for the ribs; so, they are strong and stout and, in conformity with the backward curving of the ribs, they have a backward and upward inclination. Their tips bear articular facets for the ribs except for the 11th and 12th ribs which are floating; so, the transverse processes of the 11th and 12th vertebrae are reduced in size and carry no facets.

Each *cervical* transverse process has a circular foramen, the *foramen transversarium*. The upper six foramina transmit the vertebral artery; the 7th transmits only veins.

The *lumbar* transverse processes are thin and flat. Conforming to the shape of the rounded abdominal cavity, they are directed slightly backward. Since the 4th lumbar vertebra lies at the level of the highest part of the iliac crest, it follows that the 5th lumbar vertebra must lie below the highest part of the crest.

Morphology of the Transverse Processes. Except in the thoracic region, a transverse process comprises two elements—a *costal* or rib element and a true or *morphological* transverse process, as is made clear by *figure 23.1*.

Pedicles spring from the upper half of the sides of the bodies; vertebral notches mainly lie behind the lower half. *Intervertebral foramina* generally increase in size from above downward (*figs. 18, 20*).

Laminae overlap markedly in the thoracic region and slightly in the cervical region. In the lumbar region there are **interlaminar gaps,** and also in the cervical region when the neck is bent. The largest gaps are between skull and atlas, atlas and axis, 4th and 5th lumbar verte-

Fig. 23. The line of gravity.

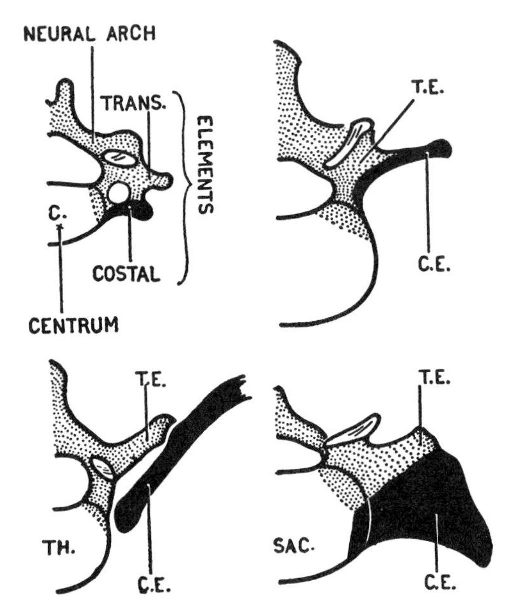

Fig. 23.1. Homologous parts of cervical, thoracic, lumbar, and sacral vertebrae. (*Clear* = centrum; *Stippled* = neural arch, transverse and spinous processes; *black* = costal element).

brae, and 5th lumbar vertebra and sacrum. Through these gaps, a physician can insert a spinal-puncture needle into the vertebral canal and the cerebrospinal fluid surrounding the spinal cord.

Vertebral Foramina and Vertebral Canal. In the thoracic region the vertebral canal is circular and of the diameter of a signet ring (*fig. 24*), circular because the spinal cord is here cylindrical; but in the regions from which the largest nerve roots for the limbs spring the canal is larger and triangular, or rather it is expanded transversely in adaptation to the more laterally expanded cord.

In the upper three cervical vertebrae, the vertebral canal is very roomy—so roomy that free movement between the head and the neck does not constrict the spinal cord.

Articular Processes. In all three regions—cervical, thoracic, lumbar—the articular processes prevent the vertebrae from slipping forwards, and they allow flexion and extension. In addition, the *cervical articular processes* allow one to look sideways and upward, because their upper facets mostly face obliquely upward, laterally, and backward. The *thoracic* processes allow rotation, but *lumbar* processes prevent it, allowing side bending.

Spinous Processes. The spinous processes become more massive as they are followed from higher to lower levels. The pull on each, as in rising from the stooping posture, is mainly a caudalward one; hence, each is directed caudalward. That of the 1st *cervical* is reduced to a tubercle. In modern man, cervical spines 2–6 are bifid. The 7th ends in a tubercle and is prominent, but not so prominent as the 1st thoracic. Both are easily felt. *Thoracic* spines are long and sloping (*figs. 20, 24*) while the *lumbar* spines are thick oblong plates with thickened ends (*fig. 16*). The 5th (and 4th) *sacral* spines and laminae are absent and the sacral canal is exposed. The sacral articular processes on each side fuse to form an irregular crest which ends below in a cornu (horn). This cornu articulates with the cornu of the coccyx.

ARTICULATIONS OR JOINTS

SKULL TYPE—*Suture and Synchondrosis; Synostosis.*

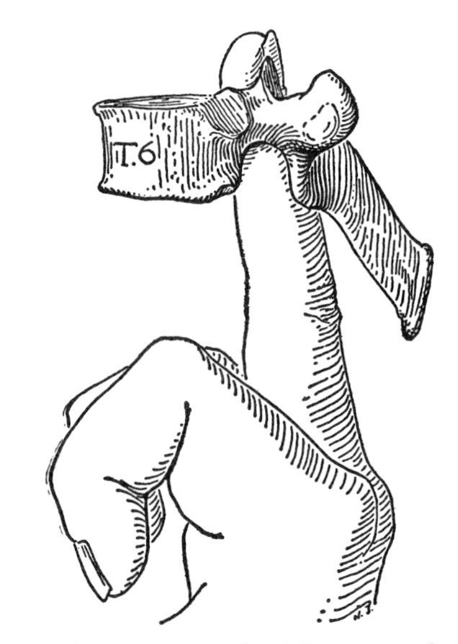

Fig. 24. The vertebral foramen of the 6th thoracic vertebra is not larger than a finger-ring.

VERTEBRAL TYPE—*Symphysis; Symphysis between Two Vertebral Bodies; Syndesmosis; Vertebral Syndesmoses.*

LIMB TYPE—*Synovial Joint: articular capsule, joint cavity, synovial membrane and folds, synovia, fat-pads, articular discs, ligaments, articular cartilage, lubrication, labra, nerves, and vessels.*

Classification of Synovial Joints.

Definition and Classifications

A joint is a junction between two or more bones. The formal classification is given below for reference—

Formal Classification

Fibrous Joints
- Syndesmoses
- Sutures
- Gomphoses

Cartilage Joints
- Synchondroses
- Symphyses

Synovial Joints

Joints also may be classified as immovable, slightly movable, and freely movable; but perhaps it is more helpful to consider them as:

1. The *skull type:* Immovable or temporary joint.
2. The *vertebral type:* Slightly movable or secure joint.
3. The *limb type:* Freely movable, insecure, or synovial joint.

"Skull Type" of Joint (Immovable)

The skull type is either a suture or a synchondrosis depending on whether the bones concerned ossify in membrane or in cartilage (*fig. 25*). Gomphoses are found at the roots of the teeth.

Suture. When the growing edges of two bones (or ossific centers) developing in membrane come together, a thin layer of fibrous tissue ("ligament") may persist unossified between them until middle age, or indefinitely. Such a union is called a suture; and sutures are confined to the skull (*fig. 25*). The edges may interlock in jig-saw fashion, or like the teeth of a saw (*fig. 667*). They may be bevelled and over-

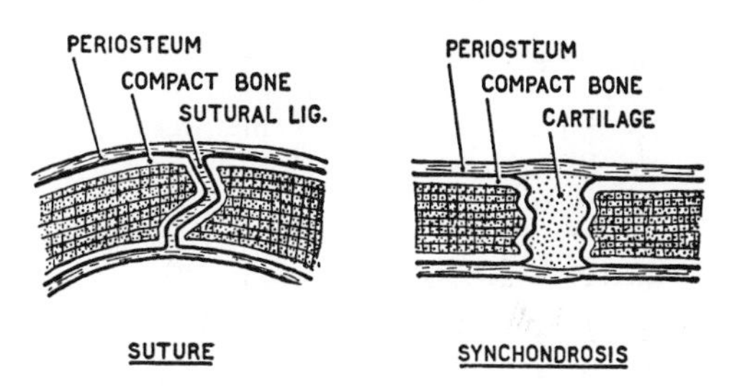

Fig. 25. A suture and a synchondrosis.

lapping or relatively flat and abutting. A ridge may fit into a groove, as it does between sphenoid and vomer.

A Gomphosis is a membranous union of a tooth to its socket (periodental membrane).

Synchondrosis. Similarly when the growing edges of two bones (or ossific centers) developing in a single mass of cartilage come together, a residual plate of cartilage may persist unossified between them for a number of years. Such a union is a synchondrosis.

Sites. Synchondroses occur at the base of the skull between the basi-occipital and basisphenoid, sphenoid and ethmoid, petrous-temporal and basi-occipital, and petrous-temporal and jugular process. Their distribution, however, is widespread, for is it not by synchrondroses (epiphyseal plates) that epiphyses are united to bones? (*fig. 13*).

Synostosis is the obliteration of a suture or a synchondrosis by bone. It is associated with cessation of growth locally.

"Vertebral Type" of Joint

The vertebral type is either a symphysis or a syndesmosis. Intervertebral joints are built for strength and security; so, their opposed bony surfaces are firmly bound together. This minimizes the risk of dislocation, which here would be disastrous.

A Symphysis is a joint where two opposed bony surfaces are coated with hyaline cartilage, are united by fibrocartilage, and are further united in front and behind by ligamentous bands. There is no joint cavity, but a small cleft may be present.

Sites. Symphyses occur (1) between the bodies of vertebrae, (2) between the pubic bones, and (3) between the manubrium and body of the sternum. They are all situated in the median plane.

The Symphysis between Two Vertebral Bodies. The upper and lower surfaces of the body of a vertebra each consist of hyaline cartilage. In the periphery of the cartilage a ring-shaped bony epiphysis appears (*fig. 26*) and finally fuses with the rest of the body in early adult life. The cartilaginous plate persists and must be regarded as helping to enclose an intervertebral disc.

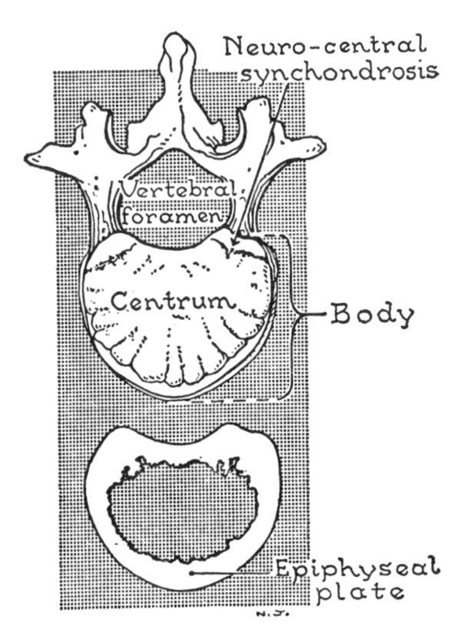

Fig. 26. A vertebra in childhood.

Intervertebral Discs (*fig. 27*). Adjacent bodies are united by a fibrocartilaginous disc whose peripheral part is composed of about a dozen concentric layers of fibers, the *anulus fibrosus*. The fibers in alternate layers cross like the limbs of the letter X. The center of the disc is filled with a fibrogelatinous pulp, the *nucleus pulposus*, which acts as a cushion or shock absorber.

Protrusion of the nucleus material through a tear in the anulus is erroneously called a "slipped" disc. The pressure on spinal nerves that may result will cause severe back pain and even localized paralysis. The hyaline plate also is apt to crack; then the nucleus herniates through it into the cancellous body of the vertebra.

Longitudinal Ligaments of the Bodies, an *anterior* and a *posterior*, extend from sacrum to base of skull: the one is attached to the intervertebral discs and adjacent margins of the vertebral bodies anteriorly; the other is attached to them posteriorly, within the vertebral canal. The anterior ligament is a broad, strong band. The posterior ligament is weak and narrow, but it widens where it is attached to the backs of the discs.

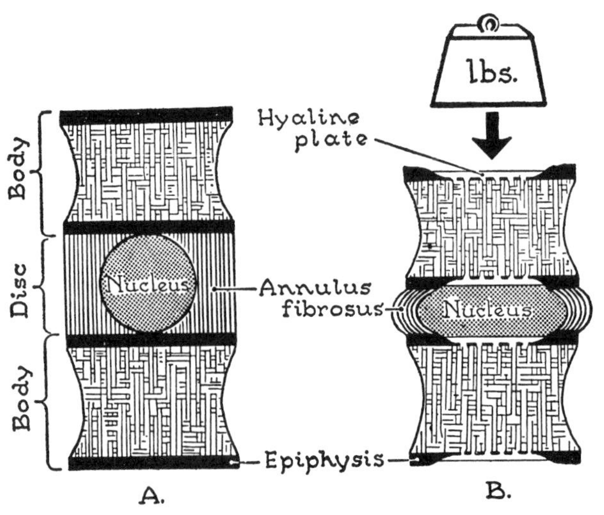

Fig. 27. Scheme of an intervertebral disc (*Annulus*, old terminology = *Anulus*, N.A.).

Vessels and Nerves. Small blood vessels from the marrow spaces pass through the hyaline plate to supply the disc until the 8th year, and some of these may persist until the 20th or 30th years. (Coventry *et al.;* and others.) Branches of the spinal nerves have been traced to the longitudinal ligaments and to the anulus (Roofe; and others).

A Syndesmosis (or Fibrous Joint) is a union by cord-like ligamentous fibers.

Sites. Syndesmoses occur between the vertebral arches, and between the lever-like processes of the vertebrae, also between the coracoid process of the scapula and the clavicle (coracoclavicular lig.), between the bones of the forearm and of the leg (e.g., interosseous membranes).

Vertebral Syndesmoses. The laminae of adjacent vertebrae are united by yellow elastic fibers, called a *ligamentum flavum* (L. = yellow). These bands unite the upper border and posterior surface of one lamina to the lower border and anterior surface of the lamina above. By virtue of their elasticity, the ligamenta flava serve as "muscle sparers," i.e., they assist in the recovery to the erect posture after bending forward and they are particularly strong in the lumbar region.

The adjacent borders of the spinous processes are united by weak *interspinous ligaments* and their tips are united by the strong *supraspinous ligament.* The transverse processes may be connected by weak *intertransverse ligaments.*

The articular processes of the vertebrae are united by articular capsules to form *Synovial Joints* (see below).

So, typical vertebrae are united to each other by three types of joints; symphyses, syndesmoses, and synovial joints.

Limb Type of Joint

Synovial Joint or Articulation

The limbs being primarily organs of locomotion have joints that permit free movement. The ends of two bones rubbing against each other are capped with hyaline cartilage and are enclosed and united by an *articular capsule (fig. 28).* This consists of a short sleeve of fibrous tissue, called the *fibrous capsule,* which extends well beyond the articular (= joint) cartilage, and it is lined with an inner sleeve of synovial membrane (the synovial capsule). The synovial membrane is reflected from the fibrous capsule onto the bone, and this it covers right up to the articular cartilage.

Synovia or Synovial Fluid, slippery like egg-white (hence its name—ova, L. = eggs), is a dialysate of blood plasma plus a mucin, called *hyaluronic acid.* To it the viscosity and lubricating properties of the synovia are due.

A Synovial or Joint Cavity develops as a cleft between the ends of the primitive bones. It is lined everywhere either with articular cartilage or with synovial membrane.

A Synovial Membrane is a thin sheet of connective tissues, characterized by its richness in blood vessels and lymphatics; it produces the synovial fluid in sufficient quantity to keep the surfaces lubricated in

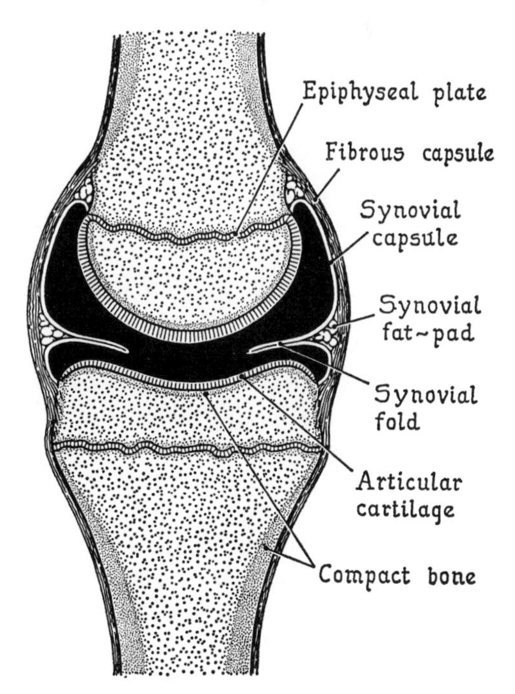

Fig. 28. A synovial joint, on section.

health. When irritated, it pours forth excess fluid (e.g., "water on the knee"). Rheumatoid arthritis inflames the membrane which is liberally supplied with pain endings of nerves.

Transparent *folds of synovial membrane* containing fat at their attached borders project into all synovial joints, commonly for a centimeter or farther (*fig. 28*).

Fat-pads (Haversian glands) are pads of fat placed between the synovial membrane and either the fibrous capsule or the bone. Fat being very pliant, the pads can accommodate themselves to changing conditions (*fig. 28*).

Articular Discs, which are pads of fibrocartilage (or of condensed fibrous tissue) interposed between the articular surfaces of two bones, are found in certain joints where double movements take place. On the proximal surface of the disc one type of movement takes place (e.g., flexion and extension); on the distal surface another type of movement (e.g., rotation or gliding). Discs are nonvascular and nonnervous, except at their peripheral attachments. They are nourished by synovial fluid.

Sites. Discs are found in the temporomandibular, sterno-clavicular, acromio-clavicular, radio-ulnar, and knee joints.

Ligaments. The fibrous capsule is thickened in parts to form cords and bands, called *ligaments*, which withstand temporary strains. Other ligaments are independent of the capsule.

Articular Cartilage, having neither blood vessels nor lymphatics (nor has it any nerves), must receive its nourishment by diffusion. The synovial fluid nourishes the cartilage from its free surface and the epiphyseal vessels from its attached surface (Ekholm). Detached pieces of articular cartilage can live, and even grow, in synovial fluid ("joint mice").

Lubrication (*fig. 29*). The articular components of a synovial joint are incongruous; that is, they do not fit each other reciprocally, as do structures made by machine. If apposing surfaces are parallel or congruous, there can be no self-lubrication. Self-lubrication seems to be dependent upon the presence of (1) a thin, convergent, or wedge-shaped space, (2) a viscous fluid, and (3) a certain speed of movement. Given these three factors, two bearing surfaces will be completely separated by a film of lubricant (MacConaill).

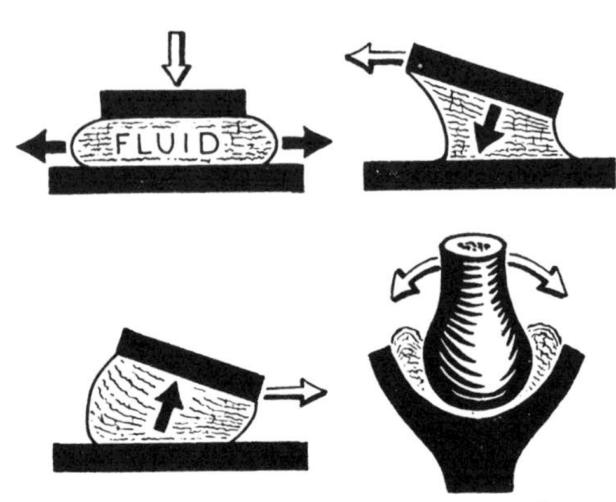

Fig. 29. Lubrication, with the use of a convergent film of viscous fluid.

Articular discs, menisci, and synovial fat-pads and folds are important aids to the formation of thin wedge-shaped films. When at rest, there is no film between weight-bearing surfaces. Indeed, Charnley holds that such a film plays no important role. He emphasizes "boundary lubrication," which depends upon entrapped hyaluronic acid in the spongy articular cartilage. (For a lucid discussion of current views, see Radin and Paul.)

Labra. The shoulder and hip joint each has its socket deepened by a pliable ring of fibrocartilage called its labrum (L., lip).

Nerve Endings are found in the fibrous capsule and synovial membrane. Those in the fibrous capsule are of a type associated with position-sense or proprioception (Ruffini corpuscles) and with pain (free endings). Those in the synovial membrane and its prolongations are believed to be pain receptors and, being associated with blood vessels, they probably supply them (E. Gardner). Articular cartilage has no nerves, nor have discs except at their attached margins.

Blood and Lymph Vessels are a feature of synovial membrane, but they are absent from articular cartilage.

Bursae. See page 24.

Classification of Synovial Joints

1. Plane. *Arthrodial or gliding joint:* The apposed bony surfaces are approximately flat, e.g., carpal joints and joints of the small tarsals.

2. Uniaxial. (*1*) *Hinge or ginglymus joint:* one surface is concave, the other convex, and movement takes place on a horizontal axis, e.g., the elbow and ankle. (*2*) *Pivot or trochoid joint:* a ring encircles a pivot set on a vertical axis, and rotation takes place as with a door on a hinge, viz., atlanto-axial and proximal radio-ulnar joints.

3. Biaxial. Circumduction is permitted; i.e., on performing the movements of flexion, abduction, extension, and adduction in sequence, a cone is described. (*1*) *Condyloid joint:* one bony surface is a ball and the other a socket, but rotation is not a conspicuous feature, e.g., the metacarpophalangeal (knuckle) joints. (*2*) *Ellipsoid joint:* one surface is an oval and the other a socket, e.g., the radio-carpal (wrist) joint.

4. Multiaxial. Movements of circumduction and of axial rotation are permitted. (*1*) *Ball and socket joint:* a ball fits into a socket and provides a universal joint; the shoulder and hip joints. (*2*) *Saddle joint:* the surfaces are reciprocally saddle-shaped; the carpo-metacarpal joint of the thumb.

MUSCLES

SKELETAL MUSCLES—*Fibers; Parts; Insertions; Synovial Bursa; Synovial Sheath.*
Internal Structure; Contraction; Investigation; Electromyography; Muscle Action.
Blood Supply; Nerves; Nomenclature; Variations and Accessory Muscles.

It is from the fancied resemblance certain muscles bear to mice, the tendons presumably being their tails, that the diminutive term "muscle" is derived (L. mus = a mouse).

There are three types of muscular tissue: (1) *skeletal* or voluntary, such as occurs in the muscles of the limbs, body wall, and face, (2) *heart* or cardiac, which is confined to the heart, and (3) *smooth*, visceral, or involuntary, such as is found in the stomach, bladder, and blood vessels.

Skeletal muscles are under the control of the will; hence, they are alternatively called voluntary muscles. Histologically their fibers possess light and dark cross-striations. Heart muscle also is striated, but neither heart muscle nor smooth muscle is under voluntary control; they are both involuntary. The accompanying table indicates that the terms striated and involuntary are comprehensive.

Appearance	Restrictive Name	Control
Striated	skeletal	or voluntary
	heart	involuntary
Nonstriated or smooth		

To contract and to relax is the function of all three types of muscle. Skeletal or voluntary muscles mostly pass from one bone across a joint (or joints) to another bone, and by contracting they approximate their sites of attachment; hence, they act upon joints. Heart muscle and smooth muscle mostly form the walls of cavities and tubes, and by contracting they expel the contents.

The distinction between voluntary and smooth muscle on the basis of their ability to be controlled by the will is not always clear. Thus, the diaphragm is structurally a voluntary muscle like the Biceps, and though it can be controlled voluntarily, as on taking a deep breath or on holding the breath, ordinarily it works automatically. Again, the upper part of the esophagus is supplied with voluntary muscle and the lower part with smooth, yet voluntary control cannot be exercised over either part.

It would appear that the distribution of voluntary and smooth muscle is determined not so much by the type of control required as by the character of the contraction required, voluntary (skeletal) muscle having the property of rapid contraction; smooth muscle, of slow sustained contraction without fatigue.

Skeletal or Voluntary Muscle

Skeletal or voluntary muscle is the subject of the remainder of this section. The red or lean of a roast of beef is voluntary muscle. Voluntary muscles form about 43 per cent of the total body weight. They are the engines or motors of the body.

Fibers

The "fibers" of a voluntary muscle are in reality thread-like cells consisting of protoplasm or sarcoplasm, multiple nuclei, and a cell membrane, the *sarcolemma* (Gk. sarx = flesh; lemma = a husk or skin; *cf.* sarcophagus, sarcoma). The fibers range from about 1 mm to 41 mm or more in length. Around each individual fiber there is some loose areolar tissue (endomysium), around a collection of fibers there is more (perimysium), and around the entire muscle still more (the muscle sheath). This areolar tissue permits swelling, gliding and, indeed, independent action of the enclosed fiber or collection of fibers. In a long muscle, such as the Sartorius, several fibers may be arranged more or less end to end.

In some animals muscle fibers are *red* or *dark*, as in the leg of a chicken; others are *white* or *pale*, as in the breast. In man, red and white fibers are thoroughly mixed in different proportions in different muscles (E. W. Walls).

Exceptions

Though skeletal muscles typically cross at least one joint and are attached at both ends to bone, certain voluntary muscles, particularly those of the face, are by one end attached to *skin*—through them we express our emotions; others, articular muscles, are by one end attached to the synovial capsules of joints—an *articular muscle*, by withdrawing the capsule, saves it from being nipped; still others form rings or *sphincters* around the entrance to the orbital cavity, mouth, and anal canal—these close the eyelids, lips, and anus, respectively. The *constrictors* of the pharynx constitute what practically is a tubular muscle; the striated muscle of the esophagus actually is tubular.

The Parts of a Muscle

The proximal attachment of a limb muscle is called the *origin*, and the distal attachment the *insertion*. The fleshy part is sometimes referred to as the *fleshy belly*. Some muscles are fleshy from end to end, but most are fibrous at one end or at both. The fibrous end has the same histological structure as ligament. When rounded it is called a *tendon*, when flattened and membranous, an *aponeurosis*, which suggests that it is nervous—but the ancients did not distinguish between nerves, ligaments, and tendinous structures.

The two chief component parts of a voluntary muscle, then, are (1) the fleshy and (2) the fibrous (tendon or aponeurosis). These have contrasting properties: *fleshy fibers* are highly specialized, contractile, vascular, expensive in upkeep, and resistant to infection, but they cannot survive pressure or friction; *tendons* are unspecialized, inelastic, nonvascular, and inexpensive in upkeep. They are designed to withstand pressure, but owing to their meager blood supply they readily die (slough) when exposed to infection. Where a muscle presses on bone, ligament, tendon, or other unyielding structure, the fleshy fibers always give place to tendon (*fig. 30*). Further, if the tendon is subjected to friction, a lubricating device—a *synovial*

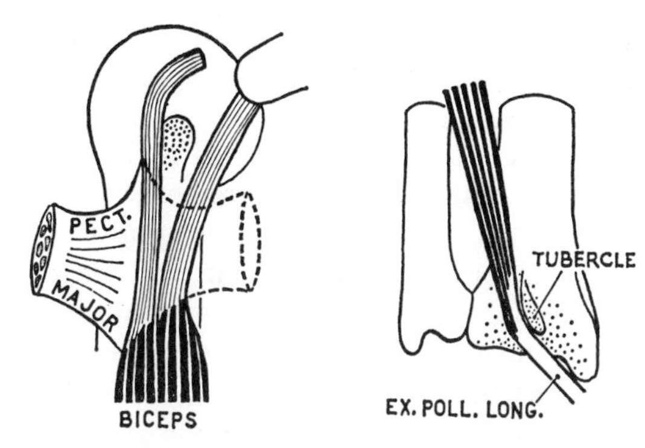

Fig. 30. Two examples demonstrating a principle: where a muscle is subjected to pressure or friction its fleshy fibers are replaced by tendon or aponeurosis.

bursa or a *synovial sheath*—is always interposed.

It is a matter of common observation that the cross-sectional area of the tendon of a muscle is much less than that of the fleshy belly (*figs. 30.1* and *61.1*). Hence, a muscle that arises by fleshy fibers and is inserted by tendon has a much more extensive origin than insertion. Hence, the precise site of attachment of a tendon, where the force of a muscle pull is concentrated and focused, is of much greater practical importance than that of a wide spread fleshy attachment. At fleshy attachments forces are dissipated and make no mark on the bone, but at tendinous attachments forces are concentrated and do. Thus, tendinous attachments create ridges, tubercles, and facets, and if large they may produce traction epiphyses.

Tendons are immensely strong; it is estimated that a tendon whose cross-sectional area is 6 cm^2 is capable of supporting a weight of from 4,000 to 8,000 kilograms (Cronkite).

The fibers of a tendon are not strictly parallel, but plaited; they twine about each other in such a manner that fibers from any given point at the fleshy end of the tendon are represented at all points at the insertional end (*fig. 31 A*); hence, the pull of the whole muscle can be transmitted to any part of the insertion. The fan-shaped manner in which most tendons are inserted into bone ensures that successive parts of the insertion shall take the full pull of the muscle as the angle of the joint changes (*fig. 31, B* and *C*).

The fibers of tendons, ligaments, and other fibrous structures commonly pass through a pad of fibrocartilage before plunging into their bony attachment. Like the rubber pad employed by the electrician at the junction of the free and the fixed point of a wire, the pads help to prevent fraying from frequent flexing.

Insertions. Muscles are usually inserted near the proximal end of a bone (or lever), close around a joint, close to an axis of movement (e.g., Biceps, Brachialis, and Triceps at the elbow). So, they help to retain in apposition the ends of the bones taking part in the joint, and thereby give it strength. By being thus inserted they produce, on contracting, rapid movement of the distal end. (MacConaill and Basmajian.)

Some muscles are inserted near the middle of the shaft of a bone (e.g., Deltoid,

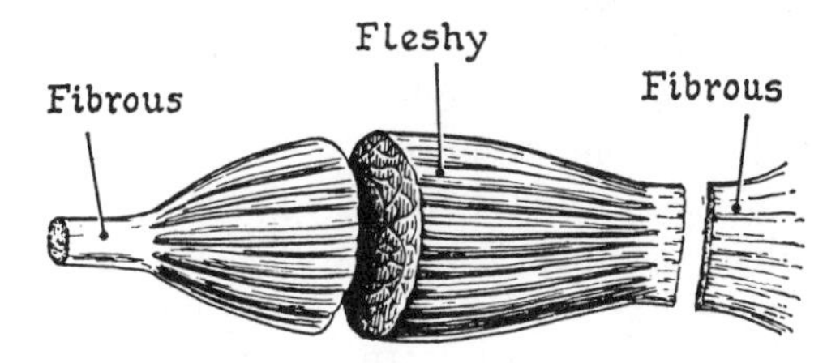

Fig. 30.1. Cross sections of tendons and fleshy belly compared.

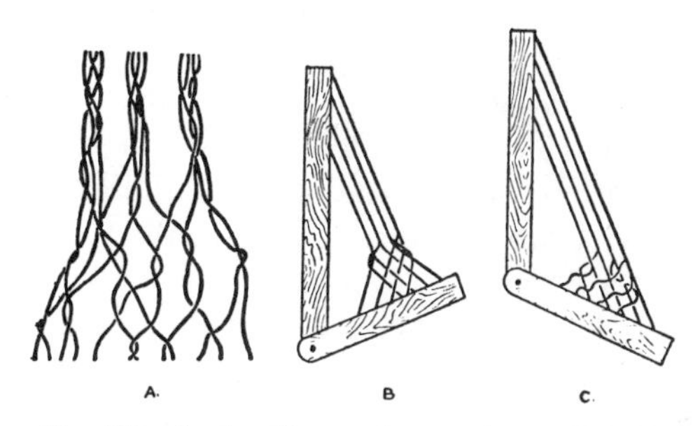

Fig. 31. *A*, the fibers of a tendon are plaited. *B* and *C*, in different positions of a joint different fibers take the strain. (After Mollier).

Coracobrachialis, Pronator Teres, Adductors of the thigh) and a few are inserted near the distal end (e.g., Brachioradialis; part of Adductor Magnus).

A Synovial Bursa (L. bursa = a purse) is a closed sac differentiated out of areolar tissue. It is roughly the size and shape of a coin. Its delicate walls separated from each other merely by a film of slippery fluid, like white of egg. As a lubricating device, diminishing friction and allowing free movement, a bursa is more effective than areolar tissue.

Bursae may be classified thus—subtendinous, articular, and subcutaneous.

Subtendinous Bursae are found wherever tendons rub against resistant structures, such as bone, cartilage, ligament, or other tendons; hence, they are commonest in the limbs (*fig. 122*). Certain subtendinous bursae constantly communicate with synovial cavities, e.g., the Biceps and Subscapularis bursae at the shoulder.

"Articular" Bursae play the part of accessory joint cavities where they separate neighboring capsules, e.g., the subacromial bursa (*fig. 183*).

Subcutaneous Bursae are present (1) at the convex surface of joints which undergo acute flexion, because here the skin requires to move very freely, e.g., behind the elbow (olecranon b.), in front of the knee (prepatellar b.); (2) over bony and ligamentous points subjected to considerable pressure and friction. Most of these, being acquired or occupational, are inconstant, e.g., the one lying superficial to the ischial tuberosity.

A Synovial Sheath is a tubular bursa that envelops a tendon. In fact, it is two tubes, one within the other. The inner or *visceral tube* adheres closely to the tendon and is separated from the outer or *parietal tube* by the synovial cavity. The visceral and parietal tubes are united longitudinally, along the surface least subjected to pressure, by a synovial fold, the *mesotendon*, which transmits vessels to the tendon (*fig. 32*). If the range of movement of the tendon is considerable, the mesotendon may disappear or be represented by a thread or *vincula* (e.g., long digital flexors).

A synovial sheath is required only where a tendon is subjected to friction or pressure on two or more surfaces (front and back). This condition obtains only at the hand, foot, and shoulder. In all instances it so happens that the friction results from the presence of bone on one surface and of a retinacular ligament on the other. In order to allow ample play such sheaths extend about 1 cm above and below the sites of friction.

Architecture of Internal Structure

The fleshy fibers of a muscle may be disposed either (1) parallel to the long axis of the muscle, (2) obliquely, like the barbs of a feather, or (3) radially like a fan.

Functional differences. The parallel type has long fibers (or chains of fibers) but relatively few of them, therefore it can lift a light weight through a long distance. The oblique type has short fibers, but they are very numerous; therefore, it can lift a heavy weight through a short distance.

1. *Fibers Parallel to the Long Axis of the Muscle* or approximately so (*fig. 33*). The fleshy fibers may be *parallel* from end to end, perhaps having a short tendon or aponeurosis at one end or at both ends. This includes many strap-like and flat muscles (e.g., Sartorius). The *fusiform type* narrows to a tendon at both ends (e.g., Biceps Brachii, *fig. 30.1*).

2. *Fibers Oblique to the Long Axis of the Muscle* (*fig. 34*). From their resemblance to

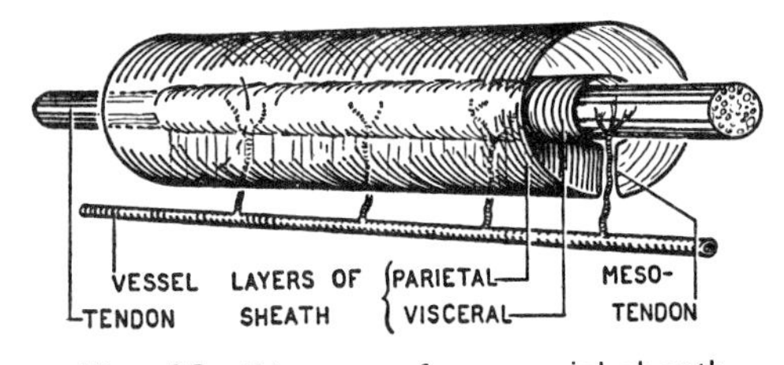

Fig. 32. Diagram of a synovial sheath.

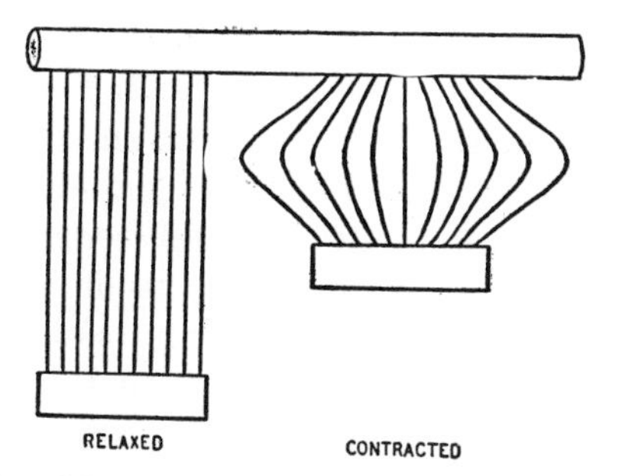

Fig. 33. On contracting, fleshy fibers shorten by one-third to one-half of their resting length and swell correspondingly.

feathers these mucles are called *pennate*, the fleshy fibers corresponding to the bars of the feather and the tendon to the shaft, for they are all inserted by tendon. They are (1) *unipennate* when the fleshy fibers have a linear or narrow origin and have the appearance of one-half of a feather; (2) *bipennate* when the arrangement of the fibers is that of a whole feather; (3) *multipennate* when septa (partitions) extend into the origin and the insertion (e.g., Deltoid, *fig. 84*), the appearance being that of many feathers; and finally, (4) muscles (e.g., Tibialis Anterior) whose fibers converge from the walls of a cylindrical space to a buried central tendon may be spoken of as "*circumpennate*".

3. In a Radial, Triangular or Fan-shaped Muscle (*fig. 34*) the fleshy fibers converge from a wide origin or base to an apex, which is necessarily fibrous. And, it creates a rough mark, line, ridge, or process on the bone.

Contraction

When muscle fibers contract or shorten, they necessarily increase in circumference (*fig. 33*)—they swell, as exemplified in gross form by the Biceps Brachii in changing from its relaxed to its contracted state.

On contracting, the fleshy fibers of a muscle shorten between a third and a half (57 per cent, Haines) of their resting length. Being, so to speak, expensive in upkeep they are never longer than necessary—those of a muscle composed of parallel fibers being 2 or 3 times the length of the distance through which the site of insertion can move. If the distance between origin and insertion is greater than the length of fiber required for full action, the surplus length is fibrous, i.e., tendinous or aponeurotic (e.g., Sartorius and Rectus Femoris, in the lower limb). A tendon can be any length.

As a corollary, knowing the range of movement of the bony point into which a muscle is inserted, the length of the fleshy fibers (or chains of fibers) of that muscle can usually be calculated.

Investigation

In investigating the action of muscles five methods are available.

1. In the cadaver a muscle may be freed

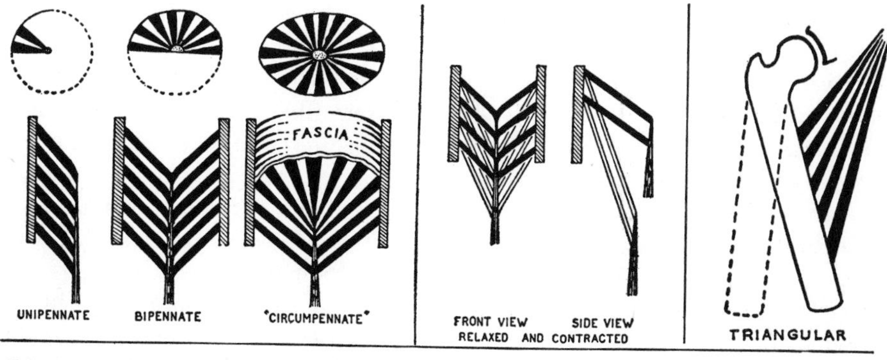

Fig. 34. Diagrams showing the architecture or internal structure of pennate and triangular muscles. Pennate muscles, obviously, are powerful muscles. (After Pfuhl).

from surrounding structures and pulled upon; this indicates the general course and action of the muscle in life.

2. In the living subject a muscle may be stimulated to contract by the suitable application of electricity. This was Duchenne's method of approach, and his work forms the basis of our knowledge of the actions of muscles (Kaplan).

3. Inspection and palpation of the muscles of the region may determine which are in action. Precautions must be taken to avoid confusing synergists with prime movers (p. 27).

4. Clinical information may be gained from the study of the effects (1) of nerve injuries—and these are plentiful during each war—and (2) of transplanting tendons surgically in cases of paralysis. These fall into the category of "natural" experiments.

5. *Electromyography* would seem to be the ultimate method. It consists in inserting electrodes in the muscle of a living human subject, and having him perform a motion. The differences in the electrical action potentials of the muscle are amplified and recorded mechanically. There is a direct relationship between the tension developed in a muscle and the electrical activity; so, by this procedure one can analyze the functions of an individual muscle during motion, noting, for example, during what stage of a movement it comes into action and during what stage it exhibits its greatest activity. Further, by using a number of amplifiers the simultaneous actions of a group of muscles can be studied.

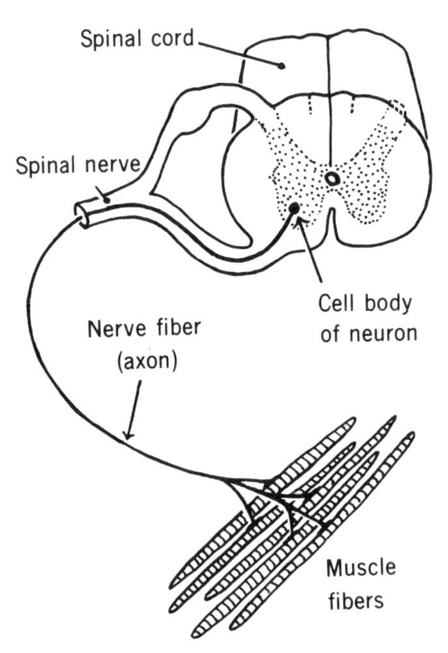

Fig. 34.1 Diagram of a motor unit. (From *Primary Anatomy*).

Methods 1 and 2 give information as to what an individual muscle acting alone might do, and have less practical importance than methods 3 and 4. Method 5 is throwing much new light on the subject. (For more information, see "Muscles Alive," 3rd Ed., Basmajian.)

Muscle Action

The structural unit of a muscle is a muscle fiber. The functional unit, known as a *motor unit*, consists of a nerve cell, situated in the anterior horn of the spinal cord, and all the muscle fibers, usually 100 or more, controlled by the nerve fiber of that cell (*fig. 34.1*).

When an impulse is carried by the nerve fiber to its muscle fibers, they all contract almost simultaneously and for a total time that is quite brief (measured in milliseconds). The result is a twitch of a tiny volume of the whole muscle.

Where great precision is required, as with the extrinsic muscles of the eyeball, one nerve fiber controls only about a dozen muscle fibers. Only a small proportion of motor units are in action at a given movement—at least they are in different phases of activity, because impulses are discharged by the central nervous system asynchronously.

Movements are produced by throwing an increasing number of motor units into action and at the same time relaxing the antagonistic muscles (reflex inhibition).

Most bodily actions, even ordinary ones, call into play few principals and more assistants. The principal muscles, *prime movers* or agonists, by actively contracting (shortening) produce the desired movement. Muscles that are so situated that they would usually produce movements in the opposite direction are called *antagonists.*

Again, when a prime mover passes over more than one joint, certain muscles are called upon to steady the unstable joints; such muscles are *synergists* (Gk. syn = together; ergon = work). When antagonists contract during a movement, their role is synergistic. Still other muscles, *fixators*, are called upon to steady the more proximal parts of the limb or trunk. Obviously, the same muscle may act as prime mover, antagonist, synergist, or fixator under different circumstances as the nervous system dictates.

Examples. (1) The long flexors of the fingers can also flex the wrist joint. So, on clenching the fist or on closing the hand, say on a broom handle, the tendons of the wrist extensors can be felt to contract. They act as synergists.

Gravity is a valuable aid to some movements, depending upon the position of the limb or of the body. *Example:* On raising the arm from the side, the Deltoid is the prime mover; gravity (certainly if there is a weight in the hand) is sufficient to lower the arm. If resistance is encountered, the Pectoral muscles and the Latissimus Dorsi come into play as prime movers. Similarly, in walking the raised limb tends of its own weight to fall to the ground. (MacConaill and Basmajian.)

Spurt and Shunt Muscles. While acceleration of a movement is an obvious function of most muscles for moving joints (spurt muscles), in other situations muscles act by pulling along the long axis of the bones and so impart no torque and so no movement (shunt muscles); nevertheless, they are important in maintaining the integrity of the joint and ensuring the "desired" action of the spurt muscles. Special architectural arrangements and accessory mechanisms (e.g., fibrous tunnels for the flexor tendons in the fingers) permit the shunting function of muscles across certain joints while simultaneously they may be providing spurt impetus across others (see MacConaill and Basmajian for details).

Blood Supply

The chief vessels enter a muscle with the nerve constantly in some muscles (Biceps Brachii and Gastrocnemius), and with varying frequencies in others; others again have multiple entries (Fl. Digit. Profundus). Accessory vessels, unaccompanied by nerves are present in many muscles, forming good anastomoses in some (Pectoralis Major) and poor in others (Sartorius and Hamstrings) (J. C. Brash). The vessels in a muscle anastomose to form a rectangular network of large and small interconnecting branches. This assures every muscle cell of adequate nutrition. *Anastomosis* (Gk. stoma = a mouth) means the junction or intercommunication of hollow tubes—usually blood vessels—by means of open mouths.

The veins in the muscles, like those in the limbs in general, have valves; so, muscular exercise, by massaging these veins, aids in circulating the blood.

Lymph Vessels run with the blood vessels.

Nerve Supply

The nerve to a muscle, called a motor nerve, contains fibers from two or more consecutive spinal cord segments (e.g., the nerve to Deltoid arises from cervical segments 5 and 6). Every motor nerve is a mixed nerve, being about three-fifths efferent (motor), two-fifths afferent (sensory) and containing sympathetic fibers (*fig. 34.2*). (1) The efferent fibers pass to *motor end-plates* (i.e., areas of sarcoplasm under

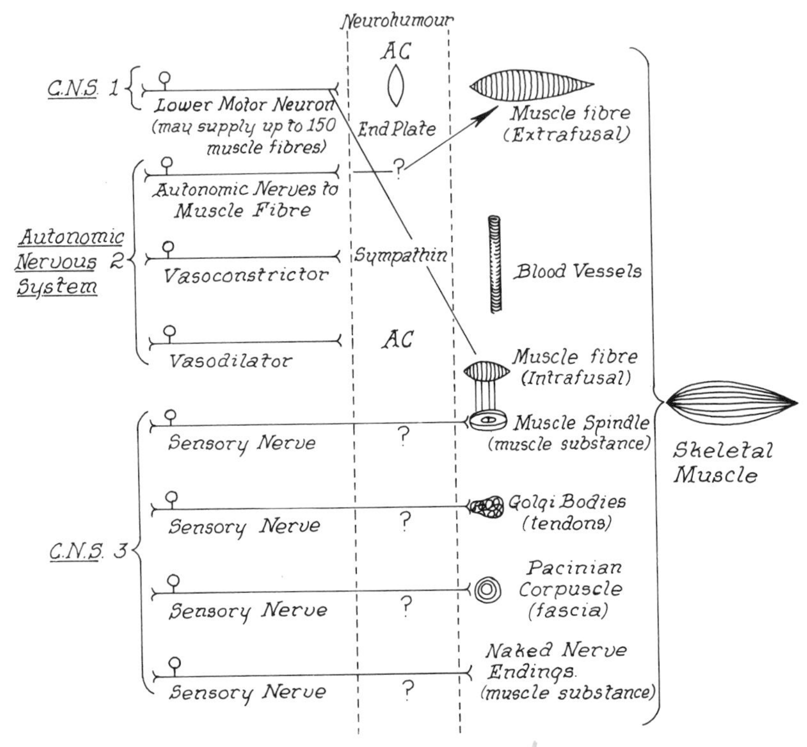

Fig. 34.2. Many different types of nerve fibers are found in a 'motor' nerve. (From *Muscles Alive: their Functions Revealed by Electromyography*).

the sarcolemma in which a nerve fibril branches like an open hand). (2) The afferent fibers begin as: *free endings* on the muscle fibers, *encapsulated endings* in the connective tissue, *muscle-spindles* (a spindle being a fusiform swelling, 1 to 4 mm long, containing poorly developed muscle fibers over which an afferent nerve fiber spreads or around which it forms a spiral) and, *tendon-spindles* at musculo-tendinous junctions. (They are similar to muscle-spindles, but collagen fibers replace muscle fibers (Bridgman).)

It is through the sensory nerves in muscles, tendons, and joints that one is kept informed of the position of the parts of one's body in space; so, even with the eyes shut, one can walk without stumbling, and know whether a joint is extended or, if flexed, the degree. (3) The sympathetic fibers supply the vessels.

Muscles placed near the surface of the body may be made to contract on applying an electrode to the skin near, but not at, the point of entry of the nerve. The most effective point of electrical stimulation for each muscle (which overlies the greatest concentration of nerve endings on muscle fibers) is called the *motor point* of that muscle.

Nomenclature

The names given to muscles are descriptive of their *shape* (e.g., Triangularis, Rhomboidei); of their *general form* (e.g., Longus, Serratus (like a saw), Vasti); of the number of heads or *bellies* (e.g., Biceps, Triceps, Quadriceps, Digastric), of their *structure* (e.g., Semitendinosus, Semimembranosus); of their *location* (e.g., Supraspinatus, Tibialis Anterior); of their *attachments* (e.g., Brachioradialis); of their *action* (e.g., Flexor and Extensor Carpi Ulnaris, Abductor and Adductor

Hallucis, Tensor Palati); of their *direction* (e.g., Transversus Linguae); of *contrasting features* (e.g., Pectoralis Major and Minor, Gluteus Maximus, Medius, and Minimus).

Variations and Accessory Muscles

Muscles are subject to variation, and books have been written about their variations (Le Double). The muscles that vary most are muscles that are either "coming" or "going", that is, muscles that are appearing in the species and muscles that are disappearing in evolution. Few have any importance. The following paragraphs are offered for interest, not for memorizing.

The *Palmaris Longus* and *Plantaris* are disappearing (the one is absent in 13.7 per cent of 771 limbs, the other in 6.6 per cent of 740 limbs) and when present the sites of origin and insertion and the size of the fleshy bellies vary greatly. The *Peroneus Tertius* is appearing (it is absent in 6.25 per cent of 400 limbs) and when present it is inserted anywhere along the dorsum of metatarsals 3, 4, or 5 or to the interosseous membrane between them. The *Peroneus Longus* has migrated across the sole of the foot to be attached to metatarsal 1. In 72.3 per cent of 213 limbs it is attached also to cuneiform 1, and in 48.2 per cent to metatarsal 2 also. The short muscles to the little finger and little toe are commonly in part suppressed (J. C. B. G.).

The wide spread fleshy origin of a muscle may spread either a little more or a little less without affecting the action of the muscle (e.g., the Pectoralis Minor, typically arising from ribs 3, 4, and 5, may extend upward to the 2nd rib or downward to the 6th; on the other hand, the lower portions of the Trapezius and Pectoralis Major are not uncommonly wanting).

Accessory Muscles mostly lend themselves to explanation on morphological grounds; they are not clinically important (Sternalis and the Axillary Arch). Sometimes the Biceps Brachii has three origins and the Coracobrachialis three insertions, as in some primates, and occasionally they have four.

2 CARDIOVASCULAR AND NERVOUS SYSTEMS

CARDIOVASCULAR SYSTEM

Blood; Definitions; Circulation of Blood; Structure of Cardiac Wall; Blood Vessels; Venous Valves; Structure of Vessels; Distribution of Blood and Its Nervous Regulation; Anastomoses and Variations; Arterio-venous Anastomoses; Sinusoids; Cavernous Tissue; Vascularity.

LYMPHATIC SYSTEM—*Capillaries; Vessels; Nodes; Main Channels; Lymphatico-venous Communications; Investigation; Exceptions; Lymphoid Tissues.*

FETAL CIRCULATION—*Changes at Birth; X-ray Findings.*

NERVOUS SYSTEM

SUBDIVISIONS.

SPINAL MEDULLA OR CORD—*Membranes; Structure; Neuron.*

PERIPHERAL NERVES—*Macroscopic or Gross Structure; Microscopic Structure; Plexuses; Blood Supply.*

AUTONOMIC NERVOUS SYSTEM—*Sympathetic System: Structure; Cerebral Control; Supply of Individual Regions and Organs (upper limb, lower limb, head and neck, thorax, abdomen and pelvic).*

PARASYMPATHETIC SYSTEM—*Oculomotor Nerve; Facial Nerve; Glossopharyngeal Nerve; Vagus Nerve; Pelvic Splanchnic Nerves.*

ENDOCRINE GLANDS

CARDIOVASCULAR SYSTEM

Heart and Blood Vessels

The blood vascular system comprises (1) the heart, and (2) the blood vessels (arteries—capillaries—veins). These form a tubular system which, with few exceptions, is closed, lined throughout with a single layer of flat cells, *endothelium*, and filled with blood.

Blood. The volume of blood in the body is about 5 liters, and it equals about 7 per cent of the total body weight. It is composed of plasma and cells. The *plasma* or fluid portion of the blood = 91 per cent water and 9 per cent solids (e.g., proteins, salts, products of digestion, and waste products), also respiratory gases (O_2 and CO_2), internal secretions, enzymes, etc. The *cells* = red corpuscles (erythrocytes), white corpuscles (leukocytes), and blood platelets; these are in suspension.

The fluid products of digestion are absorbed from the digestive tract into the bloodstream for distribution to the various tissues and organs. Oxygen from the air passes into the lungs and thence into the bloodstream, the plasma dissolving a small amount, the red cells combining with 60 times as much. This also is distributed to the tissues. The waste products of metabolism, formed in the tissues, enter the bloodstream and are brought by it to the excretory organs (kidneys, bowel, lungs, and skin). It is through the thin walls of the capillaries, which consist only of endothelium, that these various interchanges take place. Their walls are semipermeable membranes—permeable to water and crystalloids, but impermeable to the proteins of the blood plasma and to other substances of large molecular composition. Nutritive materials and O_2 pass from the blood to the tissues; and waste products and CO_2 return from the tissues to the blood.

Arteries are tubes that conduct blood from the heart to the capillaries. *Veins* are tubes that conduct it back from the capillaries to the heart.

Circulation of Blood. The blood traverses two separate circuits (*fig. 35*), the *pulmonary* and the *systemic*, each with its own pump.

1. The right side of the heart pumps the blood through the vessels of the lungs—these constitute the pulmonary or lesser circuit.

2. The left side of the heart pumps the blood through the vessels of the body generally—these constitute the systemic or greater circuit.

To serve these two circuits, the heart has four chambers, two on the right side and two on the left. One of these on each side is a thin-walled receiving chamber, the *atrium*, the other is a thick-walled distributing or pumping chamber, the *ventricle*.

The Pulmonary Circuit. The blood entering the right atrium passes to the right ventricle which pumps it through the pulmonary arteries to the capillaries of the lungs, thence through the pulmonary veins to the left atrium.

The Systemic Circuit. From the left atrium the blood passes to the left ventricle which pumps it through the aorta and its various arterial branches to the capillaries of all the rest of the body, thence through veins which become increasingly larger and fewer till finally the superior and inferior venae cavae and the cardiac veins return to the right atrium the same volume of blood as the left ventricle ejects into the aorta, and at the same rate (*fig. 36*).

Normally blood is red when the oxygen content is high and bluish when it is low. Therefore, the blood is *red* in the pulmonary veins, left side of the heart, and systemic arteries; and "*blue*" in the systemic veins, right side of the heart, and pulmonary arteries. It is customary in illustrations to color the parts red and blue accordingly.

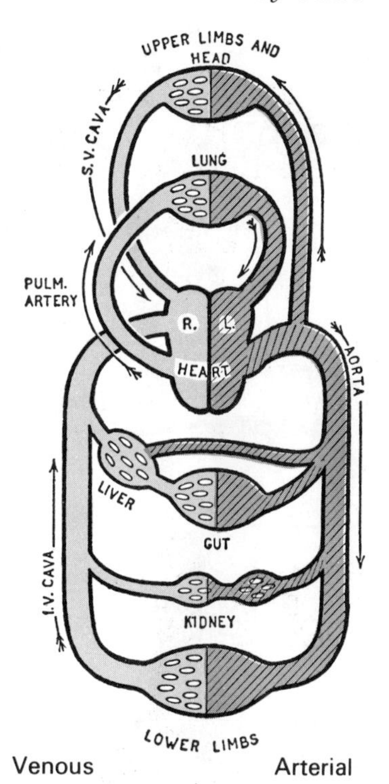

Fig. 35. Diagram of the circulatory system. Hatched vessels conduct oxygenated blood; clear vessels conduct blood laden with carbon dioxide. Note the pulmonary and systemic circuits and the double set of capillaries on the subsidiary digestive and renal circuits.

Though the capacity of each of the four heart chambers is practically the same (60–70 cc), the resistance offered by the pulmonary vessels is obviously much less than that offered by the systemic ones; accordingly, the wall of the right ventricle is much thinner than that of the left.

Structure of Cardiac Wall. (L. Cor, and Gk. Kardia = Heart.) The main layer of the cardiac wall is the middle or muscular layer, the *myocardium*. Internally, the myocardium is lined throughout with *endocardium*. This is a thin fibrous membrane, lined with flat endothelial cells continuous with those of the blood vessels. Externally the heart is coated with *epicardium* (visceral pericardium), which is similar to endocardium. Deep to the epicardium run the main cardiac vessels, often embedded in much fat, which devel-

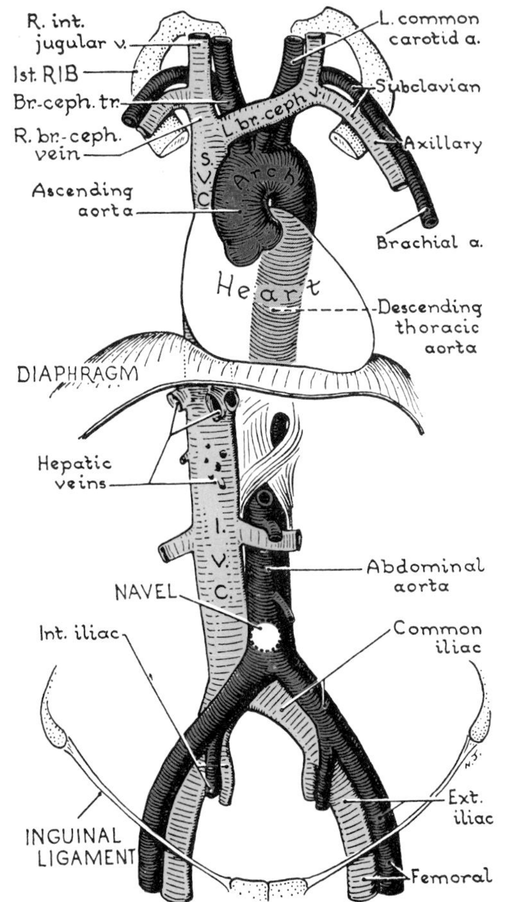

Fig. 36. The great arteries and veins of the systemic circuit.

ops in its fibrous-tissue layer with increasing age and corpulence.

For **heart valves**, see Chapter 32.

Blood Vessels. The main systemic artery (the *aorta*) and the *pulmonary trunk*, each about 30 mm in diameter, branch and rebranch like a tree, the branches becoming smaller as they become progressively more numerous. When reduced to a diameter of about 0.3 mm and just visible to the naked eye, they are called *arterioles*. These break up into a number of *capillaries* (L. capillus = a hair), each about ½–1 mm long and large enough (7 μ or 0.007 mm) to allow the passage of the red blood cells in single file.

The capillaries form an anastomosing network or *rete*. The cross-sectional area of the entire systemic rete or capillary bed is about 800 times greater than that of the aorta. In this rete the smallest veins, called *venules*, have their source.

Veins accompany arteries and have the same tree-like pattern, the branches being called *tributaries*.

Below the elbow and knee, and elsewhere, arteries are closely accompanied by paired veins, *venae comitantes*, one on each side. These veins are united to each other by short branches which form a network around their artery (*fig. 36.1*). Veins also run independently of arteries (e.g., superficial veins); veins are more numerous than arteries and their caliber is greater.

Venous Valves. The inner tunic of most medium and small veins is thrown at intervals into delicate folds, called *valvules* or *cusps*. Each cusp forms with the wall of the vein a semilunar bulging pocket or *sinus*. These pockets are mostly arranged in pairs, facing each other, to form valves. Like the gates of a canal lock, they open in one direction, namely, toward the heart (*fig. 36.2*).

Fig. 36.1. Venae Comitantes cling to their artery. (From *Primary Anatomy*.)

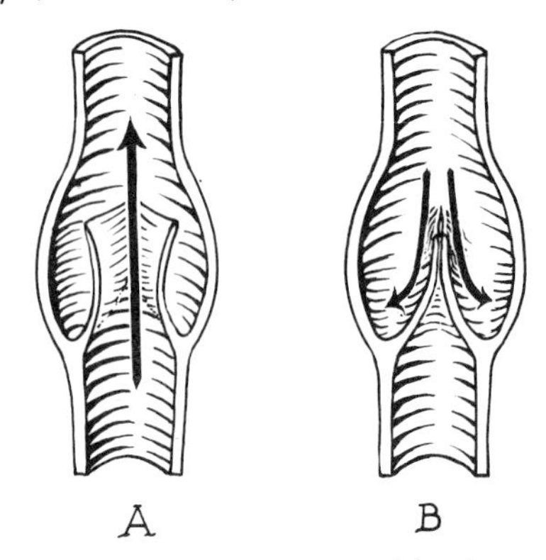

Fig. 36.2. A venous bicuspid valve open (*A*) and closed by backpressure (*B*). (From *Primary Anatomy*.)

The nearest valves to the heart lie at the ends of the internal jugular, subclavian, and femoral veins (*fig. 37*). They imprison the venous blood within the trunk, and during periods of increased intra-abdominal pressure (e.g., during defecation) and during increased intrathoracic pressure (e.g., during expiration), they prevent it from being forced back into the limbs, head, and neck.

There are no functioning valves in the *Portal System*, that is, in the veins that bring blood from the stomach and intestines (guts) to the liver; and only rarely do those (two or three) in the cardiac veins function. Valves are most numerous in the veins of the limbs, and they are commonly placed just distal to the mouth of a tributary.

To Demonstrate Valves in the superficial veins of the forearm: Circumduct your limb vigorously at the shoulder joint in order to cause the veins to fill; then, keeping the forearm below the level of the heart so that the veins shall not empty, with the tip of your index obstruct a prominent vein about the middle of the forearm and, by stroking proximally with your thumbnail, empty a long segment of the vein. Note that on removing the thumb the vein fills from above as far as the nearest valve, and that on removing the index the empty segment fills from below (*fig. 38*).

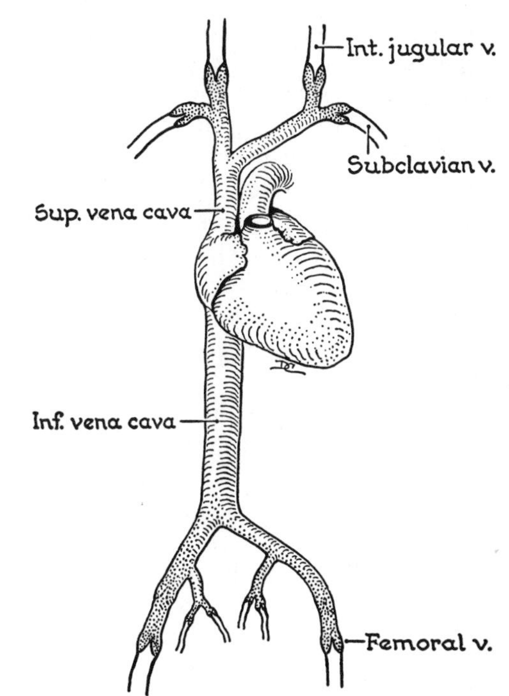

Fig. 37. The venous blood is imprisoned in the trunk by valves.

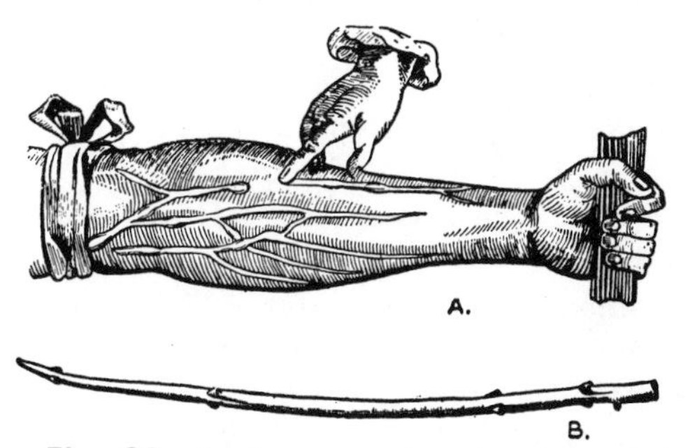

Fig. 38. *A*, demonstrating the sites of the valves in the veins of the forearm during life (after William Harvey—*The motion of the heart and blood in animals*, 1628). *B*, a vein turned inside out imperfectly showing the valves (after Fabricius, 1603).

Structure. Arteries typically have three coats or tunics. (1) The *tunica intima* or inner coat has a lining of endothelial cells, with a little subendothelial fibrous tissue and, outside this, a tube of elastic tissue (the internal elastic membrane); (2) the *tunica media* or middle coat is composed of alternate layers of smooth muscle and elastic tissue in a fibrous bedding; (3) the *tunica adventitia* or outer coat is a fibrous tube of considerable strength containing some elastic fibers.

Both muscular tissue and elastic tissue are elastic, like rubber, and after being stretched they tend to return to normal length, i.e., they contract passively. Muscles can also contract actively.

The blood ejected at each heart beat into the aorta causes it to expand. As the expanded aorta returns to normal, it continues to maintain sufficient pressure to drive the blood through the arteries and arterioles into the capillary bed. The amount escaping into the capillary bed in a given region, however, is regulated by the need of the tissue at the time. In other words, the blood is apportioned to the organs in accordance with their requirements. The mechanism controlling the distribution is the muscle in the walls of the arterioles which, so to speak, can be turned on and off like a tap. The aorta and the large arteries have much elastic tissue and relatively little muscle in their walls; whereas the arterioles are essentially muscular.

Capillaries have but one coat, namely, the endothelial lining common to all vessels and the heart.

Veins have the same general structure as their companion arteries. Their caliber, however, is greater; the blood pressure within them is lower; they have less muscle and much less elastic tissue. A vein may be described as a fibrous-tissue tube in which a muscle coat is embedded, and which possesses an endothelial lining.

Some hours after death all the muscular tissue in the body contracts; hence, the corpse becomes rigid, in *rigor mortis*, and the blood is driven from the more muscular arteries into the less muscular veins. Even after the rigor passes off and the vessels relax, the arteries are empty and the veins are filled with blood. Arteries (Gk. and L. arteria = an air tube) derive their name from the fact that they were believed to be conductors of air like the trachea, which was known as the arteria aspera or rough air tube.

Vasa Vasorum. The heart wall is supplied by the coronary arteries and the cardiac veins. Arteries and veins with a diameter greater than 1 mm are supplied by small vessels, *vasa vasorum* (= vessels of vessels).

Distribution of Blood and Its Nervous Regulation. The volume of blood pumped into the aorta in a given time equals that returning to the right atrium, which in turn equals that pumped into the pulmonary artery and returning to the left atrium. If it were otherwise, there would be temporary or permanent local congestion, which, in fact, occurs with heart failure. The blood pressure within the aorta is about 120 mm of mercury, at the arterial end of the capillaries about 30 mm, at the venous end 12 mm, and in the great veins 5 mm. This determines the direction of the blood flow.

The rate of flow in the aorta, when the subject is resting, is about 0.5 meter per second, in the capillaries about 0.5 mm per second, and in the veins it gradually increases until in the venae cavae it nearly equals the rate in the aorta.

The Heart is supplied by efferent vagal and sympathetic fibers. Stimulation of the vagus results in slowing of the heart; stimulation of the sympathetic in acceleration. When the blood pressure in the arteries rises, afferent fibers of the vagus distributed to the aortic arch (aortic nerve) and of the glossopharyngeal nerve distributed to a swelling on the internal carotid artery, called the *carotid sinus* (sinus nerve), are stimulated; this reflexly brings about slowing of the heart and vasodilatation of the vessels with consequent fall in blood pressure.

The Blood Vessels, particularly the arterioles, are supplied by efferent nerve fibers of the sympathetic nervous system—which almost always cause vasoconstriction—and perhaps in some locations by parasympathetic fibers having an opposite effect. Various degrees of vasoconstriction allow the blood to be partly shut off from one part of the body and diverted to another, e.g., during digestion it is diverted to the abdominal viscera; during hot weather it is diverted to the skin so that heat may be lost.

Anastomoses and Variations. During development, networks of vessels sprout into actively growing parts (e.g., organs and limbs) and, as the parts enlarge, the networks advance farther into them. Certain channels through these networks are chosen to be permanent arteries and veins and their branches; the others disappear (*fig. 446*). There being a wide choice of channels in the network, it is not surprising that those selected for permanency should vary somewhat from individual to individual. By way of demonstrating this fact, pull up your sleeve and compare the pattern of the superficial veins on the front of your forearm with that of your neighbor; minor if not major differences will be seen; even between your right and left limbs the patterns differ.

Although the number of channels in the network retained to form main arteries is restricted, commonly to one or two, the peripheral parts persist as capillary channels and the communications between these and neighboring capillaries are called anastomoses (i.e., furnished with stomata or mouths). Anastomoses also occur between certain large vessels (e.g., coronary arteries of the heart, arterial circle at the base of the brain) and between many small vessels and precapillary vessels. In cases of obstruction of the larger arteries, it is by the enlargement of these anastomoses that a collateral circulation is established and the vitality of distant (or neighboring) parts preserved.

An arterial channel, which under ordinary circumstances disappears, may persist as an *accessory* or *supernumerary* artery; if the normal artery disappears the persisting artery will act as a substitute for it. The same remarks apply to veins.

Arteries are *sinuous* or *tortuous* when supplying parts that are highly mobile like the cheeks and lips, protrudable like the tongue, or expansile like the uterus and colon.

Arterio-venous Anastomoses. In some regions arterioles communicate directly with venules, e.g., in the skin of the palm of the hand and of the terminal phalanges, in the nail bed, in the skin of the lips, nose, and eyelids and at the tip of the tongue. This is in addition to the regular communications by capillaries.

End-arteries are arteries that do not anastomose with neighboring arteries except through terminal capillaries. Obstruction of such an artery is likely to lead to local death, resulting in the case (1) of a cerebral artery, in paralysis, (2) of the central artery to the retina, in blindness, (3) of a branch of the renal or splenic artery, in death of a segment of the kidney or spleen, and (4) of several adjacent end arteries of the gut, in gangrene of the gut.

Sinusoids. In parts of certain organs (e.g., liver, spleen, bone marrow, suprarenal glands) the place of capillaries is taken by irregularly wide, tubular spaces, called *sinusoids.* The lining cells differ somewhat from endothelial cells, many of them being phagocytic, i.e., capable of engulfing particulate matter.

Cavernous Tissue. In the erectile tissue of the external genitals (penis in the male, clitoris in the female), there are innumerable venous spaces lined with endothelium and separated by partitions containing smooth muscle. In the mucous membrane of the nasal cavities, arterioles open into wide and abundant venous spaces. Therefore, the mucous membrane of the nasal passages also is erectile. When one has a "cold in the head," the venous spaces dilate and may obstruct the air way.

Vascularity. Cellular tissues are vascular, e.g., muscles being the cellular engines of the body require much fuel; glands (kidney, thyroid, suprarenal, and liver) are very vascular; so, obviously, are the lungs. *Connective tissues* are only slightly vascular; thus, loose and dense fibrous tissues (e.g., deep fascia, tendons, ligaments) have a very meager blood supply; fat and bone have a fair supply, but hyaline cartilage, the cornea, and the epidermis are nonvascular.

Nervous Tissue. The gray matter of the brain and spinal medulla, being cellular, is more vascular than the white matter and the peripheral nerves.

Lymphatic System

Tissue fluid is the fluid that bathes the cells (and fibers) of the tissues of the body, and it resembles blood plasma in chemical composition. From this fluid the cells get their nutritive material; to it they give their waste products; and through it they respire. Between the tissue fluid and the plasma of the circulating blood a constant interchange of fluid and dissolved substances takes place. However, some fluid transudes into the *lymph capillaries* whence it is drained by *lymph vessels* through *lymph nodes* to the great veins at the root of the neck where it rejoins the blood stream.

The fluid in the lymphatics is clear and colorless and is called *lymph* (L. lympha = pure, clear water); lymphocytes are added to it as it passes through the lymph nodes, otherwise it is free from blood cells.

Lymph Capillaries occur only where there are blood capillaries, and like blood capillaries they form a closed network, but with a larger mesh. They are especially numerous in the skin and in mucous membranes.

The Lymph Vessels draining the networks are thin-walled, ½ to 1 mm in diameter, and beaded due to numerous valves. They run in layers of areolar tissue. Though more plentiful than veins, they tend to accompany veins and to drain corresponding territories. The propensity for spread of cancer cells along lymphatics is well known.

Lymph Nodes (Glands) interrupt the flow of lymph. They vary *in size* from a pin's head to an olive, and are somewhat flattened. *In color*, they are pink in life and brownish in the embalmed cadaver; but those draining the lungs are black from inhaled carbon, and after a meal those draining the intestines are white from emulsified fat. The nodes act as filters for lymph and factories for lymphocytes. Both bacteria and cancer cells may be stopped and destroyed by phagocytic cells. Otherwise continued spread leads to more serious results.

The flow of lymph in an immobile limb is almost negligible, but during muscular activity it is very active due probably to (1) rhythmical contraction of the vessels, (2) intermittent pressure (e.g., muscular action) on the valved vessels, (3) negative pressure or suction within the thorax, and (4) positive pressure within the abdomen during inspiration. Swelling of the feet in anyone who has sat immobile for hours (as in a long airplane trip) attests to the importance of activity to overcome the effects of gravity.

Main Channels. Three paired lymph trunks (1) the *jugular* accompanying the internal jugular vein in the neck, (2) the *subclavian* accompanying the subclavian vein in the upper limb, and (3) the *bronchomediastinal* from thoracic viscera, end either separately or together near the angle of confluence of the internal jugular and subclavian veins. A fourth channel, the *thoracic duct* (*fig. 39*), drains the chest wall and the territory below the diaphragm (the liver in part excepted); it is described in Chapter 33.

Lymphatico-venous Communications exist between lymphatics and veins (Rusznyák *et al.*; Pressman and Simon).

Exceptions. Lymph capillaries are not present in epithelium (e.g., epidermis), cartilage, or other tissues devoid of blood vessels, neither are they present in the bone marrow and the main tissues of the spleen and liver. In the spleen and liver, the blood passing between the cells in irregular spaces (called sinusoids) appears to provide the sole drainage of the active tissues. However, the fibrous tissues of both organs have lymphatics. There are no lymphatics in the brain.

Lymph capillaries do not accompany the blood capillaries between the alveoli of the lungs, or in the glomeruli of the kidneys, since these capillaries are not nutritive in function; but in the lungs, lymph vessels do accompany the bronchial vessels.

Lymphoid Tissues are collections of enormous numbers of migratory and semimigratory scavenger cells called lympho-

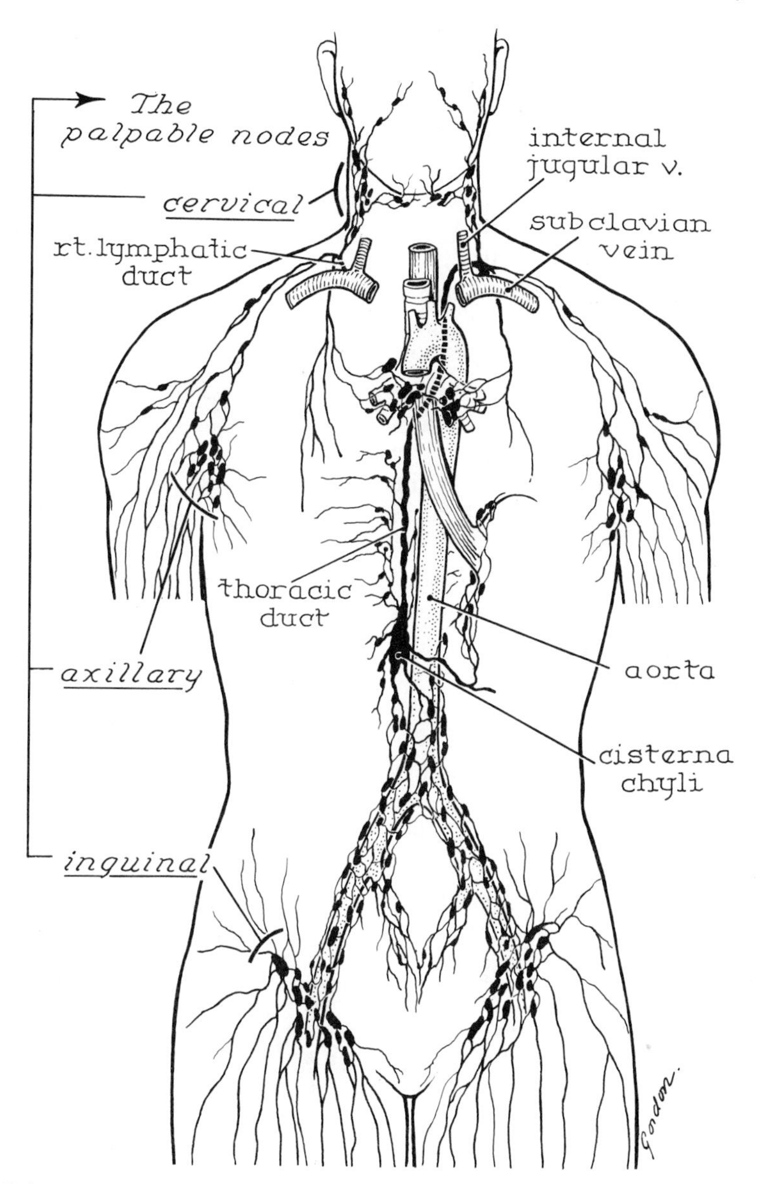

Fig. 39. Scheme of the thoracic duct and lymphatics of the body. (From *Primary Anatomy*.)

cytes. Lymphocytes are usually classed with the white blood cells for they are discharged with lymph into the blood stream. Lymphoid tissue occurs in lymph nodes, the palatine, lingual and pharyngeal tonsils, the solitary and aggregated follicles of the intestine, the appendix, the thymus and the corpuscles of the spleen. It is best developed in youth, and diseases of this tissue are commonest in youth.

Fetal Circulation

Red blood cells appear very early in the embryo, but until the 2nd month they are immature in type (containing nuclei). Nucleated red cells are found until a few days after birth, but they cannot be recovered from the maternal blood because the two circulations—fetal and maternal—are separate and closed. In the placenta, the two circulations come into close apposition with each other, being separated by semipermeable walls. Nutritive material and oxygen permeate from mother to fetus; waste products and carbon dioxide permeate from fetus to mother. The functions of the placenta, therefore, are concerned with nutrition, excretion, and respiration. In the fetus the pulmonary, portal, and renal circulations are of little account; the placental circulation is paramount.

Fetal blood, charged with nutritive material and oxygen, leaves the placenta in the *umbilical vein*. This vein traverses the umbilical cord outside the fetus and the free edge of a membrane, the falciform ligament of the liver inside, to end in the left portal vein. Some of the blood, thus brought to the left portal vein, then flows through the liver, leaves it by the hepatic veins, and enters the inferior vena cava; but most of the blood by-passes the liver, being short-circuited by the ductus venosus (*fig. 40*).

The *ductus venosus*, which occupies a sulcus behind the liver, connects the left portal vein to the inferior vena cava just below the diaphragm (*fig. 40*). Hence, the purest blood to reach the fetal heart travels through the terminal part of the i.v. cava to the right atrium.

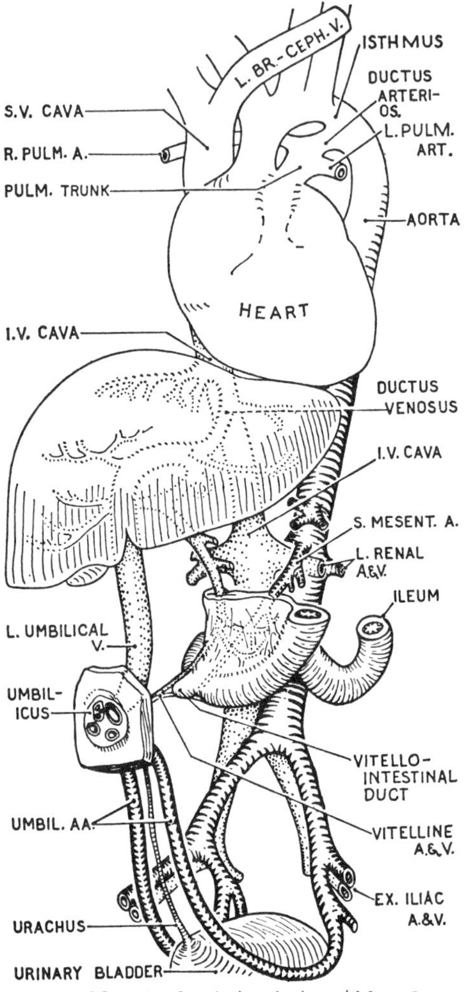

Fig. 40. The fetal circulation. (After Cullen.)

In the fetus the blood from the i.v. cava is directed across the right atrium and through an opening in the interatrial septum, the *foramen ovale*, into the left atrium, thereby short-circuiting the pulmonary circuit. Thence it passes through the left ventricle into the ascending aorta and aortic arch, and by their branches it is distributed to the walls of the heart, the head and neck (including the brain), and the upper limbs, which therefore receive "pure" blood. It is blood that has been purified in the placenta and returned by the umbilical vein and ductus venosus to the i.v. cava and so to the right atrium (*fig. 40*).

The blood from the head and neck and upper limbs returns via the superior vena cava to the right atrium, whence it passes through the right ventricle into the pulmonary trunk. Some of this blood then follows the pulmonary circuit through the lungs to the left atrium, but most of it is short-circuited to the aorta by a vessel, the *ductus arteriosus*, that connects the left pulmonary artery to the aortic arch just beyond the origin of the left subclavian artery. Beyond this connection the united streams descend through the aorta and common iliac arteries—some to be distributed to the abdomen and lower limbs, some to the placenta. The umbilical or placental arteries, one on each side, pass by the sides of the bladder and up the anterior abdominal wall to the umbilicus, thence along the umbilical cord to the placenta.

From this account it would appear (1) that the kidney has no short circuit, (2) that the liver has one (ductus venosus), and (3) that the lungs have two (foramen ovale and ductus arteriosus), and, furthermore, that the heart, head, neck, and upper limbs receive the "purest" blood.

Changes at Birth. *Lungs.* The child cries; the lungs begin to expand (Chap. 31) and assume their functions, whereupon the foramen ovale and the ductus arteriosus close—the former to be represented by the *fossa ovalis,* the latter by a fibrous cord, the *ligamentum arteriosum.*

Placenta. The cord is tied about 5 cm from the umbilicus, cut beyond, and discarded with the placenta; so, the right and left umbilical arteries become thrombosed and fibrous, and are known as the obliterated umbilical arteries; and the umbilical vein, behaving likewise, becomes the *round ligament* of the liver.

Liver. The ductus venosus also becomes a fibrous thread, the *ligamentum venosum*; so, all the blood in the portal vein must now pass through the liver.

X-ray Findings. Radio-opaque substances injected into veins of living fetal sheep reveal that (1) the superior caval blood follows the adult pattern through the heart to the ductus arteriosus into the descending aorta; (2) most of the inferior caval blood passes through the foramen ovale; (3) the ductus arteriosus closes functionally a few minutes after delivery; (4) the functional closure of the foramen ovale is apparently dependent on the onset of respiration (Barcroft *et al.*). However, permanent structural changes are more gradual (Patten), and the times of complete obliteration of the orifices are highly variable (*fig. 41*). The foramen ovale, though closed functionally, remains unobliterated (i.e., admits a probe) in about 25 per cent of adults.

NERVOUS SYSTEM

SUBDIVISIONS.

SPINAL MEDULLA OR CORD—*Membranes; Structure; Neuron.*

PERIPHERAL NERVES—*Macroscopic or Gross Structure; Microscopic Structure; Plexuses; Blood Supply.*

AUTONOMIC NERVOUS SYSTEM—*Sympathetic System: Structure; Cerebral Control; Supply of Individual Regions and Organs (upper limb, lower limb, head and neck, thorax, abdomen and pelvis).*

PARASYMPATHETIC SYSTEM—*Oculomotor Nerve; Facial Nerve; Glossopharyngeal Nerve; Vagus Nerve; Pelvic Splanchnic Nerves.*

Man is not the largest of the animals, nor are his vision, hearing, or sense of smell the most acute; he is not the strongest of the animals, nor the most fleet of foot. Even his marvellous hand is not unique—but he is superlative in the quality of his central nervous system that controls it.

The brain requires preferential treatment found in textbooks of Neuroanatomy, but the rest of the nervous system will be described.

Subdivisions of the Nervous System

Central Nervous System	Brain Spinal medulla or cord
Peripheral Nervous System	Peripheral nerves Cranial nerves—12 pairs Spinal nerves—31 pairs Autonomic nervous system Sympathetic Parasympathetic

The central nervous system controls the voluntary muscles of the body and is concerned with things both above and below the level of consciousness. The autonomic nervous system, also called the involuntary, or visceral nervous system, controls the parts over which we do not exercise voluntary control and of which, for the most part, we are unconscious (e.g., action of the heart, movements of the viscera, the state of the blood vessels, and the secretion of glands).

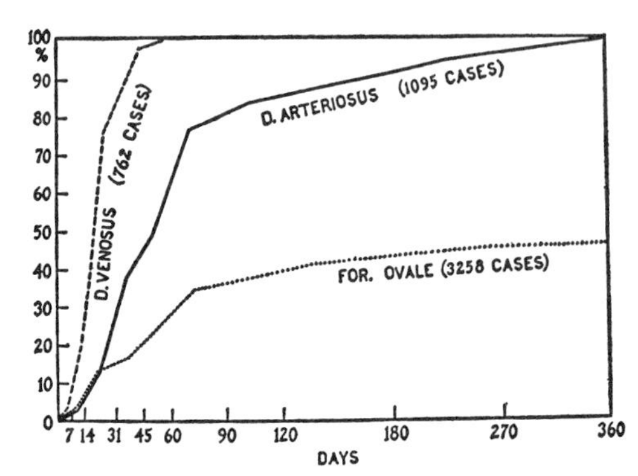

Fig. 41. Dates of permanent closure of the ductus venosus, ductus arteriosus and foramen ovale during the 1st year of life. (After Scammon and Norris.)

Neuron. The structural unit of the nervous system is a nerve cell or neuron. A neuron has the following parts: (1) the cell body, (2) a process or *dendrite* (or processes) which transmits impulses to the cell body and (3) a process or *axon* which transmits impulses from the cell body. An entire neuron may be microscopic in size; on the other hand, a process may be several feet long (e.g., a fiber of the sciatic nerve).

Spinal Medulla or Cord

The spinal cord lies within the vertebral canal. The vertebral canal (or a vertebral foramen), being too small to admit a finger (*fig. 24*), the cord obviously must be of smaller diameter than a finger, but in length (45 cm) it equals the femur.

The cord is surrounded by three **membranes:** (1) the *pia mater* (L = tender mother) which, like a delicate skin, clings tenderly to the cord, (2) the *arachnoid mater* (Gk. = like a cobweb) which is joined to the pia by threads and is flimsy, and (3) the *dura mater* which is tough (L., *dura*) and fibrous, but is easily split longitudinally, owing to the direction of its fibers. The arachnoid mater is separated from the pia by a space, the *subarachnoid space*, filled with cerebrospinal fluid. This fluid presses the arachnoid against the dura thereby obliterating a potential space, the *subdural space*.

In the fetus the spinal cord extends down to the coccyx, but as development proceeds, owing to the greater growth of the vertebral column, it is drawn upwards, so that at birth it extends only to vertebra L. 3, and in the adult to the upper part of L. 2 (*fig. 42*). Flexing the column draws the cord temporarily higher. These facts permits a physician to insert a needle between the laminae of L 3 and 4 into the subarachnoid fluid without striking the cord. A strong, glistening thread, the *filum terminale*, largely composed of pia mater, attaches the tapered end of the cord to the back of the coccyx even in the adult.

A broad band of pia, the *ligamentum denticulatum*, projects like a long lateral fin from each side of the cord and lies between the ventral and dorsal nerve roots. From the free margin of this ligament strong tooth-like processes, one for each segment, pass through the arachnoid to become firmly attached to the dura. The

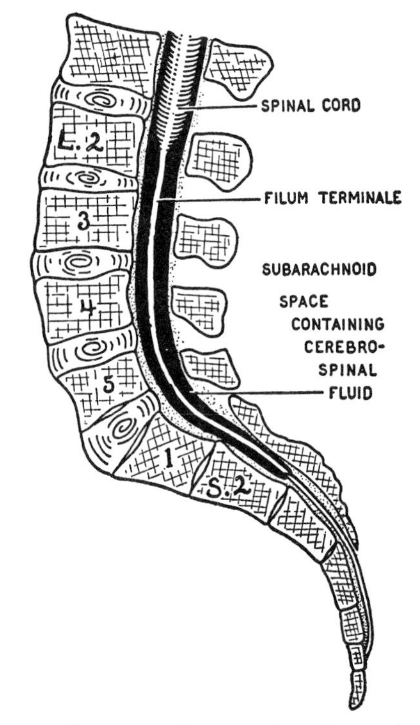

Fig. 42. The spinal cord ends at lumbar vertebra 1 or 2, the subarachnoid space at sacral vertebra 2.

highest tooth or dens of each ligament is at the level of the foramen magnum; the lowest tooth varies in level from T. 12 to L. 2 (I. B. Macdonald).

The *spinal dura* is free within the vertebral canal. It is, however, adherent to the margin of the foramen magnum and is there continuous with the cranial dura. It is fixed caudally to the coccyx by the filum terminale which, as it passes beyond the closed end of the dural sac, acquires an adherent dural covering. It is fixed on each side by the ventral and dorsal spinal nerve roots (described below). These likewise acquire adherent dural coverings which add to their thickness and give them strength.

Meningeal Nerves. At each segmental level, sensory and synpathetic nerves pass through the intervertebral foramina to the dura (Kimmel).

On *flexing the spinal column*, the dural sac stretches and nerves L. 1 and 2 are drawn from 2 to 5 mm within the intervertebral foramina. L. 3 is withdrawn less, and L. 4 negligibly. On the other hand, "*straight-leg raising*," by pulling on the sciatic nerve, can pull L. 5, S. 1, and S. 2 downward through 2 to 5 mm, but again the effect on L. 4 is negligible (Inman and Saunders). Similarly, traction on the outstretched hand pulls the roots of C. 6 and 7 (Smith).

In the adult, the subarachnoid space and the contained cerebrospinal fluid extend to the body of sacral vertebra 2. So, between L. 2 and S. 2, in place of cord there is filum terminale surrounded by the roots of the lower spinal nerves which have been drawn upwards with the cord. These loosely occupy the sac of cerebrospinal fluid, here somewhat dilated, in which they lie, and from their resemblance to a horse's tail they are called the *cauda equina*. Fortunately, the spinal-puncture needle inserted into the C. S. F. rarely if ever strikes these nerves.

Structure. The spinal cord or spinal medulla, as seen on cross section, has an H-shaped field of gray matter enveloped in a zone of white matter (*fig. 43*). The anterior and posterior limbs of the H, called the *anterior* and *posterior horns or gray columns*, divide the white matter of each side into *anterior*, *lateral*, and *posterior white columns*. A groove in front, the *antero-median fissure*, and a septum behind, the *postero-median septum*, separate the right and left sides of white matter. A canal, the *central canal of the cord*, continuous with the ventricles of the brain, runs through the gray matter.

Two continuous rows of delicate *nerve rootlets or fila* (L. filum = a thread) are attached to each side of the cord along the lines of the apices of the anterior and posterior columns of gray matter. The fila of the anterior row converge laterally in groups of a dozen or so to form *ventral or anterior nerve roots*, and the fila of the posterior row do likewise to form *dorsal or posterior nerve roots*.

The respective ventral and dorsal roots pierce the dura about 2 mm apart, one in front of the other—or rather they carry it before them. They then continue laterally to the intervertebral foramina where each

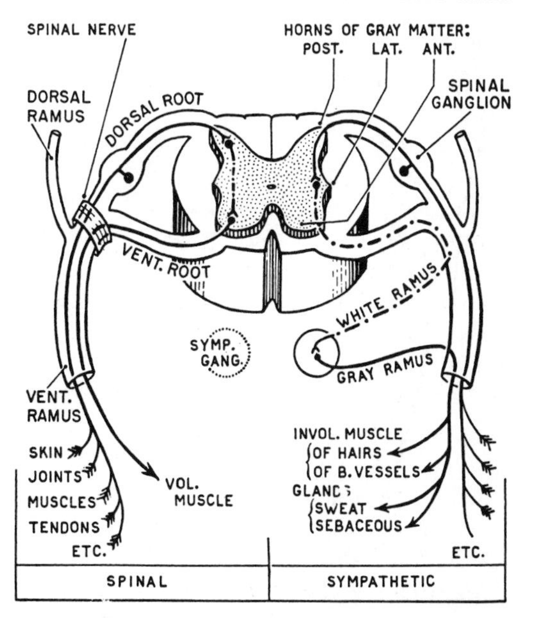

Fig. 43. The composition of a peripheral nerve.

dorsal root is swollen because it contains the bodies of (sensory) nerve cells; the swelling is a *spinal ganglion* (posterior root ganglion). Just beyond this, the two roots unite and their fibers mingle to form a *nerve trunk*. This, after a few millimeters, divides into a large *ventral* and a small *dorsal ramus* (L. ramus = branch) (*fig. 43.1*).

The length of cord to which the fila of one pair of spinal nerves is attached is called a *spinal segment*. Since there are 31 pairs of spinal nerves, the cord is said to have 31 segments.

Peripheral Nerves

Macroscopic or Gross Structure. The peripheral nerves are made up of fibrous tissue enclosing bundles of nerve fibers, both afferent and efferent (*fig. 43*). A loose sheath (epineurium), continuous with the surrounding areolar tissue, encases the entire nerve and unites the individual bundles. Each bundle has a thicker sheath, the *perineurium*, which branches when the bundle branches and which accompanies every fiber to its destination. Between the fibers are some delicate partitions (endoneurium).

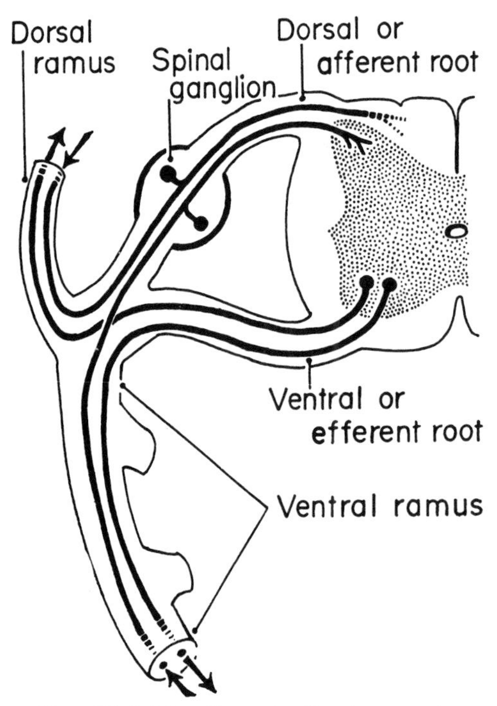

Fig. 43.1 The fibers in the two roots and two primary rami of a spinal nerve.

Microscopic Structure. A nerve fiber consists of a central thread, the *axis cylinder* (*axon*), encased in a tube of white fatty material of varying thickness, the *myelin sheath*, around which there is a thin, but tough, nucleated membrane, the *neurilemma*. If a fiber is pinched, the axon and the myelin sheath will rupture, but the neurilemma may remain intact.

White versus Gray. It is the myelin sheath that gives whiteness to nerves and to nerve tracts in the spinal cord and brain; in its absence nerve tissue looks gray.

Nerves transmit impulses quite rapidly, but the speed is variable, being about 60 meters per second in motor nerves (Smorto and Basmajian).

Functions. A dorsal nerve root transmits impulses to the cord; it is *afferent* or sensory. A ventral nerve root transmits impulses from the cord; it is *efferent* or motor. (L. ad-ferens = bringing to; ex-ferens = bringing from, the CNS here being understood.)

The cell bodies of the afferent nerve fibers lie within the spinal ganglia. Those of the efferent nerve fibers lie in the anterior horn of the gray matter of the cord (*fig. 43.1*).

Afferent or sensory spinal nerve fibers bring impulses from the endings in the skin, muscles, joints, etc. to the spinal ganglia, and thence, via the fila of the dorsal roots to the cord. Within the cord the fibers branch: some branches ascend in the cord, others descend, still others at once turn into the gray matter of the posterior horn where they synapse (i.e., make contact) with the dendrites of short *connector* (*intercalated*) cells whose axons in turn synapse with the dendrites of large multipolar cells in the anterior horn. The axons of these motor cells emerge in the anterior root fila and end in motor end-plates of voluntary muscles.

These neurons—afferent, connector, and efferent—constitute a simple spinal *reflex arc*. The arc functions thus: if one's finger is touched, even during sleep, one moves it; if it is pricked or burnt, one withdraws the limb. In the one case a simple reflex arc is invoked; in the other, afferent impulses spread more widely, even to the opposite side of the body, perhaps arousing consciousness, due to the ascending and descending and crossing fibers by which impulses spread to other parts of the cord and to the brain.

Typical Spinal Nerve. Details of the organization and ramifications are given on p. 84 and in *figure 72*.

Nerve Plexuses. Ventral nerve rami and their branches communicate with adjacent rami and branches to form plexuses, thereby widening the influence of the individual segments of the spinal cord. Close to the vertebral column the ventral rami of nerves C. 1–4 form the cervical plexus, those of C. 5–Th. 1 the brachial plexus, L. 1–4 the lumbar plexus, L. 4–S. 4 the sacral plexus, and S. 4–Co. 1 the coccygeal plexus.

Nerves communicate more peripherally also. Cutaneous nerves communicate freely with adjacent nerves and overlap them with the result that severing a cutaneous nerve diminishes sensation in its territory without as a rule abolishing it (*figs. 43.2* and *43.3*).

Peripheral nerves are themselves plexuses: their fibers do not run parallel courses but are plaited, not unlike the tendon fibers in *figure 31A*. The branch of a nerve is usually bound to the parent stem for 1–6 cm by areolar tissue, and the level at which it leaves the parent stem varies within this range. Proximal to this, the strands of the branch take part in the plait, and attempts to free the branch more proximally lead to its destruction (*fig. 43.4*).

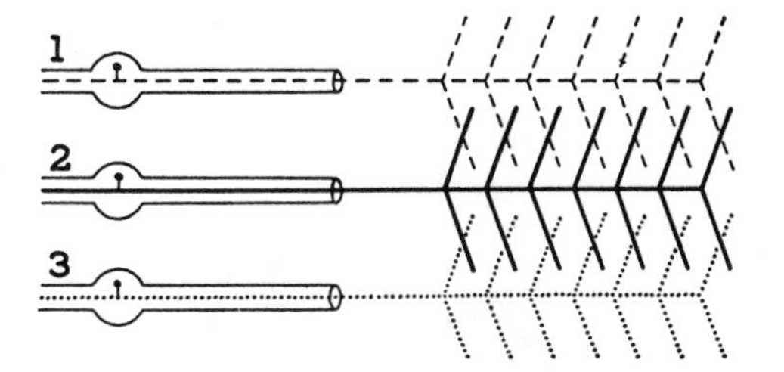

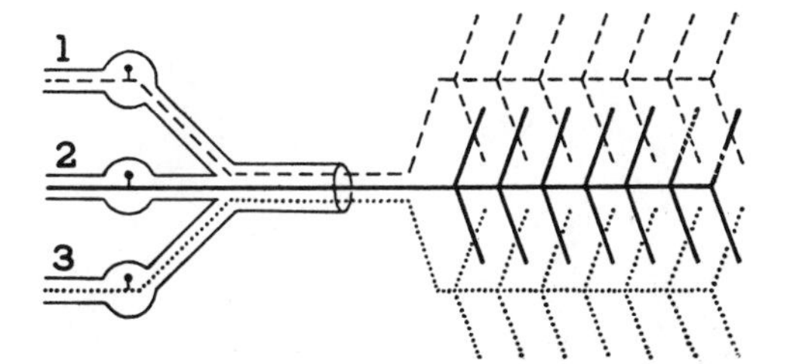

Fig. 43.2. The distribution of a series of spinal nerves overlap. (After Haymaker and Woodhall, modified.)

Blood Supply. The peripheral nerves are supplied by a succession of anastomosing nutrient vessels, *arteriae nervorum* derived from the nearest arteries, but their number, size and origin are inconstant. On reaching a nerve the artery usually divides into ascending and descending branches, which anastomose with longitudinal chains. The veins, *venae nervorum*, have a similar pattern (Sunderland).

Autonomic Nervous System

This system of nerves and ganglia distributes efferent impulses to (1) the heart, (2) smooth muscle, wherever situated, and (3) glands, and it collects afferent impulses from them. It has two parts,

(1) the sympathetic system and
(2) the parasympathetic system.

The *Sympathetic System* (*fig. 44*) has central connections with the thoracolumbar part of the spinal medulla or cord from the 1st thoracic to the 2nd (or 3rd) lumbar segment. The *Parasympathetic System*, which has (1) a cranial part and (2) a sacral part, has central connections with the brain through cranial nerves III, VII, IX,

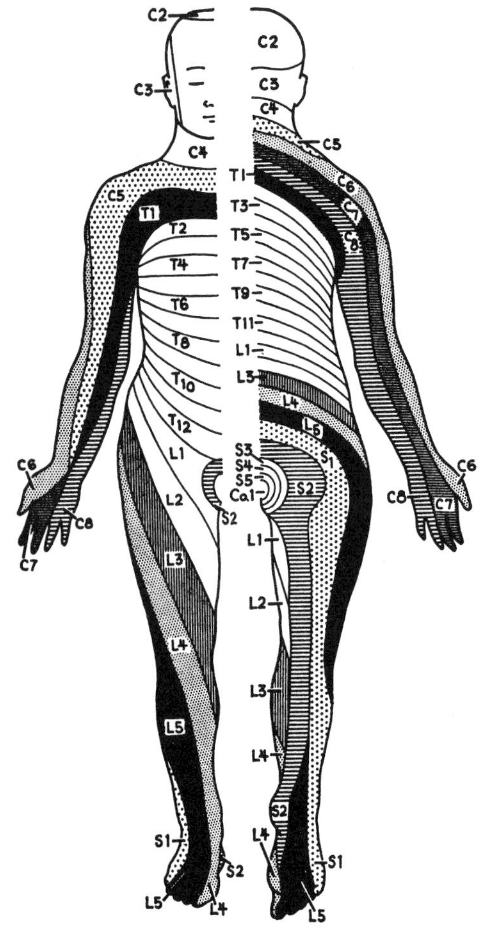

Fig. 43.3. Dermatomes: the strips of skin supplied by the various levels or segments of the spinal cord. (After Keegan, modified.)

and X, and with the spinal cord at sacral segments (2), 3, and 4.

Sympathetic System

In the *sympathetic system* there are neurons corresponding to the efferent, connector, and afferent neurons of the voluntary system (*figs. 44.1 and 44.2*). The efferent cells occur (1) in the *paravertebral ganglia* or ganglia of the sympathetic trunk, (2) in the *prevertebral ganglia* (or visceral ganglia—cardiac, celiac, intermesenteric, hypogastric, and subsidiary ganglia) which are but detached parts of paravertebral ganglia, and (3) in the medulla of the suprarenal gland.

The gray matter of the spinal cord from segments Th. 1—L. 2 possesses an intermediolateral column (horn), and it is in this intermediolateral column that the sympathetic *connector cells* are lodged (*fig. 44.2*). Their axons pass as fine, medullated *preganglionic fibers* (2.6 μ) via the ventral nerve roots and white rami communicantes (Th. 1—L. 2) to the paravertebral ganglia, and there form synapses with the excitor cells.

Some preganglionic fibers, however, ascend and descend in the sympathetic trunk to form synapses with the cells in the ganglia at various levels; whereas other preganglionic fibers pass through the paravertebral ganglia without synapsing, to form *splanchnic nerves* (*fig. 44.1*).

The *postganglionic fibers*, or axons of the excitor cells, are mostly nonmyelinated

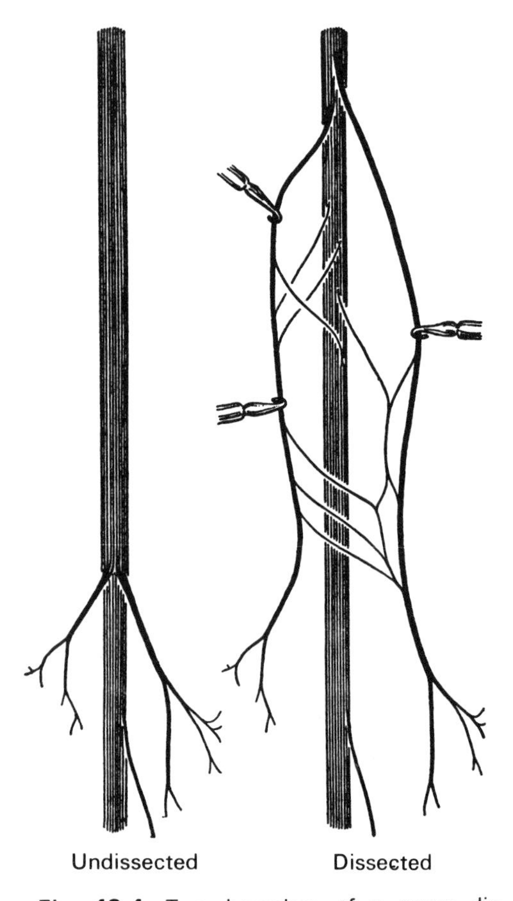

Fig. 43.4. Two branches of a nerve dissected proximally.

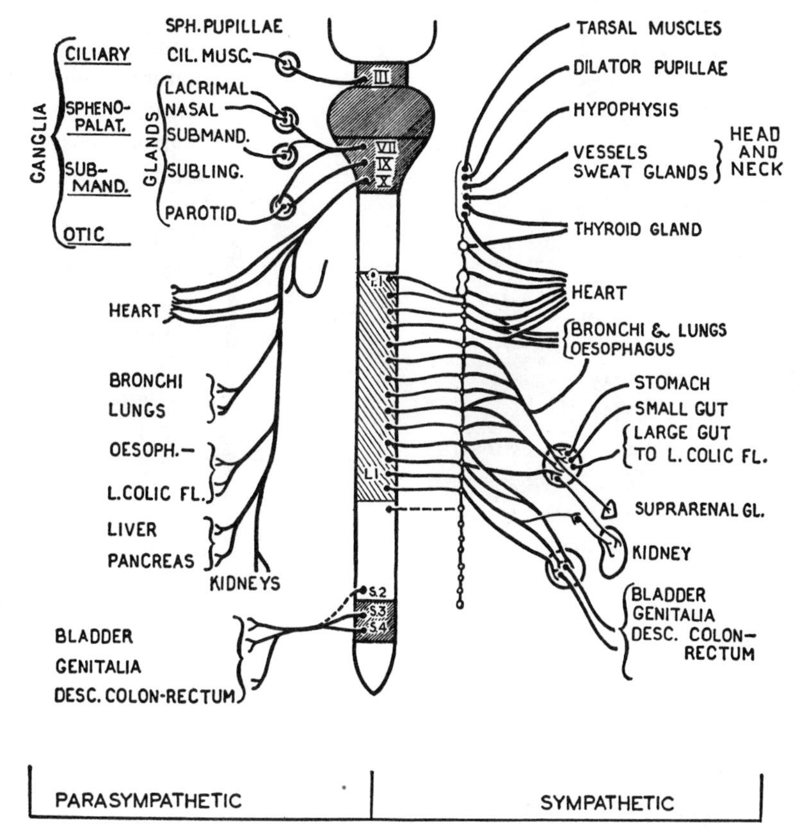

Fig. 44. General plan of the autonomic nervous system. (After Stopford.)

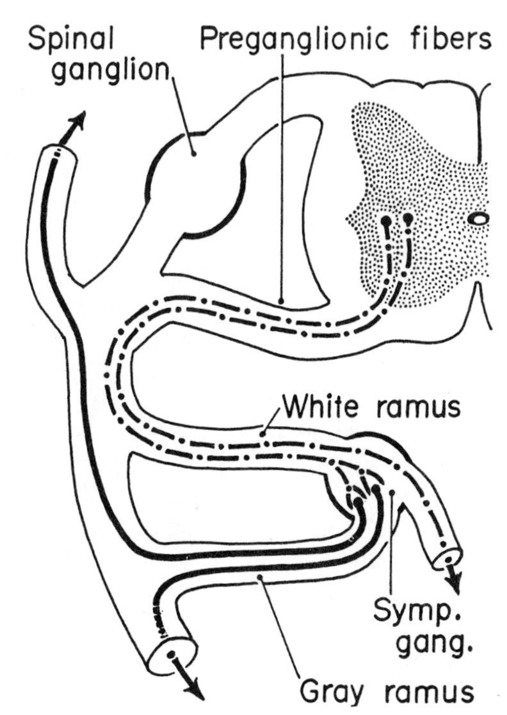

Fig. 44.1. The typical sympathetic contribution to a spinal nerve.

and therefore gray. These gray fibers, carried in a short nerve, a *ramus communicans*, pass laterally from the ganglia of the sympathetic trunk to each and every spinal nerve. They reach the blood vessels, sweat and sebaceous glands, and arrectores pilorum muscles of the entire cutaneous surface of the body and of somatic structures (limbs and body wall). Other postganglionic fibers are relayed from the superior cervical ganglion to the face, which is the territory of the trigeminal nerve (*fig. 44.3*).

The thoracic and lumbar *Splanchnic Nerves* carry elongated preganglionic fibers which have passed without interruption through paravertebral ganglia (Th. 5—L. 2). They synapse with excitor cells in prevertebral ganglia.

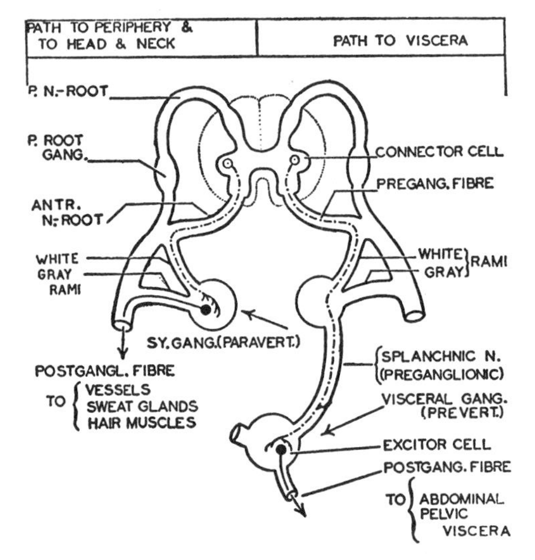

Fig. 44.2. General plan of a sympathetic ganglion.

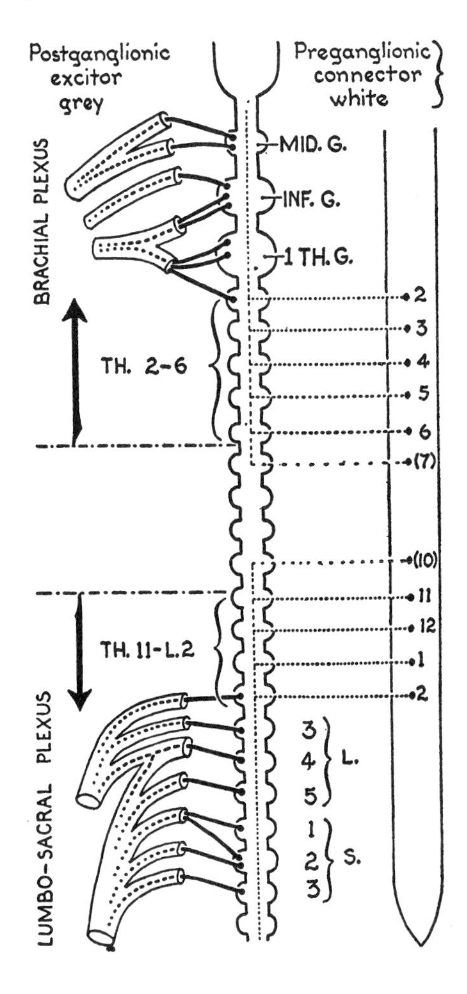

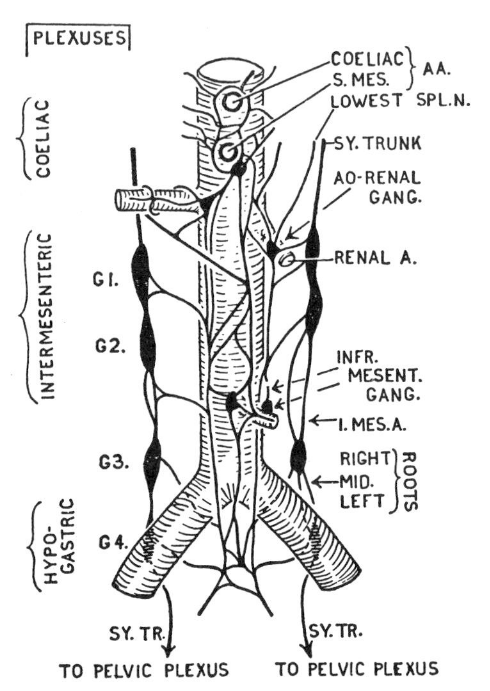

Fig. 44.4. The intermesenteric and superior hypogastric plexuses. (From a dissection by K. Baldwin.)

The greater (Th. 5—10), lesser (Th. 10, 11), and lowest (Th. 12) thoracic splanchnic nerves and the four lumbar splanchnic nerves end in the celiac and other prevertebral ganglia, whence they are relayed almost entirely as perivascular branches to the abdominal and pelvic viscera (*figs. 44.2 and 44.4*).

The *suprarenal gland* is supplied from cord segments Th. 10—L. 1, (2). Branches of the splanchnic nerves ramify among the cells of its medulla, because they develop from sympathetic neurons.

The cervical and upper thoracic sympathetic ganglia, although paravertebral in location, represent both para- and prevertebral ganglia of more caudal levels; that is to say, the postganglionic fibers of the excitor cells in these ganglia pass (1) on the one hand, in rami communicantes to the spinal nerves and, so, to somatic struc-

Fig. 44.3. The sympathetic supply to the upper and lower limbs.

tures, and (2) on the other hand, as visceral fibers to such structures as the eye, salivary glands, heart, and lungs.

The *visceral afferent* (sensory) fibers of the sympathetic system travel with the visceral efferent fibers. They pass via the white rami communicantes to the spinal ganglia (dorsal root ganglia) where, like the afferent fibers of the voluntary system, they have their cell stations. They enter the spinal cord through the dorsal roots mainly of Th. 1—L. 2, and synapse with the connector cells in the intermediolateral column of gray matter. Many, however, first ascend or descend in the sympathetic trunk.

The Sympathetic Trunk itself is composed of ascending and descending fibers some which are preganglionic efferent, postganglionic efferent, and also afferent fibers. The paravertebral ganglia (ganglia of the trunk) are formed by synapses between preganglionic efferent neurons and the cell bodies of postganglionic neurons. The influence of a single preganglionic neuron is diffused over a wide area due to the fact that each preganglionic fiber synapses with a number of postganglionic neurons.

Cerebral Control over the autonomic system is exercised by centers in the hypothalamic region, which lies toward the lower and front part of the 3rd ventricle. From here descending tracts influence the connector cells.

The Sympathetic Supply of Individual Regions and Organs will be considered by diagram and comment. *Figures 44* to *44.4* indicate for the different regions and organs (1) the probable segmental locations of the connector cells in the intermediolateral column of gray matter; (2) the ganglia in which they synapse with effector cells; and (3) the ultimate distribution of the postganglionic fibers.

Upper Limb (*fig. 44.3*). Preganglionic fibers from cord segments Th. (2), 3–6, (7) ascend in the sympathetic trunk to the upper thoracic, inferior cervical, and middle cervical ganglia. Thence a dozen or so postganglionic, gray rami pass to the roots of the brachial plexus to be distributed to the limb. Most of these fibers travel in the lower trunk of the plexus (? where they may be subjected to pressure by the 1st rib) and in the median and ulnar nerves.

Lower Limb (*fig. 44.3*). Preganglionic fibers from cord segments Th. (10), 11, 12, L. 1 and 2 descend in the sympathetic trunk to ganglia L. 2–S. 3. Thence, the postganglionic fibers pass in gray rami to the nerves of the lumbar and sacral plexuses. The upper part of the femoral artery is supplied by an extension from the aortic plexuses along the common and external iliac arteries; but as in the upper limb so in the lower, most of the femoral artery and the arteries distal to it are supplied locally.

Head and Neck (*fig. 44*). Cord segments are mainly Th. 1 and 2, i.e., connector cells are situated in segments Th. 1 and 2, though some may extend lower. The excitor cells lie mainly in the superior cervical ganglion.

Postganglionic fibers pass to the arrectores pilorum (smooth muscles that make the hairs stand up), sweat glands, and vessels of the skin; to the heart; and, via the int. carotid nerve, to the vessels of the nasal cavity (through the deep petrosal nerve), to the dura, to the cerebral vessels, and to the orbital cavity (tarsal muscles, Dilator Pupillae, and vaso-constrictors).

The orbital fibers arise mainly in cord segment Th. 1, pass to the 1st thoracic ganglion, and ascend to the excitor cells in the superior cervical ganglion.

Thorax. Cord segments are mainly Th. 2, 3, and 4. For the *heart*, the cord segments are Th. 1–4, (5). From relay stations in the three cervical and upper 4, (5) thoracic ganglia, cardiac nerves pass to the cardiac plexus (*fig. 44.5*). Pain impulses travel in the afferent fibers (but apparently not by way of the superior cervical ganglion) and have their cell stations in the upper 4, (5) thoracic spinal (dorsal root) ganglia.

For the lungs, the cord segments are Th.

2–6, (7); for the esophagus, Th. 4, 5, and 6.

Abdomen and Pelvis. Cord segments are Th. 5–L. 2. For the stomach, liver, and pancreas they are Th. 6–9, (10), and for the gall bladder Th. 4–9, (10).

For the small intestine the cord segments are Th. (8), 9, 10, (11); for the cecum and appendix Th. 10, 11 and 12; for colon to left colic flexure Th. (11), 12, and L. 1; and for left colic flexure to rectum L. 1 and 2.

For kidney the cord segments are Th. (11), 12, L. 1, (2); for ureter L. 1 and 2; and for bladder (? Th. 11, 12), L. 1 and 2. (The foregoing data are largely from Mitchell.)

From the cord segments, connector fibers pass through the thoracic and lumbar paravertebral (*trunk*) ganglia, as the thoracic and lumbar splanchnic nerves, to be relayed in the celiac, renal, mesenteric, and superior and inferior hypogastric ganglia.

For other pelvic viscera see page 302 and *figures 44.4* and *49*.

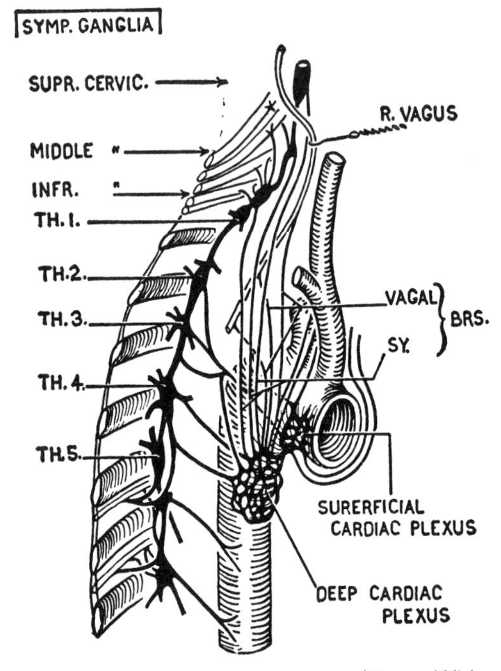

Fig. 44.5. Cardiac plexuses. (From White, after Kuntz and Morehouse.)

Parasympathetic System

Parasympathetic nerve fibers (efferent) are contained in cranial nerves III, VII, IX, and X and in sacral nerves (2), 3, and 4. Details are given here for convenience; the beginner cannot be expected to appreciate, let alone learn, them fully until he has studied the cranial nerves thoroughly.

Many branches of the trigeminal nerve (N. V) are accompanied in the terminal parts of their courses by efferent parasympathetic fibers from other nerves (e.g., the lingual nerve by the chorda tympani, see below).

Afferent (visceral sensory) nerve fibers are contained in cranial nerves IX and X and in sacral nerves (2), 3, and 4.

Oculomotor Nerve. This nerve (N. III) sends fibers to the *ciliary ganglion* in the orbit, whence they are relayed by *short ciliary nerves* to the sphincter pupillae and the ciliary muscle, mediating contraction of the pupil and accommodation of the lens to near vision (*fig. 766*).

Facial Nerve. (N. VII; see *fig. 797* and Chap. 48). *The Efferent Fibers* of the nervous intermedius (pars intermedia of the facial nerve) carry secretomotor and vasodilator impulses to the lacrimal, nasal, palatine, and salivary glands. They run thus:

(1) Via the *greater* (*greater superficial*) *petrosal nerve* to the pterygopalatine ganglion, whence as postganglionic fibers they accompany (a) the zygomatic nerve to the orbit, where, leaving it, they join the lacrimal nerve and so to the lacrimal gland (*fig. 786.1*), and (b) branches of nerve V^2 to the nasal and palatine glands (*fig. 796* and Chap. 46); and (2) via the *chorda tympani*, which, after passing through the tympanum, joins the lingual nerve which conducts it as far as the submandibular ganglion from which it is relayed to the submandibular and sublingual glands (*fig. 786* and Chap. 45, and (3) a *twig of the facial nerve* joins the lesser (lesser superficial) petrosal nerve perhaps bringing accessory secretory fibers to the parotid gland.

The *Afferent Fibers* of the nervus intermedius are: (1) mainly taste fibers coming (a) from the anterior two-thirds of the tongue in the chorda tympani, and (b) from the soft palate in the lesser palatine nerves, through the pterygopalatine ganglion and into the cranium via the greater petrosal nerve. These fibers have their cell station in the geniculate ganglion. (2) The fibers of deep sensibility in the face also have their cell station in the geniculate ganglion and travel in the nervus intermedius.

Glossopharyngeal Nerve. The tympanic branch of this nerve (N. IX, *fig. 761*) traverses the tympanum in the tympanic plexus, receives a twig from the facial nerve, and, as the lesser petrosal nerve, runs in the cranium to the otic ganglion, and thence by the auriculotemporal nerve to the parotid gland. If the twig from the facial nerve is secretory (secretomotor), then the parotid gland has a double nerve supply—from IX and from VII.

The *Afferent Fibers* are (1) fibers of taste from the posterior one-third of the tongue, (2) fibers of general sensation from the posterior one-third of the tongue, the fauces, pharynx, and tympanum, and (3) pressor fibers in the sinus nerve. Their cell station is in the ganglia on the root of nerve IX.

Vagus Nerve. This nerve (N. X) supplies the digestive and respiratory passages and the heart, as described in Chapter 43 and shown in *figure 762*. The relay stations of the efferent fibers of the vagus are in terminal ganglia, situated in (or near) the walls of the organ or part it supplies. The vagus causes hollow organs to contract and their sphincters to relax. It is secretomotor. The cell stations of its afferent fibers are in the ganglia on the root of the vagus. Its afferent fibers do not carry impulses of pain from the heart or from abdominal organs.

Pelvic Splanchnic Nerves. These nerves (S. 2, 3, and 4) behave as vagal fibers, and are described on page 302. The vagus supplies the gastro-intestinal tract as far as the left colic flexure, via a branch of the posterior vagal trunk (*fig. 306*); the pelvic splanchnic nerves supply the gut distal to that point. Their afferent fibers are conductors of pain impulses.

ENDOCRINE GLANDS

An endocrine or ductless gland is one that produces an internal secretion or *hormone*, i.e., a secretion carried off in the venous blood stream to influence the activities of another part of the body. If it also produces a secretion carried off by a duct it is both an exocrine and an endocrine gland.

Hypophysis (Pituitary Gland)

This is a small gland about the size of a pea hanging by a stalk, the infundibulum, from the hypothalamus which forms the floor of the third ventricle of the brain. It rests on the deeply concave body of the sphenoid bone in the region behind the nasal cavities (*fig. 44.6*). It is made up of two lobes, an anterior and a posterior, entirely different from one another functionally.

The *anterior lobe* forms most of the gland. It produces several hormones which, because of their influence on other endocrine glands, have caused the hypophysis cerebri to be referred to as the "master gland." Disturbances of the lobe result in dwarfism, giantism, infantilism, excessive obesity, and other manifestations associated with growth and sex.

The *posterior lobe* (*or neurohypophysis*) is a downgrowth from the brain and produces secretions which stimulate contractions of the uterus, the production of urine by the kidneys, and influence the production of insulin by the pancreas. This lobe also affects blood pressure. It is possibly associated with sympathetic and (or) parasympathetic functions because of its connections with the hypothalamus. Its portal system of veins carries the secretions first to the hypothalamus before the blood is returned to the systemic veins that drain that part of the brain.

Pineal Gland or Body

The pineal body, so called because in shape and appearance it resembles a miniature pine cone, is deeply placed within the brain. It projects backwards from the back of the roof of the third ventricle. Little is known about it and its status as an endocrine gland.

Thyroid Gland

This lies in the neck (*fig. 44.6*), and is fully described in Chapter 40. The thyroid gland consists of irregular masses of tissue (follicles) filled with a colloid containing the thyroid hormone. This hormone (thyroxin) is concerned with a great many bodily functions and, when produced in excessive quantities, disturbs the normal rate at which the cells of the body function. This disturbance of the basal metabolic rate (B.M.R.) is a notable feature of *goiter*. Congenital deficiency or absence of the thyroid gland results in a mentally defective dwarf known as a cretin. The condition is alleviated by the administration of thyroxin.

Parathyroid Glands

The parathyroid glands—two on each side—lie on the back of the thyroid gland. They are small ovoid bodies; yet, they are extremely important since they regulate the relative amounts of calcium in the blood and in the bones. Certain distressing diseases in which the bones become soft or extremely brittle are the result of malfunctioning of the parathyroids.

Thymus Gland

This gland lies behind the manubrium sterni, and in front of the great vessels above the heart. It is essentially an elongated bi-lobed gland whose period of greatest functional activity is in fetal life. After birth it gradually shrinks until, at puberty, it usually consists merely of two elongated fatty masses reaching down to the pericardium and with little thymic tissue remaining. It resembles a lymphoid organ and no

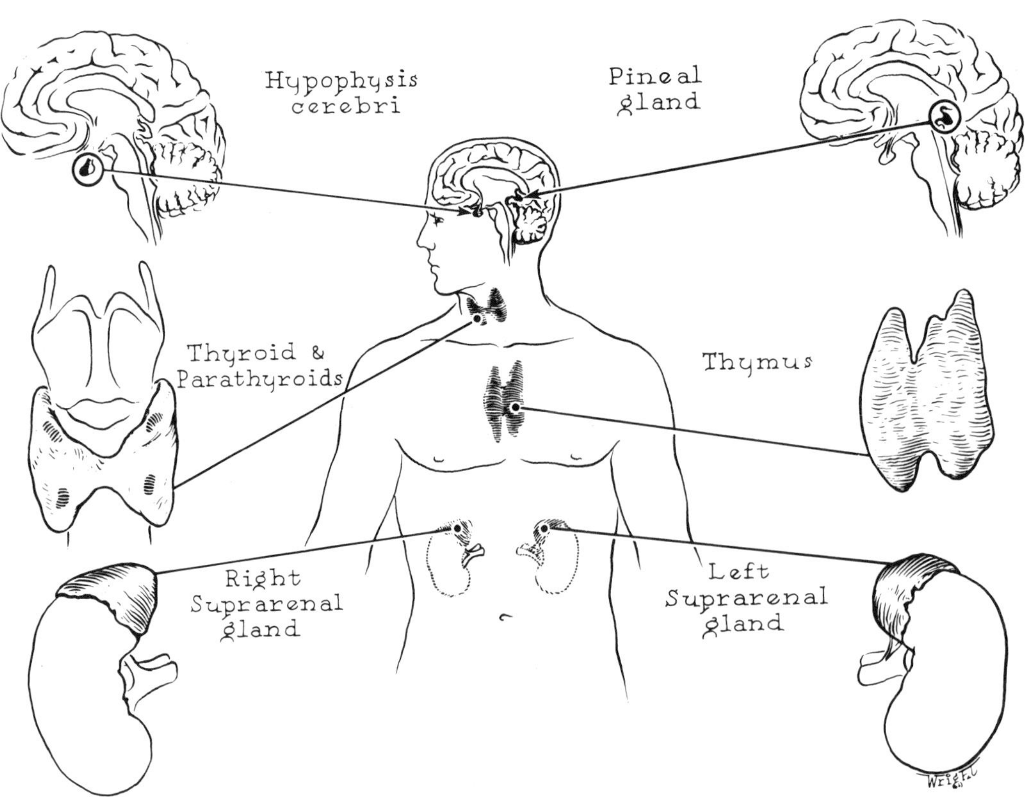

Fig. 44.6. Scheme of main endocrine glands. (From *Primary Anatomy*.)

hormone has yet been isolated from it. Though there is no general agreement on its endocrine function, there is no question that the thymus is an important production center for lymphocytes before puberty.

Suprarenal (Adrenal) Glands

As their names implies, these cap the top of the kidneys but they are not functionally related. They are described fully in Chapter 17. The cortex of each forms a thick outer layer of the gland. It produces a group of hormones—among them cortisone—which are undergoing intensive investigation. Some are essential to life because they regulate various metabolic processes, e.g., salt metabolism, production of sex hormones, and production of collagen fibers of fibrous tissue throughout the body.

The medulla (L. = marrow) of the adrenal is closely associated in developmental origin with the sympathetic ganglia. It is in the medullary part of the gland that adrenalin is produced.

Chromaffin Bodies

Scattered along the line of the sympathetic chain and the abdominal aorta are many tiny bodies identical in structure with the suprarenal medulla. They are known as para-aortic bodies or paraganglia or chromaffin bodies and like the suprarenal medulla are associated with the activity of the sympathetic system.

The two para-aortic bodies are largest at about the age of three when they measure only about one cm in length. They atrophy and disappear by the mid-teens. The paraganglia, which are close associates of the sympathetic trunk and are quite tiny, atrophy to microscopic size in the adult. The chromaffin cells (so named because they are easily stained by yellow-colored chromic acid) secrete adrenalin just as the suprarenal medulla does.

Pancreas

As noted before, the pancreas has both an exocrine and an endocrine function. Scattered among the larger clumps of glandular tissue that pour their exocrine secretions through the pancreatic duct into the duodenum are small islets of cells that release the hormone *insulin* into the blood stream. Insulin is vital to the proper metabolism of carbohydrates and its production is partly regulated by the hypophysis. The injection of the hormone into patients suffering from diabetes mellitus is a lifesaving treatment.

Testes and Ovaries

The gonads not only produce the spermatozoa and ova, which may be considered as exocrine, but they have an important endocrine function as well. Specialized cells release the male and female sex hormones. These regulate the sexual function of the adult person and determine the secondary sexual characteristics including all aspects of maleness and femaleness.

In women, the monthly cycle of ovulation followed by menstrual bleeding is regulated by a balance of hormones produced by the ovaries and other endocrine glands, especially the hypophysis. These various hormones interact in such a way as to cause: (1) a sloughing off of the uterine mucosa if the ovum is not fertilized by a sperm withing a few days of its release, or (2) an embedding and protection of a fertilized ovum in the mucosa. The placenta of a developing embryo itself produces hormones which interact with hormones of the ovaries and hypophysis to prevent permature expulsion of the child.

3 DIGESTIVE AND RESPIRATORY SYSTEMS

DIGESTIVE SYSTEM

PARTS—*Mouth; Mastication; Oral Glands: Parotid, Submandibular, Sublingual; Pharynx and Esophagus; Stomach; Intestine; Liver; Pancreas; Spleen.*

RESPIRATORY SYSTEM

PARTS—*Nasal Cavities; Pharynx; Auditory Tube; Larynx; Trachea; Bronchi; Lungs; Respiratory Act; Epithelial Surfaces.*

DIGESTIVE SYSTEM

This system is essentially (1) a long hollow tube, the digestive passage, through which food is propelled, and (2) certain accessory glands (*fig. 45*).

The *digestive passage* (alimentary canal) begins at the lips, traverses the neck, thorax, abdomen and pelvis, and ends at the anus.

Its successive **parts** are:

Mouth
Pharynx
Esophagus
Stomach
Small intestine
  duodenum, jejunum, and ileum.
Large intestine
  appendix, cecum, colon, rectum, anal canal and anus.

The stomach and intestines comprise the *gastrointestinal tract.*

The **accessory glands** are:

Glands of the mouth
  parotid,
  submandibular, and
  sublingual
Liver
Pancreas, with which is associated the Spleen.

Mouth

To survive, animals must eat. And, in search for food, they advance, mouth end foremost. Around the mouth the organs of special sense are developed: the eyes (and ears) to locate food by *sight* (and sound), the nose by *smell*, and the lips and tongue by *touch* and *taste*. Hence, the brain establishes itself at the mouth end of the body.

The mouth, or oral cavity, is bounded externally by the cheeks and lips. The cleft between the upper and lower lips is the *aperture* of the mouth. The *teeth* form two dental arches, one set in the upper jaw or maxillae and the other in the lower jaw or mandible. The horseshoe-shaped space external to the teeth and gums is the *vestibule* of the mouth. The space bounded externally by the teeth and gums is the *mouth proper*. From the floor of the mouth rises the *tongue*; the roof comprises the *hard* and *soft palates* and the median finger-like process, the *uvula*, in which the soft palate ends.

Mastication. The incisor teeth (L. incidere = to cut) bite off pieces of food, and the molar teeth (L. mola = a millstone) grind them. The food is commonly coarse; so, the mouth requires a protective lining. It is lined with an epithelium consisting of special layers of flattened (squamous) cells resembling epidermis (p. 68), but the surface cells, although flattened, retain their nuclei and do not cornify. The epithelium is lubricated and kept moist by the secretions of small glands, the size of pin-heads,

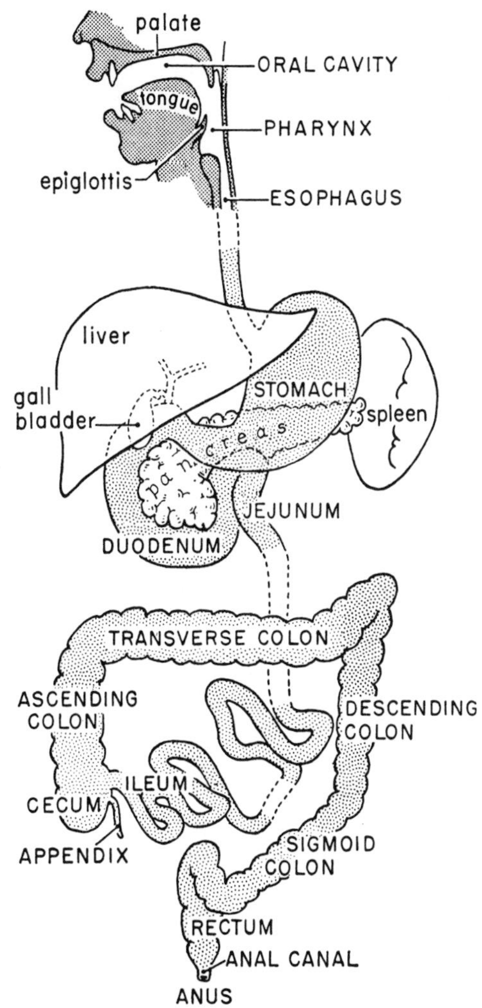

Fig. 45. Diagram of the digestive tract.

that line the palate, lips, and cheeks. These secretions augment those of the three large, paired, **Oral Glands** (Salivary Glands) (parotid, submandibular, and sublingual).

The **Parotid Gland** lies below the ear (*fig. 45.1*). Its long duct opens into the vestibule of the mouth beside the 2nd upper molar tooth.

The **Submandibular** (SUBMAXILLARY) **Gland** lies under shelter of the mandible; its duct opens on to a papilla (=nipple) beside its fellow, behind the lower incisor teeth. On opening your mouth and raising the tip of the tongue, watery secretion may be seen in a mirror to be welling up from the orifices.

Each **Sublingual Gland** produces a ridge on the side of the floor of the mouth beneath the tongue. Its many fine ducts open on to the ridge (*figs. 45.1* and *785*).

The salivary glands have either a mucous secretion or a serous secretion or both; the serous secretion (ptyalin) digests free starch but cannot break down the cellulose enclosing the starch.

The saliva moistens the food which has been ground into small particles by the teeth. This the tongue rolls into a bolus or lubricated mass easily swallowed—dry food is swallowed with difficulty. Saliva keeps the lips and mouth pliable in speaking.

Pharynx and Esophagus

The pharynx and esophagus are merely passages and, like the mouth, are protected by stratified squamous epithelium. The muscular coat of the pharynx and upper half of the esophagus, though voluntary in structure, is not under voluntary control; in the lower half of the esophagus (and therefore throughout the entire "g-i" tract to the anus) it is involuntary or smooth.

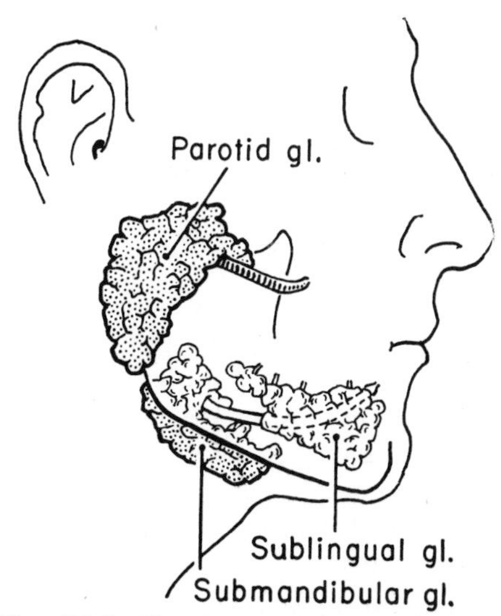

Fig. 45.1. The three oral (salivary) glands.

Stomach

Because the stomach is a receptacle for food and must expand, contract, and move about, it possesses a serous (peritoneal) coat. Its inner lining of mucosa secretes a protective mucus and hydrochloric acid which is necessary to the action of the pepsin secreted by the mucosa. These initiate the digestion of proteins. The outlet of the stomach, the **Pylorus** (Gk. = a gatekeeper) is guarded by a strong sphincter of circular fibers, the *pyloric sphincter*, which opens in response to neural control to discharge partly digested and liquefied stomach contents into the duodenum.

Intestine

The intestine or gut is divisible into two parts: (1) the small intestine, about 7 meters long, and (2) the large intestine, about 2 meters long, as measured at autopsy (B. M. L. Underhill). But during life, a tube, 3–4 meters long, swallowed by an adult may project from both mouth and anus. After death the gut lengthens greatly and progressively (W. C. Alvarez).

The partially digested food leaving the stomach is acidic and sterile. In the upper part of the small gut (10 cm beyond the pylorus) the ferments of the pancreas and the bile are added. Here digestion continues and absorption of water and digested products begins. These processes are most active in the duodenum and they diminish as the large gut is approached. The essential function of the large gut is to dehydrate the intestinal contents. The bacteria, present in enormous numbers, exemplify symbiosis, for they make an important contribution to the welfare of their host. In addition to the production of vitamin B, they convert bile salts into molecules that are resorbed and used for producing hemoglobin. They also in part break down cellulose, which is almost undigested in the small intestine.

As the intestinal contents become less fluid in the large gut more lubricating mucus is both needed and produced by special cells.

Liver (Hepar)

The liver is the largest organ in the body. Attached to its under surface is the *gall bladder*; this extends backward to a transverse fissure, the *porta hepatis* (L. = door of liver), through which the *hepatic ducts* conduct bile from the liver and through which the *portal vein* (conducting blood laden with products of digestion) and the *hepatic artery* (conducting oxygenated blood) enter the liver. After the blood has circulated through the liver, it leaves the posterior surface via the hepatic veins to enter the inferior vena cava.

The liver is a gland of compound tubular design. The cells of the tubules elaborate an *exocrine (external) secretion*, called bile, into the tiny lumina of the tubules. Each tubule, called a *bile canaliculus*, is really a series of spaces between rows or sheets of liver cells. It drains into a system of ducts, the *bile passages*, which emerge from the liver to communicate with the gall bladder and with the duodenum (*fig. 274*, p. 212).

The same liver cells also elaborate an *endocrine (internal) secretion* into the blood in the *sinusoids* (p. 35). Helping to line the sinusoids are phagocytic cells, *Kupffer's cells*, which help to dispose of effete red blood cells and are concerned with immunity mechanisms.

Blood Flow through the Liver. Traditionally, the liver is composed of hexagonal lobules (*fig. 46*) surrounding a receiving (central) radicle of the hepatic vein, placed centrally. More correctly (*fig. 46.1*), a distributing (preterminal) portal vein is placed centrally in a *liver acinus*, sending diverging sinusoids to three receiving hepatic veins, placed peripherally (Rappaport).

The *bile passages* conduct bile in the reverse direction (i.e., toward the porta). *Lymph vessels* run both with the portal vein and the hepatic veins.

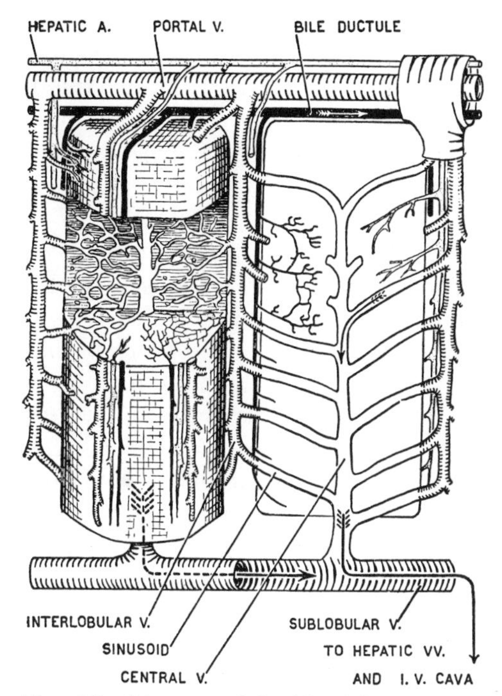

Fig. 46. Diagram of the blood flow through 2 lobules of the liver and of the course of bile, according to the traditional description.

The cells of the liver regulate the amount of glucose in the blood by removing the excess after a meal and temporarily storing it as glycogen. They also store vitamin A. They elaborate fibrinogen and heparin both vital in blood-clotting mechanisms.

Pancreas

The pancreas is known as the sweetbread in animals that are used by man for food. This long gland extends across the abdomen and lies behind the stomach. It has an external secretion which passes to the duodenum where it is changed into ferments that act upon proteins, carbohydrates, and fats. And, it has an internal secretion, called insulin, the reduction of which leads to diabetes.

Spleen (Lien)

The spleen is mentioned here for convenience; but its association with the digestive system is incidental. It is a soft sponge filled with blood, about the size of a clenched fist.

Structure and Function. See page 200.

RESPIRATORY SYSTEM

PARTS—*Nasal Cavities; Pharynx; Auditory Tube; Larynx; Trachea; Bronchi; Lungs; Respiratory Act; Epithelial Surfaces.*

The respiratory system, which buds off the primitive gut, has two portions (*fig. 47*): (1) conducting and (2) respiratory.

The *Conducting Portion*, or air passages, comprises:

Nasal cavities
Pharynx
Larynx
Trachea
Bronchi and Bronchioles

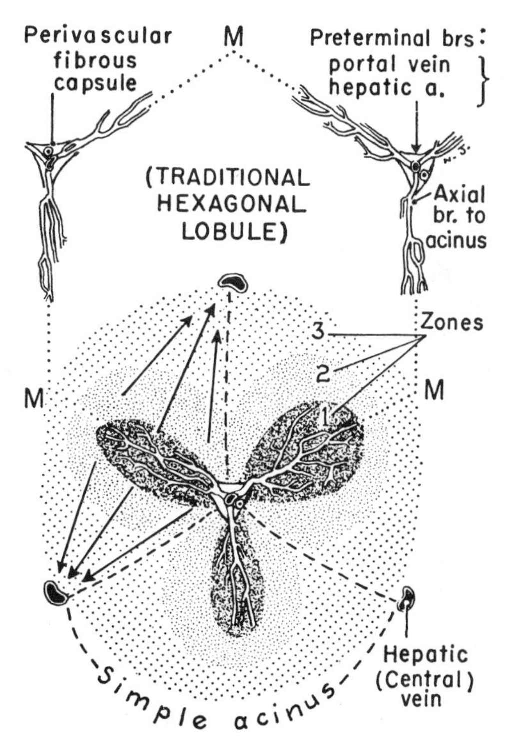

Fig. 46.1. According to the modern view, the acinus is the liver unit. It incorporates sectors of several (3) adjacent lobules, a preterminal branch of the portal vein and of the hepatic artery being central and several (3) initial hepatic veins being peripheral. *Zones 1* are supplied with better blood than *Zones 2*, and *2* than *3*. (Courtesy of Rappaport *et al.*).

The *Respiratory Portion* comprises:

Lungs (having respiratory bronchioles, alveolar ductules, alveolar sacs, and alveoli)

The lungs are sponges filled with air; their septa (partitions) consist almost entirely of blood capillaries. Thin, moist, and membranous, they allow transfusion of gases in solution from air to blood and *vice versa*.

The respiratory system absorbs oxygen, the absorption taking place in the alveoli of the lungs. Carbon dioxide is eliminated there simultaneously.

Conducting Portion

Nasal Cavities

The nasal cavities, a right and a left are separated from each other by a thin median partition, the *nasal septum*. The entrance to each cavity, called the *nostril* or naris, opens into a *vestibule* which is lined with skin.

From the side wall of each cavity three downwardly curved shelves, the *conchae*, overhang three antero-posteriorly running passages, the *meatuses*. Opening into the inferior meatus is the *tear duct* (nasolacrimal duct); opening into the other meatuses are the orifices of large *air sinuses* (air cells). These sinuses, inflated with air, invade the surrounding bones causing them to be large enough to carry the upper teeth and to form the framework of a large face in the adult, without adding great weight.

The mucous membrane covering the inferior and middle conchae, contains dilatable *venous sinuses* which warm and humidify the inhaled air. It has a ciliated epithelium and also mucous and serous glands, diffuse lymphoid tissue, and lymphoid follicles. The mucus catches inhaled dust and bacteria, and acts a sterilizing agent. The cilia, which are microscopic "hairs," waft the mucus, and the foreign matter entangled in it, backward to the pharynx.

The uppermost part (2 square cm in each nasal cavity) of the medial and lateral walls is the *olfactory area*. Here the olfactory nerves, which stream through the thin perforated roof (*cribriform plate*) of the nasal cavity, end freely among the epithelial cells.

Pharynx

The pharynx is a fibromuscular chamber, 15 cm long. It is attached above to the base of the skull; it is continuous below with the esophagus. Communicating with it in front are the nasal, oral, and laryngeal cavities. Accordingly, it is divisible into three parts (upper, middle, and lower), called the nasal, oral, and laryngeal parts, respectively (*fig. 47*).

The nasal part or *nasopharynx* is the backward extension of the nasal cavities. While it cannot be shut off from these cavities, it can be, and is, shut off from the oral part by the soft palate and uvula during the act of swallowing. Were it not so, food would be forced from the oral part into the nasal part and so to the nasal cavities—a person with a paralyzed soft palate may find that this may happen.

Auditory Tube

An air duct, the auditory tube (Eustachian) opens into each side of the nasopharynx (*fig. 47*). Each tube, by bringing the pharynx into communication with the tympanic cavity (middle ear) of the corresponding side, serves to keep the air pressure on the two sides of the tympanic membrane (ear drum) equal under changing atmospheric conditions. Normally the tubes are closed, but the act of swallowing opens them; hence, the discomfort in the ears while ascending or descending in an airplane, is relieved by swallowing. The tubes can be shut for days by inflammation of the nasal mucosa, causing temporary partial deafness; more serious, they provide a passage for infection to spread from the nasopharynx to the middle ear.

On each side of the entrance to the oral

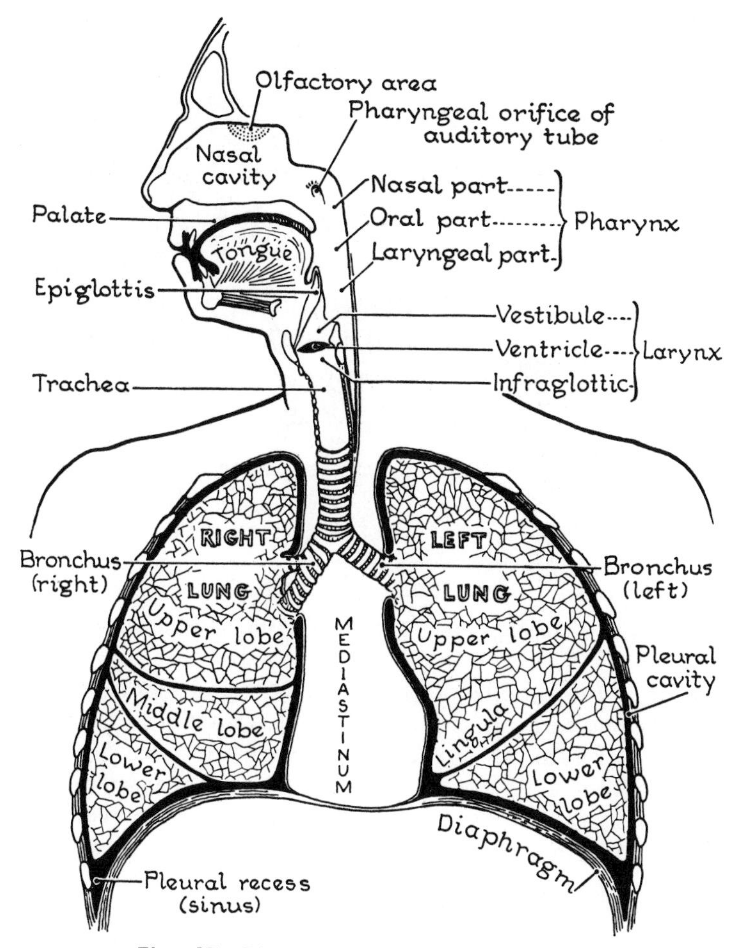

Fig. 47. Diagram of the respiratory system.

pharynx, and visible from the mouth, is a mass of lymphoid tissue, "*the tonsil*" or the palatine tonsil. Its upper pole extends upward from the side of the tongue far into the soft palate; its lower pole cannot be seen unless the tongue is depressed.

Larynx

The larynx or voice box opens off the lowest part of the pharynx and is continuous below with the trachea. This box is kept rigid by a number of hyaline and elastic cartilages which are united by membranes. It is lined with mucous membrane internally and covered with voluntary muscles externally.

The chief cartilages of the larynx are:
(1) The *thyroid cartilage*, which resembles an angular shield, has two perpendicular *laminae* which meet in front in the median plane, the prominent upper end of the angle of meeting being conspicuous as the *laryngeal prominence* (Adam's apple). Below, the thyroid cartilage grips the cricoid cartilage (*fig. 47.1*) as the knees of a horseman grip a saddle.
(2) The *cricoid cartilage* is a complete ring expanded posteriorly into a lamina or plate and so resembles a signet ring; it keeps the lower part of the larynx perpetually open.
(3) The *arytenoid cartilages* are paired, small, and pyramidal; their bases articulate with the upper border of the lamina of the cricoid cartilage. The paired *vocal cords*, the free upper edge of the *conus elasticus*, extend from the inside of the

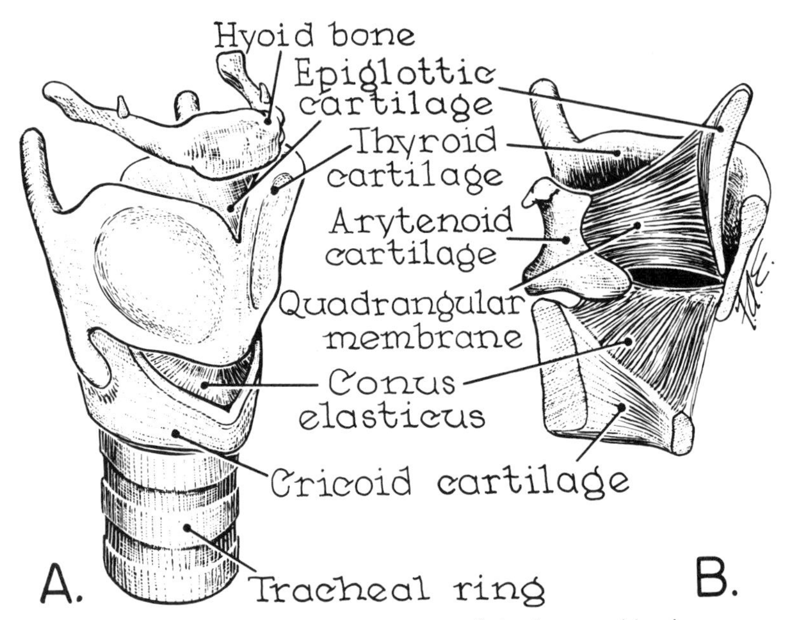

Fig. 47.1. Cartilages and ligaments of the larynx, side view.

angle of the thyroid cartilage horizontally backward to the arytenoid cartilages whose various movements control the tension and distance apart of the cords.

(4) The *epiglottic cartilage* is shaped like an elm leaf, its stalk being attached to the angle of the thyroid cartilage just above the vocal cords.

Trachea

The trachea or windpipe is an elastic tube about 10 cm long, with a caliber equal to the root of the index finger. It is kept patent by about 20 U-shaped rings of hyaline cartilage which are open posteriorly (*figs. 47, 47.1*).

At the level of the sternal angle, 5 cm below the jugular notch, the trachea bifurcates into a right and a left bronchus.

Bronchi

The bronchi have the same structure as the trachea. After an oblique course of 5 cm, each enters the respective lung at the hilus and descends toward the base, giving off branches which in turn branch and rebranch like a tree (*figs. 585–589*). Within the lung the U-shaped rings give place to flakes of hyaline cartilage which surround the tube and hold it open. When the bronchi are reduced to the diameter of 1.0 mm, they are called *bronchioles*.

The terminal bronchioles divide into a number (2 to 11) of alveolar ductules which end in dilated air sacs, *alveolar sacs*. The walls of these sacs being themselves sacculated, resemble a bunch of grapes, and hence they are called alveoli (L. alveolus = a bunch of grapes) (*fig. 590*). Adjacent *alveolar* sacs are practically contiguous—between them there is room only for a close-meshed network of capillaries.

Respiratory Portion

Lungs

The right and left lungs (*fig. 47*) are sponge-works of elastic tissue which feel like rubber sponges. In this highly elastic framework a bronchus and a pulmonary artery and pulmonary vein branch out.

The right lung is divided by two complete fissures into three separate lobes; the left is divided by one fissure into two lobes.

Pleura. Each lobe of each lung has a delicate and inseparable "skin", the *pulmonary pleura* (pulmo, L. = lung). This is a perfectly smooth, moist, serous membrane, identical in structure and in origin with peritoneum in the abdomen, i.e., a fine areolar sheet with a surface of pavement (mesothelial) cells. Another layer of pleura lines the ribs, diaphragm, and mediastinum (containing the heart); this is *parietal pleura*. Between the pulmonary and parietal layers of pleura there is a potential space, the *pleural cavity* (*fig. 47*). It allows the lung to expand and contract without friction.

The Respiratory Act has two phases—inspiration and expiration. On inspiration the diaphragm descends, thereby increasing the vertical diameter of the thorax; and the ribs rise from a sloping to a more horizontal position, thereby increasing both the anteroposterior and the side-to-side diameter, as you can determine by palpation. Air rushes down the trachea and bronchi into the lungs which must expand to avoid formation of a vacuum in the pleural cavities. Expiration is largely a matter of elastic recoil; that is, the highly elastic tissue contracts, the stretched abdominal muscles act like an elastic belt on the abdominal contents forcing the diaphragm upward; and the cartilages of the ribs, which underwent twisting during inspiration, now untwist.

As the table shows, during quiet respiration about 500 cc of air are inspired and expired. On full inspiration about an extra 2500 cc can be taken in; and on full expiration an extra 1000 cc can be forced out. Even then there remain in the alveolar sacs, trachea, and bronchi about 1000 cc which cannot be expelled.

	cc		
Total capacity	2500	Complemental air	Vital capacity
	500	Tidal air	
	1000	Supplemental air	
	1000	Residual air	

Epithelial Surfaces. The mucous membrane of the respiratory passages as far as the bronchioles of 1 mm in diameter are lined with ciliated epithelium. It has mucous and serous glands, and lymphoid tissue both diffuse and aggregated.

In the protective mucus that lines the larynx, trachea, and bronchi, inhaled dust and other foreign material are caught and entangled. The cilia cause the lining tube of mucus to move ever upward, like a moving carpet or escalator, to the entrance of the larynx where it spills over into the pharynx and is swallowed. The other method of expelling foreign material is by coughing. The lymphoid tissue also is defensive against foreign invasion.

The cilia in the nasal cavities sweep backward toward the pharynx; those in the trachea and bronchi upward toward the pharynx; those in the air sinuses spirally around the walls to the orifices.

Exceptional Areas. (1) The vestibule of the nose is lined with skin, possessing hairs, sweat glands, and sebaceous glands; (2) areas subjected to pressure or friction, where cilia could not survive, are protected by stratified squamous epithelium.

These are: (a) the parts of the upper surface of the soft palate and uvula which are applied to the pharyngeal wall during swallowing; (b) areas against which the food comes into contact, namely, the entire oral and laryngeal parts of the pharynx and also the dorsal aspect of the upper half of the epiglottis; (c) the vocal cords, which vibrate and strike each other; (3) the terminal and respiratory bronchioles, the epithelium of which is cubical; (4) the alveoli, whose continuous lining of epithelium cells is so extremely thin that its existence was doubted until it was shown by electron microscopy; (5) the olfactory (smell) epithelium.

4 UROGENITAL SYSTEM AND THE SKIN

URINARY SYSTEM

PARTS—*Kidneys: Structure, Renal Arteries; Ureters, Urinary Bladder and Urethra; Nerve Supply of Kidney, of Ureter, of Bladder and Urethra.*

GENITAL SYSTEM

Parts Compared.

MALE GENITAL SYSTEM—*Testes; Ducts; Accessory Glands: Seminal Vesicles, Prostate, Bulbo-urethral Glands; Penis; Scrotum; Semen.*

FEMALE GENITAL SYSTEM—*Ovaries; Uterine Tube; Uterus; Vagina; Accessory Glands; External Genitalia.*

SKIN AND FASCIAE

Functions; Dermis or Corium; Epidermis; Nails; Hairs; Arrectores Pilorum Muscles; Sebaceous Glands; Sweat Glands; Vessels; Nerves

SUPERFICIAL FASCIA OR TELA SUBCUTANEA —*Loose Areolar Tissue; Adipose Tissue.*

DEEP FASCIA

URINARY SYSTEM

The urinary system and the genital system develop together as the *urogenital system*. The male urethra serves as a common outlet for the products of both systems.

The urinary system comprises the following parts (*fig. 47.2*):

Kidneys, Ureters } bilateral and paired
Bladder, Urethra } median and unpaired

The *kidneys* excrete urine. This passes down two muscular tubes, one on each side, the *ureters*, into a muscular reservoir, the *urinary bladder*, where it is stored until such time as it may conveniently be discharged through the *urethra*.

Kidneys

Each of the two kidneys is of conventional kidney shape, is about 11 cm long, and weighs about 130–150 gm. Its function is to keep the composition of the blood plasma constant. This it does by removing the excess of water and the waste products, expecially those resulting from the metabolism of nitrogenous substances. It also maintains the acid-base balance of the blood by selective elimination of certain electrolytes.

Structure (*fig. 48*). Occupying the inner two-thirds of the cut surface of a kidney are dark, striated areas, the *pyramids*. The *papillae* or tips of the pyramids project into the calices of the pelvis. The outer one-third of the kidney, i.e., the part lying external to the bases of the pyramids, is *cortex*. *Renal columns*, similar to cortical tissue, extend between the pyramids. These columns and the pyramids constitute the *medulla*. (Calix, singular, Gk. = "cup".)

A kidney contains about 1,000,000 microscopical units. Each unit or *nephron* has two parts, (1) a *glomerulus* and (2) a *uriniferous tubule*.

A glomerulus (L. = a small ball) is a spherical bunch of looped capillaries which invaginates the expanded blind end of a uriniferous tubule, called a *glomerular capsule* or "*Bowman's capsule*". The surface area of all the glomeruli of each kidney is estimated to be 0.3813 sq. meter (M. H. Book). The two layers of glomerular capsule, outer and inner or invaginated, plus the glomerulus are known as a *renal* or *Mal-*

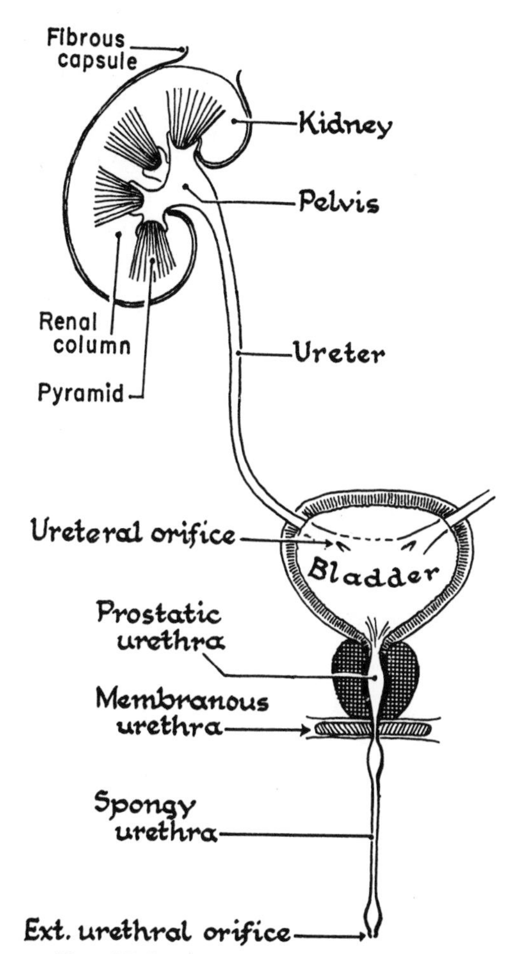

Fig. 47.2. The urinary system (male).

pighian corpuscle. But the term is variously applied. Malpighi, who first saw the corpuscles through the microscope, likened them to apples hanging from a tree.

The capsule is succeeded by the proximal convoluted tubule, the descending and ascending limbs of Henle's loop, the distal convoluted tubule and finally the junctional tubule which discharges into a system of collecting tubules. About a dozen collecting tubules open on to the papilla of each pyramid (*fig. 48*) and discharge their contents into the pelvis of the kidney.

Each named part of the tubule has a distinctive epithelium and, accordingly, a different function. The glomeruli and the convoluted tubules occupy the cortex and renal columns; the limbs of Henle's loops and the collecting tubules occupy the pyramids.

Details of Arteries and Veins. The **Renal Arteries,** a right and a left, carry far more blood to the kidneys than is needed for their nourishment, and all the blood first passes through the glomeruli.

Each renal artery typically divides into five *segmental arteries* (commonly arising irregularly) which, after entering the renal sinus, send branches up along the sides of the pyramids and, because the pyramids are spoken of as lobes of the kidney, the arteries are called *interlobar arteries*. At the junction between medulla and cortex, the interlobar arteries divide into many *arcuate arteries* which curve between these two parts, forming arcs but not arches.

From the arcuate arteries, *interlobular arteries* pass radially toward the capsule, each giving off many branches from which short arterioles, *vasa afferentia*, pass to the glomeruli (vas, L. = vessel; plural, vasa).

Each *vas afferens* enters a glomerulus and there forms capillary loops, as noted. These unite and leave the glomerulus as a *vas efferens*, which breaks up into capillaries which ramify throughout the renal substance. Thus, (1) the *vasa efferentia* from the outermost glomeruli (those nearest the capsule) provide a capillary network among the convoluted tubules; (2) the vasa from the innermost glomeruli send long

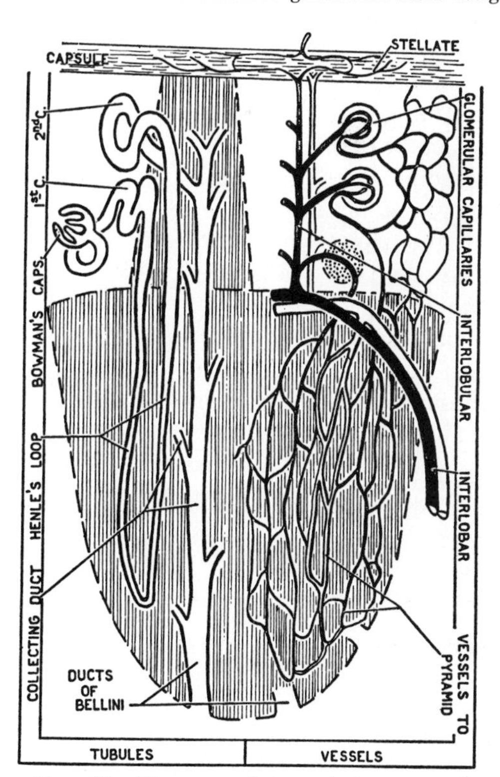

Fig. 48. Diagram of the tubules and blood supply of the kidney (see also *fig. 309*).

meshed capillaries, arteriolae rectae, to the pyramids; and (3) the intermediate vasa do both. The "arcuate' arteries are end arteries—they do not anastomose with their fellows to form arches.

The **interlobular veins** begin under the capsule as stellate veins and may be seen on stripping it off. The tributaries of the renal veins anastomose freely.

Ureters, Urinary Bladder and Urethra

From the kidney the urine is propelled by peristaltic action along a 25-cm muscular tube, the *ureter*, into a hollow muscular reservoir, the *urinary bladder*. Through a cystoscope (a lighted tube used by urologists), jets of urine are seen to squirt into the bladder from the ureteral orifices two or three times a minute.

The bladder has a widely varying capacity (average about ½ liter). From its lowest part of fibro-muscular tube, the *urethra*, leads to the exterior of the body. It is about 20 cm long in the male, 4 cm in the female.

At the junction of the bladder and urethra (i.e., at the neck of the bladder) involuntary muscle forms specialized bundles but these do not constitute a true involuntary sphincter, as formerly taught (Woodburne)—and beyond this (i.e., between the fasciae of the urogenital diaphragm) urinary control is maintained by a sphincter of voluntary muscle, the *sphincter urethrae*.

Parts of Male Urethra (*fig. 47.2*). (1) Prostatic—3 cm; (2) membranous (surrounded by a thin voluntary muscle and its membranes)—quite short but elastic; and (3) penile or spongy urethra, traversing the entire length of the spongy body of the penis—in the non-erect state, about 15–18 cm, but stretching proportionately during penile erection.

Nerve Supply

Kidney. Cutting dorsal nerve roots of T. 12, L. 1, and L. 2 relieves renal pain (White and Smithwick). A denervated kidney continues to excrete normal urine.

The Ureter. Its spinal or cord segments are L. 1 and 2. The peristalsis of the ureter is not disturbed by lumbar sympathectomy; in fact, by the withdrawal of inhibitory sympathetic influences the functions of dilated (hydronephrotic) ureter and of a dilated colon (megacolon) are improved.

The Bladder and Urethra receive both parasympathetic and sympathetic nerves (*fig. 49*).

Parasympathetic. The pelvic splanchnic nerves (S. 2, 3, 4) are the motor nerves to the bladder; when they are stimulated, the bladder empties, the blood vessels dilate, and the penis becomes erect. They are also the sensory nerves to the bladder.

Sympathetic, through its superior hypogas-

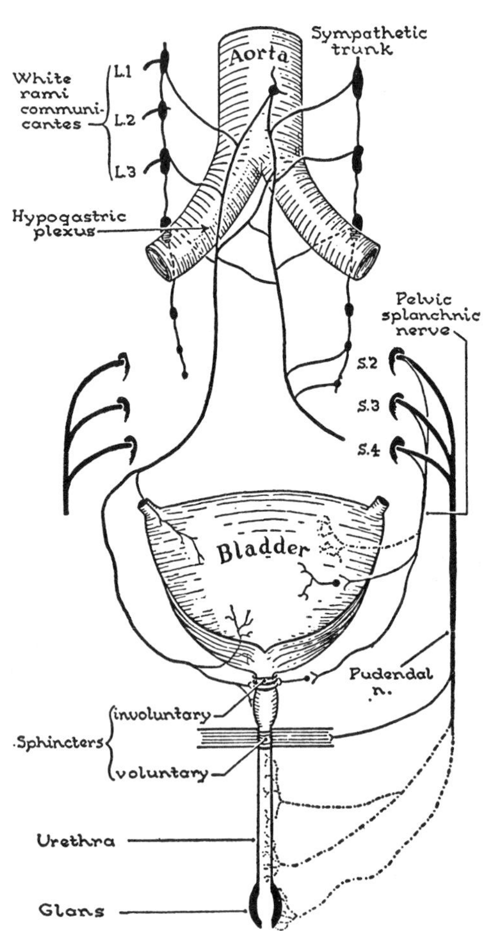

Fig. 49. Diagram of the nerve supply of the bladder and urethra.

tric plexus (lower thoracic, and lumbar 1, 2, 3), is motor to a continuous muscle sheet comprising the ureteric musculature, the trigonal muscle, and the muscle of the urethral crest. It also supplies the muscle of the epididymis, ductus (vas) deferens, seminal vesicle, and prostate. When the plexus is stimulated, the seminal fluid is ejaculated into the urethra but is hindered from entering the bladder perhaps by the muscle sheet which is drawn towards the internal urethral orifice. The sympathetic is also vaso-constrictor, and to some slight extent it is sensory to the trigonal region.

It would seem that the sympathetic supply to the bladder has a vaso-constrictor and a sexual effect and that as regards micturition it is not antagonistic to the parasympathetic supply (Learmonth; Langworthy and others).

Somatic Nerves. The pudendal nerve is motor to the sphincter urethrae and sensory to the glans penis and the urethra.

GENITAL SYSTEM

The male and female organs of reproduction are fundamentally the same, and in early fetal life are very similar. In both sexes they pass through an indifferent stage during which there is a pair of parallel ducts—mesonephric (Wolffian) and paramesonephric (Mullerian)—on each side of the body. In the male, the mesonephric ducts are utilized as genital ducts and the paramesonephric ducts largely disappear or remain vestigial; in the female the converse is true (*figs. 49.1, 50,* and *51;* and details in table on this page).

Each sex has (1) a symmetrical pair of reproductory or sex glands, which produce germ cells—spermatozoa in the male and ova in the female; (2) two different pairs of passages through which these germ cells ultimately find their way to the exterior of the body; one being well developed in each sex and largely vestigial in the opposite sex; (3) accessory glands; and (4) external genitals.

Male Genital System

The male reproductive organs are:

Testis
Epididymis
Ductus Deferens
Seminal Vesicle
Ejaculatory Duct
Prostate
Bulbo-urethral Gland

The external genital parts include:

Penis
Urethra
Scrotum

Rudimentary or vestigial structures:

Prostatic Utricle
Paradidymis
Aberrant ductules

Homologous Male and Female Parts

	Male	Female
Sexual Gland	Testis	Ovary
MALE PASSAGES Mesonephric tubules & duct	Epididymis paradidymis Ductus Deferens	Appendix (W) (?) *Epoophoron* *Paroophoron* *Duct of Gartner (of epoophoron)*
	Urethra: prostatic penile	Urethra Labia Minora, enclosing vestibule
FEMALE PASSAGES Paramesonephric duct	*Appendix (M)* (?) *Prostatic Utricle* (?)	Uterine tube Uterus Vagina
Accessory glands	Seminal vesicle Prostate Bulbo-urethral gland Urethral glands	— Para-urethral ducts Greater vestibular gland Lesser vestibular glands
External genitals	Penis bulb of penis Scrotum	Clitoris Bulbs of vestibule Labia Majora

Testes

The testes or male sex glands, one on each side, lie in the scrotum (L. = a leather bag) (*figs. 49.1, 225, 226,* and *227*). Each testis is ovoid and 4 cm long. Like the eyeball, it has a thick, white, inelastic, fibrous outer coat, the *tunica albuginea.* Within this covering are numerous, delicate, threadlike, macroscopic *seminiferous tubules*, the linings of which produce enor-

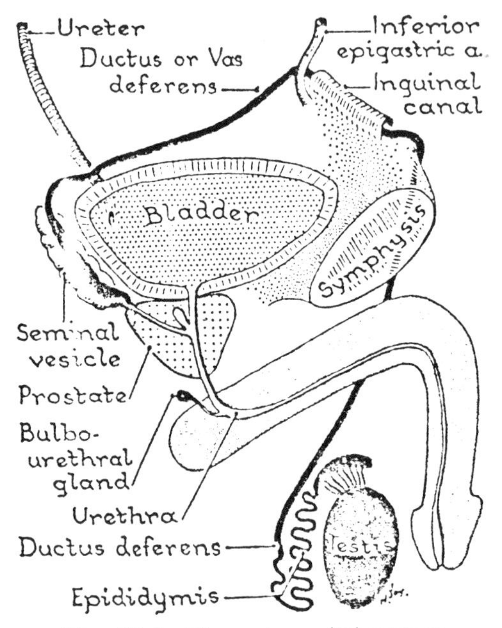

Fig. 49.1. The male genital system.

mous numbers of microscopic *spermatozoa*. These are provided with long whiplike tails which later provide propulsive power. Specialized cells in the testis produce the male sex hormone.

The testis is covered in front and at the sides with a bursal sac, the *tunica vaginalis testis*, which is peritoneum, because it was continuous with the peritoneal cavity until about the time of birth.

The Ducts of the Testis

From 6 to 12 fine *efferent ductules*, lead out of the upper part of the testis into the *duct of the epididymis*. This thread-like duct, although about 6 meters long, is so folded as to form a compact body, the *epididymis* (Gk. epi = upon; didumos = a twin; i.e., the testes are twins), which caps the upper pole of the testis and is applied to its posterior border. The cells lining the duct of the epididymis discharge a mucoid secretion to carry the spermatozoa.

The *ductus* or *vas deferens* connects the duct of the epididymis to the urethra. It is about 45 cm long. It ascends through the upper part of the scrotum to the abdominal wall. This it pierces obliquely in a tunnel, the *inguinal canal*. Continuing, it runs under cover of the peritoneum to the base of the bladder, and then descends between the bladder and the rectum. Its terminal 2 cm, the *ejaculatory duct*, pierces the prostate and opens into the urethra close to its fellow, about 3 cm beyond the bladder. From this point onward the male urethra is the common duct of the urinary and genital systems. It is also the royal road for gonorrheal infection throughout the mucous membranes of the male genital tract.

Semen

Semen (L. = seed) is composed of spermatozoa suspended in the secretions of the ducts of the testes and of the accessory glands.

Accessory Glands

1. The Seminal Vesicles, one on each side, are tubular outgrowths from the last part of the deferent ducts, which they resemble in structure. They add a yellowish sticky liquid to the semen.

2. The Prostate (Gk. pro = before, istanai = to stand), the size and shape of a large chestnut, surrounds the first 3 cm of the urethra just beyond the bladder. It is partly glandular, partly muscular and partly fibrous. It secretes into the urethra an opalescent liquid, free from mucus. It is pierced by the paired ejaculatory ducts.

3. The Bulbo-urethral Glands (of Cowper), the size of a pea, lie one on each side of the membranous urethra. Their ducts are 2–3 cm long and open into the spongy urethra. They secrete a mucus-like substance.

4. The Urethral Glands (p. 281).

External Genitalia

The Penis (L. penis = a tail) is the male organ of copulation (*see fig. 333*). It comprises three parallel fibrous tubes, two paired and one unpaired (*fig. 51.1*). These have innumerable cavernous spaces, filled with blood. They are enclosed in a single loosely fitting tube of skin and subcutane-

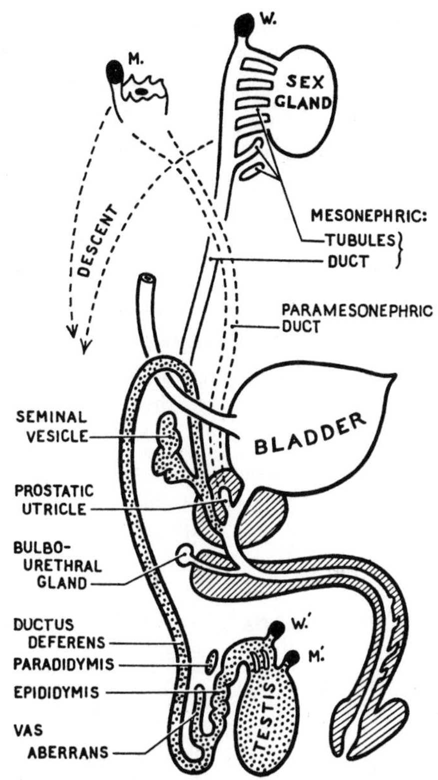

Fig. 50. Diagram of the male genital system (side view).

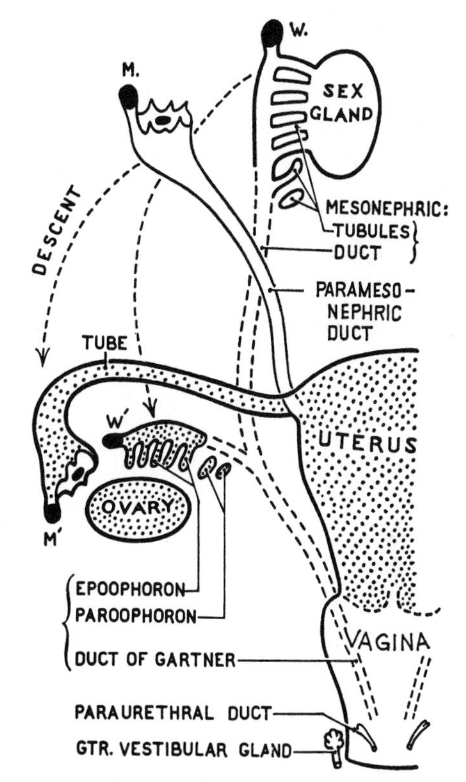

Fig. 51. Diagram of the female genital system (front view).

In the upper halves of both figures the parts are shown in their early or indifferent state. Most of the paramesonephric (Mullerian) structures disappear in the male, and of the mesonephric (Wolffian) in the female. (M = Wolffian appendage.)

ous tissue. The paired tubes, the right and left *corpora cavernosa*, are fused side by side. In front they present a rounded end; behind they diverge into right and left *crura* (L. crus = a leg) which are firmly attached to the pubic arch. The c. cavernosa are the "supporting skeleton" of the penis.

The unpaired tube, the *corpus spongiosum*, is traversed by the urethra. Its expanded hinder end is fixed to the perineal membrane which stretches between the sides of the pubic arch; its expanded anterior end, the *glans* (L. glans = an acorn), fits like a cap on the ends of the corpora cavernosa.

The redundant skin covering the glans is the foreskin or *prepuce*. The operation of

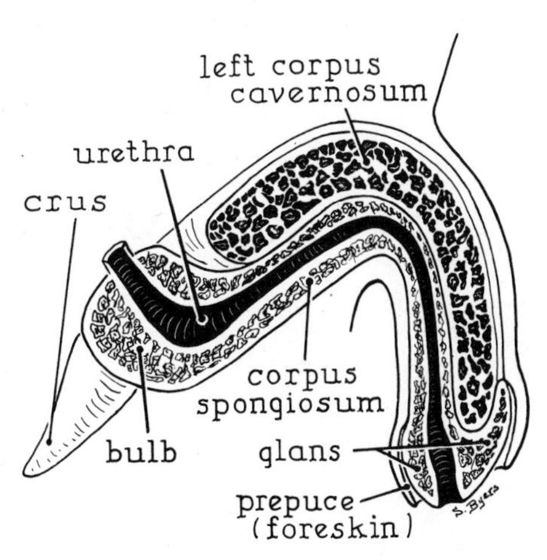

Fig. 51.1. Structure of corpora of penis cut longitudinally. (From *Primary Anatomy*.)

circumcision consists in removing the prepuce.

The Scrotum is the bag of skin and subcutaneous tissue in which the testes lie. Like the penis it is free from fat.

Female Genital System

The female reproductive organs are:

Ovary, Uterine Tube .... paired

Uterus, Vagina .... unpaired

The external genital parts include:

Clitoris

Pudenda (mons pubis, labium majus, labium minus, vestibule of vagina, bulb of vestibule, greater vestibular gland, vaginal orifice)

Rudimentary or vestigial structures:

Epoophoron and duct

Paroophoron

Ovaries

The ovaries or female sex glands, one on each side, lie on the side walls of the pelvis (*fig. 51.2*). Each ovary is a solid body about half the size of a testis. It has an attached border; otherwise it lies free in the peritoneal cavity. However, it is not covered with peritoneal cells but with cubical cells, some of which become the ova (eggs). Its surface is scarred and pitted due to the shedding of ova. At birth each ovary contains about 200,000 (immature) ova. From puberty to the end of the reproductive period (15th to 50th year) an ovum is shed into the peritoneal cavity about once a lunar month. In all about 400 ova are shed; the rest are absorbed in the ovary. Ovarian cells also release female sex hormones into the blood stream.

Uterine Tubes

The uterine tube (*fig. 51.2*) is about 10 cm long, is as large as a pencil, lies in the free edge of a fold of peritoneum, the *broad ligament*, and takes a twisted course from ovary to uterus. Its ovarian end is funnel shaped, the *infundibulum*, and at the bottom of the funnel is the abdominal mouth or *ostium* which lies open to the peritoneal

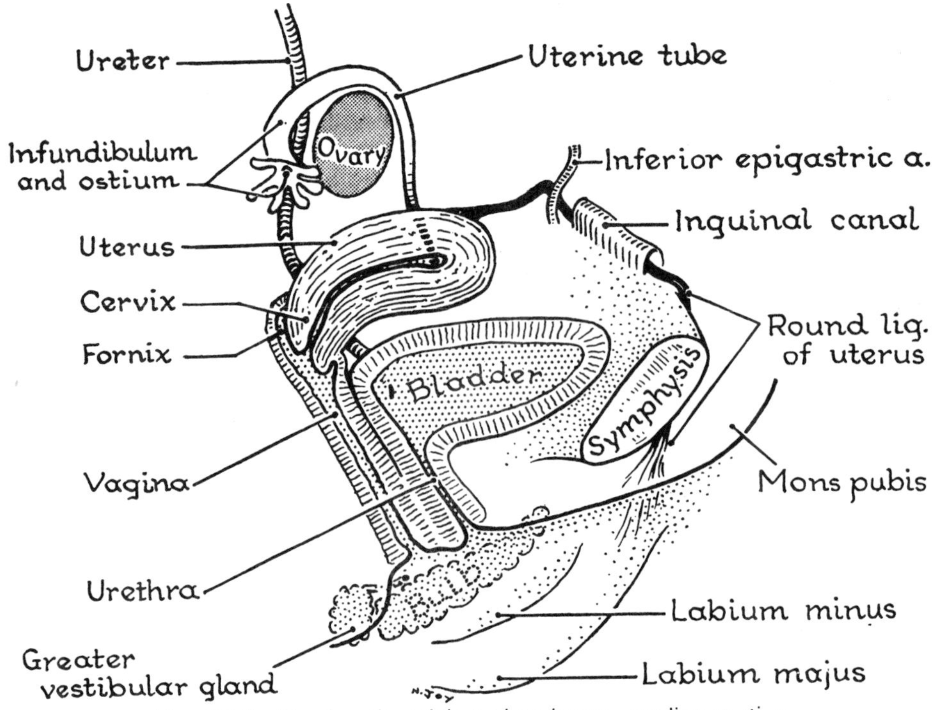

Fig. 51.2. The female pelvis and perineum, median section.

cavity, beside the ovary. Its other end opens into the cavity of the uterus.

When an ovum is about to be shed the mouth of the tube apparently lies ready to receive the ovum. Peristaltic action and cilia in the mucous membrane lining the tube propel it along. In the tube the ovum may be met and fertilized by a spermatozoon, which is able to propel itself from the vagina and through the uterus in about 6 hours.

Uterus

The uterus is a thick-walled, hollow, muscular organ placed near the center of the pelvis and projecting upward into the peritoneal cavity between the bladder and the rectum (see *fig. 379*). It is shaped like an inverted pear, somewhat flattened from before backward so that its cavity is collapsed, and is 8 cm long. Above, a uterine tube opens into it on each side; below, it opens into the vagina. The upper ⅔ (5 cm) are the *body;* the lower ⅓ is the *cervix* (= neck); the part of the body above the entrance to the tubes is the *fundus*. The peritoneum covering the body and fundus stretches from each side of the uterus as a fold, the *broad ligament*, that rises from the pelvic floor, extends to the side wall of the pelvis, and contains the uterine tube in its upper free edge. The function of the uterus is to harbor a fertilized ovum as it becomes a child for 9 months. The first 2 months the ova are in the indifferent or embryonic stage; the last 7, the formed or fetal stage.

Vagina

The vagina (L. = a sheath) is a relatively thin-walled collapsed tube, about 9 cm long. The cervix projects into its upper end; below, it opens into the *vestibule* of the external genitalia.

Accessory Glands

The Great Vestibular Glands (paired) open into the vestibule. Along with the more vital uterine tubes, they are prone to harbor chronic gonorrheal infections.

External Genitalia

These (pudenda) have their homologous parts in the male, see pages 265–267.

Nerve Supply. See page 263.

SKIN AND ITS APPENDAGES AND THE FASCIAE

Functions; Dermis or Corium; Epidermis; Nails; Hairs; Arrectores Pilorum Muscles; Sebaceous Glands; Sweat Glands; Vessels; Nerves.
SUPERFICIAL FASCIA OR TELA SUBCUTANEA —*Loose Areolar Tissue; Adipose Tissue.*
DEEP FASCIA

The skin or cutis has four appendages:
hairs and nails
sweat glands and sebaceous glands

The skin is no mere envelope wrapped around our bodies like paper around a parcel. It is, indeed, a wrapping but it is more than a wrapping—it is one of our most versatile organs.

As an *envelope* it has admirable properties: being waterproof it prevents the evaporation and escape of tissue fluids; it becomes thick where it is subject to rough treatment; it is fastened down where it is most liable to be pulled off; it has friction ridges where it is most liable to slip; "—even with our ingenious modern machinery we cannot create a tough but highly elastic fabric that will withstand heat and cold, wet and drought, acid and alkali, microbic invasion, and the wear and tear of three score years and ten, yet effect its own repairs throughout, and even present a seasonable protection of pigment against the sun's rays. It is indeed the finest fighting tissue" (Whitnall).

As an *organ* it is the regulator of the body temperature; it is an excretory organ capable of relieving the kidneys in time of need; it is a storehouse for chlorides; it is the factory for antirachitic vitamin D (ergosterol) formed by the action of the ultraviolet rays of the sun on the sterols in the skin, and necessary for the mineralization of bones and teeth; and it is the most extensive and varied of the sense organs.

In an average adult man the skin covers

a body surface of 1.7 sq. meters. At the orifices of the body it is continuous with the mucous membranes. The skin, somewhat modified, forms the conjunctiva and the outer layer of the ear drum.

The skin has two parts (*fig. 52*): (1) dermis, and (2) epidermis. (L. cutis and Gk. derma = skin; *cf.* the terms subcutaneous and hypodermic.)

Dermis or Corium (unlike epidermis) is of mesodermal origin, i.e., it develops from a middle layer of the primitive embryo, which lies between an outer layer (ectoderm) and inner layer (endoderm); from it develop the connective tissues, vessels, muscles, and skeleton. The corium (dermis) is a feltwork of bundles of white fibers and of elastic fibers. Superficially the feltwork is of fine texture; deeply it is coarse and more open and its spaces contain pellets of fat, hair follicles, sweat glands, and sebaceous glands. It is in general 1 to 2 mm thick, but is thicker on the palms and soles and back, and is thinner on the eyelids and external genital parts. The spaces in the feltwork are lozenge-shaped; hence, a puncture made with a conical instrument, such as a large needle, does not remain a round hole for the skin splits in the long axes of the lozenges. The long axes are differently directed in different parts, usually being parallel to the lines of tension of the skin. A cut across the long axes of these **Lines of Cleavage** (Langer's Lines) will gape. (Cox.)

It is due to the presence of elastic fibers that the skin, after being stretched or pinched into a fold, returns to normal. Dermis, when tanned, is called leather.

The Epidermis is a nonvascular strati-

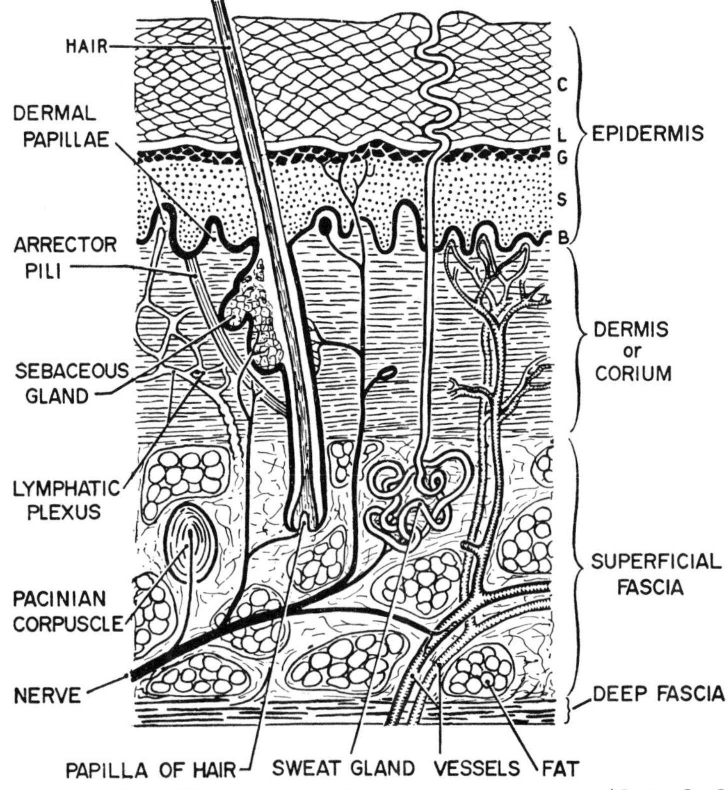

Fig. 52. Section of thick skin, superficial fascia, and deep fascia. (*C, L, G, S, B* = Strata Corneum, Lucidum, Granulosum, Spinosum, and Basal layer, respectively.) (After Appleton, Hamilton, and Tchaperoff.)

fied epithelium of ectodermal origin—the other major derivative of embryonic ectoderm being the central nervous system. The deeper layer, the "*germinative layer*," is living. It consists of several strata of polyhedral cells resting on a single stratum of columnar (basal) cells. The superficial layer, the horny layer or *stratum corneum*, is dead. It consists of several strata of dry, flattened, scaly cells without visible nuclei. The surface cells are perpetually being rubbed away and are perpetually being replaced by cells of the germinative layer.

Ridges of dermis, which on cross-section have the appearance of nipple-like processes and are therefore called *papillae*, project into the epidermis (Ham).

For *Finger Prints* see *figure 141*, and p. 131, and consult Cummins and Midlo.

The Nails are thickenings of the deeper layers of the stratum corneum. A nail has a free end which projects, a root which extends proximally deep to the overhanging nail fold, two lateral borders, a free surface, and a deep one. The white crescent appearing distal to the nail fold is the *lunule*. The deep surface rests on and adheres to the *nail bed*. This largely consists of white fibrous tissue which attaches the nail to the periosteum.

Growth takes place at the root and from the bed as far distally as the lunule; beyond this the nail probably slides distally on its bed, adhering to it. Poisons, formed during an acute illness, temporarily arrest the growth of the nails (as they do of the bones, see *fig. 15*) and a transverse ridge appearing on each nail when growth is resumed is evidence of a past illness. Seeing that the average rate of nail growth is 0.1 mm a day or 3 mm a month (i.e., about 3–5 cm a year) the date of a past illness can be estimated. The toe nails grow much more slowly than finger nails, of which the nail of digit III consistently grows fastest, that of digit V, slowest. Nails grow rapidly in "nail biters"; in immobilized limbs they grow slowly (Le Gros Clark and Buxton).

The Hairs are distributed over the entire surface of the body except the palms and soles, dorsum of the last segment of the digits of the hand and foot, red of the lips, and parts of the external genitals and the conjunctiva. They are also present in the vestibule of the nasal cavity and in the outer part of the external acoustic meatus.

Hairs may be short (a few mm) or long. Long hairs are present in the scalp, eyebrows, margins of the eyelids, vestibule of the nose, outer part of the external acoustic meatus; and at puberty they appear on the pubes, external genitals, axillae, and in the male on the face.

A hair has a *shaft* or part that projects beyond the skin surface, a *root* that lies in a follicle of the skin, and at the end of the root there is a swelling, the *bulb*, which is moulded over a dermal papilla. The life of a hair on the head is about 2 to 4 years; of an eyelash about 3 to 5 months. Like the leaves on an evergreen tree, old hairs are constantly falling out and new ones taking their place.

The Arrectores Pilorum Muscles are bundles of smooth muscle that pass obliquely from the epidermis to the slanting surface of the hair follicles deep to the sebaceous glands. By contracting they cause the hairs to stand erect. In birds, by erecting the feathers, they increase the air spaces between them thereby preserving heat; hence, sparrows look plump in cold weather. In man, spasm of the Arrectores produces "goose skin."

Sebaceous Glands are simple alveolar glands, bottle-shaped, and filled with polyhedral cells which break down into a fatty secretion, called sebum, in the hair follicles. The glands develop as outgrowths of hair follicles into the dermis, one or more being associated with each hair. Commonly, the glands are largest where the hairs are shortest (e.g., end of nose, and outer part of the external acoustic meatus). They make the skin waterproof—"like water off a duck's back" is a common expression.

Fortunately there are no hairs on the

palms and soles; neither are there sebaceous glands to make the surfaces greasy. Independently of hairs, sebaceous glands are present on the inner surface of the prepuce, on the labia minora, and on the areolae of the mammae. The tarsal glands of the eyelids also are modified sebaceous glands which waterproof the edges of the lids.

Boils (and carbuncles) start in hair follicles and sebaceous glands and are therefore possible, indeed common, in the vestibule of the nose and outer part of the external acoustic meatus.

Sweat Glands are present in the skin of all parts of the body (red of the lips and glans penis excepted), being most numerous on the palms and soles and in the axillae. They are simple tubular glands. The secretory part is coiled to form a ball (0.3 to 0.4 mm in diameter) situated in the fat deep to the dermis. The duct runs tortuously through the dermis, enters the epidermis between two ridges and proceeds spirally to the skin surface. In the stratum corneum it is represented merely by a cleft between the cells. The resemblance to the "intestines of a fairy" was fancifully suggested by Oliver Wendell Holmes.

The ceruminous (= wax) glands in the outer parts of the external acoustic meatus and the ciliary glands of the eyelids are modified sweat glands. In the axilla, about the external genitals, and around the anus are long, large (3 to 5 mm in diameter) modified sweat glands that produce an odor. They are spoken of as "sexual skin glands."

Sweating lowers the temperature. In man at rest, sweating is observed to begin abruptly when the body temperature is elevated a fraction of a degree. This is due to the action of the heated blood on the brain centers. Since the autonomic nerve fibers to sweat glands travel to the skin in the ordinary cutaneous nerves, if such a nerve is cut the area is not only deprived of sensation but it also cannot sweat.

Sweat glands are excretory organs—accessory to the kidneys. The salt taste of sweat is due to sodium chloride. The normal sweat secretion is important in keeping the thick horny layers of the palms and soles supple, and it increases the friction between the skin and an object grasped. In dogs, sweat glands are confined to the foot pads; so, being unable to sweat, dogs pant.

Vessels. The vessels for the supply of the skin run in the subcutaneous fatty tissue. From these the dermis receives two arterial plexuses; one is deeply seated near the subcutaneous tissue, and the other is in the subpapillary layer. This sends capillary loops into the papillae. The returning blood passes through several layers of thin-walled subpapillary venous plexuses, thence through a deep venous plexus, and so to the superficial veins. Arterio-venous anastomoses, which are sometimes open and sometimes closed, connect some of these arterioles and venules.

The *lymph vessels* of the skin form a plexus at the junction of the dermis and the superficial fascia. This plexus receives blind finger-like vessels (or networks) from the papillae and it drains into lymph vessels that accompany the superficial arteries and veins.

Nerves. The cutaneous nerves have (1) *efferent* autonomic fibers for the supply of:

Smooth muscle { of hairs (Arrectores)
of blood vessels

Glands .... { sweat glands
sebaceous glands

(2) *afferent* somatic fibers of general sensation, namely touch, pain, heat, cold, and pressure (*fig. 43*, p. 41).

As *figure 52* indicates, free fibers end between the cells of the germinative layer (hence, intra-epidermal injections may cause pain), and around the hair follicle and beside it (probably touch fibers); tactile corpuscles occupy occasional papillae (for touch); Pacinian corpuscles lie in the superficial fascia and are plentiful along the sides of the digits (for pressure).

Superficial Fascia or Tela Subcutanea

Superficial fascia is a subcutaneous layer of **loose areolar tissue** which unites the corium of the skin to the underlying deep fascia. It consists of (1) bundles of *white* or *collagenous fibers* which, by branching and uniting with other bundles, form an open webbing, filled with (2) *tissue fluid;* (3) a slender network of *yellow elastic fibers*, and scattered amongst all this lie (4) *connective tissue cells (fig. 53)*.

When areolar tissue is exposed to the air, held fluids rapidly evaporate with consequent drying and shrinking. Fortunately, the addition of an appropriate laboratory solution will restore to areolar tissue its original fluffy texture.

Areolar tissue is derived from those portions of mesoderm that remain after bones, ligaments, tendons, muscles, and vessels have taken form. It is, therefore, not confined to the superficial fascia but is diffusely spread; for example, it forms the sheaths of muscles, vessels, nerves, and viscera, and it fills up the spaces between them; it forms the basis of the mucous, submucous, and subserous coats of the hollow viscera; the serous membranes (i.e., peritoneum, pleura, pericardium, and tunica vaginalis testis) are but areolar membranes lined with flat mesothelial cells.

Areolar tissue is potentially **Adipose Tissue** or fat, and wherever areolar tissue occurs, there fat also may occur. The fat accumulates in the connective tissue cells. Fat is fluid at body temperature, but because each drop of fat is imprisoned in a cell, it does not "run" away; nor can it be massaged away normally.

Distribution. The superficial fascia almost everywhere contains fat—except in the eyelids, external ear, penis, and scrotum, and at the flexion creases of the digits. In the palms and soles it forms a protective cushion; here and still more so

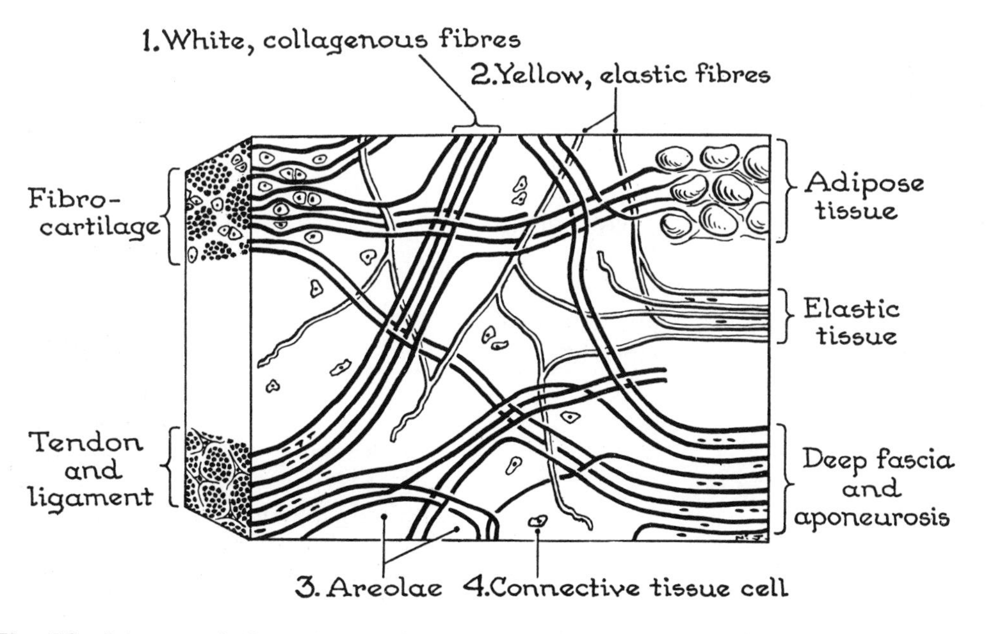

Fig. 53. Scheme to indicate that the four ingredients of areolar tissue (viz., collagenous fibers, elastic fibers, areolar spaces and cells), when blended in different proportions, form other tissues (e.g., adipose, elastic, collagenous, ligamentous, and fibrocartilaginous) and that one may merge into another.

in the breast and scalp it is loculated, i.e., imprisoned in loculi. It is most abundant in the buttocks.

In women, in whom special fat deposits are a normal secondary sexual characteristic, the typical areas are in the gluteal and lumbar regions, front of the thigh, anterior abdominal wall below the naval, mammae, postdeltoid region, and cervicothoracic regions.

Adipose or fatty tissue, as mentioned, is modified areolar tissue, and is notably present in the superficial fascia, the subserous layer of the abdominal wall and in the omenta and mesenteries. It covers parts of the urinary tract (e.g., kidney, ureter, sides of the bladder), but it leaves free the gastro-intestinal tract, liver, spleen, testis, ovary, uterus, and lung. It is odd, then, that it should be present in the sulci of the heart. There is no fat within the cranial cavity to dispute possession with the brain, nor within the dura mater covering the spinal cord. The loose fatty tissue of the orbital cavity provides the eyeball with a soft and pliant bed.

Subcutaneous Bursae. See page 24.

Deep Fascia

Deep fascia is the membranous investment of the structures deep to the superficial fascia (*fig. 53* and *54*). Like tendons, aponeuroses, and ligaments, it contains the same four ingredients as areolar tissue, but in different proportions; and, like them, being subjected to tensile strains, the white collagenous fibers form parallel bundles; the tissue fluids are at a minimum, and the cells are flattened (stellate on cross-section) from pressure.

Deep fascia is best marked in the limbs and neck, where it is wrapped around the muscles, vessels, and nerves like a bandage, the fibers being chiefly circularly arranged. Around the thorax and abdomen, which require to expand and contract, there may be a film of areolar tissue but there can be no true deep fascia.

The deep fascia sends septa, *intermuscular septa*, between various muscles and groups of muscles (*fig. 54*) and usually blends with or attaches to periosteum.

AXIOM. Where deep fascia encounters bone, it does not cross it but attaches itself to it, for the simple reason that both have a common derivation; so, unless some muscle during development intervenes and detaches the fascia from the bone, the two remain attached.

The deep fascia is thickened where muscles are attached to it, and the direction of its fibers takes the line of the pull of the muscles. It is also thickened about the wrist and ankle to form, as it were, bracelets and anklets, called *retinacula*, which prevent the tendons of muscles from bowstringing (*fig. 55*).

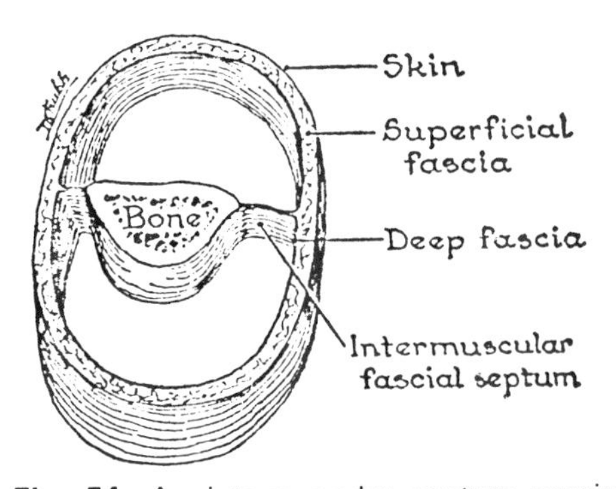

Fig. 54. An intermuscular septum passing from deep fascia to bone.

Fig. 55. Demonstrating the function of a retinaculum (anular ligament). (After Andreas Vesalius—De Humani Corporis Fabrica, 1568 edition: first published in 1543.) (By the courtesy of the Toronto Academy of Medicine.)

SECTION TWO

UPPER LIMB

5	Pectoral Region and Axilla	75
6	Scapular and Deltoid Regions	89
7	Cutaneous Nerves and Superficial Veins	103
8	Arm	109
9	Flexor Region of Forearm	116
10	Hand	125
11	Extensor Region of Forearm and Hand	142
12	Joints of Upper Limb	152

5 PECTORAL REGION AND AXILLA

INTRODUCTION

Anatomical Position; Mooring Muscles; Lines of Force Transmission.

Definitions and Boundaries; Landmarks; Sternum; Coracoid Process; Clavicle; Pectoralis Minor; Two Triangles.

WALLS OF AXILLA

ANTERIOR WALL—*Pectoralis Major: parts, origins, insertions, actions; Cephalic Vein; Costocoracoid Membrane and Clavipectoral Fascia.*

POSTERIOR WALL—*Muscles; Quadrangular and Triangular Spaces.*

LATERAL WALL

MEDIAL WALL—*Serratus Anterior.*

VESSELS AND NERVES OF AXILLA

Axillary Sheath; Great Arterial Stem; Axillary Artery and Branches; Axillary Vein and Tributaries.

Brachial Plexus (of Nerves); Relation to Artery; Identification of Parts; Motor Nerves.

Typical Spinal Nerve; Cutaneous Nerves.

FASCIA, MAMMARY GLAND, LYMPH NODES, AND ANOMALIES

INTRODUCTION

Human upper limbs move freely, being adapted to purposes of prehension. Each articulates with the trunk at one small joint, the *sternoclavicular joint*. The lower limbs have no such freedom. They have to bear the weight of the body during locomotion and standing. Accordingly, they are united behind to the vertebral column at the sacro-iliac joints, and in front to each other at the symphysis pubis. (See table 1, p. 76).

In prenatal life the thumb and the radius, and the big toe and the tibia, are situated on the cranial or head side of the central axis of the respective limbs and are said to be *preaxial*. However, the 5th finger and ulna, and the 5th toe and fibula lie on the caudal or tail side of this axis and are said to be *postaxial*.

On assuming the **Anatomical Posture** (*fig. 1*), the upper and lower limbs undergo rotation but in opposite directions; the thumb by rotating laterally brings the palm of the hand to the front, while the big toe by rotating medially brings the sole of the foot to the ground.

Mooring Muscles (*fig. 56*). The upper limb is moored to the head, neck and trunk by muscles, which may be likened to dynamic (rather than static) guy ropes.

Lines of Force Transmission. The following fundamental points should be verified by reference to the skeleton and *figure 57*.

The *clavicle* is a strut that, through the medium of a strong ligament, the *coracoclavicular*, holds the scapula (and the rest of the limb with it) away from the median plane. The clavicle has an enlarged medial end which articulates with a shallow socket formed by the sternum and 1st rib cartilage. A strong ligament, in the form of an *articular disc*, prevents the clavicle from being driven medially on to the sternum. The lateral end of the *clavicle* is not enlarged, doing little more than making contact with the *acromion*.

At the shoulder joint the rounded head of the humerus makes side-to-side contact

TABLE 1

Bones of Upper Limb

Region	Bones
Shoulder or Pectoral Girdle	Clavicle Scapula
Arm or Brachium	Humerus
Forearm or Antebrachium	Radius Ulna
Hand or Manus	Carpal bones Metacarpal bones Phalanges 1st or Proximal 2nd or Middle 3rd or Distal

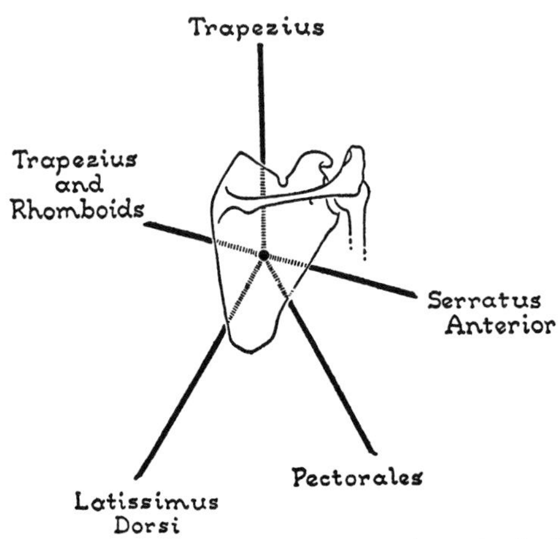

Fig. 56. Upper limb is moored to body by muscles.

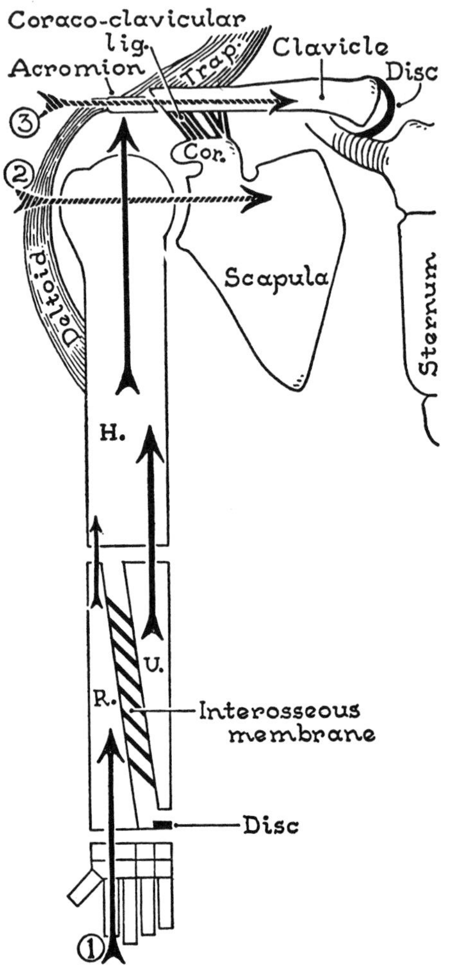

Fig. 57. Scheme of skeleton of limb showing lines of force transmission and ligaments that may transfer force as the result of a fall:

1. on the hand—interosseous membrane,
2. on the shoulder—coracoclavicular ligament,
3. on the acromion—articular disc.

with the shallow glenoid cavity of the scapula. The acromion and related coraco-acromial ligament form a hood-like roof or arch, which prevents upward dislocation of the humerus. The lower end of the humerus articulates with the ulna and radius.

The ulna and radius are united by a strong *interosseous membrane*. Clearly, at the elbow joint the ulna is more important than the radius; at the wrist joint the contrary is true. The enlarged lower end of the radius articulates with the carpal bones, but the ulna does not. *Figure 57* summarizes the lines of force transmission.

Definition and Boundaries of Axilla. The **axilla** is the pyramidal space above the arm pit. It has four walls, an apex, and a base (*fig. 58*). The *anterior wall* is fleshy and is formed by the Pectoralis Major and two muscles that lie behind it enclosed in a sheet of fascia, called the *clavipectoral fascia*. [This wall is practically synonymous with the Pectoral Region, which includes the breast or mamma.]

The *posterior wall* is formed by the scapula overlaid by the Subcapularis, and below this by the Latissimus Dorsi and Teres Major. The thick fleshy lower borders of the anterior and posterior walls can be grasped between the fingers and the thumb.

The *medial wall* is formed by the upper ribs (2nd to 6th) covered with Serratus Anterior. The narrow *lateral wall* is the intertubercular sulcus (bicipital groove) of the humerus. It lodges the long tendon of the Biceps, and its lips give attachment to muscles of the anterior and posterior walls.

The base is the skin and fascia of the arm pit. *The truncated apex* is the triangular space bounded by three bones—the clavicle, the upper border of the scapula, and the 1st rib.

The Contents of the Axilla (*fig. 65*). The great vessels and nerves of the upper limb pass through the axilla on their way to the distant parts of the limb—they are the chief contents. The other contents are: the two heads of the Biceps, the Coracobrachialis, and lymph nodes.

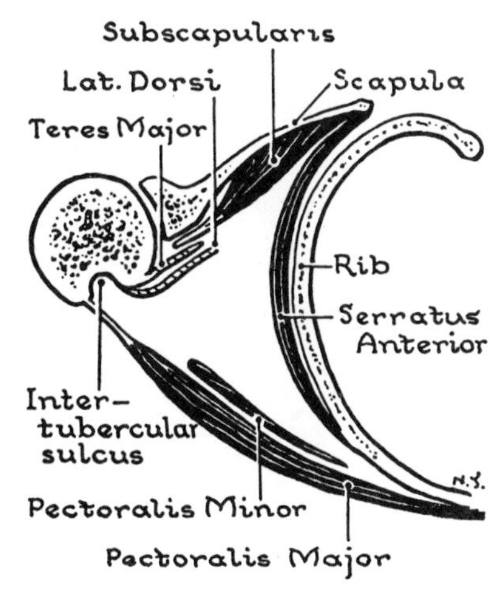

Fig. 58. Walls of axilla, on cross section.

Landmarks, Bony and Muscular (*fig. 59*). The *Sternum*, or breast bone, is a flat bone, shaped like a short flat sword. It consists of three parts: manubrium, body, and xiphoid process. The manubrium, or handpiece, meets the body at a slight angle, called the *sternal angle*. On each side the sternum articulates with a clavicle and seven costal (rib) cartilages. (For details see Chap. 29.)

The *Coracoid Process* of the scapula (Gk. Korax = a crow, i.e., a crow's beak) is misnamed (*fig. 60*). Its *tip* lies below the

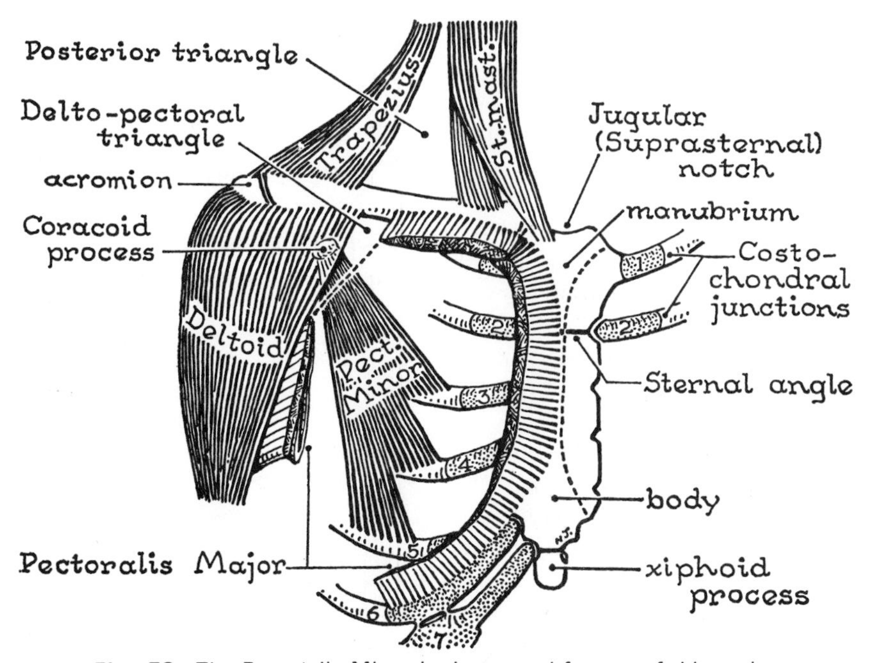

Fig. 59. The Pectoralis Minor is the central feature of this region.

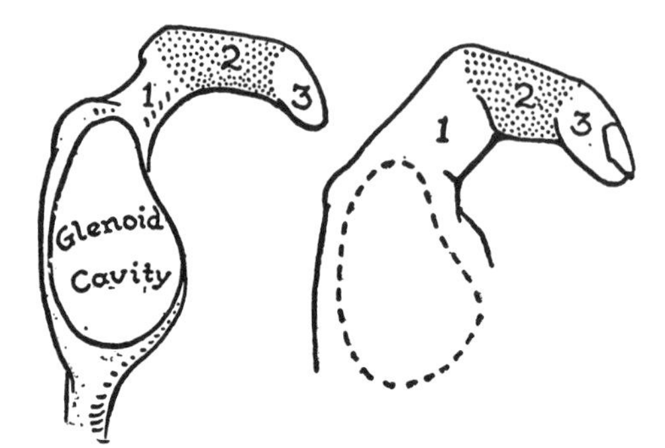

Fig. 60. The coracoid process of the scapula resembles a bent finger.

clavicle under shelter of the anterior edge of the Deltoid muscle (*fig. 59*). To palpate this essential landmark in the living subject, press firmly not backward, but laterally.

The *Clavicle*, or collar bone (*figs. 59* and *60.1*), is a long bone with a double curve and set nearly horizontally. It has two parts: the part medial to the underlying coracoid process (medial three-fourths) is triangular on cross-section like long bones in general; but the part lateral to the process is flattened and is rough below for the ligaments that bind the scapula, via its coracoid process, to the clavicle (*fig. 60.2*). When the clavicle is fractured ("broken collar bone") the shoulder region falls medially, forward and downward.

The clavicle, although almost visible through the skin, is not merely subcutaneous but also subplatysmal because a thin broad sheet of muscle, the *Platysma*, descends from the neck to the level of the 2nd or 3rd rib and intervenes between the clavicle and the skin. The Platysma is superficial even to the cutaneous nerves, the *supraclavicular nerves*, that cross in front of the clavicle (*fig. 61*).

Along the superficial aspect of the clavicle, a bare strip lies between the attachments of the Sternomastoid and Trapezius above and of the Pectoralis Major and Deltoid below (*figs. 59 and 63*).

(The clavicle is described in fuller detail for specialists on p. 98.)

The *Pectoralis Minor* (*fig. 59*), completely concealed by the Pectoralis Major, is the central landmark. It arises from the (2nd), 3rd, 4th, 5th, (and 6th) ribs, where bone and cartilage join, and it is inserted into the medial border of the coracoid process near its tip. Of course, the insertion or apex of this triangular muscle is tendinous (*fig. 61.1*).

***AXIOM**: When the site of insertion of a muscle is restricted, the insertion must be fibrous.*

Surface Anatomy. The only certain way of identifying any rib lies in reckoning from the *sternal angle* which is at the level of the 2nd costal cartilages. About 5 cm below the *jugular notch* (suprasternal notch), you

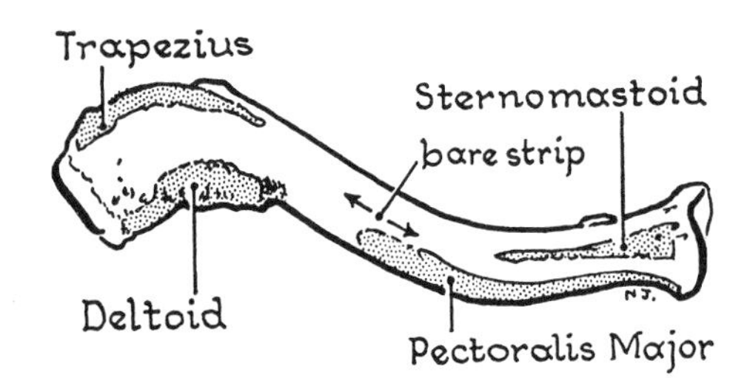

Fig. 60.1. Clavicle and attachments of muscles.

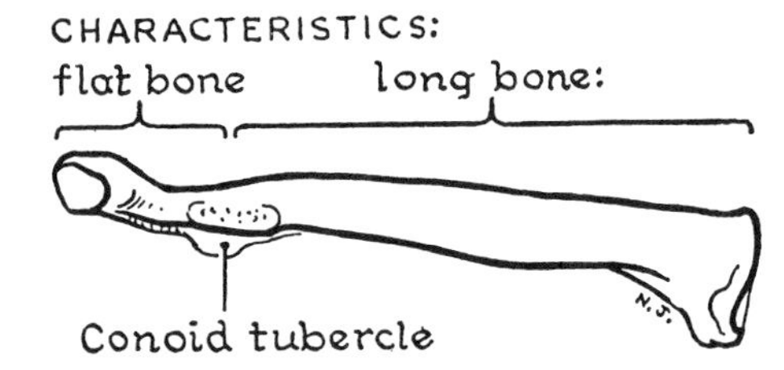

Fig. 60.2 The two parts of the clavicle.

Fig. 61. The clavicle is occasionally pierced by a branch of the supraclavicular nerves.

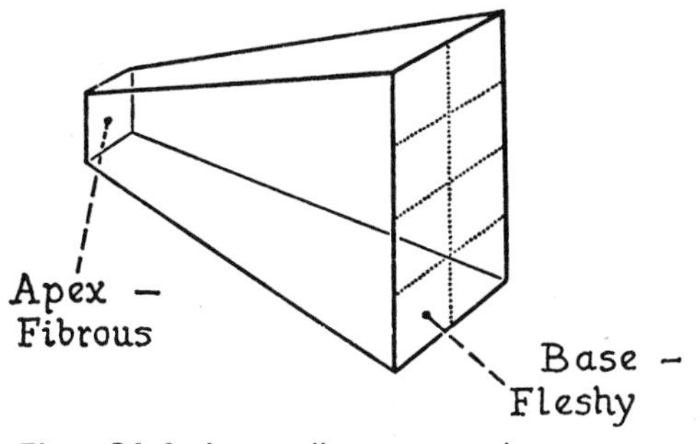

Fig. 61.1 A tendinous attachment conserves space.

encounter the transverse ridge which indicates where the manubrium and body of the sternum articulate with each other. This is the sternal angle. Follow the ridge laterally to the 2nd costal cartilage; you may then readily palpate the 3rd, 4th, and 5th cartilages through the substance of the Pectoralis Major. By joining a point on the 3rd rib about 5–7 cm from the midline to the coracoid process you define the upper border of the Pectoralis Minor; similarly by joining the 5th rib near its cartilage, say 10 cm from the median plane, to the coracoid process you map out the lower border of this triangular muscle.

WALLS OF AXILLA

Anterior Wall of Axilla

The anterior wall of the axilla is formed throughout by Pectoralis Major, and behind this by Pectoralis Minor and Subclavius within their sheaths of clavipectoral fascia (*fig. 63*).

Pectoralis Major. The *clavicular head* arises in line with the Deltoid from the anterior aspect of the clavicle. The *sternal head* arises from the sternum: its origin curves downward and laterally along the 5th (or 6th) costal cartilage.

The two heads have, therefore, a continuous curved fleshy origin from clavicle, sternoclavicular joint, sternum, and one or two ribs, and even the aponeurosis of the External Oblique muscle of the abdomen.

The Pectoralis Major is inserted by means of a folded (3-layer) aponeurosis into the crest of the greater tubercle of the humerus (lateral lip of the bicipital groove) (*figs. 62, 64*).

The clavicular part is inserted by means of an anterior lamina; the upper sternal fibers pass to the middle lamina; the lower fibers pass to the posterior lamina. In birds of flight that have need of powerful wing muscles, a bony keel develops from the sternum to increase the origin for the Pectoralis Major.

Actions. Obviously all parts of the Pectoralis Major adduct the humerus (i.e., move it toward the median plane of the body) and rotate it medially (on its own long axis). If, while palpating your Pectoralis Major, you put the radial side of your closed fist under the edge of a heavy table and try to raise it, you will find that the clavicular part of the muscle comes into action. If, now, you put the ulnar side of your closed fist on the table and press downward, you will find that the sternal part comes into action. This indicates that the clavicular part aids in flexing the shoulder joint (i.e., raises the humerus forward, as in pushing), and that the sternal part aids in extending the joint (i.e., brings the flexed humerus downward and backward to the side).

The Pectoralis Major and Deltoid are continuous with each other below at their humeral attachments, but above they diverge slight (1 cm) from each other and form with the clavicle the *Deltopectoral Triangle*. It allows the passage of the cephalic vein, its accompanying arterial twig (the deltoid), and lymph vessels.

The cephalic vein (*fig. 62*), which occupies the furrow between the Deltoid and the Pectoralis Major, may be followed through the deltopectoral triangle and through the costocoracoid membrane to its termination in the axillary vein.

The Practical Significance of the cephalic vein is that a surgeon may pass a fine, pliable tube through it and the axillary and subclavian veins and onward into the heart.

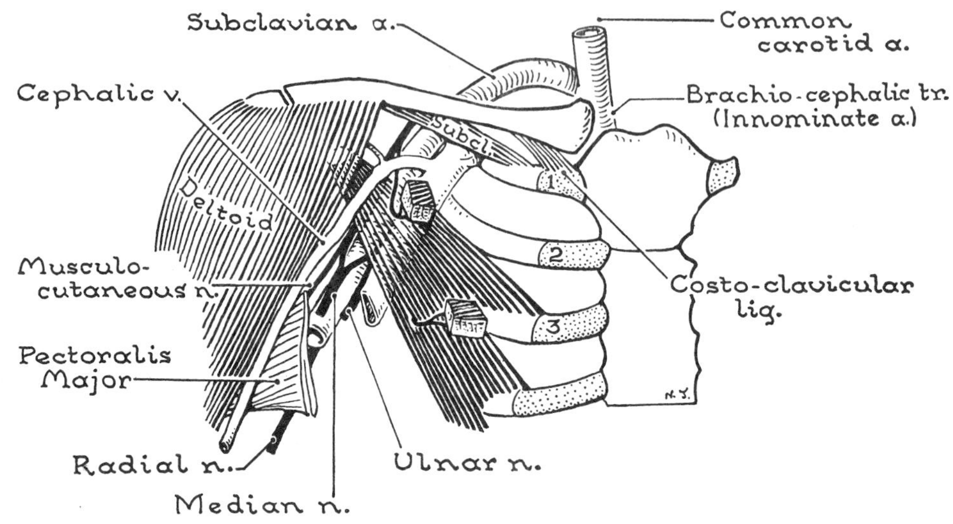

Fig. 62. Cephalic vein. Two buttons of Pectoralis Major hang from its nerves.

The costocoracoid membrane extends laterally from the 1st and 2nd costal cartilages medially, to the coracoid process. It is part of a larger fascial sheet, the **clavipectoral fascia** (*fig. 63*), which splits to enclose the Pectoralis Minor and Subclavius muscles.

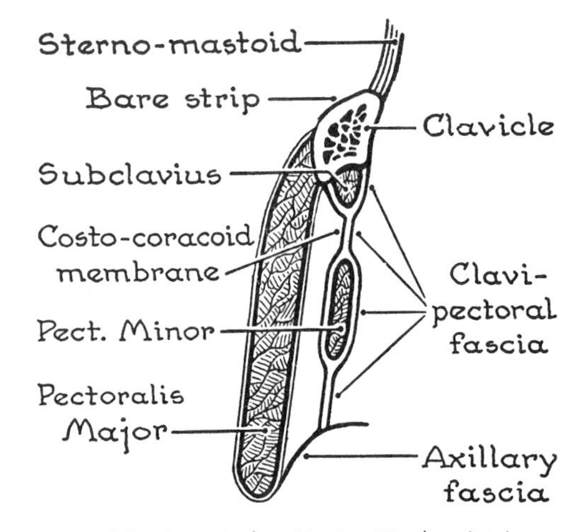

Fig. 63. Anterior wall of axilla (sagittal section).

Posterior Wall of Axilla

The posterior wall is formed by the Subscapularis, Teres Major, and Latissimus Dorsi; the scapula being the background (*fig. 64*). Their tendons insert near each other medial to the intertubercular sulcus (lesser tubercle and medial lip of the sulcus).

(For Origin of Latissimus from vertebral spines and iliac crest, see p. 90).

If the Subscapularis is separated from the Teres Major, a long triangular space, whose lateral side or base is the surgical neck of the humerus, will be opened up. In its depths the long head of the Triceps Brachii will be seen. This head of the Triceps subdivides this long triangular space into a small, unimportant, medial *Triangular Space* and an important, lateral *Quadrangular Space* (*fig. 64* and p. 95).

Lateral and Medial Walls of Axilla

The **lateral wall** (*figs. 64, 65*) is the narrow interval between these two crests, called the *intertubercular sulcus*.

The **medial wall** is formed by ribs and intercostal muscles covered with Serratus Anterior.

Serratus Anterior is the chief muscle to protract or pull forward the scapula, as when pushing (*fig. 66*). If it is paralyzed, attempts to raise the arm in front of the body largely result in causing the inferior angle of the scapula to project from the back. Its nerve (long thoracic n.) runs vertically a little behind the midaxillary line (*fig. 65*).

Origin. The Serratus arises from the

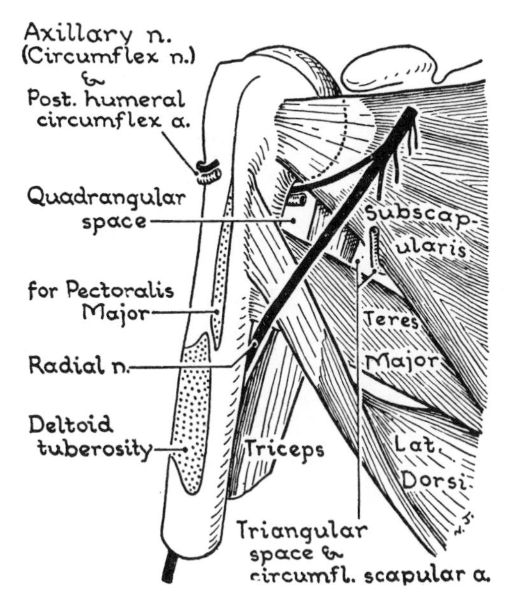

Fig. 64. The posterior wall of axilla.

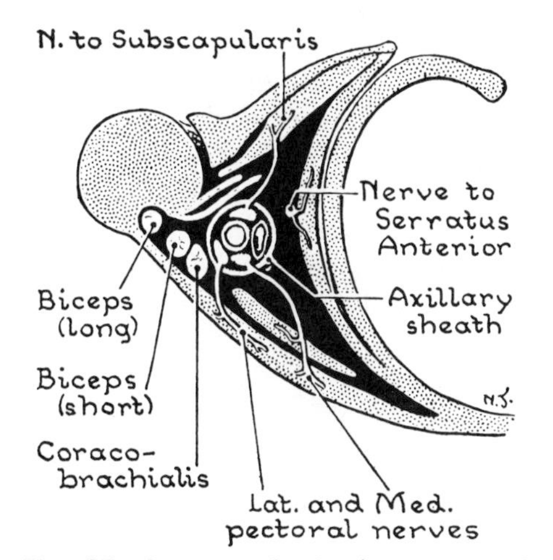

Fig. 65. Contents of axilla (cross section).

outer surfaces of the upper eight ribs by a series of fleshy digitations (*fig. 66*).

Insertion. The Serratus is inserted (*fig. 66.1*) into the costal aspect of the medial border of the scapula with concentration on the inferior angle. The usefulness of this is apparent; the fibers are concentrated where they act to best advantage, at the end of a lever (*fig. 66*).

VESSELS AND NERVES OF AXILLA

The great axillary vessels and nerves are enveloped in a thin tube, the **axillary sheath,** continuous with fascia in the neck. They enter the axilla through the triangular space formed by three bones—clavicle, 1st rib, and upper border of the scapula. They curve downward to the humerus.

The Great Arterial Stem of the limb (*fig. 67*) is called the *subclavian artery* until it reaches the lower border of the 1st rib. From there to the lower border of the Teres Major it lies within the axilla and is known as the *axillary artery*. On leaving the axilla to enter the arm or brachium it becomes

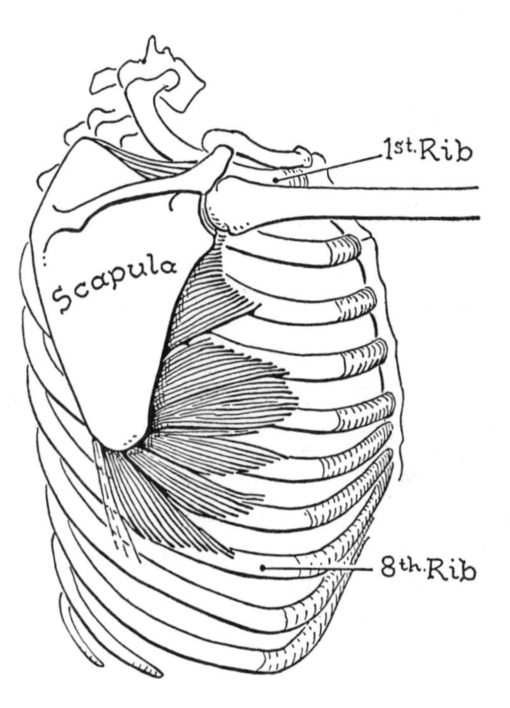

Fig. 66. The Serratus Anterior.

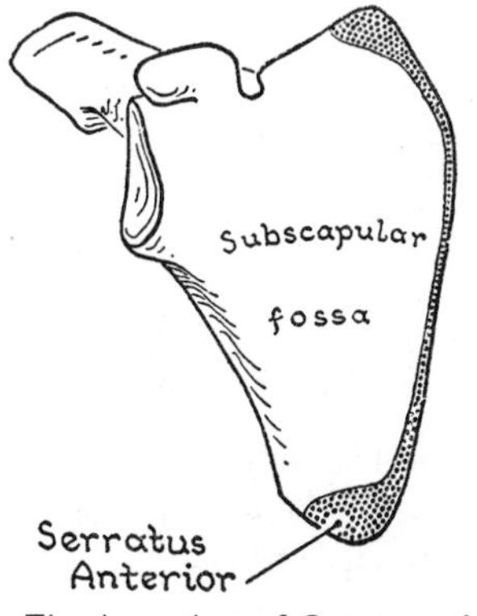

Fig. 66.1 The insertion of Serratus Anterior.

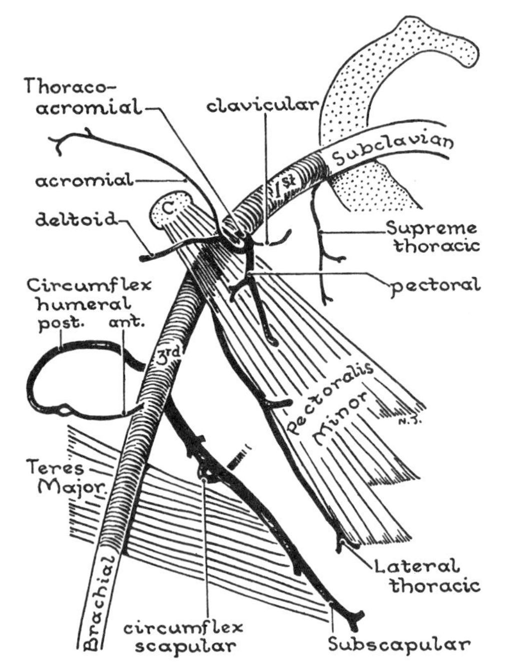

Fig. 67. Three parts of axillary artery and its branches.

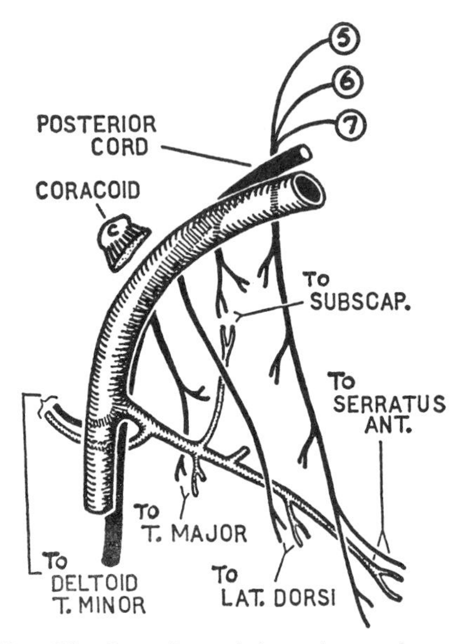

Fig. 68. Branches of the subscapular artery accompanying 5 motor nerves.

the *brachial artery* (p. 112).

The Axillary Artery is conveniently divided into three parts by the Pectoralis Minor (*fig. 67*). The 2nd part lies close behind the Pectoralis Minor a finger's breadth from the tip of the coracoid process—this is important. The relations of the brachial artery to the nerves are considered on page 83.

On voluntarily forcing your arm backward and downward you compress the artery between the 1st rib and the clavicle and so arrest or diminish the radial pulse at the wrist—a trick known to old soldiers. In the event of the clavicle being fractured the Subclavius serves as a buffer to protect the great vessels and nerves from the ragged ends of the bone. This is the only significant "use" of this rather insignificant muscle.

Branches. One arises from the first part, two from the second, and three from the third. Details are shown in *fig. 67*. Of these 6 branches, the *subscapular* is much the largest; it is the artery of the posterior wall (*fig. 68*), and it follows the lower border of the Subscapularis to the medial wall. It sends a large branch, the *Circumflex scapular a.*, to the dorsum of the scapula.

It is through the anastomoses that this, the largest of all the branches of the axillary artery, makes on the chest wall and around the scapula that the circulation in the limb is maintained after the distal part of the subclavian a. or the proximal part of the axillary artery has been ligated (*fig. 79*).

The main duty of the thoraco-acromial *a.* is to supply the Pectoral Muscles.

The lateral thoracic a. is part of the blood supply of the female breast, and therefore important in surgery.

The three branches of the 3rd part of the axillary artery arise either singly or together about the level of the surgical neck of the humerus, in front of the quadrangular space (*fig. 64*).

Emboli or blood clots expelled from the heart during life are apt to lodge at this site, because beyond it the main vessel is much reduced in caliber.

The circumflex humeral arteries encircle the surgical neck of the humerus. The posterior artery accompanies the axillary nerve, and is large.

The Axillary Vein lies on the medial or concave side of its artery, but it overlaps and conceals the artery when the arm is abducted. It is the continuation of the basilic vein, and at the 1st rib becomes the **subclavian vein;** it has two or three bicuspid valves.

Tributaries. In addition to receiving tributaries corresponding to the six branches of the axillary artery, it receives the two **venae comitantes** of the brachial artery and the **cephalic vein.** (Venae comitantes, p. 32 and *fig. 36.1.*)

The Brachial Plexus is formed by the alternate union and bifurcation of nerves, thus: five ventral nerve *rami* unite to form three *trunks*, which bifurcate to form six *divisions*, which unite to form three *cords*, which bifurcate to form six *terminal branches* (*fig. 69*). Of these six terminal branches, two soon unite to form the median nerve; hence, the plexus may be said to begin as five ventral rami and to end as five nerves.

The rami and the trunks of the plexus lie in the neck, the divisions behind the clavicle, the cords above and behind the Pectoralis Minor, and the terminal branches distal to it.

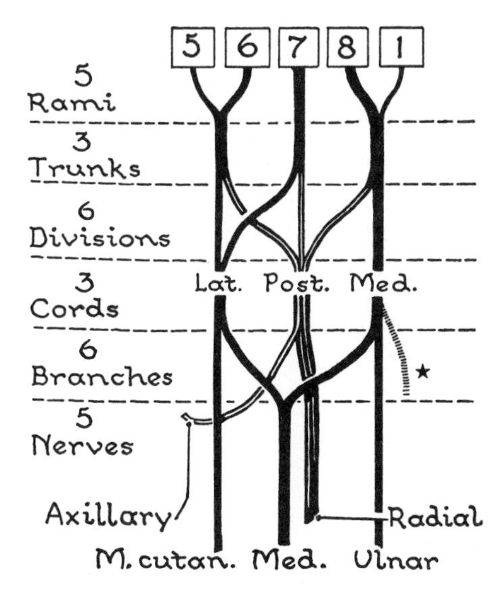

Fig. 69. Scheme of the brachial plexus. *Indicates the medial cutaneous nerve of the forearm.

Each of the three trunks divides into an anterior and a posterior division. The three posterior divisions unite together to form a single *posterior cord.*

Of the three anterior divisions, the lateral and intermediate unite to form the *lateral cord*, while the medial continues its course as the *medial cord.*

AXIOM: ***It is a fundamental truth that the nerves to the muscles on the original ventral and dorsal surfaces of the upper and lower limbs are derived from the anterior and posterior divisions respectively of the brachial and lumbosacral plexuses.***

Each of the three cords gives off one or more *collateral branches* and ends by dividing into two *terminal branches* (as shown in table 2 (*next page*), the lateral cord dividing into the musculocutaneous nerve and the lateral root of the median nerve, the medial cord into the ulnar nerve and the medial root of the median nerve, and the posterior cord into the radial and the axillary nerve (circumflex n.).

From a glance at *figure 69* it should be clear that the musculocutaneous nerve and the lateral root of the median nerve can derive fibers from the 5th, 6th, and 7th nerve segments; the ulnar nerve and medial root of the median nerve from the 8th and 1st; the median nerve itself and the posterior cord from each of the 5 segments.

In point of fact their origins are: musculocutaneous 5, 6, 7; ulnar (7) 8, 1; median (5) 6, 7, 8, 1; radial 5, 6, 7, 8 (1); axillary 5, 6.

The medial and lateral cords might better have been called the "antero-medial" and "antero-lateral" cords. The posterior cord is well named because it is destined to supply all the muscles on the posterior or extensor aspect of the limb.

Relation of Plexus to Artery. The brachial plexus (*fig. 70*) emerges from the cervical portion of the vertebral column; thus it should at first be above, behind, and lateral to the artery; and it is.

The three cords are disposed around the second part of the axillary artery—the part behind the Pectoralis Minor—as their names suggest.

Identification. The medial and lateral cords and their terminal branches have the form of a capital "M" (*fig. 70*). This fact should be used in identifying them.

Two Cutaneous Branches also can be found. The larger of these is the *medial cutaneous nerve of the forearm.* It and the much smaller *medial cutaneous nerve of the arm* arise from the medial cord just before it bifurcates.

If the derivatives of the lateral and medial cords are held aside, the only remaining large nerve, namely, the *radial nerve,* will be identified by this process of exclusion.

Unless the foregoing procedure is followed, the radial nerve may easily be mistaken for the ulnar nerve. (*See fig. 71.*)

The *radial nerve* and the *axillary nerve* will be seen to be the two terminal branches of the posterior cord. Of the five terminal nerves of the plexus, the axillary nerve alone does not run longitudinally, but disappears into the quadrangular space.

Motor Nerves of the Axilla in Review—*See figs. 171 and 172.*

A Typical Spinal Nerve (*fig. 72*) is a serially segmental structure which is attached to the spinal cord by *two roots*—a ventral (anterior) which is motor or efferent, and a dorsal (posterior) which is sensory or afferent. The two roots leave the vertebral canal through an intervertebral

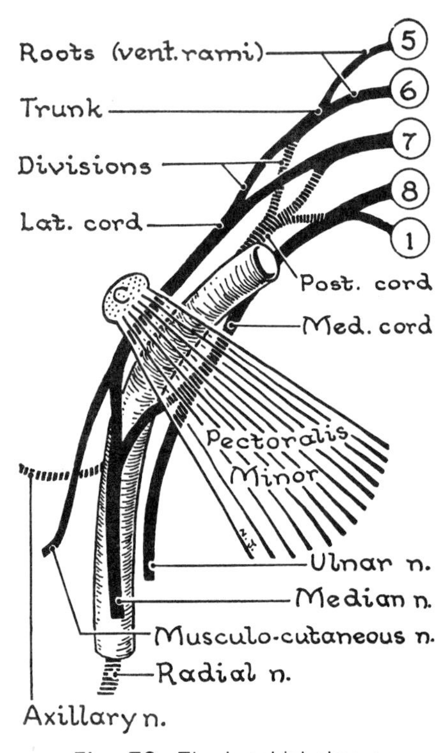

Fig. 70. The brachial plexus.

TABLE 2

Derivatives of the Brachial Plexus within the Limits of the Axilla

Cords	Terminal Branches	Collateral Branches	
	Mixed (motor and sensory)	Motor	Cutaneous
Lateral	1. Musculocutaneous 2. Lateral root of median	1. Clavicular head of Pectoralis Major (and upper part of sterno-costal head)	
Medial	1. Medial root of median 2. Ulnar	1. Sterno-costal head of Pectoralis Major 2. Pectoralis Minor	1. Med. cutan. n. of arm 2. Med. cutan. n. of forearm
Posterior	1. Axillary 2. Radial	1. Subscapularis 2. Latissimus Dorsi 3. Teres Major	
From the musculocutaneous		1. Coracobrachialis	
From the radial		1. Triceps (long head) and	Post. cutan. n. of arm
From the 5, 6, and 7 roots of the plexus		1. Serratus Anterior	

foramen and join immediately beyond it to form a *spinal nerve*. Having both motor and sensory fibers, the nerve is said to be mixed. After the course of a few millimeters the spinal nerve divides into a *ventral* and a *dorsal ramus*. Roughly, the dorsal rami supply the muscles of the back that act on the vertebral column and the skin covering them; whereas ventral rami supply the muscles and skin of the anterior three-quarters of the body wall.

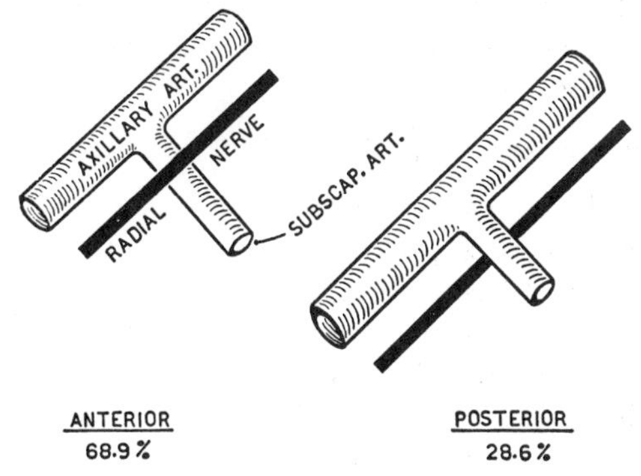

Fig. 71. Illustrating that the relationship of nerves to arteries is subject to variation (e.g., radial n. to subscapular a.). (Based on 360 limbs.)

There is an enlargement on the dorsal root as it lies in the intervertebral foramen. This enlargement, known as a *spinal ganglion* (post. root ganglion), consists of the cell bodies of all the afferent (sensory) fibers in that spinal nerve.

Spinal Ganglia (Post. Root Ganglia). Ganglion C. 1, when present, lies on the posterior arch of the atlas; C. 2 on the lamina of the axis; C. 3—L. 4 in the intervertebral foramina; L. 5—Co. 1 in the vertebral and sacral canals.

Spinal nerves and their branches also carry efferent fibers of the sympathetic nervous system. Sympathetic ganglia lie on the sides of the vertebral bodies in front of the spinal nerves to which they are connected by white and gray rami communicantes. White rami carry fibers from the spinal cord to the ganglia. Gray rami carry fibers from the ganglia, where their cell bodies are located, to the spinal nerves and thence to their branches. Thus, the white rami contain preganglionic fibers, the gray contain postganglionic fibers, and

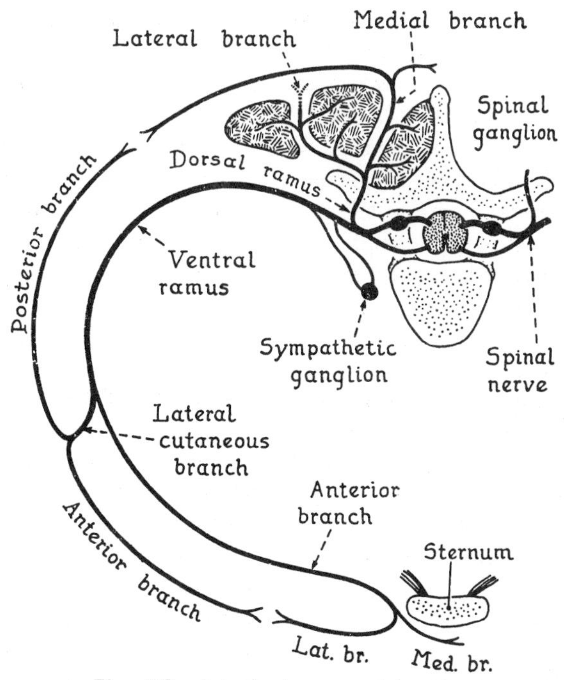

Fig. 72. A typical segmental nerve.

the ganglia are the site of transmission of impulses from one to the other. The fibers traveling along the peripheral nerves confine their attention to involuntary muscle and to glands (*fig. 43*).

The Cutaneous Nerves. The ventral rami of the thoracic nerves are usually called *intercostal nerves*. A typical intercostal nerve gives off a *lateral cutaneous branch* in the midaxillary line and ends, after piercing the Pectoralis Major at the side of the sternum, as an *anterior cutaneous branch* (*fig. 72*).

The *first three intercostal nerves* are atypical: the 1st in having neither a lateral nor an anterior cutaneous branch; the 2nd in sending its lateral branch across the dome of the axilla in the laminated fascia and then descending on the posteromedial aspect of the arm as the *intercostobrachial nerve* (*fig. 98*); and the 3rd in sending a small lateral branch to the medial side of the arm.

Development. In prenatal life, the ventral rami of nerves C. 5, 6, 7, 8, and Th. 1 are drawn out from the trunk into the developing limb to supply it with both motor and sensory fibers. They form the brachial plexus and their simple segmental arrangement is lost. Since these five rami (5, 6, 7, 8, and 1) send no cutaneous branches to the pectoral region, it becomes the duty of the supraclavicular branches of C. 3 and 4 to descend in front of the clavicle to the level of Th. 2 thereby closing the gap or hiatus thus occasioned. In consequence, a person whose spinal cord is injured due to a fracture of the 5th cervical vertebra might retain sensation to pinprick as low as the 2nd intercostal space (*fig. 97*).

The cutaneous supply to the limb draws upon more segments than the motor supply, for branches from C. 3 and 4 descend to the deltoid region; branches from Th. 2 and 3 descend to the medial side of the arm. This may be represented thus:

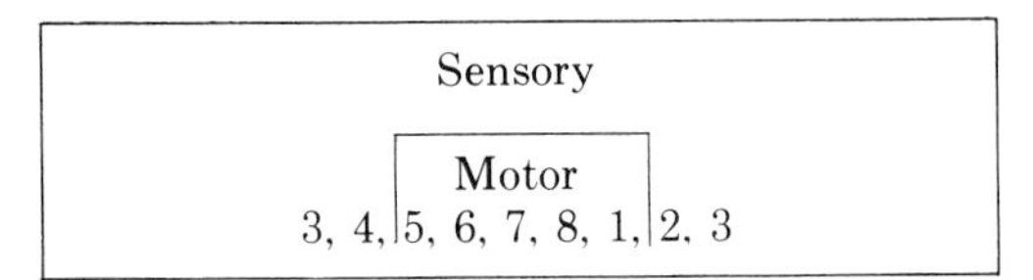

FASCIA, MAMMARY GLAND, LYMPH NODES, AND ANOMALIES

Fascia

The large quantities of very loose, moist, areolar tissue found between the various muscles of this region are necessary to allow the limb a wide range of movements where it forms the dome-shaped floor of the axilla, it is irregularly laminated.

The *clavipectoral fascia*, which is the sheath of Pectoralis Minor and Subclavius, acts as a suspensory ligament for the dome-shaped floor of the axilla.

Mammary Gland

The mammary gland or breast (*fig. 73*) lies in the superficial fascia. It is made up of 15–20 units of glandular tissue, whose lobules, enclosed in a fibro-areolar stroma, radiate from the nipple into the surrounding superficial fat. The periphery of the gland extends from the 2nd to the 6th rib in the vertical plane and from the side of the sternum to near the midaxillary line in the horizontal plane.

About two-thirds of the gland overlies the Pectoralis Major; one-third the Serratus Anterior. Although easily separated from the fascia covering these muscles, the gland is firmly connected to the true skin by *suspensory ligaments* (*of Cooper*) that pass from its stroma between lobules of fat.

In the pinkish *areola* surrounding the nipple there is a number of nodular rudimentary milk glands, *the areolar glands*, and deep to the areola is some involuntary muscle, a lymph plexus, and an absence of fat.

Vessels and Nerves. *Nerves:* Intercostal nerves (2nd–6th), via lateral and anterior cutaneous branches. These or the vessels convey sympathetic fibers.

Arteries. Perforating branches (especially the 2nd and 3rd) of the *internal thoracic a.* (internal mammary a.) and two branches of the *lateral thoracic a.* approach the gland from the sides, ramify on its superficial surface, send branches into it, and anastomose around the nipple. Twigs

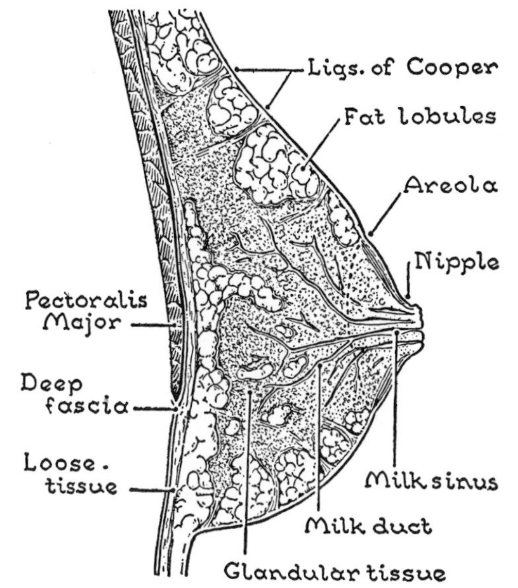

Fig. 73. The mamma on vertical section. (After Testut.)

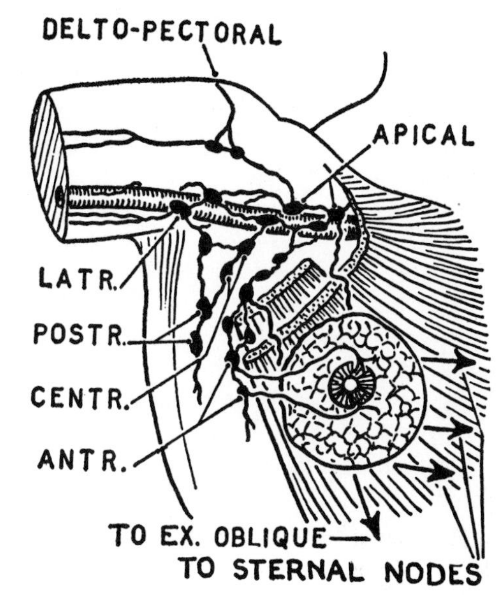

Fig. 74. Lymphatics of breast and axilla (*sternal* nodes = *parasternal*). (Poirier and Charpy.)

from the *intercostal aa.* may enter the deep surface of the gland.

For *Venous Connections* see end of Chap. 36.

Lymphatics. The main lymph vessels of the mamma, like the ducts, converge on the nipple. Deep to the areola they form a subareolar lymph plexus (*fig. 74*). (1) From this plexus two or three distinct vessels course superficially to the pectoral lymph nodes. (2) From the medial border of the mamma, lymph vessels pass to the parasternal lymph nodes, which lie along the internal thoracic a. (3) From the deep surface of the mamma, lymph vessels pass through the Pectoralis Major to the interpectoral nodes, which lie superficial to the Pectoralis Minor and costocoracoid membrane, or passing through these end in the apical nodes.

Axillary Lymph Nodes

The axillary lymph nodes (glands) (*fig. 73*) are arranged in several main groups: (1) *the lateral* nodes lie along the lower parts of the axillary vein. They receive the lymph vessels that ascend along the medial side of the arm and they empty into (2) *the apical* (*infraclavicular*) nodes that lie along the upper part of the axillary vein between the Pectoralis Minor and the clavicle. All the vessels of the limb, including those following the cephalic vein, drain either directly or indirectly into this group, and it in turn drains into *the subclavian lymph trunk* which ends in the right lymph duct or (on the left) the *thoracic duct.* (3) *The pectoral* (*anterior*) nodes lie along the lower border of the Pectoralis Minor with the lateral thoracic vein. (4) *The subscapular* (*posterior*) nodes lie along the subscapular veins. (5) *The central* nodes lie between the layers of fascia at the base of the axilla or in the fat deep to it. (6) Occasionally one or two small nodes occur along the cephalic vein in the *deltopectoral triangle.*

Anomalies

Accessory Nipples or even *Mammae* are occasionally found along the "milk line" (*fig. 75*).

Sternalis muscle. A narrow band of muscle, lying in front of Pectoralis Major and in line with Sternomastoid and Rectus Abdominis, is commonly found.

Axillary Arch: a variable band of voluntary muscle that stretches across the base

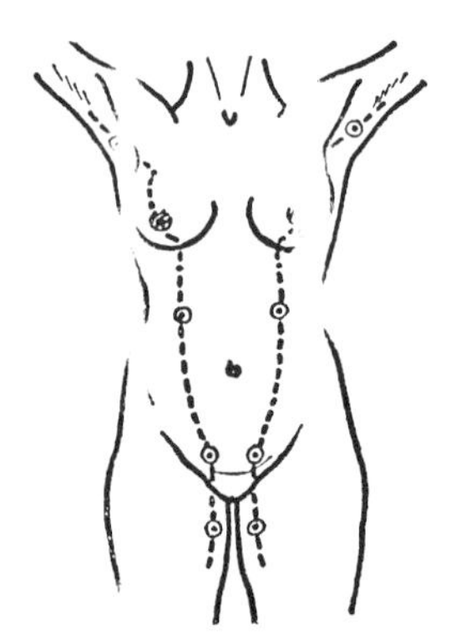

Fig. 75. Accessory nipples may appear on the milk line.

of the axilla from Latissimus Dorsi to Pectoralis Major or coracoid process.

Pectoralis Major: its sternal head may be absent.

The lateral root of the median nerve (or part of it) may travel with the musculocutaneous nerve far into the arm before joining the median nerve.

6 SCAPULAR AND DELTOID REGIONS

SCAPULA

Parts, relations to ribs and vertebrae.
To Measure the Limb.

CUTANEOUS NERVES OF BACK

Dermatomes.

FIRST LAYER OF MUSCLES OF BACK

Latissimus Dorsi—origin, insertion, nerve supply, actions and functions, two associated triangles.
Trapezius—origin, insertion, functions, nerve supply.
Upper Border of Scapula: Suprascapular Vessels and Nerve.

SECOND LAYER OF MUSCLES OF BACK

Levator Scapulae and Rhomboids: origins, insertions, nerve supplies.
Transversa Colli Artery; Anastomoses around scapula.

UPPER HALF OF HUMERUS AND SHOULDER REGION

Upper Half of Humerus—Upper End; Lesser and Greater Tubercles; Surgical Neck; Upper Half of Body.
Deltoideus—Attachments; Tuberosity; Structure.
Quadrangular Space; Axillary Nerve; Acromion; Subacromial Bursa.
Subscapular Fossa and Subscapularis.
Inferior Angle and Lateral Border of Scapula.
Teres and Spinatus Muscles.

BONES OF PECTORAL GIRDLE: IN DETAIL

Clavicle—Flattened Lateral One-Fourth; Prismatic Medial Three-Fourths; Medial (Sternal) End; Variations; Ossification.
Scapula—Orientation; Angles; Borders; Surfaces; Spine.
Acromion; Coracoid Process; Muscle Attachments.
Medullary Foramina; Ossification; Anastomoses; Variations.

SCAPULA

The scapula is a flat bone with two surfaces, a *costal* and a *dorsal*; it is triangular in shape, having three borders and three angles; and it has two processes, the *coracoid process* and the *spine* which ends as the *acromion* (*fig. 76*).

The *spine* of the scapula crosses obliquely the medial four-fifths of the dorsal surface, dividing it into a smaller *supraspinous* and a larger *infraspinous fossa*. These two fossae communicate with each other at the "*spinoglenoid notch*," which lies between the lateral border of the spine and the *glenoid cavity*. The posterior border or *crest of the spine*—often for the sake of brevity referred to as "the spine" —ends above the shoulder joint in a free, flattened and expanded piece of bone, the *acromion*. The acromion has on its medial border, very close to its tip, a small oval bevelled facet for articulation with the flattened acromial end of the clavicle.

Details of the scapula are given on p. 99—but only for those who especially require it.

Palpable Parts. The crest of the spine, the acromion, and the clavicle form the limbs of a V which you can palpate from end to end through the skin.

The *acromioclavicular joint* is likewise subcutaneous and can be easily located. Sometimes it is a visible bump.

The *medial* (*vertebral*) *border* of the scapula is covered in its upper half with Trapezius, but you can easily palpate it

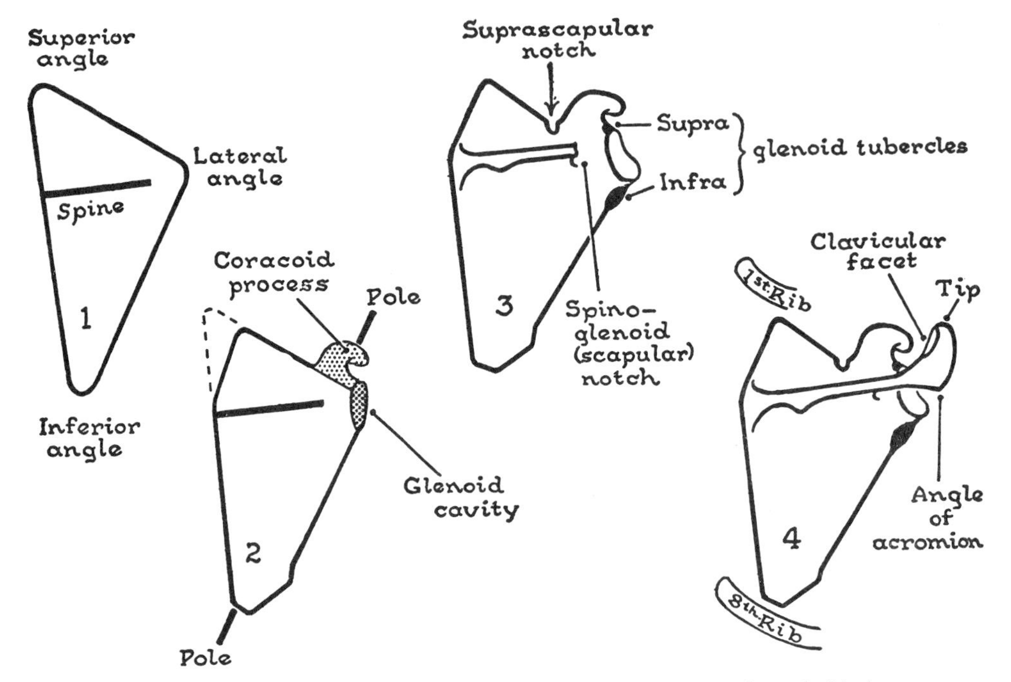

Fig. 76. The progressive stages in sketching a scapula, from behind.

from the *inferior angle* to the *superior angle*. The *lateral (axillary) border* you can trace with some difficulty. The inferior angle and the tip of the coracoid are, so to speak, at opposite "*poles*" of the scapula. By grasping the inferior angle with one hand and palpating the coracoid process with two fingers of the other, you can manipulate the scapula in the same manner as a surgeon does to reveal a fracture between the "*two poles*".

The *medial border* is almost parallel to the spines of the vertebrae and is about 5 cm distant from them. It crosses ribs 2 to 7.

To Measure the Limb. The *angle of the acromion* is the usual point from which to measure the length of the limb, since it is readily palpated between the thumb and index.

CUTANEOUS NERVES OF THE BACK

A typical dorsal nerve ramus divides into a medial and a lateral branch (*fig. 72*). Both branches supply muscles, and one or the other of the branches becomes cutaneous. Above the midthoracic region it is the medial branches that become cutaneous, and they do so close to the median plane; below, the lateral branches become cutaneous at some distance from the median plane. The cutaneous branch of Th. 2 extends to the acromion and is the longest of the dorsal rami.

The area of skin supplied by a single dorsal (i.e., sensory) nerve root is called a *dermatome*. On the trunk the dermatomes form obliquely encircling bands (*fig. 43.3*).

FIRST LAYER OF MUSCLES THE BACK

Latissimus Dorsi and Trapezius

Latissimus Dorsi (*fig. 77*). *Origin.* This fan-shaped muscle arises by aponeurosis from the 7th thoracic spine to the middle of the outer lip of the iliac crest (specifically lower 6 thoracic, lumbar, and sacral spines, and the outer lip of the iliac crest) through the medium of the *thoracolumbar fascia* (lumbar fascia). It also arises from the lower three ribs by fleshy fibers.

Insertion and Nerve Supply: see pages 80 and 84 (Table 2).

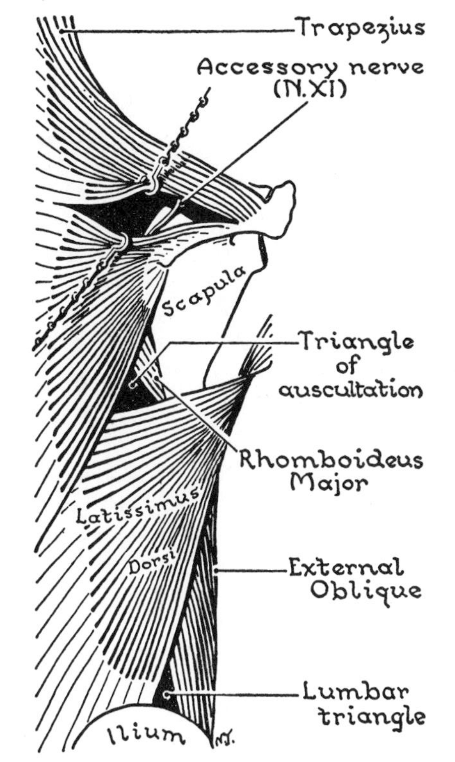

Fig. 77. Exposure of superior border of scapula. Two triangles.

Actions and Functions. The Latissimus extends the humerus, rotates it medially, and adducts it. It brings the outstretched arm from above the head to behind the back. It is used in swimming, in raising the body when one hangs from a horizontal bar, in rowing, in pulling, in elbowing one's way through a crowd.

Two Triangles with some minor interest are associated with Latissimus Dorsi (*fig. 77*).

Trapezius. This, too, is a triangular muscle. The Trapezii of the two sides together form a trapeze or table.

The *Origin* of the Trapezius extends from the skull above to the spinous process of the last thoracic vertebra below. As will be seen in Chapter 35 it arises from the skull, the spines of all the cervical vertebrae (*via* the ligamentum nuchae) and thoracic vertebrae.

Insertion. The upper fibers of the muscle pass inferolaterally to the lateral third of the clavicle; the intermediate fibers pass horizontally to the acromion, as well as to the upper lip of the crest of the spine of the scapula; the lower fibers converge as they pass superolaterally and form an aponeurosis which is inserted into and accounts for the well marked tubercle on the lower lip of the crest of the spine.

Functions. The Trapezii are the suspensory muscles of the shoulder girdles. In health they hold the shoulders back and up, thus giving a "military carriage." The Trapezius comes into play, raising or steadying the shoulder girdle, whenever a weight is either supported on the shoulder or carried in the hand (*fig. 176*). The lower fibers assist the upper fibers to rotate the scapula by pulling on their tubercle of insertion to depress the medial border.

Nerve Supply. The Trapezius is supplied on its deep surface by (1) the accessory nerve (external branch), which arises from the upper five cervical nerve segments and takes a devious course (*fig. 764*), and (2) by the ventral rami of C. 3 and 4, which take a direct course, and are not purely sensory in man as they are in some species (Corbin and Harrison; Wookey).

The Upper Border of the Scapula lies deep to Trapezius and is the key to the suprascapular region. It extends from the superior angle, where the Levator Scapulae is inserted, to the upper part of the glenoid cavity where the long head of the Biceps arises from the *supraglenoid tubercle* (*fig. 76*). Laterally, it is abruptly deeper—the *scapular notch* (suprascapular notch). This notch is bridged by a sharp, taut band, the *transverse scapular ligament* (suprascapular lig.). Between the notch and the supraglenoid tubercle the border is drawn out into the stout *coracoid process.*

Suprascapular Vessels and Nerve. The vessels cross above the ligament; the nerve, below. On the dorsum of the scapula the suprascapular vessels and nerve lie in contact with the bone. They pass from the supraspinous fossa through the spinoglenoid notch to the infraspinous fossa.

Distribution. The suprascapular nerve supplies the Supraspinatus and Infraspinatus, and it sends twigs to the shoulder joint. It has no cutaneous branches.

SECOND LAYER OF MUSCLES OF THE BACK

Levator Scapulae and Rhomboidei. When the Trapezius is reflected, the **Levator Scapulae** and the **two Rhomboids** are exposed. Together they occupy the whole length of the medial border of the scapula (*fig. 78*).

Origins. The *Levator Scapulae* arises from the upper four cervical vertebrae (posterior tubercles of the transverse processes). The *two Rhomboids* arise from the lower part of the ligamentum nuchae and the spines of cervical and upper four thoracic vertebrae.

Insertions—from the superior angle and along the whole length of the medial border, as seen in *figure 78*.

Nerve Supply. Levator—C. 3, 4 and twigs from 5. Rhomboids—C. 5 (as dorsal scapular nerve) (*fig. 78*).

Transversa Colli Artery (*figs. 78, 79*). This artery [transverse cervical] runs backward in the neck to the anterior border of the Levator Scapulae where it divides. A *superficial branch* accompanies the *accessory nerve* while the *deep branch* accompanies the *nerve to the Rhomboids.*

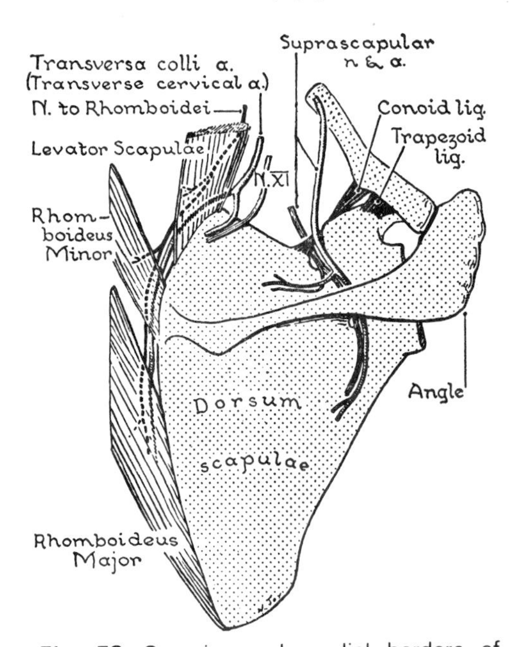

Fig. 78. Superior and medial borders of scapula.

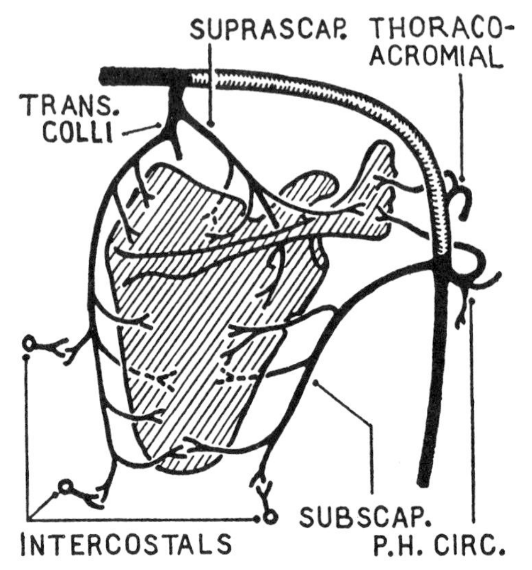

Fig. 79. Scheme of anastomoses around scapula.

The Anastomoses on and around the Scapula are very free. They bring notably the 1st part of the subclavian artery into communication with the 3rd part of the axillary artery on the body of the scapula, on the acromion, and with intercostal arteries around the scapula, as shown in *figure 79.*

Serratus Anterior (p. 80) is interposed between the Subscapularis and the chest wall (*fig. 80*). Rhomboids and the Serratus Anterior together keep the scapula applied to the thoracic wall. If either is paralyzed, the medial border will project from the back—a "winged scapula."

UPPER HALF OF HUMERUS AND SHOULDER REGION

Upper Half of the Humerus

The upper half of the humerus (*figs. 81, 82*), being devoted to the shoulder, is quite different from the lower half, devoted to the elbow. Thus, a description of the whole

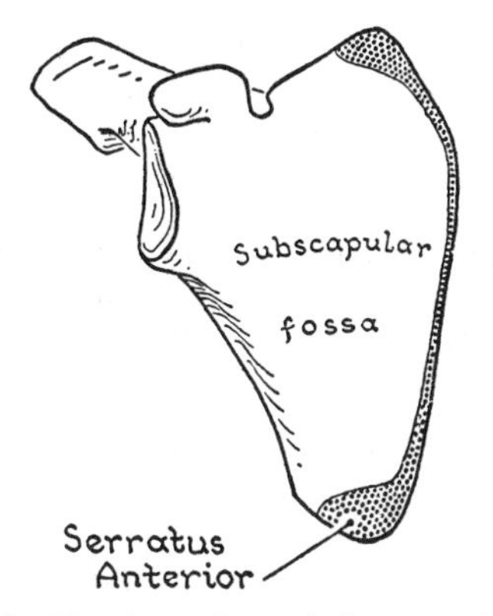

Fig. 80. The insertion of Seratus Anterior. More than half the muscle inserts on the medial surface of the inferior angle.

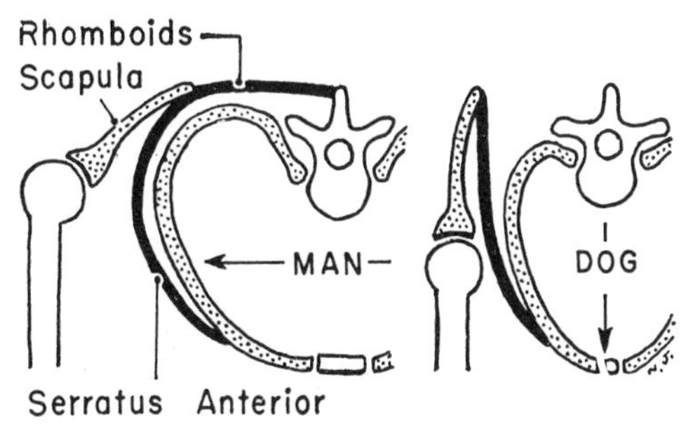

Fig. 80.1. To show that the Rhomboids and the Serratus together hold the medial border of the scapula applied to the thoracic wall. In the quadruped, the thorax is suspended by the Serrati.

bone here would be artificial and useless. The lower half is described with the arm on p. 109.

The **upper end** consists of: (a) an articular portion, the *head*, which is covered with cartilage, directed medially and upward, and delineated by (b) the *anatomical neck* (to which the fibrous capsule of the joint is attached) and (c) and (d) two tubercles, the *lesser and greater tubercles* (tuberosities). These are separated by a groove, the *intertubercular sulcus* (bicipital groove) wherein lodges the long tendon of the Biceps (*fig. 81*).

The prominent **lesser tubercle**—the insertion for Subscapularis—is directed straight forward; it can be palpated (through the Deltoid) about 4 cm inferolateral to the coracoid process.

The less prominent, although larger, **greater tubercle** projects laterally beyond the acromion and so gives the shoulders its roundness. It too can be palpated through the Deltoid. It possesses three flat contiguous facets (*fig. 82*) for the Supraspinatus, Infraspinatus, and Teres Minor.

The **surgical neck** is the zone between the head and tubercles above and the body below. It is completely encircled by the circumflex humeral vessels and partly encircled by the axillary nerve.

The **upper half of the body** is cylindrical in shape. Its features are *two crests* and a *sulcus* on the anterior aspect, and a *tuberosity* for the insertion of the Deltoid on the lateral aspect.

Into the rugged *crest of the greater tubercle* (lateral lip of the bicipital groove)

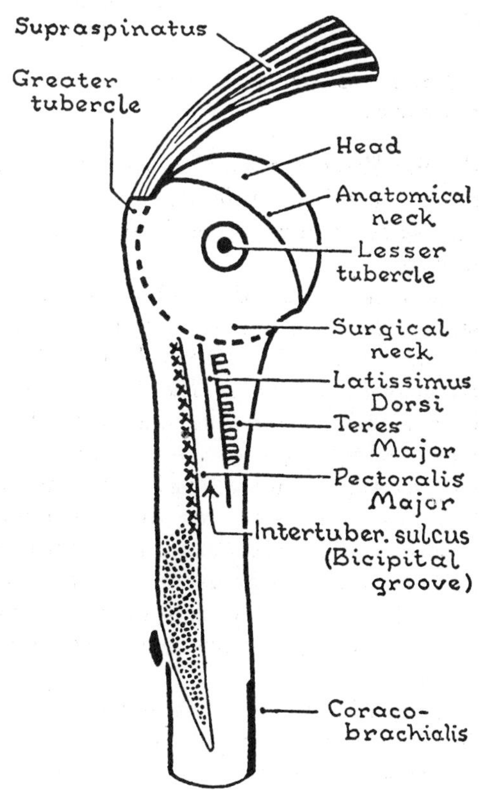

Fig. 81. Upper half of humerus (anterior view, schematic).

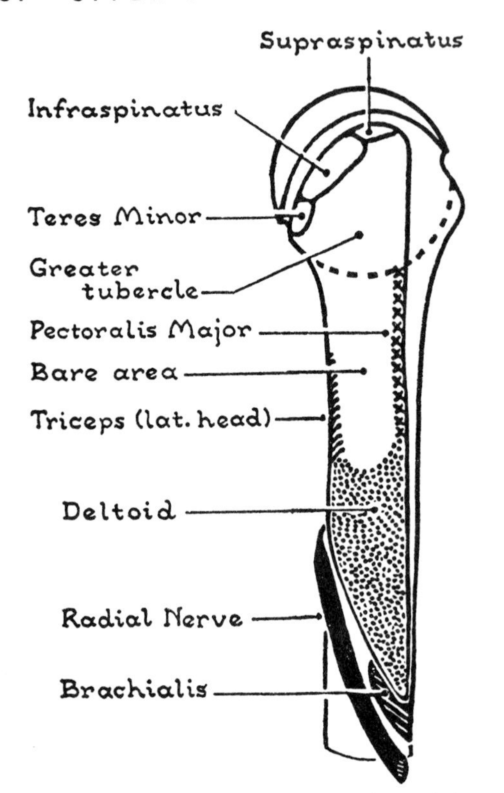

Fig. 82. Upper half of humerus (lateral view, schematic).

the Pectoralis Major is inserted; into the *crest of the lesser tubercle* (medial lip of the bicipital groove) the Teres Major is inserted.

Deltoideus (*fig. 83*). This muscle is shaped like the Greek letter delta inverted.

Attachments. The Deltoid arises from the lateral third of the clavicle, the lateral border of the acromion, and the whole length of the spine of the scapula. From this extensive linear origin the muscle descends to its restricted and, therefore, fibrous insertion into the *deltoid tuberosity* halfway down the humerus (*figs. 82, 83*). Behind the tuberosity is a broad shallow groove, the (*spiral*) *groove* for the radial nerve.

Abductors. Since the shoulder joint has but two muscles lying above it, it can evidently possess but *two abductors:* they are the Supraspinatus and the intermediate fibers of the Deltoid (*fig. 83.1*). Each is supplied by the 5th and 6th cervical nerve segments, the one via the suprascapular nerve; the other via the axillary nerve.

Internal Structure of Deltoid (*fig. 84*). The intermediate part of the Deltoid, is multipennate, the muscle fibers being very numerous but very short. On this account it is very powerful, but its range of action is very short. The anterior and posterior parts have a different internal structure; they are composed of long parallel fibers because they take part in the more extensive movements of flexion and extension of the shoulder.

Deep to Deltoid. If you detach its origin and turn it downward you expose the axillary nerve (circumflex n.), accompanied by the posterior humeral circumflex artery, which adheres to the

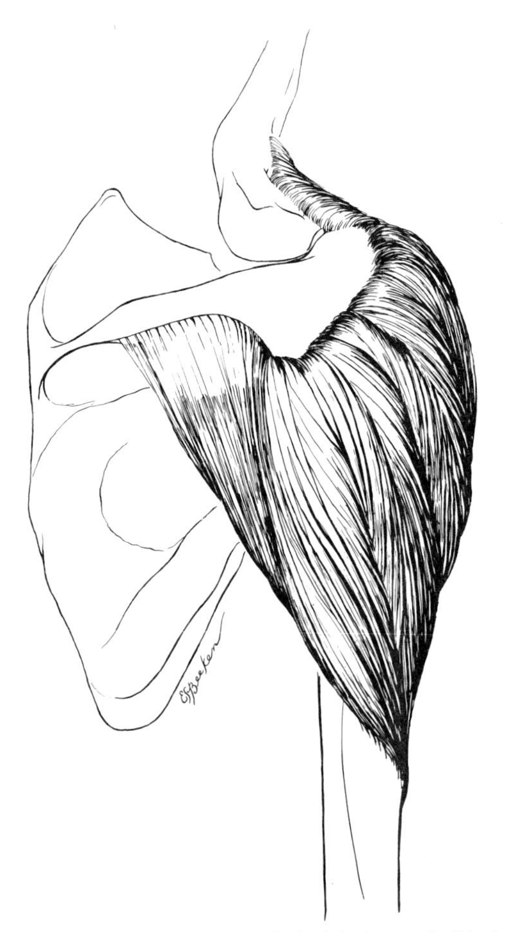

Fig. 83. The right Deltoid from behind. (From *Primary Anatomy*.)

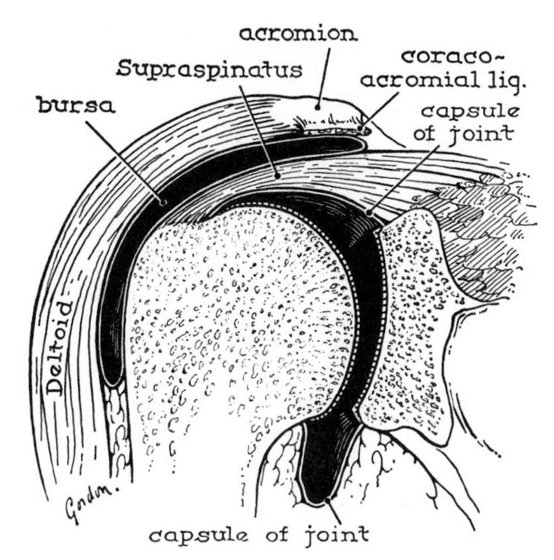

Fig. 83.1. Deltoid and Supraspinatus in coronal section. (From Primary Anatomy.)

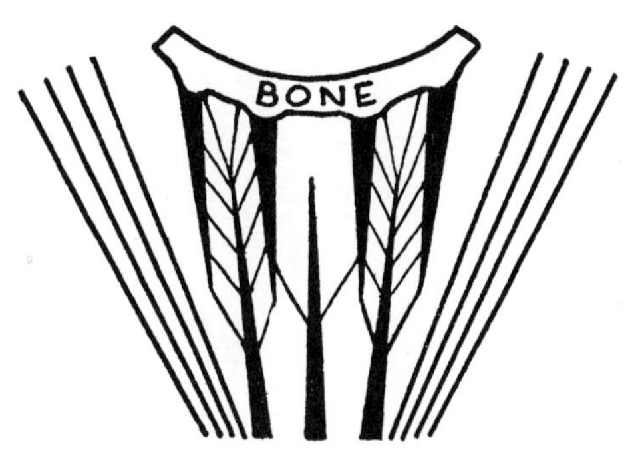

Fig. 84. Architecture or internal structure of Deltoid.

deep surface of the Deltoid and supplies it (*fig. 86*). Follow the nerve and vessel to the quadrangular space. Also exposed is the greater tubercle and the bare area on the lateral aspect of the humerus.

The Quadrangular Space is bounded *above* by the lateral border of the scapula and the capsule of the shoulder joint; *laterally*, the surgical neck of the humerus; *below*, the Teres Major (*figs. 85 and 86*) and *medially*, the *long or scapular head of the Triceps* (*fig. 86*). The last arises from the rough triangular infraglenoid tubercle of the scapula. Passing through the quadrangular space are the *axillary nerve* and the *posterior humeral circumflex artery.*

Medial to the long head of Triceps (in an unimportant triangular space), a large anastomotic artery, the *circumflex scapular branch* of the subscapular artery, which, on its way to the infraspinous fossa, grooves the axillary border of the scapula (*fig. 86*).

Axillary Nerve (Circumflex N.) (C. 5 and 6). This nerve has two important relations:

1. The capsule of the joint above.
2. The surgical neck of the humerus laterally.

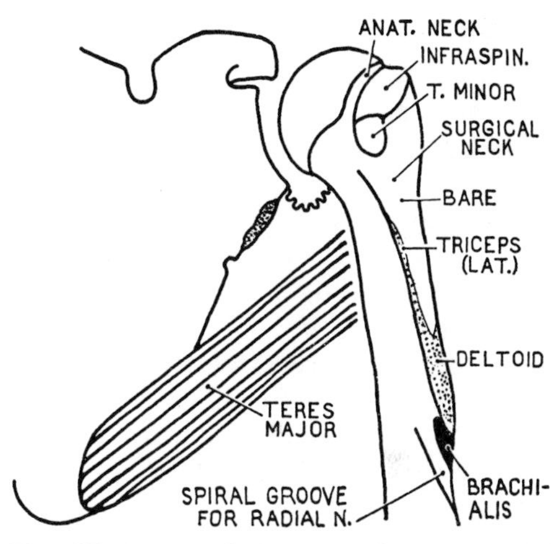

Fig. 85. Upper half of humerus (posterior view).

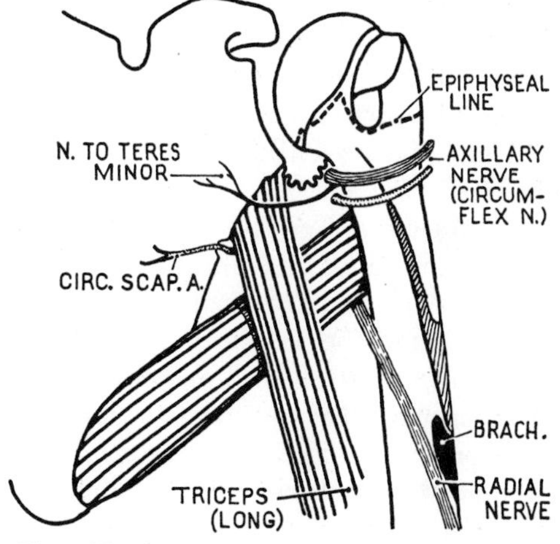

Fig. 86. The triangular and quadrangular spaces.

As the axillary nerve passes through the quadrangular space it sends twigs to the capsule of the joint; and, on leaving the space it supplies the Teres Minor. Thereafter, it supplies the Deltoid and the skin covering the Deltoid (*fig. 98*).

Surface Anatomy. The position of the axillary nerve is indicated on the skin surface by a horizontal line drawn 5 cm below the angle of the acromion.

Parts Covered by Acromion. When the Deltoid and acromion are removed, it is not the fibrous capsule of the joint that you see but the tough tendons of three dorsal scapular muscles—Supraspinatus, Infraspinatus, and Teres Minor—which conceal and adhere to the capsule.

The **subacromial bursa** lies between the acromion and the Supraspinatus tendon. It plays the part of an accessory joint cavity and is large, extending beyond the acromion laterally deep to the Deltoid as the subdeltoid bursa (*fig. 83.1*). When the arm is abducted, the greater tubercle passes completely under cover of the acromion; hence, the necessity for an extensive subacromial (plus subdeltoid) bursa.

Epiphyses. During the 15th year the *coracoid process* is fusing with the scapula (*fig. 87*), and the acromial epiphysis begins to ossify (*fig. 88*). The acromial epiphysis commonly fails to fuse and on X-ray examination may simulate a fracture. (Age of maturation or complete fusion, see p. 102.)

Four Short Muscles that surround the shoulder joint—called **"rotator cuff muscles"** by surgeons—along with the **Teres Major,** are of great surgical importance in the cause and treatment of shoulder dislocations. The four "rotator cuff muscles" are the *Subscapularis*, *Supra-* and *Infraspinatus*, and *Teres minor*.

Subscapular Fossa and Subscapularis. The concave *subscapular fossa* provides for the **Subscapularis** a multipennate origin (*fig. 89*). The tendon of insertion (to the lesser tubercle) grooves the anterior margin of the glenoid cavity and helps to give it a pear-shaped appearance (*fig. 90*).

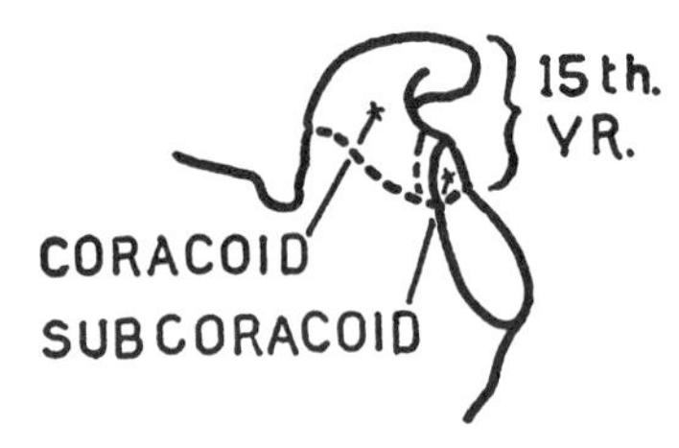

Fig. 87. Coracoid epiphyses.

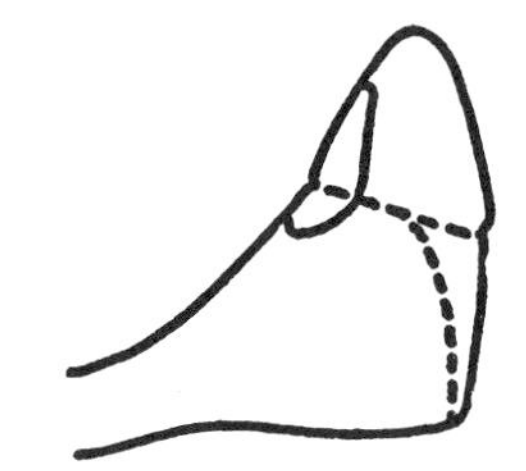

Fig. 88. Acromial epiphyses.

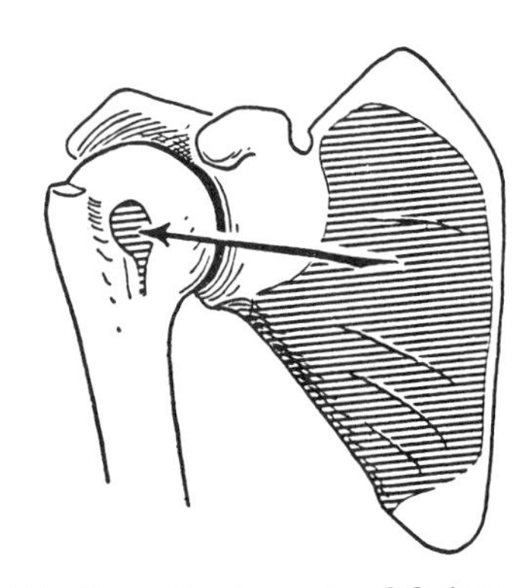

Fig. 89. The attachments of Subscapularis.

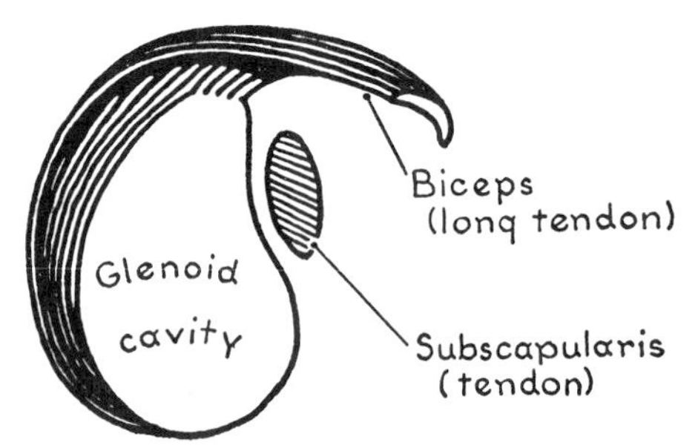

Fig. 90. Glenoid cavity, Biceps, and Subscapularis.

The Inferior Angle and Lateral Border of the Scapula are thick and strong because they form a power lever (*figs. 91, 92, 93*). The *infraglenoid tubercle* gives origin to the long or scapular head of the Triceps.

Teres Muscles (teres, L. = round) (*fig.*

94). *Origins.* The *Teres Minor* arises from the lateral border of the scapula. The *Teres Major*, which is a very large muscle, arises chiefly from the large impression on the dorsum of the inferior angle of the scapula.

Insertions. Both muscles are inserted into the humerus by tendon—the Minor into the lowest facet on the greater tubercle; the Major into the crest of the lesser tubercle.

Nerve Supply. Both muscles are supplied by nerve segments C. 5, 6, but via different nerves.

Actions. The Minor is a lateral rotator of the humerus; the Major is a medial rotator.

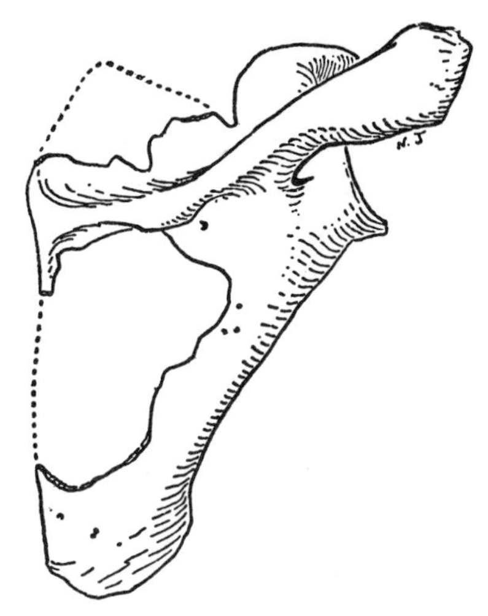

Fig. 93. The lateral or axillary border is the strong border; the others can be broken away with the fingers.

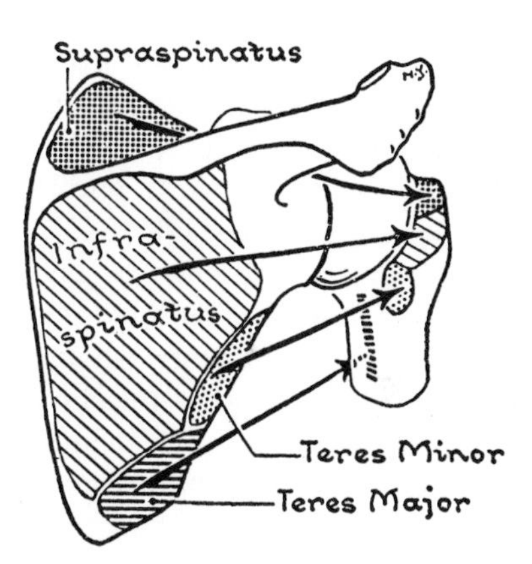

Fig. 94. Attachments of Teres and Spinatus muscles.

Fig. 91. The three muscles attached to the end of the lever.

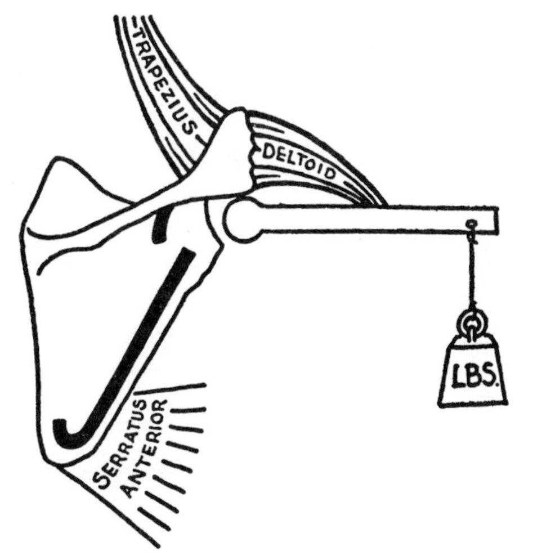

Fig. 92. Three chief muscles concerned in elevation of limb: Two strengthening bars on scapula.

***AXIOM:** When muscle fibers contract they shorten by about a third or a half of their relaxed length, i.e., if a muscle is to move its insertion through 1 cm, its fleshy fibers must be 2 to 3 cm long.*

Spinatus Muscles. *Attachments* (*fig. 94*). The Supraspinatus and Infraspinatus *arise* by fleshy fibers from the respective supraspinous and infraspinous fossae of the scapula. They are *inserted* by tendon into the upper and middle facets on the greater tubercle of the humerus, the tendons fusing with the fibrous capsule of the shoulder joint.

Nerve Supply. They are supplied by the suprascapular nerve (C. 5 and 6).

Actions. See Muscle Force Couples, page 161.

BONES OF PECTORAL GIRDLE: IN DETAIL

Additional details are offered here for either those few readers who are required to commit them to memory or those who seek clarification. Normally, these details can be skipped over.

Clavicle

Its *function* is to act as a strut holding the scapula and therefore the upper limb laterally, backward, and slightly upward. When the clavicle is fractured, the shoulder falls medially, forward, and slightly downward, as might be expected.

Animals such as the dog, ox, and horse that use their forelimbs merely for support and locomotion (i.e., for forward and backward motions) have either no clavicles or only rudimentary ones; but primates, rodents, guinea pigs and bats that employ them for grasping, climbing, burrowing, or flying (i.e., for side to side motion as well) have clavicles.

To *identify the side* to which a detached clavicle belongs, so hold it (1) that the flattened part is lateral, (2) that the aspect with rough markings at both ends is inferior, and (3) that the forward convexity is medial and the forward concavity lateral (*fig. 94.1*).

The clavicle has two functionally distinct parts—one lateral to the coracoid process, the other medial. The lateral part is flattened and forms one-fourth to one-third of the bone; the medial part is prismatic, being triangular on cross-section, and forms two-thirds to three-fourths. The flattened part affords anchorage for the scapula; the prismatic part plays the role of a long bone.

Flattened Lateral One-Fourth. It is compressed from above downward and is united to the scapula both terminally and inferiorly, thus: (1) *Terminally*—there is a small, oval *articular facet* for the acromion of the scapula, so bevelled that in cases of dislocation the acromion is driven under the clavicle. The articular end is not enlarged like the ends of long bones in general, so the articulation depends for security upon the conoid and trapezoid portions of the coracoclavicular ligament. (2) *Inferiorly*—these unite the coracoid process of the scapula to the clavicle and are responsible for the following markings: a tubercle, the *conoid tubercle*, which is placed below the posterior border of the bone at the junction of the flattened and prismatic parts, and a rough line, the *trapezoid ridge*, which extends diagonally laterally and forward across the inferior surface from the conoid tubercle to the anterior end of the articular facet (*fig. 94.2*).

Prismatic Medial Three-Fourths. Triangular on cross section like a typical long bone,

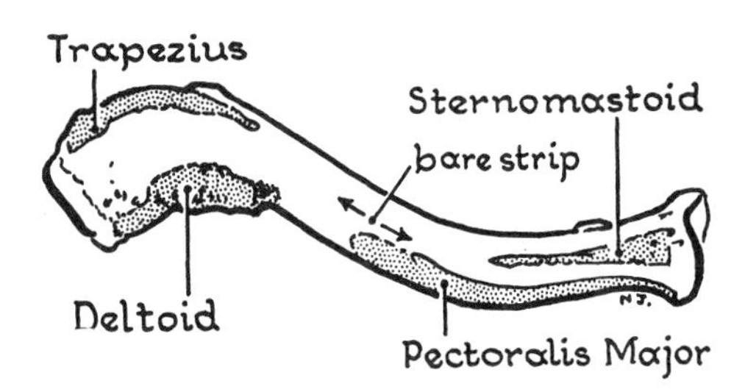

Fig. 94.1 Right clavicle from above—muscle attachments.

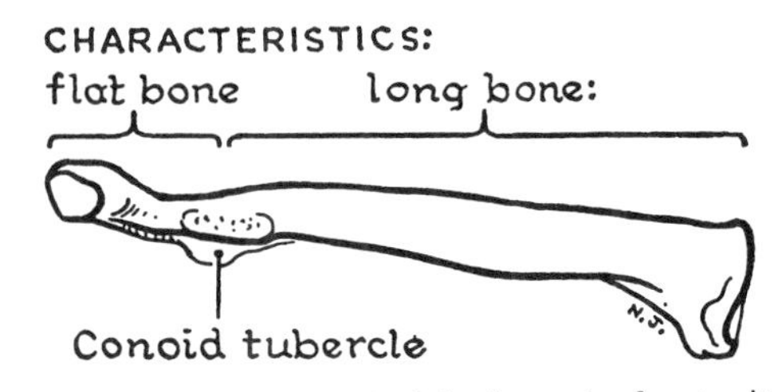

Fig. 94.2. Right clavicle from in front—its two parts.

accordingly, it has three surfaces separated by three borders. The *posterior surface*, continuous with the posterior border of the flattened lateral quarter, is concave, nearly vertical, and perfectly smooth; and so it protects the subclavian vessels and the brachial plexus which pass behind it. A nutrient foramen opens on to this surface and by its direction indicates the sternal end to be the more actively growing one.

The *anterior surface*, continuous with the upper surface of the flattened part, faces anterosuperiorly.

The *inferior surface*, continuous with the inferior surface of the flattened part, extends from the conoid tubercle to the medial end where there is a rough, centimeter-long depression (sometimes tubercular) for the costoclavicular ligament. The area between these points is fusiform, smooth, and gives insertion to the Subclavius.

The inferior surface of the clavicle, then, is mainly for ligamentous attachments; the anterior (or anterosuperior) is subplatysmal and palpable between 4 muscles; the posterior protects the great vessels and nerves entering the upper limb, and it also can readily be palpated, provided the Trapezius and Sternomastoid are relaxed as when leaning heavily on the arms of a chair.

THE MEDIAL (STERNAL) END of the bone is enlarged and is covered with articular cartilage, and this extends on to the inferior surface; it is, however, separated from its socket, formed by the sternum and first costal cartilage, by a diagonally set articular disc. Crossing in front of this end of the bone is the flattened sternal head of the Sternomastoid.

Variations. The clavicle varies more in shape than most other long bones. It is peculiar in having no medullary cavity. The right clavicle, though stronger than the left, is usually shorter. It may be perforated by one of the supraclavicular nerves (*fig. 61*).

Ossification. The clavicle is the first bone in the body to start to ossify, beginning during the 5th fetal week in membrane from two centers which are placed close together and which soon fuse. One represents the Trapezius-Deltoid end, the other the Sternomastoid-Pectoralis Major end. The two ends pass through a cartilaginous phase. A secondary center, which appears at the sternal end, forms a scale-like epiphysis which begins to fuse with the diaphysis between the 18th and 25th years, and is completely fused to it between the 25th and 31st years. (McKern and Stewart.) This is the last of the epiphyses of the long bones to fuse. An even smaller scale may be present at the acromial end.

Scapula

Orientation (i.e., position in space): Since the scapula is applied to the upper part of the barrel-shaped thorax, its inferior angle lies behind the plane of the glenoid cavity, therefore the lateral (axillary) border is directed obliquely downwards and backwards as well as downwards and medially. The glenoid cavity faces laterally and slightly upwards and forwards.

Angles. The *lateral angle* is truncated (*fig. 94.3*). It is enlarged to form an articular socket for the head of the humerus. Because the socket is shallow it is called the *glenoid cavity* (Gk. glene = shallow). Its anterior margin is grooved for the Subscapularis tendon; this helps to give it a pear-shaped appearance. The coracoid process, which rises above it, represents the stalk of the pear. In the quadruped mammal the cavity faces the ground and rests upon the head of the humerus; in erect man the scapula is thrust backwards by the clavicle, so the cavity faces laterally —making side to side contact with the humerus—but with a slight upward and forward tilt.

The *inferior angle* is thick and rounded and gives attachment to three strong mus-

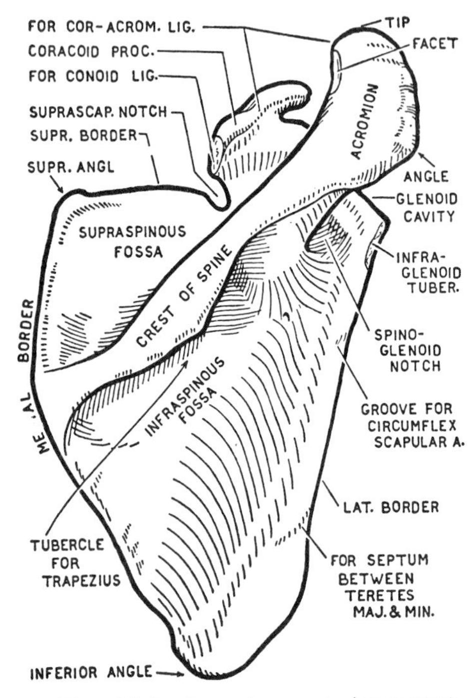

Fig. 94.3. Scapula—posterior aspect.

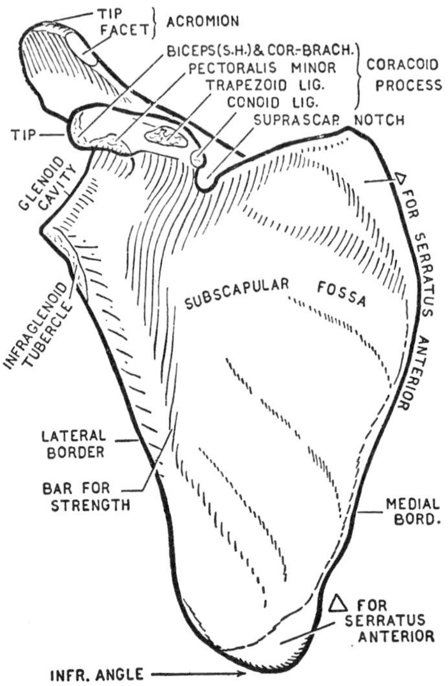

Fig. 94.4. Scapula—anterior aspect.

cles (*fig. 91*). The *superior angle* is relatively thin and acute.

Borders. The *superior border* inclines laterally and downwards from the superior angle, where the Levator Scapulae is inserted, to the upper end of the glenoid cavity, where the long head of the Biceps arises from the *supraglenoid tubercle*. This border, not being subjected to stress or strain and not giving attachment to muscles—save the delicate Omohyoid—is thin and sharp. Laterally it becomes abruptly deeper, the (*supra-*)*scapular* notch, which is bridged in life by a ligamentous band, the suprascapular ligament, which converts the notch into a foramen. Between the notch and the supraglenoid tubercle the border is, so to speak, drawn out to form the *coracoid process* (*fig. 94.4*).

The *medial* (vertebral) *border* is arched and is thicker than the superior one, because it affords insertion to the Rhomboid Minor at the apex of the spine, the Levator Scapulae between the apex and the superior angle, and the Rhomboid Major between the apex and the inferior angle.

The *lateral* (axillary) *border* is thick and smooth. It extends from the inferior angle upwards, laterally, and forwards to the glenoid cavity. At its upper end it has a rough, triangular impression, the *infraglenoid tubercle*, for the long head of the Triceps.

Surfaces. The *costal* (ventral) *surface* is monopolized by the Serratus Anterior and Subscapularis. Thus: converging fibers of the Serratus pass to raised triangular areas at the superior and inferior angles, the inferior being the larger; diverging fibers pass to a line along the medial border which connects these triangles. The Subscapularis arises by fleshy fibers from the remainder of this surface, except near the glenoid cavity. It also arises from four fibrous septa which create sharp lines that run from the medial border obliquely upwards and laterally towards the glenoid cavity.

A smooth *strengthening bar*, running close to the lateral border, connects the glenoid cavity to the inferior angle (*fig. 92*). Between this bar and the Serratus insertion the surface is concave, the *subscapular fossa*, the concavity being deepest at the level of the glenoid cavity.

The *dorsal surface* (dorsum scapulae) is slightly arched from above downward, and near the lateral border it is corrugated longitudinally. It is divided by a triangular plate of bone, the *spine*, into a smaller, deep *supraspinous fossa* and a larger, shallow *infraspinous fossa*.

Spine. Its *apex* lies at the medial border of the bone and presents a smooth, triangular area across which aponeurotic fibers of the Trapezius play. The base or *lateral border* of the spine lies a finger's breadth from the midpoint of the posterior margin of the glenoid cavity. It is very stout and rounded and forms the medial border of the *spinoglenoid notch* through which the infraspinous branches of the suprascapular vessels and nerve pass from the superior to the inferior fossa. The *anterior border* is attached to the dorsum scapulae. The posterior border or *crest* is broad and flat, and has an upper and a lower sharp lip with a subcutaneous strip between them. The lower is for the Deltoid throughout. The lowest fibers of the Trapezius converge on the lower lip superficial to the Deltoid, and there create a downwardly directed *tubercle*. The upper lip lateral to the tubercle is also for the Trapezius.

Acromion (Gk. akros = a point; omos = the shoulder; cf. acropolis, acromegaly). The spine of the scapula is drawn out laterally into a flat, stout, triangular plate, the *acromion*, which overhangs the glenoid cavity. It is greatly strengthened by the lateral border of the spine which fades away on its under surface. Due to this it can withstand the lateral thrust of the clavicle, the upward pull of the Trapezius, the downward pull of the Deltoid, and impacts from the humerus below.

The *upper surface* of the acromion is subcutaneous and faces posterosuperiorly. The *lateral border* of the acromion meets the lower lip of the spine at a right angle, the *acromial angle*. This border gives origin to the Deltoid and possesses four tubercles for the attachment of septa that descend into the Deltoid (*fig. 84*).

The *medial border* has an oval, *articular facet* for the clavicle; behind this the Trapezius is attached, in front the Deltoid; and the *tip* gives attachment to the apex of the coraco-acromial ligament under cover of the Deltoid.

Coracoid Process (Gk. korax = a crow). Shaped like a bent finger, it has two parts, a vertical and a horizontal. The *vertical part* is the upward continuation of the most lateral part of the superior border of the scapula. It is smooth and flattened from before backward.

The *horizontal part* is rounded and is directed forwards, laterally and downwards. Its tip is rough for the common tendon of the Biceps (short head) and Coracobrachialis; the anterior half of the medial border is marked for the tendon of the Pectoralis Minor, which occasionally crosses the upper surface (*fig. 178*), so only the posterior half of the upper surface is marked for the trapezoid ligament; a tubercle at the junction of the vertical and horizontal parts is for the conoid ligament, and also for the suprascapular ligament. The lateral border is marked throughout its length for the base of the coraco-acromial ligament; and approximately in line with the Pectoralis Minor the coracohumeral ligament arises from this border.

Attachments. *Muscle Attachments.* Levator Scapulae and Rhomboidei to the whole length of the medial border; long heads of the Biceps and Triceps to the supra- and infraglenoid tubercles, the Biceps also gaining attachment to the posterior margin of the glenoid cavity (*fig. 90*); Serratus Anterior and Subscapularis monopolize the entire costal surface; Spinati and Teres muscles engage all the dorsal surface; Trapezius and Deltoid are attached to opposite lips of the crest of the spine and to opposite borders of the acromion; Omohyoid to the upper border beside the notch; Pectoralis Minor, Coracobrachialis and Biceps (short head) to the coracoid process; Latissimus Dorsi may receive a few fibers from the inferior angle.

Ligamentous Attachments: In addition to the capsular ligaments of the shoulder and acromioclavicular joints there are four scapular ligaments; and all four are attached to the coracoid process. They are: the coracoclavicular (conoid and trapezoid), coraco-acromial, coracohumeral and suprascapular.

Medullary Foramina are small and numerous; a large one occurs on the dorsum just below the spine, and another large one at about the corresponding point on the costal surface.

Ossification. The primary center appears at the 8th fetal week. Secondary centers: for the coracoid in the 1st year, and subcoracoid, including the upper end of the glenoid cavity, in the 10th year; these fuse with the scapula during the 15th year. About puberty two centers appear in the acromion and others in the medial border and inferior angle; these usually fuse with the scapula between the 18th and 20th years, and in all cases by the 23rd year. (McKern and Stewart.)

Anastomoses around the scapula (*fig. 79*).

Variations. The terminal *epiphysis* (metaphysis) of the acromion commonly fails to unite; the failure is usually unilateral; the epiphyseal line passing transversely through the hinder part of the oval, articular facet or just behind it (*fig. 88*). The portion of the *medial border* of the scapula below the level of the spine may be convex, straight, or concave. The *suprascapular ligament* may be ossified.

7 CUTANEOUS NERVES AND SUPERFICIAL VEINS

CUTANEOUS NERVES

General; Dermatomes; Regional Supplies: pectoral, axilla, shoulder, brachium or arm, seeking nerves, antebrachium or forearm, palm of hand, dorsum of hand; Variations in Pattern in Hand.

SUPERFICIAL VEINS

Basilic Vein; Cephalic Vein; Median Cubital Vein; Relations; Anomaly.

LYMPH NODES

CUTANEOUS NERVES

General. The muscles of the upper limb are supplied by nerve segments C. 5, 6, 7, 8 and Th. 1 by way of the brachial plexus. The cutaneous supply is more extensive: it draws not only on the plexus, but also on two additional cephalic and two additional caudal segments. It comes from segments 3, 4——5, 6, 7, 8, 1——2, 3.

The upper limbs make their appearance in the embryo on a level with the segments from which they derive their nerves, and like many other organs (e.g., heart, diaphragm, stomach, and sex glands) they descend, dragging their nerves after them.

The limbs sprout and grow from the anterior or ventral half of the body; so, the brachial plexus is derived from ventral nerve rami (*fig. 95*).

Dermatomes is the name given to the areas of skin supplied by individual dorsal nerve roots in orderly numerical sequence from the shoulder down the preaxial border of the limb to the thumb, from the thumb to the little finger, and from the little finger up the postaxial border to the axilla (*figs. 96, 97*); *figure 97.1* summarizes the arrangement of the nerves themselves.

Axilla. The *intercostobrachial* (Th. 2) and Th. 3 supply the skin of the armpit.

Pectoral Region. Above the 2nd rib this region is supplied by the *supraclavicular nerves* (C. 3, 4); and below the 2nd rib by the *anterior cutaneous nerves* and anterior branches of the *lateral cutaneous nerves* (Th. 2–6). (*See fig. 98.*)

Shoulder. *Supraclavicular nerves* (C. 3, 4) also supply the upper half of the deltoid region. The cutaneous branch of the axillary nerve (C. 5, 6), known as the *upper lateral cutaneous nerve of the arm*, appears at the posterior border of the Deltoid and spreads forward over the lower half of the Deltoid. Cutaneous branches of the dorsal rami of spinal nerves, especially of Th. 2, reach the back of the shoulder (*fig. 98*).

Brachium or Arm. *The posterior cutaneous nerve of the arm* (C. 5, 6, 7, 8) arises from the radial nerve while it is still in the axilla by a stem common to it and the motor nerve to the long head of the Triceps. *The intercostobrachial nerve* (Th. 2), accompanied by a branch of the 3rd intercostal nerve, pierces the fascial floor of the axilla and becomes cutaneous. *The medial cutaneous nerve of the arm* (C. 8 and Th. 1) springs from the medial cord of the plexus and becomes cutaneous in the upper third of the arm, medial to the brachial artery. *The lower lateral cutaneous nerve of the arm* (C. 5, 6), a branch of the radial nerve, becomes cutaneous where the spiral groove for the radial nerve crosses the lateral intermuscular septum.

AXIOM: When seeking cutaneous nerves in the limbs, one should think vertically and act vertically—not horizontally—because the border is constant, whereas the level is inconstant.

Antebrachium or Forearm. It is supplied by *medial, lateral,* and *posterior cutaneous nerves of the forearm.* Each of these three cutaneous nerves is derived either directly or indirectly from the similarly named medial, lateral, and posterior cords of the brachial plexus. Each of the three outcrops on the surface above the elbow.

The medial cutaneous nerve of the forearm, a direct branch of the medial cord ((7) 8, 1), lying medial to the brachial artery, becomes cutaneous half way down the arm in the company of the basilic vein. It divides into an anterior and a posterior branch, both of which pass in front of the elbow and then descend on the front and back of the ulnar side of the forearm as far as the wrist. Its branches pass either superficial or deep to the median cubital vein and other cutaneous veins at the elbow.

The lateral cutaneous nerve of the forearm is the end branch of the musculocutaneous nerve (C. 5, 6 (7)), and therefore of the lateral cord. It appears at the lateral border of the Biceps and divides into ante-

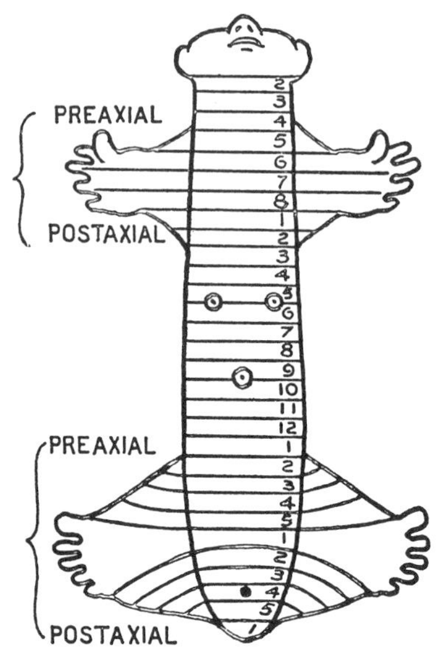

Fig. 96. Scheme of primitive segmental nerve distribution. (After Purves-Stewart.)

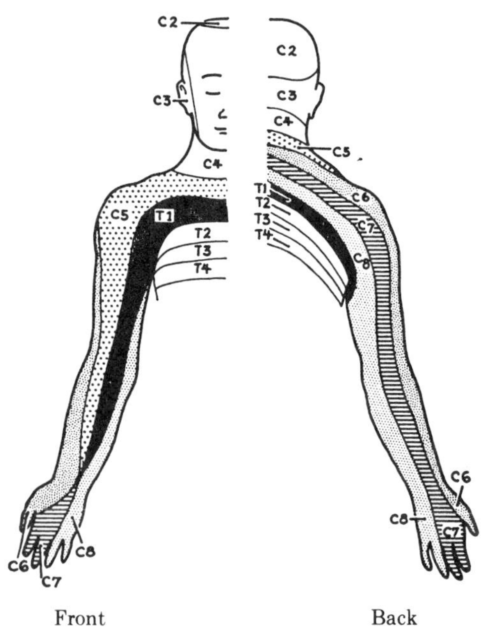

Fig. 97. The cutaneous distribution of the spinal nerves—dermatomes. (After Keegan and Garrett.)

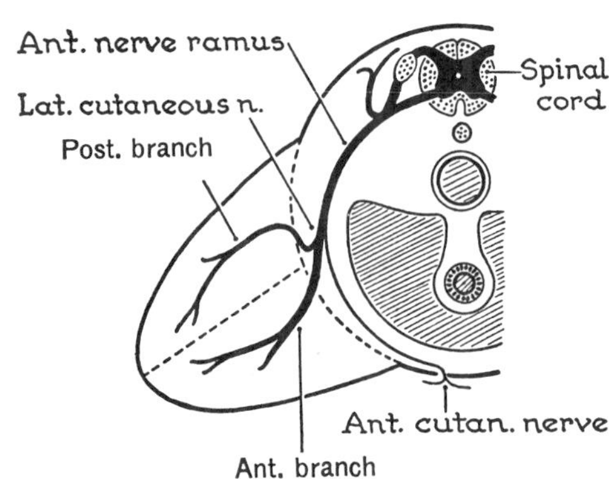

Fig. 95. Scheme of the developing limb bud. (After Patterson.)

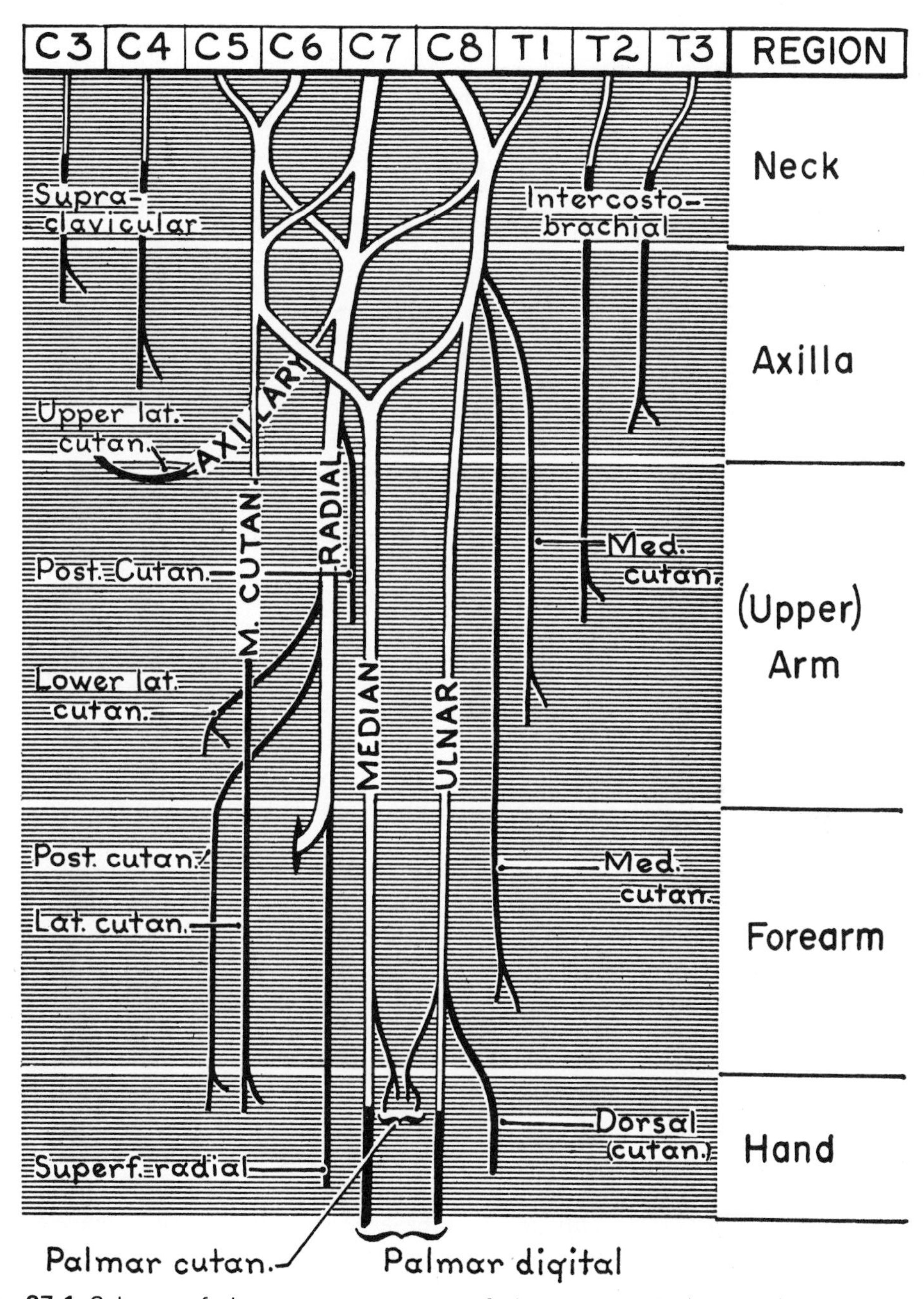

Fig. 97.1 Scheme of the cutaneous nerves of the upper limb (in black). It shows three features—their source, level of origin, and level of termination.

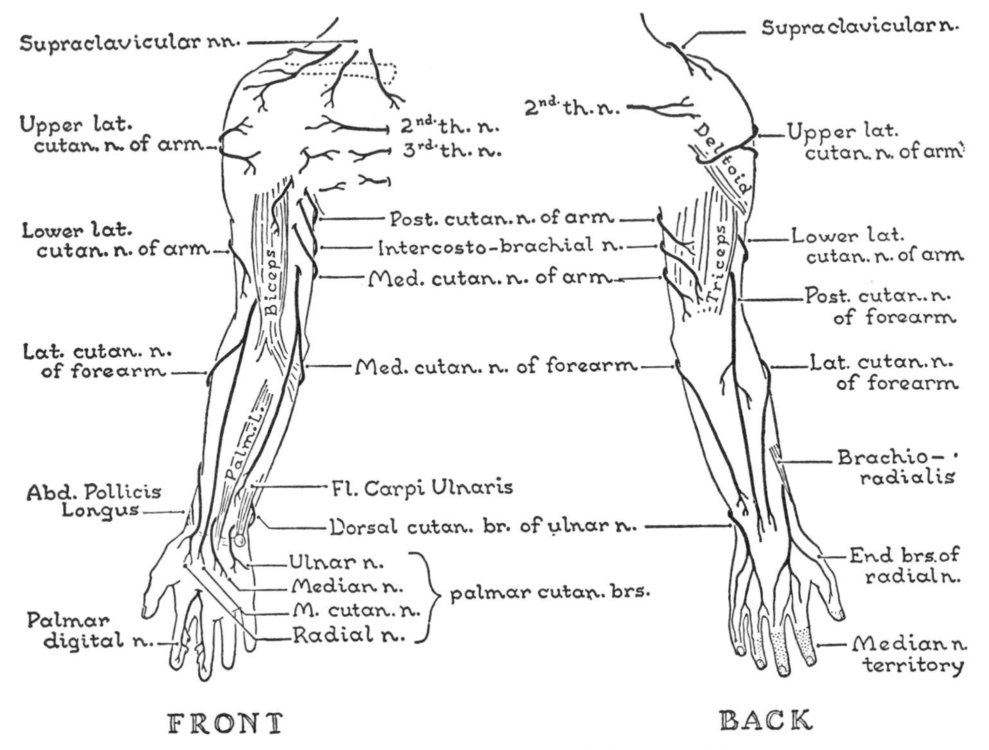

Fig. 98. The cutaneous nerves of the upper limb.

rior and posterior branches which pass in front of the elbow before descending on the front and back of the radial side of the forearm; the anterior branch to reach the ball of the thumb, the posterior branch to end at or beyond the wrist.

The posterior cutaneous nerve of the forearm, a branch of the radial nerve and therefore of the posterior cord (C. 5, 6, 7, 8), becomes cutaneous along the line of the lateral intermuscular septum and descends in the midline of the back of the forearm to the wrist or beyond it.

Palm of the Hand. Of the five terminal branches of the brachial plexus all, except the axillary nerve, pass into the arm and each ultimately contributes a *palmar cutaneous branch* to the hand (*fig. 98*).

Perhaps the most important cutaneous nerves of the upper limb are the **Palmar Digital Nerves.** There are 10 palmar digital nerves—one for each side of each digit. The median nerve supplied 3½ digits; the ulnar nerve 1½ (*fig. 100*). The palmar digital nerves furnish branches to the entire palmar surfaces of the digits, and to the distal parts of the dorsal surfaces including the subungual regions (i.e., deep to the nail) and to the local joints (*fig. 99*).

Dorsum of the Hand. The dorsum is supplied by the radial, ulnar, and median nerves. *The dorsal cutaneous branch of the ulnar nerve* (C. 7, 8) and *the superficial branch of the radial nerve* (C. 6, 7), after passing deep to the Flexor Carpi Ulnaris and Brachioradialis, respectively, outcrop on the surface at the posterior borders of these muscles from 2 to 10 cm above the corresponding styloid process. The dorsal branch of the ulnar nerve reaches the dorsum of the hand by crossing the medial (ulnar collateral) ligament of the wrist; the superficial radial nerve, by crossing the "snuff-box." As at the front so at the back, the ulnar nerve supplies one and one-half digits (*fig. 100*). The radial supplies the

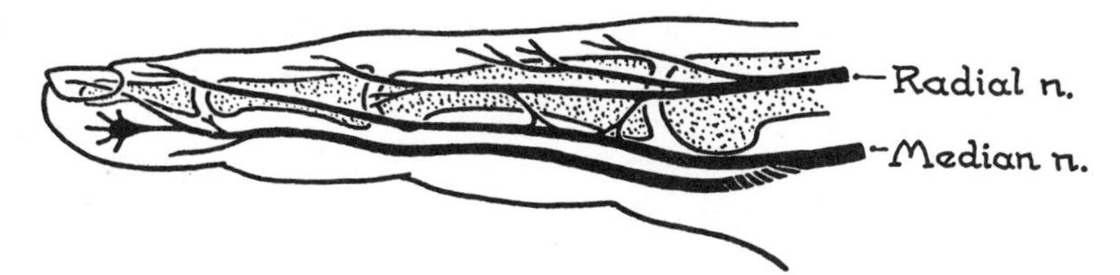

Fig. 99. Digital nerves of the index. (From a dissection by M. Wellman.)

remainder, except for the distal parts of the lateral three and one-half digits, which the median nerve supplies.

Variations in Pattern in the Hand (*fig. 101*) are common.

SUPERFICIAL VEINS

The superficial veins of the upper limb (*fig. 102*) form many patterns. *The Dorsal Venous Arch*, usually seen with ease on the back of your own hand, receives *digital branches*. It ends medially as the basilic vein and laterally as the cephalic vein. The **Basilic Vein** ascends on the ulnar side of

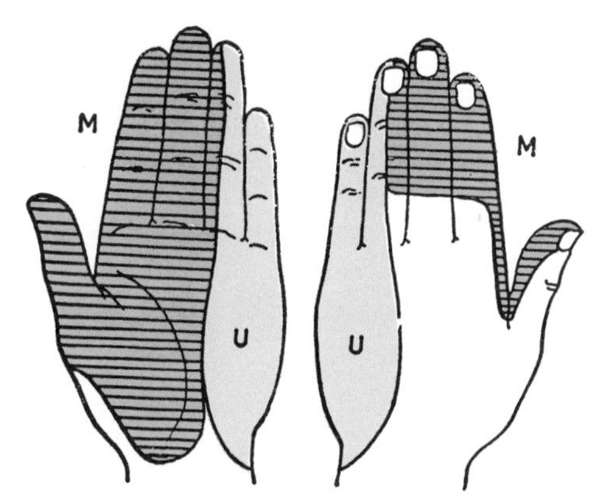

Fig. 100. Distribution of median and ulnar nerves in the hand. (After Stopford.)

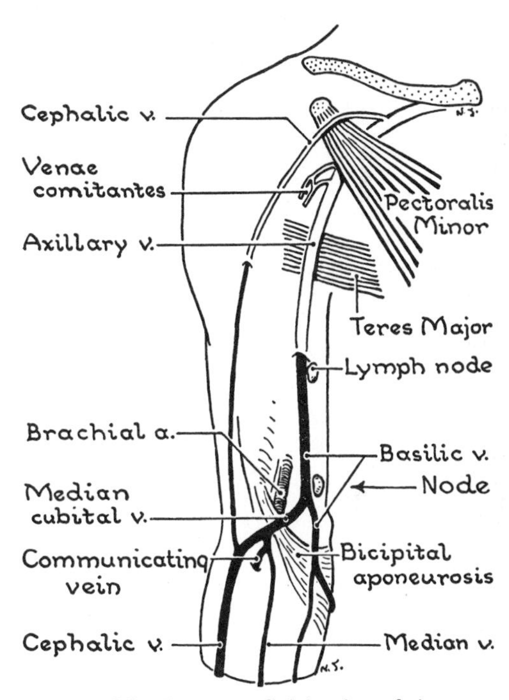

Fig. 102. The superficial veins of the arm.

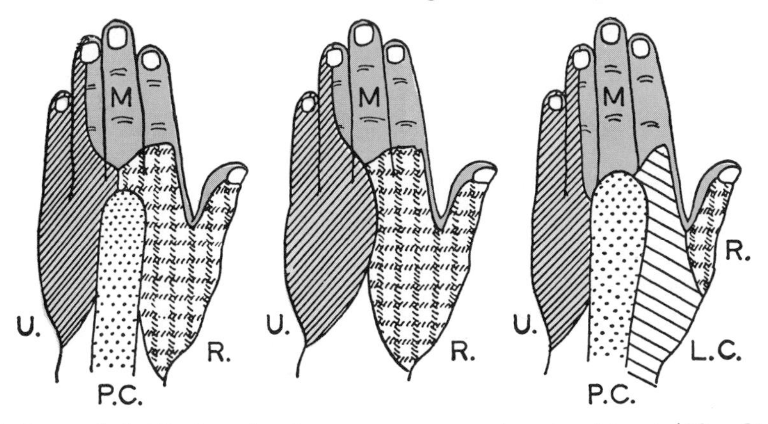

Fig. 101. Patterns of distribution of cutaneous nerves on dorsum of hand. (After Stopford.)

the forearm to the elbow and then in the medial bicipital furrow to the middle of the arm where it pierces the deep fascia. It then accompanies the brachial artery and its venae comitantes to the axilla and becomes the axillary vein.

The **Cephalic Vein** crosses "the snuffbox" (that is, the depression at the side of the wrist proximal to the base of the metacarpal of the thumb) superficial to the branches of the radial nerve. It ascends on the radial border of the forearm; in the lateral bicipital furrow of the arm; and in the cleft between the Deltoid and Pectoralis Major at the shoulder. It pierces the deltopectoral triangle and ends in the axillary vein (*fig. 102*).

[In the embryo, the cephalic vein crosses in front of the clavicle and ends in the external jugular vein; it may continue to do so in postnatal life.]

A small vein, the *median vein*, runs up the front of the forearm and, after communicating with a *deep vein*, bifurcates into a medial and a lateral branch, which join the basilic and cephalic veins respectively. This is the simple M-like pattern. More commonly, however, a large oblique vein, the **Median Cubital Vein,** placed in front of the elbow, joins the cephalic vein to the basilic vein. The median cubital vein passes in front of the bicipital aponeurosis (medial to the tendon). That alone separates it from the brachial artery and median nerve; and it passes in front of (or between) the branches of the medial cutaneous nerve of the forearm. Beware of confusing an anomalous *superficial ulnar artery* for a vein. This can be disastrous in clinical practice.

The practitioner of today employs the median cubital vein in blood transfusion; the barber of former days employed it in blood-letting. Hence the barber's sign—red and white wound spirally on a pole and a brass dish hanging from the pole. The pole for the patient to grasp; white for the bandage; red for blood; the dish to catch the blood.

Lymph Nodes

The most distal superficial node in the upper limb is the *cubital* (*supratrochlear*) *node*, placed 5 cm above the medial epicondyle. There may be a node where the basilic vein pierces the deep fascia (*fig. 102*). Though not palpable, these are tender when inflamed.

8 ARM

LOWER HALF OF HUMERUS

Humerus (cont'd from p. 93): Lower End; Lower Half of Body.

FASCIAE AND MUSCLES

Fascia; Coracobrachialis; Biceps; Brachialis; Triceps.

PALPABLE STRUCTURES AROUND ELBOW

CUBITAL FOSSA

Borders; Covering; Contents; Floor.

ARTERIES AND NERVES OF ARM

Brachial Artery and Branches.
Nerves: Musculocutaneous; Ulnar; Median; Radial.
Profunda Brachii Artery.

Segmental Innervation of Muscles

LOWER HALF OF HUMERUS

The **Lower End** of the humerus has articular and nonarticular parts. The former is divided into two areas: the *capitulum* for the head of the radius and the *trochlea* for the trochlear (semilunar) notch of the ulna. One is spheroidal; the other is shaped like a spool (*fig. 103*). Immediately above and wide of these two condylar or knuckle-like articular areas are two projections, the *medial* and *lateral epicondyles*. The medial epicondyle can easily be grasped between the finger and thumb; the posterior surface of the less prominent lateral epicondyle is smooth, subcutaneous, and palpable.

There are also three fossae; a large triangular one behind, the *olecranon fossa*, receives the olecranon of the ulna when the elbow is extended.

The **Lower Half of the Body of the Humerus** is flattened from before backward and is divided by the medial and lateral supracondylar ridges into an *anterior* and a *posterior aspect*. These surfaces are entirely devoted to fleshy fibers of the Brachialis in front and medial head of the Triceps behind; so, they are smooth.

The medial and lateral *supracondylar ridges* ascend from the epicondyles and afford attachment to the medial and lateral intermuscular septa, which give additional attachment to the muscles. The lateral and more prominent ridge extends to a broad shallow groove, the *groove for the radial nerve* (spiral groove), which intervenes between the ridge and the *deltoid tuberosity*. A rounded strengthening bar of bone (*figs. 104, 105*) causes the lower half of the humerus to be triangular on cross-section.

FASCIAE AND MUSCLES

Fascia

The muscles of the arm or brachium are enveloped in a sleeve of tough deep fascia whose fibers take a more or less circular course. This sleeve is divided into an anterior and a posterior compartment by the medial and lateral intermuscular septa which pass from the enveloping deep fascia to the supracondylar ridges (*fig. 106*).

Muscles

In the anterior compartment there are three muscles—Coracobrachialis, Biceps

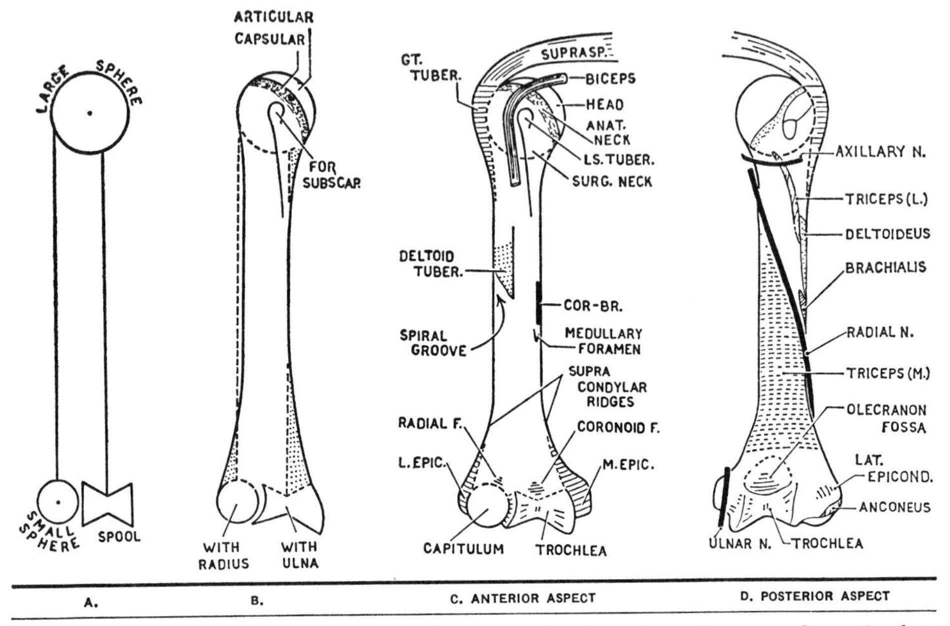

Fig. 103. *A, B,* and *C* represent progressive stages in sketching a humerus from the front. *D,* posterior aspect showing attachments of muscles and contacts with nerves.

Fig. 104. Cross-sections of humerus: the borders are indicated by letters.

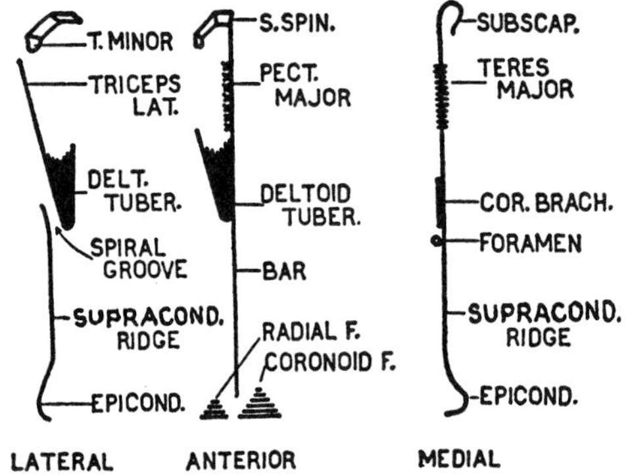

Fig. 105. The borders of the humerus.

Brachii, and Brachialis. In the posterior compartment there is one muscle—Triceps Brachii.

The Long Head of the Biceps descends in the intertubercular sulcus, encased in a *synovial sheath* (i.e., a tubular bursa), continuous with the synovial cavity of the shoulder joint.

The Coracobrachialis shares the tip of the coracoid process with the *short head of the Biceps;* so, both have tendinous origins. The roughness for the Coracobrachialis is of little significance.

Anterior to the Brachialis, the two heads of the Biceps unite. Their common tendon is inserted into the posterior, rough half of the tuberosity of the radius, which is situated just below the medial side of the neck of the radius (*fig. 107*); so, when the Biceps contracts, as when driving in a screw or corkscrew with the right hand, it supinates the radius. A *bursa* is required between the tendon and the smooth anterior part of the olive-shaped tuberosity. A secondary insertion of the Biceps, the *bicipital aponeurosis* is shown in fig. 108.

The Brachialis arises from the entire breadth of the anterior aspect of the lower half of the humerus and from the intermuscular septa. Its restricted and therefore fibrous insertion produces a rough elevation (tuberosity of the ulna) on the anterior aspect of the coronoid process of the ulna.

Actions. Brachialis is the main flexor of the elbow joint. The Biceps is not only a powerful flexor, but it also is a powerful supinator of the forearm. At the shoulder, its actions (in the expected directions) are much less powerful. Coracobrachialis is relatively unimportant.

Nerve Supply. These three muscles, and these three only, are supplied by the musculocutaneous nerve (*fig. 172*).

The Triceps Brachii: Its *long* or scapular head springs from the *infraglenoid tubercle*.

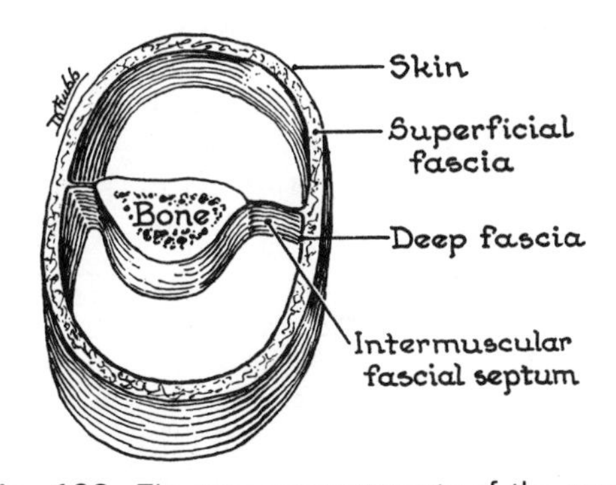

Fig. 106. The two compartments of the arm.

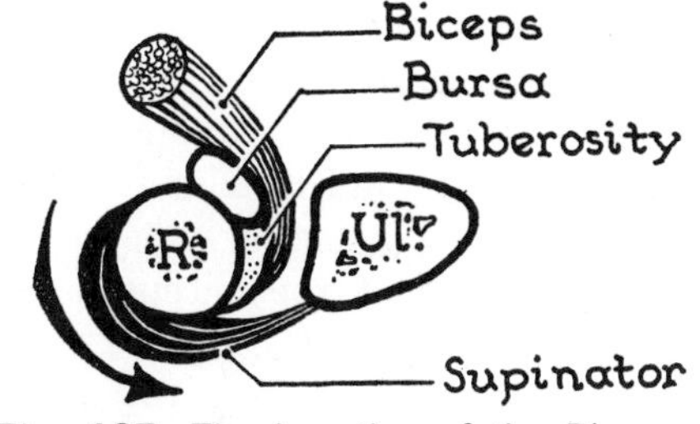

Fig. 107. The insertion of the Biceps. The Biceps supinates.

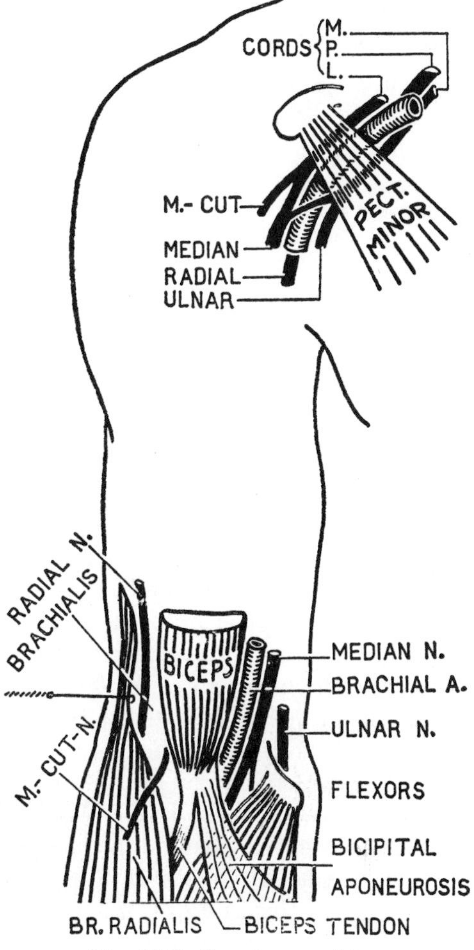

Fig. 108. Two key positions.

The Triceps has two humeral heads, a *medial* and a *lateral;* of these the medial head arises by fleshy fibers from the entire posterior surface of the humerus below the level of the (spiral) groove for the radial nerve and from the intermuscular septa (*fig. 103*). The lateral head is largely tendinous and is responsible for the line that ascends from the posterior margin of the deltoid tuberosity.

The common tendon of insertion is attached to the olecranon process of the ulna.

Action and Function. The Triceps is the extensor of the elbow joint. Its chief function or use is to keep the extended elbow extended (i.e. to prevent it from flexing) when one is pushing an object.

Nerve Supply. All three heads are supplied by the radial nerve.

PALPABLE STRUCTURES AROUND ELBOW

Before studying the vessels and nerves of the arm, become familiar with the disposition of the structures around the elbow, for they assume a key position both to the front of the arm and to the front of the forearm (*figs. 108 and 109*). You can, therefore, at any time roll up your sleeve, palpate, and thereby refresh your memory.

Note that when the elbow is flexed to a right angle and the forearm forcibly supinated (the palm of the hand then faces upward), the *Biceps tendon* and the *bicipital aponeurosis* stand out at the middle of the front of the elbow as a prominent central landmark, 2 cm wide. They can be grasped between the index finger and the thumb. When the Biceps is relaxed (rest forearm on the table) the pulsations of the *brachial artery* can be felt just medial to the Biceps tendon; and with the tips of the fingers the *median nerve*, which lies just medial to the artery, can be rolled on the Brachialis.

The *ulnar nerve*, lying behind the medial epicondyle, can be felt to slip from under the finger tips as they are drawn across it and tingling feelings in the little finger may be felt. It can be traced half way up the arm. Brachioradialis is seen and felt along the lateral edge of the elbow.

CUBITAL FOSSA

The cubital fossa is the triangular space at the front of the elbow, bounded laterally

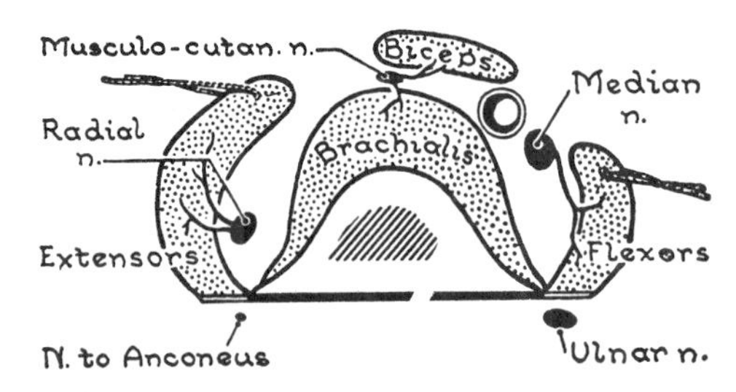

Fig. 109. Section showing nerves just above the elbow. Note their "sides of safety."

by Bracnioradialis and medially by Pronator Teres.

Covering the fossa is deep fascia reinforced with Bicipital aponeurosis, superficial to which are cutaneous nerves (*fig. 98*) and superficial veins (*fig. 102*).

The Contents are: Biceps tendon, brachial art. and its two terminal branches, median nerve, and (on retracting Brachioradialis) radial nerve (*figs. 108* and *112*).

The Floor is chiefly Brachialis.

ARTERIES AND NERVES OF THE ARM

Brachial Artery (*fig. 110*). This is the largest artery whose pulsation and whose walls can be felt satisfactorily in the living subject. It may be palpated along the medial bicipital furrow throughout the length of the arm to the point where it disappears behind the bicipital aponeurosis. At the level of the neck of the radius, 2 to 3 cm below the crease of the elbow, it divides into its two terminal branches, a larger *ulnar artery* and smaller *radial artery*. Triceps and Brachialis lie behind it through its course in the arm.

Its *Collateral Branches* (*fig. 128*) are:
Profunda brachii a. (p. 114),
Superior ulnar collateral a. and
Inferior ulnar collateral a.
whose special interest lies in the anastomoses they effect at the shoulder and elbow.
Muscular branches, and
Nutrient a. to the humerus.

Venae comitantes accompany the brachial artery and make a very open network around it. They join the axillary vein.

The Musculocutaneous Nerve arises from the lateral cord of the plexus (5, 6, 7) and must course distally and laterally. It pierces the Coracobrachialis and continues between the Biceps and Brachialis; and it supplies these three muscles. It emerges at the lateral margin of the Biceps as the *lateral cutaneous nerve of the forearm*.

The Ulnar Nerve arises from the medial cord of the plexus ((7), 8, 1) and, without

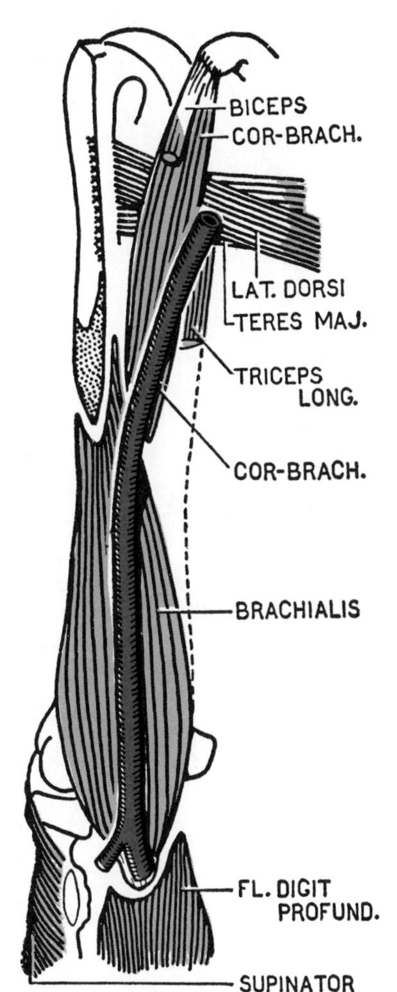

Fig. 110. Posterior relations of the brachial artery.

any branching, extends to the back of the medial epicondyle, and onward into the forearm. It, therefore, must pass distally and medially. It is applied to the medial side of the great arterial stem till it reaches the middle of the arm. There it leaves the artery and passes behind the medial intermuscular septum into the posterior compartment of the arm, where it lies subfascially, applied to the medial head of the Triceps.

[Accompanying it behind the medial septum are the superior ulnar collateral artery and a nerve, called the *ulnar collateral nerve*, sent by the radial nerve to the medial head of the Triceps.]

The Median Nerve. Its two roots ((5), 6, 7, 8, and 1) unite in the axilla on the lateral side of the artery. It slowly crosses the brachial artery, lying medial to it at the elbow (*fig. 111*). The median nerve gives off no branches in the arm. It commonly receives a large communication from the musculocutaneous nerve and occasionally it gives one to it.

The Radial Nerve being a terminal branch of the posterior cord (5, 6, 7, 8 (1)), lies behind the axillary artery. On entering the arm, it still lies behind the main arterial stem and is in front of the long head of the Triceps, which conducts it into the posterior compartment of the arm. On leaving the long head of the Triceps, the nerve enters the spiral groove which is converted into a tunnel by the lateral head of the Triceps. Within the tunnel, the radial nerve descends almost vertically along the origin of the medial head of the Triceps, which lies medially (*fig. 112*).

On escaping from the tunnel, the radial nerve enters the anterior compartment of the arm and continues its vertical descent in the depths between the Brachioradialis and the Brachialis (*fig. 113*). Distal to this it lies on the capsule of the elbow joint and on the Supinator. At a variable point it

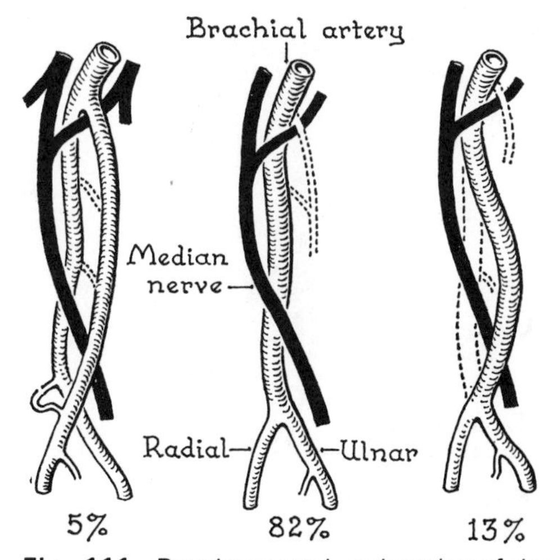

Fig. 111. Developmental explanation of the variable relationship of median nerve to brachial artery. (Based on 307 limbs.)

divides into two terminal branches: (1) a sensory branch, the *superficial radial nerve*, and (2) a branch that is essentially motor, the *posterior interosseous nerve* (deep radial n.). The latter disappears into the substance of the Supinator just below the elbow (*fig. 113*).

Collateral Branches in the axilla and arm are shown in *figures 112* and *113*. At *the lateral side of the arm and front of the elbow*, branches pass from the radial and posterior interosseous nerves to four muscles—Brachioradialis, Ex. Carpi Radialis Longus, Ex. Carpi Radialis Brevis, and Supinator. The radial nerve usually supplies the proximal two muscles and the posterior interosseous nerve the distal two muscles.

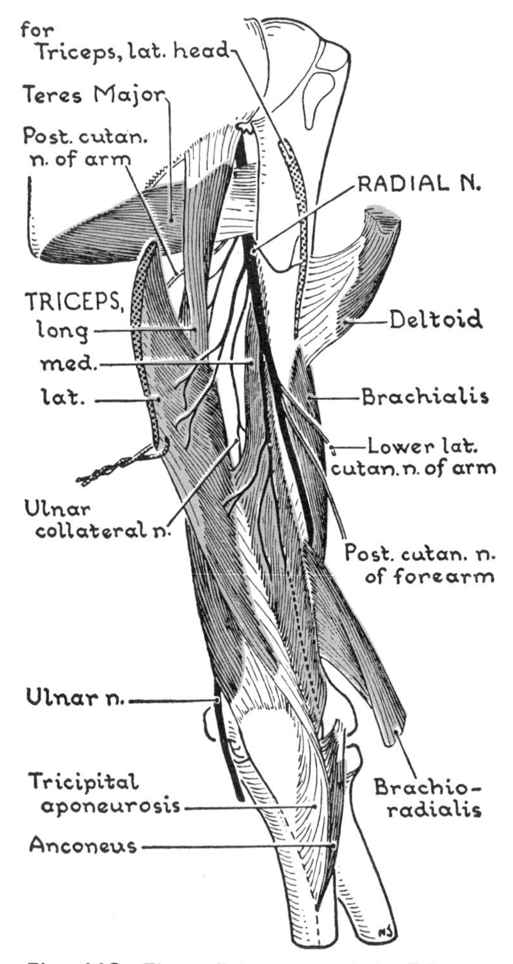

Fig. 112. The radial nerve and the Triceps.

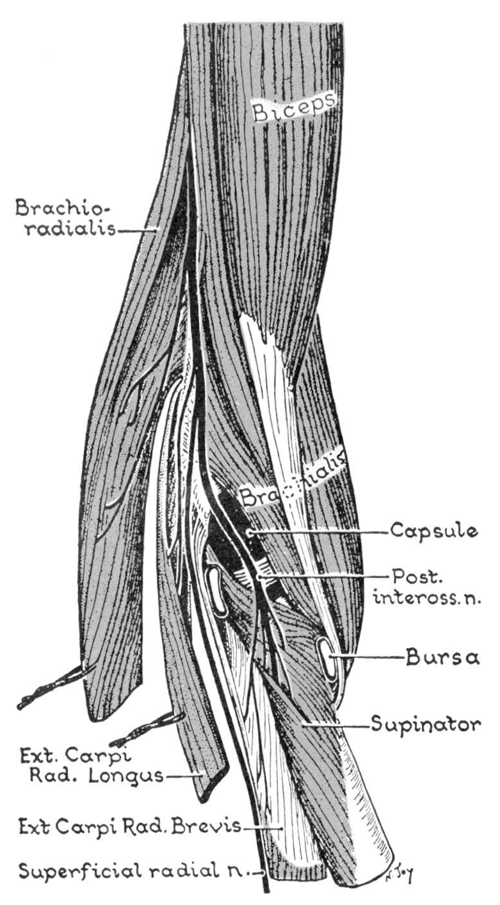

Fig. 113. The radial nerve at the elbow.

***AXIOM:** The side from which motor branches leave a nerve is the side nearest the muscles to which they are distributed. Certain nerves have sides of safety and sides of danger; sides on which it is safe to dissect and sides on which it is dangerous.*

The Profunda Artery arises from the brachial artery just below the Teres Major. It is the companion artery of the radial nerve. By its anastomoses it brings the axillary artery above into communication with the radial and ulnar arteries below.

Thus (for those who need details):

Above, a *recurrent branch* (commonly present) anastomoses with the posterior humeral circumflex art. (*fig. 114* and p. 92).

Below, it has *two terminal branches*, an anterior and a posterior, which form anastomoses at the elbow (*fig. 128*).

A *nutrient branch* is commonly given off.

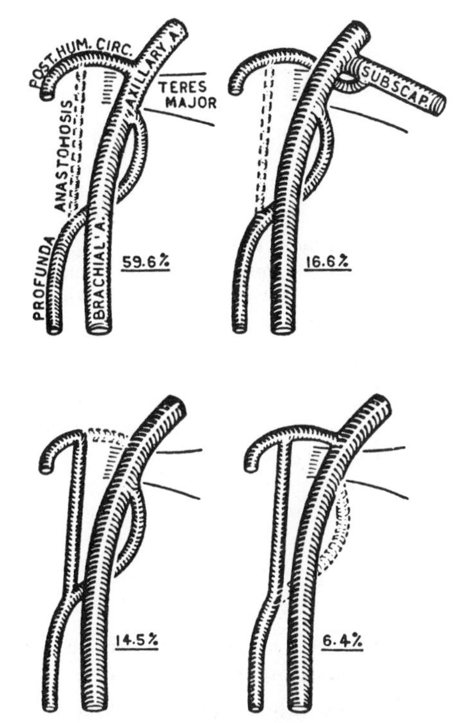

Fig. 114. Four types of variations in origin of the posterior humeral circumflex and profunda brachii arteries; in 2.9 per cent the arteries were otherwise irregular. Percentages are based on 235 specimens.

In 7.3 per cent of 123 specimens the profunda artery arose from the stem of the subscapular artery.

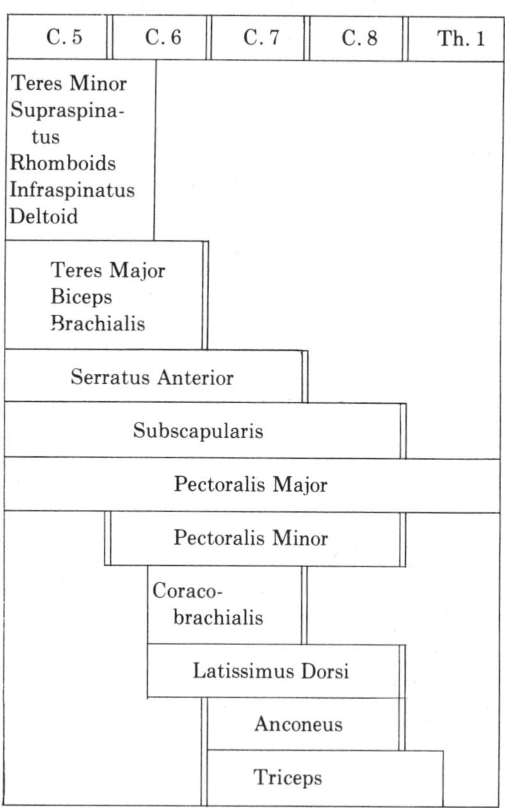

Segmental Innervation of Muscles of Shoulder and Upper Arm Supplied by Brachial Plexus*

<table>
<tr><th>C. 5</th><th>C. 6</th><th>C. 7</th><th>C. 8</th><th>Th. 1</th></tr>
<tr><td colspan="2">Teres Minor
Supraspinatus
Rhomboids
Infraspinatus
Deltoid</td><td></td><td></td><td></td></tr>
<tr><td colspan="2">Teres Major
Biceps
Brachialis</td><td></td><td></td><td></td></tr>
<tr><td colspan="3">Serratus Anterior</td><td></td><td></td></tr>
<tr><td colspan="4">Subscapularis</td><td></td></tr>
<tr><td colspan="5">Pectoralis Major</td></tr>
<tr><td></td><td colspan="3">Pectoralis Minor</td><td></td></tr>
<tr><td></td><td colspan="2">Coraco-
brachialis</td><td></td><td></td></tr>
<tr><td></td><td colspan="3">Latissimus Dorsi</td><td></td></tr>
<tr><td></td><td></td><td colspan="2">Anconeus</td><td></td></tr>
<tr><td></td><td></td><td colspan="3">Triceps</td></tr>
</table>

* Modified after Bing; and Haymaker and Woodhall.

9 FLEXOR REGION OF FOREARM

RADIUS AND ULNA

Ends and Flexor Surfaces; Terminology.
Boundaries of Region.

MUSCLES, VESSELS, AND NERVES

Muscles Clothing Bones—Pronator Quadratus; Flexor Pollicis Longus; Flexores Digitorum Superficialis and Profundus; Observations and Axiom.
Anterior Interosseous Nerve and Artery.
Flexor Digitorum Superficialis.
Surface Anatomy of Wrist.
Four Superficial Flexors—Pronator Teres; Flexor Carpi Radialis; Palmaris Longus; Flexor Carpi Ulnaris.

ARTERIES AND NERVES

Arteries—ulnar, (median), radial; Branches.
Nerves—ulnar, superficial radial, median; Branches; Internervous Lines.

RADIUS AND ULNA

The proximal end of the radius consists of: *Head, neck, and radial tuberosity* for the insertion of the Biceps (*fig. 115*).

The distal end of the radius consists of: *Lower articular surface, ulnar notch, styloid process*, anterior and posterior aspects. On the last are the *dorsal radial tubercle (of Lister)* and several grooves.

The proximal end of the **ulna** consists of: *Olecranon, coronoid process, trochlear notch, radial notch*, and *ulnar tuberosity* for the insertion of the Brachialis.

The distal end of the ulna consists of: *Head, styloid process, pit* for the attachment of the articular disc, and *groove* for Ex. Carpi Ulnaris.

The **body of the radius** is cylindrical and somewhat tapering above; the **body of the ulna** is cylindrical and somewhat tapering below. The expanded lower two-thirds of the radius and the expanded upper two-thirds of the ulna are triangular on cross-section.

The facing interosseous borders of the two bones are united by an **interosseous membrane,** which, on account of the inferomedial direction of its fibers would seem able to transfer from radius to ulna the force of an impact received by the hand (*fig. 116*)—though recently Halls and Travill have shown experimentally that this is highly improbable.

Conventions in Terminology. Typically, the bodies of long bones are triangular on cross-section, and their surfaces and borders are named by opposites (*fig. 117*).

At the summit of the convexity of the lateral surface of the radius there is a rough *area for the Pronator Teres.* From it the *anterior* and *posterior oblique lines* ascend across the anterior and posterior aspects of the bone to the radial tuberosity. The area of the body between the oblique lines is appropriated by the *Supinator.*

The posterior border of the ulna extends from the apex of the posterior *subcutaneous surface of the olecranon,* on which lies the olecranon bursa, to the back of the styloid process. This border is sharp, subcutaneous, and always palpable from end to end; and when the forearm is raised in self-protection, it is liable to be struck.

The anterior border is smooth and rounded (*fig. 118*), except in its lower quarter where it forms a rough ridge, the

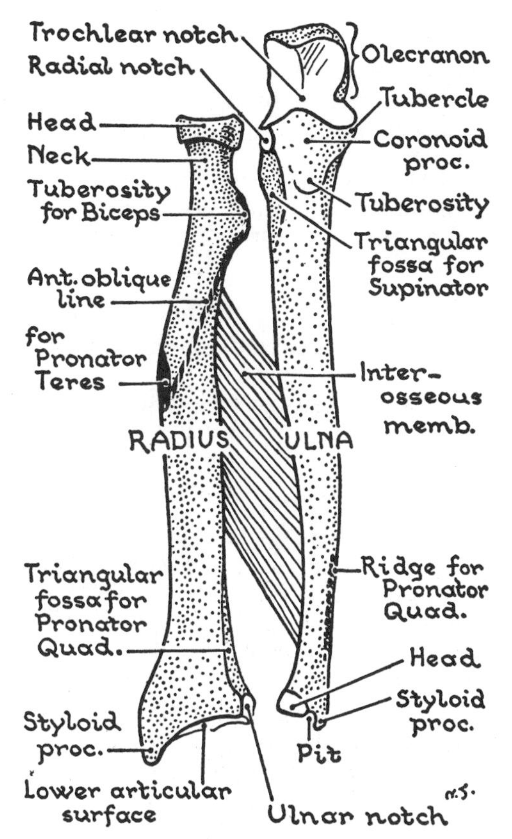

Fig. 115. The radius and ulna, anterior aspect.

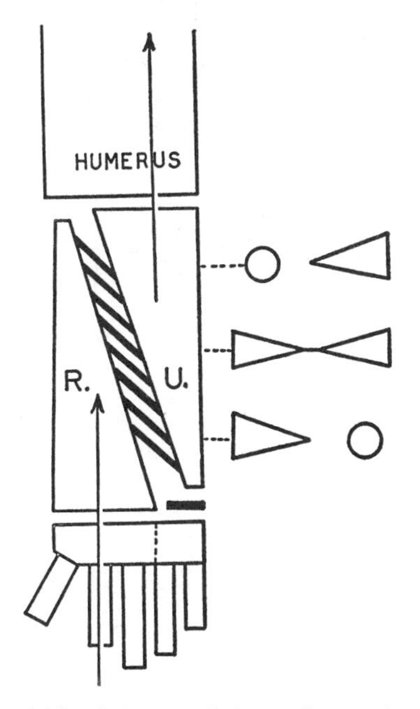

Fig. 116. Scheme of the radius and ulna.

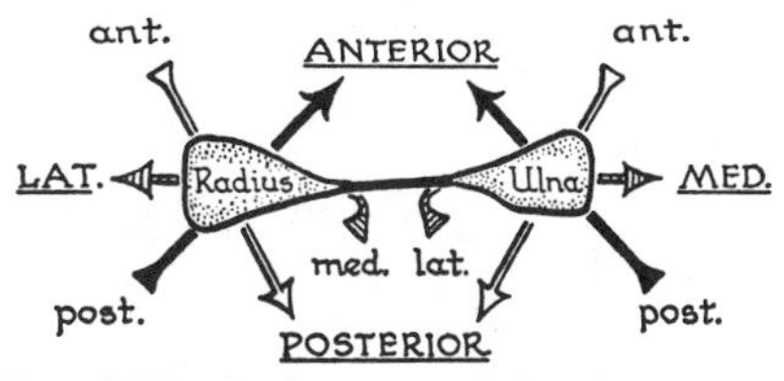

Fig. 117. Surfaces and borders illustrated on cross-section. Surfaces in large type; borders in small type.

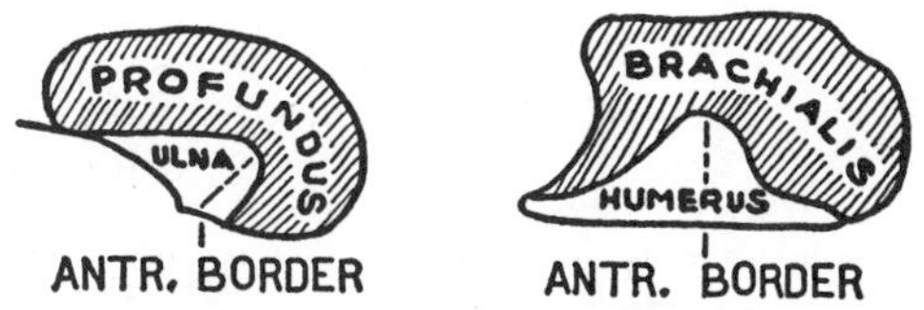

Fig. 118. Rounded borders are usually strengthening bars; sharp and rough borders give attachment to fibrous tissue.

pronator ridge, for the fibrous origin of the Pronator Quadratus (*figs. 115, 121*).

Boundaries of Flexor Region (*fig. 119*). The flexor region of the forearm includes the medial surface of the ulna. Medially, it is marked off from the extensor region by the sharp, palpable, posterior border of the ulna. Laterally, it is marked off from the extensor region by the impalpable anterior border of the radius (*fig. 120*). More superficially, the course of the radial artery serves as a practical guide to the lateral boundary.

Since neither the medial nor the lateral boundary line is crossed by a motor nerve, the deep parts of the limb may be explored by the surgeon through these internervous lines.

MUSCLES OF FLEXOR REGION

Muscles Clothing the Flexor Aspects of the Bones (*fig. 121*). The *Pronator Quadratus* arises from the pronator ridge of the ulna. It is inserted by fleshy fibers into the smooth, lower quarter of the anterior surface of the radius (*fig. 122*). Bridging it are the long tendons.

The *Flexor Pollicis Longus* arises from the smooth, anterior surface of the radius and adjacent part of the interosseous mem-

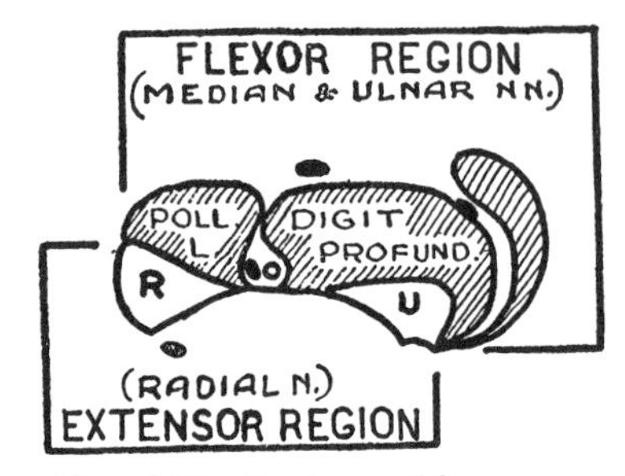

Fig. 119. Regions of forearm.

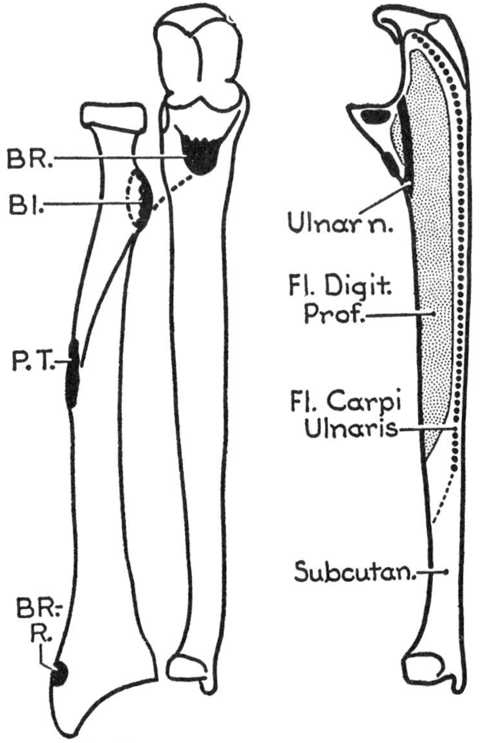

Fig. 120. Flexor aspect of the bones of the forearm. The four flexors of the elbow (Brachialis, Biceps, Pronator Teres, and Brachioradialis) like sentinels guard the boundary between flexor and extensor territories.

brane from the level of the Pronator Quadratus below right up to the anterior oblique line from which the aponeurotic radial head of the *Flexor Digitorum Superficialis* (for the 3rd digit) arises. Above this level the *Supinator* has been seen to occupy the bone up to and beyond the insertion of the *Biceps* into the posterior, rough half of the radial tuberosity.

The *Flexor Digitorum Profundus* arises from the smooth anterior and medial surfaces of the ulna, adjacent part of the interosseous membrane and the medial surface of the olecranon (*fig. 120*).

The vessels and nerves that pass through the flexor region of the forearm are separated from the bones by a carpet formed by the above 7 muscles.

Tendons. The five tendons of the deep digital flexors converge on the midline of

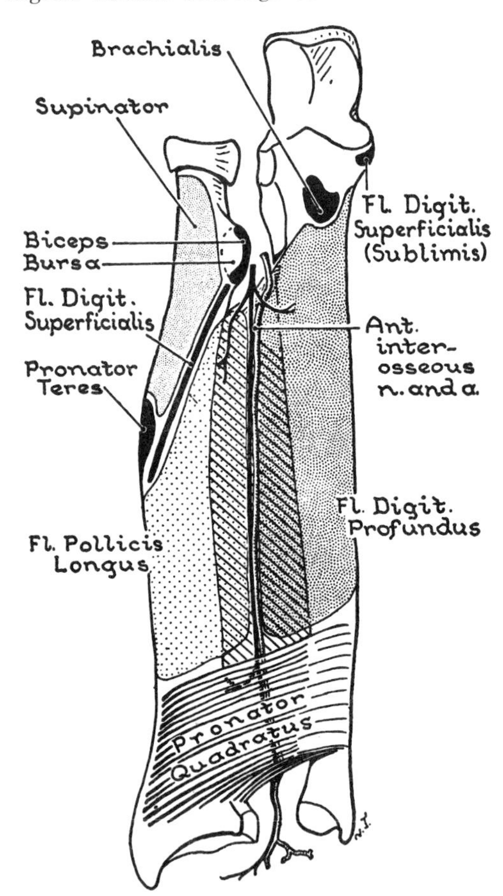

Fig. 121. Radius, ulna, and interosseous membrane, showing attachment of muscles; front view.

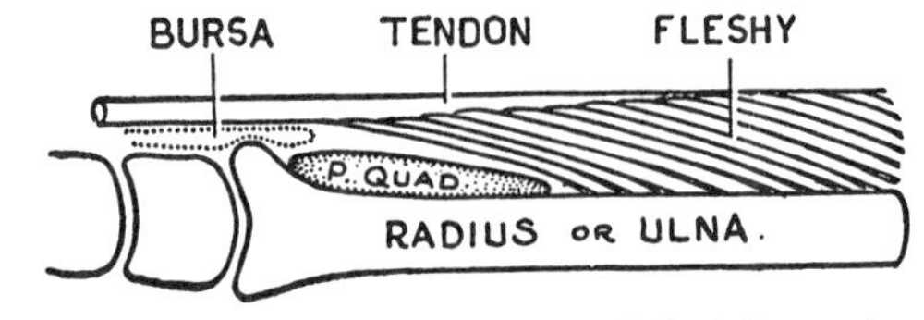

Fig. 122. Tendon of deep digital flexor (*see Axiom*).

the limb in order to enter the carpal tunnel side by side (*fig. 144*). So, the tendon of the Flexor Pollicis Longus must leave the lateral part of the Pronator Quadratus uncovered, thereby allowing the radial artery to come to lie on it (*fig. 123*).

AXIOM: Fleshy fibers cannot survive severe pressure or friction, and even a tough tendon requires a bursa or a synovial sheath where it rubs against bone or other resistant structure, e.g., fibrous sheath, other tendons.

In lower primates there is a common deep digital flexor for the five digits, but in man the Flexor Pollicis Longus is an independent muscle and the deep tendon to the index acquires a certain independence in the forearm. The designer of the work glove (*fig. 123.1*) apparently recognized that fact.

The Anterior Interosseous Nerve and Artery arise from the median nerve and common interosseous artery 2 or 3 cm below the elbow. They cling to the interosseous membrane. The *nerve* supplies the three deep muscles (Fl. Pollicis Longus, ½ of Fl. Digitorum Profundus for digits 2 and 3 and Pronator Quadratus)—the remainder of the Flexor Digitorum Profundus is supplied by the ulnar nerve.

Intermediate Layer

Flexor Digitorum Superficialis (or Sublimis) (*fig. 124*). This muscle occupies an intermediate position between the superficial and deep flexors. It arises from the humerus, ulna and radius, and from a fibrous bridge between the latter two. Under the bridge run the median nerve and the ulnar artery (*fig. 124*). The median nerve supplies the Superficialis, clinging closely to its deep surface.

Unlike the tendons of the Fl. Digitorum Profundus which lie side by side, those of the Fl. Digitorum Superficialis are "two deep" (*fig. 124*). As the tendons pass behind the flexor retinaculum they come to lie four abreast.

Superficial Layer

The Four Superficial Flexors are the Pronator Teres, Flexor Carpi Radialis, Pal-

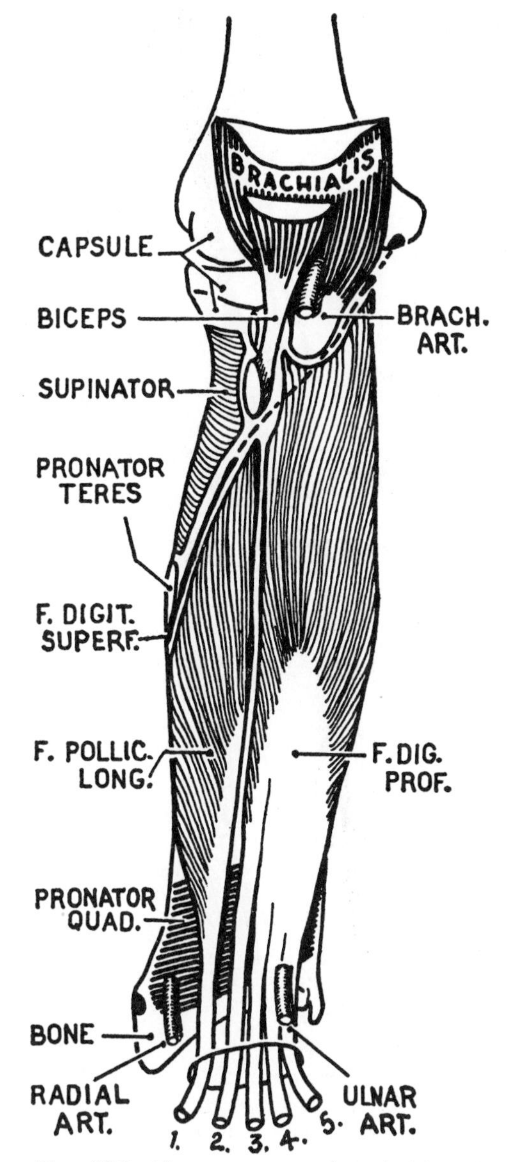

Fig. 123. The seven muscles clothing the flexor aspect of the radius and ulna.

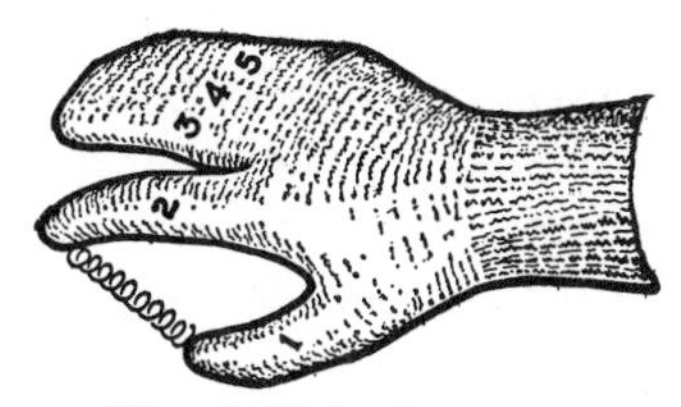

Fig. 123.1. See text.

maris Longus, and Flexor Carpi Ulnaris (*fig. 125*). They have a common and therefore a fibrous origin from the front of the

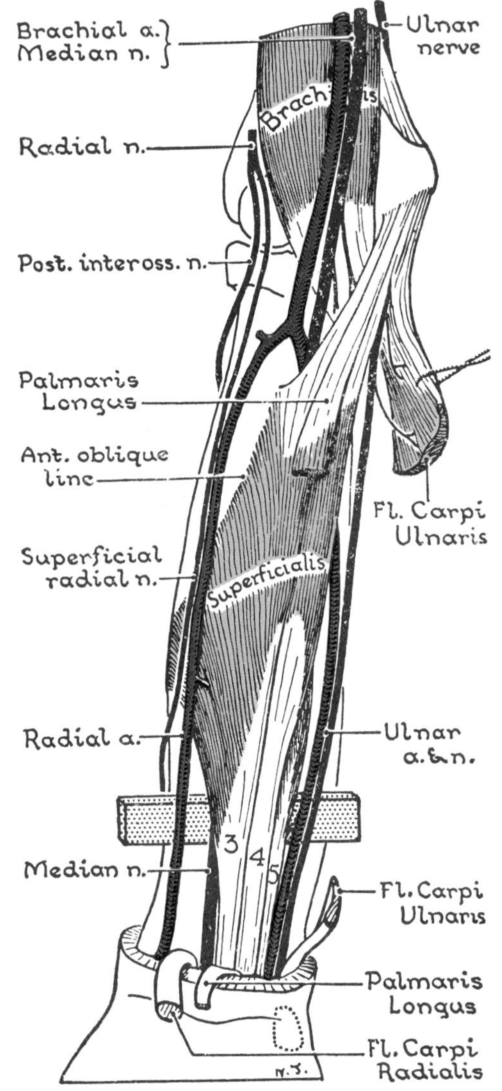

Fig. 124. The key position of the Flexor Digitorum Superficialis (Sublimis) with reference to the nerves and arteries.

medial epicondyle of the humerus, from the investing deep fascia, and from the septa between adjacent muscles. The *Flexor Carpi Ulnaris* bridges the ulnar nerve and increases its origin down the sharp posterior border to the ulna (*fig. 120*).

Surface Anatomy of Front of Wrist

The *distal skin crease* at the wrist (*fig. 126*), slightly convex toward the palm, corresponds to the upper border of the **flexor retinaculum** (transverse carpal ligament). It crosses the two prominent proximal bony pillars to which the retinaculum is attached, namely: the *tubercle of the scaphoid bone* laterally, the *pisiform bone* medially.

Where the prominent tendon of the *Flexor Carpi Radialis* crosses this skin crease at the junction of its lateral one-third with its medial two-thirds, it passes in front of the scaphoid tubercle and so serves as a guide to it. The Flexor Carpi Radialis is rendered very prominent when the clenched fist is fully flexed against resistance.

The tendon of the *Flexor Carpi Ulnaris*, the most medial of all the structures at the

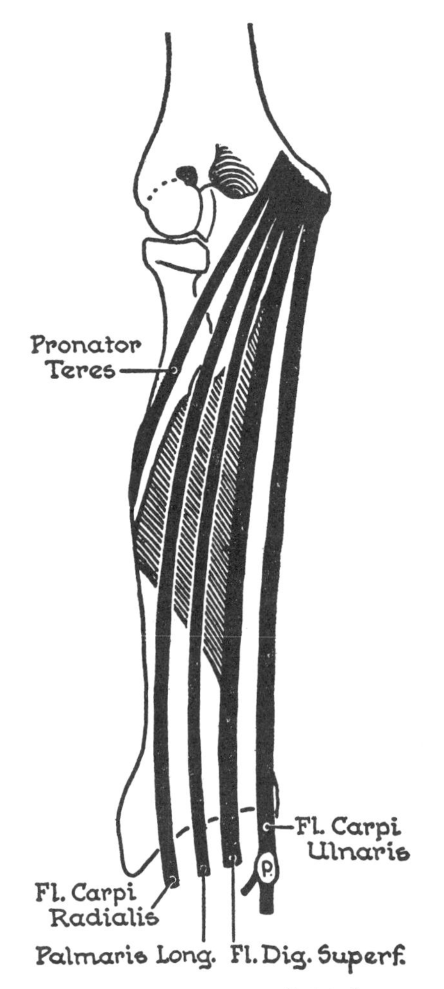

Fig. 125. The four superificial flexors and the Flexor Digitorum Superficialis.

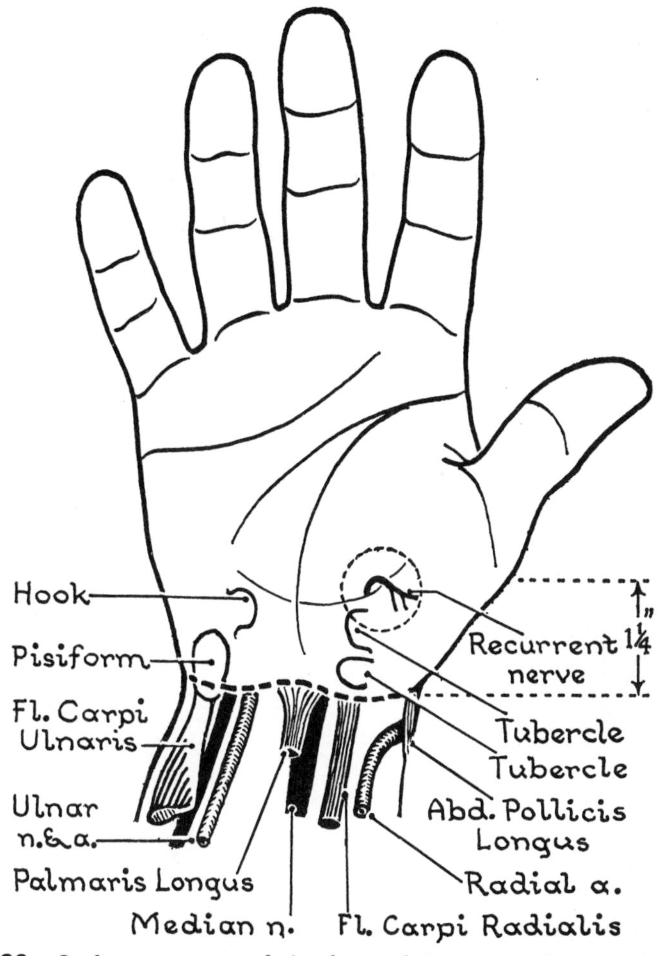

Fig. 126. Surface anatomy of the front of the wrist: a key position.

wrist, can be traced to the pisiform at the medial end of the crease.

Bisecting the distal skin crease (i.e., at the middle of the front of the wrist) is the most important of all the structures—the *median nerve*. It cannot be palpated for layers of fascia and the *Palmaris Longus* tendon covers it. The latter is sometimes absent.

Palmaris Longus is absent on both sides in 7 per cent, absent on the right only in 7 per cent, and absent on the left only in 6 per cent (R. K. George).

The *radial artery* is easily felt pulsating lateral to the Fl. Carpi Radialis. The *ulnar artery* and *nerve* pass into the palm immediately lateral to the pisiform, which is protective to them (*fig. 127*). As, however, a strong band of fascia covers them, the pulse of this artery is not easily felt.

Finally, when the wrist is extended, the *Fl. Digitorum Superficialis* bulges between the Fl. Carpi Ulnaris and the Palmaris Longus. The *lower end of the radius* can be grasped between the fingers and thumb. The anterior margin of its articular surface is surprisingly prominent and rugged.

ARTERIES AND NERVES OF FRONT OF FOREARM

Lines connecting the bifurcation of the brachial artery below the skin crease at the elbow, to the points where the ulnar and radial arteries were identified at the wrist, indicate the general directions of the two arteries (*fig. 127*).

Ulnar Artery. In its course through the forearm, the ulnar artery passes deep to the fibrous arch of the Fl. Digitorum Superficialis. Under this arch the median nerve must cross the ulnar artery (*fig. 124*). Behind the ulnar artery are the muscles clothing the anterior aspect of the ulna (*fig. 123*).

Its pulse is not readily felt at the wrist because it is there bridged by deep fascia (the *volar carpal lig.*).

AXIOM: Most branches of arteries are muscular branches and they are usually unnamed. Certain muscular branches anastomose, especially about joints; so do a few nonmuscular branches, and they are named.

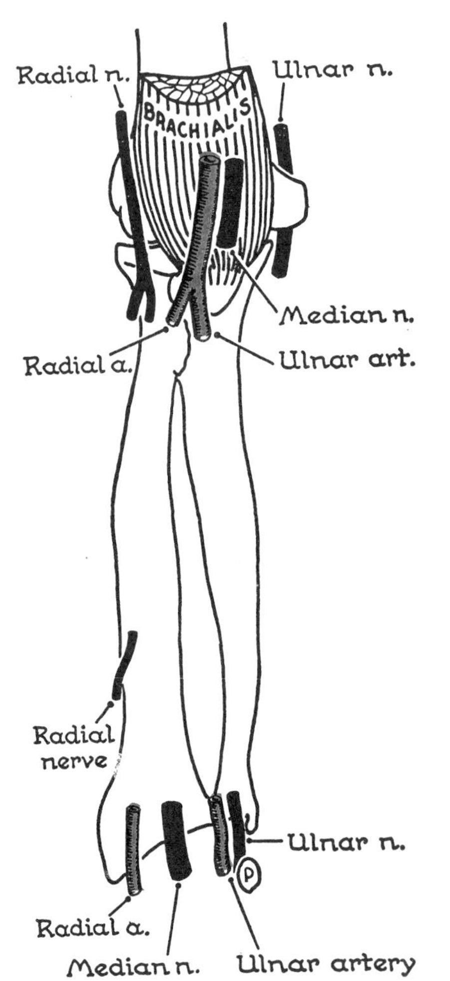

Fig. 127. The vessels and nerves of the forearm at key positions. Complete this picture by filling in the missing segments.

Branches of Ulnar Artery in the forearm include *recurrent aa.* that anastomose around the medial epicondyle; *common interosseous a.*, which supplies the bulk of the forearm muscles and bones, and has anastomotic branches around the elbow and wrist; *muscular aa.;* and *carpal aa.*, at the wrist.

Details of Branches for advanced specialists and reference:

1 and 2. *Anterior and posterior ulnar recurrent aa.* take part in the anastomoses around the medial epicondyle (*fig. 128*).

3. *Common interosseous a.* is a short stem that passes to the upper border of the interosseous membrane and divides there into the anterior and posterior interosseous aa. Their branches are:

Anterior interosseous artery:
- Median, companion to median nerve.
- Nutrient, to radius and ulna.
- Muscular.
- Anterior communicating, anastomoses in front of the wrist.
- Terminal, anastomoses behind the wrist (*figs. 155* and *157*).

Posterior interosseous artery:
- Recurrent, anastomoses behind the lateral epicondyle.
- Muscular.
- Terminal, anastomoses behind wrist.

4. *Muscular branches.*

5, 6. *Palmar and dorsal ulnar* carpal aa. take part in the anastomoses around the wrist (*fig. 155*).

The *median artery* is of historical interest. For a brief period in embryonic life it accompanies the median nerve.

Radial Artery. This trunk is not crossed by any muscle, but the Brachioradialis is lateral and overlaps it; the Flexor Carpi Radialis is medial in its lower two-thirds.

Immediately behind it are six muscles that clothe the anterior aspect of the radius (*fig. 123*) and beyond these are the lower end of the radius and the capsule of the wrist joint. The artery gains the back of the

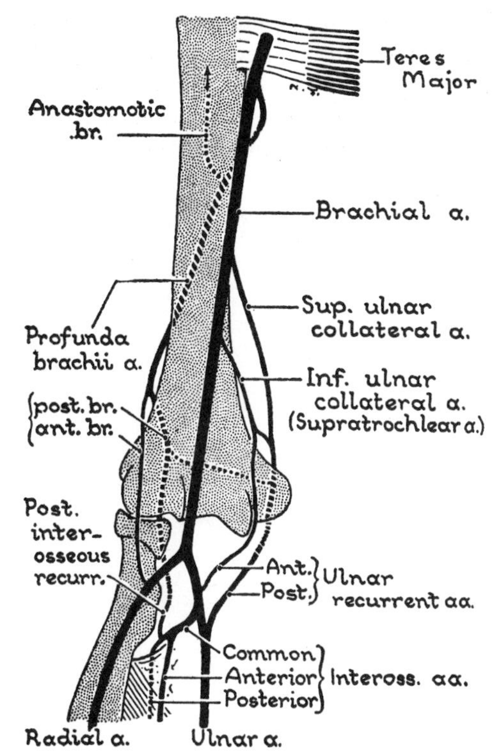

Fig. 128. The arterial anastomoses around the elbow (details). Their existence, not their details, is the essential point for students to remember.

wrist by passing below the styloid process and deep to the Abductor Pollicis Longus.

The superficial radial nerve accompanies the radial artery only briefly in the middle third of the forearm.

Branches of the Radial Artery given off in the forearm are *muscular, recurrent* (*fig. 128*), *palmar* (*anterior*) *carpal* (p. 141), and *superficial palmar* (*fig. 148*).

Ulnar Nerve. The ulnar nerve takes a straight course from the back of the medial epicondyle to the lateral side of the protective pisiform. It lies throughout on the Fl. Digitorum Profundus. The Fl. Carpi Ulnaris covers it proximally and overlaps it distally. Its artery approaches it, obviously from the lateral side, and is in contact with it in the lower two-thirds of its course.

Branches (*fig. 172*). At the elbow the ulnar nerve supplies one and one-half muscles—the Fl. Carpi Ulnaris and the medial half of the Fl. Digitorum Profundus—and it sends twigs to the elbow joint. Somewhat above the wrist its palmar and dorsal cutaneous branches arise (*fig. 98*).

Plastic surgeons take advantage of the fact that if you sever the humeral origin of the Fl. Carpi Ulnaris, you can bring the ulnar nerve to the front of the elbow, shortening its course by several centimeters.

Superficial Radial Nerve. This is the sensory continuation of the radial nerve distal to the origin of the motor (posterior interosseous) nerve. To expose it, peel the Brachioradialis laterally (*fig. 113*).

After crossing the capsule of the elbow joint, the nerve descends with the Brachioradialis, to which it passes deep to become cutaneous about 5 cm above the styloid process of the radius.

The radial artery approaches the nerve from the medial side and accompanies it through the middle third of the forearm.

Median Nerve. It plunges with the ulnar artery deep to the "Superficialis bridge," running a straight course to the midpoint of the wrist (*fig. 127*), where it passes deep to the flexor retinaculum. Behind it are the muscles clothing the front of the ulna, especially the Fl. Digitorum Profundus.

Branches (*fig. 172*). The median nerve supplies all the flexor muscles of the forearm, except the one and one-half supplied by the ulnar nerve. It adheres to the deep surface of the Fl. Digitorum Superficialis, giving it branches in its course.

It also sends articular twigs to the elbow and wrist joints.

Internervous Lines. The Fl. Digitorum Superficialis is supplied by the median nerve and the Fl. Carpi Ulnaris by the ulnar nerve; so, the septum between them marks an internervous line. This may safely be opened up, the two muscles pulled apart, and access gained to the deeper parts of the forearm without fear of damaging a motor nerve—until a point is reached about 6 cm below the medial epicondyle, for at this level motor branches of the ulnar nerve are encountered.

Other Internervous Lines. The posterior border of the ulna marks the internervous line between the motor territories of the ulnar and radial nerves; similarly, the course of the radial artery marks the internervous line between the motor territories of the median and radial nerves.

10 HAND

CARPAL BONES AND JOINTS

Carpal bones; Drawing Carpus; Distal Surface of Carpus and Bases of Metacarpals; Dorsal and Palmar Aspects; Flexor Retinaculum.
Midcarpal Joint; Joint Surfaces.
Intercarpal Ligaments; Striking a Blow.

METACARPAL BONES AND JOINTS

Bases; Bodies; Metacarpal of Thumb; Metacarpophalangeal Joints; Collateral Ligaments; Palmar Ligaments; Deep Transverse Metacarpal Ligaments; M-P Joint of Thumb.

PHALANGES AND INTERPHALANGEAL JOINTS

OSSIFICATION OF BONES OF HAND

PALM OF THE HAND

Finger prints; Skin Creases; Surface Anatomy; Digital Formula.
Definitions: Movements of Digits; Axial Line of Hand.
Carpometacarpal Joint of Thumb.
Flexor Retinaculum; Fl. Carpi Ulnaris.
Thenar Muscles—Abductor Pollicis; Opponens Pollicis and Flexor Pollicis Brevis; Nerve Supply.
Hypothenar Muscles—Nerve Supply.
Abductor Pollicis; Opposition.
Ulnar Nerve in Hand; Palmaris Brevis; Median Nerve in Hand; Variations in Motor Distribution.
Palmar Aponeurosis; Structures Crossing Skin Crease at Wrist; Fibrous Sheaths of Digits; Insertions of Long Flexors; Lumbricals; Synovial Sheaths; Mesotendons; Common Carpal Synovial Sheath.
Palmar Spaces.
Arteries of Hand—Superficial Palmar Arch; Radial Artery; Deep Palmar Arch; Palmar Carpal Arch; Dorsal Carpal Arch.

There is little purpose in studying the bones of the hand individually before you are familiar with them collectively; so, we shall consider first the hand as a whole, and later attend to the notable features of the individual bones. And, because considerable areas of the different bones take parts in joints, we should find it difficult to avoid discussing the joints and ligaments with the bones.

When reading the following remarks, have beside you the bones of the hand, preferably strung on catgut. Do not omit also to make the suggested observations on your own hand.

The following comprise the skeleton of the hand:

1. The carpal bones, or bones of the wrist, or carpus.
2. The metacarpal bones, or bones of the palm, or metacarpus.
3. The phalanges, or bones of the digits. The digits are numbered I, II, III, IV, V. The first digit is the thumb or pollex; the second, the forefinger or index; the third, the middle finger or digitus medius; the fourth, the ring finger or digitus anularis; and the fifth is the little finger, or digitus minimus.

CARPAL BONES AND JOINTS

The carpal bones are short or cubical bones (p. 5). Two surfaces, the *anterior* and *posterior*, are rough for the attachment of ligaments, while four surfaces articulate with adjacent bones and are, therefore, covered with cartilage. Like all short bones, they retain red marrow for some years after it has disappeared from the bodies of long bones. Carpal bones do not start to ossify until shortly before birth or within a few years after it. The sequence provides the radiologist with an index of "bone age" in children.

The carpal bones, eight in number, are arranged in a proximal and a distal row, each of four bones. Their names express their general appearance. From radial to ulnar side they are:

Proximal row—*scaphoid, lunate, triquetrum,* and *pisiform.*

Distal row—*trapezium, trapezoid, capitate,* and *hamate.*

(Scaphoid = navicular, in B.N.A.
Trapezium = greater multangular,
Trapezoid = lesser multangular.)

It may be helpful first to picture these as eight uniform cubes arranged as in *figure 129,* but a glance at an articulated skeleton shows a more complex arrangement. The series of diagrams in *figure 130* illustrates the plan. Observe that: (1) the pisiform articulates with only the triquetrum and lies entirely in front of it; therefore, (2) to the wrist joint the remaining three proximal bones present a common articular surface, which is a convex ovoid that allows movements in all directions (except rotation); (3) the capitate is the largest bone; it is centrally located and around it the others are organized; its caput or head fits into a concavity provided by the proximal row; (4) on the lateral side the scaphoid is elongated distally while the trapezium and trapezoid are relatively small; (5) the hamate reaches the lunate between the capitate and triquetrum, and so (6) the line of articulation between proximal and distal rows—called the **midcarpal joint**—is seen to be sinuous, having a very sharp "jog" laterally; this limits abduction and adduction between the two rows but allows flexion; (7) the trapezium has a saddle-shaped articular surface for the base of the first metacarpal, giving it great mobility (*fig. 131*); (8) the second metacarpal is deeply wedged and immobilized between the trapezium, trapezoid, and capitate; (9) the broad distal surface of the capitate articulates with the flat base of the third metacarpal; (10) the hamate has two adjoining concave surfaces for the convex

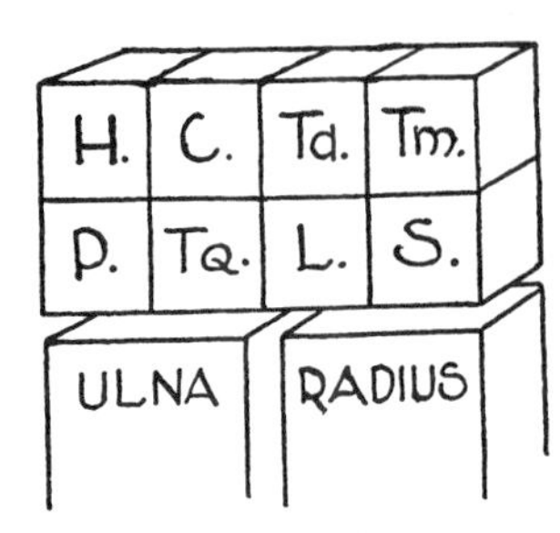

Fig. 129. The carpal bones as cubes.

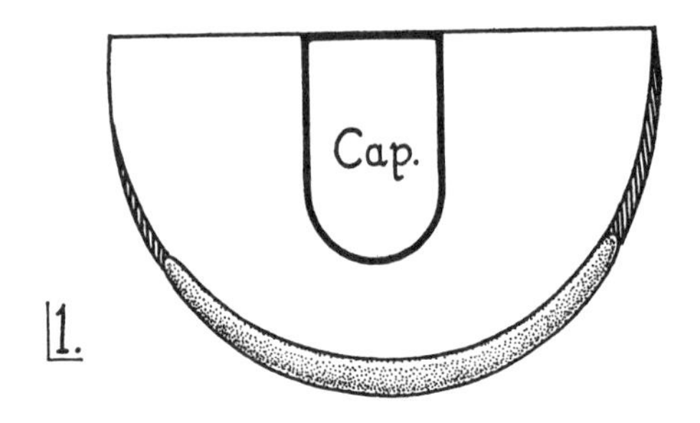

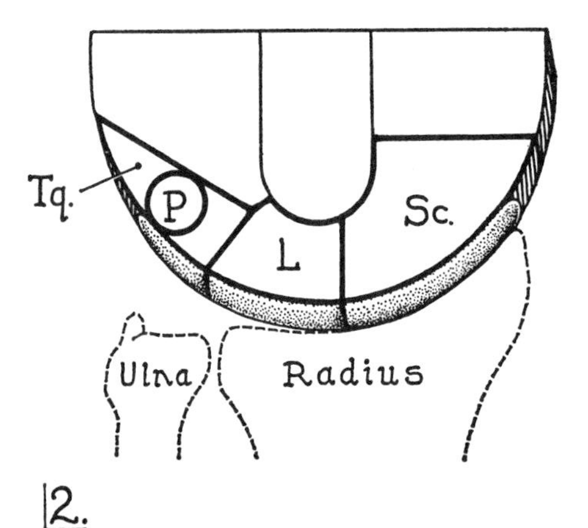

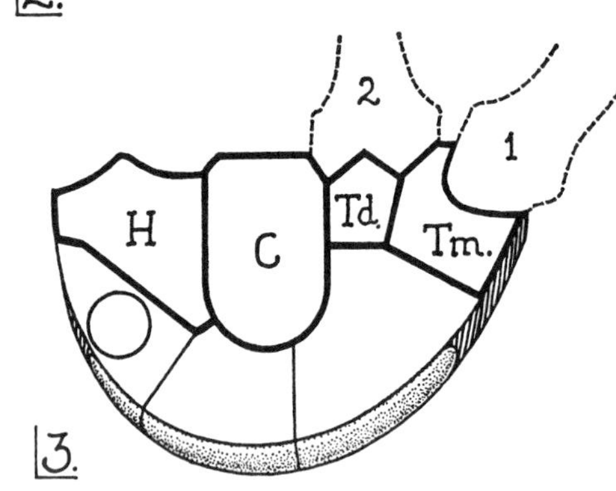

Fig. 130. Steps in drawing carpal bones (see text).

base of the 4th and 5th metacarpals; these permit a slight degree of flexion; and (11) the sides of the bases of the metacarpals 2, 3, 4, and 5 articulate with each other.

Distal Surface of Carpus and Bases of Metacarpals. Here we do not find plane gliding surfaces, such as would allow the metacarpals to shift from side to side, and backward and forward. They are very irregular (*fig. 131*). Although the metacarpal of the thumb or pollex can be flexed and extended, abducted and adducted, and rotated, it does not possess the spherical base you naturally look for, but a saddle-shaped surface which fits on to the distal surface of the trapezium, which is reciprocally saddle-shaped.

Details. The metacarpal bones of the five digits (thumb and four fingers) have only four carpal bones with which to articulate. The articulated hand reveals that the bases of the 4th and 5th metacarpals articulate with the hamate. [In the more primitive carpus, e.g., that of the turtle, the *hamate* like its homologue in the foot, the *cuboid*, is represented by two bones.]

Grasp in turn the knuckles or heads of the metacarpals of your own fingers, noting that the 5th metacarpal can be moved freely backward and forward, that the 4th metacarpal can be moved to a less degree, and that the 3rd and 2nd metacarpals are almost immobile. Again, view the back of your knuckles as you tightly close your fist (*fig. 206*) and see illustrated the same fact: that the 5th and 4th metacarpals flex at their carpometacarpal joints while the 3rd and 2nd remain rigid and immobile. An examination of the carpus gives the reason: the 5th and 4th carpometacarpal joints are clearly hinge joints, the hamate presenting two concavities for the convex bases of the 5th and 4th metacarpals. [In the foot the corresponding articular surfaces of the cuboid for the 5th and 4th metatarsals are similarly fashioned.]

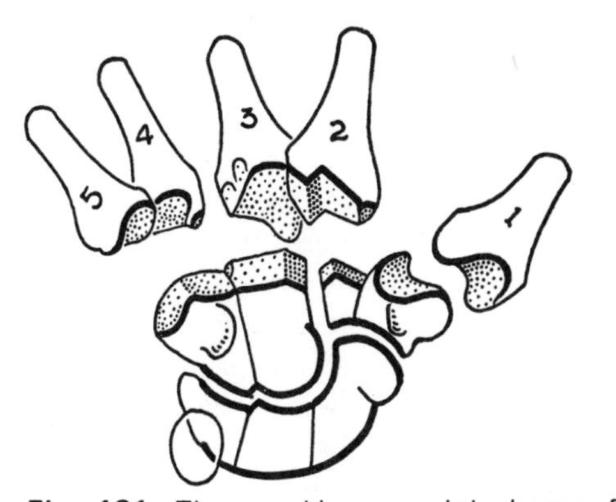

Fig. 131. The carpal bones and the bases of the metacarpal bones.

The capitate has an expansive plane surface for the base of the 3rd metacarpal. The trapezoid (lesser multangular) does not project so far distally as the carpals on each side of it; and in consequence, the base of the 2nd metacarpal is mortised between the capitate and trapezium (greater multangular), and in part articulates with them. The trapezoid, moreover, possesses an anteroposteriorly placed ridge which fits like a wedge into the guttered base of the 2nd metacarpal (*fig. 132*).

Dorsal and Palmar Aspects of Carpus. Observe that the dorsum of your wrist is transversely arched or convex, and that the dorsum of an articulated carpus likewise is transversely arched or convex. Its palmar surface is hollow or concave. The arched condition is maintained by a tie beam, the **flexor retinaculum** (transverse carpal ligament), which unites the marginal bones of the carpus. Its proximal part extends between two rounded prominences, namely, the *pisiform* and the *tubercle of the scaphoid* (*fig. 133*), whereas its distal part stretches between two crests, namely, the *hook of the hamate* and the *tubercle of the trapezium*.

The tubercle of the trapezium is also the lateral lip of a groove in which the tendon of the Flexor Carpi Radialis runs on its way to the base of the 2nd (and 3rd) metacarpal, where it is inserted.

Like the stones in an arch of masonry the carpal bones are broad on their convex or dorsal aspects, and narrower on their concave or palmar aspects, with the single exception of the lunate which is more expansive ventrally than dorsally. When the lunate becomes dislocated or dislodged, the displacement is usually forward.

Midcarpal Joint (Transverse Carpal Joint). This joint lies between the proximal and distal rows of carpal bones (*fig. 134*) and is sinuous. The greater part of the

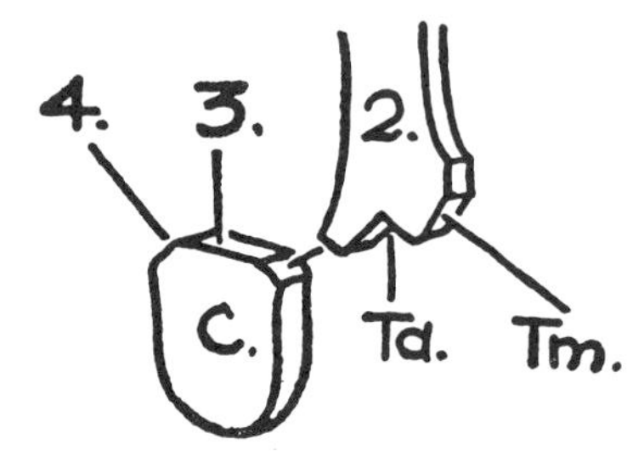

Fig. 132. The capitate articulates with three metacarpals. The 2nd metacarpal articulates with three carpals.

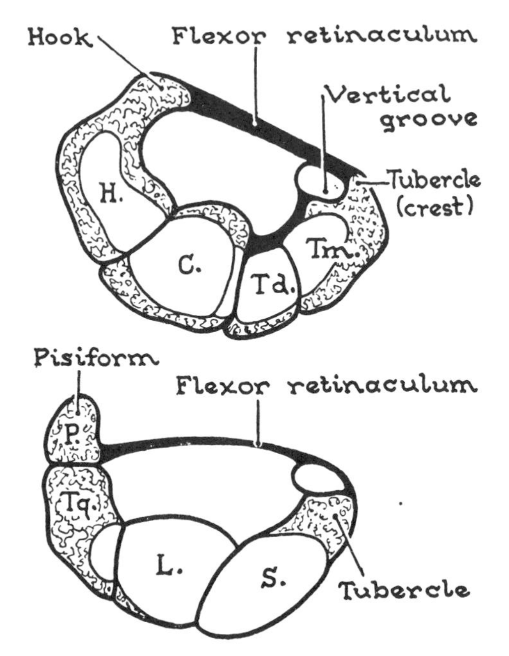

Fig. 133. The four marginal bones of the carpus give attachment to the flexor retinaculum.

flexion that appears to take place at the radiocarpal joint actually takes place at the midcarpal joint.

Intercarpal Ligaments. The various carpal bones are bound to each other by dorsal and palmar bands which largely radiate from the capitate, and by interosseous ligaments. Between the individual bones of each row a slight amount of gliding can take place; that is to say, the joints are plane (arthrodial).

Striking a Blow. You may crystallize your knowledge of the carpals by considering the mechanism involved in striking a blow. When you strike a blow with your fist, you instinctively employ the knuckles of the 2nd and 3rd metacarpals in preference to those of the 4th and 5th. Why is this? Because: (1) The 2nd and 3rd metacarpals are long, stout, and strong; whereas the 4th and 5th are shorter and more slender. (2) The bases of the 2nd and 3rd are individually equal in surface area to the combined bases of the 4th and 5th. (3) The 2nd and 3rd are rigid and immobile; whereas the 4th and 5th are not. (4) The line of force traveling along the 2nd and 3rd metacarpals is transmitted directly to the radius via the trapezoid and scaphoid, and via the capitate and lunate (*figs. 134* and *143*); whereas that traveling along the 4th and 5th is transmitted via the apex of the hamate to a linear facet on the lunate and so to the radius. Further, (5) the upper articular facet of the triquetrum, such as it is (*fig. 133*), is applied to the ulnar collateral ligament (medial lig.) of the wrist, except during adduction of the wrist when it moves into contact with the articular disc of the radiocarpal joint; so the triquetrum and pisiform are not force transmitters.

Ligaments. On striking with the 4th and 5th knuckles, the direction taken by the strong interosseous ligaments uniting the 4th to the 3rd metacarpal, and the hamate to the capitate, is such as to relieve the linear facet on the lunate of the brunt of the impact.

METACARPAL BONES AND THEIR JOINTS

Proximal Ends or Bases. These have articular facets which are counterparts of the distal surfaces of the carpals (*fig. 131*). Further, the apposed surfaces of the bases of the 2nd, 3rd, 4th, and 5th metacarpals articulate with each other and, therefore, carry articular facets.

Bodies of the Metacarpal Bones of the Four Fingers (*fig. 135*). Each has a rounded, anterior border which separates an anterolateral from an anteromedial surface. As these surfaces, which give origin to Dorsal Interossei approach the base of the bone they wind dorsally. In consequence, the dorsal palpable surface is bare and flat.

Metacarpal of the Thumb. This is short, stout, and rounded posteriorly; it provides surfaces for the fleshy attachment of muscles (*fig. 136*). The base is saddle-shaped. The head is compressed anteroposteriorly; and, in front, two sesamoid bones play.

The Metacarpophalangeal Joint. Observe

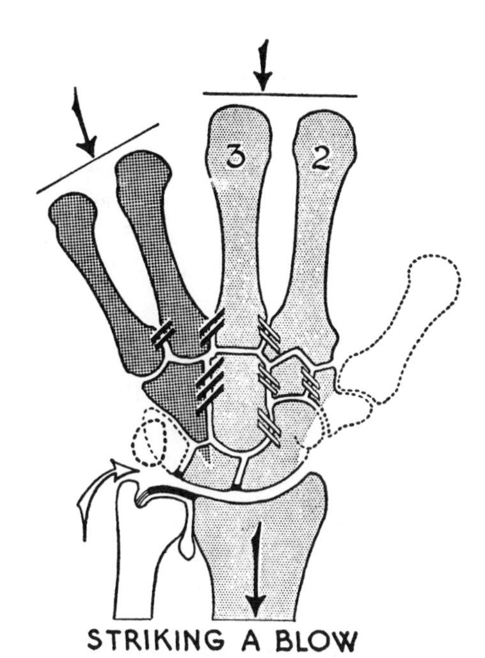

Fig. 134. When striking a blow use the 2nd and 3rd knuckles. Observe that the ligaments take the directions of usefulness.

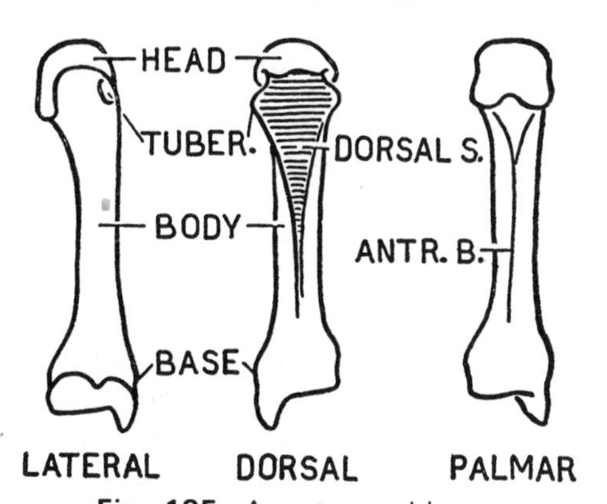

Fig. 135. A metacarpal bone.

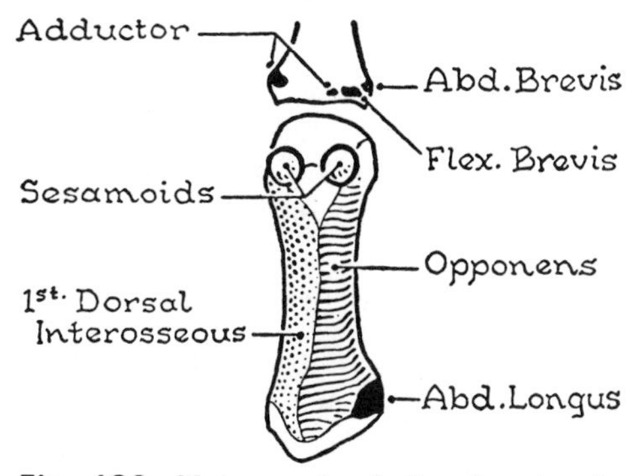

Fig. 136. Metacarpal of thumb, showing attachments of muscles, palmar aspect.

on your knuckles, i.e., the heads of the metacarpals of the fingers, that they are spheroidal. The bases of the proximal phalanges fit the metacarpal heads and, so, are concave.

The metacarpophalangeal articulations of the four fingers can be flexed and extended; and, when the hand is open, they can also be abducted and adducted; that is to say, they can perform the four component movements of circumduction. They, therefore, are condyloid joints (Kondulos, Gk. = a knuckle).

When the joints are flexed, neither abduction nor adduction is possible because (1) the heads of the metacarpals are flattened in front; and (2) the **collateral ligaments,** though slack on extension, are taut on flexion, due to their eccentric attachments to the sides of the heads of the metacarpals (*fig. 137*).

Posteriorly there are no ligaments to these joints: the extensor (dorsal) expansions of the extensor muscles effectively serve the part. Anteriorly, the capsule is thickened to form the **palmar ligament** or **plate** (volar accessory lig.). Fibers of the collateral ligaments radiate to the sides of this plate and keep it firmly applied to the front of the head of its metacarpal, visor-fashion (*fig. 137*).

The sides of the palmar ligaments of the fingers are united to each other by three

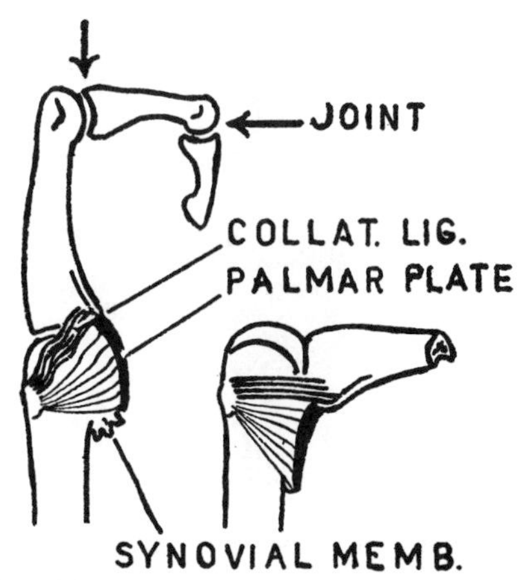

Fig. 137. Metacarpophalangeal and interphalangeal joints.

ligamentous bands, the **deep transverse metacarpal ligaments,** which help to prevent the metacarpals from spreading.

In front of the transverse ligaments pass the digital vessels and nerves and the Lumbricals; behind them pass the Interossei. Slips from the four digital bands of the palmar aponeurosis are attached to the transverse ligs. (*fig. 149*) in front; slips from the extensor expansions are attached to them behind.

The *Metacarpophalangeal Joint of the Thumb*. The head of the metacarpal is rounded and the base of the phalanx is concave. The movements allowed are: flexion and extension, some rocking from side to side, and some rotation.

PHALANGES AND THEIR JOINTS

The phalanges also are long bones, two for the thumb and three for each of the other digits. They are known as: the proximal or 1st; the middle or 2nd; and the distal or 3rd (*fig. 138*).

The distal ends of the distal phalanges have a smooth surface which the fingernail overlies; the area for the finger-pad is rough owing to the attachment of fibrous bands that bind the skin to it.

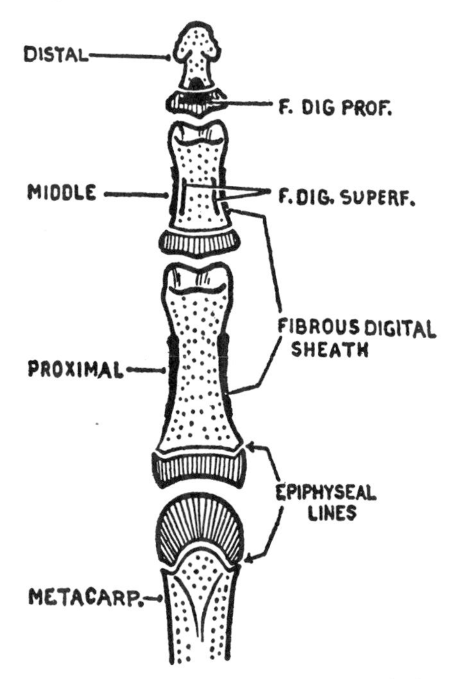

Fig. 138. Phalanges (palmar aspect) showing epiphyseal lines and attachments of fibrous sheath and flexor tendons.

The dorsal aspects of the proximal and middle phalanges are smooth, rounded and covered with the extensor expansion. The palmar surfaces take part in the floor of the osseofibrous tunnels (*fig. 150*) in which the flexor tendons glide. They are smooth and flat.

Each phalanx has two borders. The borders of the proximal and middle phalanges provide attachment for the fibers of the fibrous digital sheath. Those of the middle phalanx also receive the slips of insertion of the Fl. Digitorum Superficialis.

When your hand is closed, the heads of the middle and proximal phalanges are uncovered. They have two little condyles. The bases of the distal and middle phalanges have two little depressions.

The Interphalangeal Joints are hinge joints with their movements restricted to flexion and extension.

OSSIFICATION OF BONES OF HAND

Primary Centers. The bodies of the metacarpals and phalanges start to ossify during the 3rd prenatal month. The carpal bones, unlike the tarsal bones, have not started to ossify at the time of birth, although, in the female, centers may have appeared in the capitate and hamate. The carpals proceed to ossify in orderly spiral sequence, approximately in the following years: capitate and hamate 1st, triquetrum 3rd, lunate 4th, scaphoid 5th, trapezoid and trapezium 6th—and the pisiform 12th (*fig. 139*).

Epiphyses. Each of the long bones has one epiphysis: in the metacarpals these occur at the heads; in the phalanges at the bases—the metacarpal of the thumb is the exception, for it resembles a phalanx in that its epiphysis is at its base.

The epiphyses start to ossify in the 2nd to 3rd year and fuse in the 17th to 19th year. Epiphyses have been found at both ends of the 1st and 2nd metacarpals.

Radiograms (*fig. 140*). At any given time

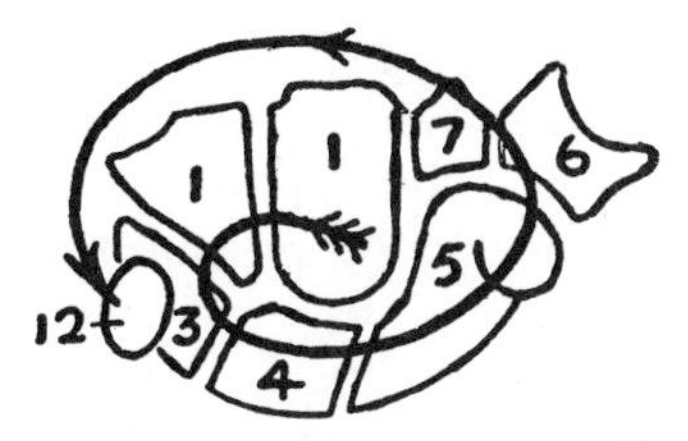

Fig. 139. Spiral sequence of ossification of carpals; approximate ages in years.

during the period of physical growth, which is from birth to about the 17th year in girls and to about the 18th or 19th in boys, a radiogram of the hand and wrist will reveal the *skeletal age* and therefore the progress made toward physical maturity—which may be either accelerated or retarded—and it may reveal much else. The skeletal state of a girl aged 13½ years is not achieved by a boy until he is 15½ years old (Greulich and Pyle).

Garn *et al.* find that the hand alone cannot be used as a standard of skeletal development for the whole body, that all ossific centers do not have equal predictive value, and that a small number of centers of high predictive value in the hand and foot provide more useful information than does the entire number. (For details on the reliability of the data, see Johnson *et al.*)

PALM OF THE HAND

The skin of the grasping or palmar surface of the hand is very thick. It rests on a protective pliable layer of fat but is anchored to the underlying tissues by fibrous bands, which are most dense at the pads of the fingers, and in front of the palmar aponeurosis. The fat, accordingly, tends to be imprisoned in loculi, as in the breast.

In order that the grip shall not slip, the skin is corrugated; it is ridged and furrowed, and there is a convenient absence of greasy, sebaceous glands. On the summits of the ridges the mouths of numerous sweat glands open. The disposition of the ridges in arches, loops, and whorls differs in detail from person to person; so does the spacing, the shape, and the size of the mouths of the sweat glands. The impressions left by the ridges and gland mouths are known as **Finger Prints** (*figs. 141* and *142*).

Permanent **Skin Creases** occur in the hand in response to its movements. On the digits they are transverse; in the palm they have the form of the letter M. At the creases there is an absence of fat. The relation of the creases to the joints is shown in *figures 143* and *125*.

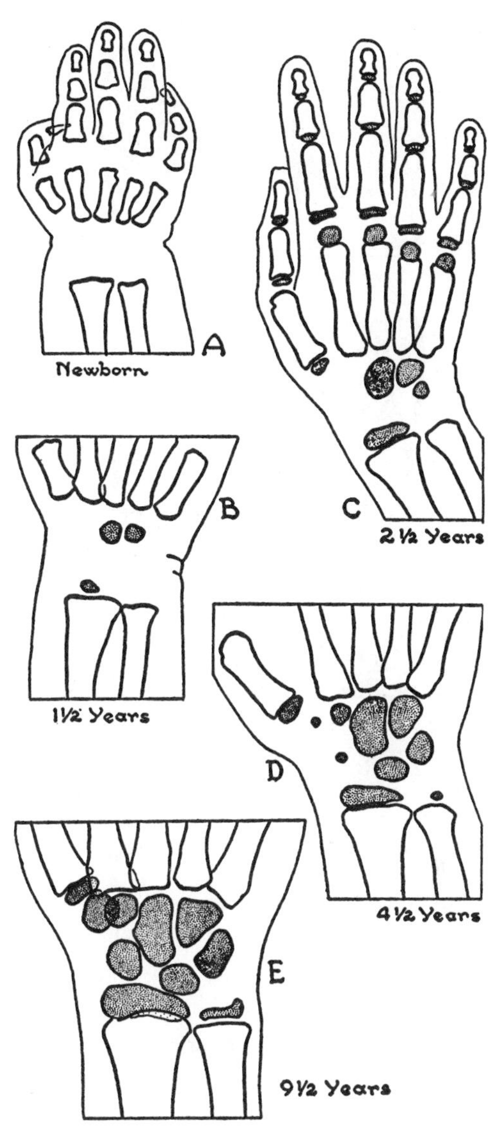

Fig. 140. Progressive ossification of the bones of the hand. (Courtesy of Dr. J. D. Munn, Hospital for Sick Children, Toronto.)

The most projecting digit is the middle finger; next in order come the ring finger, index, little finger, and thumb, so the **Digital Formula** reads, 3 > 4 > 2 > 5 > 1 (Wood Jones). This is the primitive arrangement. It is common to apes and man including the N. American Indian. In 19 per cent of white people the index

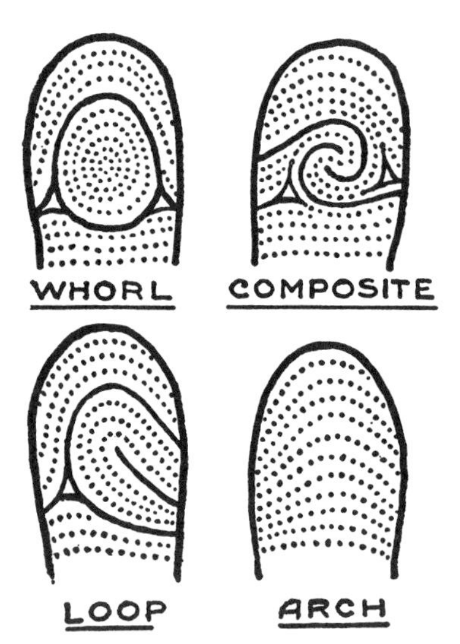

Fig. 141. Types of finger prints (After Wilder.)

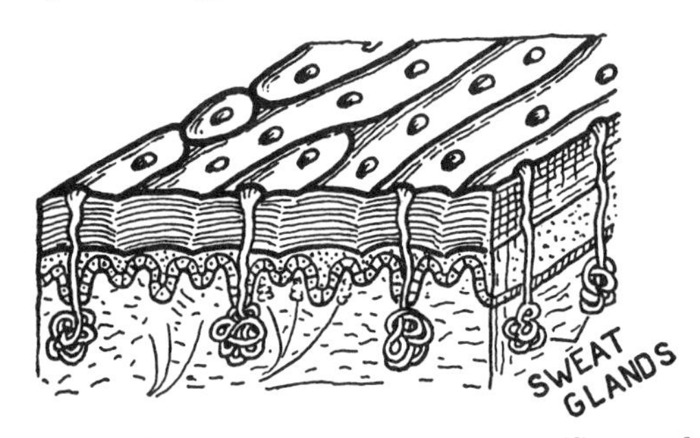

Fig. 142. Friction ridges and orifices of ducts of sweat glands on the fingers. (After Wilder.)

and ring project equally (3 > 4 = 2 > 5 > 1) and in 33 per cent the index exceeds the ring finger (3 > 2 > 4 > 5 > 1). Among white females the percentages are higher than among the males (George).

Movement of Digits. A line drawn through the middle finger, middle metacarpal and capitate is called the *Axial Line of the Hand* (*fig. 143.1*). Movement of a finger away from this axial line is called abduction; movement toward it, adduction; and the movements take place at the metacarpophalangeal joints. The thumb is set at right angles to the other digits. It also can be abducted and adducted; the movements take place, however, not at its metacarpophalangeal joint, but at its carpometacarpal joint; the movement forward (anteriorly) being abduction, and backward (posteriorly) adduction; that is to say, in the anteroposterior (sagittal) plane you flex and extend your fingers, abduct and adduct your thumb; whereas in the side-to-side (coronal) plane you abduct and adduct your fingers and flex and extend your thumb.

Carpometacarpal Joint of Thumb. It is to movements of medial and lateral rotation that the thumb owes its peculiar value, a value that exalts it above many fingers. The metacarpal of the thumb sits on the trapezium as though astride a saddle, enjoying all the movements of a ball and socket joint; it is a multiaxial joint of the saddle variety (p. 21).

Flexor Retinaculum. A tie-beam, the *flexor retinaculum* (*fig. 144*), forms with the carpal bones an osseofibrous tunnel, the *carpal tunnel.* The retinaculum stretches between two rounded prominences, the *tubercle of the scaphoid* and the *pisiform*, and between two crests, the

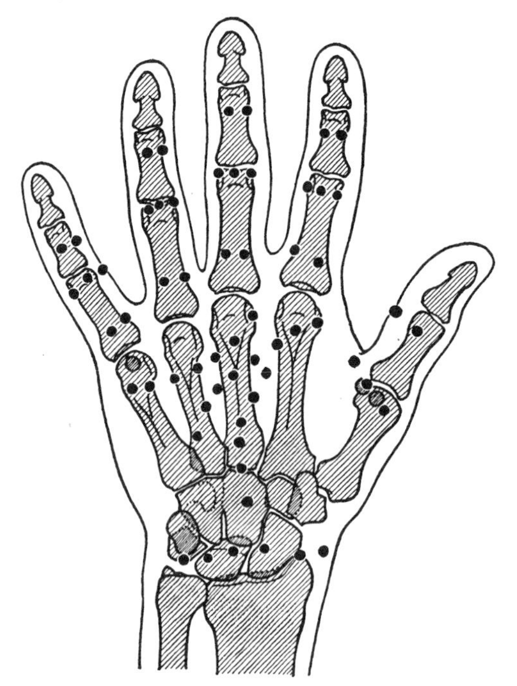

Fig. 143. Tracing of X-ray of hand. Shot was placed on the skin creases to show their relations to the joints (*see fig. 126*).

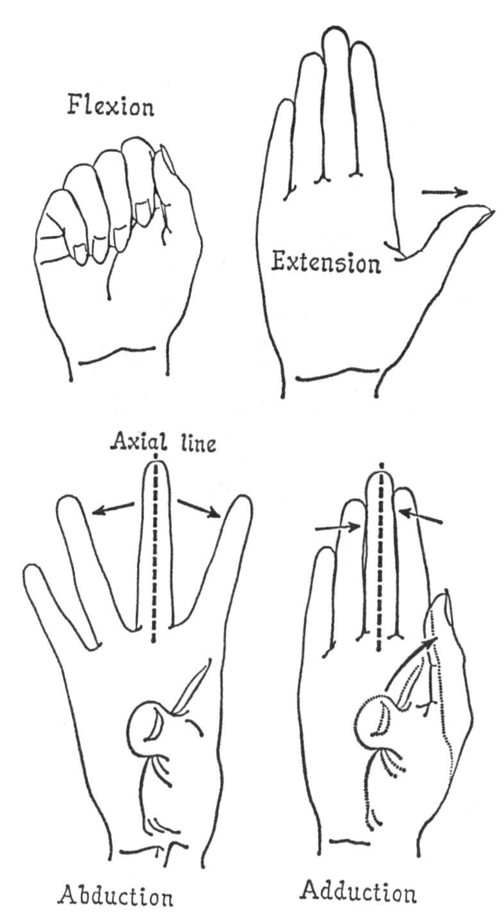

Fig. 143.1. Movements of fingers and thumb defined.

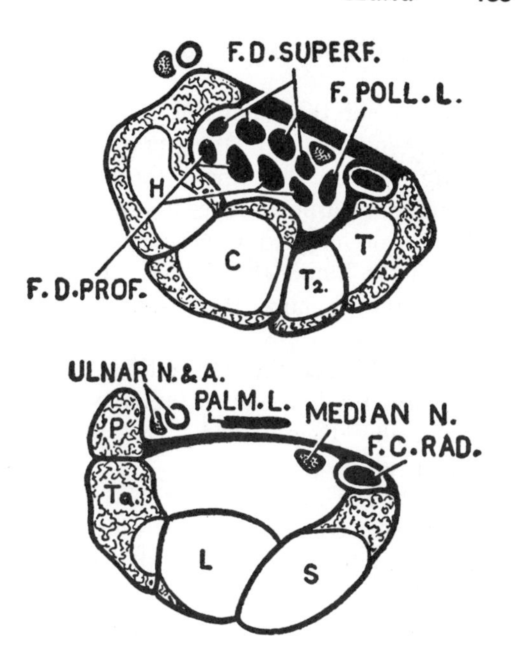

Fig. 144. Proximal and distal rows of carpal bones viewed from above (i.e., distal surfaces of the radiocarpal and midcarpal joints). The flexor retinaculum forming the osseofibrous carpal tunnel.

tubercle of the trapezium and the *hook of the hamate* (*fig. 145*).

The flexor retinaculum not only acts as a restraining band that prevents the long flexors of the digits from "bow-stringing," it also affords chief origin to the thenar and hypothenar muscles.

The Three Thenar Muscles (*Muscles of the Ball of the Thumb*) arise together from the flexor retinaculum and its two lateral bony pillars (tubercle of scaphoid and tubercle of trapezium).

The most superficial of them is the *Abductor Pollicis Brevis*. It is self-evident from its position and attachments (*fig. 146*) that it draws the thumb forward (i.e., abducts it); and the movement takes place

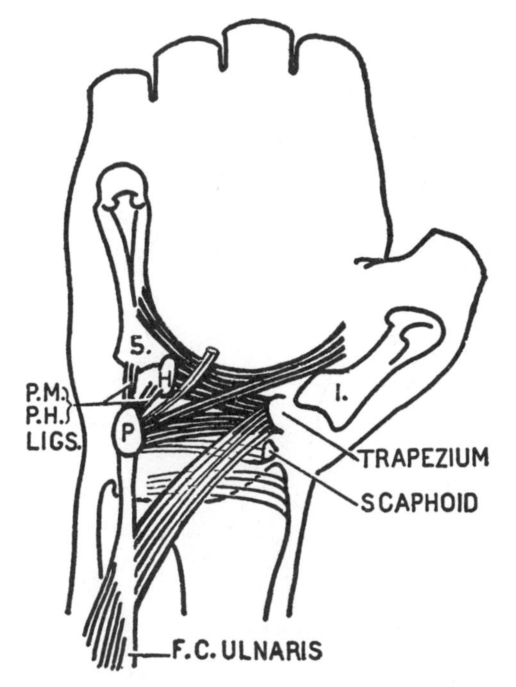

Fig. 145. The flexor retinaculum (details) (P.M. = pisometacarpal; P.H. = pisohamate).

at the saddle-shaped carpometacarpal joint.

Reflection of the Abductor uncovers a fleshy sheet (*fig. 146*) that pronates or medially rotates the whole thumb and flexes its metacarpophalangeal joint.

The portion inserted into the metacarpal is named the *Opponens Pollicis;* the portion into the phalanx, the *Flexor Pollicis Brevis* (*fig. 146*).

The insertion of the Fl. Pollicis Brevis into the proximal phalanx is largely via the radial sesamoid embedded in the palmar ligament (plate).

Nerve to the Three Thenar Muscles. A branch of the median, it takes a recurrent course around the lower border of the retinaculum. It is to be found lying 3 cm vertically below the palpable tubercle of the scaphoid (*figs. 147* and *125*). A small coin centered on this point will cover the nerve, whose importance is bound up with the importance of the thumb itself.

The deep fascia covering the thenar eminence is too thin to afford the nerve much protection, and for practical purposes the nerve is subcutaneous.

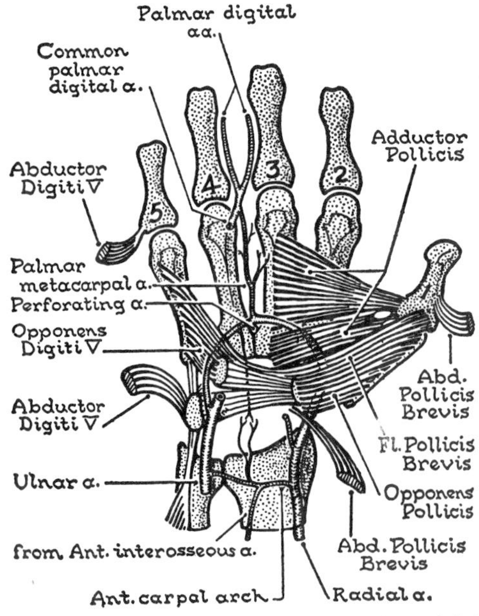

Fig. 146. Short muscles of thumb and 5th digit. Arteries of hand (semischematic).

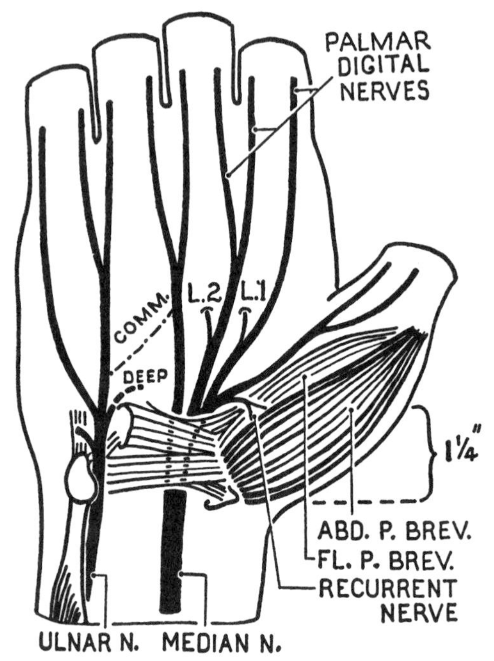

Fig. 147. Median and ulnar nerves and thenar muscles. (L.1 and L.2 = Lumbrical branches of median n.)

The Three Hypothenar Muscles (*Muscles of the Ball of the Little Finger*) are in most respects mirror images of the muscles of the thenar eminence. The common origin is from the flexor retinaculum and its two medial pillars—viz., pisiform and hook of hamate. They act on a metacarpophalangeal joint, but, of course, opposition is impossible.

Nerve Supply. The hypothenar muscles are supplied by the ulnar nerve (deep branch) as it runs between the pisiform and the hook of the hamate.

Adductor Pollicis. The thumb possesses a fourth short muscle, the Adductor Pollicis. It lies in the depths of the palm. A transverse head arises from the palmar border of the middle metacarpal, and an *oblique head* from the corresponding carpal bone (the capitate). It is inserted into the base of the first phalanx of the thumb (*fig. 146*). Obviously the Adductor draws the thumb to the palm.

Nerve Supply. Contrary, perhaps, to

expectation, it is supplied by the ulnar nerve.

Opposition is bringing of the pad of the thumb to the pad of a finger and holding it there, as in pinching, writing, holding a cup by its handle, or fastening a button. Most delicate actions performed by the hand involve opposition.

As regards the thumb, the movements executed almost simultaneously are circumduction, rotation, and flexion. Rotation takes place principally at the metacarpophalangeal joint but also at the carpometacarpal joint, while the trapezium moves on the scaphoid and the scaphoid angulates forward (Bunnell).

The three joints of the opposing finger (or fingers) are flexed by the Profundus, the Superficialis, a Lumbrical, and two Interossei.

Note. Without two muscles supplied by the median nerve (Opponens and Flexor Pollicis Brevis), there would be no rotation. Without two muscles supplied by the ulnar nerve (Adductor Pollicis and an Interosseus), the grip would be weak. An Interosseus is required to steady (i.e., to prevent ulnar deviation of) the finger against which the Adductor Pollicis is exerting pressure. In the case of the index it is the 1st Dorsal Interosseus. Hence, ulnar paralysis results in weak opposition.

Ulnar Nerve. It enters the hand by passing vertically between the pisiform and the hook of the hamate (*figs. 147* and *148*) in front of the flexor retinaculum and the pisohamate lig. It is covered first by a slip of deep fascia, and then by the **Palmaris Brevis** (*fig. 148*), which is a superficial sheet of muscle that passes from the retinaculum to the skin at the ulnar border of the hand. Between pisiform and hamate the nerve divides into a *deep* and a *superficial branch*.

Distribution. The superficial branch supplies cutaneous branches to the medial one and one-half fingers and the motor branch to the Palmaris Brevis, and it communicates with the median nerve.

The deep branch supplies the three muscles of the hypothenar eminence, and then curves around the lower edge of the hook of the hamate into the depths of the palm where it supplies all the short muscles of the hand, except the five usually supplied by the median nerve.

Median Nerve. Crossing the midpoint of the skin crease of the wrist, the median nerve enters the palm through the carpal tunnel adhering to the deep surface of the flexor retinaculum (*fig. 144*). It appears in the palm deep to the prolongation of the Palmaris Longus called the palmar aponeurosis, so it is fairly superficially placed. At once it begins to break up into "recurrent" and digital branches. These are distributed to five muscles and to the skin of the lateral three and one-half digits, to the joints of these digits and to the local vessels.

Motor Branches. The 5 muscles supplied by the median nerve are the three thenar muscles and the two lateral lumbrical muscles:

Abductor Pollicis Brevis }
Opponens Pollicis } by the recurrent branch
Flexor Pollicis Brevis }
and
1st Lumbrical }
2nd Lumbrical } by the palmar digital branches

These are shown in *figure 147*, and the "recurrent" branch is described on page 134. Variations are common.

Details of Variations in Motor Distribution. The standard, or textbook, pattern of motor innervation to the hand just described, is commonly departed from either by the *ulnar nerve* encroaching on median nerve territory or vice versa. Thus, in 226 hands the ulnar nerve supplied Fl. Pollicis Brevis partially in 15 per cent and completely in 32 per cent; Abd. Pollicis Brevis in 3 per cent and all 3 thenar muscles in 2 per cent. Conversely, the *median nerve* supplied Add. Pollicis in 3 per cent, and Add. Pollicis + first Dorsal Interosseus in 1 per cent. Neither the musculocutaneous nor the radial nerve has been proved to supply a thenar muscle (T. Rowntree, clinical observations). Forrest confirmed the variability of thenar nerve supply, but using electrical stimulation and electromyography revealed that 85

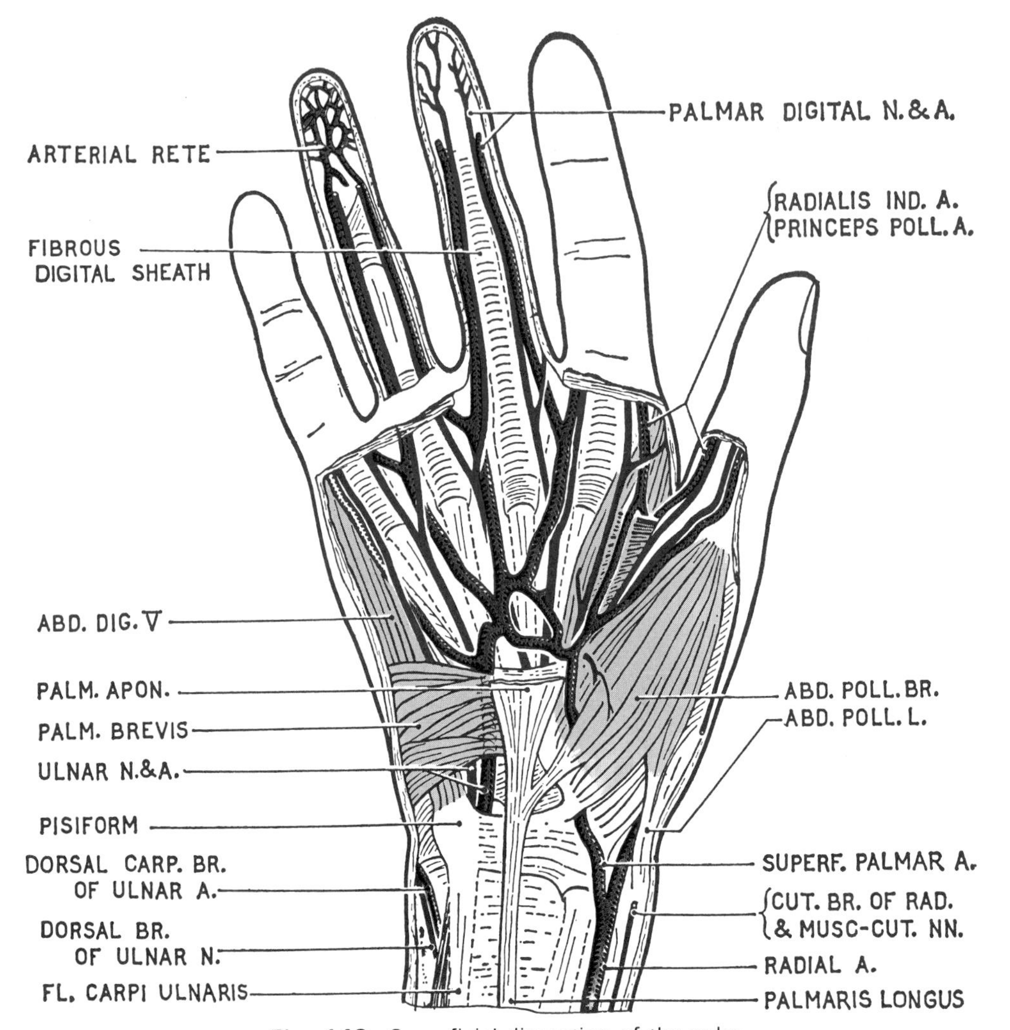

Fig. 148. Superficial dissection of the palm.

per cent of Fl. Pollicis Brevis muscles in 25 hands are supplied by both median and ulnar nerves. Morphological studies by Harness and Sekeles provide overwhelming confirmation.

The Palmar Digital Branches of the Median and Ulnar Nerves lie on the sides of the fibrous digital sheaths (*fig. 148*).

The three common palmar digital nerves descend to the three interdigital clefts (*fig. 148*), protected by the tough palmar aponeurosis and crossed by the superficial palmar arch. The two digital branches to the thumb accompany the Flexor Pollicis Longus tendon (*fig. 148*). The ulnar nerve commonly extends its influence to median nerve territory and vice versa, through communicating branches to the palm.

Palmar Aponeurosis. The Palmaris Longus tendon adheres to the front of the flexor retinaculum, enters the palm, and divides into four broad, diverging bands. These descend to the roots of the four fingers, there to blend with subcutaneous tissues (*fig. 149*).

In the distal half of the palm it sends fibrous septa dorsally to the palmar ligs. (plates) and to the deep fascia (*fig. 154*).

Fibrous Sheaths of the Digits. A pair of tendons, a *Superficialis* and a *Profundus*, descends deep to the palmar aponeurosis. In front of the head of a metacarpal bone each pair enters a *fibrous digital sheath* (*fig. 150*). Each sheath extends from the palmar lig. (plate) of a metacarpophalangeal joint to the insertion of a Profundus tendon into the base of a distal phalanx. Each sheath, therefore, crosses 3 joints. In front of the joints the sheath, for mechanical reasons, must be pliable and thin.

Insertions of Long Digital Flexors. Each Profundus tendon is inserted into the anterior aspect of the base of a distal phalanx (*fig. 150*). Each Superficialis tendon splits in front of a proximal phalanx into medial and lateral halves (*fig. 150–B*), which ultimately insert on the margins of a middle phalanx. Each Profundus tendon passes through a perforation in a Superficialis.

The intricate perforation remains so widely open that you can easily thread a cut Profundus tendon through it again.

The Lumbricals (*fig. 153*) are four muscles which in shape, size, and color resemble earthworms. They arise in the palm from the Profundus tendons. They lie behind the

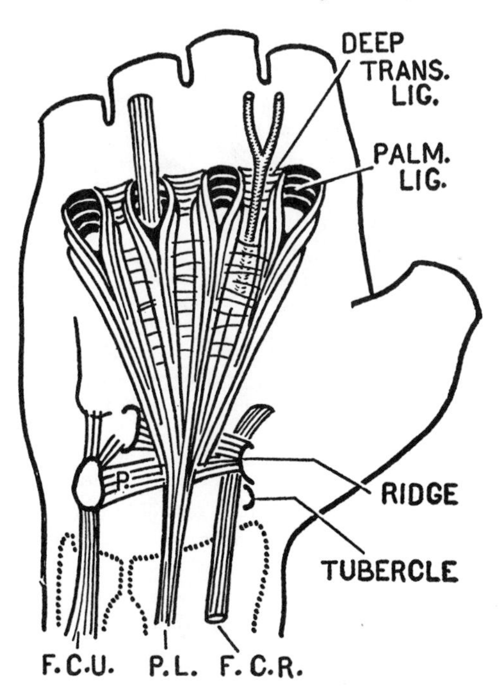

Fig. 149. The three flexors of the wrist. Palmar aponeurosis. Flexor retinaculum. (*ridge of trapezium = tubercle*)

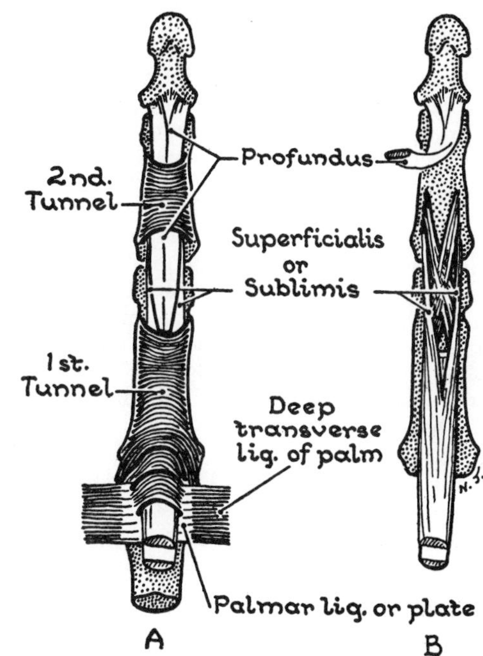

Fig. 150. *A*, a fibrous digital flexor sheath showing the two osseofibrous tunnels. *B*, mode of insertion of the long digital flexors.

digital vessels and nerves, and they accompany them in front of the deep transverse ligaments of the palm to the radial side of the fingers, where they join the extensor (dorsal) expansions distal to the attachments of the Interossei (*figs. 165* and *170*). The medial two are supplied by the ulnar nerve; the others by the median.

Action (see page 147).

Synovial Sheaths or Vaginae Synoviales. A lubricating device, a tubular bursa envelops or ensheaths any tendon subject to friction on more than one side. It is a tube within a tube, and the potential cavity between the inner and the outer tube is closed at both ends (*fig. 151*).

Every tendon lying within a synovial sheath has, or must at one time have had, a **mesotendon** (*fig. 152*)—a double layer of synovial membrane that attaches a tendon to the wall of its sheath and conveys vessels to it.

A synovial sheath may extend 2 or 3 cm proximal to and distal to the site of friction. But this depends upon the excursion the tendon makes, as does the probability of disappearance of the mesotendon. The very end portions of the original mesotendons of the long flexors remain as triangular folds, the *vincula brevia*, and several thread-like portions persist in front of the proximal phalanges as *vincula longa*. The vincula convey blood vessels to the tendons. (The singular of vincula is *vinculum*.)

Infections as a result of perforations entering the sheaths are serious because synovial fluid is an excellent culture medium for bacteria—and death of the enclosed tendons is a real possibility. Before the days of antibiotics, heroic surgical opening and drainage of the sheaths often became necessary.

The long flexor tendons require synovial sheaths where they pass first through the carpal tunnel and then through the digital tunnels. They have, therefore, *carpal synovial sheaths* and *digital synovial sheaths*. The thumb has obviously the freest range of movement. This and the shortness of the metacarpals of the thumb and of digit V result in their carpal and digital sheaths being continuous (*fig. 153*). Those of the thumb probably always unite; those of the little finger fail to unite in about 10 per cent of persons.

The carpal sheaths of the four Superficialis and four Profundus tendons usually become one, the **Common Synovial Sheath** of the digital flexors. The carpal sheath of the Flexor Pollicis Longus commonly joins, too. Then an infection starting in the sheath of the little finger may spread by this route to the thumb.

Palmar Spaces (*fig. 154*). There are in the palm four closed fascial spaces, also important in surgery because of potential closed infections from piercing wounds. The thenar muscles occupy one, the *thenar space;* the hypothenar muscles occupy another, the *hypothenar space*. Between these two there is a large triangular *central space* that contains the tendons of the fingers. Its anterior wall is the palmar aponeurosis. Its posterior wall is formed by the three medial metacarpals, the palmar and deep trans-

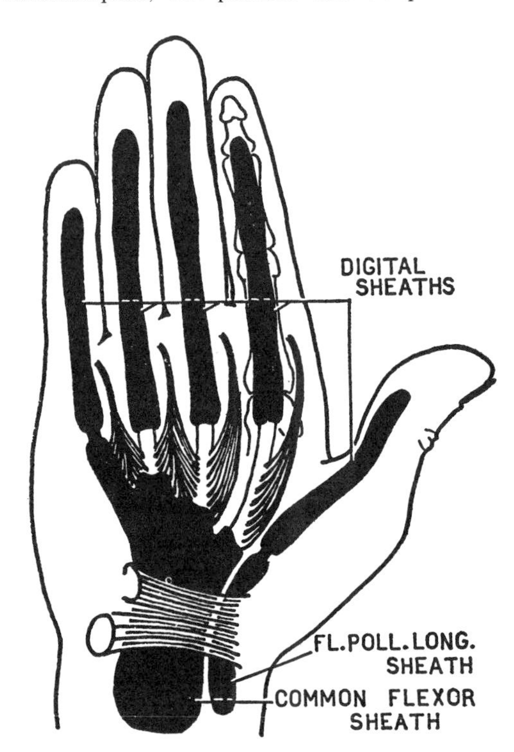

Fig. 153. Synovial sheaths or tubular bursae are required for the long digital flexor tendons at the osseofibrous carpal tunnel and at the osseofibrous digital tunnels. The Lumbricals are shown.

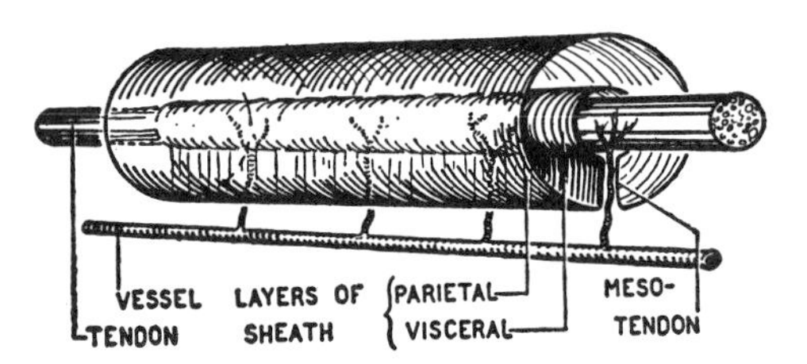

Fig. 151. Diagram of a synovial sheath.

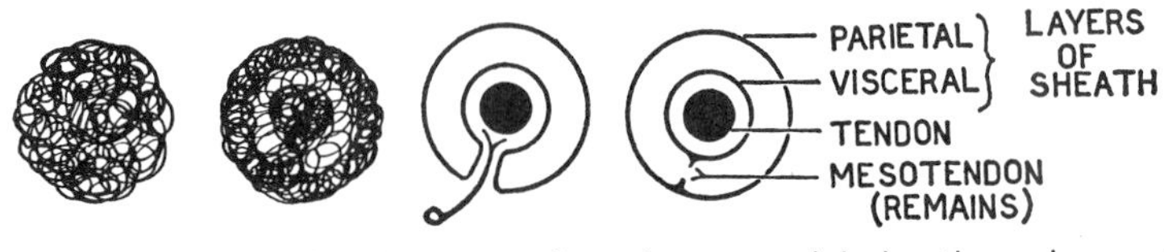

Fig. 152. Stages in the development of tendon, synovial sheath, and mesotendon.

verse ligs. (*fig. 149*), and the fascia covering the medial Interossei and the Adductor Pollicis. Its side walls are the backwardly turned edges of the palmar aponeurosis which fuse with the thenar and hypothenar fasciae. The fourth space is placed between the Adductor Pollicis in front and the two lateral intermetacarpal spaces behind.

In the distal half of the palm, the central space has eight subdivisions or tunnels (*fig. 154*). The septa separating the tunnels are derived from the palmar aponeurosis. The tendons of the Lumbricals prolong the spaces downwards on to the dorsum of the digits. It is by this lumbrical route that infection in the central palmar space may spread to the dorsum of the hand.

Arteries of the Hand, Palmar and Dorsal

The blood supply from the ulnar and radial arteries is good and the anastomoses are excellent through four transversely placed arterial arches (*fig. 155*):

1. The superficial palmar arch lying deep to the palmar aponeurosis (*fig. 148*).
2. The deep palmar arch. } on
3. The dorsal carpal arch. } skeletal
4. The palmar carpal arch. } plane

The Superficial Palmar Arch is the largest and most distal. It is the continuation of the ulnar artery completed by one or other of the palmar or digital branches of the radial artery. (The deep branch of the

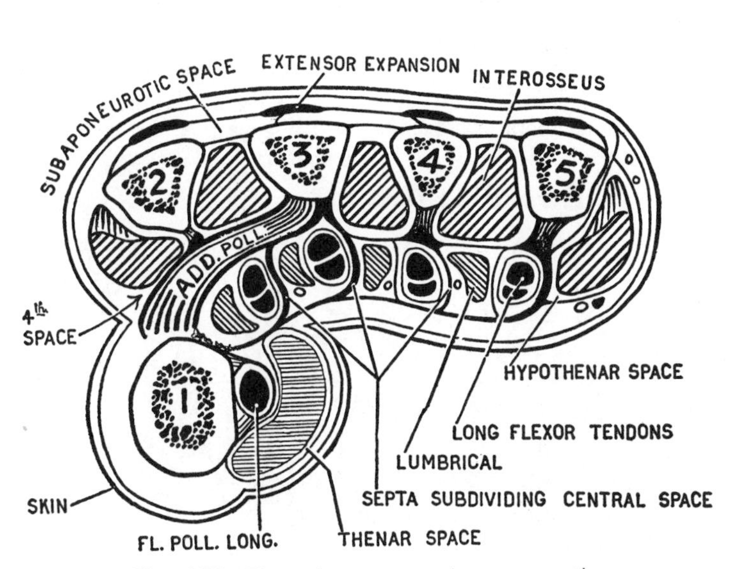

Fig. 154. The palmar spaces in cross section.

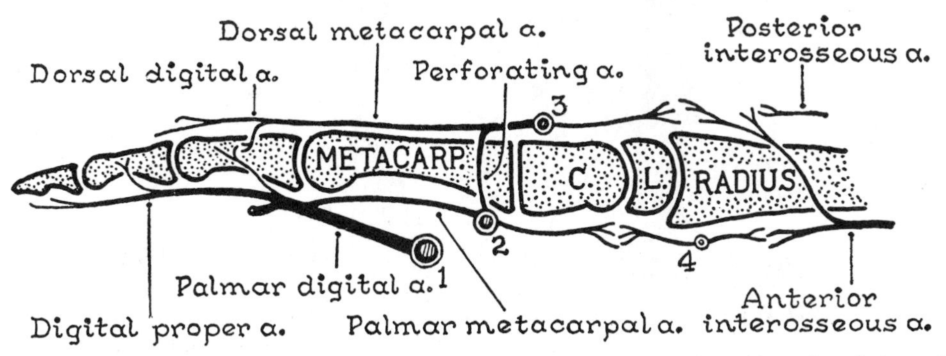

Fig. 155. Scheme of arteries of the hand: The four arches are numbered in order of size—three cling to the skeletal plane.

ulnar artery accompanies the deep branch of the ulnar nerve and completes the deep arch.)

Palmar Digital Arteries. The superficial palmar arch supplies the medial three and one-half digits, leaving the lateral one and one-half to the care of the deep palmar arch.

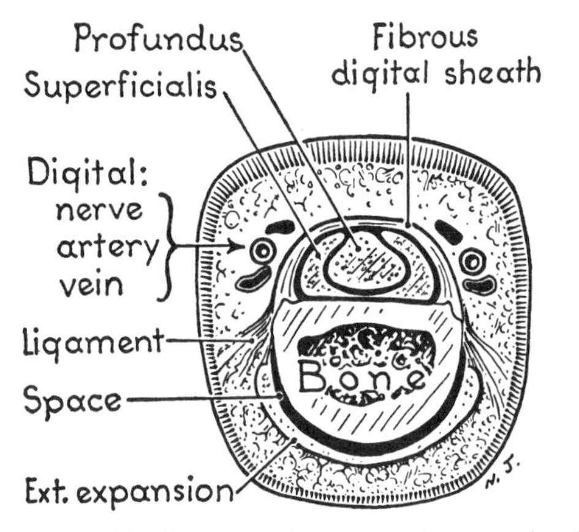

Fig. 156. Cross section through a proximal phalanx.

Palmar digital arteries and nerves run on the sides of the flexor tendons in their fibrous sheaths, the nerve being anteromedial to the artery (*fig. 156*). To expose them, you should feel with your finger nail for the edge of a phalanx and make a longitudinal incision in front of it.

Radial Artery. After giving off the *palmar radial carpal* and *superficial palmar arteries*, the radial artery turns round the lateral border of the wrist and descends vertically through the anatomical snuffbox to reach the proximal end of the first intermetacarpal space (*fig. 157*), where it passes between the two heads of the First Dorsal Interosseus, and enters the palm to become the *deep palmar arch*.

The radial artery crosses the radial collateral ligament of the wrist, the scaphoid and trapezium; and, in turn it is crossed by the three

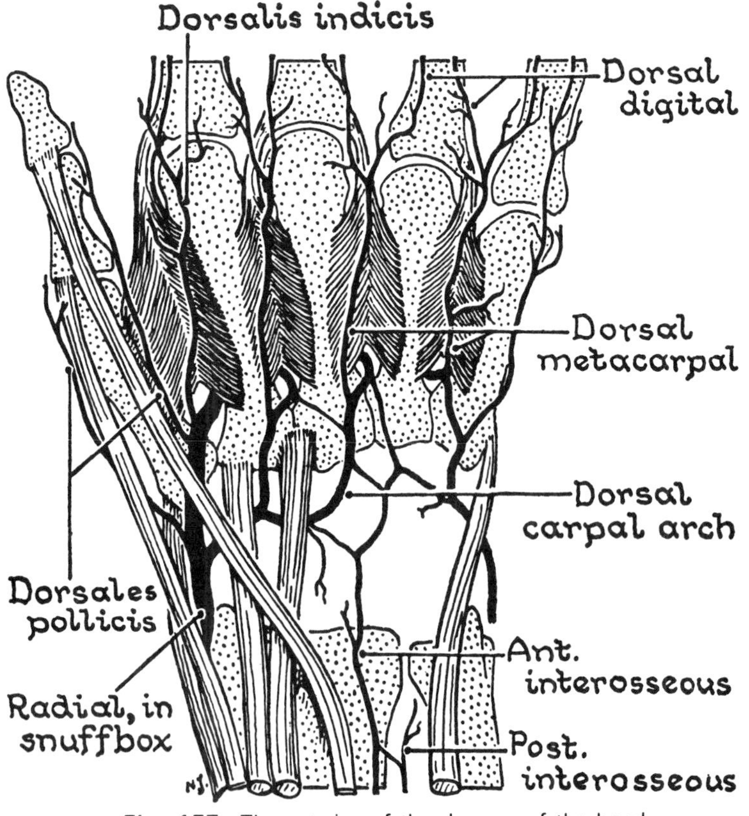

Fig. 157. The arteries of the dorsum of the hand.

tendons that bound the snuff-box, branches of the radial nerve to the thumb, and the dorsal venous arch (*fig. 157*).

While in the snuff-box the radial artery gives off the *dorsal radial carpal artery*, and sends small branches, *dorsal digital aa.*, to the sides of the lateral one and one-half digits.

The Deep Palmar Arch. (*fig. 146*) the radial artery continued into the palm, is completed by the deep branch of the ulnar artery.

Branches. (*1*) *Palmar digital branches* to the lateral one and one-half digits (viz., *princeps pollicis* and *radialis indicis aa.*). (*2*) *three palmar metacarpal arteries;* (*3*) *three perforating arteries; and* (*4*) *several recurrent branches.*

The Palmar Carpal Arch, more accurately a **rete** or network, is formed by the union of the *palmar carpal branch of the ulnar artery* and the *palmar carpal branch of the radial artery.*

It receives twigs of the anterior interosseous artery and recurrent branches of the deep palmar arch (*fig. 146*).

The Dorsal Carpal Arch is applied to the dorsal surface of the carpal bones and its branches have many anastomoses (*fig. 157*).

11 EXTENSOR REGION OF FOREARM AND HAND

FOREARM

Boundaries; Tendons; Palpable Parts of Ulna and Radius.

SNUFF-BOX AND THREE DORSAL TENDONS

Internervous Line.

MUSCLES, VESSELS, AND NERVES

Supinator; Posterior Interosseous Nerve; Muscles of the Extensor Region—Lateral Superficial Group; Posterior Superficial Group; Deep Group.

Posterior Surfaces of Ulna and Radius; Arteries; Cutaneous Nerves; Deep Fascia; Extensor Retinaculum.

DORSUM OF WRIST AND HAND

Tendons at Back of Wrist; Extensor (Dorsal) Expansions: Lumbricals and Interossei.

Interossei—attachments, course, insertions, nerve supply.

Nerve supply to components of expansion and nerve lesions.

FOREARM

Boundaries. No motor nerve crosses the boundary lines, which are: *medially*, the subcutaneous border of the ulna and the ulnar border of the hand, and *laterally*, the course of the radial artery in the forearm and the radial border of the hand (*fig. 158*).

Tendons (*figs. 158.1, 163, 164*). All tendons passing from the forearm to the back of the hand span the carpus and reach the bases of metacarpals or phalanges. There are no fleshy muscles and so no motor nerves.

Palpable Parts of Ulna and Radius (*fig. 159*). While the whole length of the ulna is palpable, only the upper and lower ends of the radius are easily felt. The head of the radius lies immediately below the smooth posterior aspect of the lateral epicondyle. When the elbow is extended, the head lies at the bottom of a visible hollow in which its upper margin is easily felt on pressing firmly downward. When the elbow is flexed and the forearm alternately pronated and supinated, the head can be felt to revolve under the palpating fingers. The *styloid process* of the radius lies in the anatomical snuff-box and is more than 1 cm below the level of the ulnar styloid process. In order to palpate the tips of these processes, grasp the sides of the wrist between your thumb and index, and press upward.

SNUFF-BOX AND THE THREE DORSAL TENDONS OF THE THUMB

When the thumb is fully extended, a hollow, called the **"snuff-box,"** can be seen on the dorsum of the wrist, at the root of the thumb. Bounding the snuff-box medially is the tendon running to the dorsum of the thumb (*fig. 160*) and easily followed in the living subject.

Crossing the snuff-box superficially are the *cephalic vein* and *superficial radial nerve*. The *radial artery* crosses on the skeletal plane.

The Three Dorsal Tendons of Thumb Traced Proximally (*fig. 160*). The line of the

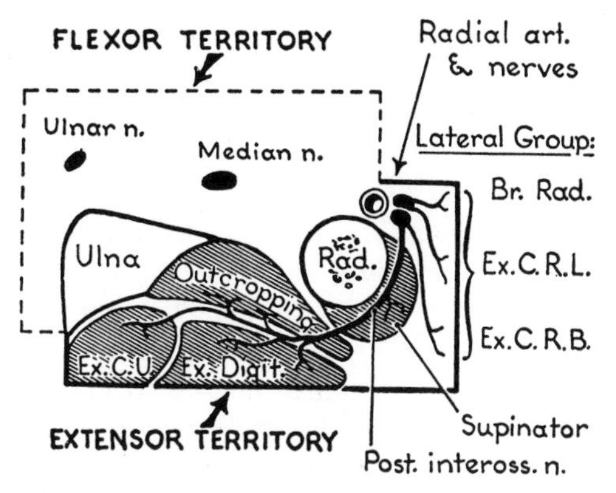

Fig. 158. The extensor region of the forearm, on cross-section.

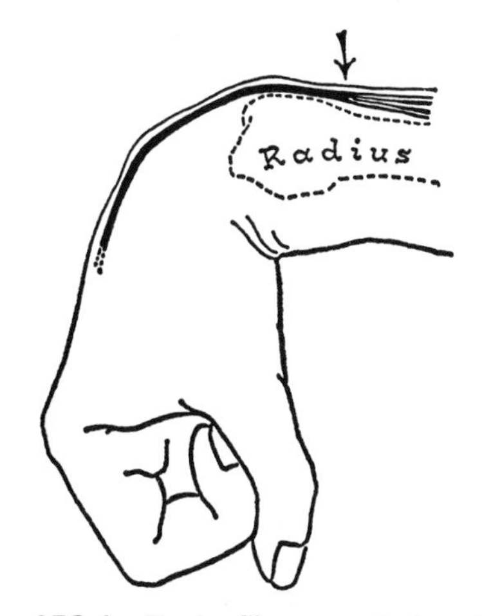

Fig. 158.1. Fleshy fibers must give place to tendon beyond the arrow at the lower end of the radius.

"*three outcropping muscles*"of the thumb divides the superficial muscles of the extensor region into a lateral and a posterior group, each with its own nerve supply. It is an **internervous line.** This makes it the safest line of approach and entry to the deeper parts of the back of the forearm. On separating the two groups of muscles, the Supinator, which is one of the deep muscles, is seen wrapped around the upper third of the radius.

MUSCLES, VESSELS, AND NERVES

Supinator

This, the primary supinator—the other being the Biceps—arises from the lateral

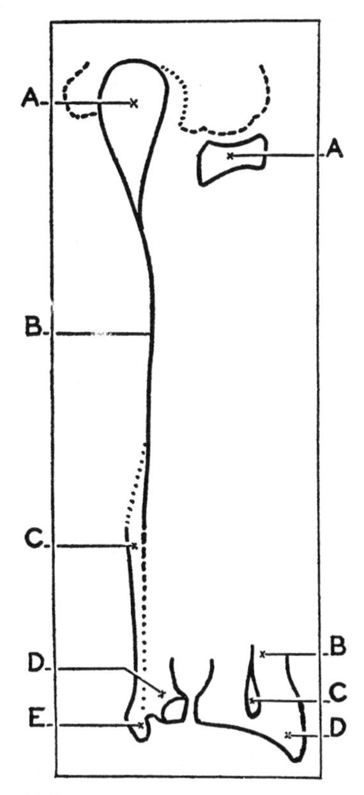

Fig. 159. Palpable parts of ulna and radius (from behind). *Ulna: A,* olecranon; *B,* posterior border; *C,* area between flexor and extensor tendons; *D,* head; and *E,* styloid process. *Radius: A,* head; *B,* lower end dorsally; *C,* dorsal tubercle; and *D,* styloid process.

ligaments of the elbow (*fig. 160*) and an adjacent triangular area of the ulna distal to its radial notch (*figs. 107, 189*). It inserts into the body of the radius between the anterior and posterior oblique lines (*fig. 161*).

Posterior Interosseous Nerve

This motor nerve (also called *Deep Radial Nerve*) starts as one of the two terminal branches of the radial nerve—the other branch being the superficial radial nerve —in front of the capsule of the elbow joint and under cover of the Brachioradialis. It winds round the radius in the substance of the Supinator, which forms for it a fleshy tunnel. It emerges from the muscle some 7 cm below the head of the radius and finds

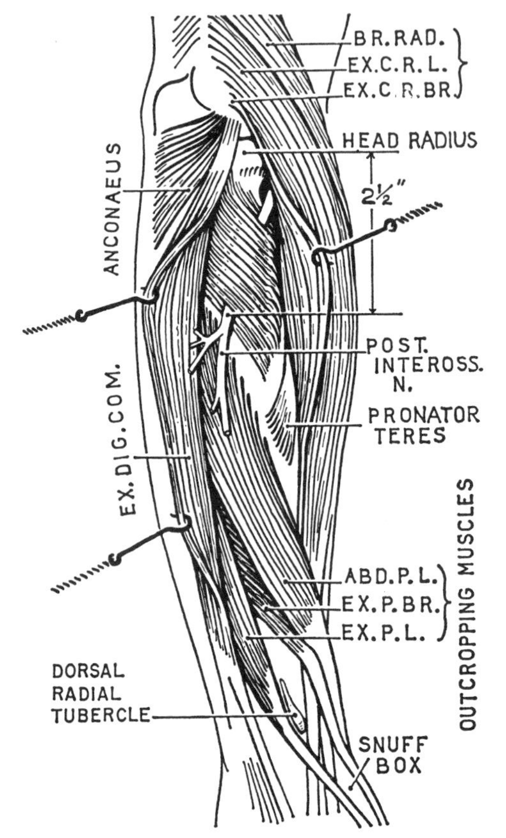

Fig. 160. The furrow of "the 3 outcropping muscles of thumb" opened up—line of relative safety.

itself under cover of the posterior groups of superficial muscles.

The radial and posterior interosseous nerves together supply the three muscles of the lateral group (Brachioradialis, Ex. Carpi Radialis Longus, Ex. Carpi Radialis Brevis) and the Supinator before the latter nerve enters the Supinator. On emerging from the Supinator, it proceeds to supply all the remaining muscles on the back of the forearm.

Muscles of Extensor Region of Forearm

These Extensors are arranged in two layers—a *superficial* and a *deep*. The superficial layer is subdivided into two groups—a *lateral* and a *posterior*—by the line of the three outcropping muscles of the thumb.

The origin of the superficial muscles is from a flattened, *common extensor tendon*, which is attached to the front of the lateral epicondyle and neighboring fascia and supracondylar ridge.

Lateral Group of Superficial Extensors:

1. *Brachioradialis.*
2. *Extensor Carpi Radialis Longus.*
3. *Extensor Carpi Radialis Brevis.*

The Brachioradialis arises from the upper part of the lateral supracondylar ridge and intermuscular septum. It bounds the cubital fossa laterally and shelters the radial nerve. Its tendon is inserted into the base of the styloid process of the radius.

It is peculiar in having both origin and insertion at the distal ends of bones. Clearly, it can act as a flexor, though

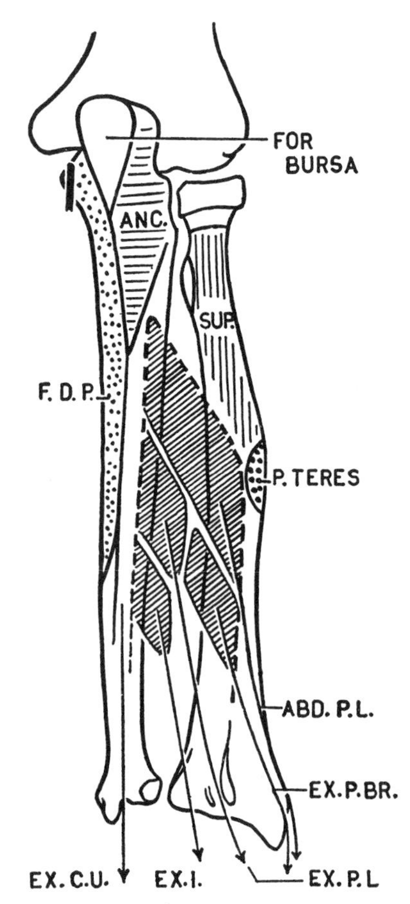

Fig. 161. Posterior aspect of ulna and radius, clothed.

developmentally it belongs to the extensor group of muscles and is supplied by an "extensor" nerve. Functionally, it acts during both fast extension and fast flexion, apparently as a "shunt muscle" (p. 27).

The Extensor Carpi Radialis Longus arises from the lower one-third of the supracondylar ridge and lateral intermuscular septum. *E. C. R. Brevis* arises from the common tendon and adjacent capsule and septa. Their tendons occupy the broad sulcus on the lower end of the radius lateral to the dorsal radial tubercle (*fig. 164*). They cross the snuff-box and pass to the bases of the 2nd and 3rd metacarpals where they may be palpated while dorsiflexing the wrist (*fig. 157*).

A *bursa* exists at each end of the Ex. Carpi Radialis Brevis. The one at its tendon of origin is said to be inflamed in "tennis elbow."

Posterior Group of Superficial Extensors:

1. *Extensor Digitorum.*
2. *Extensor Digiti Minimi or V.*
3. *Extensor Carpi Ulnaris.*
4. *Anconeus.*

The Three Extensors arise from the common tendon of origin, the deep fascia covering them, and the intermuscular septa. Their *grooves* on the lower end of the radius and ulna and *insertions* are given on p. 146.

The Anconeus arises by tendon from the back of the lateral epicondyle. It is inserted by fleshy fibers into the posterior surface of the ulna above the oblique line and into the lateral surface of the olecranon (*figs. 160, 161*). Developmentally, it is a part of the medial head of the Triceps.

Deep Group of Extensor Muscles of Forearm:

1. *Abductor Pollicis Longus.*
2. *Extensor Pollicis Brevis.*
3. *Extensor Pollicis Longus.*

} *"Outcropping Muscles"*

4. *Extensor Indicis.*
5. *Supinator* (considered on p. 143).

The exact origin of the three tendons to the thumb (*fig. 161*) is unimportant. After outcropping (*fig. 160*) they run to the thumb to insert on the bases of the three long bones of the thumb: Abd. Longus to the metacarpal, and Ex. Brevis and Ex. Longus to the proximal and distal phalanges (*fig. 157*). The E.P. Longus angles around a pulley (dorsal radial tubercle) and so is separated from the other two—hence the snuff-box.

Extensor Indicis joins the extensor expansion to the index.

Details of Posterior Surfaces of Ulna and Radius (*fig. 161*). The posterior surface of the ulna is crossed by an *oblique line* that passes from the radial notch to the bend on the sharp posterior border at the junction of the upper one-third and lower two-thirds of the bone. The area above this line belongs to the Anconeus.

The posterior surface of the radius also is crossed by an oblique line, the *posterior oblique line*. The area above this belongs to the Supinator.

From the oblique line of the ulna a *vertical line* descends and divides the posterior surface into medial and lateral halves. The Ex. Carpi Ulnaris overlies the medial half and its tendon plays in the groove.

The lateral half of the posterior surface of the ulna below the oblique line and the whole width of the posterior surface of the radius below the oblique line, plus the intervening interosseous membrane, are utilized by the origins of four deep muscles, three being for the thumb, one for the index (*fig. 161*).

Arteries. The *Posterior Interosseous Artery* is the smaller of the two terminal branches of the common interosseous artery (*fig. 128*). It does not enter this region with the nerve of the same name, but by passing over the upper border of the interosseous membrane. It descends between the superficial and deep muscles, supplies them, and takes part in the anastomoses at the elbow and wrist.

Cutaneous Nerves (*figs. 98 and 101*).

Enclosing Deep Fascia. For some distance *below the elbow* the deep fascia gives origin to the extensor muscles; so, its fibers are strong and run vertically.

At the lower end of the forearm the fibers are required to retain the extensor tendons in place, forming an **extensor retinaculum.** The more proximal fibers, violating the rule that deep fascia must be attached to all the exposed, subcutaneous, bony points it crosses, turn round the head of the ulna and become continuous with the deep fascia on the front of the forearm (*fig. 162*); and the more distal fibers pass obliquely downward and medially from the radius to the medial carpal bones. Septa attach the retinaculum to the radius and medial carpals thereby forming osseofascial tunnels for the tendons. There are six of these, each lined with a synovial sheath (*fig. 163*).

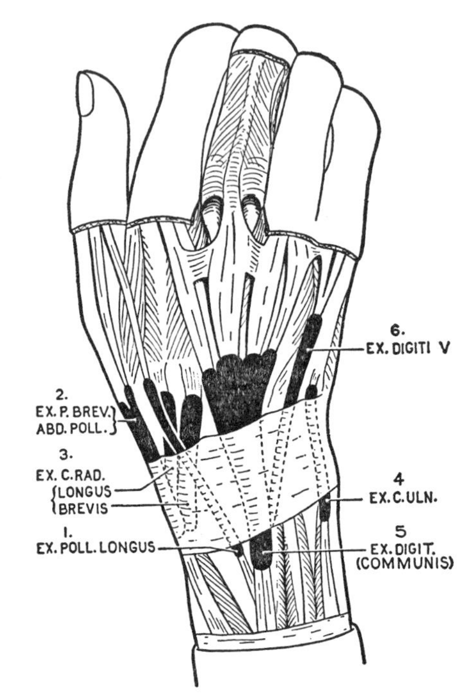

Fig. 163. The six synovial sheaths on the dorsum of the hand and a dorsal expansion—see also *figure 164*. (Dissection by C. P. Rance and J. W. Rogers.)

DORSUM OF WRIST AND HAND

Tendons at Back of Wrist. The nine tendons occupy six grooves (*figs. 163, 164*).

Extensor Expansions (*Dorsal Expansions*). The four flat tendons of the Ex. Digitorum (Communis) traverse the most medial tunnel at the back of the lower end of the radius and, diverging, pass to the four fingers. The common tendons of the index and little fingers are joined on their medial sides near the knuckles by their respective proper tendons—Ex. Indicis and Ex. Digiti V (*figs. 163, 164*).

Three oblique bands unite the four tendons proximal to the knuckles (*fig. 163*). Hence, independent action of the fingers is restricted.

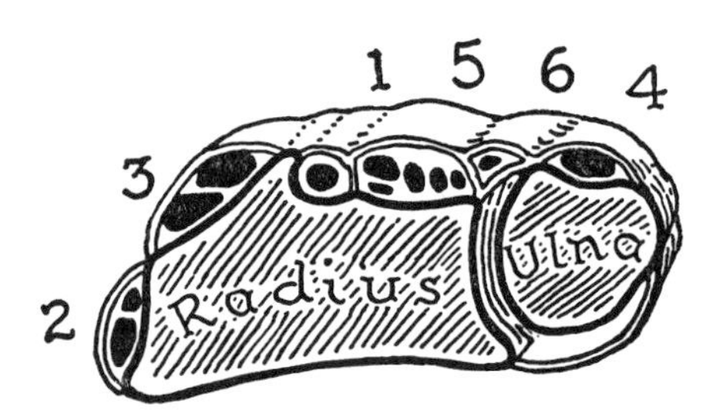

Fig. 164. Figure 163 on cross-section.

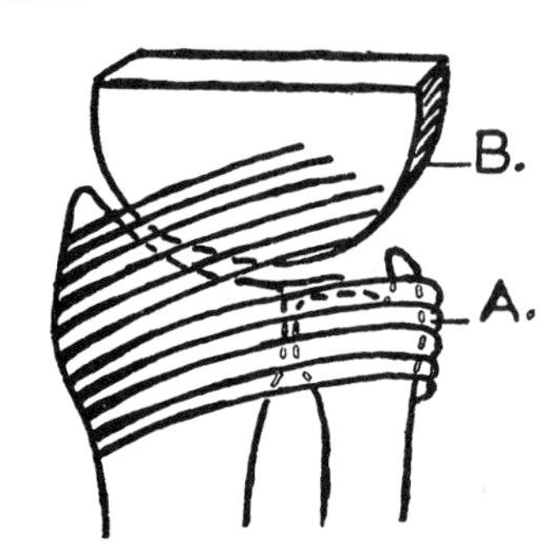

Fig. 162. The extensor retinaculum.

Structure and Attachments (*fig. 165*). On the distal ends of the metacarpals and on the digits the extensor tendons become still further flattened, to the thickness of deep fascia and are called *extensor* or *dorsal expansions*.

Each expansion is wrapped around the dorsum and sides of a metacarpal head and of a proximal phalanx. The *visor-like hood*, thus thrown over the metacarpal head, is anchored on each side to the palmar lig. or plate and thereby it serves to retain the

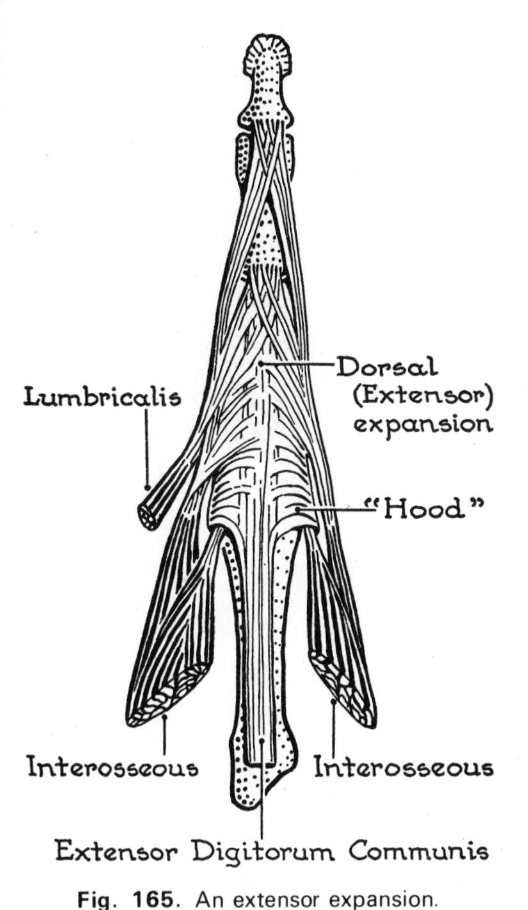

Fig. 165. An extensor expansion.

extensor tendon in the midline of the digit. A broad *fibro-areolar ribbon* passes from the hood to the base of the proximal phalanx.

On the proximal phalanx the expansion divides into a *median band*, which passes to the base of the middle phalanx, and into *two side bands* which pass to the base of the distal phalanx.

The extensor tendon pulls mostly via the median band and its attachment to the proximal phalanx, and only slightly via the side bands.

Each side band is joined by half an Interosseus tendon and more distally, on the radial side, by an entire Lumbrical tendon. These tendons are united across the dorsum of the proximal phalanx by a sling of transverse fibers; and via the side bands they run to the bases of the 2nd and 3rd phalanges (*fig. 166*).

The Interossei and Lumbricals flex the metacarpophalangeal (m-p) joints by means of the arched fibers thrown across the proximal phalanges (*fig. 165*). In addition, they extend the interphalangeal (i-p) joints (*fig. 167*) and impart side motion (abduction and adduction) to the fingers, provided the m-p joints are stabilized or steadied (by the long extensor tendons).

Origin of Lumbricals (see page 137).

Interossei. Each of the five digits can be abducted and adducted, that is, moved away from and moved toward a line passing through the middle finger, and called the *axial line of the hand*. For this, 10 muscles are required. The adductors and the 2 abductors of the thumb attend to its

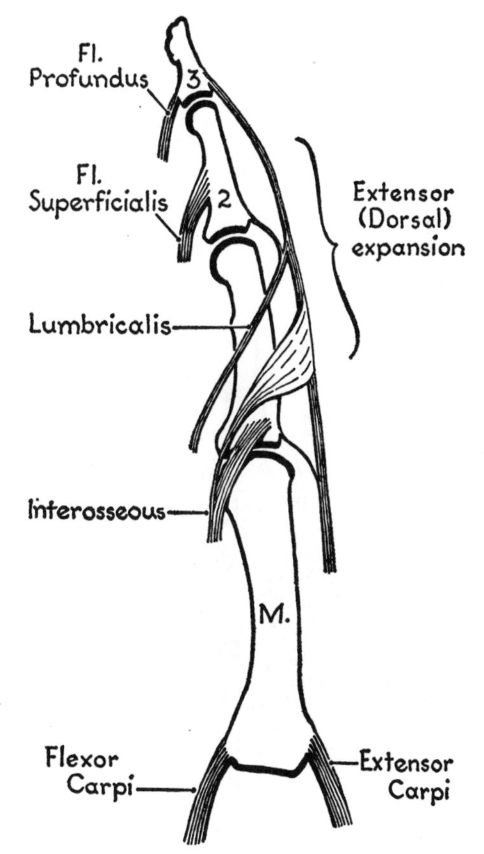

Fig. 166. The insertions of the tendons of a finger, side view.

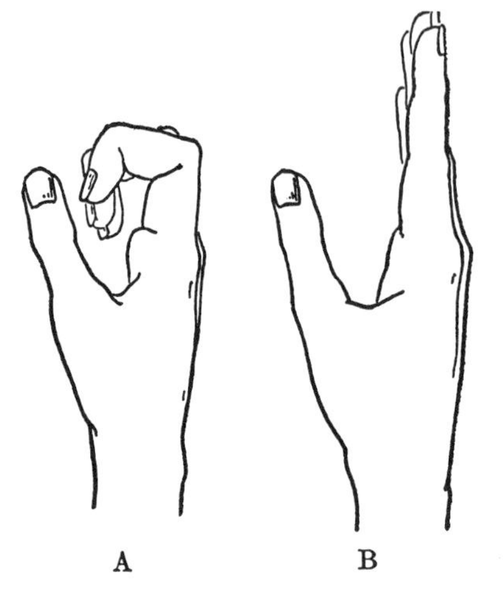

Fig. 167. *A*, long digital extensors extend metacarpophalangeal joints. *B*, Lumbricals and Interossei extend interphalangeal joints.

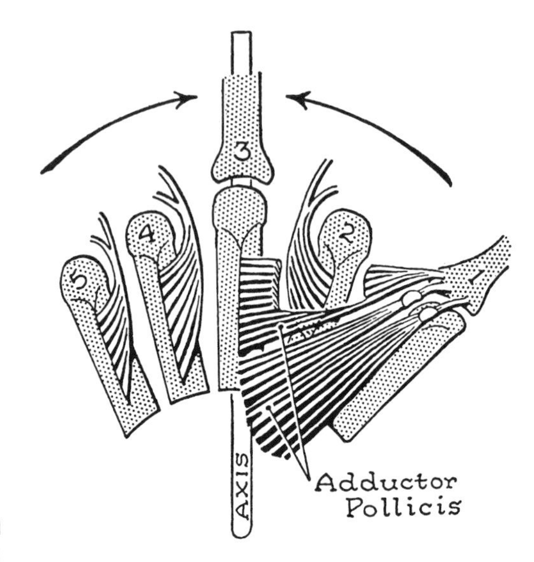

Fig. 169. Three Palmar Interossei.

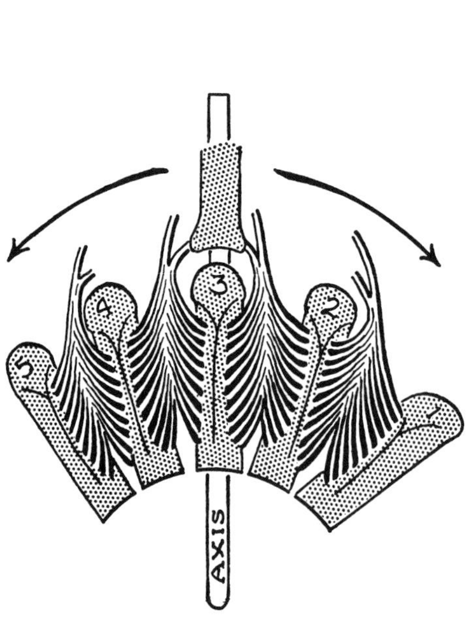

Fig. 168. Four Dorsal Interossei.

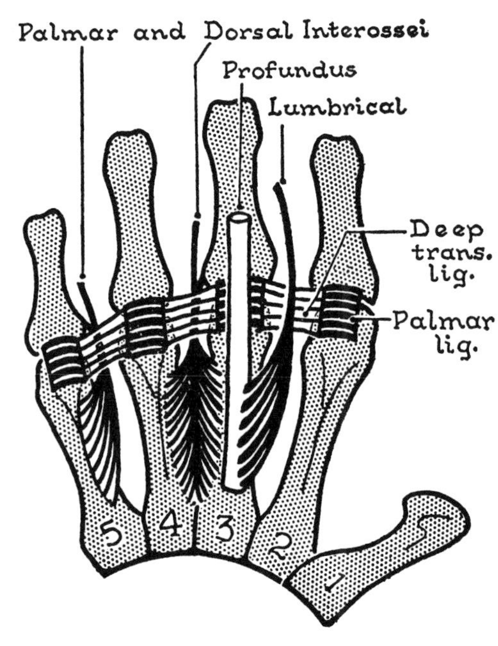

Fig. 170. The different relationships of Lumbricals and Interossei to the deep transverse metacarpal ligaments, palmar view.

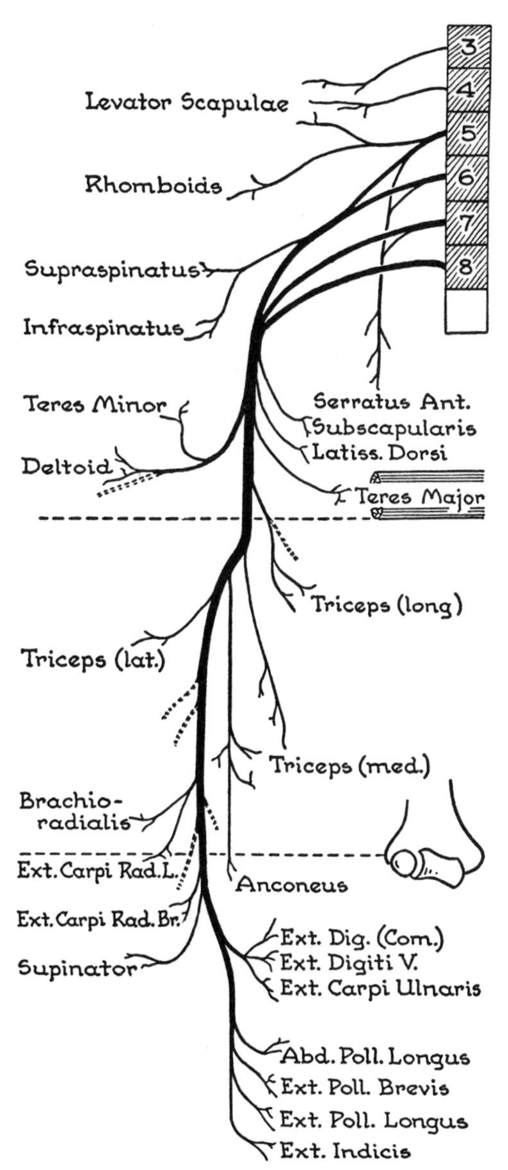

Fig. 171. The motor distribution of the nerves of the back of the limb.

requirements, and the Abductor Digiti V abducts the little finger. The remaining 7 movements are performed by the 7 Interossei (*figs. 168, 169*).

Four Dorsal Interossei (*fig. 168*) are conspicuous from the dorsum of the hand filling the four intermetacarpal spaces and arising by double heads from the adjacent sides of the bodies of the five metacarpals. They are abductors.

The three Palmar Interossei arise by single heads from the anterior borders of the metacarpals of the fingers with available borders, namely, 2nd, 4th, and 5th (*fig. 169*)—the 3rd anterior border is monopolized by the transverse head of the Adductor Pollicis. Moreover, each Palmar Interosseus arises from the metacarpal of the digit on which it acts—its action being adduction.

Course. The Interossei pass behind the deep transverse metacarpal ligaments (*fig. 170*); the Lumbricals and the palmar digital vessels and nerves pass in front.

Insertions. The sides of the fingers to which the Interossei must pass to perform these movements are apparent. Their insertions are partly into the bases of the proximal phalanges and partly into the extensor expansions.

Nerve Supply. All Interossei are supplied by the ulnar nerve.

Nerve Lesions and Extension of the Fingers. Since the ulnar nerve supplies the Interossei and at least two of the Lumbricals, its loss produces inability to extend the i-p joints. There is also a reactive hyperextension of the m-p joints through the pull of the radial-nerve-supplied Extensor Digitorum. Radial nerve loss results in failure to extend the m-p joints, but the i-p joints can be extended by the Interossei and Lumbricals.

Review of Nerve Supplies. The student will find *figures 171* and *172* and the table on p. 150 useful for review.

*Segmental Innervation of Muscles of Forearm and Hand**

Muscles	C.5	C.6	C.7	C.8	Th.1
Brachioradialis Supinator	X	X			
Pronator Teres		X	X		
Ext. Carpi Radialis longus and brevis Flexor Carpi Ulnaris Flexor Carpi Radialis		X	X	X	
Ext. Digitorum Ext. Carpi Uln. Ext. Indicis Ext. Digiti V Ext. Poll. Longus Ext. Poll. Brevis			X	X	
Abd. Poll. Longus			X	X	
Palmaris Longus			X	X	X
Pronator Quadratus Fl. Digitorum Profundus Fl. Digitorum Superficialis Fl. Pollicis Longus Lumbricals			X	X	X
Opponens Poll. Abd. Poll. Br. Flexor Poll. Br.				X	X
Palmaris Brevis Add. Pollicis Fl. Digiti V Abd. Digiti V Opponens Digiti V Interossei				X	X

* Modified after Bing; and Haymaker and Woodhall.

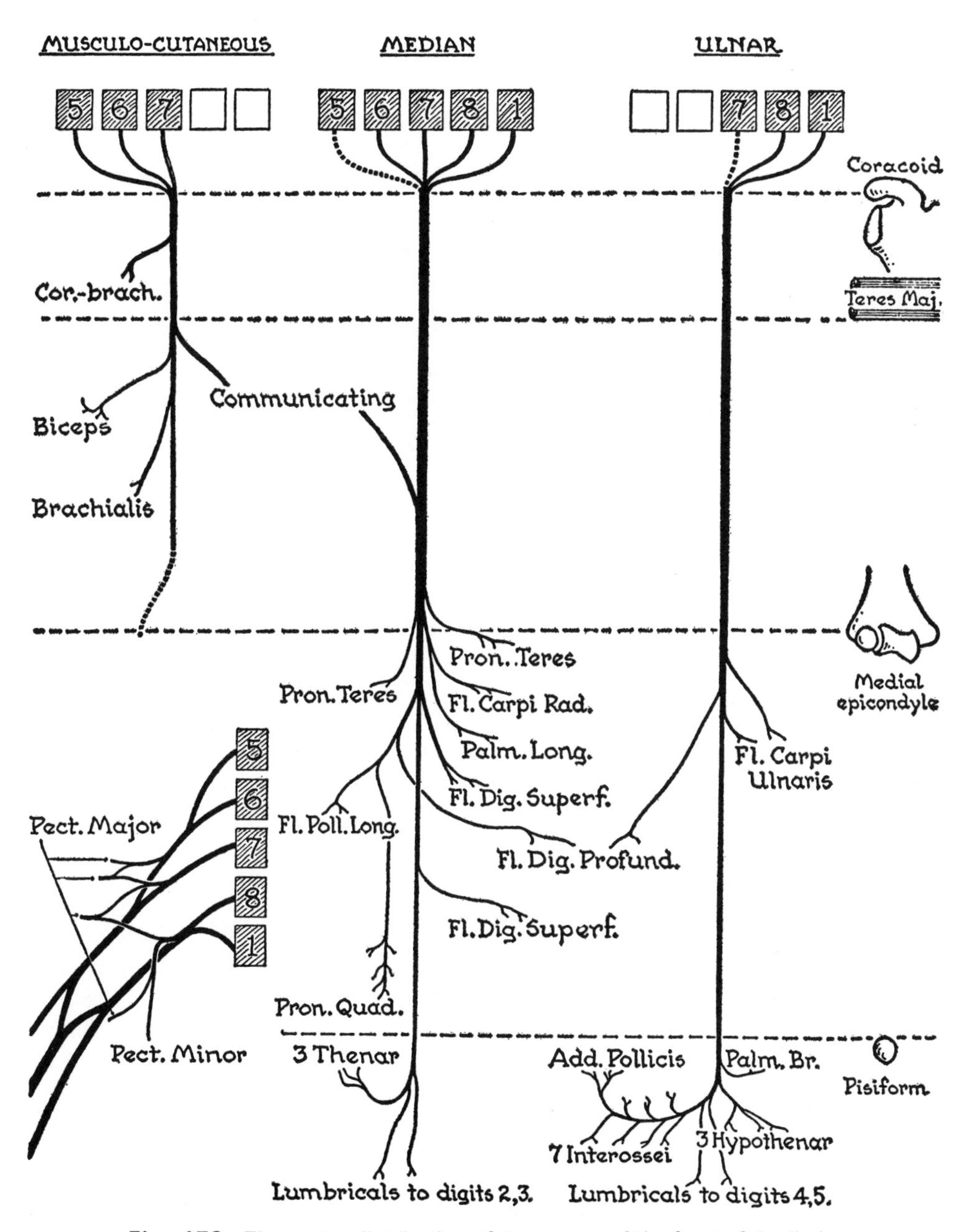

Fig. 172. The motor distribution of the nerves of the front of the limb.

12 JOINTS OF UPPER LIMB

JOINTS OF THE SHOULDER GIRDLE

Sternoclavicular Joint; Coracoclavicular Ligament; Acromioclavicular Joint; Movements of Shoulder Girdle.

SHOULDER JOINT

The Ball and Socket; Movements; Fibrous Capsule and Ligaments: Coracohumeral and Glenohumeral; Long Head of Biceps; Strength of a Joint.
Four Short Muscles; Bursae: Subacromial Bursa; Coraco-acromial Arch.
Epiphyses; Muscles Acting on Shoulder Joint; Nerve Supply.

Elevation of Upper Limb

Muscle Force Couples

ELBOW JOINT

Bones Concerned: humerus, ulna, head of radius.
Proximal Radio-Ulnar Joint; Anular Ligament.
Ligaments and Capsule; Movements; Relations; Anastomoses.
Epiphyses (around elbow); Nerves

RADIO-ULNAR JOINTS

Intermediate Radio-Ulnar Joint

Interosseous Membrane; Oblique Cord.

Distal Radio-Ulnar Joint

HEAD OF ULNA; EPIPHYSES.

WRIST OR RADIOCARPAL JOINT

Joint Surfaces; Movements of Wrist; Attrition of Disc and Ligaments.
Ligaments; Muscles; Relations; Nerve Supply.

INTERCARPAL, MIDCARPAL, CARPOMETACARPAL AND INTERMETACARPAL JOINTS

METACARPOPHALANGEAL AND INTERPHALANGEAL JOINTS

Movements of Fingers; Ligamentous Functions; Epiphyses of Upper Limb: Table of Chronology.

JOINTS OF THE SHOULDER GIRDLE

(Joints in which the clavicle takes part)

1. Sternoclavicular joint.
2. Coracoclavicular ligament.
3. Acromioclavicular joint.

Unity of Function. Almost the sole duty of the clavicle is to thrust the scapula, and with it the arm, laterally and to prevent it from being driven medially. Since muscles can render the clavicle little aid in this, the function is performed by two structures, the *coracoclavicular ligament* laterally and the *articular disc* medially (*fig. 173*). A fall on the side of the shoulder must put a strain on them (*fig. 174*).

Sternoclavicular Joint

Structure. The enlarged sternal end of the clavicle articulates in the shallow socket at the upper angle of the manubrium and adjacent part of the 1st costal cartilage. The end of the clavicle rises above the manubrium, the two making an ill fit; but the strong, thick, *articular disc* of fibrocartilage divides the joint cavity into two and prevents medial displacement of the clavicle. Its attachments must be to the clavicle above and to the 1st costal cartilage below. In older persons, the disc may be perforated centrally, yet its hold is not lost. Strong *anterior* and *posterior ligaments*, to which the margins of the disc are attached, strengthen the joint.

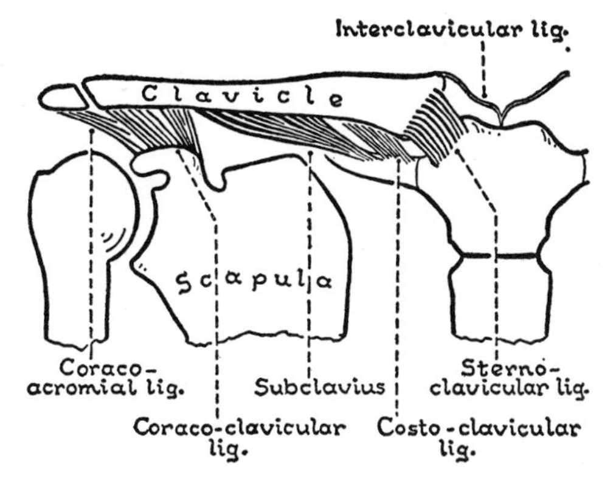

Fig. 173. Structures having unity of direction. The feeble interclavicular lig. may be homologous with the wishbone of birds.

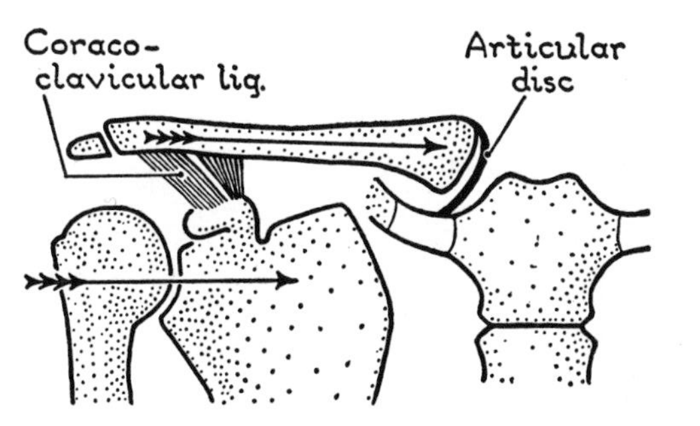

Fig. 174. Structures having unity of function.

Movements and Function. As you may discover by palpation, movements at this joint allow the scapula considerable mobility. Also, it allows the clavicle to undergo axial rotation during elevation (p. 161).

The Costoclavicular Ligament passes from the 1st costal cartilage to a rough impression below the sternal end of the clavicle (*fig. 173*).

Coracoclavicular Ligament

The joint has no articular surfaces —junction is effected by a powerful ligament; so, the joint is a syndesmosis. The coracoclavicular ligament is in two parts, a conoid and a trapezoid, of interest only to a few orthopedists and other specialists (*fig. 174*).

Functions. (1) Owing to their medial (and downward) direction, the ligament prevents the scapula from being driven medially; in this, only the Serratus Anterior gives it help. (2) It is the mainstay of the acromioclavicular joint, and so long as it is intact, the joint may, indeed, undergo subluxation (partial dislocation), but the acromion cannot be driven under the clavicle (*fig. 175*). (3) With the aid of muscles it suspends the scapula.

Acromioclavicular Joint

The medial border of the acromion has near its tip a small oval facet which articulates with a similar small facet on the lateral end of the clavicle (*fig. 175*). Strong parallel fibers form a complete capsule for the joint. A small *articular disc* hangs into the cavity from above.

Function. (1) This joint, while not vital to force transmission, does permit the scapula to move vertically on the chest wall when the pectoral girdle rises and falls (e.g., as when shrugging the shoulders). (2) It also permits the scapula (and with it the glenoid cavity) to glide forward and backward on the clavicle and so to face directions convenient to the head of the humerus (e.g., forward when striking a blow). (3) Also, its freedom is essential to free elevation of the limb (p. 161).

Relations. The joint is subcutaneous and is easily felt on pressing medially.

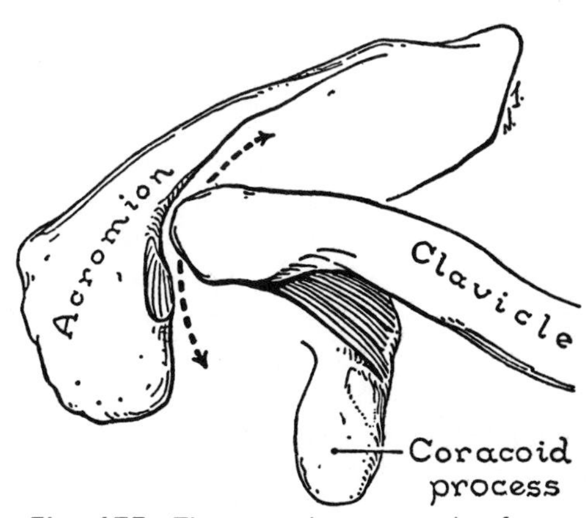

Fig. 175. The acromion may swing forward or backward, but, so long as the coracoclavicular ligament is intact, it cannot be driven under the clavicle.

Movements of the Shoulder Girdle

1. *Simple elevation* of scapula: i.e., the scapula moves vertically upward. A low level of activity in Trapezius (upper), Levator Scapulae, and Serratus Anterior (upper) is sufficient to suspend the girdle; but when a weight is either supported on the shoulder or carried in the hand, these muscles contract vigorously (*fig. 176*). (See table 3.)

2. *Simple depression* is brought about by the weight of the limb. But as an active movement, e.g., pressing downward or resting on parallel bars, it calls into action the Pectoralis Minor, which acts on the girdle, and the Pectoralis Major and Latissimus Dorsi, which act on the humerus (*fig. 177*). The timely contraction of these muscles saves the clavicle from fracture, when one falls on one's outstretched hand (p. 76).

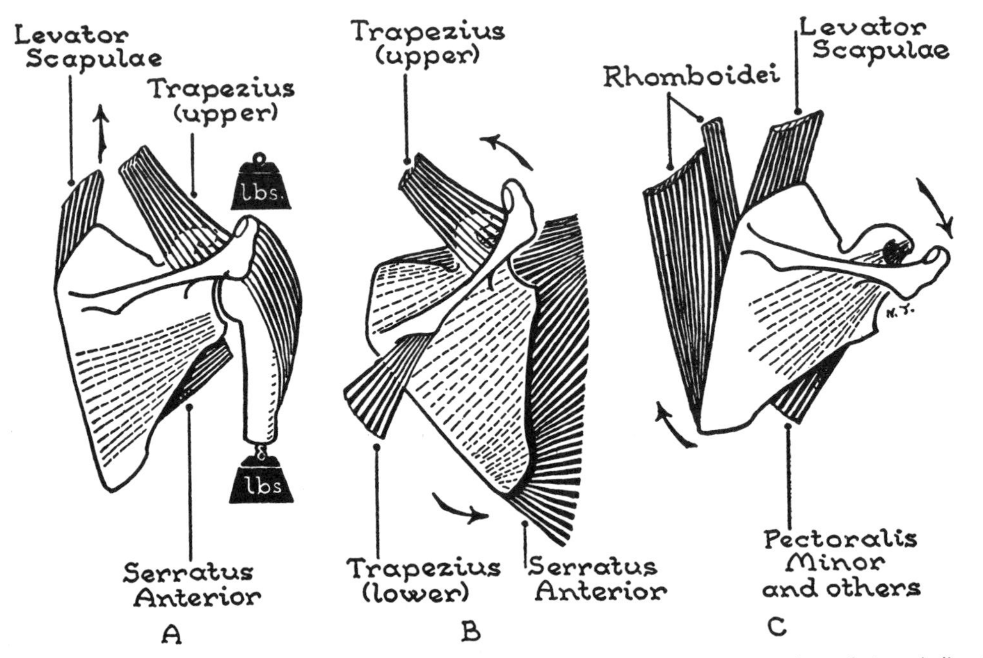

Fig. 176. *A*, simple elevation of the shoulder girdle. The suspensory muscles of the girdle. *B*, elevation of the girdle with upward rotation of the glenoid cavity. *C*, depression of the girdle with downward rotation of the glenoid cavity.

TABLE 3

*Muscles Acting upon the Shoulder Girdle**

Simple Elevation	Simple Depression	Elevation with Upward Rotation of Glenoid Cavity	Depression with Downward Rotation of Glenoid Cavity	Protraction or Forward Movement	Retraction or Backward Movement
Trapezius (upper) Lev. Scapulae Serratus Anterior (upper)	Pect. Minor Subclavius Pect. Major and Lat. Dorsi	Trapezius (upper) Trapezius (lower) Serratus Anterior	Lev. Scapulae Rhomboids Pect. Minor Trapezius (mid.) Pect. Major and Lat. Dorsi	Pect. Minor Lev. Scap. Serratus Anterior and Pect. Major	Trapezius (mid.) Rhomboids and Lat. Dorsi

* *Note.* Pectoralis Major and Latissimus Dorsi act on the girdle indirectly, through the humerus.

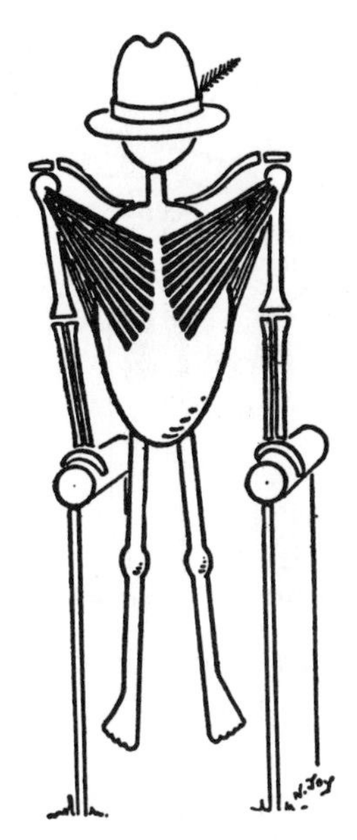

Fig. 177. Pressing downward or resting on parallel bars calls into action the Pectoralis Major and the Latissimus Dorsi.

3. *Elevation with upward rotation of glenoid cavity.* In this movement the acromion rises, the superior angle of the scapula descends, and the inferior angle swings laterally. The Trapezius (upper), Trapezius (lower), and the Serratus Anterior combine in this rotation. This movement is almost always part of a larger movement involving either abduction or flexion of the shoulder joint, as when the hand reaches for some object above the head, i.e., the entire limb is elevated—see page 161.

4. *Depression with downward rotation of glenoid cavity,* that is, recovering from the last movement or overstepping the recovery, e.g. chopping wood (*fig. 176*). The Pectoralis Minor, Rhomboids, Levator Scapulae, and Trapezius (especially the middle portion) are called into play; and the Pectoralis Major and Latissimus Dorsi, which act indirectly through the humerus, give them powerful assistance.

5. *Protraction of the scapula* or forward movement, e.g., pushing. The Serratus Anterior, Pectoralis Minor, and Levator Scapulae act together with the Pectoralis Major.

6. *Retraction of the scapula* or backward movement, that is, recovering from the last movement or overstepping the recovery, e.g., pulling. The Trapezius (middle portion) and the Rhomboids act with the Latissimus Dorsi.

SHOULDER JOINT

(*Articulatio Humeri*)

Bones

The shoulder joint is a ball-and-socket or multiaxial joint (p. 21).

The **Ball** is the head of the humerus. It forms one-third of a sphere, and faces medially, upward, and backward.

The **Socket** is the shallow, pear-shaped glenoid cavity of the scapula. At its upper end, that is, at the root of the coracoid process, is the supraglenoid tubercle for the long head of the Biceps. At its lower end, that is, on the lateral border of the scapula, is the rough infraglenoid tubercle for the long head of the Triceps. The playing of the tendon of the Subscapularis across the front of the socket is responsible for the concavity that contributes largely to its pear shape. A strip of fibrocartilage, the *glenoid labrum*, runs round the rim of the socket, deepens it somewhat, and makes a pliable elastic cushion for the ball to roll against, (*cf.* the cushion of a billiard table).

Movements

It is obvious that there is more freedom at the shoulder joint than at any other joint in the body. It is also obvious that the movements permitted are flexion and extension, and abduction and adduction—and, therefore, circumduction which combines them. The humerus can rotate medially and laterally on its own long axis. (To demonstrate these movements of axial rotation, bend your elbow to a right angle so that movements of the bones of the forearm shall not confuse.)

Fibrous Capsule and Ligaments

To allow such free movement, taut ligaments are impossible and the capsule of the joint is very loose. In fact, when the shoulder muscles are removed, leaving the humerus attached to the scapula only by the capsule and ligaments, the head of the humerus can be pulled 1 cm away from its socket. However, in the adducted position the capsule becomes very tight superiorly, especially a part called the coracohumeral ligament, effectively preventing downward displacement of the humerus.

The *Fibrous Capsule* stretches as a loose tube from just proximal to the margin of the glenoid cavity of the scapula to the anatomical neck of the humerus. Inferiorly, however, it passes well down on to the surgical neck. When the arm is adducted, this lower part lies in folds (*fig. 183*).

The *Synovial Membrane* lines the fibrous capsule and is reflected along the anatomical (and surgical) neck as far as the hyaline cartilage of the head of the humerus.

The Coracohumeral Ligament extends from the lateral border of the process to the anatomical neck of the humerus beside the lesser and greater tubercles—which are situated anteriorly and laterally—and its posterior fibers so blend with the capsule that the ligament can only be distinguished from the capsule when viewed from the front. The ligament prevents downward dislocation of the adducted humerus and also limits lateral rotation (*fig. 178*).

Downward dislocation of the shoulder joint is prevented by a **locking mechanism** dependent on three factors: (1) the slope of the glenoid fossa, leading to (2) the tightening of the superior part of the capsule (including the coracohumeral ligament), and (3) the activity of the Supraspinatus muscle (Basmajian and Bazant).

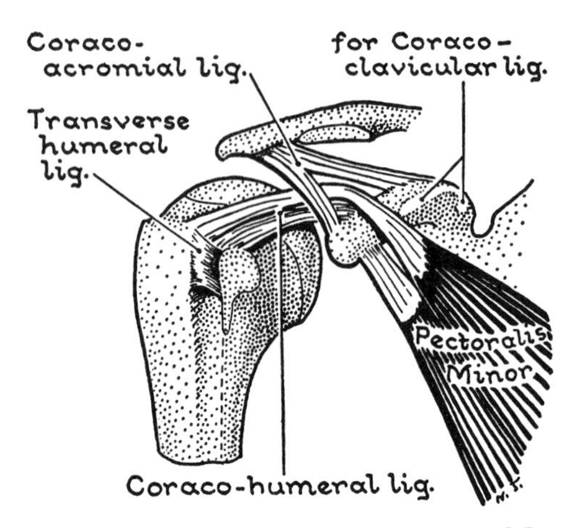

Fig. 178. *Variation:* Part of tendon of Pectoralis Minor augmenting the coracohumeral lig.

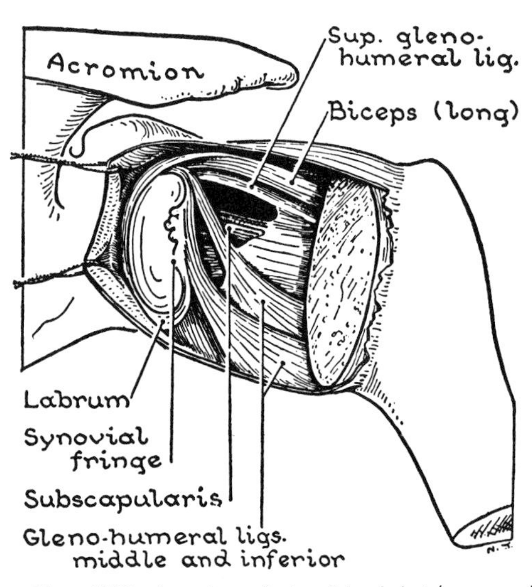

Fig. 179. Interior of shoulder joint (exposed from behind).

Interior of the Shoulder Joint

Figure 179 shows three unimportant bands, the superior, middle, and inferior glenohumeral ligaments; these are slight thickenings of the fibrous capsule, which may stand out in relief when viewed from within the joint. Unfortunately, their importance is greatly exaggerated by some authors; but their thickness and functions are quite insignificant. Undue attention has fallen on them because the middle band often stands out distinctly; this is due to the perforation of the capsule both above and below it. Thus, the subscapularis bursa usually communicates with the joint cavity, and so the *Subscapularis tendon* enters into this picture (*fig. 180*).

The Long Head of the Biceps may be followed from the intertubercular sulcus (which faces forward, *fig. 181*) across the front of the head of the humerus to its origin from the supraglenoid tubercle and posterior lip of the glenoid cavity where it constitutes the glenoid labrum (*figs. 90, 179*).

The long tendon is retained in the intertubercular sulcus by a band, the *transverse humeral ligament* (*fig. 181*); its synovial bursa obviously communicates with the joint space (*fig. 184*).

Strength of a Joint

The strength of a joint depends on three main factors: (1) Bony formation; (2) Ligaments; (3) Muscles. It is evident that for strength the shoulder joint depends on ligaments some times and on muscles at other times. Now, of the muscles around the shoulder joint some are long; others are short. The long muscles perform movements; the short muscles—disposed closely around the head—have also the important function of retaining the head in its socket as "dynamic ligaments." In this they have the assistance of the overhinging coracoacromial arch, which obviously prevents upward displacement of the humerus, and the coracohumeral ligament which prevents direct downward dislocation.

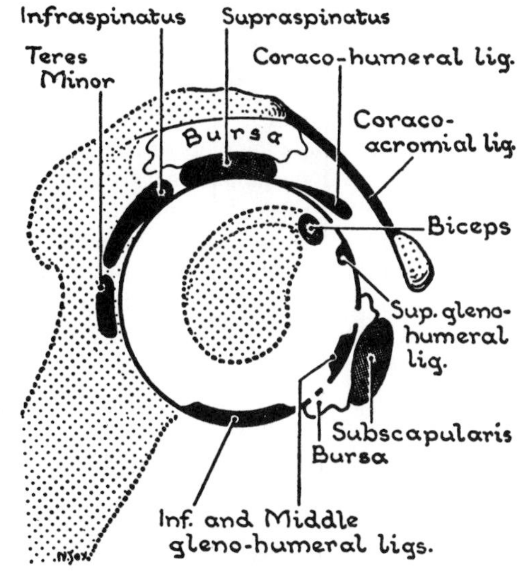

Fig. 180. Scheme of shoulder joint on sagittal section (lateral view).

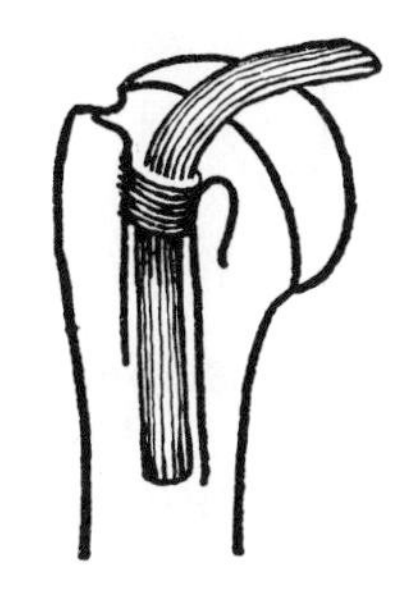

Fig. 181. Upper end of humerus (front view). Long tendon of Biceps and transverse ligament.

Four Short Muscles and Immediate Relations of the Joint

The short muscles round the joint (*figs. 180* and *182*) are:

Supraspinatus—above
Infraspinatus and Teres Minor—behind
Subscapularis—in front

The tendon of Supraspinatus fuses intimately and extensively (2–3 cm) with the underlying fibrous capsule; posteriorly, the tendons of Infraspinatus and Teres Minor, and anteriorly, the tendon of Subscapularis fuse progressively less from above downward.

The tendons of these four short muscles are in a sense *accessory dynamic ligaments*.

Below, there is no supporting short muscle—no alert ligament, but loose unsupported capsule and the quadrangular space. And through this space pass the axillary nerve (which supplies Deltoid and

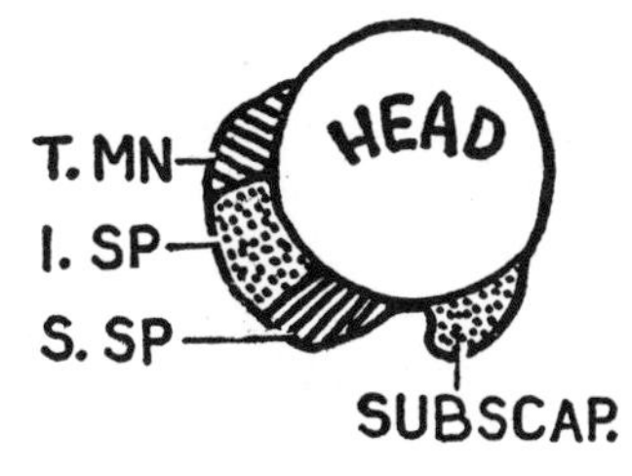

Fig. 182. Insertions of the four short muscles that act as "accessory ligaments" of the shoulder joint viewed from above.

Teres Minor) and the posterior humeral circumflex artery.

Lying side by side on the insertion of the Subscapularis (*fig. 65*), close to the joint, are (1) the brachial plexus and great vessels, (2) the Coracobrachialis, and (3) the short head of the Biceps, which in turn lies side by side with the long head.

Bursae Associated with the Shoulder Joint

The Subscapularis tendon plays in its groove on the anterior border of the glenoid cavity, where it has the **subscapularis bursa.** Usually bursal wall and joint capsule break down so that their cavities come into communication (*fig. 179*).

The *biceps synovial sheath,* where it plays in the upper 5 cm of the intertubercular sulcus (*fig. 184*) has been mentioned above.

The Subacromial Bursa (*fig. 183*) lies between the acromion and the Supraspinatus tendon, and it extends downward between the Deltoid and the greater tubercle. How far down? As far as is necessary; i.e., to cover the part of the greater tubercle that passes under the acromion during abduction of the humerus. *Bursitis*—probably beginning with degeneration and inflammation of an irritated Supraspinatus tendon—is a common affliction of middle age.

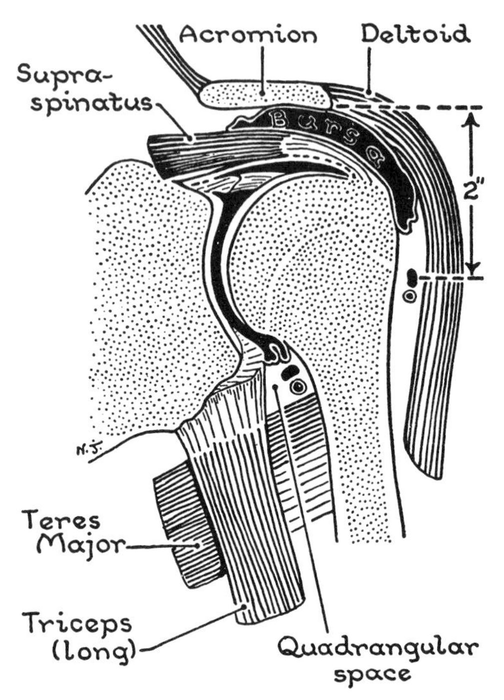

Fig. 183. Shoulder joint, on coronal section. Observe: (1) the upper epiphyseal plate of the humerus; (2) the axillary (circumflex) nerve in the quadrangular space and 5 cm below the acromion; (3) the capsule in folds during adduction; and (4) the bursa which is both subacromial and subdeltoid.

As the result of breakdown of the Supraspinatus tendon and underlying capsule, it is by no means uncommon for the subacromial bursa to be in wide open communication with the synovial cavity of the shoulder joint (*fig. 185*). In our series, none (of 16) under 50 years was perforated; but 3 (of 17) between 50 and 60 years, and 16 of the 46 over 60 were perforated.

The Coraco-Acromial Arch is formed by the coracoid, the coraco-acromial ligament, and the acromion (*figs. 184* and *186*). The ligament joins the tip of the acromion to the lateral border of the coracoid process (*fig. 187*).

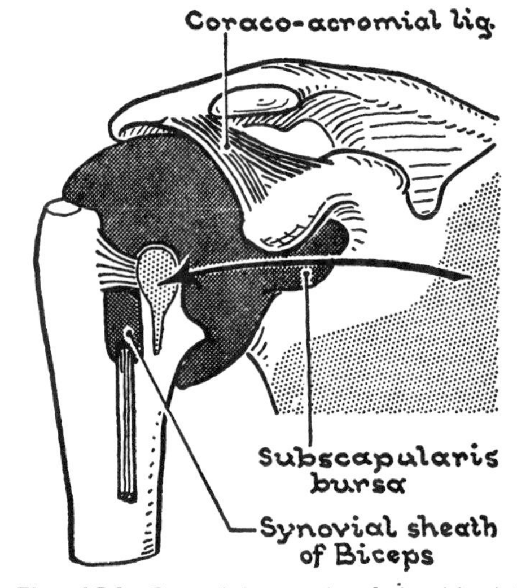

Fig. 184. Synovial capsule of shoulder joint and communicating bursae, distended.

The arch, with the subjacent subacromial bursa, forms a resilient **secondary**

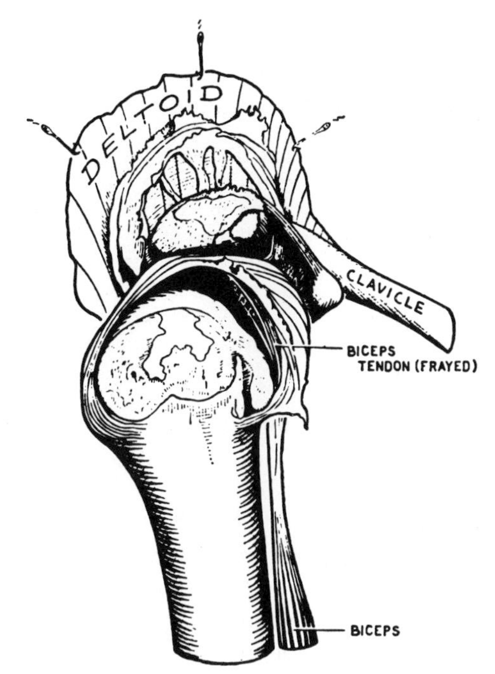

Fig. 185. The supraspinatus tendon and the underlying capsule are commonly worn through. This is an advanced stage with eburnation (ivory-like polishing) of the contact surfaces of humerus and acromion.

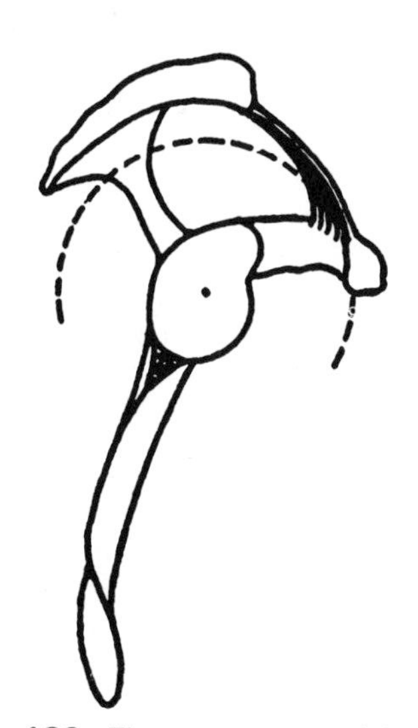

Fig. 186. The coraco-acromial arch.

socket for the head of the humerus, preventing its upward displacement. When the arm is abducted, the greater tubercle of the humerus would impinge on the lateral edge of the coraco-acromial arch were it not for the automatic lateral rotation of the humerus permitting the greater tubercle to slip under the acromion.

Epiphyses Near Shoulder (*fig. 188*)

The upper epiphysis of the **humerus** rests on the spike-like end of the diaphysis. The epiphyseal line lies at the upper limit

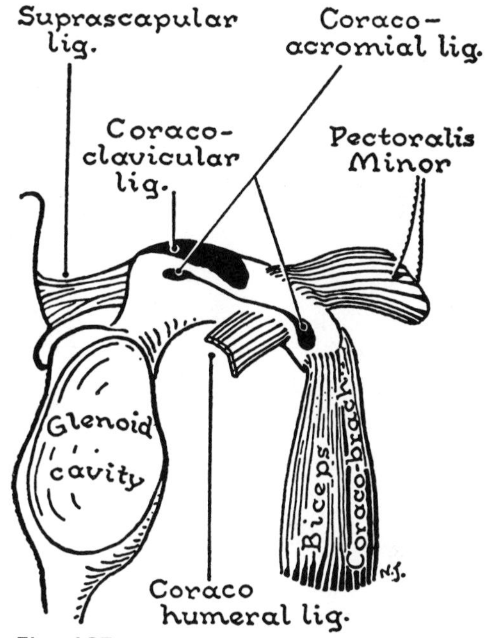

Fig. 187. The structures attached to the finger-shaped coracoid process. (See *fig. 60.*)

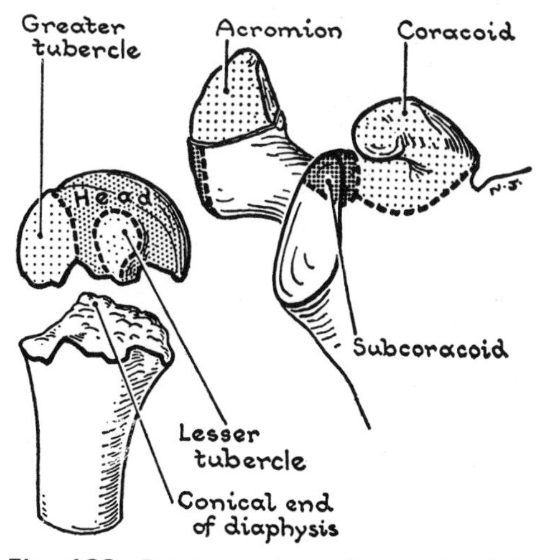

Fig. 188. Epiphyses about the shoulder joint.

of the surgical neck. The upper epiphysis of the humerus is an amalgamation of three smaller epiphyses: a pressure epiphysis for the head which appears during the 1st year, and two traction epiphyses, one for the greater tubercle (3rd year), the other for the lesser tubercle (5th year). All three fuse together before the 7th year, and the resulting single mass fuses with the shaft completely by the 18th year in some cases and by the 24th year in all cases (McKern and Stewart).

The epiphyseal line of the **coracoid process** crosses the upper part of the glenoid cavity. Union occurs during the 15th year. The epiphyseal line of the acromion crosses the clavicular facet. If the epiphysis has not fused by the 23rd year, it will remain separate (os acromiale).

Table 8 (on p. 171) summarizes the times of appearance and fusion of all epiphyses in the upper limb. It is intended for reference, not memorization.

Muscles Acting on Shoulder Joint

All muscles passing from the clavicle and scapula to the humerus must act upon the shoulder joint; and those passing from the trunk to the humerus must act on the shoulder joint and also on the joints of the clavicle or shoulder girdle (table 4).

Note. The sternal fibers of the Pectoralis Major bring the humerus from the raised to the dependent position; the Teres Major is a powerful muscle—the sectional area of its fleshy belly is almost as large as that of the Biceps. The long head of the Triceps can extend the shoulder only when the arm is abducted. The Supraspinatus can barely raise the arm when the Deltoid is paralyzed.

Reference to table 4 makes it evident that the 5th and 6th cervical nerve segments control flexion, abduction, and lateral rotation, and that the 5th, 6th, 7th, 8th, and Th. 1 control extension, adduction, and medial rotation. Therefore, if the 5th and 6th segments are paralyzed, as when these roots of the brachial plexus are excessively stretched by forceful downward pulls, the arm will come to occupy a typical posture—the extended, adducted, and medially rotated position—described as the "position of a waiter taking a tip" (*fig. 188.1*). C. 5–6 root damage can result from falling on the shoulder off a moving motorcycle, or being born shoulder first.

TABLE 4

Muscles Acting on the Shoulder Joint with Their Approximate Spinal Nerve Segments

Flexion		Extension		Abduction	
Deltoid (clav.)	5, 6	Deltoid (postr.)	5, 6	Deltoid (mid.)	5, 6
Supraspinatus	5, 6	Pect. Major (st.)	7, 8, 1	Supraspinatus	5, 6
Pect. Major (clav.)	5, 6	Teres Major	5, 6		
Biceps	5, 6	Lat. Dorsi	7, 8		
Coracobrachialis	7	Triceps (long.)	7, 8		
5, 6 (7)		5, 6, 7, 8, 1		5, 6	
Adduction		**Med. Rotate**		**Lat. Rotate**	
Deltoid (postr.)	5, 6	Deltoid (clav.)	5, 6	Deltoid (postr.)	5, 6
Pect. Major (cl.)	5, 6	Pect. Major (cl.)	5, 6	Infraspinatus	5, 6
Pect. Major (st.)	7, 8, 1	Pect. Major (st.)	7, 8, 1	Teres Minor	5, 6
Coracobrachialis	7	Subscapularis	5, 6		
Teres Major	5, 6	Teres Major	5, 6		
Lat. Dorsi	7, 8	Lat. Dorsi	7, 8		
Triceps (long.)	7, 8				
5, 6, 7, 8, 1		5, 6, 7, 8, 1		5, 6	

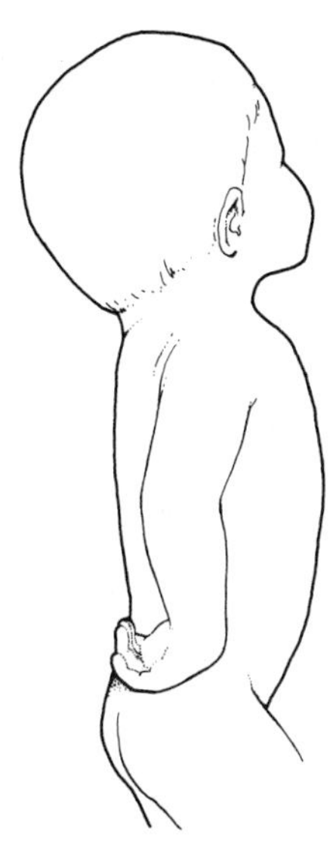

Fig. 188.1 The "tip-taking" position of the upper limb of a baby with a damaged upper trunk of the brachial plexus.

Nerve Supply of the Shoulder Joint

From C. 5 and 6 via the suprascapular, axillary, and lateral pectoral nerves, and posterior cord; perhaps also pain fibers from the adventitia of the axillary artery (E. Gardner).

Elevation of the Upper Limb

The vertical position can be attained either through forward flexion or through abduction. In this movement the sternoclavicular, acromioclavicular, and shoulder joints take part simultaneously. During the elevation through 180 degrees, the humerus and scapula move in the ratio of 2:1 from almost the beginning of the movement to the termination. Thus, in general, for every 15 degrees of elevation 10 occur at the shoulder joint and 5 are due to movement of the scapula (Inman *et al.*).

As elevation progresses beyond shoulder level (90 degrees), the clavicle rotates backward on its own long axis and its lateral end rises. Without this clavicular movement, elevation above shoulder level is greatly restricted. If the scapula is held fixed, the humerus can move only through a right angle and the power of movement is greatly diminished. Therefore, in elevation of the limb the scapula must be free to rotate and it does so with the permission of the clavicular joints (Inman *et al.*).

During the early phases of elevation, the sternoclavicular joint passes through its greatest range of movement, and in the terminal phase the acromioclavicular joint does so.

Muscle Force Couples. (*1*) *Movements of the Humerus.* The Supraspinatus and the Deltoid act together and progressively in elevating the arm. The Supraspinatus alone is unable to initiate abduction.

Electromyography reveals that both muscles act together throughout the entire range of the movement. These elevator muscles are assisted by the three short depressors—Subscapularis, Infraspinatus, and Teres Minor. The two groups act as a force couple, the one elevating and the other depressing (*cf.* prime mover and synergists).

(2) *Movements of the Scapula.* The upper parts of the Trapezius and Serratus Anterior (and the Levator Scapulae) constitute the upper component of the force couple necessary to the rotation "up" of the scapula, whereas the lower parts of the Trapezius and Serratus Anterior constitute the lower component.

ELBOW JOINT

(*Articulatio Cubiti*)

The elbow joint is a hinge or *ginglymus* joint, continuous with the proximal radio-ulnar joint. A hinge demands that: (1) the capsule will be loose in front and loose behind in order to permit flexion and extension; and (2) collateral ligaments will be required to prevent medial and lateral movements.

Bones

The Humero-Ulnar Parts. *Lower end of Humerus.* This part has a spool-shaped pulley with sharp edges, the *trochlea.* The trochlea leads in front to a depression, the *coronoid fossa* and behind to a broad triangular hollow, the *olecranon fossa.*

Upper end of Ulna. Here a triangular bracket, the *coronoid process*, projects forward (*fig. 189*). The portion of the ulna continued upward beyond the level of the coronoid process is cubical, the *olecranon*. Together they embrace a concavity, the *trochlear notch* (semilunar notch). This notch is reciprocally saddle-shaped for the trochlea of the humerus with which it articulates (*fig. 192*).

Posteriorly, the olecranon presents a triangular, subcutaneous surface, overlaid by the subcutaneous **olecranon bursa** (*fig. 190*).

The lateral surface of the coronoid process is a concave facet, called the *radial notch* of the ulna (*fig. 189*).

The Brachialis tendon is inserted into the anterior surface of the coronoid process, where it produces a rough area, the *tuberosity of the ulna*.

The Humero-Radial Parts. The upper concave surface of the *disc-shaped* **head of the radius** rotates on the lower and anterior aspects of the rounded articular **capitulum** or little head. Above and laterally, the capitulum merges with the lateral epicondyle. The rim of the upper surface of the disc-like head plays upon the lateral lip of the spool-shaped **trochlea** (*fig. 191*).

The Proximal Radio-Ulnar Joint

The head of the radius is held in position by the **anular ligament**, which is attached to the ends of the radial notch of the ulna. The notch forms one-fourth of a circle; the ligament three-fourths.

The ligament is not strictly speaking ring-shaped, but rather it is cup-shaped, being of smaller circumference below than above (*fig. 192*). In consequence, the head of the radius cannot be withdrawn from the cup.

"*Pulled Elbow.*" Before the age of seven, the grip of the anular ligament on the head may not be as firm as in later life; so sudden traction on a child's hand or forearm, as when pulling it out of the way of a passing motor car, may result in partial dislocation of the radius downward.

Ligaments and Capsule

The *Radial Collateral Ligament* (Lateral Lig.) of the elbow joint is fan-shaped. It

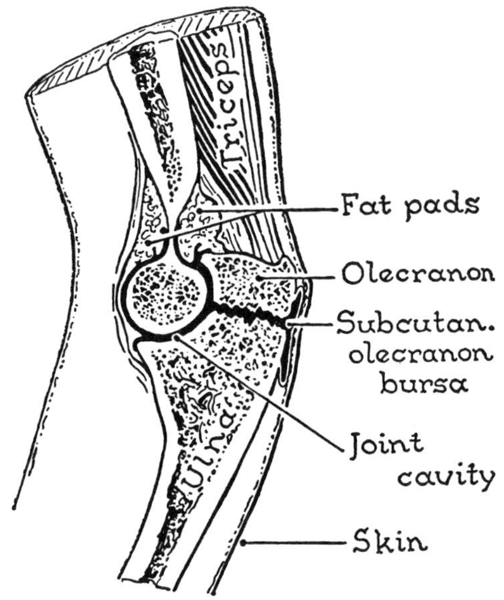

Fig. 190. The joint cavity and the olecranon bursa united by a fracture (uncommon).

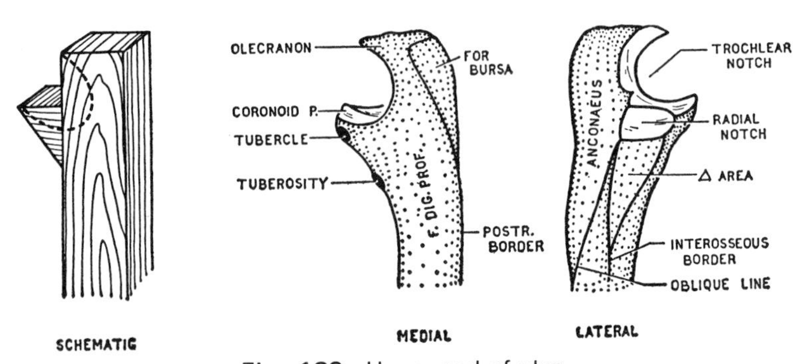

Fig. 189. Upper end of ulna.

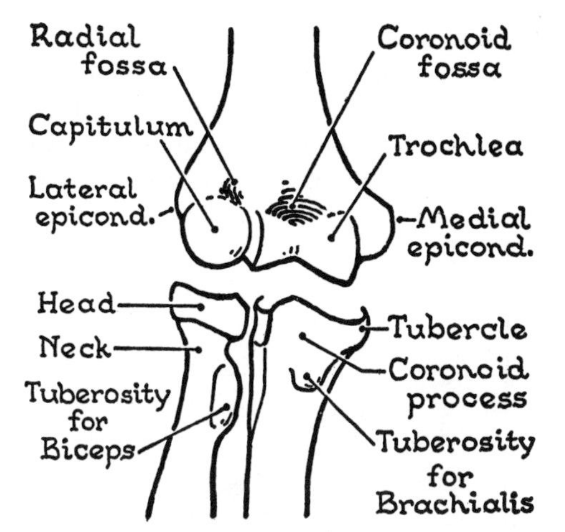

Fig. 191. The bony parts concerned in the elbow joint (front view).

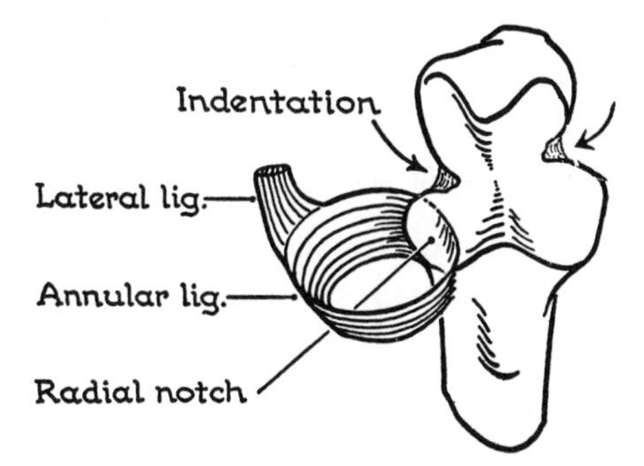

Fig. 192. Socket for head of radius (from above). The indentations are occupied by fat pads.

extends from the lateral epicondyle to the side of the anular ligament; so, indirectly it helps to retain the head of the radius in position. From it the Supinator and the Ex. Carpi Radialis Brevis in part arise.

The *Ulnar Collateral Ligament* (Medial Lig.) of the elbow joint extends in a fan-shaped manner from the lower part of the medial epicondyle to the medial margin of the trochlear notch (*fig. 193*).

The anterior fibers form a thick cord that is attached to a tubercle on the medial side of the coronoid. From this cord, the Fl. Digitorum Superficialis in part arises.

The Fibrous Capsule extends to the upper margins of the coronoid and radial fossae in front, but not quite to the top of the olecranon fossa behind. Below, it is attached to the margins of the trochlear notch and anular ligament.

The Synovial Capsule does not reach so high in the radial, coronoid, and olecranon fossae as the fibrous capsule. The intervals between the two capsules in these fossae are filled with fat, fluid at body temperature, and known as fat pads ("Haversian glands").

Below, the synovial capsule bulges a short distance (½ cm) below the lower free margin of the anular ligament and surrounds the neck of the radius in a sac-like manner (*fig. 194*). This device obviously allows the radius to rotate without tearing the synovial membrane. Redundant *folds of synovial membrane* project into this joint—as they do into other joints—and assist the fat pads to fill the unoccupied spaces.

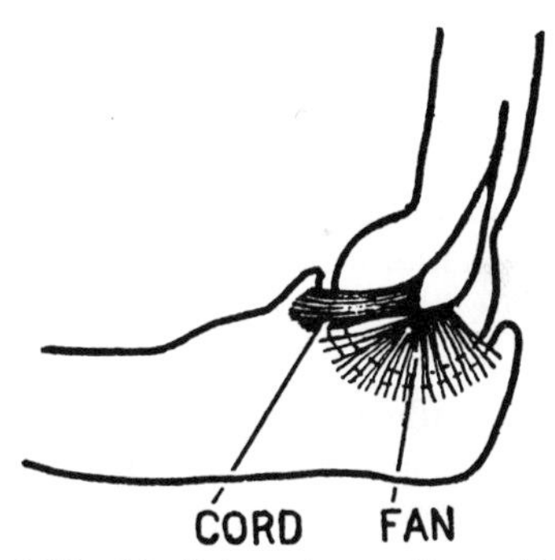

Fig. 193. Medial or ulnar collateral ligament of elbow joint.

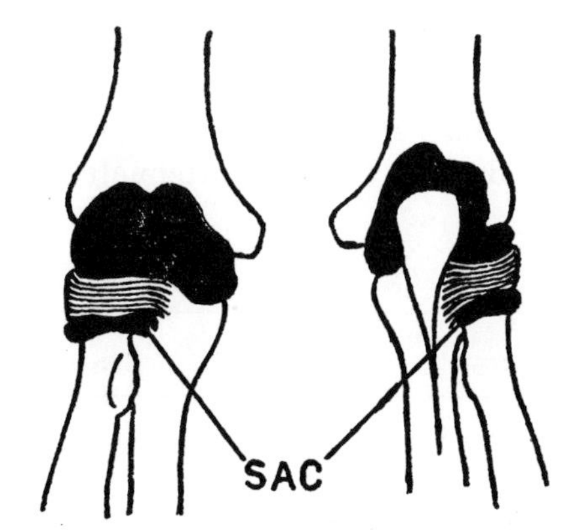

Fig. 194. Synovial capsule of elbow joint, distended.

"*Carrying Angle.*" The medial lip of the trochlea of the humerus descends 5 to 6 mm lower than the lateral lip; for the lower end of the humerus is oblique. On account of this obliquity you will observe that when your elbow is extended the humerus and ulna do not lie in line with one another but meet at an angle, especially obvious in women.

The medial lip of the trochlea also projects forward about 5 mm in advance of the lateral lip; therefore, though the humero-ulnar joint may be called a hinge joint, the ulna does not move on the trochlea like a door on a hinge, but rather it revolves on a cone. As a result the distal end of the ulna remains lateral to the axial line of the humerus in the flexed as well as in the extended position.

Muscles

Flexion is more powerful than extension in the ratio of 14:9.

The chief flexor of the elbow joint is the *Brachialis*. The Biceps, as a flexor of the elbow, acts best when the forearm is supinated, less well when semipronated, and not when pronated. As a supinator of the forearm, it does not act when the elbow is extended, unless against resistance. (Basmajian and Latif.) (See table 5.)

The Brachioradialis creeps up the humerus (*fig. 195*), and after the Brachialis and Biceps are paralyzed, the Brachioradialis is a useful flexor.

The common use of the *Triceps* is not so much to extend the elbow as to prevent flexion of the elbow, or to regulate flexion, as in pushing a wheelbarrow. It is also an aggressive muscle, as in boxing.

Relations

The Brachialis (*fig. 196*), though thick in the middle, is thin and attenuated at its edges. In consequence, the musculocutaneous nerve, which lies anterior to the middle of the muscle, is far removed from the joint; but the median nerve is separated from the capsule of the joint merely by the

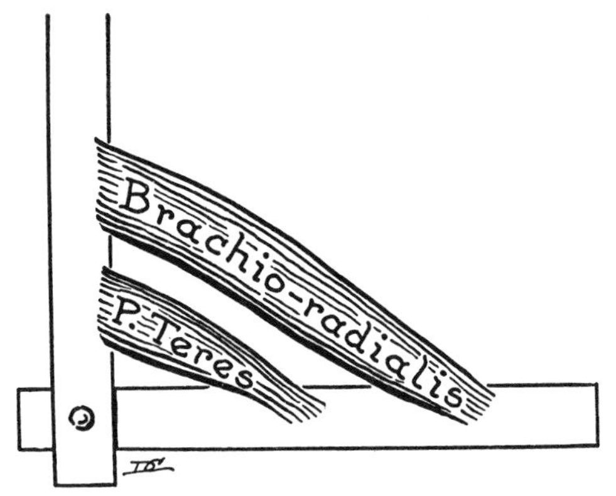

Fig. 195. As a flexor of the elbow joint, the Brachioradialis is more advantageously situated than the Pronator Teres.

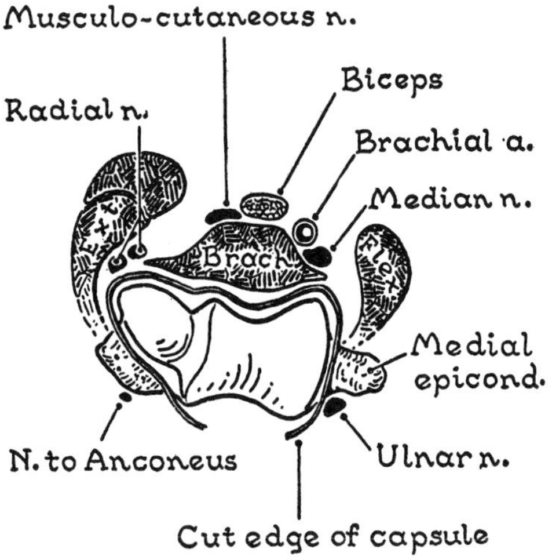

Musculo-cutaneous n.
Biceps
Radial n.
Brachial a.
Median n.
Brach.
Medial epicond.
N. to Anconeus
Ulnar n.
Cut edge of capsule

Fig. 196. Cross-section through elbow joint showing important relations.

TABLE 5

Muscles Acting on the Elbow Joint

Subdivisions of the Joint	Flexors	Nerve Segments	Nerve	Extensors	Nerve Segments	Nerve
Humero-ulnar	Brachialis	5, 6	Musculocutan.	Triceps Anconeus	7, 8	Radial
Humero-radial	Biceps Brachioradialis Pronator Teres	5, 6 (5), 6 6	Musculocutan. Radial Median			

thinness of the medial edge of the Brachialis; the radial nerve (or its two terminal branches) is in direct contact with the capsule (*fig. 113*). The ulnar nerve is always in immediate contact with the ulnar collateral ligament and it is covered by Fl. Carpi Ulnaris. The nerve to the Anconeus crosses the capsule behind the lateral epicondyle.

Anastomoses around the joint. These are shown in *figure 128*.

Epiphyses

Humerus: The epiphyseal line separating the trochlea, capitulum, and lateral epicondyle from the diaphysis runs transversely just above the articular cartilage (*fig. 197*).

This distal epiphysis is the first of all long bone epiphyses to fuse, synostosis being complete by the 17th or 18th year. The medial epicondyle fuses with a spur of bone that descends from the diaphysis and separates it from the lower epiphysis proper.

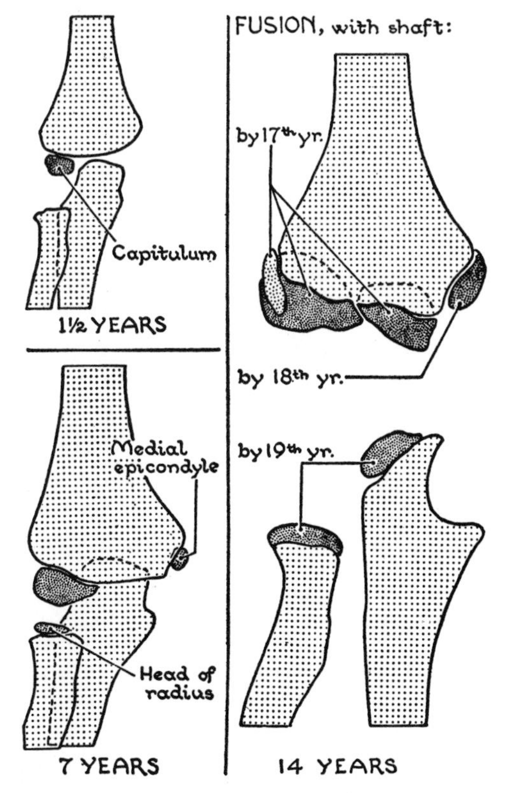

Fig. 197. Epiphyses about the elbow joint. (Courtesy of Dr. J. D. Munn, The Hospital for Sick Children, Toronto.)

Ulna: The upper epiphysis is a traction epiphysis of the Triceps. It may be a mere scale.

Radius: The upper epiphyseal line lies just below the head.

Nerve Supply

Each of the five nerves shown in *figure 196*, including the nerve to the Anconeus, sends a twig or twigs to the joints.

RADIO-ULNAR JOINTS

The two bones of the forearm are united at the proximal, "intermediate," and distal radio-ulnar joints.

Proximal Radio-Ulnar Joint

Considered with the elbow joint (p. 162).

Intermediate Radio-Ulnar Joint

The bodies of the radius and ulna are united to each other by the *interosseous membrane*.

The Interosseous Membrane is best marked at the middle two-fourths of both bones. It is responsible for the sharpness of their interosseous borders. It is about equally taut (or lax) in all positions—pronation, semipronation, and supination.

The Oblique Cord. This unimportant thickening in the fascia overlying the Supinator (Martin) extends from the tuberosity of the ulna inferolaterally to the radius, below its tuberosity.

Distal Radio-Ulnar Joint

This joint to some degree duplicates the design of the proximal joint. These similarities are illustrated in *figure 198*.

The Head of the Ulna, that is the articular part of the lower end (*fig. 115*), has a semicircular margin around which the ulnar notch of the radius rotates, and a semilunar or crescentic distal surface on which the articular disc plays. Within the concavity of the semilune, at the root of the

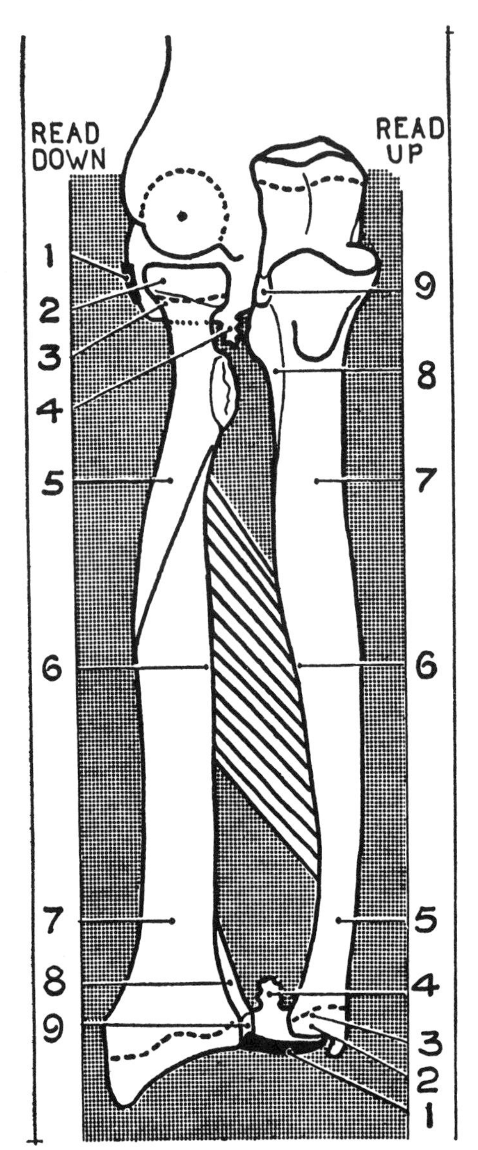

Fig. 198. Points of similarity between the proximal and distal radio-ulnar joints.

1. Anular ligament———Articular disc.
2. Circular head———Semicircular head.
3. Epiphyseal line enters the joint cavity.
4. Sacciform recess of synovial membrane.
5. The bone here circular on cross-section.
6. The interosseous border is here sharp.
7. The bone here triangular on cross-section.
8. Triangular fossa between border and notch.
9. Ulnar notch of radius———radial notch of ulna.

styloid process, there is a *fovea or pit*. The apex of the disc is attached to the pit. The base is attached to the lower margin of the ulnar notch of the radius.

The **disc** is subjected to pressure and friction; therefore, it is fibrocartilaginous, but its two margins and its apex, not being so subjected, are ligamentous and pliable, and they are strong. The disc closes the joint cavity and separates it from the radiocarpal joint.

Muscles. See table 6.

Observation. When your elbow is flexed to a right angle and applied closely to the side of your body, so as to eliminate shoulder movements, your hand can be rotated through nearly two right angles. When your shoulder is abducted and your elbow extended, as in fencing, your hand can be rotated through nearly three right angles. The movements of pronation and supination of the radio-ulnar joints under these circumstances augment the axial rotation of the shoulder joint.

Epiphyses (*fig. 198*). The lower or distal **radial epiphysis** appears before that of the **ulna** (*fig. 140*). Radiographically, these two epiphyses unite with the diaphyses about the 16th or 17th year in the female and the 18th or 19th in the male (see Greulich and Pyle); it may be delayed until the 23rd year (McKern and Stewart).

WRIST JOINT OR RADIOCARPAL JOINT

Observations. You can easily observe on your own limb that the movements permitted at the radiocarpal joint are: flexion, extension, adduction, and abduction. These together comprise the movement of circumduction. This, then, is a biaxial or ellipsoid articulation (*fig. 199*).

It is at the midcarpal joint that the greater part of the flexion often attributed to the wrist joint actually takes place. Because the radial styloid process descends farther than the ulnar styloid process, abduction is more restricted than adduction.

Joint Surfaces. The socket is formed by the lower articular surface of the radius

TABLE 6

Muscles Acting upon the Radio-Ulnar Joints and Their Nerves and Nerve Segments

Pronators	Segment	Nerve	Supinators	Segment	Nerve
Pronator Quadratus	(7), 8, 1	Anterior Interosseous	Supinator	5, 6	Posterior Interosseous
Pronator Teres	6	Median	Biceps	5, 6	Musculocutaneous
Fl. Carpi Radialis	6	Median	2* dorsal muscles of the thumb	(6), 7, (8)	Posterior Interosseous

* The Ex. Poll. Brevis arises from the radius and therefore cannot assist in rotation.

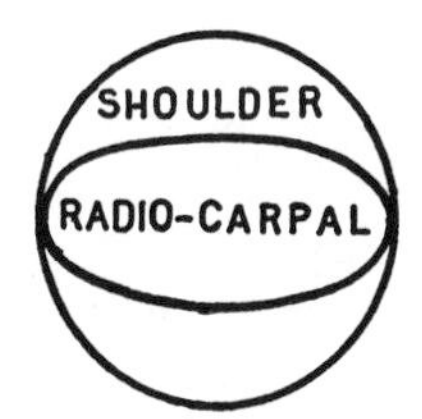

Fig. 199. Circle versus ellipse.

plus the articular disc. The egg-shaped or ellipsoidal convex surface is formed by the proximal articular surfaces of the scaphoid, lunate, and triquetrum, plus the interosseous ligaments binding these three bones together (*fig. 200*).

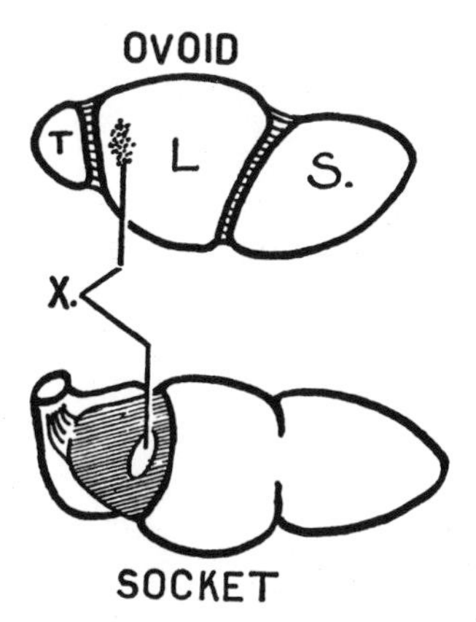

Fig. 200. "Ovoid" and socket of wrist joint. "X" marks the sites at which the disc is sometimes perforated and the lunate softened.

Details Revealed by X-rays. In *Abduction* (radial deviation, *fig. 201*), the scaphoid rotates forwards, the lunate moves on to the articular disc, and the space between the triquetrum and the ulna is increased.

In *Adduction* (ulnar deviation, *fig. 202*) the scaphoid, lunate, and triquetrum move laterally, the lunate coming entirely on to the radius and the triquetrum making contact with the articular disc (*fig. 203*).

As to the *Distal Row of Carpals*, you may observe that in adduction the apex of the hamate comes into articulation with the lunate; the long axis of the capitate changes with reference to lunate and scaphoid; and the trapezoid and trapezium move medially on the scaphoid.

Attrition. Commonly the articular disc and also both interosseous carpal ligs., that take part in the "ovoid and socket" of the wrist joint, shown in *figure 200*, are perforated. In consequence, the distal radio-ulnar joint is brought into communication with the wrist joint, and the wrist joint with the intercarpal joints.

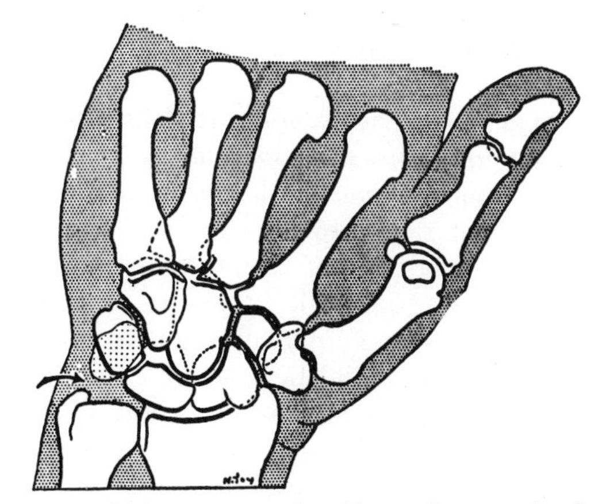

Fig. 201. Tracing of an X-ray photograph of the wrist in abduction. (*Arrow* indicates a space, proximal to the triquetrum, present in abduction.)

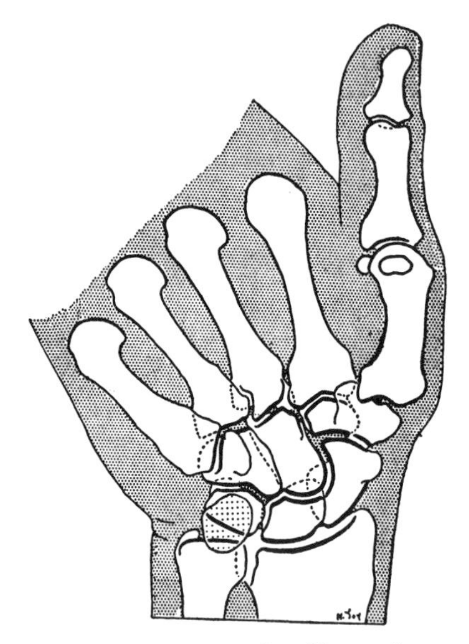

Fig. 202. Tracing of an X-ray photograph of the wrist in adduction.

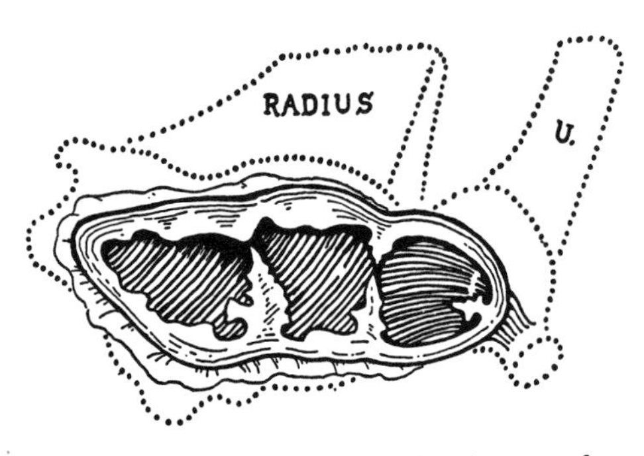

Fig. 203. Socket for proximal row of carpals. Note the transparent synovial fringes.

Ligaments. The *palmar and dorsal radiocarpal ligaments* pass obliquely inferomedially from the front and back of the lower end of the radius to the front and back of the proximal row of carpal bones and to the capitate (*fig. 204*).

The radial and ulnar collateral ligaments (lateral and medial ligs.) extend from the styloid processes of the radius and ulna to the scaphoid and triquetrum, respectively.

Muscles. The muscles acting on the radiocarpal and midcarpal joints and their nerve segments are given in table 7.

Actions. The flexors and extensors of these joints are indicated in *figure 205*. The three important flexors are more advantageously situated than the three extensors. Flexion is more powerful than extension in the ratio of 13:5 (Fick).

Relations of the Joint are shown thus: *Anterior—figs. 123* and *126* which is a key for tendons, nerves, and arteries. *Posterior—figs. 163* and *164* for tendons; *Fig. 157* for arteries; and *fig. 98* for cutaneous nerves.

Nerve Supply: Anterior interosseous branch of the median, posterior interosseous branch of the radial, and deep palmar and dorsal cutaneous branches of the ulnar.

INTERCARPAL, MIDCARPAL, CARPOMETACARPAL, AND INTERMETACARPAL JOINTS

These are described with the carpal bones on pages 125–130. It should be noted that the 4th and 5th metacarpal bones have slight hinge movements; the 5th especially can be flexed, and it is the forward or flexor movement of the 5th metacarpal that largely prevents a cylindrical object, such as the handle of a wheelbarrow or a garden rake, from slipping through the closed hand (*fig. 206*); similarly, it helps when climbing or pulling a rope.

Each head of a metacarpal is separated from its neighbor not only by adjacent collateral ligaments and Interosseous tendons but also by areolar tissue which commonly contains a **bursa.**

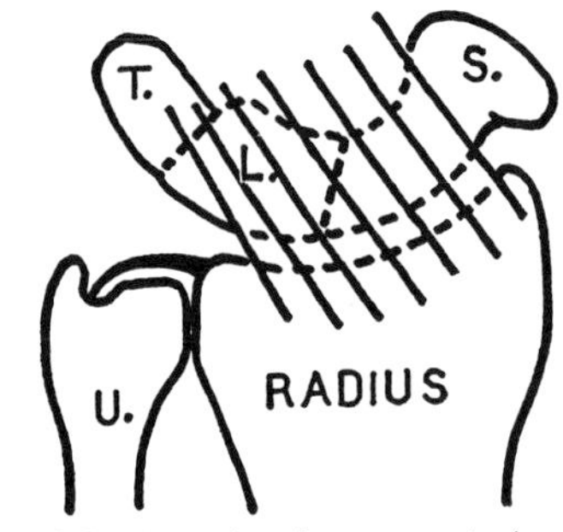

Fig. 204. Anterior (or posterior) radiocarpal ligament.

TABLE 7
Muscles Acting on the Radiocarpal Joint and Their Nerve Segments

Flexion		Extension		Abduction		Adduction	
Fl. C. Ulnaris	8, 1	Ex. C. Ulnaris	7	Ex. C. Radialis Longus	6, 7	Ex. C. Ulnaris	7
Fl. C. Radialis	6	Ex. C. Radialis Longus	6, 7	Ex. C. Radialis Brevis	6, 7, 8	Fl. C. Ulnaris	8, 1
Palm. Longus	7, 8, 1	Ex. C. Radialis Brevis	7	Fl. C. Radialis	6		
Abd. Poll. Long.	7			Abd. Poll. Long.	7		

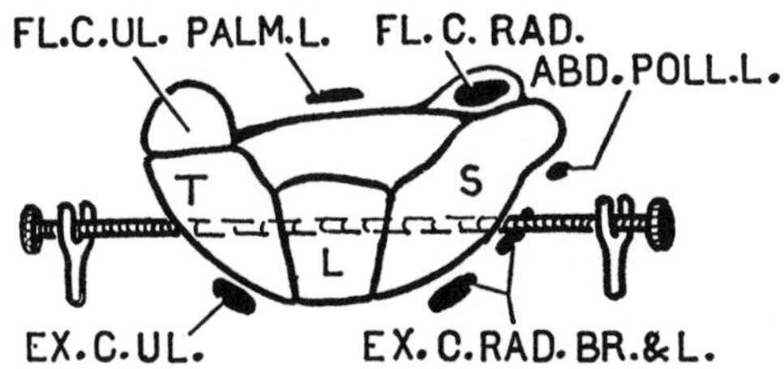

Fig. 205. The flexors and extensors of the wrist, on transverse section.

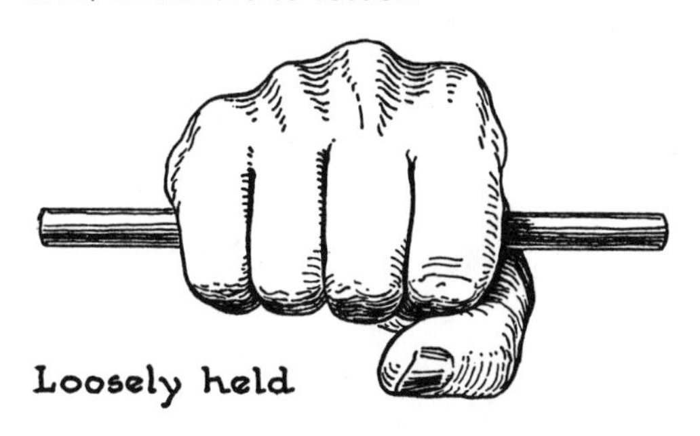

Fig. 206. Because the 4th and 5th carpometacarpal joints are hinge joints, the grip on a rod is more secure.

METACARPOPHALANGEAL AND INTERPHALANGEAL JOINTS

Described with the metacarpal bones and phalanges on pages 128–130.

Movements of Fingers. The flexed fingers are adducted so that, when flexed and adducted, they steady each other (*fig. 207*).

The Ex. Digitorum, Ex. Indicis, and Ex. Digiti V extend the metacarpophalangeal joints. The interphalangeal joints are extended by the Lumbricals, Interossei, and Abd. Digiti V through their attachments to the dorsal expansions.

Fine Interplay of Muscles and Ligaments. The following *details*—of great importance in specialized hand surgery—were revealed by F. Braithwaite *et al.*, J. M. F. Landsmeer, and R. W. Haines.

The Interossei and Abd. Digiti V abduct and adduct the fingers at the metacarpophalangeal joints; this they can do while the palm lies flat on a table. Also, they can move each finger separately.

The Lumbricals, Interossei, and Abd. Digiti V flex the metacarpophalangeal joints of the fingers and they can, either separately or at the same time, extend the interphalangeal joints, as in making the upstroke in writing (*fig. 208*).

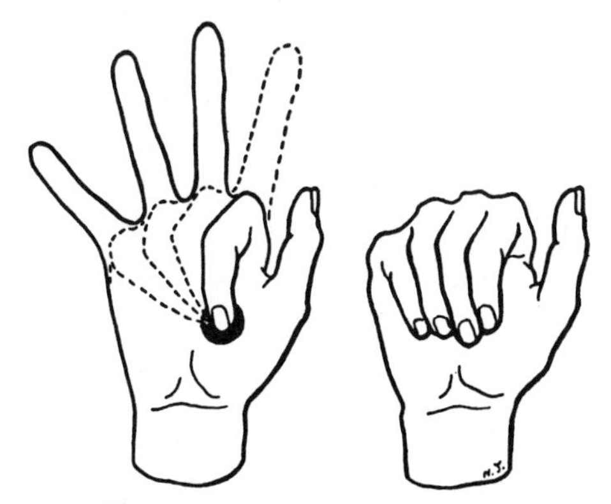
Fig. 207. To demonstrate that flexed fingers buttress each other.

Synergists. The four flexors of the wrist acting as synergists, steady the wrist when the extensors of the fingers are in action; similarly, the three extensors of the wrist act synergically when the flexors of the fingers are in action (*fig. 209*). Indeed, you cannot clench your fist or grasp an object firmly unless you extend your wrist. And, it follows that the hand is rendered relatively useless if the extensors of the wrist are paralyzed (e.g., in lesions of the radial nerve).

Shortness of the Long Digital Tendons. You cannot fully flex your wrist while the hand is closed, but only when the fingers are extended; this is owing to the relative shortness of the extensor tendons. Similarly, it is not easy fully to to extend your wrist while the fingers are extended; this is owing to the relative shortness of the flexor tendons.

Ligamentous Functions. If, beginning at the base of the 3rd or distal phalanx, you trace the free border (medial or lateral) of an extensor expansion proximally, you will be conducted by a narrow band of fibers to the corresponding border of the 1st phalanx and to its fibrous digital sheath (*fig. 210*). This component of the expansion is a ligamentous band, called the **retinacular lig.** (Landsmeer) or the *link lig.* (Haines). It spans the 2nd phalanx, crossing dorsal to the axis of the distal interphalangeal joint and palmar to the axis of the proximal

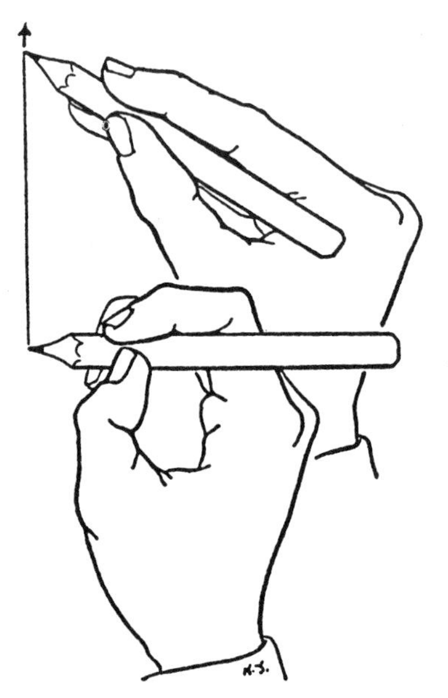

Fig. 208. The Interossei and Lumbricals making the upstroke in writing.

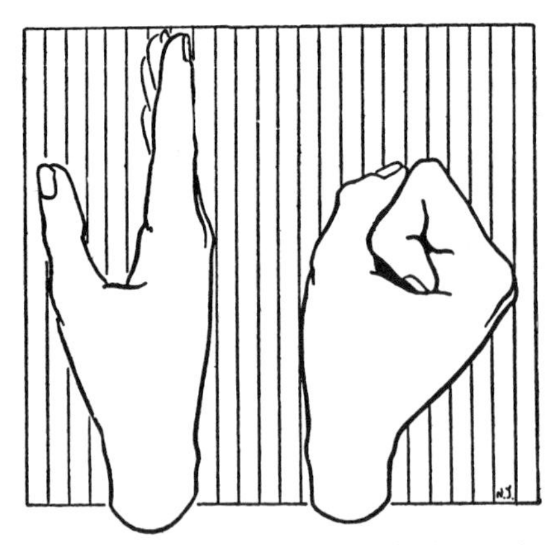

Fig. 209. The extensors of the wrist act synergically with the flexors of the fingers.

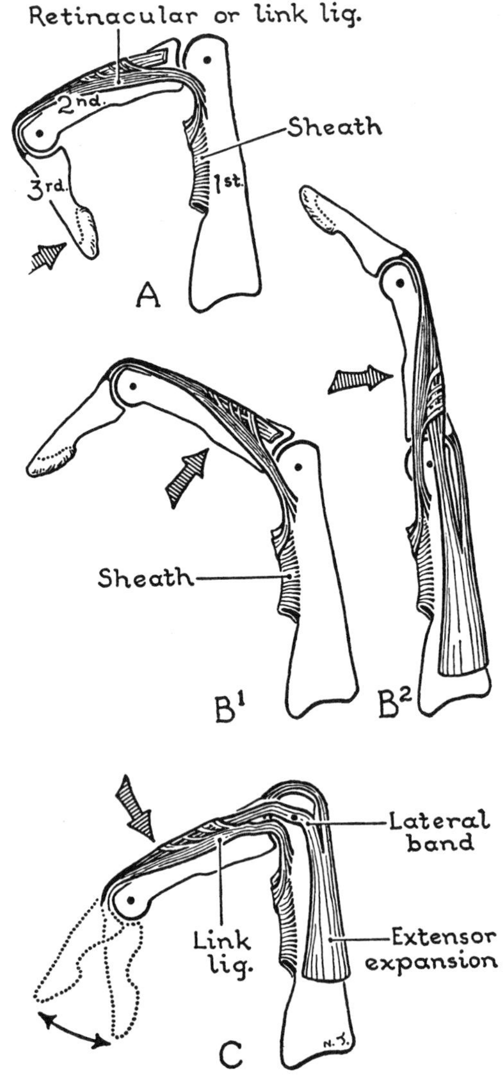

Fig. 210. To demonstrate the functions of the retinacular or link ligaments.

joint.

Observations. 1. On flexing the distal joint (either passively in the cadaver or voluntarily in your own hand, through the action of the Fl. Digitorum Profundus) (*fig. 210A*) the retinacular lig. becomes taut and pulls the proximal joint into flexion.

2. Similarly, on extending the proximal joint (either in the cadaver or voluntarily), the distal joint is pulled by the retinacular lig. into nearly complete extension (*fig. 210B*).

3. While the proximal joint is fully flexed passively (e.g., by pressing on the 2nd phalanx) the 3rd phalanx cannot be extended voluntarily (actively), though it offers no resistance to passive extension. Indeed, it is flail. This is due to the fact that the lateral bands of the extensor expansion slip forward and hence are slack and ineffective (*fig. 210C*).

You will note, then, that the metacarpophalangeal joint of a finger can move independently of the interphalangeal joints whereas the interphalangeal joints are compelled by the retinacular or link ligs. to move together.

TABLE 8

Table of Appearance and Fusion Times of Epiphyses in Upper Limb Including Primary Centers for Carpus, in Years Unless Otherwise Stated

	Appears	Fuses
Scapula, Acromion	15	19
Coracoid	1 }	15*
Subcoracoid	10 }	
Inf. Angle	15	22
Vertebral Border	15	22
Clavicle Sternal end	18+	22+
Humerus, Head	Birth }	
Greater Tubercle	3 }	6 } 19*
Lesser Tubercle	5 }	
Medial Epicondyle	6	18
Lateral Epicondyle	12 }	
Capitulum	1 }	15 } 17
Trochlea	10 }	
Radius Upper end	7	19
Lower end	1	19*
Ulna Upper end	11	19
Lower end	6	19*
Scaphoid	6	
Lunate	5	
Triquetrum	4	
Hamate	½	
Capitate	½	
Trapezoid	6	
Trapezium	6	
Pisiform	12	
1st Metacarpal base	−3	19
2nd–5th Metacarpal heads	−3	19
Thumb, Proximal Phalanx base	2	18
Distal Phalanx base	−3	18
2nd–5th Fingers, Proximal Phalanx base	2	18
Middle Phalanx base	−3	18
Distal Phalanx base	−3	−18
Sesamoids of thumb	−13	

Since the student will be concerned with the age periods at which he may reasonably be sure that fusion is complete, the latest times of fusion are given. Female bones fuse distinctly earlier.

* Asterisks denote times that are found to be quite constant. The table is a compilation from several authorities and is based on X-ray findings. It has been deemed advisable to give definite ages rather than a spread. Figures denote years unless otherwise stated.

+ = a later tendency. − = an earlier tendency.

SECTION THREE

ABDOMEN

13 Anterior Abdominal Wall and Scrotum 175

14 Abdominopelvic Cavity 190

15 Stomach, Liver, and Related Structures 212

16 Mesenteric Vessels, Duodenum, and Pancreas 223

17 Three Paired Glands 231

18 Posterior Abdominal Structures 238

13 ANTERIOR ABDOMINAL WALL AND SCROTUM

ANTERIOR ABDOMINAL WALL

Boundaries; Superficial Fascia.

Muscles and Relations

Rectus Abdominis and its Sheath—Surface Anatomy; Linea Alba; Actions; Posterior Wall of Sheath; Anterior Wall; Contents of Sheath; Nerves. Epigastric Arteries.
Other Vessels of Wall.
Three Flat Muscles
External Oblique—origin, insertion, inguinal ligament, lacunar ligament, superficial inguinal ring.
Internal Oblique.
Transversus Abdominis.
Falx Inguinalis or Conjoint Tendon.
Cremaster Muscle.
Functions of Three Flat Muscles; Features above the Level of the Anterior Superior Spine and below.
Layers of Abdominal Wall below Ant. Sup. Spine and Layers of Scrotum.
Spermatic Cord; Inguinal Canal.
Inferior Epigastric and Deep Circumflex Iliac Arteries.

TESTIS, SPERMATIC CORD, AND SCROTUM

Descent of Testis; Gubernaculum Testis; Processus Vaginalis Peritonei.
Scrotum; Dartos Muscle; Three Coverings of Testis and Spermatic Cord; Spermatic Cord (cont'd); Constituents of Cord; Ductus Deferens.
Testis—Epididymis; Structure of Testis and Epididymis; Vessels and Nerves; Development; Rudimentary Structures.

ANTERIOR ABDOMINAL WALL

To examine a patient's abdomen, one inspects and palpates the anterior abdominal wall, and when the contents must be explored, cuts through it. Its essential component is muscular to allow for expansion and compression of the abdominal cavity and its contained hollow viscera (= organs).

The different parts of the gastro-intestinal canal dilate, contract, and move about. This is permitted by what is virtually an enormous bursa, called the *peritoneal cavity*, into which they project freely. The interior of the wall is lined with peritoneum.

Boundaries

The anterior abdominal wall is bounded above by the xiphoid process and the costal cartilages 7—10; 11 and 12 costal arches are not long enough to reach the front (*fig. 211*). It is bounded below on each side by the iliac crest, anterior superior spine, inguinal ligament, pubic tubercle, pubic crest, and pubic symphysis.

The **xiphoid process** is tucked between the 7th costal cartilages and is not easily palpated. Vigorous attempts to do so cause discomfort. Hence, the easily palpated lower end of the body of the sternum, at the xiphisternal joint, is preferred as a landmark.

Superficial Fascia

In the lower part of the abdominal wall, fat or adipose tissue is layed down in varying—sometimes horrifying—layers. Throughout the fat, membranous sheets of collagenous fibers are found. When the anatomist locates one that is denser than the others, he reverts to an old, still widespread, but erroneous concept (disproved by Tobin and Benjamin)—calling it "Scarpa's fascia." The several layers of fat he

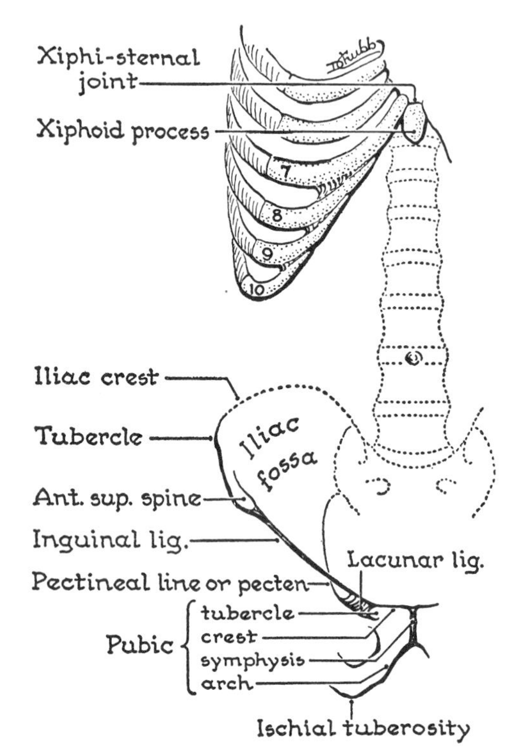

Fig. 211. Boundaries of the anterior abdominal wall.

collectively calls "Camper's fascia or layer."

A finger passed downward between the membranous layer and the underlying aponeurosis of the External Oblique will be arrested by the attachment of the membranous layers to the fascia lata (i.e., deep fascia of the thigh) a finger's breadth below the inguinal ligament (*fig. 212*); medial to the pubic tubercle it can be passed downward deep to "Colles fascia," in the scrotum.

Muscles and Relations

On each side, they are:

Rectus Abdominis (and Pyramidalis)

and

Three flat muscles:
- Obliquus Externus Abdominis
- Obliquus Internus Abdominis
- Transversus Abdominis

***AXIOM.** The areolar tissue found between muscles affords them freedom of movement, and its quantity increases with an increase of independent contractions.*

The Rectus Abdominis and Its Sheath. The Rectus Abdominis (*fig. 213*) is a long strap-like muscle. It lies within a sheath (*fig. 214*) formed by the splitting of the aponeuroses of the three flat muscles.

Three or four transverse *tendinous intersections* in the Rectus, indicative of its segmental origin, adhere to the anterior wall of its sheath.

Attachments. The Rectus is attached horizontally above to the front of the xiphoid process and neighboring costal cartilages. Its lower attachment is by tendon to the pubic body.

Surface Anatomy. The *lateral border* of the Rectus curves from the pubic tubercle, across the midpoint of the line joining the anterior superior spine to the umbilicus, across the chest margin about 8 cm from the midline—where (by chance) it overlies the gall bladder on the right side.

Its *medial border* is separated from that of its fellow by the **linea alba;** that is, the line extending between the xiphoid process and the symphysis pubis in which the aponeurotic fibers of the three flat muscles decussate.

Small gaps may be present in the linea alba through which herniae of extraperitoneal fat protrude and sometimes cause symptoms.

Actions. Recti Abdominis approximate chest to pubis and flex the thoracic and lumbar vertebrae. By tensing them, we break the force of a blow and so protect the abdominal viscera.

Posterior Wall of Rectus Sheath. At four different levels four different tissues lie in contact with the posterior surface of the Rectus (*fig. 214*). From below upward they are: (*a*) areolar, (*b*) aponeurotic, (*c*) muscular, and (*d*) cartilaginous. This much every student should know; the details are important to specialists.

Anterior Wall of Rectus Sheath. This wall is aponeurotic throughout (*fig. 214, C*). Again, the details should not concern the student.

Pyramidalis. This thin triangular muscle (commonly absent) lies in front of the lower part

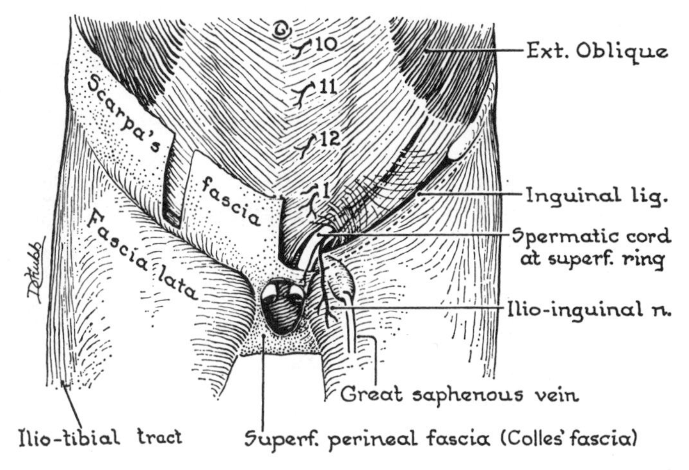

Fig. 212. The fasciae of Scarpa and Colles.
(The penis and scrotum are cut away and the spermatic cords are cut across.)

of the Rectus Abdominis. It is unimportant.

The *Three Flat Muscles* will be discussed after the Nerves and Vessels. They are disposed in three layers, the Transversus being the deepest.

Nerves and Vessels

Nerves. *The Ventral Rami* of the *lower six thoracic nerves* run with their accompanying vessels between the Internal Oblique and the Transversus to the back of the sheath of the Rectus (*figs. 215, 216*). Thereupon, they pierce every structure, branches reaching the skin as anterior cutaneous nerves (*fig. 211*).

The ventral ramus of the *1st lumbar nerve* divides far back into two: the *iliohypogastric* and *ilio-inguinal*. They run forward between the Internal Oblique and the Transversus and near the ant. superior spine they pierce the Internal Oblique. They continue between the Internal and External Obliques, and finally reach the skin.

Distribution (*fig. 216*). Th.10 supplies a strip of the skin leading to the umbilicus, leaving three nerves (9, 8 and 7) to supply the region above the umbilicus, and three (11, 12, and 1) to supply the region below the umbilicus (*fig. 212*). This distribution should be memorized!

The ilio-inguinal nerve supplies the skin at the root of the penis, anterior part of the scrotum, and adjacent part of the thigh.

While these nerves (Th.7–L.1) are running between the Internal Oblique and the Transversus, they supply them, 12 and 1 being the most important because they control the vital lower part of the abdominal wall (*fig. 218*).

Those nerves that pierce the Rectus (Th.7 to 12, but not L.1) also supply it; and to sever any one of them results in paralysis of that segment of the Rectus.

The Lateral Cutaneous Branches of the lower six thoracic and first lumbar ventral rami arise and become cutaneous in front of the midlateral line half way around the body (*fig. 72*). As they pierce the External Oblique, they supply it.

Epigastric Arteries (*fig. 216*). *The Superior Epigastric Artery* (a continuation of the internal thoracic a.) enters the Rectus sheath behind the 7th costal cartilage and anastomoses with the much larger *Inferior Epigastric Artery* which enters from below. Thus, vessels to the upper and lower limbs are brought into communication with each other—for the internal thoracic (int. mammary) is a branch of the subclavian artery, and the inferior epigastric artery is one of the two collateral branches of the external iliac artery. The two epigastric arteries must supply everything in the Rectus Sheath and its coverings.

Branches. Their branches accordingly are *cutaneous, muscular,* and *anastomotic.*

The inferior epigastric artery has in addition two branches of some importance: (1) a *cremasteric branch*—discussed later (*fig. 227.1*), and (2) a *pubic branch* which anastomoses on the back of the pubic bone (see p. 286).

The lateral border of the Rectus is a nearly bloodless line, because very few branches anastomose across it (see below).

Other Vessels in the Anterior Wall: *Intercostal and Lumbar Arteries* accompany nerves T.11, 12, and L.1 and their lateral branches.

The Deep Circumflex Iliac Artery ramifies between the Internal Oblique and the Transversus (p. 184), as do twigs of the musculophrenic artery (a branch of the internal thoracic).

The Three Superficial Inguinal Arteries are cutaneous. They spring from the femoral artery and run in the superficial fascia, thus: (1) the *superficial epigastric a.* runs toward the navel; (2) the *external pudendal a.* crosses in front of the spermatic cord to supply the scrotal wall; and (3) the *superficial circumflex iliac a.* runs below the lateral half of the inguinal ligament.

The Three Superficial Inguinal Veins end in the great saphenous vein of the thigh.

The superficial epigastric and lateral thoracic veins anastomose, thereby uniting the veins of the upper and lower halves of the body. Everyone is familiar with this conspicuous anastomosis in the flank of the horse.

The Superficial Lymph Vessels of the anterior abdominal wall above the level of the navel pass to the axillary nodes, those below to the inguinal nodes.

The Three Flat Muscles of the Abdomen

The three flat muscles of the abdomen —Obliquus Externus, Obliquus Internus, and Transversus—are the prototype of "three ply" wood, for the general directions of their fibers are, respectively, downward and medially, upward and medially, and transversely. Though the majority of the fibers of the three muscles take these *three general directions*, you will find that below the level of the anterior superior iliac spine all the fibers of all three muscles take *one particular direction*—downward, forward, and medially. These may be the most important fibers of the muscles; and, their attachments are complicated by the fact that they are traversed by the inguinal canal (*fig. 217*).

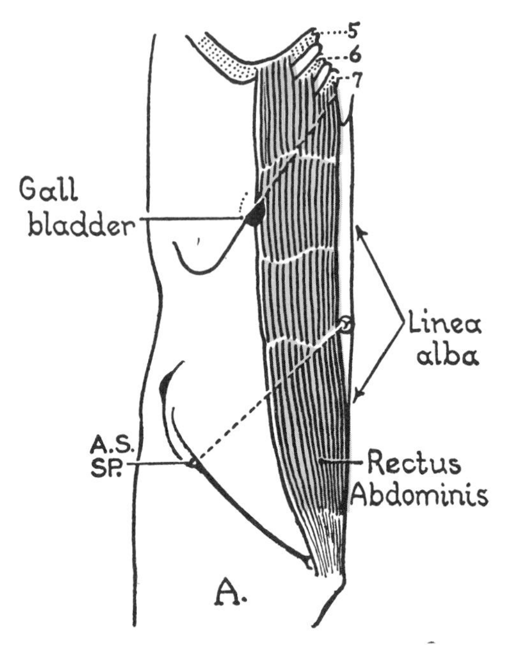

Fig. 213. *A*, The Rectus Abdominis.

Obliquus Externus Abdominis. The External Oblique *arises from* the lower eight ribs a hand's breadth from the costal margin by means of fleshy digitations. It is fleshy above and laterally; elsewhere it is aponeurotic (*fig. 217*). Its uppermost fibers are nearly horizontal. Its most posterior fibers are fleshy and nearly vertical.

Insertion. The fibers of the External Oblique spread out fanwise:

A. The *fibers of the upper half* of the muscle help to form the front of the Rectus sheath, decussating in the whole of the linea alba.

B. The *fibers of the lower half* of the muscle find attachment: (1) to the anterior ½ of the iliac crest by fleshy fibers. (2) From the anterior superior spine to the pubic tubercle, the muscle has a free, lower, aponeurotic border, known as the **inguinal ligament** (Poupart's lig.). Doyle has shown that the lateral part is an insertion of the aponeurosis into a toughened line of iliopsoas fascia. (3) Some fibers of the inguinal ligament pass horizontally backward to attach to the pecten pubis (pectineal line)

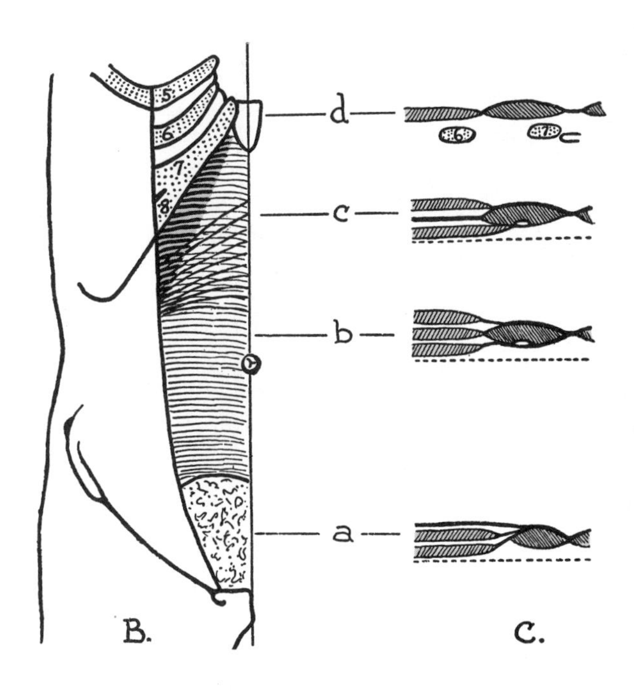

Fig. 214. *B*, The posterior wall of the Rectus sheath. *C*, Transverse sections of the sheath at four levels.

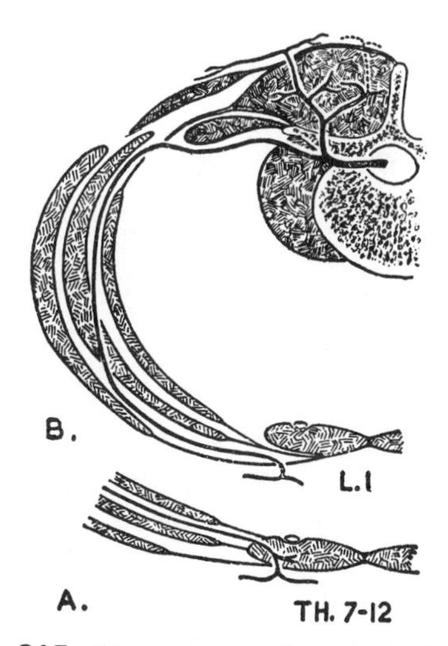

Fig. 215. The course of a ventral nerve ramus in the abdominal wall. *A*, lower thoracic. *B*, 1st lumbar.

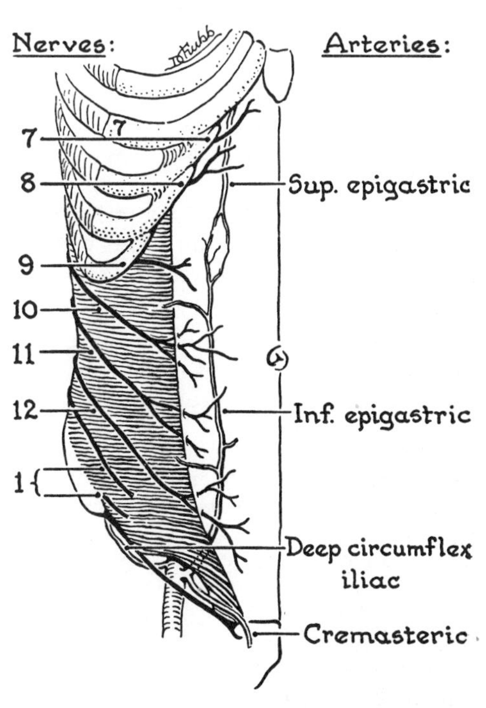

Fig. 216. The nerves and arteries within the rectus sheath.

as the **lacunar ligament** (*fig. 211*) in infants, but falls short of it in adults (Doyle). (4) Between tubercle and symphysis the aponeurosis is thin, areolar and

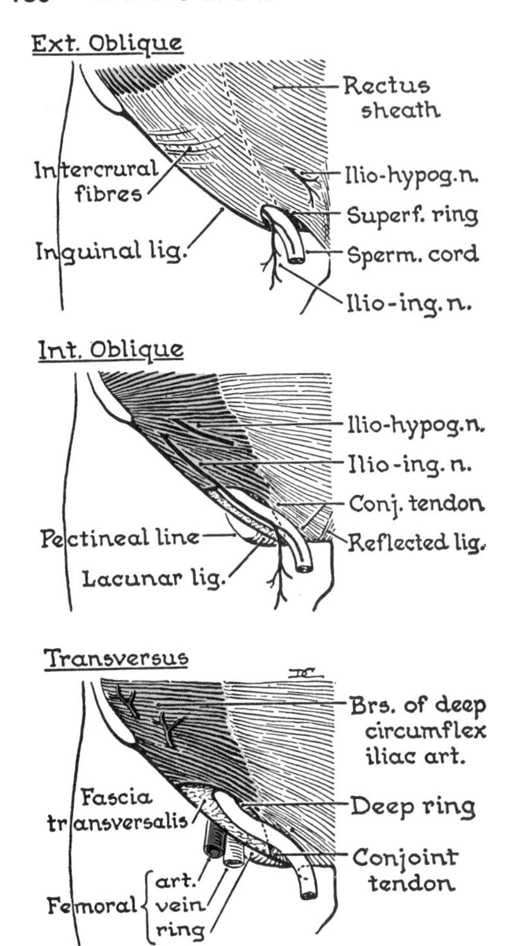

Fig. 217. Three flat muscles below level of anterior superior spine. Walls of inguinal canal.

ballooned out to cover the testis as the *external spermatic fascia.* The mouth of the balloon is the *superficial inguinal ring.*

Superficial Inguinal Ring. This oblique, triangular opening, 2 to 3 cm long, is the "mouth" of the inguinal canal (p. 183). Its center lies above the pubic tubercle.

Its apex is superolateral to the tubercle; its base is formed by the lateral half of the pubic crest. The two sides of the ring, known as the *upper and lower crura* (L. crus = a leg) are diverging portions of the External Oblique aponeurosis. Scattered curved fibers, the *intercrural fibers,* lying lateral to the ring, blend with the aponeurosis and prevent the crura from spreading (*fig. 212*).

Obliquus Internus Abdominis. The direction of the muscle fibers is anterosuperior, at right angles to those of the External Oblique. Its origin from the iliac crest extends dorsally beyond the External Oblique.

Origin. The Internal Oblique, then, arises from more than the anterior ½ of the iliac crest, from more than the lateral ½ of the inguinal ligament (and commonly from the posterior aponeurosis of the Transversus).

Insertion. (1) Its most posterior fibers ascend to the cartilages of the lower four ribs (12, 11, 10, and 9). The next fibers become aponeurotic and continuing obliquely superomedially or horizontally, they split at the lateral border of the Rectus, contribute to the formation of the anterior and posterior walls of its sheath, and so reach the linea alba; (2) those from the inguinal ligament curve downward, some to contribute to the falx inguinalis or conjoint tendon (described below) but most to form the Cremaster muscle, which loops down into the scrotum to form a covering of the testis.

Transversus Abdominis. Its *pelvic origin* is from the inner lip of the iliac crest and from the lateral third of the inguinal ligament. The *costal origin* is by fleshy slips that spring from the inner surface of the lower six costal cartilages where they interdigitate with fleshy slips of the Diaphragm. Between its iliac and costal origins, it arises from the *lumbar vertebrae* through the medium of an aponeurosis (*fig. 323,* and p. 246).

Its fibers pass forward to decussate in the linea alba. Those from the lateral third of the inguinal ligament form the falx inguinalis.

Falx Inguinalis or Conjoint Tendon. The inguinal parts of the Internal Oblique and Transversus join to form a common aponeurosis, the lowest part of which becomes a delicate membranous sickle (= L., falx), which is attached to the medial centimeter or more of the pecten pubis (*fig. 217.1;* also p. 273). *The Nerve Supply* to the fleshy fibers of this region is from L. 1, via the

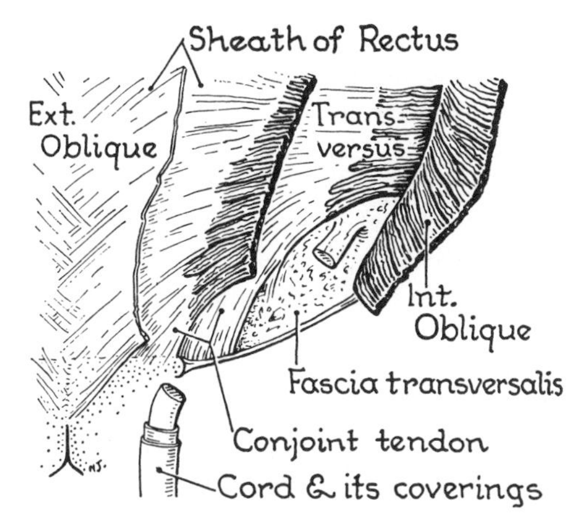

Fig. 217.1 Dissection of conjoint tendon.

ilio-inguinal and genitofemoral nerves (*fig. 218*).

On pulling on the fleshy fibers of the Internal Oblique and Transversus, it becomes evident that both contribute to the aponeurosis, and that the Transversus alone becomes the sickle-shaped band. The band usually blends with the underlying fascia transversalis (Anson and McVay; and Chandler).

Cremaster Muscle. This is the covering that the Internal Oblique and Transversus together give to the testis and spermatic cord. It is a "balloon," like the external spermatic fascia (p. 180). It is, therefore, attached to the medial half of the inguinal ligament and to the pubis. It fills the concavity formed by the arching fibers of the two flat muscles, and in the scrotum its individual fleshy fibers form elongated loops connected to each other by areolar tissue.

Its Nerve is the genital branch of the genitofemoral (L. 1, 2).

Its Action is to retract or draw up the testis, and stroking the skin of the thigh where it is supplied by the ilio-inguinal nerve (L. 1) reflexly brings about this retraction.

Functions of the Three Flat Muscles. They are the chief lateral flexors of the trunk. The External Oblique fibers of one side and the Internal Oblique fibers of the other, being parallel, act together in rotation of the trunk.

All three flat muscles help to maintain the intra-abdominal pressure. This pressure is much more important than ligaments and mesenteries in retaining the viscera in place.

The three flat muscles (1) by contracting alternately with the diaphragm as antagonist, help to bring about the expiratory act during forced breathing, and they are necessary to the abdominal form of respiration; (2) by contracting simultaneously with the diaphragm, they aid in expelling the contents of the abdominal organs (in micturition, defecation, vomiting, and parturition) and in driving on venous blood to the heart. They contract vigorously during coughing.

Layers of Abdominal Wall. A surgeon requires to know what depth he has reached in entering the peritoneal cavity.

AXIOM. Above the level of the ant. sup. spine, the direction of muscle fibers, the presence of areolar tissue, the presence of nerves and of branches of the deep circumflex iliac artery all indicate the depth or layer at which you have arrived.

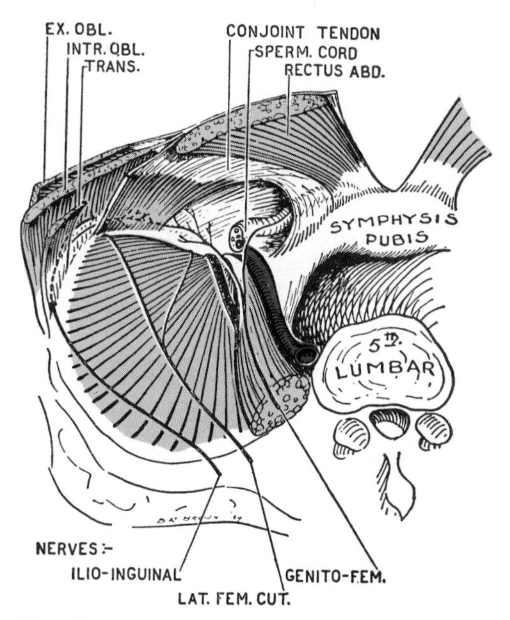

Fig. 218. Branches of ilio-inguinal and genitofemoral nerves to muscle fibers controlling conjoint tendon. (Dissection by Dr. R. G. MacKenzie.).

The lower six thoracic nerves cling to the deep surface of the Internal Oblique. The ascending and terminal branches of the *deep circumflex iliac artery* pierces the Transversus near the ant. sup. spine to ascend between Internal Oblique and Transversus (*fig. 217*).

Below the Level of the Anterior Superior Spine the abdominal wall consists of eight layers, each of which has a tubular evagination; of these, the outer two form the scrotum, whereas the inner six are contained within the scrotum (*fig. 219*). In table 9 the names of the layers of the abdominal wall and the corresponding layers found in the scrotum are set out in parallel columns.

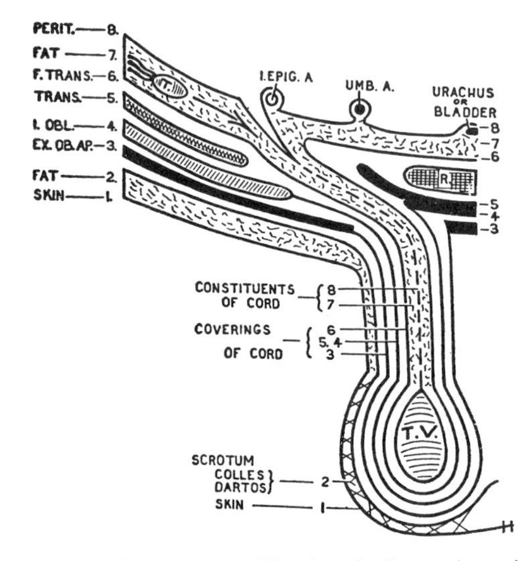

Fig. 219. Scheme: The inguinal canal, and the layers of the anterior abdominal wall prolonged into the scrotum. (See table 9.) (In this cross section the scrotum is raised to the horizontal position.)

The testis, like other abdominal organs (e.g., kidney and ureter; bladder and urachus; the great vessels—but not the spinal nerves), develops in layer 7, that is to say, between the peritoneum and the fascia lining the abdominal muscles (e.g., Transversus, Ilio-Psoas, Diaphragm) (*fig. 220*). A month or so before birth the testis makes a descent through the abdominal wall. In descending it encounters the outer six layers of the wall and carries before it a tubular prolongation from each. Layers 1 and 2 form the *scrotum;* they do not envelop the testis and cord so snugly as layers 3–6, which are known as the *coverings of the cord.*

The testis drags after it its duct, arteries, veins, lymph vessels, and nerves, a quantity of the extraperitoneal tissue from layer 7 in which it develops, and a tube of peritoneum, called the *processus vaginalis,* belonging to layer 8. These constitute the **spermatic cord.** The passage through the abdominal wall is called the *Inguinal Canal.*

The foregoing is a simplified description. In reality, the testis is not the active agent in producing the tubular prolongations within which it is later contained. The scrotum and the coverings of the cord are formed early, even before the abdominal muscles differentiate, and, as Curl and Tromly properly point out, there is a definite inguinal canal in the male

TABLE 9

The Layers of the Abdominal Wall	The Corresponding Layers in the Scrotum	
1. Skin	1. Skin	Scrotum
2. Superficial fascia {(a) fatty (Camper) (b) membranous (Scarpa)	2. Dartos muscle and fascia	
3. External Oblique (aponeurotic)	3. External spermatic fascia	Coverings of the cord
4. Internal Oblique } (fleshy)	4. } Cremaster muscle	
5. Transversus } (fleshy)	5. } Cremaster muscle	
6. Fascia Transversalis (lining abdominal cavity in this region)	6. Internal spermatic fascia	
7. Extraperitoneal fatty tissue (layer inhabited by organs)	7. Areolar tissue with localized collections of fat	Two of the constituents of the cord
8. Peritoneum	8. Processus vaginalis	

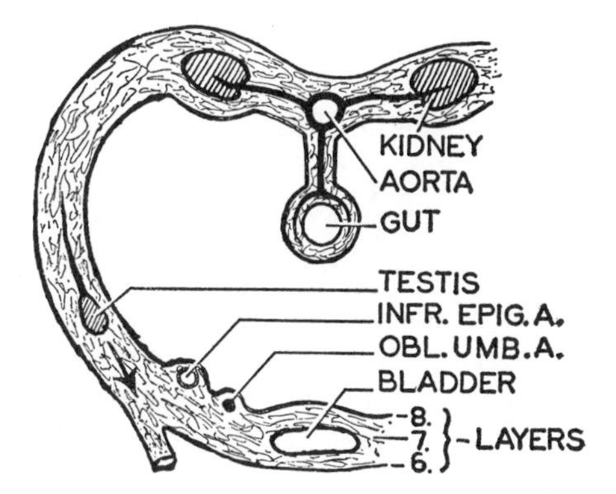

Fig. 220. The testis making its descent in layer number 7—i.e., the layer in which the organs and great vessels reside.

fetus before the testis passes through the abdominal wall, and in the female it is just as definite. Originally, the inguinal canal is an almost straight anteroposterior passage through the abdominal muscles; subsequently, during the process of growth it becomes more oblique.

Inguinal Canal

The first of the six layers of the abdominal wall that the testis encounters during its descent has to be the fascia lining the Transversus, the *fascia transversalis.* The tubular covering it provides is the *internal spermatic fascia.* The mouth of this tube, the *deep inguinal ring,* is the entrance to the inguinal canal. It lies a finger's breadth above the midpoint of the inguinal ligament and immediately lateral to the stem of the inf. epigastric artery (*fig. 217*).

The tube derived from the Exp. Oblique aponeurosis is the *external spermatic fascia.* Its mouth, the *superficial* (*subcutaneous*) *inguinal ring,* is the exit from the inguinal canal.

Between the deep and the superficial inguinal ring the Transversus and Internal Oblique together contribute a tubular muscular covering, the *Cremaster*, which, as you would expect, lies in a layer of *cremasteric fascia.*

The Inguinal Canal extends from the deep inguinal ring to the superficial inguinal ring and is about 4 cm long. (1) The *anterior wall* is formed throughout by the External Oblique aponeurosis, as *figures 217* and *219* show. (2) The *posterior wall* is formed throughout by the fascia transversalis, which is quite thick here, reinforced medially by the falx inguinalis and the lateral edge of the Rectus sheath.

(3) Between "1" and "2," are the Internal Oblique and the Transversus. Their fleshy fibers lie in front of the most lateral part of the canal and arch over it, and as the conjoint tendon or falx, they lie behind its most medial part, i.e., they lie in front of the deep ring and behind the superficial ring.

More precisely, the Internal Oblique almost always arises from more than the lateral half of the inguinal lig., and therefore covers the deep ring and takes part in the anterior wall of the canal. But the origin of the Transversus is very variable; rarely does it extend to the deep ring to take part in the anterior wall; indeed in one in four cases it fails to arise from the inguinal ligament (Anson and McVay).

Running behind the posterior wall in the extraperitoneal fat are the inferior epigastric and the obliterated umbilical arteries. The inferior epigastric artery lies at the medial boundary of the deep ring and here the ductus deferens takes a recurrent course lateral to it. In thin persons the three innermost layers of the abdominal wall—namely, the fascia transversalis, the areolar framework devoid of fat, and the peritoneum—blend (*fig. 221*).

(4) *The floor of the canal* is formed by the grooved surface of the inguinal liga-

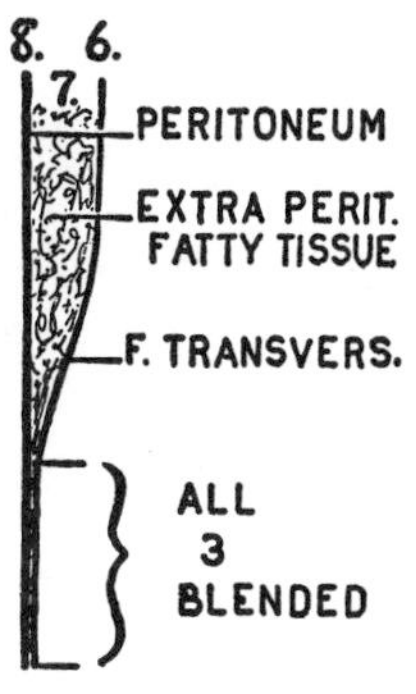

Fig. 221. In the absence of fat the inner three layers of the abdominal wall blend to form a single layer.

ment and by the lacunar ligament (p. 179).

What control has one over the canal: How is it closed? Your conception of the canal may be helped by likening it to an arcade of three arches formed by the Transverse and the two Oblique muscles (*fig. 222*). During standing there is continuous contraction of Internal Oblique and Transversus in the inguinal region. During coughing and straining, when the raised intra-abdominal pressure threatens to force a hernia through the canal, vigorous contraction of the arched fleshy fibers of the Internal Oblique and Transversus "clamp down," without damaging the cord. The action is that of a half-sphincter.

Lytle, and later Patey, concluded from dissections and stimulation studies that active pulling by Transversalis on the deep ring (lateralward) closes the canal.

Nerves. After piercing the Internal Oblique, the ilio-inguinal nerve, now purely sensory, runs less than a finger's breadth above the inguinal lig., and emerges through the superficial ring lateral to the spermatic cord, but it may be anterior, or medial, or posterior to it. The iliohypogastric nerve runs at a higher level (*fig. 217*). The genitofemoral nerve runs with the motor part of the Cremaster, which it supplies.

Inferior Epigastric and Deep Circumflex Iliac Arteries

These arise from the external iliac artery just before it passes behind the inguinal ligament to become the femoral artery. The one ends within the Rectus sheath; the other between the Transversus and the Internal Oblique.

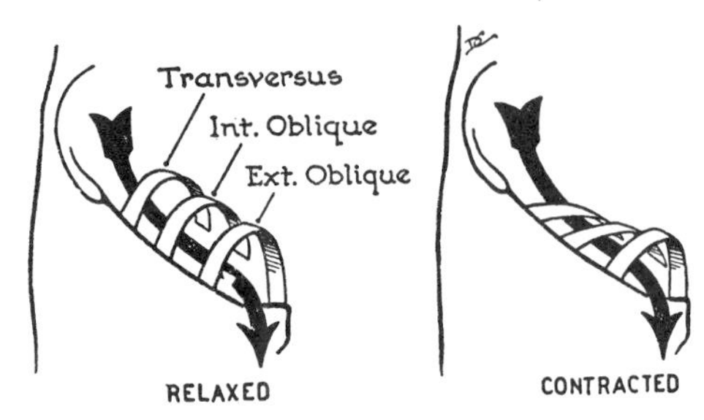

Fig. 222. "The inguinal arcade." The inguinal canal likened to an arcade of three arches traversed by the spermatic cord.

TESTIS, SPERMATIC CORD, AND SCROTUM

Descent of the Testis. The testis (like the ovary) develops from the cells covering the medial part a primitive kidney (mesonephros), which otherwise atrophies in man. In early fetal life a cord of spindle-shaped cells, called the **gubernaculum testis,** pushed before it into the scrotum the layers of the growing abdominal wall, and evaginated them as though they were so many sheets of rubber.

The upper end of the gubernaculum was attached (1) to the testis, (2) to the adjacent part of the peritoneum, and (3) to the mesonephric duct (later the epididymis and ductus deferens) (*figs. 219* and *223*). These it dragged after it, or at least it constrained them to follow in its wake. As a result, the peritoneum was drawn out into a blind tube, the **processus vaginalis peritonei,** and the testis, which was adherent to the outer surface of the tube, was drawn with it into the scrotum. The lower end of the epididymis was drawn down too.

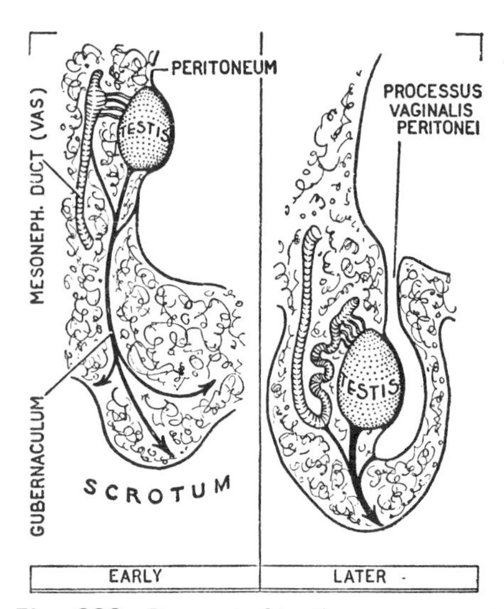

Fig. 223. Descent of testis, processus vaginalis peritonei, and epididymis (diagrammatic).

The testis descended to the iliac fossa during the 3rd prenatal month, traversed the inguinal canal during the 7th prenatal month, and reached the bottom of the scrotum after birth.

Variations. The testis sometimes fails to leave the abdomen, an *undescended testis*, or it may be out of place, an *ectopic testis*, and the scrotum empty. This may be the result of hormonal insufficiency (L. J. Wells). Hormonal, surgical, and conservative-waiting therapy may be appropriate in specific cases, but none can be ignored because of later complications (e.g., an intra-abdominal testis may produce no spermatozoa in adult life).

Comparative Anatomy. In the elephant the testis is retained in the abdomen; in certain rodents it descends periodically and then returns to the abdomen; in the pig it descends to the perineum; and in the marsupials it becomes prepenile.

Scrotum is the name of the bag of skin and subcutaneous tissues in which the testes lie. The scrotum has a bilateral origin. It is derived from the right and left labioscrotal folds. In the female these folds remain discrete as the labia majora; in the male they fuse behind the penis to form the scrotum (*fig. 224*).

The skin of the scrotum forms a single pouch; the subcutaneous tissues of the scrotum form a right and a left pouch with a common median partition. The subcutaneous tissue contains a sheet of involuntary muscle supplied by sympathetic nerves, the **Dartos muscle.** The fibers of the Dartos adhere to the skin and cause it to wrinkle when cold.

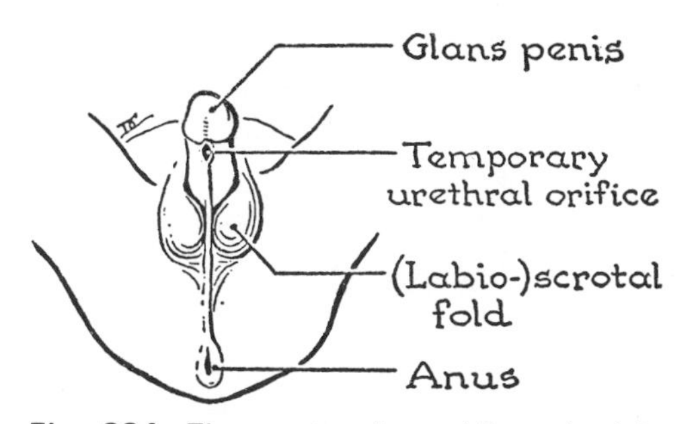

Fig. 224. The scrotum has a bilateral origin.

Nerves. The 1st lumbar segment, via the ilio-inguinal nerve, supplies the ventral part of the scrotum and the root of the penis; the (2nd), 3rd, and 4th sacral segments, via the posterior scrotal nerves and perineal branch of the posterior femoral cutaneous nerve, supply the perineal part of the scrotum (Perineum, p. 257).

Vessels. The scrotum is supplied by the ext. and int. pudendal arteries and veins. The lymph vessels pass to the superficial inguinal nodes (*fig. 399*).

Three Coverings of Testis and Spermatic Cord. The outer and inner coverings are the external and internal spermatic fasciae, and between them is the middle covering or Cremaster muscle. These tubular coverings tend to blend into a single areolar membrane in which the fibers of the Cremaster are spread out in loops.

Supply. The Cremaster is supplied by segments L. 1 and 2 (via genitofemoral n.) and by the cremasteric branch of the inf. epigastric artery.

Spermatic Cord (cont'd). The spermatic cord consists of the structures running to and from the testis, and representatives of the inner two layers of the abdominal wall, viz., peritoneum and extraperitoneal fatty-areolar tissue.

These constituents of the cord assemble at the deep inguinal ring lateral to the inferior epigastric artery, pass through the inguinal canal, and descend in the scrotum to the testis. As the cord emerges from the superficial inguinal ring, it rolls over the pubic tubercle and acquires its covering of external spermatic fascia.

The Constituents of the Cord are:

1. Representatives of the inner two layers of the abdominal wall:

   a. Processus vaginalis peritonei.

   b. Areolar tissue continuous with the extraperitoneal fatty areolar tissue.

2. Structures pertaining to the testis:

   a. Ductus deferens.

b. Vessels (artery, veins, and lymphatics) and nerves of the testis.

c. Vessels and nerves of the ductus deferens and epididymis.

The Processus Vaginalis Peritonei is the tube of peritoneum behind which the testis follows the gubernaculum into the scrotum. Its upper part lies in front of the ductus deferens, and is normally obliterated before birth (or within a month after birth) and becomes a fibrous thread, the *funicular process of peritoneum*. When it remains patent, it leads to congenital inguinal hernia. Its lower part remains patent, is invaginated from behind by the testis, and is known as the **tunica vaginalis testis.**

Areolar Tissue surrounds the structures passing to and from the testis. In it are commonly found circumscribed *areas of fat*, which may attain the size of a pigeon's egg.

The Ductus Deferens (Vas Deferens) (*fig. 225*) is the duct that conveys spermatozoa from the testis to the urethra. It is the continuation of the canal of the epididymis. In length it equals the femur or the spinal cord, i.e., 45 cm.

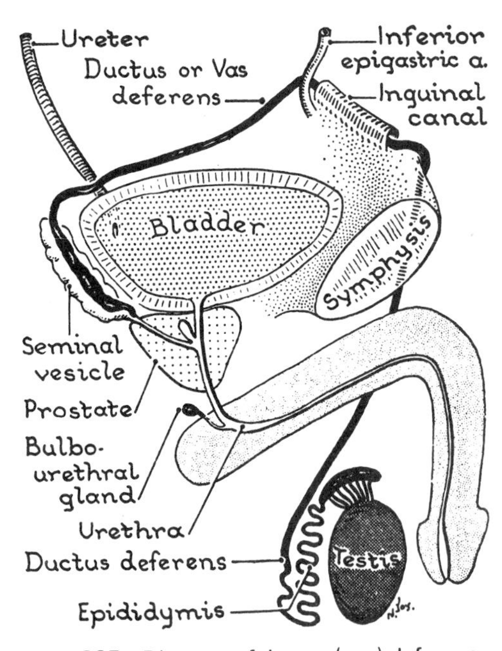

Fig. 225. Diagram of ductus (vas) deferens.

Except at its two extremities, which are dilated and thin walled, it has a thick muscular coat and a capillary lumen; hence, it feels firm like a whip cord. It first ascends behind the testis along the medial side of the epididymis; it continues through the scrotum and inguinal canal as the posterior constituent of the spermatic cord; it then hooks around the lateral side of the inferior epigastric artery and descends subperitoneally to the posterolateral angle of the bladder and thence to the urethra (*fig. 225*).

Testis (*fig. 226*). *The Testis* is an ovoid gland measuring 4 × 3 × 2 cm. Its normal size is quite variable (Farkas). It is enveloped in the *tunica vaginalis testis* except where the epididymis and the structures of the spermatic cord are attached to its upper pole and posterior border. It has a tough, fibrous, white, outer coat, the *tunica albuginea*, which is comparable to the sclerotic, white, outer coat of the eyeball. Posteriorly the outer coat is thicker and less dense and is known as the *mediastinum testis.*

Areolar *septa* extend from the mediastinum to the tunica albuginea and divide the testis into about 250 elongated pyramidal compartments. Each contains a lobule

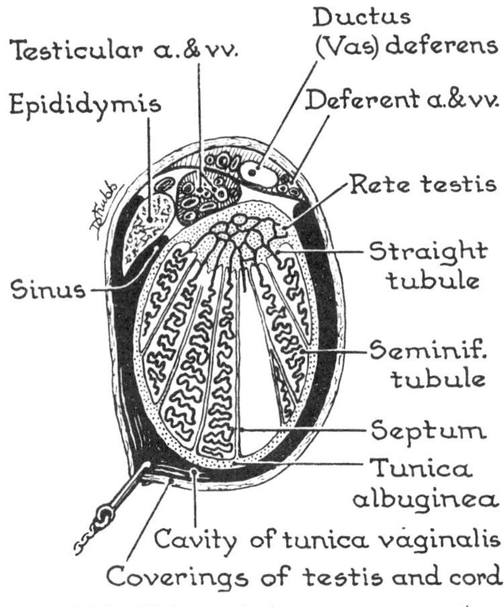

Fig. 226. Right testis in transverse section.

formed from two or more *seminiferous tubules* which are each a half meter long; they are closely packed and convoluted, except at the apex of the compartment where they join together and take a short, straight course, the *straight tubules.* In the mediastinum the straight tubules anastomose to form a network, the *rete testis.*

From six to a dozen fine ducts, the *efferent ductules*, connect the upper part of the rete testis with the head of the epididymis (*fig. 227*).

Epididymis. The epididymis is applied to the upper pole and posterior border of the testis. It is somewhat larger in diameter than a lead pencil. It tapers from above downward. It is subdivided into a *head* or upper part, a *body* or intermediate part, and a *tail* (which turns upward as the ductus deferens). The *sinus of the epididymis* is a small recess of the tunica vaginalis lateral to it (*fig. 227*). The ductus deferens is medial.

Structure. In the head, the efferent ductules become coiled to form lobules; they empty into the duct of the epididymis. The duct of the epididymis forms the body and tail of the organ. It is greatly twisted and folded, and when unraveled it is many meters long, rivalling the small gut in length.

Vessels and Nerves of Testis, Epididymis, and Ductus

Arteries: The *testicular artery*, a branch of the aorta, pierces the mediastinum testis in a variety of ways, branches and anastomoses with the *deferent artery* (which is a companion of the ductus from the pelvis) and the *cremasteric a.*, a twig of the inf. epigastric a. (*figs. 227.1, 227.2*).

Veins: Up to a dozen veins from the region form an anastomosing plexus, the *pampiniform plexus*, which ascends, reducing to two or three veins and ultimately to one. On the right side it joins the inf. vena cava and on the left side, the left renal

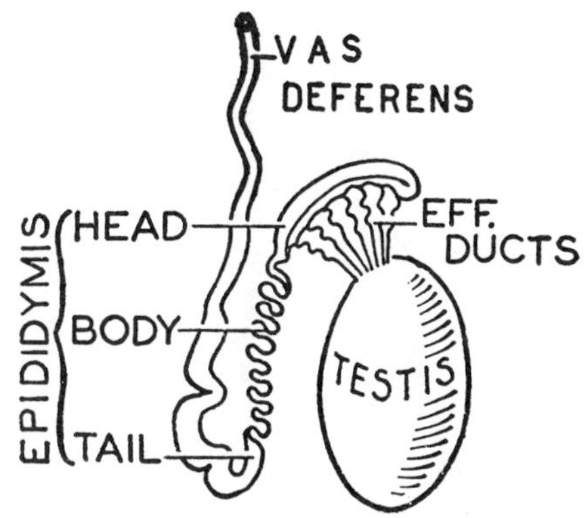

Fig. 227. The testis and epididymis.

vein. The testicular veins may become dilated and tortuous, a clinical condition called varicocele.

Lymphatics: The lymph vessels of the testis, like those of the ovary, end in nodes situated between the common iliac vein and the renal vein (*fig. 228*). In this they sharply differ from the lymph vessels of the scrotum and penis, which drain into the inguinal nodes, a matter of importance in cancer-spread.

Nerves: Of testis—Th. (6, 7, 8, 9, and) 10. Of epididymis—Th. 11, 12, and L. 1, via the inf. hypogastric plexus (Mitchell).

Development (*fig. 229*). *The duct system of the testis* is of three-fold origin: (1) The seminiferous and straight tubules and the rete testis are developed from anastomosing cords of cells in the genital ridge between vertebral segments L. 4——S. 2. (2). The efferent ductules and the lobules in the head of the epididymis are formed from the six or more mesonephric tubules that succeed in establishing connections between the rete testis and the duct of the epididymis. (3) The duct of the epididymis and the ductus (vas) deferens are derived from the mesonephric duct, the duct of the primitive kidney (mesonephros), while the modern mammalian model (metanephros) is forming by its side (*fig. 229*).

Rudimentary Structures about the testis and epididymis are five in number. Of these, two are brought into view when the tunica vaginalis is opened. They are little bodies attached to the upper pole of the testis and head of the epididymis. They probably represent the cranial ends of the paramesonephric (Mullerian) and mesonephric (Wolffian) ducts and are known as the *appendix of the testis* and the *appendix of the*

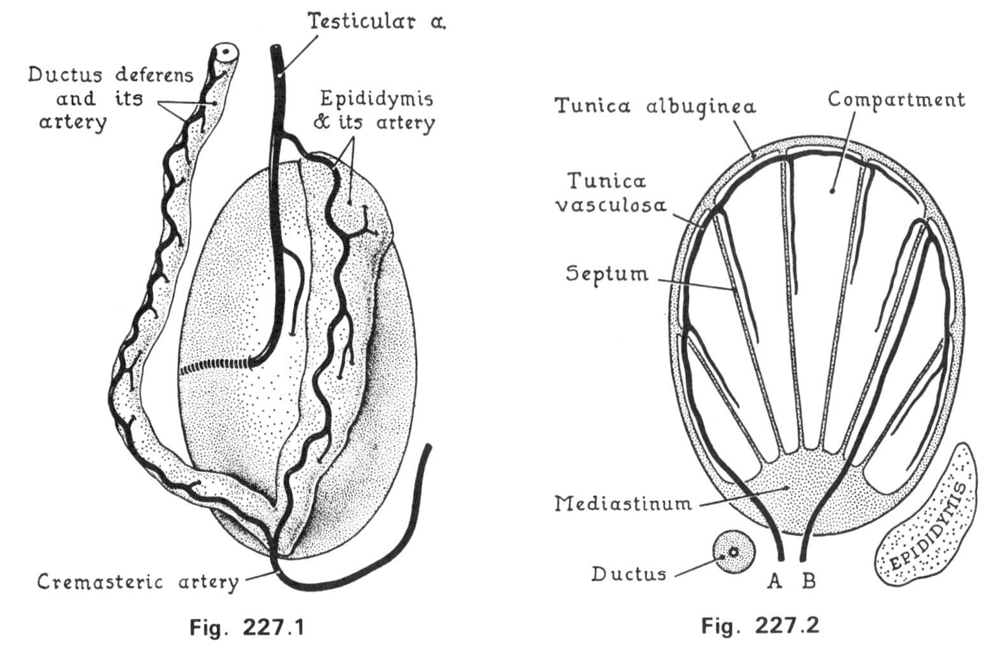

Fig. 227.1

Fig. 227.2

Figs. 227.1 and 227.2. The testicular artery and its anastomoses. (Injection and dissection by Dr. Neil Watters.)

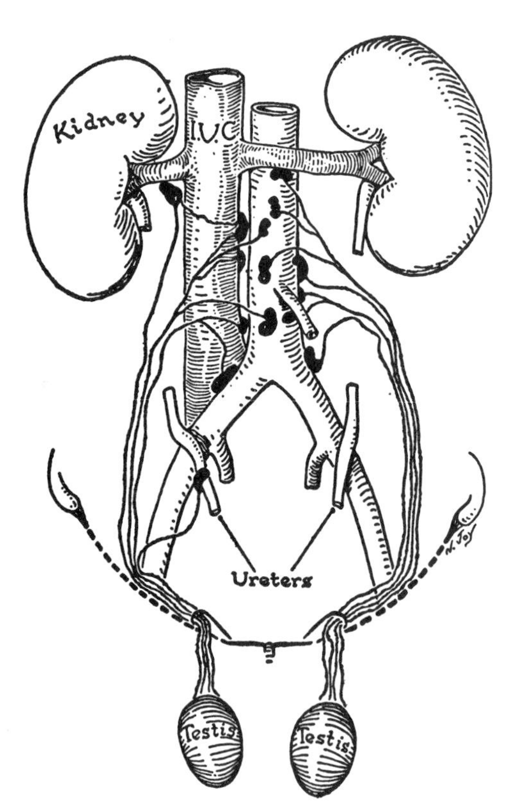

Fig. 228. The lymph drainage of the testes. (After Jamieson and Dobson.)

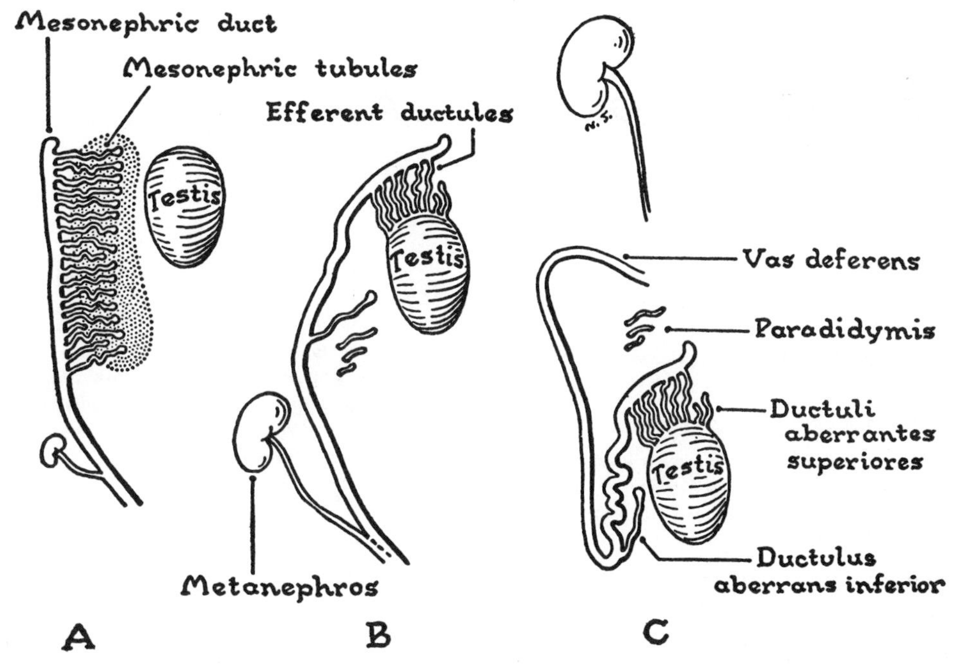

Fig. 229. Some mesonephric tubules make connections; others persist as vestigial structures.

epididymis, respectively.

The three other rudiments are revealed only when the epididymis is unraveled. They are: *ductulus aberrans superior*, *ductulus aberrans inferior*, and *paradidymis* (*fig. 229*). These vestiges may rarely become cystic and require diagnosis and treatment.

When the testis descends, a fragment of spleen is rarely, a fragment of adrenal gland is occasionally, carried down with it.

14 ABDOMINOPELVIC CAVITY

LANDMARKS

Subdivisions; Planes: Transpyloric and Transumbilical; Midinguinal Point.
Protection of Viscera; Definitions (of parts of peritoneum).

PARTS OF GASTRO-INTESTINAL CANAL

General Disposition; Large Intestine (parts).
Examination of G.I. Tract—Stomach; Duodenum; Jejunum and Ileum; Features of Large Intestine; Cecum; Appendix; Colon.

UPPER ABDOMINAL VISCERA AND THEIR CONNECTIONS

Liver; Gall Bladder; Epiploic Foramen (Mouth of Lesser Sac).
Peritoneal Attachments of Spleen; Pedicle of Spleen; Omental Bursa (exploration).
Peritoneal Attachments of Liver: *Falciform Ligament; Round Ligament; Ligamentum Venosum.*
Lesser Omentum; Triangular Ligaments; Coronary Ligaments; Epiploic Foramen (cont'd); Caudate Lobe of Liver.
Spleen (cont'd)—Surfaces; Borders; Surface Anatomy; Development; Variations.
Ever Changing Positions of Viscera: *Body Types or Bodily Habitus; Details of Changing Positions of Various Organs; Features of Hypersthenic and Asthenic Habitus.*

UMBILICUS

Development: Urachus; Umbilical Artery and Vein; Changes of Birth; Meckel's Diverticulum.

PERITONEUM

Peritoneal Folds—Those Containing and Not Containing Tubes; Fossae, Recesses, and Gutters;

Notes on Development Explaining Relationships of Abdominal Structures

Afferent Nerves of Peritoneum.
Obliteration of Peritoneal Cavity

LANDMARKS

Subdivisions. Without landmarks and measurements, you are lost. But instead of thinking in inches and centimeters you must *think in terms of vertebral heights.*

The transpyloric plane bisects the line joining the top of the sternum to the top of the symphysis pubis, and lies at the level of the disc between the 1st and 2nd lumbar vertebrae (*fig. 230*). **The transumbilical plane** passes through the umbilicus, or navel, and lies at the level of the disc between the 3rd and 4th lumbar vertebrae. It is true that the level of the umbilicus varies somewhat with age, sex, obesity, and posture, but it is, for all that, a valuable landmark.

For purposes of elaborate topographical work it is customary to divide the abdomen proper into nine regions in all (*fig. 231*).

Protection to Viscera. The abdominal viscera lie largely "within" the thorax and pelvis. Of course, the viscera are not within the thoracic cavity, for they are situated below the diaphragm; but, each dome of the diaphragm rises to the level of the 5th rib in the midclavicular line. The lower abdominal viscera are protected behind and at the sides by the ilia. In the flanks only the breadth of two fingers separates the 11th ribs from the iliac crests.

Definitions. By *abdominopelvic cavity*" is meant the space enclosed by the bones, muscles and fasciae of the abdominal and pelvic walls from the diaphragm cranially or above to the pelvic diaphragm caudally or below.

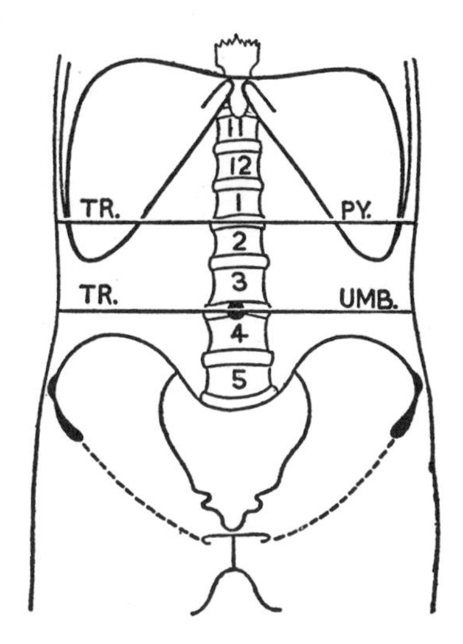

Fig. 230. Horizontal planes with the vertebral column as a measuring rod. To remember 1:2 and 3:4 should not tax the memory.

This cavity is divided at the pelvic inlet into the abdominal cavity (proper) and the pelvic cavity (*fig. 232*).

The Peritoneum is an areolar membrane covered with a single layer of squamous cells, like a wall supporting its wall paper.

The Peritoneal Cavity is a complicated sac, lined everywhere with peritoneum, and moistened with serous (watery) fluid. The cavity is a potential cavity; it is completely empty, except when air is admitted at operation or when fluid collects in a number of conditions. Accordingly, each part of its free surface is in contact with, and rubs against, some other part of its free surface. In the female, the peritoneal cavity communicates with the exterior through the uterine tubes, uterus, and vagina.

The portion of peritoneum lining the walls or parietes of the peritoneal cavity is called the *parietal layer* of peritoneum; that covering the organs or viscera is the *visceral layer*. The peritoneum forms certain folds and double layers. A double layer connecting the stomach to another structure is called an *omentum*. Of these there are two: (1) *the lesser omentum* which is attached to the lesser curvature of the stomach, and (2) *the greater omentum* which is attached to the greater curvature (*fig. 233*). A fold connecting the intestine to the posterior abdominal wall and conveying vessels and nerves to it is a *mesentery*. All other folds are called *ligaments*. The distinction, however, between the three terms is not one of importance.

PARTS OF THE GASTRO-INTESTINAL CANAL

General Disposition (*fig. 234*). The digestive passage extends from the mouth to the anus. It is divisible into the following parts: mouth, pharynx, esophagus, stomach, small intestine, and large intestine.

The stomach and intestines (i.e., from stomach to anus inclusive) are collectively called the **gastro-intestinal canal.** The last 2 cm of the esophagus and the gastro-intestinal canal are situated within the abdomen and pelvis.

The *esophagus* pierces the diaphragm about 2 cm to the left of the median plane and ends at the cardiac or esophageal

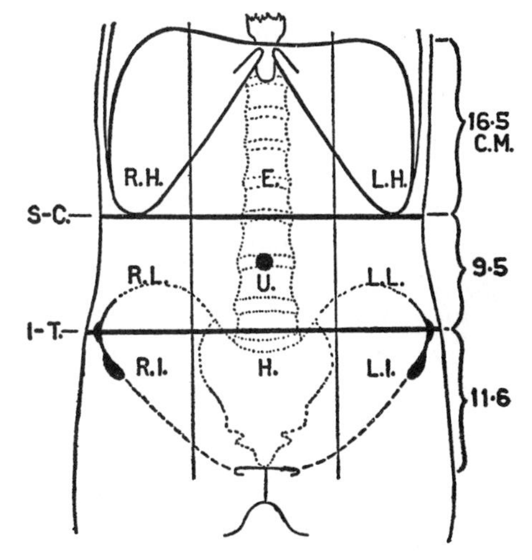

Fig. 231. The nine regions of the abdomen. The right and left sagittal or "vertical" planes are erected on the midpoint of the line joining the corresponding anterior superior spine of the ilium to the top of the symphysis pubis, the **midinguinal point.** *RH, LH* = right and left hypochondriac. *RL, LL* = right and left lateral or lumbar. *RI, LI* = right and left inguinal or iliac. *E* = epigastric. *U* = umbilical. *H* = suprapubic or hypogastric.

Fig. 232. The abdominal and pelvic cavities are almost set at a right angle with each other.

orifice of the stomach. The *stomach* lies to the left of the median plane. Its exit, the *pylorus*, lies about 2 cm to the right of the median plane in the transpyloric plane. It is succeeded by the *small intestine or gut*, which is subdivided into three parts: duodenum, jejunum, and ileum.

The *duodenum* is the horseshoe-shaped portion of the small intestine that can be defined loosely as that portion that once had and then lost its primitive mesentery. It begins at the pylorus and ends at the *duodenojejunal junction*, about 2 or 3 cm to the left of the median plane just below the transpyloric plane, i.e., several centimeters from where it begins. Having no mesentery, the duodenum adheres to the structures on the posterior wall of the abdomen. On this account and because the transverse colon crosses in front of it and adheres to it, it is not conspicuous. The *jejunum* and the *ileum* are the two parts of the small gut that retain their mesenteries. They extend from the duodenojejunal junction to the right iliac fossa where the ileum opens into the large gut. The opening is called the *ileocecal orifice*.

The large intestine is subdivided into: vermiform appendix, cecum, ascending colon, right colic flexure, transverse colon, left colic flexure, descending colon, sigmoid colon, rectum, anal canal, and anus.

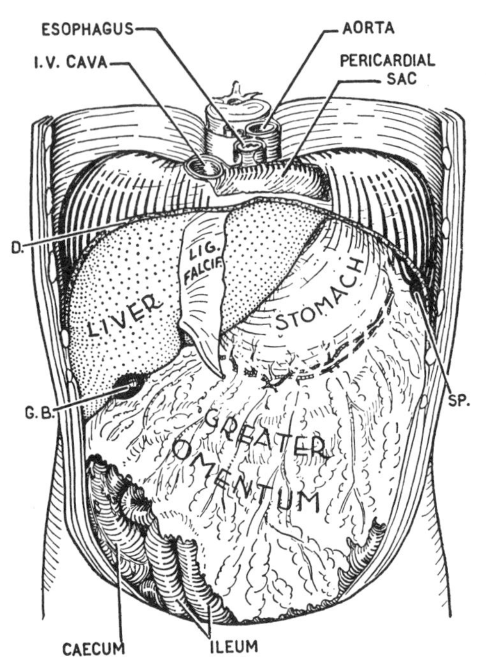

Fig. 233. Abdominal contents undisturbed. *G.B.* = gall bladder; *Sp* = spleen; *D* = cut edge of diaphragm.

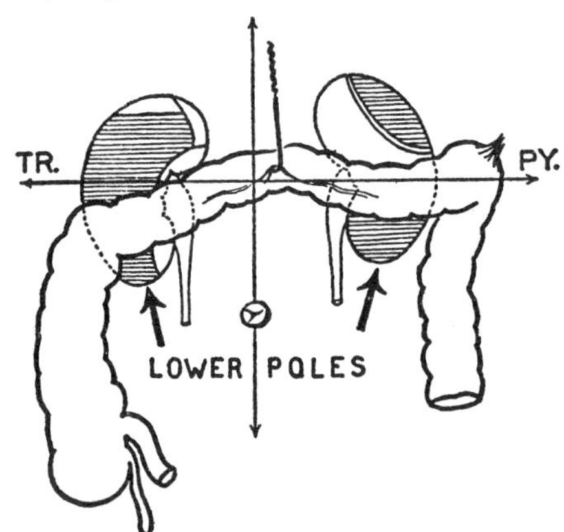

Fig. 234. Contacts of the colon with the kidneys. The shaded parts of the kidneys are covered with peritoneum and are palpable when the peritoneal cavity is open.

The *cecum* is the blind cul-de-sac situated below the ileocecal orifice. The worm-shaped *appendix* opens into the cecum about 2 or 3 cm below the ileocecal orifice. The *ascending colon* ascends from the right iliac fossa, across the iliac crest, to the under surface of the liver. There, in front of the kidney (*fig. 234*), it makes a bend, the *right colic (or hepatic) flexure,* and becomes the *transverse colon,* which extends across the abdomen to the under surface of the spleen (*fig. 235*). In front of the left kidney, it bends again to become the *left colic (or splenic) flexure.* From here the *descending colon* descends to the pelvic brim where it becomes the *sigmoid* (pelvic) *colon.* As sigmoid colon it passes to the middle of the sacrum where it becomes the *rectum.* The lower portion of the rectum and the anal canal passes through the floor of the pelvis and in the region known as the perineum opens on to the surface at the *anus.*

Examination of Gastro-intestinal Canal

By inspection and by handling one may examine the abdominal portions of the alimentary canal.

The last part of the *esophagus* lies in a groove on the posterior aspect of the attenuated left lobe of the liver.

The Stomach has entrance and exit: the one, the *cardiac* (or esophageal) *orifice,* is situated near the heart 2–3 cm to the left of the median plane behind the 7th costal cartilage; the other, the *pyloric orifice,* is situated about 2–3 cm to the right of the median plane on the transpyloric plane. The two borders of the stomach extend between these two orifices. The *lesser curvature* is short and concave and, with the first 3 cm of the duodenum, gives attachment to the *lesser omentum.* The *greater curvature* is long and convex and, with the first 3 cm of the duodenum, gives attachment to an extensive double layer of peritoneum, the *greater omentum.* The greater omentum is divisible into three parts: (1) a lower apron-like part, the *gastrocolic lig.,* (2) a left part, the *gastrolienal lig.,* and (3) an upper part, the *gastrophrenic lig.* (*fig. 235.1*). These are attached to the transverse colon, lien (spleen), and diaphragm, respectively.

The stomach is subdivided thus: a line drawn horizontally at the level of the cardiac orifice separates the *fundus* from the *body* (*fig. 236*). An oblique line joining an *incisura* (notch) on the lesser curvature (incisura angularis) to the greater curva-

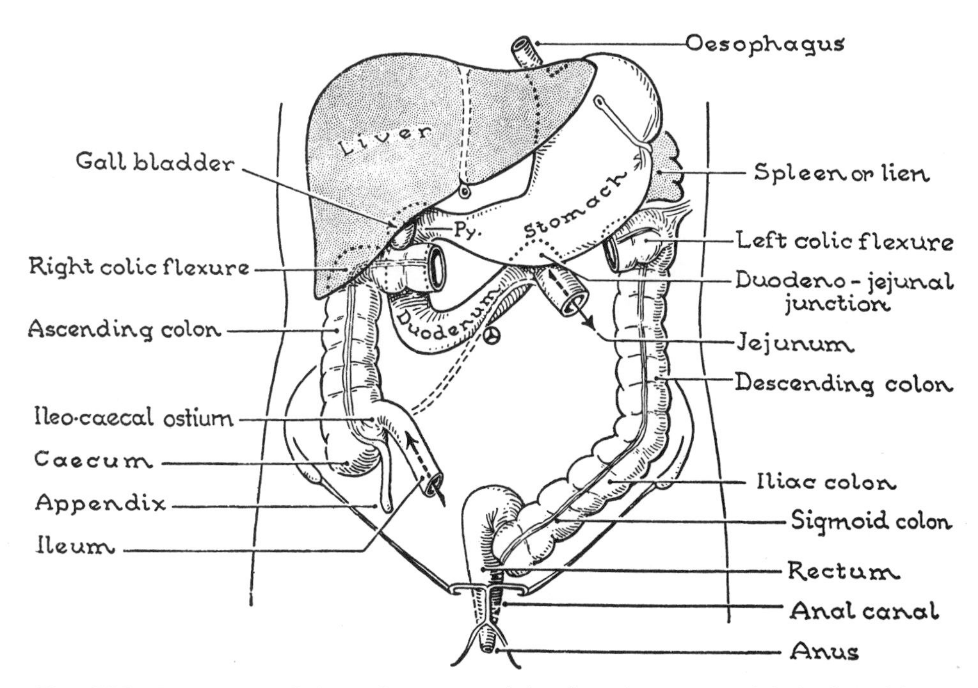

Fig. 235. The names of the various parts of the digestive tract and their dispositions.

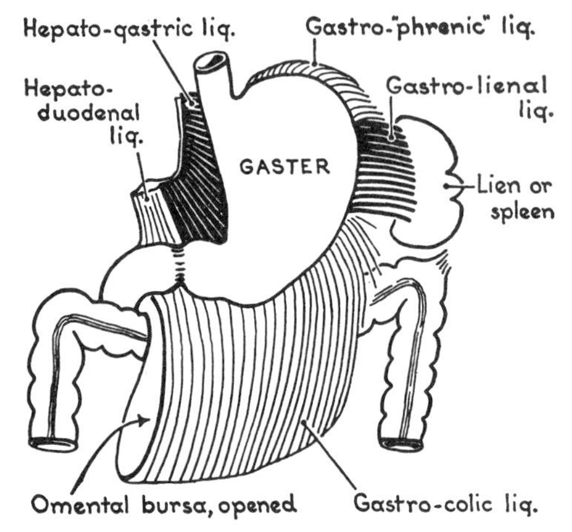

Fig. 235.1. Diagram of the two parts of the lesser omentum and of the three parts of the greater omentum.

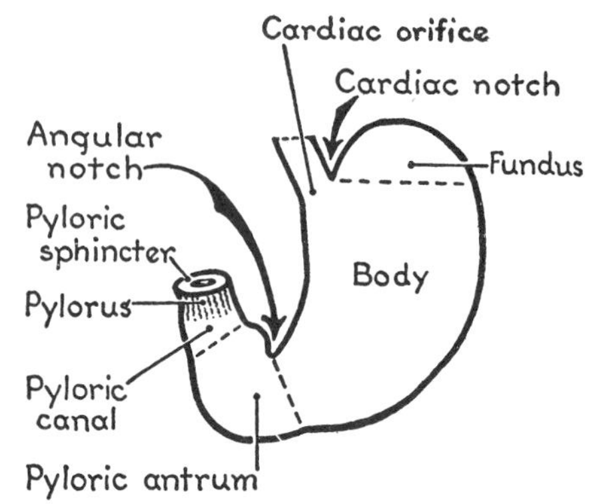

Fig. 236. The parts of the stomach.

ture separates the body from the pyloric part of the stomach. Another oblique line passing from an indentation on the greater curvature (sulcus intermedius) to the lesser curvature subdivides the pyloric part into a large chamber, the *pyloric antrum*, and a more tubular portion, the *pyloric canal*, which ends at the *pylorus*.

The Duodenum is the fixed part of the small gut or the part that has lost its primitive mesentery. Though 25 cm in length its two ends are but 5 cm apart, for it begins at the pylorus, which lies on the transpyloric plane about 2 or 3 cm to the right of the median plane, and ends at the duodenojejunal junction, where it is continuous with the jejunum, about 2 or 3 cm to the left of the median plane and a little below the transpyloric plane. It is molded around the head of the pancreas and is horseshoe-shaped (*fig. 291*). Later (p. 226), we shall see that it is divided into four parts, and that the 2nd part is largely concealed by the transverse colon which crosses it and adheres to it.

The **remainder of the small gut** retains its mesentery and extends from the duodenojejunal junction to the ileocecal orifice, which is situated in the right iliac fossa where intertubercular and "vertical" planes intersect. Though a distance of about 15 cm separates these two points, the gut, under dissecting room conditions, steers a varying course of 5 to 7 meters between them. The root of the mesentery of the gut is attached diagonally across the posterior abdominal wall between the two points; accordingly, it likewise is about 15 cm long, whereas its intestinal border is elaborately ruffled and frilled to accommodate the gut.

The small intestine is so convoluted and mobile that you can pass many coils of it through your hands without being able to decide whether it is leading you to its duodenal end or to its cecal end. But, by the simple device of placing a hand on each side of the mesentery and drawing the fingers forward from root to intestinal border, the convolutions are locally untwisted and the direction of the gut or intestine rendered quite obvious.

Jejunum Contrasted with Ileum. The upper two-fifths of the free part of the small gut are called jejunum, and the lower three-fifths ileum. The jejunum has a greater digestive surface than the ileum, because (1) its *diameter* is greater; (2) its spirally arranged folds of mucous membrane, called *plicae circulares* (p. 229), are bigger and more closely packed; and (3) the minute finger-like projections of its mucous membrane, called *villi*, are larger and more numerous. Hence, its wall feels thick and velvety, whereas the wall of the smaller calibered ileum with its fewer plicae and villi may be almost parchment-like in thinness.

The extraperitoneal *fat*, normally present in the mesentery, creeps along the vessels on to the ileal wall but fails to reach the jejunal wall; hence, there are translucent "windows" in the mesentery at the edge of the jejunum. Further, the disposition of the *vessels* from which the *vasa recta* spring (Arcades, p. 223) to become progressively more complex from the jejunum to the ileum (Michels). Also the vasa recta become progressively shorter (*figs. 237, 238*).

Features of the Large Intestine. The *large gut* forms three and one-half sides of a "picture-frame." Its outer longitudinal muscle coat does not form a complete coat, as in the small gut, but it is arranged in three narrow bands, the *teniae coli*, which, being shorter than the gut itself, cause it to be gathered up into *sacculations*. These teniae lead directly to the root of the appendix.

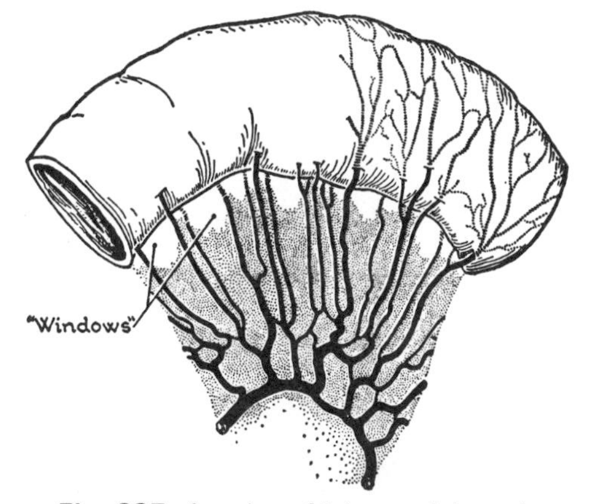

Fig. 237. Arteries of jejunum injected.

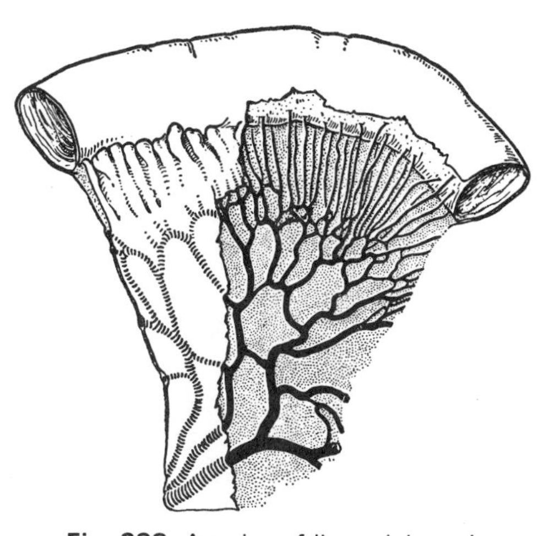

Fig. 238. Arteries of ileum injected.

Peritoneal bags of fat, *appendices epiploicae*, hang from the large gut throughout its whole length (*fig. 239*). Those from the appendix, cecum, and rectum generally contain no fat.

Size alone does not necessarily distinguish large gut from small gut, the descending colon commonly having a *caliber* less than that of the small gut.

The primitive mesentery, possessed by the large gut during prenatal life, is constantly retained by the transverse and sigmoid colons, while the appendix acquires a mesentery, and the cecum is free. The extent to which the ascending and descending colons lose their primitive mesenteries varies (*fig. 240*).

The **cecum** is free and lies in the right iliac fossa, commonly "hanging" over the pelvic brim. The cecum may have a short mesentery or even two mesenteries, with a *retrocecal fossa* extending upward between them, a favorite hiding-place of the appendix. An extensive retrocecal fossa is, of course, a *retrocolic fossa*.

The **vermiform appendix** in fetal life opened into the apex of the cecum; now it opens into the cecum about 2 cm below the ileocecal junction. It may be long or short

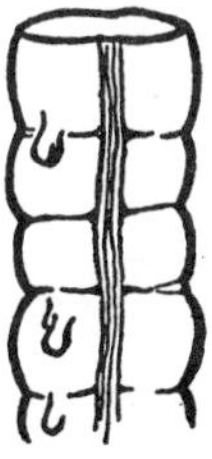

Fig. 239. Segment of large intestine showing one of the three teniae, sacculations, and appendices.

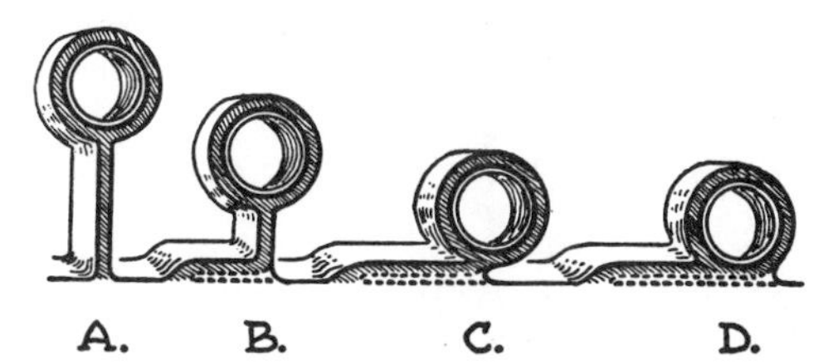

Fig. 240. Primitive mesentery of the large gut in various stages of absorption.

(4 to 20 cm in Solanke's series), and may occupy any position consistent with its length (*fig. 241*).

A triangular fold of peritoneum, known as the *meso-appendix*, attaches the appendix to the terminal part of the left (lower) layer of the mesentery of the ileum.

The appendix has a uniform external coat of longitudinal muscle fibers continuous with the teniae coli. On the cecum and colon, the teniae coli remain discrete till they reach the rectum when they form for it a nearly uniform coat again.

Colon. The *ascending colon* (*fig. 242*) crosses the iliac crest and ascends in front of the muscular posterior wall and lower pole of the right kidney to the under surface of the liver. Here, in front of the lower pole of the right kidney, it makes a right angle bend, the *right colic flexure*, and becomes the *transverse colon*.

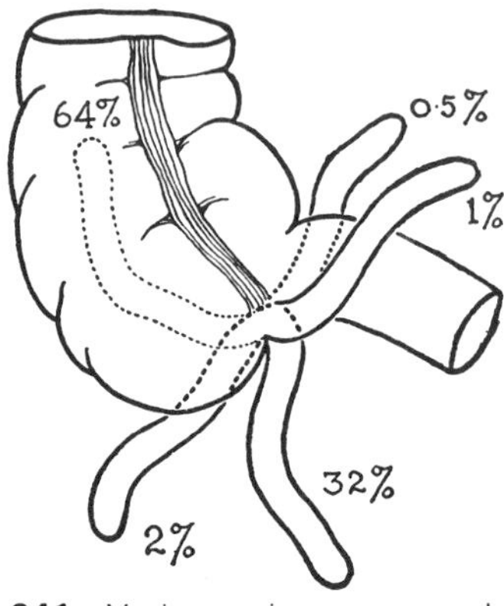

Fig. 241. Various sites assumed by vermiform appendix and approximate frequencies. (After Wakeley.)

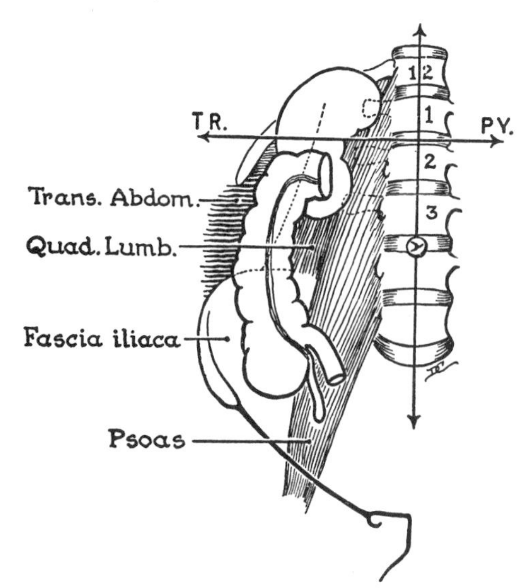

Fig. 242. The ascending colon.

Resting on the transverse colon are (*fig. 235*): the liver, gall bladder, and stomach. It is attached to the greater curvature of the stomach by the gastrocolic lig. (part of the greater omentum) (*figs. 248, 282*) and to the pancreas by the *transverse mesocolon* (*fig. 243*).

The right extremity of the transverse colon crosses and adheres to the front of: the right kidney, the second part of the duodenum, and head of the pancreas (*fig. 243*). The remainder hangs down for a varying distance but ascends again in front of the descending colon, and makes with it an acute angle at the left colic flexure.

The *left colic flexure* is attached to the diaphragm below the spleen (*fig. 254*) by a bloodless fold of peritoneum, the *phrenicocolic ligament*.

The *descending colon*, often much reduced in caliber, descends, crosses the iliac crest, and proceeds across the iliac fossa to the pelvic brim where it becomes the sigmoid (pelvic) colon (*fig. 235*).

The *sigmoid colon* has a mesentery, the *sigmoid mesocolon*, whose root runs a ∧-shaped course: upward and then downward in front of the sacrum as far as its middle or third piece. Its appendices epiploicae are very long fatty tags. The sig-

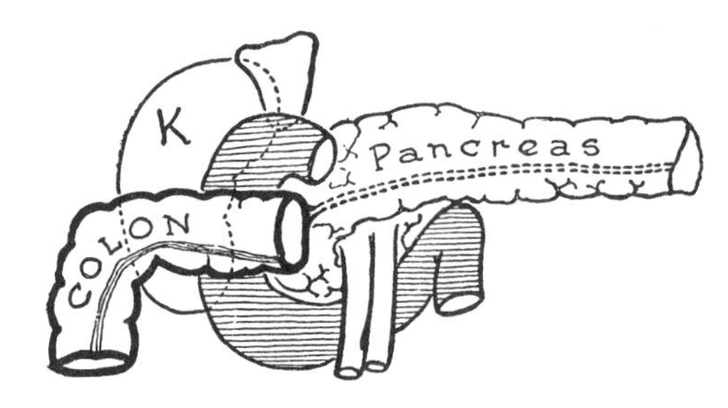

Fig. 243. The attachment of the transverse mesocolon (shown by *dotted lines*).

moid colon may be of the short or long type—20 to 45 cm.

The long type forms quite a long loop which may become twisted on itself, thereby causing an intestinal obstruction.

The *rectum* begins at the 3rd piece of the sacrum, and since it lies in the pelvis, it is described with the pelvic organs (Chapter 20).

UPPER ABDOMINAL VISCERA AND THEIR CONNECTIONS

Liver (hepar, L.). The liver is a soft, pliable organ, molded by its surroundings and weighing almost 1½ kg. When hardened by embalming, its *inferior* or *visceral surface* (*fig. 244*) faces downward, to the left, and backward. It is coated with peritoneum.

The remainder of the liver is in contact with the anterior abdominal wall and the diaphragm, which make it smooth and round. It is called the *diaphragmatic surface*. Most of it, too, is coated with peritoneum except for a small "bare area" on the posterior aspect which is in contact with the diaphragm (p. 220).

The visceral and diaphragmatic surfaces are separated from each other (except behind) by the sharp, *inferior border*.

Far back on the visceral surface there is a deep transverse fissure, 5 cm long, the **porta hepatis.** It is the door through which vessels, nerves, and ducts enter and leave the liver. To its lips a part of the lesser omentum is attached.

(Cont'd on pp. 198–199 and 219–221.)

Gall Bladder. This pear-shaped vesicle, about 8 cm long and holding about 40 to 50 cc of bile, is divided indefinitely into—*fundus*, *body*, and *neck*. The fundus of the gall bladder projects beyond the sharp, inferior border of the liver and comes into contact with the anterior abdominal wall where the lateral border of the Rectus crosses the costal margin (*fig. 213*). The body and neck adhere to the sloping inferior surface and run to the porta hepatis (*fig. 245*).

Relations (*fig. 245*). A gall stone could penetrate its way through the walls of the gall bladder (1) upward into the liver substance, or (2) downward into the duodenum, or (3) into the colon, or (4) forward through the anterior abdominal wall.

Comparative Anatomy. A gall bladder is present in most fish and higher orders of vertebrates. It is present in most birds but absent in the pigeon; present in the ox, sheep, goat, and pig, but absent in the horse and deer; present in the guinea pig and rabbit but absent in the white rat.

Epiploic Foramen (Mouth of the Lesser Sac). The gall bladder serves as a guide to the mouth of a diverticulum of the general peritoneal cavity, called the **omental bursa** (*lesser sac of peritoneum*). This is because the *cystic duct*, which drains the gall bladder, lies in the free edge of the lesser omentum; and the mouth of the sac lies behind this free edge (*fig. 246*). This sac is not unlike an empty, rubber hot water bottle (*fig. 246.1*).

Peritoneal Attachments of Spleen (Lien, L. = Spleen). A double layer of peritoneum, the *lienorenal ligament*, passes from spleen to kidney (*fig. 247*). A second double

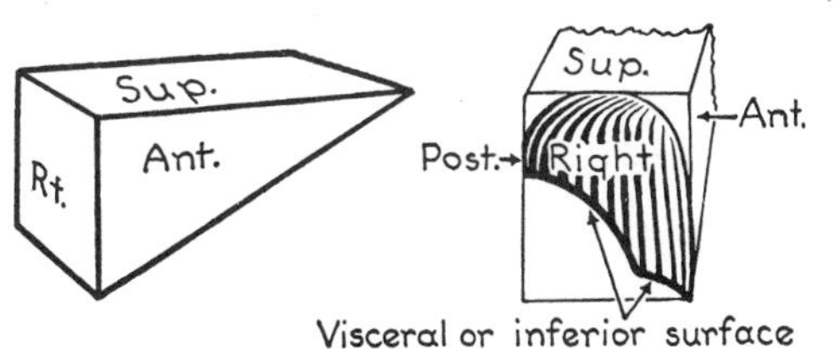

Fig. 244. The general shape of the liver.

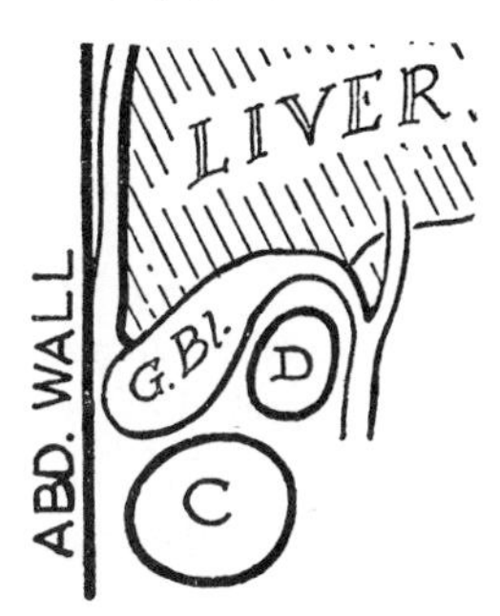

Fig. 245. The four relations of the gall bladder (sagittal section).

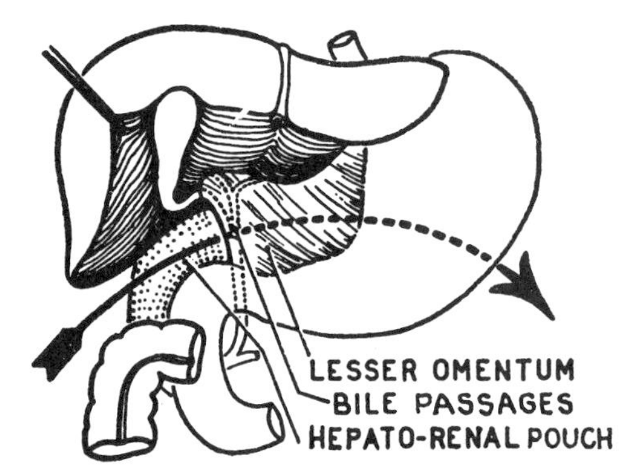

Fig. 246. Showing why the gall bladder serves as a guide to the epiploic foramen.

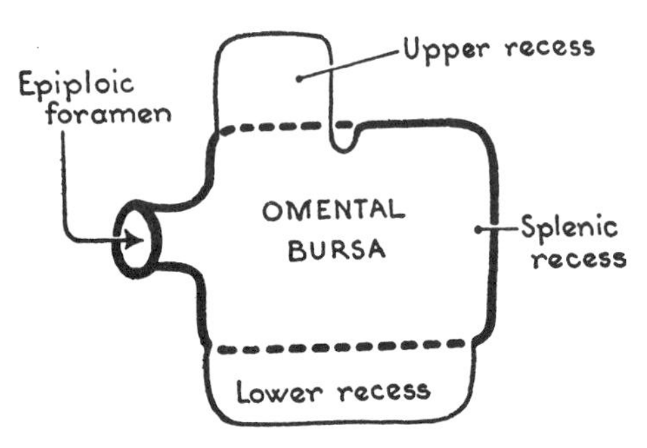

Fig. 246.1. Scheme of omental bursa.

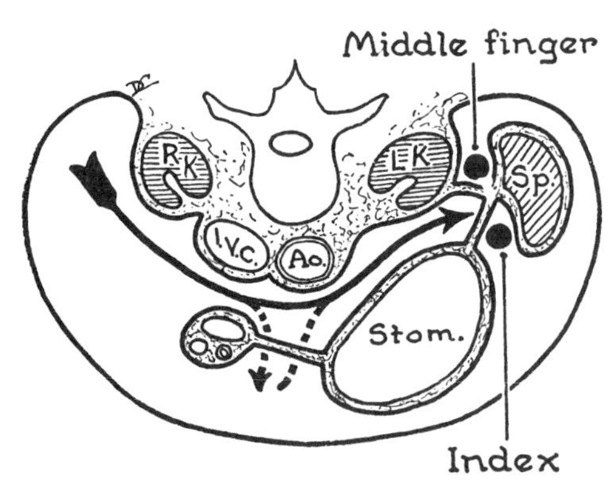

Fig. 247. Palpating the hilus of the spleen while its pedicle is clamped between two fingers of the right hand. If you pass a finger to the left (as shown by the *arrow*) it will reach the hilus at the left limit of the omental bursa.

layer, continuous with the gastrocolic lig., passes from the greater curvature of the stomach to the spleen; this is the *gastrolienal (gastrosplenic) lig.* These two ligaments form a stalk or pedicle in which the vessels run to and from the spleen.

Pedicle of the Spleen; Omental Bursa (The Lesser Sac of Peritoneum). The "pedicle or stalk" of the spleen may be clamped between two fingers (*fig. 247*). It has a free lower border (and a free upper border) or you could not grasp it. Its linear site of attachment to the spleen is around the hilus. Clamped between your index and middle fingers are four layers of peritoneum (*fig. 247*).

The Lower Recess of this bursal sac—*see figure 248.*

The Upper Recess lies in the median plane between the liver and the diaphragm (*fig. 251*).

Peritoneal Attachments of the Liver

The free edge of the **falciform ligament** contains the **round ligament,** which extends from the umbilicus to the sharp inferior border of the liver. Before birth, the round ligament was the *umbilical vein* that returned purified blood from the placenta of the mother to the liver of the fetus. After birth, it becomes a fibrous cord, the *round ligament of the liver or ligamentum teres* (*figs. 249 and 269*).

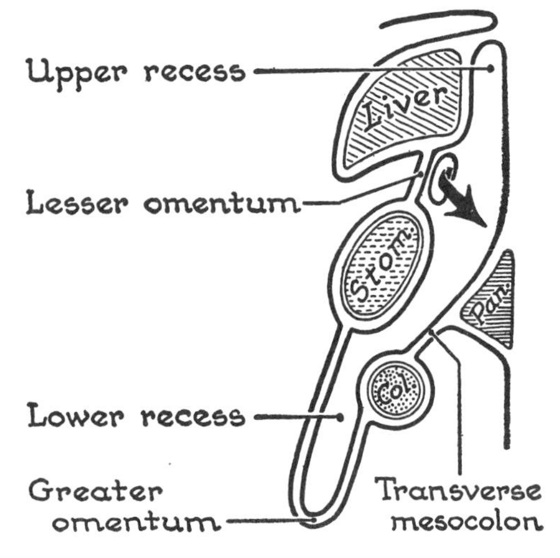

Fig. 248. Showing the vertical extent of the omental bursa (lesser sac). The *arrow* passes through the epiploic foramen (mouth of the lesser sac).

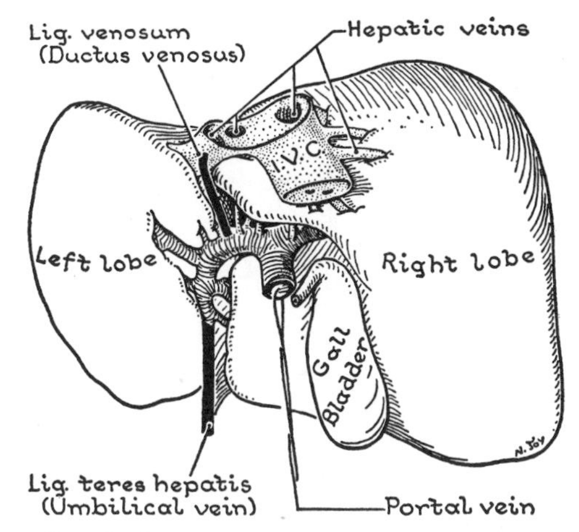

Fig. 249. Portal vein, hepatic veins, and two obliterated veins, called ligaments. (Postero-inferior view.)

The falciform lig. is attached to the anterior abdominal wall and diaphragm in the median plane; and thence to the convex surface of the liver (to the right of the median plane). On the inferior surface of the liver to the left end of the porta, the round ligament occupies the *fissure for the round lig.*

Before birth the left umbilical vein opened for a short time into the left portal vein and, so, poured its blood into the liver. But the *ductus venosus* develops to connect the left portal vein to the inferior vena cava (*fig. 249*).

The ductus venosus is obliterated after birth and becomes the **ligamentum venosum.** It continues the sagittal course of the umbilical vein, at the bottom of the *fissure for the lig. venosum*, on the posterior aspect of the liver.

For descriptive purposes, these three ligaments—falciform, round, and venosum—divide the liver into a right and left lobe.

The **lesser omentum** extends from the lesser curvature of the stomach and first 3 cm of the duodenum to the fissure for the lig. venosum and to the porta hepatis (*fig. 250*).

The **triangular ligaments** are the sharp, bloodless, peritoneal folds at the extreme right and left limits of the attachment of the liver to the diaphragm (*fig. 251*).

The **coronary ligament** is the peritoneal reflections from the edges of the bare area of the liver to those of the contacting "bare" area of the diaphragm. Generations of students have wasted many hours in fruitlessly learning its intricacies. A brief review of *figure 251* is sufficient.

The inf. vena cava occupies the left or basal part of the bare area of the liver, and the peritoneum, there applied to the i.v. cava (*fig. 251*), is the left side or base of the coronary ligament, the right triangular lig. being the apex.

The **caudate lobe** is the lobe with a tail, the tail or **caudate process** being the narrow isthmus of liver that bounds the epiploic foramen above and connects the caudate lobe to the visceral surface of the right lobe.

(The liver is continued on p. 219.)

Spleen or Lien (cont'd. from p. 198). A thin peritoneal-covered, and easily torn capsule encases the soft, vascular spleen, which is molded by the structures in contact with it. The size of a small fist after death, it is much larger in life. *Notches* on its superior border are a feature of note.

It has a convex diaphragmatic surface and a visceral surface shared unequally by *stomach*, *kidney*, and *colon* (*figs. 253, 253.1*).

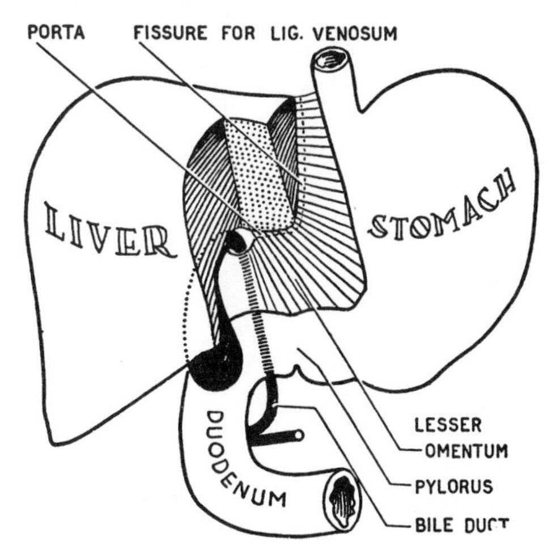

Fig. 250. The attachments of the lesser omentum. (A section has been taken from the liver to show the fissure for the lig. venosum.)

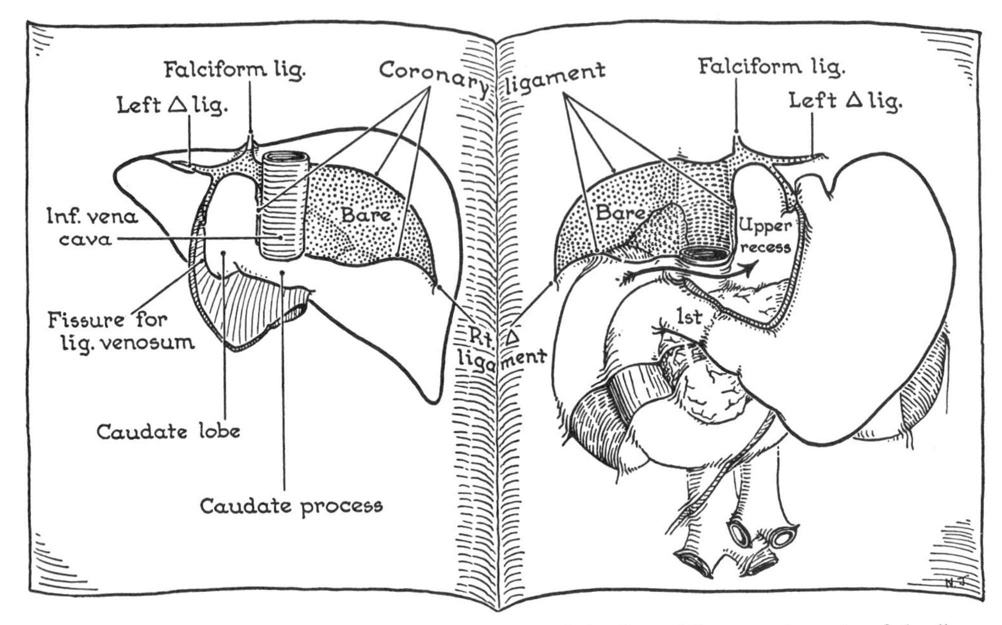

Fig. 251. The coronary and triangular ligaments of the liver. (The attachments of the liver are cut through, and the liver is turned to the left, as you would turn the page of a book. Hence, the posterior aspect of the liver is revealed on the left page and its posterior relations on the right page. The *arrow* passes through the epiploic foramen.)

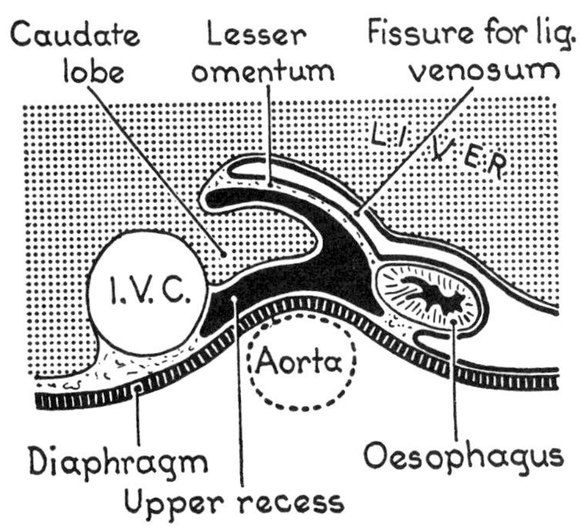

Fig. 252. The boundaries of the upper recess, viewed from below.

The *hilus* admits the vessels and nerves running in the pedicle along with the *tail of the pancreas* which abuts against the hilus.

Development (*figs. 269–271*).

Surface Anatomy. The spleen lies deep to the 9th, 10th, and 11th left ribs. Its long axis follows the 10th rib and extends from, or almost from, the suprarenal gland to the

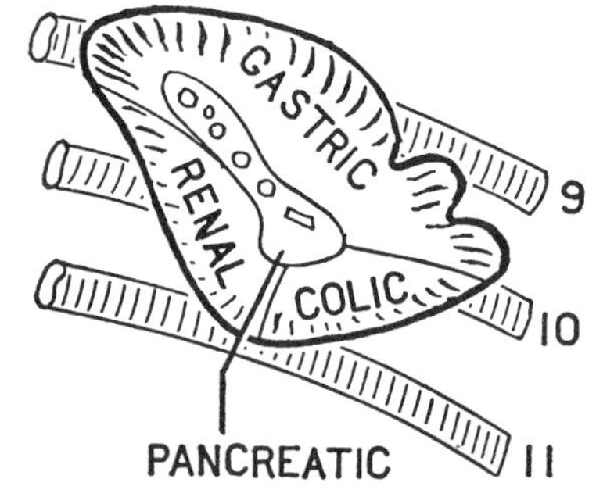

Fig. 253. Visceral surface of spleen and its "circumferential" border.

midaxillary line. Separating it from the ribs are the peritoneal cavity, the diaphragm, and the pleural cavity; in its upper half the left lung also intervenes (*fig. 254*).

Structure. The spleen has a capsule of white fibers, elastic fibers and smooth muscle fibers which allow it to expand and contract. Supporting *trabeculae* of the same materials spread inward from the capsule. The spaces between the trabeculae contain a supporting *sponge work* of reticular fibers and reticulo-

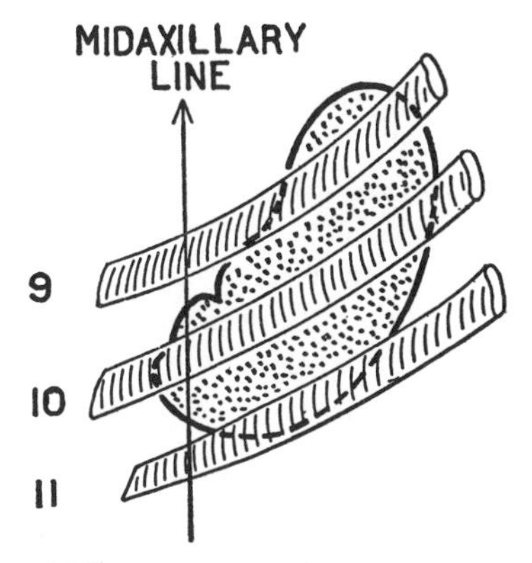

Fig. 253.1. Parietal (diaphragmatic) surface of spleen.

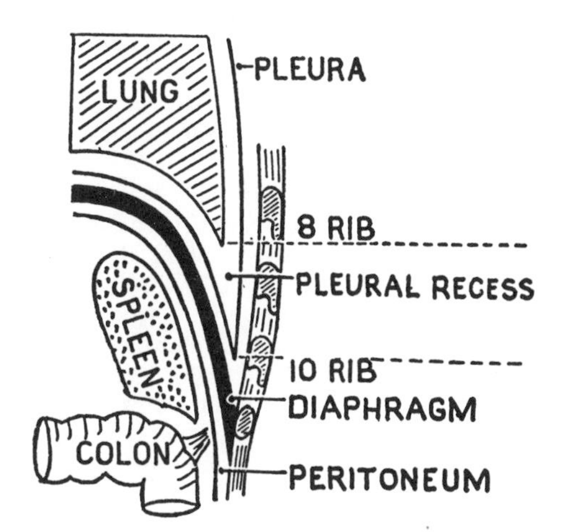

Fig. 254. Coronal section in the midaxillary line, to show the parietal relation of the spleen.

endothelial cells and are filled with blood. About one-sixth of the total volume of blood in the body can be stored in the spleen, which accordingly varies greatly in size.

Function. The spleen is the largest of the lymphocyte-producing organs. It is the main storehouse of blood. It is the chief depot of reticulo-endothelial cells, which break down the hemoglobin of effete red cells and in so doing produce bile pigment, they also rid the blood of other debris; they also are concerned with resistance to disease and with immunity.

The spleen is not essential to life; in fact, in certain conditions its removal improves health.

Rupture of the Spleen, fortunately rare, can be caused by severe abdominal blows. Surgical removal of the organ seems to be easily compensated for by the body.

Accessory Spleens, the size of large lymph nodes (and very small ones), may lie on the course of the splenic artery and its left gastro-epiploic branch, and elsewhere. In performing splenectomy to relieve certain disorders of the blood, all accessory spleens must be removed if recurrence of the disorder is to be avoided (Curtis and Movitz; Halpert and Eaton).

EVER CHANGING POSITIONS OF VISCERA

The positions of the various abdominal viscera vary considerably from subject to subject, depending largely upon the body build—upon whether the subject is of the broad type (when, characteristically, they are placed high in the abdomen) or of the intermediate and slender types (when they are placed lower).

Body Types or Bodily Habitus. Healthy human beings differ from each other not only in outward appearance, form, and size, but also inwardly. Mills described two extreme physical types, the *hypersthenic* and the *asthenic* (*figs. 254.1 and 254.2*) and two intermediate types, the *sthenic* and the *hyposthenic;* and these have their subdivisions.

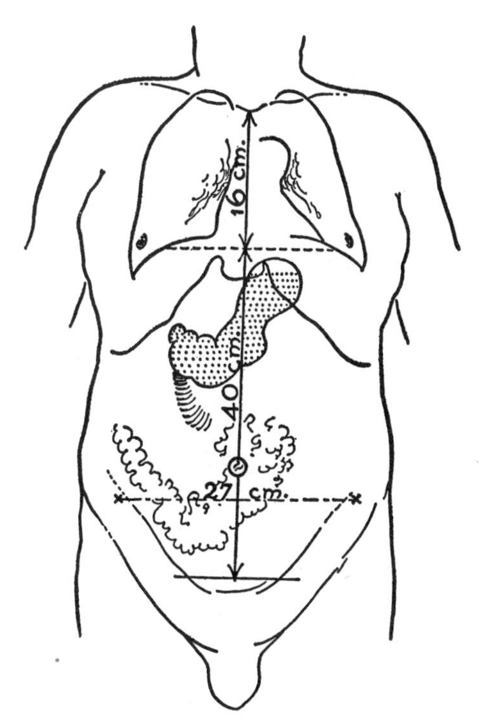

Fig. 254.1. Hypersthenic habitus.

There are: (1) variations in general bodily physique and in the relative capacities of the thorax, upper abdomen, lower abdomen, and pelvis; (2) variations in the form, position, tone, and mobility of the viscera; and (3) constant relationships between (1) and (2). A powerful, heavily built hypersthenic individual with short thorax and long abdomen has the gastro-intestinal tract placed high (*fig. 254.3*), rare in those of slender physique (asthenic).

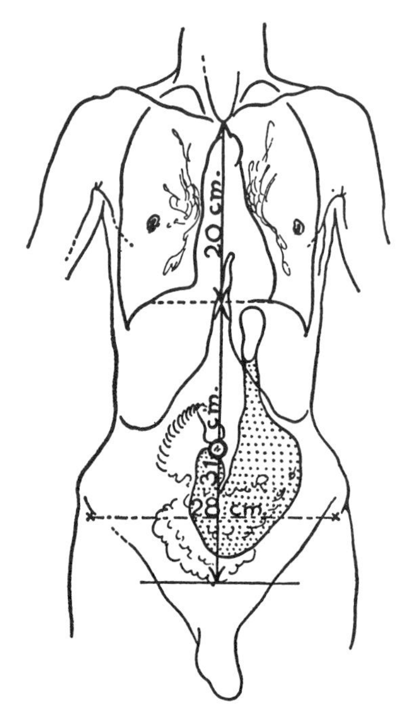

Fig. 254.2. Asthenic habitus.

The viscera are ever changing both their shapes and their positions. They move with the movements of the diaphragm and of the anterior abdominal wall. They move when the posture alters, being highest when the subject is recumbent, lower when he stands, and still lower when he sits (*fig. 255*). They rise when the anterior abdominal wall is voluntarily retracted.

The hollow organs (e.g., stomach, intestines, bladder, and uterus) vary as they fill and empty, and they vary with the tone of their muscle coats (e.g., fear and other emotions result in relaxation of the stomach so that the greater curvature suddenly falls).

The following data by Moody and van Nuys are submitted for your appreciation—*not for you to memorize*.

The Stomach. The *Cardiac Orifice* of the stomach is nearly stationary. The *Greater Curvature* of the empty stomach when erect, varies in level from vertebra L. 1 to S. 1, and when supine from Th. 12 to L. 5 (*fig. 256*). In any subject it is lowest in the erect posture, and on an average it rises the height of a vertebra on assuming the prone posture (face down), and of another vertebra on assuming the supine (face up). The position of the stomach when lying down is no indication of its position when standing up; it moves from 1 to 16 cm.

Similarly, the position of the *Pylorus* ranges from the level of L. 2 to L. 5 in the erect posture, and from Th. 12 to L. 4 in the supine (*fig. 256*), and horizontally, it ranges between 7 cm to the right of the median plane and 5 cm to the left.

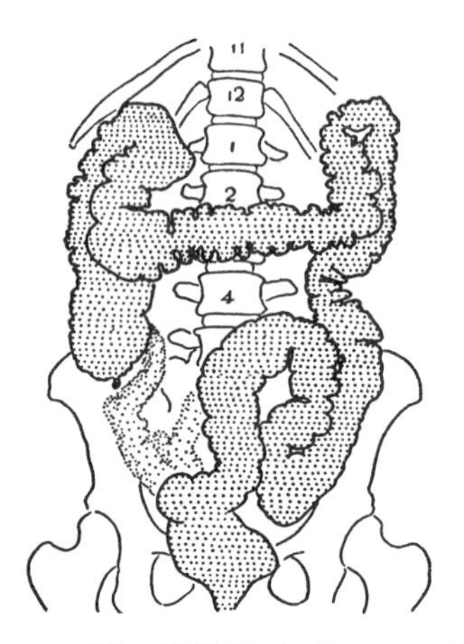

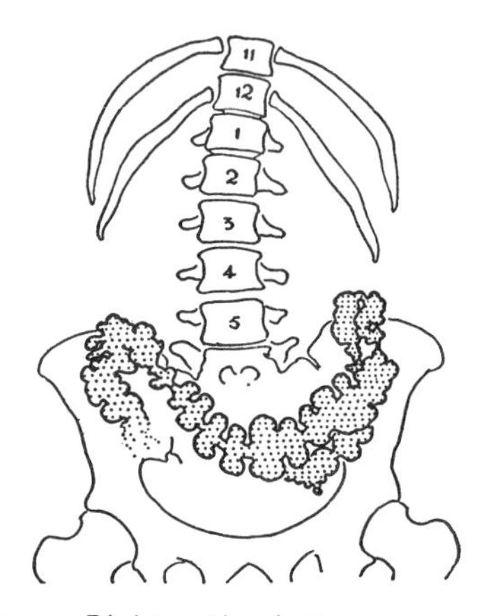

Fig. 254.3. *Left*, hypersthenic type. *Right*, asthenic type.

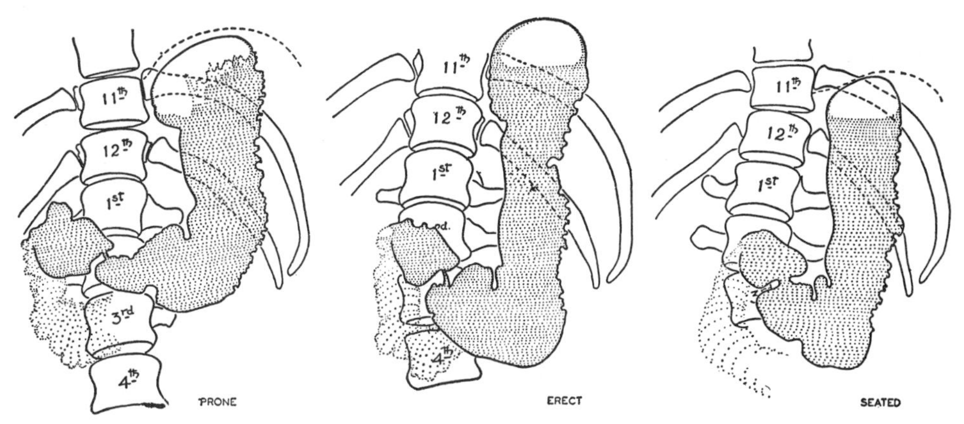

Fig. 255. Tracings of radiograms of the stomach of a healthy female, aged 30 years (Radiograms by Dr. Keith Bonner).

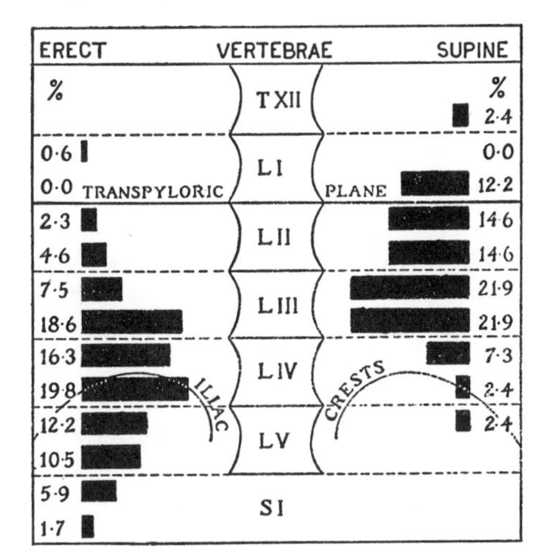

Fig. 256. *Left*, frequency distribution (in percentages) of the relation of the lowest part of the greater curvature of the empty stomach to the vertebral column in healthy adult males (172 erect, 41 supine). *Right*, frequency distribution (in percentages) of the relation of the pylorus to the vertebral column in healthy adult males (86 erect, 41 supine). (Based on the work of Moody, van Nuys, and Kidder.)

In any subject, on changing from the erect posture to the supine, it moves from 2 to 10 cm (usually 6 to 8 cm). On changing from the erect posture to the prone, and again from the prone to the supine, it moves upward and to the right (e.g., in the erect posture 34 per cent of pylori are to the left of the median plane, but in the supine only 8 per cent). The greater curvature and the pylorus are usually lower in the female than in the male.

The Duodenum and the *Head of the Pancreas* slide downward on the areolar bed behind them, provided the liver adjusts its shape (*fig. 302*).

The Transverse Colon, in the erect posture, is most commonly U- or V-shaped, its lowest point most often being 10 cm below the interiliac line (i.e., a line joining the highest points on the iliac crests).

The Liver and Spleen also vary in position. The greatest height of the right lobe of the *liver* in the erect posture varies from 15.5 to 25.5 cm (average 20 cm). In 50.2 per cent of males and 34 per cent of females the lower end of the right lobe lies below the interiliac line, as much as 5.0 cm. The average excursion made by this lower end, on changing from the erect posture to the supine, varies from 0 to 10.9 cm (average 2.5 cm). The lower limit of the *spleen* reaches L. 1 to L. 5 (average L. 3).

The Kidneys. The commonest position of

both kidneys in both sexes is opposite the upper four lumbar vertebrae, when the subject is erect; and opposite the 12th thoracic and upper three lumbar vertebrae, when the subject is supine.

With the subject supine, the caudal pole of the right kidney is below L. 3 in 38 per cent of the men and in 48 per cent of the women; whereas this pole of the left kidney is below L. 3 in 17 per cent of the men and only 9 per cent of the women. The excursion of the kidneys due to forced respiration varied from 0.1 to 6.5 cm.

Features of the Hypersthenic Habitus:

1. A powerful and massive *physique*, great body *weight*, and heavy *bony framework*.

2. The *thorax* is short, deep, and wide; the *abdomen* is long and of great capacity in its upper zone. The *subcostal angle* is very obtuse, and the xiphoid process is broad.

3. The *lungs* are wide at their bases, and contracted at their apices which project but little above the clavicles.

4. The long axis of the *heart* is nearly transverse.

5. The *gastro-intestinal tract* is high. The stomach is of the bull horn type, the pylorus being the lowest, or nearly the lowest, part of the stomach. The entire colon is short; the cecum is well above the iliac basin even when the subject is standing; the transverse colon is short, actually transverse, and high; consequently the descending colon is long, it is also straight. The relative proportions of the colon are characteristic, so are its fine, numerous haustrations. The gastric motility is fast; there is marked tone and rapid motility of the colon. Defecation takes place 2 to 3 times a day.

Features of the Asthenic Habitus:

1. Frail and slender *physique*, light body *weight*, and delicate *bony structure*.

2. The *thorax* is long and narrow; the *abdomen* is short. There is disproportion between the *pelvic capacity* and that of the upper abdomen, the false pelvis being often as wide and capacious as that of a hypersthenic subject of twice the weight. The *subcostal angle* is narrow.

3. The *lungs* are widest above, and their apices reach well above the clavicles.

4. The long axis of the *heart* is approximately in the median plane.

5. The *gastro-intestinal* tract is low. The stomach is atonic and largely pelvic when the subject is standing. The entire colon is long; the cecum is capacious and low in the pelvis; the transverse colon dips down toward, or into, the pelvis; the haustrations of the colon are coarse. The tone of the gastro-intestinal tract is poor and its motility slow.

UMBILICUS

When the umbilicus or navel is examined from its peritoneal aspect, four fibrous cords are seen radiating from it. They are the obliterated remains of four tubes which in fetal life traversed the umbilical cord (*fig. 257*). The tubes are: the urachus, the right and left umbilical arteries, and the umbilical vein. Each of the four may produce for itself a peritoneal fold or mesentery; but whether occupying peritoneal folds or not, they are all situated in the extraperitoneal fatty-areolar layer (*figs. 219, 260*).

The Obliterated Allantoic Duct or *Urachus* ascends in the median plane from the apex of the urinary bladder to the umbilicus. [*Cf.* the *allantois* (*fig. 258*), which lines the egg shell.]

On each side of the urachus an *Obliterated Umbilical Artery* proceeds from the internal iliac artery to the umbilicus—equivalent of the allantoic arteries (*fig. 259*). After birth the umbilical cord is cut and the arteries thereafter become the obliterated umbilical arteries. (Wide of these on each side an inf. epigastric artery passes from the ext. iliac artery to the rectus sheath; occasionally each creates a pronounced fold.)

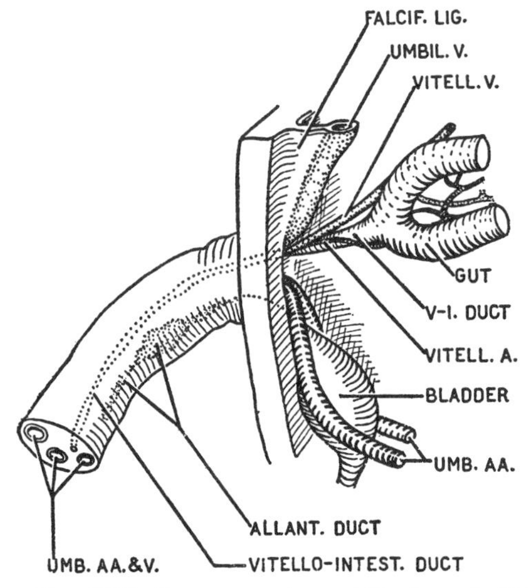

Fig. 257. Structures in umbilical cord. (After Cullen.)

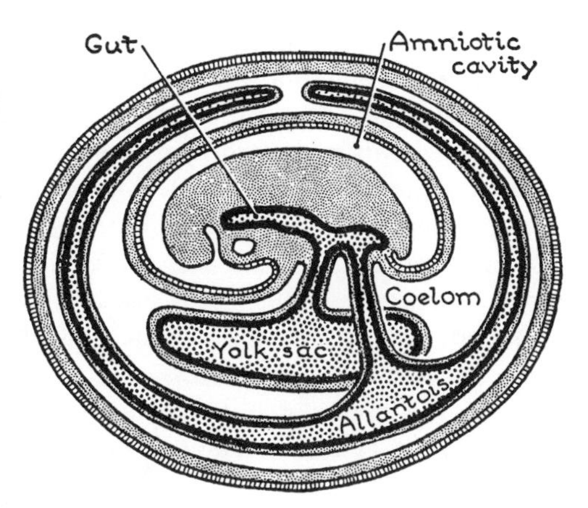

Fig. 258. In the chick embryo the allantois is an enveloping respiratory sac. (After Paterson.)

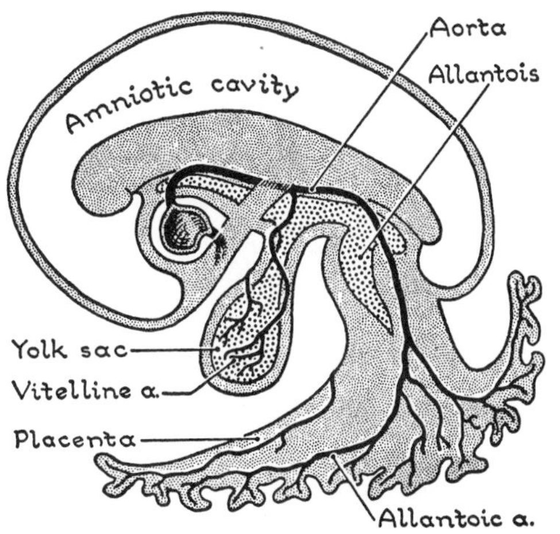

Fig. 259. In the human embryo the allantois is superseded by the placenta. (After Paterson.)

The Obliterated Umbilical Vein is, of course, the *round ligament of the liver* (*fig. 249*).

If at birth the cord is cut very short: (1) urine will escape from the umbilicus if the urachus is patent; (2) feces, if the vitello-intestinal duct is patent; and (3) the peritoneal cavity will be opened if the extra-embryonic celom is patent.

Meckel's Diverticulum (*fig. 259.1*). The vitello-intestinal duct persists in 2 per cent of persons as a patent appendage of the gut, springing from the ileum within a meter or so of the cecum (Jay *et al.*).

PERITONEUM

Peritoneal Folds. Peritoneal folds are often commonly "*the mesenteries*" of tubes:

SIXTEEN FOLDS CONTAINING TUBES are represented in *figure 260* and table 10.

FOLDS NOT CONTAINING TUBES are indicated alphabetically in *figure 260*.

A. Left triangular lig. of the liver.
B. Right triangular lig. of the liver.
C. Phrenicocolic lig.
D. "Supporting lig. of liver."
E. Acquired folds lateral to the ascending and descending colons.
F. Fold guarding inf. duodenal fossa.
G. Inf. ileocecal fold (bloodless fold) (*fig. 261*).
H. Ligament of the ovary.

Peritoneal Fossae, Recesses, and Gutters occur as follows:

1. *Omental Bursa* (p. 198).

2. *Above the greater omentum:* The right and left *Subphrenic Spaces* lie between diaphragm and liver, one on each side of the falciform ligament. The *Hepatorenal Recess* or *Pouch* lies between the right lobe of liver, right kidney, and right colic flexure (*fig. 262*).

3. *Below the greater omentum* are the following fossae and gutters:

Duodenal Fossae: The (inconstant) superior duodenal, inferior duodenal, paraduodenal, and retroduodenal fossae are

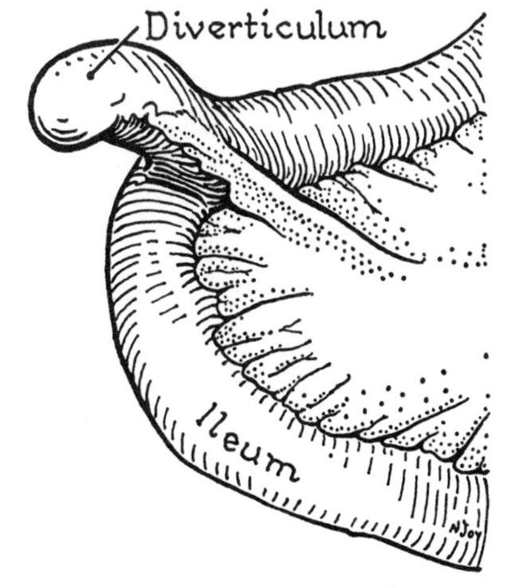

Fig. 259.1. Diverticulum ilei (of Meckel).

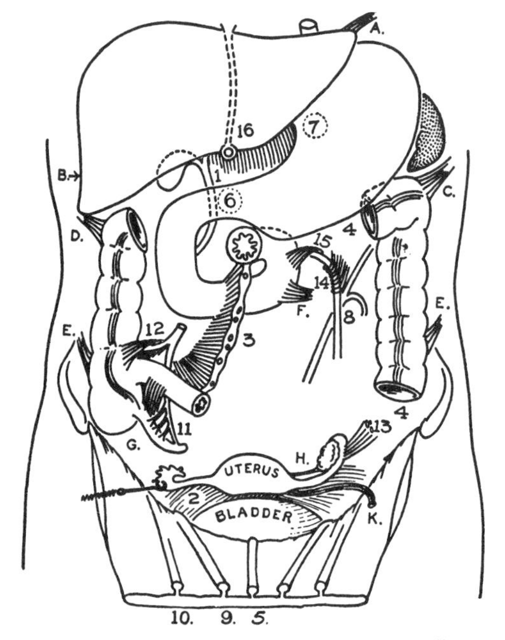

Fig. 260. Peritoneal folds acting as "mesenteries" of tubes (see table 10).

little pockets whose mouths face each other near the duodenojejunal junction.

Cecal Fossae: The superior ileocecal, inferior ileocecal, and retrocecal fossae are related to the cecum (*fig. 261*). An extensive retrocecal fossa is a retrocolic fossa.

An Intersigmoid Fossa is sometimes present. Its mouth opens at the apex of the Λ-shaped root of the sigmoid mesocolon, where the left ureter crosses the common iliac vessels. A pencil can be pushed up the fossa for several centimeters in front of the ureter.

4. *Pelvic Fossae:* In the male the rectovesical fossa lies between rectum and bladder. In the female the uterus and its broad ligaments divide the rectovesical fossa for several centimeters in front of the fossae.

5. *The "Retro-Omental" or Paracolic Gutters (fig. 262.1).*

TABLE 10

Peritoneal Folds Acting as "Mesenteries" of Tubes

Nature of Tube	Patent or Obliterated	Name of Fold	Name of Tube of Which the Fold is a Mesentery	Number on Figure 260
Duct	Patent	The lesser omentum	The bile passages	1
		The broad ligament of uterus	The uterine tube	2
		The mesentery	The small intestine	3
		The mesocolon	The large intestine	4
	Obliterated	The median umbilical ligament	The urachus (Allantoic duct)	5
Artery	Patent	The right gastropancreatic fold	The hepatic a.	6
		The left gastropancreatic fold	The left gastric a.	7
		The fold of the paraduodenal fossa	The asc. branch of left colic a. sometimes	8
	Obliterated	The medial umbilical ligament	Obliterated umbilical a.	9
Artery and Vein	Patent	The lateral umbilical ligament	Inferior epigastric vessels	10
		The mesentery of the appendix	Appendicular vessels	11
		The superior ileocecal fold	Anterior cecal vessels	12
		The suspensory lig. of ovary	Ovarian vessels	13
Vein	Patent	The fold of paraduodenal fossa	Inferior mesenteric vein	14
		The fold of sup. duodenal fossa	Inferior mesenteric vein	15
	Obliterated	Falciform ligament of the liver	Lig. teres hepatis (Obliterated umbilical vein)	16

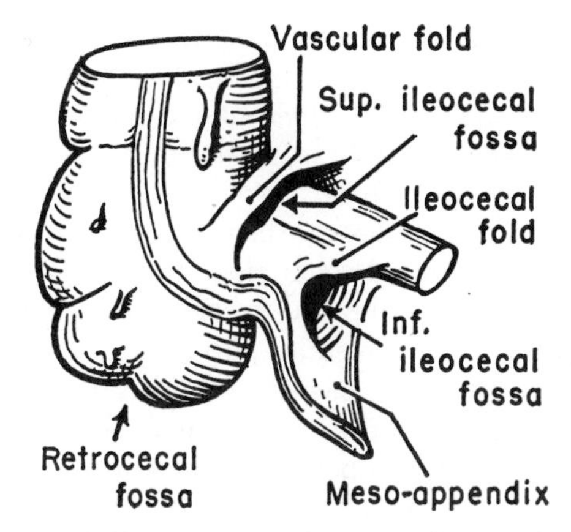

Fig. 261. Peritoneal folds and fossae about the cecum.

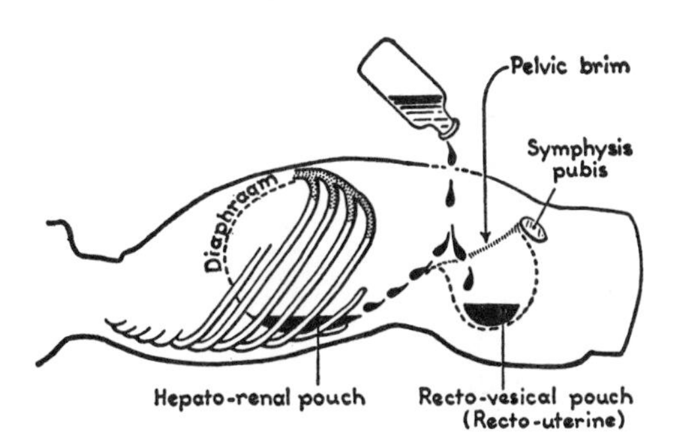

Fig. 262. Two pouches which are the lowest parts of the peritoneal cavity when the subject is supine. Fluid (e.g., blood) from torn organs follow these paths.

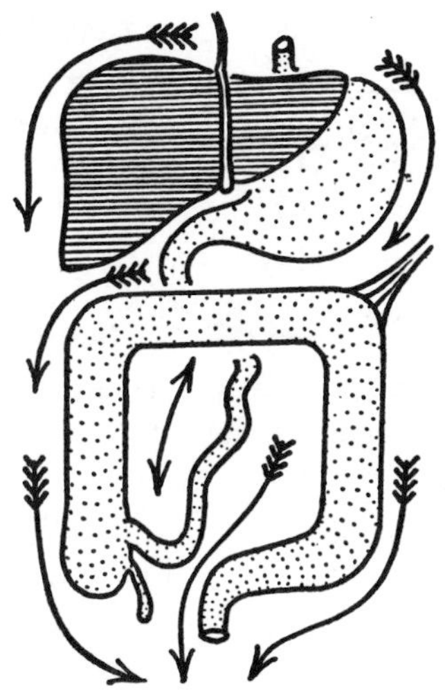

Fig. 262.1. The four retro-omental or "paracolic" gutters and three "supra-omental spaces." The right lateral gutter is the only gutter open above. It would conduct fluid from the hepatorenal pouch and right subphrenic space past the appendix and into the pelvis—if the subject is sitting.

NOTES ON DEVELOPMENT EXPLAINING RELATIONSHIPS OF ABDOMINAL STRUCTURES

1. To explain relationships, *the Gastro-intestinal Tract* and its two derivatives, the *liver* and *pancreas*, with which is associated the *spleen*, may be called "*The G. I. Tract & Co.*" Then the *Right and Left Urogenital Tracts*, and their two paired glands, the *kidneys* and *sex glands* (testes or ovaries) with which are associated the right and left *suprarenal glands* are "*The U. G. Tracts & Co.*"

2. During embryonic life the "three paired glands" lie on each side of the aorta, covered with peritoneum of the posterior abdominal wall. [*Cf.* the adult frog.].

3. At the same early period the gastro-intestinal canal was a straight tube of uniform caliber, slung from the front of the vertebral column and aorta by a mesentery, the *primitive dorsal mesentery* (*fig. 263*).

4a. Three unpaired branches of the abdominal aorta, named the celiac trunk and superior mesenteric and inferior mesenteric arteries, supplied the gastro-intestinal tract and its three unpaired glands (*figs. 263* and *264*). The superior mesenteric artery continued as the *vitelline artery* through the umbilicus to supply the yolk sac (*fig. 259*).

4b. The portal vein is formed by three unpaired veins, named the splenic (= the celiac), superior mesenteric, and inferior mesenteric veins; so, it returns to the liver the blood that the celiac, superior mesenteric, and inferior mesenteric arteries conveyed to the "G. I. Tract & Co.," the liver, of course, excepted. The portal vein receives no other blood (*fig. 265*).

5. The diaphragm is supplied by cervical segments 3, 4, and 5 by way of the phrenic nerves. It developed in the neck. With the advent of lungs, the diaphragm

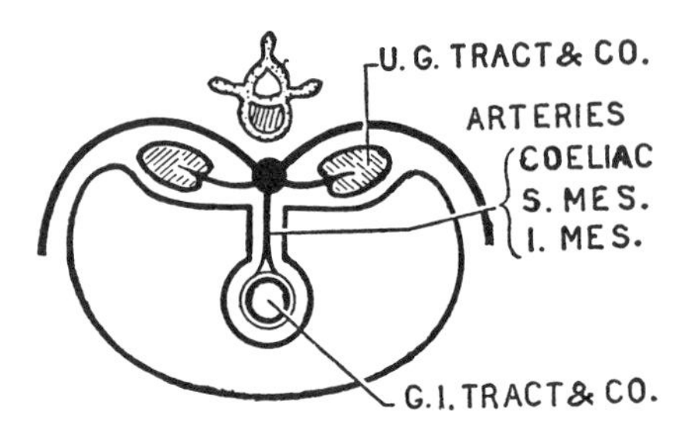

Fig. 263. Transverse section of the abdomen of an embryo (schematic).

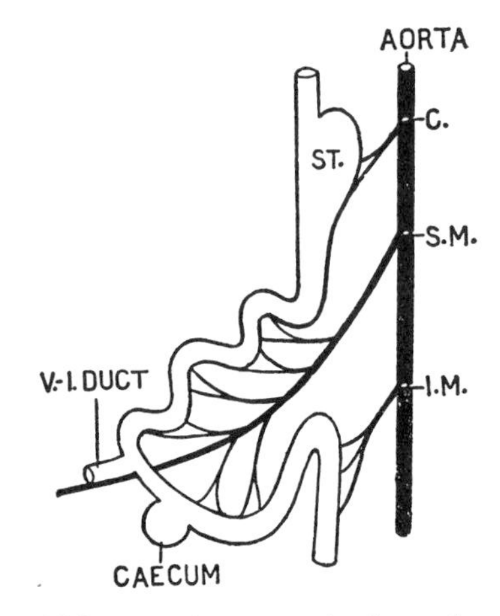

Fig. 264. The three unpaired arteries of the "G. I. Tract & Co." in the primitive dorsal mesentery.

descended, pushing the stomach and celiac trunk before it and dragging the phrenic nerves after it.

6. The adult intestinal canal may reach 6 or 7 meters in length, but the abdominal cavity is less than ½ meter long. Therefore, the gut ceases to be a straight tube confined to the median plane. The small intestine becomes convoluted; and a long loop of gut, involving the cecum and adjacent parts of the small and large intestine, taking the superior mesenteric artery as an axis, rotates counterclockwise around it. This brings the cecum temporarily to the under surface of the liver. From there it ultimately descends into the right iliac fossa (93 per cent of adults) and so helps to encircle the small gut (*fig. 266*).

7. Thereafter, the mesenteries of the ascending, the beginning of the transverse, and the descending colon, together with any branches of the mesenteric vessels they may contain, adhere to the posterior abdominal wall (*fig. 267*).

The remainder of the transverse colon and the sigmoid colon alone retain mesenteries. The rectum and the anal canal retain their primitive median positions.

8. Just as the pages of an open book, on falling to the left or right, must come to lie in front of pages before and after (*fig. 268*), so every part of "the G. I. Tract & Co." that falls to the left or right must come to occupy a position ventral to anything and everything pertaining to "the U. G. Tracts & Co." it happens to cross, i.e., suprarenals, kidneys, testes or ovaries; suprare-

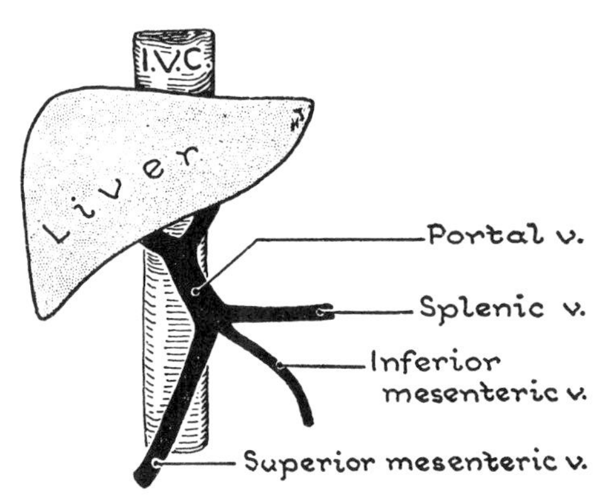

Fig. 265. The three unpaired veins of the "G. I. Tract & Co." form the portal vein.

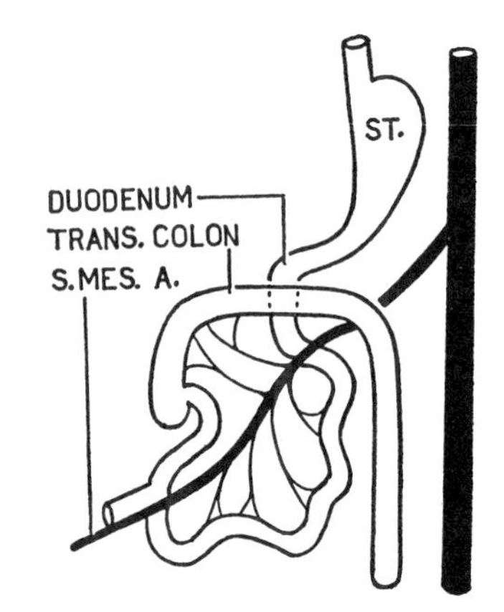

Fig. 266. Rotation of gut around the superior mesenteric artery.

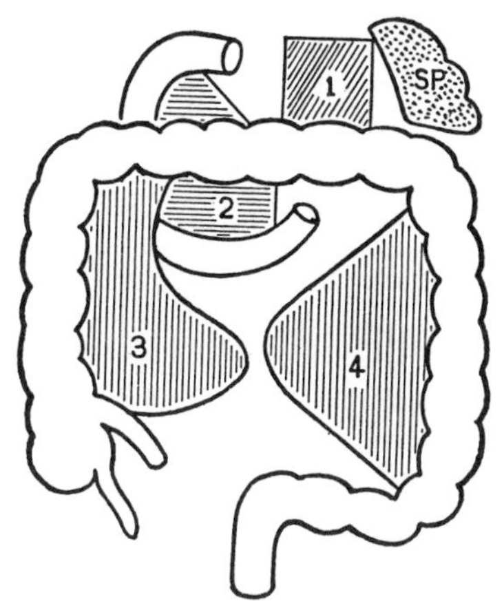

Fig. 267. Sites where primitive mesenteries adhere to the posterior abdominal wall, obliterating the peritoneal cavity: (*1*) dorsal mesogastrium, (*2*) mesoduodenum, (*3*) mesocolon of ascending colon and right colic flexure, (*4*) mesocolon of descending colon.

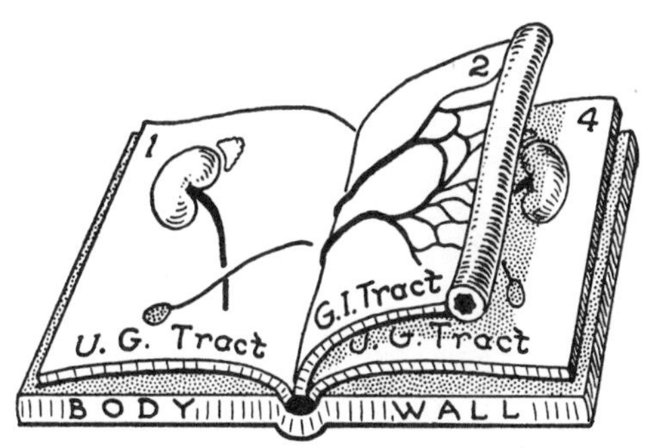

Fig. 268. The pages of a book serve to demonstrate that the "G. I. Tract & Co." is necessarily ventral to the "U. G. Tract & Co."; and the "U. G. Tract & Co." to the body wall.

nal, renal, testicular, or ovarian vessels; and ureters, deferent ducts or uterine tubes (*fig. 268*).

9. The counterclockwise rotation of the gut must bring some part of the large intestine in front of the superior mesenteric artery—the transverse colon—and some part of the small intestine behind it—the 3rd part of the duodenum (*fig. 266*).

10. The transverse colon would seem to have forced the entire duodenum (save its first 2 cm) and the pancreas, which lay in the mesoduodenum, against the posterior abdominal wall, for the duodenum has lost its mesentery.

A surgeon uses the above principles in order to free or mobilize a portion of the colon or duodenum without damage to vessels.

11. The *primitive ventral mesentery* exists only above the umbilicus and first 2 cm of duodenum, so it is called the *ventral mesogastrium*. The liver divides it into two portions, the falciform ligament and the lesser omentum (*fig. 269*).

The spleen divides the *dorsal mesogasterium* into (1) Gastrolienal ligament, and (2) Lienorenal ligament (*fig. 269*).

12. The liver comes to occupy especially the right side of the body, and relegates the stomach and spleen to the left (*figs. 270* and *271*). In accordance with principles

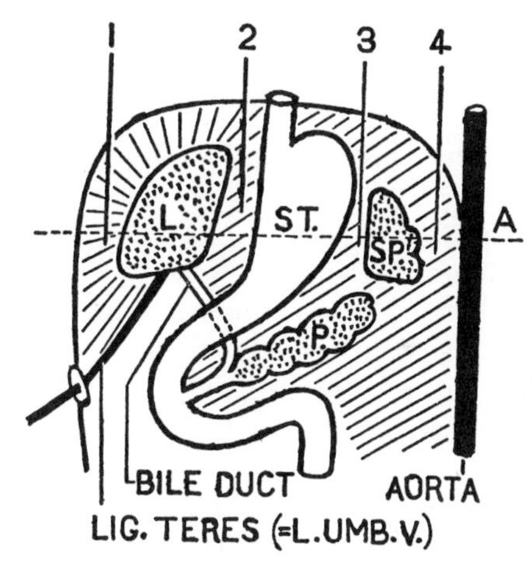

Fig. 269. Primitive ventral and dorsal mesogastria give rise to: (*1*) falciform lig., (*2*) lesser omentum, (*3*) gastrolienal lig., (*4*) lienorenal lig.

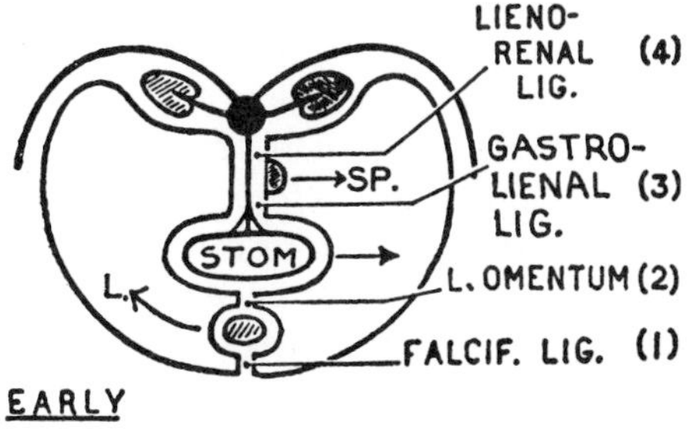

Fig. 270. Transverse section of abdomen of embryo at *level A, figure 269*, indicating that the liver moves to the right; the stomach and spleen to the left.

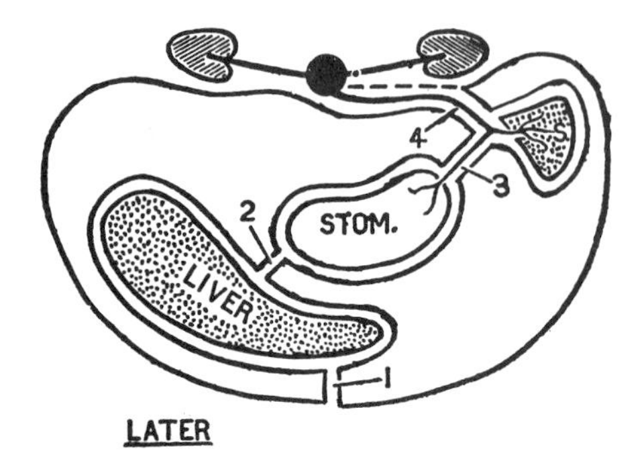

Fig. 271. A later stage than *figure 270.* Partial absorption of dorsal mesogastrium; the unabsorbed part is the lienorenal lig. You may restore the primitive state and expose the left kidney and its vessels along the broken line without damaging any structures.

stated above, these three organs, belonging to the G. I. Tract and Co., must lie in front of the right and left three paired gland system; the liver being in front of right kidney and suprarenal; the stomach and spleen in front of the left kidney and suprarenal.

13. When the stomach moved to the left, it underwent a rotation on its long axis. As a result, its original left surface became the ventral surface; the original right surface, the dorsal surface.

14. The portion of the dorsal mesogastrium passing between spleen and aorta is forced against the posterior abdominal wall; and absorbed between the median plane and the front of the left kidney. The new attachment becomes the *lienorenal ligament.*

15. The duodenum, by losing its mesentery and adhering to the posterior abdominal wall, limited the epiploic foramen (mouth of the omental bursa) below; the liver, by enlarging, encroached on the mouth, narrowing it from above; and the bile passages, passing from liver to duodenum in the free edge of the lesser omentum, limited the mouth in front.

16. The primitive omental bursa was at first limited below by the dorsal mesogastrium, but in time it bulged downward in front of the transverse mesocolon and transverse colon, and adhered to them, thereby forming the gastrocolic portion of the greater omentum. Developmentally, therefore, the transverse mesocolon is four layers thick (*fig. 272*).

17. The portion of the peritoneal cavity between the base of the bladder and prostate anteriorly and the rectum posteriorly underwent obliteration. And, it is safe to open it up (*fig. 273*).

In summary, the chief *sites of obliteration* of the peritoneal cavity are:

1. The portion of the dorsal mesogastrium between the aorta and the middle of the left kidney (*fig. 271*).

2. The mesoduodenum, including the part containing the pancreas (*fig. 272*).

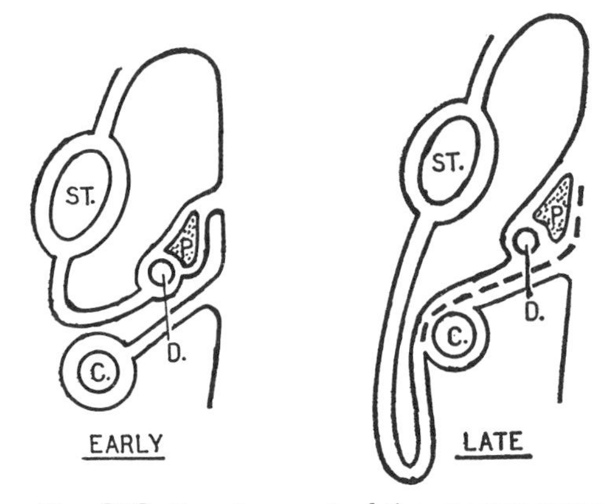

Fig. 272. Development of the greater omentum.

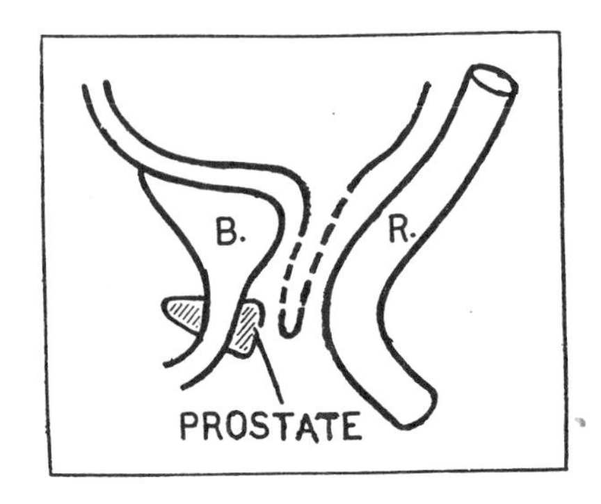

Fig. 273. The posterior aspect of the prostate was formerly subperitoneal. Obliteration of the "rectoprostatic" peritoneal fossa takes place along the broken line.

3. The ascending and descending mesocolons (*fig. 267*).

4. The right portion of the transverse mesocolon adheres to the right kidney, 2nd part of the duodenum, and head of the pancreas (*fig. 307*).

5. The greater omentum adheres to transverse colon and mesocolon (*fig. 272*).

6. The walls of the greater omentum commonly cohere, thereby obliterating the lower recess of the omental bursa.

7. The pouch between the prostate and rectum is obliterated (*fig. 273*).

8. The processus vaginalis peritonei is obliterated in part (*fig. 223*).

Afferent Nerves of the Peritoneum. These travel as follows: (1) from the *central parts of the diaphragm* via the phrenic nerve (C. 3, 4, 5); direct mechanical stimulation of this area causes pain referred by the supraclavicular nerves (C. 3, 4) to the lower part of the anterior border of the Trapezius; (2) from the *peripheral parts of the diaphragm* via the intercostal and subcostal nerves (Th. 7–12); here stimulation causes pain referred through these same nerves to the skin of the abdominal wall; (3) from the *parietal peritoneum* again via these same nerves (Th. 7–12) and L. 1; here stimulation is correctly localized at the point stimulated; (4) the mesenteries of the small and large intestines are sensitive from their roots to near the intestine, whereas the *greater omentum* and the *visceral peritoneum* are insensitive to mechanical stimulation (Morley).

15 STOMACH, LIVER, AND RELATED STRUCTURES

LESSER OMENTUM, BILE PASSAGES, CELIAC TRUNK

Lesser Omentum and Bile Passages

Cystic Duct, Common Hepatic Duct.

Celiac Trunk

Access to it: Surface Anatomy and Relations; Distribution and Branches.
Left Gastric Artery—branches, variations.
Splenic Artery—branches.
Hepatic Artery—terminal branches; Cystic Artery: variations important to surgeon; Collateral Branches of Hepatic Artery; Collateral Circulation about Liver.

PORTAL VEIN

Course; Relations; Tributaries; Portacaval Anastomoses; Development.

STOMACH

(Cont'd); Structure (Coats); Arteries; Lymphatics; Nerves; Relations; Stomach Bed.

LIVER

(Cont'd); H-Shaped Group of Fissures and Fossae; Posterior Aspect; Visceral Surface and Contacts; Bare Areas.
Surface Anatomy.
Lobes and Segments.

BILE PASSAGES AND GALL BLADDER

Functions; Vessels and Nerves; Variations.

LESSER OMENTUM, BILE PASSAGES, CELIAC TRUNK

Lesser Omentum and Bile Passages

The **lesser omentum** extends from the lesser curvature of the stomach and first 2 cm of the duodenum to the fissure for the ligamentum venosum and to the porta hepatis (*fig. 250*), and in its free edge runs the cystic duct.

The **cystic duct** is to be traced from the neck of the gall bladder to a point 1 cm above the first part of the duodenum where it unites at an acute angle with the common hepatic duct to form the bile duct. Later the bile duct may be followed behind the first part of the duodenum and the head of the pancreas to the second part of the duodenum which it enters 8 cm from the pylorus (*fig. 274*).

The **common hepatic duct** is formed in the porta hepatis by the union of the right and left hepatic ducts, and is bound to the cystic duct by a tough areolar web. Medial to these ducts lies the hepatic artery; and behind the ducts and artery lies the thin-walled portal vein. Accompanying these are lymph vessels and branches of the vagal trunks and of the celiac plexus.

The **epiploic foramen** separates them from the i.v. cava which is the immediate posterior relation when they pass behind the first part of the duodenum.

Lying between the two layers of the lesser omentum along the lesser curvature of the stomach are the right and left gastric arteries with their accompanying veins, lymph vessels, lymph nodes, and the anterior and posterior vagal trunks.

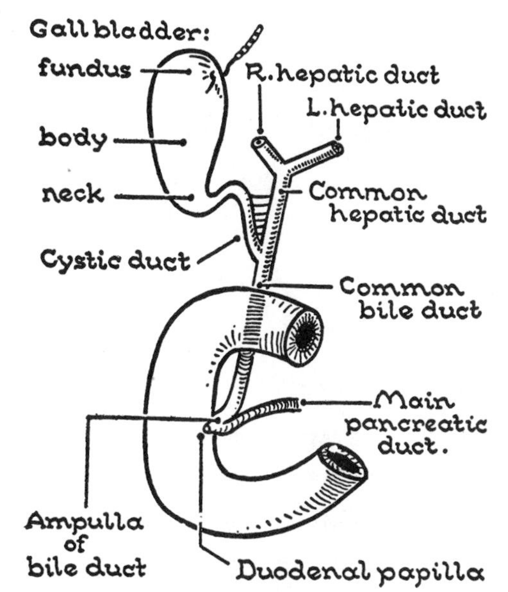

Fig. 274. The bile passages. (The average lengths of the cystic, common hepatic, and (common) bile ducts are 3.4, 3.2, and 6.5 cm, respectively.)

Celiac Trunk

The celiac trunk is the first of the three unpaired arteries that supply the gastrointestinal tract (*fig. 264*).

Access to the celiac trunk, pancreas, and other structures behind the omental bursa may be had via 3 routes—by cutting through: (1) the lesser omentum or (2) the transverse mesocolon or (3) the greater omentum (*fig. 248*).

Surface Anatomy and Relations. The celiac trunk springs from the aorta between the right and left crura of the diaphragm which straddle it (*fig. 274*). The pancreas lies below it. In the median plane, the arch that joins the two crura crosses the aorta at the disc between the last thoracic and first lumbar vertebrae, i.e., the depth of a vertebral disc above the transpyloric plane.

The celiac trunk or artery is only about 1 cm long. It is surrounded by the celiac plexus of nerves, wide of which are the tough, nodular celiac ganglia; wide of these are the suprarenal glands (*fig. 274.1*).

A stout and strong branch of the posterior vagal trunk, containing fibers from both vagi, descends along the left gastric a. to the celiac plexus.

Later, the greater and lesser splanchnic nerves will be seen piercing the crura of the diaphragm to reach the celiac ganglia (pp. 243–244).

Distribution and Branches

The celiac trunk supplies the stomach, the adjacent parts of the esophagus and duodenum, and the three unpaired glands —liver, pancreas, and spleen (*figs. 275 and 276*). It has three branches:

1. Left Gastric:
    - esophageal
    - gastric
    - (aberrant left hepatic)
2. Splenic:
    - pancreatic
    - splenic
    - short gastric
    - left gastro-epiploic
3. Hepatic:
    - gastroduodenal:
        - supra- and retroduodenal
        - post. sup. pancreaticoduodenal
        - ant. sup. pancreaticoduodenal
        - right gastro-epiploic

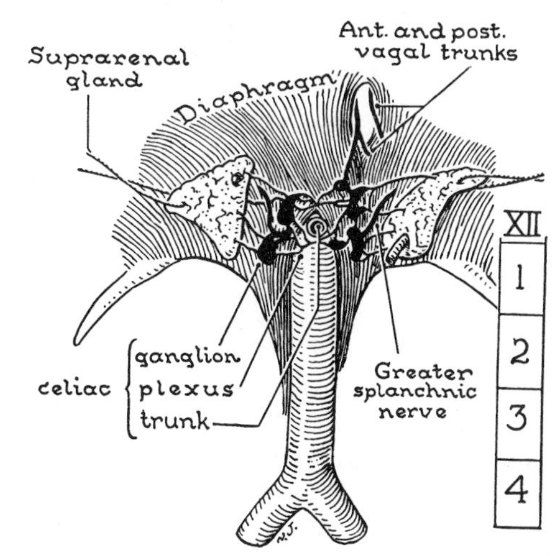

Fig. 274.1 The relations of the celiac trunk and its vertebral level. (Glands are retracted.)

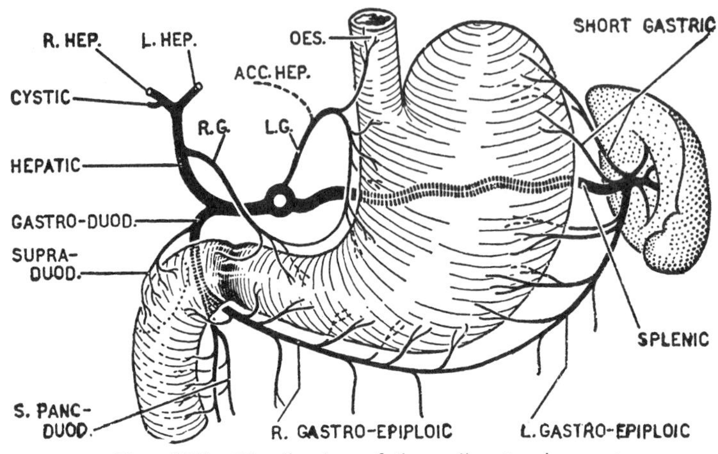

Fig. 275. Distribution of the celiac trunk or artery.

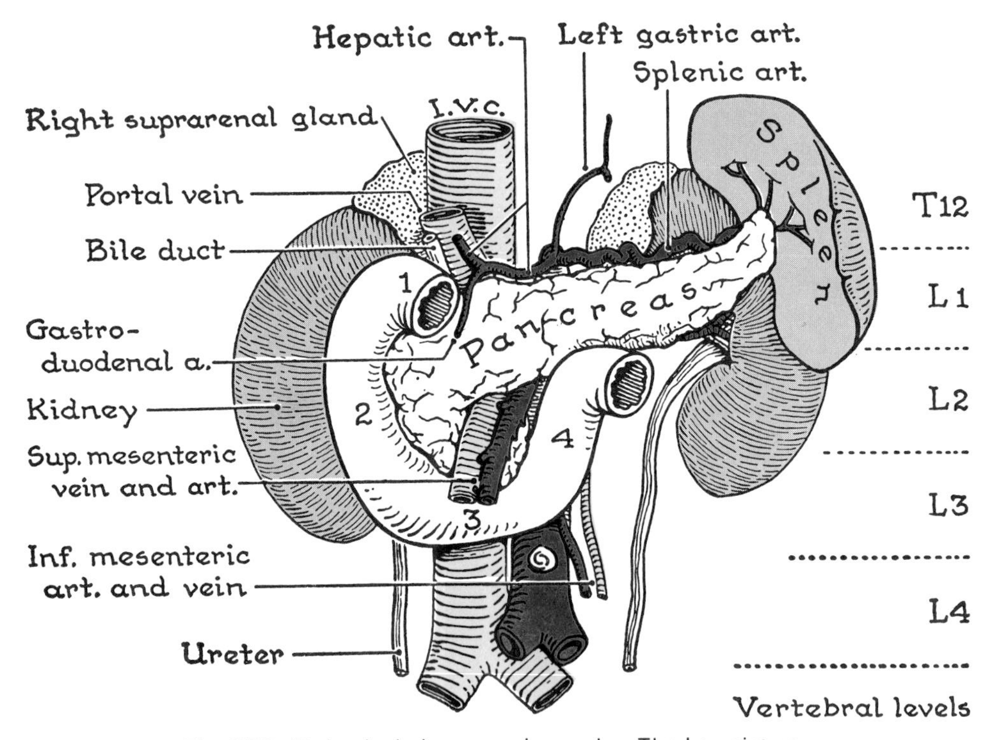

Fig. 276. Abdominal viscera and vessels—The key picture.

right gastric
right hepatic and cystic branch
left hepatic

The Left Gastric Artery is the smallest of the three branches of the celiac trunk but the largest of the five arteries to the stomach. It runs along the lesser curvature of the stomach between the layers of the lesser omentum and anastomoses with the right gastric artery.

Branches. Its branches are *gastric* and *esophageal*, which anastomoses with esophageal branches of the thoracic aorta.

Variation. Aberrant left hepatic arteries arise from the left gastric a. in one subject in four, many completely replacing normal left hepatic aa. (Michels).

The Splenic Artery is the largest branch. It takes a serpentine course to the left along the upper border of the pancreas. It crosses the left suprarenal gland and half the breadth of the left kidney to run between the layers of the lienorenal ligament to the hilus of the spleen.

Branches. Pancreatic branches to the body and tail of the pancreas; several *short gastric branches*, via the gastrolienal ligament to the fundus and greater curvature of the stomach; and the *left gastro-epiploic branch*, via the gastrolienal lig. into the gastrocolic lig. to the greater curvature of the stomach (never closer than a finger's breadth to it). As its name indicates, it sends *gastric* branches to the stomach and *epiploic* (omental) branches to the greater omentum.

The Hepatic Artery swings to the right along the upper border of the pancreas to the front of the portal vein. There it divides into two limbs. The ascending limb is the continuation of the main artery upward within the lesser omentum, in front of the portal vein, and on the left side of the bile passages (*fig. 276*). The descending branch is the *gastro-duodenal a.*

Terminal Branches. In the porta, the hepatic artery ends by dividing into the right and left hepatic aa., which enter the corresponding lobes. The right hepatic a. crosses the common hepatic duct (*fig. 277*).

The Cystic Artery arises from the right hepatic a. and is an end-artery to the gall bladder.

Details Important Only to Surgeons. The cystic a. usually arises in the angle between the common hepatic duct and the cystic duct. Its two branches anastomose with each other, supply the gall bladder, and send twigs to the liver. In about 25 per cent of cases the two branches arise independently, i.e., there are two cystic arteries. The cystic a. may arise from any nearby artery. When it arises to the left of the common hepatic duct, it must cross that duct to reach the gall bladder, and the crossing usually takes place in front of the duct. (Daseler, Anson, Hambley, and Reimann.)

Collateral Branches of the Hepatic A. (1) The *right gastric artery* (*fig. 275*) descends to the lesser curvature of the stomach. (2) The *gastroduodenal artery* passes downward between the first part of the duodenum and the pancreas, where, after a course of about 2 cm, it divides into the *right gastro-epiploic* and the *ant. sup. pancreaticoduodenal artery* (*fig. 275*).

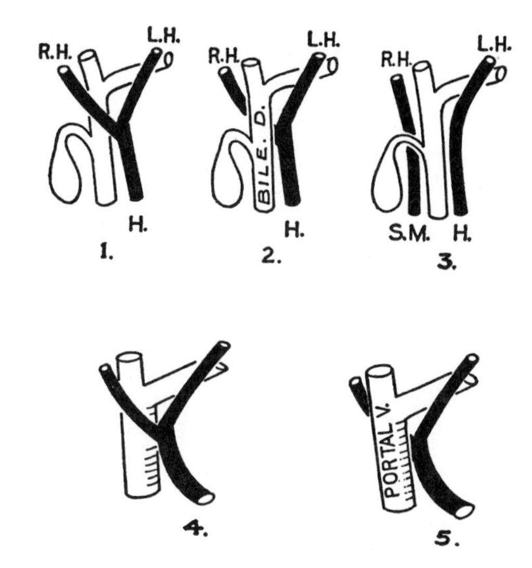

Fig. 277. Variations in the course of the right hepatic artery in 165 specimens:

1. anterior to bile duct 24%
2. posterior to bile duct ...... 64% } 100%
3. arising from superior mesenteric artery 12%
4. anterior to portal vein 91%
5. posterior to portal vein ..... 9% } 100%

The gastroduodenal a. gives off an end-artery, the "*Supraduodenal a.*" (Wilkie) (*fig. 275*), *retroduodenal twigs* to the back of the duodenum, and the *sup. pancreaticoduodenal aa.* (*post. and ant.*). These last effect a double arch with the inferior pancreaticoduodenal branches of the superior mesenteric a. (*fig. 300*), one arch lying in front of the head of the pancreas and the other behind it; both arches supply pancreatic and duodenal branches.

The right gastro-epiploic a. runs between the two layers of the greater omentum, a finger's breadth from the greater curvature of the stomach, and commonly anastomoses with the left gastro-epiploic a. Both vessels have *gastric branches* and long slender epiploic branches, which descend in the omentum.

Collateral Circulation about the Liver. To deprive the liver altogether of its arterial blood is usually fatal. There is however a *collateral anastomosis.* Thus: (1) the larger branches and the precapillaries of the right and left hepatic aa. anastomose so well both in the fissures of the liver and deep to the capsule that fluid injected into the one artery flows from the cut end of the other. The deep intrahepatic aa., however, are end-arteries. (2) If the common hepatic a. is ligated, the arterial supply to the liver may yet be assured in those 12 per cent of persons in whom the right hepatic a. arises from the superior mesenteric a. (*fig. 277*). (3) It may also be assured in the 11.5 per cent in whom a "replaced left hepatic a." springs from the left gastric a.; and perhaps in some of those in whom an accessory left hepatic a. does so (p. 214). (4) If the hepatic a. is obstructed gradually on the aortic side of the origin of the right gastric a., the circulation is maintained by the anastomosis the right gastric a. effects with the left gastric a. (5) The inferior phrenic, the cystic, and the superior epigastric aa. send fine twigs to the liver, the last *via* the falciform lig.

PORTAL VEIN

The portal vein drains all, and drains only, the gastro-intestinal tract and its unpaired glands, the liver of course excepted. It returns to the liver the blood delivered by the celiac, superior mesenteric, and inferior mesenteric arteries to these parts (*fig. 277.1*).

Course. The portal vein is formed between the head and neck of the pancreas by the union of the splenic (which represents the celiac artery), the sup. mesenteric, and the inf. mesenteric veins. It ascends to the right end of the porta where it divides into the right and left portal veins. The right vein enters the right lobe; the left vein passes transversely to the left end of the porta and supplies the caudate, quadrate, and left lobes (*fig. 249*, p. 199).

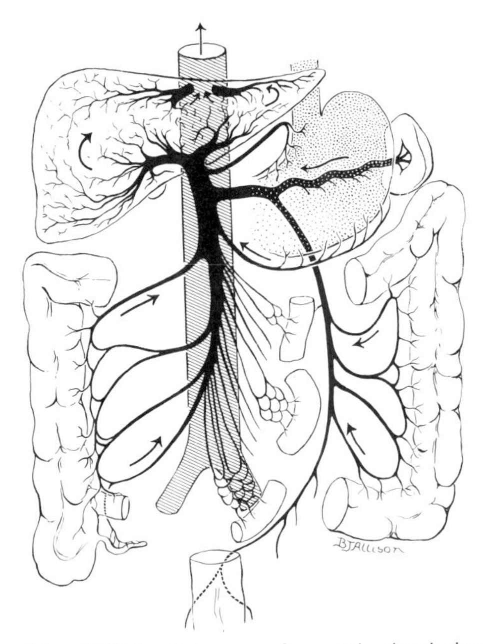

Fig. 277.1. Scheme of portal circulation. (From *Primary Anatomy*.)

There are no functioning valves in the portal system.

Relations. The portal vein ascends behind the neck of the pancreas, first part of the duodenum, and the gastroduodenal a. It then enters the lesser omentum, where the bile passages and hepatic artery lie in front of it.

Behind the portal vein lies the i. v. cava (fig. 277.2). Intervening are: (1) at the epiploic foramen (*fig. 247*), (2) below the foramen—two areolar membranes, one covering the i. v. cava, the other covering the portal vein (*fig 302*), and (3) above the foramen—the caudate process of the liver (*fig. 249*).

The ligamentum teres and l. venosum are attached to the left portal vein at the left end of the porta (*fig. 249*).

Tributaries (*fig. 277.1*).

Portacaval Venous Anastomoses. When the portal vein is slowly obstructed, as a result of disease of the liver or from other causes, the portal blood may enter the inf. vena cava by way of certain anastomotic veins, which then become dilated and varicose—and they may burst.

They are as follows (*fig. 278*):

a. At the upper end of the gastro-intestinal tract: the esophageal branches of the left gastric vein anastomose with esophageal branches of the azygos veins in the thorax. (When varicose they are called esophageal varix.)

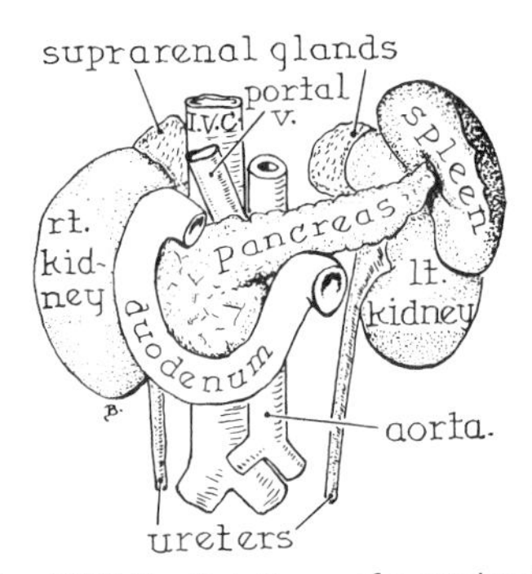

Fig. 277.2. Relations of portal vein.

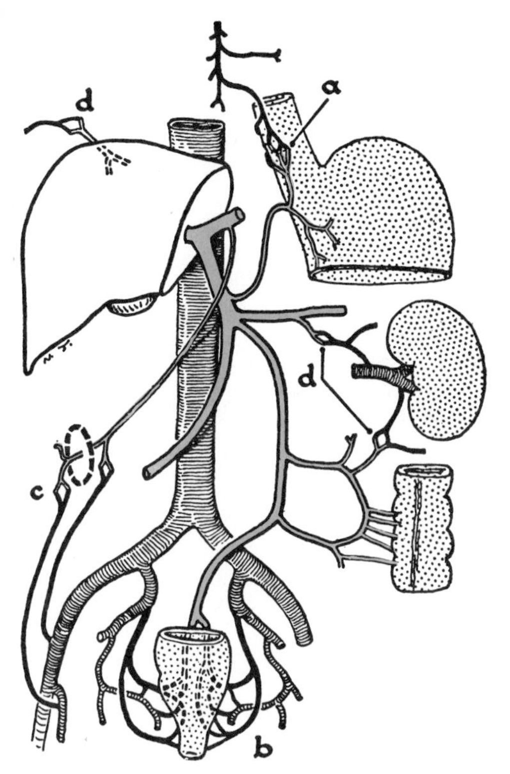

Fig. 278. Diagram of the chief portacaval anastomoses (see text).

b. At the lower end of the gastro-intestinal tract: the superior rectal vein anastomoses with the middle and inferior rectal veins and, most important of all, with the *pelvic venous plexuses* (p. 286). (When varicose, the rectal or hemorrhoidal veins are called hemorrhoids.)

c. Fine *para-umbilical veins* run with the round lig. of the liver from the left portal vein to the umbilicus where they anastomose with the superficial and deep epigastric veins.

d. Twigs of the colic and splenic veins anastomose in the extraperitoneal fat with twigs of the renal vein and with veins of the body wall. Here may be included twigs from the bare area of the liver.

Since the anastomotic veins seldom possess valves, they can conduct blood equally readily in either direction depending on whether the obstruction is in the portal vein or in the inferior vena cava (E. A. Edwards).

Development of the Portal System. In prenatal life the *right* and *left vitelline veins* from the yolk sac and the *right* and *left umbilical veins*, originally from the allantois but later from the placenta, opened independently into the common sinus venosus of the heart (*fig. 279*) until the developing liver intercepted them and broke them up into the anastomosing sinusoids of the liver. Thereafter, the prehepatic parts of the right umbilical, left vitelline, and left umbilical veins disappeared, leaving only the prehepatic part of the right vitelline to conduct blood from the liver to the heart. [Prehepatic = cephalad to the liver; posthepatic = caudad to the liver.] Definitively, this prehepatic part of the right vitelline vein becomes the terminal segment of the i. v. cava (*fig. 316*).

The posthepatic part of the right umbilical vein disappeared, leaving the corresponding part of the left umbilical vein to bring blood from the placenta to the liver. A short cut develops, the *ductus venosus*, connecting the left umbilical vein with the prehepatic part of the right vitelline vein.

The *right and left vitelline veins* made a figure-of-8 anastomosis around the first and third parts of the duodenum. Out of this the portal vein took form by the disappearance of the posterior (right) limb of the 8 below and of the anterior (left) limb of the 8 above. It is joined by the superior mesenteric, inferior mesenteric, and splenic veins (*fig. 292*).

STOMACH

Surface Anatomy. The stomach (Gk. gaster; L. venter) has been described on p. 193.

Structure. The coats of the stomach are: serous, subserous, muscular, submucous, and mucous.

Muscular coat. It has three layers—an outer longitudinal, a middle circular, and an inner oblique (*fig. 280*). The *longitudinal fibers* are continuous with those of the esophagus; they are best marked along the curvatures; at the pylorus they dip in to join the sphincter and only a few are continuous with those of the duodenum.

The *circular fibers* are present everywhere

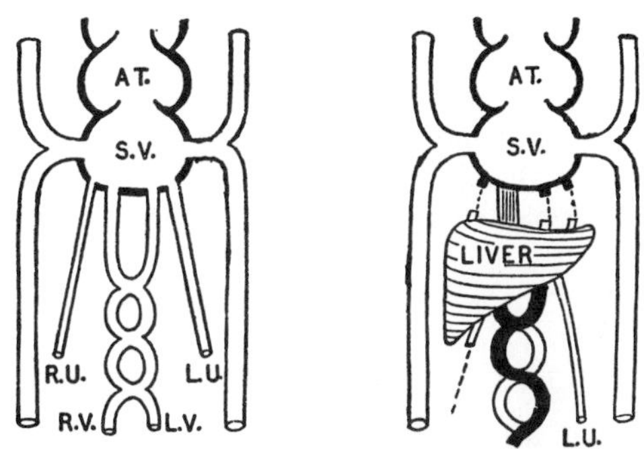

Fig. 279. Development of portal vein and terminal part of i. v. cava (see text). (See also *fig. 316*.)

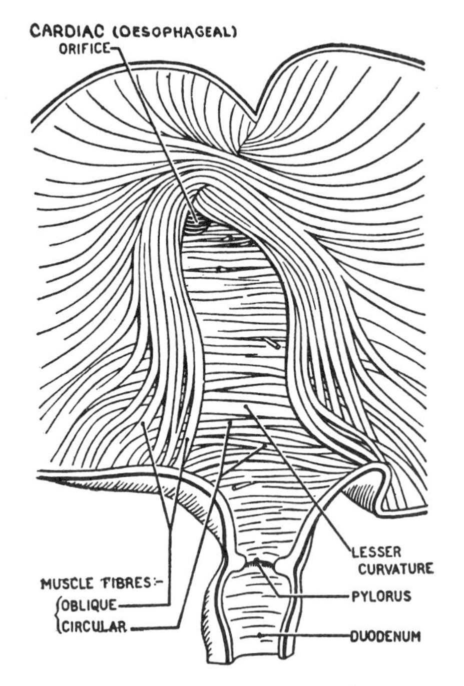

Fig. 280. The muscular coat of the stomach seen from within. (The stomach was opened along the greater curvature and the mucous and submucous coats were removed.)

except at the fundus, and they are greatly increased at the pylorus to form a sphincter. At the pylorus and anus powerful sphincters are required to keep the contents from escaping from the stomach and rectum.

Anatomically the presence of a sphincter at the esophageal or cardiac end of the stomach is disputed; perhaps the diaphragm suffices. Physiologically, however, there is a cardiac sphincter which opens on stimulation of the vagus and closes on stimulation of the sympathetic.

The *oblique* fibers form ∩-shaped loops that extend over the fundus and down both surfaces of the stomach to the pyloric antrum, the cardiac notch forming their medial limit.

Mucous membrane is rugose when the stomach is empty and three or four uninterrupted ridges lie along the lesser curvature from esophagus to pylorus forming a gutter (*fig. 281*).

Arteries

The arteries of the stomach are derived from the celiac trunk. The right and left gastric aa. form an arch close to the lesser curvature; the right and left gastro-epiploic aa. usually form a feeble arch at some distance from the greater curvature, the left artery receding from the stomach as it is traced to the left. Three or four short gastric aa. pass to the greater curvature at the fundus via the gastrolienal ligament (*fig. 275*).

All arterial branches supplying the stomach penetrate the muscular coats and enter the submucosa where they form a very extensive network of comparatively large vessels. From this network in the submucosa two systems of branches are given off. Of these, one turns back to the muscular coats; the other continues to the mucosa. The branches to the mucosa usually divide twice, run spirally toward the muscularis mucosae, and pierce it to enter the mucosa where they suddenly become smaller by giving off end-branches (i.e., vessels connected only by means of a capillary network). Each end-artery continues to run a spiral course, and supplies an area of mucosa of about 2.5 mm in diameter. The submucous network on the lesser curvature is made up of long parallel vessels which are smaller, make fewer anastomoses, and run more than twice the distance of similarly sized vessels in other parts of the stomach (Reeves).

Lymphatics of the Stomach, p. 252 and *fig. 327.1*.

Nerves of the Stomach

Distribution of Vagal Trunks within the Abdomen. Due to the anticlockwise rotation undergone by the stomach during development, the anterior vagal trunk (left vagus) enters the abdomen in front of the

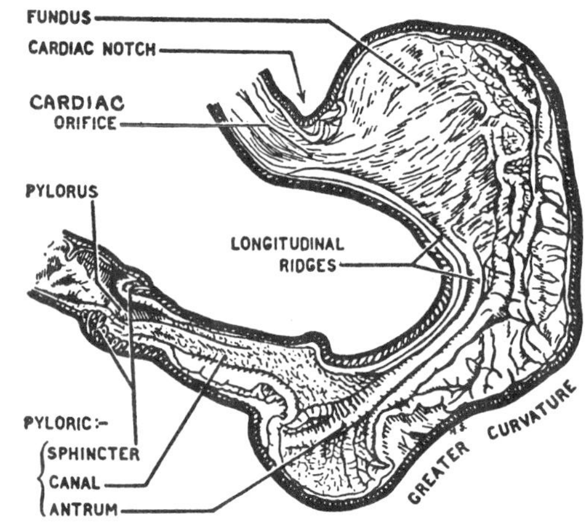

Fig. 281. The mucous coat of the posterior half of the stomach.

esophagus and the posterior vagal trunk (right vagus) behind it. Either trunk is occasionally in two or three branches. Both trunks, each carrying fibers from both vagi, run close to the lesser curvature and send gastric branches to the respective anterior and posterior surfaces of the stomach as far as the pyloric antrum.

The *Anterior Vagal Trunk* sends *hepatic branches* curving upward in the lower part of the lesser omentum to the porta hepatis, and it is through *pyloric branches* descending from these that the pyloric antrum, pylorus and first part of the duodenum are supplied.

From the *Posterior Vagal Trunk* a branch descends along the stem of the left gastric artery to the *celiac plexus* whence its fibers, in company with sympathetic fibers, are distributed along the branches of the aorta to the abdominal viscera, e.g., intestines (proximal to the left colic flexure), pancreas, and spleen. This is the only connection these organs have with parasympathetic nerves (*figs. 274.1* and *282*).

Sympathetic fibers, both afferent and efferent, from cord segments Th. 6, 7, 8, 9, (and 10) via the splanchnic nerves relayed in the celiac ganglia, pass to the stomach along the blood vessels.

Relations of the Stomach

Antero-superiorly are: the left lobe of the liver, diaphragm, and anterior abdominal wall. The diaphragm separates it from the left lung and pleura and apex of the heart.

Postero-inferiorly, the omental bursa intervening, is **the stomach bed** formed to the extent shown in *figure 282*.

LIVER

(Continued from pp. 197, 198–199.)

An **H-Shaped Group of Fissures and Fossae** subdivides the posterior and inferior surfaces into four combined areas. The crossbar of the H is the *porta hepatis;* the area in front of it is the *quadrate lobe;* the area behind, the *caudate lobe* (*fig. 283*).

The left sagittal limbs of the H are deep fissures containing the lig. teres and the lig. venosum.

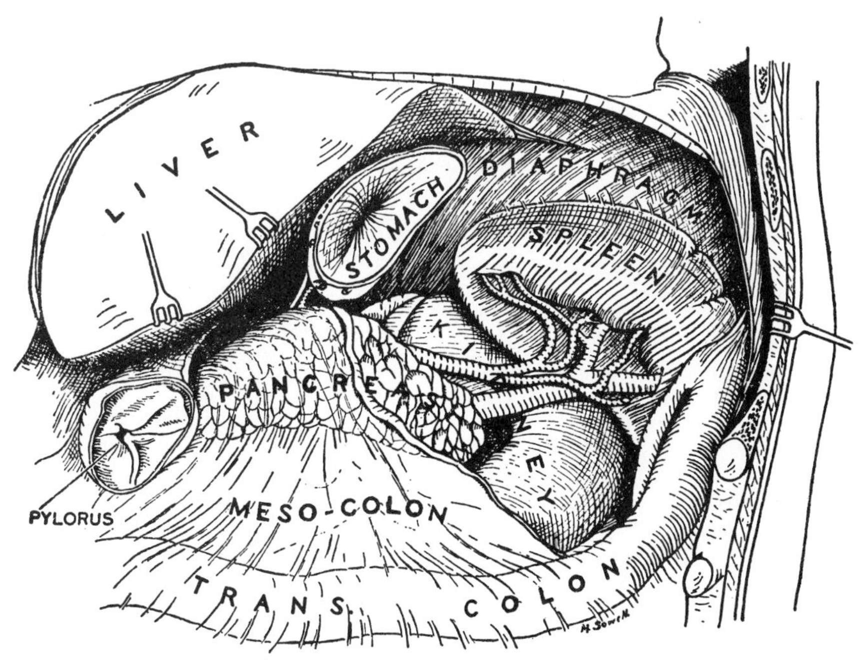

Fig. 282. The stomach bed. (This pancreas is unusually short; the left suprarenal gland, the left gastric artery and the branch of the posterior vagal trunk to the celiac plexus are not labeled.)

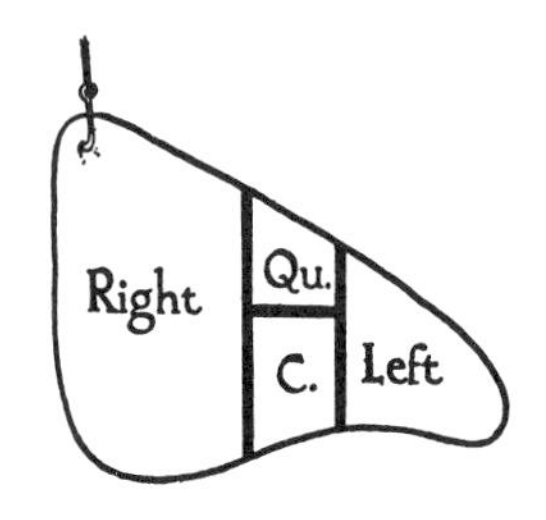

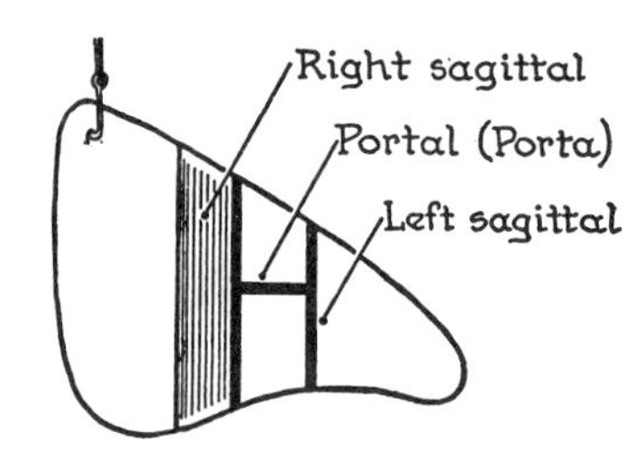

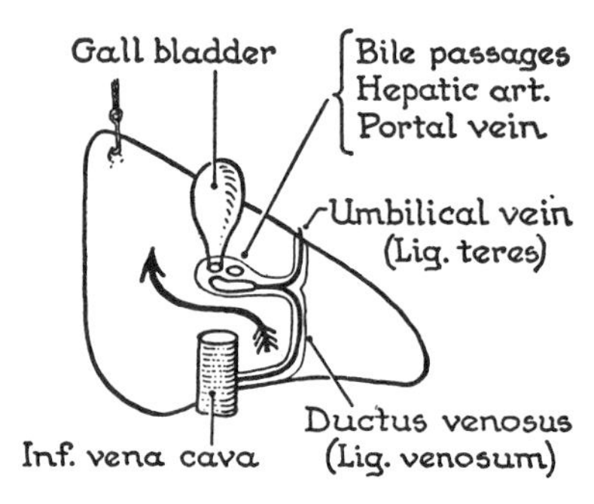

Fig. 283. Diagrams of the liver, hooked upward to show (1) the H-shaped fissure on the inferior and posterior surfaces, (2) the subdivisions of this fissure, and (3) their contents.

The right sagittal limbs of the H are fossae for the lodgment of the gall bladder and i. v. cava.

The **Posterior Aspect** cannot be seen until the liver is removed. On the left, it is covered with peritoneum and is grooved by the esophagus.

On the right, it is in naked contact with the diaphragm. This *bare area* is bounded by the coronary ligament where peritoneum is reflected from diaphragm to liver (*fig. 251*). The i.v. cava occupies the leftmost portion of the bare area; the right kidney and suprarenal gland encroach on the bare area from below (*fig. 283.1*).

Visceral Surface of the Liver. Aided by *figures 283.1* and *251*, place the viscera methodically, thus:

Parts of G.I. Tract & Co.
esophagus—duodenum—stomach—colon

Parts of Three Paired Glands
right kidney—right suprarenal gland

Details. The transverse colon, which crosses in front of the 2nd part of the duodenum (*fig. 243*), runs from right to left behind the sharp, inferior margin of the liver as far as the median plane, and leaves its impress on the right lobe, gall bladder and quadrate lobe.

Behind these intestinal areas the right lobe is hollowed for the right kidney and suprarenal,

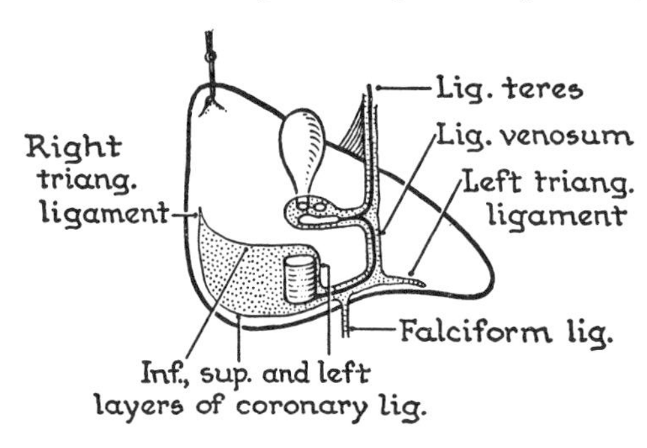

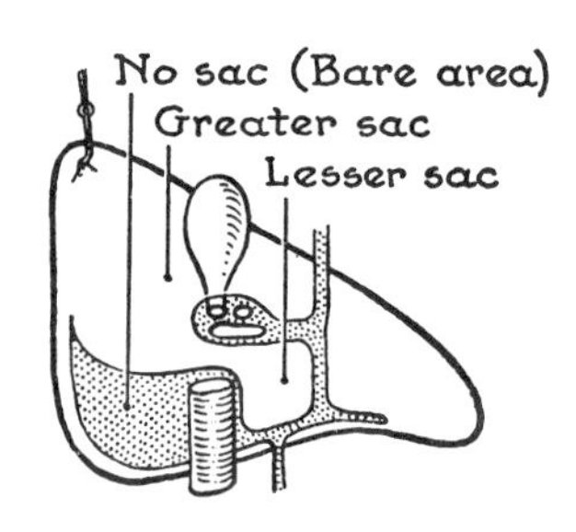

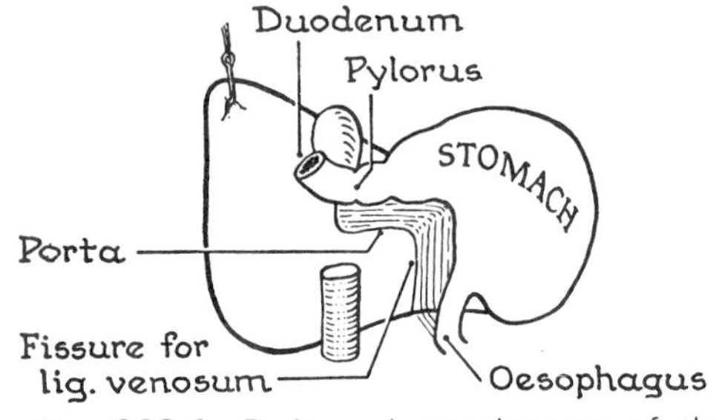

Fig. 283.1. Peritoneal attachments of the inferior and posterior surfaces of the liver.

but is separated from these two glands by the peritoneum of the hepatorenal pouch (*fig. 246*). The upper ends of these two glands usually extend above the hepatorenal ligament (i.e., lower layer of the coronary lig.) and so, come into direct contact with the bare area of the liver.

Surface Anatomy of the Liver (*fig. 233*).

1. *The base* or right lateral aspect extends from near the right iliac crest (above or below) in the midlateral line, across ribs (11), 10, 9, and 8, up to rib 7.

2. *The upper limit*, of course, is the same as the upper limit of the diaphragm: it crosses the xiphisternal joint in the median plane and rises to the 5th rib in the right midclavicular line; *the apex* or leftmost part fails by about 2 cm to reach the left midclavicular line and lies 2–3 cm inferomedial to the left nipple.

3. *The sharp inferior border* crosses the transpyloric plane about 2 to 3 cm to the right of the midline.

The position of the **gall bladder** varies with that of the liver. Its fundus typically lies at the lateral border of the Rectus Abdominis somewhat below the costal margin (*fig. 213*).

Lobes and Segments of Liver (*fig. 284*). *Descriptively*, the liver is divided into unequal right and left lobes by the falciform lig. *Structurally*, however, the right and left portal veins, hepatic arteries, and hepatic ducts serve approximately equal halves of the liver. The plane between the two halves (structural lobes) run anteroposteriorly through the fossa for the gall bladder and the fossa for the inf. vena cava; i.e., through the right sagittal fossae, and between the right and left groups of vessels and ducts in the porta.

The entire quadrate lobe and half the caudate lobe are served by left vessels and ducts. And, according to the prevailing pattern, no branches of the portal vein, hepatic artery or hepatic duct cross this right sagittal plane.

According to the biliary drainage, the left half of the liver is divided by the plane of the left sagittal fissures into a medial and a lateral segment; whereas the right half is divided by an oblique plane into an anterior and a posterior segment. Each of these four segments is subdivisible into an upper and a lower area, as in *figure 284*. (Healey, Schroy and Sorensen; and Hjortsjo.)

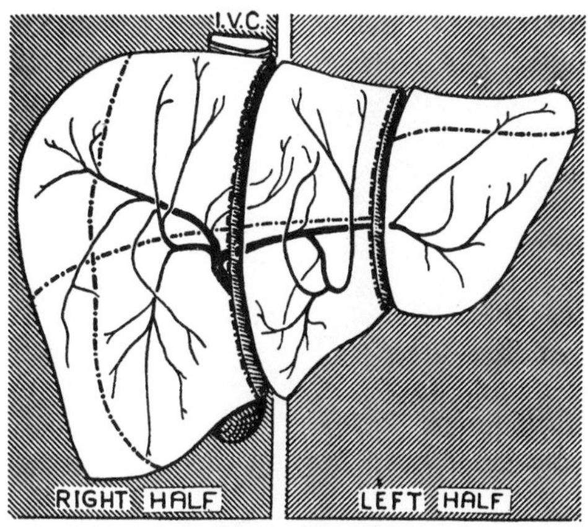

Fig. 284. The segments of the liver, according to the biliary drainage. (After Healey.)

BILE PASSAGES AND GALL BLADDER

These structures have been described with the liver (*figs. 274, 284.1*).

Function. The function of the gall bladder is to concentrate and store the bile brought to it from the liver via the cystic duct between meals and to discharge it into the intestine via the cystic duct during meals.

The Vessels and Nerves of the gall bladder. *The Cystic Artery* (see p. 215). *The Cystic Veins* mostly plunge through the fossa for the gall bladder into the liver substance and behave like branches of the portal vein. *The Lymph Vessels* pass to the cystic node at the neck of the bladder and thence downward along the biliary chain. *The Nerves* are derived from the anterior vagal trunk (p. 219), and from cord segments Th. (6), 7, 8, 9, (10) via the celiac plexus.

Variations. The bile duct developed as an outgrowth from the duodenum (*fig. 293*). It branched and rebranched more or less dichotomously. One of the main branches, the *cystic duct*, instead of branching gave rise to a blind vesicle, the *gall bladder*.

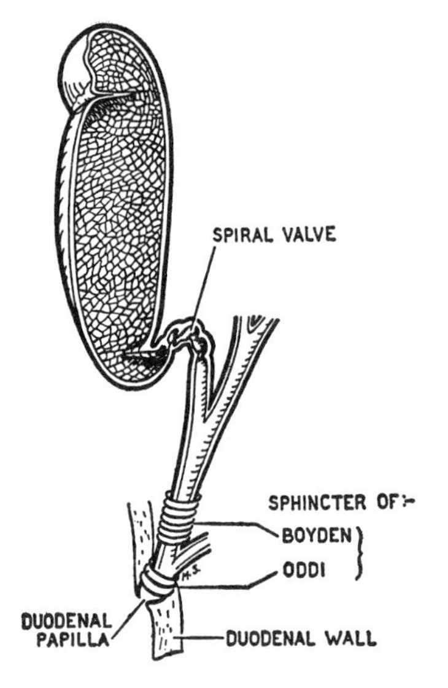

Fig. 284.1. The mucous membrane of the gall bladder and extrahepatic bile passages. The two sphincters are shown diagrammatically. This bladder happened to have a folded fundus.

Irregular branching of the bile passages is common, and when the branch is in surgical danger, the fact is of importance. Thus (*fig. 285 A*) the cystic duct may end much lower than usual or much higher, sometimes joining the r. hepatic duct (*B*), and it may swerve across the common hepatic duct (*C*). When there are two right hepatic ducts, the lower is erroneously called an *accessory hepatic duct*—it is not an additional or supplementary duct, but merely one that arises unusually early (*fig. 285, D, E.* and *F*).

The gall bladder is rarely *absent* (unless like an offending appendix, it has been removed). It is rarely *buried* in the liver, but occasionally it is suspended from the liver by an *acquired mesentery*. It is commonly attached to the transverse colon or to the duodenum by a *peritoneal fold*. Occasionally, it has a *sacculation* at its neck (Hartman's pouch), and occasionally the fundus is *congenitally folded* upon itself within its serous or peritoneal coat (*fig. 284.1*), or the folding may include the serous coat. Although *bilobed* gall bladders are very common in domestic animals, in man they are rare [17 cases of double bladder, however, each with a separate cystic duct, have been reported (Boyden)].

STRUCTURE. *The Intrahepatic Ducts* are described on p. 54. *The Extrahepatic Ducts* (viz., right and left hepatic, common hepatic, cystic, and bile) are fibrous tubes containing many elastic fibers and lined with columnar epithelium, but without a muscular coat, except at the lower 5 to 6 mm of the bile duct where there is a strong and effective submucous sphincter; the duodenal papilla (or the ampulla) also has a sphincter. In the cystic duct there is a spiral fold (valve) which probably serves to keep the duct patent (*fig. 284.1*).

The inner surface of the *Gall Bladder* is covered with small polygonal compartments, opening on to the interior and resembling the cut surface of a honeycomb. Like villi, these greatly increase the absorptive surface. The wall of the gall bladder has: a single, inner layer of columnar cells, a tunica propria, a muscular coat of decussating fibers, a subserous coat, and a serous coat (except where the bladder is applied to the liver).

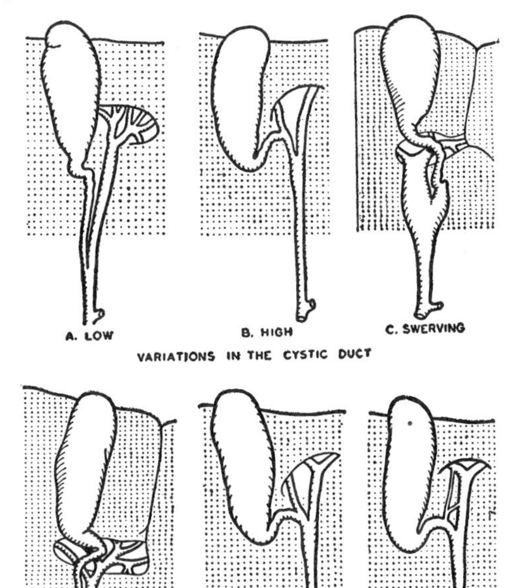

Fig. 285. Variations of the bile passages. *A, B,* and *C* = variations in length and course of cystic duct. *D, E, F* = accessory right hepatic ducts. (Of 95 gall bladders and bile passages injected in situ with melted wax and then dissected, seven had accessory hepatic ducts in positions of surgical danger. Of these, four joined the common hepatic duct near the cystic duct (*D*), two joined the cystic duct (*E*), and one was an anastomosing duct (*F*).

16 MESENTERIC VESSELS, DUODENUM, AND PANCREAS

MESENTERIC VESSELS

Superior Mesenteric Artery—Branches: To Small Intestine; To Large Intestine; Arcades; Colic Arteries.
Inferior Mesenteric Artery—Marginal Artery (of Drummond); Variations.
Superior and Inferior Mesenteric Veins—Marginal Vein.

STRUCTURE OF INTESTINES

DUODENUM AND PANCREAS

Surface Anatomy and Vertebral Levels of Duodenum.
Pancreas—Structure and Function; Surface Anatomy and Vertebral Levels; Notes Explaining Relationships.
Anterior Relations of Duodenum and Pancreas; Posterior Relations; Blood Supply.
Interior of Duodenum; Main Duct of Pancreas; Displaying Common Bile Duct.

MESENTERIC VESSELS

Superior Mesenteric Artery

The superior mesenteric artery arises from the aorta just below the celiac trunk and supplies the small and large intestines from the 2nd part of the duodenum to the left colic flexure. For embryological reasons (*fig. 294*), it passes between the head and neck of the pancreas.

"Clamped" between it and the aorta, like a nut in a nutcracker, are the left renal vein and the third part of the duodenum (*figs. 285.1, 291*). Both—at least theoretically—may be obstructed by the "clamp." Entering the root of the mesentery applied to the inferior vena cava, it runs to the ileum 10 or 12 cm from the ileocecal junction. There it ends by forming an arch with one of its own branches (*fig. 286*).

Branches:

1. To the small intestine:
   inferior pancreaticoduodenal, jejunal and ileal.
2. To the large intestine:
   ileocolic, right colic, middle colic.

The **inferior pancreaticoduodenal arteries** (*fig. 300*), anterior and posterior, give branches to the head of the pancreas and the 2nd and 3rd parts of the duodenum.

The **jejunal and ileal arteries,** 18 or so, fan out from the left side of the artery into the mesentery where they unite to form loops or arches from which arise the *vasa recta*.

Arcades. The number of tiers of arcades varies from subject to subject and from area to area, the heaviest concentration generally being in the second quarter of the length of the small intestine. In the first quarter there are 2–4 tiers (average, 2); in the second quarter, 3–5 (average, 4); in the third quarter, 2–4 (average, 2); and in the last quarter, 0–4 (only one in the majority of cases) (Michels *et al.*).

The vasa recta do not themselves anastomose (*fig 238*). In the submucosa their twigs anastomose freely. (Benjamin and Becker.)

The Three Colic Branches of the s. mesenteric a. arise from its right border (*fig. 286*), commonly from only 2 stems, one of which bifurcates. Each in turn bifurcates and forms loops or arches at a very variable distance (0.5–8 cm) from the colon. Similarly, the colic branches of the inferior mesenteric a. form loops, and the result is a series of anastomosing links, "the **mar-**

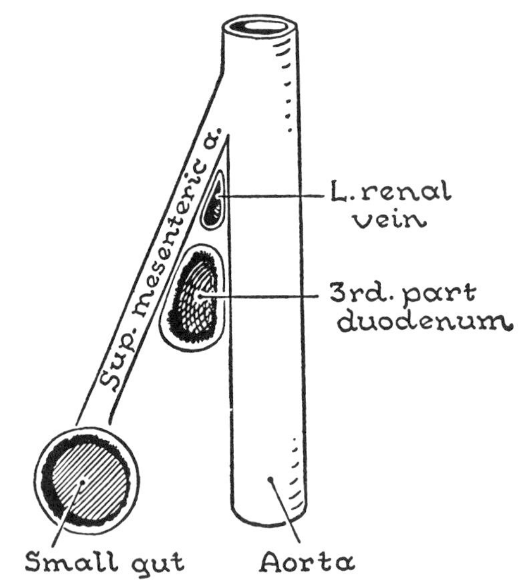

Fig. 285.1. Compression of the left renal vein and the duodenum—as in a nutcracker.

ginal artery", which parallels the entire colon (*fig. 286* and p. 225).

The **ileocolic artery** descends subperitoneally toward the ileocecal region, crossing the i. v. cava and ureter, and then bifurcates. A descending branch divides into—anterior cecal, posterior cecal, appendicular, and ileal branches (*fig. 287*).

The *appendicular branch* descends behind the end of the ileum and runs in the free edge of the mesentery of the appendix.

Variations. Commonly the appendix is supplied by two arteries (Shah and Shah, Solanke).

The **right colic a.** crosses the same structures as the ileocolic a. and divides into a descending and an ascending branch: the latter crosses the lower pole of the right kidney.

The **middle colic a.** arises at the lower border of the pancreas. It curves downwards in the right half of the transverse mesocolon and divides into a right and a left branch (*fig. 286*).

Inferior Mesenteric Artery

The inferior mesenteric artery supplies the large gut from the left end of the transverse colon to the lower end of the rectum. It arises from the front of the aorta 4 cm above its bifurcation (therefore, 2 cm above the umbilicus, and therefore in front of the third lumbar vertebra), where the lower border of the duodenum overlaps it (*fig. 291*). It descends retroperitoneally on the aorta and the psoas fascia, crosses the left common iliac artery, and enters the pelvis as the superior rectal (sup. hemorrhoidal) a. (*fig. 288*).

Branches:

1. left colic (upper left colic),
2. sigmoid (lower left colic).

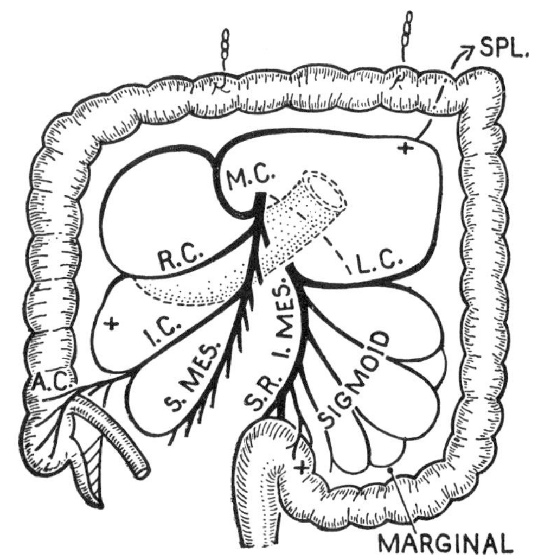

Fig. 286. The superior and inferior mesenteric arteries. (+ denotes three weak points in the marginal anastomosis: between ileocolic and right colic, between middle colic and left colic, between lowest sigmoid and superior rectal.)

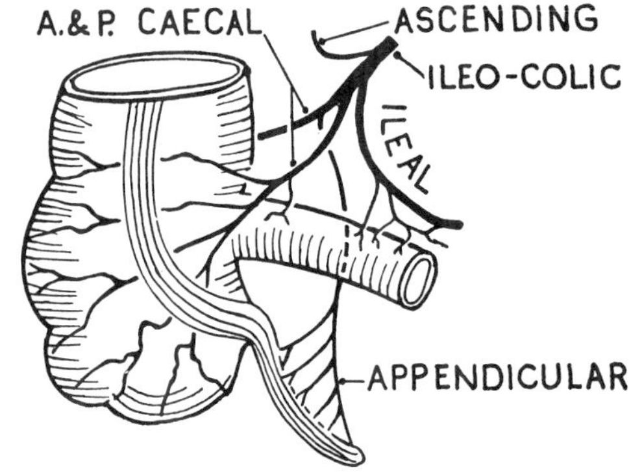

Fig. 287. The branches of the ileocolic artery.

The **left colic artery** passes retroperitoneally to the left across the inf. mesenteric vein, ureter, and testicular vessels, and divides into an ascending and a descending branch.

The **sigmoid arteries** are generally 2 to 4 in number (*fig. 288*). The upper branches cross the structures in front of the Psoas (*viz.*, inf. mesenteric vein, ureter, and testicular vessels). The lower branches cross the common iliac vessels and enter the sigmoid mesocolon to form two or three tiers of arches.

Marginal Artery (of Drummond)

The anastomosing loops of colic branches result in a continuous marginal artery situated from 0.5 to 8 cm from the wall of the large gut. It is closest along the descending and sigmoid colons.

Long and short terminal branches proceed from "the marginal artery" to the colon (*fig. 289*). The anastomoses they effect across the antimesenteric border are meager (Ross).

The short branches are 4 to 5 times as numerous as the long branches (*fig. 289*). Like the long branches they pass to the submucous plexus after a short tortuous subserous course. The muscular coats are mainly supplied by recurrent branches from the submucous plexus.

VARIATIONS. (1) In about 5 per cent of 100 specimens the marginal artery is discontinuous, due to the ileocolic a. failing to anastomose with the right colic. (2) The right colic a. very commonly takes origin either from the middle colic or the ileocolic artery. (3) The middle colic a. commonly has an accessory left branch. (4) The left colic a. supplies the left end of the transverse colon in about two-thirds of cases, and in one-third it fails to reach the left colic flexure. (5) The middle colic and left colic aa. probably always anastomose, although the portion of the marginal artery so-formed is long and usually single. (6) A large branch (the arc of Riolan) not rarely connects the stem of the s. mesenteric a. with the left colic a. on the posterior abdominal wall. (7) Occasionally a branch connects the left colic a. with the splenic a. (8) The marginal artery does not link up the lowest sigmoid a. with the superior rectal a. —except occasionally and feebly—so, if the superior rectal a. is obstructed beyond the origin of the lowest sigmoid a., there is little chance of an effective collateral circulation

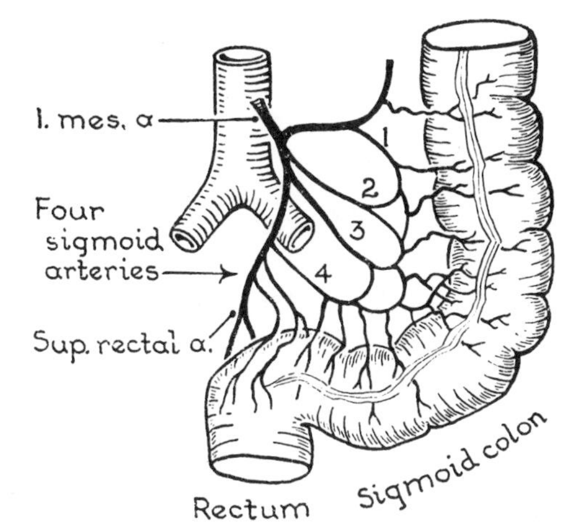

Fig. 288. Branches of inferior mesenteric artery.

being established (*fig. 286*) (Steward and Rankin; Basmajian; and Michels *et al.*).

Mesenteric Veins

The *superior mesenteric vein* lies on the right side of its artery and ends in the portal vein. The *inferior mesenteric vein* is the continuation of the superior rectal vein (sup. hemorrhoidal vein). It ends behind the neck of the pancreas by joining the angle of union between the sup. mesenteric and splenic veins as they form the portal vein.

STRUCTURE OF INTESTINES

When the gut is opened, folds of mucous membrane, the *plicae circulares*, are seen running transversely for variable distances around the gut wall. They begin in the duodenum, 5 cm from the pylorus, and end beyond the middle of the ileum. In the duodenum and upper part of the jejunum they are high (about 6 mm) and closely set; lower down they gradually become smaller and more widely separated.

Finger-like projections, *villi* (0.5 to 1.5 mm long) cover the mucous surface from pylorus to ileocecal orifice.

The *Ileocecal Orifice and Valve*. Here the circular muscle of the small gut, covered with mucous membrane, pouts into the large gut (*fig. 290*).

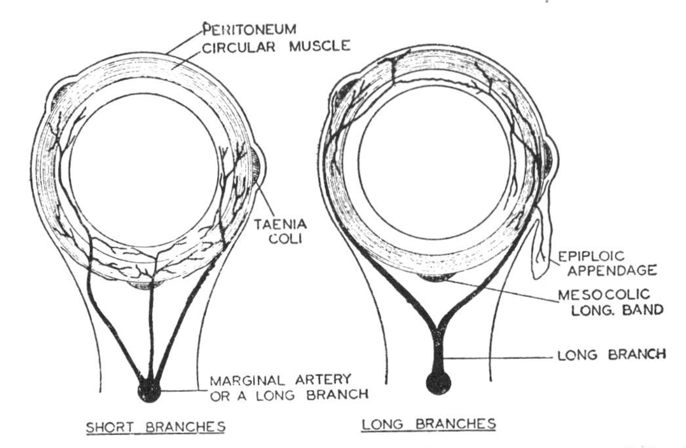

Fig. 289. The blood supply of the colon. (After Steward and Rankin.)

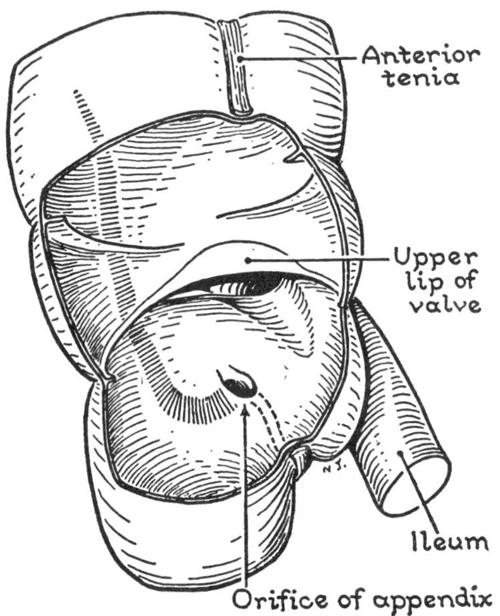

Fig. 290. The ileocecal valve, as seen on opening a dried cecum. In life, the orifice is circular.

Lymph follicles. From the middle of the ileum to the colon, small collections of lymphoid tissue, *solitary lymph follicles*, 2 to 3 mm in diameter, are present. Also in the ileum, at the antimesenteric border, are 20 or more *aggregated lymph* follicles (Peyer's patches); they are oblong, 1 cm wide by 3 cm or more long, the long axis being in the long axis of the gut.

DUODENUM AND PANCREAS

The word *Duodenum* is a Latin corruption of the Greek word dodekadaktulos, meaning 12 fingers (*cf.* the 12 islands in the Levant called the Dodecanese Islands). About 300 B.C., Herophilus of Alexandria introduced the name dodekadaktulos from its being as long as 12 fingers are broad in certain animals (Finlayson).

In man the duodenum is the part of the small intestine that has lost its dorsal mesentery. About 25 cm long, it is molded around the head of the pancreas in horseshoe fashion and is divided into four parts (*fig. 291*):

1st or superior—5 cm long.
2nd or descending—7½ cm long.
3rd or horizontal—10 cm long.
4th or ascending—2½ cm long.

Surface Anatomy of Duodenum: Vertebral Levels. The duodenum begins at the pylorus in the transpyloric plane 2 to 3 cm to the right of the median plane, and ends at the duodenojejunal junction slightly below the transpyloric plane 2 to 3 cm to the left of the median plane. The ends of the horseshoe are, therefore, only about 5 cm apart (*fig. 291*).

The Pancreas (Gk. Pan = all; kreas = flesh) weighs only about 170 grams and resembles the letter J, or a chemist's retort, set obliquely. The bowl of the retort, known as the *head* of the pancreas, lies within the concavity of the duodenum, whilst the stem of the retort, divided indefinitely into *neck*, *body*, and *tail*, slants obliquely across the abdomen. The tail abuts against the spleen. The head projects medially, behind the superior mesenteric vessels as the *uncinate process* (*fig. 291*).

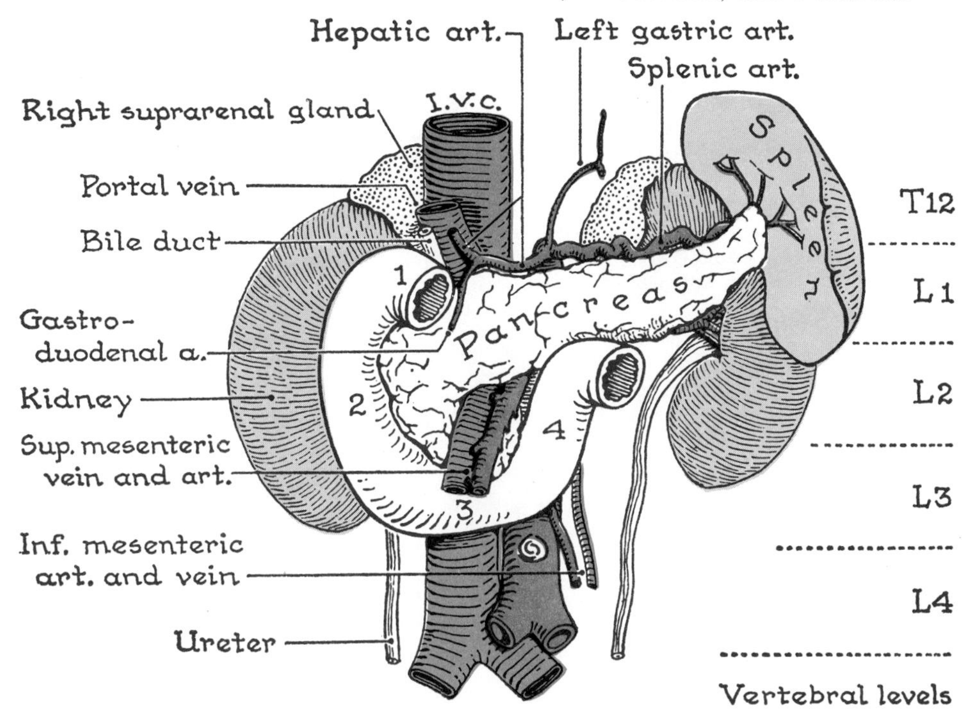

Fig. 291. Abdominal viscera and vessels.

The celiac trunk lies at the upper border of the pancreas and the splenic branch runs along the upper border of the body and tail.

The **main duct** of the pancreas resembles a herring bone, in that small ducts spring from the main duct which is straight. This duct empties the *exocrine secretion* of the pancreas into the second part of the duodenum (*fig. 301*).

The *endocrine secretion*, insulin, is formed by tiny clusters of cells, the *islets of Langerhans*. Its deficiency is the disease, diabetes mellitus.

Surface Anatomy of Pancreas (*fig. 291*): *Vertebral Levels*. The head lying within the concavity of the duodenum lies in front of vertebra L. 2; the body rises to the level of the celiac trunk and, therefore, is in front of vertebra L. 1.

Notes Explaining Relationships:

1. During prenatal intestinal rotation (*fig. 266*), the *duodenum* is thrust by the transverse colon against the structures lying on the posterior abdominal wall, and it loses its dorsal mesentery. The *pancreas* likewise loses the layer of peritoneum that formerly clothed its right surface—now its posterior surface.

2. The right and left vitelline veins return blood from the yolk sac (*fig. 292*). The veins twine about the duodenum in figure-of-8 fashion. The portions anterior to the 1st part of the duodenum and posterior to the 3rd part disappear with the result that the superior mesenteric and portal veins of adult anatomy take form.

3. The ducts of the liver and pancreas arise as two outpouchings or hollow buds, a dorsal and a ventral, of the endoderm of the duodenal wall (*fig. 293*). It is the rudiment of the bile passages and of the liver and also of the pancreatic duct around which the head of the pancreas develops.

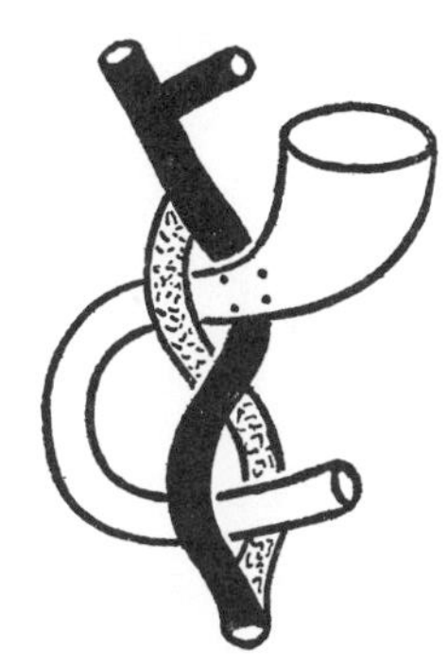

Fig. 292. Showing how the portal vein develops from a figure-of-8 anastomosis around the duodenum.

4. The 2nd part of the duodenum undergoes partial rotation to the right on its own long axis (*fig. 294*). This explains why the (common) bile duct passes upward behind the accessory pancreatic duct and 1st part of the duodenum.

For descriptive purposes, the head of the pancreas may be regarded as swinging around behind the junction of the splenic, superior mesenteric, and portal veins, thereby causing them to occupy a position between it and the neck of the pancreas.

5. The primitive dorsal and ventral pancreatic ducts remain separate as one of several options (*fig. 295*). The (common) bile duct and the main pancreatic duct open separately into the duodenum, one above the other in 5 per cent of 200 cases. (Millbourn.)

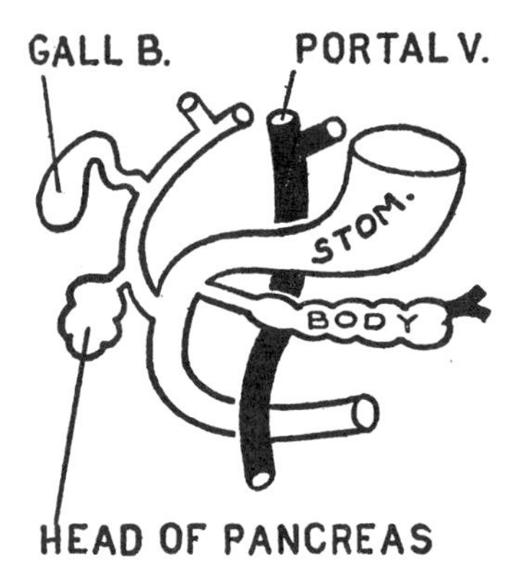

Fig. 293. The rudiments of the liver and pancreas arise as outpouchings of the duodenum into the ventral and dorsal mesogastria.

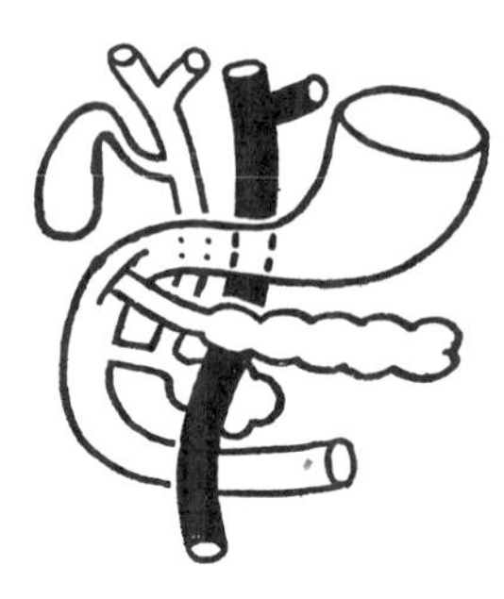

Fig. 294. The two rudiments of the pancreas close on the portal vein (or s. mesenteric vessels) like a book on a bookmark.

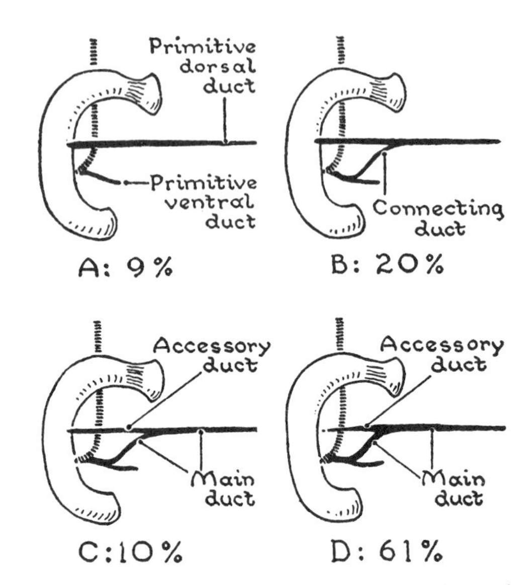

Fig. 295. Varieties of pancreatic ducts in 200 specimens.

Anterior Relations of Duodenum and Pancreas: (a) gall bladder and liver on each side of gall bladder; (b) transverse colon (direct anterior relation of the 2nd part of the duodenum) (*fig. 297*); (c) superior mesenteric vein and artery cross anterior to the 3rd part of the duodenum—*fig. 300*); (d) coils of jejunum.

The **Pancreas** has two aspects, an *anterior* and a *posterior*. The transverse colon is attached to the anterior aspect of the head by areolar tissue and is suspended from the anterior aspect of the body and tail (*fig. 296*) by the transverse mesocolon. Above and below the attachment, the pancreas is obviously covered with peritoneum. The tip of the tail extends into the lienorenal ligament and abuts against the spleen.

The pancreas is surrounded by various portions of the gastro-intestinal tract (*figs. 297, 298*).

Posterior Relations of Duodenum and Pancreas (*fig. 299*). These structures belong to the "three paired gland system" and to the great vessels.

Blood Supply of Duodenum and Pancreas (*fig. 300*). The *blood supply of the 1st part*

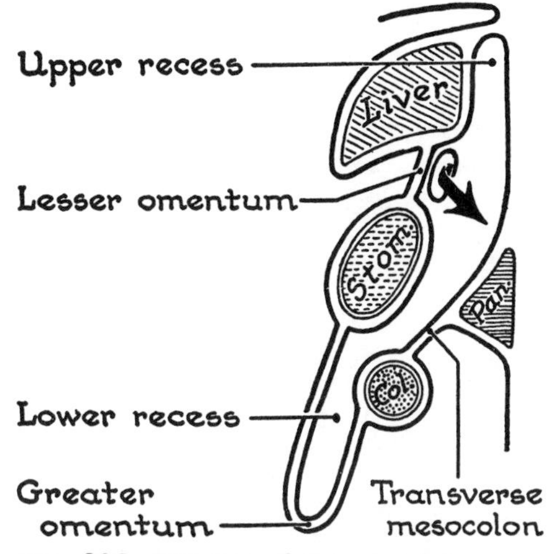

Fig. 296. Diagram of the stomach, omental bursa, and stomach bed, on sagittal section.

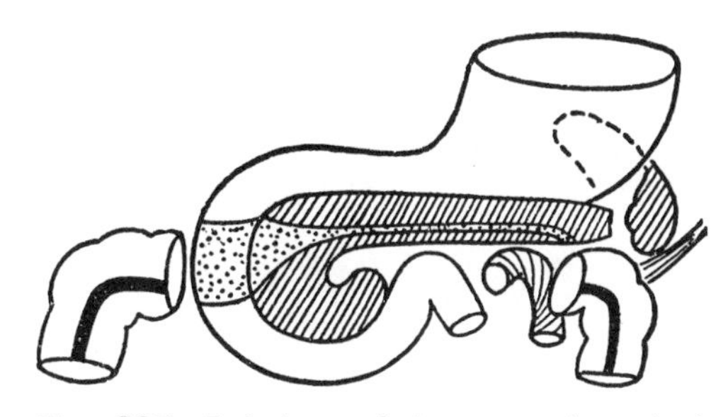

Fig. 297. Relations of the gastro-intestinal apparatus to the pancreas.

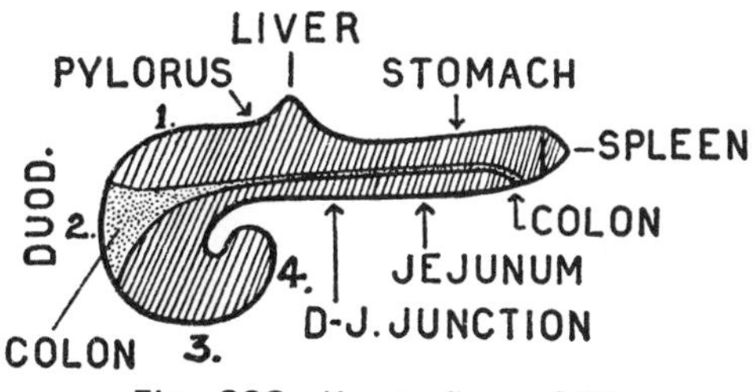

Fig. 298. Key to figure 297.

of the duodenum is of special interest because this is a common site of ulcers. An independent twig (or twigs), the "*retroduodenal*" *branch* of the gastroduodenal a. (not seen in *fig. 300*) helps to supply the posterior wall.

The body and tail of the pancreas are supplied by two constant arteries: (1) the *splenic artery*, which runs behind the upper border of the gland, and (2) the *inferior pancreatic artery*, commonly derived from the celiac a., which runs behind the lower border to the tail. Inconstant arteries also supply the gland. The various arteries anastomose freely to form a network around the lobules of the gland. (Pierson; Wharton; Woodburne and Olsen; and Michels.)

Interior of the Duodenum (*fig. 301*): (1) The first part of the duodenum is smooth; beyond this, *plicae circulares* of mucous membrane are large and numerous; (2) the conjoint bile and pancreatic duct (some-

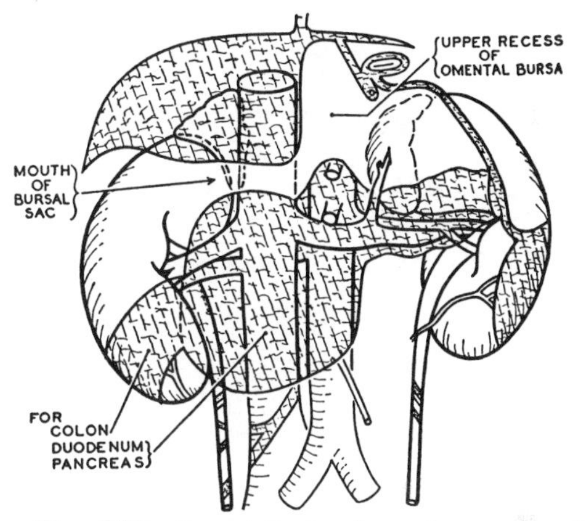

Fig. 299. Especially to show the extent of the duodenum and pancreas.

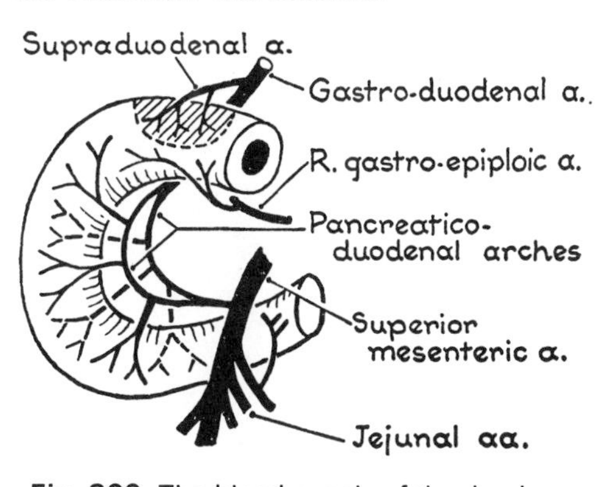

Fig. 300. The blood supply of the duodenum.

times dilated to form an *ampulla*, 5 mm long, as it traverses the duodenal wall) opens onto the (*major*) *duodenal papilla*, which is situated on the concave side of the duodenum about 8 cm from the pylorus;

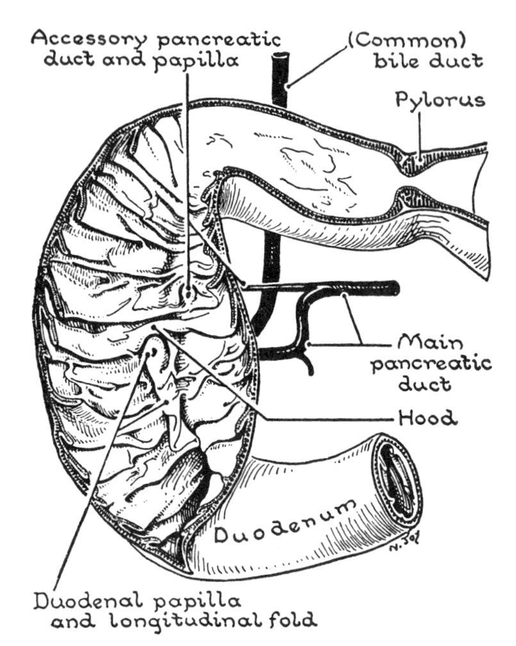

Fig. 301. Interior of duodenum. Pancreatic ducts.

(3) from the papilla a *plica longitudinalis* descends, and over the papilla a semicircular *hood-like fold* is commonly thrown; (4) the accessory pancreatic duct opens into the duodenum on a *minor or accessory papilla*, 2 cm anterosuperior to the major duodenal papilla.

The **main duct of the pancreas,** which runs near the posterior surface of the gland, resembles a herring bone, in that small ducts spring from the main duct which is straight.

Application of Embryology. *To display the (common) bile duct (fig. 302).*

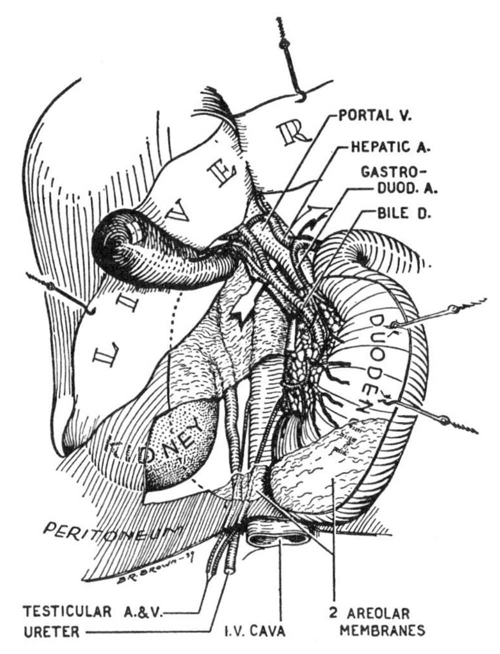

Fig. 302. Display of bile duct by embryological approach. (*Arrow* passes through epiploic foramen.)

17 THREE PAIRED GLANDS

SUPRARENAL, RENAL, AND GENERATIVE OR SEX GLANDS

Migrations during Development; Testicular Vessels —posterior relations, anterior relations, testicular veins contrasted; Ovarian Vessels; Ductus (Vas) Deferens.

Suprarenal Glands—Structure and Development; Surface Anatomy; Relations; Vessels; Nerves.

KIDNEY AND URETER

General; Surface Anatomy; Anterior Relations; Posterior Relations; Structure; Vessels and Nerves; Arterial Segments; To Explain Variations; Ureter; Anomalies.

SUPRARENAL, RENAL, AND GENERATIVE OR SEX GLANDS

Migrations. All three glands developed in the retroperitoneal tissue. The kidney ascends, dragging its duct after it and picking up new vessels. The descending testis (or ovary) dragged after it its duct, artery, vein, lymph vessels, and nerves in front of the path of the kidney and ureter (*figs. 303–305*).

If the kidney fails to let go its hold of one or more of the lower arteries it acquires and loses in its ascent, then the adult kidney will possess more than one renal artery. It is on this basis that accessory renal arteries are explained (*fig. 311B*).

The lymph nodes that drain the testis (and ovary) are to be sought not in the groin but in the abdomen.

Testicular or Ovarian Vessels. The gonadal (of either sex) artery, one on each side, arises from the aorta just below the renal artery. It descends in the retroperitoneal tissue on the psoas fascia to the deep inguinal ring (male). In the female it enters the pelvis, crossing the ureter. The right artery in addition crosses the i. v. cava (*fig. 313*). It adheres to the peritoneum, except where crossed by parts of the gastro-intestinal system; the *details* are:

On the right, by the 3rd part of duodenum, the right colic and ileocolic vessels, and the end of the ileum or the cecum; *on the left*, by the 4th part of the duodenum, the inf. mesenteric vein, the left colic and sigmoid vessels and the sigmoid colon.

The right testicular or ovarian vein ends in the i. v. cava; *the left vein* in the left renal vein.

Ductus Deferens (Vas Deferens). This duct runs a short subperitoneal course in the abdomen. From the deep inguinal ring it curves round the lateral border of the inf. epigastric artery and, crossing the ext. iliac vessels, enters the pelvis.

Suprarenal Glands (Adrenal Glands) (*fig. 306*)

These paired endocrine glands overlap the upper ends of the kidneys. They are crescentic in shape and easily torn. The glands are situated one on each side of the celiac trunk, and, as the celiac ganglion alone intervenes, they are less than about 2 to 3 cm from the trunk; indeed, their medial borders are exactly 5 cm apart. A

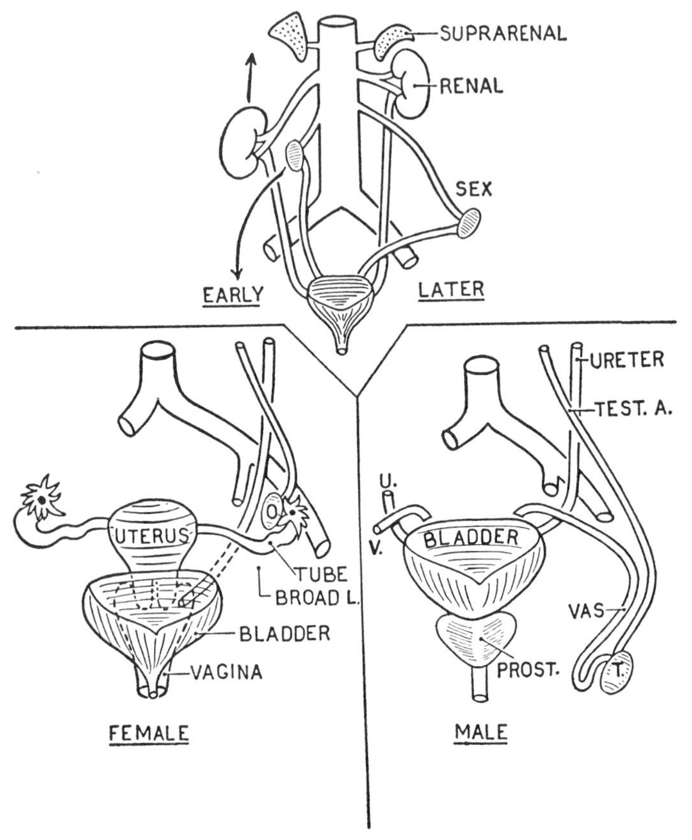

Figs. 303, 304, and 305. The migrations of the kidneys and sex glands. Definitively the ovarian vessels and the uterine tube in the female bear the same relationships to the ureter as do the testicular vessels and the ductus (vas) deferens in the male.

peak added to the right gland converts its crescentic form into a triangular one.

Structure and Development. Each gland is composed of two parts: an outer golden cortex and an inner vascular medulla which develops from nerve cells.

At birth the suprarenal gland covers nearly the upper half of the kidney; in the adult it lies at the upper pole and along the medial border above the hilus, some fat intervening. Its association with the kidney is accidental.

Surface Anatomy and Relation. Its vertebral level is shown in *figure 306*. It lies in front of the crura of the diaphragm. *In front*, lie portions of the gastro-intestinal canal and its three unpaired glands (*fig. 307*). *Left*—pancreas, splenic artery, and stomach (separated by the omental bursa). *Right*—liver and, embedded in the bare area, the i. v. cava (*fig. 307*). *Details*—

Vessels. Numerous tenuous twigs from three arteries converge on each gland—(1) from the suprarenal artery proper, which springs from the aorta, (2) from the (inferior) phrenic artery above, and (3) from the renal artery below. A single large suprarenal vein leaves the anterior surface of each gland: the right vein ascends to the i.v. cava; the left vein descends to the left renal vein.

Nerves. The greater and lesser splanchnic nerves pierce the crus of the diaphragm and pass to the celiac ganglion or an off-shoot of it. A branch of the posterior vagal trunk, containing fibers from both vagi, divides and passes to the ganglion (*fig. 306*).

From the ganglion, the half dozen nerve filaments that pass to each suprarenal gland are probably derived from the lesser and lowest splanchnics; other filaments from the 1st (and

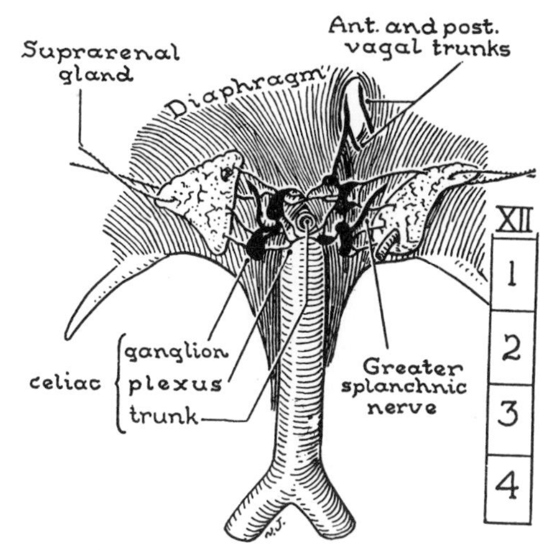

Fig. 306. The celiac plexus and the suprarenal glands (retracted). Note vertebral levels.

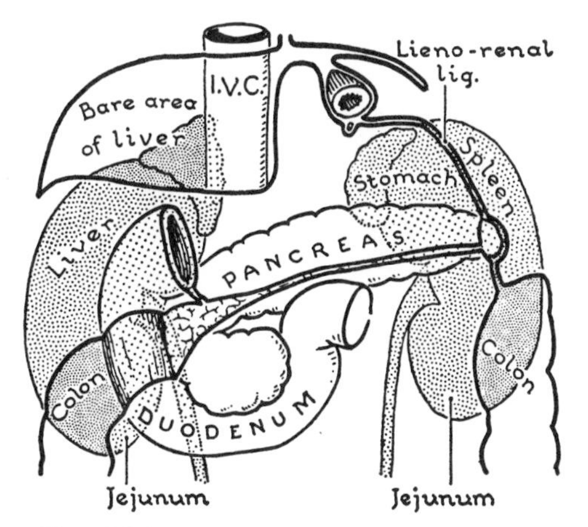

Fig. 307. The anterior relations of the kidneys and suprarenal glands. (The duodenum and pancreas are drawn in situ; the other relations are indicated by name.)

2nd) lumbar sympathetic ganglia pass to the gland.

KIDNEY AND URETER

The kidneys (L. ren; Gk. nephros) are paired (*fig. 313*). Each kidney is about 11 cm long and weighs about 130 to 150 gm.

The intermediate third of the medial border presents a cleft, the *hilus* or door, which leads into a cavity, the *renal sinus*. Passing through the hilus into the sinus are the *pelvis* or expanded upper end of the ureter, the renal vessels and nerves, and some fat.

Above the hilus, the suprarenal gland is in contact with the medial border and upper pole; below the hilus, the ureter is close to the medial border.

Like other abdominal organs, the kidney lies in the extraperitoneal fatty tissue. In the midst of this fatty tissue there is a tough areolar membrane, the *renal fascia*, which splits to close the kidney and a certain quantity of fat, the *perinephric fat* or *fatty capsule*.

The two layers of fascia do not blend below, so, the kidneys can move downward with the diaphragm during inspiration; neither do they blend medially but pass in front of and behind the renal vessels, aorta, and inferior vena cava to unite indefinitely with the respective layers of the opposite side. Above and laterally the two layers of fascia blend.

Surface Anatomy. The *lower poles* of the kidneys lie about 2 cm above the transumbilical plane (level of 3–4 intervertebral disc); their *upper poles* reach the level of the 12th thoracic vertebra and, therefore, beyond the 12th rib.

Its *medial border* is about 5 cm from the aorta. The hilus is at or near the transpyloric plane (*fig. 307*).

Considering the subject from behind, the kidney extends from a point one or two fingers' breadth above the highest part of the iliac crest, which is on a level with the spine of the fourth lumbar vertebra, upward to or almost to the 11th rib.

(For vertebral levels see *fig. 291.*)

Anterior Relations (*fig. 307*). The kidneys are in direct contact with some structures and indirectly in contact with others. These are given below for reference only.

The Direct Contacts are:

1. The 2nd part of the duodenum and the tail of the pancreas at the hili.
2. The right and left colic flexures.
3. Suprarenal glands.
4. Bare area of liver (on right side only).

The Indirect Contacts are:

1. Coils of jejunum.

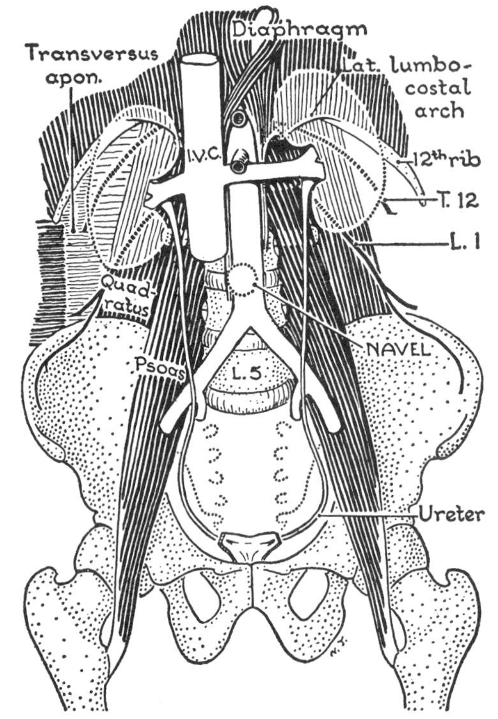

Fig. 308. The posterior relations of the kidneys. The course of the ureters.

2. Liver at the hepatorenal pouch on the right side.
3. Spleen on the left side.
4. Stomach on the left side.

Posterior Relations. Behind the kidney are parts of the roof and posterior wall of the abdomen (*fig. 308*). Thus, for reference, not memory work:

1. Muscles and Bones: The diaphragm, together with the medial and lateral arcuate ligs. (lumbocostal arches) from which it arises; the Psoas and Quadratus Lumborum and, in the angle between them, the uncovered tips of the transverse processes of the 1st, 2nd, and (3rd) lumbar vertebrae; and the posterior aponeurosis of the Transversus Abdominis, which arises from these transverse processes (*fig. 323*).

2. Nerves and Vessels: Th. 12 (subcostal nerve) and L. 1 (iliohypogastric and ilio-inguinal nerves).

Structure of the Kidney. The kidney has a *fibrous capsule* which is easily stripped off. When a kidney is divided with a sharp knife into anterior and posterior halves, its cut surface is seen to possess six or more smooth, darkish, longitudinally striated, triangular areas known as the pyramids (*fig. 309*).

Each *pyramid* is seen to have a free, rounded apex or *papilla* (L. = nipple) which projects into the renal sinus, and a *body* or main portion. From the *base* of the body radiations, the *medullary rays*, occupy the cortex of the kidney and extend to the surface. A papilla, a body, and a series of medullary rays constitute a complete pyramid.

The outer or surface layer of the kidney is the *cortex*. It is the part that lies superficial to the bases of the pyramids and it comprises the entire outer one-third of the kidney substance. It looks granular. Cortical tissue also fills the areas between the pyramids and is there known as *renal columns*. *Details of fine structure* are given for reference on p. 60.

Renal Vessels and Nerves. The blood supply of the kidney is peculiar in that all, or almost all, of the blood passes through the glomerular capillaries where it is purified before it passes through a second set of capillaries from which it nourishes the

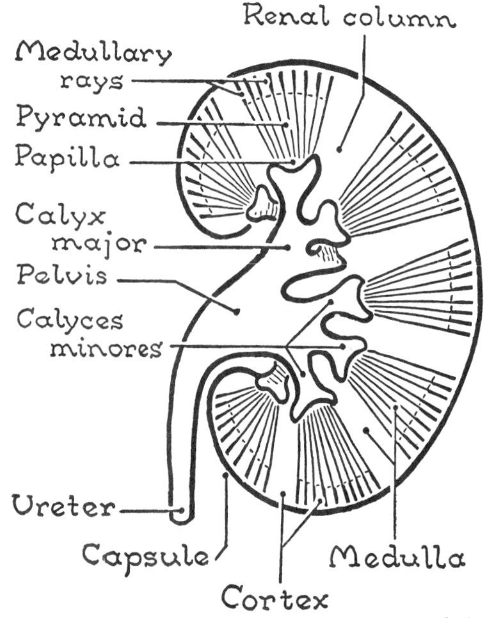

Fig. 309. The macroscopic structure of the kidney (seen on longitudinal section).

kidney substance (see p. 61).

The Renal Arteries, one for each kidney, arise from the sides of the aorta 1 cm below the superior mesenteric a. (See also p. 61).

Arterial Segments of the Kidney. According to the distribution of the branches of the renal artery, the kidney has five segments (*figs. 309.1, 309.2*); these branches do not anastomose. The veins, however, anastomose freely (R. T. Graves).

The Renal Nerves are probably derived from segments T. 12, L. 1 and 2. (For details see p. 62).

Pelvis of the Kidney (Pelvis of the Ureter) (*fig. 310*). The pelvis is the expanded, funnel-shaped, upper end of the ureter. It lies partly within the renal sinus and partly outside it. Traced toward the kidney it is seen to divide into two stalks, the *cranial and caudal calices majores*, with sometimes a third or middle calix. Each major calix divides into several goblet-shaped *calices minores* into each of which one or more papillae project. A dozen or more papillary ducts open on to each papilla.

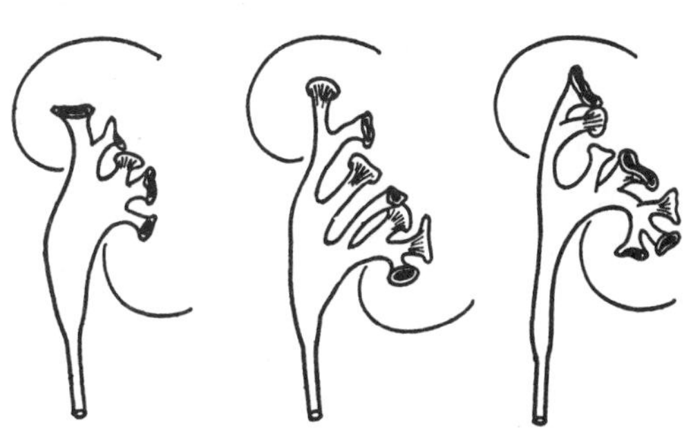

Fig. 310. Sketches of casts of the renal pelvis.

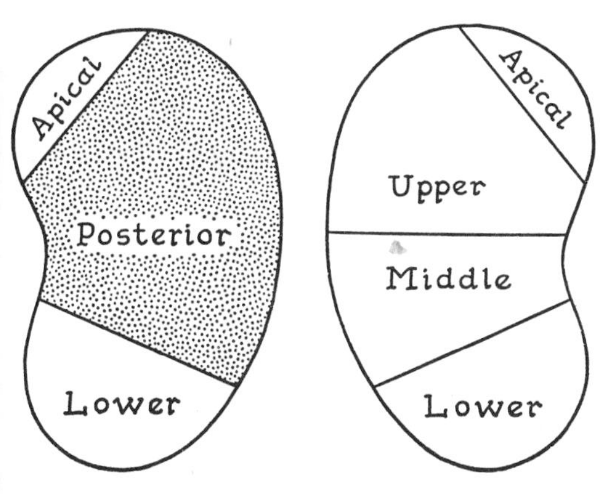

Fig. 309.1. The five segments of the kidney, according to its arterial supply. (After Graves.)

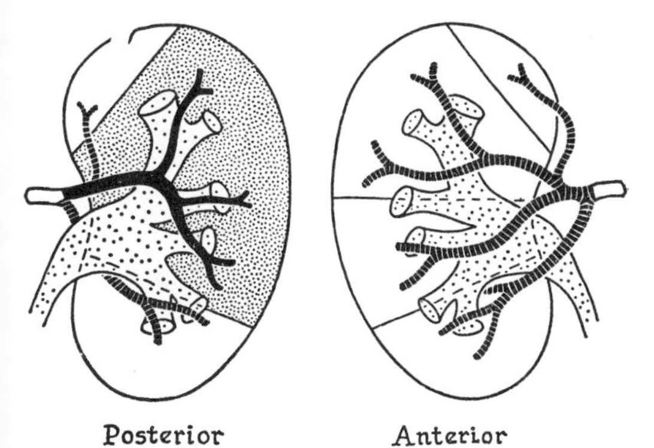

Fig. 309.2. The five segmental branches of the renal artery. (After Graves.)

To Explain Variations (Details). The primitive ureter bifurcates to form a cranial and a caudal calix major. These continue to bud and divide progressively, the terminal branches being the collecting tubules in the pyramids of the renal medulla.

The medulla is composed of seven pairs of pyramids (seven being ventral and seven dorsal) of which three pairs are connected with the cranial calix and four with the caudal.

The papillae of the pyramids become crowded at the two poles of the kidney (especially the upper pole), with the result that groups of two or more papillae fuse to form compound papillae. Accordingly, the maximal number of papillae is 14 and the average is 9.

Commonly there is a third calix major, the *middle calix*, which receives the fourth and fifth pairs of papillae, leaving only the sixth and seventh for the caudal calix (F. Lofgren).

Ureter

The ureter or duct of the kidney is 25 cm long. Its upper half is in the abdomen; its lower half is in the pelvis; its terminal part pierces the posterolateral angle of the bladder.

Its abdominal part extends almost vertically from the lower part of the hilus of the kidney (less than 5 cm from the median plane) to the bifurcation of the common iliac artery (*fig. 308*).

The ureter lies in the subperitoneal areolar tissue and adheres to the peritoneum.

When the peritoneum is mobilized, the ureter is in danger of injury, for it moves with it.

The uterer descends on the Psoas fascia and crosses the genitofemoral nerve. The i. v. cava is close to the medial side of the right ureter; the inferior mesenteric vein is close to the medial side of the left ureter.

Otherwise its *anterior relations* are the vessels of the testis or ovary, and parts of the gastro-intestinal canal. [On the *right side* they are: the 2nd part of the duodenum, the right colic and ileocolic arteries, the root of the mesentery, and the end of the ileum. On the *left side* they are: the left colic and sigmoid arteries and the sigmoid mesocolon.]

Arteries. The pelvis and the ureter possess a longitudinal anastomosing network of arteries derived from the renal artery above and the vesical artery below. This is reinforced along its length by an aortic or testicular (ovarian) or common iliac branch.

Nerves. Like the arteries, the nerves are derived from nearby sources, i.e., the renal and intermesenteric plexuses above, the inf. hypogastric plexus (pelvic plexus) below; and the testicular and sup. hypogastric plexuses in between. The cord segments for the kidney are Th. 12, L. 1 and 2.

Anomalies (*fig. 311*). Much the com-

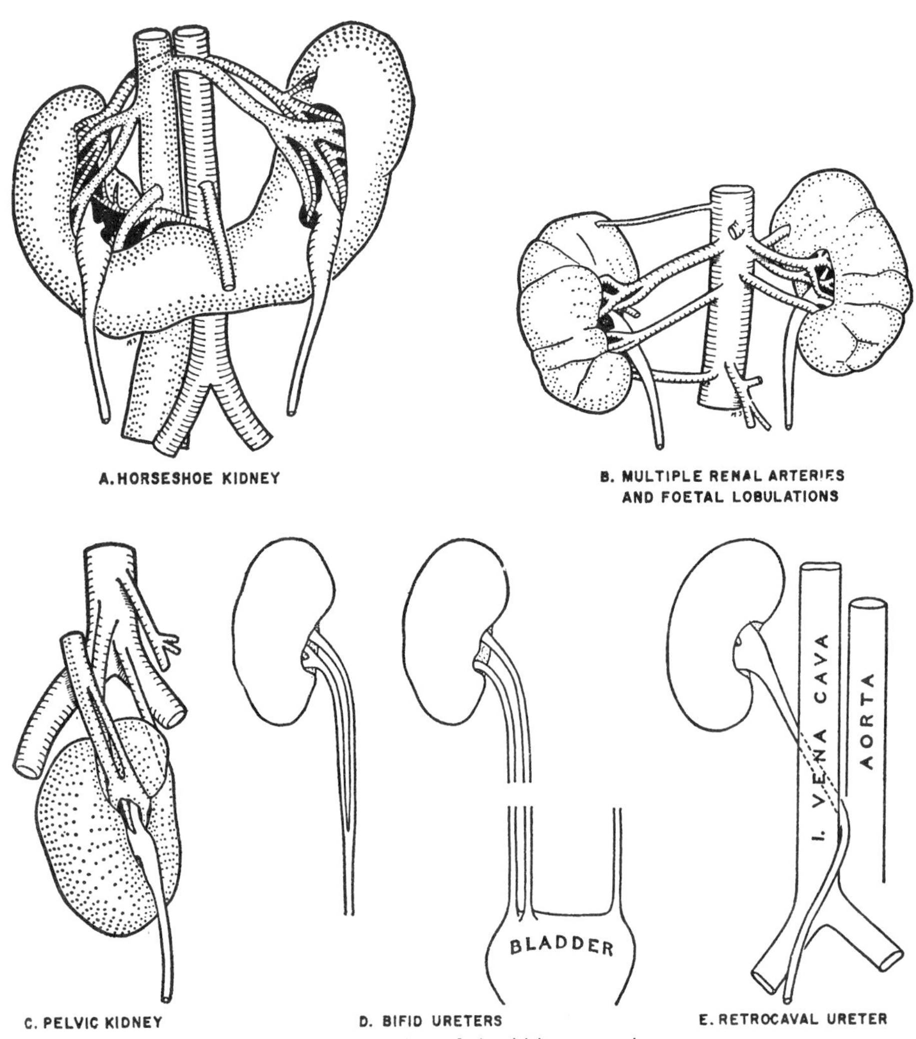

Fig. 311. Anomalies of the kidneys and ureters.

monest gross anomaly of the urinary tract is a *bifid ureter* and pelvis, the result of premature division of the ureteric bud in the fetus. The condition is generally incomplete and unilateral and the ureter is commonly constricted at the point of fission. When completely bifid, one of the ipsilateral ureters may open into other parts of the U. G. tract, e.g., the floor of the urethra, or roof of the vagina or seminal vesicle (rare). The *fetal lobulation* (*fig. 311, B*) present at birth (and well seen in the cow) is commonly retained.

About 3 per cent of kidneys have *two renal arteries* arising from the aorta, of which one usually goes to the upper or lower pole. *Fused kidneys:* in 1 in 700 persons the right and left kidneys are fused, generally at their lower poles, and form a horseshoe kidney (*fig. 311, A*). The ascent may be arrested by the inferior mesenteric a. crossing in front of the isthmus. *Congenital absence of kidney and ureter* or *rudimentary kidney with ureter* occur with the same frequency as horseshoe kidney.

Other uncommon to rare anomalies. The kidney on one side may be *double.* One or other kidney may migrate across the median plane. A *pelvic ectopic kidney* (*fig. 311, C*) is one that, failing to ascend, remains lodged in the pelvis and is supplied by arteries of low origin. A *congenital cystic kidney* results when the secreting parts of a kidney and their ducts fail to unite.

A retrocaval ureter (Pick and Anson) is rare (*fig. 311, E*).

18 POSTERIOR ABDOMINAL STRUCTURES

GREAT VESSELS OF ABDOMEN

Abdominal Aorta and Iliac Arteries.
Collateral Branches of Aorta.
Inferior Vena Cava—Course; Development; Tributaries; Varicocele of Testicular Veins.
Ascending Lumbar Vein.
Relationship of Branches of Aorta to I.V.C.
Relationships of Great Vessels in General.

ABDOMINAL AUTONOMIC NERVES

Sympathetic Trunk; Celiac Plexus; Intermesenteric and Superior Hypogastric Plexus.
Sympathetic Connections; Parasympathetic Connections. Sensory Nerves.

POSTERIOR WALL OF ABDOMEN PROPER

Bony Parts—Lumbar Vertebrae; Variations of Vertebrae.
Muscles—Iliacus; Psoas Major and Minor; Quadratus Lumborum; Transversus Abdominis; Intertransversarii.
Fascia Iliaca; Psoas Fascia.
Lumbar Nerves.
Lumbar Plexus—Lumbosacral Trunk; Obturator Nerve; Femoral Nerve; Other Branches of Plexus; Variations.

DIAPHRAGM

Surface Anatomy; Attachments and Marginal Gaps; Anomaly; Vertebrocostal Trigone; Structure.
Structures Piercing Diaphragm; Nerve and Blood Supply; Relations.
Development.

ABDOMINAL LYMPHATICS

Chains of Nodes; Cysterna Chyli.
Drainage and Nodes of Various Organs and Areas: intestine, pancreatic group of nodes, stomach, liver, pancreas, spleen.

GREAT VESSELS OF THE ABDOMEN

The great vessels are the abdominal aorta, the inferior vena cava, and the common, external, and internal iliac arteries and veins.

Abdominal Aorta and Iliac Arteries

The aorta enters the abdomen in the median plane at the disc between vertebrae T. 12 and L. 1. It ends in front of L. 4 by bifurcating into the right and the left common iliac artery (*fig. 312*).

Each common iliac artery bifurcates into an internal iliac and an external iliac artery. The main trunk to the thigh curves and runs on the Psoas to the midinguinal point where the name changes to femoral artery. Two branches arise from the external iliac artery just before it passes behind the inguinal lig. (*fig. 312*)—the *inferior epigastric* and *deep circumflex iliac arteries*.

The Veins lie within the bifurcation of the arteries, as shown in *figure 313*. The common iliac veins join to form the i. v. cava behind the right common iliac artery, in front of vertebra L. 5.

Collateral Branches of the Abdominal Aorta:

	Branches	Distribution
A.	Celiac trunk S. mesenteric a. I. mesenteric a.	To the G.I. canal and the 3 unpaired glands.
B.	Suprarenal aa. Renal aa. Testicular aa.	To the 3 paired glands.
C.	(Inf.) Phrenic aa. Lumbar aa. Median sacral a.	To the roof and walls of the abdomen.

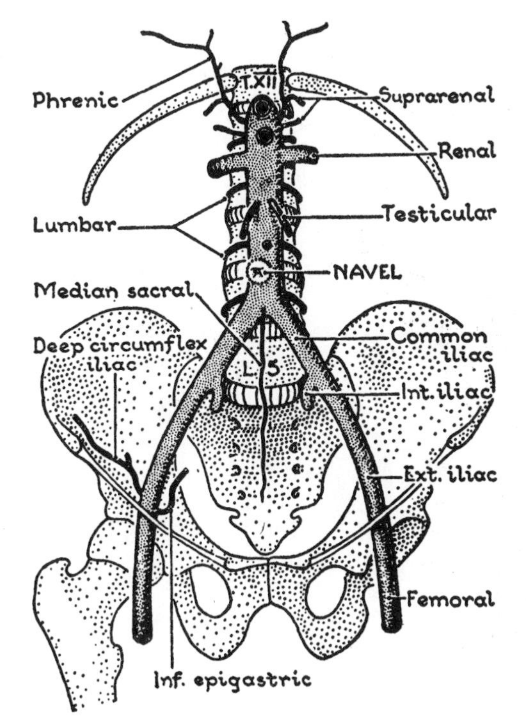

Fig. 312. The abdominal aorta and its branches—collateral and terminal.

These arteries occupy three planes (*fig. 314*).

The **3 unpaired arteries** to the G.I. plane arise from the front of the aorta (*fig. 315 A*).

The Celiac Trunk or Artery. The median arcuate ligament, which unites the two crura of the diaphragm, rests upon the celiac trunk (*fig. 315*). Its level is, therefore, the disc between 12th thoracic and 1st lumbar vertebrae.

The Superior Mesenteric Artery takes origin just below the celiac trunk, and therefore behind the neck of the pancreas, and therefore behind the splenic vein, which is embedded in the pancreas. It "clamps" and prevents ascent of the left renal vein, which crosses the aorta close below it (*fig. 314 A*).

The Inferior Mesenteric Artery takes origin 4 cm above the aortic bifurcation, and 2 cm above the umbilicus, and therefore in front of the 3rd lumbar vertebra. It would arrest the ascent of a horseshoe kidney (*fig. 314 A*).

The **arteries to the 3 paired glands** arise close together (*fig. 315 B*).

From their inception the **phrenic, lumbar, and median sacral arteries** are part either of the roof or of the wall of the abdomen (*fig. 315 C*).

The (Inferior) Phrenic Artery arises beside the celiac trunk and passes at once to the diaphragm.

The Lumbar Arteries. Four pairs of lumbar arteries hug the bodies of the vertebrae 1–4, and pass dorsally medial to the sympathetic trunk and Psoas. A 5th pair, the *iliolumbar arteries*, spring from internal iliac arteries.

The Median Sacral Artery is the tiny "continuation" of the aorta (*fig. 312*).

Inferior Vena Cava

This, the largest vein in the body, begins in front of the 5th lumbar vertebra (*fig. 313*), below the aortic bifurcation and behind the right common iliac artery. At the level of the 8th thoracic vertebra, it pierces the central tendon of the diaphragm to join the right atrium of the heart. It extends, therefore, across eight vertebrae and is about twice the length of the abdominal aorta.

Development (*figs. 316* and *317*). The inferior vena cava is a composite vein of complex origin. But the elementary facts of its development are needed to make its adult anatomy meaningful. A longitudinal vein, the *posterior cardinal vein*, appears on each side of the vertebral column and travels through the abdomen and thorax to join with the similar vein, the *anterior cardinal vein*, from the head, neck, and upper limb, to form the *common cardinal vein* (duct of Cuvier) which ends in the sinus venarum (the receiving chamber) of the heart. In addition to receiving the *somatic segmental veins* (lumbar and intercostal) from the body wall, the posterior cardinal vein receives the *veins from the three paired glands*, but *not* from the

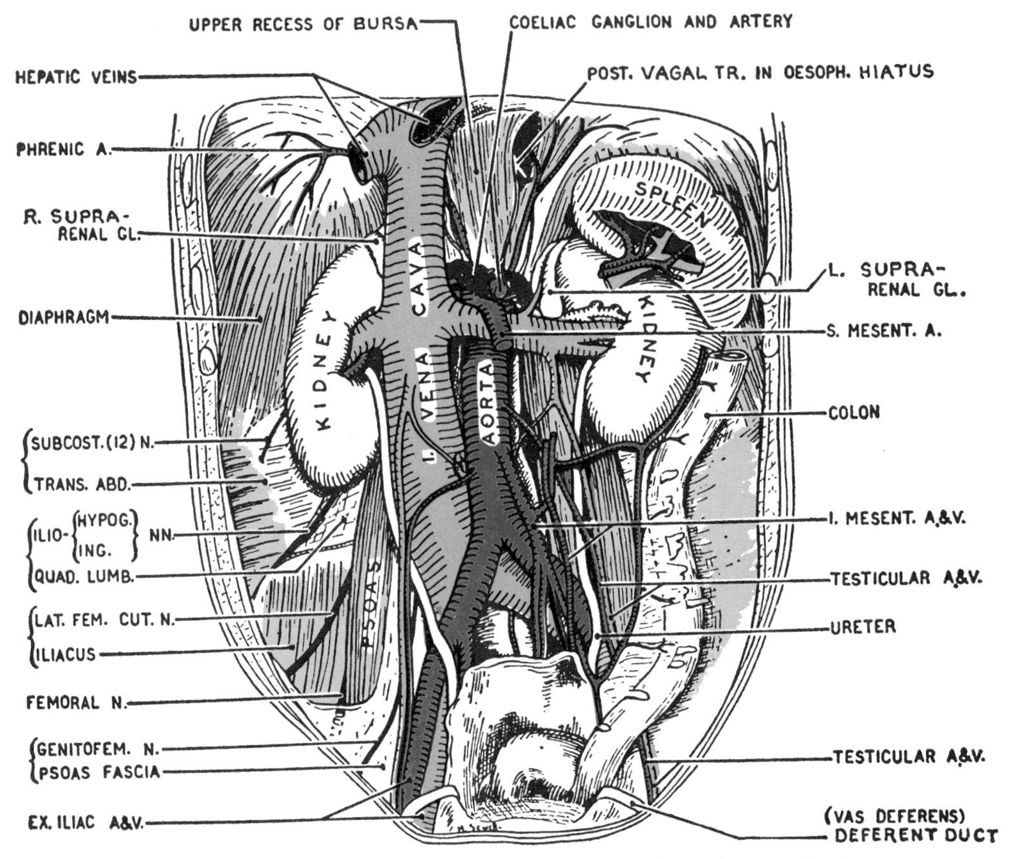

Fig. 313. Primary retroperitoneal structures, also spleen and colon.

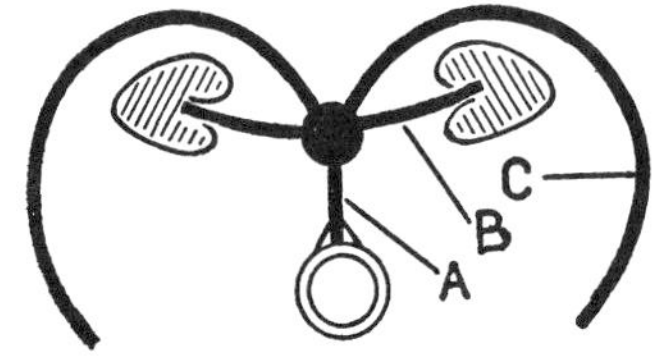

Fig. 314. "The three vascular planes." (Cross section, see *fig. 315*.)

gastro-intestinal tract and its three unpaired glands. They return their blood via the portal vein to the liver, and thence the heart by the persisting right vitelline vein, which was destined to become the terminal part of the inferior vena cava (*fig. 279*, p. 217).

Two new cross-communications, of which one is the *left common iliac vein*, the other the *left renal vein*, divert their blood to the right posterior cardinal vein, while the left posterior cardinal vein disappears (*fig. 316 B*). (A left inferior vena cava is an anomaly which reminds us of this history.)

A new vessel sprouts from the posterior cardinal vein near the renal veins, and connects it with the right vitelline vein behind the liver (*fig. 316 B, broken lines*).

Now the blood returns to the heart from below the diaphragm through the adult i.v. cava formed (*fig. 317*) from—(1) the right cardinal vein as far as the entrance of the renal veins, (2) the new connection, and (3) the terminal or prehepatic portion of the right vitelline vein.

If you will pull forward the right kidney, you will see a small vein leave the back of the i. v. cava near the renal veins, and pass headward through the right crus of the diaphragm. This is a vestige of the prerenal portion of the right post. cardinal vein. In the thorax it joins intercostal veins which form the azygos vein.

Tributaries of the Inferior Vena Cava. These fall into the same three groups as

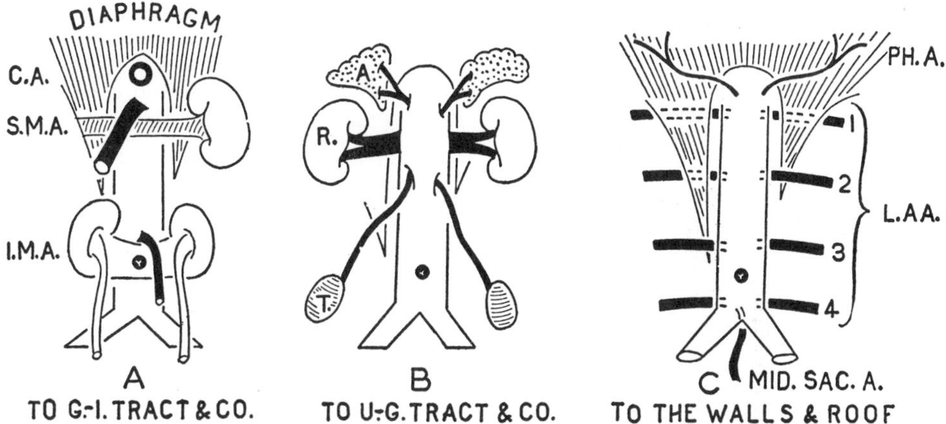

Fig. 315. The branches of the abdominal aorta arranged according to the planes they occupy.

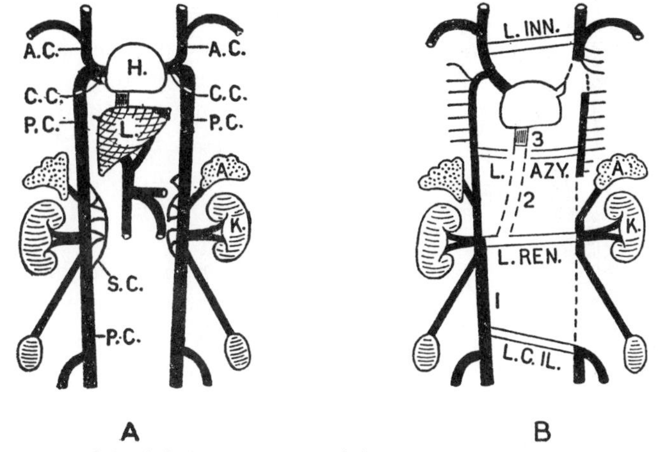

Fig. 316. Development of the inferior vena cava. (*A*): *A.C., P.C., S.C., C.C.* = anterior, posterior, sub-, and common cardinal veins. (*B*): *1* = postrenal segment of inf. vena cava; *2* = new connection; *3* = prehepatic segment.

the branches of the abdominal aorta:

A. *The Blood from the Gastro-intestinal Canal*, after circulating in the liver, leaves it via three large hepatic veins and six or more small hepatic veins to enter the last part of the i.v. cava.

B. *Veins of the Three Paired Glands:* The right suprarenal, renal, and testicular (or ovarian) veins enter the i.v. cava; the left renal vein drains all three left organs (*fig. 316B*).

C. *Veins of the Body Wall:* (1) right and left (inferior) phrenic veins.

(2) The lumbar veins.

The lumbar veins, on each side, are irregularly linked together by an **ascending lumbar vein,** which lies in front of the lumbar transverse processes (*fig. 677*). This vein begins caudally at the common iliac vein. The right and left ascending lumbar veins are (1) accessory to the i.v. cava, and (2) they are to be classified with the vertebral venous system (Chapter 36).

The right and left iliolumbar veins, which are the veins of the 5th lumbar segment, end in the right and left common iliac veins.

The median sacral vein ends in the left common iliac vein.

Relationships of the Branches of the Aorta to the I. V. Cava are explained on a developmental basis in *figures 317* and *318*.

Relationships. Not only is it an unbearable burden, but it also is folly to commit to rote memory the relationships of all the great vessels. They can be arrived at by applying the logical principles described in these pages.

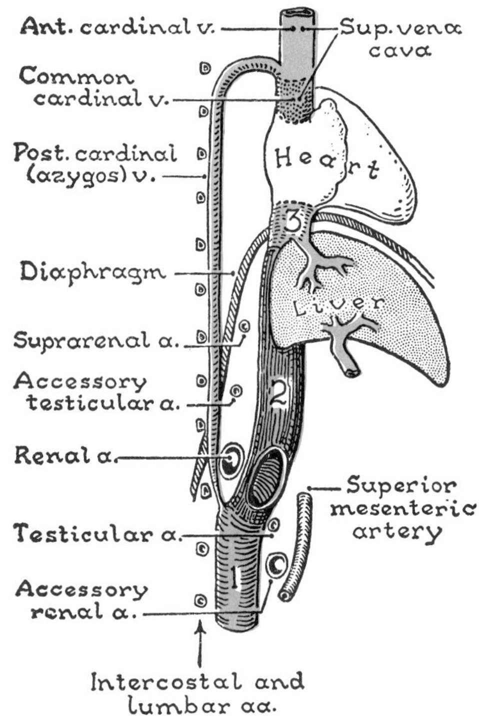

Fig. 317. Scheme of the development of the inferior vena cava, explaining its relationship to the arteries that cross it, side view.

ABDOMINAL AUTONOMIC NERVES

The autonomic nervous system within the abdomen is represented by:

1. The sympathetic trunks, white and gray rami communicantes, lumbar splanchnic nerves, and thoracic splanchnic nerves.
2. The prevertebral plexuses—celiac, intermesenteric, and superior hypogastric.
3. Parasympathetic nerves—vagus and pelvic splanchnic nerves.

Sympathetic Trunk (*fig. 318.1*). In the abdomen the sympathetic trunk follows faithfully the anterior border of the Psoas. It descends, therefore, on the bodies of the vertebrae and the intervetebral discs, the transversely running lumbar vessels alone intervening. It enters from the thorax with the Psoas behind the medial arcuate lig. (lumbocostal arch) and it passes into the pelvis behind the common iliac vessels (*fig. 324*).

The *right trunk* is concealed by the i.v. cava, and crossed by the right renal artery. The *left trunk* is crossed by the left renal vessels, the left testicular and the inf. mesenteric artery.

As a rule each trunk has four ganglia—not five—two probably having fused.

Connections. Each trunk receives a

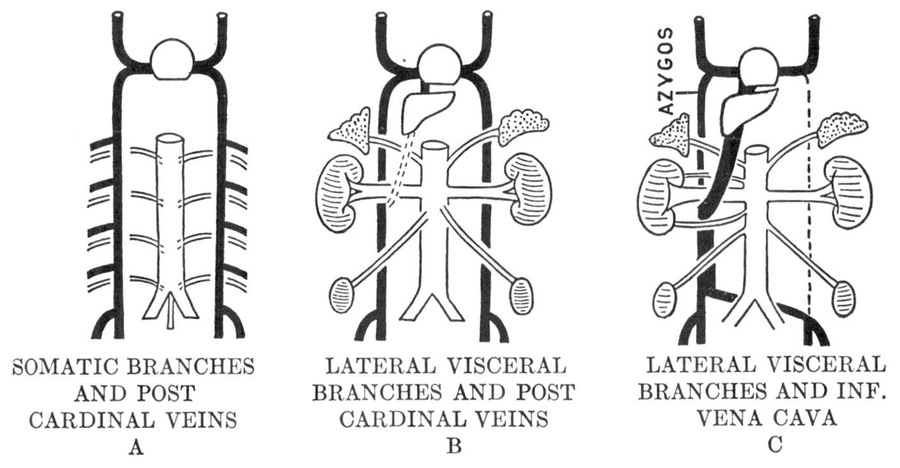

Fig. 318. Explaining the relationships of the branches of the aorta to the cardinal veins and subsequently to the i. v. cava. In *C* note the relations of the accessory renal arteries to the i. v. cava.

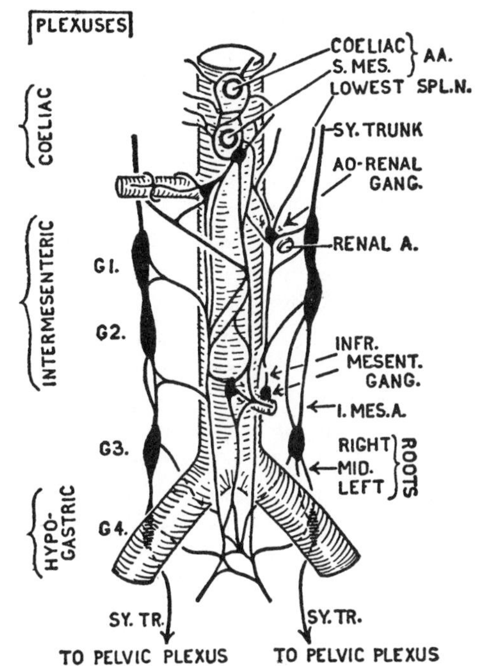

Fig. 318.1. The intermesenteric and sup. hypogastric plexuses. (Dissection by K. Baldwin.)

white ramus from each of the upper two (or three) lumbar nerves and sends one or more *gray rami* to each of the five lumbar nerves to be distributed with nerves to somatic structures. These white and gray rami curve backward and laterally on the sides of the vertebrae either with the lumbar vessels or independently (*fig. 319*). Four rami, the *lumbar splanchnic nerves*, run medially to the intermesenteric and superior hypogastric plexuses to be distributed largely with blood vessels to viscera.

Celiac Plexus. The *celiac ganglia* (*fig. 319.1*) are tough, nodular masses connected to each other by fibers that encircle the celiac trunk, the whole being known as the *celiac plexus*, or as the *solar plexus*, because its branches radiate like the rays of the sun.

Each ganglion lies behind the peritoneum, between the celiac trunk and the suprarenal gland and, therefore, on the crus of the diaphragm. The i.v. cava largely conceals the right ganglion; the pancreas and splenic artery the left one.

The celiac plexus extends down the front of the aorta and is reinforced by the lumbar splanchnic nerves to form the **intermesenteric and superior hypogastric plexuses.** The intermesenteric lies in front of the aorta; the hypogastric lies within the bifurcation of the aorta and therefore in front of the left common iliac vein and the 5th lumbar vertebra and disc (*fig. 319*).

From the celiac plexus fine branches stream into the suprarenal gland (*fig. 319.1*).

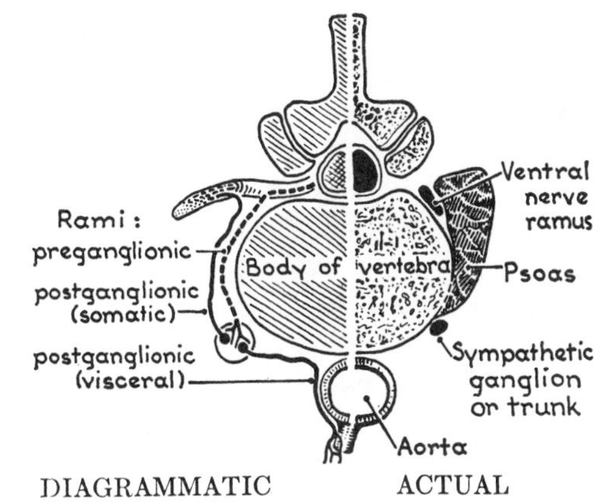

Fig. 319. Pre- and postganglionic fibers of a sympathetic ganglion.

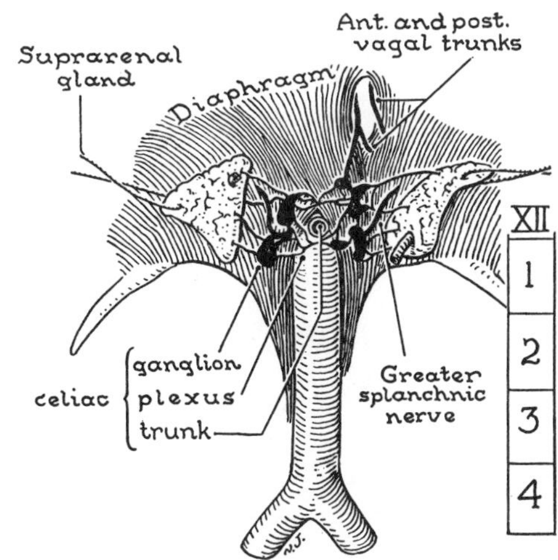

Fig. 319.1. The greater splanchnic nerves, celiac ganglia, and plexus; also suprarenal glands, and anterior and posterior vagal trunks.

Sympathetic Connections. The greater splanchnic nerve pierces the crus abreast of the celiac trunk and at once joins the celiac ganglion; the lesser splanchnic nerve pierces the crus just below and laterally and also joins the ganglion; the lowest splanchnic nerve also pierces the crus and joins an offshoot of the celiac plexus, called the *renal plexus*.

Parasympathetic Connections. Fibers of both *vagus nerves*, via a large branch of the posterior vagal trunk, pass through the celiac plexus (*fig. 319.1*) to be distributed with arteries (celiac, sup. mesenteric, and ? renal) to the abdominal viscera.

Fibers of both **pelvic splanchnic nerves** ascend from the pelvis in the hypogastric plexuses to be distributed with the inferior mesenteric artery to the gut.

The vagi control the gut as far as the left colic flexure; the pelvic splanchnics beyond it.

Sensory Nerves. Accompanying fibers of the autonomic nervous system run sensory (afferent) fibers which ultimately enter the spinal cord or brainstem along sensory roots of spinal or cranial nerves. Sensation from internal organs is diffuse and often "referred" at the level of consciousness to cutaneous areas (*fig. 319.2*).

POSTERIOR WALL OF ABDOMEN PROPER

Bony Parts

In the median plane are the bodies, intervertebral discs, and transverse processes of the five lumbar vertebrae; laterally the wall extends from the 12th rib above to the pelvic brim below, and it is divided into upper and lower parts by the iliac crest. The *bodies* of the lumbar vertebrae increase in height and width from 1st to 5th; so do the discs (*fig. 320*).

The *transverse processes* project laterally at the levels of the upper halves of the bodies. The 3rd projects farthest; those above and below it project progressively less. The 5th is stout and conical and it projects upward, backward, and laterally. The *12th rib*, variable in length, curves downward and laterally to about the level of the 2nd lumbar disc. The *iliac crest* curves upward and laterally to the level of the middle or lower part of the 4th lumbar vertebra (*fig. 320*).

The region of the abdomen between the iliac crests and the pelvic brim is the *pelvis major*. It includes the alae of the sacrum and the iliac fossae described on page 271.

The Lumbar Vertebrae (*fig. 321*). The

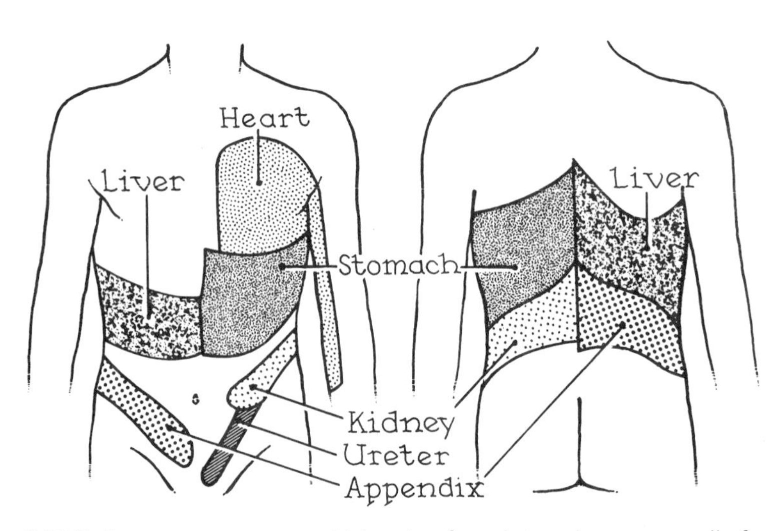

Fig. 319.2 Cutaneous areas to which pains from internal organs are "referred".

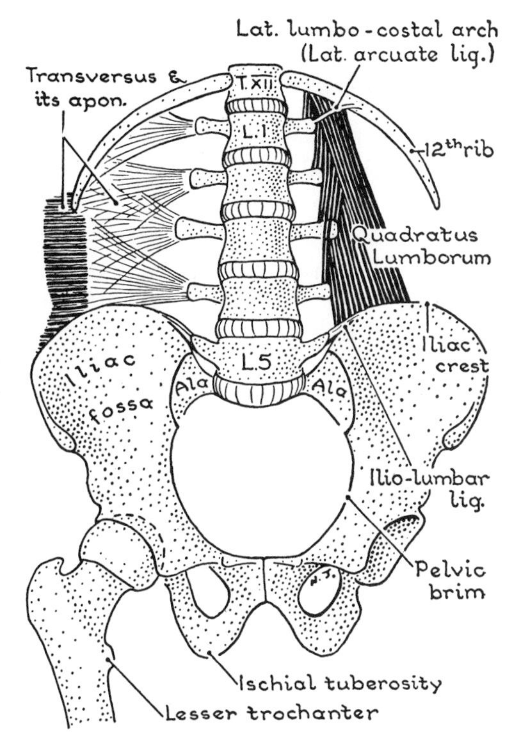

Fig. 320. The skeleton of the posterior abdominal wall. Note that the discs bulge.

The Transversus Abdominis and its aponeurosis.

The Quadratus Lumborum.

body is large, kidney-shaped, and flat above and below. In view of the lumbar curve, it is surprising that only L. 4 and 5 are deeper in front than behind.

The *pedicles* are directed backward and laterally. Above and below each pedicle, there is a small superior *vertebral notch* and a large inferior one. The *laminae* are thick and slope downward and backward, enclosing a triangular *vertebral foramen.*

The *spine* is a thick oblong plate that projects nearly horizontally backward and ends in a thickened posterior border.

The *transverse processes* each spring from the junction of pedicles and laminae.

Their ends give attachment to the Transversus Abdominis aponeurosis and to the Quadratus Lumborum. Owing to the topographical position of the 5th vertebra, its processes extend onto the body and are conical.

The superior *articular processes* spring from the pedicles and, facing medially, grasp the inferior processes of the vertebra above, which spring from laminae and face laterally. The directions, however, gradually change, the inferior articular processes of the 5th lumbar vertebra facing nearly forward.

Variations (Details):

1. In about 6 per cent of adult white males, and perhaps in twice this number in blacks, one of the lower vertebrae (usually L. 5) is a *Bipartite Vertebra,* having a so-called "separate neural arch"—the spine, laminae, and inferior articular processes being detached (*fig. 322*). This is probably the result of an unhealed fracture.

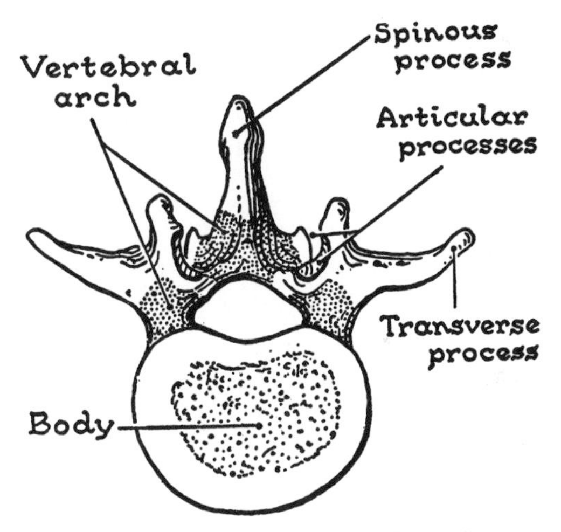

Fig. 321. A lumbar vertebra from above.

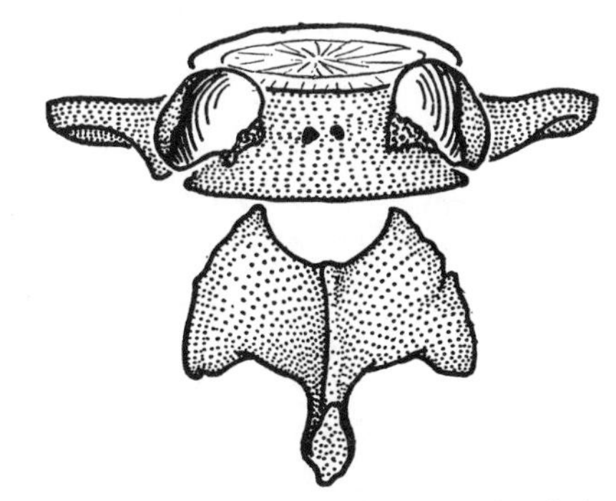

Fig. 322. The 5th lumbar vertebra is in two pieces in 6 per cent of individuals, rendering them liable to a deformity called spondylolisthesis.

The bodies, losing the restraining influence of the inferior articular processes, tend to slip forward (spondylolisthesis). Now, this clinical condition increases with age and is uncommon in women. It was found in 10 per cent of 162 Japanese and in 27 per cent of 350 Eskimos (T. D. Stewart). (See also Roche and Rowe.)

2. The 5th lumbar vertebra is commonly partly sacralized (*fig. 323*).

3. The two sides of the vertebral arch of the lower (or of any) vertebrae may fail to meet (spina bifida).

Muscles and Fascia

Muscles. Iliacus, Psoas, Psoas Minor, Quadratus Lumborum, Transversus Abdominis, and Intertransversarii (*fig. 324*).

The **Iliacus,** like the iliac fossa from which it arises, is fan-shaped. Its fleshy fibers are inserted mostly into the lateral and anterior aspects of the Psoas tendon. The Iliacus and Psoas are referred to collectively as the **Iliopsoas.**

The **Psoas** takes fleshy origin from the sides of the bodies and intervertebral discs of all the lumbar vertebrae and Th. 12, forward as far as the sympathetic trunk.

The ventral rami of the upper four lumbar nerves plunge into the substance of the Psoas when they emerge from the intervertebral foramina, but the 5th ventral ramus escapes under the medial margin of the muscle. On the sides of the vertebral bodies, which are constricted like a waist, the lumbar vessels and the rami communicantes of the sympathetic trunk run dorsally protected by sheets of fascia which bridge them and afford the Psoas an uninterrupted origin (*fig. 324*).

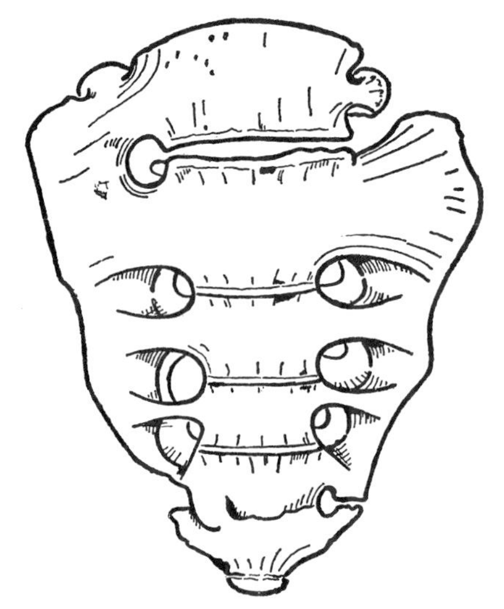

Fig. 323. The 5th lumbar vertebra is commonly partly sacralized.

The Psoas tendon inserts into the lesser trochanter of the femur. It plays in the groove between the anterior inferior iliac spine and the iliopubic eminence prior to crossing the middle of the front of the hip joint, of which it is a powerful flexor. The Psoas has a minor effect on the lumbar segments of the column (increasing the lumbar curvature).

The **Psoas Minor** is either absent (50 per cent) or insignificant. In some lower mammals the Psoas Minor is larger than the Psoas itself. As a flexor of the pelvis on the spine, its value to the rabbit when running and to the ape when brachiating is apparent.

The **Quadratus Lumborum** is quadrate but not rectangular, for its lateral border is oblique (*fig. 324*); and this oblique border is a landmark when exposing the kidney from behind (*fig. 308*).

The Quadratus attachments to the iliac crest, lumbar transverse processes, and 12th rib are well shown in *figure 324*.

Its fascia is slightly thickened above to form the *lateral arcuate lig.* (lumbocostal arch)—which gives origin to the diaphragm—and greatly thickened below to form the important *iliolumbar ligament* (p. 275).

The **Transversus Abdominis** helps to form a sheath for the deep muscles of the back. Near the lateral border of the Quadratus it becomes aponeurotic. This posterior aponeurosis divides into two layers, the *anterior and posterior layers of the thoracolumbar fascia.* Of these, the posterior passes to the tips of the lumbar spines and supraspinous lig., enclosing back muscles; the anterior passes behind the Quadratus and attaches itself to the tips of the lumbar transverse processes (*fig. 320*), the last rib above and the iliac crest below.

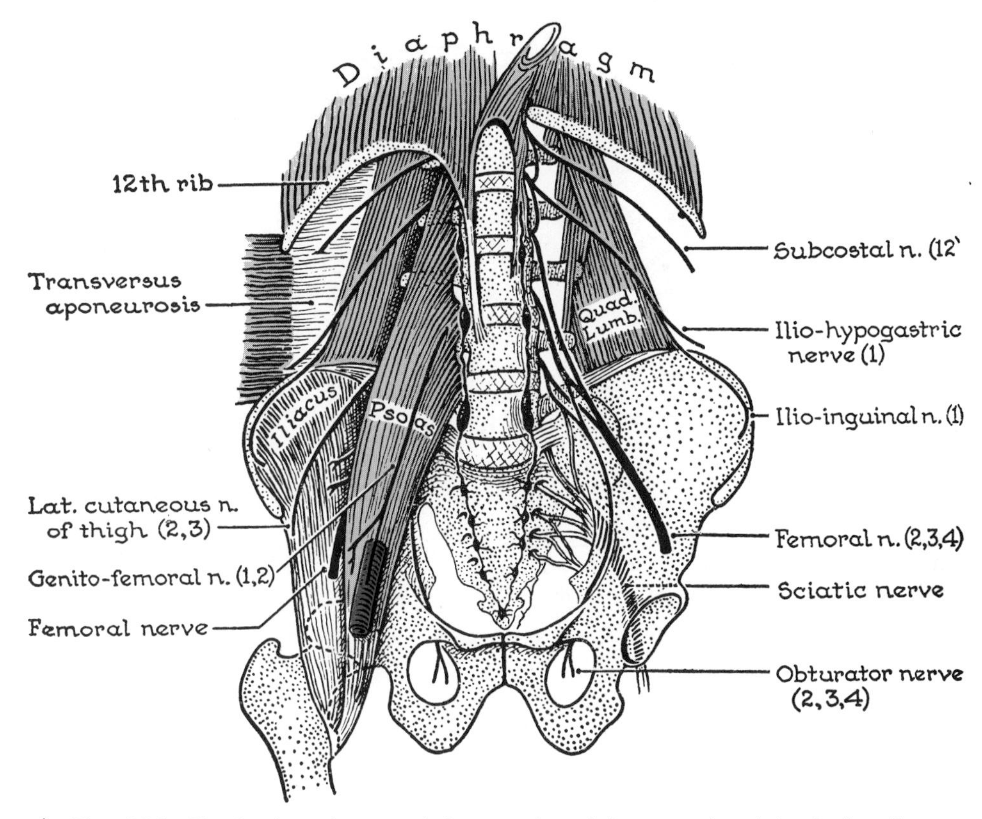

Fig. 324. The lumbar plexus and the muscles of the posterior abdominal wall.

The **Intertransversarii** are fleshy bands that pass between adjacent borders of transverse processes.

Fascia Iliaca and Psoas Fascia. These are part of the general fascia that lines the muscles that enclose the abdominal cavity. This strong fascial sheet covers the Iliacus and the Psoas. The psoas fascia is thickened above to form the medial arcuate lig., which gives origin to the diaphragm (*fig. 325*).

It is carried downward in front of the Iliacus and Psoas into the thigh behind the inguinal lig.; it adheres to it and to the fascia transversalis; and in the thigh it lies deep to the fascia lata. But, the part covering the Psoas (and also the pectineal fascia) is separated from the inguinal ligament by the femoral artery, femoral vein, and deep inguinal lymph vessels wrapped in extraperitoneal areolar tissue, the femoral sheath (*fig. 401*).

Nerves

The Lumbar Nerves. The *ventral rami* are large, and they increase in size from the 1st to the 5th. Each of the five receives one or two gray rami communicantes from the sympathetic trunk; each of the upper two (or three) sends a white ramus communicans to the sympathetic trunk; most of the lumbar nerves supply the Psoas, Quadratus Lumborum, and Intertransversarii. Each of the five continues downward across the front of the root of the transverse process of the vertebra next below, the 5th crossing the ala of the sacrum (*fig. 324*). The ventral rami form the *lumbar plexus* (see below) and *lumbosacral trunk*.

Lumbar Plexus

The lumbar plexus is formed by the ventral rami of the upper three and one-half lumbar nerves. The first ramus is joined by a branch of the 12th thoracic

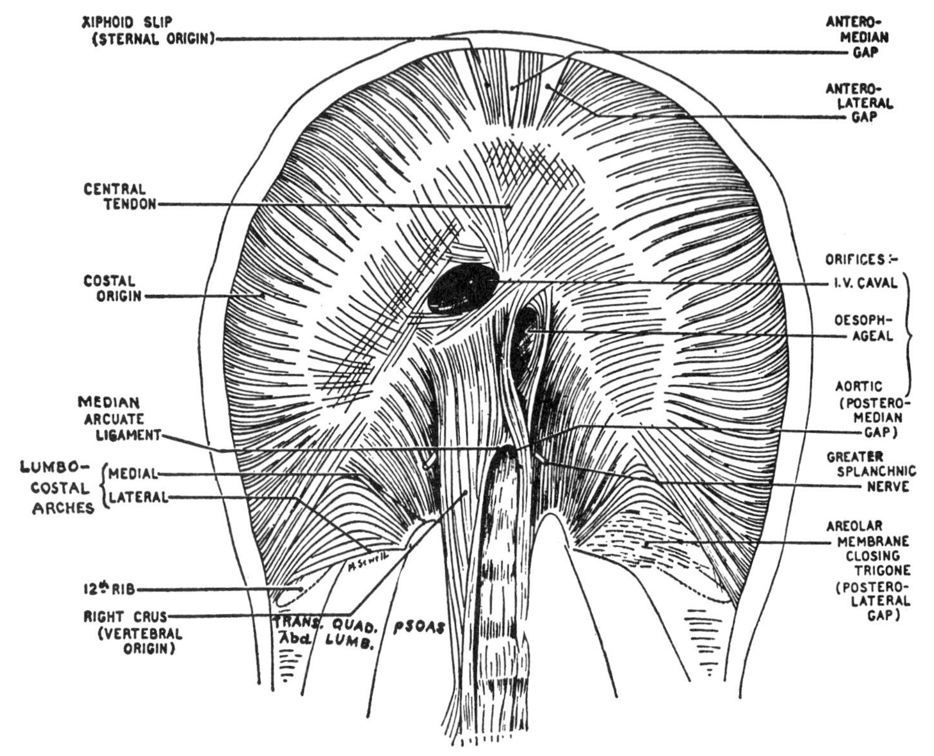

Fig. 325. The abdominal surface of the diaphragm. (*Lumbocostal arches* = *Arcuate ligaments.*)

ramus. [The lower half of the 4th ramus joins the 5th ramus near the anterior border of the ala of the sacrum to form the *Lumbosacral Trunk.*] The branches of the plexus run through the Psoas. Its largest and most important branches are the *femoral* and *obturator nerves*, both of which spring from the segments L. 2, 3, and 4.

The **Obturator Nerve** courses to the upper part of the obturator foramen on the side wall of the pelvis. It appears from under cover of the medial border of the Psoas, pierces the psoas fascia, crosses the sacroiliac joint, passes lateral to the internal iliac vessels and ureter, and enters the pelvic cavity.

The **Femoral Nerve** lies lateral to the femoral sheath and enters the thigh behind the inguinal ligament. It appears at the lateral border of the Psoas, runs downward in the angle between the Psoas and Iliacus. It does not enter the femoral sheath, which encloses the femoral vessels (*fig. 412*). In the false pelvis, it supplies the Iliacus.

It is of interest to observe that though the femoral nerve supplies the muscles on the front of the thigh, its three roots arise behind the three roots of the obturator nerve, which supplies the muscles on the medial aspect of the thigh. The explanation is that during development the limb undergoes medial rotation whereby the femoral nerve region, originally behind, is brought to the front; and the obturator nerve region is carried from the front to the medial side.

In addition to the femoral nerve, four other nerves appear at the lateral border of the Psoas. In ascending order they are:

1. *The Lateral (femoral) Cutaneous Nerve* (from L. 2 and 3 either directly or from the femoral nerve) enters the thigh behind the inguinal ligament lateral to the femoral nerve.

2. and 3. *The Ilio-inguinal and Ilio-hypogastric Nerves* (from L. 1) enter the abdomen behind the medial arcuate lig., and cross in front of the Quadratus, pierce the Transversus near the ant. sup. spine and then the Int. Oblique. Piercing the

Ext. Oblique aponeurosis, they supply the skin of the suprapubic and inguinal regions.

The lateral cutaneous branch of the iliohypogastric nerve crosses the iliac crest and descends to the level of the greater trochanter of the femur (*fig. 425*).

4. *The Subcostal Nerve* (*ventral ramus of Th. 12*) is not a branch of the lumbar plexus, but it enters the abdomen behind the lateral arcuate lig. and pierces the Transversus aponeurosis, and then runs to the Rectus sheath which it enters (*fig. 216*). Its lateral cutaneous branch crosses the iliac crest and descends to the level of the greater trochanter.

The Genitofemoral Nerve (L. 1 and 2) comes to lie within the extraperitoneal fatty layer, and so, pierces the Psoas and the Psoas fascia. It divides at a very variable level into two branches, *femoral* and *genital*, which descend in front of the Psoas.

The *femoral branch* supplies the skin of the femoral triangle, piercing the fascia lata. The *genital branch* supplies the cremaster muscle and traverses the inguinal canal to end in the skin of the scrotum. In its course it pierces the coverings of the spermatic cord.

Variations. Considerable interchange takes place between the above nerves, and so their territories are quite variable.

An *Accessory Obturator Nerve* (from L. 3 and 4) is a small nerve, when present, along the medial border of the Psoas. It crosses the superior ramus of the pubis to the Pectineus.

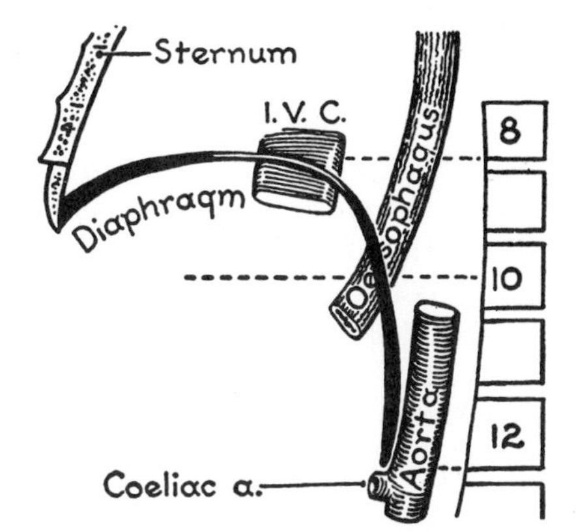

Fig. 325.1 The higher the vertebral level, the more ventral is the hiatus in the diaphragm.

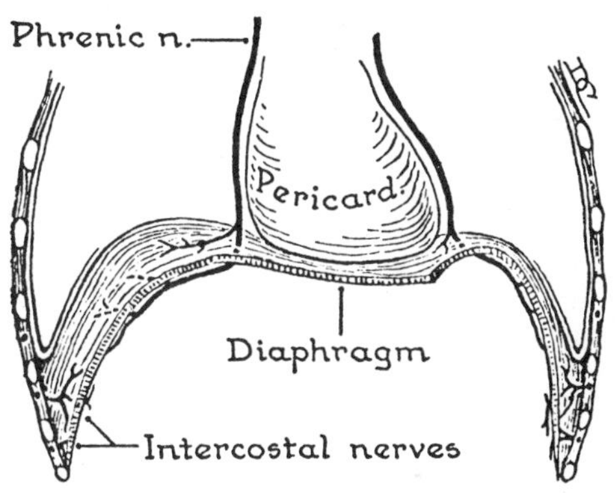

Fig. 326. The diaphragm and its nerves.

DIAPHRAGM

The diaphragm (Gk. dia = through, across; phragma = a partition) has a rounded cupola on each side below the lungs, and a depressed median portion on which the heart lies at the level of the xiphisternal joint (*figs. 325.1, 326*).

The right cupola, occupied by the liver, rises to the 5th rib, just below the right nipple; the left cupola falls short of the left nipple by 3 cm. (The nipples correspond to the inferior angles of the scapulae on the back.)

Attachments and Marginal Gaps. The diaphragm arises by fleshy digitations from the back of the xiphoid process, from the inner surfaces of the 7th to 12th costal cartilages (interdigitating with the Transversus Abdominis) and from the vertebrae L.1–3 by the *crurae* (*fig. 325*).

Anteriorly, there are slight gaps both between the right and left xiphoid slips, and lateral to them. The superior epigastric vessels pass into the rectus sheath through the lateral gaps.

Posteriorly, in the median plane, there is a large gap between the two crura; through this the aorta passes. The medial parts of the crura are fibrous, and they join in front of the aorta immediately above the celiac trunk to form the fibrous *median arcuate ligament.*

Posteriorly, on each side, between the crus and the 12th rib, the pleural and peritoneal cavities are continuous with each other in prenatal life. Then crural fibers migrate across and generally separate the two cavities (*fig. 325*). Sometimes they fail, leaving the *vertebrocostal trigone* (lumbocostal trigone), above the 12th rib. The areolar and muscular fibers closing the space find attachment to the fascia covering the Psoas and Quadratus Lumborum, which in response becomes thickened and strengthened to form the *medial* and *lateral arcuate ligs.* (lumbocostal arches). Of these, the medial ligament bridges the Psoas and extends from the lateral border of the crus to the transverse process of the 1st lumbar vertebra; the lateral ligament bridges the Quadratus and extends from the latter point to the middle of the 12th rib.

Anomaly. Rarely a baby is born in whom the pleural and peritoneal cavities remain in open communication through the vertebrocostal trigone, a *congenital diaphragmatic hernia.* Some of the abdominal contents may come to occupy the pleural cavity and jeoparidize the child's life.

Structure. The central part of the diaphragm is called the *central tendon.* It is composed of decussating and interwoven tendinous fibers, and it has the shape of a trefoil or clover leaf. The peripheral fleshy fibers converge on it.

Structures Piercing (*fig. 325.1*). *The I. V. Cava* pierces the central tendon at the level of the 8th Th. vertebra. Its orifice enlarges during inspiration.[1]

The Esophagus is circled by fleshy fibers of the right crus (which swing across the midline) at the level of the 10th vertebra. These fibers appear to act as a sphincter for the cardiac end of the stomach and prevent its contents from returning to the esophagus, but some studies appear to deny this role (Mann, Greenwood, and Ellis).

The Aorta does not pierce the diaphragm but passes behind the median arcuate ligament at the level of the 12th vertebra.

[1] This is not agreed upon for man; in some animals there is a certain degree of constriction of the i. v. cava during strong contraction of the diaphragm (Franklin).

The vertebral levels, then, are 8, 10, and 12.

Other Structures Piercing. *Through the caval foramen* pass some branches of the *right phrenic nerve.* The left phrenic and other branches of the right phrenic pierce the diaphragm independently to spread out on its abdominal surface (*fig. 326*).

Through the esophageal hiatus pass the *anterior and posterior vagal trunks* and the *esophageal branches of the left gastric artery and vein.* The vein is of special importance, because it anastomoses with esophageal branches of the azygos veins to connect the portal and systemic venous systems (p. 216; *fig. 278*).

Through the aortic hiatus passes the *thoracic duct* [also a vein connecting the right ascending lumbar vein to the azygos system] (*fig. 317*).

The three *splanchnic nerves* (greater, lesser, and lowest) pierce the crura to end in the celiac (and aorticorenal) ganglia.

Nerve Supply (*fig. 326*). The diaphragm is supplied by (1) the phrenic nerve (C. 3, 4, and 5), its only motor nerve—but it is also sensory: and by (2) the lower intercostal nerves, which are sensory to the peripheral parts.

Arteries. Pericardiacophrenic and musculophrenic aa. (from int. thoracic a.) and intercostal, sup. phrenic and inf. phrenic aa. (from the aorta).

Relations. *The abdominal relations* are: the liver, stomach, and spleen; the celiac ganglion, suprarenal glands, and kidneys. *The thoracic relations* are: the heart and pericardium, the lungs and pleurae, and pleural recesses and below the recesses are the lower intercostal spaces and ribs, the thoracic aorta and the esophagus.

Development. The fully formed diaphragm is derived from a mesodermal partition of composite origin (*fig. 326.1*). The anteromedian part arose from the septum transversum; the posteromedian part, from the primitive dorsal mesentery; the lateral parts from the body wall. The *pleuroperitoneal canal*, on each side is closed by a membrane, the *pleuroperitoneal membrane.*

In the young embryo the hinder part of this composite partition lies at the level of vertebra C. 2. On passing vertebrae C. 3, 4, and 5, portions of the myotomes of these segments, supplied by the phrenic nerve, extend into it and pervade it, thereby forming the muscular diaphragm (Wells).

THE SEPTUM TRANSVERSUM is the thick mesodermal mass that surrounds the vitelline veins (also the common cardinal veins) prior to their entering the sinus venarum of the heart. At a certain period this septum projects horizontally backward from the anterior body wall to meet the dorsal mesentery at the level where the duct system of the liver (of endodermal origin) buds from the duodenum.

Ultimately, the septum separates into: (1) a part of the pericardium; (2) part of the diaphragm; (3) the connective tissue stroma of the liver.

The enlarging peritoneal cavity, by extending into the septum transversum between layers 2 and 3, "dissects" the liver from the diaphragm—except for the falciform ligament and the bare area bounded by the coronary ligament—here liver and diaphragm remain loosely connected.

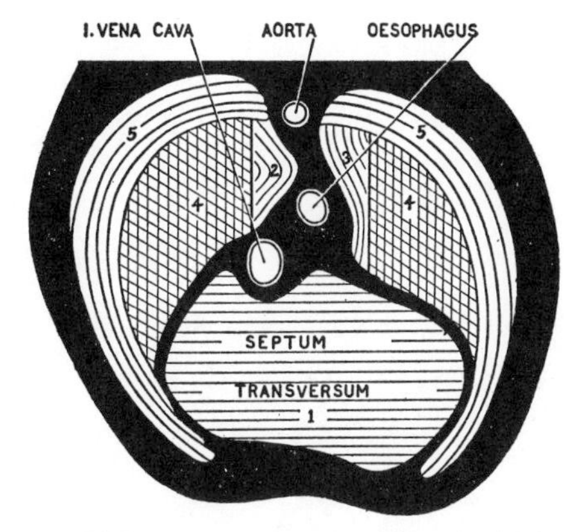

Fig. 326.1. The elements from which the diaphragm is developed.

1 = septum transversum; *2, 3* = dorsal mesentery;

4 = pleuroperitoneal membrane; *5* = body wall. (After Broman.)

ABDOMINAL LYMPHATICS

Lymph capillaries, lymph vessels, and lymph nodes occupy areolar and fascial planes, e.g., the deep fascia, and the submucous and subserous coats of viscera. Lymph capillaries form networks which drain by the nearest issuing lymph vessels. Retrograde flow is prohibited by numerous valves.

After a meal the intestinal lymph vessels contain emulsified fat; and, since this is white like milk, the vessels are called *lacteals*.

The external and common iliac chains of nodes (*fig. 396*) continue upward along the sides of the aorta and around it as the *right* and *left aortic* or *lumbar chains of nodes*. They open by means of a *right* and a *left lumbar lymph trunk* into a thin-walled tubular sac, the *cisterna chyli*. The inf. vena cava runs through the right chain making it less accessible than the left.

These chains receive (1) lymph already filtered through nodes and coming from the lower limbs, lower part of the anterior abdominal wall, external genitals, perineum, and pelvis; (2) lymph vessels that follow the lumbar arteries and drain the posterior abdominal wall; (3) the vessels from the three paired glands—suprarenal, kidney, and testis (the ovary, upper part of the uterus, and the uterine tube in the female); and (4) the part of the gastrointestinal tract supplied by the inferior mesenteric artery draining into the left aortic chain, but the parts supplied by the celiac and superior mesenteric arteries draining by means of a gastro-intestinal trunk into the cisterna chyli.

The Cisterna Chyli (*fig. 326.2*) resembles a segment of a vein about 5 cm long. Its diameter is less than that of a lead pencil, but it may be irregularly dilated. It lies between the aorta and the right crus of the diaphragm. It receives five or more trunks, namely, the *right* and *left lumbar trunks*, the *gastro-intestinal trunk*, and a *pair of vessels* that descend from the lower intercostal spaces. On passing through the aortic hiatus it becomes the thoracic duct.

Lymphatic Drainage

Intestine. The nodes of the *large intestine* are numerous and are roughly divisible into three groups: (1) *paracolic nodes* on the

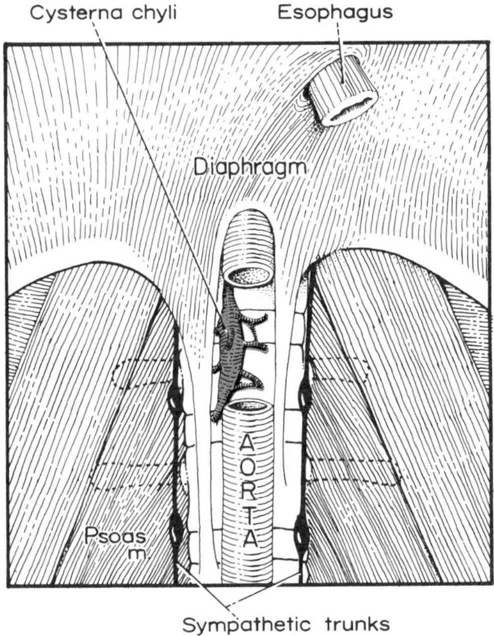

Fig. 326.2. The cysterna chyli.

marginal artery close to the gut wall, (2) *intermediate nodes* on the stems of the colic arteries, and (3) *main nodes* near the roots of the colic arteries, beside the aorta. Also, there are a few very small *epicolic nodes* applied to the surface of the colon.

The lymph vessels from the intestine follow the blood vessels fairly closely and each vessel is interrupted by one or more groups of nodes (*fig. 327*). The lymph vessels from the segment of the large gut between the appendix and the left colic flexure follow the branches and stem of the superior mesenteric artery and, joining those of the small gut, form the *intestinal trunk*. Those from the left colic flexure and remainder of the large gut follow the branches of the inferior mesenteric artery and end among the nodes of the left lumbar (aortic) chain.

There may be an appendicular node in the mesentery of the appendix; there are many nodes clustered in the ileocolic angle; the vessels from the transverse colon have the longest distance to travel; those at the left colic flexure communicate with the splenic nodes, traveling probably along the route of the occasional artery described on page 225.

The Small Intestine is drained through numerous nodes divisible like those of the large intestine into three groups. Ultimately a channel, the *intestinal trunk*, emerges and, by joining the *gastric trunk*, forms the *gastro-intestinal trunk* which opens into the cisterna chyli.

Pancreatic Group of Nodes. (Celiac and Superior Mesenteric Nodes, *figure 327.1*). Along the upper border of the pancreas there are *middle, right, and left suprapancreatic groups* of nodes related to the celiac artery and to its hepatic and splenic branches. A *subpyloric* group is applied to the front of the head of the pancreas below the pylorus. A *left gastric chain* lies on the course of the left gastric artery. A *biliary chain* extends along the bile passages from the porta hepatis above, through the lesser omentum, and behind the first part of the duodenum and head of the pancreas, to the second part of the duodenum below. And, there is the *main group of submesenteric nodes* at the root of the mesentery.

Stomach. The lymph vessels from the part of the stomach that lies to the left of a vertical line dropped through the esophagus pass with the left gastro-epiploic and short gastric arteries through the gastrolienal and the lienorenal ligaments to the suprapancreatic nodes; some, however, pass to a necklace of nodes that encircle the cardiac orifice.

Of the lymph vessels to the right of this vertical line, (1) those from the upper two-thirds of both surfaces of the stomach run to the left to nodes placed on the left gastric artery at the left end of the lesser curvature of the stomach and are there in part intercepted; but some vessels pass by these nodes to more distant ones on the stem of the left gastric artery, and so to suprapancreatic nodes. (2) The lower vessels run to the right to nodes placed on the right gastro-epiploic artery at the right end of the greater curvature, thence to the subpyloric group, from which they are

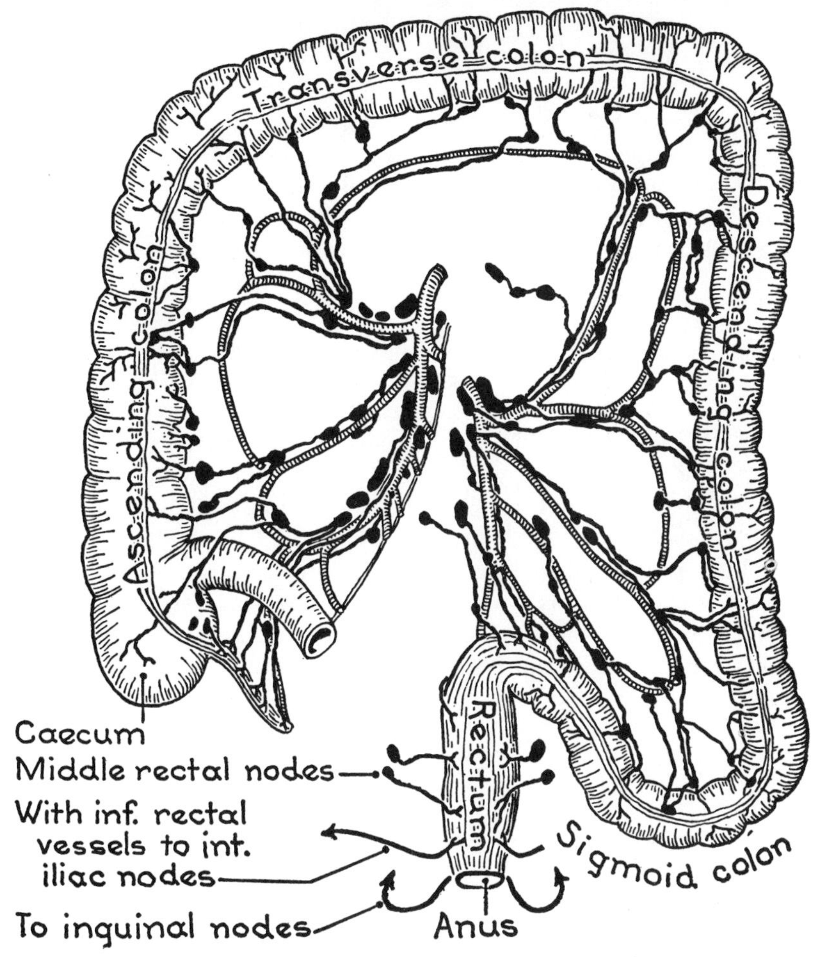

Fig. 327. The lymphatics of the large intestine. (After Jamieson and Dobson.)

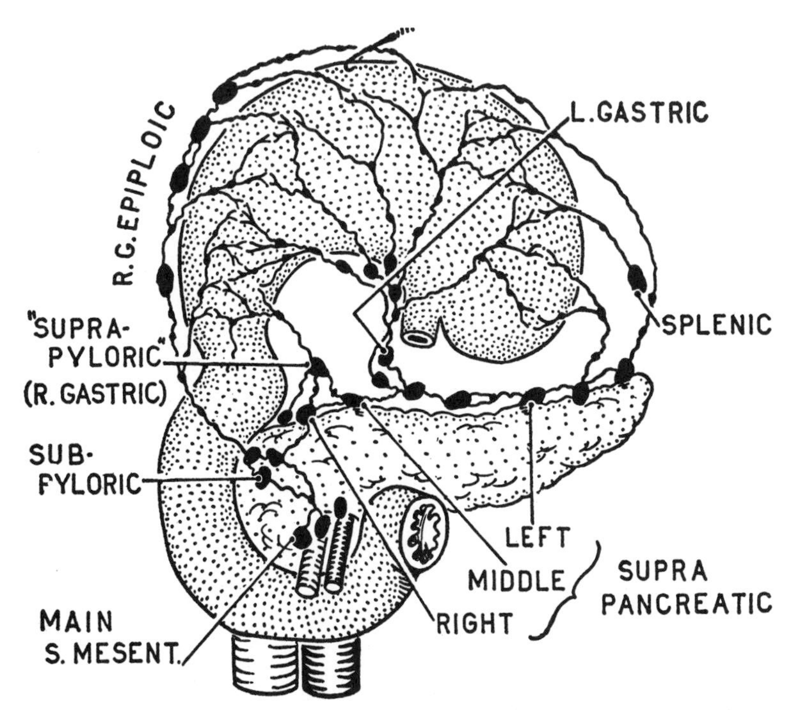

Fig. 327.1. The lymphatics of the upper abdomen and stomach.

dispersed to suprapancreatic nodes and to the main group of superior mesenteric nodes. (3) At the extreme pyloric end of the lesser curvature several lymph vessels follow the right gastric artery.

The lymph plexuses of the stomach communicate freely with those of the esophagus; but only feebly, if at all, with those of the duodenum. This is due to: the connective tissue septum in the submucous coat at the pyloric sphincter, to discontinuity of the circular muscle fibers, and to indipping of the longitudinal muscle fibers.

Liver. Lymph vessels from the upper surface of the liver pass through the falciform ligament to retrosternal nodes, which lie on the diaphragm and discharge into the parasternal (int. mammary) chain; some from the interior, following hepatic veins, pass with i. v. cava through the diaphragm to diaphragmatic nodes, thence to the thoracic duct; others, following the branches of the portal vein, emerge at the porta and travel down the biliary chain to be distributed to the various pancreatic nodes. There is an intercepting gland, the *cystic node*, at the neck of the gall bladder, and there are other "hepatic nodes" in the porta hepatis.

Pancreas. The lymphatics of the pancreas drain into adjacent nodes.

Spleen. The lymphatics of the spleen drain into splenic nodes situated where the tip of the pancreas abuts against the spleen; these in turn drain into the suprapancreatic group.

SECTION FOUR

PERINEUM AND PELVIS

19 Perineum 257

20 Male Pelvis 268

21 Female Pelvis 290

22 Pelvic Autonomic Nerves and Lymphatics 302

19 PERINEUM

Definition; Developmental Considerations; Definitions of: urogenital diaphragm, perineal membrane or inferior fascia of u.g. diaphragm, superficial perineal fascia and pouch, deep pouch, perineal body, anococcygeal raphe.

ANAL REGION

Sphincter Ani Externus; Ischiorectal Fossae; Contents; Vessels and Nerves.

UROGENITAL REGION IN THE MALE

Superficial Perineal Pouch and Contents.

Penis—Parts; Skin; Suspensory Ligament; Vessels and Nerves; Erectile Tissue; Lymphatics.

Superficial Perineal Muscles.

Deep Perineal Pouch and Contents.

Internal Pudendal Vessels and Nerves—Vasomotor Nerves; Erection; Ejaculation.

Exposure of Prostate.

FEMALE PERINEUM

Comparisons (between female and male).

The perineum is the diamond-shaped region at whose angles are the arcuate (inferior) pubic ligament, the tip of the coccyx, and the ischial tuberosities (*fig. 328*). The pubic arch and the sacrotuberous ligaments form its sides. The sacrotuberous lig., however, is hidden by the lower border of the Gluteus Maximus.

The anterior half of this diamond is the *urogenital region (triangle)*; the posterior half is the *anal region (triangle)*.

Developmental Considerations. In the embryo the endodermal-lined alimentary canal ends in a blind receptacle, the *cloaca*, shaped somewhat like a coffee pot, the spout being the *allantoic diverticulum*, definitively the *urachus* (*figs. 329 and 330*).

The *mesonephric duct* (definitively the *ductus deferens* in the male) grows caudally and opens into the anterior part of the cloaca. The *ureter* develops as an outgrowth from the mesonephric duct, and the two have for a period a common terminal duct. This common duct is absorbed subsequently into the posterior wall of the bladder and prostatic urethra with the result that the ureter and ductus deferens come to have independent openings.

In reptiles and birds the cloaca opens on the skin surface through an orifice, guarded by a sphincter of striated muscle, the *cloacal sphincter*. In mammals, including man, a septum of mesoderm, the *urorectal septum*, divides the cloaca into (1) an anterior or urogenital part, and (2) a posterior or intestinal part. The cloacal sphincter also divides into anterior and posterior parts: the posterior part becomes the Sphincter Ani Externus; the anterior part becomes the other perineal muscles (p. 262).

From these considerations you will understand why a single nerve, the *pudendal n.*, serves the muscles as well as the skin of the region, and why its companion artery, the *internal pudendal a.*, nourishes the entire territory; also why the bladder and rectum have a common nerve supply (pelvic splanchnic n. and hypogastric plexus).

Definitions:

1. The Urogenital Diaphragm (*fig. 331*) is a thin sheet of striated muscle which stretches between the two sides of the pubic arch. Its most anterior fibers and its

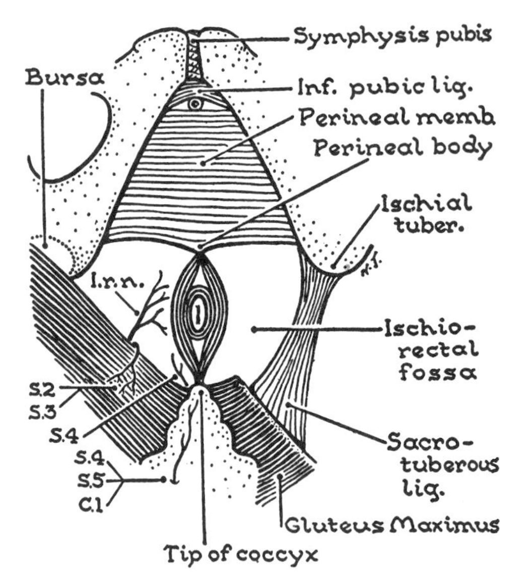

Fig. 328. Boundaries and subdivisions of the perineum. (*Inf. pubic lig. = Arcuate lig.*)

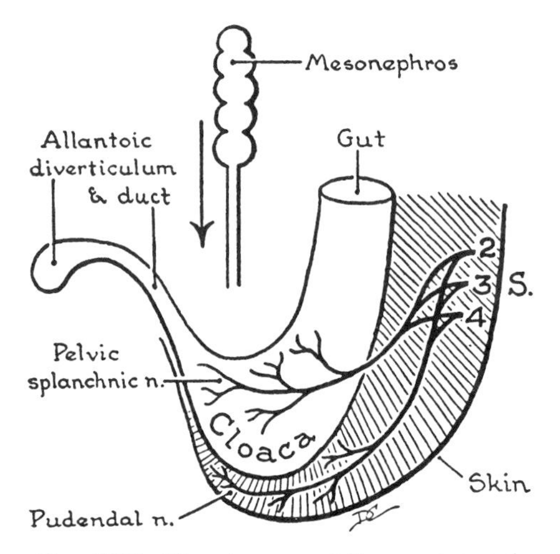

Fig. 329. The cloaca and its nerve supply.

most posterior fibers (Transversus Perinei Profundus) run transversely. Its middle fibers (Sphincter Urethrae) encircle the urethra.

Like other muscles, the urogenital diaphragm is enveloped in areolar tissue, and because the diaphragm is flat, its envelope forms two sheets, the *inferior and superior fasciae of the urogenital diaphragm. Perineal Membrane* is a shorter name for the inferior fascia.

2. *The Superficial Perineal Fascia of Colles* is the "fascia of Scarpa" continued into the perineum and is attached to: the fascia lata, the pubic arch, and the base of the perineal membrane (*fig. 331*). In front it is prolonged over the penis and scrotum.

3. *The Superficial Perineal Pouch* is the fascial space deep to the superficial perineal fascia. Should the urethra rupture into this space, the attachments of the fascia will determine the direction of flow of the extravasated urine—not to the anal triangle nor the thigh, but into the scrotum, around the penis, and upward into the abdominal wall.

4. *The Deep Perineal Pouch* is the space enclosed by the superior and inferior fasciae of the u.g. diaphragm. Among its contents are the membranous urethra and the Sphincter Urethrae (*fig. 331*).

5. *The Perineal Body* is a small fibrous mass at the center of the perineum (*fig. 328*). Attached here are the base of the perineal membrane and several muscles that converge from 5 directions: Sphincter Ani Externus, Transversus Perinei Superficialis (R. and L.), Bulbospongiosus, and in part the Levator Ani (*fig. 336*).

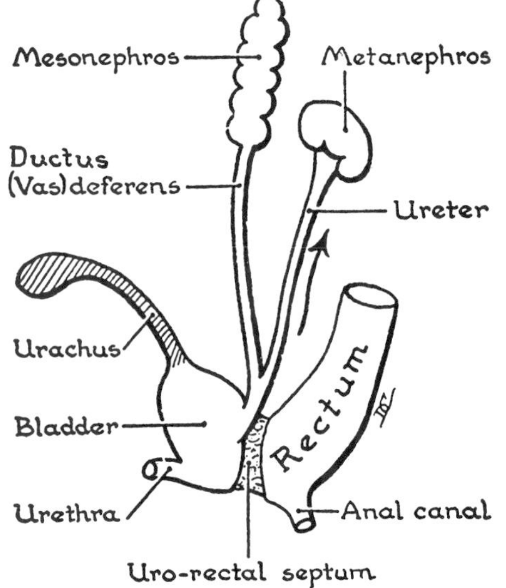

Fig. 330. Connections and subdivisions of cloaca.

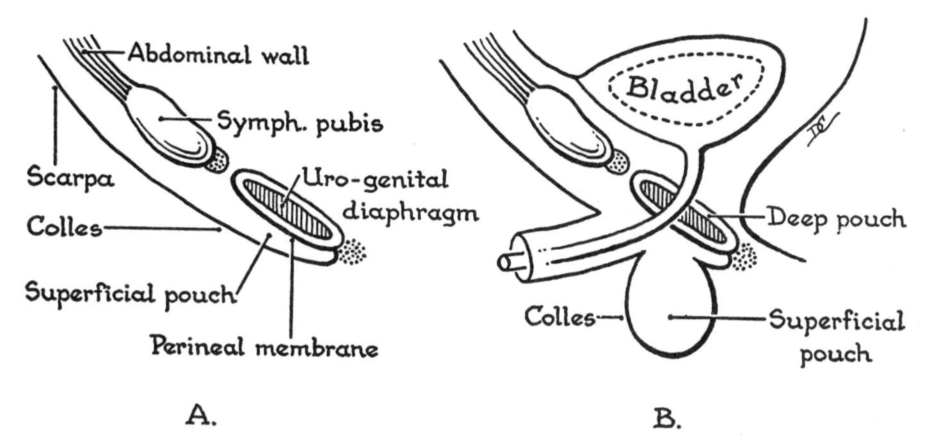

Fig. 331. To explain the urogenital diaphragm and the perineal pouches (schematic).

6. *The Anococcygeal Raphe* is a fibrous band between the anus and coccyx.

ANAL REGION (TRIANGLE)

Sphincter Ani Externus (Anus, L. = a ring)

This is a sphincter of voluntary muscle, 2 to 3 cm deep, placed around the anal canal. It has three parts—subcutaneous, superficial, and deep (*figs. 336, 370*).

The Subcutaneous Part is slender and encircles the anal orifice. *The Superficial Part* is elliptical and extends from the tip of the coccyx and anococcygeal raphe to the perineal body (*fig. 336*). It moors the circular anus to the median plane, but offers it little support fore and aft.

The Deep Part encircles the anal canal like a collar; in front, however, some fibers decussate and join the opposite Superficial Transverse Perineal muscle. Above, it blends with the Levator Ani. It is supplied by many branches of the inferior rectal (hemorrhoidal) vessels and nerve (*fig. 372*). Its nerve is the inf. rectal branch of the pudendal nerve, chiefly from S. 4 (*fig. 336*).

Ischiorectal Fossa

The ischiorectal fossae are the fascia-lined, wedge-shaped spaces, one on each side of the anal canal and rectum. Filled with fat, they allow the rectum to become distended and to empty. Each fossa is bounded laterally by the ischium, from which the Obturator Internus arises (*fig. 332*); medially, by the rectum and anal canal, to which the Levator Ani and External Sphincter are applied; posteriorly, by the sacrotuberous ligament and the overlying Gluteus Maximus; anteriorly, by the base of the urogenital diaphragm and its fasciae.

The fascia covering the Obturator Internus is fairly strong and it extends upwards beyond the fossa in the pelvis minor, reaching the pelvic brim (*fig. 346*). The fascia covering the Levator Ani is weak. The apex or roof of this fascia-lined, wedge-shaped fossa is formed by the Levator Ani arising from the Obturator Internus fascia less than the length of an index finger above the ischial tuberosity.

A finger tip passed forward enters a short cul-de-sac above the base of the u.g. diaphragm. Here the superior fascia of the diaphragm and the fasciae of the medial and lateral walls and of the roof adhere.

The fascial linings of the right and left fossae blend posteriorly, between anal canal and coccyx, to form a weak, median, areolar partition whose lower free border is the **anococcygeal raphe.**

Contents of the Fossa:

1. Fat.
2. Internal pudendal vessels and pudendal nerve (pudere, L. = to be ashamed; *cf.* impudent) run forward in an areolar

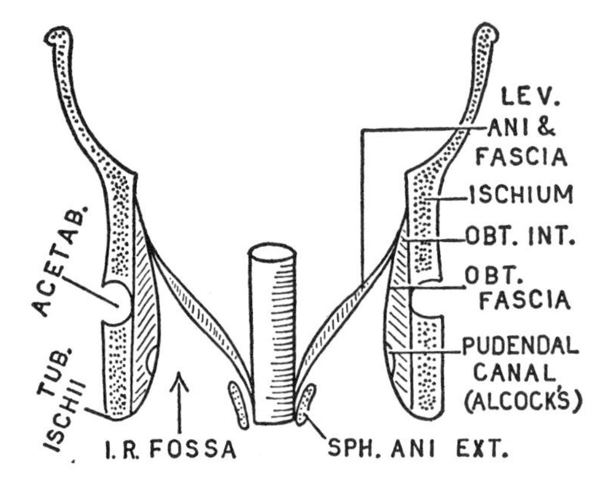

Fig. 332. The pelvis in coronal sections to show the ischiorectal fossa (schematic).

sheath, the *pudendal canal*, in the Obturator Internus fascia, 2 to 3 cm above the tuberosity. Far back they give off the inferior rectal (hemorrhoidal) vessels and nerve, which pass forward and medially through the ischiorectal fossa toward the surface to supply: External Sphincter Ani, skin around the anus, and mucous membrane of the lower canal.

3. Two other cutaneous nerves, the *perforating cutaneous* branch of the 2nd and 3rd sacral and the *perineal branch* of the 4th sacral are shown in *figure 328*.

UROGENITAL REGION IN THE MALE

Superficial Perineal Pouch

The superficial perineal pouch has been defined on p. 258.

The *two posterior scrotal branches of the perineal nerve* pierce the base of the perineal membrane, and the *perineal branch of the posterior cutaneous nerve of the thigh* pierces the attachment of the superficial perineal fascia to the pubic arch; thereafter these three sensory nerves can be followed forward to the scrotum accompanied by the two *posterior scrotal arteries*. A small artery, called the *transverse perineal a.*, runs medially along the base of the pouch to meet its fellow.

The Contents of the Superficial Perineal Pouch are:

1. The root of the penis.
2. Three superficial nerves (paired).
3. Three superficial arteries (paired).
4. Three superficial muscles (paired).

Penis (penis, L. = a tail)

The penis is composed of three fibroelastic cylinders, the right and left *corpora cavernosa penis* and the *corpus spongiosum penis* (*fig. 333*), which are filled with erectile tissue and are enveloped in fasciae and skin (corpus, L. = body; pl., corpora).

The corpora cavernosa fuse with each other in the median plane, except behind where, as two diverging *crura*, they separate to find attachment on each side to 2 cm of the pubic arch. They support the corpus spongiosum, which lies below and between them.

The corpus spongiosum is traversed by the urethra; it is swollen in front where, as the *glans penis*, it fits on to the blunt end of the united corpora cavernosa, and swollen behind where, as the *bulb of the penis*, it is fixed to the perineal membrane. This membrane moors the bulb to the pubic arch.

The crura and the bulb are the attached parts or *root of the penis* (*fig. 334*).

Skin and Fasciae (*fig. 334.1*) of the penis are prolonged loosely over the glans as the *foreskin* or *prepuce*. The skin is devoid of hairs and the fascia of fat. Deep to these, a closed tube of denser and more tightly fitting fascia, *the deep fascia of the penis*,

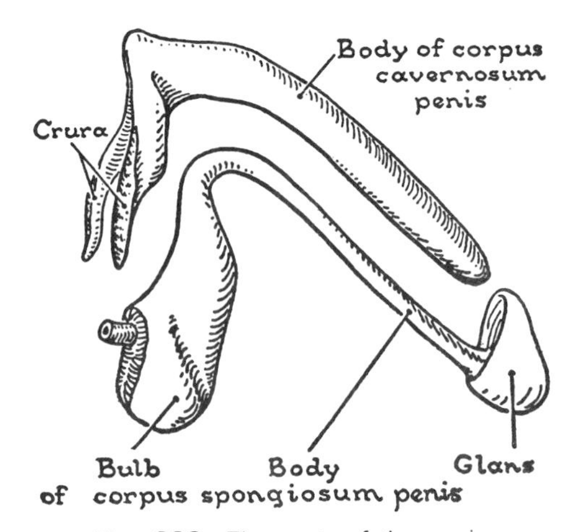

Fig. 333. The parts of the penis.

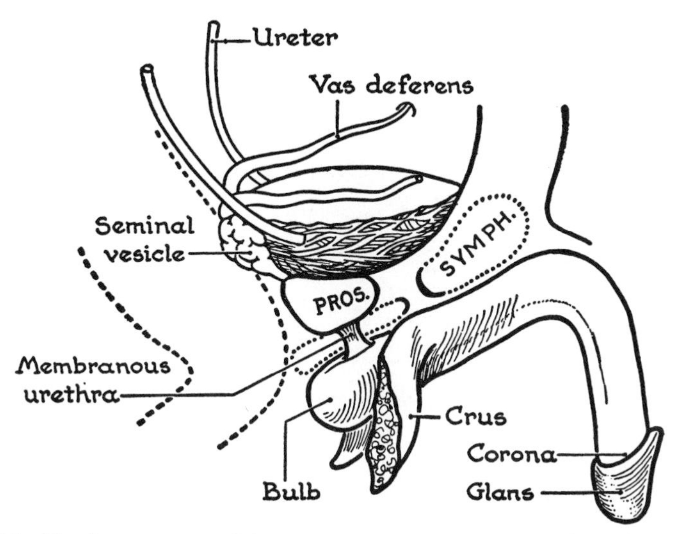

Fig. 334. The lower parts of the genital and urinary tracts and their relations.

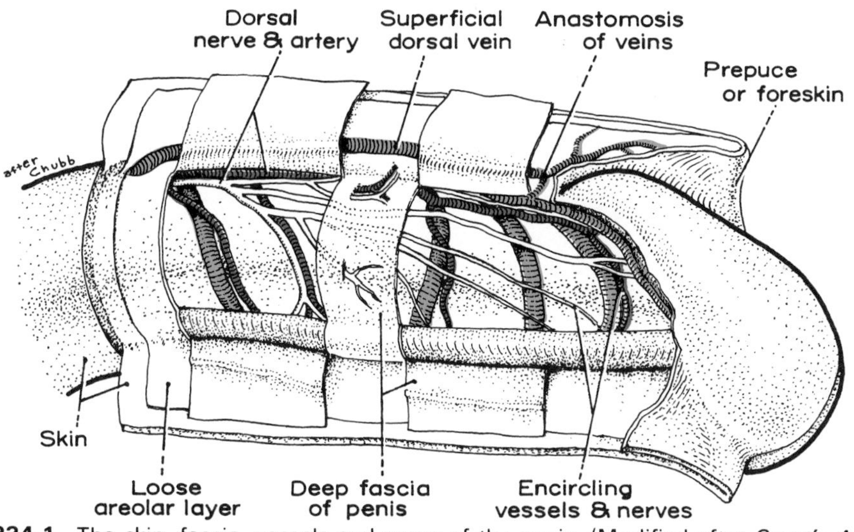

Fig. 334.1. The skin, fascia, vessels and nerve of the penis. (Modified after *Grant's Atlas*.)

envelops the body of the penis from its root to the corona of the glans (*fig. 335*).

The Suspensory Ligament of the Penis is a thick, triangular fibro-elastic band. Above, it is fixed to the lower part of the linea alba and upper part of the symphysis pubis; below, it splits to form a sling for the penis at the junction of its fixed and mobile parts (i.e., where the organ is bent), and here it blends with the fascia penis.

Vessels and Nerves of the Penis (*fig. 335*). There is a superficial and a deep dorsal vein of the penis. Each is single and occupies the median plane. The superficial dorsal vein is accompanied by lymph vessels, but it has no companion artery. It drains a cutaneous plexus which bulges through the thin penile skin. The deep dorsal vein, on the other hand, has a companion dorsal artery and nerve on each side, and lymph vessels. They run along the dorsum penis deep to the deep fascia of the penis. Branches of the arteries conduct blood to the corpora cavernosa and corpus spongiosum, to be returned by branches of the vein (*fig. 335*). The nerves end mainly in the very sensitive glans.

The Deep Dorsal Vein (unpaired) passes

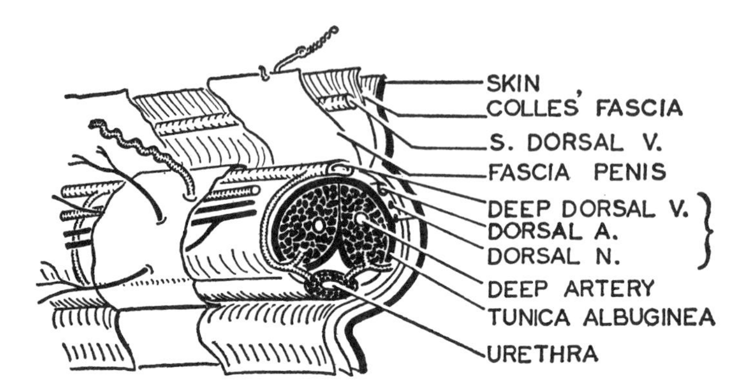

Fig. 335. The penis on cross-section, its coverings and its vessels.

below the symphysis pubis to end in the prostatic plexus of veins.

The Lymph Vessels end in the deep inguinal nodes (p. 303).

The Erectile Tissue of the penis is supplied by three paired arteries—artery to the bulb, artery to the crus (deep artery, p. 264), and dorsal artery. The last named artery sends encircling twigs to assist the former two arteries.

The Vasomotor Nerves are derived from the pelvic splanchnics (p. 264 and *fig. 49*).

The Coverings of the Penis (skin and fasciae) are supplied by the dorsal aa. of the penis and by the ext. pudendal branches of the femoral aa. They are drained by the *superficial dorsal vein*, which divides; thence the right and left branches pass via the ext. pudendal veins to the great saphenous veins of the thigh.

The Superficial Lymph Vessels anastomose in the prepuce with branches of the deep vessels and end in the superficial inguinal nodes.

The Cutaneous Nerves are branches of the dorsal nerves. The genital branch of the genitofemoral n. and the ilio-inguinal n. supply the parts near the pubis.

Superficial Perineal Muscles

On each side three muscles of the superficial perineal pouch cover the perineal membrane. They are: the slender *Transversus Perinei Superficialis* which extends from the ischial tuberosity to the perineal body; the *Ischiocavernosus*, which is applied to the crus; and the **Bulbospongiosus,** which arises from the perineal body and a median raphe below the bulb of the penis (*fig. 336*).

The most posterior fibers of each Bulbospongiosus pass to the perineal membrane; the intermediate fibers of the two sides meet on the dorsum of the corpus spongiosum; and the most anterior fibers meet on the dorsum of the penis, where they blend with the fascia penis. The Bulbospongiosus is a compressor that empties the bulb and the hinder part of the spongy urethra. There is evidence that it is essential in maintaining erection.

Deep Perineal Pouch (*fig. 331*)

This pouch is the narrow space enclosed by the fasciae of the urogenital diaphragm.

Contents: (1) the u.g. diaphragm (Transversus Perinei Profundus and Sphincter Urethrae), (2) the membranous urethra which, thin walled and 1 cm long, perforates the diaphragm, (3) two small glands, the *bulbo-urethral glands* (of Cowper), each the size of a pea, lie deep to the diaphragm and alongside the urethra. Their long ducts travel in the wall of the urethra for 2–3 cm before opening into the spongy urethra (*figs. 339 and 347*), and (4) vessels and nerves to be described now.

Internal Pudendal Vessels and Pudendal Nerve

The artery, which is a branch of the internal iliac artery, and the nerve, which

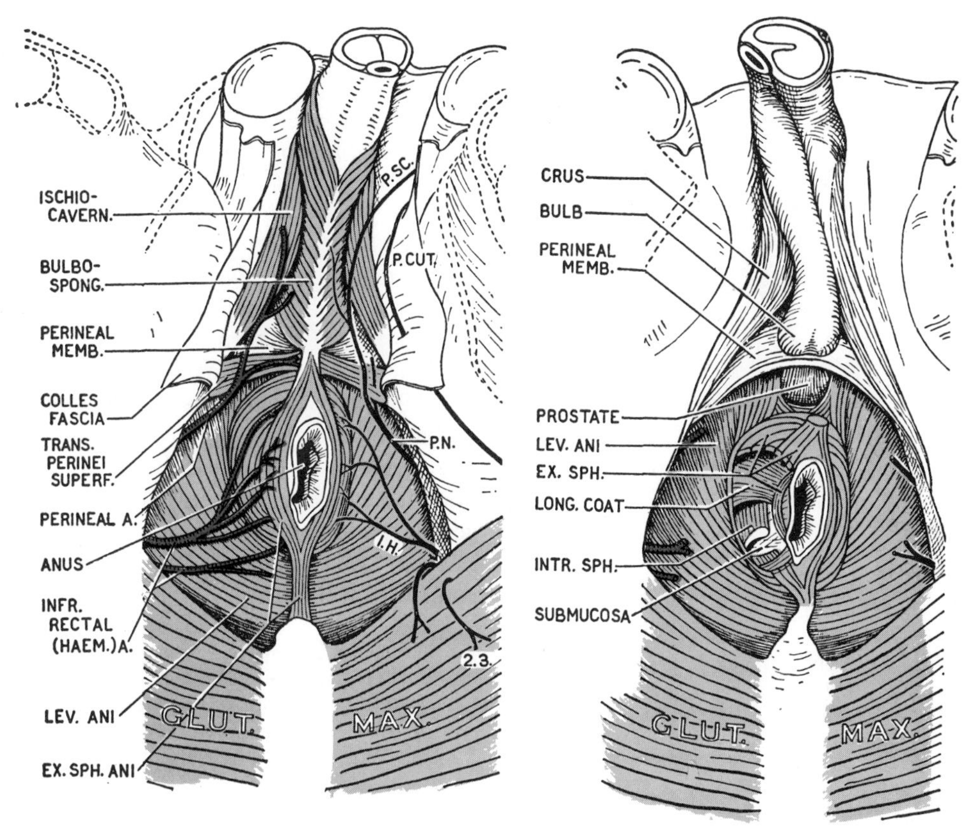

Fig. 336. *Left,* superficial dissection of the male perineum. (By Dr. H. C. Hair.) *Right,* exposure of the prostate. Dissection of the anal canal. (By Dr. V. P. Collins.) *I.H.* = inferior rectal (hemorrhoidal) n.; *P.N.* = perineal n.; *P.SC.* = posterior scrotal n.; *P.CUT* = perineal branch of posterior cutaneous n. of thigh.

arises from sacral segments 2, 3, and 4, together leave the pelvis through the greater sciatic foramen, flit through the gluteal region (*fig. 337*), crossing the ischial spine and passing through the lesser sciatic foramen to enter the pudendal canal on the side of the ischiorectal fossa. They have three territories to supply: (1) anal triangle, (2) urogenital triangle and scrotum (labium majus in female), and (3) penis (clitoris in the female).

Dermatomes: see *fig. 336.1.*

The nerve divides while entering the pudendal canal into three terminal branches (*figs. 337, 338*)—(1) inf. rectal, (2) perineal, and (3) dorsal n. of penis. *Details:*

1. *The inferior rectal (hemorrhoidal) nerve* supplies the External Sphincter Ani, the skin about the anus, and the lining of the canal below the anal valves. It may enter the region independently by piercing the sacrospinous ligament (Roberts and Taylor).

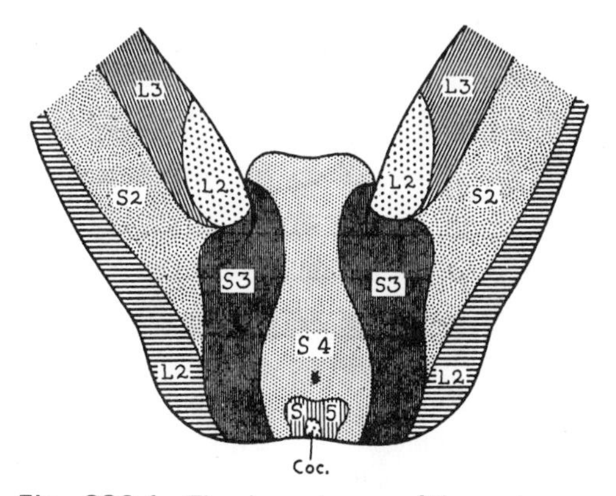

Fig. 336.1. The dermatomes of the perineum.

2. *The perineal nerve* runs through the pudendal canal to the perineal membrane where

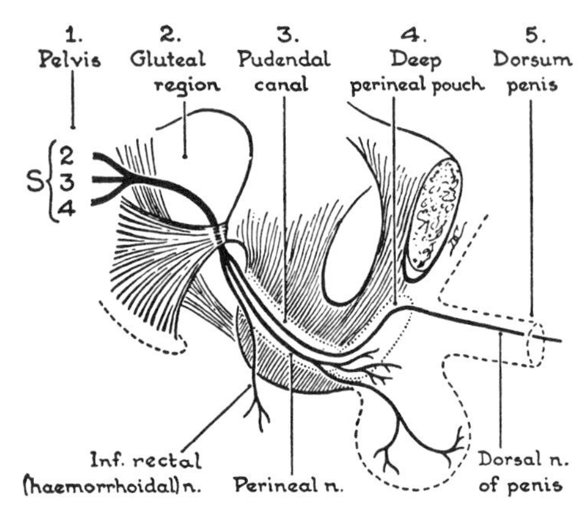

Fig. 337. The course of the pudendal nerve: Its three divisions and five regions traversed.

its two cutaneous branches, the *posterior scrotal nerves*, become superficial and continue to the scrotum. Its motor branch, the *deep perineal nerve*, innervates the diaphragm and the three superficial muscles, and sends twigs to Levator Ani and Ext. Sphincter Ani.

3. *The dorsal nerve of the penis* burrows through to the dorsum of the penis, where it lies deep to the fascia penis. It supplies the glans, the prepuce, the skin of the penis, and the spongy urethra (*fig. 49*, p. 62).

The **Internal Pudendal Artery** travels through the pudendal canal and deep perineal pouch to become the *dorsal artery of the penis*, Deep to the fascia penis, on the dorsum of the penis, it lies between its companion vein and nerve.

Its Branches are: (1) *The inferior rectal a.* (inf. hemorrhoidal a.) which supplies the anal triangle; (2) *The perineal a.* which gives off the slender *transverse perineal a.* and *posterior scrotal aa.* (3) *The artery to the bulb*, a large vessel to the bulb. (4) *The artery to the crus* (deep artery of the penis), which arises deep to the crus, and at once plunges into it.

The Vasomotor Nerves to the cavernous tissue are derived from the pelvic splanchnic nerves and the hypogastric plexus. They pass below the symphysis pubis to the erectile tissue either directly or indirectly via the pudendal nerve.

Erection. Stimulation of the pelvic splanchnic nerves produces erection of the penis by causing dilatation of the arteries and cavernous tissue. In the normal reflex act the afferent limb is the pudendal nerve (S. 2, 3, 4); the efferent is the pelvic splanchnic (S. 2, 3, 4) (see *fig. 49*). By causing the Bulbospongiosus and Ischiocavernosus to contract, the pudendal nerve is responsible for increasing the erection.

Ejaculation. The impulse spreading, sympathetic nerves are stimulated to cause closure of the internal urethral orifice, to set up peristaltic waves which empty the epididymis and propel its contents through the vas to the urethra, and to empty the seminal vesicles and prostate of their secretions. The cord segment is L. 1. The path is probably via the intermesenteric and hypogastric plexuses, because the usual operation for removal of the lumbar sympathetic does not impair ejaculation, but removal of the hypogastric plexus does so permanently (Learmonth).

Exposure of the Prostate from the Perineum (*fig. 336*). Between the borders of the Levatores Ani, pushing the anal canal and rectum backward exposes the tough fascia covering the posterior surface of the prostate.

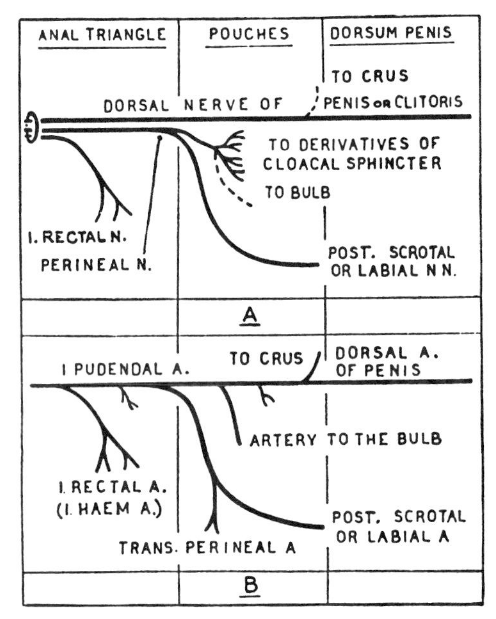

Fig. 338. *A*, the three *divisions* of the pudendal nerve: each for a region. *B*, The three *branches* of the pudendal artery. A necessary difference in terminology: an essential similarity in anatomy.

FEMALE PERINEUM

An incision into the male urethra, entering it on the under surface just behind the glans penis, and carried back to the prostatic urethra, restores it to its fetal condition, which is similar in both sexes. The incision (*fig. 339*) divides everything encountered including the urethra, scrotum, Bulbospongiosus, bulb of the penis and u.g. diaphragm. However, the female penis, called the clitoris, is diminutive and is not traversed by the urethra. It comprises two *corpora cavernosa clitoridis* and a *glans clitoridis* which caps the conjoint corpora cavernosa (*fig. 340*).

The incision suggested above would bring about the following changes and emulate homologous female parts:

1. The edges of the incised urethra = the right and left *labia minora*.

Each *labium minus* is a thin cutaneous fold, devoid of fat and lying alongside the orifice of the vagina. The posterior end is free. The anterior end divides into two lesser folds which unite with their fellows across the median plane, the upper folds forming a hood, the *prepuce of the clitoris*, over the glans, the lower joining to form a band, the *frenulum of the clitoris*, which is attached to the under surface of the glans.

2. The scrotum is split into—right and left *labia majora* (*fig. 341*).

Each *labium majus* is a broad, rounded, cutaneous ridge lying lateral to the labium minus and covering a long finger-like process of fat. This process extends backward from a median skin-covered mound of fat, the *mons pubis* situated in front of the pubis. Entering the fat of the labium from behind and running forward are: the medial and lateral labial (*cf.* scrotal) nerves, arteries, and veins, and also the perineal branch of the posterior cutaneous nerve of the thigh. Entering it from the front are: branches of the ilio-inguinal nerve and external pudendal artery and vein, and a fibrous band, called the *round ligament of the uterus* (*fig. 382*).

Sebaceous glands open onto both surfaces of the labia minus and majus; hair covers the mons and the lateral surface of the labium majus.

3. The superficial perineal fascia and pouch into—r. and l. parts.

4. Bulbospongiosus into—r. and l. parts (Sphincter Vaginae).

5. The bulb of the penis into—the r. and l. *bulbs of the vestibule*.

Each *bulb* is a loosely encapsuled mass of erectile tissue, shaped like a half pear.

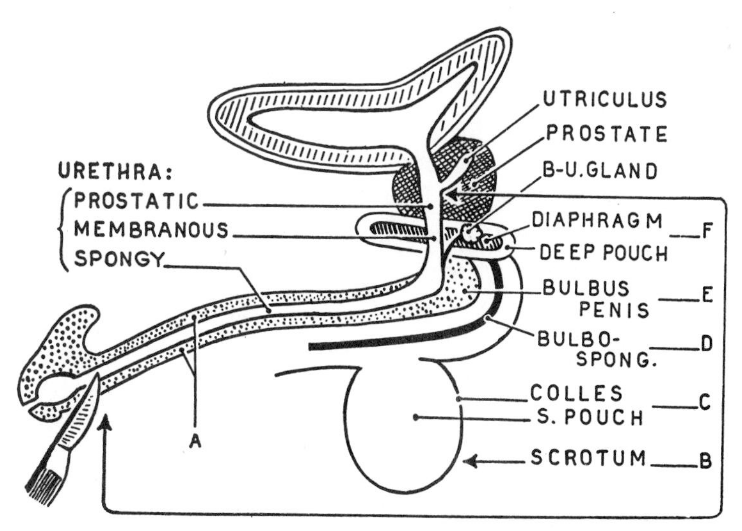

INCISION EXTENDS FROM CORONA OF GLANS TO UTRICULUS

Fig. 339. Incision converting male perineum into female.

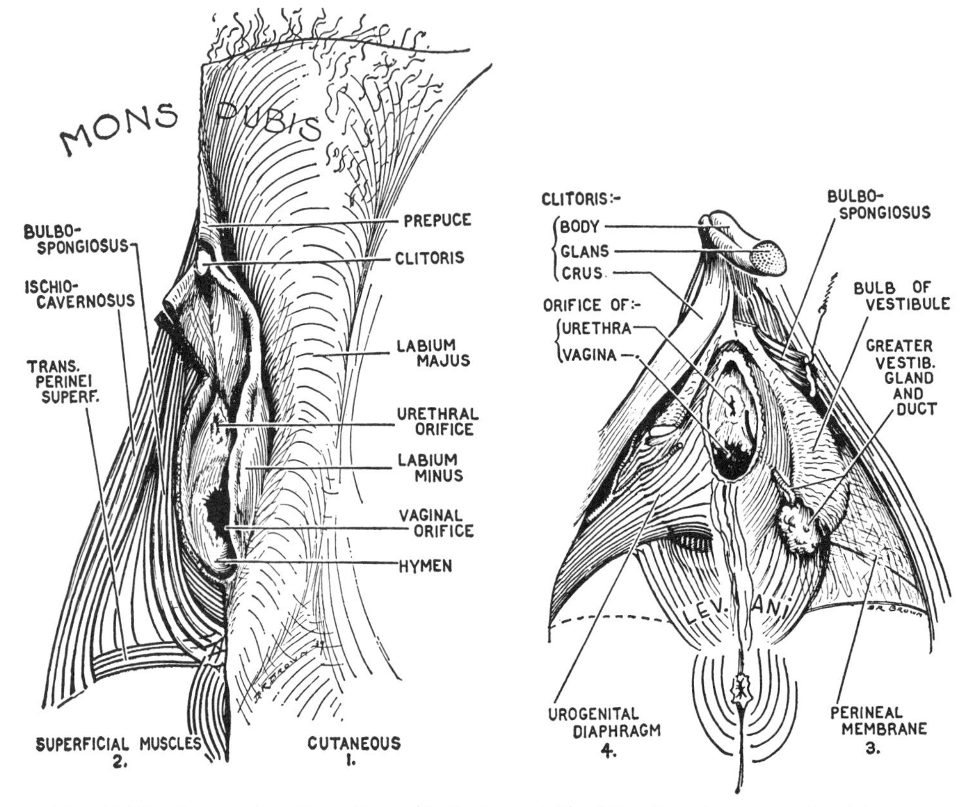

Fig. 340. Successive dissections (1, 2, 3, and 4) of the female urogenital triangle.

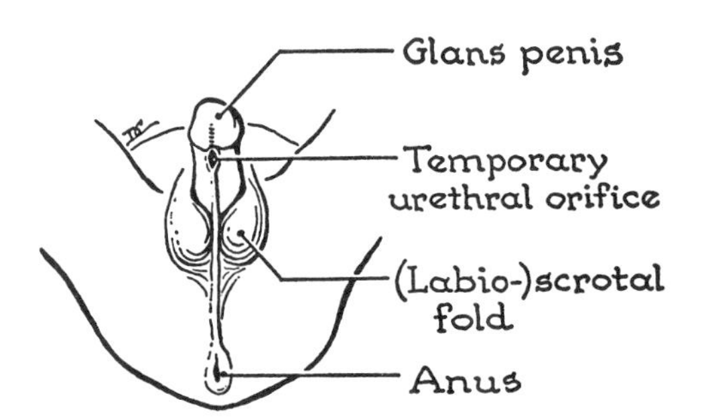

Fig. 341. Showing (1) the paired, right and left, labioscrotal (genital) folds about to unite below the penis to form the scrotum, and (2) the urethral orifice moved forward.

The convex surface is lateral and is covered with Bulbospongiosus; the flat or concave surface is medial and is applied to the perineal membrane which intervenes between it and the wall of the vagina; the enlarged end of the pear is posterior; the narrow stalk is anterior and, after joining its fellow, ends as the *glans clitoridis*.

6. The urogenital diaphragm, its inferior fascia (perineal membrane), and its superior fascia into—right and left parts. These blend with the vaginal wall.

7. The prostatic utricle comes to open on to the skin surface—the *vagina* (? and uterus).

The *vestibule* of the vagina is the cleft between the labia minora. It represents the spongy urethra of the male. The *hymen* is a thin membranous fold of irregular outline that surrounds the vaginal orifice like the ruptured membrane of a drum. Opening into the vestibule are: the urethra, vagina, para-urethral glands, and greater and lesser vestibular glands. (The lesser glands are the many small mucous glands that open here.)

8. The bulbo-urethral glands come to open on to the skin surface—the *greater vestibular glands* (Bartholin's glands).

Each *gland* is larger than a pea (but

smaller in the aged) and is situated at the hinder end of the bulb. Its duct, 2 cm long, opens into the hinder part of the vestibule.

9. The remaining part of the urethra (i.e., the part above the utricle) is homologous with the entire female urethra. In the male this part is 2 cm long; in the female it is 3 cm. The female urethra lies immediately in front of the anterior wall of the vagina and is intimately adherent to it (*fig. 379*).

The *urethral orifice* opens just in front of the vaginal orifice and is 2 to 3 cm behind the glans.

10. Though a prostate is not found in the female, the *para-urethral glands*, whose ducts open one on each side of the female urethra, are probably homologous with prostatic glands.

11. The ejaculatory ducts, which in the male open on to the lips of the prostatic utricle (*fig. 364*), generally disappear in the female, but, as the ducts of Gartner (*fig. 51*, p. 65), they may persist as blind tubes on the anterior wall of the vagina, and become cystic; rarely they open on to the skin surface.

12. In the male, the primitive urethral orifice opened in the perineum. Later, when the lips of the genital folds met and fused, a secondary orifice was formed behind the glans penis (*fig. 341*); subsequently, the urethra traversed the glans; so, a third and permanent orifice opens at the end of the glans. In the female, and as a rare anomaly in the male, the primary perineal orifice is the permanent one. The glans clitoridis is not canalized.

13. In the male, the gubernaculum testis passed to the scrotum, and the processus vaginalis peritonei and the testis followed it. In the female, the gubernaculum ovarii and the processus vaginalis (of Nuck) entered the labium majus but, owing to a side attachment that the gubernaculum makes with the uterus, the *ovary* enters the pelvis; only rarely does it descend into the labium (*fig. 382*).

20 MALE PELVIS

Developmental Considerations.
Male Pelvis Viewed from above: Peritoneum; Bladder; Retropubic Space; Ureter and Ductus Deferens.
Organs that Developed in Primitive Urorectal Septum.

Side Wall of Pelvic Cavity

Obturator Nerve; Levator Ani; Pelvic Fascia.

Interior of Bony Pelvis

Sacrum and Coccyx—Os Sacrum: Base; Pelvic Surface; Sides; Posterior Surface; Coccyx; Ossification.
Pelvic Brim—Pelvis Major vs. Pelvis Minor
Pelvis Minor
Side Wall; Foramina; Muscles of Walls.
Pelvic Diaphragm—Levator Ani, Coccygeus.
Orientation of Pelvis; Mechanism.

ARTICULATIONS

Lumbosacral Joint: intervertebral disc, ligaments.
Mechanisms of Pelvis and Joints—Sacro-iliac Joint and Symphysis Pubis; Ligaments Resisting Rotations of Sacrum; Movements; Relations, Surface Anatomy; Development and Variations.

UROGENITAL SYSTEM IN PELVIS

Urinary Bladder—Its Bed; Ureter; Interior of Bladder; Vessels and Nerves.
Urethra—Prostatic, Membranous, and Spongy Parts; Palpation; Nerve Supply.
Structure of Urinary Tract

INTERNAL PARTS OF MALE GENITAL SYSTEM

Prostate.
Ductus (Vas) Deferens, Seminal Vesicles, Ejaculatory Ducts; Vessels and Nerves.

RECTUM AND ANAL CANAL

Parts; Structure; Arteries; Veins; Lymph Vessels and Nerves; Relations.

VESSELS AND NERVES OF PELVIS

Arteries (in detail)—Branches of Internal Iliac Artery: Visceral, to Limb and Perineum, Somatic Segmental.
Veins of Pelvis.
Sacral and Coccygeal Nerve Plexuses (in detail); Arteries Piercing Plexus; Branches of Sacral Plexus; Sciatic Nerve and Pudendal Nerve.
Coccygeal Plexus.

Developmental Considerations. In the mesodermal partition, the *urorectal septum* (*fig. 330*), certain reproductive organs develop:

In the Male: deferent ducts, seminal vesicles, and prostate.
In the Female: uterine tubes, uterus, and vagina.

Male Pelvis Viewed from Above. The following *peritoneal fossae* are seen: on each side of the rectum, the *pararectal fossae;* on each side of the partly filled bladder, the *paravesical fossae;* and, between them, the *rectovesical pouch.*

Follow the Peritoneum in the Median Plane (fig. 342).

Four "visceral tubes" adhere to the peritoneum even when it is detached and mobilized (*fig. 343*).

The Retropubic Space (*fig. 342* and *344*) is a continuous bursa-like cleft in the areolar tissue at the sides and front of the bladder which allows the bladder to fill and empty without hindrance. The space is limited *posteriorly* by a broad areolar sheet enclosing a leash of vessels that pass from the internal iliac artery and vein to the posterolateral border of the bladder (*fig. 375*). Two taut cord-like thickenings, one

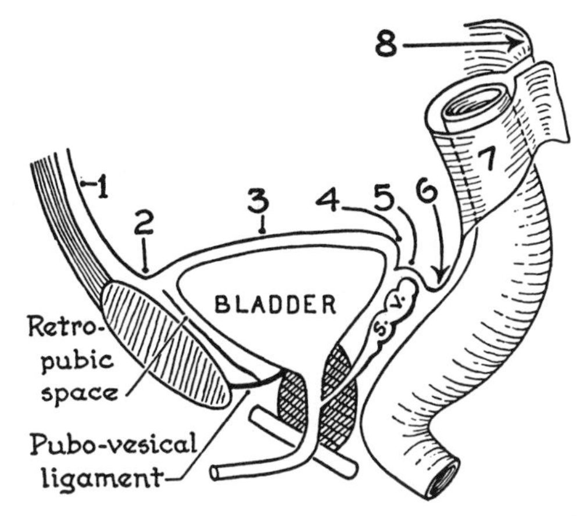

Fig. 342. The peritoneum of the male pelvis in paramedian section (see *fig. 380*). (1) anterior abdominal wall, (2) back of the pubis, (3) superior surface of the empty bladder, (4) posterior surface for 1 cm, (5) upper ends of the seminal vesicles on each side of the median plane, (6) across the bottom of the recto-vesical pouch, (7) the rectum, (8) the mesentery of the sigmoid colon.

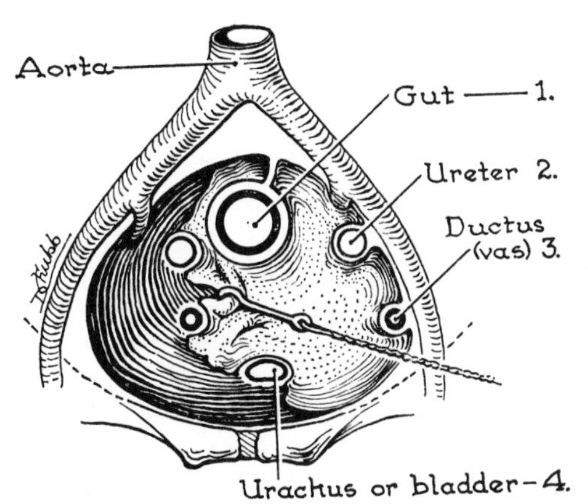

Fig. 343. Four visceral tubes adhere to the peritoneum. (The peritoneum has been detached from the side wall of the pelvis).

on each side of the median plane, attach the neck of the bladder to the lower end of the symphysis; these are the *puboprostatic* or *pubovesical ligaments* (*fig. 348*).

Ureter and Ductus (*Vas*) *Deferens.* The *ureter* crosses the external iliac artery, just in front of the bifurcation of the common iliac artery. It enters the lateral angle of the bladder where its last 2–3 cm is enveloped in the leash of vessels just mentioned, and is crossed by the ductus deferens (*fig. 345*).

The *ductus deferens* runs from the deep inguinal ring, where it turns round the inferior epigastric vessels.

The seminal vesicles are situated lateral to the deferent ducts behind the bladder (*fig. 346*).

Side Wall of the Pelvic Cavity

The side wall may be divided into two

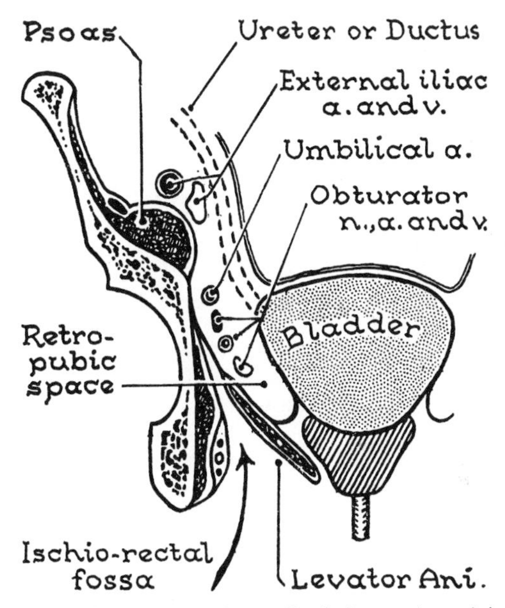

Fig. 344. The side wall of the male pelvis on coronal section (diagrammatic).

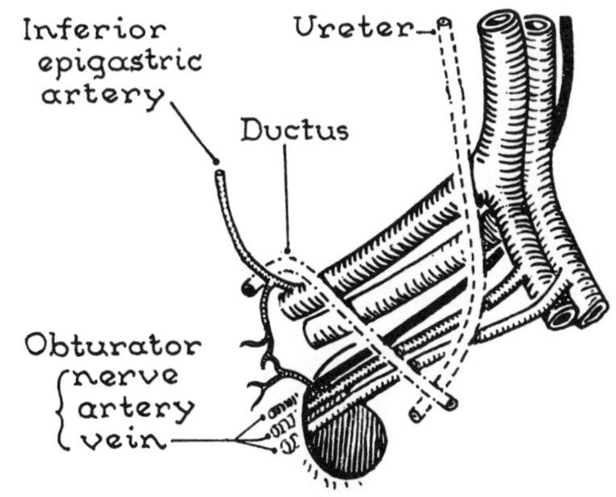

Fig. 345. The structures on the side wall of the male pelvis. Note the medial positions of the ureter and vas or ductus deferens.

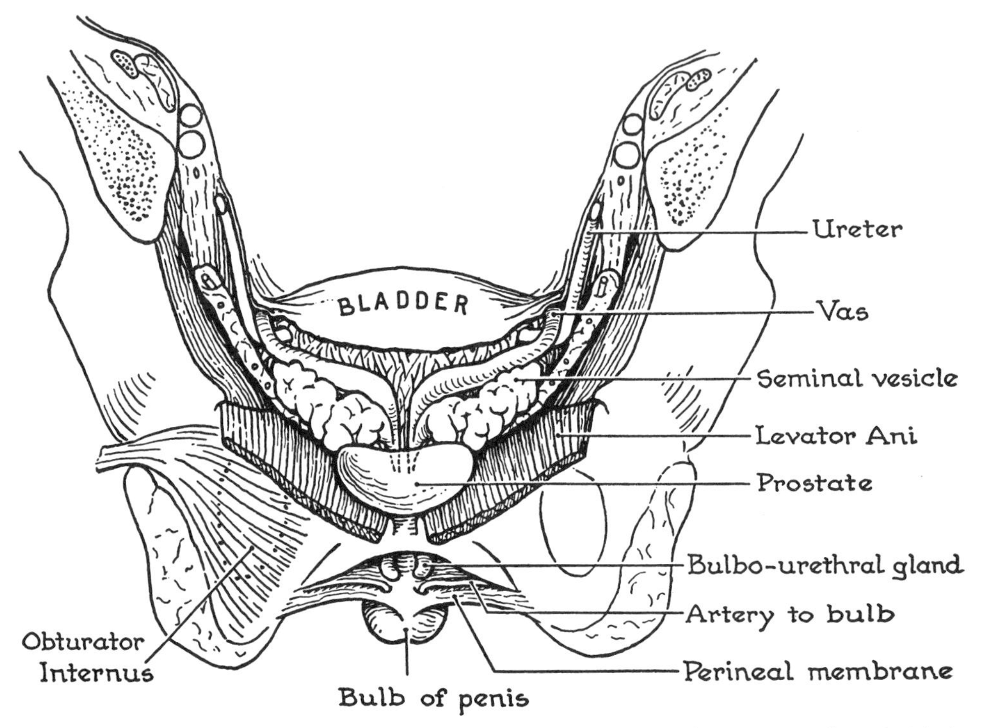

Fig. 346. A coronal section of the pelvis to show the genito-urinary organs from behind.

parts: the part above the obturator nerve and the part below it (*fig. 347*).

The Obturator Internus, covered with its fascia (*fig. 347*), rises posteriorly to the pelvic brim, but falls anteriorly below the upper part of the obturator foramen leaving the pubic bone exposed and allowing the obturator nerve, artery, and vein to escape from the pelvic cavity unhindered.

The Levator Ani forms the greater part of the pelvic floor or diaphragm. It arises from the inner surface of the body of the pubis, from the inner surface of the ischial spine, and, between these two points, from the obturator fascia, which is commonly thickened to form a *tendinous arch* of origin (*fig. 347*).

The structures encountered on the side wall are shown in *figures 344 and 345*. Also running forwards from the internal iliac artery to the umbilicus is the (obliterated) umbilical artery (p. 204).

Pelvic Fascia (*fig. 348*).

All the contents of the pelvis are covered

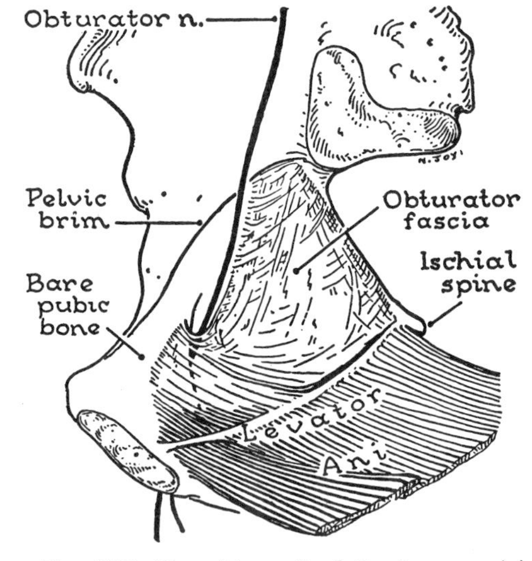

Fig. 347. The side wall of the lesser pelvis divided into upper and lower (anterior and posterior) parts by the obturator nerve.

all over with areolar tissue. The covering of organs that expand and contract, notably the rectum and bladder, is necessarily loose; that given to organs that do not expand may be dense, as in the case of the prostatic fascia, or thin as in the case of the fascia covering the

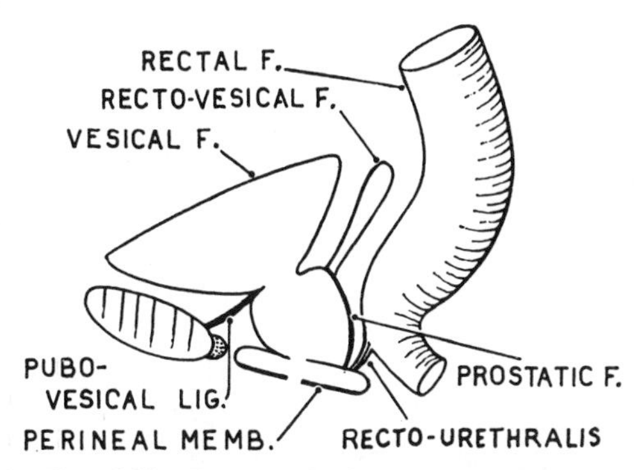

Fig. 348. The pelvic fascia of the male in median sagittal section. Recto-urethralis consists of strands of smooth muscles.

seminal vesicles and deferent ducts. Its texture depends upon the strains put upon it.

Subdivisions of the Pelvic Fascia. This fascia is divided into *parietal, diaphragmatic* (for the pelvic diaphragm or floor), and *visceral* layers. Note:

1. The puboprostatic ligaments, which anchor the prostate (and bladder) to the pubis, are thickenings of this fascia.
2. The handle of the knife passed through the roof of the ischiorectal fossa enters the retropubic space.
3. The fascia invests the numerous veins and few arteries that are passing from the internal iliac vessels to the base of the bladder and internal genital organs (within the rectovesical fascia) (*fig. 375*). This limits the retropubic space posteriorly.
4. Four layers of fascia separate the bladder from the rectum (*fig. 348*).
5. Dense fascia clothes the back of the prostate (p. 264).

6. The potential "retrorectal space" is bounded on each side by an areolar fold containing the pelvic splanchnic nerves, which run from the 2nd, 3rd, and 4th sacral foramina to the side of the rectum (*fig. 348.1*).

INTERIOR OF BONY PELVIS

The pelvis (L. = basin) is formed by the right and left hip bones, the sacrum, and the coccyx. Each hip bone (*L., os coxae*) articulates with the other in front at the symphysis pubis; behind they articulate with the first three sacral vertebrae at the sacro-iliac joints. The right and left hip bones constitute the pelvic girdle.

Sacrum and Coccyx

The Os Sacrum is composed of five fused vertebrae, and the **os coccygis** of 3–5, though before birth it has 7–11 cartilaginous caudal rudiments. Both sacrum and coccyx are triangular with base above and apex below (*fig. 349*).

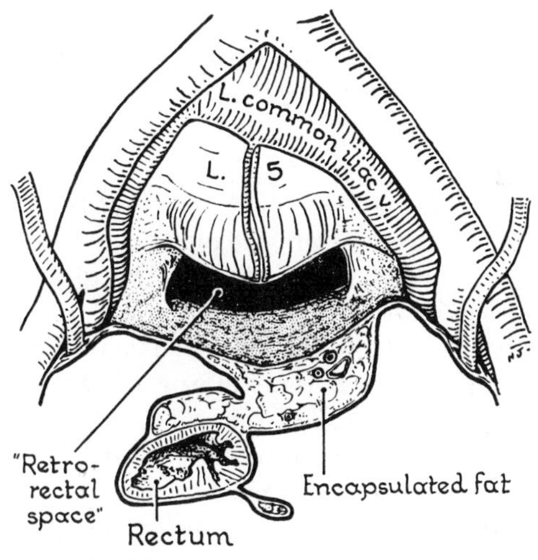

Fig. 348.1 The "retrorectal space."

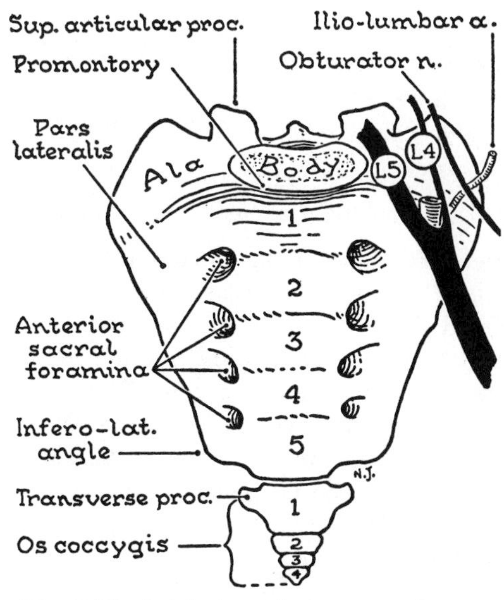

Fig. 349. Pelvic (anterior) aspect of sacrum and coccyx.

The Base of the Sacrum is divided into three parts. The median part is the oval upper surface of the body of the first sacral vertebra. Its anterior border is an important landmark named the *promontory* of the sacrum. Behind its posterior border is situated the somewhat compressed triangular entrance to the *sacral canal*, flanked by prominent superior articular processes. The right and left lateral parts, called the *alae*, are fan-shaped and represent fused costal and transverse elements (*fig. 23.1*).

Each ala is crossed by the constituents of the lumbosacral trunk, the iliolumbar artery, the obturator nerve, and the Psoas. Only the 5th lumbar ventral ramus is so taut that it grooves the surface (*fig. 349*).

The Pelvic Surface of the sacrum is concave and is crossed by ridges at the sites of obliterated discs (*fig. 349*). Lateral to them are the four *pelvic* (or *anterior*) *sacral foramina*. Their margins are shaped by the emerging ventral rami of the upper four sacral nerves which pass laterally. The mass of bone lateral to the foramina is the *pars lateralis* (lateral mass).

The **sides** of the upper three pieces of the sacrum form a large *auricular* or *ear-shaped facet* which articulates with a corresponding auricular facet on the ilium (*fig. 350*); the sides of the lower two pieces, as well as the sides of the coccyx, are thin for the sacrotuberous and sacrospinous ligaments (*fig. 358*).

(Posterior Surface. *See* Chapter 36.)

The First Piece of the Coccyx possesses a pair of *transverse processes* each of which is joined by a ligament to the sacrum (inferolateral angle) thereby making a foramen through which the ventral ramus of the 5th sacral nerve enters the pelvis (*fig. 376*). The remaining pieces of the coccyx are nodular. The 1st piece commonly fuses with the sacrum.

Ossification. The 5th, 4th, and 3rd pieces of the sacrum are always completely fused together by the 23rd year, the 3rd and 2nd by the 24th year, and the 2nd and 1st by the 25th year or even later (McKern and Stewart).

The Pelvic Brim is the boundary line between the greater or *major pelvis* above and the lesser or *minor pelvis* below. Its component parts are: the promontory of the sacrum, the anterior border of the ala of the sacrum, the iliopectineal line, the pubic crest, and the upper end of the symphysis pubis (*fig. 351*). The *iliopubic eminence* (iliopectineal eminence) marks the site of union of ilium and pubis.

The pubic part of the iliopectineal line,

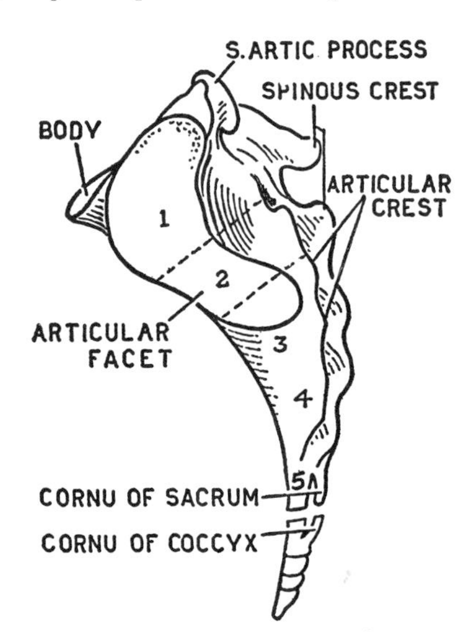

Fig. 350. The lateral aspect of sacrum and coccyx.

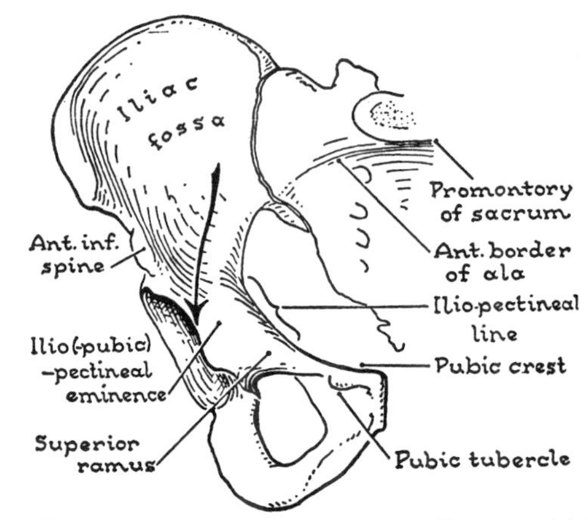

Fig. 351. The pelvis major (false pelvis) and the pelvic brim.

called the **pecten pubis,** gives attachment in its whole length to the fascia covering the Pectineus, a muscle of the thigh, and to the lacunar ligament and conjoint tendon. A strong fibrocartilaginous band, of periosteum, the *pectineal ligament* (Cooper's lig.), through which a surgeon's needle and thread can get a good grip, raises the line into a ridge.

The Pelvis Major (False or Greater Pelvis) (*fig. 351*) is formed on each side by the ala of the sacrum and the concave, fan-shaped iliac fossa. The handle of the fan immediately above the acetabulum is a broad groove for the Iliopsoas tendon.

Pelvis Minor

The pelvis minor (*true or lesser pelvis*) is formed by the sacrum and coccyx, inner surface of the ischium and pubis, and a small part of the ilium (*fig. 352*). The main features posteriorly are the greater and lesser sciatic notches and the ischial spine which separates them. The oval *obturator foramen* separates the pubis and ischium.

Foramina in the Walls of the Pelvis Minor. The posterior wall is perforated by the four *anterior or pelvic sacral foramina*. Two ligaments, the *sacrotuberous* and *sacrospinous*, so unite the posterior wall to the side wall as to leave two gaps, the *greater* and *lesser sciatic foramina* (*fig. 358* and *361*). The large *obturator foramen* is closed by the obturator membrane except above where a gap transmits the obturator nerve and vessels, and may be the site of a rare type of hernia.

Muscles of the Walls of the Pelvis Minor. *The Obturator Internus* arises by fleshy fibers from almost the entire inner surface of the side wall of the pelvis minor (*fig. 353*). It leaves the pelvis through the lesser sciatic foramen to be inserted into the greater trochanter of the femur.

The Piriformis arises by fleshy fibers from the anterior surface of the sacrum and enters the gluteal region through the greater sciatic foramen and is inserted into the greater trochanter (*fig. 420*). *The obturator fascia—See fig. 353.*

The Pelvic Diaphragm

The Levator Ani (and Coccygeus) of opposite sides stretch across the pelvis, like a hammock, separating the pelvic cavity from the perineum. This pelvic "floor" is perforated by the urethra and the anal canal and, in the female, by the vagina

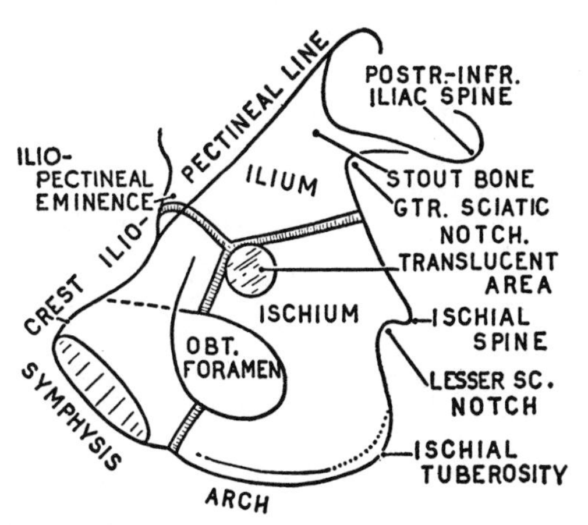

Fig. 352. Lateral wall of pelvis minor (*Eminence: iliopectineal = iliopubic.*)

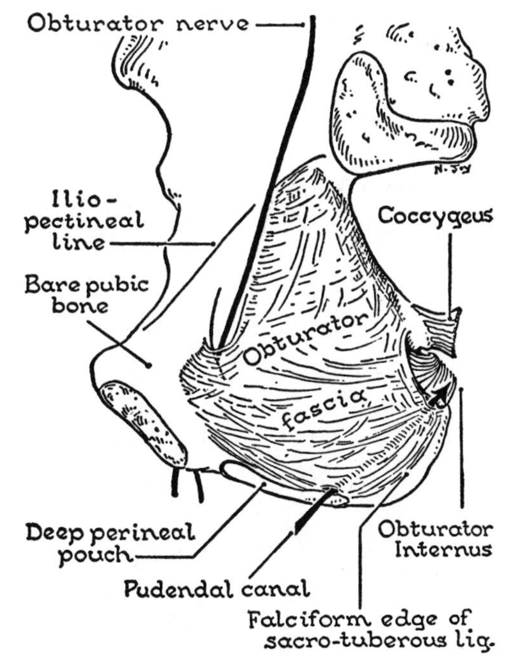

Fig. 353. The Obturator Internus Fascia forms a pocket. In it runs the pudendal canal and from it arises the Levator Ani (*cf. fig. 332*).

also. More specifically, the Levator Ani has a free anterior border which is separated from its fellow by the urethra, vagina, and anal canal.

The pelvic diaphragm arises in front from the body of the pubis, behind from the spine of the ischium, and between these two points from an arched thickening of the obturator fascia, called the *tendinous arch* (*fig. 347*).

Functions. The pelvic diaphragm supports the pelvic viscera. The integrity of the pelvic floor depends especially on the Levator Ani, including the part known as the puborectal sling, since these hold forward the lower part of the rectum, which in turn helps to support the bladder, prostate and seminal vesicles in the male, and (which is more important) the bladder and vagina in the female (*fig. 354*).

The puborectal sling keeps the anorectal angle closed, but during defecation it relaxes and allows the anorectal junction to straighten, while other fibers draw the anal canal over the feces that are being expelled.

Nerve Supply. Branches of S. (2), 3, and 4 supply the muscle on its pelvic surface; twigs of the perineal n. (S. 2, 3, and 4) supply it on its perineal surface.

Tailed Animals		*Man*
1. Pubococcygeus	=	Levator Ani
2. Iliococcygeus		
3. Ischiococcygeus	=	Coccygeus & sacrospinous lig.

The Coccygeus (*figs. 353, 355*) stretches like a fan from the ischial spine to the sacrum and coccyx, lining the sacrospinous ligament.

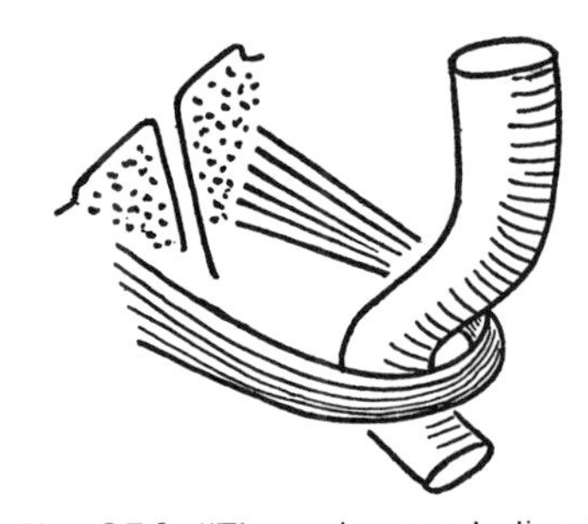

Fig. 354. "The puborectal sling."

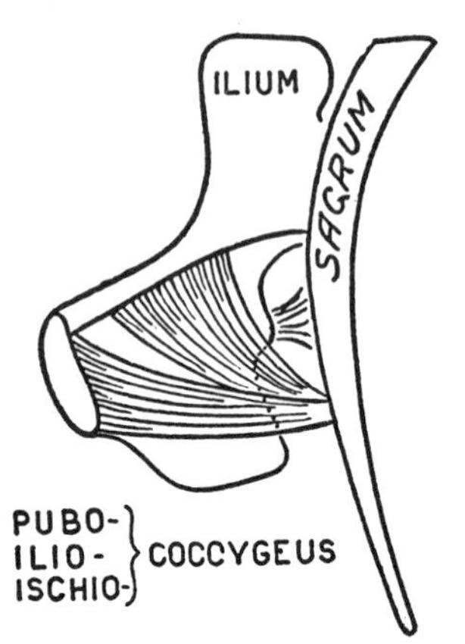

Fig. 355. Pelvic diaphragm of monkey (After Keith.)

The Pubococcygeus, which is the thickest and most important part of the Levator Ani, runs downward and backward from the pubis and meets it fellow (a) in the perineal body in front of the anus and (b) in the anococcygeal raphe behind the anus; between the two (c) it is carried down by the rectum and anal canal, which perforate it and with which it blends. On the pelvic surface of the muscle many fibers pass backward from pubis to coccyx; others meet those of the opposite side in the angle between the rectum and anal canal and so form a U-shaped puborectal sling that maintains this angle (*fig. 354*). The different parts of the diaphragm, though overlapping somewhat, blend to form a single sheet.

Orientation of the Pelvis

In the "neutral" position of the pelvis the anterior superior iliac spines and the top of the symphysis are in a vertical plane. This is approximated by applying the pelvis to a wall.

In the female the ant. sup. spines overstep the vertical slightly. In compensation for the forward tilting of the pelvis in the female, the lumbar curvature of the spine is increased. Hence, the lower part of the back is concave in the female, and the buttocks more prominent.

Mechanism of the Pelvis (*fig. 356*)

The weight of the body superimposed on the 5th lumbar vertebra is transferred to

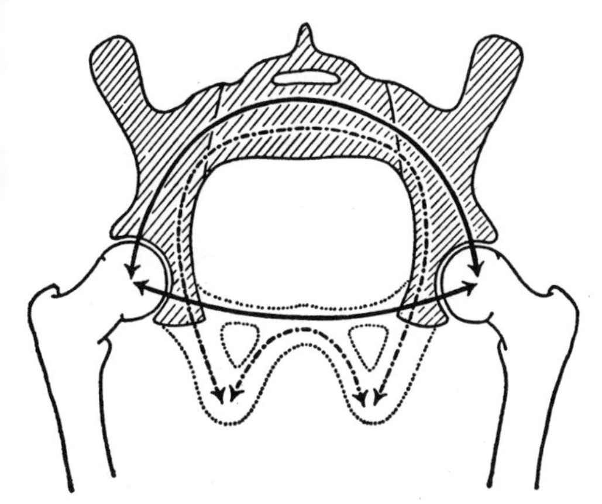

Fig. 356. The mechanism of the pelvis. *Solid lines* = the standing arch and its tie beam or counter arch which, like the clavicle, resists compressive forces. *Broken lines* = the sitting arch and its tie beam or counter arch which resists spreading forces. (After Braus.)

the base of the sacrum, thence to the upper three pieces of the sacrum, across the sacro-iliac joints to the ilia, and thence (1) when standing—to the acetabula and so to the femora (pl. of femur) or (2) when sitting down—to the ischial tuberosities. Along these lines the bony parts are thickened. In the standing posture, the acetabula and the side walls of the pelvis tend to be forced together, but the pubic bones, acting as struts, prevent this from happening (*fig. 357*).

Many mammals possess a symphysis ischii as well as a symphysis pubis and so have a powerful strut. In birds the pubic bones are wide apart and, so, offer no obstruction to the laying of eggs.

ARTICULATIONS

Lumbosacral Joint

The Intervertebral Disc is much thicker than other intervertebral discs, permitting more movement. It is so much deeper in front than behind that it contributes to the lumbar curve.

The Iliolumbar Ligament suspends the 5th lumbar transverse process from the iliac crest (*fig. 361*). The iliolumbar ligs. are important for they limit axial rotation of the 5th vertebra on the sacrum, and they assist the articular processes in preventing forward gliding of the 5th vertebra on the sacrum.

Joints of the Pelvis

The joints of the pelvis: (1) sacrococcygeal joint, (2) symphysis pubis, and (3) right and left sacro-iliac joints.

Sacrococcygeal Joint. This is an atypical intervertebral joint. Movement backward of the coccyx takes place on defecation and on parturition.

The bodies of the last sacral and 1st coccygeal vertebrae are united by an intervertebral disc, and the transverse processes and the cornua by ligamentous bands. And, a very tough membrane, which is a downward prolongation of the supraspinous and interspinous ligaments, closes the sacral canal posteriorly and extends to the posterior surface of the coccyx.

Symphysis Pubis. Here as between the bodies of two vertebrae, the opposed bony surfaces are coated with hyaline cartilage and are united by fibrocartilage. Dense anterior decussating fibers and a strong arcuate ligament (inf. pubic lig.) unite the pubic bones.

Aging. The surface and margins of the pubic symphysis undergo small changes, especially between the 20th and 40th years, which serve as reliable criteria of age

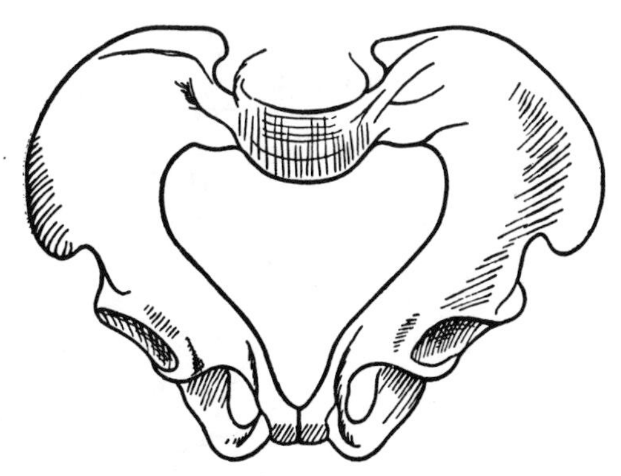

Fig. 357. Pelvis from case of osteomalacia. The femora have driven in the softened bones, narrowing the pelvis.

(Todd). This symphysis normally never fuses.

Sacro-Iliac Joint

The Bony Surfaces of this synovial joint are:

(*1*) *The Internal Surface of the Ilium* behind the iliac fossa. This part is bounded above by the posterior one-third of the iliac crest, below by the greater sciatic notch, and behind by the posterior superior and posterior inferior iliac spines and the slight notch between them (*fig. 358*).

It is subdivided into two parts: a lower, the *auricular surface;* and, an upper, the *tuberosity*. The auricular or ear-shaped part articulates with the sacrum, is covered with cartilage, and is traversed by a longitudinal sinuous ridge. The iliopectineal line begins at its most anterior part. The tuberosity is rough and tubercular for the numerous short fibers of the strong interosseous sacro-iliac ligament.

(*2*) *The Sacrum* possesses the counterpart: thus, on the side of the pars lateralis there is an *auricular surface* with a sinuous furrow. Posterosuperior to the auricular surface is the *sacral tuberosity* (*fig. 350*).

Structural Requirements. Weight on the sacrum by the superimposed column will tend to cause its upper end to rotate forward and its lower end with the coccyx to rotate backward (*fig. 359*). Ligaments are so disposed as to resist this tendency. Further, the articular surfaces of the sacrum are farther apart in front than behind, so the sacrum behaves not as a keystone, but as the reverse of a keystone, and tends therefore to sink forward into the pelvis. As it does so, the posterior ligaments become taut and draw the ilia closer together with the result that the interlocking ridge and furrow engage more closely. Here is an **automatic locking device** (*fig. 360*).

Ligaments Resisting Forward Rotation of Upper End of Sacrum:

1. *The Interosseous Sacro-iliac Ligament* (*fig. 360*), which is a ligament of great strength, unites the iliac and sacral tuberosities. It lies posterosuperior to the joint.

2. *The Dorsal Sacro-iliac Ligaments,* very strong bands, unite the transverse

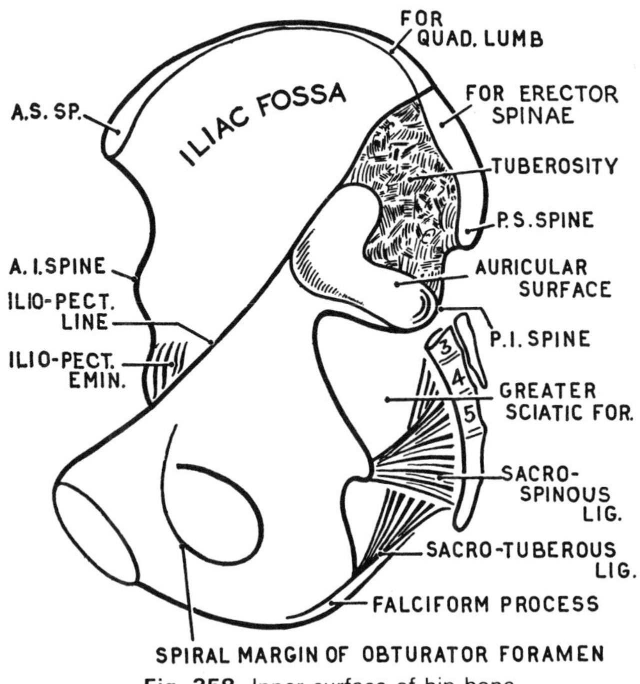

Fig. 358. Inner surface of hip bone.

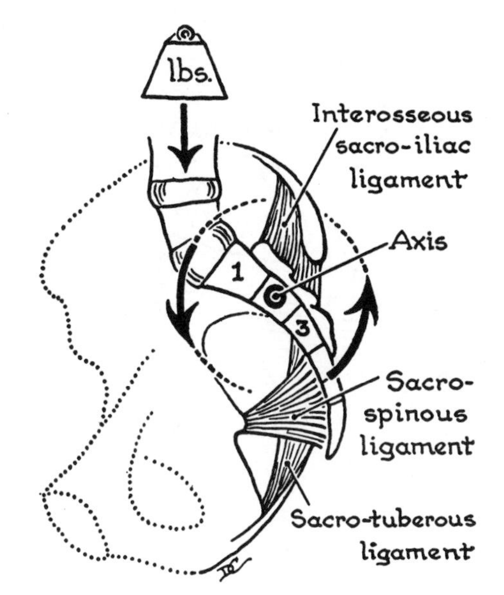

Fig. 359. Ligaments resisting rotation of sacrum.

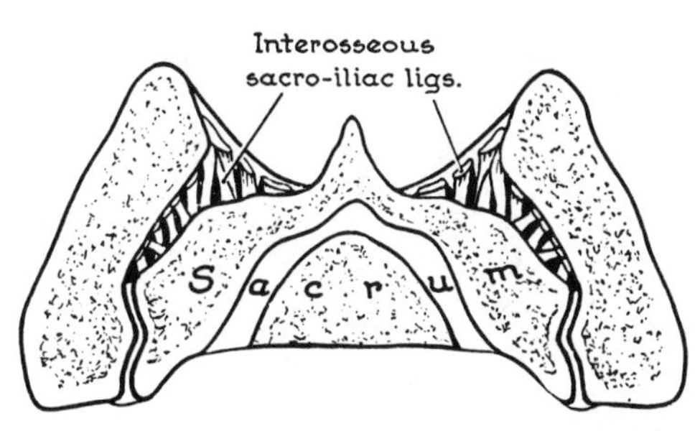

Fig. 360. The sacro-iliac joint on transverse section. Note the locking device.

tubercles of the sacrum to the post. sup. iliac spine.

3. The Illiolumbar Ligament (p. 275).

Ligaments Resisting Backward Rotation of Lower End of Sacrum: the *sacrotuberous* and *sacrospinous ligaments* (*figs. 359* and *361*). The one passes from the tuberosity of the ischium, the other from the spine of the ischium, to the available parts of the side of the sacrum and coccyx, i.e., to the lateral border of the sacrum and coccyx below the articular facet.

The Sacrotuberous Lig. is a broad band from the medial part of the tuber ischii to the sides of the sacrum and coccyx and adjacent dorsal surfaces.

[The lower end of the ligament curves forwards as a falciform process of which one edge creates a line (or ridge) on the ischium, while the other is continuous with the Obturator Internus fascia. The lateral part of the ligament arches from ischial tuberosity to posterior iliac spines (inf. and sup.); its only function is to afford origin to the Gluteus Maximus.]

The Sacrospinous Ligament is a triangular sheet, co-extensive with the Coccygeus; it extends from the ischial spine to the side and dorsum of the sàcrum and coccyx.

The sacrotuberous and the sacrospinous ligaments convert the greater and lesser sciatic notches into foramina. They are no doubt responsible for the curvature of the sacrum, which is a characteristic of man, who alone walks erect.

Movements. In a fresh specimen it is easily demonstrated (1) that the sacrum can rotate backward and forward between the hip bones; and, after division of the disc and ligaments of the symphysis pubis, (2) that the pubic bones easily spread 1 cm, and (3) allow the hip bones to rock fairly freely on the sacrum. (See Frigerio *et al.*)

During pregnancy the ligaments of the pelvis are relaxed and movements are more free.

Anterior Relations (*fig. 376*). The lumbosacral trunk, the superior gluteal a., and

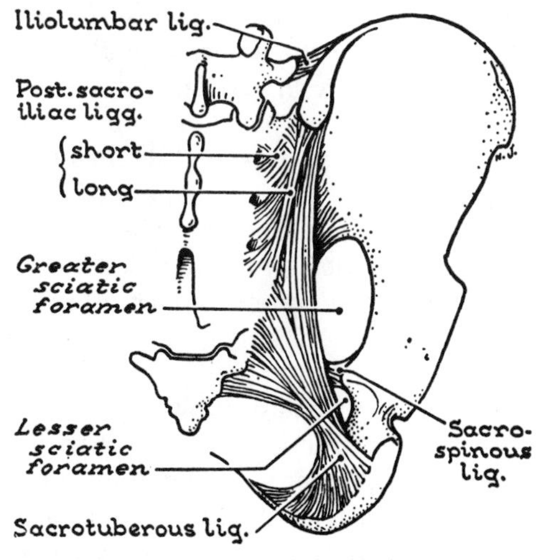

Fig. 361. The ligaments of pelvis (posterior).

the 1st sacral nerve cross the pelvic surface of the capsule.

Surface Anatomy. The post. sup. iliac spine, which lies at the level of the 2nd sacral vertebra is readily palpated. It lies in a prominent skin dimple. The joint lies lateral to this (Chapter 24 and *fig. 419*).

The anterior part of the joint cavity is closed by a strong fibrous capsule called the *anterior sacro-iliac ligament*. After middle life, particularly in males, this ligament may ossify, forming a crust of bone that prevents movement of the joint (synostosis).

DEVELOPMENT AND VARIATIONS. The sacro-iliac joint and the symphysis pubis do not develop, like other joints, as clefts in a continuous rod of condensed mesenchyme but by the coming into apposition of the ilium and sacrum posteriorly and of the pubic bones of opposite sides anteriorly. The auricular surface of the sacrum is usually covered with hyaline cartilage, that of the ilium with fibrocartilage, and between them there is a joint cavity which is present before birth (Schunke).

Accessory articular facets, 1 to 2 cm in diameter, are commonly found between the opposed tuberosities of the sacrum and ilium.

MALE UROGENITAL SYSTEM IN PELVIS

(continued)

Urinary Bladder (*Vesica Urinaria*). The empty and contracted urinary bladder, shaped not unlike the forepart of a ship, has four surfaces and four angles (*fig. 362*). (*Note:* vesical is the adjective and should not be confused with vesicle, = little bladder.)

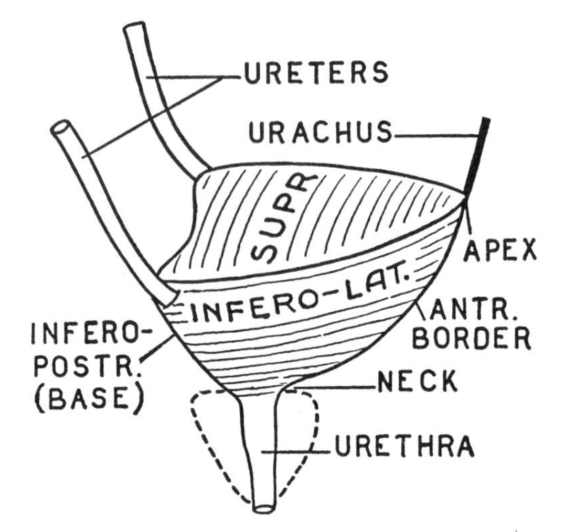

Fig. 362. The 4 surfaces, 4 angles, and 4 ducts of the urinary bladder.

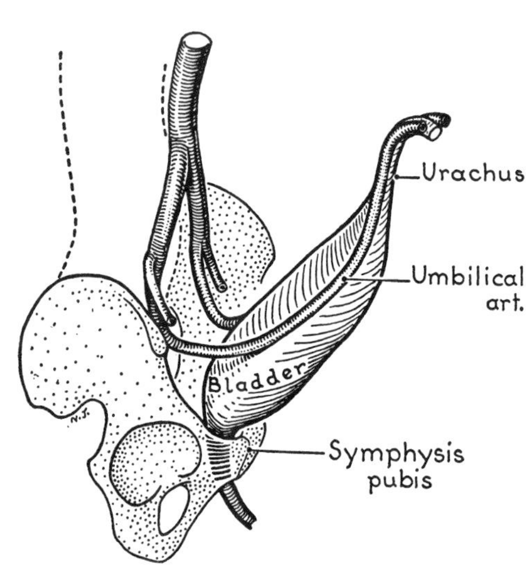

Fig. 362.1. The bladder is abdominal at birth.

The superior surface and 1 cm of the inferoposterior surface are the only parts covered with peritoneum. The superior surface is bounded by the rounded borders that connect the ureters to each other and to the urachus. It supports the sigmoid colon and ileum.

The urinary bladder owes its shape largely to the structures in contact with it. At birth the bladder is an abdominal organ (*fig. 362.1*) fusiform in shape, lying in the extraperitoneal tissue of the anterior abdominal wall; only by about the 6th year has a pelvis enlarged sufficiently to allow it to sink to its permanent pelvic position and tetrahedral shape.

Bed of Bladder. The bed or mould in which the bladder lies is formed on each side by bare pubic bone, the Obturator Internus, and the Levator Ani, and behind by the rectum, hence the inferolateral and inferoposterior surfaces (*figs. 346* and *363*).

The entire organ is enveloped in areolar tissue, the *vesical fascia*. A dense plexus of veins, the *vesical plexus*, lies in this fascia on the side of the bladder, clings to the bladder, and separates it from the retropubic space.

The base or *inferoposterior surface* of the bladder is separated from the rectum by the seminal vesicles and ampullated ends of the deferent ducts, which are enclosed between the two layers of rectovesical fascia; above these the peritoneum of the rectovesical pouch separates the upper 1 cm of the bladder from the rectum (*figs. 342* and *348*).

Pelvic Portion of the Ureter. Each ureter enters the pelvis at the bifurcation of the common iliac artery and descends immediately in front of the internal iliac artery. It lies subperitoneally, is crossed by the deferent duct near the bladder, and it crosses the apex of the seminal vesicle (*fig. 363*).

The Interior of the Bladder. The mucous membrane lies in folds, except over the trigone where it is smooth. *The Trigone* (Gr. = triangle) is a relatively fixed part of the base of the bladder (*fig. 364*). The two ureters and the urethra open at its angles about 2 cm apart. The internal urethral orifice lies at the lowest point of the bladder and, therefore, is situated advantageously for drainage.

The orifice of each ureter is guarded by a flap of mucous membrane and is collapsed. A ridge, the *interureteric fold*, connects the two ureters at the upper border of the trigone.

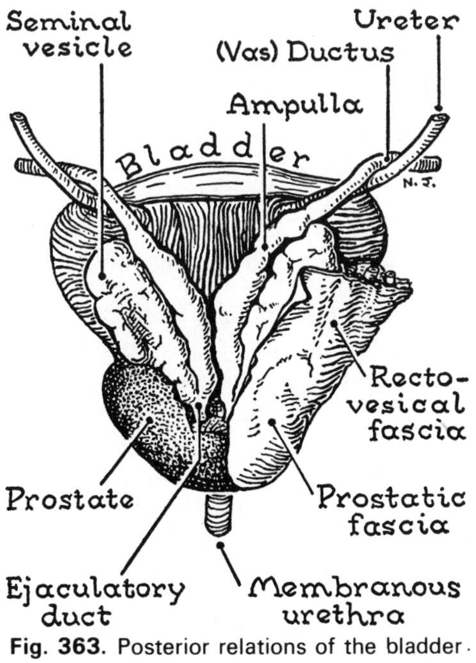

Fig. 363. Posterior relations of the bladder.

An elevation, the *uvula*, overlying the middle lobe of the prostate, is sometimes to be seen at the apex of the trigone behind the internal urethral orifice.

The Trigonal Muscle is a submucous sheet distinct from the muscular wall proper. It is continuous with the muscular wall of the ureters above; its apex descends in the posterior wall of the urethra to the utricle. This trigonal area of the bladder and urethra is derived from the Wolffian or mesonephric ducts. It is supplied by the hypogastric plexus (L. 1, 2).

Though 500 cc can be injected into the bladder without causing discomfort, urine is usually voided when little more than half this volume has accumulated, the exact volume depending largely on acquired habit.

As the bladder fills, the trigone enlarges but little, so the three orifices are only slightly displaced; but the rest of the bladder stretches. When the fundus reaches the level of the umbilicus, the peritoneum is stripped from the anterior abdominal wall for several centimeters above the symphysis pubis.

Vessels and Nerves of the Bladder.

Arteries: superior and inferior vesical.

Veins pass to the vesical plexus and thence to the internal iliac veins.

Lymph Vessels (p. 304).

Nerves. Pelvic splanchnic nn. and hypogastric plexus (p. 62).

Male Urethra. The male urethra is a fibro-elastic tube, 20 cm long (*fig. 365*). It is divided by the superior and the inferior fascia of the urogenital diaphragm into three parts: *prostatic*, *membranous*, and *spongy*. The part above the superior fascia traverses the prostate; the part between the two fasciae has no covering; the part beyond the inferior fascia or perineal membrane traverses the corpus spongiosum penis.

The Prostatic Urethra (*fig. 364*) is the widest and most dilatable part of the urethra. It runs almost vertically and is the length of the prostate—2 cm or more. In transverse section it is crescentic, owing to

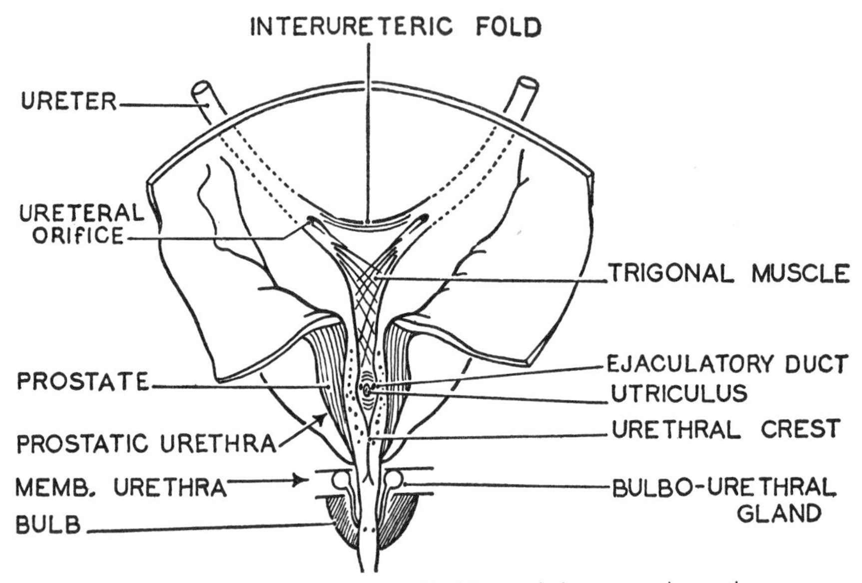

Fig. 364. The trigone of the bladder and the prostatic urethra.

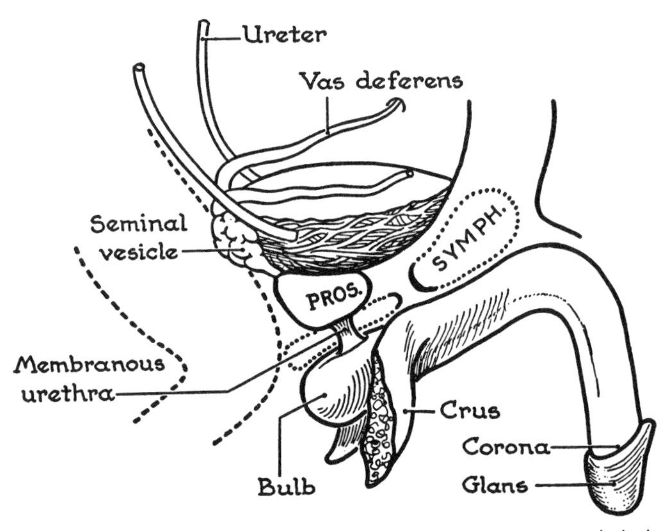

Fig. 365. The lower parts of the genital and urinary tracts and their relations.

a ridge on the posterior wall, called the *urethral crest*. The gutter on each side of the crest is called the *prostatic sinus*. The urethral crest rises to a summit, the *colliculus*. Opening on it is a diverticulum within the prostate the *prostatic utricle* (= little uterus), up which a probe can be passed about 1 cm. Onto the lips of the utricle open the pinpoint orifices of the right and the left *ejaculatory duct*. Into the prostatic sinus several dozen *prostatic ducts* open (*fig. 364*).

The Membranous Urethra passes through the middle of the urogenital diaphragm and its two fasciae and is only 1 cm long. Behind it on each side lies a bulbourethral gland. The *Sphincter Urethrae* part of the diaphragm forms its sphincter of voluntary muscle, supplied by the perineal branch of the pudendal nerve. Like the Sphincter Ani Externus, also supplied by the pudendal nerve, it is constantly on guard and relaxes only during micturition.

The Spongy or Penile Urethra traverses

the bulb, body, and glans of the corpus spongiosum penis. It enters the bulb on its upper surface and ends near the lower part of the apex of the glans at the external urethral orifice. Its lumen is dilated both in the bulb and in the glans. The dilatation in the glans is known as the navicular (terminal) fossa (*fig. 339*).

The external urethral orifice—like the opening of most ducts—is the narrowest part of the urethra. A small kidney stone may stick at the vesical orifice of the ureter, or, having passed, it may yet stick at the external urethral orifice.

Palpation. A catheter passed into the bladder can be palpated in the *spongy urethra* from the under surface of the penis, in the *membranous urethra* from the perineum, and in the *prostatic urethra* by a finger in the rectum.

Nerve Supply (p. 302 and *fig. 395*).

Structure of the Urinary Tract

Although the membranous and spongy urethrae are lined with stratified columnar epithelium, the navicular fossa is lined with stratified squamous epithelium, as are other orifices opening on to the skin surface, such as the nostrils, mouth, mammary ducts, sebaceous ducts (but not the sweat ducts), anal canal, and vagina.

The Tunica Propria is thick and loose in the ureter and bladder, and it falls into folds when they are empty.

Muscle. The urinary tract has both an inner longitudinal and an outer circular coat of muscle. The lower part of the ureter and the bladder have an additional outer longitudinal coat; the layers are interwoven in the bladder. Fibers from the longitudinal and circular muscle coats loop from behind forward around the front and sides of the upper part of the prostatic urethra. These slings are optimistically called the *Sphincter Vesicae*.

The smooth muscle fibers in the urethra are not numerous.

The intramural part of the ureter has longitudinal fibers, but no circular ones. Of these, some join their fellow in the interureteric fold, others radiate into the trigonal muscle (*fig. 364*).

Urethral Glands. Mucous cells, singly and in clusters lining recesses (lacunae urethrales), as well as in branching outpouchings (urethral glands) in the tunica propria, occur especially on the dorsum of the anterior two-thirds of the spongy urethra. They may extend to the neck of the bladder.

Common Direction. The mouths of all ducts and tubes opening into the urethra open forward, in the direction in which the urine flows. Hence, urine does not enter them during micturition, but fine instruments passed in the reverse direction may do so.

INTERNAL PARTS OF MALE GENITAL SYSTEM

Prostate (*fig. 365*). The prostate surrounds the urethra between the bladder and the u.g. diaphragm. It occupies the same bed as the bladder and it resembles the bladder in shape, its surfaces being superior, inferolateral, and inferoposterior. Of these, the superior blends with the bladder; the two inferolateral ones lie on the Levatores Ani; and the inferoposterior lies on the rectum. Its apex, from which the urethra emerges, abuts against the superior fascia of the diaphragm between the anterior borders of the Levatores Ani (*fig. 346*).

The prostate is encased in a strong envelope of pelvic fascia, which is continuous below with the superior fascia of the u.g. diaphragm, and which is anchored to the pubes by the puboprostatic (-vesical) ligaments. The *prostatic fascia* is distinct from the outermost part of the gland proper, which is called the *capsule of the prostate* (*figs. 348* and *363*).

On each side of the prostate the capsule is separated from the fascia by a venous plexus, called the *prostatic plexus of veins*. This plexus receives the deep dorsal vein of the penis in front, communicates with the vesical plexus above, and drains into the internal iliac veins behind.

The posterior part of the prostatic fascia forms a broad strong sheet, called by the surgeon the *fascia of Denonvilliers*. It is easily separated from the loose rectal fascia behind it.

The prostate can be palpated per rectum and exposed readily from the perineum (p. 264).

The urethra passes vertically through the forepart of the prostate; the prostatic utricle projects into the hinder part, and the ejaculatory ducts pierce its upper surface and open on to the lips of the utricle.

STRUCTURE. The prostate is a modified portion of the urethral wall. In composition it is one-half glandular, one-fourth involuntary muscle, and one-fourth fibrous tissue. The glands are arranged in three concentric groups (*fig. 366*). The innermost or *mucosal* are short and simple. They open all round the urethra above the level of the colliculus. "All hypertrophies of the prostate arise from these mucosal or subu-rethral glands" (Young). The intermediate or *submucosal* glands open into the prostatic sinus at the level of the colliculus. The outermost or *prostatic glands proper* are long and branching. They envelop the other two groups, except in front where those of opposite sides are joined by a non-glandular isthmus. Their ducts open into the prostatic sinus.

The Ductus (Vas) Deferens, like the femur, is about 45 cm long. It has a course in the scrotum, in the inguinal canal, on the side wall of the pelvis, and between the bladder and rectum where it rounds the ureter. Medial to the seminal vesicle, it is ampullated and thinwalled (*fig. 367*).

Seminal Vesicles. Each vesicle is a tortuous, branching diverticulum developed from the ampullated end of the ductus deferens (*fig. 368*). Below, it joins the ductus deferens to form the ejaculatory duct.

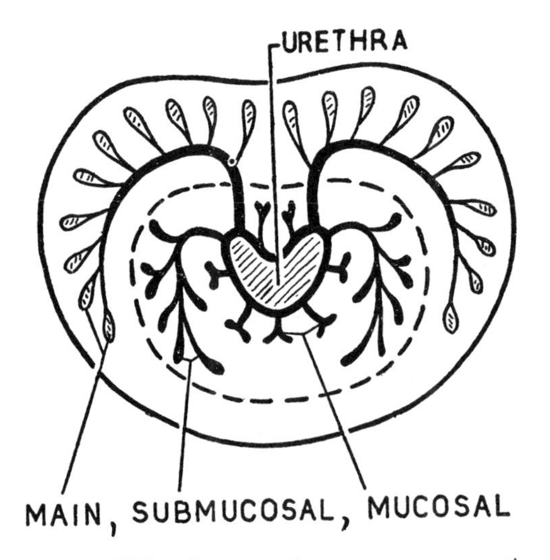

Fig. 366. Prostate in transverse section showing three concentric groups of tubules (schematic). (After Adrion.)

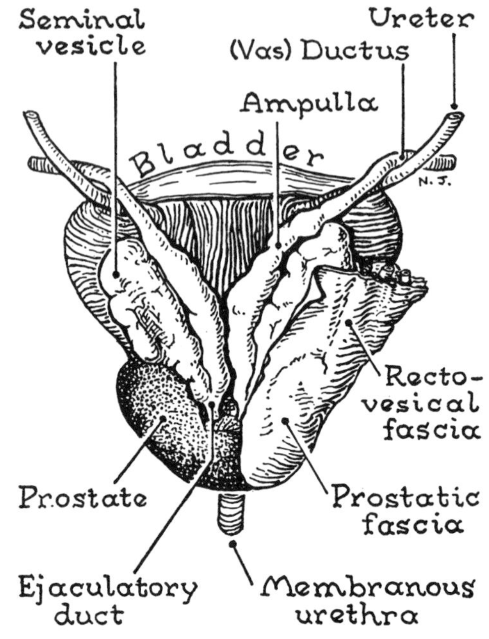

Fig. 367. The seminal vesicles and the deferent ducts (vasa deferentia).

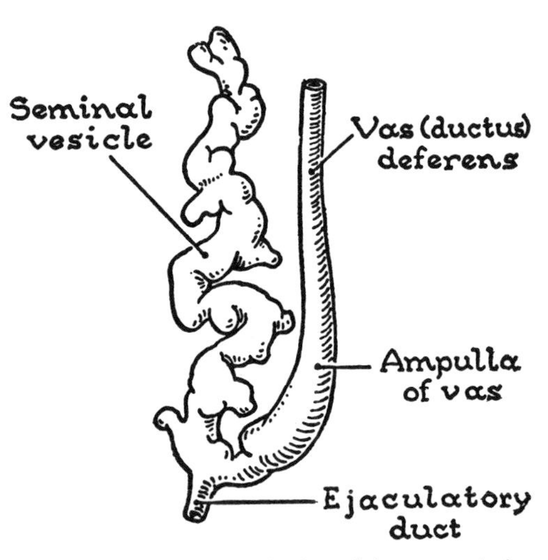

Fig. 368. The seminal vesicle unraveled.

The vesicles, prostate, and bulbo-urethral glands each add a distinctive secretion to the semen, but seminal vesicles are absent in dogs and other carnivora.

Ejaculatory Duct (paired). This duct is common to the ductus deferens and the seminal vesicle. It has the diameter of the lead of a pencil. It is easily torn away from the prostate, the upper half of which it pierces obliquely to open beside the prostatic utricle.

Vessels and Nerves. *The inferior vesical aa.*, with a little help from the middle *rectal aa.*, supply all the structures in between the bladder and the rectum, viz., prostate, seminal vesicles, ampullae of the deferent ducts, and the ends of the ureters. *The deferent a.* itself springs from an inferior vesical a.

Nerves. (*See* p. 302.)

RECTUM AND ANAL CANAL

The intestinum rectum and anal canal are the terminal parts of the large intestine. The rectum begins where the colon ceases to have a mesentery, which is in front of the 3rd piece of the sacrum, and it is about 12 cm long. The rectum continues the curvature of the sacrum and coccyx downward and forward for 4 cm beyond the coccyx, and there, at the apex of the prostate, makes a right angled bend and becomes the anal canal (*fig. 365*). The anal canal passes downward and backward for about 3 cm or more to its orifice, the *anus*.

Rectum (*fig. 342*). The upper part of the rectum is covered with peritoneum in front and at the sides; the middle part is covered in front only; and the lower part, which is commonly dilated to form the *ampulla of the rectum*, lies below the level of the rectovesical pouch and, therefore, is devoid of peritoneal covering. The distance from the skin surface to the peritoneal cavity (i.e., to the rectovesical pouch), measured in front of the anal canal and rectum, is 8 to 10 cm (*fig. 369*).

The rectum in man is not straight as its name would imply; it has the anteroposterior curve and bend just described; it has also three lateral curvatures: thus, at the bottom of the rectovesical pouch, the right wall is indented with the result that a transverse fold or shelf projects into the lumen; and similar but less pronounced indentations from the left occur about 2 to 3 cm below and above this. The shelves within the gut are the *plicae transversales* (*fig. 369*). They consist of mucous membrane and circular muscle. Their form is maintained by the prolongations of the three teniae coli, which spread out to form a continuous outer longitudinal muscle coat, thick ventrally and dorsally but thin at the sides.

The plicae must be avoided by tubes and instruments introduced into the rectum.

Anal Canal (*figs. 370* and *371*). The anal canal extends from the anorectal junction,

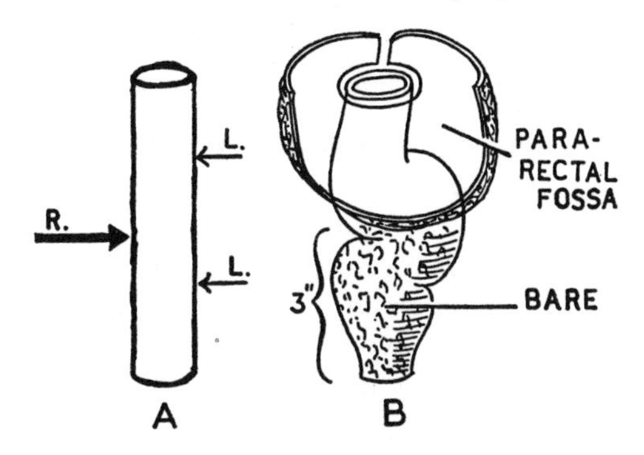

Fig. 369. The peritoneal coverings and lateral flexures of the rectum, front view. (3″ = 8 cm.)

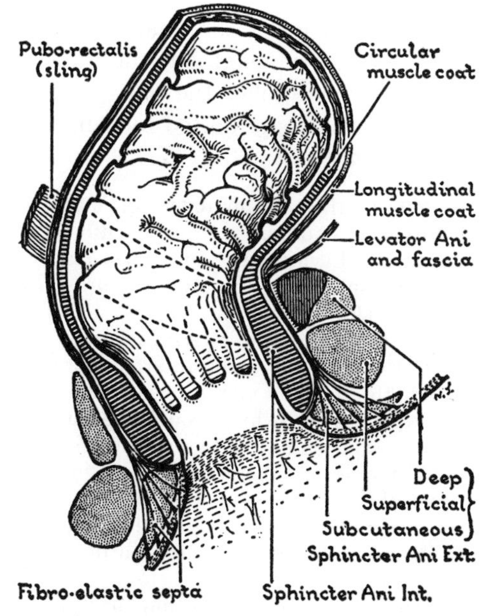

Fig. 370. The sphincters of the anus, and the puborectal sling.

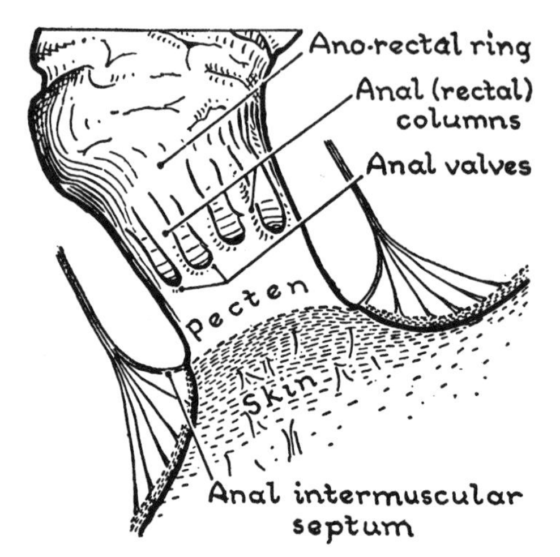

Fig. 371. Interior of anal canal.

which lies above the level of the puborectal sling (*fig. 370*) and the sphincters, to the anus below. It remains closed, except during defecation. Accordingly, it is surrounded throughout by two sphincters, (1) and *External* (voluntary), described on page 259 and (2) an *Internal* (involuntary), which is merely the much thickened lower end of the circular muscle coat of the gut (*fig. 370*). Of the two, the voluntary sphincter is the more important (Hardy; Stephens; Varma).

The upper part of the anal canal possesses 5 to 10 permanent longitudinal folds of mucous membrane, the *anal columns* whose lower ends are united by semilunar folds, the *anal valves* (*fig. 371*).

Between the two sphincters, the longitudinal muscle coat of the rectum, re-inforced by fibers of the Levator Ani and its fasciae, descends and splits into a number of *fibro-elastic septa* (*fig. 370*).

One of these, the *anal intermuscular septum* (*fig. 371*), passes below the Internal Sphincter to reach the mucous membrane of the canal, whereas the others swing through the Subcutaneous External Sphincter to reach the skin (*fig. 370*). Not only are there differences in detail between the two sexes, but even different parts of the circumference of the sphincter in individuals show significant differences in detail (Oh and Kark).

Above the anal valves the canal has columnar epithelium containing goblet cells. The 1 cm of canal below the valves is smooth, is lined with stratified squamous epithelium, and is known to the surgeon as the "pecten." The remainder is lined with skin.

Anteroposterior versus side-to-side flattening. When the rectum is empty, its anterior and posterior walls are in apposition; when the anal canal is empty, its lateral walls are in apposition. The same is true of the vagina and the vaginal orifice, and of the male spongy urethra and the external urethral orifice.

Arteries (*fig. 372*). The superior, middle, and inferior rectal (*hemorrhoidal*) aa. supply the rectum and anal canal, making five arteries in all—for the superior a. is unpaired.

The Superior Rectal Artery is the continuation of the inferior mesenteric a. It divides into a right and a left branch.

Terminal branches, given off irregularly from these, pierce the muscle coat, ramify in the submucosa, and descend in the anal columns (*fig. 371*). They anastomose with each other and with the middle and inferior rectal arteries lower down.

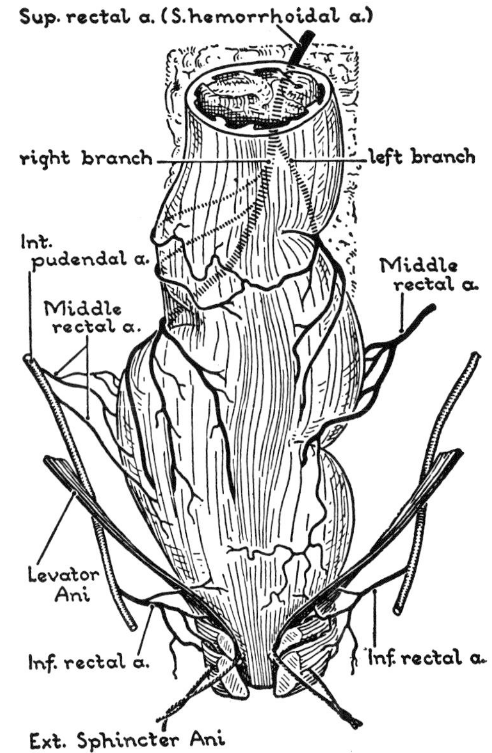

Fig. 372. The arterial supply of the rectum and anal canal, front view.

Other Arteries (e.g., median sacral, inf. gluteal, int. pudendal) send twigs to the lower part of the rectum.

Veins. The superior, middle, and inferior rectal or hemorrhoidal veins accompany their arteries and drain corresponding parts of the rectum and anal canal. The superior vein becomes the inferior mesenteric vein and, therefore, belongs to the portal system. It is valveless. The middle and inferior veins are paired and belong to the caval system. They have valves.

The Superior Vein begins in the anal columns. It has extensive mucous and submucous plexuses, and it receives branches from the circumrectal tissues.

The Middle Rectal Vein is a much more important vessel than the corresponding artery. It drains the rectum above the Internal Sphincter and communicates both submucously and in the rectal fascia with the inferior rectal vein, and it makes free anastomoses submucously with the superior rectal vein. Its branches communicate with the prostatic (vaginal and uterine) plexus. It is the chief link between the portal and caval systems—a point of clinical significance, and it ends in the internal iliac vein.

The Inferior Vein begins in the Sphincter Ani Externus, the walls of the anal canal, and the subcutaneous veins at the anal margin. These subcutaneous veins are continuous above with the submucous veins of the anal canal and rectum.

Lymph Vessels. *See* p. 305.

Nerves. *See* p. 302.

Relations. Behind the rectum (*figs. 373, 376*): three pieces of the sacrum, the coccyx, and sacrospinous lig.; Piriformis, Coccygeus, Levator Ani; the median sacral a. and v., the sympathetic trunks and ganglion impar, parts of the last three sacral and the coccygeal nerves, the lateral sacral vessels and in the rectal fascia, which is usually thick and fatty posteriorly, are branches of the superior rectal vessels and lymph nodes.

On each side are: the pararectal fossa containing sigmoid colon or ileum; the middle rectal vessels, the pelvic splanchnic nerves and the inferior hypogastric plexus, the Levator Ani, and the ischiorectal fossa.

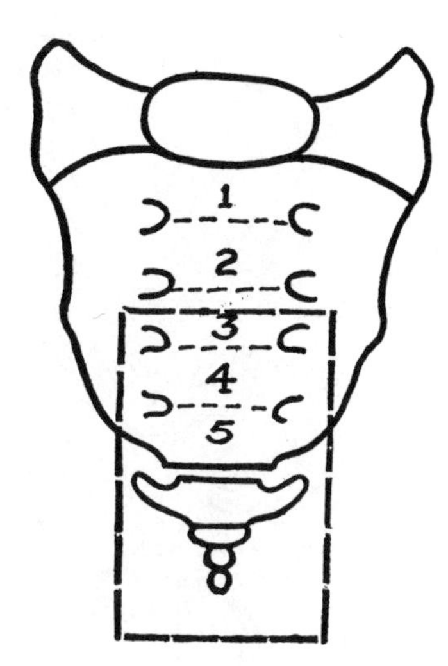

Fig. 373. The parts of the sacrum and coccyx covered by the rectum (see *fig. 376*).

In front are: the bladder separated by rectovesical pouch, seminal vesicles, and ampullae of the deferent ducts, and below these is the prostate (*fig. 365*).

At the anorectal junction are: the puborectal sling, the base of the u.g. diaphragm, and the membranous urethra.

In front of the anal canal are the perineal body and the bulb of the urethra.

VESSELS AND NERVES OF PELVIS

Arteries of the Pelvis

The arteries of the pelvis are:

internal iliac artery—paired.
median sacral artery } unpaired; previously
superior rectal artery } described.

The Internal Iliac (Hypogastric) **Artery** (*fig. 374*) takes origin from the common iliac artery one-third of the way, i.e., 5 cm, along the line joining the aortic bifurcation to the midinguinal point. It descends subperitoneally and crosses medial to the external iliac vein, the Psoas, and the obturator nerve. In front of it runs the ureter; behind it lies its vein.

Branches of Internal Iliac Artery. These arise erratically. They may be grouped thus:

1. Visceral branches:

Umbilical	*Superior vesical*
Inferior vesical	*Middle rectal*

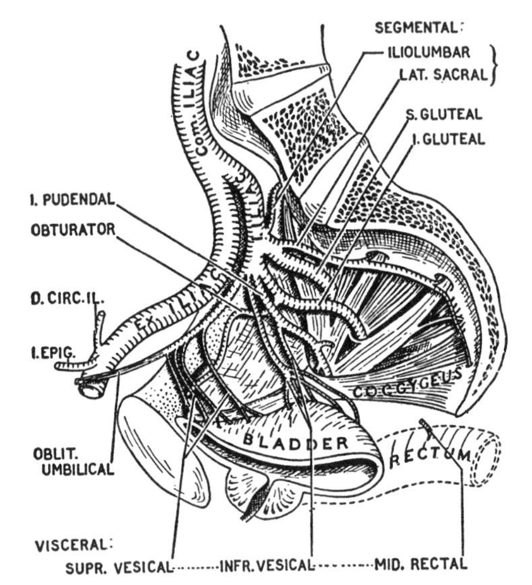

Fig. 374. Internal iliac artery and branches.

(also, *Uterine* and *Vaginal* in the female)

2. Branches to the limb and perineum:
   Superior gluteal — *Inferior gluteal*
   Obturator — *Internal pudendal*
3. Somatic segmental branches:
   Iliolumbar — *Lateral sacral*

1. The Visceral Branches:

(a) *The umbilical artery* (*fig. 259*, p. 205) is obliterated as far back as the branches to the bladder, i.e., the *superior vesical aa.*

(b) *The inferior vesical arteries* and (c) *the middle rectal artery* run in the leash of veins that form the posterior limit of the retropubic space.

Of the two inferior vesical aa., one (the *vesiculodeferential a.*) is constant in arising from the umbilical a. at its origin (Braithwaite). The other has a variable origin. On each side, these arteries supply the side and base of the bladder, the prostate, seminal vesicle, ampulla of the deferent duct, and end of the ureter. One gives off the deferent a. *The middle rectal a.* varies in origin and in size (*fig. 372*).

2. The Branches to the Limb and Perineum leave the pelvis through the greater sciatic and obturator foramina. Thus: (a) *the superior gluteal artery*, which is much the largest branch of the internal iliac artery, makes a U-shaped turn round the angle of the greater sciatic notch into the gluteal region. Its vein and the superior gluteal nerve accompany it (*fig. 421*).

(b) *The inferior gluteal artery* and (c) *the internal pudendal artery* descend in front of the sacral plexus and pass between the borders of the Piriformis and Coccygeus into the gluteal region. There the int. pudendal artery, which is the smaller and more anterior, crosses the ischial spine and enters the perineum through the lesser sciatic foramen in company of its own nerve and the nerve of the Obturator Internus.

(d) *The obturator artery* runs forward on the side wall of the pelvis to the obturator foramen between its nerve and vein.

The obturator and inferior epigastric arteries both supply branches to the back of the pubis. These pubic branches anastomose, and the anastomotic channel is commonly (33 per cent) so large that the obturator artery derives its blood from the epigastric artery. This is known as an *accessory obturator artery*.

The obturator vein likewise is commonly "abnormal."

3. The Somatic Segmental Branches.

(a) *The iliolumbar artery*, being the artery of the 5th lumbar segment, ascends in front of the ala of the sacrum, and divides into iliac and lumbar branches. The iliac branches anastomose in the iliac fossa.

(b) *The lateral sacral artery* descends in front of the roots of the sacral plexus.

Veins of Pelvis (*fig. 375*)

The pelvic viscera may be said to lie within a basket woven out of large thin-walled veins among which the arteries thread their way. The basket is divided into vesical, prostatic (or uterine and vaginal), and rectal venous plexuses, which drain largely into the internal iliac vein, but partly via the superior rectal (hemorrhoidal) vein into the inferior mesenteric vein and so to the portal vein.

The *middle rectal vein* is relatively large. It emerges from the lower part of the side of the rectum and passes to the internal iliac

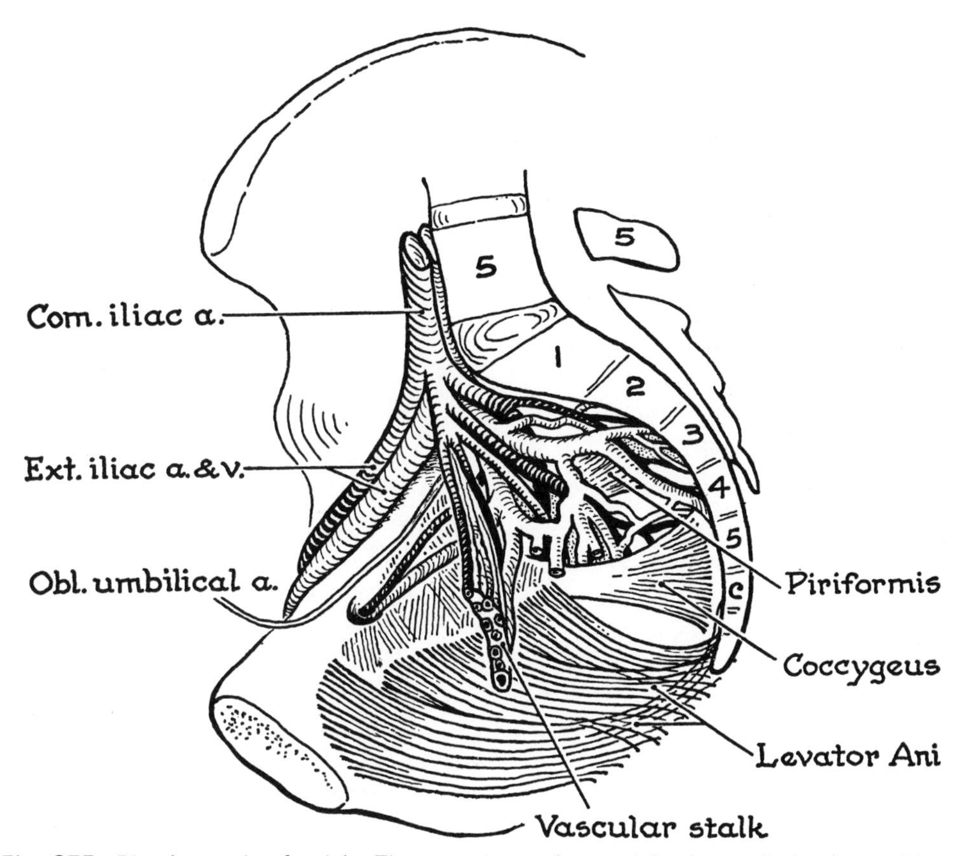

Fig. 375. Blood vessels of pelvis. The vascular stalk to pelvic viscera limits retropubic space posteriorly.

vein. It anastomoses with the superior and inferior rectal veins and also with the other plexuses of pelvic veins, thereby taking a prominent part in the portacaval anastomoses (p. 216).

Nerves of Pelvis

Sacral and Coccygeal Nerves. The lumbar, sacral, and coccygeal plexuses are derived from the ventral rami of spinal nerves Th. 12—Co. 1, as in table 11.

Note that rami L. 4 and S. 4 both contribute to two plexuses.

The ventral rami of the sacral and coccygeal nerves, like all other ventral rami, receive gray sympathetic rami communicantes, which they conduct to the blood vessels, sweat and sebaceous glands, and arrectores pilorum in their territory. No white sympathetic rami communicantes arise caudal to L. 3, but parasympathetic fibers, called the *pelvic splanchnic nerves*, arise from the ventral rami of S. (2), 3 and 4 (*figs. 44.2* and *377*).

TABLE 11

Ventral rami of	Th.	L.	S.	C.
Lumbar plexus	12	1, 2, 3, 4		
Sacral plexus		4, 5,	1, 2, 3, 4	
Coccygeal plexus			4, 5,	1

Sacral Plexus (*figs. 376* and *377*). Of the six **roots** of the sacral plexus, L. 4 and L. 5 cross and groove the ala of the sacrum as the **lumbosacral trunk,** which joins S. 1 to pass in front of the sacro-iliac joint.

Covered with fascia, they unite with S. 2, 3, and 4 in front of the Piriformis to form the sacral plexus. The plexus has many

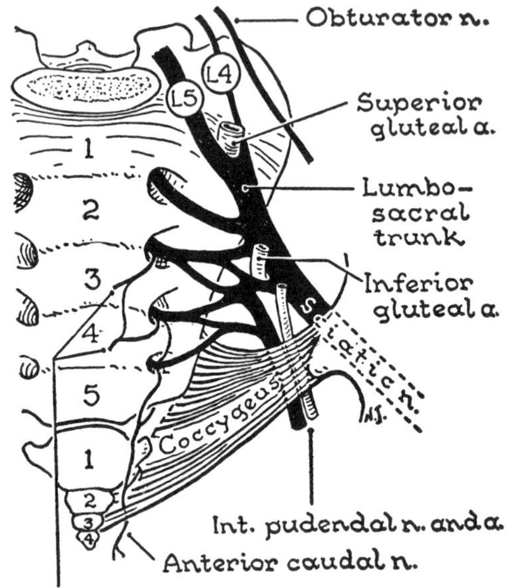

Fig. 376. The (left) sacral plexus.

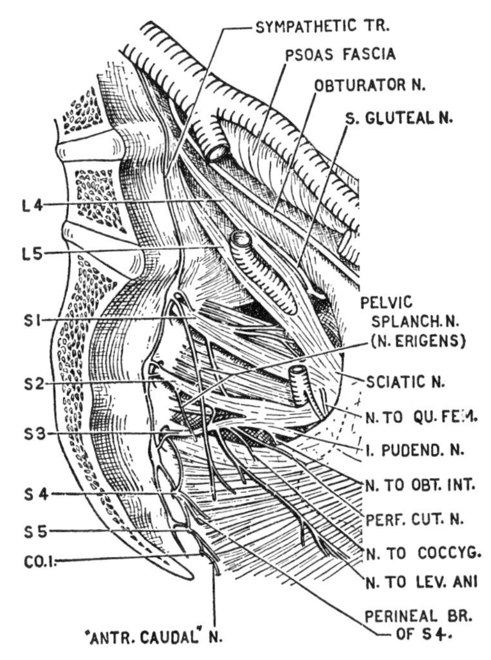

Fig. 377. The (left) sacral and coccygeal plexuses.

collateral branches and it ends as two terminal branches, the *sciatic* and *pudendal nerves*.

Arteries Piercing the Plexus (fig. 376). Four branches of the internal iliac artery pierce the plexus: the *iliolumbar*, the *superior gluteal*, the *inferior gluteal*, and the *internal pudendal aa*.

The Branches of the Sacral Plexus (table 12) may be grouped thus:

1. Branches from roots of plexus:
   (a) *Muscular* (to Piriformis, Levator Ani, and Coccygeus)
   (b) *The pelvic splanchnic nerves.*
2. Branches that pass through the greater sciatic foramen:
   (a) *Two terminal* (sciatic and pudendal)
   (b) *Five collateral.*
3. Branches that emulate the coccygeal plexus in piercing the structures attached to the side of the coccyx in order to become cutaneous:
   (a) *Perforating cutaneous of S. 2, 3*
   (b) *Perineal of S. 4.*

1. The branches arising from the roots of the sacral plexus are short twigs to Piriformis (S. 1, 2), long branches to Levator Ani and Coccygeus (S. 3, 4), and the pelvic splanchnic nerves (S. 2, 3, 4).

The *pelvic splanchnic nerves*, like the pudendal nerve, arise from S. 2, 3, 4, but are "mixed" parasympathetic nerves. They supply the involuntary muscles derived from the cloaca that cause the involuntary sphincters guarding the rectum and bladder (Sphincter Ani Internus and Sphincter Vesicae) to relax while the organs contract, and they are sensory to them. They also cause dilatation of the arteries of the erectile tissue of the penis or clitoris and thereby produce erection, hence the alternative name *nervi erigentes*.

2. The branches that pass through the greater sciatic foramen enter the gluteal region and they should be studied with it. They are seven in number, two being terminal and five collateral. The exact segments of origin of the collateral branches are unimportant.

The sciatic nerve is the largest nerve in the body and it forms the greatest part of the sacral plexus. It arises by five roots (L.

TABLE 12

Branches of the Sacral Plexus

Terminal	Collateral		
	From the back	From the front	From front and back
1. Sciatic	3. Superior Gluteal	5. N. to Quadratus Femoris	7. Posterior cutaneous of the thigh
2. Pudendal	4. Inferior Gluteal	6. N. to Obturator Internus	

4, 5, S. 1, 2 and 3), and leaves the pelvis between the Piriformis and the ischial border of the greater sciatic notch.

The pudendal nerve (S. 2, 3, 4) already seen in the perineum, escapes between the Piriformis and Coccygeus just medial to the sciatic nerve.

The superior gluteal nerve (L. 4, 5, S. 1) and *the inferior gluteal nerve* (L. 5, S. 1, 2) arise from the back of the plexus. The nerves to *Quadratus Femoris* (L. 4, 5, S. 1) and *Obturator Internus* (L. 5, S. 1, 2) arise from the front of the plexus; the former descends in front of the sciatic nerve, the latter on the medial side of the sciatic nerve. The *posterior cutaneous nerve of the thigh* (S. 1, 2, 3) arises from the back and front of the plexus.

The superior gluteal nerve and its companion artery and vein escape from the pelvis above the Piriformis at the angle of the greater sciatic foramen. All other structures passing through the foramen pass below the Piriformis. The inferior gluteal nerve and the posterior cutaneous nerve of the thigh escape below the Piriformis and behind the sciatic nerve.

3. The two branches that emulate the terminal branch of the coccygeal plexus are the *perforating cutaneous branch* of S. 2 and 3 and the *perineal branch* of S. 4. These descend in front of the Coccygeus and then pass through it, reaching the skin.

The Coccygeal Plexus is formed by the ventral rami of S. 5 and C. 1 which emerge from the sacral hiatus onto the dorsum of the sacrum and supply skin as the *anococcygeal nerve*, which corresponds to the ventral *caudal nerve* of tailed mammals. A similarly formed nerve, derived from the dorsal rami of S. 4, 5, and C. 1, corresponds to the dorsal *caudal nerve* of tailed mammals. In man they supply a circular area of skin around the coccyx—the area on which one sits down (*fig. 336.1*).

21 FEMALE PELVIS

Female Pelvis in Median Section: How to Sketch it; Female Pelvis from above.

INTERNAL GENITAL ORGANS

Ovary; Gubernaculum Ovarii; Derivative of Mullerian or Paramesonephric Duct; Broad Ligament; Uterine Tube.
Uterus: Parts; Structure.
Vagina: Walls; Fornices; Relations; Structure.
Development; Remnants of Mesonephros; Unique Relations of Female Ureter.
Vessels Peculiar to Female Pelvis (Ovarian; Uterine; Vaginal Arteries and Veins).
Nerves.
Support of Female Pelvic Viscera

DIFFERENCES BETWEEN MALE AND FEMALE PELVIS

Functions Related to Childbirth Peculiar to Female Pelvis: Shape of Pelvic Cavity; Diameters; Attachment of Crura of Clitoris; Other Features; Ischium-pubis Index.
Types of Female Pelvis

Female Pelvis in Median Section. Because of its obstetrical importance you should be able to sketch the skeletal parts, then insert the viscera, and lastly put on the peritoneal covering.

1. The symphysis pubis is 3 to 4 cm (or about 1½ inches) deep. Its posterior surface is flat and faces more upward than backward (*fig. 378*).

2. Erect a vertical line 9 cm (or about 3½ inches) high on the symphysis. Draw two parallel horizontal lines, one at the level of the top of the symphysis, the other at the top of the vertical line.

3. The sacral promontory lies on the upper horizontal line, three times the depth of the symphysis (about 11 cm) from the upper end of the symphysis.

4. The tip of the coccyx lies on the lower horizontal line, three times the depth of the symphysis more or less (about 11 cm) from the lower end of the symphysis.

5. The upper part of the sacrum is straight and parallel to the symphysis; its lower part and the coccyx are often much curved.

6. A line joining the upper end of the symphysis to the promontory of the sacrum indicates the plane of the pelvic brim, which bounds the pelvic inlet. The inlet makes an angle of 60° with the horizontal.

7. A line joining the lower end of the symphysis to the tip of the coccyx is the plane of the outlet of the pelvis. It makes an angle of 15° with the horizontal.

The soft parts are now to be inserted:

8. Insert the urogenital diaphragm and its upper and lower fasciae.

9. The bladder and urethra, as in the male: The urethra pierces the urogenital diaphragm about 2 or 3 cm from the symphysis. It is about 4 cm long (*fig. 379*).

10. The rectum and anal canal, as in the male.

The uterus and vagina occupy the positions taken by the seminal vesicles, ampullae of the deferent ducts, and prostate in the male.

11. The vagina is over 9 to 10 cm long. It lies nearly parallel to the pelvic brim. Its anterior wall is in structural continuity with the urethra in its lower part; it is in

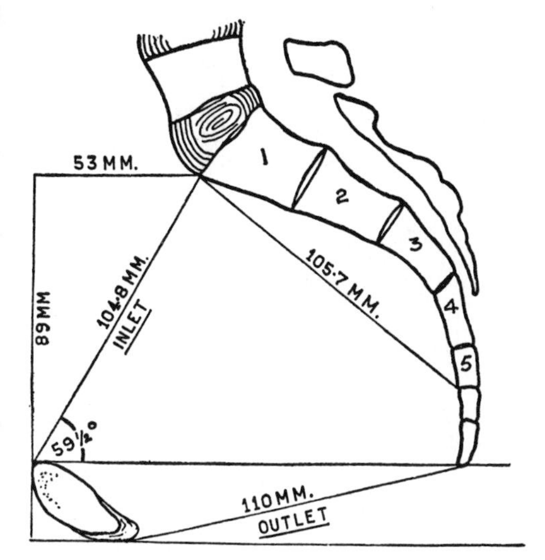

Fig. 378. Outlines of female pelvis in median section (drawn to scale).

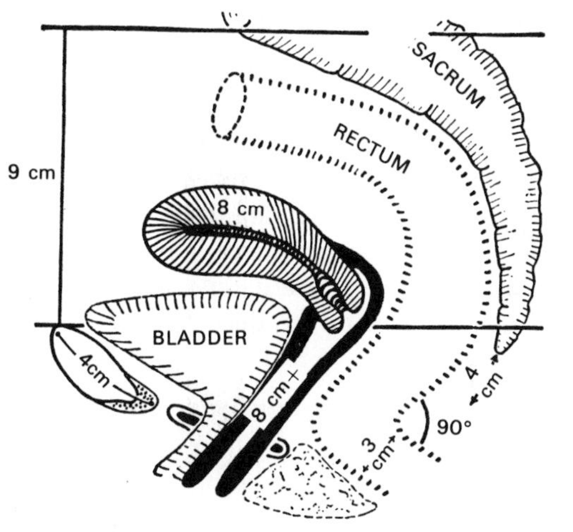

Fig. 379. The soft parts inserted in *figure 378* (scale in cm).

contact with the bladder in its middle part; and it is pierced by the cervix of the uterus in its upper part. Like the urethra, it pierces the urogenital diaphragm.

12. The posterior wall of the vagina is separated from the rectum by the rectal and the vaginal layer of fascia (rectovaginal septum), and from the anal canal by the triangular mass of fibromuscular tissue called the *perineal body*, much larger than in the male.

13. The uterus is 8 cm long. It lies nearly at right angles to the pelvic brim and to the vagina. Its upper 5 cm (or fundus and body) are 2 to 3 cm thick; its lower 3 cm (or cervix) is 2 cm thick. The external orifice of the uterus lies on (or below) the lower of the two parallel horizontal lines. The fundus does not reach to the pelvic inlet. The body and cervix meet at a slight angle, so the uterus is said to be anteflexed.

14. The anterior and posterior fornices (fornix L. = an arch) of the vagina are the shallow depression in front of the anterior lip of the cervix and the deeper depression behind the posterior lip. The depression runs like a gutter all round the cervix, so there are also a right and a left lateral fornix, but they are not seen in sagittal section.

15. The peritoneum passes from the symphysis onto the upper surface of the bladder, and from the third piece of sacrum onto the rectum, as in the male. It falls at least 2 cm short of the anterior fornix of the vagina in the vesico-uterine pouch (item 4 in *fig. 380*) but it clothes 1 cm or more of the posterior fornix (item 5 in *fig. 380*).

Female Pelvis Seen from Above (*fig. 381*). The bladder and rectum have the same disposition as in the male. However: (1) the uterus and vagina situated medianly replace the seminal vesicles, ampullae of the deferent ducts, and prostate of the male;

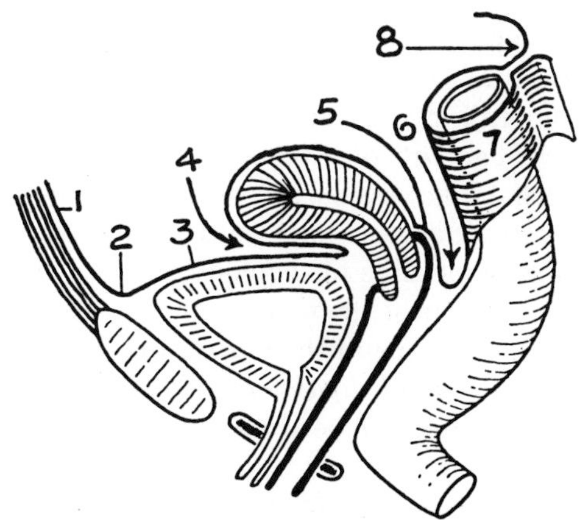

Fig. 380. The peritoneum of the female pelvis in median section (*see* text and *fig. 342*).

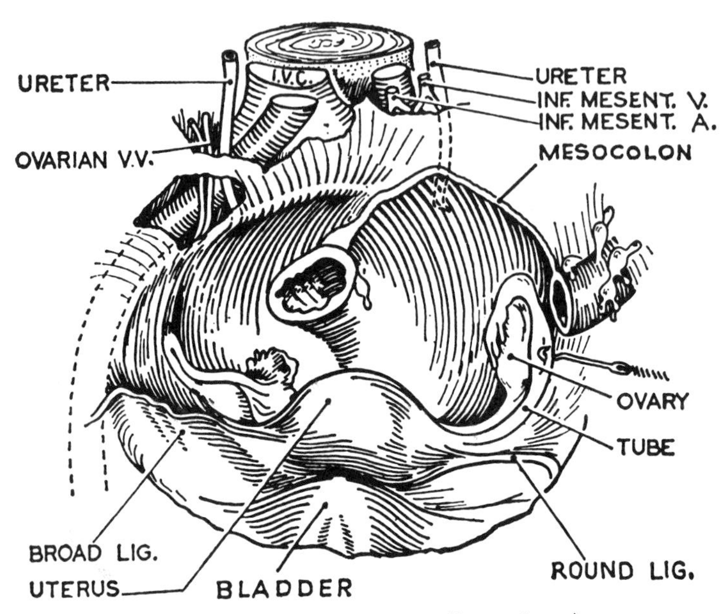

Fig. 381. Female pelvis (from above).

(2) the ovaries and their ducts (the uterine tubes) lie on either side, with the *broad ligaments of the uterus*, which pass from the lateral margins of the uterus to the side walls of the pelvis.

The anterior subdivision is the *vesico-uterine pouch;* the posterior is the *recto-uterine pouch.* The uterus rests on the empty bladder and the shallow vesico-uterine pouch is empty; the deeper *recto-uterine pouch* (item 6 in *fig. 380*) is occupied by sigmoid (pelvic) colon and ileum.

The ovary is attached to the back of the broad ligament. The uterine tube occupies the upper border of the broad ligament except at its lateral end. There the broad ligament is continued as a fold, the *suspensory ligament of the ovary*, across the external iliac vessels. A cord of fibromuscular tissue curves backward and upward from the junction of the body and cervix of the uterus, past the rectum, to the sacrum. It helps to suspend the uterus and is called the *uterosacral ligament.* The overlying crescentic fold of peritoneum is the *recto-uterine fold.*

A round fibromuscular cord, the *ligament of the ovary*, stands out in relief from the back of the broad ligament. It joins the lower pole of the ovary to the angle between the side of the uterus and the uterine tube. A similar cord, the *round ligament of the uterus* (lig. teres uteri), stands out from the front of the broad ligament. It passes from the angle between uterus and tube across the pelvic brim to the deep inguinal ring (*fig. 382*).

FEMALE INTERNAL GENITAL ORGANS

These comprise: ovaries, uterine tubes, uterus, and vagina.

Ovary (*fig. 383*)

The ovary has the shape of a testis but is only half the size. It is covered with cubical epithelium—not peritoneum. The pits and scars on its surface mark the sites of the absorbed corpora lutea which result monthly from the shedding of ova. The *mesovarium* attaches its anterior border to the back of the broad ligament. The *suspensory lig. of the ovary* suspends the tubal (upper) pole of the ovary from the external iliac vessels and conducts the ovarian artery, vein, lymph vessels, and nerves to the ovary. The uterine (lower) pole of the ovary

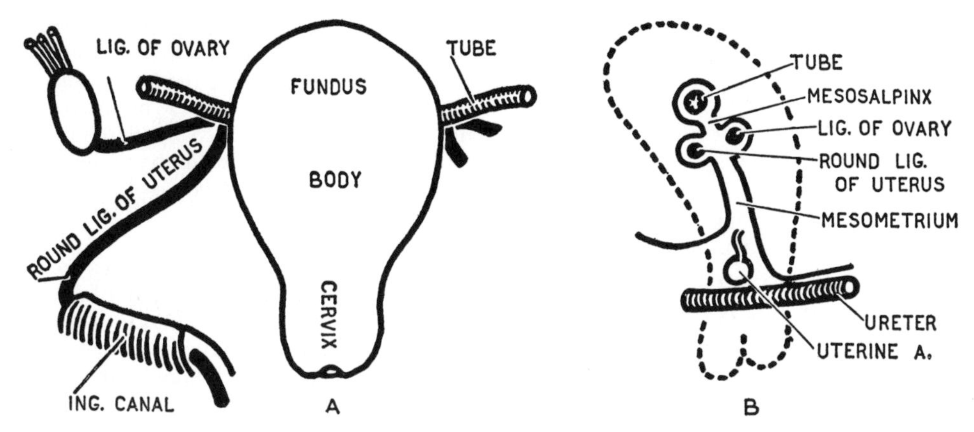

Fig. 382. *A*, derivatives of the gubernaculum ovarii—lig. of ovary and round lig. of uterus. *B*, sagittal section through broad lig. of uterus.

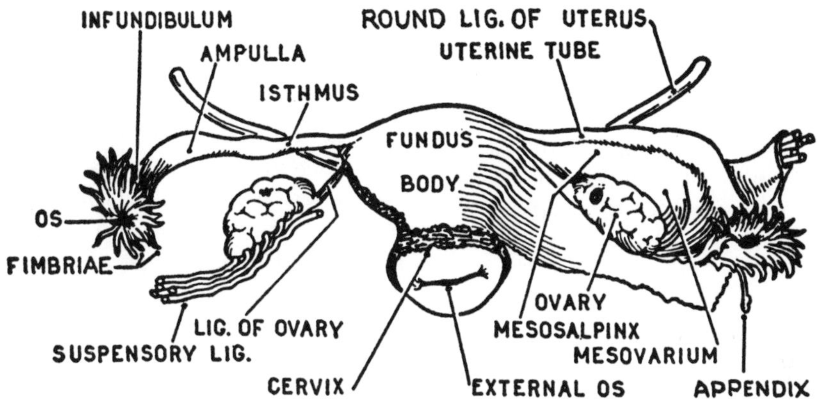

Fig. 383. The uterus and appendages (from behind).

is attached to the lateral margin of the uterus by the *ligament of the ovary*.

The typical position of the ovary is on the side wall of the pelvis, behind the broad ligament, and in the angle between the external iliac vein and the ureter. It is hidden by the uterine tube which falls over it medially. But its position is variable.

Uterine Tube (of Fallopius) (*fig. 383*)

The uterus is 5 cm across at its widest part, which is where the tubes enter it. Each tube is about 11 cm long. The spread, therefore, of the uterus and its two tubes greatly exceeds the diameter of the pelvic inlet (13 cm). Evidently the tubes cannot lie in a straight line. Each tube occupies the free edge of the broad ligament, its "mesentery," and runs upward, laterally, and backward to the side wall of the pelvis and there curves backward over the ovary.

Each tube has the following parts: *infundibulum*, *abdominal orifice*, *ampulla*, *isthmus*, *uterine part*, and *uterine orifice*.

The *abdominal orifice*, 2–3 mm in diameter, lies at the bottom of a funnel-shaped depression, the *infundibulum*. Fringes or fimbriae lined with ciliated epithelium, project from the infundibulum and encourage ova, when shed, into the tube. One fimbria is attached to the ovary.

The main part of the tube is long, irregular, and dilated, and is called the *ampulla*. The succeeding shorter part is straight and narrow and is called the *isthmus*. The *uterine part* passes through the uterine wall, which is at least 1 cm thick. Four longitudinal folds, bearing secondary folds, project into the lumen of the tube, which is lined with columnar ciliated epithelium.

Uterus

The uterus is thick walled and muscular. It is pear-shaped, 8 cm long, 5 cm at its widest part, and 2 or 3 cm at its thickest part. It is flattened in front where it rests on the bladder, and convex behind. The uterine tubes enter it at its widest part. The broad ligaments are attached to its margins; the ligament of the ovary and the round ligament of the uterus are attached just below the tube (*figs. 382, 384*).

The uterus is divisible into three parts—*fundus, body*, and *cervix*. The fundus and body form the upper 5 cm; the cervix the lower 3 cm. The fundus is the part that rises above the tubes. The uterine artery runs tortuously up the side of the uterus between the layers of the broad ligament.

The *external os* of the uterus is round until the birth of the first child; thereafter, it becomes a transverse slit guarded by an anterior and a posterior lip. There is a slight angle at the junction of the body and cervix, so the uterus is said to be anteflexed. The long axis of the uterus seldom lies in the median plane, but is deflected to one side or other.

The potential *cavity of the body* of the uterus is triangular; the uterine orifices of the tubes, which are about 1 mm in diameter, open at the upper lateral angles, the *internal os of the uterus* at the lower angle. The anterior and posterior walls are applied to each other. The *cervical canal* extends from the internal os to the external os of the uterus. It is spindle-shaped and 3 cm long.

STRUCTURE: The uterus has three coats—serous, muscular, and mucous. *In the fundus and body* the muscular coat (myometrium) is more than 1 cm thick and consists of interlacing bundles of smooth muscle, an arrangement which, after the birth of a child, brings about the natural arrest of hemorrhage by constricting the penetrating vessels. The serous (peritoneal) coat is adherent to the muscular coat except at the sides, where it passes onto the broad ligaments. The mucous coat (endometrium) is thick and is lined with columnar cells, many of which are ciliated, and it possesses numerous tubular glands which extend to, or even into, the muscle coat.

In the cervix the muscular bundles are largely circularly arranged, as at a sphincter. The mucous membrane is thrown into branching folds, is lined with columnar mucus-secreting cells, and possesses both simple and branched tubular mucus-secreting glands.

Vagina (L. = a sheath or scabbard)

The vagina is about 7 to 9 cm long and is approximately parallel with the pelvic brim. It extends from the vestibule of the vagina, which is guarded by the labia minora and where its orifice opens, through the urogenital diaphragm to reach the recto-uterine peritoneal pouch (item 6 in *fig. 380*). Its anterior and posterior walls are applied to each other.

Its anterior wall is structurally continuous with the urethra in its lower third; in contact with the bladder in its intermediate third; and is pierced by the cervix in its upper third. Around the cervix there is a circular gutter, described as the anterior, posterior, and lateral **fornices of the vagina.** Since the anterior aspect of the cervix is not covered with peritoneum, the anterior fornix is 2 cm from the vesico-uterine pouch of peritoneum; but the posterior fornix is clothed with peritoneum of the recto-uterine pouch.

Note that an instrument, forced upward through the posterior fornix, would enter the recto-uterine pouch, which is the lowest part of the peritoneal cavity and, therefore, well placed for surgical drainage.

Each lateral fornix is crossed by the base of the broad ligament, the uterine artery, and the ureter. Because the uterus is seldom median in position, the ureter is almost apposed to the cervix on one side (generally the left) and 1 to 2 cm away on the other (*figs. 384* and *385*).

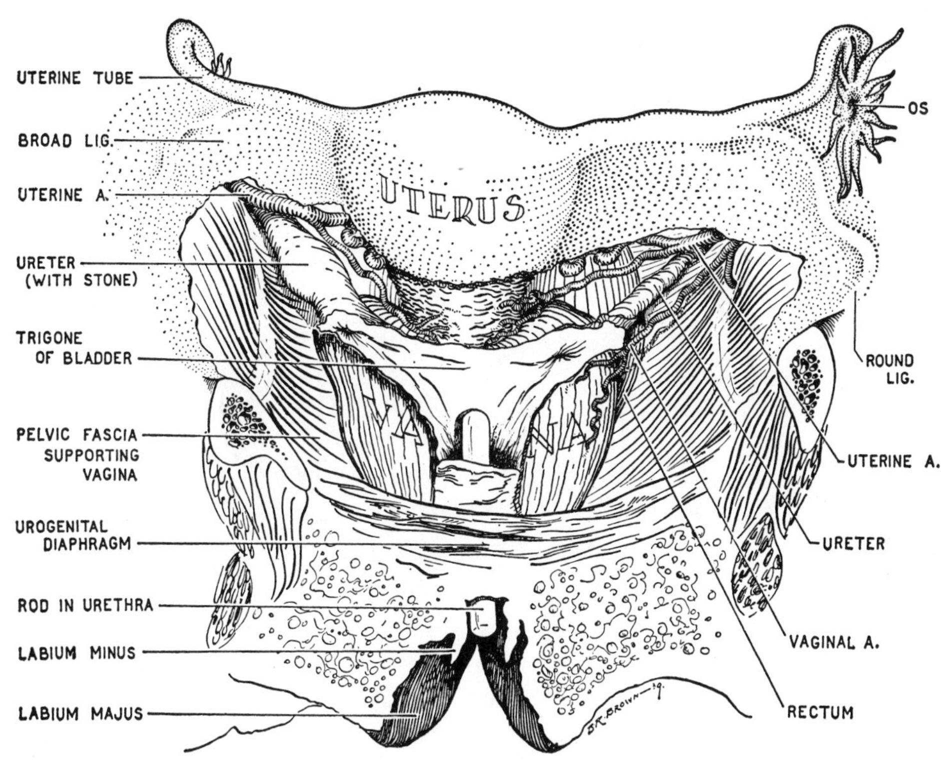

Fig. 384. The female internal genitalia. (Parts of the pubic bones and all of the bladder, the trigone excepted, have been removed.) The uterus is here symmetrically placed; hence, the ureters are nearly equidistant from the cervix. (Note the stone in the right ureter. Dissection by Dr. B. L. Guyatt.)

Relations. The posterior wall of the vagina is separated from the rectum by the vaginal and the rectal layer of pelvic fascia, and from the anal canal by the perineal body.

In the pelvic cavity, the side walls of the vagina rest on the Levatores Ani. On passing between the free anterior borders of the Levatores Ani, from which it receives fibers, the vagina enters the perineum and at once encounters, carries before it, and perforates, the u.g. diaphragm and its fasciae. These fasciae and the Levator Ani fascia and the superficial perineal fascia all fuse and blend with the laminated outer surface of the vaginal wall. To a great extent these form the support of the female pelvic viscera (*fig. 386*). Applied to this wall are the bulbs of the vestibule, the bulbo-urethral glands, and the Bulbospongiosus (*fig. 340*).

STRUCTURE: The vagina is lined with stratified squamous epithelium. This epithelium lines a *tunica propria* which presents numerous transverse folds (rugae). Outside the tunica propria there is a thin *muscular coat* of longitudinal fibers and some interlacing circular ones, and a thick fibro-areolar *adventitious coat.*

The stratified squamous epithelium is reflected from the vagina onto the cervix, clothing it as far as the external os. Though the vagina possesses no glands, its epithelial surface is moist from the secretion received from the cervical canal.

Development. In the female the *gubernaculum ovarii* passed through the inguinal canal into the labium majus followed only by a short *processus vaginalis peritonei* (the canal of Nuck). But, the gubernaculum ovarii acquired a side attachment to the uterus. As a result, the ovary passed into the pelvis drawing its vessels and nerves after it.

The gubernaculum in the female be-

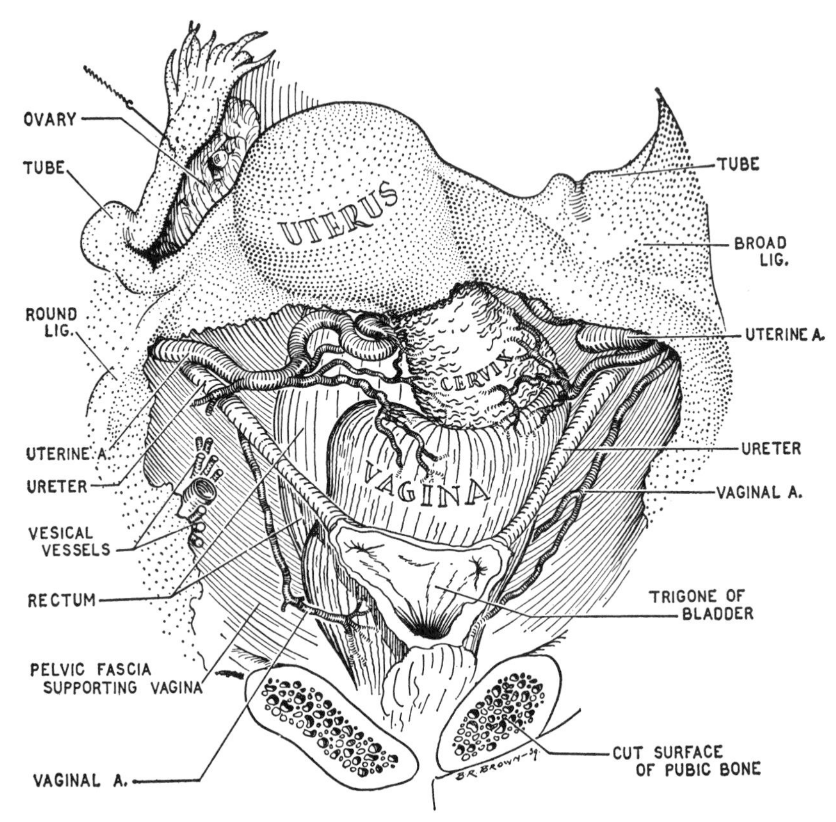

Fig. 385. The female internal genitalia. (Parts of the pubic bones and all of the bladder, the trigone excepted, have been removed.) The uterus is asymmetrically placed; hence, one ureter is close to the cervix; the other is far removed from it.

comes the ligament of the ovary and the round ligament of the uterus; and these two ligaments are all but continuous at their attachment to the side of the uterus just below the uterine tube (*fig. 382*). The round ligament in the female practically repeats the course taken by the deferent duct in the male, i.e., it is subperitoneal; it crosses the side wall of the pelvis and the external iliac vessels; and it turns round the inferior epigastric artery, passes through the inguinal canal, and ends in the labium majus, the homologue of the scrotum.

Very rarely the ovary does follow the gubernaculum into the labium majus.

Derivatives of the Mullerian or Paramesonephric Duct. Before the sex of the embryo is apparent, two vertical ducts appear on each side of the posterior abdominal wall. They run to (but do not empty into) the ventral subdivision of the cloaca, the *urogenital sinus*. They are the Wolffian or *mesonephric ducts* (which predominate in the male) and the Mullerian or paramesonephric ducts (which predominate in the female).

The mesonephric ducts in the male serve as sperm ducts and on each side become the duct of the epididymis, deferent duct, and ejaculatory duct. The paramesonephric ducts in the female serve as ducts for ova (*fig. 386*), their cranial parts becoming the uterine tubes; their intermediate parts fusing to form the uterus (drawing their mesenteries, the broad ligaments, off the side walls of the pelvis); and their caudal ends fusing to form the upper part of the vagina. As growth proceeds a solid plate of cells, the *vaginal plate*, extends downward between the urogenital sinus and the rectum. Eventually this plate becomes canalized and joined by a right and a left upwardly growing diverticulum of the sinus to form the lower part of the vagina.

Remnants of the (Wolffian) Mesonephric Duct and Tubules (*fig. 386*) persists in the female as

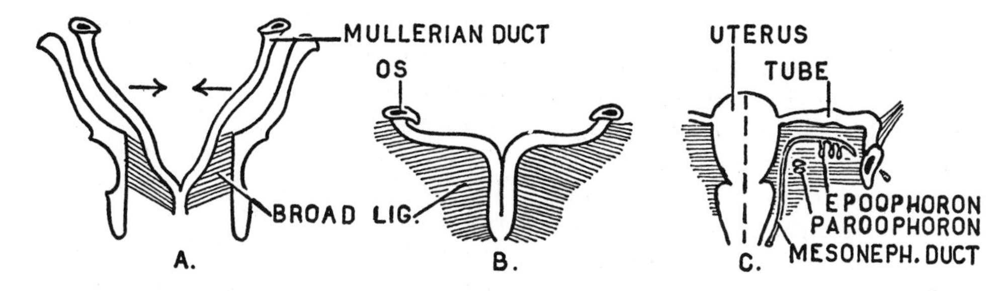

Fig. 386. Scheme of the development of the broad lig. of uterus as the "mesentery of the paramesonephric or Mullerian duct."

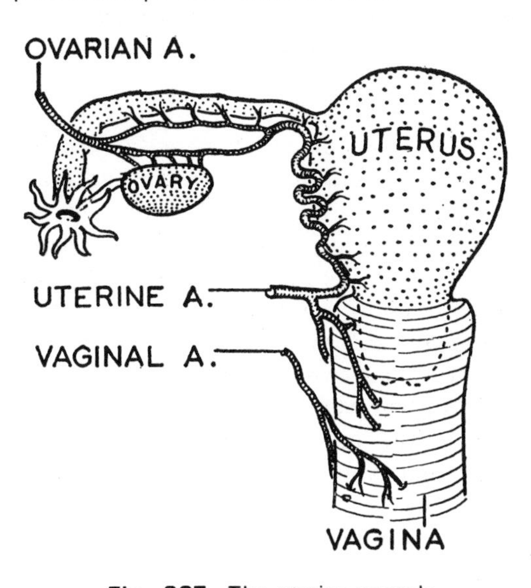

Fig. 387. The uterine vessels.

the *epoophoron*, *paroophoron*, and *duct of Gartner* (*of the epoophoron*). Because they commonly become cystic and cause trouble they are of clinical importance. Naturally, they are to be sought for at the sides of the (Mullerian) paramesonephric ducts, now converted into the uterine tubes, uterus, and vagina.

The epoophoron (= above the egg basket) lies between the layers of the broad ligament, above the ovary. It is a vestigial part of the mesonephric duct and tubules, and corresponds in the male to the duct of the epididymis and the efferent ductules of the testis (*fig. 229*, p. 189).

The paroophoron (= beside the egg basket) lies between the layers of the broad ligament, medial to the ovary. It is formed from mesonephric tubules and corresponds in the male to the paradidymis. *The duct of Gartner* is the segment of the mesonephric duct that lies in front of the anterior wall of the vagina. It corresponds to the end of the ductus deferens.

Unique Relations of Female Ureter

The ureter (*figs. 384* and *385*) (a) crosses the lateral fornix of the vagina, (b) below the broad ligament, and (c) below the uterine artery, and because of the obliquity of the uterine axis, (d) it lies closer to the cervix on one side (generally the left), and (e) enters the bladder in front of the vagina.

Special Vessels

The Ovarian Artery (*fig. 387*) arises from the aorta, like the testicular a., but differs from it in crossing the external iliac vessels, about 1 cm in front of the ureter (*fig. 381*), to enter the suspensory lig. of the ovary. It anastomoses freely with the uterine a., which may practically replace it.

The **Uterine and Vaginal Arteries** spring from the internal iliac artery (*figs. 384, 385, 387*). The uterine a. is large. It descends in front of the ureter to the base of the broad ligament and, at the lateral fornix of the vagina, it crosses above the ureter. After sending branches to the vagina and cervix, it continues tortuously up the side of the uterus between the layers of the broad ligament, supplies the uterus and the tubes, and anastomoses freely.

Veins. *The Ovarian Vein* opens into the i.v. cava or left renal vein according to the side.

The uterine and vaginal *venous plexuses* join the vesical plexus and pass as several large branches to the internal iliac vein. These plexuses communicate with the rectal plexus.

Nerve Supply

The ovaries, like the testes, are supplied by (?) T. 10; the fibers travel with the ovarian vessels and some of them supply the *uterine tubes*. The *uterus* is supplied by the hypogastric plexus and the pelvic splanchnic nerves. The afferent fibers from the fundus and body pass through the hypogastric plexus to T. 11, 12; those from the cervix and vagina via the pelvic splanchnic (parasympathetic) nerves (S. 2, 3, 4); but the lowest part of the vagina is supplied by the (somatic) pudendal nerve (also S. 2, 3, 4).

Support of the Female Pelvic Viscera

Muscular. As in the male, so in the female, the thick pubic parts of the Levatores Ani form a puborectal sling for the rectum, drawing it forward until it forms a sloping shelf (*fig. 355*). Upon this shelf the vagina rests, and on the vagina rests the bladder. This is the essential support.

The pubic parts of the Levatores Ani are also inserted into the perineal body and thereby they act as a sling for the lower part of the posterior wall of the vagina.

The urogenital diaphragm and its fasciae blend with the lower third of the vagina and assist the Levatores Ani to support it.

Fascial. The bladder and the rectum are clothed with vesical and rectal fasciae, as in the male (*fig. 348*). The uterus and the vagina likewise have their own fasciae, which are thick and tough around the cervix and vagina.

The apposed layers of rectal and vaginal fasciae are so loosely connected that a rectovaginal areolar space may be said to exist between them, whereas the apposed layers of the vesical and vaginal fasciae, being more closely blended, constitute a vesicovaginal fascial septum.

There are three paired suspensory structures:

1. *The perivascular stalk* is the first and most important of these. In effect, it includes (1) those branches of the internal iliac vessels that have been referred to previously as the leash of vessels (uterine, vaginal, vesical) that limit the retropubic space posteriorly (*fig. 375*), and (2) the fascia in which these vessels are imbedded. This fascia is conducted by its contained vessels to the side of the junction of the cervix and vagina and to the side of the bladder, and there it blends firmly with the fasciae of these organs, and with that of the rectum as well. This mass of fascia and vessels plus an accession of fibers from the region of the ischial spine is commonly referred to as the *lateral cervical lig.* or the *cardinal lig.*

2. *The uterosacral ligament* lies in the recto-uterine fold of peritoneum and may be regarded as the posterior free curved margin of the lateral cervical ligament. It extends from the middle of the sacrum to the junction of the body and cervix of the uterus where it meets its fellow. It moors the cervix posterosuperiorly.

Slips of involuntary muscle everywhere pervade the pelvic fascia and give it a supporting value not appreciated after death.

3. *The round ligament* of the uterus, by being attached along the side of the uterus as far caudally as the cervix, perhaps helps to moor the uterus and vagina anterosuperiorly.

DIFFERENCES BETWEEN THE MALE AND FEMALE PELVIS

The female pelvis may be contrasted with the male pelvis under the following headings:

1. Features dependent on the fact that woman is smaller and weaker.
2. Features related to the unique function of the female pelvis.
3. Markings on the pubic arch for the crura clitoridis.
4. Other features.

Women are less muscular than men; they are shorter and lighter. For these and other reasons their bones, including the bones of the pelvis, are lighter and the ridges and markings for tendons, aponeuroses, and fasciae are less pronounced.

Joints. The articular surfaces are relatively and absolutely smaller in the female. Thus, the oval articular facet on the base of the sacrum and the acetabulum are both strikingly small (*fig. 388*). The symphysis pubis is short. The auricular facet of the sacrum barely encroaches on the 3rd vertebra.

Functions Related to Childbirth. Note:

1. The cavity (*figs. 389, 389.1, 389.2*). All **diameters** are absolutely greater in the female than in the male (*fig. 390*), each of the three diameters of the inlet being about ½ cm greater: and the anteroposterior and transverse diameters of the outlet being about 2½ cm greater.

2. The Transverse Diameter or Breadth is greater in the female because:

a. The body and crest of the pubic bone are wider in the female.

b. The acetabulum is approximately its own diameter distant from the symphysis in the male, but in the female it is relatively much more.

c. The base of the female sacrum is relatively wide.

d. The female pubic arch is almost a right angle; it equals the angle between the outstretched thumb and the index finger

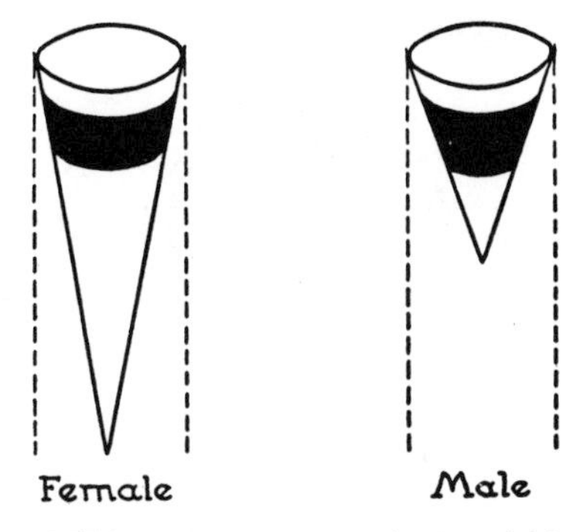

Fig. 389. The true pelves (colored *black*) as segments of cones. Female—short segment of long cone; male, the reverse.

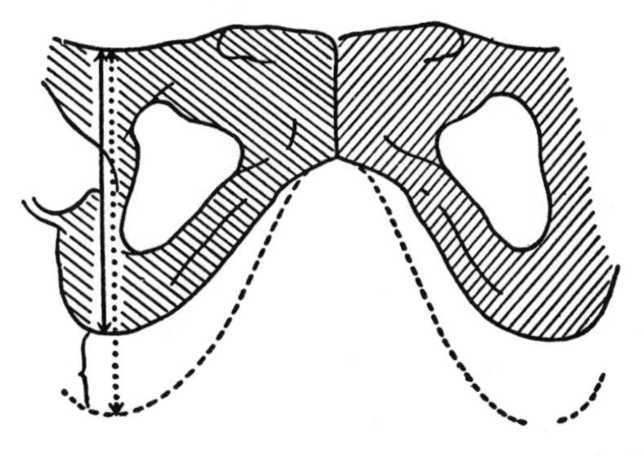

Fig. 389.1. The pelvic cavity is shallower in the female, and the subpubic angle is greater.

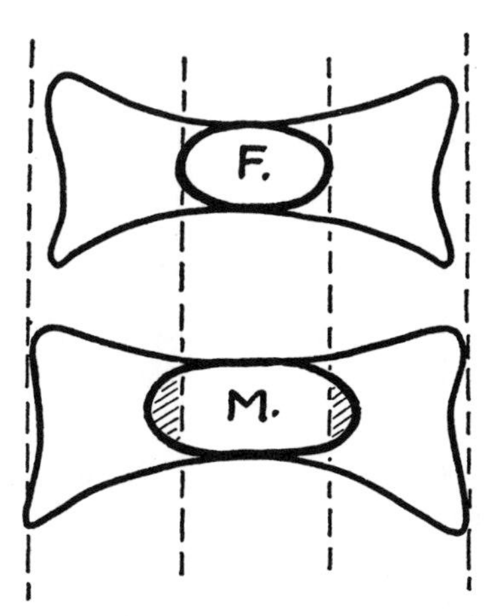

Fig. 388. Base of sacrum: male and female compared.

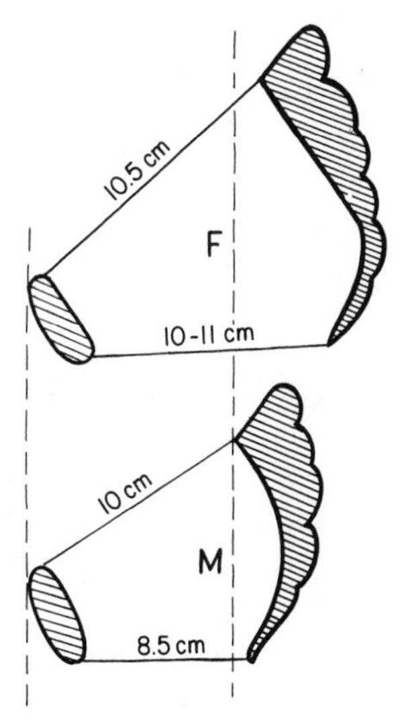

Fig. 389.2. Pelvic cavities on median section: male and female compared.

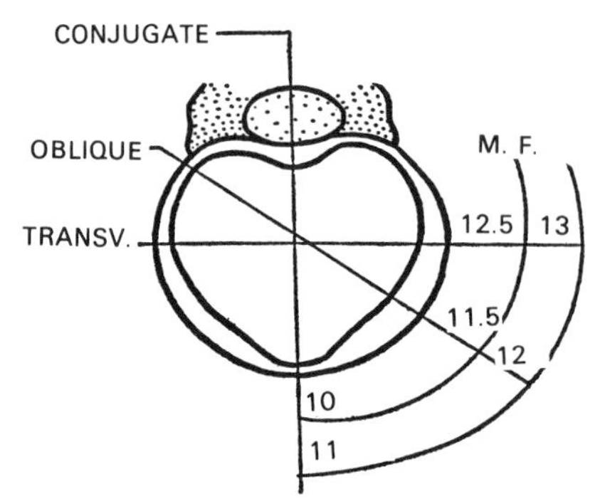

Fig. 390. Diameter of pelvic brim (in cm) male and female compared.

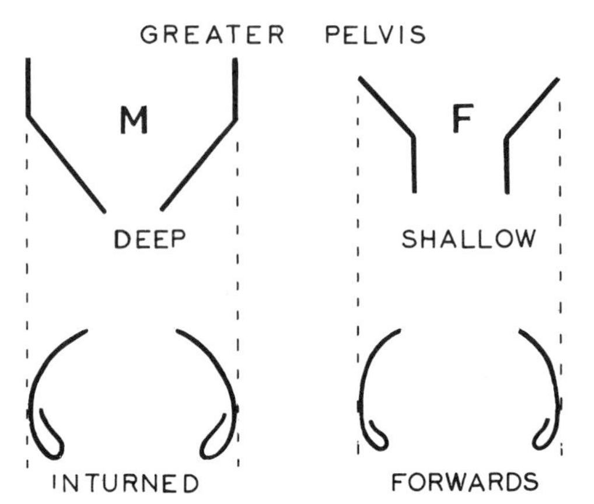

Fig. 392. Illiac crests: male and female compared.

(*fig. 389.1*). In the male it is an acute angle, equal to the angle between the index and middle fingers when spread.

e. The female ischial tuberosities are everted.

3. The Anteroposterior Diameter is greater in the female than in the male:

a. *At the inlet* the promontory is less prominent, so the female inlet tends to be round; the male to be heart-shaped.

b. *At the outlet* the coccyx is carried backward in the female by increasing the angle of the greater sciatic notch to approximately a right angle in the female (*fig. 391*).

Other Features. In the female the area for the crus clitoridis is narrow and the arch is thin. The pelvis major is shallow and the anterior superior spines rather point forward (*fig. 392*).

Skirting the anterior margin of the auricular

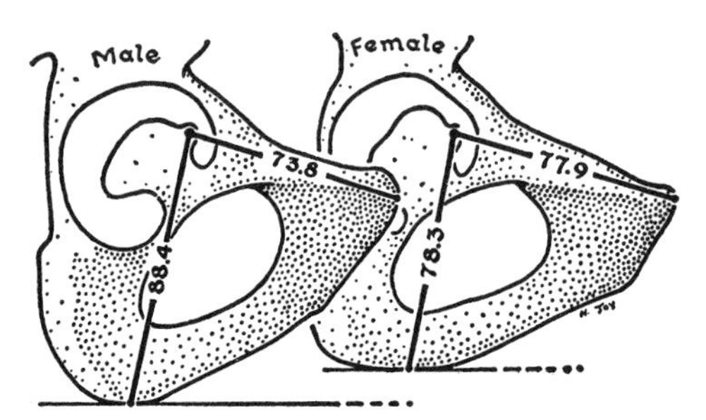

Fig. 393. The ischium-pubis index.

$$\frac{\text{length of pubis} \times 100}{\text{length of ischium}}$$

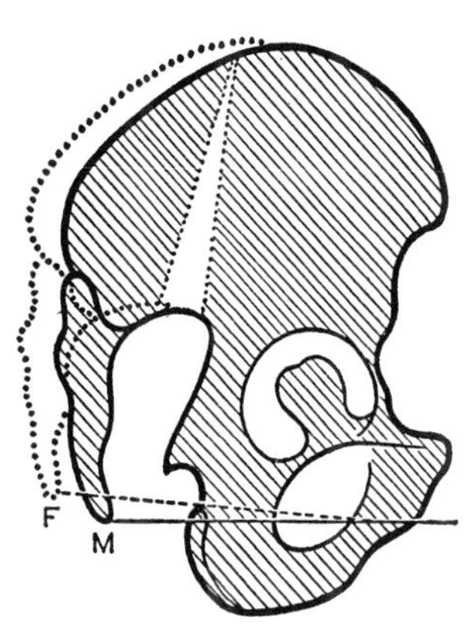

Fig. 391. Conversion of the male outlet into the female enlarges the angle of the greater sciatic notch.

TABLE 13

Percentage Incidence of Pelvic Types

(After Greulich and Thoms)

Type	300 Clinic Patients	100 Nurses	107 Girls Age 5–15
Dolichopellic	16.3	37	57.9
Mesatipellic	44.0	46	33.6
Brachypellic	36.3	17	8.3
Platypellic	3.3		

Dolichopellic (long): a-p. diameter > transverse diameter.

Mesatipellic (round): transverse > a-p. diameter by 0–1.0 cm.

Brachypellic (oval): transverse > a-p. diameter by 1.1–2.9 cm.

Platypellic (flat): transverse > a-p. diameter by 3 cm. or more.

TABLE 14
*Average Anteroposterior (Conjugate) and Transverse Diameters of the Pelvic Inlet in White Females, Classified According to the Four Pelvic Types of Caldwell and Moloy**

Pelvic Type	No. of Cases	Conjugate Diameter	Transverse Diameter
		cm.	
Gynecoid	26	10.86	13.76
Android	25	10.59	13.56
Anthropoid	19	11.75	12.94
Platypelloid	3	8.55	14.45
Mixed average for females	73	10.90	13.51
Average for males	43	10.10	13.00

* After T. W. Todd.

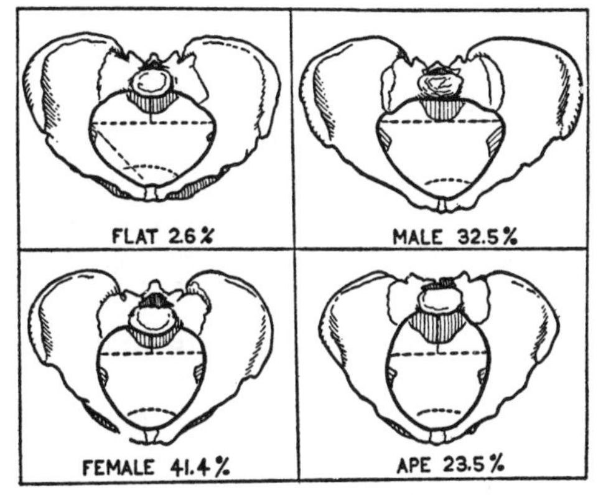

Fig. 394. The inlets of the four major types of female pelvis. (After Caldwell and Moloy.)

facet on the ilium there may be an identifying deep groove, the *pre-auricular sulcus* (Derry).

Identification. In cases where it is difficult to arrive at a decision as to the sex of a given pelvis, greatest weight should be placed upon: (1) the area for the attachment of the crus penis or clitoridis; (2) the angle of the pubic arch; (3) the size of the acetabulum; (4) the size of the facet on the base of the sacrum relative to the alae; (5) the distance of the acetabulum from the symphysis pubis; and (6) the size of the greater sciatic notch.

The medicolegal expert also uses the *Ischium-Pubis Index (fig. 393)*, which reveals the sex of an adult pelvis in more than 90 per cent of instances; that is because the pubic bone is relatively and absolutely longer in the female, and the ischium relatively and absolutely longer in the male. The index is less than 90 in adult males and more than 90 in adult females (Washburn). Other features of the pubis are valuable in "sexing" adult pelves (Phenice).

TYPES OF FEMALE PELVIS. Female pelves by no means all conform to the description given above; and since for the obstetrician it is of practical importance to know the dimensions and shapes of the pelves of his patients, various classifications have been proposed (*Tables 13, 14; fig. 394*).

The size and shape of the pelvis are influenced by various factors (hormonal, environmental, hereditary, and mechanical). At puberty, over a period of 1½ years, the predominantly dolichopellic pelvis of childhood changes to one of the adult types. (Greulich and Thoms).

22 PELVIC AUTONOMIC NERVES AND LYMPHATICS

PELVIC AUTONOMIC NERVES

Sympathetic Trunk; Superior Hypogastric Plexus; Inferior Hypogastric Plexus. Pelvic Splanchnic Nerves.

PELVIC LYMPHATICS

Pelvic Nodes; Structures Drained (area by area).

PELVIC AUTONOMIC NERVES

Sympathetic Trunk. *Within the Abdomen* the sympathetic trunk descends on the bodies of the lumbar vertebra, crosses behind the common iliac vessels and enters the pelvis.

Within the Pelvis, the trunks of the two sides, each having four ganglia, converge to the front of the coccyx to end in the unpaired unimportant *ganglion impar*. They lie medial to the pelvic sacral foramina (*figs. 324, 377*) and send gray rami laterally to each of the sacral and coccygeal nerves and a few visceral twigs to the inferior hypogastric plexus.

Superior Hypogastric Plexus [Presacral Nerve]. This plexus lies below the bifurcation of the aorta (*fig. 318.1*). It is a downward prolongation of the pre-aortic (intermesenteric) plexus, joined by the 3rd and 4th lumbar splanchnic nerves (*fig. 395*).

Branches enter the pelvis and descend in front of the sacrum as *right and left hypogastric nerves*. As these descend, they become plexiform and are joined by *twigs from the sacral sympathetic ganglia* and the pelvic splanchnic nerves to form the **inferior hypogastric plexus** [pelvic plexus].

Pelvic Splanchnic Nerves [Nervi Erigentes]. These threads run forward on each side from sacrum to rectum (*fig. 395*). Parasympathetic nerves, they spring from S. (2), 3, 4. They join the corresponding inf. hypogastric plexus which then becomes mixed sympathetic and parasympathetic.

Mixed branches are distributed with these vessels to the various pelvic viscera, the parasympathetic constituents being the more important.

The pelvic splanchnic nerves (both right and left) send ascending fibers across the left common iliac artery to the inferior mesenteric artery and are distributed with it to the descending and sigmoid colons (Stopford).

Functions: It is generally believed that the parasympathetic has exclusive control on the muscular walls of the bladder, urethra, and rectum (Sheehan). The pelvic splanchnics also cause relaxation of the arteries to the erectile tissue, producing erection of the penis (or clitoris), hence the alternative name—*nervi erigentes* (*fig. 395*).

The hypogastrics, when stimulated, cause the epididymis, deferent duct, seminal vesicle, and prostate to contract and empty their contents; at the same time the region of the neck of the bladder is shut off (Learmonth). Hence, on ejaculation, the seminal fluid is hindered from entering the bladder.

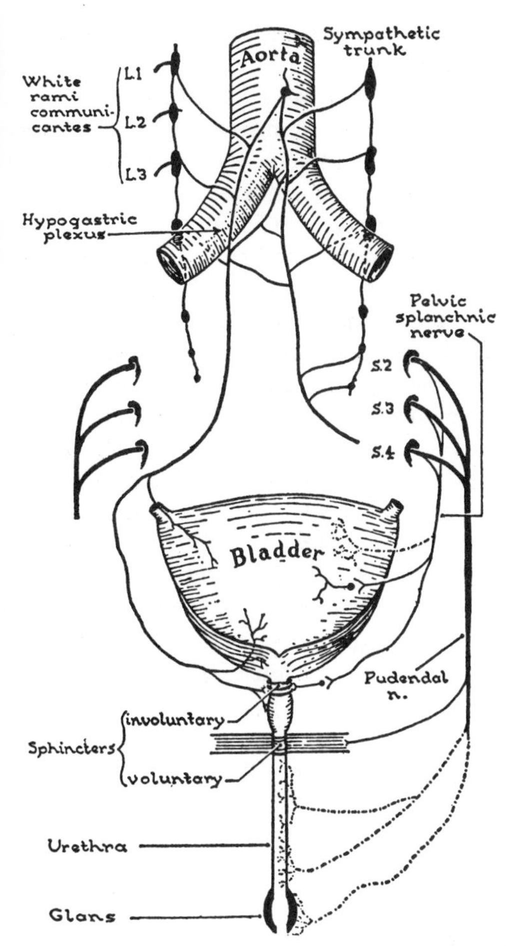

Fig. 395. Diagram of the nerve supply of the bladder and urethra.

The pelvic splanchnic nerves also carry many afferent fibers from the pelvis (e.g., pain, distention). But, the sympathetic is sensory to the body and fundus of the uterus (Mitchell).

PELVIC LYMPHATICS

Many lymphatics leave the superficial and deep inguinal nodes (p. 311), pass behind the inguinal ligament, and end in iliac nodes (*fig. 396*). These lymph vessels surround the femoral artery and vein. The medial vessels pass through the femoral canal and may be interrupted by a node (of Cloquet) that lies in the canal.

The Pelvic Nodes are in two groups:

(*1*) *nodes near the pelvic brim*—(a) the *external and common iliac nodes*, arranged as several intercommunicating chains around the respective blood vessels, and (b) the nodes above the *sacral promontory*.

(*2*) *nodes within the cavity* (a) the *internal iliac*, *lateral sacral*, and *median sacral nodes* arranged on the respective blood vessels; and (b) others in the *vesical fascia*, in the *rectal fascia* mainly behind the rectum (pararectal nodes) and on the course of the superior rectal artery, *in the broad ligament* near the cervix uteri, and between the prostate and rectum.

Structures Drained. The following structures are drained by lymph vessels traveling to lymph nodes, thus:

The Skin of the Penis and the Prepuce→ with the superficial dorsal vein in the subcutaneous tissues to the superficial inguinal nodes of both sides.

The Glans Penis and the Penile Urethra →accompany the deep dorsal vein deep to the fascia penis to the deep inguinal nodes

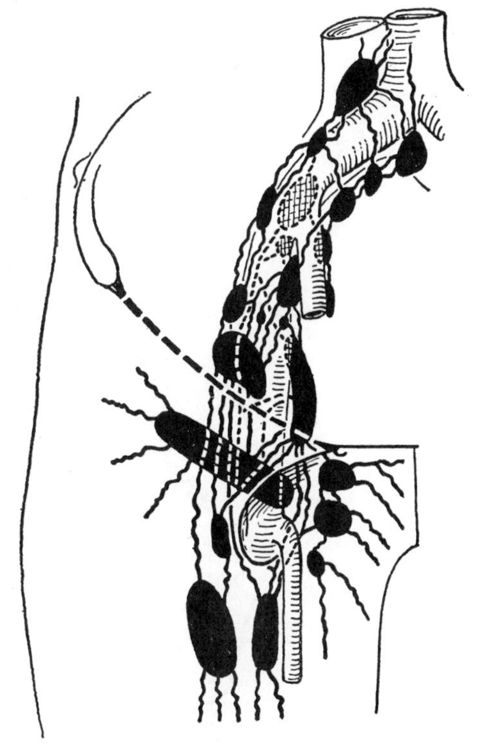

Fig. 396. Dissection of the inguinal lymphatics and those at the pelvic brim.

of both sides, thence to the external iliac nodes. Some vessels pass without interruption through the femoral and inguinal canals to the external iliac nodes. The lymph

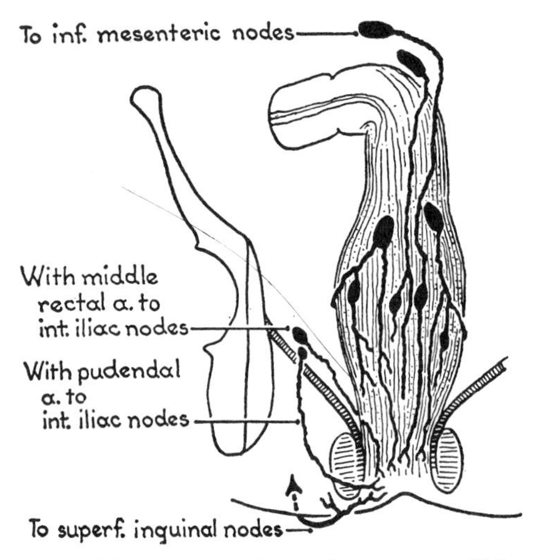

Fig. 397. Lymphatics of rectum. (After Rouviere.)

vessels of the glans and of the prepuce anastomose.

Bulbar Urethra → follows the internal pudendal artery to internal iliac nodes and the deep dorsal vein below the infrapubic ligament, thence to external iliac nodes.

Membranous and Prostatic Urethrae (*or Whole Female Urethra*) → internal iliac nodes.

Bladder (superior and inferolateral surfaces) → follow the general course of the branches of the superior vesical artery, deferent duct, and ureter to external iliac nodes lying along the medial side of the external iliac vein, some being interrupted by anterior and lateral vesical nodes.

Base of the Bladder and the Male Internal Genital Organs (prostate, seminal vesicles, ampullae of the deferent ducts) → internal iliac nodes; also to the external iliac nodes; and on the Levator Ani to the sacral nodes.

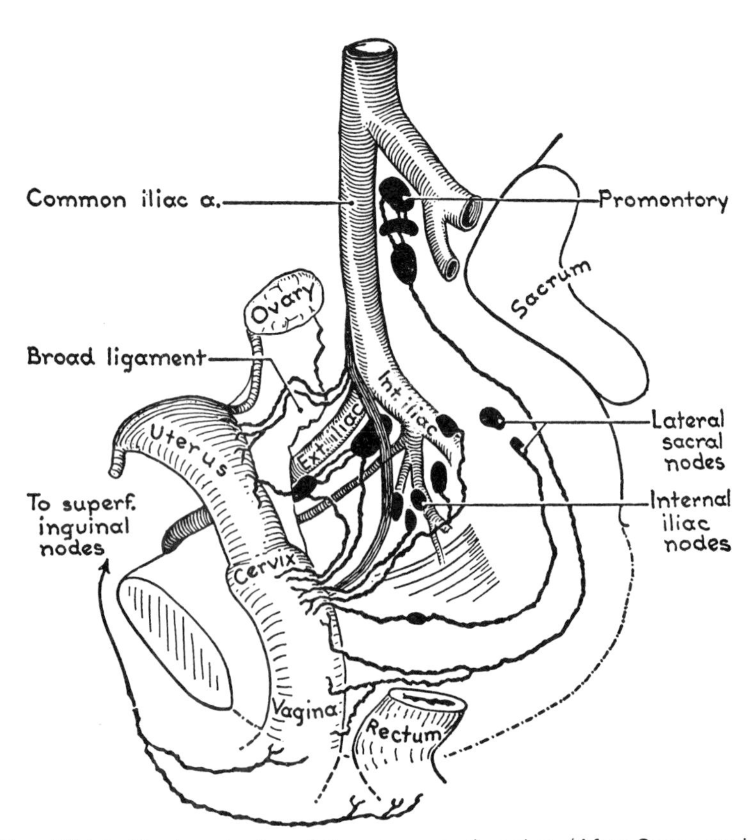

Fig. 397.1 The lymphatics of the uterus and vagina. (After Cunco and Marcille).

Anus and Lowest Part of the Anal Canal → by cutaneous vessels to the superficial inguinal nodes.

Anal Canal → internal iliac nodes by vessels piercing the Levator Ani (others crossing the ischiorectal fossa with branches of the pudendal vessels) and by others following the middle rectal vessels.

Rectum, above the anal valves (*fig. 397*), → (1) pararectal nodes, which lie in the fatty tissue behind the rectum enclosed within the rectal fascia, thence to the superior rectal and inferior mesenteric nodes; (2) lateral and median sacral nodes; and, (3) with the middle rectal artery to internal iliac nodes.

Ovary, like the testis (*fig. 228*), → lateral aortic and pre-aortic nodes, between the levels of the common iliac vessels below, and the renal vessels above.

Uterine Tube and Fundus of the Uterus → with those of the ovary, ending in the lower pre-aortic and lateral aortic nodes. They traverse the broad ligament and cross the pelvic brim.

Body of Uterus (*fig. 397.1*) → via broad ligament to the external and common iliac nodes, and by a single vessel running in the round ligament to the superomedial group of superficial inguinal nodes.

Cervix of Uterus and Upper End of Vagina → external iliac nodes; with uterine and vaginal arteries to internal iliac nodes; and, by vessels passing close to the rectum in the uterosacral fold, to lateral sacral nodes and nodes of the promontory.

Lower End of Vagina and labium majus, like the scrotum → to superficial inguinal nodes.

SECTION FIVE

LOWER LIMB

23	Femur and Front of Thigh	309
24	Hip Bone and Gluteal Region	320
25	Back and Medial Region of Thigh	327
26	Leg and Dorsum of Foot	336
27	Bones and Sole of Foot	350
28	Joints of Lower Limb	364

23 FEMUR AND FRONT OF THIGH

FRONT OF THIGH

Bony Landmarks.

SUPERFICIAL STRUCTURES

Fascia Lata and Iliotibial Tract.
Great (Long) Saphenous Vein—Saphenous Hiatus.
Lymph Nodes–Inguinal Nodes.
Cutaneous Nerves: Femoral Sheath; Canal and Ring.

FEMUR

Proximal End; Shaft or Body; Distal End.
Epiphyses; Variations.

FEMORAL TRIANGLE

Outline; Contents; Femoral Artery and Vein; Profunda Femoris Artery and Veins; Venous Valves.
Femoral Nerve.
Psoas Tendon.

ADDUCTOR CANAL

Sartorius; Contents of Canal; Internervous Lines Regions of Thigh—Iliotibial Tract.

QUADRICEPS FEMORIS

Rectus and Vasti; Internervous Line.
Distribution of Femoral Nerve.

FRONT OF THIGH

Bony Landmarks. The bony parts labelled in *figure 398* should be located on the skeleton and palpated on the living model, which will likely be yourself or a fellow student; also feel the *ischial tuberosity* by sitting on your fingertips.

The greater trochanter lies about 10 cm below the tubercle on the iliac crest. Grasp it. It is subaponeurotic and, so, difficult to feel unless the parts are relaxed (e.g., by lying down).

The *body of the femur* is buried in muscles and runs obliquely to its lower end at the knee.

SUPERFICIAL STRUCTURES

The Skin, both dermis or true skin and epidermis, is thick and tough where it needs to be; on the back, buttocks, lateral aspects of both limbs, palm, and sole.

The Superficial Fascia over bony points, such as the acromion, olecranon, knuckles, and patella, and under skin creases is generally reduced; over the ischial tuberosity and the heel the fat is imprisoned in much fibrous tissue. On the buttocks large quantities of fat may be deposited; in the Bushmen of South Africa, the quantity is enormous.

The Deep Fascia is especially strong; in the thigh it is called the **fascia lata**. According to rule, it is attached to bone where they meet.

It is extremely strong laterally because in between two thin layers of circularly disposed fibers there runs a broad band of coarse vertical fibers called the **iliotibial tract** (p. 323). This tract is the conjoint aponeurosis of the Tensor Fasciae Latae and the Gluteus Maximus.

The Great (Long) Saphenous Vein ascends throughout the length of the limb, in the subcutaneous fat (*fig. 398.1*).

It begins at the medial end of the dorsal venous arch of the foot, passing in front of the medial malleolus, crossing the lower third of the medial surface of the tibia, and following 1 cm behind its medial border as

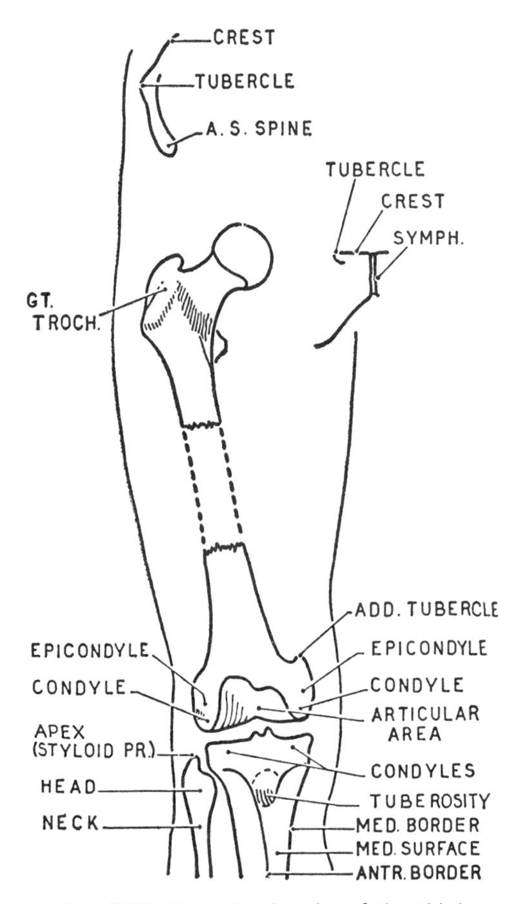

Fig. 398. Bony landmarks of the thigh.

far as the knee. At the knee it is found on incising a hand's breadth (or slightly more) behind the medial border of the patella. From there it takes a straight, but oblique, course up the thigh to the femoral vein which it joins 4 cm inferolateral to the pubic tubercle.

It anastomoses freely with the short saphenous vein; it communicates along intermuscular septa with the deep veins; and it receives numerous tributaries including three *superficial inguinal veins*.

These veins accompany the three **superficial inguinal arteries**, which are branches of the femoral artery—(1) the *(superficial) external pudendal artery* which crosses in front of the spermatic cord to supply the scrotum; (2) the *superficial epigastric artery* which passes towards the navel; and (3) the *superficial circumflex iliac artery* which passes laterally below the inguinal ligament.

Terminology. Saphenous, sesamoid, nuchal, and *retina* are among the few **terms of Arabic origin** that remain in our anatomical vocabulary (Singer).

Saphenous Opening or Hiatus (Fossa Ovalis) (*fig. 399*). The great saphenous vein has to pass through the fascia lata in order to reach the femoral vein 4 cm below the pubic tubercle; it appears responsible for the large hole that exists, called the *saphenous opening*.

Only in man is there such an opening in the fascia and only in man does the saphenous vein ascend far above the knee. From the pubic tubercle, its lateral border is crescentic or falciform, and it blends with some areolar tissue (cribriform fascia) that covers the opening.

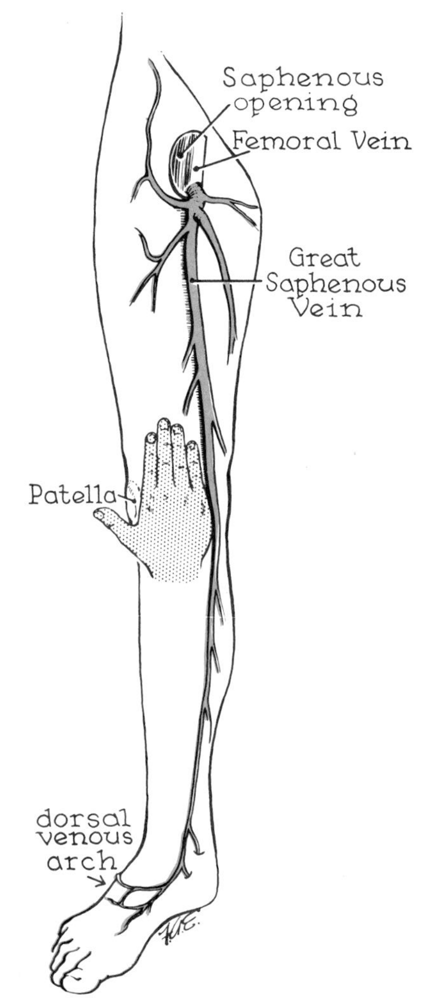

Fig. 398.1. The great saphenous vein.

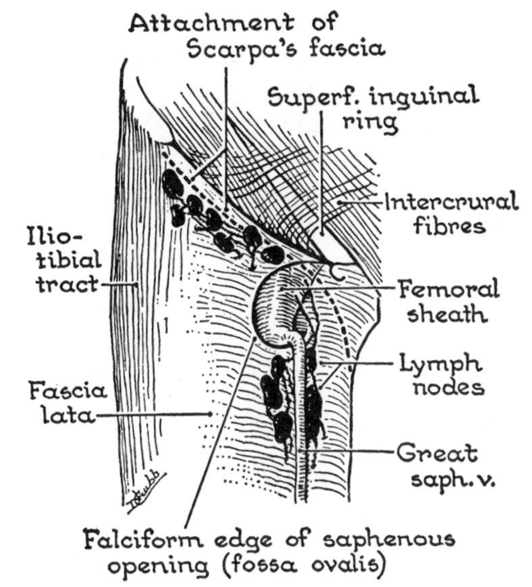

Fig. 399. Saphenous opening; great saphenous vein joining the femoral vein within the femoral sheath; lymph nodes; and the attachment of Scarpa's fascia (*broken line*).

Lymph Nodes. In the lower limb, the lymph nodes are situated at the knee and the groin; those at the knee (popliteal) being deep and impalpable; those at the groin (inguinal) being superficial and palpable.

The Inguinal Nodes (*fig. 399*) are in two groups, a superficial and a deep. The *superficial nodes* are subdivided into (1) an upper horizontal group and (2) a lower vertical group along the great saphenous vein.

The lower superficial inguinal nodes, receive all the superficial lymph vessels of the lower limb except the few that, following the small saphenous vein, end in the popliteal nodes. *The upper superficial inguinal nodes* drain the regions supplied by the three superficial inguinal blood vessels; namely, the subcutaneous tissues of the anterior abdominal wall below the navel, of the penis and scrotum (vulva in the female and lowest part of the vagina) and of the gluteal region, perineum, and lower part of the anal canal, but not of the testis (nor of the ovary).

The *deep nodes*, one to three in number, lie on the medial side of the femoral vein, in and below the femoral canal. They receive the deep lymph vessels of the limb and of the glans penis (or clitoridis) and spongy urethra.

Cutaneous Nerves (*fig. 400*). The cutaneous nerves of the front of the thigh are derived from the ventral rami of L. 1, 2, 3, 4. The *lateral*, *intermediate*, and *medial cutaneous nerves of the thigh* and the *saphenous nerve* are branches of the femoral nerve. They pierce the deep fascia along an oblique line that roughly marks the Sartorius and their exact disposition is unimportant. *Details* are given below for specialists—

The *lateral cutaneous nerve* commonly arises independently from the lumbar plexus and enters the thigh close to the anterior superior spine. When it springs from the femo-

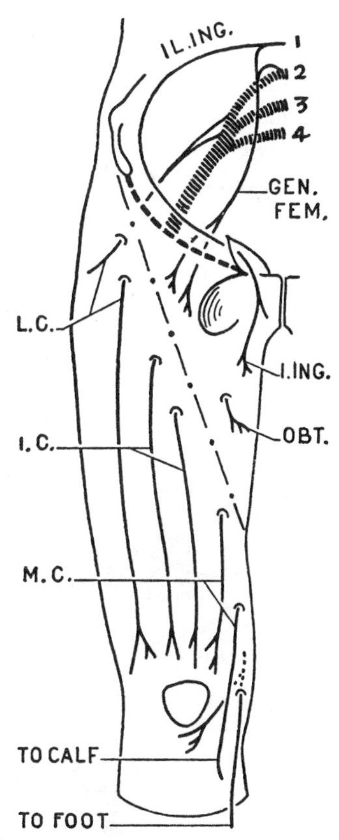

Fig. 400. Cutaneous nerves of the front of the thigh, above and below the line of the Sartorius.

ral nerve, it enters at some distance from the spine. A posterior branch passes to the gluteal region; the posterior branch of the medial cutaneous nerve extends to the calf; and the saphenous nerve extends half way along the medial border of the foot. The other branches remain above the knee.

The *saphenous nerve* comes to the surface between the Sartorius and Gracilis behind the adductor tubercle. It gives off a *patellar branch* below the patella.

The skin overlying the femoral triangle (p. 314) is supplied by (1) the *ilio-inguinal nerve*, and (2) the femoral branch of the *genitofemoral nerve*.

A branch of the *obturator nerve* becomes cutaneous at the middle of the thigh and may extend to the calf.

Femoral Sheath (*fig. 401*). The femoral artery, vein, and some lymph vessels are wrapped up in a prolongation of the extraperitoneal areolar tissue that envelops the external iliac vessels in the abdomen. This wrapping, the femoral sheath, is funnel-shaped; it has three compartments—a lateral one for the artery, a middle one for the vein, and a medial one, the **femoral canal**, partly occupied by lymph vessels. It is into the femoral canal that a femoral hernia may bulge from above.

The Femoral Ring is the mouth of the femoral canal. It is bounded *laterally* by the femoral vein; *posteriorly* by the superior ramus of the pubic bone covered with a coating of Pectineus and pectineus fascia; *medially* at a variable distance (Doyle) by the lacunar ligament and the conjoint tendon, both of which are attached to the pecten pubis (pectineal line); and *anteriorly* by the inguinal ligament and the spermatic cord.

To enlarge the ring in an operation for a femoral hernia, you dare not cut laterally into the vein; it would be useless to cut posteriorly on to the bone; you would require to cut either medially or anteriorly (*fig. 412*).

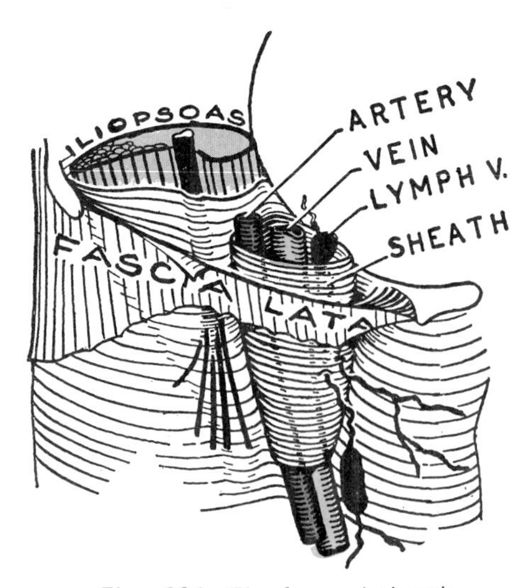

Fig. 401. The femoral sheath.

FEMUR

The femur is the longest bone in the body, being a quarter of the stature, or about 45 cm in an average man.

The Proximal End presents for examination, a head, neck, greater trochanter, and less trochanter (*figs. 402* and *403*).

The Head forms two-thirds of a sphere, and is directed medially, upward, and

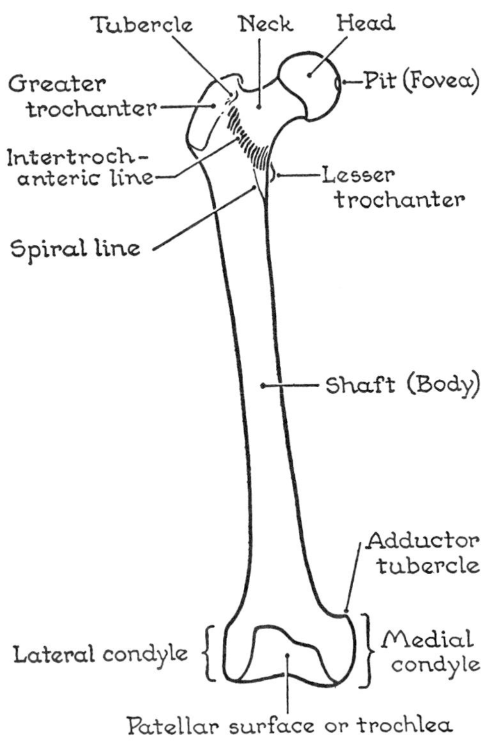

Fig. 402. The femur, front view.

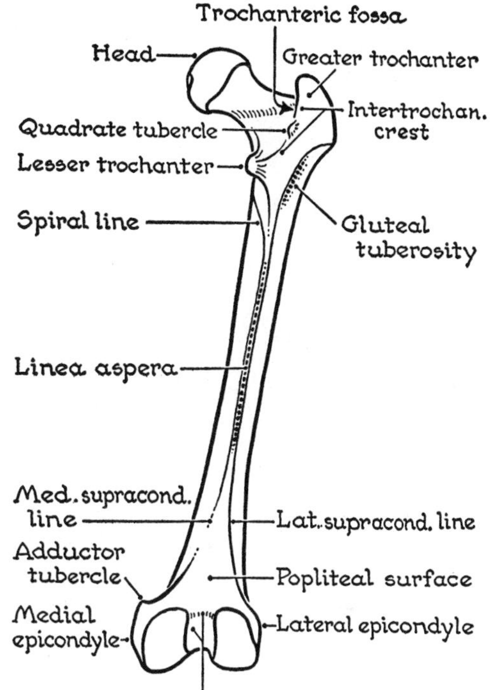

Fig. 403. The femur, posterior view.

foward. It is much more secure in its socket than the head of the humerus.

The Neck is obliquely placed. It is limited laterally by the greater trochanter and really is the medially curved upper end of the shaft (*fig. 404*). A broad, very rough, oblique line, the *intertrochanteric line*, is due to the attachment of the massive iliofemoral ligament.

The posterior aspect of the neck is separated from the shaft by a prominent, rounded ridge, the *intertrochanteric crest.*

In the child the pelvis is narrow before the bladder descends into the pelvis; so, the neck and shaft of the femur are nearly in line with each other. As the pelvis widens the neck becomes more horizontal and the angle between neck and shaft becomes smaller (125° male).

The Lesser Trochanter is the traction epiphysis of the Iliopsoas. It is conical; it projects from the posterior surface of the bone; and it points medially (*fig. 403*).

The Greater Trochanter is a traction epiphysis especially for the Gluteus Medius, which draws it upward, medially, and backward. Its highest point is its postero-superior angle.

Some Details. The *quadrate tubercle* marks the site where the epiphyseal line crosses the intertrochanteric crest. On the medial side of the greater trochanter there is a circular depression, the *trochanteric fossa*, in which the Obturator Externus is inserted. From the fossa a shallow groove for the tendon of the Obturator Externus runs horizontally across the back of the neck (*fig. 435*).

The Shaft or Body is slightly bowed forward. Its middle two quarters are approximately circular on cross-section; from its back a broad rough line, the **linea aspera** stands out prominently. The linea aspera bifurcates above and below into diverging lines that bound triangular areas (*fig. 403*).

The anterior and lateral aspects of the shaft give origin to the fleshy fibers of the Vastus Intermedius; the medial aspect is bare. The shaft, therefore, is smooth except for the linea aspera. Many muscles and three intermuscular septa crowd onto this line.

The Distal End of the femur is divided into two large knuckles, the *medial* and *lateral condyles*. The hinder parts of the condyles project backward like thick discs beyond the popliteal surface (*fig. 405*). Between the opposed surfaces of these discs is a U-shaped notch, the *intercondylar fossa*, the width and depth of the thumb. At the center or hub of the nonop-

Fig. 404. The neck is the incurved shaft to which the trochanters are added.

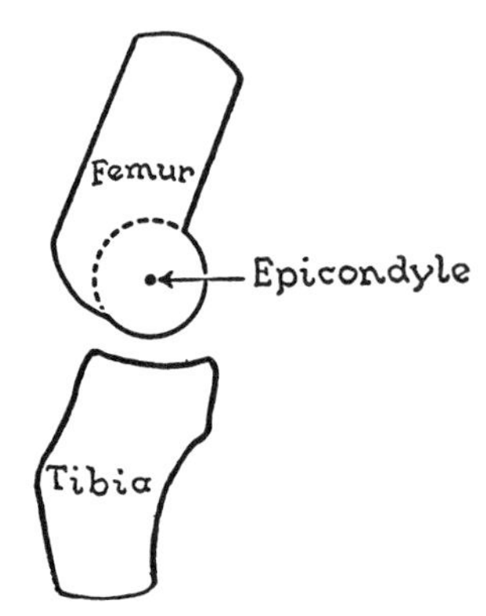

Fig. 405. A disc-shaped condyle and an epicondyle.

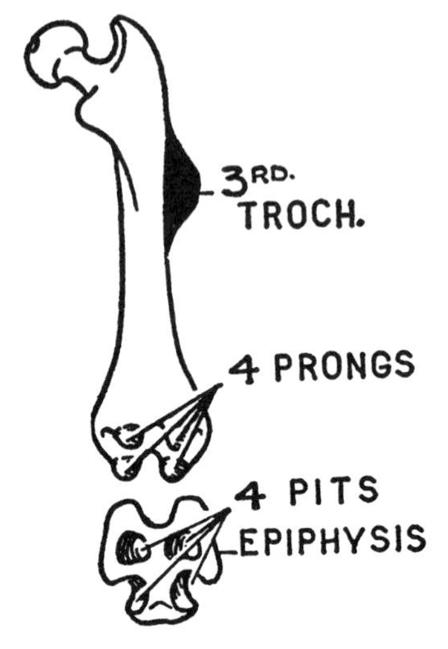

Fig. 406. Femur of muskrat showing two primitive mammalian characters: the 3rd trochanter for gluteus maximus and the exaggerated interlocking of the inferior epiphysis with the shaft.

posed surface of each disc there is a fullness, the *epicondyle*, to which the medial and lateral ligaments of the knee joint are attached (*fig. 507*). The femoral condyles are covered with cartilage below and behind for articulation with the tibia, and the cartilage of the two sides meets in front in a V-shaped trough, the *patellar surface* or *trochlea* in which the patella plays. The lateral lip of this pulley projects further forward and further upward than the medial lip.

The **epiphyseal line** of the *lower end* of the femur runs irregularly through the adductor tubercle and along the intercondylar line. (*fig. 406*). *At birth* the lower epiphysis is an ossific nodule about ½ cm in diameter.

Variations. A *third trochanter* (*fig. 406*), is constant in certain rodents and the horse, and may appear in any mammal including man as an enlarged gluteal tuberosity. *Platymeria* or marked anteroposterior flattening of the upper part of the shaft of the femur, occurs in some races, but is not common in Europeans.

FEMORAL TRIANGLE

The femoral triangle (*fig. 407*) is really a shallow trough. Its base is formed by the inguinal ligament. Its apex lies about 10 cm below the inguinal ligament, where the Sartorius crosses the lateral border of the Adductor Longus.

The Central and Dominant Structure within this triangular frame is the *femoral artery*. It begins at the *midinguinal point*.

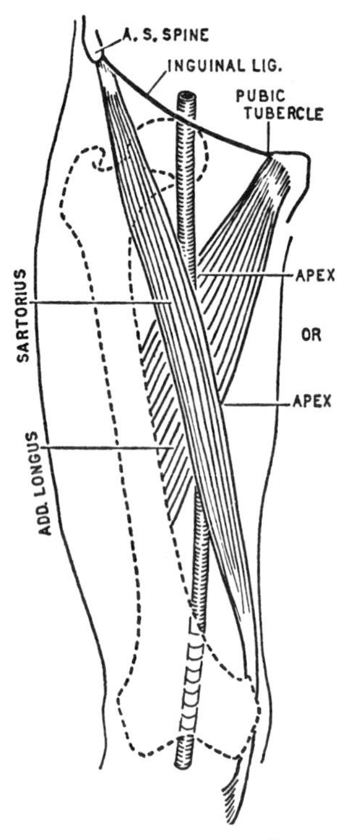

Fig. 407. The sides of the femoral triangle. The course of the femoral artery.

It leaves the triangle at its apex and enters the adductor canal (subsartorial canal) and so only its upper 10 cm lie in the triangle; its remaining 15 cm travel through the adductor canal.

The Floor of the Triangle is muscular (see *fig. 408*).

Contents of the Triangle (*fig. 409, 410*):

Femoral artery, vein, and nerve, certain of their branches, and the deep inguinal lymph nodes; and—

Profunda femoris artery and vein, and their circumflex branches (p. 333).

The Profunda Femoris Artery usually takes origin several centimeters below the ligament. It is only slightly smaller than the continuation of the femoral artery itself, and is therefore no mean vessel. For good reason, earlier authors considered this a bifurcation of a common artery into its deep and superficial branches.

The Femoral Vein lies medial to the femoral artery in the femoral sheath, but lies behind the artery at the apex of the triangle; so do also the profunda artery and vein, all four vessels being almost insepara-

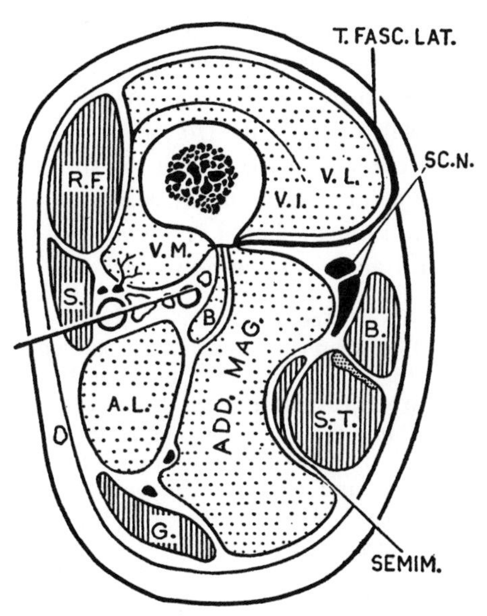

Fig. 409. Cross-section of thigh near apex of the femoral triangle showing (1) attachments of muscles to linea aspera; (2) six muscles (*hatched*) that span the femur, and (3) that a bullet might penetrate the femoral and profunda femoris vessels.

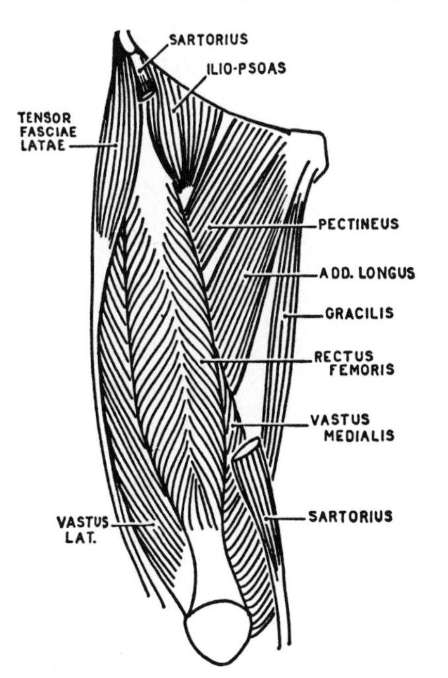

Fig. 408. The floor of the femoral triangle. The walls of the adductor canal.

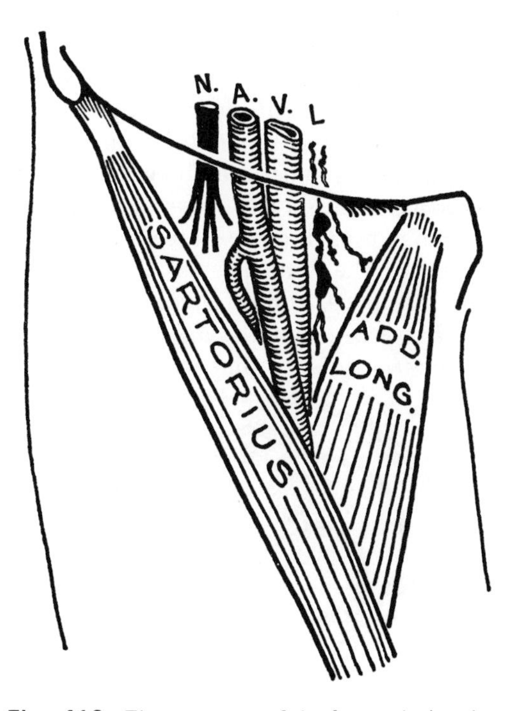

Fig. 410. The contents of the femoral triangle.

bly united in a tough areolar sheath. A stab or bullet wound at the apex of the triangle would penetrate in succession the four great vessels of the limb (*fig. 409*).

The great *saphenous vein* is the last tributary to join the femoral vein.

Venous Valves. The i.v. cava and the common iliac vein have no valves.

Indeed, in 24 per cent of 200 limbs no valve was found between the mouth of the saphenous vein and the heart. This may be important in relation to varicose veins. In 76 per cent one valve was found above the saphenous vein, usually in the femoral vein but sometimes in the external iliac, and sometimes in both. Usually (90 per cent) a valve occurs just below the opening of the profunda vein and there may be two or three others (Basmajian). The saphenous vein is well supplied with valves.

The Femoral Nerve enters the thigh slightly lateral to the artery (*fig. 411*), deep to the fascia lata and also to the fascia iliaca. Soon after crossing the inguinal ligament it breaks up into numerous motor and sensory branches. Two of these follow the artery, closely applied to its lateral side, into the adductor canal; one, the *nerve to the Vastus Medialis* is motor; the other, the *saphenous nerve*, is sensory.

The *medial femoral circumflex* branch of the profunda artery passes backward, leaving the triangle between Pectineus and Iliopsoas, while the *lateral femoral circumflex* branch passes laterally through or behind the branches of the femoral nerve, disappears under cover of the muscles that arise from the anterior superior and anterior inferior iliac spines (namely, Sartorius and Rectus Femoris), and breaks up into three terminal branches.

Relationships. The femoral artery lies in front of the **Psoas tendon**; the nerve, in front of the Iliacus; and the vein and most lymph vessels, in front of the Pectineus (*fig. 412*).

No motor nerve, except the *twig to the Pectineus*, crosses the femoral artery.

ADDUCTOR CANAL

The adductor canal of Hunter (the Subsartorial Canal) is the narrow outlet of the femoral triangle roofed in by the Sartorius and thereby converted into an intermuscular tunnel, triangular on cross-section and 15 cm long. It begins (less than) 10 cm below the inguinal ligament and ends 10 cm above the adductor tubercle (*fig. 413*). Only the femoral vessels and the saphenous nerve run right through it.

The Sartorius arises side by side with the inguinal ligament from the anterior superior iliac spine. It is inserted into the medial surface of the tibia below the level of the tuberosity (*fig. 428.2*).

Nerve Supply—femoral nerve. It is commonly pierced by the femoral cutaneous nerves in transit (*fig. 400*).

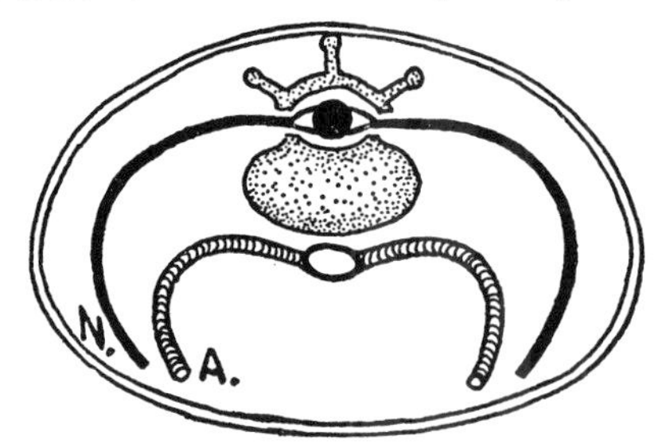

Fig. 411. The origins of the femoral nerve and artery account for their relative positions.

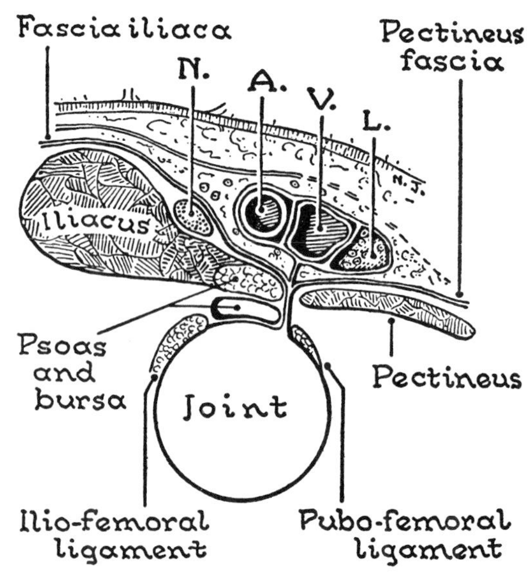

Fig. 412. The relationship of the femoral vessels and nerve to the hip joint.

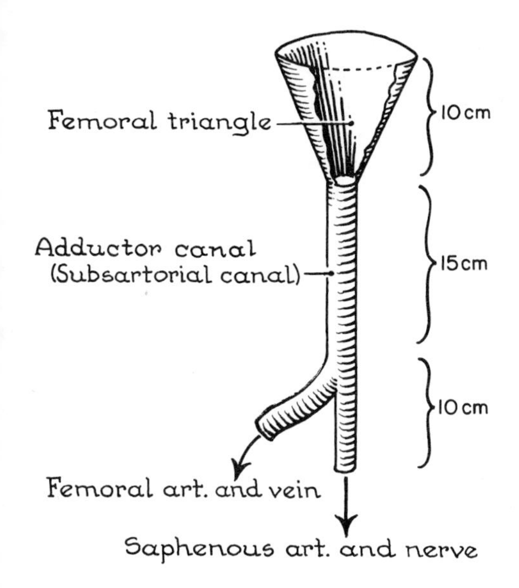

Fig. 413. The femoral "trough" and subsartorial "tunnel" and its two outlets.

When origin and insertion are approximated, the limbs are brought into the position that the tailor (L. sartor = a tailor) traditionally assumes when at work—hence its name, Sartorius.

Walls of the Adductor Canal. The adductor canal is a trough closed over with Sartorius (*fig. 437*).

The muscles of the lateral side of this long trough are: Iliopsoas and Vastus Medialis. The posteromedial ones are five **adductors,** which run from the pubis and ischium to the linea aspera (*fig. 414*). They spread out fanwise as they descend to their aponeurotic insertions. Add. Brevis appears in the interval between Pectineus and Longus; the Magnus appears between Longus and Gracilis (*fig. 414*). The Gracilis is inserted into the medial surface of the tibia deep to the Sartorius, and it lies immediately behind the Sartorius as the two cross the medial femoral condyle. The Magnus descends as far as the adductor tubercle.

Contents of the Adductor Canal. *The Femoral Artery and its (Small) Branch, the Saphenous Artery.* Their course is mapped out by a line joining the midinguinal point to the adductor tubercle. Its upper two-thirds maps out the course of the femoral artery, the lower third that of the saphenous artery. The femoral artery becomes the popliteal artery by passing through a gap or hiatus in the insertion of the Magnus 10 cm above the adductor tubercle. The saphenous artery becomes cutaneous (*fig. 414*).

The femoral artery and vein are so firmly bound together in a common sleeve that one cannot move without the other.

The *Nerve to Vastus Medialis* accompanies the femoral artery far into the canal, clinging to its muscle (*fig. 409*).

The Saphenous Nerve accompanies the femoral artery through the femoral triangle and adductor canal; then it accompanies the saphenous artery to the surface; here it joins and accompanies the great saphenous vein through the leg, ending half way along the medial border of the foot.

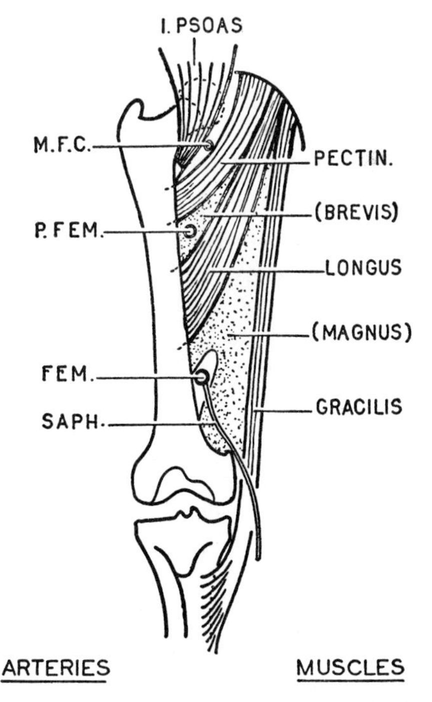

Fig. 414. The muscles posterior to the femoral and saphenous arteries. Adductor muscles spread fanwise to their insertions. Arteries course between them, the femoral artery piercing the Adductor Magnus.

Note on Comparative Anatomy. The vein is peculiar in man in not following the artery and nerve into the adductor canal. True, it still sends a communicating branch via the old route to the lower end of the femoral vein.

Regions of the Thigh (Details)

THEORETICALLY, the thigh may be regarded as divided into four regions, each separated from its neighbor by an intermuscular septum, and each containing a group of muscles and their nerve (*fig. 415*).

ACTUALLY, this scheme is departed from in the following main respects: Of the three abductors, all of which are supplied by the sup. gluteal nerve, two (Glutei Medius et Minimus) do not descend beyond the greater trochanter, while the third, the Tensor Fasciae Latae, spans the femur and gains attachment to the front of the lateral condyle of the tibia and to the side of the patella by means of the **iliotibial tract.** It falls to the three remaining groups to surround the femur.

The anterior muscles envelop and monopolize the shaft of the femur, except along its posterior border which alone is free and available to the medial and posterior muscles and to all intermuscular septa. The attachments of the muscles to this restricted border are necessarily aponeurotic; hence, the roughness of the border and its name–the *linea aspera*.

QUADRICEPS FEMORIS

The four heads of this very powerful knee extensor are the Rectus Femoris and the 3 Vasti (Medialis, Lateralis, and Intermedius). Of these the Rectus arises from the ilium and so also flexes the hip joint; the 3 Vasti arise from the shaft of the femur (*figs. 408, 409*).

This enormous muscle is inserted into a small area of bone, namely, to the tuberosity of the tibia—so the insertion must be tendinous. Where the tendon plays across the front of the lower end of the femur a sesamoid bone, the *patella*, is developed. The portion of the tendon distal to the patella is called the *ligamentum patellae.*

The Rectus Femoris (*fig. 408*) has a tendinous origin from the anterior inferior iliac spine (*straight head*) and to the acetabular margin for about 3 cm behind it (*reflected head*).

Internervous Line between the motor territories of the femoral and gluteal nerves: If a surgeon makes an incision vertically downward from the anterior superior spine, the Sartorius may be pulled medially, the Tensor Fasciae Latae laterally, and the anterior inferior spine exposed; thereafter the Glutei Medius et Minimus may be pulled backward, the Rectus and Iliacus forward, and the hip joint exposed. No motor nerve crosses this line.

The Three Vasti clothe the shaft of the femur.

Origins. The Vastus Intermedius arises by fleshy fibers from the anterior and lateral aspects of the femur. No muscle arises from the medial aspect, but the Vastus Medialis overlies it. The Vasti Lateralis et Medialis arise largely by aponeuroses from the lateral and medial lips, respectively, of the linea aspera as well as from the upward and downward continuation of these lips.

The most distal fibers of origin of the Intermedius, called the *Articularis Genu*, are attached to the synovial capsule of the knee joint to retract the capsule during extension of the knee.

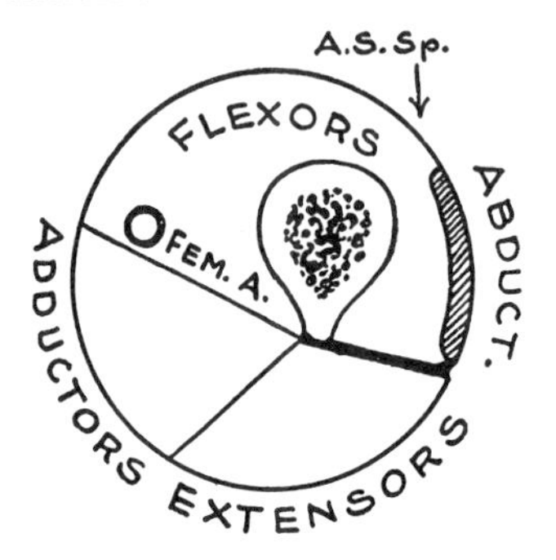

Fig. 415. The regions of the thigh; theoretical and actual.

Insertions on Patella (Detail). The Vasti Medialis et Lateralis are continuous at their insertions, and they occupy a plane between the Rectus Femoris and Vastus Intermedius, and all four are inserted into the base of the patella (*fig. 501*).

Distribution of the Femoral Nerve (L. 2, 3, 4). *Motor Distribution* (*fig. 491* on p. 362). *The Psoas* is supplied by the roots of the femoral nerve. *The Iliacus* is supplied while in the iliac fossa. *The Pectineus* is supplied near its lateral border by a twig that passes behind the femoral sheath. *The Sartorius* and *Rectus Femoris* are both supplied by one or two branches which enter them from 8 to 15 cm from the anterior superior spine. *The Three Vasti* receive short, stout branches at their upper ends. The Vasti Medialis and Lateralis both receive one long branch also, which enters their respective anterior borders near their middles.

Cutaneous Distribution (*fig. 400* and p. 311).

ARTICULAR DISTRIBUTION. The femoral nerve is distributed to the hip joint and to the knee joint, thus:

To the capsule of the *hip joint* via the nerve to Pectineus and either the nerve to Rectus Femoris or the nerve to Vastus Lateralis or directly from the femoral nerve itself.

To the capsule of the *knee joint* via the nerves to the three Vasti and the saphenous nerve (E. Gardner).

VASCULAR DISTRIBUTION. According to rule, sympathetic fibers run to all companion vessels.

24 HIP BONE AND GLUTEAL REGION

HIP BONE (OS COXAE)

Component Parts; Gluteal Aspect: Adductor Aspect; Acetabulum.

ILIUM

Dorsum Ilii.

ISCHIUM

Sciatic Notches; Body; Tuber Ischii; Ramus of Ischium.

PUBIS

Body; Superior Ramus; Pectineal Surface.

GLUTEAL REGION

Bony Landmarks.

Gluteus Maximus—Origin; Insertion; Iliotibial Tract; Functions.

Structures Deep to Gluteus Maximus—Door to Gluteal Region; At Lower Border of Piriformis; At Upper Border.

Sciatic Nerve; Greater Trochanter; Quadrate Tubercle; Superior Gluteal Nerve.

Gluteus Medius et Minimus; Tensor Fasciae Latae.

Quadratus Femoris; Obturator Internus and Gemelli; Two Gluteal Arteries; Posterior Cutaneous Nerve.

Five Regions of Back of Lower Limb.

HIP BONE (OS COXAE)

Component Parts. The hip bone is composed of three elements—ilium, ischium, and pubis. These three meet at the cup-shaped cavity for the head of the femur, the *acetabulum*, and are there united by a tri-radiate cartilage until the 16th year, when fusion takes place (*fig. 416*).

The *ilium* is a flat bone; the *ischium* and the *pubis* are irregular V-shaped bones. The ilium lies above the acetabulum; the ischium behind and below; the pubis in front and below. The site of fusion of ilium with pubis is the *iliopubic* (iliopectineal) *eminence* (*fig. 352*). The pubis and ischium surround the *obturator foramen*.

Gluteal Aspect (*Posterolateral Aspect*) *of the Hip Bone* (*fig. 417*). This aspect is wide above and is there bounded by the whole length of the iliac crest. Below, it is narrow and forms the lower part of the ischial tuberosity. It can be subdivided into three obviously different areas: (1) above is the fan-shaped *dorsum ilii*, (2) below is the rough, oval *ischial tuberosity*, and (3) between these is a large quadrate area, the *dorsum ischii*, behind the acetabulum.

Adductor Aspect (*Anterior Aspect*) *of the Hip Bone* (*fig. 418*). It affords origin to the adductor muscles.

Acetabulum, described with the hip joint on page 364.

ILIUM

The Ilium is fan-shaped (*figs. 416, 417*). It provides areas of origin for the Gluteus muscles (Minimus, Medius and Maximus on its **dorsum ilii**.

The upper border of the dorsum ilii, the *iliac crest*, is palpable throughout. In front it ends in the rounded *anterior superior iliac spine*, behind in the sharp *posterior superior iliac spine;* and from its most lateral point (about 6 cm behind ant. sup. spine) a broad *tubercle* projects downward. The pull of the abdominal muscles creates a traction epiphysis along the whole length of the crest (*fig. 416*).

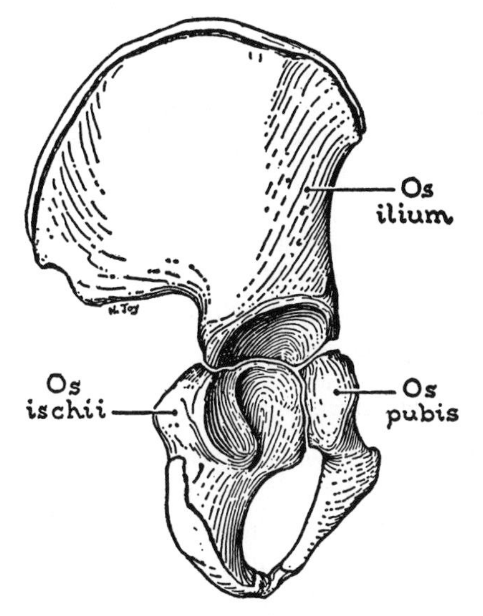

Fig. 416. A young hip bone.

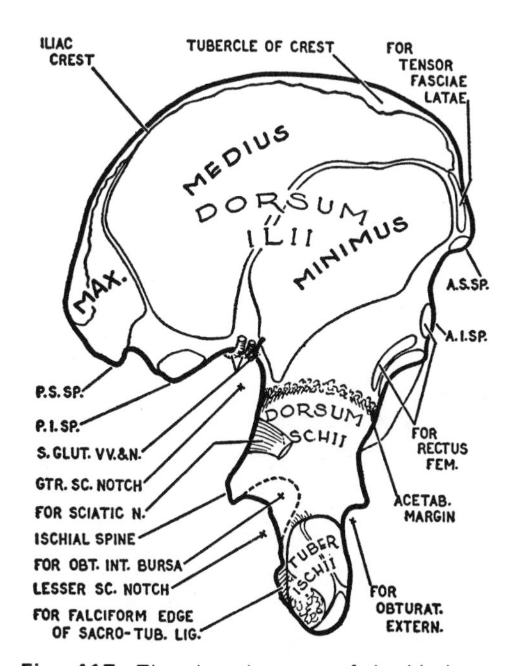

Fig. 417. The gluteal aspect of the hip bone.

The *anterior inferior spine* for the straight head of the Rectus Femoris and the *posterior inferior spine* (at the posterior limit of the sacro-iliac joint) lie below their respective superior spines. The upper part of the V-shaped *greater sciatic notch* completes the outline.

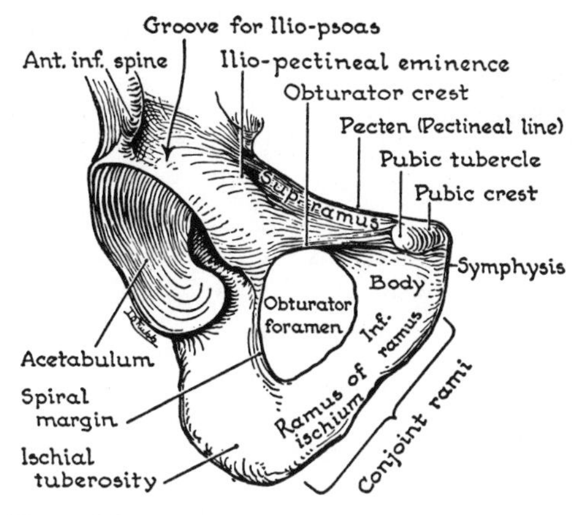

Fig. 418. The adductor aspect of the hip bone.

ISCHIUM

The ischium (os ischii) has three parts—a *body* adjoining the ilium, a *tuberosity projecting downwards from the body*, and *a ramus* passing from the tuberosity upward and forward below the obturator foramen.

The Sciatic Notches, greater and lesser, are separated from each other by a beak-shaped process, the *ischial spine*. This gives attachment to the sacrospinous ligament which, with the sacrotuberous ligament convert the sciatic notches into sciatic foramina (*fig. 361*).

The **Dorsum Ilii** (or Gluteal Surface of the Body) lies between the acetabular margin and the greater sciatic notch. It is crossed by the sciatic nerve (*fig. 417*).

The **Tuberosity of Tuber Ischii** is the oval mass of bone from which the hamstrings arise. Its chief function is to support the body when sitting.

The **Ramus of the Ischium** (*fig. 418*) is a flattened bar that unites with the inferior ramus of the pubis, fusing in the 8th year to form the *conjoint ramus* of the ischium and pubis. The conjoint rami of the two sides constitute the *pubic arch*.

PUBIS

The pubis (os pubis) (*fig. 418*) has three parts—a *body* lying medially, a *superior ramus* passing upwards and laterally from the body, and an *inferior ramus*, descending from the body and forming part of the pubic arch.

Body. Its *symphyseal surface* is elliptical, covered with cartilage, and 4 cm long. The pubic crest ends laterally at the *pubic tubercle*.

The Superior Ramus is a three-sided pyramid whose base forms a fifth of the articular part of the acetabulum.

The three borders are formed by two lines that diverge from the pubic tubercle: (1) the *pecten pubis* (pectineal line); (2) the other line is the spiral *margin of the obturator foramen*. The two ends of the spiral bound the inferior grooved surface of the ramus and account for two borders. In it run the obturator vessels and nerve.

The *pectineal surface* faces anterosuperiorly and gives origin to the Pectineus.

GLUTEAL REGION

Bony Landmarks (*fig. 419*). The whole length of the *iliac crest* and the *anterior superior spine* are easily palpated. The *posterior superior spine* lies deep to a visible, and rather pretty, dimple. A line joining them crosses the *2nd sacral spine* and marks the bottom of the dural and arachnoid sacs with the contained cerebrospinal fluid. Produced laterally, this line obviously must cross about the *center of the sacro-iliac articulation* (*fig. 419*).

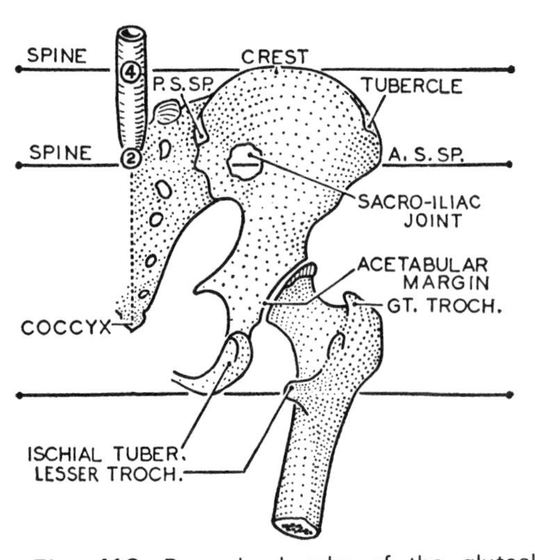

Fig. 419. Bony landmarks of the gluteal region.

A line joining the highest points on the iliac crests crosses the *4th lumbar spine*, and is the guide to that spine for physicians performing "spinal punctures" of the subarachnoid space.

The tip of the *greater trochanter* is palpable at the greatest width of the "hips"—as is the tip of the coccyx in the furrow between the buttocks about 3 to 4 cm behind the anus.

Gluteus Maximus

This very coarse-grained muscle, with a thin fascial covering, is rhomboidal (*fig. 420*). It arises from available surfaces of bones and ligaments along a strip between the posterior superior spine and the tip of the coccyx.

The lower border of the Gluteus Maximus extends from the tip of the coccyx across the tuber ischii and onwards to the shaft of the femur, but the tuberosity is uncovered when you sit down.

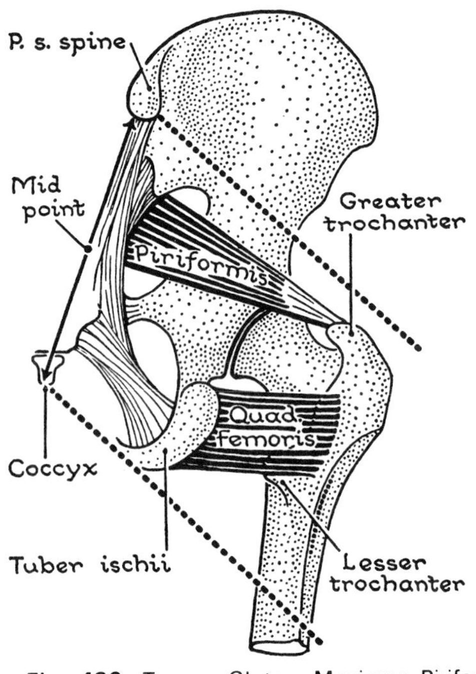

Fig. 420. To map Gluteus Maximus, Piriformis, and Quadratus Femoris. The lower border of the Piriformis is the key line of the gluteal region.

The upper parallel border runs from the post. sup. spine to pass above the greater trochanter.

Insertion. One-quarter of the Gluteus Maximus is inserted into the gluteal tuberosity of the femur, and the rest joins the Tensor Fasciae Latae to form the **iliotibial tract.** The tract descends between the circularly disposed layers of the fascia lata to be attached to the lateral condyle of the tibia in front of the axis of the knee joint.

Owing to this attachment (*fig. 507*), it assists in maintaining the extended knee joint in the extended position. You can feel it become prominent and taut on the side of the knee.

The Maximus attaches to the whole length of the linea aspera because the iliotibial tract also runs into the lateral intermuscular septum. Where Gl. Maximus plays across the greater trochanter, there must be a bursa (*fig. 427*).

Nerve and Vessels (*fig. 421*). The inferior gluteal nerve and vessels enter the middle of its deep surface, and other large neighboring arteries give it branches.

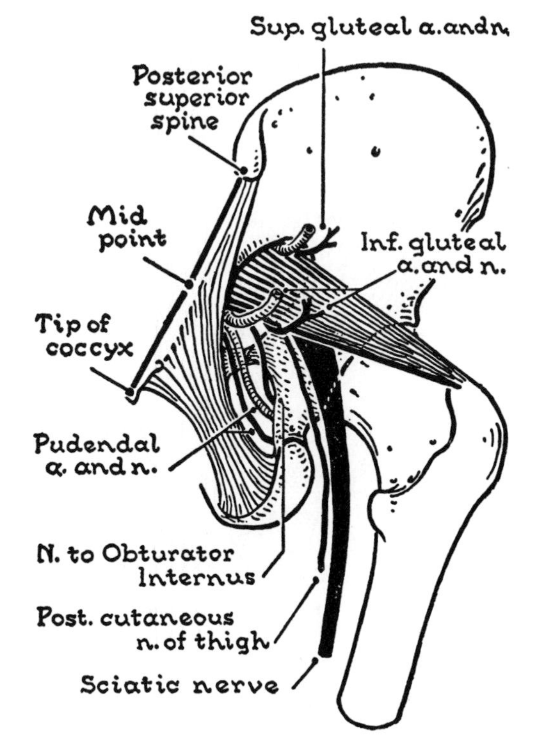

Fig. 421. Structures passing through the "door" to the gluteal region.

Functions. The Gluteus Maximus is the great extensor of the hip joint acting in rising from the sitting position, straightening, walking upstairs, and running; but very little in gentle walking. It is also a lateral rotator against resistance.

Structures Deep to Gluteus Maximus

The main door from the pelvis to the gluteal region is the greater sciatic foramen.

The **Piriformis** itself enters here and occupies a *key position.* Only the superior gluteal vessels and nerve enters *above* the muscle (*fig. 421*).

The Obturator Internus is the only structure that enters by the lesser foramen.

At the Lower Border of Piriformis these nerves and vessels enter (*fig. 421*):

5 {
- Sciatic nerve, which hides nerve to Quadratus Femoris.
- Inferior gluteal nerve.
- Inferior gluteal vessels.
- Posterior cutaneous nerve of thigh.

3 {
- Nerve to Obturator Internus.
- Internal pudendal vessels.
- Pudendal nerve.

The group of 3 cross the ischial spine (or the sacrospinous lig.), and swing forward through the lesser sciatic foramen into the ischiorectal fossa.

Various surgical approaches through the Gluteus Maximus to the region of the hip joint and sciatic nerve depend on a knowledge of the location of the lower border of Piriformis (*fig. 422*).

The Sciatic Nerve, from the lower border of the Piriformis, curves downward midway between the tuber ischii and the greater trochanter, covered by the Gluteus Maximus (*fig. 423.1*). It has a side of danger and a side of safety—its lateral side (*fig. 423*).

Greater Trochanter (*fig. 424*). The ten-

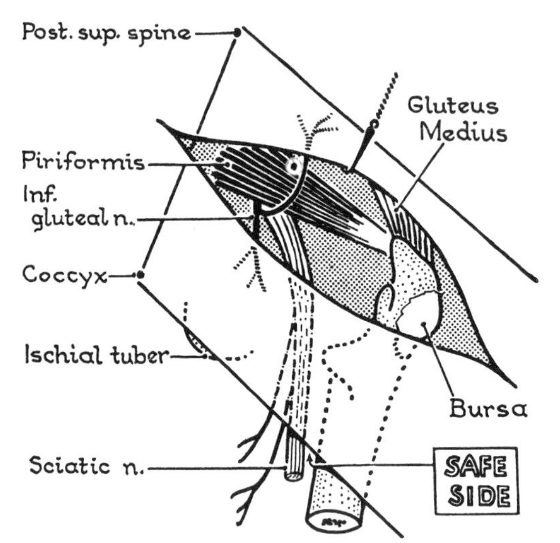

Fig. 422. Incision to be made through the Gluteus Maximus to expose the "key line" of the region, the lower border of Piriformis, indicated on the skin by joining a point midway between the posterior superior spine and the tip of the coccyx to the top of greater trochanter (*cf. fig. 420*).

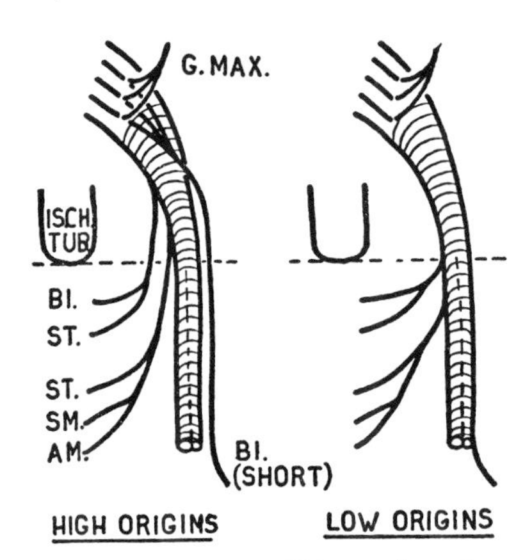

Fig. 423. Principle: Sides of safety and sides of danger.

dons of five muscles are inserted into this great traction epiphysis, and of these the two greatest are (1) the Gluteus Minimus, and (2) the Gluteus Medius. The others are (3) the Piriformis, (4) the Obturator Internus and its Gemelli, and (5) the Obturator Externus (to a pit on the medial surface).

The Superior Gluteal Nerve, accompanied by the superior gluteal vessels, passes through the greater sciatic foramen, above the Piriformis, in contact with the bony angle of the foramen, and it runs between the Gluteus Medius and Gluteus Minimus. It supplies the 3 abductors of the hip joint, viz., the Gluteus Medius, Gluteus Minimus, and Tensor Fasciae Latae.

Glutei Medius et Minimus. Their anterior borders are commonly fused. The Medius receives many fibers from the overlying fascia (*fig. 424.1*).

The Tensor Fasciae Latae arises from the ant. sup. iliac spine and adjacent outer lip

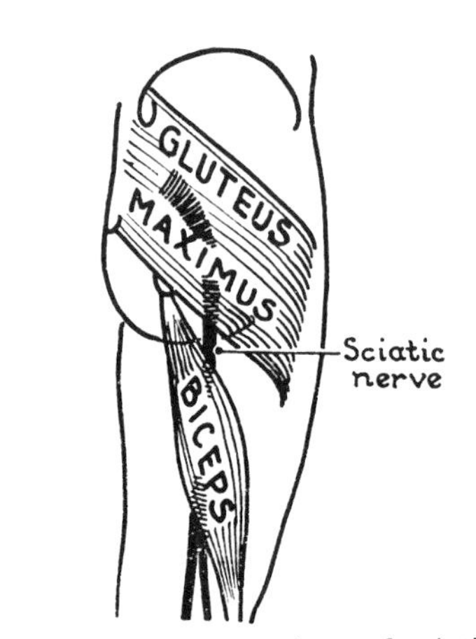

Fig. 423.1. Posterior relations of sciatic nerve.

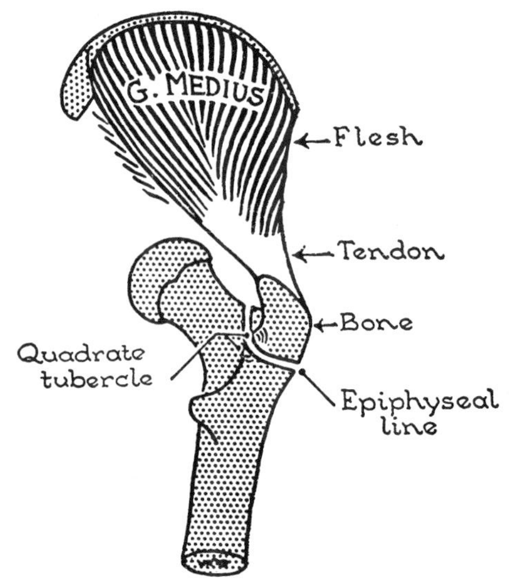

Fig. 424. The fleshy, tendinous, and bony portions of the Gluteus Medius.

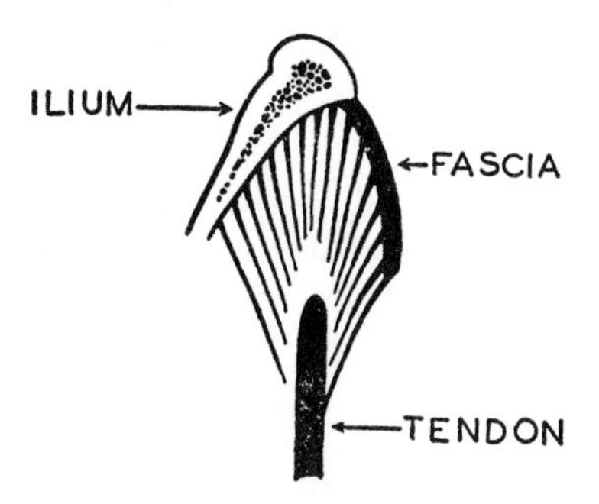

Fig. 424.1. The anterior part of the Gluteus Medius arises largely from the fascia covering it; hence, the thickness of this fascia.

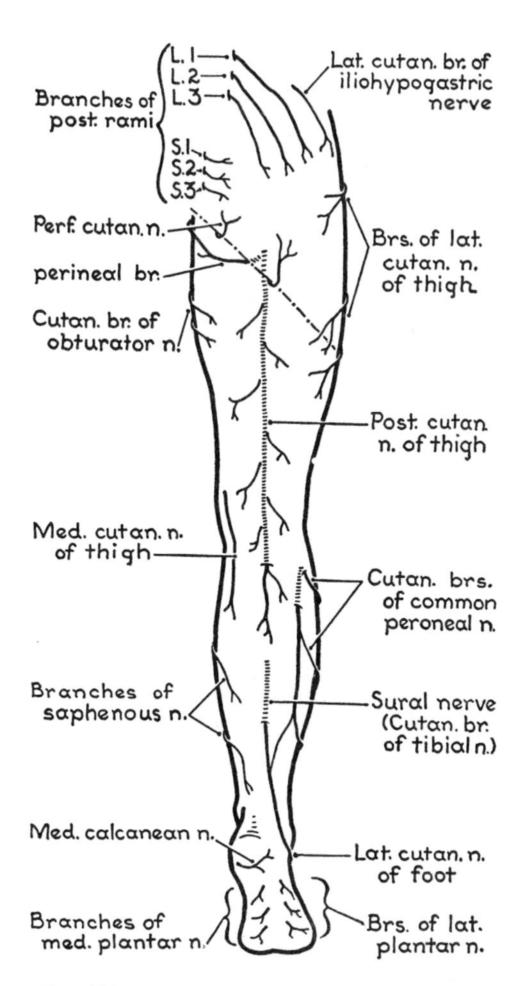

Fig. 425. Cutaneous nerves of back of lower limb.

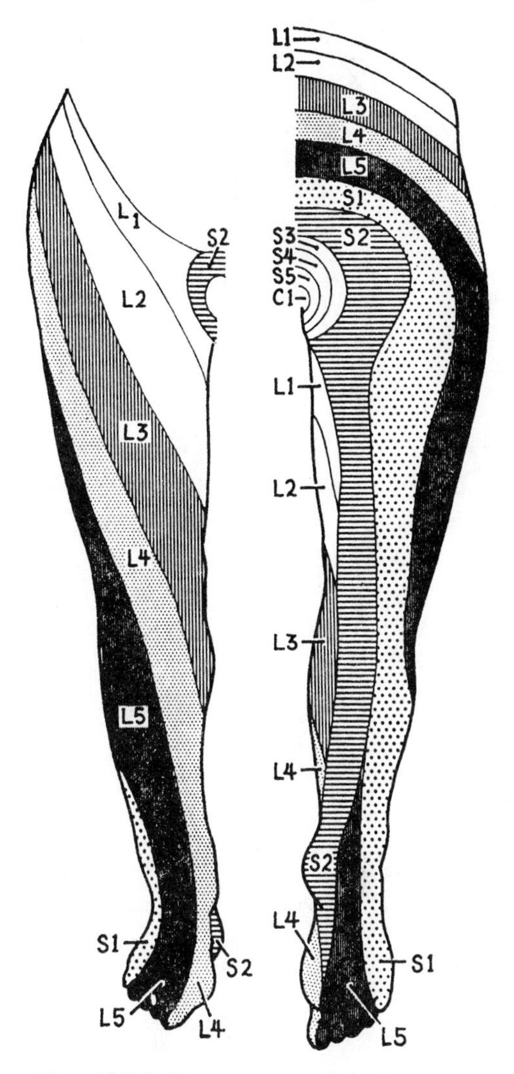

Fig. 425.1 Dermatomes of the lower limb.

of the iliac crest. The iliotibial tract is its aponeurosis of insertion (p. 323).

The 6 Small Lateral Rotators

(1) The Quadratus Femoris (*figs. 420, 427*). This unimportant muscle is part of the "red carpet" for the sciatic nerve as it runs down into the thigh. Its nerve arises from the front of the sciatic nerve and supplies a twig to the hip joint. **(2) The Obturator Internus and (3, 4)** its **Gemelli** form the carpet above the Quadratus. The *Obturator Internus* arises from the pelvic surface of the obturator membrane and most of its periphery of bone. It makes a sharp turn as it

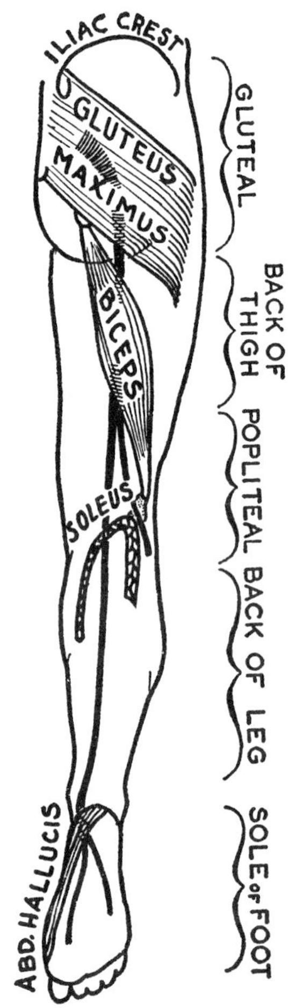

Fig. 426. The five regions of the back of the limb, the four muscles separating them, and the tibial division of the sciatic nerve.

passes through the lesser sciatic foramen (*fig. 353*) separated from the bone by a *bursa*. Its tendon inserts on the upper border of the greater trochanter.

The Superior Gemellus arises from the upper margin of the lesser foramen (ischial spine); the Inferior from the lower margin of the lesser foramen (ischial tuberosity); and their fleshy fibers joins the tendon of the Obturator Internus.

(5) Piriformis also is a lateral rotator of the hip joint, as is the deeply placed **(6) Obturator Externus** which spirals from its origin on the obturator membrane to the back of the femoral neck and is deeply situated.

The Gluteal Arteries, superior and inferior, are responsible for supplying the entire region, their largest branches being *muscular*.

The Posterior Cutaneous Nerve of the Thigh runs vertically behind the line of the sciatic nerve (*fig. 425*) and has two large branches, the *gluteal* and *perineal*. It ends on the calf.

Dermatomes, see figure 425.1.

Five Regions of the Back of the Limb (*fig. 426*). Four muscles delineate 5 regions: Gluteus Maximus, Biceps, upper border of Soleus, and Abductor Hallucis. Passing deep to all four is the tibial division of the sciatic nerve.

25 BACK AND MEDIAL REGION OF THIGH

BACK OF THIGH AND POPLITEAL FOSSA

Floor of Region; Contents of Back of Thigh.
Hamstring Muscles—Definition; Insertions; Origins; Short Head of Biceps; Hybrid Muscles.
Nerves and Arteries
Sciatic Nerve: Posterior Relations; Termination; Anterior Relations; Posterior Cutaneous Nerve; Arteries.
Popliteal Fossa (Back of Knee)
Definition; Palpable Tendons; Contents.
Tibial Nerve: Branches.
Relationships.
Popliteal Artery and Its Branches.
Common Peroneal Nerve: Branches.
Small Saphenous Vein; Lymph Nodes.

MEDIAL REGION OF THIGH

Muscles—Pectineus; Adductors Longus, Gracilis, Brevis, Magnus; and Obturator Externus.
Origins; Insertions.
Three Hernial Sites.
Obturator Nerve and Its Branches; Obturator Artery; Profunda Femoris Artery and Its Distribution; Lateral Circumflex Artery; Four Perforating Arteries.
Branches of Popliteal Artery; Primary Routes of Arteries.

BACK OF THIGH AND POPLITEAL FOSSA

The Floor of the Region is flat (*fig. 427*). It is formed by the Adductor Magnus and Vastus Lateralis, which is covered with the strong lateral intermuscular septum.

Contents of the Back of the Thigh:

1. The hamstring muscles—
   Long head of Biceps Femoris
   Semitendinosus
   Semimembranosus
2. The short head of Biceps Femoris.
3. The sciatic nerve.
4. The posterior cutaneous nerve.
5. Certain vessels.

The Hamstring Muscles. A hamstring is one that (1) arises from the tuber ischii (2) is inserted into one or other of the two bones of the leg, and (3) is supplied by the tibial division of the sciatic nerve. They extend the hip joint and flex the knee joint, but you cannot do both fully at the same time—try it!

The Short Head of Biceps arises by fleshy fibers from the linea aspera and from the adjacent part of the lateral supracondylar line and lateral intermuscular septum (*fig. 427*). It receives a branch of the lateral or peroneal division of the sciatic nerve.

When the knee is flexed, the Biceps, being attached to the head of the fibula, rotates the leg laterally; the Semimembranosus and Semitendinosus, being attached to the medial side of the tibia, rotate the leg medially. (Leg = crus, L. = the limb from knee to ankle.)

Insertions. See figs. 428.1, 428.2. In addition to the direct bony attachments, the tendons send aponeurotic expansions to the fascia around and below the knee. *Origins*—see *figure 429*.

Hybrid Muscles. The Biceps clearly is an amalgamation of two muscles. The long head belongs developmentally to the front of the limb; the short head, to the back.

The Adductor Magnus also is hybrid. The part arising from the pubic arch is supplied by

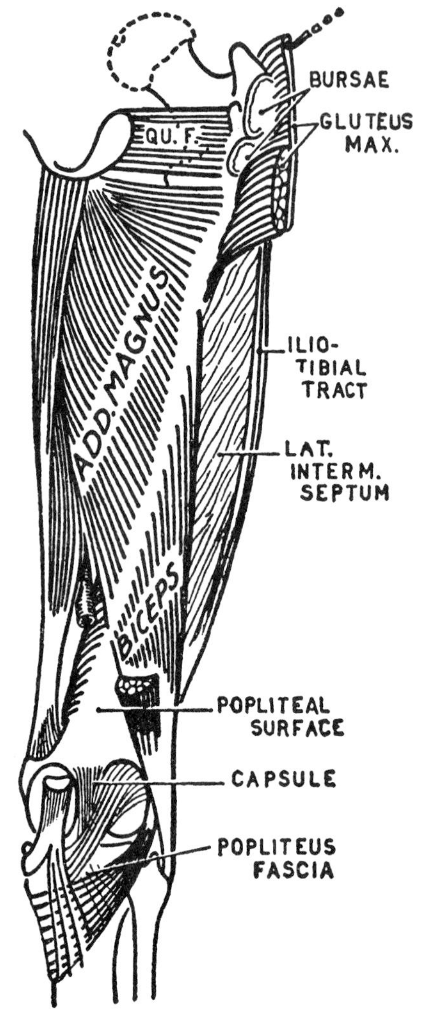

Fig. 427. The "floor" of the back of the thigh and politeal fossa.

the obturator nerve, whereas the part arising from the tuber ischii is supplied by the tibial division of the sciatic nerve (*fig. 428*).

The Pectineus is supplied by both femoral and obturator nerves.

Nerves and Arteries

The Sciatic Nerve (*Nervus Ischiadicus*) is only accessible deep in the angle between the Gluteus Maximus and the long head of the Biceps (*fig. 426*).

Posterior Relations. Above, the Maximus covers it; below, the Biceps (long head).

Termination. Deep to the Biceps, the sciatic nerve divides into its two terminal branches—*tibial and common peroneal nerves* (med. and lat. popliteal nerves). (Gk. perone = L. fibula = a pin or skewer.) They are merely bound together by loose areolar tissue right up to the sacral plexus (*fig. 430*).

The nerve to the short head of the Biceps springs from the lateral side of the sciatic nerve (peroneal division). The branches to the hamstrings spring from its ischial or medial side (*fig. 423*).

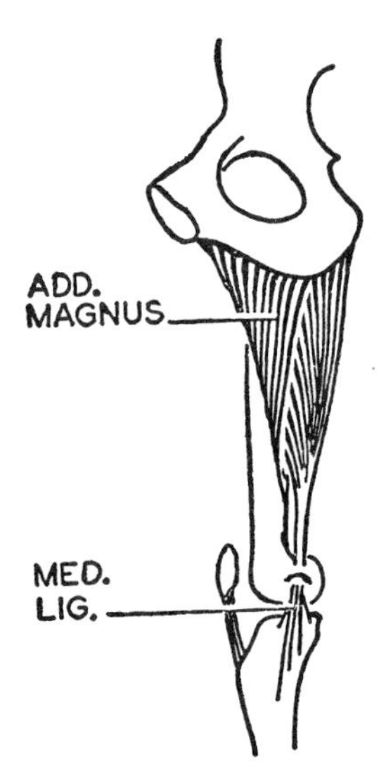

Fig. 428. The ischial part of the Adductor Magnus almost "made it" as a hamstring.

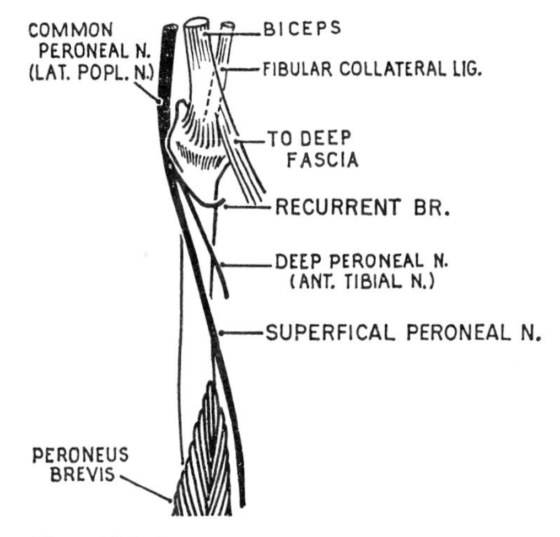

Fig. 428.1. Insertion of Biceps Femoris.

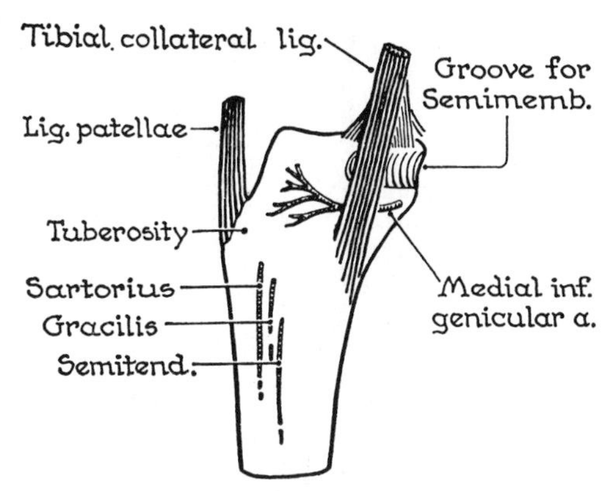

Fig. 428.2. Insertion of four tendons into tibia.

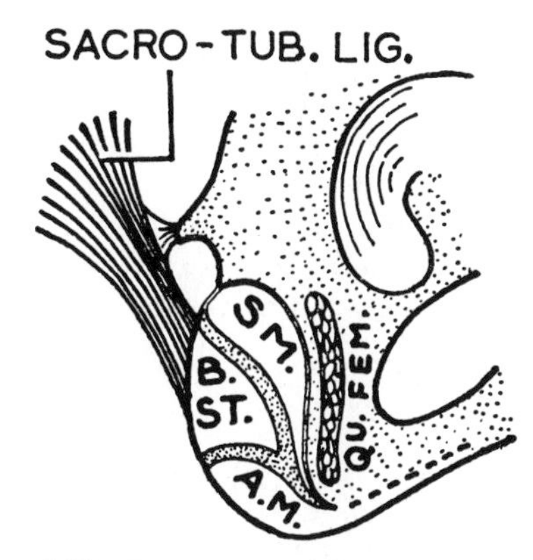

Fig. 429. The tuber ischii and the structures attached to it. (For reference, not memorization.)

AXIOM: The side from which a motor nerve leaves its parent stem is constant—it leaves from the side nearest the muscle it supplies —but the level at which it leaves is variable.

Posterior Cutaneous Nerve of the Thigh (*fig. 425*). The posterior cutaneous nerve (S. 1, 2) gives cutaneous twigs to right and to left, ending on the calf.

Arteries to back of thigh (see page 334).

Popliteal Fossa (*Back of the Knee*)

The Floor of the Popliteal Fossa (*fig. 427*) is formed by: (1) the popliteal surface of the femur; (2) the capsule of the knee joint, and (3) the fascia covering Popliteus.

The fossa is a potential one because the circular fibers in the deep fascia bandage the structures together. Dissection converts it into a diamond-shaped recess (*fig. 431*).

Palpable Tendons. While sitting down and using both hands to feel, intermittently press the heel backward against a chair leg, noting that: (1) laterally the *Biceps tendon* becomes taut and is easily followed to the head of the fibula; do not mistake it for the prominent posterior border of the *iliotibial tract* which runs a full finger's breadth in front of it. The *Biceps tendon* also becomes taut on lateral rotation of the leg (knee being flexed); it is the only lateral rotator. (2) medially, the *Semi-*

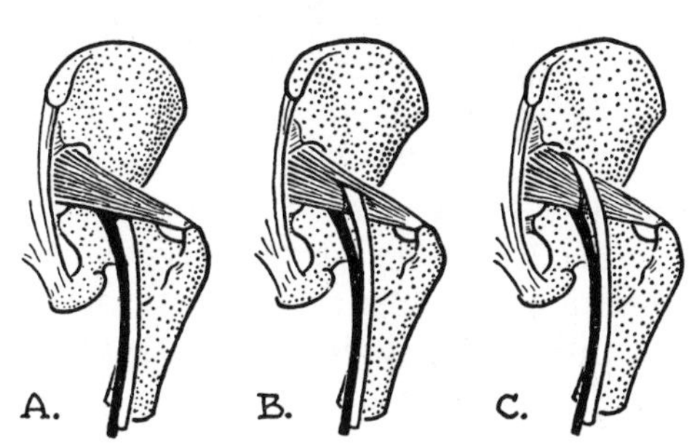

Fig. 430. The relationship of the sciatic nerve to the Piriformis: *A.*, in 87.5 per cent of 420 limbs; *B.*, in 12 per cent the peroneal division passed through the Piriformis; *C.*, in 0.5 per cent (i.e., in both limbs of one subject) it passed above.

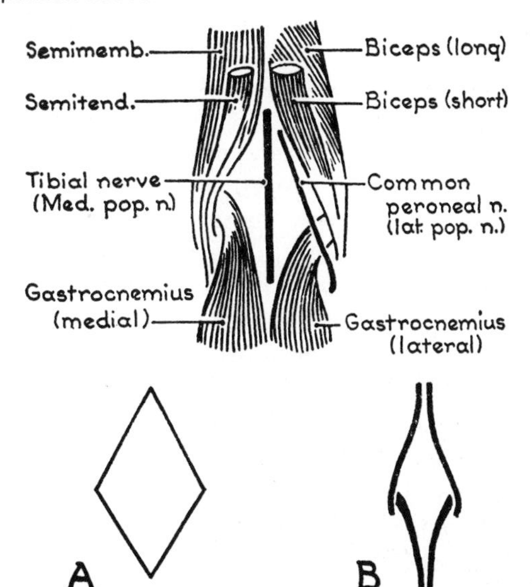

Fig. 431. Boundaries of the popliteal fossa.

tendinosus tendon is felt to spring, like a bow string, backward from the rounder *Semimembranosus tendon.*

Contents. The fossa contains four important structures:

1. tibial nerve (med. popliteal n.).
2. and 3. popliteal artery and vein.
4. common peroneal nerve (lat. popliteal n.).

Relationships. As the tibial nerve and the popliteal vessels are passing through the narrow ravine bounded on each side by a femoral condyle and a head of the Gastrocnemius, they are crowded one behind the other; the order from the surface to the floor of the fossa being retained—nerve, vein, artery. The customary rule—nerve, artery, vein—does not hold here.

The Tibial Nerve (*Med. Popliteal N.*) (L. 4, 5, S. 1, 2, 3), which is the larger of the two terminal branches of the sciatic nerve, passes from the upper to the lower angle of the popliteal fossa, and then descends to the ankle where it enters the medial side of the sole.

Branches.

1. *Motor:* to local muscles—Plantaris, medial and lateral heads of Gastrocnemius, Soleus, and Popliteus (*fig. 432*). Deep to it are the popliteal vessels (*fig. 432*).

2. *Articular* } See below under common peroneal nerve.
3. *Cutaneous* }

The Popliteal Artery and Vein are intimately bound together in a fascial sheath. The artery divides into its two terminal branches, the *anterior and posterior tibial arteries* behind the tibia (*fig. 432*); soon after, the latter gives off the peroneal a. The arteries below the knee are accompanied by venae comitantes; so, the *popliteal vein* begins as an assembly of veins.

Anterior Relations (*fig. 433*).

Branches: Terminal: ant. and post. tibial arteries.

Collateral: (1) Cutaneous (2) muscular, and (3) articular (*fig. 438*).

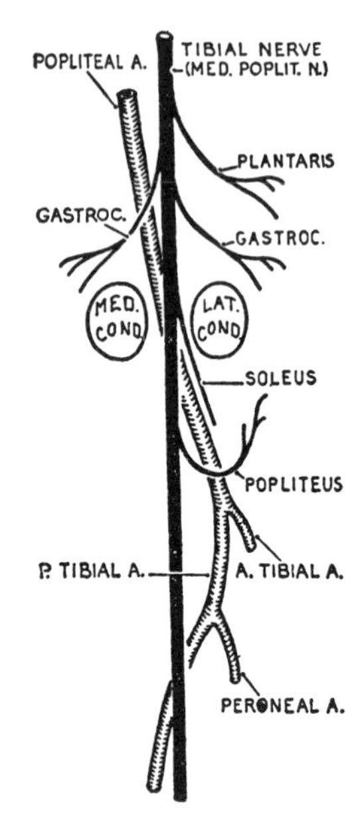

Fig. 432. The tibial nerve (med. popliteal n.)—its side of safety and side of danger and its relation to the bowed arterial stem.

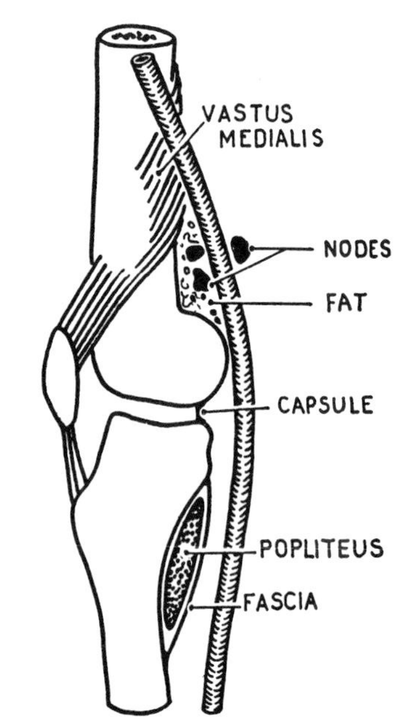

Fig. 433. Anterior relations of popliteal artery, diagrammatic, medial view.

The Common Peroneal Nerve (Lat. popliteal n.) (L. 4, 5, S. 1, 2), separates from the tibial nerve about the middle of the thigh and ends lateral to the neck of the fibula by dividing into two terminal branches—*deep* *peroneal* and *superficial peroneal nerves* (ant. tibial and musculocutaneous nn.).

It follows the Biceps tendon, crossing in turn: the Plantaris, Gastrocnemius (lateral head), and the back of the head of the fibula coated with a thin veneer of Soleus. Here it is readily palpated by fingertips drawn horizontally across it.

Branches. Terminal: Deep and superficial peroneal nerves (*fig. 428.1* and p. 339).

Collateral: Articular and *cutaneous branches* spring from both tibial and peroneal nerves.

The cutaneous branch of the tibial nerve, called the *sural nerve*, lies with the small saphenous vein in the furrow between the two bellies of the Gastrocnemius. It is joined by a branch of the common peroneal nerve (called the *communicating peroneal nerve*) to travels distally on the Gastrocnemius.

The Small (short) **Saphenous Vein** is the lateral continuation of the dorsal venous arch of the foot. It passes below and then behind the lateral malleolus and is accompanied in its course by the sural nerve. It pierces the popliteal fascia and, after dividing, ends in the popliteal and profunda femoris veins.

A large superficial branch connects it to the upper end of the great saphenous vein.

Lymph Nodes. The most distal nodes drain into the popliteal nodes, and have little practical value.

The popliteal nodes, five or six in number, lie in the fat around the popliteal vessels. The efferents from the popliteal nodes follow the femoral vessels to the deep inguinal nodes.

MEDIAL REGION OF THE THIGH

Muscles

The muscles of the medial or adductor region of the thigh arise collectively from the anterior aspect of the hip bone and the obturator membrane (*fig. 434*).

From this compact area of origin—the details of which may be safely ignored—they spread out fanwise to a linear insertion that extends the length of the femur and beyond to the tibia (*fig. 435*). They

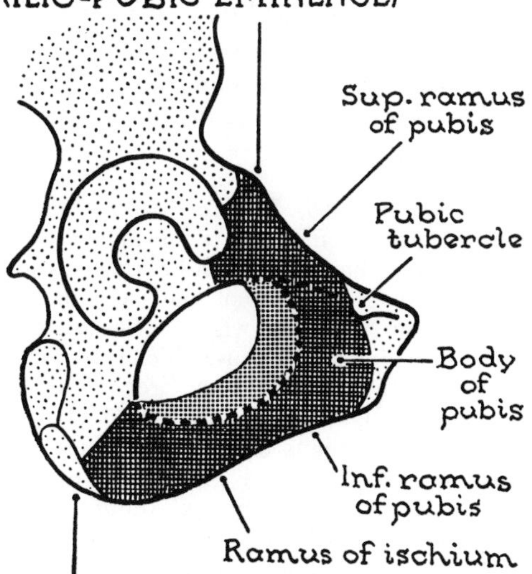

Fig. 434. Compact area of origin of Adductors.

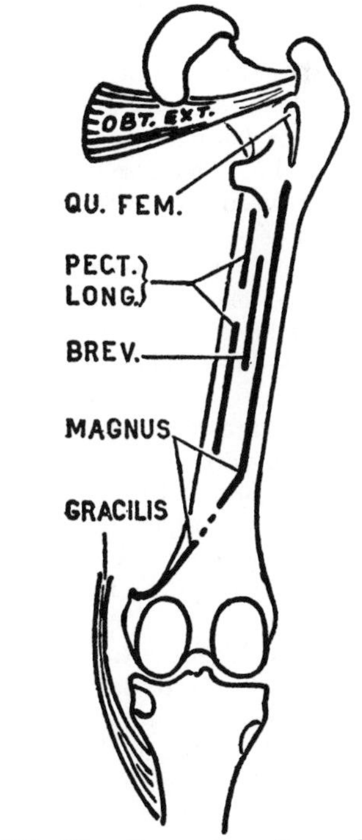

Fig. 435. The linear insertion of the Adductors.

share their chief actions—adduction, medial rotation of the hip joint, and their nerve—obturator. The profunda femoris and obturator arteries nourish them.

The large muscle mass consists of the following six individual muscles: Pectineus, Adductor Longus, Gracilis, Adductor Brevis, Adductor Magnus, and Obturator Externus. Read but do not memorize the following details.

Origins (*fig. 495*). The *Pectineus, Adductor Longus,* and *Gracilis* have a continuous curvilinear origin that extends from the iliopubic eminence to the ramus of the ischium. The origin of the Pectineus, which is fleshy, meets that of the Adductor Longus, which is tendinous at the public tubercle; the Gracilis has an aponeurotic origin. If the thigh is abducted, the tendon of the Longus becomes prominent and palpable and acts as a guide to the pubic tubercle. The *Adductor Brevis, Adductor Magnus,* and *Obturator Externus* arise by fleshy fibers in successively deeper, overlapping planes. (Some of the upper fibers of Adductor Magnus occasionally form a separate muscle, *Adductor Minimus.*)

Insertions. The restricted and therefore fibrous insertions of these six muscles are (*fig. 435*):

The *Pectineus:* to the line running from the lesser trochanter toward the linea aspera.

The *Adductor Longus:* to almost the whole length of the linea aspera in line with the Pectineus. (Between the Pectineus and Longus the Brevis can be seen.)

The *Gracilis:* to the medial surface of the tibia below the level of the tuberosity and between the insertions of the Sartorius and Semitendinosus. The Gracilis is the only muscle of the adductor group to cross the knee joint. (Between the Longus and Gracilis the Magnus can be seen) (*fig. 414*).

The *Adductor Brevis:* to the lower part of the pectineal line and upper part of the linea aspera. Its upper part is overlapped by the Pectineus; its lower part by the Adductor Longus.

The *Adductor Magnus:* to the linea aspera, extending upward onto its lateral continuation (the gluteal tuberosity), and downward onto its medial continuation (the medial supracondylar line). In fact, it extends from the level of the lesser trochanter above, where it is continuous with the Quadratus Femoris, to the adductor tubercle below. The portion of the Adductor Magnus that arises from the tuber ischii does so by tendon, belongs developmentally to the hamstring muscles, is supplied by the tibial nerve, and is inserted mainly by a palpable tendon into the adductor tubercle and into the supracondylar line just above it.

The *Obturator Externus:* to the trochanteric fossa. It passes below the head of the femur and grooves the back of the neck. It will be studied with the hip joint.

Three Hernial Sites

Three hernial sites of clinical significance are illustrated in *figure 436.*

Nerves and Vessels

The Obturator Nerve (*fig. 491* on page 362), like the femoral nerve, is derived from L. 2, 3, and 4, and has motor, cutaneous, articular, and vascular distribution.

On passing through the obturator foramen the nerve divides into an anterior and a posterior division which supply the six adductors (*fig. 437*).

The sole *Cutaneous Branch* is the continuation of the nerve to the Gracilis. It reaches the surface about the middle of the

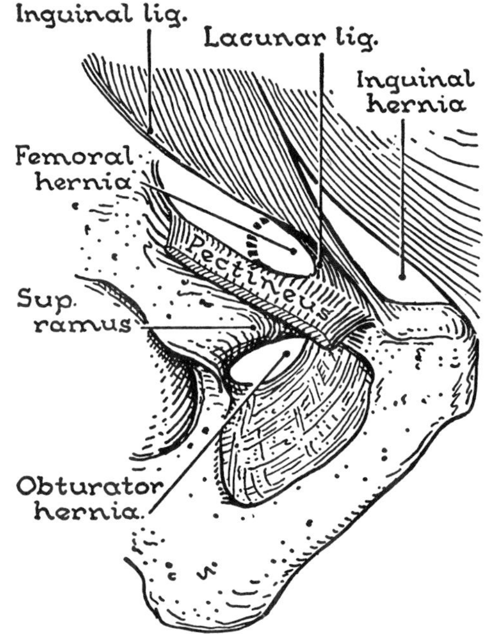

Fig. 436. Three hernial sites, and the structures separating them.

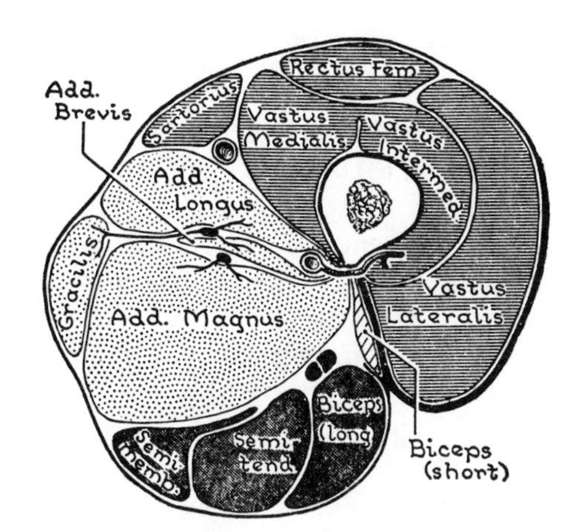

Fig. 437. Cross section of the thigh. (1) Adductor Longus separates the femoral and profunda femoris arteries; (2) Adductor Brevis separates the two divisions of the obturator nerve; and (3) the perforating arteries hug the bone.

thigh, where it supplies a restricted area, but it may extend to the calf.

Articular Branches supply both the hip joint and the knee joint. The following details are for reference only.

The Branch to the Hip springs from the main nerve within the obturator canal. Its numerous twigs ramify in the pubofemoral lig. Some reach the synovial membrane and also run along the lig. of the head of the femur.

The Branch to the Knee is the continuation of the n. to Adductor Magnus. It runs down on the medial side of the popliteal vessels, gives *vascular branches* to them and ramifies in the posteromedial part of the fibrous capsule. An articular branch of the saphenous nerve may reach the knee joint. (E. Gardner.)

An *Accessory Obturator Nerve* sometimes follows the medial border of the Psoas over the superior ramus and rejoins the main nerve deep to the Pectineus. It may supply a twig to the hip joint and to the Pectineus.

The Obturator Artery assists the profunda artery to supply the adductors.

Articular twigs run through the acetabular foramen to the acetabular fossa and often a branch traverses the lig. of the head of the femur (lig. teres, *fig. 496*, and p. 366).

The Profunda Femoris Artery usually arises from the lateral side of the femoral artery about 4 cm below the inguinal lig. At the apex of the femoral triangle it lies behind the femoral vessels with its own vein. It descends among the adductors (*figs. 437* and *438*).

Distribution. Its various branches supply most of the *muscles* of the thigh; *articular branches* to the hip and knee joints; and a *nutrient* branch (or two) to the femur; and they effect numerous *anastomoses.* Its named branches are:

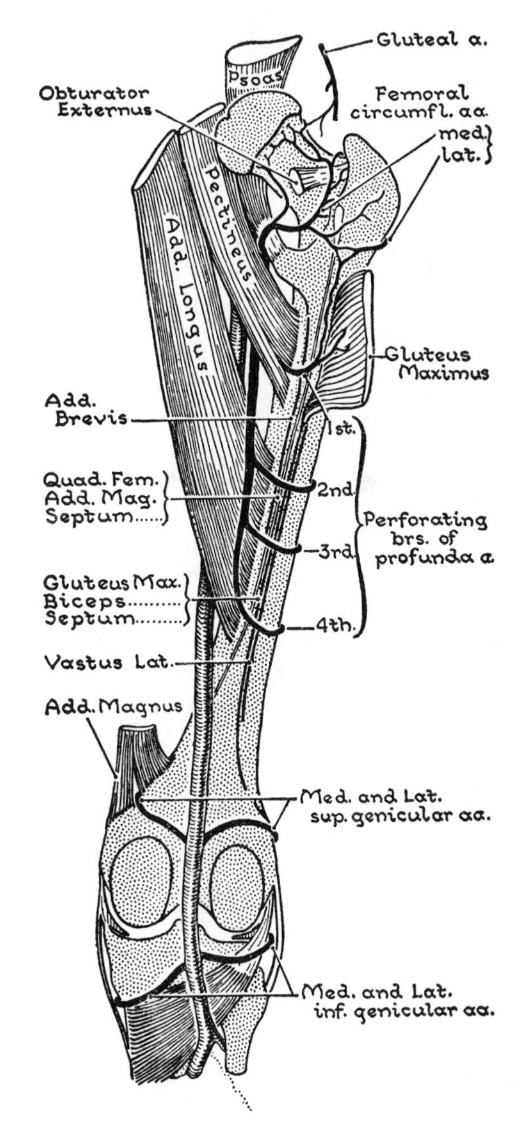

Fig. 438. Arteries of thigh and knee from behind.

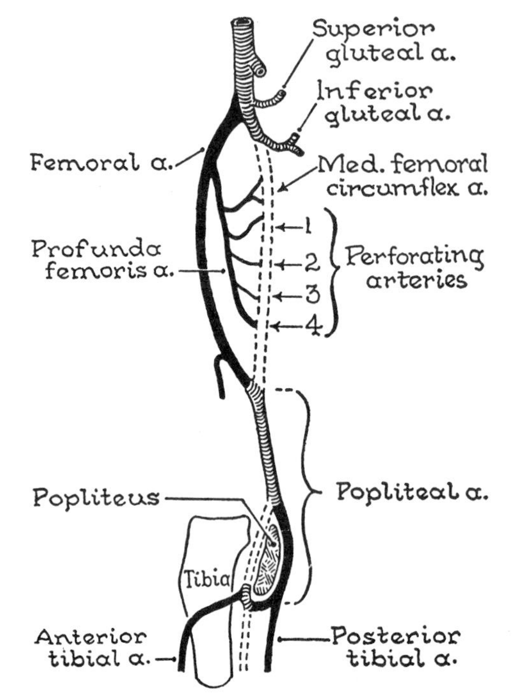

Fig. 439. Development of the main arterial trunk: (*hatched segments* are primary; *solid black* are secondary; *clear* disappear). (After Senior.)

1. Lateral femoral circumflex,
2. Medial femoral circumflex,
3. 1st, 2nd, 3rd and 4th perforating,
4. Muscular (unnamed branches).

The *Lateral Femoral Circumflex Artery* is a large artery that runs laterally between the branches of the femoral nerve, and divides into three branches.

The *ascending* and *transverse* branches anastomose in the gluteal region, and they send branches along the front of the neck of the femur to the head. The large *descending* branch follows the anterior border of the Vastus Lateralis and anastomoses at the knee.

The *Medial Femoral Circumflex Artery* (*fig. 438*). The chief duty of the medial circumflex a. is to supply the neck and head of the femur. An *articular* branch passes through the acetabular foramen with a branch of the obturator a. (p. 333). Other branches are *muscular*, and *anastomotic*.

The medial circumflex a. divides into (1) a transverse branch, and (2) an ascending branch, which anastomoses behind the hip.

The *Four Perforating Arteries* encircle the shaft of the femur, hugging it so closely that they must surely be torn if the shaft is fractured. They perforate any muscle they encounter (*figs. 437–439*).

Distribution. The perforating arteries are essentially muscular in this most muscular of regions, but one also supplies the nutrient artery to the femur.

Anastomoses. The term *cruciate anastomosis* is applied to the union of the medial and lateral femoral circumflex arteries with the inferior gluteal artery above and the 1st perforating artery below (*fig. 438*).

Primary Route of Arteries. In the embryo the primary arterial trunk arises as a branch of the internal iliac a., and passes down the back of the limb, accompanying the sciatic nerve. A vessel, which becomes the external iliac and femoral arteries, grows down the front of the thigh and joins the primary trunk above the knee (*fig. 439*). The proximal part of the primary trunk is resorbed, except in the gluteal region.

*Segmental Innervation of Muscles of Hip and Thigh**

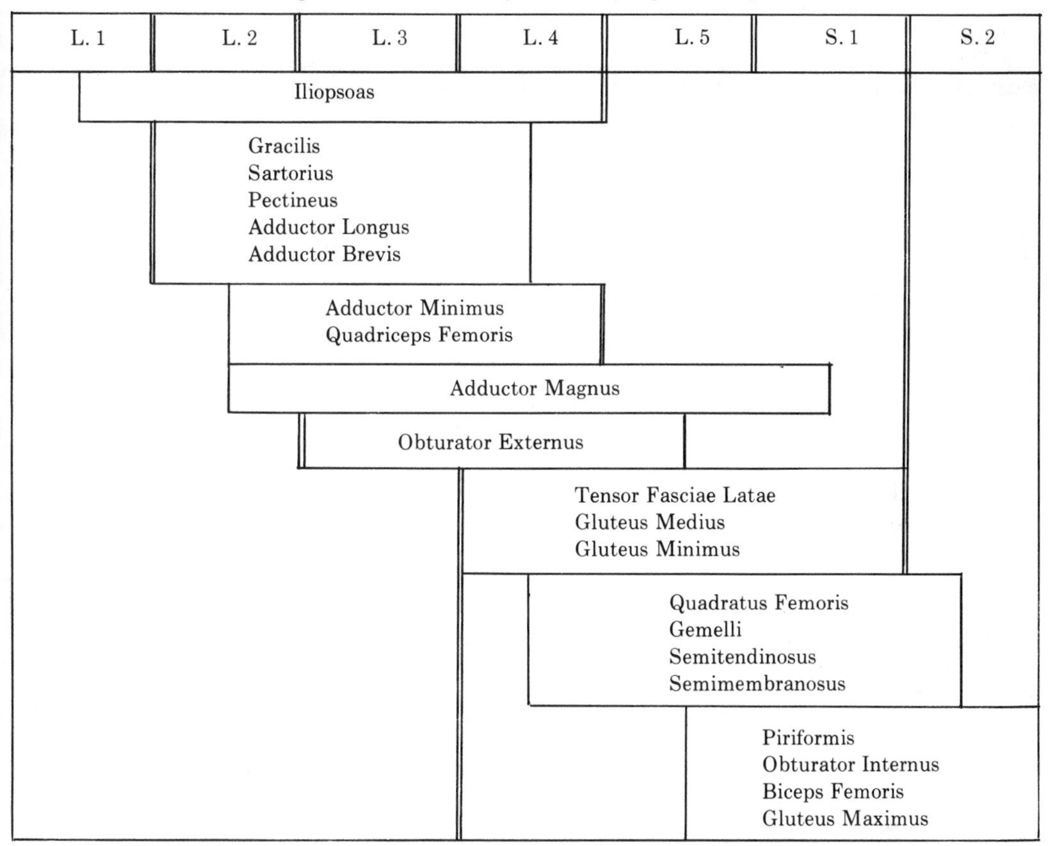

* Modified after Bing; and Haymaker and Woodhall. (For Leg and Foot, page 361.)

26 LEG AND DORSUM OF FOOT

MEDIAL SURFACE AND BODY OF TIBIA

Definitions and Boundaries.
Medial Crural Region and Body of Tibia; Tendinous Expansions of Sartorius, Gracilis and Semitendinosus.
Tibial Collateral Ligament; Medial Malleolar Vessels; Great Saphenous Vein.
Terminology of Surfaces and Borders of Tibia.
Interosseous Border of Tibia; Lower End; Lower Surface.

ANTERIOR CRURAL REGION AND DORSUM OF FOOT

Great Arterial Trunk and Its Branches.
Common Peroneal Nerve; Deep Peroneal Nerve; Superficial Peroneal Nerve.
Muscles of Anterior Crural Region—Tibialis Anterior; Extensor Digitorum Longus and Peroneus Tertius; Extensor Hallucis Longus; Extensor Digitorum Longus and Peroneus Tertius; Extensor Hallucis Longus; Extensor Digitorum Brevis and Hallucis Brevis.
Deep Fascia; Extensor Retinaculum; Peroneal Retinaculum.

PERONEAL (FIBULAR) REGION

Palpation; Boundaries and Contents.
Muscles—Peronei Longus et Brevis: Origins; Insertions; Comparative Anatomy.
Functions of Fibula.
Tibiofibular Joints.
Body of Fibula: Surfaces and Borders.
Ossification.

POSTERIOR CRURAL REGION

Bony Framework
Muscles—Gastrocnemius; Plantaris; Soleus; Popliteus; Tendons of Ankle; Tibialis Posterior; Flexor Digitorum Longus; Flexor Hallucis Longus.
Arteries and Nerves—Tibial Nerve and Its Branches; Posterior Tibial Artery; Peroneal Artery.
Small Saphenous Vein; Cutaneous Nerves.

The leg or *crus* is the segment of the lower limb between the knee and the ankle. Its main regions (fascial compartments) are (*figs. 440, 441*):

1. Anterior crural,
2. Posterior crural (much larger),
3. Lateral or "fibular" or "peroneal."

MEDIAL SURFACE AND BODY OF TIBIA

By palpation note that the medial surface of your own tibia is subcutaneous and smooth from the medial malleolus below to the medial condyle above; that it is bounded in front by the subcutaneous *anterior border*, the "shin," throughout; and that it is bounded behind by the subcutaneous *medial border* of the tibia (*fig. 442*).

Already you have felt 2 of the 3 tendons at the medial side of the knee (p. 329). Sartorius tendon is much more difficult to feel.

The Tendinous Expansions of the Sartorius, Gracilis, and Semitendinosus find attachment to the medial surface of the tibia below the level of the tuberosity. Each represents a different region of the thigh—a curious but insignificant detail; each is supplied by a different nerve (femoral, obturator, or sciatic); each passes across the medial ligament of the knee to reach its insertion, a *bursa* intervening. Much more important to surgeons is the spread of fibers from these tendons which reinforce the medial side of the knee capsule.

The Tibial Collateral Ligament (*medial lig. of the knee*) attaches to the tibia as shown in *figures 442* and *443*.

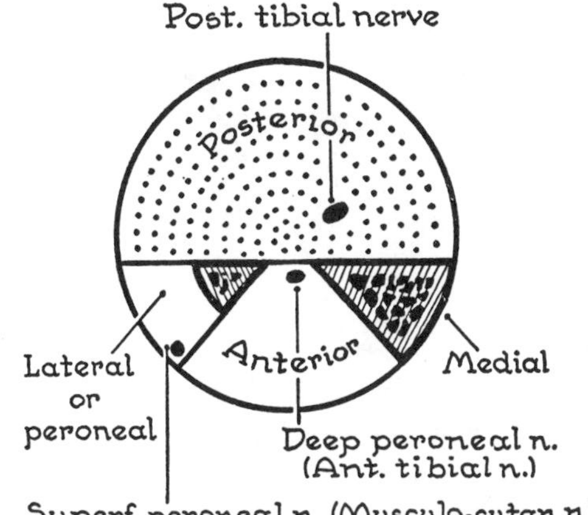

Fig. 440. Scheme of regions of the leg on transverse section, showing their relative sizes and nerves.

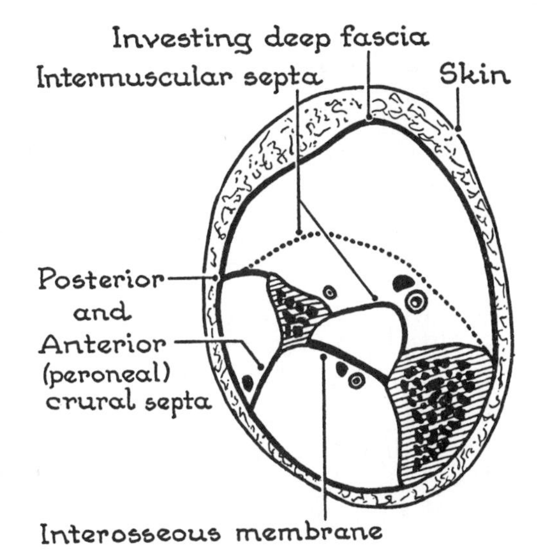

Fig. 441. The leg on cross section, showing the interosseous membrane and the various intermuscular fascial septa enclosing the fascial compartment.

The **great saphenous vein** crosses the lowest third of the medial surface of the tibia obliquely and continues proximally. At the knee it lies a hand's breadth behind the medial border of the patella (*fig. 442*).

TERMINOLOGY OF SURFACES AND BORDER is clarified in *figure 444*.

Interosseous Border of Tibia. Because the

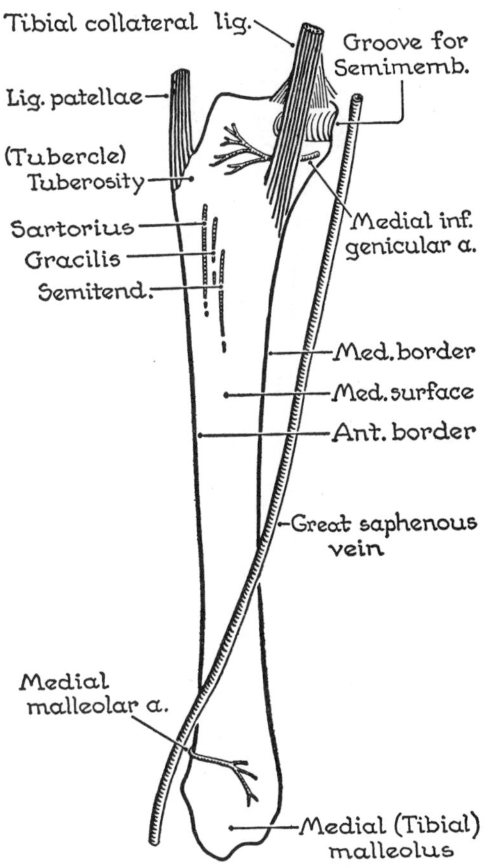

Fig. 442. The subcutaneous or medial aspect of the tibia.

lateral border gives attachment to the interosseous membrane, which unites the fibula to the tibia, it is sharp and is named the *interosseous border* (*fig. 443*). It runs to the apex of the deep, rough, triangular notch for the reception of the lower end of the fibula. The fibula requires to be firmly united to the tibia at its lower end, otherwise the talus would prize the two bones apart, so the *fibular notch* is filled with strong interosseous bands, and is rough.

The Lower End of the Tibia is expanded and, so, offers a large quadrangular bearing surface that sits on the talus.

The Lateral Surface lies deep to and is devoted to Tibialis Anterior.

The Posterior Surface (see p. 344).

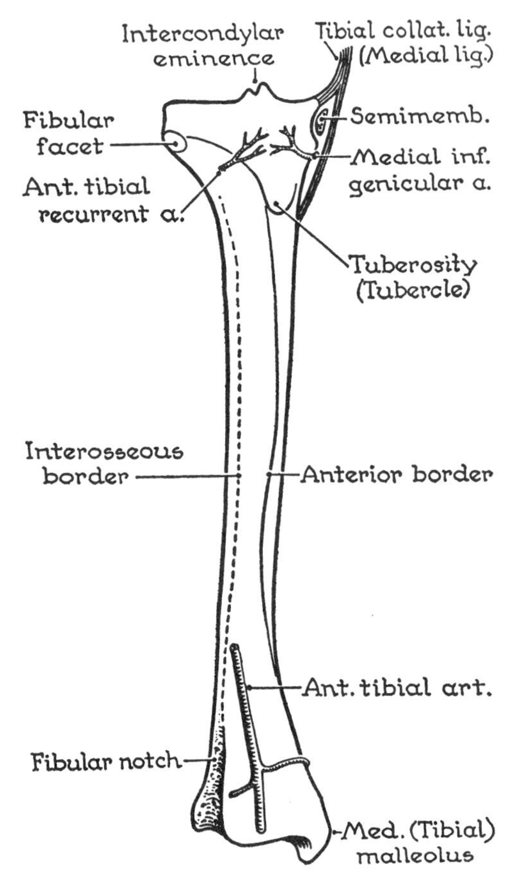

Fig. 443. Anterolateral aspect of the tibia.

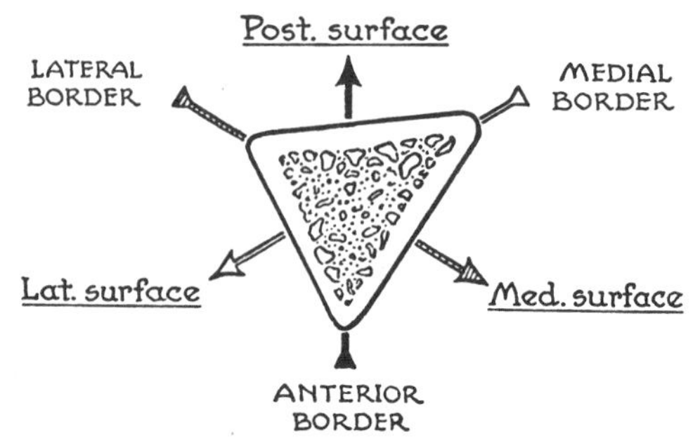

Fig. 444. Terminology: surfaces and borders are named by opposites.

ANTERIOR CRURAL REGION AND DORSUM OF FOOT

The Arterial Trunk of the front of the leg and dorsum of the foot, with its venae comitantes, enters the anterior osseofascial compartment as the **anterior tibial a.** in contact with the medial side of the neck of the fibula. After entering the foot and changing its name to **dorsalis pedis a.,** it ends near the web between the great and 2nd toes by dividing into dorsal digital branches for them. A line joining these two points and crossing the middle of the front of the ankle gives the course of the great arterial trunk (*fig. 445*).

The arteries hug the "skeletal plane," crossing in turn the interosseous membrane, the lowest third of the lateral surface of the tibia, the ankle joint, the tarsal

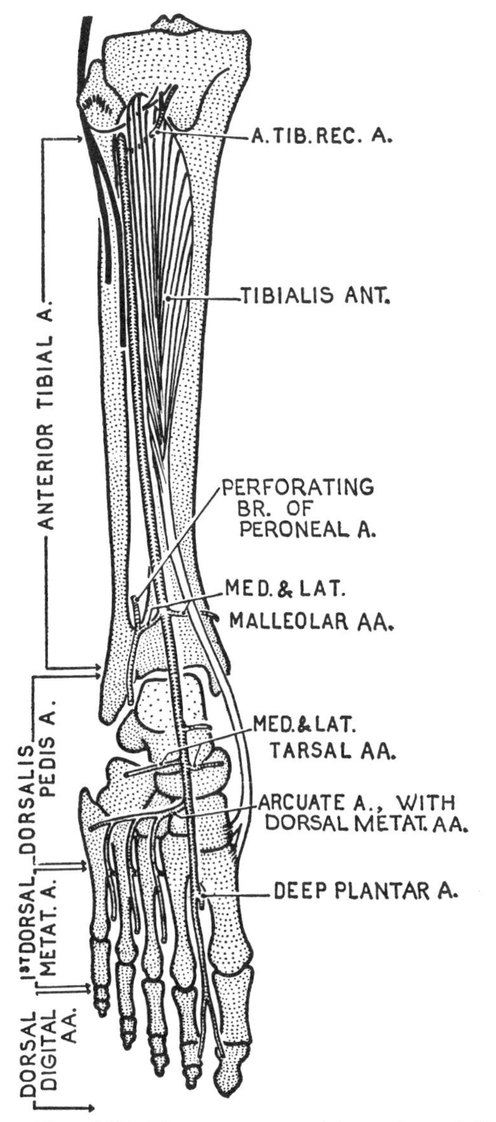

Fig. 445. The great arterial trunk and its named branches lie on the skeletal plane.

bones, and the fascia covering the First Dorsal Interosseus muscle.

At the proximal end of the first intermetatarsal space, a *deep plantar branch* plunges into the sole of the foot.

Branches (*fig. 445*) are:

anterior tibial recurrent artery.
branch accompanying the superficial peroneal nerve.
medial and lateral malleolar aa.
medial and lateral tarsal aa.
arcuate artery.
dorsal metatarsal aa.
dorsal digital aa.

Variation. The anterior tibial a. may fail to grow more than a short way down the leg, in which case the dorsalis pedis artery springs from the perforating branch of the peroneal artery (*figs. 445, 446*).

The following details should be appreciated but not memorized.

The *Anterior Tibial Recurrent Artery* (*fig. 445*), with the anterior tibial recurrent nerve, runs upward in the Tibialis Anterior to the lateral condyle to take part in the genicular anastomoses. The *lateral malleolar a.* joins the perforating branch of the peroneal a. The common stem thus-formed runs downward in front of the inferior tibiofibular joint to take part in the anastomoses on the lateral side of the ankle. It may become the chief source of the dorsalis pedis a., as noted above.

The *lateral tarsal a.* and the *arcuate a.* run laterally on the dorsum of the foot deep to the Extensor Digitorum Brevis. The *dorsal metatarsal branches* to the 2nd, 3rd, and 4th spaces arise from the arcuate a.; each is joined by a perforating branch of the deep plantar arch; and each in turn divides into two *dorsal digital aa.*

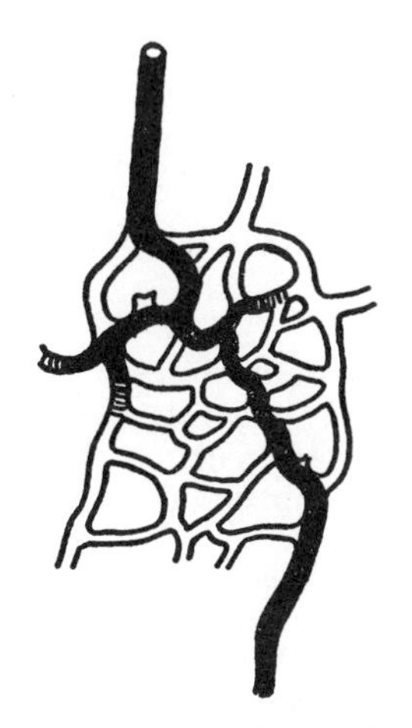

Fig. 446. Arteries are formed from channels through a network.

The medial side of the 1st toe and the lateral side of the 5th toe receive digital branches from the 1st and 4th metatarsal arteries respectively. Hence, the arrangement of vessels on the dorsum of the foot is almost identical with that on the dorsum of the hand.

Common Peroneal Nerve (Lateral Popliteal N.). This subfascial nerve, after following the posterior border of the Biceps tendon and crossing the Plantaris, lateral head of the Gastrocnemius, and back of the head of the fibula, from which it is separated by a film of Soleus, comes finally into direct contact with the lateral side of the neck of the fibula where it divides into its two terminal branches (*fig. 451*):

1. Deep peroneal nerve.
2. Superficial peroneal nerve.

These two nerves take origin on and, except for their terminal cutaneous branches, literally never leave the skeletal plane (*fig. 451*); and therefore they burrow through various structures.

These are: the posterior crural septum, Peroneus Longus, anterior crural septum, and Ext. Digitorum Longus.

The *anterior tibial recurrent nerve* (really only a twig) joins an artery of the same name to the knee joint and Tibialis Anterior.

The Deep Peroneal Nerve (Anterior Tibial N.) approaches the anterior tibial artery from the lateral side and accompanies it through the leg; thereafter, it accompanies the dorsalis pedis a. through the foot and, becoming cutaneous, divides into *two dorsal digital nerves* (*fig. 447*).

It supplies the 4 muscles in the anterior crural region, and on the dorsum of the foot it sends a lateral branch to supply the Ext. Digitorum Brevis and the various joints. In 22 per cent of limbs, an accessory deep peroneal n. runs behind the lateral mal-

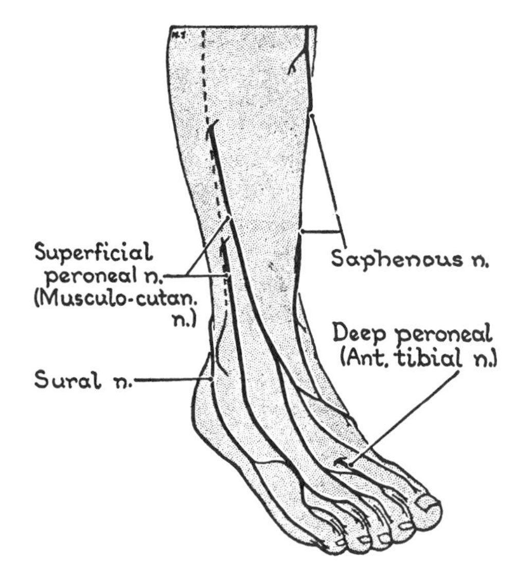

Fig. 447. Cutaneous nerves of dorsum of foot.

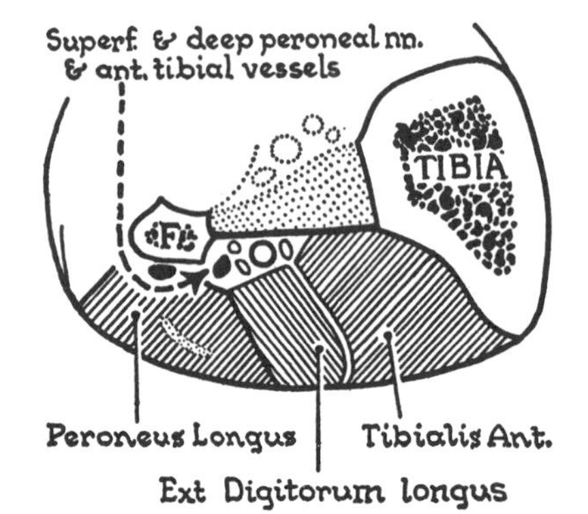

Fig. 448. Section through the upper third of the front of the leg. There are two muscles in the anterior region; one in the lateral. The vessels and nerves cling to the skeletal plane.

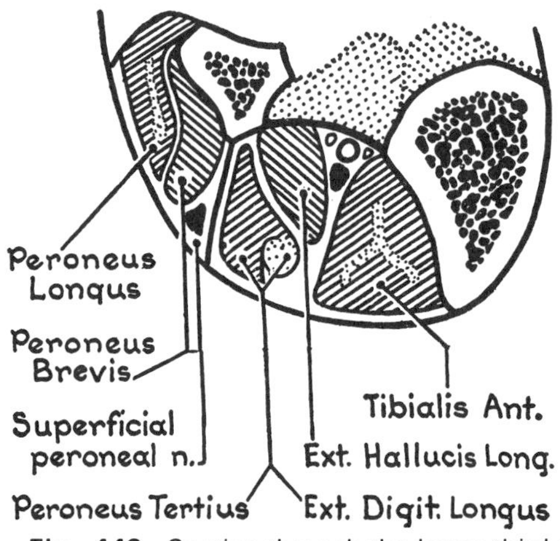

Fig. 449. Section through the lower third of the front of the leg. There are four muscles in the anterior region; two in the lateral.

leolus and is motor to the Extensor Digitorum Brevis (Lambert).

The Superficial Peroneal Nerve (Musculocutaneous N.) runs downward and forward, in contact with the shaft of the fibula and covered by the Peroneus muscles (Longus and Brevis) (*fig. 451*). It becomes superficial along the line of the anterior crural septum, a variable distance above the ankle (*fig. 447*).

It supplies Peroneus Longus and Peroneus Brevis and sends digital branches to all the toes, except adjacent sides of the 1st and 2nd (*deep peroneal n.*) and the lateral side of the 5th (*sural n.*). To these it sends communicating twigs (*fig. 447*).

Muscles of Anterior Crural Region

As may be seen in the cross-sections (*figs. 448* and *449*), there are two fleshy muscles: *Tibialis Anterior and Extensor Digitorum Longus*, in the upper part of the region; and there are two more, *Extensor Hallucis Longus* and *Peroneus Tertius*, in the lower part. At the ankle these are represented by their four tendons.

The Tibialis Anterior takes origin from the tibia; the other three crowd onto the fibula. Only *Extensor Digitorum Brevis* arises from, and confines itself to, the dorsum of the foot. The exact origins of these muscles are not important but they are given below for reference.

Details of Origins

The Tibialis Anterior arises from the upper two-thirds of the lateral surface of the tibia, from the adjacent part of the interosseous membrane and from its own covering of deep fascia.

The Extensor Hallucis Longus arises from the middle two-quarters of the anterior surface of the fibula and from the interosseous membrane.

The Extensor Digitorum Longus is a thin, unipennate muscle. It arises from the entire length of the narrow anterior surface of the fibula and from adjacent parts of the interosseous membrane, anterior crural septum, and deep fascia.

The Extensor Digitorum Brevis arises from the anterior part of the calcaneus and extensor retinaculum. Its four tendons pass to the medial four toes. The section serving the 1st toe is the *Extensor Hallucis Brevis.* Its tendon is inserted into the base of its proximal phalanx; the three other tendons join the dorsal expansions of the Extensor Digitorum Longus to the 2nd, 3rd, and 4th toes.

The fleshy belly of the Ext. Dig. Brevis is responsible for the soft swelling seen in life on the dorsum of the foot in front of the fibular malleolus.

Insertions. The stout tendon of the *Tibialis Anterior* turns medially to be inserted into the medial surface of the 1st metatarsal and 1st cuneiform, a *bursa* intervening. *Ext. Digitorum Longus* inserts by means of dorsal expansions into the distal two phalanges of the lateral four toes. Its lowest quarter, known as the *Peroneus Tertius,* fails to reach the toes, but it gains attachment anywhere along the dorsum of the (4th or) 5th metatarsal. It is a special evertor of the foot, and is almost peculiar to man. *Ext. Hallucis Longus* inserts into the base of the distal phalanx of the great toe.

Deep Fascia

In the uppermost part of the front of the leg the deep fascia gives origin to muscles, so its fibers run longitudinally and are strong. In the ankle region it acts as retinacula (*fig. 55,* p. 72); so, the fibers are disposed circularly.

The Extensor Retinaculum is in two parts —superior and inferior (*fig. 450*).

The *inferior part* is placed in front of the ankle and has the appearance of a Y-shaped band. The stem of the Y is attached to the anterior part of the upper surface of the calcaneus; the fibers of the Y form loops or slings—especially for Peroneus Tertius, Ext. Digitorum Longus, and Ext. Hallucis Longus—that prevent the tendons

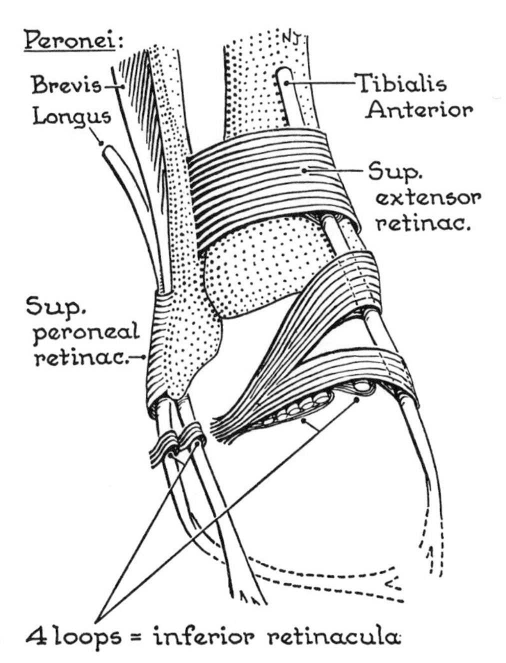

Fig. 450. Extensor and peroneal retinacula.

from bowstringing forward and also from bowstringing medially.

The Inferior Peroneal Retinaculum is attached to the calcaneus; it consists of two loops for the Peroneus Brevis and the Peroneus Longus. (Superior peroneal retinaculum, *fig. 450.*)

Synovial Sheaths that extend about 2 or 3 cm proximal and distal to the points of friction envelop the tendons.

PERONEAL (FIBULAR) REGION

By **palpation** of your own limb determine the following points: only the upper and lower ends of the fibula are subcutaneous; the *head* is rounded and the *common peroneal nerve* can be rolled behind; the *malleolus* is triangular; its anterior and posterior borders are conspicuous and palpable; its lateral surface is continuous with a subcutaneous isosceles triangle on the body (*fig. 450*) leading to the anterior crural septum and anterior border of the fibula (*fig. 441*).

The cord-like *fibular collateral lig.* can be felt (when the knee is flexed) where it runs obliquely downwards and backwards to be attached to the head of the fibula, just in front of its apex.

The cord-like *calcaneofibular lig.* runs downward and back from the malleolus just in front of its tip. It cannot be felt because the peroneal tendons cross it.

Boundaries and Contents. (*See figs. 448, 449*, and *451.*)

Muscles

Peronei Longus et Brevis fill the peroneal (fibular) compartment.

Origins. They arise from the lateral aspect of the fibula and septa, one high and the other low, with considerable overlap. They are bipennate above and unipennate below (*fig. 451*).

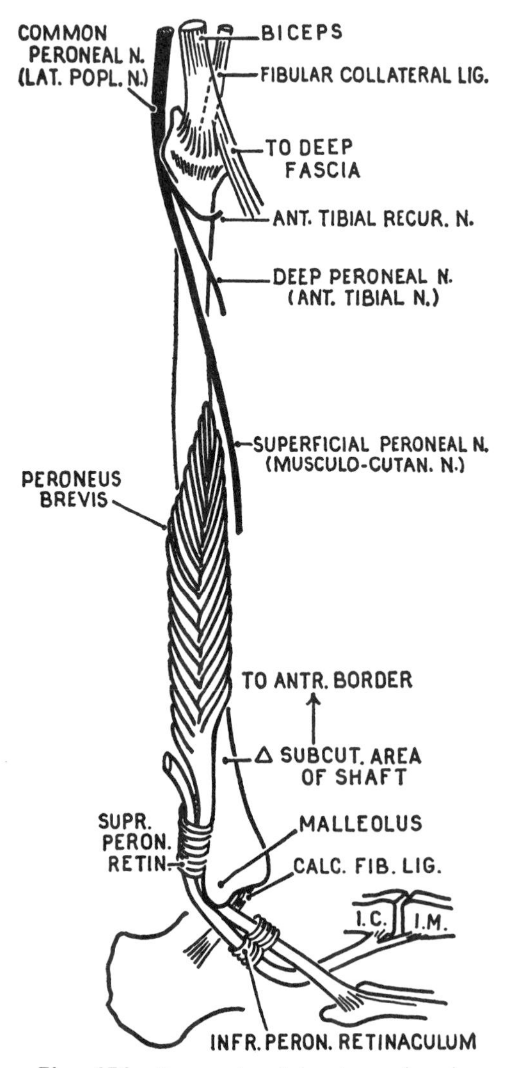

Fig. 451. Peroneal or lateral crural region.

Insertions. The Brevis inserts on the base of the 5th metatarsal, but the Longus enters the sole behind the 5th metatarsal, runs in a groove on the under surfaces of the cuboid, and grasps the same two bones as Tibialis Anterior (i.e., medial cuneiform and 1st metatarsal).

The two tendons use the posterior aspect of the fibular malleolus as a pulley, and both cross the calcaneofibular ligament. They are bound down by the *superior* and *inferior peroneal retinacula* (*fig. 451*), and "use" a single synovial sheath, which bifurcates below.

Fibula and Its Joints

The human fibula carries no weight to the ground, but it holds the talus in its socket and it does this in a resilient manner. A main function of the bone is to give origin to muscles. Of the nine muscles attached to it, only the Biceps pulls upward; the others all pull downward (*fig. 452*). All ligamentous connections between tibia and fibula are so directed as to resist this downward pull (see *fig. 453*). Weinert *et al.* have shown that, contrary to expectations, the fibula moves down (not up) to deepen the ankle socket at the strike phase of running.

Comparative Anatomy. In amphibia the fibula is as large as the tibia. It articulates with the femur above and with the tarsus below and it is weight supporting. In reptiles it is smaller than the tibia and it bears less weight. In monotremes and marsupials it is still further reduced. In horses and ruminants the shaft of the fibula either disappears or is represented by a fibrous band that connects the upper and lower ends; and the ends are either incorporated with the tibia or in articulation with it. In carnivora and primates the complete fibula exists, but it does not bear weight.

Tibiofibular Joints. The fibula is moored to the tibia at its upper end, along its shaft,

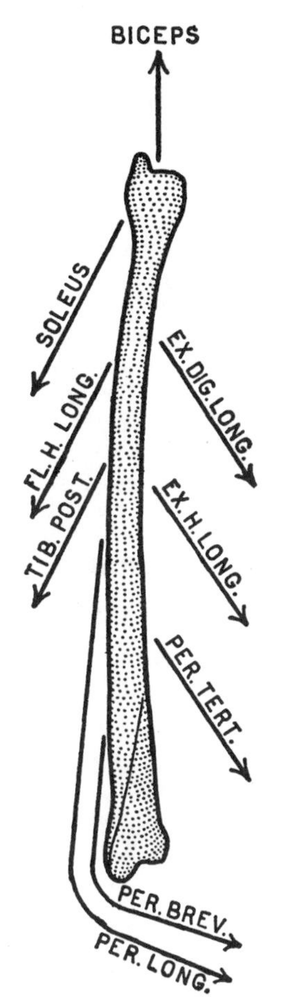

Fig. 452. The muscles attached to the fibula pull downward, except the Biceps.

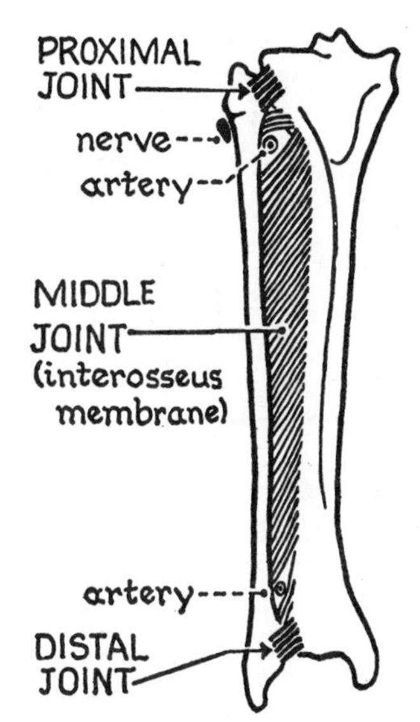

Fig. 453. The tibiofibular articulations. Note unity of direction of ligamentous fibers.

and at its lower end—at proximal, middle, and distal joints.

The Proximal Tibiofibular Joint has a synovial cavity and is of the plane or gliding variety. A small, flat, round facet on the head of the fibula articulates with a similar facet on the posterolateral part of the lateral condyle of the tibia. Behind the joint is the Popliteus tendon, separated by the Popliteus *bursa*, which may communicate with the joint (p. 375 and *fig. 517*).

The Middle and Distal Tibiofibular Joints are syndesmoses. The *interosseous membrane* (middle joint) stretches between the "opposing" sides of the tibia and fibula, producing a sharp line on each. Distally, each line expands into a large, rough, triangular area (*fig. 454*). A strong *interosseous ligament* binds the opposed areas together. Additional ligaments are are associated with the socket of the ankle joint (p. 378).

Body of the Fibula (*fig. 454*). The shape of the body depends largely on the muscles and septa attached to it. In general terms: the peroneal surface is dedicated to the peronei; the flexor (posterior) surface to the Soleus and Fl. Hallucis Longus, with a special area just behind the interosseous membrane for the Tibialis Posterior; the narrow anterior surface gives origin to the extensors of the toes.

If you must learn details for a special purpose, begin with the important peroneal surface. What follows should be appreciated but *not memorized* by most students.

The Peroneal Surface is found by placing a finger behind the malleolus, which is the pulley for the Peronei, and letting it run up the shaft to the head of the bone. This surface is broad and spiral, facing posteriorly below and laterally above. It is bounded by lines (the *anterior* and *posterior borders*) which give attachment to the anterior and posterior crural septa. To make doubly sure of the anterior border, which sepa-

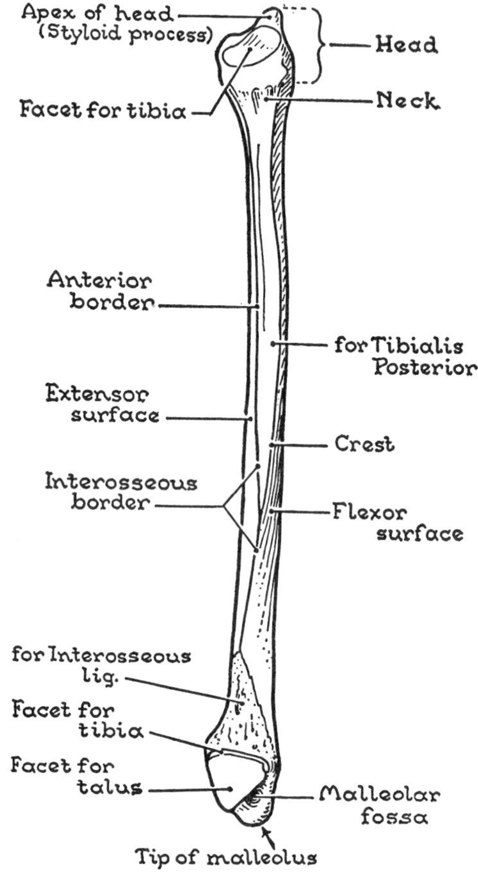

Fig. 454. Medial aspect of the fibula.

rates the peroneal from the extensor surface, place a finger on the subcutaneous, isosceles triangle (*fig. 451*) above the malleolus and run the finger straight up to the head of the bone.

The Flexor or Posterior Surface is broad and it gives origin to the Soleus in its upper third and to the Fl. Hallucis Longus in its lower two-thirds. It is spiral, like the peroneal surface next to it.

The Surface for the Tibialis Posterior is the enigma; it is fusiform and is to be found thus: put a finger on the rough area for the interosseous ligament—i.e., the area above the smooth, triangular facet on the malleolus for articulation with the talus—and follow it upwards. It becomes a line which splits, one-third to one-half of the way up the shaft, into an *anterior line* and a *prominent posterior crest*. These enclose a fusiform area for the Tibialis Posterior. The anterior line is the *interosseous border* for the interosseous membrane; the prominent crest is for the intermuscular septum behind the Tibialis Posterior (*fig. 441*). It is common to mistake the crest for the interosseous border.

There is a deep hollow, the *malleolar fossa*, between the grooved pulley behind the malleolus and the triangular facet for the talus.

The Extensor or Anterior Surface faces consistently forward and is almost linear, because it gives origin to the unipennate Ext. Digitorum Longus in its upper three quarters and to the Peroneus Tertius in its lower quarter. It broadens somewhat in its middle two quarters to afford origin to the Ext. Hallucis Longus.

Ossification. The shaft of the fibula begins to ossify about the 8th prenatal week, like the shafts of other long bones. The upper end begins to ossify about the 5th year; fusion is always complete by the 22nd. The lower end begins to ossify about the 2nd year; fusion is always complete by the 20th.

POSTERIOR CRURAL REGION

(The Back of the Leg)

Bony Framework, viewed from behind (*fig. 455*).

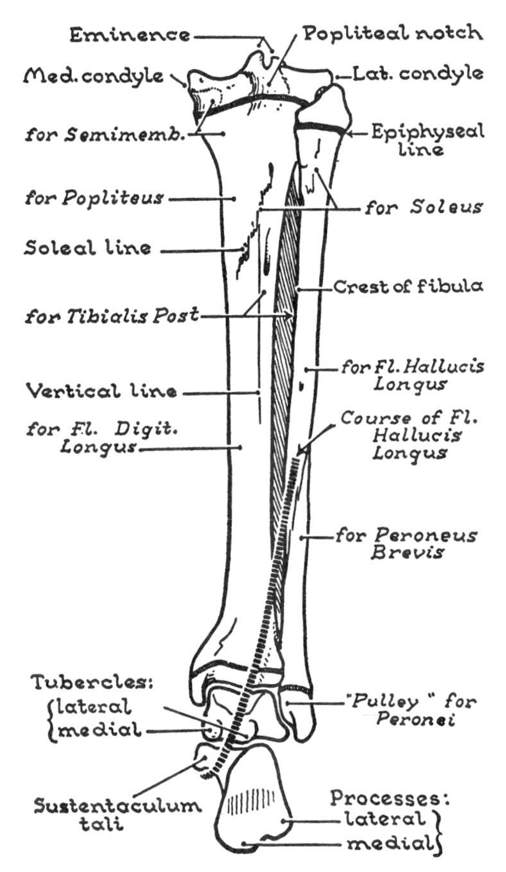

Fig. 455. Posterior aspect of the bones of the leg and foot.

Above, the *femoral condyles* project backward.

The *tibial condyles* overhang the shaft of the tibia posteriorly and at the sides. The rounded *head of the fibula*, supported by its neck, is well below the knee joint.

The posterior surface of the body of the tibia is crossed obliquely by the *soleal line*. *The tibial malleolus* is 2 cm shorter than the fibular malleolus (*fig. 455*).

In view are: the back of the talus including the *upper articular surface* and a *deep groove* between the *medial* and *lateral tubercles* for the Fl. Hallucis Longus tendon which runs downward and medially to a groove on the under surface of the sustentaculum tali. The *posterior third of the calcaneus* projects backwards beyond the talus, forming the heel. It ends in a large medial process which rests on the ground (*fig. 455*).

Muscles of Posterior Crural Region

Superficial group (*fig. 456*):

1. Gastrocnemius,
2. Soleus, and
3. Plantaris, between them.

Deep group:

1. Popliteus, a rotator of the knee
2. Flexor Hallucis Longus and
3. Flexor Digitorum Longus to the end phalanges of the toes.
4. Tibialis Posterior to every small tarsal and most metatarsals.

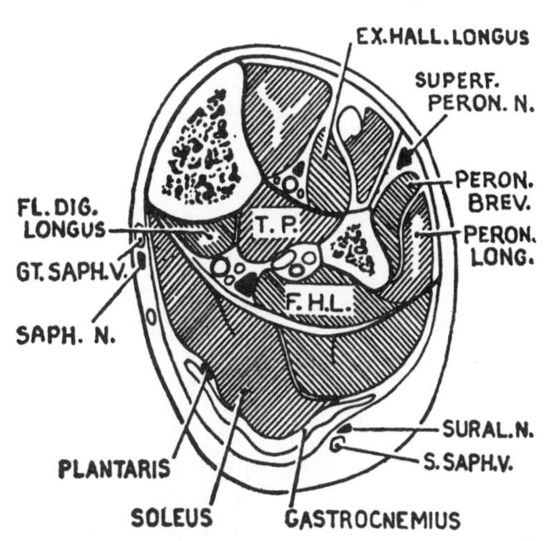

Fig. 456. Transverse section through middle of leg.

The Gastrocnemius has two flattened heads of origin which are largely tendinous. The lateral head arises from the epicondyle of the femur (just above the impressions for the Popliteus and fibular collateral lig.); the medial head arises from the popliteal surface of the femur, above the medial condyle (*fig. 456.1*).

The two resulting fleshy bellies, which form the fullness of the calf, unite and then end at the middle of the leg in a broad aponeurosis that blends with the aponeuro-

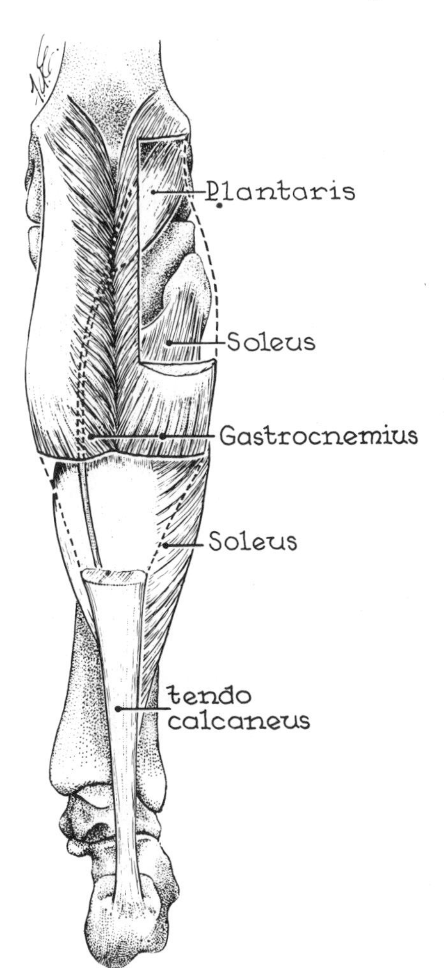

Fig. 456.1. The Gastrocnemius, Soleus, and Plantaris muscles form the tendo calcaneus.

sis of the Soleus to form the *tendo calcaneus* (tendon of Achilles). The Gastrocnemius and Soleus are referred to as the *Triceps Surae* (Sura, L. = the calf) (*fig. 456.1*).

A constant *bursa* between the tendon of the Semimembranosus and the medial head of the Gastrocnemius may communicate with a *bursa* between the latter and the capsule of the knee joint. This in turn may communicate with the knee joint (in 42.4 per cent of 528 dissecting room limbs). So, fluid in the joint, resulting say from a sprain, can pass from the knee joint via the Gastrocnemius bursa into the Semimembranosus bursa and cause a swelling at the back of the knee. A *sesamoid bone* is commonly found in the lateral head of the Gastrocnemius, and less commonly in the medial head.

The Tendo Calcaneus [Achillis]. See *figure 457.*

The Soleus is shaped like the sole of a boot. In man, its origin from the upper part of the back of the fibula is greatly increased by crossing to the soleal line of the tibia, and even lower (*fig. 458*). So, its *origin is horseshoe shaped* (*fig. 459*). Short fleshy fibers of this powerful muscle join the tendo calcaneus.

The Plantaris is placed between the Gastrocnemius and the Soleus. It has a small fleshy part and a very long ribbon-like tendon (*fig. 456.1*). It arises near the lateral head of the Gastrocnemius and it is inserted along the medial side of the tendo calcaneus.

In lower mammals it is a large muscle whose tendon runs in a groove behind the calcaneus on its way to the toes, which it flexes. In man its sural part is vestigial and, like its homologue, the Palmaris, is variable and sometimes absent (p. 29). The plantar part has become the plantar aponeurosis.

Popliteus, inserted into the posterior surface of the tibia by fleshy fibers (*figs. 460* and *461*), arises by tendon just below the lateral epicondyle of the femur. It runs behind the lateral meniscus and the proximal tibiofibular joint.

A variable bundle of fleshy fibers from its upper border gains a fibrous attachment to the hinder part of the lateral meniscus, thereby forming an *articular muscle* (Last; Lovejoy *et al.*).

Deep Muscles

Below the horseshoe-shaped origin of the Soleus there are three bipennate muscles—Tibialis Posterior, Fl. Digitorum Longus and Fl. Hall. Longus (*fig. 459*). Their tendons are the quills of the feathers. They pass downward and medially to enter the sole of the foot.

The Tibialis Posterior is the deepest, *arising* from the interosseous membrane and adjacent bones. Its tendon clings faith-

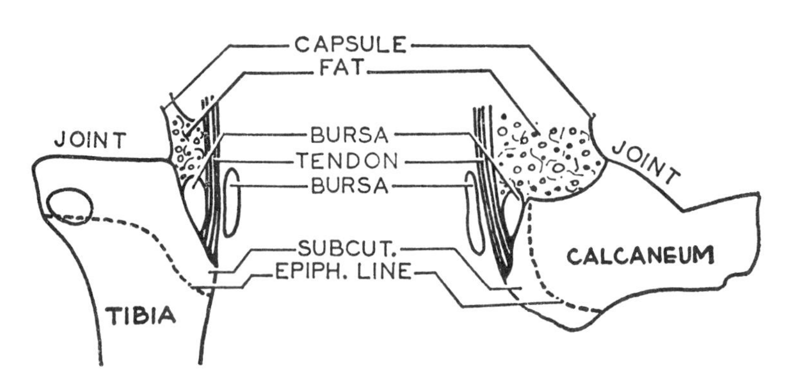

Fig. 457. The tendo calcaneus and ligamentum patellae compared. About equal in length, each is inserted into an epiphysis, and has a bursa deep to it with a large pad of fat between the bursa and the capsule of the nearest joint. Overlying the lig. patellae is the superficial infrapatellar bursa; in the skin over the tendo calcaneus there may be a bursa.

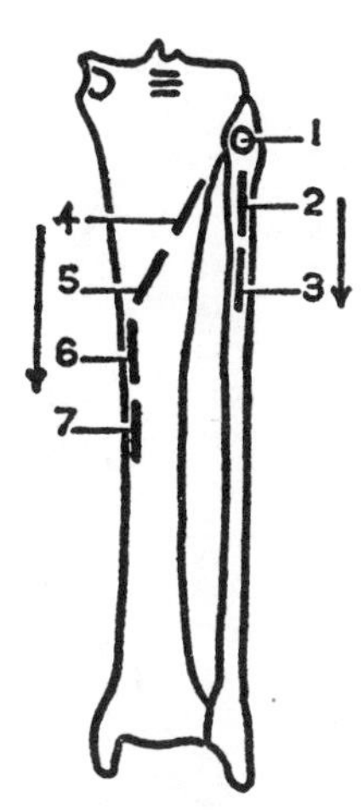

Fig. 458. The increasing origin of the Soleus is horseshoe-shaped.

Fig. 459. Three bipennate muscles resembling three feathers.

fully to the skeletal plane, using the medial malleolus as its pulley.

Insertion—see fig. 484 and p. 359.

The *Flexor Digitorum Longus* arises from the posterior surface of the tibia and the fascia covering the Tibialis Posterior.

The *Flexor Hallucis Longus* is the largest of the three. Its origin from the fibula below the Soleus overflows to the posterior crural septum, the fascia covering the Tibialis

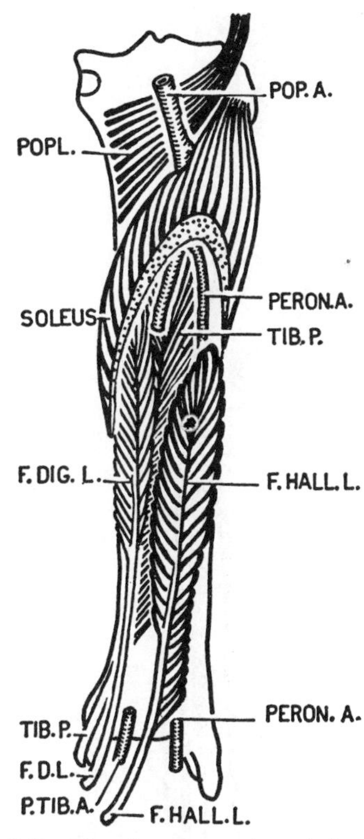

Fig. 460. The deep muscles of the back of the leg—one above and three below Soleus.

Posterior, and even the interosseous membrane and tibia inferiorly (*fig. 460*).

(Continued into the foot on p. 358).

The Tendons behind the Ankle (*fig. 462*) are the tendo calcaneus and 5 others. Of these, two, the Peronei Brevis et Longus, groove the back of the fibular malleolus; two, the Tibialis Posterior and Fl. Digitorum Longus, groove the back of the tibial malleolus; and one, the Fl. Hallucis Longus, grooves the lower end of the tibia midway between the malleoli (*fig. 460*).

Arteries and Nerves

There are two large arteries and one nerve in this region (*fig. 461*), namely:

Posterior tibial artery
Peroneal artery
Tibial nerve (post. tibial n.)

Tibial Nerve. The tibial nerve takes a straight course to the back of the lower end of the tibia deep to the flexor retinaculum, where it divides into the *medial* and *lateral*

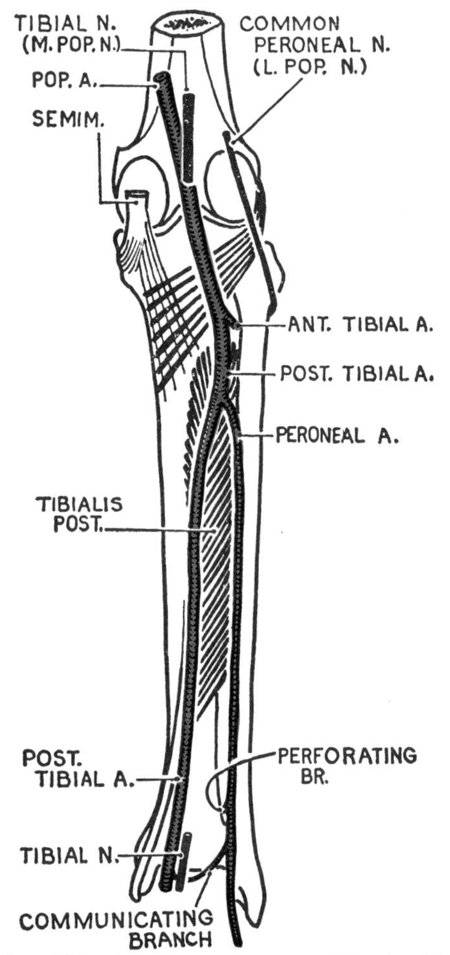

Fig. 461. Arteries and nerves of back of leg.

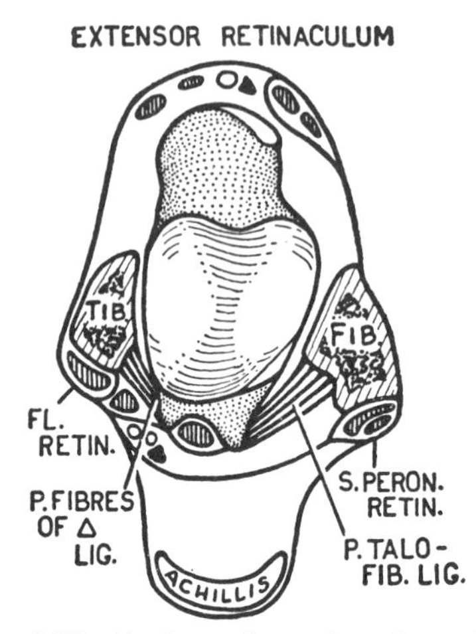

Fig. 462. Horizontal section through ankle joint showing (1) wedge-shaped socket, (2) direction of ligaments, (3) five posterior tendons, and (4) the investing and intermuscular deep fascia.

plantar nerves. It is applied to the lateral side of the posterior tibial artery (*fig. 461*), on the fascia covering tibialis posterior and then the tibia.

Branches: (1) *muscular;* (2) *cutaneous*, the medial calcanean nerves, to the heel (*fig. 489*); (3) *articular*, to the ankle joint; (4) *vascular*, branches to the local vessels; and *terminal*, the medial and lateral plantar nerves (p. 361).

Cutaneous Nerves are summarized in *figures 425* and *447*. The patterns formed by these nerves are numerous.

The Posterior Tibial Artery begins at the upper border of the Soleus, and ends deep to the flexor retinaculum by dividing into the *medial* and *lateral plantar artery*. Near the ankle two layers of fascia cover it (*fig. 462*). When this fascia is relaxed by inverting the foot, the pulsations of the artery can be felt behind the medial malleolus.

The Peroneal Artery arises from the posterior tibial artery and descends behind the fibula, distal tibiofibular joint, and ankle joint. It ends on the lateral surface of the calcaneus as the *lateral calcanean artery*.

Branches. Both the posterior tibial artery and its peroneal branch have *muscular*, *cutaneous*, *nutrient*, *communicating*, and *calcanean* branches. The peroneal artery gives off a *perforating branch* which runs down in front of the distal tibiofibular joint, and anastomoses on the lateral side of the ankle.

The Small Saphenous Vein drains the lateral end of the dorsal venous arch of the foot. It runs from below the lateral malleolus to the groove between the two heads of the Gastrocnemius, and, after piercing the popliteal fascia, ends partly in the popliteal vein and partly in the profunda femoris vein.

Communicating Veins. The cutaneous veins form an open network between the great and small saphenous veins, and they communicate along the intermuscular septa with the deep veins. Above the ankle the valves in the communicating veins are so directed that the blood flows from the superficial veins into the deep veins, and from the small saphenous vein into the great saphenous vein.

27 BONES AND SOLE OF FOOT

BONES OF ARTICULATED FOOT

Organization (How to Draw); Talus; Calcaneus; Middle Unit (Five Small Tarsals); Metatarsals.
Arches.
Ossification.

IDENTIFIABLE PARTS IN LIVING FOOT

SOLE OF FOOT

Articulated Foot from Below; Plantar Fascia and Aponeurosis; Cutaneous Nerves.
Plantar Muscles—First, Second, Third, and Fourth Layers; Long Tendons on Skeletal Plane: Peroneus Longus, Tibialis Posterior et Anterior.
Lateral Plantar Artery—(Deep) Plantar Arch.
Lateral Plantar Nerve.
Medial Plantar Artery and Nerve.
"Door" of Foot; Re-arrangement of Tendons.

Comment on Vessels and Nerves.

BONES OF ARTICULATED FOOT

Study of the bones of the foot need not be as forbidding as most books make it. If the student follows the remarks describing the accompanying diagrams, verifies them by referring to an articulated foot and his own foot, he will find himself familiar in a practical way with the essential facts and more important details.

Examine the dorsum of an articulated foot and make the following observations:

1. *Outline*. The medial border of the foot is almost straight; the most projecting toe is usually the big toe or *hallux* (*fig. 463*).

2. Draw or imagine a line joining the midpoints of the medial and lateral borders. In front of it lie the long bones—metatarsal bones and phalanges; behind it lie the tarsal bones.

The middle and distal phalanges are in reality nodular and rudimentary. The hallux has two stout phalanges.

Metatarsals. The 1st is the stoutest and strongest; the 2nd is the longest; the 5th has a palpable tuberosity at its base (*fig. 464*).

3. *The Transverse Tarsal Joint*. You can only flex and extend your ankle joint. To obtain a view of the sole, you invert your foot by rotating joints other than the ankle. The *head of the talus*, which is globular, fits into the posterior surface of the *navicular*, which is cup-shaped; and the anterior surface of the *calcaneus*, which is sinuous, articulates with the reciprocal posterior surface of the cuboid. At these two joints, collectively known as the transverse tarsal joint, part of the movement of *inversion* and *eversion* takes place; it augments the more important movements at the joints between the *talus* and the supporting *calcaneus* (fig. 466).

4. *The transverse tarsal joint* of inversion and eversion serve to divide the foot into three units; anterior, middle, and posterior (*fig. 465*). The "middle unit" comprises 5 small tarsal bones; the "posterior unit," two large tarsal bones—talus and calcaneus (*figs. 464, 466 and 467*).

The **Talus** rests on the anterior two-thirds of the calcaneus and projects slightly in front of it. It may remind you of a tortoise with a body, neck, and head. The upper surface of the body supports the tibia, is entirely articular, and is saddle-

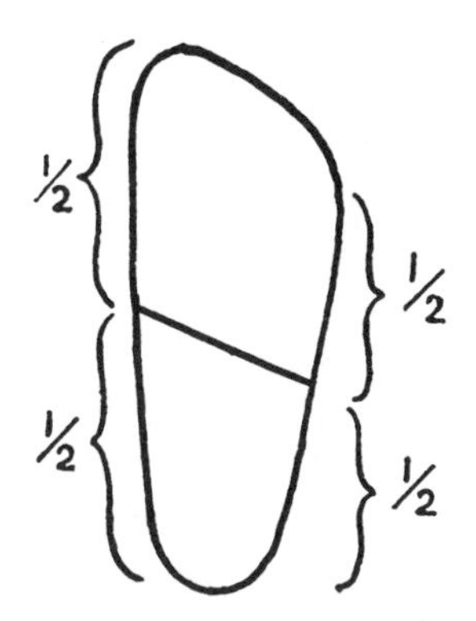

Fig. 463. The outline of the foot.

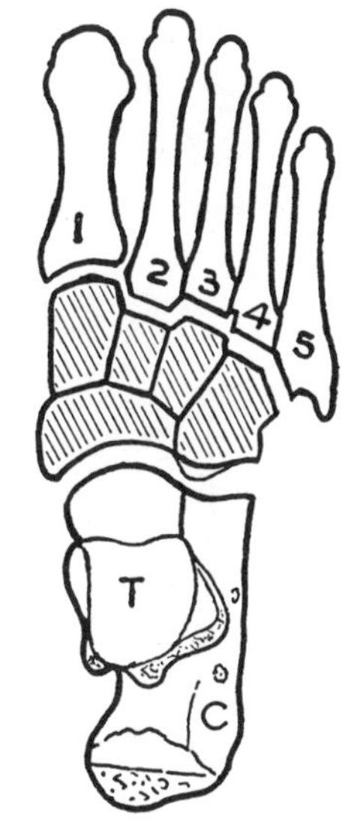

Fig. 464. The dorsum of the foot.

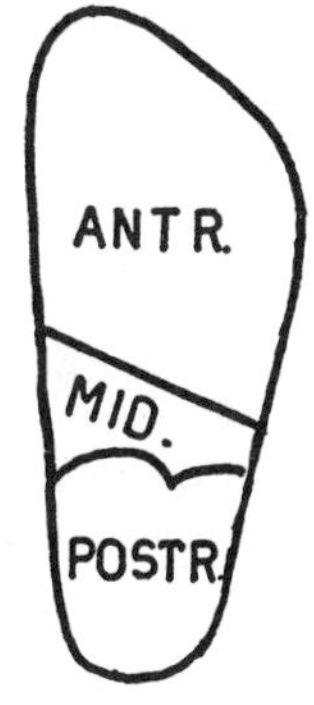

Fig. 465. The three "units" of the foot.

shaped; its lower surface forms the large *subtalar joint*. The sides of the body are grasped by the malleoli, and the facets that result correspond in length and shape with those of the malleoli. The posterior surface of the body tapers to two tubercles, a *medial* and a *lateral*, separated by a groove for Fl. Hallucis Longus tendon on its way into the sole (*fig. 467*). The head is rounded in front to articulate with the navicular bone. Below, it partly rests on the calcaneus.

The **Calcaneus** is by far the largest tarsal bone. It is divided into three thirds. The anterior two-thirds supports the talus; the posterior one-third forms the prominence of the heel and rests on the ground

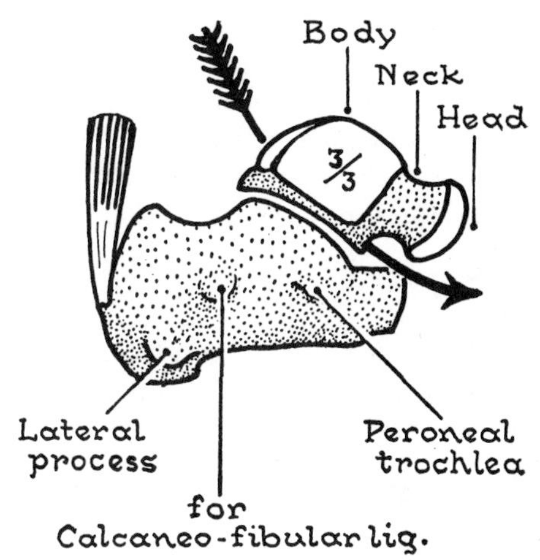

Fig. 466. Lateral aspects of talus and calcaneus. (A probe may be passed like the *arrow*, traversing the tarsal tunnel, which is filled with an interosseous ligament.)

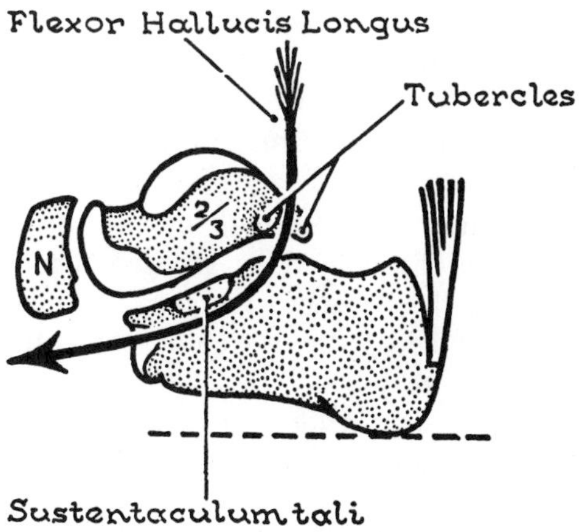

Flexor Hallucis Longus
Tubercles
2/3
N
Sustentaculum tali

Fig. 467. The medial aspects of the talus and calcaneus.

(*fig. 467*). The anterior surface is wholly in articulation with the *cuboid* (*fig. 468*). The posterior third of the upper surface is saddle-shaped, horizontal, and free; the intermediate third supports the body of the talus and is convex, articular, and oblique; the anterior third forms a nonarticular horizontal platform laterally and a projecting shelf, the *sustentaculum tali*, medially; each has a small facet, which may be continuous with the other, for the head of the talus (*fig. 468*).

The lateral surface (*fig. 466*) is palpable and presents features associated with the ankle joint and peroneal tendons.

The posterior surface [tuber calcanei] is wider below than above. It is continuous, on the plantar surface, with a large medial process or "tubercle" (weight-bearing) and a small lateral process (*fig. 466*).

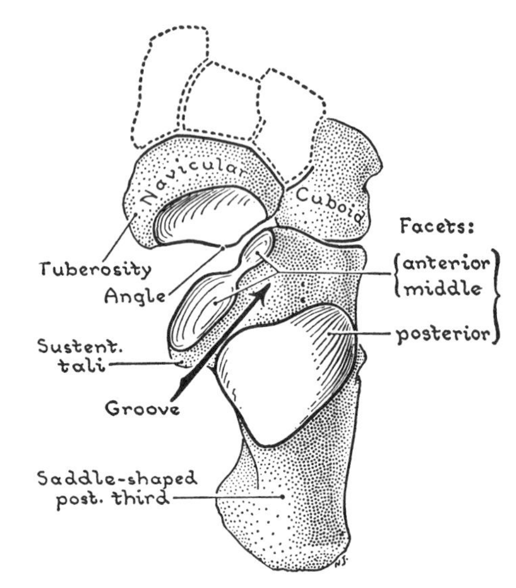

Fig. 468. The upper surface of the calcaneus.

As an aid for beginners to the names and relative positions of the 7 tarsal bones—calcaneus, cuboid, and 3 cuneiforms, navicular and talus—the accompanying design has been found useful. (Right foot, dorsal aspect, after G. H. Paff.)

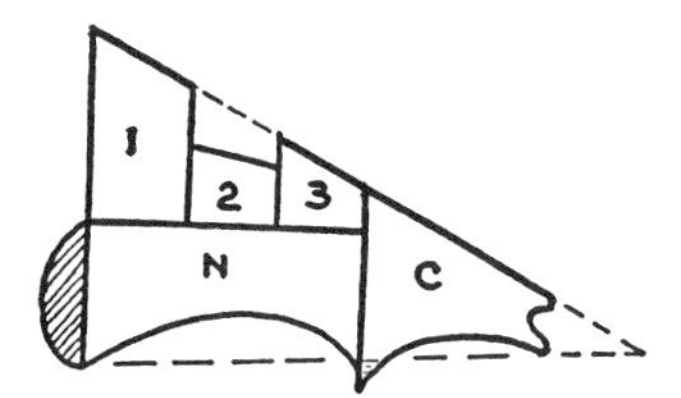

Fig. 469. The middle "unit," comprising the five small tarsal bones, is a modified wedge.

The Middle Unit comprises the five small tarsal bones and roughly resembles a triangle or a wedge whose apex is lateral and base medial (*figs. 468* and *469*). The anterior side is indented between the 1st and 3rd cuneiforms to receive the base of the 2nd metatarsal which is morticed between them, preventing side to side shifting.

The blunt apex, formed by the lateral surface of the cuboid, is grooved for the tendon of the Peroneus Longus. The base, where formed by the navicular, presents a tuberosity.

5. *Two continuous articular facets never, or hardly ever, lie in precisely the same plane; even when two facets on one bone are continuous with one another, they always, or almost always, meet at an angle, forming inclined planes. See if this is not so* (*fig. 468*).

The *posterolateral angles of the bases* of the 2nd, 3rd, 4th, and 5th metatarsals "point" to the sides to which they belong (*fig. 470*).

The *contiguous sides of the bases of the lateral four metatarsals have articular facets* (*fig. 471*). But there are no such facets between the bases of the 1st and 2nd metatarsals, thus indicating, as does its curved surface, that the great toe was once a free member like the thumb.

The 2nd metatarsal is morticed between the medial and lateral cuneiforms. The interlocking of the various bones prevents side-shifting of the bones (*fig. 472*).

6. *Arches.* The foot has two longitudinal arches. It is designed for walking and so is not a piece of masonry. It is a spring rather than an arch (*fig. 474*).

The three medial digits, their metatarsals, and cuneiforms, the navicular and

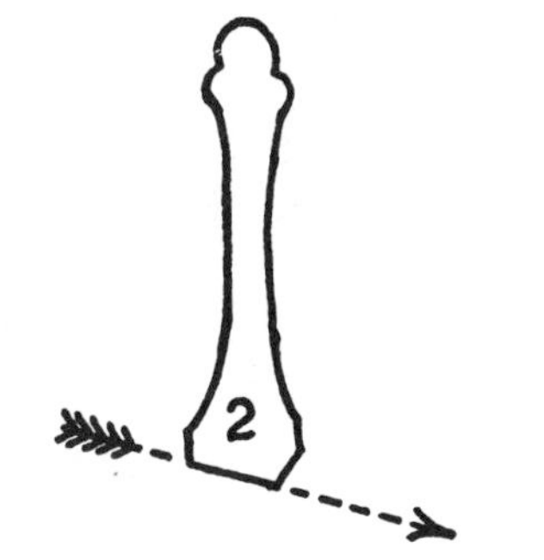

Fig. 470. The base tells the side.

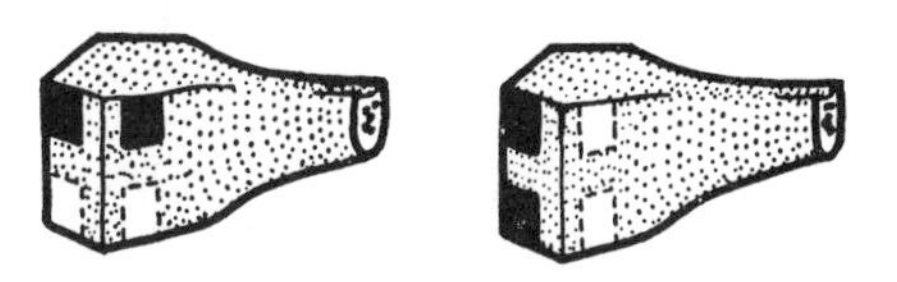

Fig. 471. Variations in metatarsal facets.

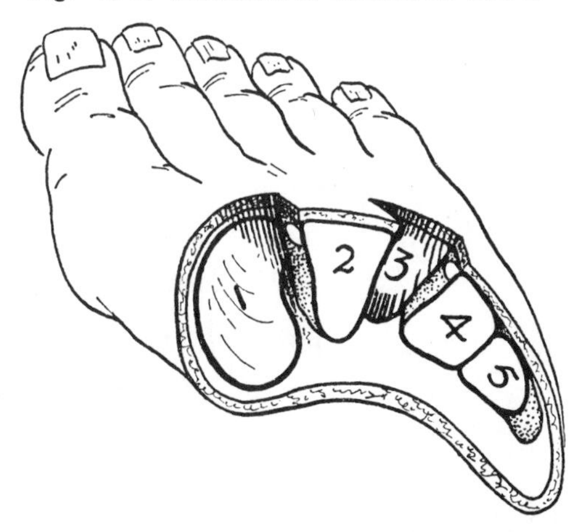

Fig. 472. The bases of the metatarsals interlock with the small tarsals and so prevent side-to-side shifting.

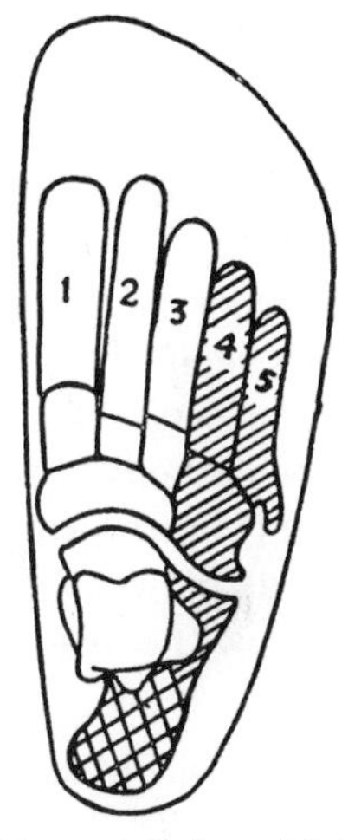

Fig. 473. The medial and lateral longitudinal arches. The calcaneus is common to both.

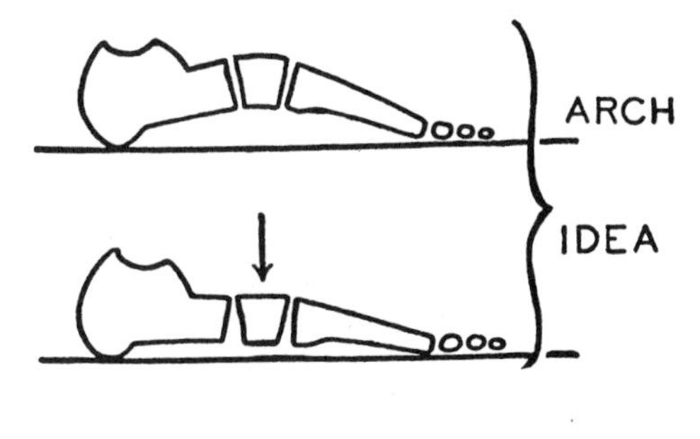

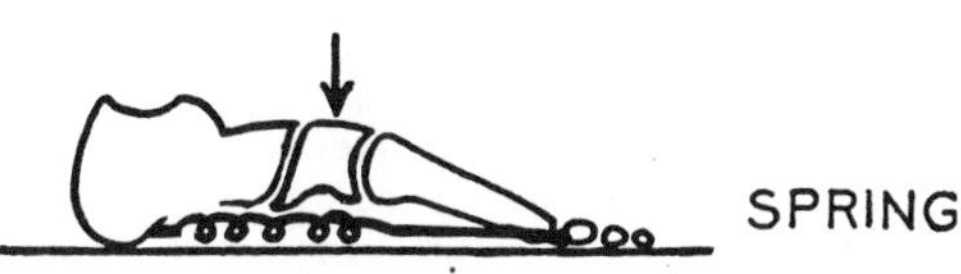

Fig. 474. The foot is not an inert masonry arch; it is a spring.

talus are collectively known as the *medial longitudinal arch*—it is the vital one. The two lateral digits, their metatarsals, the cuboid and the calcaneus, collectively known as the *lateral longitudinal arch* (spring) (*fig. 473*) is a very low arch and not much of a spring (*fig. 475*).

The foot also forms a transverse half-arch at its midpoint because the medial arch is so much higher than the lateral one (*fig. 472*). To call this an arch as some authors do is poor biomechanics, and the theory will no more stand up than will a half-arch.

7. *Viewed from the Lateral Side* the foot is a low, flat arch, which flattens under your weight.

8. *Viewed from the Medial Side* the foot is highly arched. The medial process of the calcaneus forms the posterior pillar; the heads of the 1st, 2nd, and 3rd metatarsals

and the sesamoid bones under the head of the 1st form the anterior pillar (*fig. 475*). The summit of this arch lies at the junction of its posterior one-third and anterior two-thirds.

The round head of the talus is placed at the summit, between the sustentaculum tali and the tuberosity of the navicular, and here receives no bony support. But the *spring ligament* (plantar calcaneonavicular), which extends between these two bony processes, lies below the head and supports it (*fig. 533*).

Ossification of the Bones of the Foot. X-ray photographs (*fig. 475.1*) show that at the time of birth the calcaneus and talus are well ossified and the cuboid is starting to ossify. The sequence is calcaneus, talus, cuboid, 3rd, 1st, 2nd cuneiform, navicular, epiphysis of calcaneus (table 17 on p. 387).

The calcaneus is the only short bone (either tarsal or carpal) that regularly has an epiphysis. This epiphysis gives attachment to the tendo calcaneus and the plantar aponeurosis; hence, it includes the posterior surface and the processes of the calcaneus (*fig. 457*). It appears about the 11th year and fuses about the 17th year.

The metatarsals and phalanges have each a primary center for the body, which appears about the 3rd prenatal month, and secondary centers for the heads of the 2nd–5th metatarsals and for the bases of the 1st metatarsal and all the phalanages. These appear about the 3rd year and fuse about the 18th year (*cf.* the hand, *fig. 140*).

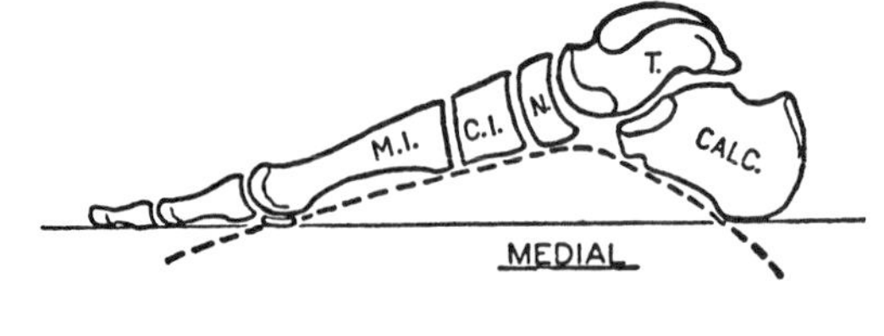

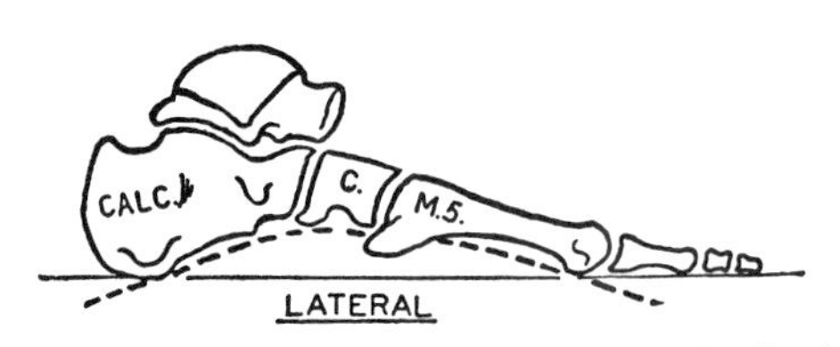

Fig. 475. Medial and lateral longitudinal arches.

IDENTIFIABLE PARTS IN LIVING FOOT

Now you can and should review with your eyes and your fingertips:

1. The blunt end of the medial malleolus.

2. The sustentaculum tali more than a finger's-breadth below the malleolus.

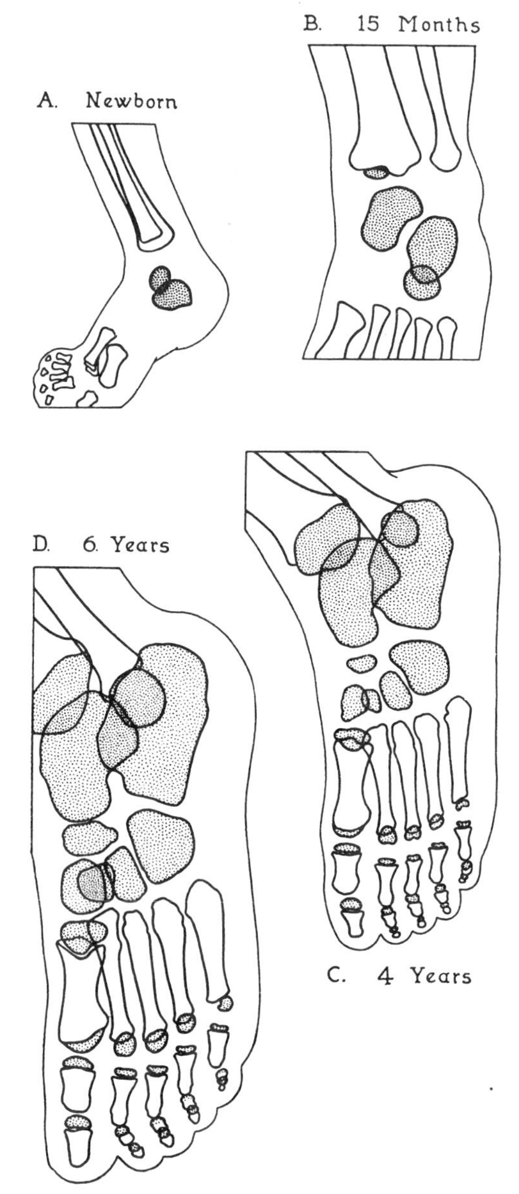

Fig. 475.1 Progressive ossification of the bones of the foot. (Courtesy of Radiology Dept., Kingston General Hospital).

3. The tuberosity of the navicular two finger's-breadths in front of the sustentaculum.

4. The head of the talus, occupying the space between these two. All four are easily palpated, and may indeed be rendered visible through the skin. In palpating the tip of the malleolus, the sustentaculum, and the tuberosity, *approach them from below*.

5. The sesamoid bones, which play on the under surface of the head of the 1st metatarsal, may be felt under the ball of the big toe (*figs. 476, 481*, and *483*).

6. The lateral malleolus is visible; its pointed end is 2 cm lower than the medial malleolus and more posterior.

7. The projecting base of the 5th metatarsal is easily felt about half way along the lateral border of the foot.

8. The calcaneocuboid joint lies two-thirds of the way between the tip of the lateral malleolus and the projecting base of the 5th metatarsal. Verify this by inverting the foot and palpating the superolateral portion of the anterior surface of the calcaneus which is thereby uncovered. This is easily done.

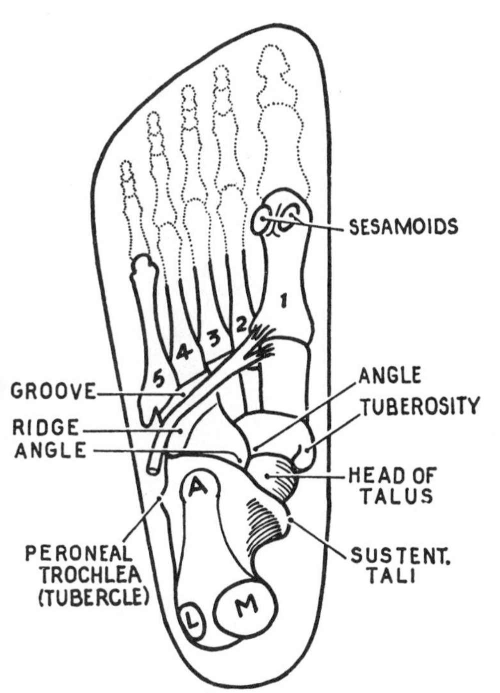

Fig. 476. The chief features of the plantar aspect of the articulated foot.

9. With the foot still inverted, palpate the upper part of the front of the round head of the talus, just superomedial.

10. Of necessity the muscles of inversion and eversion, the *2 Tibiales* and the *3 Peronei*, must be inserted anterior to the transverse tarsal joint.

(a) With the foot inverted and the ankle dorsiflexed, the tendon of the Tibialis Anterior may be seen easily as it juts out anterior to the medial malleolus, on its way to its insertion (the medial surface of the 1st cuneiform and adjacent part of the 1st metatarsal—*fig. 484*).

(b) The Tibialis Posterior is not so obvious. With the foot inverted and ankle plantarflexed, its tendon may be felt as it passes from its pulley (the medial malleolus) to its insertion—the tuberosity of the navicular.

(c) Now, with the foot everted, trace the tendons of the Peronei as they turn forward below the lateral malleolus; the Brevis to the base of the 5th metatarsal, the Longus to pass into the sole of the foot just behind the base of the 5th metatarsal (*fig. 476*).

11. The tendons of the Ex. Hallucis Longus and Ex. Digitorum Longus can generally be made to stand out on the dorsum of the foot.

12. The soft cushion-like mass in front of the lateral malleolus, caused by the relaxed Ex. Digitorum Brevis, should not be mistaken for a swelling due to a sprain.

13. The pulsation of the anterior tibial and dorsalis pedis arteries may be felt midway between the two malleoli where two tendons (Tibialis Anterior and Ex. Hallucis Longus) lie medial to it, and two tendons (Ex. Digitorum Longus and its Peroneus Tertius) lie lateral.

14. The posterior tibial artery may be felt to pulsate behind the medial malleolus, but the foot should be slightly

inverted (passively) in order to relax the fascia, and you should palpate firmly forward and laterally (*fig. 489*).

THE SOLE OF THE FOOT

In the following pages on the foot, you will recognize many structures as homologues of those already met in the hand.

Articulated Foot from Below (*fig. 476*)

1. *The Bearing Points of the Foot are the medial process of the calcaneus* posteriorly, and the *heads of all five metatarsal bones* anteriorly—the *two sesamoid bones* that play one on each side of the V-shaped ridge below the head of the 1st metatarsal raise that head off the ground (*figs. 476, 477*). There is no transverse arch.

2. *The Flexor Hallucis Longus* grooves the under surface of the sustentaculum (*fig. 467*) and subsequently passes between the two sesamoid bones. The *Peroneus Longus* grooves the cuboid as it passes obliquely across the sole to the adjacent parts of the 1st metatarsal and 1st cuneiform.

3. *The Transverse Tarsal Joint.* The inferior surface of the cuboid is large and triangular. It ends behind in an *angle*, which lies below the calcaneus, which is cut away to receive it. The under surface of the navicular also has an *angle* opposite its *tuberosity*. The head of the talus is without bony support between the sustentaculum tali behind and the tuberosity and angle of the navicular in front.

Coverings

The Plantar Fascia (*fig. 478*). The *medial* and *lateral portions* of the plantar fascia cover the abductors of the great and little toes and are thin. The central portion, the **plantar aponeurosis,** has become detached from the tendon of the Plantaris and modified to act as a strong tie for the longitudinal arches of the foot. Posteriorly, it is attached to the medial process of the calcaneus; anteriorly, it splits into five bands similar to the palmar fascia (*fig. 478*).

Observe that dorsiflexion of the toes, as in walking, renders the plantar aponeurosis taut and increases the arches.

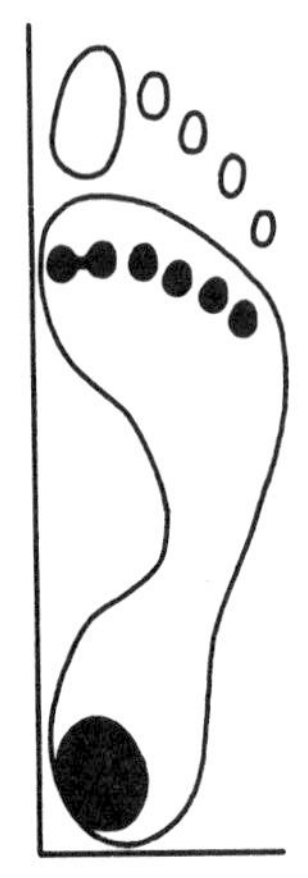

Fig. 477. The bearing points of the foot.

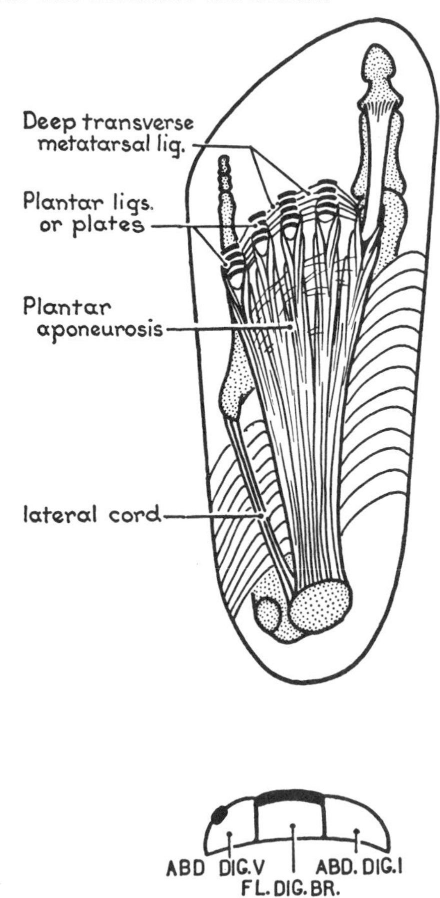

Fig. 478. The plantar fascia.

The **epidermis** and **dermis** are both much thicker on the palms and soles than elsewhere—even at birth. With pressure and friction the thickness increases. The subcutaneous fat is bound down within small fibrous compartments that serve as cushions.

Function of the Toes. In man they press into the ground and form a friction surface during the act of walking. They afford a purchase for the forward thrust of the body during running and pushing.

Cutaneous Nerves: The medial plantar nerve supplies the 3½ digits on the hallux side of the foot, just as the median nerve supplies the 3½ digits on the pollex side of the hand, leaving the lateral plantar nerve, which corresponds to the ulnar nerve in the hand, to supply 1½ digits (*fig. 479*).

The skin under the heel is supplied by the medial calcanean branches of the posterior tibial nerve which pierce the flexor retinaculum with their companion arteries (*fig. 489*).

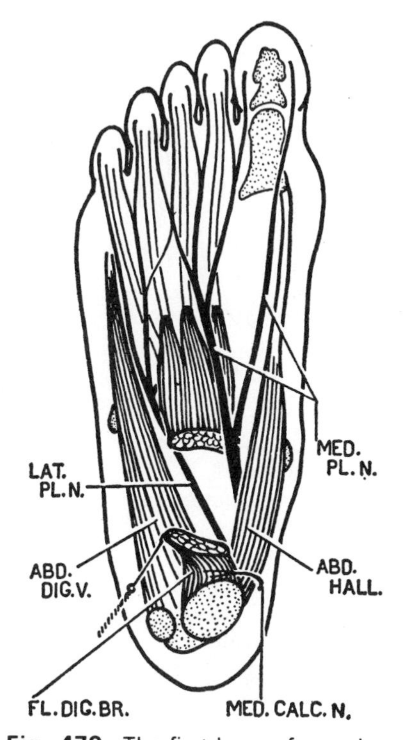

Fig. 479. The first layer of muscles.

Plantar Muscles

These are arranged in four layers.

First or Superficial Layer:

These arise from the calcaneus (*fig. 479*).

1. Abductor Hallucis and } inserted into proximal phalanges
2. Abductor Digiti Quinti }
3. Flexor Digitorum Brevis—inserted into middle phalanges.

The abductors of the 1st and 5th digits have a continuous fleshy origin from the heel. The insertions are into the bases of the proximal phalanges. Though these two muscles are "abductors," they find no occasion to abduct the toes; they flex them and act as elastic ties for their respective arches (*fig. 480*).

The Flexor Digitorum Brevis lies deep to the plantar aponeurosis. It arises mainly from the aponeurosis and slightly from the medial process of the calcaneus. Its mode of insertion into the middle phalanges is similar to that of its homologue in the hand, Fl. Digitorum Superficialis (*fig. 150*).

Second Layer of Muscles (*fig. 481*).

Long flexor tendons and short fleshy muscles attached to them form this layer.

1. Flexor Digitorum Longus tendon.
2. Quadratus Plantae or Flexor Accessorius.
3. Lumbricals.
4. Flexor Hallucis Longus tendon.

The tendon of the *Fl. Digitorum Longus* appears from under cover of the Abductor Hallucis, crossing the tendon of the Fl. Hallucis Longus. Its *insertion*, as for Fl.

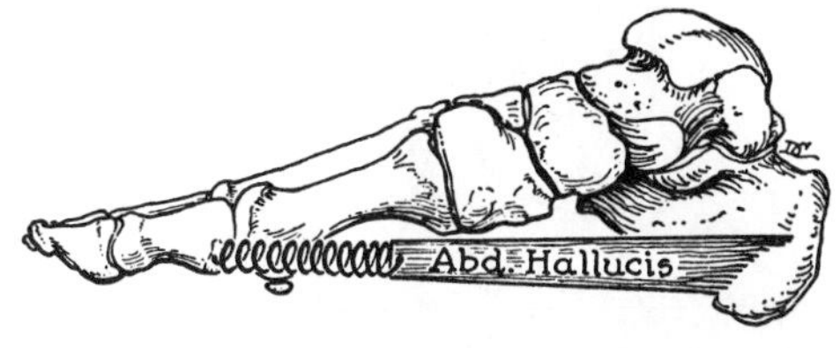

Fig. 480. The short plantar muscles act as elastic springs or ties for the arches of the foot.

Digitorum Profundus (*fig. 150*), is into the bases of distal phalanges.

The Quadratus Plantae [Fl. Accessorius] arises from the medial side of the calcaneus (*fig. 488*), and it is inserted into the tendon of the Fl. Digitorum Longus apparently to redirect the line of pull of the oblique tendons (*fig. 481*).

The four Lumbricals arise from the tendons of the Fl. Digitorum Longus and are inserted into the dorsal digital expansions.

The Flexor Hallucis Longus tendon passes onward between the sesamoid bones developed in the tendons of insertion of Fl. Hallucis Brevis, and finally is inserted into the distal phalanx of the hallux. It also sends fibrous slips to those tendons of the Fl. Digitorum Longus that pass to the 2nd (and 3rd) digits, a reflection of its phylogeny (*fig. 482*).

Third Layer of Muscles—all short (*fig. 483*).

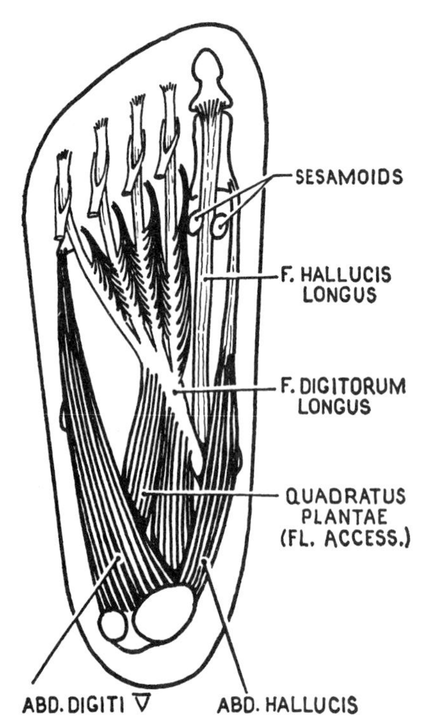

Fig. 481. The second layer of muscles displayed by removal of the Flexor Digitorum Brevis.

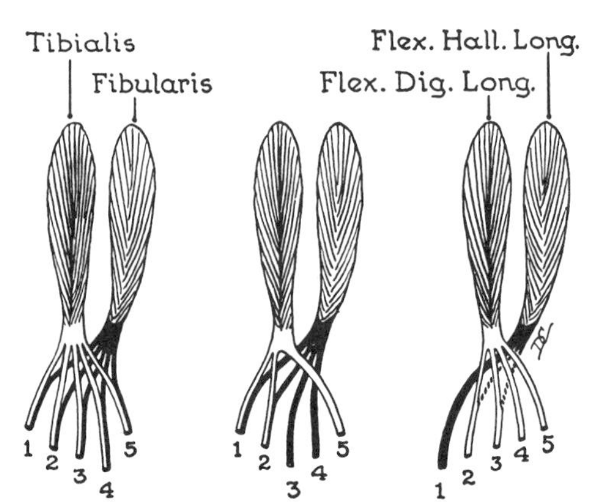

Fig. 482. Scheme explaining the crossing and variable union of long digital flexors in various mammals and man. (After Straus.)

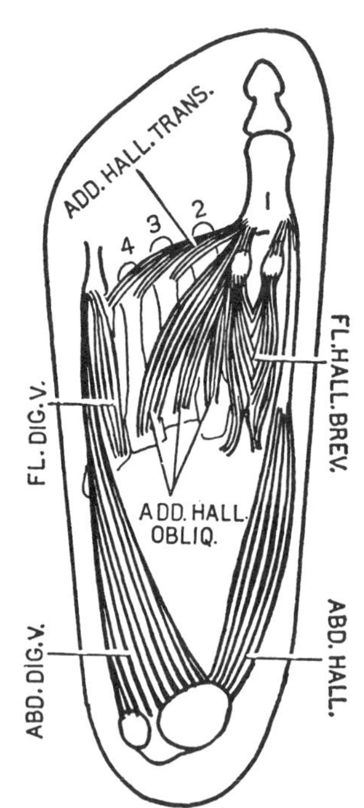

Fig. 483. The 3rd layer of muscles displayed by excision of the Flexor Digitorum Longus and the Quadratus Plantae and Lumbricals which are attached to it.

1. Flexor Hallucis Brevis.
2. Flexor Digiti Quinti and
3. Adductor Hallucis Transversus (form three sides of a square open posteriorly).
4. Adductor Hallucis Obliquus (largely fills the square).

The muscles of both third and fourth layers practically confine themselves to the anterior half of the foot. Details of the third layer of small muscles—*see figure 483.*

The Abductor Hallucis is inserted in conjunction with the medial head of Fl. Hallucis Brevis; the *Adductores Hallucis Obliquus et Transversus* with the lateral head. Two sesamoid bones develop in the tendons of insertion (*figs. 483, 483.1*).

Fourth Layer of Muscles:

1. Seven Interossei.
2. Two tendons on the skeletal plane.

Interossei, 4 Dorsal and 3 Plantar. These seven muscles are like their homologues in the hand, except that the axis runs through the 2nd metatarsal, not the 3rd. The dorsal muscles would abduct and the plantar adduct if they could, but their real *function* emerges during walking. The Interossei flex the metatarsophalangeal joints and thereby draw the heads of the metatarsals together (i.e., they keep the foot from spreading), and they extend the interphalangeal joints and thereby keep the toes from curling up.

Two Long Tendons on Skeletal Plane. The Tibialis Posterior and the Peroneus Longus belong to the ligamentous or deepest layer of the posterior half of the foot. One being an invertor, the other an evertor, they must find attachment in front of the transverse tarsal articulation.

1. The **Tibialis Posterior** (*fig. 484*) sends two-thirds of its tendon to insert into the tuberosity of the navicular. One-third divides into a number of finger-like bands that pass below the spring ligament and diverge to reach the cuboid, cuneiforms, and 2nd, 3rd and 4th metatarsal bases.

2. The **Peroneus Longus** enters the sole deep to the Abductor Digiti V, passes obliquely across the sole in the groove on the cuboid to its insertion into the lateral side of the base of the first metatarsal and adjacent part of the first cuneiform (*fig. 476*). The **Tibialis Anterior** is inserted into the same two bones as the Peroneus Longus, but on their medial side (*fig. 484*).

Plantar Vessels and Nerves

The Lateral Plantar Artery (*fig. 487*) is much larger than the *medial plantar artery* and is, in fact, the continuation of the posterior tibial artery. It enters the sole deep to the Abductor Hallucis and, with its companion nerve, runs forward and laterally between the 1st and 2nd layers of muscles. Then, dipping deeper, it runs medially between the 3rd and 4th layers (*fig. 485*), forming the (deep) **plantar arch.**

The arteries of the anterior half of the foot are arranged like the arteries of the hand (*figs. 486, 487*). There are similar anastomotic channels. The names are not important.

BRANCHES. The deep branch of the dorsalis pedis a. is but an enlarged (1st) perforating a.

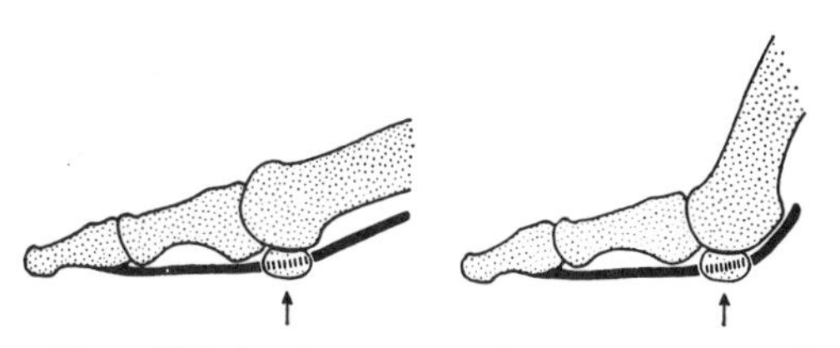

Fig. 483.1. The sesamoids are always the bearing points for the head of the 1st metatarsal, allowing free play for the long flexor tendon.

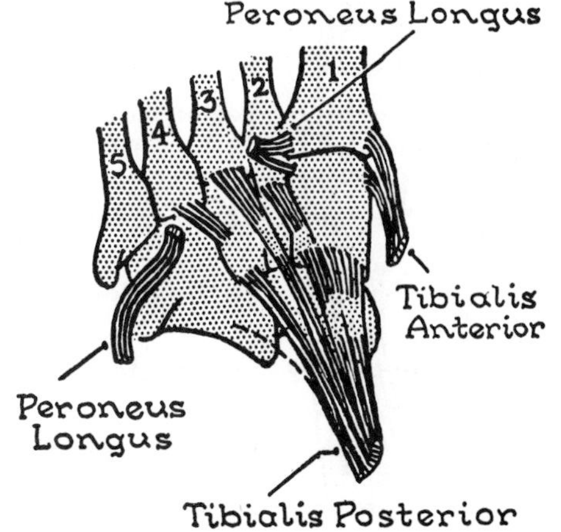

Fig. 484. The grasp of the Tibialis Posterior extends to the small tarsals and to the metatarsals.

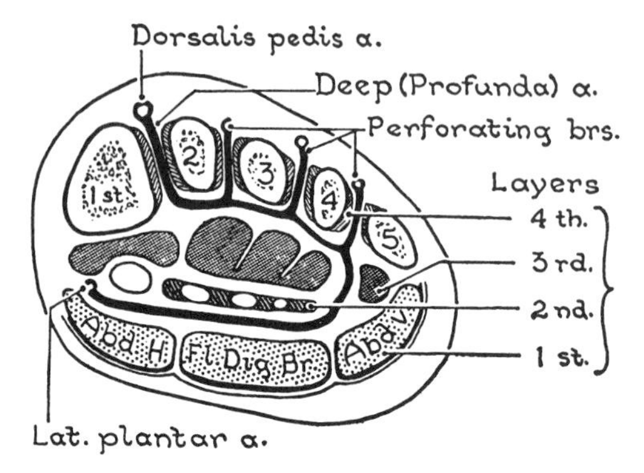

Fig. 485. Scheme to show the lateral plantar artery coursing first between the 1st and 2nd layers of muscles, then between the 3rd and 4th layers.

Fig. 486. Scheme of arteries in the anterior part of the foot (compare with *fig. 155*).

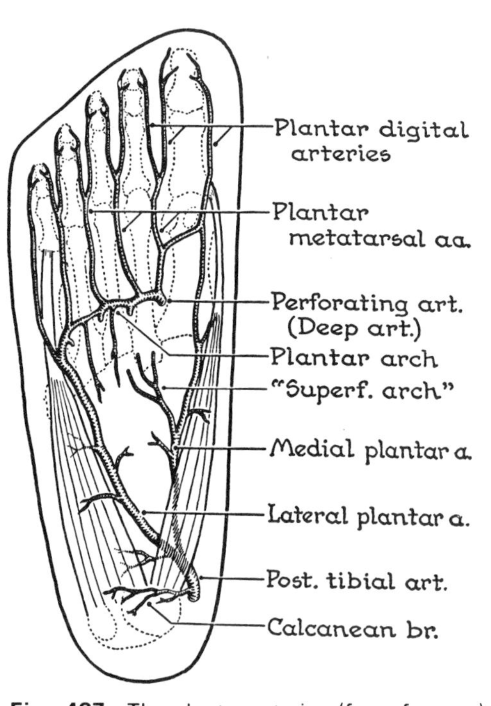

Fig. 487. The plantar arteries (for reference).

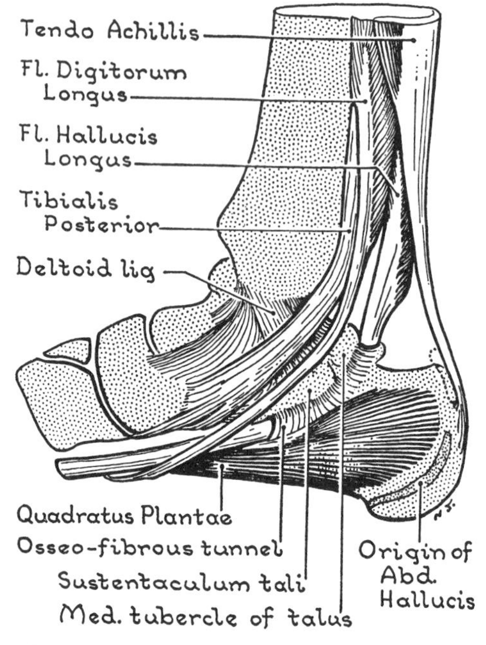

Fig. 488. The tendons of the 3 deep muscles, at the ankle.

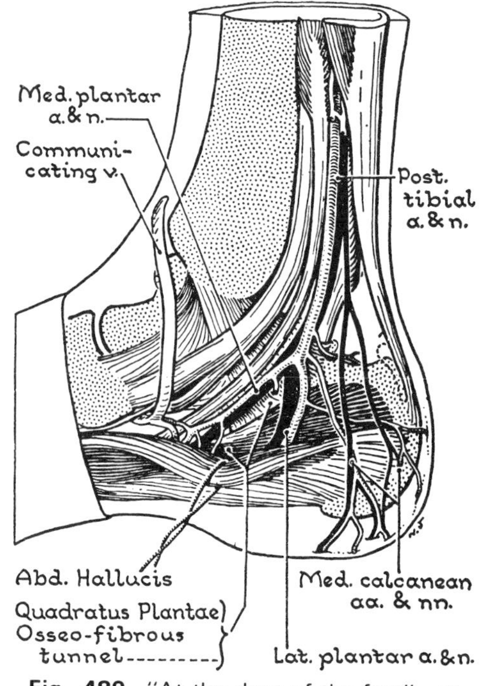

Fig. 489. "At the door of the foot"—structures passing deep to the Abductor Hallucis.

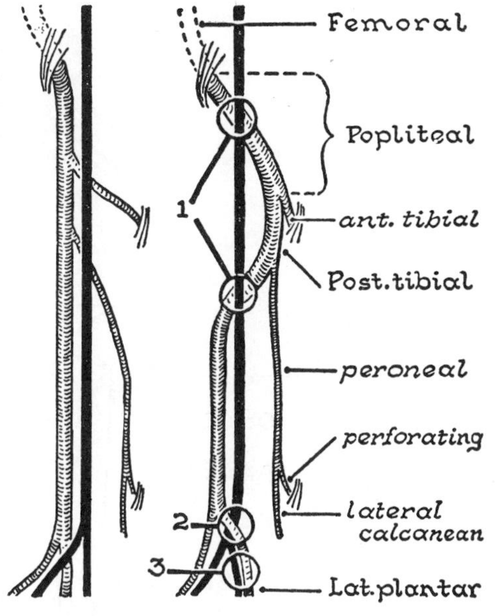

Fig. 490. Unexpected relationships of arteries to nerves.

(*fig. 485*). From the deep plantar arch spring: (1) the three other *perforating aa.*, which join the dorsal metatarsal aa.; (2) four *plantar metatarsal aa.*, which bifurcate into *plantar digital aa.*; and (3) a branch to the medial side of the big toe and one to the lateral side of the little toe.

The Lateral Plantar Nerve follows its artery closely and sends: (1) cutaneous branches to the lateral 1½ digits, and (2) motor branches to all the muscles of the sole not supplied by the medial plantar nerve. This includes all seven Interossei and the Adductor Hallucis (*cf.* ulnar nerve).

The Medial Plantar Nerve, like the median nerve of the hand, sends:

1. Cutaneous branches to 3½ digits which fan out subjacent to the plantar fascia.

2. Motor branches to the four muscles between which the nerve runs, namely:

   a. Abductor Hallucis
   b. Flexor Digitorum Brevis

*Segmental Innervation of Muscles of Leg and Foot**

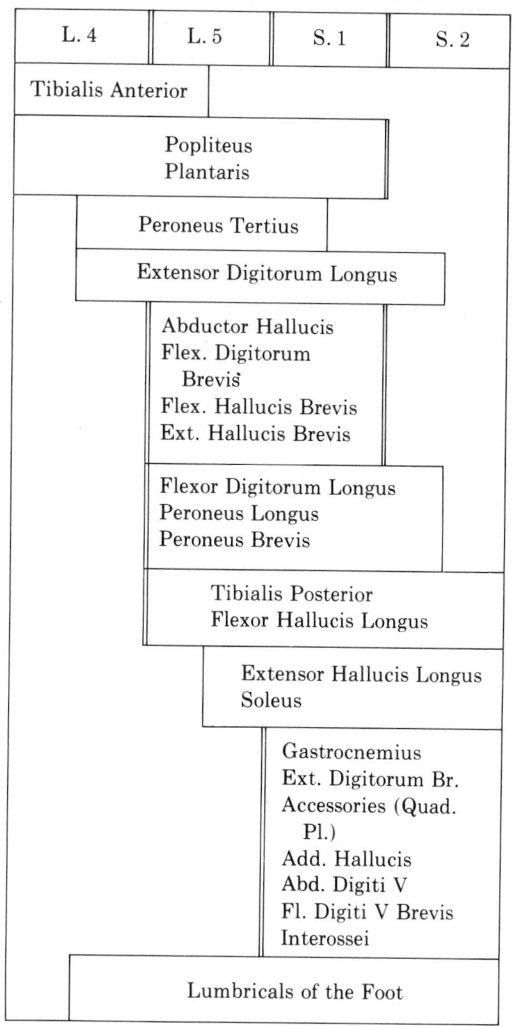

L. 4	L. 5	S. 1	S. 2
Tibialis Anterior			
Popliteus Plantaris			
Peroneus Tertius			
Extensor Digitorum Longus			
	Abductor Hallucis Flex. Digitorum Brevis Flex. Hallucis Brevis Ext. Hallucis Brevis		
	Flexor Digitorum Longus Peroneus Longus Peroneus Brevis		
	Tibialis Posterior Flexor Hallucis Longus		
	Extensor Hallucis Longus Soleus		
		Gastrocnemius Ext. Digitorum Br. Accessories (Quad. Pl.) Add. Hallucis Abd. Digiti V Fl. Digiti V Brevis Interossei	
Lumbricals of the Foot			

* Modified after Bing; and Haymaker and Woodhall. (For Hip and Thigh, p. 335.)

   c. Flexor Hallucis Brevis
   d. Lumbricalis I.

"The Door of the Foot". (1) Since the lateral and medial plantar vessels and nerves and three tendons enter the sole on the medial side by passing deep to the Abductor Hallucis, this entrance may be thought of as the door of the sole of the foot. (2) The Peroneus Longus enters on the lateral side deep to the Abductor Digiti V, through the back door, so to speak. (3)

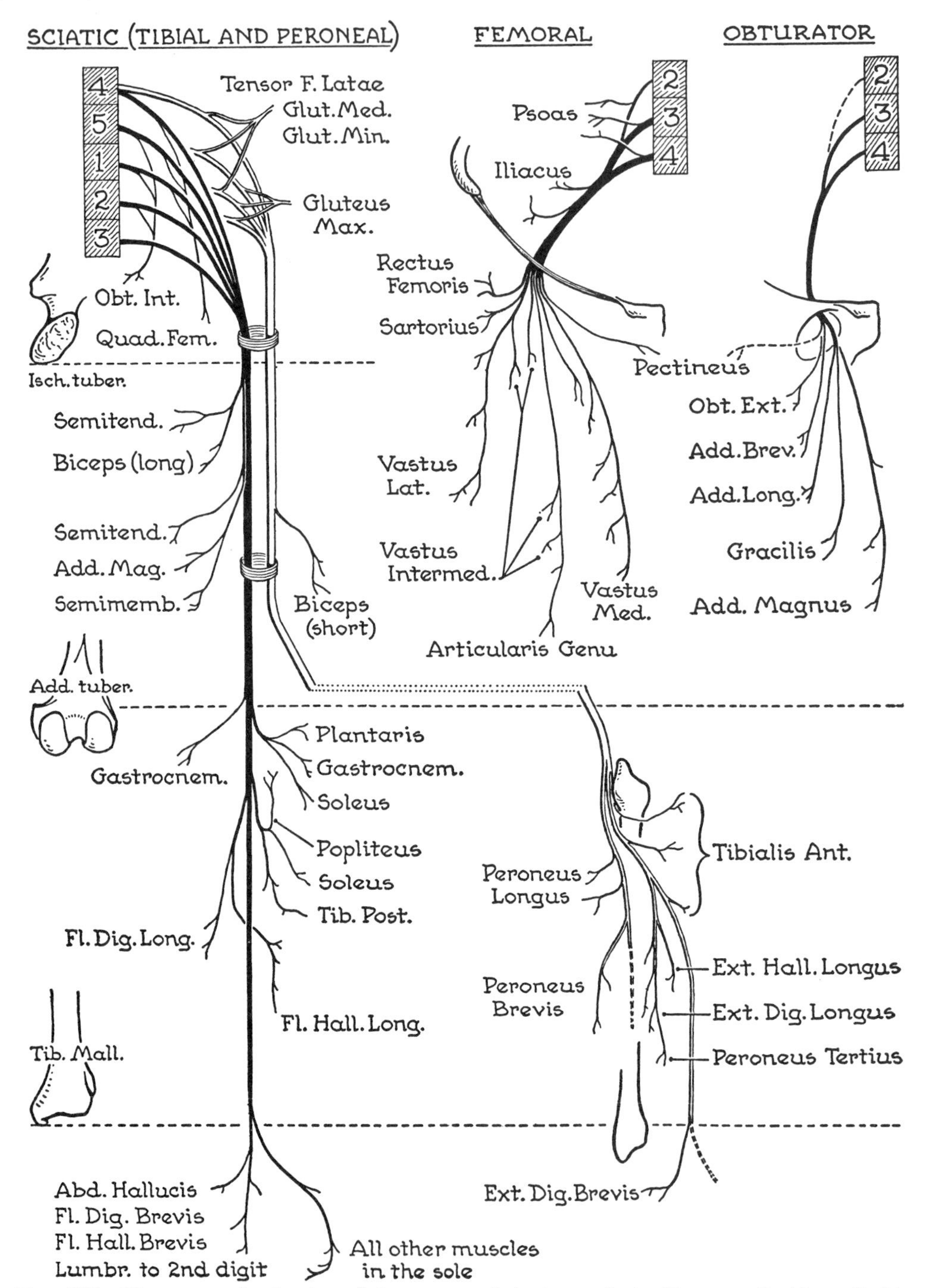

Fig. 491. The motor distribution of the nerves of the lower limb. (You may find it profitable to compare the levels of origin of the branches in the limb you are dissecting with these average levels.)

The perforating vessels pass through the intermetatarsal spaces, as it were, through windows. The Abductor Hallucis, then, guards the door, where *Quadratus Plantae* provides a soft door-mat on the bone (*figs. 488, 489*).

Comment on Vessels and Nerves. Recall that *an artery accompanies its companion nerve on the side from which it approaches it*. In the lower limb, the rule is tested by several exceptions, which, happily, have little other importance (*fig. 490*).

Review of Nerve Supplies. See the tables on pages 335 and 361 and *figure 491*.

28 JOINTS OF LOWER LIMB

HIP JOINT

The Ball and Socket; Epiphyses; Fibrous Capsule and Ligaments: iliofemoral, pubofemoral, and ischiofemoral.
Sequences; Tendon of Psoas; Relationships; Synovial Membrane; Ligament of Head; Dislocation; Relations.
Blood Supply; Nerve Supply; Movements (and Muscles).

KNEE JOINT

General Observations; Bursae; Divisions
Patella and Patellar Articulation–Facets; Palpation; Functions.
Tibiofemoral Articulations–Fibrous Capsule; Collateral Ligaments: Tibial; Fibular; Sequences: 13 in all.
Menisci.
Intercondylar Septum; Infrapatellar Synovial Fold.
Cruciate Ligaments
Synovial Cavity and Communicating Bursae; Extent of Cavity; Relations; Blood Supply; Nerve Supply.
Bursae about the Knee; Epiphyses.

ANKLE JOINT

Nine Basic Observations; Bony Parts.
Ligaments; Synovial Membrane; Relations; Movements.
Blood Supply; Nerve Supply; Other Observations; Epiphyses.

JOINTS OF FOOT

Talocalcanean Joint—Bony Parts Reviewed; Ligaments; Five Tendons; Subdivisions.
Spring Ligament; Bifurcate Ligament.
Transverse Tarsal Joint—Talonavicular and Calcaneocuboid.
Long and Short Plantar Ligaments.

Arches of the Foot

Joints Distal to Transverse Tarsal Joints

DYNAMICS OF FOOT

Standing—Weight Distribution; Arch Support; Muscles.
Pentadactyl Hand and Foot.
Supernumerary Ossicles.

THE HIP JOINT

The Ball and Socket. The *head of the femur* forms two-thirds of a sphere, slightly flattened above where the acetabulum rests most heavily upon it. It points medially, upward, and forward, so its anterior part is not engaged in the socket when the body is in the anatomical position.

The *acetabulum* (L. = vinegar cup) is deficient inferiorly at the *acetabular notch*. The remainder is horseshoe-shaped and covered with cartilage (*fig. 492*). Within the horseshoe, the *acetabular fossa* is nonarticular and thin.

The acetabular notch is converted into a foramen by the *transverse acetabular ligament* (*fig. 492*). The acetabulum is deepened by a complete ring of pliable fibrocartilage, the *acetabular labrum*, which is attached to its brim and to the transverse ligament. The labrum grasps the head of the femur beyond its "equator."

Epiphyses. The epiphysis of the head of the femur fits like a cap on a spike, and the epiphyseal line encircles the articular margin. It lies entirely within the synovial capsule (*fig. 496*). *Ossification* of the head, which starts during the 1st year, is always complete before the 20th year (McKern and Stewart).

The ilium, ischium, and pubis each contribute to the acetabulum (*fig. 492*). The epiphyseal line is triradiate; synostosis is always complete by the 17th year.

Fibrous Capsule and Ligaments. The fibrous capsule is a very strong, thick sleeve attached around the brim of the acetabuum and to the labrum and transverse ligament. Distally, its femoral attachment is along the whole length of the intertrochanteric line and to a line on the surface of the neck below, but not posteriorly or above. The fibers run an oblique or spiral course.

Assuming the fibers were originally parallel, then their oblique or spiral course can be attributed to the assumption of the erect posture. It has 3 parts.

The Iliofemoral Ligament (*fig. 493*) is a broad, strong band, shaped like an inverted Y. Above, it is attached to the ant. inf. iliac spine and to the acetabular margin, so it lies deep to the two heads of origin of the Rectus Femoris and is co-extensive with them. Below, it creates the broad, rough intertrochanteric line of the femur.

The Pubofemoral and Ischiofemoral Ligaments are attached to the pubic and ischial parts, respectively, of the acetabular margin and they pass across the back of the neck of the femur to the upper end of the intertrochanteric line (*fig. 494*).

The orbicular zone is a collection of deep fibers that run circularly and cause an hourglass constriction within the capsule (*fig. 496*).

Sequences. When you stand erect, your line of gravity passes behind the hip joints. Your trunk tends to fall backward, or rather to rotate backward. It is to check this backward rotation that the anterior part of the capsule is thickened to form the iliofemoral ligament.

Between the iliofemoral and pubofemoral ligaments, an area needing support, there is no ligament; indeed, this part of the capsule is commonly perforated. However, guarding this critical area is the **tendon of the Psoas**—active and alert. Its *bursa* communicates with the joint through the perforation (*fig. 495*)—rarely in the young, but commonly in the adult (20 per cent of 478 limbs, aged 20 to 92 years).

Synovial Membrane lines all parts of the interior of the joint, except where there is

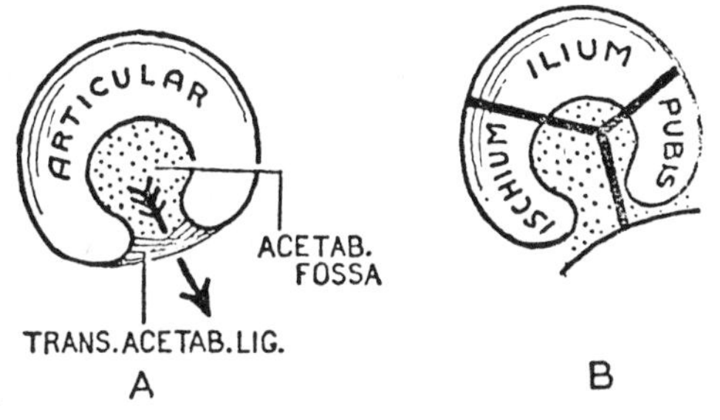

Fig. 492. *A*, the acetabulum shown in correct orientation and the transverse acetabular ligament. The *arrow* passes through the acetabular foramen. *B*, triradiate "epiphyseal" cartilage.

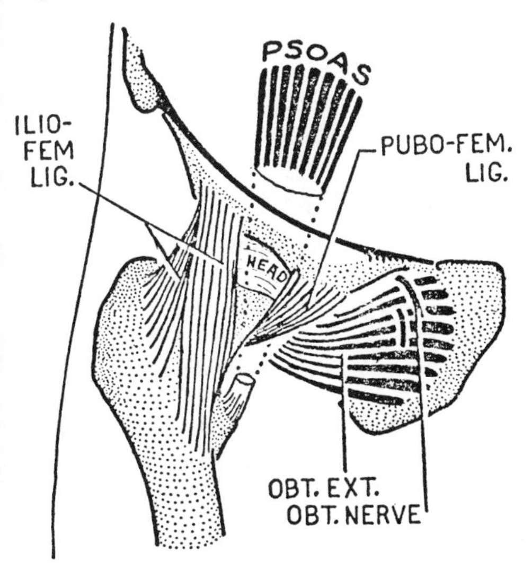

Fig. 493. Capsule of hip joint (from the front). The Psoas guards the weak point.

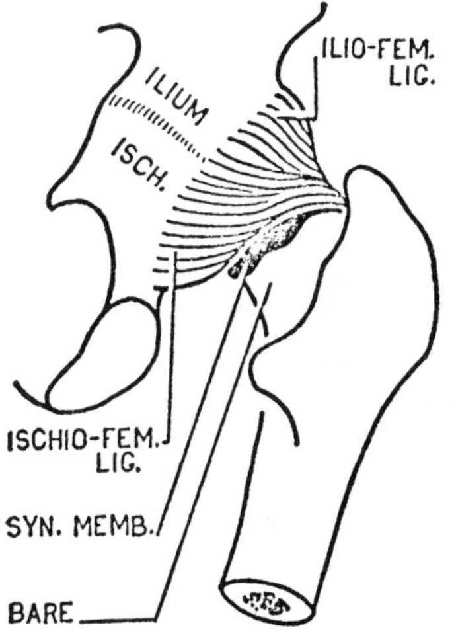

Fig. 494. Capsule of hip joint (from behind).

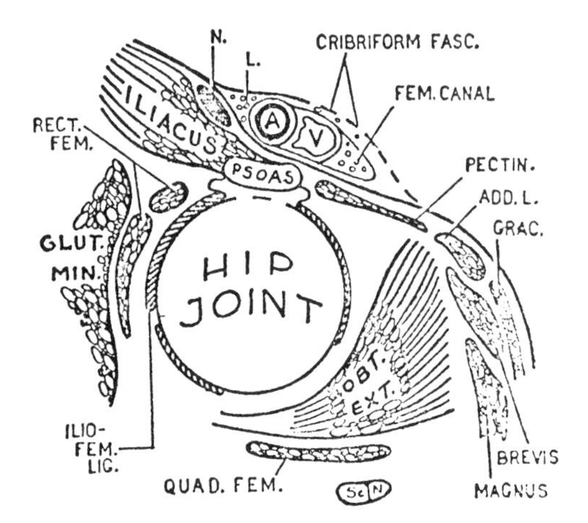

Fig. 495. Sequences: The hip joint and its relations; coronal section, semidiagrammatic.

cartilage. It is so with all synovial joints. The membrane lines the neck of the femur completely in front, and as far as the Obturator Externus tendon behind, and it stretches across the acetabular fossa.

The synovial membrane protrudes posteriorly and acts as a bursa for the Obturator Externus tendon (*fig. 494*). The synovial membrane around the neck is raised into several loose longitudinal folds, in which arteries ascend to the head.

The fat in the obturator region sends a prolongation under the transverse ligament into the acetabular fossa where it forms an extrasynovial fat pad (*fig. 496*).

Ligament of the Head. Within the joint there is a hollow cone of synovial membrane that transmits blood vessels to the head of the femur. This lies between the pit or *fovea* on the head of the femur and the acetabular fossa, so it is flattened and triangular like a flattened megaphone. The base is continuous with the sheet of synovial membrane that covers the fat pad in the acetabular fossa—this sheet is attached to the articular margins of the fossa.

A probe pushed through the acetabular foramen will pass either into the acetabular fossa, or into the synovial cone (*fig. 496*). Branches of the medial femoral circumflex and obturator arteries follow both courses.

Observations:

1. When you stand erect or bend forward over a basin, your ankle and knee joints are locked, and your foot, leg, and thigh bones become temporarily a rigid unit (*fig. 497*). A rotary force then applied to the foot is greatly amplified at the upper end of the femur—as at the end of a long screw driver—and may possibly result in fracture of the neck of the femur.

2. A pillow placed under your knee, when you lie down, automatically results in flexion of the hip, knee and ankle.

3. The posture conducive to **dislocation of the hip joint** is one in which the joint is fully flexed and medially rotated, as on bending forward with toes turned in. Flexion brings the shallow part of the acetabulum to rest on the femur; medial rotation

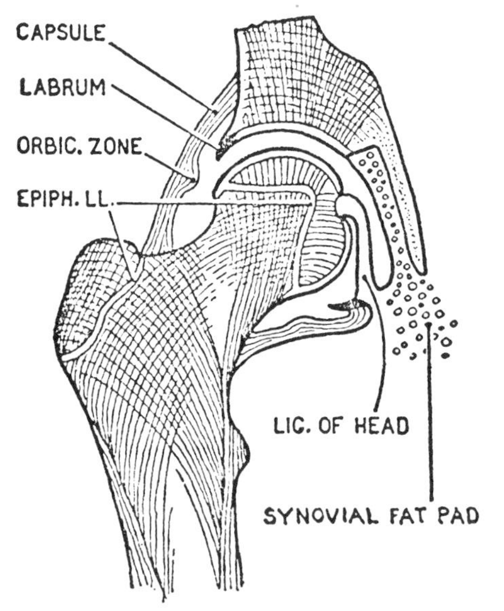

Fig. 496. The hip joint in coronal section. Note the lines of strain and stress.

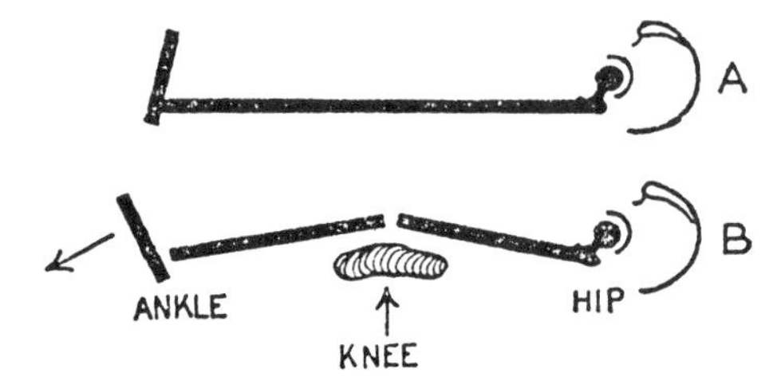

Fig. 497. The lower limb. *A*, rigid; *B*, flail, due to pillow flexing the knee.

brings the head of the femur to the back. A weight (such as a sack of potatoes) then falling on the back will dislocate the head of the femur onto the dorsum ilii.

Relations are shown in *figure 495*. The relationship of the sciatic nerve is appreciated only when the obliquity of the acetabular margin is kept in mind (*figs. 419–421*).

Blood Supply. Medial femoral circumflex, lateral femoral circumflex, obturator, and gluteal arteries supply large twigs.

The head of the femur receives three sets of arteries. Details of clinical significance are:

(1) The never failing and main set of three or four arteries that ascends in the synovial retinacula on the posterosuperior and postero-inferior parts of the neck, perforate the neck just distal to the head, and bend at 45° toward the center of the head where they anastomose freely with (2) terminal branches of the nutrient artery of the shaft, and in 80 per cent of instances with (3) the artery of the lig. of the head (lig. teres). This last artery has established a precarious anastomosis in the epiphysis of the head by the 10th year, further improvement occurring during adolescence (Trueta). This anastomosis persists even in those of advanced age, but in 20 per cent it is never established (Wolcott).

Nerve Supply. *Femoral nerve* (via n. to Rectus Femoris) p. 319. *Obturator nerve* (via one or two branches that pass laterally) p. 333. *Sciatic nerve* (via n. to Quadratus Femoris) to the back of the capsule, and sometimes twigs from the *Superior gluteal nerve* (E. Gardner).

Movements. The movements permitted are circumduction and rotation. *Flexion* is arrested by the hams, when the knee is extended. *Extension* "winds up" the spirally running fibers of the capsule and thereby forces the head deeper into the socket, and is self-arresting. When the joint is slightly hyperextended, the articular surfaces are completely congruous (T. Walmsley). *Medial rotation* also winds up the fibers; *lateral rotation* unwinds them and is more free.

To understand the *Actions* of the abductors and medial rotators, the pelvis and femur must be held in correct orientation. Thus, (1) the anterior superior spines and the upper end of the symphysis lie on the same coronal plane, and (2) the head of the femur is directed forward as well as medially and upwards, i.e., the greater trochanter lies behind the plane of the head. The *function of the abductors* is to prevent the pelvis from becoming adducted, that is, to prevent the body from falling to the unsupported side when one foot is off the ground, as in walking (*fig. 498*). The *function of the medial rotators* is to rotate the unsupported, or opposite, side of the pelvis forward and thereby to increase the stride. (See *fig. 499* and Table 15.)

The Gluteus Maximus is necessary for rising from the sitting position, for climbing stairs, and for running and jumping, but it is hardly required in walking on the flat, which is essentially movement for flexors, and there the hams suffice as extensors.

THE KNEE JOINT

General Observations. It is apparent that the chief movements occurring at the knee joint are *flexion* and *extension* and that the joint is of the hinge variety. Some degree of *axial rotation* is also permitted while the knee is in the position of flexion and semiflexion. Later you will see that a slight degree of medial rotation of the femur is necessary to the completion of the act of extension.

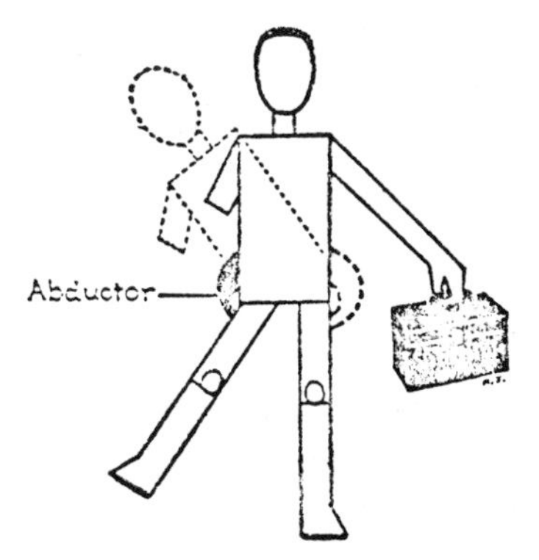

Fig. 498. Weight, substituting for paralyzed Abductors, demonstrates the chief function of Gluteus Medius and Gluteus Minimus.

TABLE 15

Muscles Acting on the Hip Joint

Circumductors			
Flexors	Extensors	Abductors	Adductors
Iliopsoas Tensor Fasciae Latae Sartorius Pectineus Rectus Femoris Adductor Longus Adductor Brevis Adductor Magnus (Obt. part)	Gluteus Maximus The three hams Adductor Magnus (Ham part)	Gluteus Medius Gluteus Minimus (Tensor Fasciae Latae) Piriformis Sartorius	Adductor Magnus Adductor Brevis Adductor Longus Pectineus Gluteus Maximus Short Muscles Obturator Internus Gemelli Obturator Externus Quadratus Femoris
		Rotators	
		Medial	Lateral
		Gluteus Medius Gluteus Minimus (Tensor Fasciae Latae) Adductor Magnus (Ham part) Pectineus Upper adductor mass	Gluteus Maximus Short Muscles Piriformis Obturator Internus Gemelli Obturator Externus Quadratus Femoris Iliopsoas

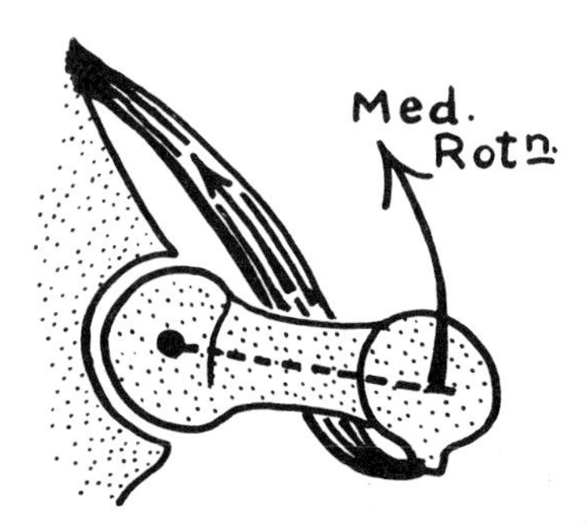

Fig. 499. Explaining why the upper adductor muscles, although inserting on the linea aspera (on the back of the femur), are medial rotators of the femur.

When standing up and leaning forward, as when washing your face, you can, using your hands, *move your patella from side to side* because the Quadriceps Femoris is then relaxed. Hence, an unexpected blow on the back of the knee may cause you to fall. The Quadriceps is not required to be in action when you stand erect, because the *line of gravity* passes in front of the axis of the knee joint (*fig. 23*). A considerable economy in muscular effort is effected thereby.

Bursae. A bursa, the *prepatellar bursa*, lies between the skin and patella, permitting free movement of the skin.

In front of the lig. patellae, there is another bursa, the *superficial infrapatellar bursa*.

The precise depth of the *prepatellar bursa* varies: it may be either in the subcutaneous areolar tissue, or deep to the fascia lata, or actually in the substance of those fibers of the Quadriceps tendon that pass across the front of the patella. Commonly, two bursae are present, one in front of the other, with the intervening partition largely broken down.

Divisions of the Joint. The bones taking part in the knee joint are: femur, tibia, and patella; the fibula is only indirectly associated. Primitively, there were *three joint cavities* now merged into one. One is situated between the medial condyles of the femur and tibia; one between the lateral condyles of the femur and tibia; and one between the patella and the femur. They may be referred to as the *medial and lateral condylar articulations* and the *patellar articulation.* The condylar articulations are partly subdivided by medial and lateral menisci (semilunar cartilages) into upper and lower parts.

The Patella and the Patellar Articulation

The femora are set obliquely. There is, accordingly, an open angle at the lateral side of the knee, toward which the patella tends to become dislocated, when the Quadriceps contracts. Dislocations are, however, not common, largely because of two factors:

1. The forward projection of the lateral condyle of the femur (*fig. 500*).
2. The low attachment to the patella of the Vastus Medialis (*fig. 501*), which draws the patella medially.

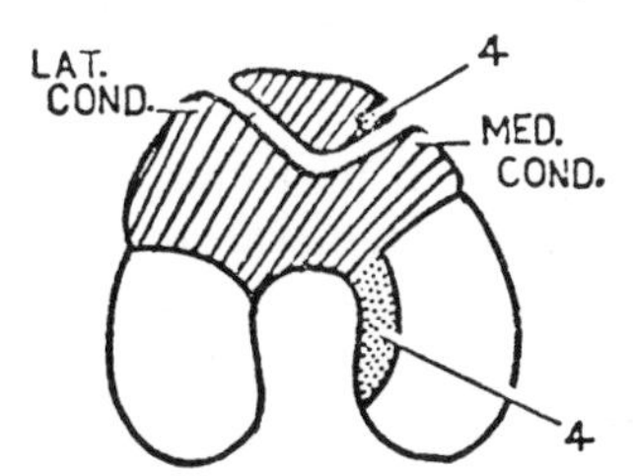

Fig. 500. The patella in its trochlea.

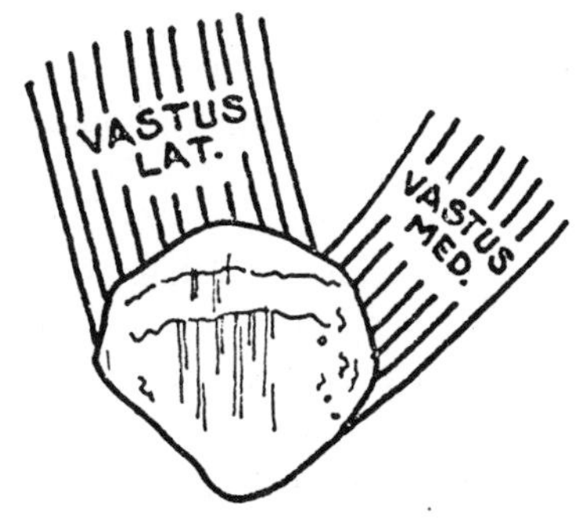

Fig. 501. The patellar attachments of the Vasti Medialis and Lateralis.

The Posterior Surface of the patella (*fig. 502*) has paired facets that articulate in turn during extension, slight flexion, flexion, and full flexion (*fig. 503*).

The human patella would be less liable to fracture if it possessed a single concave facet as in lower mammals (*fig. 504*).

Palpation. A considerable amount of the articular surfaces of the knee joint can be palpated.

Functions of the Patella. Experimental work has shown that when the knee joint is flexed, the patella is a mechanical hindrance to extension. Indeed, the removal or excision of the patella results in increased efficiency—but in the later degrees of extension (say 150°–180°) the patella improves the efficiency by holding the patellar tendon away from the axis and thereby increasing the extending momentum of the Quadriceps pull (H. Haxton) (*fig. 505*).

Tibiofemoral or "Condylar" Articulations

On the upper surface of each tibial condyle there is an oval articular area for the corresponding femoral condyle. The articular areas are separated from each other by a narrow nonarticular area, which widens in front and behind into an *anterior* and a *posterior intercondylar area* (*fig. 509*).

The Fibrous Capsule. The lig. patellae and the coronary and medial ligaments, but not the lateral lig., are part of the fibrous capsule. The fibrous capsule is reinforced in front and at the sides by

Fig. 502. The three paired facets on the posterior surface of the patella articulate with the femur as shown in *figure 503*. The medial vertical facet [4] articulates along the margin of the intercondylar notch during full flexion (*see fig. 500*).

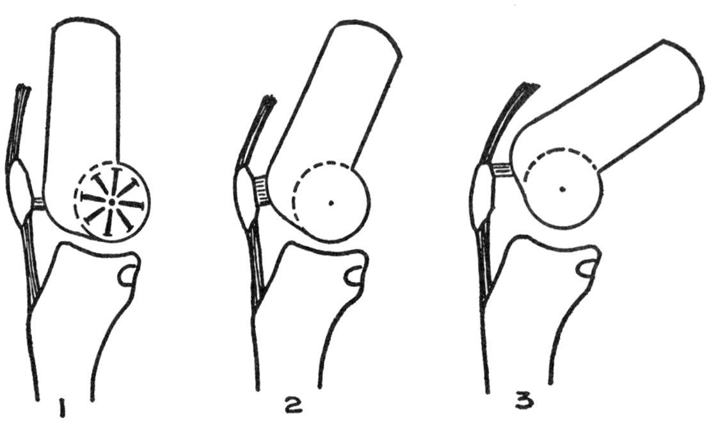

Fig. 503. The knee joint during [*1*] *extension,* [*2*] slight flexion, [*3*] flexion. At the hub of the wheel, or center of the disc, lies the epicondyle.

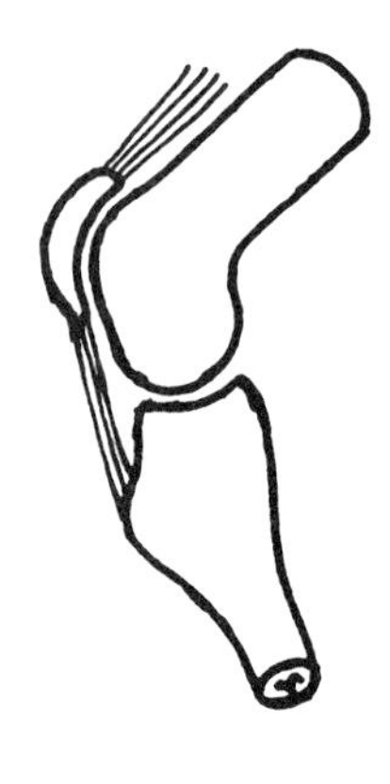

Fig. 504. Knee joint of a sheep with concave patellar facet.

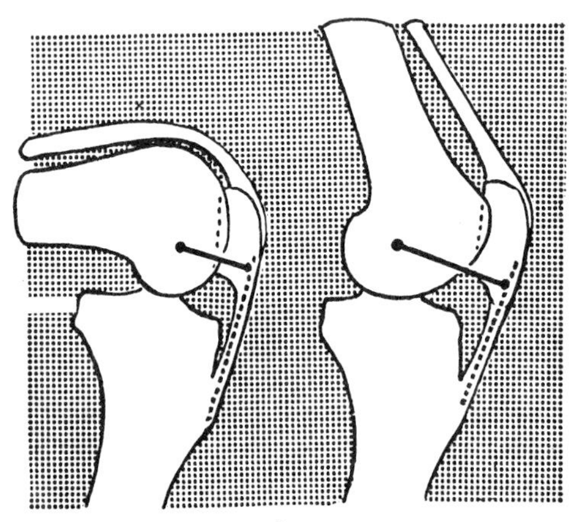

Fig. 505. The patella hindering (*left*) and assisting (*right*) extension. (After Haxton.)

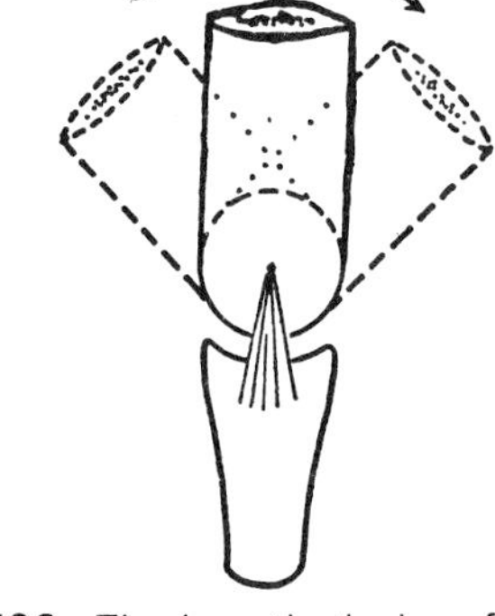

Fig. 506. The hypothetical perfect hinge.

expansions from the Vasti, Sartorius, Semimembranosus, Biceps, and iliotibial tract. (For *details*, see Harty; Hughston and Eilers; Marshall *et al.*; Noyes and Sonstegard; and Robichon and Romero.) Behind, the capsule is composed of fibers that run parallel with the Popliteus, i.e., obliquely downward and medially from femur to tibia; and one band, known as the *oblique popliteal ligament*, is attached to the Semimembranosus tendon.

Collateral Ligaments. If the femoral condyles were round and their collateral ligaments arranged as in *figure 506*, the joint could be flexed both forward and backward, because the ligaments would be equally taut in all positions. But: (1) the medial and lateral femoral condyles project backward like discs or wheels; (2) the collateral ligaments, attached above to the epicondyles at the centers of the superadded wheels; and (3) below they are attached far back (*figs. 507* and *508*). As a result, when the knee is flexed, the collateral ligaments are slack and permit medial and lateral rotation. When the knee is extended, they are taut.

The Tibial Collateral Ligament (Medial Lig.) has a superficial and a deep part. The superficial part is a long band that bridges the tibial condyle and the hollow below (*fig. 511*). The deep part is deltoid and is attached to the margin of the tibial condyle.

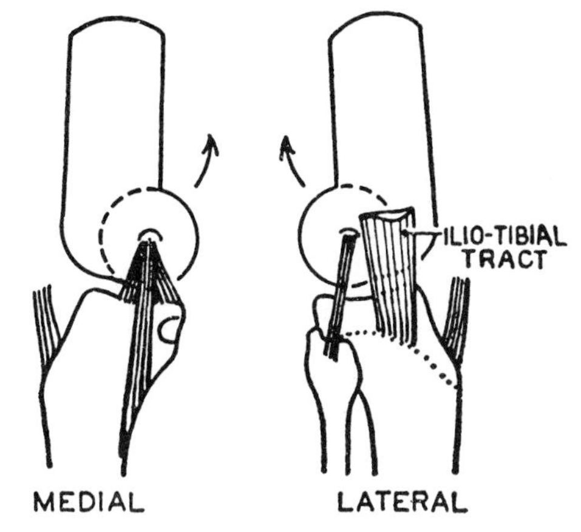

Fig. 507. The eccentric attachments of collateral ligaments. The insertion of the iliotibial tract in front of the transverse axis of the joint helps to keep the extended joint extended.

Fig. 508. The collateral ligaments are slack during the flexion and permit rotation.

One or more *bursae* lie deep to the long band. The band is crossed by three tendons—Sartorius, Gracilis, and Semitendinosus.

The Fibular Collateral Ligament (Lateral Lig.) is a cord that extends from the lateral epicondyle downward and backward to the head of the fibúla (*figs. 507, 517*). This cord is crossed by one tendon—the Biceps tendon—a bursa intervening (*fig. 428.1*).

Menisci, Cruciate Ligaments, Condyles, and Movements

Sequences. The upper ends of the tibia and fibula may be arbitrarily fitted into *an oblong frame* in which the head of the fibula occupies the posterolateral angle (*fig. 509*) From this fact the following sequences might by surmised:

1. The lateral condyle of the tibia is shorter from front to back.

2. Therefore, of the two **menisci (semilunar fibrocartilages)** that rest on and fit the tibial condyles, the lateral is shorter than the medial (*fig. 510*). The shorter lateral meniscus is shaped like a small "o," the longer medial meniscus like a capital "C". The ligamentous horns of the "C" are attached far apart, embracing those of the "o" on the nonarticular part of the tibial plateau (*fig. 510*).

3. Therefore, the portion of the lateral condyle of the femur that articulates with the lateral condyle of the tibia (and its

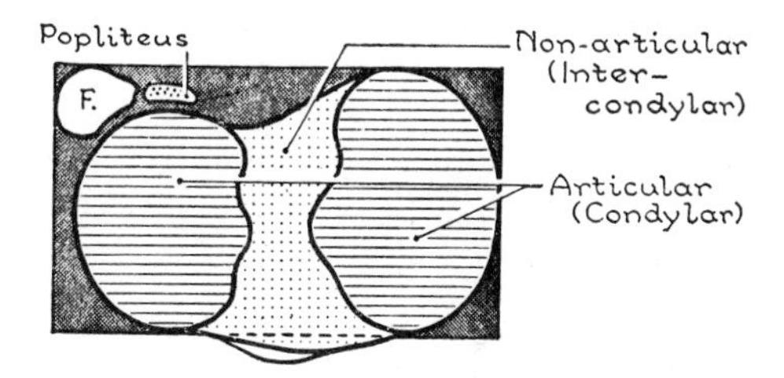

Fig. 509. Upper ends of tibia and fibula in oblong frame.

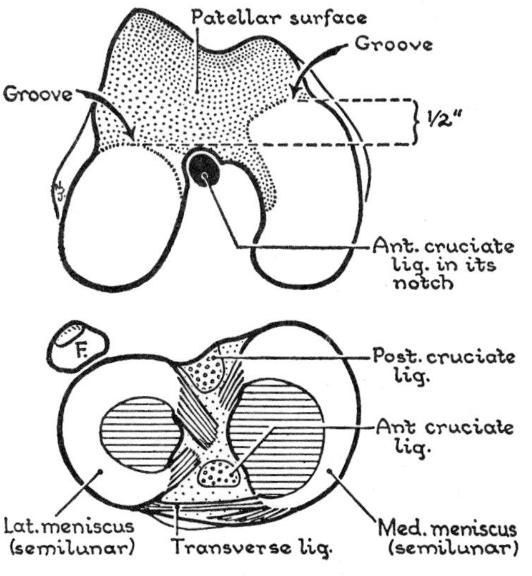

Fig. 510. The articular surfaces of the knee joint.

meniscus) is shorter anteroposteriorly than the corresponding part of the medial condyle of the femur (*fig. 510*). (Note, the condyles of the femur also have patellar areas. Just now reference is to the tibiomeniscal areas.)

4. Therefore, as the joint passes from full flexion to full extension, the medial condyle of the femur can roll farther than the lateral condyle.

5. During extension, the two femoral condyles revolve or "spin" on the tibia and its menisci. The *posterior cruciate lig.*, acting as a drag, greatly restricts the forward roll of the condyles and causes them to spin. In effect, the movement is a *hinge movement* with the addition of some true *forward roll.*

When the lateral femoral condyle can revolve no more, the longer medial condyle still has surface available (1 cm) to revolve (*fig. 510*).

6. During extension the turning of the lateral femoral condyle is arrested by two obstacles. One, the anterior margin of the lateral meniscus, creates and fits into the *curved groove* on the femur (running from the anterior part of the intercondylar notch to the lateral margin of the condyle); the other, the taut *anterior cruciate ligament*, creates and fits into its own secondary notch in the intercondylar notch (*fig. 510*).

7. Extension continues while the medial femoral condyle is completing its final spin and roll, but the femur must rotate medially on its long axis to permit this. The pivot around which it rotates is the anterior cruciate ligament.

8. At the same time the lateral femoral condyle, which is no longer rolling, and the lateral meniscus, whose sharp anterior margin is locked to it by its groove, slide forward together on the tibia, moving as one structure.

It is a matter of interest and significance that the posterior end of the lateral meniscus is attached to the femur by an oblique band that passes either in front of or behind the posterior cruciate lig. In many mammals this oblique band, greatly enlarged, is almost the sole posterior attachment of the lateral meniscus, indicating thereby that its allegiance is to the femur rather than to the tibia.

Part of the Popliteus is attached to the posterior end of the lateral meniscus and may serve as a retractor muscle (Basmajian and Lovejoy; Last).

9. The medial femoral condyle is, then, completing the process of extension at the same time as the whole femur is undergoing axial rotation. No special rotator muscles are provided. When these movements are completed, the anterior border of the medial meniscus also fits into its *curved groove* on its femoral condyle (*fig. 510*).

And finally some *details*—

10. It follows that the upper surface of the lateral tibial condyle is flat to allow forward gliding; and the lateral meniscus is broad and expansive in order to act as a carriage or toboggan for the lateral femoral condyle (*fig. 511*).

11. The upper aspect of the medial tibial condyl is flat but allows flexion, extension, and rotation. And its meniscus, having but restricted sliding action, is narrow; indeed, it tapers anteriorly.

12. The fibular collateral lig. passes to the lateral margin of the head of the fibula. Therefore, it lies wide of the lateral tibial condyle and lateral meniscus. In fact, the space between them is wide enough to afford a passage for the Popliteus tendon (*fig. 511*).

13. But with the tibial collateral lig. it is different; its deeper deltoid part is attached to the margin of the medial condyle of the tibia, and therefore does come into contact with the medial meniscus and blends with it.

Prehistoric Man. The anterior cruciate lig. has been seen to occupy, on full extension, a notch of the femur. Its presence in a femur affords evidence that the particular knee joint was capable of being fully extended, and therefore that its owner walked erect. The femora of certain prehistoric men present such well marked notches for the anterior cruciate ligament that they unquestionably walked erect; but Neanderthal Man whose femora have no suggestion of the notch, probably walked with a crouching gait.

The Intercondylar Septum (*figs. 512, 513*). In prenatal life a vertical septum separates

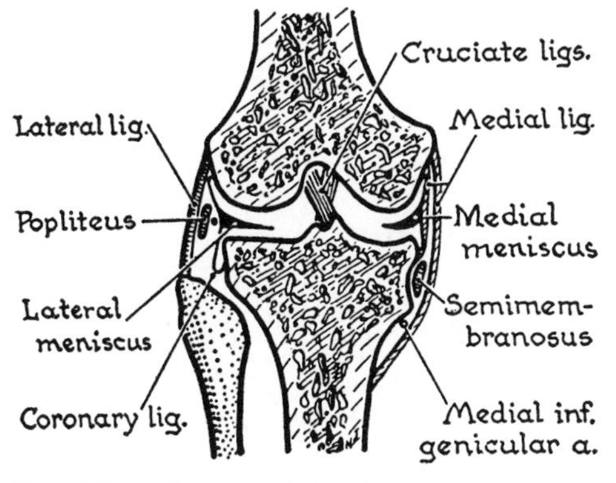

Fig. 511. The knee joint, in coronal section.

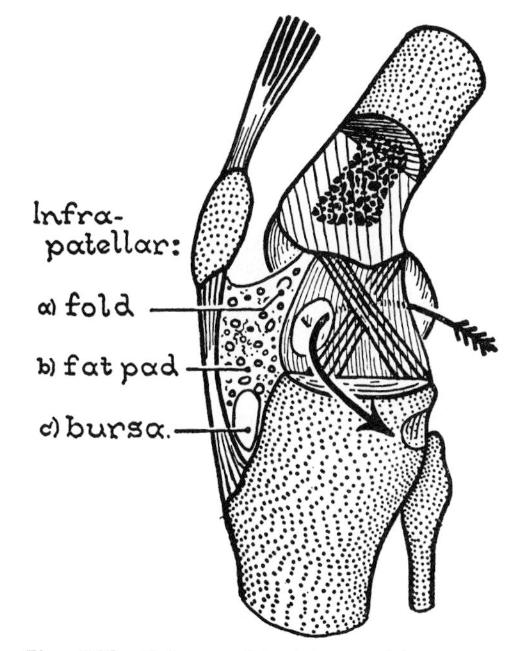

Fig. 512. Scheme of the intercondylar septum.

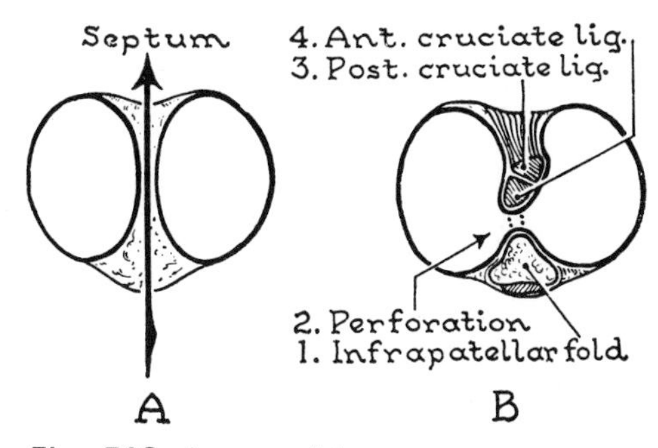

Fig. 513. Intercondylar septum as seen from above on the tibia. *A*, early; *B*, later. Four derivatives are seen.

the medial and lateral condylar joints from each other. Its lower border remains attached to the intercondylar area of the tibial plateau. The posterior half of its upper border is attached to the intercondylar notch of the femur; the anterior half is free and extends from the intercondylar notch of the femur to the patella just below its articular surface.

A *perforation* usually appears in the septum and extends backward to the anterior cruciate lig. (*fig. 513*). It divides the septum into an anterior part, the *infrapatellar synovial fold*, and a posterior part in which the *anterior* and *posterior cruciate ligaments* develop.

The Cruciate Ligaments develop in the hinder part of the septum and cross each other obliquely, like the limbs of a St. Andrew's cross. Thus, the limb attached to the tibia anteroinferiorly is attached to the femur posterosuperiorly, and contrariwise. It is from their tibial attachments that they take their names. *Figure 514* makes clear their functions.

The articular surfaces of the tibial condyles rise gently to two peaks, the *intercondylar eminence*, in the nonarticular area (*fig. 443*). A sideways slide of the femur would cause one or other condyle to

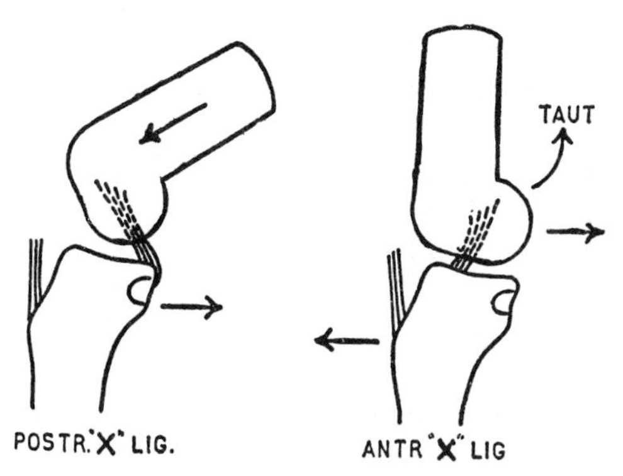

Fig. 514. The posterior cruciate ligament prevents forward displacement of the femur or backward displacement of the tibia. The anterior "X" ligament prevents backward displacement of the femur and hyperextension.

mount an incline. This the cruciate and collateral ligaments resist.

The Infrapatellar Synovial Fold is a flattened hollow cone of synovial membrane. Its apex remains attached to the most anterior point of the intercondylar notch of the femur (*fig. 512*).

Its open base extends from just below the articular cartilage of the patella to the anterior intercondylar area of the tibia; its two sides are free, and from each a small wing, the alar fold, projects.

An *infrapatellar pad of fat* (i.e., the fat behind the lig. patellae) is continued upward into the infrapatellar fold and, in stout subjects, into the alar folds also.

Synovial Cavity and Communicating Bursae

1. Developmentally, the joint possesses *three synovial cavities:* a patellar and two condylar, which later communicate freely.

2. Each condylar cavity is divided into an *upper* and a *lower part* by the meniscus. The coronary ligaments—a part of the capsule—attach the convex margins of the menisci to the upper end of the tibia just below the articular margin (*fig. 515*).

3. **Three bursae** communicate with the knee joint (*figs. 515–517*). These bursae lie deep to the tendons of Quadriceps Femoris, Popliteus, and Gastrocnemius.

a. The bursa deep to the **Quadriceps**

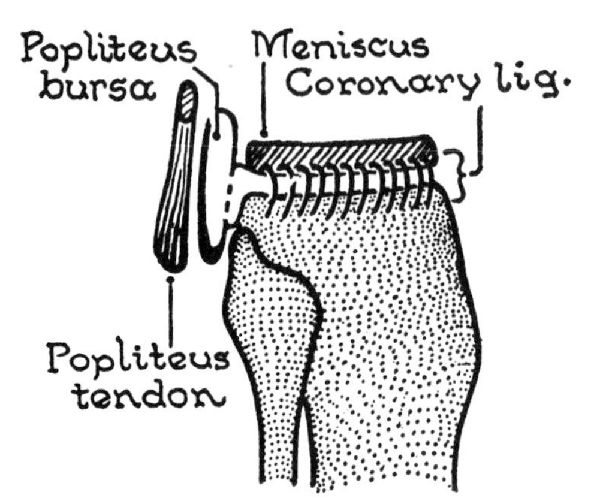

Fig. 515. Popliteus bursa communicating with the joint cavity, and coronary ligament attaching convex border of meniscus to tibia.

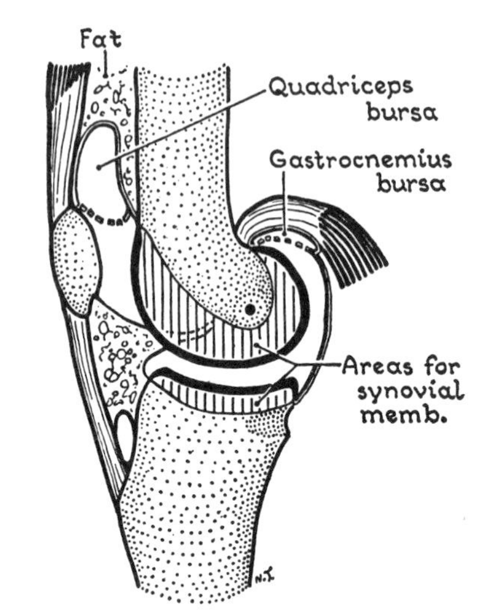

Fig. 516. The Quadriceps and Gastrocnemius bursae communicating with the synovial cavity.

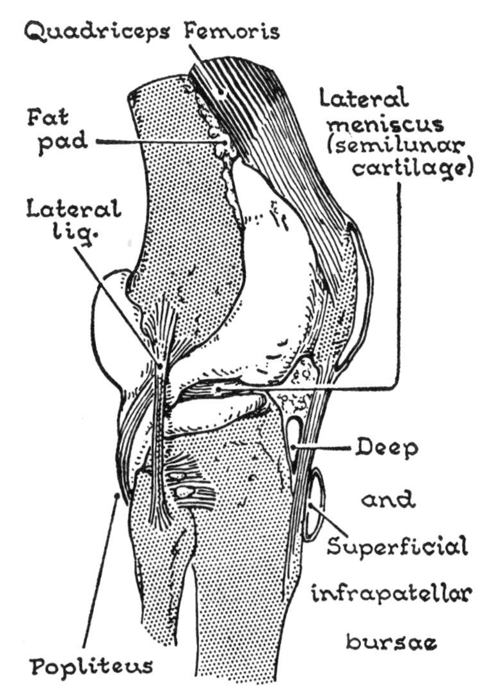

Fig. 517. Knee joint filled with latex to show the extent of the synovial cavity, lateral view.

Femoris tendon, known as the **suprapatellar bursa**, almost always opens into the patellar cavity (*figs. 516, 517*). During extension the *Articularis Genu* retracts it.

b. The *Popliteus bursa* opens into the lateral condylar cavity below the meniscus (*figs. 515, 517*).

Sometimes the partition between bursa and tibiofibular joint gives way, bringing it and the knee joint into communication.

c. The *Gastrocnemius bursa* (deep to the medial head) commonly communicates. This bursa may also communicate with a bursa deep to the Semimembranosus.

Extent of the Synovial Cavity (*figs. 516, 517*). How far may an incision be carried downward on the femur without invading the joint cavity? In front, to the top of the suprapatellar bursa, that is two fingers' breadth above the patella; at the sides the level must fall below the epicondyles, for they give attachment to the collateral ligaments.

Details of Bursae about the Knee, for reference. There are 11 or more:

—3 communicate with the joint—Quadriceps (suprapatellar), Popliteus, and Medial Gastrocnemius.

—3 related to the patella and lig. patellae—prepatellar, superficial infrapatellar, and deep infrapatellar (*fig. 517*).

—2 Semimembranosus bursae—the one between Semimembranosus and Gastrocnemius tendons may communicate with the Gastrocnemius b. and so may communicate indirectly with the knee joint; the other lies between Semimembranosus tendon and the tibial condyle.

—2 superficial to the collateral ligaments—the one between the fibular collateral lig. and the overlying Biceps tendon; the other between the tibial collateral lig. and the three overlying tendons (Sartorius, Gracilis, and Semitendinosus). The latter is commonly continuous with a bursa between Sartorius superficially and Gracilis and Semitendinosus deeply.

—1 bursa between the superficial and deep parts of the tibial collateral lig.

Relations. All muscles crossing the joint are relations of the joint. The **iliotibial tract** is to be noted especially, on account of its great protective value to the exposed lateral side of the knee. About 2–3 cm (or more) wide and placed between the lig. patellae and the Biceps tendon, it alone separates the skin from the synovial membrane. The common peroneal nerve follows the posterior border of the Biceps. The tibial nerve is behind the popliteal vein, which in turn is behind the popliteal artery (*fig. 438*).

Muscles and Movements: see table 16.

Epiphyses. The more actively growing ends of the femur and tibia are at the knee (p. 8).

The lower epiphysis of the femur begins to ossify about the 9th intra-uterine month and fuses with the diaphysis about the 19th year. The epiphyseal line runs through the adductor tubercle.

The upper epiphysis of the tibia includes the tibial tuberosity. Ossification begins before birth or soon after birth; fusion occurs about the 19th year.

The patella begins to ossify about the 3rd year, probably from several centers. The superolateral angle may remain unossified (emarginate patella) or may ossify independently (*fig. 518*).

Blood Supply. Of the five articular branches of the popliteal a., the middle genicular a. passes forward and supplies the structure in the intercondylar septum. The two medial and two lateral genicular aa. course deep to all muscles and ligaments they encounter, and embrace either the femur or the tibia. The lat. inf. genicular a. is an exception in as much as it passes (1) behind the Popliteus tendon (*fig. 438*), and then (2) runs along the margin of the lateral meniscus.

These arteries anastomose with each other, with the lateral circumflex (descending branch), the descending genicular, and the ant. tibial recurrent artery.

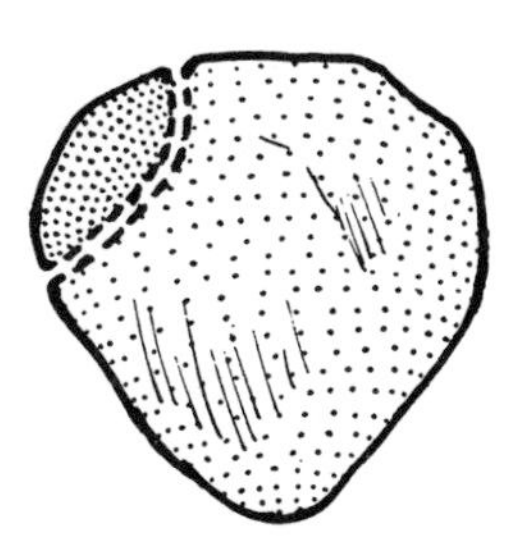

Fig. 518. Bipartite patella.

TABLE 16

Muscles Acting upon the Knee Joint (All the Muscles That Cross It)

Nerve Supply	Muscles	Accessory Actions	Main Actions
Sup. Gluteal Inf. Gluteal	Iliotibial Tract T. Fasciae Latae Gluteus Max. (part)	Retain knee in the extended position	Extensors
Femoral	Quadriceps Femoris Rectus Femoris V. Intermedius V. Lateralis V. Medialis		
	Sartorius	Rotate leg medially	Flexors
Obturator	Add. Gracilis		
Tibial division of sciatic	Semitendinosus Semimembranosus Popliteus		
	Gastrocnemius Plantaris		
	Biceps (Long)	Rotate leg laterally	
Peroneal division of sciatic	Biceps (Short)		

Nerve Supply. This is derived from the *femoral n.*, via branches to the Vasti, and the saphenous nerve; the *obturator n.* via the branch to the Adductor Magnus; and the tibial and common peroneal branches of the *sciatic n.* via the six genicular branches that accompany the corresponding arteries (Details on pp. 319, 333, and 331).

Tibiofibular Joints. (p. 342).

THE ANKLE JOINT

[Talocrural Joint]

Nine Basic Observations. 1. In walking, the Triceps Surae (i.e., two heads of the Gastrocnemius and the Soleus) raises the heel from the ground—it causes *plantarflexion* of the ankle joint at the end of the "stance phase." During the act of advancing the limb or "swing phase," the four anterior crural muscles cause the foot to clear the ground—they cause *dorsiflexion.*

2. The weight of the body is transmitted to the talus through the tibia. The malleoli grasp the sides of the talus (*fig. 519*).

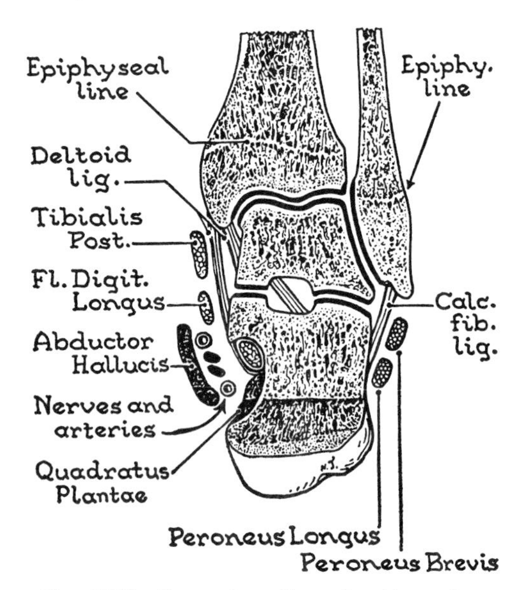

Fig. 519. Coronal section of ankle region.

3. The sharp tip of the fibular malleolus is felt 2 cm below the level of the blunt end of the tibial malleolus. Hence, the lateral side of the body of the talus is three-thirds

articular, and the medial side is one-third articular (*figs. 520, 521*).

4. The sides of the malleoli and of the shafts of the bones above them are subcutaneous—there are no muscles at the sides of the ankle for this is an ideal hinge joint.

5. Because of the hinge movements, the upper surface of the body of the talus is articular and convex from before backward.

Comparative Anatomy. In quadrupeds, a pronounced anteroposterior flange projects from the tibia into a slot in the talus (*fig. 522*). In man, this feature is reduced (*fig. 519*).

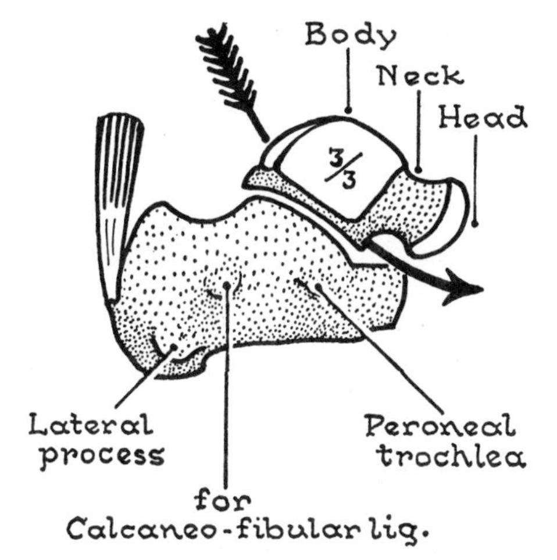

Fig. 520. Lateral aspect of calcaneus and talus. The *arrow* traverses the tarsal tunnel (tarsal sinus).

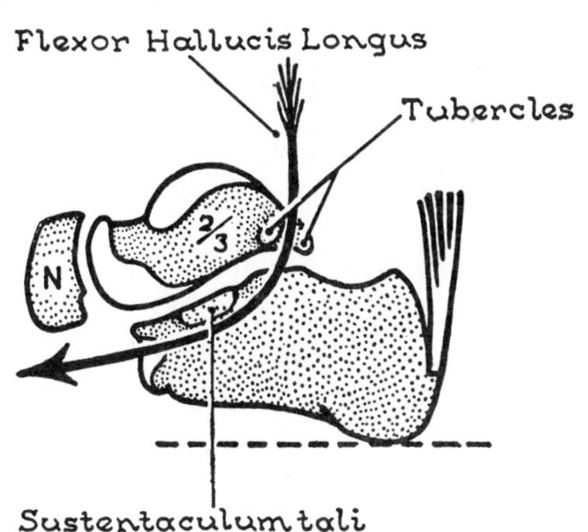

Fig. 521. Medial aspect of calcaneus and talus.

6. With the exception of the tendo calcaneus, all tendons cross the ankle and transverse tarsal joints, acting on both.

7. While the integrity of a joint depends upon four factors—bones, ligaments, muscles, and gravity—here muscles and gravity will tend toward forward displacement of the leg bones, so it falls to the bony parts and the ligaments to resist (*fig. 523*). This demands—

8. that the socket shall be so fashioned that it cannot slide forward on the talus; and—

9. and that the ligaments shall pass downward and backward.

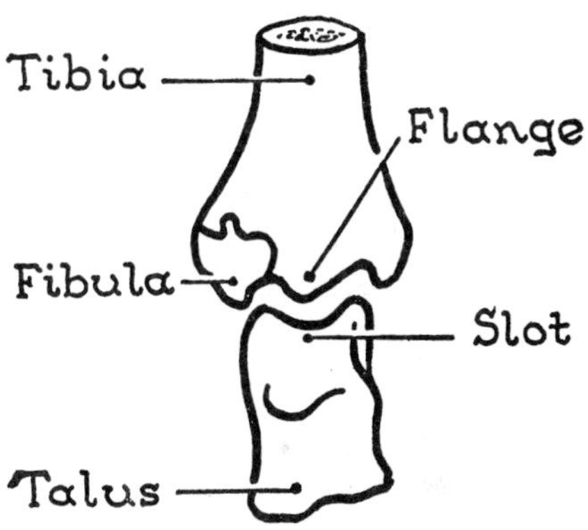

Fig. 522. The ankle joint of the sheep.

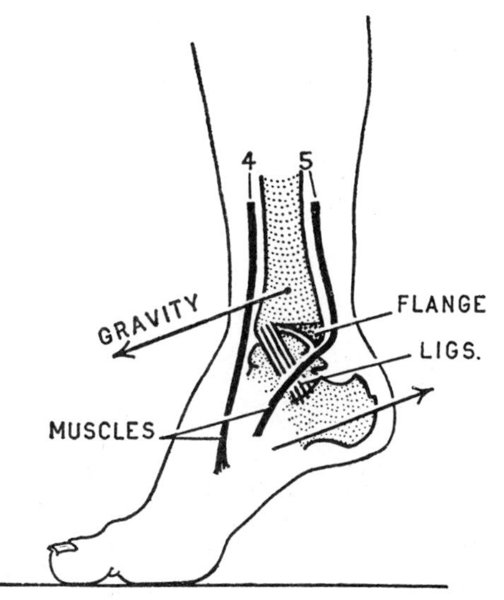

Fig. 523. At the ankle joint the ligaments and bones resist the forces of the muscles and of gravity.

Bony Parts of the Ankle Joint. These include (1) parts of the talus and (2) its socket.

The Talus. The whole of the upper surface of the body is articular, and it is continuous with the large facet for the fibular malleolus and restricted facet for the tibial malleolus (*figs. 520, 521*). The body is wedge shaped (*fig. 524*), the broad end being in front.

The Socket comprises (1) the malleoli, (2) the lower articular surface of the tibia, and (3) a transverse tibiofibular lig. which deepens the socket behind.

Ligaments. The **medial** or **deltoid ligament** (*fig. 525*) arises from the blunt end of the tibial malleolus and has two parts, a *superficial* and a *deep*, like the medial ligament of the knee: (1) The superficial band passes downward and backward to the sustentaculum tali. (2) The deep or *deltoid part* spreads out to the nonarticular part of the medial aspect of the talus. It even reaches the navicular bone. Between the sustentaculum and the navicular it is continuous with the spring ligament, forming part of the socket for the head of the talus (*fig. 525*).

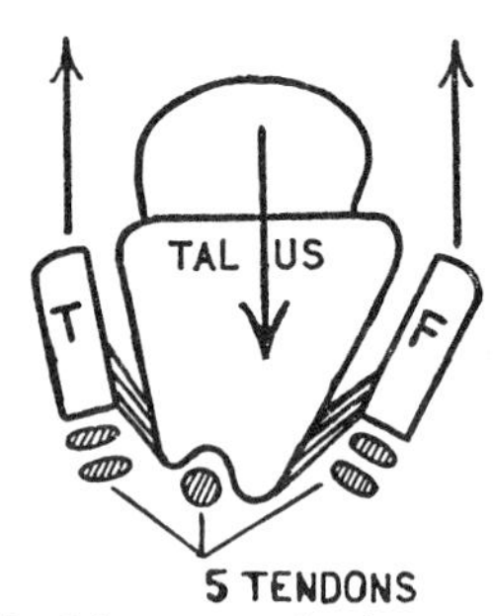

Fig. 524. Diagram of the ankle joint in horizontal section to show that the direction of the ligaments and the converging malleoli prevent backward displacement of the talus.

The **lateral ligament** (*fig. 526*) has three parts: (1) The *calcaneofibular ligament* is a cord that runs down and back. (2) The *posterior talofibular ligament* (*fig. 528*) passes medially and backward (from the malleolar fossa of the fibula—*fig. 454*) to the lateral tubercle of the talus. (3) The *anterior talofibular ligament* is a thin, weak band in the anterior part of the capsule—it is easily "sprained."

The **synovial membrane** extends well forward onto the neck of the talus (*fig. 527*) and has a pad of fat.

Movements. *Dorsiflexion* (*Flexion*)—Tibialis Anterior, Peroneus Tertius, and the long extensors of the toes. *Plantarflexion* (*Extension*)—Gastrocnemius and Soleus.

The 5 tendons passing behind the ankle are too close to the axis of the joint to act to advantage on the joint. In fact, if the tendo calcaneus is cut, the power to plantarflex is lost.

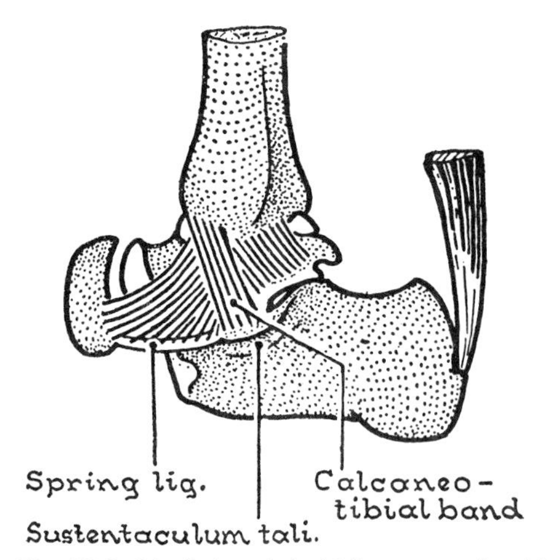

Fig. 525. Medial or deltoid ligament of ankle joint.

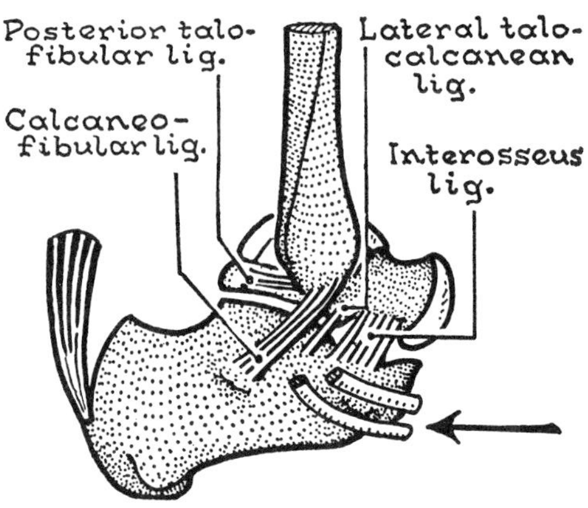

Fig. 526. Lateral view of the ligaments about the ankle, so directed as to prevent backward displacement of the bones of the foot.

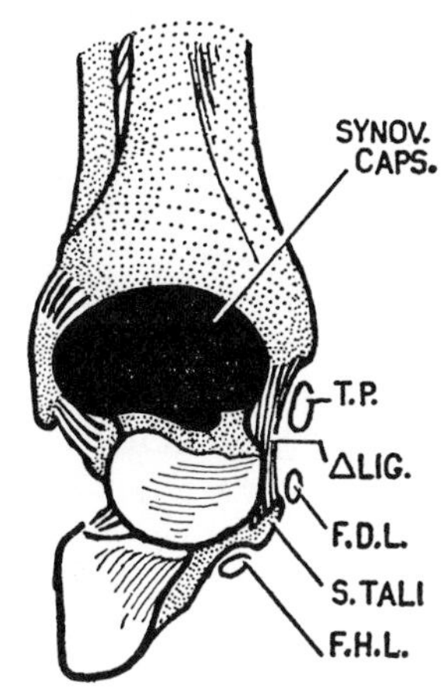

Fig. 527. The synovial cavity of the ankle joint, distended. The anterior articular surfaces of the talus and calcaneus: sustentaculum tali and related tendons.

Relations Summarized. In front, the anterior tibial vessels and deep peroneal nerve lie midway between the malleoli with 2 tendons on each side of them. Behind, the Fl. Hallucis Longus tendon lies midway between the malleoli and there are 2 tendons behind each malleolus; the posterior tibial vessels and tibial nerve lie between Fl. Digitorum Longus and Fl. Hallucis Longus (*fig. 528*). The deltoid ligament is crossed by Tibialis Posterior and Fl. Digitorum Longus, whereas the calcaneofibular ligament is crossed by Peronei Longus and Brevis.

The suppleness of the ankle socket is illustrated in *figure 529*. It probably saves the lateral malleolus from being snapped off when forces in the direction of the white arrow occur. But the medial malleolus is not so fortunate because of the unyielding deltoid ligament (*fig. 525*)—Pott's fracture results and may include snapping of the fibular shaft.

The most unstable position the joint can assume is plantar flexion, as when you rise on your toes. On going down hill you instinctively dig your heels in (1) because in this position of dorsiflexion the broad edge of the wedge is closely grasped by the malleoli, and (2) because the heel is at the short end of a lever; the toes being at the long end.

Epiphyses. The lower epiphyseal plate of the fibula is at the level of the upper surface of the joint; that of the tibia is a centimeter above the joint (*fig. 519*). The distal fibular and tibial epiphyses are always completely fused to their diaphyses by the 20th year.

Blood Supply. From the anastomoses around the joint.

Nerve Supply. From the deep peroneal and tibial nerves.

JOINTS OF THE FOOT

Joints of Inversion and Eversion:

1. Talocalcanean Joint:
   Post. talocalcanean or Subtalar Joint.
   Ant. talocalcanean Joint.
2. Talonavicular Joint.
3. Calcaneocuboid Joint, which is auxiliary to the talonavicular joint, providing it extra freedom.

Definitions of Movements. Most of the **inversion,** the movement of turning the sole of the foot to face the opposite sole, is a combination of *supination* and *adduction.* Conversely, **eversion** is a combination of

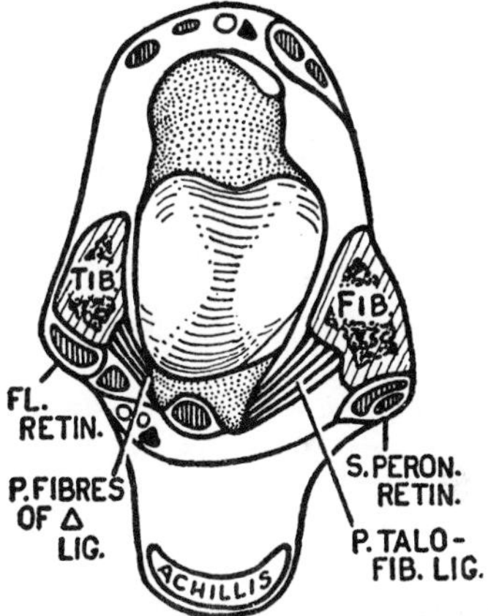

Fig. 528. Horizontal section through ankle joint showing (1) wedge-shaped socket, (2) direction of ligaments, (3) five posterior tendons, and (4) the investing and intermuscular deep fascia.

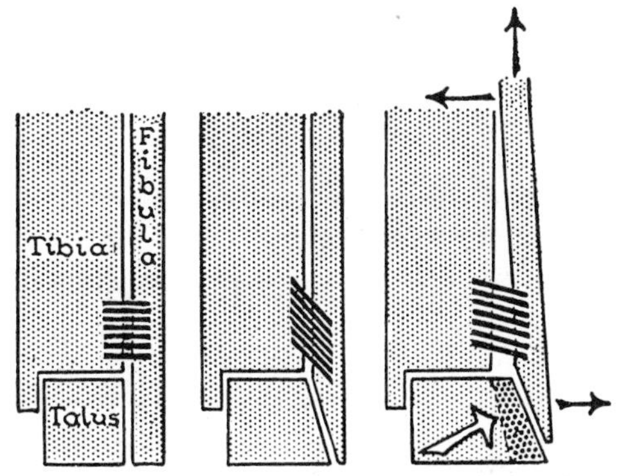

Fig. 529. Transverse fibers would make a rigid ankle socket; oblique fibers make a supple socket.

pronation and *abduction*, and it results in the sole facing slightly laterally. But the movements are complex and the terms are often used loosely, e.g., inversion as synonymous with supination, and eversion with pronation.

Inversion almost automatically recruits plantarflexion of the ankle joint and eversion is uncomfortable without dorsiflexion.

Inversion and eversion are movements: (1) *Of the entire foot* (except the talus) *about the talus;* here the articular surfaces involved are all the facets below and in front of the talus and (as seen in *figure 531*) all 3 facets above the calcaneus and the concave facet behind the navicular; and (2) *Of the forepart of the foot on the hindpart* at the transverse tarsal joint.

Talocalcanean Joints

Much of the movement of inversion and eversion occurs here (*fig. 530, 531*).

Bony Parts Reviewed. The entire body and part of the head of the talus rest upon the anterior two-thirds of the calcaneus and project slightly in front of it (*fig. 530*). The *Upper Surface of the Calcaneus* presents a very large posterior talar facet and a small anterior facet, which is usually continuous posteromedially with a larger facet (middle talar facet) on the sustentaculum tali. The combined anterior and middle talar facets are concave, the posterior talar facet is convex. A deep groove separates the posterior facet from the middle facet, and when the talus and calcaneus are in articulation, the groove is converted into a tunnel, the *tarsal sinus*. This sinus lodges some fat and the variable *interosseous talocalcanean ligament*, which unite the two bones.

The Ligaments. The two ligaments uniting the calcaneus to the talus are:

1. The interosseous talocalcanean—not very strong (*figs. 519, 526*).

2. The lateral talocalcanean—a mere slip (*fig. 526*).

The two ligaments uniting the calcaneus to the bones of the leg—and therefore spanning the talus—are:

3. The calcaneofibular portion of the lateral ligament of the ankle (*fig. 526*).

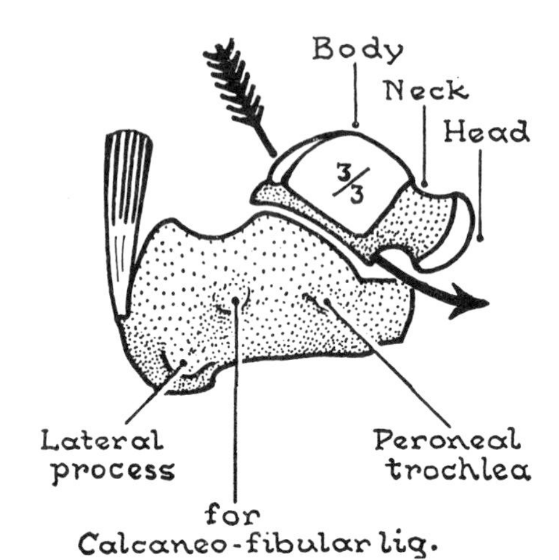

Fig. 530. The lateral aspect of the talus and calcaneus. *Arrow* traverses the tarsal sinus.

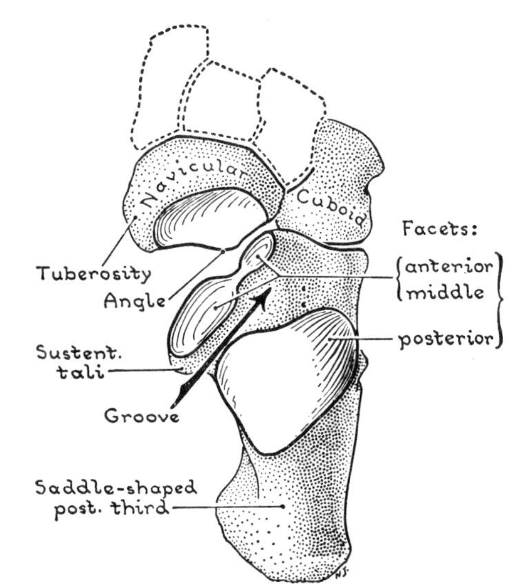

Fig. 531. The bony socket, or bed, for the talus formed by the calcaneus and the navicular.

4. The calcaneotibial portion of the medial or deltoid ligament of the ankle (*fig. 525*).

The **five tendons**, so important as stabilizers of the ankle joint, play a similar role here, just below.

Subdivisions of Talocalcanean Joint. These are the **Subtalar Joint,** below the body of the talus immediately in line with the ankle joint, and the **Talocalcaneonavicular Joint** for the head of the talus (*fig. 532*).

The Socket of the Talocalcaneonavicular Joint is well seen in *figure 532*. It is completed by the *plantar calcaneonavicular ligament*, commonly called "*the spring ligament*" (*fig. 533*).

The Spring Ligament is continuous medially with the deltoid ligament and supports the head of the talus. The portion of the head of the talus that rests upon the "spring ligament" is at the summit of the medial arch of the foot (*fig. 475*). If the Tibialis Posterior, which is a dynamic assistant to the ligament, is paralyzed, the spring ligament may stretch. This is one type of flat foot (paralytic); but much commoner is the type resulting from primary weakness of the ligaments, which become overstrained and let the head of the talus sag to the ground.

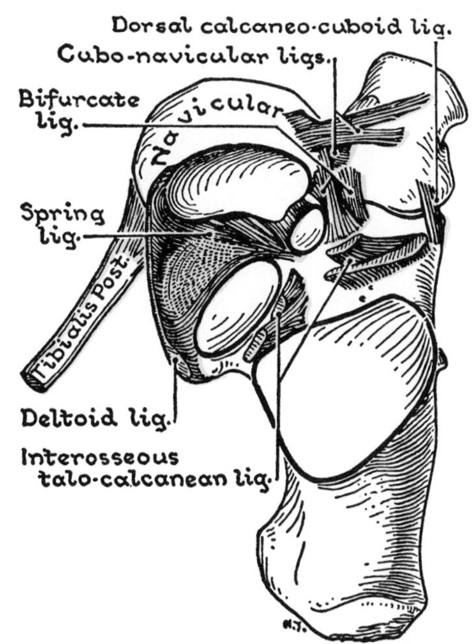

Fig. 532. The talus has been removed in order to show its bed and the ligaments of the joints of inversion and eversion.

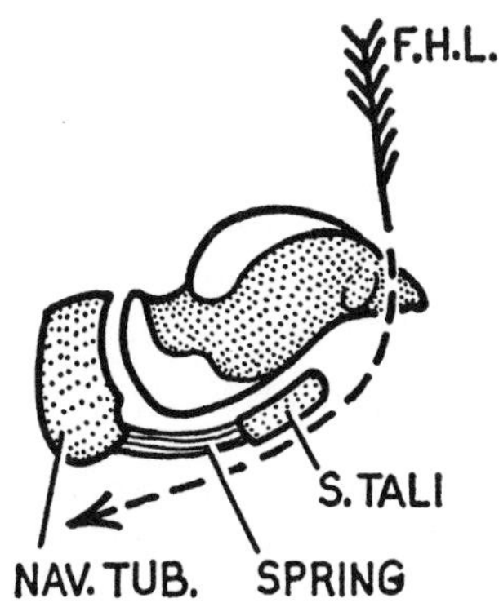

Fig. 533. The spring ligament. The *arrow* indicates the position of the Fl. Hallucis Longus tendon.

The Bifurcate Ligament (*fig. 532*) is really the collateral ligaments for the talonavicular and calcaneocuboid joints.

Transverse Tarsal Joint

The transverse tarsal joint has two component parts; together they account for about half of all inversion and eversion of the foot:

1. Talonavicular Joint,
2. Calcaneocuboid Joint.

The talonavicular joint has been described above with the talocalcaneonavicular joint, but it must be considered here also. If you nail or fuse the calcaneus and talus (and tibia) into one mass, inversion and eversion, though restricted, are not abolished. The two muscles of inversion and three of eversion are inserted in front of the transverse tarsal joint. Dorsal, interosseous, and plantar ligaments unite the cuboid to the navicular and cuneiforms, so when the two tibial muscles cause inversion, the cuboid cannot choose but follow.

Calcaneocuboid Joint

The calcaneocuboid joint is an accessory joint of inversion and eversion. The anterior surface of the calcaneus is rounded off medially (i.e., convex); the posterior surface of the cuboid is concave from side to side, thus allowing inversion.

A process (the calcanean angle) projects from the inferomedial part of the cuboid backward below the calcaneus, perhaps to "support" it.

The **short plantar ligament** [plantar calcaneocuboid lig.] stretches from the inferior surface of the calcaneus to a large area behind the ridge on the cuboid. It belongs solely to the calcaneocuboid joint, and is on the same plane as the spring lig. (*fig. 535*).

The **long plantar ligament** (*fig. 534*) bridges the short one (fig. 534) and converts the groove on the cuboid into a tunnel for the Peroneus Longus tendon.

The short plantar and spring ligaments form an almost continuous sheet of thick parallel fibers that run anteromedially from calcaneus to cuboid and navicular (*fig. 535*). They are, in fact though not in name, the deep components of the *plantar lig. of the transverse tarsal joint*, the long plantar lig. being the superficial component.

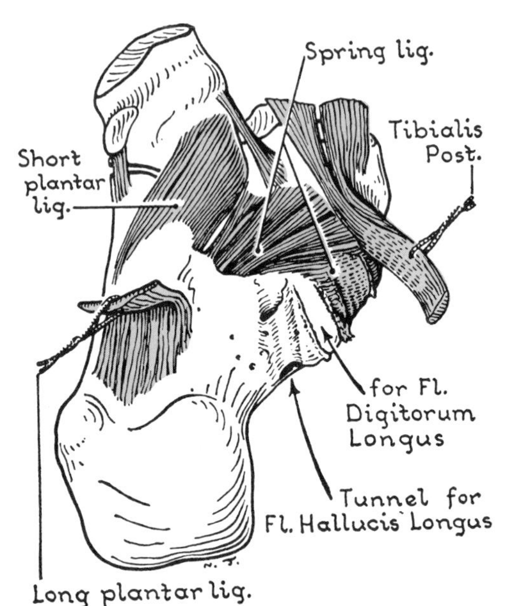

Fig. 535. Calcaneonavicular and calcaneocuboid ligaments (i.e., spring lig. and short plantar lig).

Arches of the Foot

The Lateral Longitudinal Arch is low (*fig. 535.1*). During walking it receives and bears the weight of the body before the medial arch comes into play. It yields or flattens at the hinge surfaces between the cuboid and metatarsals IV and V.

The Medial Longitudinal Arch is formed by the calcaneus, talus, navicular, three cuneiforms, three medial metatarsals, and the two sesamoid bones. It is a high arch. At its summit, placed at the junction of its posterior one-third with its anterior two-thirds, lies the head of the talus with the sustentaculum tali behind it and the navicular in front. The Tibialis Posterior is largely (two-thirds) inserted into the tuberosity of the navicular bone, but the remainder passes below the spring ligament and may act as a sling for the arch. At the hinge surfaces between the talus and the navicular and also between the navicular and the three cuneiforms, the spring or arch can flatten and recoil.

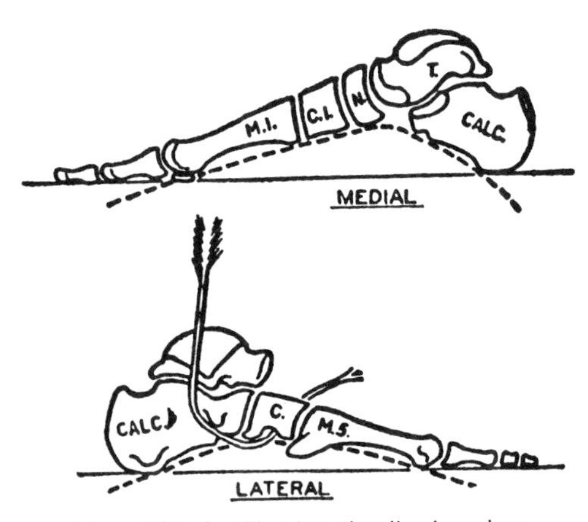

Fig. 535.1. The longitudinal arches.

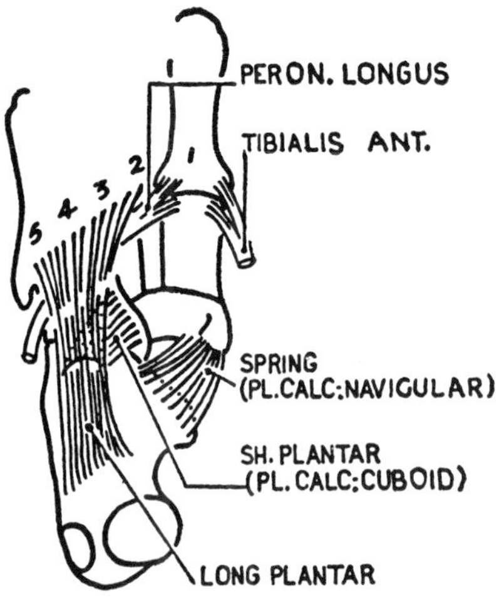

Fig. 534. Three plantar ligaments, and the two tendons that form a stirrup.

A true "transverse arch" does not exist. The half-arch of the cuboid and cuneiforms (*fig. 536*) cannot act as an arch for it has no medial pillar.

Joints Distal to Transverse Tarsal

These include: small intertarsal, tarsometatarsal, intermetatarsal, metatarsophalangeal, and interphalangeal joints.

The *Metatarsophalangel and Interphalangeal Joints* are fashioned and supplied with ligaments like the corresponding joints of the hand.

The *deep transverse metatarsal ligs.* (ligs. of metatarsal heads) extend to the hallux; in the hand the corresponding ligaments leave the pollex free (*fig. 149*).

The *Small Intertarsal Joints* (i.e., between cuboid, navicular, and cuneiforms), the *Tarsometatarsal Joints*, and the joints between the *Bases of the Metatarsals* may be considered together under the headings: (1) *side-to-side joints* and (2) *end-to-end joints*.

The Side-to-Side Joints have strong *plantar ligaments* which act as transverse ties (*fig. 536*). These ties are not limited to the plantar aspect but, as *interosseous ligaments*, extend dorsalward between the individual bones, being absent only between the bases of metatarsals I and II. The *articular facets* are generally situated near the dorsal part of the surface, where ligaments would have least value. The *dorsal ligaments* are weak.

The End-to-End Joints, or those concerned in the longitudinal arches. Observe that the apposed anterior and posterior surfaces of all the bones of the foot are completely covered with cartilage, and that the joints, accordingly, do not possess interosseous ligaments (*fig. 536*). They have, however, *strong plantar ligaments*, *weak dorsal ligaments*, and in some instances, *collateral ligaments* (*fig. 537*). The side-to-side facets are commonly continuous with the end-to-end facets.

Arrangement of Ligaments. The ligaments of the foot are so disposed as to resist certain forces. Accordingly, (1) the dorsal, interosseous, and plantar ligaments between any two bones have a common direction. Further, from *figure 538* it may be noted that (2) of the ligamentous bands uniting the bones of the lateral arch to each other and to the bones of the medial arch, all—or practically all—take a common direction; this is the *direction of resistance* (a) to the backward pull of the muscles inserted into the lateral arch, and (b) to backward thrusts applied to the 4th and 5th toes (as in walking and kicking); and similarly that (3) the ligamentous bands of the medial arch are so directed that the backward thrust given to the 1st metatarsal (as

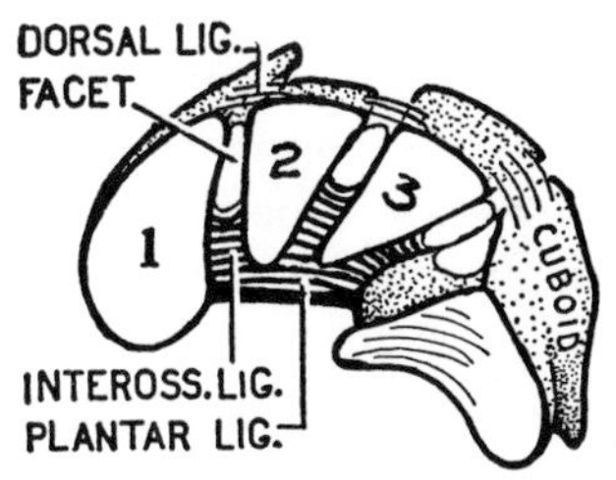

Fig. 536. The ligaments or ties of the "side-to-side" joints. Note that the cuboid supports the cuneiforms and the navicular.

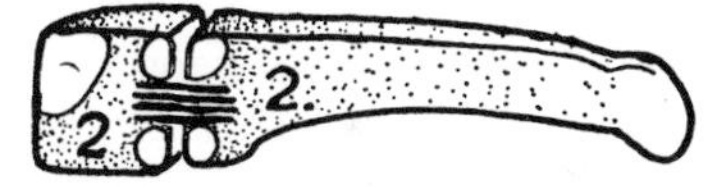

Fig. 537. The collateral ligament of an "end-to-end" joint.

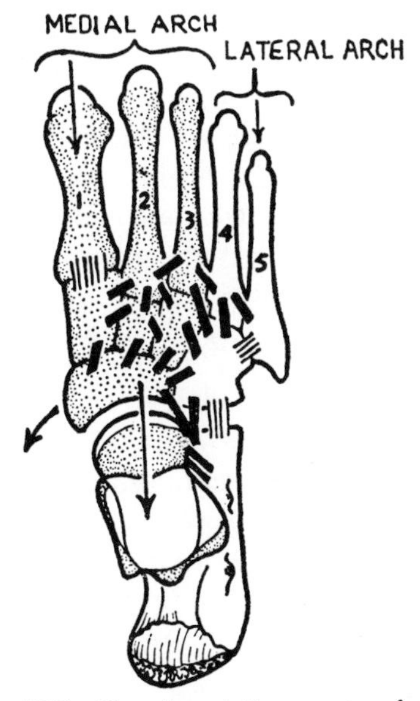

Fig. 538. The dorsal ligaments of the foot are not scattered in a haphazard way (*see text*).

in rising on the ball of the big toe in walking) is dispersed laterally to the 2nd and 3rd metatarsals which, therefore, share in conducting the thrust via the cuneiforms and navicular to the talus and so to the tibia.

DYNAMICS OF FOOT

Standing. *Weight Distribution.* The body weight of a person weighing say 72 kg and "standing relaxed in a naturally held position" is distributed through the feet as follows: 36 kg to each foot; of this, 18 kg is through the hindpart (calcaneus) and 18 through the forepart.

The forepart of the foot has six points of contact with the ground, namely, the two seasamoids under the head of the 1st metatarsal and the heads of the lateral four metatarsals, each supporting approximately 3 kg; the 1st metatarsal through its sesamoids supports a double load (D. J. Morton). (*See fig. 477.*)

If the ligaments at the base of a metatarsal or at the proximal end of a cuneiform became lax, that metatarsal, being mobile, will largely cease to be a weight-distributing bone (*figs. 538.1* to *538.3*).

Arch Support. It has long been the teaching that muscles are all-important for the support of the main arch of the foot. It would appear that this extreme view must be abandoned. R. L. Jones calculated that the plantar ligaments and aponeurosis bear the greatest stress. The chief function of the invertor and evertor muscles is to preserve a relative constancy in the ratio of the weight distribution among the heads of the metatarsals.

The height of the arch varies with your posture. Babies have no arch at all; their arches develop with maturity; indeed there is great controversy as to whether treatment is either needed or effective—probably the truth lies between.

Any failure of the arch is related to the duration of the stress to which it is subjected rather than to the severity of the stress; e.g., athletes and those who walk much subject their arches to great stress intermittently; whereas those who stand immobile subject their arches to relatively continuous stress, and it is these latter who develop trouble with their arches.

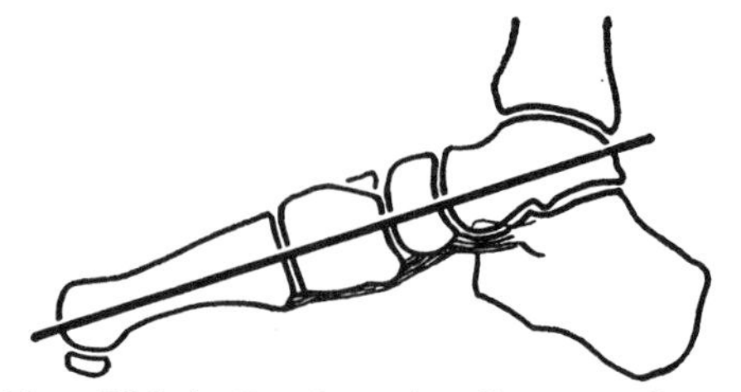

Fig. 538.1. Tracing of radiogram of a normal foot, standing. (After Ewen A. Jack.)

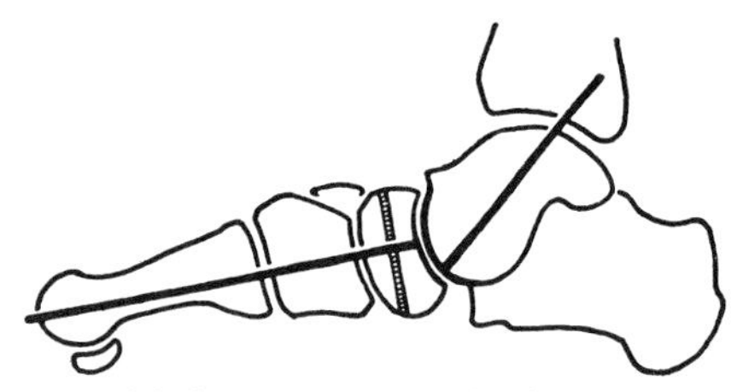

Fig. 538.2. One form of flatfoot, talonavicular joint, standing. (After Ewen A. Jack.)

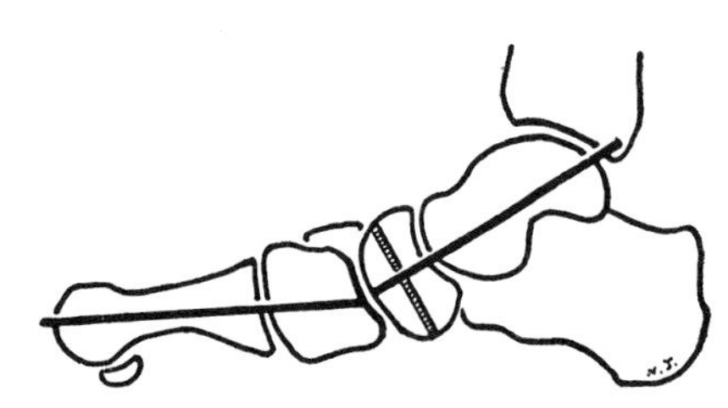

Fig. 538.3. Another form of flatfoot, naviculocuneiform joint, standing. (After Ewen A. Jack.)

It has been demonstrated electromyographically that Tibialis Anterior, Peroneus Longus, and the intrinsic muscles of the foot play no important role in the normal static support of the long arches of the foot. They are commonly completely inactive. Even with the addition of abnormally great weights to the static arches, the muscles are relatively inactive (Basmajian and Stecko). During locomotion it is different, the intrinsic muscles of the foot being then very active when one rises on the toes even to the slightest degree (Basmajian and Bentzon) apparently because this is a greater stress requiring muscular reinforcement. Calculation of the mechanical stresses confirms this view.

Other points have been clarified also (*fig. 538.4*). Passive dorsiflexion of the big toe (as in rising on the ball of the foot), even without direct aid from muscles, through the "windlass action" of the plantar aponeurosis approximates the heads of the metatarsals to the heel, thereby shortening the foot and heightening the arch (J. H. Hicks).

Distribution of Weight. Contrast *figure 539 A with 539 B.*

If, while standing on one foot (knee flexed, if you wish), you overbalance laterally, in order to compensate, the arch rises and the leg rotates laterally; and vice-versa when overbalancing medially. Indeed, when the foot is fixed, these two movements (supination of the foot and lateral rotation of the leg) are invariably associated. If, however the foot is free and the leg stationary, rising of the arch (supination) is associated with adduction of the foot (inversion).

Explanation. Of two joints (1) the subtalar (post. talocalcanean) and the talocalcaneonavicular and (2) the transverse tarsal, the former provides for hinge action between the leg + talus and the foot; the latter between the hind and fore parts of the foot. The axes of rotation of these two joints are so similar that they may be thought of as a single "oblique hinge" of which the axis runs obliquely downward, backward, and laterally from the superomedial aspect of the head of the talus to the inferolateral aspect of the heel (*fig. 538.5*).

Function of These Two Joints. This oblique hinge allows side-to-side swinging of the body, walking on sloping surfaces, and, above all, control of side-to-side balance. Indeed, without this hinge you cannot maintain side-to-side balance while standing on one leg.

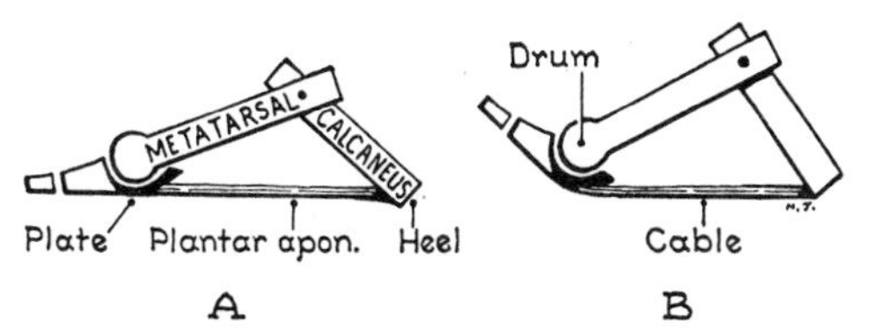

Fig. 538.4. Traction through the windlass effect on the plantar aponeurosis causes heightening of the arch. (After J. H. Hicks.)

The Pentadactyl Hand and Foot. The limbs of amphibia are paddles, all built to the same design. Hence, we may speak of homologous parts of the upper and lower limbs (*fig. 540*). The hand and foot of man are seen to retain the primitive generalized plan.

Supernumerary Ossicles which may show up in X-ray pictures. The commonest ossicles are:

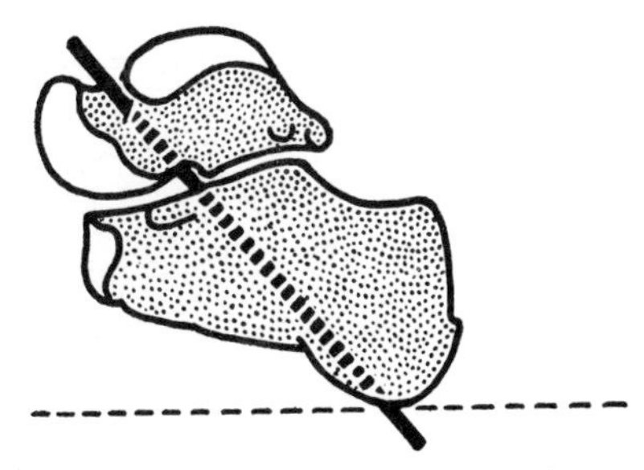

Fig. 538.5. Axis of oblique hinge. (After J. H. Hicks.)

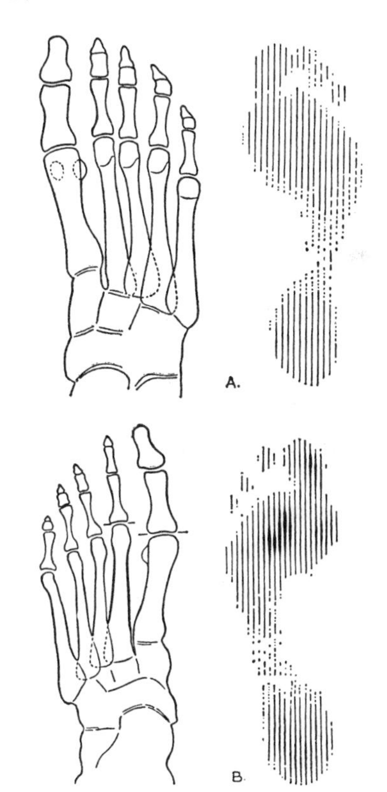

Fig. 539. *A*, tracing of a radiogram and an accompanying print of a normal right foot made by walking on a corrugated mat. *B*, tracing, and a print of a left foot with a short 1st metatarsal bone. (After Morton.)

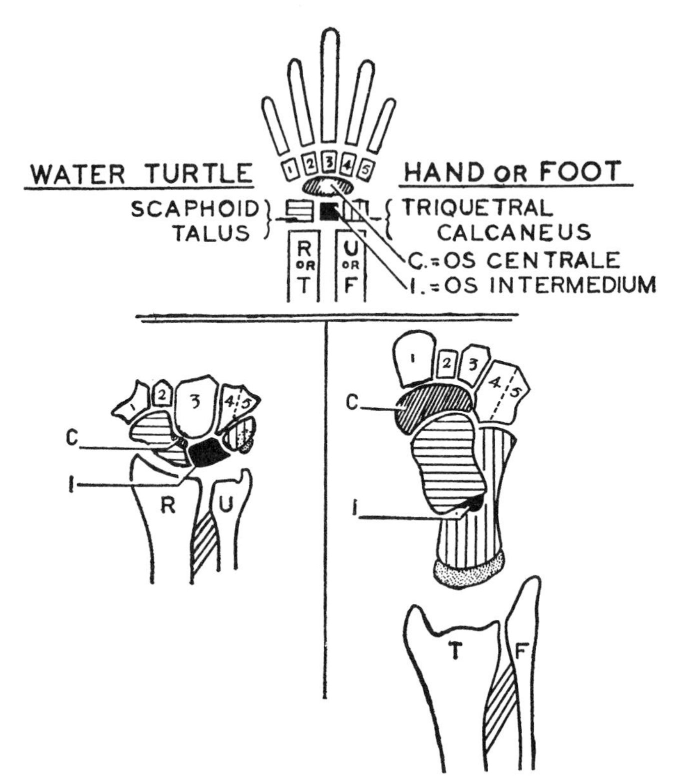

Fig. 540. Homologies in the bones of the hand and foot.

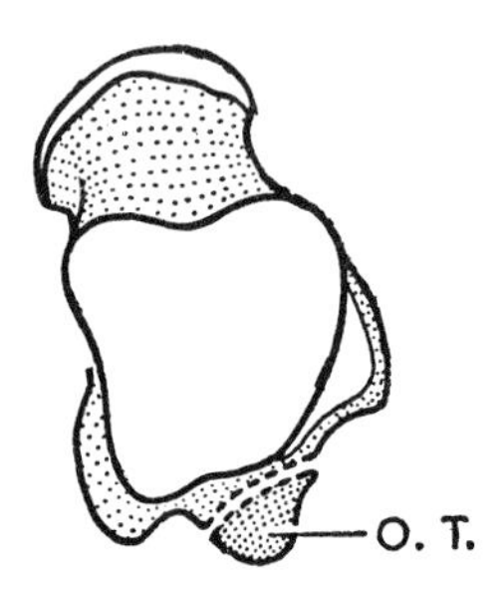

Fig. 541. Os trigonum. [In 558 paired lower limbs (i.e., 279 adult cadavers) 43 or 7.7 per cent had ossa trigona. Of these, 22 occurred bilaterally (in 11 subjects); whereas 21 occurred unilaterally (in 21 subjects), 10 being in the right foot and 11 in the left. In other words, it occurred twice as commonly unilaterally as bilaterally. (C. Storton.)]

(1) The *os trigonum* or separate lateral tubercle of the talus (*fig. 541*). Phylogenetically it is an ununited os intermedium (*fig. 540*). (2) The *tibiale externum* or separate navicular tuberosity. (3) A *bipartite medial cuneiform*, in upper and lower halves. (4) A fibrocartilaginous nodule in the Peroneus Longus tendon lateral to the cuboid is common and commonly it is ossified. (5) A sesamoid bone in the Tibialis Posterior tendon—not to be confused with a separate navicular tuberosity. (6) The tuberosity of metatarsal V, existing as a separate bone (Os Vesalianum), is rare.

TABLE 17

Table of Appearance and Fusion Times of Epiphyses in Lower Limb including Primary Centers for Tarsus, in Years Unless Otherwise Stated

	Appears	Fuses
Hip bone, three primary parts	Before birth	16*
Ischium and Pubis	Before birth	9
Ischial tuberosity	19	20
Iliac crest	16	22
Femur, Head	1	18*
Greater Trochanter	–5	18*
Lesser Trochanter	–14	18*
Lower end	Birth	19*
Tibia-Upper end	Birth	19
Lower end	2`	18
Fibula-Upper end	5	19
Lower end	2	18
Calcaneus—before birth	–30 week	
Tuberosity	11	17
Talus—before birth	–30 week	
Navicular	3	
Cuboid—before birth	+30 week	
1st Cuneiform	3	
2nd Cuneiform	–4	
3rd Cuneiform	1	
Great Toe Metatarsal base	–3	18
2nd–5th Metatarsal heads	–5	18
Great Toe, Proximal Phalanx base	–4	19
Distal Phalanx base	2	19
2nd–5th toes, Proximal Phalanx base	–5	19
Middle Phalanx base	–5	19
Distal Phalanx	–5	19
Patella	–5	
Sesamoids of great toe	–12	22+
Rib Head	15	As it ossifies
Tubercle	18	

Since the student will be concerned with the age periods at which he may reasonably be sure that fusion is complete, the *latest* times of fusion are given. Female bones fuse distinctly earlier.

* Asterisks denote times that are found to be quite constant. The table is a compilation from several authorities and is based on X-ray findings. It has been deemed advisable to give definite ages rather than a spread. Figures denote years unless otherwise stated.

+ = a later tendency. – = an earlier tendency.

SECTION SIX

THORAX

29 Walls of Thorax 391

30 Pleurae 400

31 Lungs 411

32 Heart and Pericardium 420

33 Superior and Posterior Mediastina 437

29 WALLS OF THORAX

BONY THORAX AND ITS LIGAMENTS

Apertures of Thorax; Cavity of Thorax; Chest of Child.
Thoracic Vertebrae.
Body; Transverse, Spinous, and Articular Processes; Vertebral Notches.
Sternum.
Manubrium; Body; Xiphoid Process.
Vertebral Levels.
Muscles; Posterior Relations; Ossification; Common Anomalies.
Manubriosternal Joint.
Ribs and Their Cartilages—Costae.
Classification; Typical Rib.
Joints—Articulations of Costae: of Head of Rib. Sternocostal, Interchondral, of Tubercle (Costotransverse); Ligaments.
Angles of Ribs; Ossification; Variations; 1st Rib.

INTERCOSTAL SPACES

Muscles.
Intercostales Externi, Interni, and Intimi.
Transversus Thoracis.
Internal Thoracic Artery and Veins; Parasternal Lymph Nodes.
Posterior Intercostal Artery and Vein; Intercostal Nerves.

BONY THORAX AND ITS LIGAMENTS

The bony thorax comprises: 12 vertebrae, 12 pairs of costae (ribs and costal cartilages), and 1 sternum.

Its Upper Aperture (*fig. 547*) is formed by the body of the 1st thoracic vertebra, the 1st ribs, 1st costal cartilages, and the thick upper margin of the sternum. If one makes a kidney-shaped outline with the thumbs and fingers of the two hands, it simulates the size and shape of one's own upper thoracic aperture.

The Lower Aperture of the Thorax is formed by the lower six costal cartilages and the 12th ribs, the xiphoid process in front, and the body of the 12th thoracic vertebra behind. The 11th rib is much longer than the 12th rib and reaches to within two or three fingers' breadth of the iliac crest.

The Cavity of the Thorax also is kidney-shaped on transverse section (*fig. 542*) because the ribs are carried backward beyond the vertebral bodies. They bend to form angles near the transverse processes, which act as buttresses. At birth, the chest is almost circular.

The dome of the diaphragm rises to the level of the 5th or 6th rib; so the bony thorax affords protection not only to the heart and lungs, but also to the upper abdominal viscera—notably the liver, stomach, and spleen.

Thoracic Vertebrae

Typical features (*figs. 543, 544*):

The Body has far back on each side an upper and a lower demifacet for articulation with the head of ribs. But some bodies have only one (*fig. 543*).

The bodies in the middle of the thoracic series are heart-shaped (*fig. 545*). Each body contributes to making the thoracic portion of the vertebral column concave forward (*fig. 543*).

The Transverse Processes act as buttresses. They are stout and conform with the backward sweep of the ribs, having on their tips facets for the tubercles of the ribs (*figs. 554, 555*). The transverse proc-

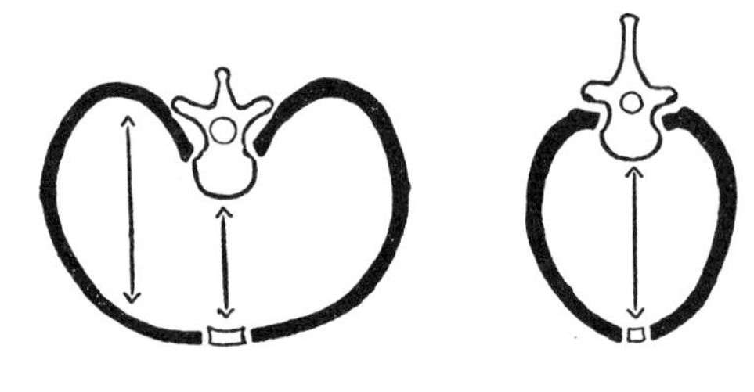

Fig. 542. Transverse section of thorax. Note anteroposterior diameters in man.

esses become progressively shorter from the 1st to the 12th.

The Spinous Processes (Spines) of the intermediate four (5, 6, 7, and 8) are long and vertical, but those above and below are progressively more horizontal (*fig. 543*).

The Articular Processes are set almost vertically on the arc of a circle (*fig. 545*). This decides that such movements as take place between adjacent thoracic vertebrae shall be mainly rotary. The facets on the superior articular processes face posterolaterally; those on the inferior articular processes, anteromedially (*fig. 544*).

The Vertebral Foramen is small and circular (*fig. 546*).

The Inferior Vertebral Notches are large; the superior ones are absent (*fig. 544*).

Transitions. As the thoracic vertebrae are followed from the middle of the series up and down, they are found gradually to assume the characteristics of cervical and lumbar vertebrae. Th. 1 is very much like C. 7, and Th. 12, like L. 1.

Sternum

The sternum (*fig. 547*), or breast bone, likened to a broad sword, is composed of three parts:

1. Manubrium sterni or handle.
2. Corpus sterni or body.
3. Xiphoid process.

Manubrium Sterni. Feel: (1) the very thick concave upper border, called the *jugular notch*, much deepened by (2) the proximal sternal ends of the clavicles,

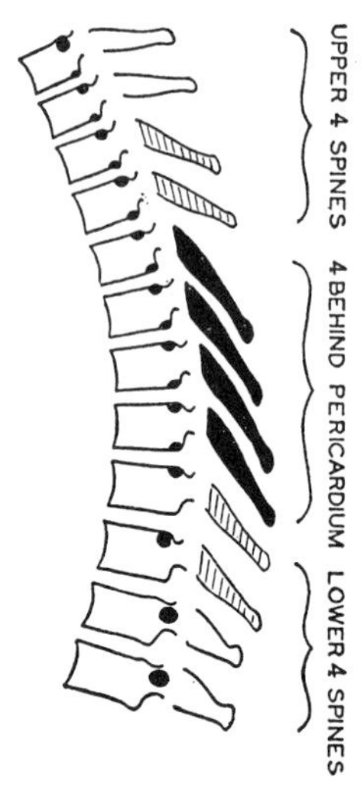

Fig. 543. Thoracic vertebrae, showing costal facets (for heads of ribs) and the inclination of the spinous processes.

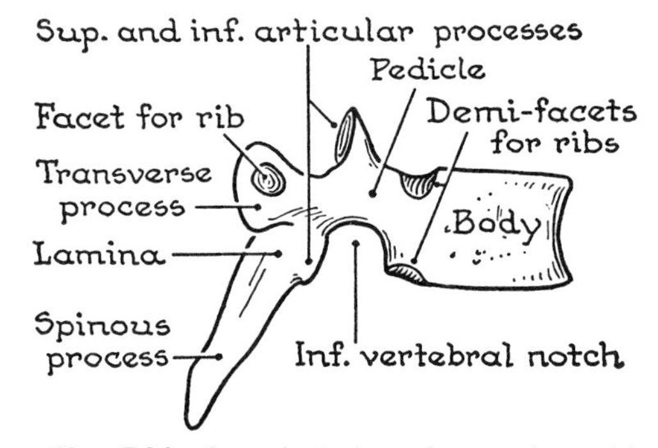

Fig. 544. A typical thoracic vertebra, side view.

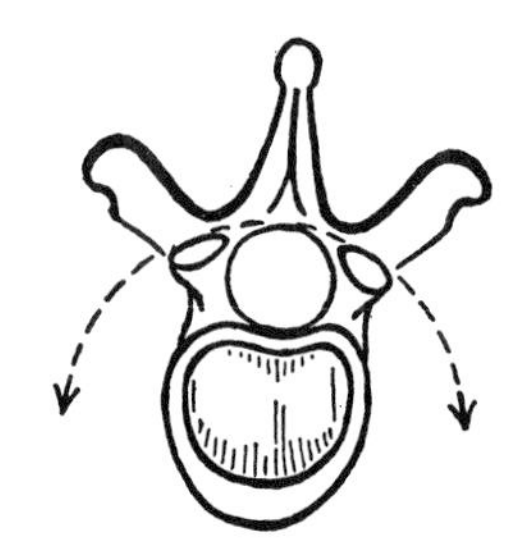

Fig. 545. The thoracic articular processes are set on an arc, so they permit rotation.

which are too large for the notches provided for them at the superolateral angles. Feel also the tendons of the Sternomastoids crossing the sternoclavicular joints (*fig. 547*).

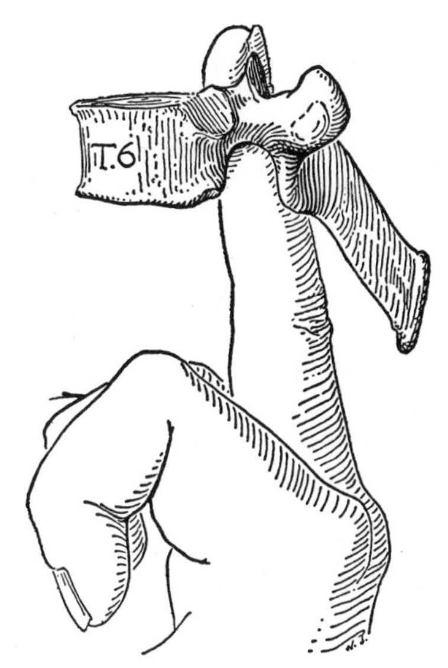

Fig. 546. A vertebral foramen barely admits a finger.

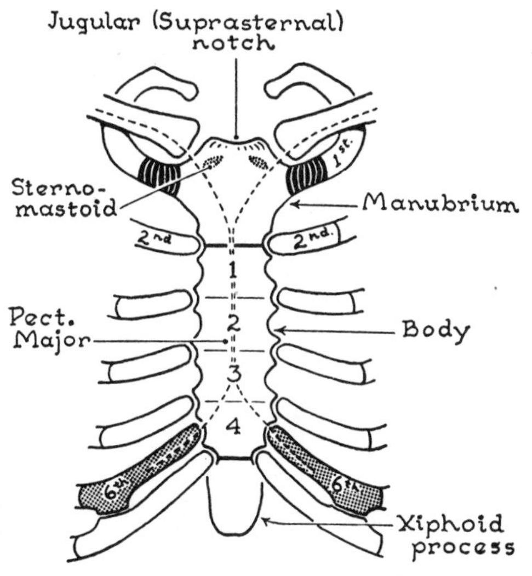

Fig. 547. The anterior surface of the sternum. The *broken lines:* origins of the Pectorales Majores.

Immediately below the clavicle, the 1st costal cartilage joins the manubrium to the rib (*fig. 547*)—a synchondrosis (p. 17).

The manubrium is 5 cm long—as long as two vertebrae. Its upper border lies at the level of the lower border of Th. vertebra 2, and only 5 cm from it (*fig. 548*). ~~Its lower border of Th. vertebra 2, and only 5 cm~~

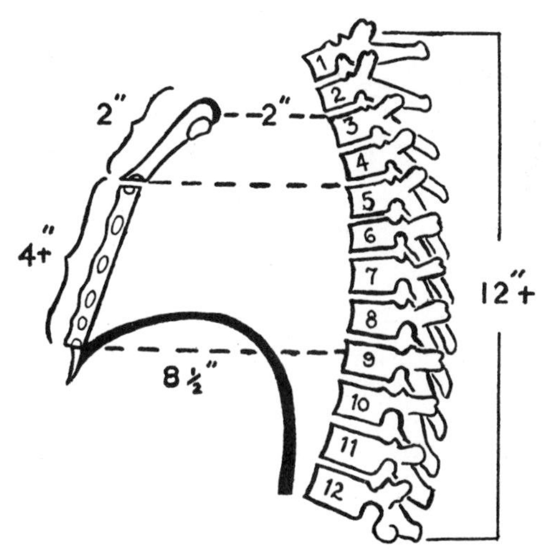

Fig. 548. The bony thorax in median section—levels and lengths. (1″ = 2.5 cm.)

~~from it~~ (*fig. 548*). Its lower border articulates with the body at an angle, the *sternal angle*. Traced laterally the transverse ridge that indicates the angle conducts the palpating fingers to the 2nd costal cartilage, the starting point from which all ribs should be counted (*fig. 547*).

The Body of the Sternum is composed of four fused pieces or *sternebrae*, which together are slightly more than twice the length of the manubrium.

The 2nd rib cartilage articulates in a notch on the side of the sternal angle; the 7th cartilage articulates with the angle between the body and the front of the xiphoid.

The Manubriosternal Joint plays a part in respiration, because it allows hinge-like movements of the body of the sternum forward and backward. It is a symphysis type of joint (p. 17).

Synostosis of the manubrium and body of the sternum is found in at least 10 per cent of adults after the age of 30 years (Ashley; Mildred Trotter).

Comparative Anatomy. In birds a keel or crest projects from the sternum, providing the powerful pectoral muscles an increased origin.

The Xiphoid Process extends downwards for a variable distance into the posterior wall of the sheath of the Rectus Abdominis. It is only half as thick as the body of the sternum, and its posterior surface is flush with the posterior surface of the body.

The xiphoid is not only variable in length, but efforts to palpate it cause discomfort; so use the sharp easily palpated edge of the lower end of the body at the *xiphisternal synchrondrosis* as your landmark.

Vertebral Levels of the jugular notch, sternal angle, and xiphisternal joint are 2nd, 4th, 8th+ (*fig. 548*).

Muscles attached to posterior aspect:
of manubrium— Sternothyroid, Sternohyoid } Neck muscles
of body—Transversus Thoracis—unimportant.
of xiphoid process—Diaphragm.

Posterior Relations: Pleurae and lungs, heart and great vessels, thymus.

Ossification: The manubrium ossifies about the 6th intra-uterine month. The sternebrae and xiphoid develop from right and left mesenchymal bars, which chondrify and fuse in the median plane. The four sternebrae, then, ossify from above downward about the 6th, 7th, 8th, and 9th intra-uterine month or later; fusion takes place from below upward about the 15th, 20th and 25th years. The xiphoid process starts to ossify in youth. The xiphisternal synchondrosis commonly becomes a synostosis in middle age.

Common Anomalies (1) The lower two or three sternebrae commonly ossify separately from right and left centers. If these fail to fuse, a perforation, suggestive of a bullet wound, will appear in an X-ray photograph (*fig. 549*). (2) The sternomanubrial joint may occur between the 1st and 2nd sternebrae, as in the gibbon. The sternal angle is then situated about halfway down the sternum.

Ribs and Their Cartilages—Costae

Classification. A rib bone and its cartilage constitute a costa. In all, there are 12 pairs of costae. Every rib articulates posteriorly with the vertebral column. The cartilages of the upper 7 pairs of ribs articulate directly with the sternum; hence, they are known as *true* or *vertebrosternal ribs*. The remaining 5 pairs are false ribs: of these the cartilages of three pairs (8th, 9th, 10th) articulate with the cartilages immediately above them and, so, form a subgroup of *vertebrochondral ribs*. Their connection with the sternum is indirect. The cartilaginous ends of the last two pairs (11th, 12th) are free; hence, they form a subgroup of *floating* or *vertebral ribs*.

Ribs are flattened, have a very thin outer compact layer, and are highly resilient.

A Typical Rib (*fig. 550*) consists of the following parts:

1. *Body:* Internal and external surfaces; superior and inferior borders; an angle and a costal groove. The posterior $\frac{1}{4}$ of the body is cylindrical; the anterior $\frac{3}{4}$ is compressed.

2. *Vertebral End:* head, neck, and tubercle.

3. *Sternal End:* pit for costal cartilage.

Examination of Rib Cage. *With the articulated skeleton before you, confirm the following facts*, because they have an impor-

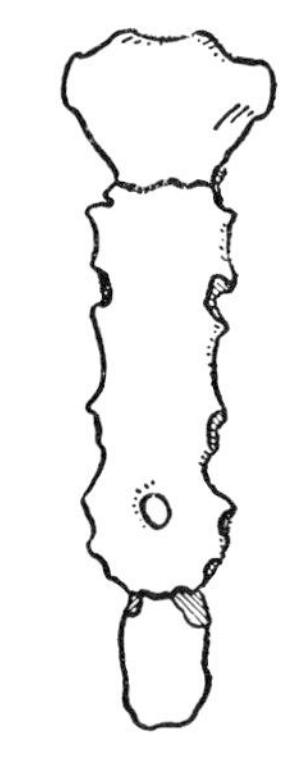

Fig. 549. A perforated sternum, the result of faulty ossification.

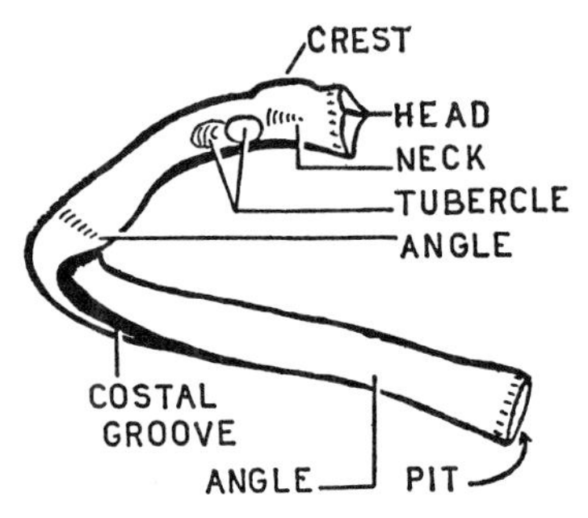

Fig. 550. A typical rib viewed obliquely from behind.

tant bearing on the mechanism of respiration (*fig. 550.1*):

1. The typical rib takes a downward slope; the cartilage, upward (*fig. 551*).
2. The 1st arch slopes downward throughout (*fig. 566*).
3. The sternal end of each arch lies at a lower level than the vertebral end (*fig. 551*).
4. The middle of each arch (except the 1st) lies at a lower level than a straight line joining its two ends (*fig. 551*).
5. Both ribs and cartilages increase in length progressively from 1st to 7th.
6. The transverse diameter of the thorax increases progressively from 1st to 8th rib, the 8th rib having the greatest lateral projection.
7. The ribs increase in obliquity progressively from 1st to 9th; the 9th rib being the most obliquely placed.
8. The anterior ends of the 11th and 12th ribs, not being subjected to terminal pressure, are tapering.

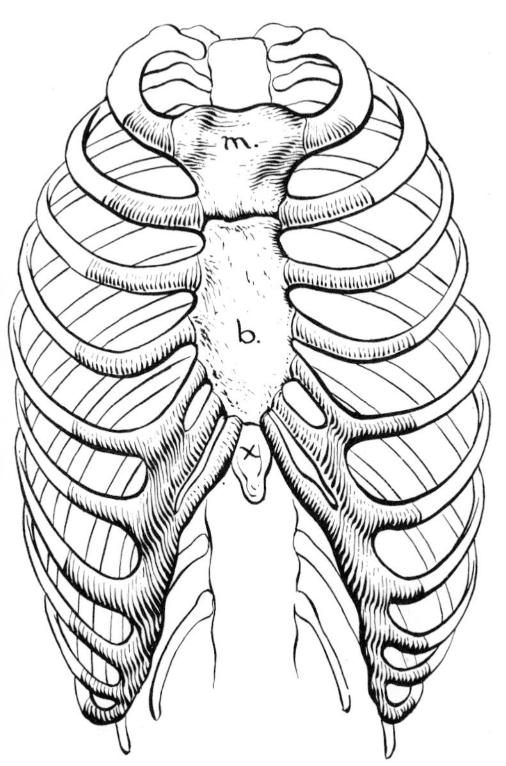

Fig. 550.1. The thoracic cage. The costal cartilages are shaded; *m*, *b*, and *x* are the manubrium, body, and xiphoid process of the sternum. (From *Primary Anatomy*.)

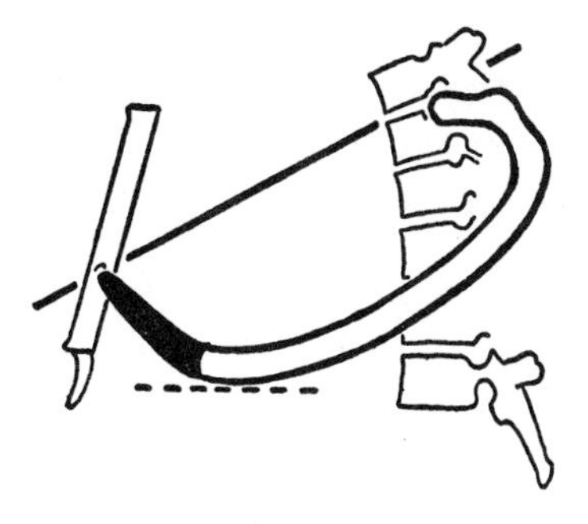

Fig. 551. A costal arch (side view).

Costal Joints

Typically the head of a rib articulates with the sides of the bodies of two vertebrae; the tubercle of a rib articulates with the tip of a transverse process; and the costal cartilage articulates with the sides of two sternebrae. Hence, the following:

1. Costovertebral articulation:
   a. Joint of the head of a rib.
   b. Joint of the tubercle of a rib.
2. Sternocostal articulation.
3. Interchondral joints, between costal cartilages 7, 8, and 9.

Joint of the Head of a Rib. The head of each typical rib (2nd to 10th) articulates with the demifacet of two adjacent vertebrae and with the intervertebral disc between these vertebrae (*fig. 552*). It is attached to the intervertebral disc by a transversely placed *intra-articular ligament*. The *capsule* is strongest in front where its fibers *radiate* from the anterior margin of the head.

The heads of ribs 1, (10), 11, and 12, being confined to single vertebrae, are rounded and their joints have no intra-articular ligaments.

Sternocostal Articulations (*fig. 552*). Like the posterior joints, each joint cavity is divided into two by an intra-articular ligament, and it is closed ventrally by a ligament that radiates from the perichondrium to the sternum. The cavity may be obliterated by fibrous union (Gray and Gardner).

Interchondral Articulations. By means of upward and downward projections, the lower cartilages (except 12) form joints —synovial, or fibrous, or even complete union—with each other (Bristoe).

Joint of the Tubercle of a Rib. *The tubercle of a rib* articulates with the facet at the tip of the transverse process of its own vertebra (except 11 and 12) to form a synovial joint, called a **costotransverse joint.** (See *figs. 553, 554.*)

Ligaments. Figure 555 shows that the strong ligamentous fibers that bind a rib to a transverse process are divided into a medial and a lateral group by the cavity of the joint. They are the *(medial) costotransverse ligament* (lig. of the neck) and the *lateral costotransverse ligament* (lig. of the tubercle).

A band, the *superior costotransverse lig.*, descends from a transverse process to the upper border of the neck of the rib next below, producing the sharp *crest of the neck.*

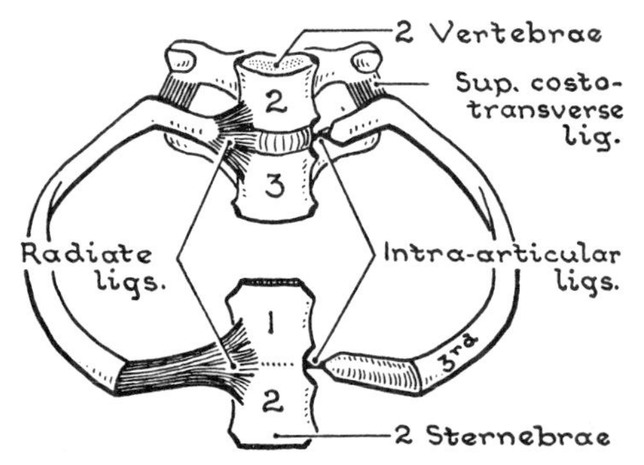

Fig. 552. The articulations at the dorsal and ventral ends of a costal arch, compared.

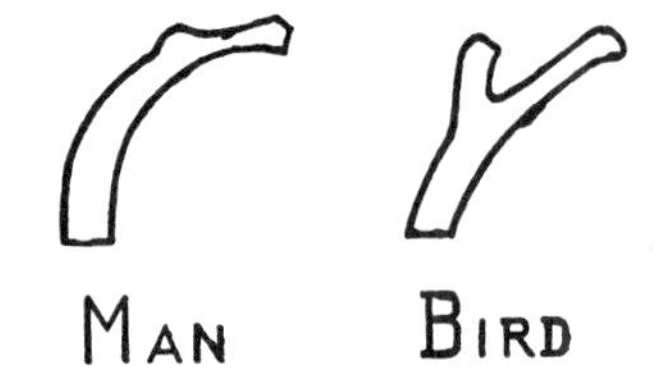

Fig. 553. Tubercle of rib is a reduced second head.

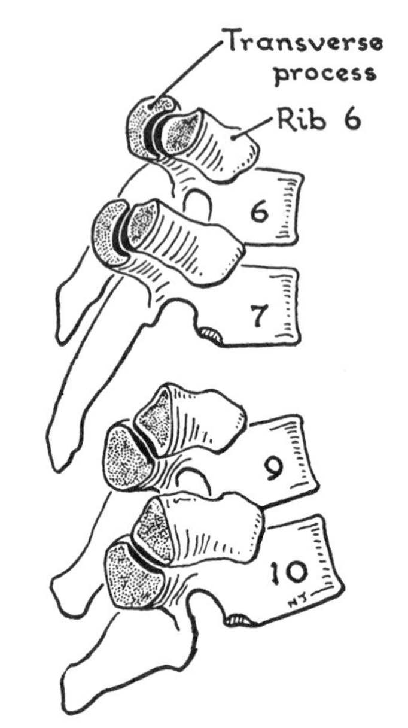

Fig. 554. To demonstrate why upper ribs rotate on transverse processes and lower ribs glide.

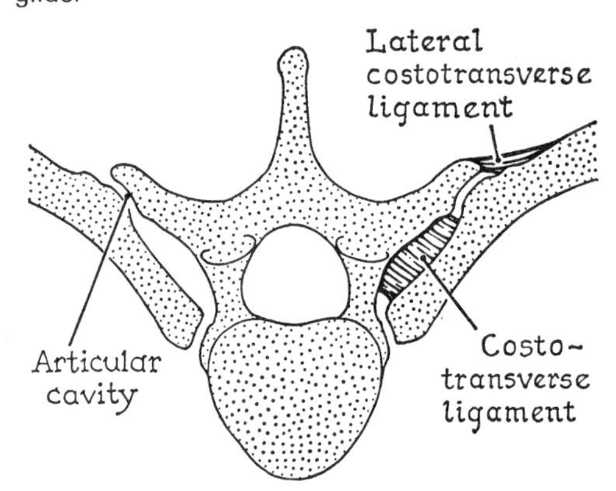

Fig. 555. Costotransverse articulation.

Angles of the Ribs. The vertical series of insertions of the Iliocostalis (the lateral column of the deep muscles of the back), are at the most backwardly projecting part of the outer surface of the ribs (*fig. 671*). Anterior to this the ribs are twisted downward, forward, and medially.

Ossification begins near the angle (about the 9th prenatal week) and spreads in both directions but fails to reach the sternal end;

hence, the costal cartilages. Scale-like epiphyses, which cap the head and tubercle, are in all cases fused by the 24th year.

Variations. *See figures 556* and *557.*

The 1st Rib is a very superlative rib, being the highest, shortest, strongest, flattest, and most curved of all the ribs. *Details—*

To a small triangular area on its outer surface, *the scalene tubercle*, the Scalenus Anterior is attached. It separates the groove for the subclavian vein in front from the groove for the subclavian artery and lowest trunk of the brachial plexus behind. Between the latter groove and the tubercle of the rib the Scalenus Medius, Levator Costae, and the first digitation of the Serratus Anterior are attached. The fascia, the *suprapleural membrane* (Sibson's fascia), clothing the deep surface of the Scalenes is attached to the sharp upper border of this rib.

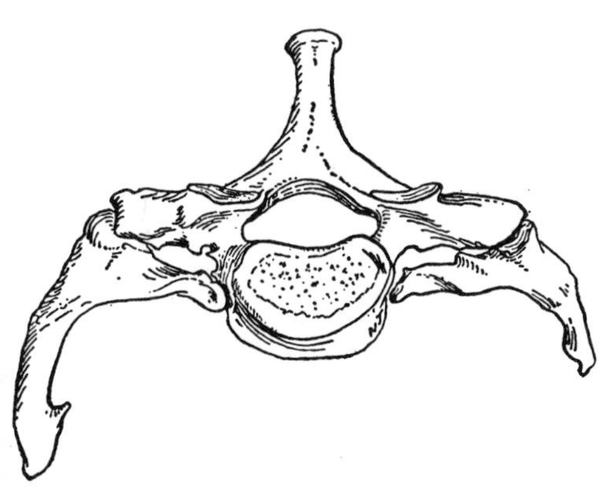

Fig. 556. Cervical rib.

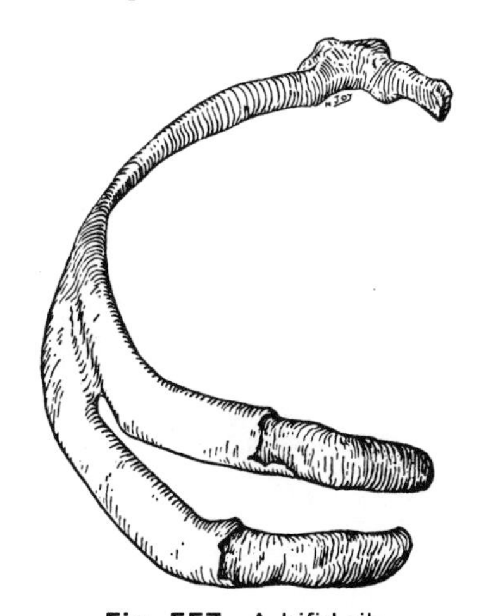

Fig. 557. A bifid rib.

To the outer surface of the *1st costal cartilage* are attached: the intra-articular disc of the sternoclavicular joint, the costoclavicular lig., and the origin of Subclavius.

INTERCOSTAL SPACES

Muscles Covering the Thorax. Except at the *triangle of auscultation*, bounded by Trapezius, Latissimus Dorsi, and Rhomboideus Major (*fig. 77*), the costae and intercostal spaces are completely covered with muscles.

These muscles are: Pectorales Major et Minor, Rectus Abdominis, Obliquus Externus Abdominis, Serratus Anterior, Latissimus Dorsi, Trapezius, Rhomboidei Major et Minor, Levator Scapulae, Serratus Posterior and Erector Spinae. All but four of these are inserted into the bones of the limb.

Muscles of the Thoracic Wall. These correspond to the flat muscles of the abdominal wall (*fig. 558*).

Abdominal	*Thoracic*
1. *Obliquus Externus*	*Intercostales Externi*
2. *Obliquus Internus*	*Intercostales Interni*
Nerves and Vessels in the Interval	
3. Transversus	Intercostales Intimi or Innermost Subcostalis Transversus

The fibers of the **Intercostales Externi** run downward and forward between the adjacent borders of two costae. Their intercartilaginous parts are the *external intercostal membranes.* **Levator Costae** arises from the tip of each transverse process and extends fanwise to the rib below.

The fibers of the **Intercostales Interni** and **Intercostales Intimi** run obliquely downward and backward (*fig. 559*).

The posterior part of an Internal Intercostal, an *internal intercostal membrane*, merges with a superior costotransverse ligament.

The **Transversus Thoracis** is the upward continuation of the Transversus Abdominis. It arises from the back of the xiphoid and sternal body and fans out to be inserted into the 3rd to 6th costochondral junctions.

The Intercostal Nerves. The 1st differs notably from the others because of the very

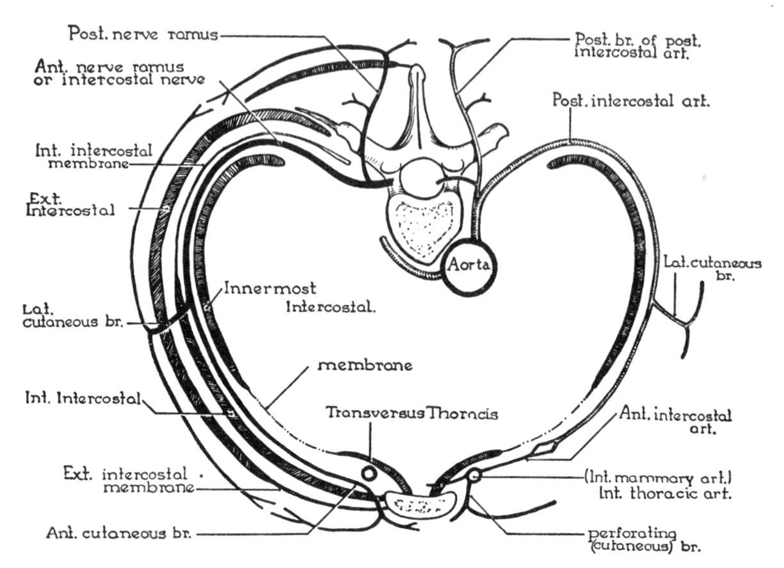

Fig. 558. The contents of an intercostal space (horizontal section). (*Nerve rami: posterior and anterior = dorsal and ventral.*)

large contribution it sends across the 1st rib to the brachial plexus.

The intercostal and subcostal nerves are typical, serially segmental nerves. About the midlateral line they give off *lateral cutaneous* branches, and they end as *anterior cutaneous branches* (*fig. 558*).

A Posterior Intercostal Artery and Vein—under the shelter of a costal groove, accompany each intercostal nerve (*fig. 559*), and provide a *lateral cutaneous* and a *dorsal* (posterior) *branch* (*fig. 558*).

The upper two posterior intercostal arteries arise from the supreme intercostal artery; the lower nine from the aorta (p. 446, and *fig. 642*).

The Internal Thoracic Artery (Internal Mammary A.) (*fig. 558*), a branch of the subclavian artery, descends a finger's breadth from the sternum, behind the upper six costal cartilages. It divides into two terminal branches, the *superior epigastric* and *musculophrenic arteries.*

The venae comitantes of the internal thoracic artery pass to the brachiocephalic (innominate) vein. A **parasternal lymph node** is found in most spaces, beside the vessels.

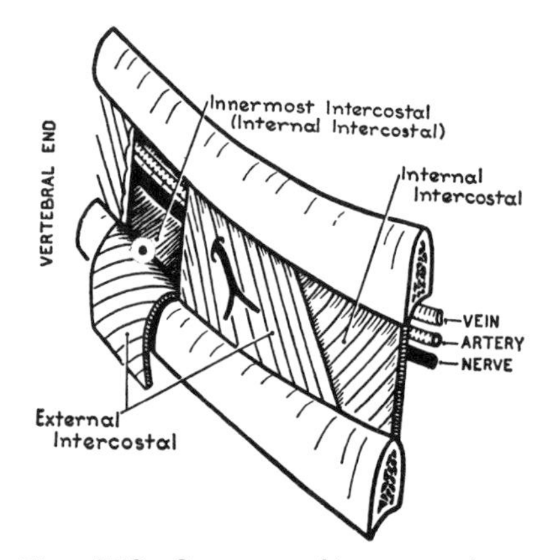

Fig. 559. Contents of intercostal space.

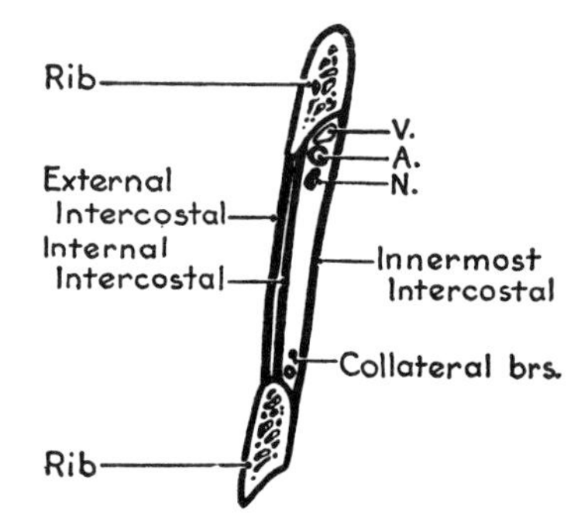

Fig. 559.1. Intercostal Muscles, coronal section.

Being the sole artery in the neighborhood, the internal thoracic art. supplies the entire neighborhood.

Branches.

1. *Perforating* (or cutaneous) branches supply the cutaneous structures, including the breast, hence its old name, "int. mammary".

2. *Anterior intercostal branches* run laterally (*fig. 559.1*).

3. The *musculophrenic art.* sends corresponding branches to the lower spaces and to the abdominal muscles and diaphragm.

4. The *pericardiacophrenic art.* follows the phrenic nerve on the surface of the pericardium.

5. *Mediastinal branches* or *twigs*—to thymus, bronchi, and pericardium.

6. The *superior epigastric art.* descends behind the 7th costal cartilage and Rectus Abdominis to anastomose with the inferior epigastric branch of the external iliac art. thereby bringing the great vessels of the upper and lower limbs into communication (*fig. 216*).

Variant. Occasionally, a *lateral costal artery*, arises above the 1st rib, and descends near the midlateral line, linking the upper five or six intercostal arteries.

30 PLEURAE

Subdivisions of Thoracic Cavity; Pleurae: parietal; visceral; and root of lung.
Lines of Pleural Reflexion—Surface Anatomy; Variations; Relationship of Costal Reflexion to Costal Margin; Three Pleural Recesses.
Mediastinum.
Subdivisions: Superior, Anterior, Posterior, and Middle; Mediastinal Pleura.
Right Aspect of Mediastinum—Structures Seen; Meso-esophagus.
Root of Lung and Pulmonary Ligament; Bronchi; Nerves and Vessels.
Left Aspect of Mediastinum—Structures Seen.
Ligamentum Arteriosum; Great Vessels; Nerves and Viscera.
Cupola of Pleura.
Just Outside Mediastinum: Sympathetic Trunk—Course; Relationships; Connections; Distribution to Viscera; Splanchnic Nerves.

Subdivisions of the Thoracic Cavity

The thoracic cavity is divided into:

1. Right and left pleural cavities.

2. Mediastinum, or thick partition between the two pleural cavities.

The contents of the mediastinum are briefly: (1) the heart within its pericardium, (2) the vessels proceeding to and from the heart, (3) the trachea, and (4) the structures in transit from neck to abdomen, e.g., esophagus, the vagus nerves, phrenic nerves, and thoracic duct.

The pleural, pericardial, and peritoneal cavities are closed potential cavities within thin walled sacs of serous membrane (*fig. 561*). In prenatal life these cavities are continuous and together make up the embryonic celom. The lung invaginates the pleural cavity as it is forming (*fig. 560*).

The Pleurae

Each pleura has three parts—parietal, visceral, and connecting.

1. The Parietal Layer lines each pleural cavity, which possesses two walls (costal and mediastinal), a base, and an apex (*figs. 561, 563*). Therefore its parts are: (1) The *costal pleura*, which lines the costae; (2) The *mediastinal pleura*, applied to the mediastinum. (3) The *diaphragmatic pleura*; (4) The *cupola* (cervical pleura) which rises into the neck.

2. The Pulmonary Pleura (i.e., *Visceral Layer*) coats the lung precisely as the peritoneum invests the liver.

3. The connecting portion, or *root*, connects the pulmonary pleura to the mediastinal layer of parietal pleura. It is a tube or sleeve of pleura in whose upper half lie all the structures that pass to and from the lung; its lower half, being empty (except for a few lymph vessels), is collapsed and is known as the *pulmonary ligament* (lig. pulmonale) (*fig. 562*).

Lines of Pleural Reflexion (*fig. 563*). The costal pleura is continuous with (1) the mediastinal pleura in front of the vertebral column. (2) It is also continuous with the mediastinal pleura behind the sternum. (3) And, it is continuous with the diaphragmatic pleura near the chest margin—the *costal reflexion.*

Surface Anatomy. The sternal and costal reflexions are of high clinical importance. To plot them on the surface of the body you

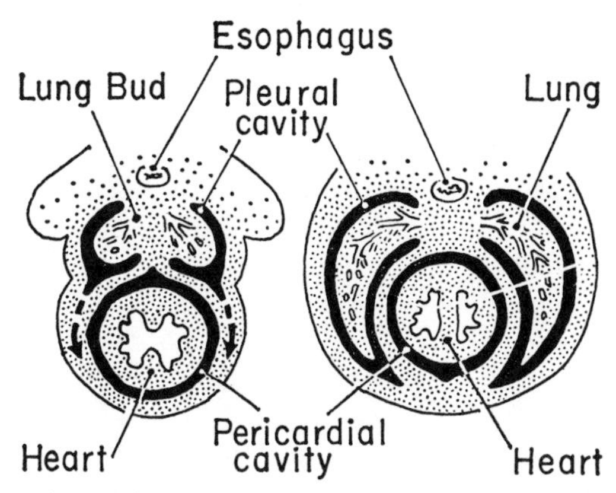

Fig. 560. The lung buds expand in their pleural cavities to embrace the heart in its pericardial sac. (After Patten.)

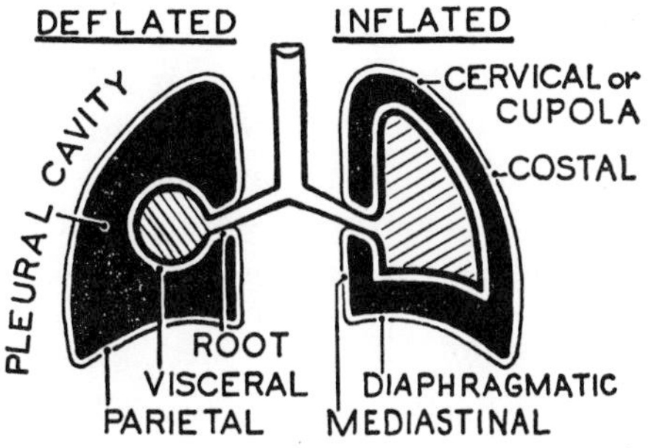

Fig. 561. The pleura: The lung represented as a balloon with a stalk (highly schematic).

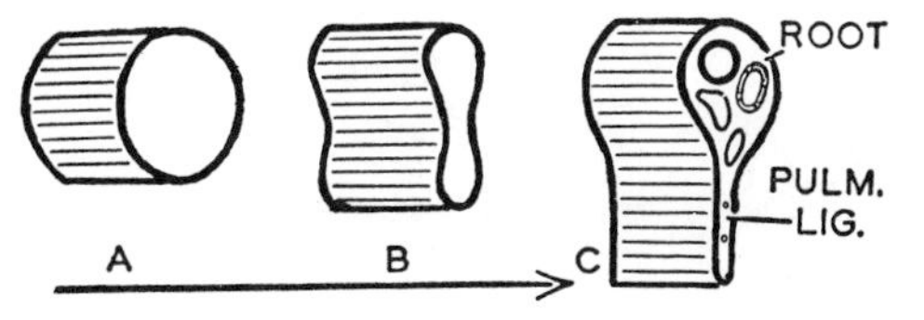

Fig. 562. Diagram of pleura at root of lung.

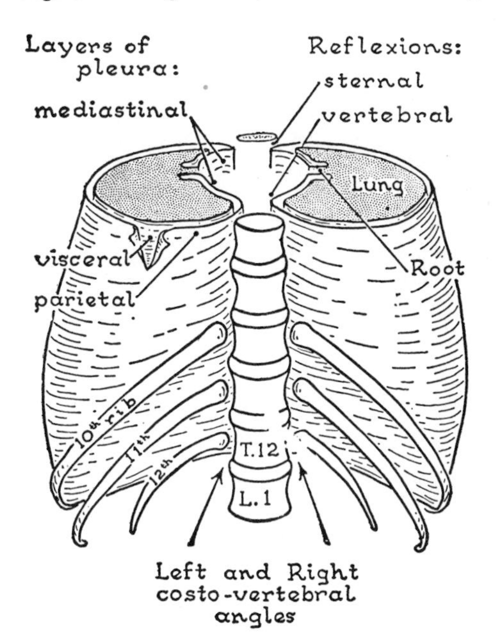

Fig. 563. Pleural reflexions, from behind.

employ the even numbers—2, 4, 6, 8, 10, and 12 ribs (*figs. 563, 564*). Thus:

The right and left *sternal reflexions* pass behind the sternoclavicular joints, meet each other in the median plane at (or above) the sternal angle, which lies at the level of the 2nd costal cartilages.

Now the right reflexion continues downward in the midline to the back of the xiphoid process; the left parts from the right at the level of the 4th cartilage, where it is deflected to about the margin of the sternum, along which it is continued downward to the 6th cartilage. Enlarged hearts seem not to effect the line (Woodburne).

Thereafter, as the left costal reflexion, it passes obliquely across the 8th rib in the midclavicular line, the 10th in the midlateral line, and the 12th at its neck or even lower (Lachman).

The right costal reflexion differs from the left in descending to a lower level anteriorly and it passes from the back of the xiphoid across the xiphicostal angle to the 7th costal cartilage.

The pleurae descend below the costal margin at three angles where an incision for abdominal surgery might inadvertently enter a pleural sac (*figs. 563, 564*).

1. Right xiphicostal angle.
2. Right costovertebral angle.
3. Left costovertebral angle.

Pleural Recesses. There are three:

1. Right costodiaphragmatic recess.
2. Left costodiaphragmatic recess.
3. (Left) costomediastinal recess.

Below the level of the costal reflexion (*fig. 565*), the diaphragmatic pleura lies in direct contact with the costal pleura because the sides and back of the lower border of the lung do not descend to the level of the costal reflexion. The potential

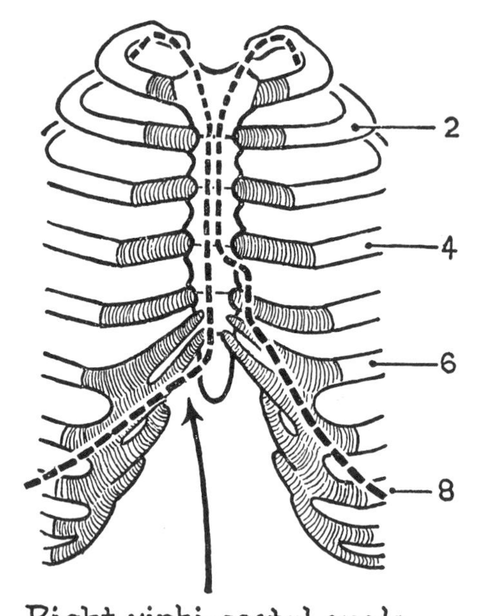

Fig. 564. Sternocostal reflexion of pleura.

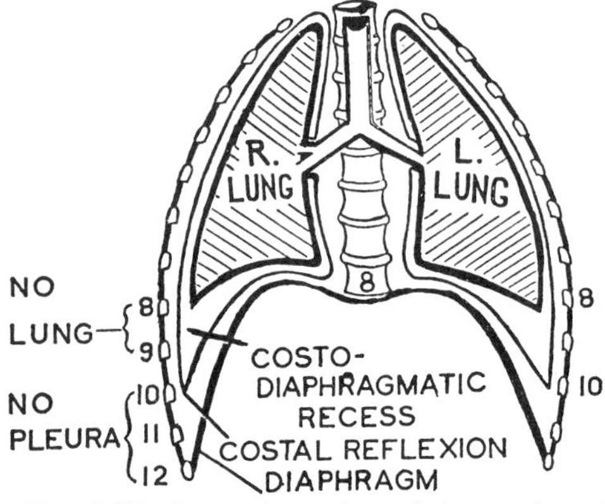

Fig. 565. Coronal section of thorax (semi-schematic).

space unoccupied by lung is known as the *costodiaphragmatic recess*. This recess becomes alternately smaller and larger as the lung advances into it and recedes from it during inspiration and expiration (*fig. 565*).

Instruments passed through the anterior parts of the (7), 8, 9, 10, and 11 intercostal spaces miss the pleural cavity but penetrate intercostal muscles, diaphragm, and enter the peritoneal cavity. At a slightly higher level they penetrate intercostal muscles, costodiaphragmatic recess, diaphragm, and enter the peritoneal cavity. Still higher they penetrate the lung also.

The third region of a pleural cavity not occupied by lung lies at the anterior ends of the 4th and 5th left interspaces. Here, owing to a deficiency in the anterior border of the left lung, where it overlies the heart, and hence called the *cardiac notch of the lung*, the costal and mediastinal layers of the left pleura come into direct contact with each other and form the *costomediastinal recess*. A thin edge of the left lung slides in and out of it in front of the heart.

Cupola (*fig. 566*). The cupola rises to, but not above, the neck of the 1st rib, which therefore protects it from injury from behind. But, since there is a drop of 3 to 4 cm between the vertebral and sternal ends of the 1st rib, it follows that the pleura rises that far above the sternal end. Here the clavicle offers some protection.

Mediastinum

The mediastinum is covered on both sides with mediastinal pleura (*fig. 563*).

Subdivisions (*fig. 567*). The central structure within the mediastinum is the heart. It is contained within a fibrous sac, the *fibrous pericardium*. The areas above, in front of, and behind the pericardium are known, respectively, as the superior, anterior, and posterior mediastina, while the area *within* the pericardium is the middle mediastinum.

In man, who walks erect, but not in quadrupeds, the fibrous pericardial sac is fused below with the central tendon of the

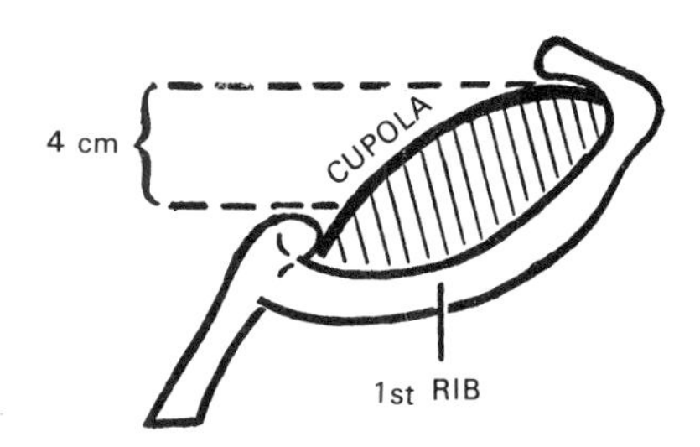

Fig. 566. The cupola of the pleura.

diaphragm; above, it reaches to the level of the sternal angle. In fact, it happens to be co-extensive in the median plane with the body of the sternum; behind it are the middle 4 thoracic vertebrae (*fig. 567*).

The Superior Mediastinum is the region above the plane joining the sternal angle to the intervertebral disc between the 4th and 5th thoracic vertebrae. At or near this plane, many structures end, or begin, or arch, or bend. Hence, it is a critical plane.

The Anterior Mediastinum is an insignificant area in front of the pericardium where the sternal reflexion of the left pleura fails to meet the right pleura in the median plane (*fig. 564*). It contains only a little fat and some lymph nodes.

The Posterior Mediastinum lies behind the pericardium and also extends downward behind the diaphragm to Th. 12.

The Middle Mediastinum or pericardial sac contains not only the heart but also the roots of eight great vessels. Lateral to the pericardium on each side run the phrenic nerves and their companion vessels.

The structures covered with the right and left sheets of mediastinal pleura are readily displayed as there is but little fat within the thorax to conceal them. For reasons that will become apparent, they should be considered under the groupings given below, and not in a haphazard manner.

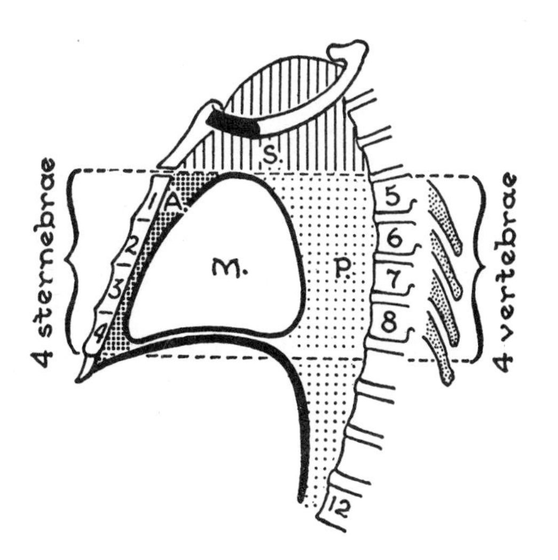

Fig. 567. Subdivisions of the mediastinum.

The thoracic contents were originally disposed symmetrically on the two sides of the body, but this initial symmetry was soon lost, largely in consequence of the disappearance of certain veins from the left side and of certain arteries from the right, as explained on page 437. The right side is the simpler; so, it will be examined first.

Right aspect of Mediastinum

The following structures are displayed:

1. Pericardial sac.
2. Two sites devoid of contents.
3. Great veins } Right phrenic nerve
   - Inferior vena cava
   - Superior vena cava
     - Right brachiocephalic (innominate) vein
     - Right jugular vein
     - Right subclavian vein
4. Esophagus, Trachea, and Right Vagus Nerve.
5. Root of lung and pulmonary ligament: Bronchus, Pulmonary artery, Pulmonary veins, and Vessels and Nerves.
6. Arch of the azygos vein.

The Pericardial Sac is separated from the right half of the body of the sternum by the thickness (or rather thinness) of the anterior border of the right lung and pleura. The sac is separated from the middle four thoracic vertebrae by the thickness of the aorta and the collapsed esophagus (*fig. 568*).

Two Sites Devoid of Contents. In two places the mediastinum has no contents; so, the right and left layers of mediastinal pleura come into apposition: (1) between the sternum and pericardium from the level of the second to the fourth costal cartilages, and (2) between the lower part of the esophagus and the aorta. Here the two layers form a dorsal **meso-esophagus** (*fig. 568, 578*).

Great Veins. A long stick or probe passed upward through the i. v. cava will enter and traverse the right atrium, the s. v. cava and the right brachiocephalic vein and, if continued into the neck, will enter the internal jugular vein. The venae cavae are like two rivers that flow due north and south to empty into a lake, the *right atrium* (*fig. 569*).

On each side of the body the internal jugular and subclavian veins unite behind the sternal end of the clavicle to form the corresponding brachiocephalic (innominate) vein. The left brachiocephalic vein crosses behind the upper half of the manubrium and joins the right brachiocephalic vein at the right margin of the sternum to form the s. v. cava.

The s. v. cava descends from the 1st to the 3rd right costal cartilage and there opens into the right atrium. Its upper half is outside the pericardial sac; its lower half is inside. The i. v. cava, which likewise is partly outside and partly inside the pericardial sac, pierces the diaphragm and enters the heart at the levels of the xiphisternal joint and 6th costosternal joint, respectively.

The Right Phrenic Nerve enters the thorax behind the junction of the subclavian and the internal jugular vein (*fig. 576*), and descends subpleurally *along the line of the great venous channel*, i.e., on the right side of the right brachiocephalic vein, superior vena cava, pericardium, and inferior vena cava—the pericardium separating it from the lower part of the s. v. cava, right atrium, and upper part of the i. v. cava. It then pierces the diaphragm and spreads out on its abdominal surface.

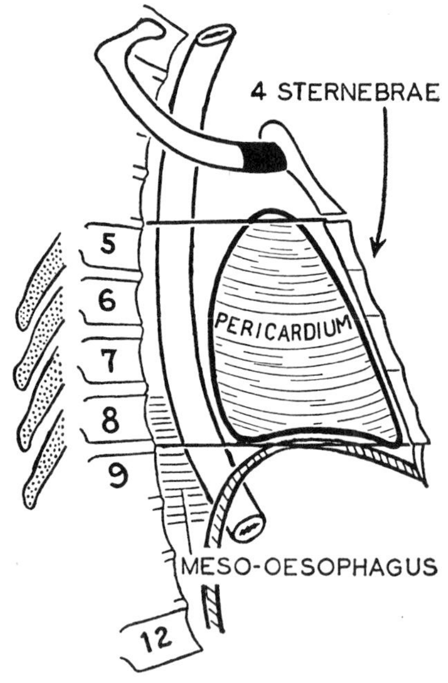

Fig. 568. The pericardial sac has 4 sternebrae in front of it and 4 vertebrae behind.

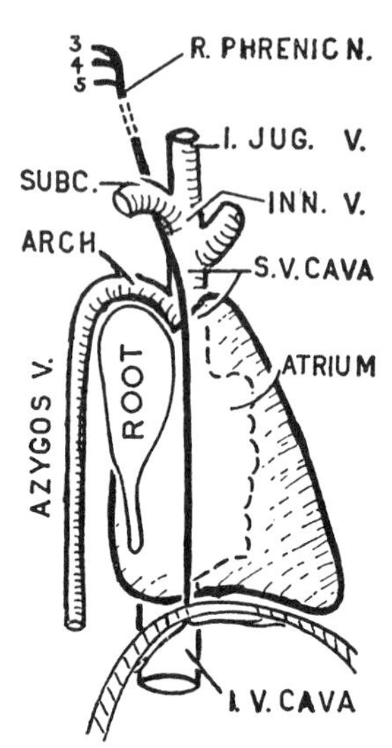

Fig. 569. The right phrenic nerve runs subpleurally along a great venous channel. (*Innominate = brachiocephalic.*)

The phrenic nerve is accompanied by the *pericardiacophrenic artery*—a branch of the internal thoracic artery.

Esophagus, Trachea, and Right Vagus Nerve. The *esophagus* lies in front of the vertebral column except below, where the aorta gains the median plane and interposes itself. The *trachea* lies immediately in front of the esophagus throughout the superior mediastinum, and on reaching the plane between the superior and posterior mediastina it bifurcates into a right and a left bronchus.

In man (i.e., mammals) lungs, bronchi, and a trachea make their appearance as an outgrowth from the upper part of the food passage and therefore, the vagus nerves, being the nerves of the food passage, are called upon to supply them (*fig. 570*).

The right vagus, after entering the thorax between the subclavian artery and brachiocephalic vein, passes obliquely downward and backward first on the side of the brachiocephalic trunk (*fig. 571*), then on the side of the trachea to the back of the root of the lung where it takes part in the posterior pulmonary plexus. From this plexus the main trunk then passes to the esophagus and adheres to it thereafter.

Root of Lung and Pulmonary Ligament. The three chief structures in the root of a lung are:

1. *The pulmonary artery* which brings blood, charged with carbon dioxide from the heart to the lungs.
2. *The pulmonary veins* (two on each side, an upper and a lower) which return oxygenated blood to the heart.
3. *The bronchus or* air passage.

The order is the same on both sides; namely, bronchus behind, artery above, veins below (*fig. 573*). On the right side, however, the bronchus to the upper lobe of the lung is higher even than the artery—hence, it was formerly called the *eparterial bronchus.*

Of the vessels, the two pulmonary veins on each side pierce the pericardium to end in the left atrium and are, therefore, below the corresponding right or left pulmonary artery, which lies along the upper border of the atria (*fig. 599*).

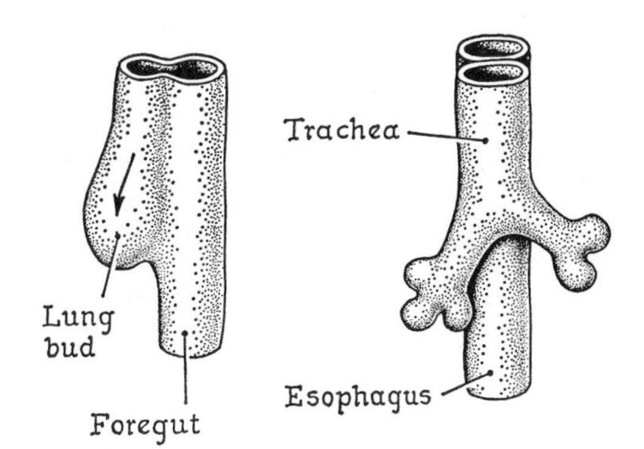

Fig. 570. The air passage buds downward from, and remains an intimate relation of, the front of the food passage. (Modified after Langman.)

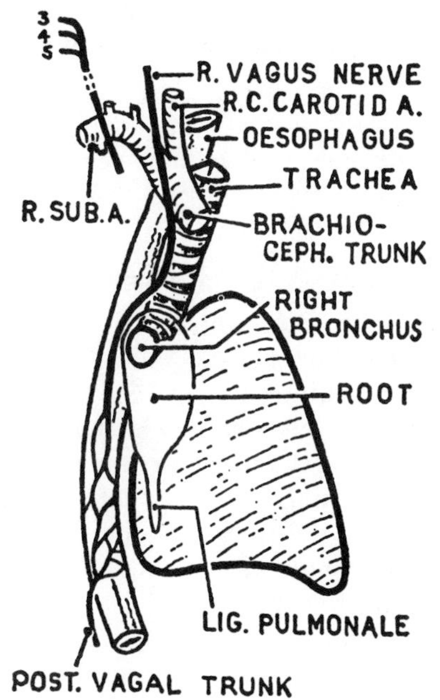

Fig. 571. The right vagus nerve is applied to the trachea and esophagus.

Also in the root of the lung are:

4. Nerve plexuses.
5. Bronchial vessels.
6. Lymph nodes.

Branches of the vagus behind the root of each lung, joined by sympathetic fibers (Th. 2, 3) from the *posterior pulmonary plexus;* branches enter the lung. Some fibers from the vagus pass to the front of the root where, with others from the cardiac plexus, they constitute the *anterior pulmonary plexus.* From it, fibers enter the lung.

The stroma of the lung derives pure arterial blood from the *bronchial artery,* which springs either from the aorta or from an intercostal artery and runs with the bronchi. In the root of the lung there are many *lymph nodes,* black from inhaled pigment.

The Arch of the Azygos Vein. The azygos vein runs upward in front of the vertebral column to the junction of the posterior and superior mediastina and there arches forwards above the root of the lung to end in the s. v. cava before the latter pierces the pericardium. The arch crosses lateral to the esophagus, trachea, and right vagus.

The brachiocephalic trunk, the right end of the left brachiocephalic vein, the beginning of the aortic arch, and the fatty remains of the thymus are also in contact with the right mediastinal pleura.

Left Aspect of Mediastinum

The structures covered with left mediastinal pleura are:

1. Pericardial sac.
2. Root of left lung and pulmonary lig.
3. Aortic arch and descending aorta.
4. Esophagus, trachea, left recurrent nerve, and thoracic duct.
5. Left common carotid and left subclavian arteries.
6. Left phrenic and vagus nerves and left superior intercostal vein.
7. Esophagus.

The Pericardial Sac, the Root of the Left Lung, and the Pulmonary Ligament are disposed as on the right side (*fig. 573*).

The Aortic Arch (*fig. 572*) is the portion of the aorta lying in the superior mediastinum. It begins where the ascending aorta leaves the pericardium, which is at the level of the sternal angle. It arches backward and to the left above the root of the left lung, rising half the height of the manubrium sterni and reaching the vertebral column at the 5th thoracic vertebra.

Continued as the *descending aorta*, it traverses the posterior mediastinum in contact with the bodies of the lower eight vertebrae, first on their left sides; but, gradually gaining the median plane, it intervenes between the esophagus and the vertebrae. Finally, at the disc between the last thoracic and the first lumbar vertebra, it passes behind the median arcuate ligament of the diaphragm to become the abdominal aorta.

(*Anomalies of the Arch, fig. 627*).

The Trachea and Esophagus in the Superior Mediastinum (*figs. 573, 574*). In the angle between them runs the *left recurrent*

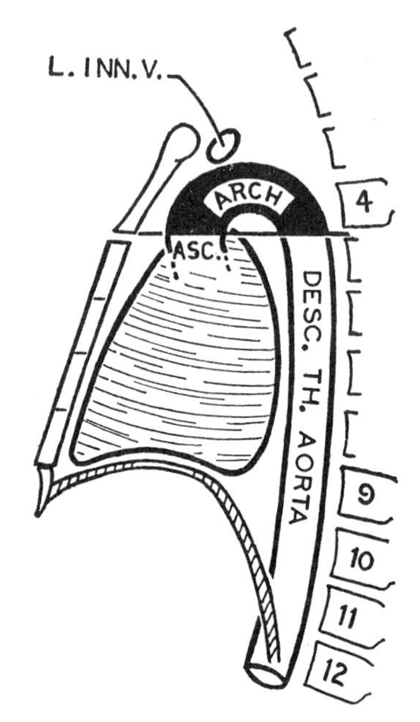

Fig. 572. The thoracic aorta.

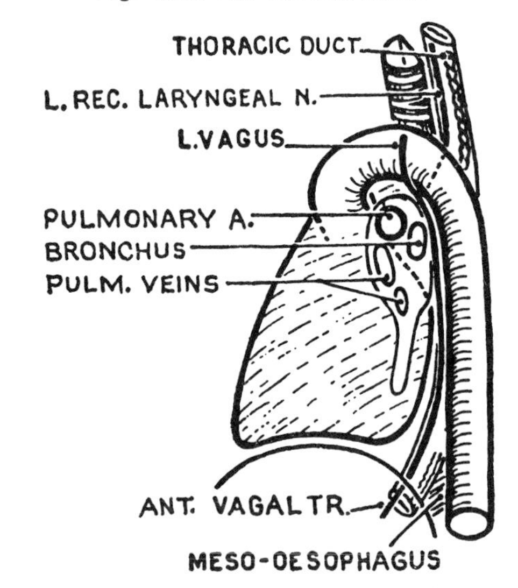

Fig. 573. Trachea and esophagus, left side.

laryngeal nerve. On the side of the esophagus, the thin-walled *thoracic duct* ascends embedded in a film of fat. The aortic arch crosses these 4 structures as a unit.

(*The Thoracic Duct* is described on page 446.)

The Left Common Carotid and Left Subclavian Arteries (*fig. 575*) arise from the convexity of the aortic arch, where it crosses the trachea, and pass obliquely upward and backward in semispiral fashion round the "*unit of 4*," (*fig. 574*) crossing in turn the trachea, recurrent nerve, esophagus, and thoracic duct.

The Left Phrenic Nerve (*fig. 576*) enters the thorax between the subclavian artery and vein (at the beginning of the brachiocephalic) and runs a subpleural course, being covered throughout with mediastinal pleura.

After crossing the subclavian a., it at once crosses the internal thoracic artery and from it acquires a companion, the *pericardiacophrenic artery*. In its course it runs just in front of the root of the lung.

(*Origin* in the neck, page 518.)

(*Distribution*, page 417.)

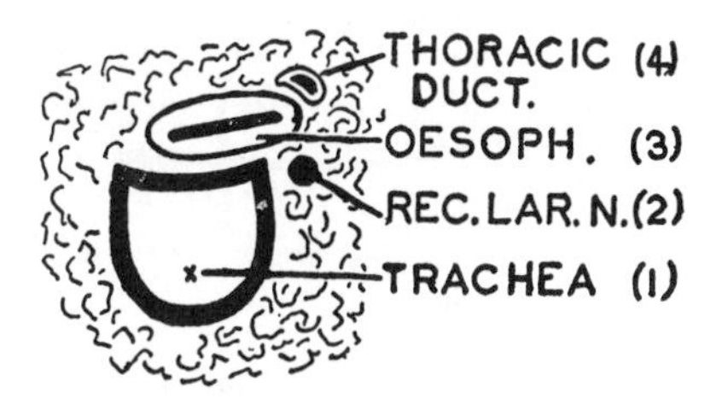

Fig. 574. Four parallel structures—a "unit of 4" (transverse section).

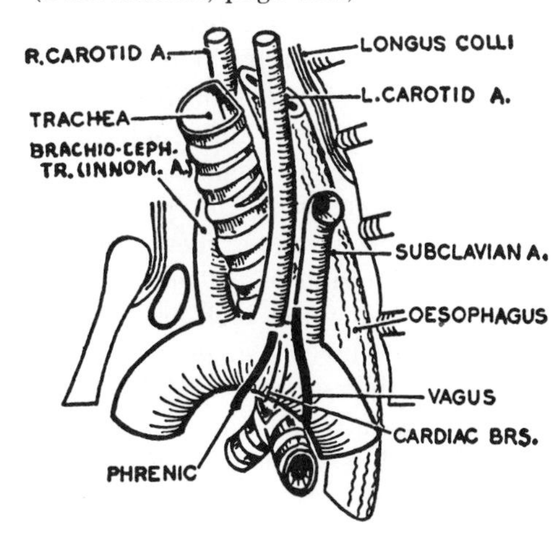

Fig. 575. Arteries ascend the "unit" in semispiral fashion.

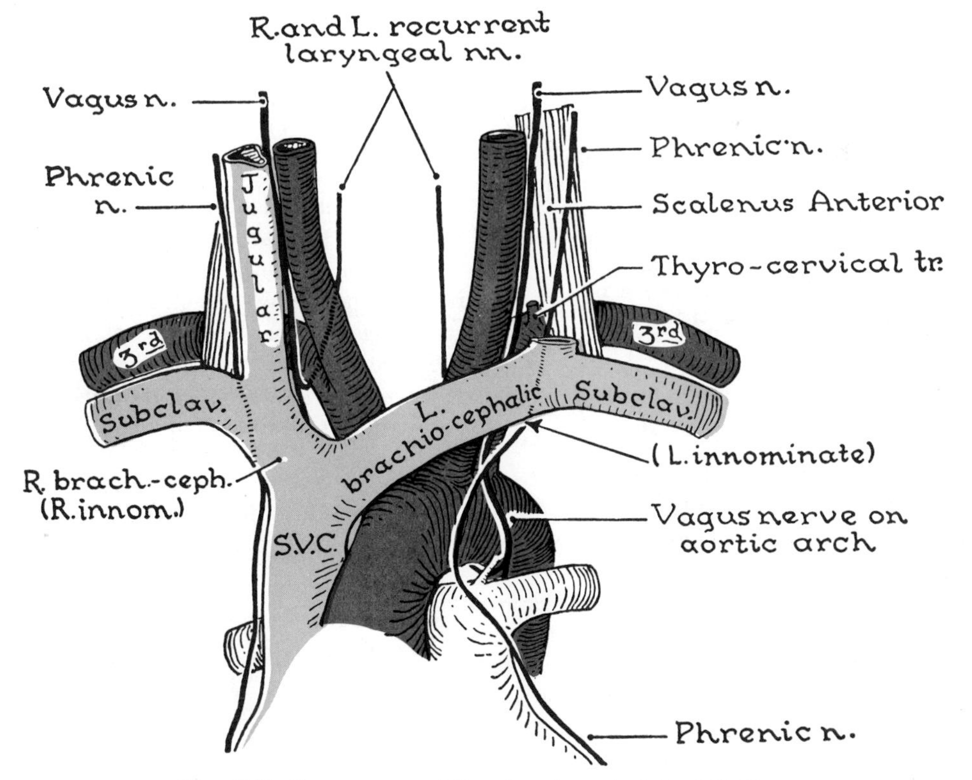

Fig. 576. The courses of the phrenic and vagus nerves.

The Left Vagus Nerve, after traversing the neck within the carotid sheath on the lateral side of the common carotid a. (*fig. 576*), continues through the superior mediastinum on the lateral side of the same artery, and, therefore, between it and the subclavian artery. These two arteries conduct it to the aortic arch which it crosses to gain the back of the root of the lung. Thereafter, it passes to the esophagus and behaves like the right vagus.

The Left Recurrent Laryngeal Nerve springs from the vagus where the latter crosses the left side of the aortic arch. It then passes below the arch and up on its right side to gain the angle between the trachea and the esophagus (*fig. 577*).

Embryologically speaking, the left recurrent laryngeal nerve recurs around the lig. arteriosum rather than around the aortic arch. **The ligamentum arteriosum** is the obliterated posterior half of the primitive VI left aortic arch (*fig. 625*).

In prenatal life the lig. arteriosum was a patent vessel, the *ductus arteriosus* (*fig. 40*), which discharged into the aorta beyond the origin of the left subclavian artery.

Two slender *Cardiac Nerves* arising in the neck, one from the vagus, the other from the sympathetic trunk, cross the aortic arch between the left phrenic nerve in front and the left vagus behind (*fig. 575*), and join the (*superficial*) *cardiac plexus*, which lies on the immediate right of the lig. arteriosum.

The Left Superior Intercostal Vein (*fig. 641*) drains the 2nd, 3rd, and 4th intercostal veins across the aortic arch into the brachiocephalic vein.

The Esophagus in the Posterior Mediastinum. In the upper half, the left side of the esophagus is concealed by the descending aorta. In the lower half, the esophagus, in the meso-esophagus, crosses the aorta very obliquely to gain the left side of the thorax and is clothed with left mediastinal pleura down to the level of the 10th thoracic vertebra where it pierces the diaphragm (*fig. 577*).

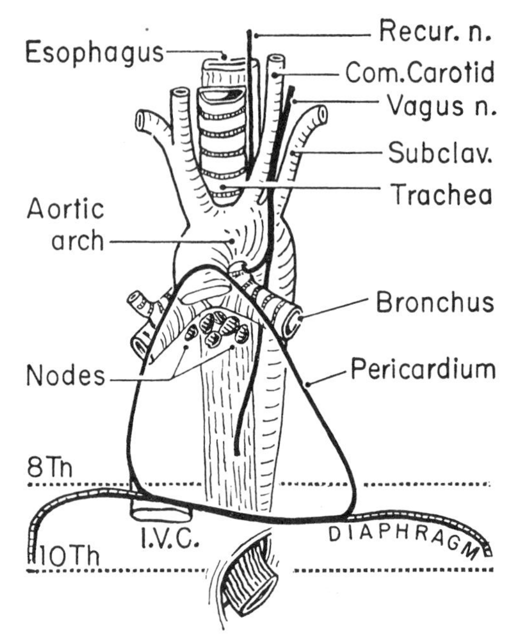

Fig. 577. The anterior relations of the esophagus.

This meso-esophagus permits the esophagus to advance several centimeters from the vertebral column. The lungs then approach each other behind the esophagus, particularly on inspiration and with the subject erect. A lateral X-ray picture reveals this *retro-esophageal "window."*

The Cupola of the Pleura (*fig. 579*) rises to the neck of the 1st rib. It forms the floor of the root of the neck (*fig. 723*). *Figure 579* reveals the details of structures seen on stripping off the cupola (*fig. 579*). The *right vagus* gives off the *right recurrent laryngeal nerve* which winds tightly below and behind the subclavian artery to ascend in the neck.

Just Outside the Mediastinum

The Sympathetic Trunk (thoracic portion) (*figs. 578, 580*) is covered throughout with costal pleura, nothing intervening, and it lies a little wide of the mediastinum. Traced caudally, the trunks of the two sides get progressively closer together.

At the superior aperture of the thorax, the trunk lies on the neck of the 1st rib. In the thorax, it crosses the heads of the 2nd to 9th ribs, the 10th costovertebral joint, and the bodies of the 11th and 12th vertebrae. It continues into the abdomen along the anterior border of the Psoas. The intercostal arteries and veins cross the trunk posteriorly (*fig. 580*).

Typically, a ganglion is to be found in front of each rib. Since *ganglia* are segmental structures, there is developmentally one for each of the 31 spinal nerves.

The inferior cervical and the 1st thoracic ganglion are usually amalgamated to form the *cervicothoracic or stellate ganglion*, which lies behind the origin of the vertebral artery.

Owing to their tendency to amalgamate, there are with fair constancy only 3 (or 4) cervical, 11 thoracic, 4 lumbar, 4 sacral pairs, and 1 coccygeal (unpaired).

Connections. From the ventral rami of a limited number of spinal nerves (all the thoracic and the upper two or three lumbar) a white ramus communicans, carrying

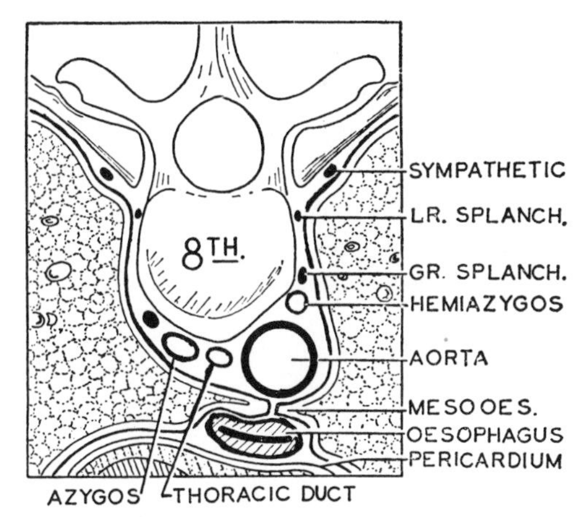

Fig. 578. Transverse section through posterior mediastinum, showing meso-esophagus and general relations.

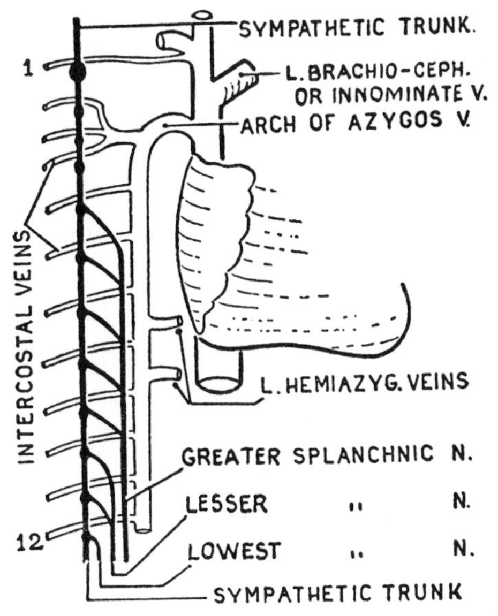

Fig. 580. Sympathetic trunk and splanchnic nerves.

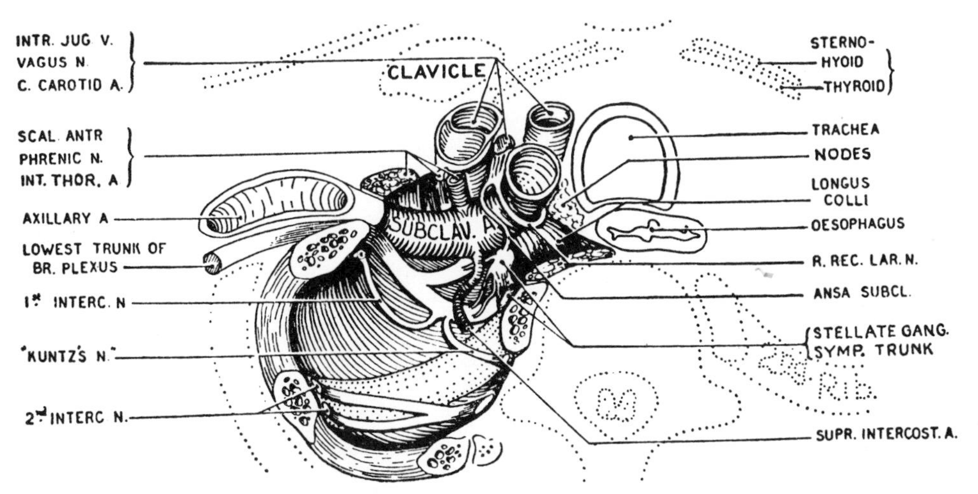

Fig. 579. Cupola of pleura has been removed in order to display immediate relations (from below).

preganglionic fibers, passes anteromedially to join a sympathetic ganglion; whereas the ventral ramus of every spinal nerve, without exception, receives from a sympathetic ganglion one or more gray rami communicantes, carrying postganglionic fibers.

Distribution to Viscera. From the upper five thoracic ganglia, postganglionic fibers pass to the cardiac plexus, posterior pulmonary plexus, and upper thoracic parts of the esophagus and aorta. Fibers either from the lower thoracic ganglia or from the splanchnic nerves contribute to the supply of the lower thoracic portions of the esophagus and aorta.

The Splanchnic Nerves (*fig. 580*) are preganglionic fibers coming from white rami communicantes and making "nonstop journeys" through sympathetic ganglia. They end in the celiac and renal ganglia, whence they are relayed as unmyelinated postganglionic fibers (*fig. 44.1*).

There are three (paired) splanchnic nerves—greater, lesser, and lowest. Their sources vary widely; they are rarely bilaterally symmetrical; and the lowest may be absent (A. F. Reed). Thus:

1. The *greater splanchnic nerve*, larger than the sympathetic trunk itself, springs from the (4th), (5th), (6th), (7th), **8th, 9th, and 10th** ganglia, runs caudally just lateral to the azygos (or hemiazygos) vein, pierces the crus of the diaphragm, and ends in the celiac ganglion.

2. The *lesser splanchnic nerve* springs from the (9th), **10th, 11th,** and (12th) ganglia, runs caudally—lateral to the greater nerve—pierces the crus, and ends in the lower part of the celiac ganglion (specifically, the aorticorenal ganglion).

3. The *lowest splanchnic nerve*, when present, springs from either the **11th** or **12th** ganglion or from both ganglia, takes a similar course, and ends in the renal plexus.

31 LUNGS

Shape and Parts; Borders and Surface Anatomy; Cardiac Notch.
Lobes and Fissures; Variations.
Bronchial Tree; Bronchopulmonary Segments; Removal of Mucus and Foreign Matter.
Structure of Bronchi and Lungs; Lung Units.
Vessels and Nerves: Pulmonary Artery and Veins; Bronchial Arteries and Veins; Lymphatics; Nerves.
Movements and Shifts; Surface Anatomy.

MECHANISM OF RESPIRATION

Inspiration and Expiration; During Quiet Inspiration; On Deeper Inspiration; Expiration; Posture.

Lungs (Pulmones)

The shape of the lungs, like that of the liver and spleen, depends largely upon the surrounding structures, and the impressions these make are best observed in formalin-hardened specimens. A full description of the form and relations of the lungs is in large measure redundant, since it is a description of the counterpart of the parietal pleura already given.

The weight of the lungs varies with their content of blood. With the vessels empty, the right lung weighs 240 gm; the left, being smaller, weighs less. Filled with blood, they double their weights (Gradwohl).

The lungs are conical (*fig. 581*). Each has an apex and base, costal and medial surfaces, anterior and inferior borders, and a hilus.

Surfaces. *The Apex* rises to the neck of the 1st rib.

The Base or diaphragmatic surface is concave. Because the right dome of the diaphragm rises higher and is more convex than the left dome, the right lung is shorter. Its base overlies the liver; the left base overlies the liver, stomach, and spleen.

The Costal Surface is marked by the ribs.

The Medial Surface has two parts: (1) vertebral and (2) mediastinal. *The vertebral part* ("posterior border") is full and rounded; it passes imperceptibly into the costal surface; and it occupies the gutter at the side of the vertebral column.

The mediastinal part (*figs. 582, 583*) bears the impress of the structures covered with the mediastinal pleura. The most noticeable feature of this surface is the *hilus* of the lung and the line of the attachment of the pulmonary lig. which descends from the hilus (*fig. 582*). In front of these, an excavation, the *cardiac impression*, is deeper on the left side than on the right, because two-thirds of the heart lie on the left of the median plane.

Borders of Lungs: Surface Anatomy. As stated already (p. 401), the lungs fill the *pleural cavities* except at three sites: right and left costodiaphragmatic and left costomediastinal recesses. The lower limits of the costodiaphragmatic recesses are related to ribs 8, 10, and 12; the sharp *lower borders* of the lungs are related to ribs 6, 8, and 10, so they are two ribs higher. They are highest in the erect posture, and lowest when a person lies prone.

The **cardiac notch,** or "bite" that the heart takes out of the anterior border of the left lung, is greater than the bite it takes out of the left pleura—hence the left costomediastinal recess. During inspiration the lungs invade, but by no means fill, the pleural recesses.

The anterior borders of the lungs are insinuated between the body of the sternum and the pericardium, so they are thin and sharp.

Lobes and Fissures. In each lung a complete fissure, the *oblique fissure*, cuts through the costal, diaphragmatic, and mediastinal surfaces as far as the root. It follows the line of the 6th rib, being much lower during life than one sees in the dead. When the arm is raised above the head, the medial border of the scapula practically overlies this oblique fissure. The left one is a little higher and more vertical (Brock).

On the right side, the *horizontal fissure*, lying at the level of the 4th costal cartilage runs backward from the anterior border to meet the oblique fissure in the midlateral line. The right lung has, therefore, three lobes—upper, middle, and lower; the left has two—upper and lower.

The antero-inferior part of the left upper lobe is called the *lingula*. The lingula (lingular process) and the cardiac notch above it correspond to the middle lobe of the right lung. The middle lobe lies at the front of the chest, posteriorly reaching the midlateral line. You can cover it with your own hand placed on your chest.

Variations. A *lobe of the azygos vein* results when the apex of a developing right lung encounters the arch of the azygos vein and is cleft by it (*fig. 584*). The vein is suspended in a

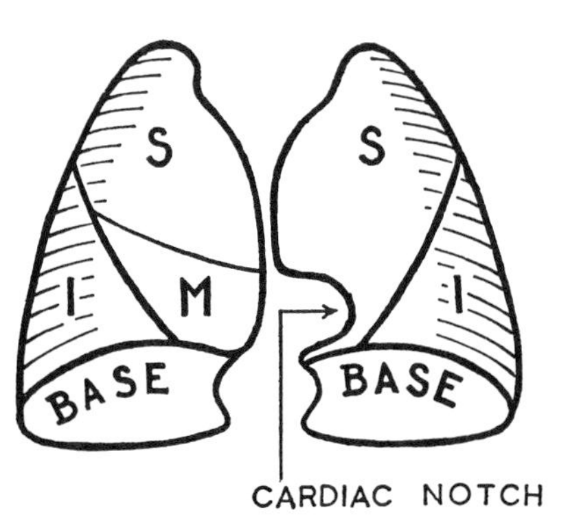

Fig. 581. The lungs, anterior aspect.

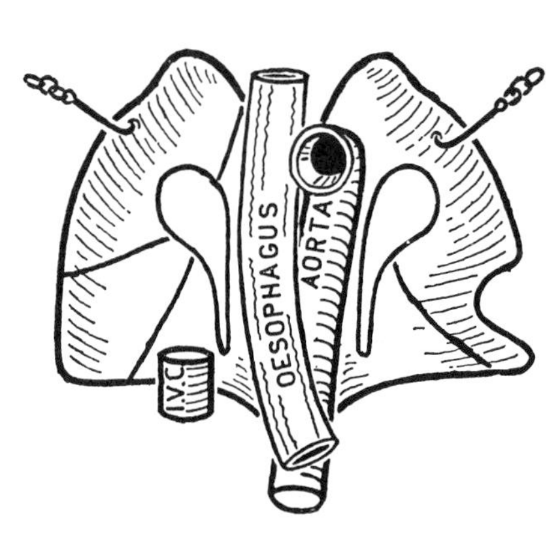

Fig. 582. The mediastinal surfaces of the lungs. This picture reveals the posterior, lateral, and anterior relations of the pericardium.

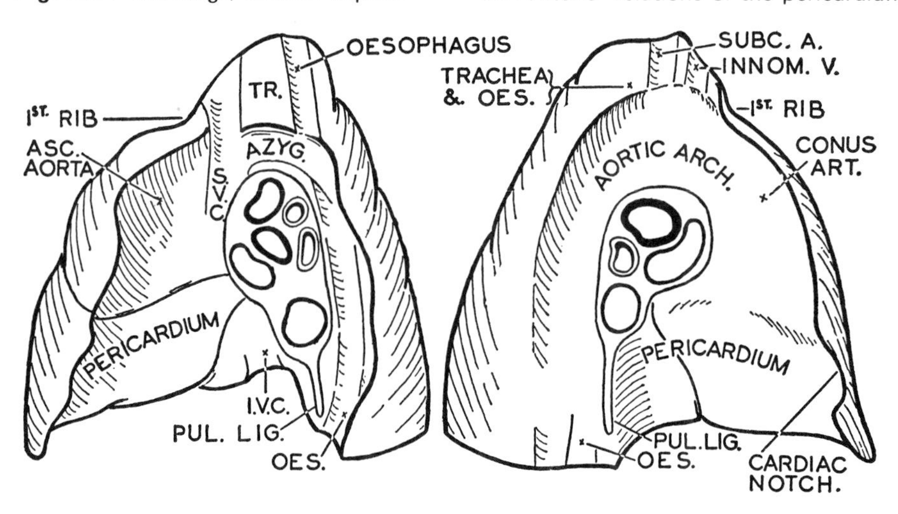

Fig. 583. Impressions commonly made on the mediastinal surfaces of the lungs. Of course, some intimate relations (e.g., phrenic n.) make no impressions.

pleural "mesentery," and it may cause a shadow by X-ray.

Fissures. The horizontal fissure (r. lung) is very commonly either completely or partially absent. The upper part of the oblique fissure (r. or l.) is commonly absent, and lower part sometimes is absent.

Additional fissures have been described between most bronchopulmonary segments (see below) and between many subsegments (Foster-Carter). Notably, they commonly define the superior segment (r. or l. lower lobe), the lingula (l. upper lobe), and a slight fissure commonly defines the right cardiac, or medial basal, segment.

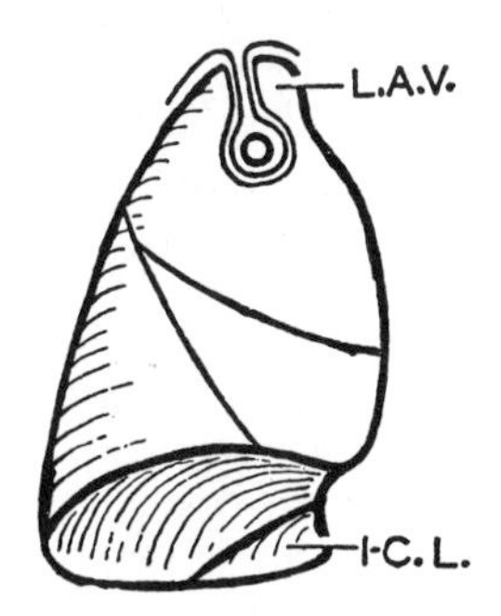

Fig. 584. Accessory lobes of the right lung: lobe of the azygos vein; infracardiac lobe.

The Bronchial Tree (*figs. 585* to *589*). The trachea bifurcates on the plane between the superior and posterior mediastina (4–5 disc) into a right and a left *primary* (main) *bronchus* for the supply of the respective lungs. Each primary bronchus descends to the hilus of its own lung where it lies behind the pulmonary vessels and on a level with the 5th and 6th vertebral bodies. On the right side it gives off three *secondary* (lobar) *bronchi*, and on the left side, two, for the corresponding lobes of the lungs. The secondary or lobar bronchi divide into *tertiary* (segmental) *bronchi*. Each segmental bronchus, together with the portion of the lobe it supplies, is called a **bronchopulmonary segment** (*fig. 586*). As seen from *figure 585*, there are usually 10 segmental bronchi on the right side, and 8 on the left.

As the bronchi continue to branch and rebranch, the bronchopulmonary segments subdivide into smaller and smaller subsegments. If you are required to learn *details*, then—

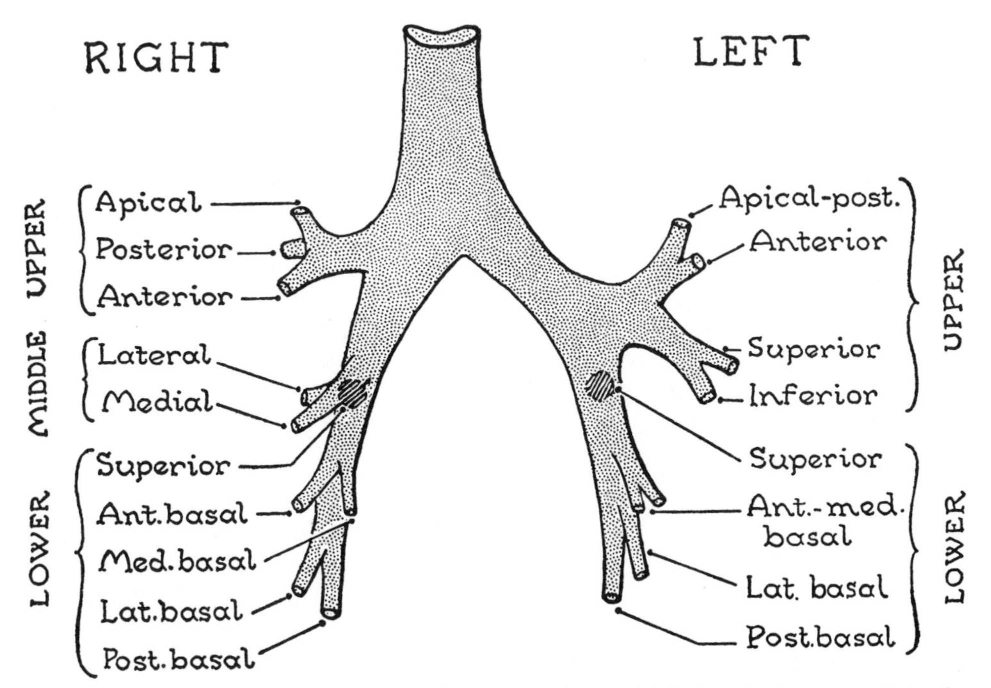

Fig. 585. The 10 right and 8 left segmental bronchi. (After Jackson and Huber.)

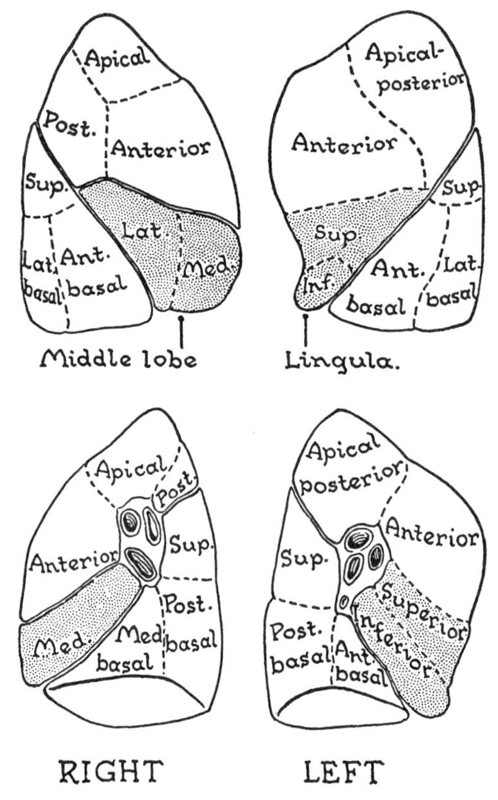

Fig. 586. The 10 right and 8 left bronchopulmonary segments. (After Jackson and Huber.)

Study Figs. 585 to 589. The right upper lobe bronchus (eparterial bronchus) arises 2–5 cm from the tracheal bifurcation. After the course of 1 cm it divides into its three segmental bronchi.

The right middle lobe bronchus arises about 2 cm below the upper lobe bronchus.

The left upper lobe bronchus arises 5 cm from the tracheal bifurcation. After the course of less than 1 cm, it bifurcates and then both of the forked branches bifurcate again into four segmental bronchi. Owing to its low origin, the apical posterior bronchus and its branches make a steep ascent.

The superior (apical) lower lobe bronchi of both sides arise almost opposite the mouths of the right middle and left upper lobe bronchi, respectively.

The posterior basal bronchus is the largest of the lower lobe segmental bronchi; it may be regarded as the continuation of the stem of the bronchial tree (*figs. 587–589*).

A right **tracheal bronchus** (present in the sheep and pig) occurs occasionally in man. It is usually a right apical bronchus displaced onto the trachea (Boyden).

Removal of Mucus and Foreign Material. The tracheobronchial tree is cleared by ciliary action and by coughing. But, by assuming various recumbent postures, the assistance of gravity may be obtained for drainage of accumulated fluids. The bronchoscopist's approach (e.g., removing an aspirated foreign body) is via the larynx and trachea.

Structure. The C-shaped bars of hyaline cartilage found in the trachea and extrapulmonary bronchi give place in the intrapulmonary bronchi to plates of cartilage which are scattered irregularly around a circular lumen. The *bronchi* branch and rebranch until their diameter is reduced to 1 mm or less—bronchioles.

Microscopic Structures

In the bronchioles the cartilage and the mucous glands cease. The noncartilaginous terminal bronchioles divide several times until the cilia cease and alveolar outpouching appear on their walls, whereupon they are aptly called *respiratory* bronchioles, as they are the beginning of the respiratory part of the lung. Each respiratory bronchiole opens into a number of *alveolar ducts* [*ductules*]. From the ducts arise the alveolar sacs and alveoli (*air cells*) (*fig. 590*).

Lung Units. On the surface of the lung dark lines are seen to enclose *polygonal fields*, 10 to 20 mm in diameter. They are the bases of pyramidal portions of lung, supplied by bronchioles 1 mm or less in diameter, and enclosed by areolar septa rendered black by inhaled pigment contained in lymphatics. These portions of lung are anatomical or secondary lobules (*fig. 591*). Finer lines subdivide the bases into smaller areas representing primary lobules or *lung units*, each served by a respiratory bronchiole (*fig. 590*).

Vessels and Nerves of the Lungs

Pulmonary Artery (*fig. 592*). The arterial tree, in general, follows the bronchial pattern.

Pulmonary Veins. On each side two pulmonary veins, an upper and lower, enter the left atrium of the heart; the right upper vein drains the upper and middle lobes.

Arteries. After giving off branches to the upper lobe, each pulmonary artery descends on the posterolateral side of the "stem" bronchus. The segmental and subsegmental arteries radiate from the hilus *medial* to ascending bronchi, *cranial* to transverse bronchi, and *lateral* to descending bronchi.

Segmental arteries vary in origin, and may,

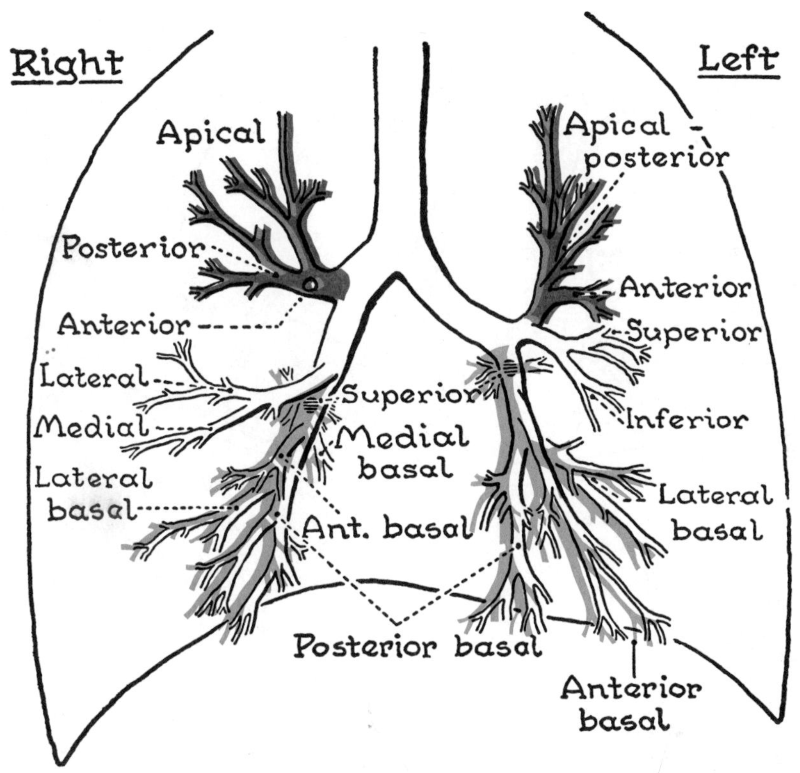

Fig. 587. The distribution of the bronchi, front view. (After Nelson, modified.)

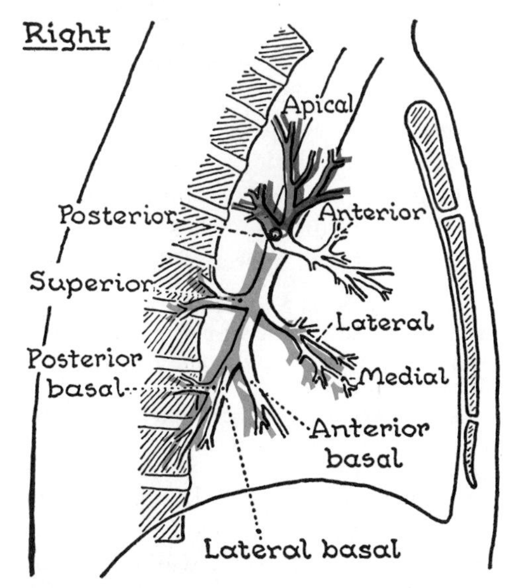

Fig. 588. The distribution of the right bronchus, side view. (After Nelson, modified.)

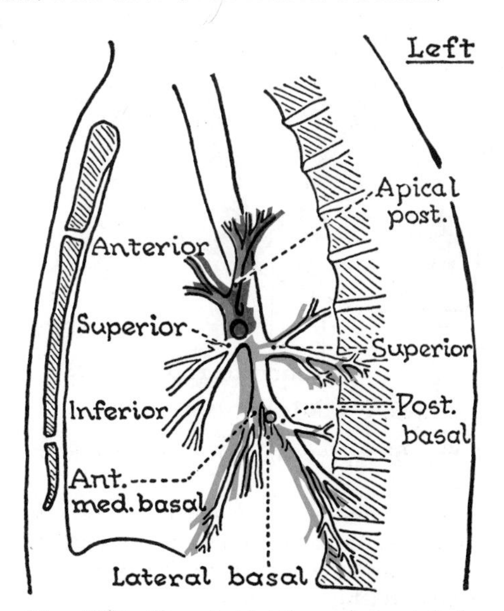

Fig. 589. The distribution of the left bronchus, side view. (After Nelson, modified.)

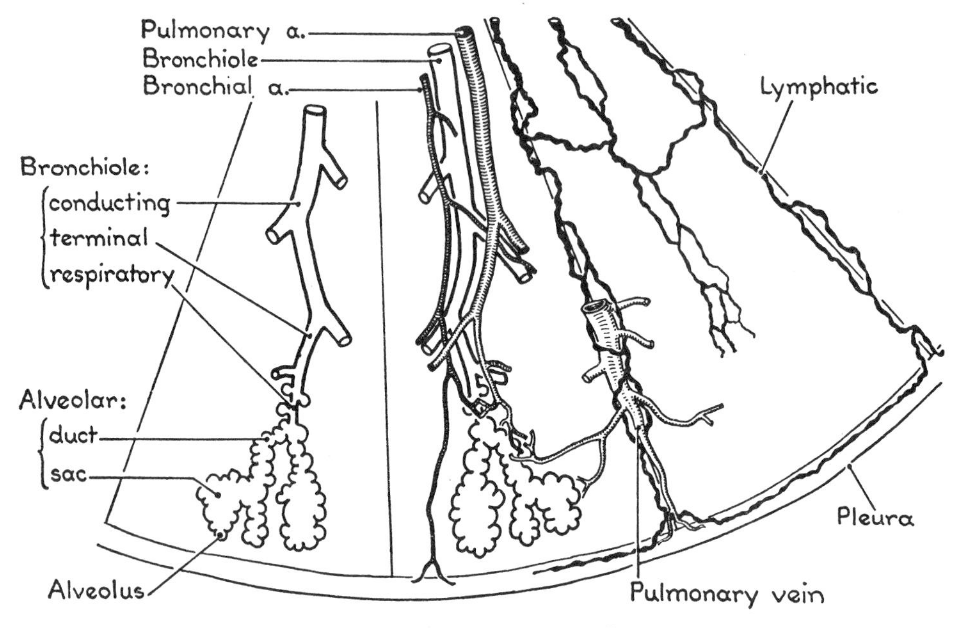

Fig. 590. Structure of a lobule of the lung.

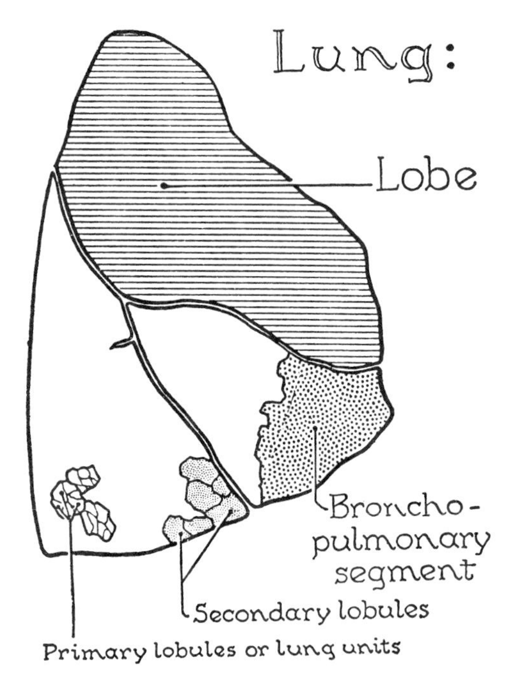

Fig. 591. The subdivisions of the lung.

indeed, be absent, in which case the subsegmental arteries arise independently from the "stem" arteries or elsewhere (Boyden).

Veins. Within the lung, they lie in areolar intrasegmental and intersegmental septa; they alternate with the arteries, each vein receiving a share of blood delivered by two arteries (including arteries from contiguous segments) and each artery contributes blood to two veins.

The Bronchial Arteries arise by a stem from the aorta, but the right may arise from an intercostal artery. Each supplies the bronchi and related structures, and runs through interlobular septa to supply the pulmonary pleura (*fig. 590*).

The pressure in the bronchial arteries is high; in the pulmonary arteries it is low. Nevertheless, anastomoses between them occur—at capillaries of the respiratory bronchioles and even earlier. In the newborn and in disease, these anastomoses may achieve a diameter of 0.5 mm.

Arteriovenous Shunts between the pulmonary artery and vein occur both in the lobules of the lung and in the pulmonary pleura (Tobin and Zariquiey).

As a rare *anomaly* the bronchial artery replaces the pulmonary artery.

Lymphatics (*fig. 590*). Lymph channels, accompanying small blood vessels, occur adjacent to the alveolar walls, but not in the interalveolar partitions. Lymph drainage is into the bronchopulmonary nodes in

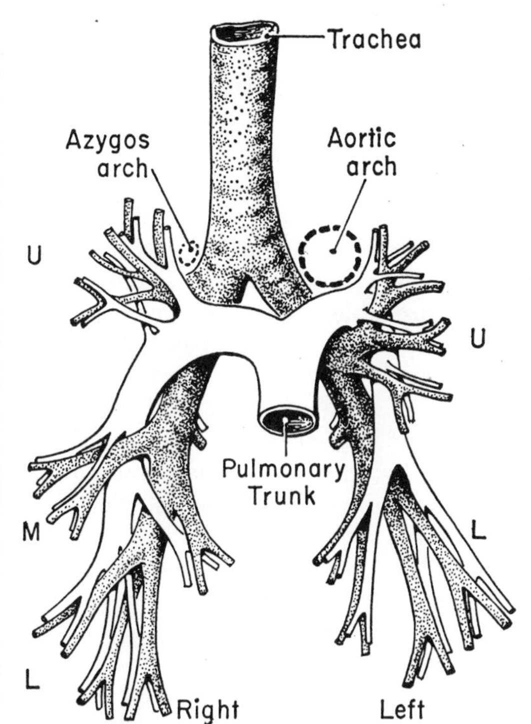

Fig. 592. To show the relationship of pulmonary arteries to bronchi, front view. *U, M, L* = branches to upper, middle, and lower lobes. (After Hayek, translated by V. E. Krahl.)

the hilus. These drain into the tracheobronchial nodes. A few vessels may drain into posterior mediastinal nodes.

Nerves of Lungs and Pleura. Branches of the vagus and of the thoracic sympathetic ganglia 1 to 5 (mainly 2, 3, and 4) form the pulmonary plexuses and these supply the lungs. Sensory vagal fibers constitute the afferent limb of the respiratory reflex arc.

Efferent vagal fibers are bronchoconstrictor and secretomotor. Efferent sympathetic fibers are bronchodilator; hence, spasms of the bronchi, as in asthma, are relieved by adrenalin (G. A. G. Mitchell).

Each *phrenic nerve* is not only motor to its own half of the diaphragm, it is also the sensory nerve to the central part of the diaphragmatic pleura, and the adjacent part of the mediastinal pleura. The *intercostal nerves* are sensory to the costal pleura and to a broad marginal strip of diaphragmatic pleura (*fig. 326*).

The visceral pleura is insensitive to mechanical stimulation and therein it resembles the visceral peritoneum, apparently most of the mediastinal pleura, and both the visceral and the parietal pericardium.

Movements and Shifts

The alterations in form of the bronchial tree are shown in *figure 593*. Elastic fibers form a longitudinal network throughout the entire tracheobronchial tree, providing the recoil mechanism of the entire lung (Macklin).

The Tracheal Bifurcation, both in the cadaver and in the supine living subject, lies at T. 4–5; but when erect it is commonly at T. 6 or even lower. Moreover, it descends during inspiration (*fig. 593*). At T. 3 or 4 during the 1st year of life, it gradually lowers over the next dozen years.

The Domes of the Diaphragm fall 2 to 3 cm when the erect posture is assumed; similarly the *Aortic Arch* may sink below

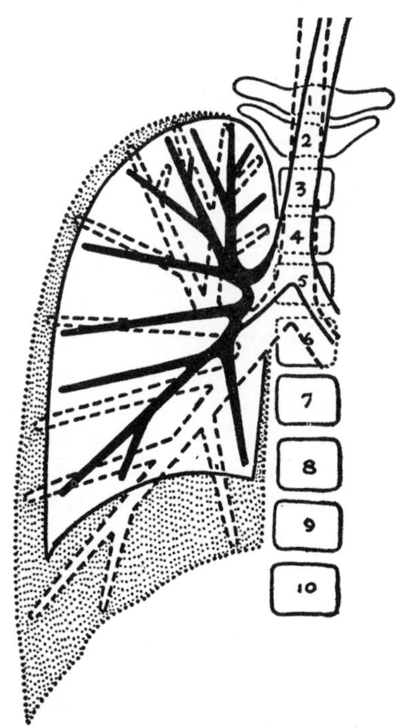

Fig. 593. Excursion of bronchial tree in forced inhalation (*dotted lines*) and forced exhalation (*solid black*) from X-ray pictures. (After C. C. Macklin.)

the superior mediastinum; and the *Heart* likewise descends (p. 422).

MECHANISM OF RESPIRATION

The two phases of respiration, *inspiration* and *expiration*, are brought about by the alternate increase and decrease in the three dimensions of the thoracic cavity. As the dimensions increase, air is drawn through the trachea and bronchi into the lungs, and blood is sucked into the thoracic veins, and thereby the formation of a vacuum is avoided. At the same time the capillaries in the lung dilate and, so, facilitate the pulmonary circulation.

At birth the ribs are horizontal and are, therefore, in the position of full inspiration; movement either upward or downward would be an expiratory act. At this age respiration is performed by the upward and downward piston-like action of the diaphragm and is said to be *abdominal in type*. By the end of the 2nd year the ribs are oblique, and by the 7th year respiration is shared by the ribs and is said to be *thoracic in type*.

During Quiet Inspiration the kidney-shaped unit (manubrium and right and left 1st costal arches) are raised slightly by neck muscles pulling from above.

The Intercostals cause the 2nd to 7th costal arches, each of which hangs like the handle of an inverted bucket, to rotate slightly at their ends so that their middle parts rise. Hence, the transverse diameter of the chest increases.

The intercostals at the same time cause the sternal ends of the arches to rise; in consequence, the body of the sternum lifts forward and the anteroposterior diameter of the chest increases.

The Diaphragm and Lower Ribs. Owing to the intra-abdominal pressure, the right and left domes of the diaphragm are rounded when relaxed, and they rise above the level of the central tendon. When the diaphragm contracts, its fibers shorten and straighten, and thereby enlarge the costodiaphragmatic recesses and cause the domes to descend. Thus is the vertical diameter increased.

The three pairs of vertebrochondral ribs (8, 9, and 10) do not imitate the movements of the vertebrosternal (2 to 7) ribs. Owing to the shape and position of the costotransverse facets they cannot rotate, but they can and do glide backward and upward (*fig. 554*). By this movement, which resembles a pair of curved spreading calipers opening, the transverse diameter of the lower part of the thorax and upper part of the abdomen increases. For this, the diaphragm, acting against the resistance offered by the abdominal muscles, is largely responsible. It forces the upper abdominal contents laterally, and this causes the lower ribs to spread. (In an animal from which the abdominal viscera have been removed, the lower ribs are drawn inward.)

During inspiration the intercostal spaces widen, and during expiration they diminish. The tone and elasticity of the Intercostals prevents sucking in of the spaces; a fibrous membrane would not suffice here.

On Deeper Inspiration the movements described are amplified. The Scalene muscles raise the first and second costal arches and the sternal heads of the Sternomastoids raise the manubrium (Jones, Beargie, and Pauly). Perhaps the Levatores Costarum and the Serratus Posterior Superior assist. The Serratus Posterior Inferior and Quadratus Lumborum steady the lower ribs.

On still deeper or labored breathing associated with shortness of breath (whether from exertion or disease) and when sneezing and coughing, the Pectoralis Minor (perhaps also the Pectoralis Major, and Serratus Anterior) assist in elevating the ribs. For the Pectoralis Minor to act, the scapula must first be fixed. In the quadruped standing on all fours, fixation is already achieved. In man, in whom the forelimbs are free, the scapulae must be fixed either (1) by finding a purchase for the upper limbs, for example by grasping the arms of the chair in which one is sitting, or (2) by the muscular action of the Trapezius, Serratus Anterior, Levator Scapulae, and Rhomboidei. The Erector

Spinae and deep muscles of the back, by straightening the thoracic curvature, help still further to cause the ribs to open out. Even the nostrils and glottis dilate rhythmically to allow of easier entrance of air.

Expiration is brought about by the elastic recoil of the lungs, Transversus Abdominis, and costal cartilages. The rotation that the ribs undergo during inspiration involves twisting of the costal cartilages and widening of the costochondral angles. It is from this twisting and widening that the cartilages recoil. Deep or forced expiration brings into play the Oblique and Transverse Abdominal muscles, and perhaps the Iliocostalis and Latissimus Dorsi.

Interesting Details. (1) Thoracic and abdominal types of respiration are usually not sharply demarcated but merge into each other, one or the other type predominating. By practice and exercise the type can be modified. (2) In quiet respiration the domes of the diaphragm move about one centimeter. The region of the caval foramen remains stationary. (3) A certain intra-abdominal pressure is necessary for the upstroke of the piston-like action of the diaphragm. This is supplied by the muscles of the anterior abdominal wall. (4) In expressing the contents of the hollow abdominal viscera (i.e., during micturition, defecation, vomiting, and parturition) a deep inspiration is taken and is held by closing the glottis, while the abdominal muscles and the diaphragm act in concert.

Posture. Gravity may work with the diaphragm or against it, thus: the diaphragm rises highest and its excursion is greatest when the subject lies flat on his back with the foot of the bed raised; it is lower and the excursion is less when horizontal; lower still when erect, and even lower when sitting down, because then the abdominal muscles are relaxed; and finally in persons whose abdominal muscles have lost their tone (e.g., cases of poliomyelitis, large umbilical herniae, visceroptosis), the diaphragm tends to remain relaxed and the respiration becomes thoracic. When the subject lies (horizontally) on one side, the dome of the diaphragm of that side is higher and makes a greater excursion than the dome of the upper side.

32 HEART AND PERICARDIUM

PERICARDIUM

Fibrous and Serous.
Contents of Pericardial Sac.

HEART

Parts; Surfaces.
Sternocostal Surface of Heart and Great Vessels, and how to draw them.
Surface Anatomy of Heart. Radiographic Anatomy.
Serous Pericardium; Oblique Pericardial Sinus; Transverse Pericardial Sinus.
Sulci of Heart: Coronary, and Anterior and Posterior Interventricular.
Ascending Aorta and Pulmonary Trunk.
BLOOD SUPPLY OF HEART—*Coronary Arteries; Cardiac Veins; Coronary Sinus; and Myocardial Circulation.*
Development of Heart; general notes.

Chambers of Heart

Right Atrium—Exterior; Interior; History of Foramen Ovale.
Left Atrium—Exterior; Interior; History.
Ventricles—Walls; Interior; Septomarginal [Moderator] Band; Interventricular Septum; Atrioventicular Valves. Surface Anatomy of Four Cardiac Orifices; Heart Sounds.

Structure of Walls of Heart

Skeleton of Heart.
Musculature of Heart.
Impulse Conducting System; Nerve Supply.

PERICARDIUM

Pericardium (Gk. Peri = around; Kardia = the heart) consists of an outer *fibrous sac*, lined with an inner *serous sac*. The heart and the roots of the great vessels lie inside the fibrous sac and invaginate the serous sac from behind (*fig. 594*). Hence, the serous pericardium has both a visceral and a parietal layer, the fibrous pericardium being the parietes (L. = walls). The visceral layer is known as *epicardium*.

The Fibrous Pericardium blends with the central tendon of the diaphragm. The ascending aorta carries the pericardium upward beyond the heart to the level of the sternal angle (*fig. 594.1*).

The Serous Pericardium (see p. 424).

Contents of Pericardial Sac. The heart and its great vessels and related structures are the contents. They are dealt with below in detail.

HEART

(L. Cor; Gk. Kardia)

This muscular pump is somewhat larger than a closed fist. It has four chambers, the *right* and *left atria* and the *right* and *left ventricles*. The atria are separated from the ventricles by a constriction that completely encircles the heart and is appropriately called the *coronary* (*atrioventricular*) *sulcus*. The ventricles are separated from each other by the *anterior* and *posterior interventricular* (*longitudinal*) *sulci*. The notched anterosuperior part of each atrium resembles a dog's ear and is called the *auricle* (L. auris = an ear) (*Fig. 594.2*).

The heart has 3 surfaces: *sternocostal* (anterior), *diaphragmatic* (inferior), and *base* (posterior); it also has an *apex* (i.e., the lowest and leftmost point).

Sternocostal Surface of Heart and Great Vessels and Skin-Surface Anatomy

Students and physicians must be able to plot the heart and great vessels on the skin of the living person. You should also practice drawing them on paper (*fig. 595*).

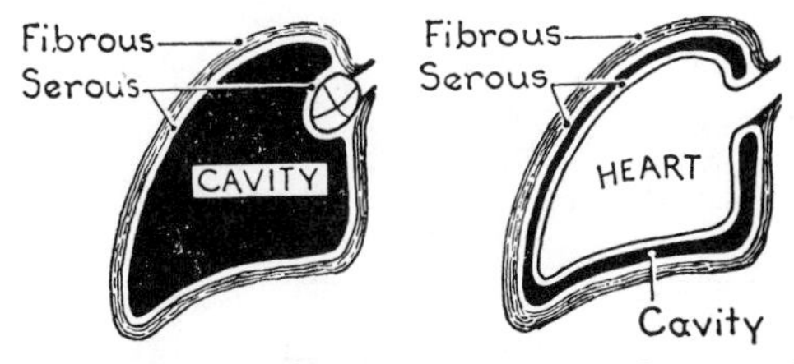

Fig. 594. To explain the layers of the pericardium.

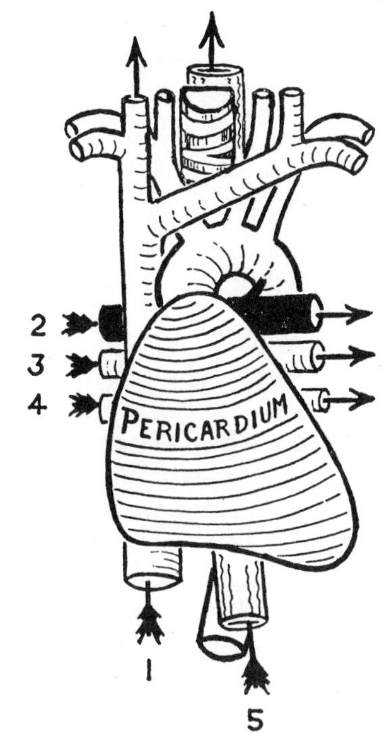

Fig. 594.1. The fibrous pericardium and the channels through which five arrows are passed without obstruction—*1*, through the I.V.C., right atrium, S.V.C., right brachiocephalic and int. jugular vv.; *2*, from right pulmonary art. to the left one; *3* and *4*, through pulmonary vv. and the intervening left atrium; *5*, up the esophagus.

Thus:

1. A vertical line drawn from the neck to the abdomen, less than a finger's breadth from the right margin of the sternum (*fig. 595B*), represents from above downward the *right borders* of:
   a. The right internal jugular vein.
   b. The right brachiocephalic vein.
   c. The superior vena cava (s. v. c.).
   d. The right atrium.
   e. The inferior vena cava (i. v. c.).
2. On each side the *internal jugular* and the *subclavian vein* unite behind the sternal end of the clavicle to form the brachiocephalic or innominate vein; but—
3. The *left brachiocephalic vein* passes obliquely behind the upper half of the manubrium and joins the right brachiocephalic vein to form the superior vena cava (*fig. 595C*).
4. The *superior vena cava* ends in the right atrium at the 3rd costal cartilage.
5. The *inferior vena cava* pierces the diaphragm at the level of the xiphisternal joint, and after a course of 1 cm enters the right atrium, at the level of the 6th costal cartilage.
6. A slight bulge or convexity in the line between the 3rd and 6th cartilages represents the right atrium, and so the *right border* of the heart.
7. An appropriately wrinkled, oblique line will define the *left margin of the right atrium* and at the same time the right part of the coronary (atrioventricular) sulcus.
8. The *inferior margin* of the heart extends from the i.v. cava, across the xiphisternal joint, and then slightly downward,

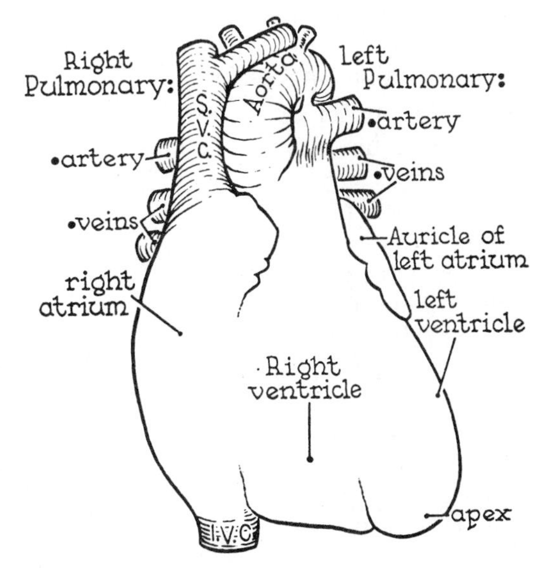

Fig. 594.2. Anterior view of undissected heart.

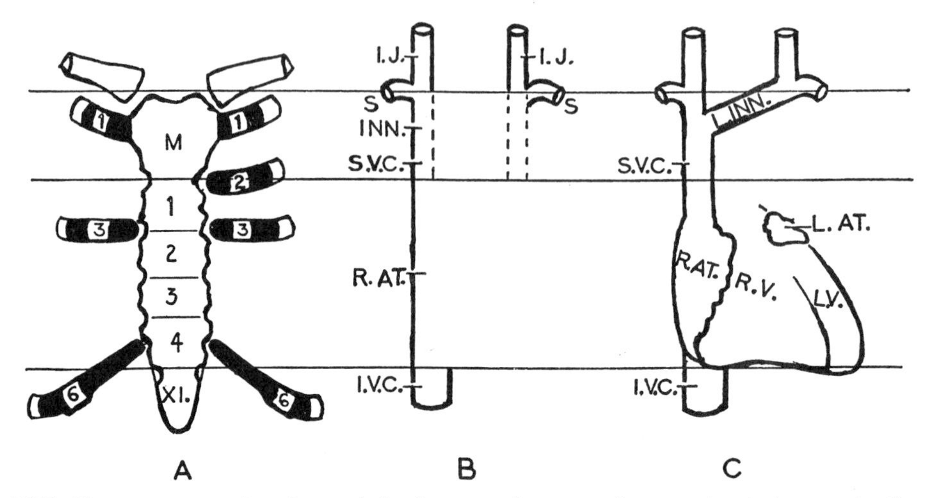

Fig. 595. The sternocostal surface of the heart and great veins constructed on projection lines.

to the apex of the heart, which is situated in the 5th left intercostal space or behind the 6th rib, 10 cm from the median plane. Here the **apex beat** is palpable. One-third of the heart lies to the right of the median plane; two-thirds to the left.

9. The *left margin* is rounded, and passes from the apex to the 2nd left interspace a finger's breadth from the sternal border.

10. The tip of the *left auricle* peeps round this margin in the 2nd interspace.

11. A line parallel to the left margin, and about one-third of the distance from it to the right margin, separates the right and left ventricles and represents the *anterior interventricular sulcus*. Portions of all four chambers are now outlined on the sternocostal surface. The portion of left ventricle appearing on this surface can be covered by a finger or two.

12. The *ascending aorta* (*fig. 596*) is the segment of the aorta lying within the pericardium, and therefore below the level of the sternal angle. The continuation of the left ventricle, it passes upward and *to the right* (overlapping the s. v. cava).

The *arch of the aorta* passes backward and to the left behind the lower half of the manubrium, and therefore below the left brachiocephalic vein, to reach the left side of the thoracic vertebra. There, as it enters the plane between the superior and posterior mediastina, it becomes the *descending aorta* (*fig. 596A*).

13. *The pulmonary trunk* (*fig. 596*), the continuation of the right ventricle, passes upward and *to the left* between the right and left auricles, which embrace it. It lies in front of and conceals the root of the aorta. As it pierces the parietal pericardium below the aortic arch, it divides, like the letter T, into *right* and *left pulmonary arteries;* they pass to the roots of the lungs (*fig. 596B*)—the *left* passes anterior to the descending aorta; the *right* posterior to the ascending aorta and s. v. cava (*fig. 594.1*).

14. The *ligamentum arteriosum*, lies outside the pericardium and passes from the *left* pulmonary artery to the "aortic arch" beyond the origin of the subclavian artery. Indeed, it continues the direction of the stem of the pulmonary trunk because of its fetal development as the ductus arteriosus (*fig. 40*).

Variations in Radiographic Anatomy (*fig. 597*). In life, the inferior border of the heart crosses the median plane 5 or 6 cm below the xiphisternal joint. While recumbent, the cadaveric position is almost achieved, except for the inferior border which is now 3.5 cm below the xiphisternal joint (Mainland and Gordon). Deep inspiration and expiration have most effect on the position and shape of the heart (*fig. 598*).

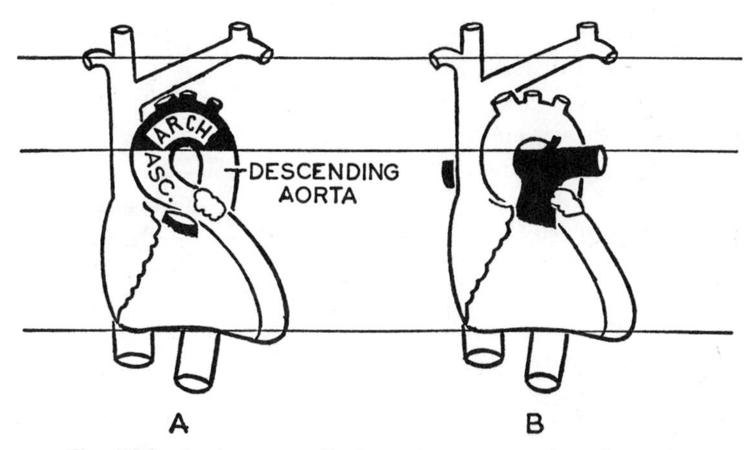

Fig. 596. *A*, the aorta. *B*, the pulmonary trunk and arteries.

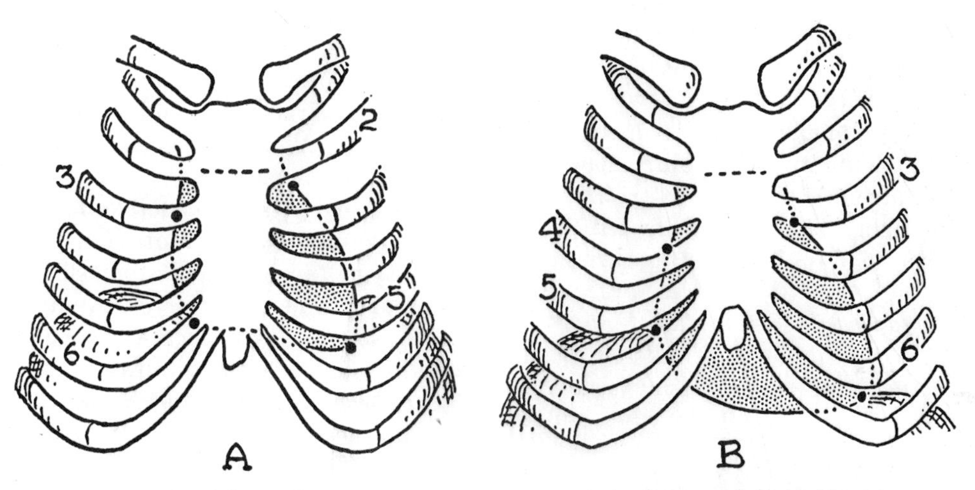

Fig. 597. The surface anatomy of the heart. (From Lachman, after Main and Gordon.)

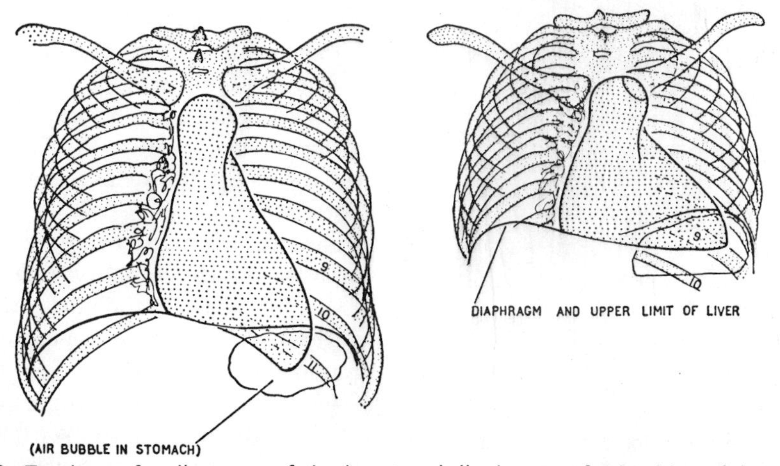

Fig. 598. Tracings of radiograms of the heart and diaphragm of a healthy adult male, showing their ever changing shapes and positions.

Further, (1) a broad stocky build, (2) the recumbent position, and (3) a distended abdomen (e.g., pregnancy, gas, fat, and the large liver of childhood) are associated with a high diaphragm and, therefore, with transverse hearts (Lachman).

Serous Pericardium. Observe these three points about the serous pericardium:

1. *The S. V. Cava and I. V. Cava* are clothed in serous pericardium in front and at the sides, but they are bare posteriorly (*fig. 599*).

2. *Oblique Pericardial Sinus*. Two fingers, passed upward behind the heart, on the left side of the i.v. cava, enter an empty recess, the *oblique pericardial sinus* (*fig. 600*). This sinus is significant because it (a) is limited *at the sides* by the two right and two left pulmonary veins (which having pierced the fibrous pericardium, soon enter the left atrium) as well as by the i. v. cava on the right; (b) is behind the left atrium (*fig. 599*); and (c) is in front of the esophagus.

3. *Transverse Pericardial Sinus*. The stems of the pulmonary trunk and ascending aorta, lying within a single sleeve of serous pericardium, form the *anterior boundary* of a potential space, the transverse pericardial sinus. Its *posterior and side boundaries* are the upper portions of the atria, their auricles, and the end of the s. v. cava. A finger can be insinuated through this sinus.

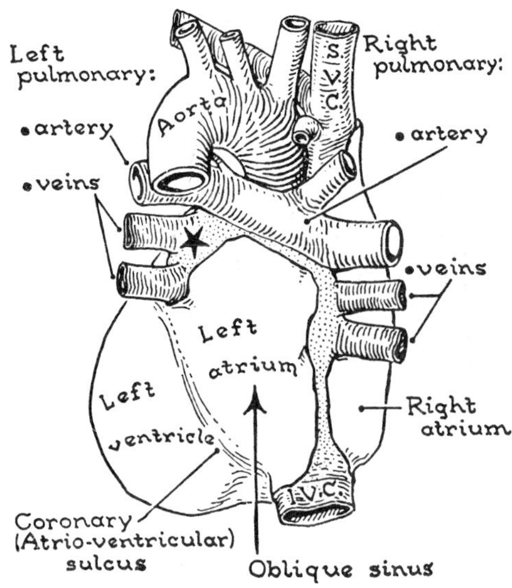

Fig. 599. The posterior aspect of the heart. * Indicates site of contact of left bronchus with left atrium.

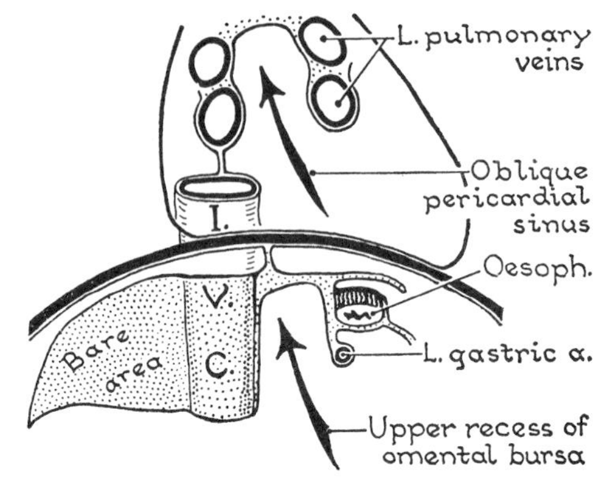

Fig. 600. Two empty inverted pockets lie in the midline of the body.

The Diaphragmatic (Inferior) Surface of the Heart—formed by the ventricles, is separated from the liver and stomach by the diaphragm. The posterior interventricular sulcus divides this surface into a right one-third and a left two-thirds.

The Base (Posterior Surface or "Back")—of the heart is formed by the atria and slightly by the left ventricle (*fig. 599*). The coronary venous sinus, lying in the coronary sulcus, runs along, or just above, its lower border. The heart does not rest on its base; the term derives from the cone shape of the heart, the base being opposite the anteriorly pointing apex.

Sulci (*fig. 602*). The **coronary sulcus** (atrioventricular sulcus) completely encircles the heart between the atria and the ventricles.

The **anterior interventricular sulcus** extends from the coronary sulcus at the left of the root of the pulmonary trunk, downwards across the sternocostal surface, around the inferior border 2 cm to the right of the apex. There it becomes continuous with the **posterior interventricular sulcus,** which continues backward on the diaphragmatic surface to meet the coronary sulcus. Cardiac vessels, including the coronary sinus, occupy the sulci and are often embedded in fat.

Ascending Aorta and Pulmonary Trunk

These lie within the fibrous pericardium below the level of the sternal angle, the pulmonary trunk extending beyond the left margin of the sternum to the 2nd intercostal space. The upper parts of the atria, their auricles, and the entering s. v. cava embrace the two great arteries from behind but fail to meet in front of them (*fig. 594.2*).

The **pulmonary trunk**, 5 cm long, begins in front of the aorta and passes semispirally (upward, backward, and to the left) around it until it reaches the concavity of the aortic arch where it bifurcates into the right and the left pulmonary artery. The stems of the right and left coronary arteries, arising from the ascending aorta, pass forwards on each side of the root of the pulmonary trunk (*fig. 602*).

The *right* and *left pulmonary arteries* lie along the upper borders of the atria and of the upper pulmonary veins (*fig. 599*) like the cross-stroke of the letter T, set slightly obliquely (*fig. 596*).

Behind the right and left arteries lie the bronchi and the inferior tracheobronchial lymph nodes (*fig. 633*).

The **ascending aorta** begins behind the pulmonary trunk and passes obliquely upward, forward, and to the right to reach the right margin of the sternum within the pericardial sac. The right pulmonary artery crosses behind it (*fig. 599*). Its right wall is dilated and is known as the **bulb of the aorta.**

Aortic Valve and Valve of the Pulmonary Trunk (Pulmonary Valve). Both valves prevent back flow of blood into the ventricles, and both have three semilunar valvules or cusps, the aortic cusps naturally being stronger than the pulmonary cusps.

Each *valvule* or *cusp* has a fibrous basis covered on both surfaces with endothelium. At the middle of the free edge of each valvule there is a *fibrous nodule*, and on each side of the nodule there is a thin, crescentic area, the *lunule* (*fig. 601*). When the valve closes, the nodules meet in the center of the lumen and the ventricular surfaces of the lunules of contiguous valvules are applied to each other.

At the root of both the arteries there are three dilatations, the *sinuses;* one is placed external to each valvule to prevent it from sticking to the wall of the artery when the valve is open.

The aortic orifice and about the first 5 to 10 mm (¼″) of the aorta are fibrous and not dilatable, so the valve remains competent. The three valvules of the aortic valve are the right valvule, the left valvule, and the posterior or noncoronary valvule. The disposition of the valvules is explained on page 429 (*figs. 610, 619*).

The aortic valve lies slightly lower than the valve of the pulmonary trunk, is posteromedial to it, and faces a different direction (*fig. 602*). The aortic valve faces upward, forward, and to the right; the valve of the pulmonary trunk faces upward, backward, and to the left.

Blood Supply of the Heart

1. Coronary arteries,
2. Cardiac veins,
3. Collateral circulation—
   a. cardiac,
   b. extracardiac.

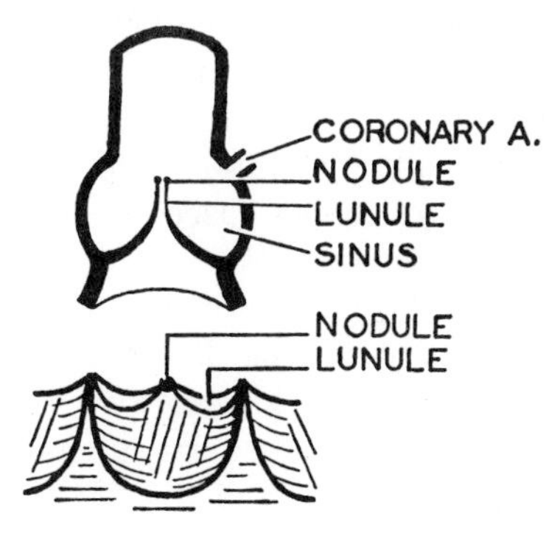

Fig. 601. The aortic valve: on sagittal section, opened up.

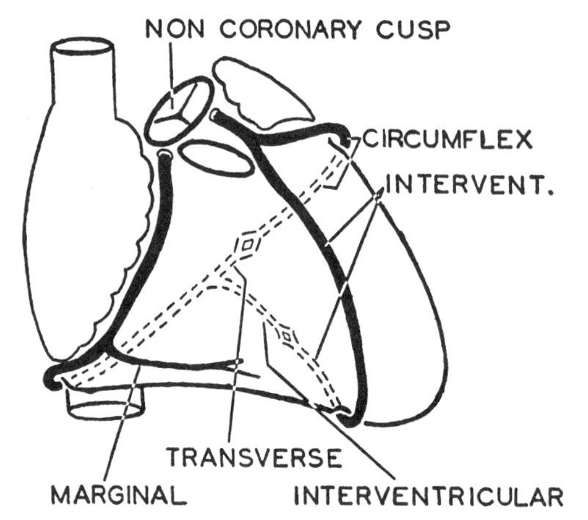

Fig. 602. The coronary arteries. The left marginal is not shown.

The Right and Left Coronary Arteries supply the heart. They spring from two of the three aortic sinuses (*fig. 610*) usually just below the level of the free edges of the cusps. They run forward, one on each side of the root of the pulmonary trunk, sheltered by the corresponding auricle (*fig. 602*).

The **left coronary artery** divides into: (1) an *anterior interventricular branch* which descends in the anterior interventricular sulcus to the inferior margin of the heart, where it turns round into the posterior interventricular sulcus, and (2) a *circumflex branch* which runs in the coronary sulcus round the left margin of the heart and gives off a *left marginal branch*, which runs down the left margin of the left ventricle.

The **right coronary artery** descends in the right part of the coronary sulcus, turns round the inferior margin of the heart in it, and divides into: (1) a *posterior interventricular branch* which descends in the posterior interventricular sulcus to meet the interventricular branch of the left coronary artery, usually in the lower ⅓ of that sulcus, and (2) a *transverse branch*, which, continuing in the coronary sulcus, meets the circumflex branch of the left coronary artery. It sends (3) a large *right marginal branch* along the inferior margin of the right ventricle.

The *left coronary artery* supplies—the anterior part of the interventricular septum and the adjacent part of the right ventricle, as well as the anterior surface, rounded left margin and a small part of the inferior surface of the left ventricle (*fig. 603*).

The *right coronary artery* supplies—the remainder of the right ventricle, the posterior part of the interventricular septum, and the extensive remaining part of the left ventricle.

Variations. (1) The posterior interventricular artery springs from the left coronary art. via its circumflex branch in about 10 per cent of hearts. (2) Accessory coronary arteries are not uncommon, but most of them are very fine. (3) For a single coronary art. to supply the entire heart is rare.

The Cardiac Veins mostly accompany the arteries in the sulci and tend to lie superficial to them (*fig. 604*). Five of the seven cardiac veins mentioned below end in the coronary sinus.

The Coronary Sinus is derived from the left horn of the primitive receiving chamber of the heart, the sinus venarum; it is about 3 to 4 cm long, lies in the coronary sulcus on the back of the heart and opens into the right atrium at the left of the orifice of the i. v. cava. At its left end it receives the companion of the left coronary artery, called the *great cardiac vein*, a companion of the interventricular branch

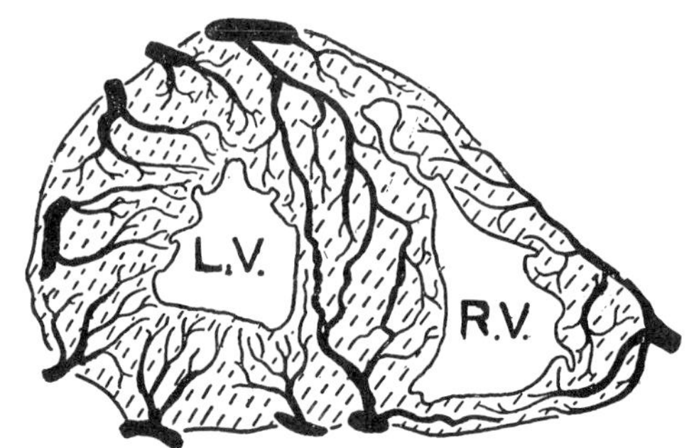

Fig. 603. Transverse section of ventricles showing branches of coronary arteries plunging into the heart substance. (After Gross and Kugel.)

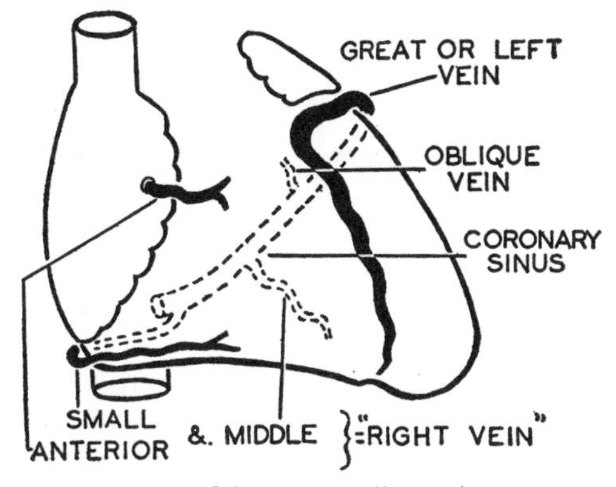

Fig. 604. The cardiac veins.

of the left coronary artery and then its circumflex branch. The companion of the right marginal artery, called the *small cardiac vein*, enters the right end of the coronary sinus. The companion of the posterior interventricular artery, called the *middle cardiac vein*, ends in the coronary sinus; so do the *posterior ventricular veins* from the diaphragmatic surface of the left ventricle. The *oblique vein* is a twig that lies behind the left atrium and ends in the coronary sinus. It was once a left-sided vein running to the primitive heart and corresponding to the S. V. C. (*fig. 611*).

One or two large *anterior cardiac veins* pass from the front of the right ventricle, across the coronary sulcus, and open directly into the right atrium.

The mouths of the great and small cardiac veins commonly have single-cusped valves, but these are rarely competent. Tiny veins, *venae cordis minimae* (*Thebesian veins*) begin in the heart wall and open directly into the chambers of the heart.

Myocardial Circulation

Figure 605 summarizes the selective pathways available. Note that:

(1) When physiological saline solution or India ink is perfused through a coronary artery most (about 90 per cent) escapes into the lumen of the heart via the venae minimae and very little passes through the capillaries to the coronary sinus.

(2) Similarly, when a vein is perfused most escapes by venae minimae and a little passes through the capillaries to the arteries.

(3) A fluid too viscous to pass through the capillary bed will, when injected into an artery (or vein), escape into the lumen of the heart via the venae minimae.

(4) A fluid too viscous to flow out through the arteries may yet, when injected into a coronary vein, enter the heart via the venae minimae. Evidently, then, the pathway from cardiac vein via venae minimae to the lumen of the heart is wider than from coronary artery via venae minimae to the lumen, and this in turn is wider than the passage through the capillary bed.

(5) When particles (e.g., carborundum), too large to traverse either the capillaries or venae minimae of the heart or the capillaries of the lung, are injected during life into the external jugular vein of the dog, they accumulate in large numbers in the coronary sinus and in the veins on the surface of the heart. To get there they must have taken a retrograde course in the veins (Batson).

Comparative and Developmental Anatomy. The primitive vertebrate heart is a nonvascular heart—there are no coronary vessels, and the myocardium is spongy or trabecular. Reptiles acquire coronary vessels and venae minimae as well.

The developing mammalian heart is spongy, and early is nourished by blood in the spaces of the myocardium. Coronary veins spread over the surface of the myocardium, enter it, and communicate with the intertrabecular sinusoidal spaces. Outgrowths from the endocardium

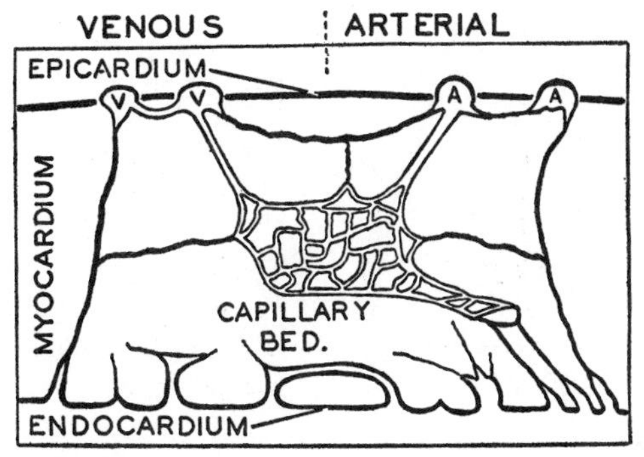

Fig. 605. Scheme of the myocardial circulation. (After Batson.)

give rise to the venae minimae of adult anatomy; they also connect with each other. The coronary arteries then sprout from the future aorta, spread over the heart, and unite with the capillary network already formed by the veins and intertrabecular spaces in the developing myocardium. (R. T. Grant and L. E. Viko.)

Collateral Circulation. (*1*) *Cardiac.* Branches of the coronary arteries seldom anastomose on the epicardiac surface of the heart. In the myocardium indeed there are anastomoses, but the vessels taking part are small. Challenged, the anastomoses can enlarge.

Thus, intercoronary arterial anastomoses were found in less than 10 per cent of apparently normal hearts, in 40 per cent of anemic hearts, and in 100 per cent of hearts with old coronary occlusion (Zoll, Wessler, and Schlesinger).

(*2*) *Extracardiac Anastomoses.* If both coronary arteries are obstructed, there is an extracardiac collateral circulation to be called upon, but, unless it has had a long preparation, it cannot answer the call.

All are twigs—the vasa vasorum in the tunica adventitia of the aorta and pulmonary arteries, and branches of the internal thoracic, bronchial and phrenic arteries.

Development of the Heart

The following notes may assist in the appreciation of the inter-relationships of the various parts of the heart and explain certain anomalies.

Elongation of the Tubular Heart. The primitive tubular heart received blood at its caudal end and discharged it from its cephalic end. This tubular heart had *five sacculations*—sinus venarum, primitive atrium, primitive ventricle, bulbus cordis, and truncus arteriosus (*fig. 606*). The constriction between the primitive atrium and ventricle becomes the *coronary* or *atrioventricular sulcus.*

Just as the intestine has a (dorsal) mesentery, so the tubular heart had a *dorsal mesocardium.* When the cardiac tube became too long for the pericardial cavity, it formed an S-shaped loop (*fig. 607*). Its two caudal segments (sinus venarum and atrium) and the entering veins came to lie dorsal to the three cephalic segments of which the last (truncus arteriosus) divided to form the ascending aorta and the pulmonary trunk. Hence, the atria and the entering veins of adult anatomy lie posterior to the ventricles and the emerging arteries (*fig. 608*).

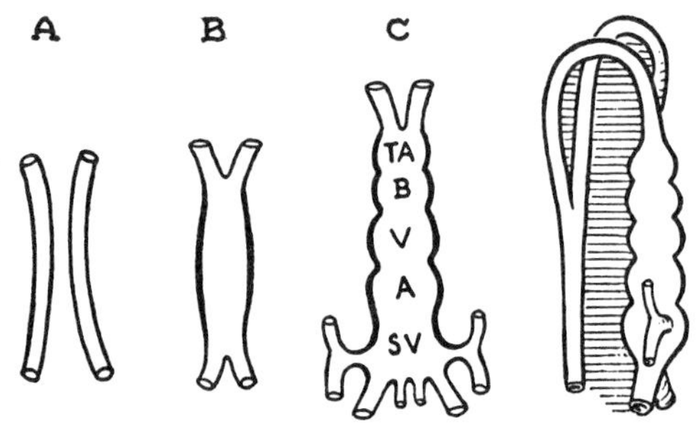

Fig. 606. Scheme of development of tubular heart.

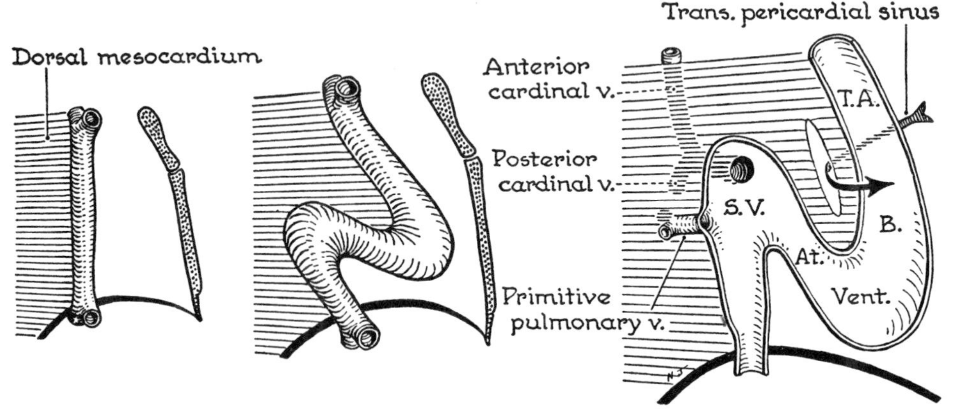

Fig. 607. The elongated tubular heart becomes "S-shaped."

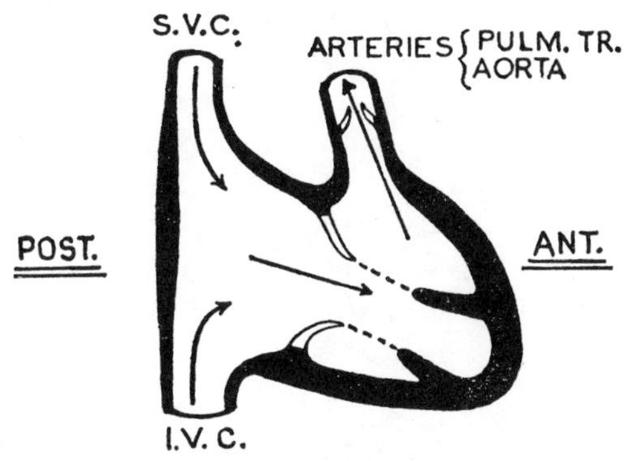

Fig. 608. Diagram of heart in sagittal section to show that the atria and entering veins are posterior to the ventricles and emerging arteries.

Perforation of Dorsal Mesocardium. In consequence of a perforation in the mesocardium, the *transverse pericardial sinus* appears and the truncus arteriosus finds itself enveloped in a tube or sleeve of visceral pericardium. Hence, when the truncus splits into two, both resulting vessels (aorta and pulmonary trunk) lie within a single pericardial sleeve.

Oblique Pericardial Sinus (p. 424).

Spiral Septum in Truncus Arteriosus. The fact that a spiral septum develops within the truncus explains the twisted courses of the pulmonary trunk and ascending aorta around each other (*fig. 609*). The course of the spiral septum was such that its cardiac end bisected the right and left lateral cusps of the primitive four-cusped valve of the truncus arteriosus; hence, the relative positions of the definitive cusps of the aortic valve and valve of the pulmonary trunk (*fig. 610*). The primitive aortic arches and their anomalies are described on p. 437; the septa dividing the primitive atria and ventricles (each into two), on p. 431.

Axial Rotation of the Heart. The heart undergoes a slight rotation to the left on its long axis, and so—(1) the right atrium is conspicuous at the right margin of the heart and anteriorly; the left atrium is conspicuous posteriorly, (2) the right ventricle is largely in front and slightly inferior; the left ventricle is largely inferior and slightly in front; and as will be seen later, (3) the interatrial and interventricular septa come to face forward and to the right (and backward and to the left); so do both cusps of the mitral valve and the septal cusp of the tricuspid valve, for these three cusps are approximately parallel to the septum; and lastly (4) the cusps of the valves of the two great arteries, which were originally disposed anteriorly, posteriorly, to right, and to left (*fig. 610*), share in the rotation.

Terminology of the Valvulae or Cusps. Confusing terms are often applied to the cusps of the aortic valve. It is simplest to relate the cusps to the coronary arteries. *Figure 610* provides the key. (For *details*, see Merklin.)

Anomalies, both life-threatening and minor, arise from imperfect separations of the arteries and their valves and the associated parts of the ventricles, e.g., transposition of the great arteries and interventricular septal defects.

Primitive Veins (*fig. 611*). The early arrangement is symmetrical until cross-communicating veins (left brachiocephalic and

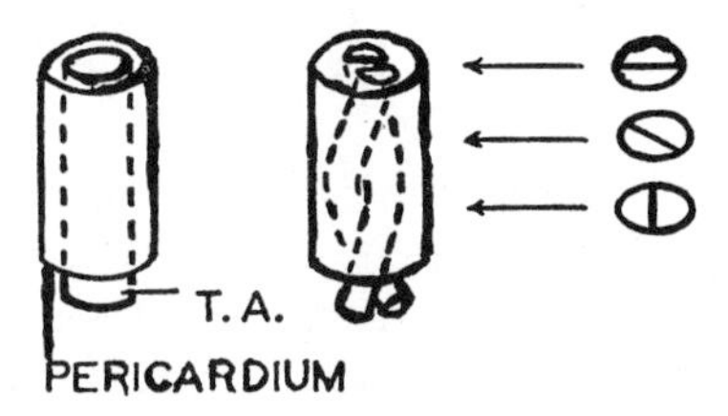

Fig. 609. The spiral septum within the truncus arteriosus explains the twisted courses of the aorta and pulmonary trunk.

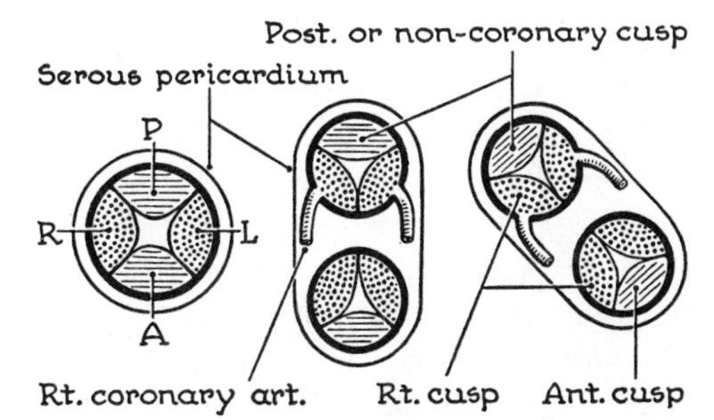

Fig. 610. The four-cusped valve of the truncus splits to form two valves, each with three cusps or valvules. Axial rotation occurs, but it need not affect the nomenclature.

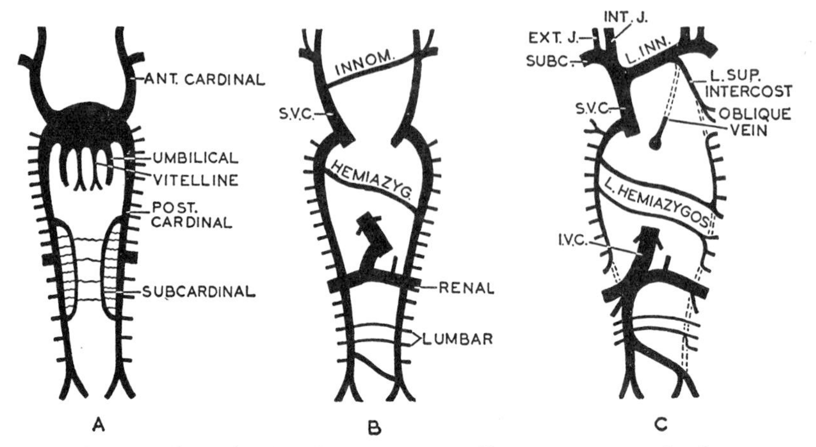

Fig. 611. Veins. *A*, six veins join the sinus venarum. *B*, cross-communications appear. *C*, only two of the six veins survive. They become caval veins. (After Arey.)

hemiazygos veins) develop and the left common cardinal vein disappears.

The left horn of the sinus venarum becomes the *coronary sinus*, now playing the part of a vein. When, as a rare anomaly, the left s. v. cava persists, it ends—as expected—in the coronary sinus.

CHAMBERS OF THE HEART

Right Atrium (*fig. 612, 613*). The right atrium developed from the right half of the sinus venarum and right half of the primitive atrium which merged to form a single chamber. The rotation of the heart, which took the other chambers away from their original orientation, leaves the right atrium forming the right margin of the heart.

Interior (*fig. 613*). On opening the rather cubical atrium, a ridge, the *crista terminalis*, is seen to correspond in position to a faint *sulcus terminalis* on the exterior, indicating where the two primitive chambers merge. The portion of the atrium behind the crista is smooth; it developed from the sinus venarum. The portion in front is trabeculated; it developed from the primitive atrium. The parallel ridges running forward from the crista terminalis toward the auricle are the *musculi pectinati* (L. pecten = a comb).

From the lower end of the crista terminalis a prominent fold of endocardium, the (imperfect) *valve of the i. v. cava*, passes in front of the orifice of the i. v. cava to become continuous with the crescentic

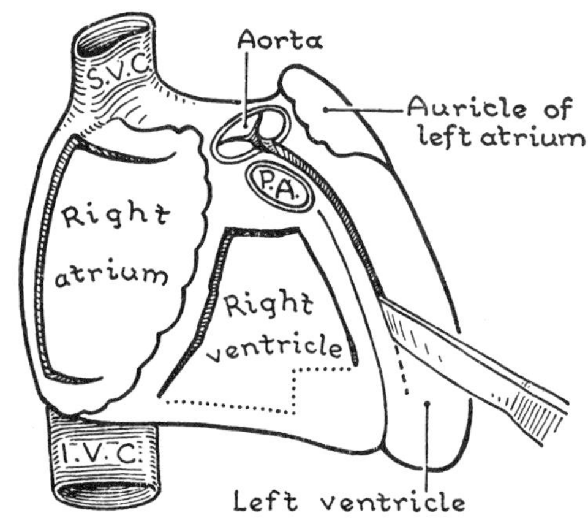

Fig. 612. Incisions for opening chambers of heart.

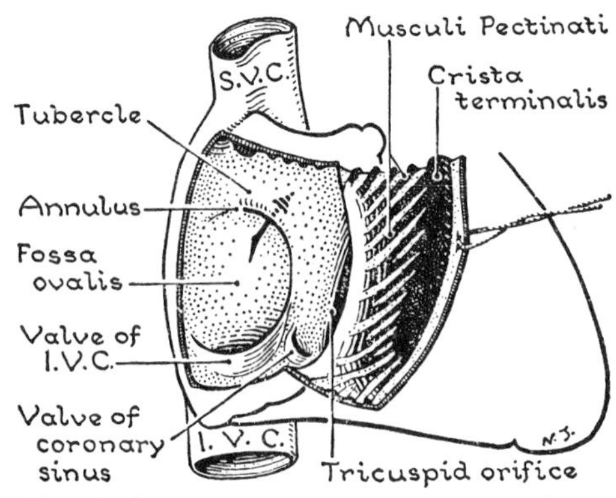

Fig. 613. The interior of the right atrium.

margin of a depression, the *fossa ovalis*. The fossa ovalis is the size of a thumb nail and is situated on the interatrial septum, which forms the medial wall of the right atrium. The right atrioventricular or *tricuspid orifice* takes the place of an anterior wall (*fig. 608*). The *orifice of the coronary sinus* opens between the orifice of the i. v. cava and the tricuspid orifice. It is "guarded" by a fold, detached from the valve of the i.v. cava. This fold, often perforated like a piece of lace, is the *valve of the coronary sinus*.

The History of the Foramen Ovale (*fig. 614*). The septum primum grew downward, dividing the primitive common atrial chamber. Before its lower end fused with the tissues that divided the common primitive atrioventricular orifice into the right and left orifices, its connection with the roof dissolved. A *septum secundum* then grew downward on the right till its lower edge overlapped the upper edge of the septum primum.

Until birth the blood from the i.v. cava is largely directed through this valve-like foramen. After birth the pulmonary circulation is established and the flap valve is closed, leaving the fossa ovalis.

Variations. In about 75 per cent of individuals the opposed surfaces fuse. In 25 per cent of individuals the edges of the primary and secondary septa overlap but fail to fuse, so the foramen is patent anatomically, though closed physiologically (*fig. 41*). Rarely the edges of the septa fail to meet; the result is a foramen patent both anatomically and physiologically. In consequence, the pulmonary circulation is disturbed—this is one type of "blue baby" condition.

Left Atrium. *Exterior* (*fig. 599*). The left atrium forms two-thirds of the back of the heart and its auricle peeps round the left border (*fig. 612*). It is demarcated from the left ventricle below by the coronary sulcus. The right and left pulmonary veins open into it near its right and left margins (*fig. 599*).

Interior. There is little to see inside this atrium. The auricle is trabeculated; the rest of the cubical cavity is smooth. The mouths of the four pulmonary veins open on the posterior wall; the left atrioventricular or *mitral orifice* replaces the anterior wall. The interatrial septum is set obliquely because the left atrium was rotated to lie posterior to the right atrium.

History. The auricle of the left atrium is derived from the left half of the primitive atrium. The smooth part is formed anew with the developing pulmonary veins (*fig. 614*). Nothing is obtained from the sinus venarum.

Ventricles. The ventricles, right and left, lie in front of their atria (*fig. 608*). They form the apex of the heart, the entire inferior margin and diaphragmatic surface (*fig. 599*), most of the left margin and

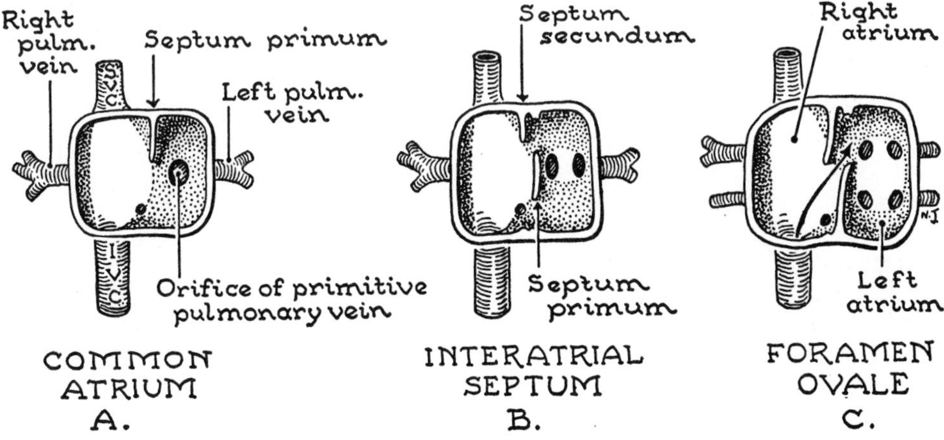

Fig. 614. Development of the left atrium. Incorporation of the stem of the primitive "common pulmonary vein." The history of the foramen ovale.

sternocostal surface, and a trivial part of the back (or base).

Walls. On cross-section it is seen that the thickness of the walls of the two ventricles is proportional to the amount of work each has to do (*fig. 615*). After birth, the left ventricle is the pump of the systemic system and the right ventricle of the pulmonary system, and so the ratio of their thickness is 3:1. The rotation of the heart to the left causes one-third of the left ventricle to face anteriorly, and two-thirds of the left ventricle and one-third of the right ventricle to face inferiorly. This is indicated by the positions of the anterior and posterior interventricular sulci and explains the obliquity of the i-v. septum which joins these sulci (*fig. 615*).

Interiors. The *cavity* of the right ventricle is triangular (*fig. 616*); the cavity of the left ventricle is conical (as it narrows to the aorta). The *entrances* or atrioventricular orifices are posterior; the *exits* or orifices of the aorta and pulmonary trunk are superior, so the blood pursues a V-shaped course within the ventricles. In each ventricle the exit is on the septal side of the entrance.

Except near the exits, the ventricular walls are lined with muscular bundles, *trabeculae carneae* (L. carnea = flesh; *cf.* carnal). Some of these bundles are merely elevated *ridges*, others are attached at both ends like bridges—the *septomarginal band* in the right atrium being important—and others form fingerlike projections, the *papillary muscles*. In each ventricle an *anterior and a posterior papillary muscle* rise from the corresponding walls. Those on the left side are larger than those on the right. In the right ventricle small *septal papillary muscles* rise from the septum also. From the apices of the papillary muscles fibrous cords, *chordae tendineae*, pass to the cusps of the atrioventricular valves.

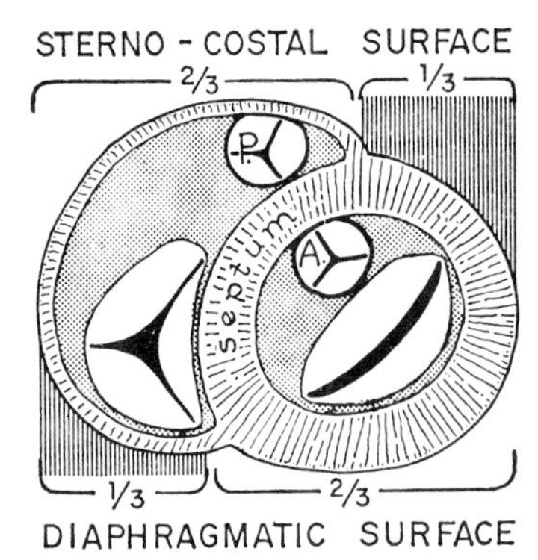

Fig. 615. Ventricles on cross-section, front-view.

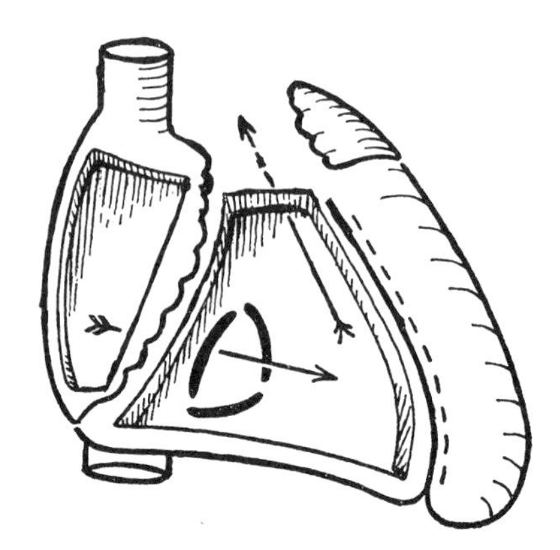

Fig. 616. Interior of the right ventricle showing relative positions of orifices.

The *septomarginal band* (Moderator Band) (present in 60 per cent of 500 human hearts) transports a fascicle of the cardiac impulse conducting system (p. 435).

The portion of the right ventricular cavity preceding the pulmonary trunk is smooth and is called the *conus arteriosus* or *infundibulum* (L. = funnel). The corresponding portion of the left ventricle is smooth, largely fibrous and nondistensible, and called the *aortic vestibule*. It can neither contract nor dilate.

The interventricular septum is fleshy except at its uppermost part where an area, the size of a thumbnail, is membranous. The *fleshy part* is an upgrowth from the apex; the *membranous part* is a downgrowth from the interatrial septum and right side of the root of the aorta (*fig. 620*). Failure of the fleshy and membranous parts to fuse results in an *interventricular septal defect* with subsequent leakage into the right ventricle from the high pressured left ventricle. The thin *pars membranacea* can be felt between the finger and thumb placed one in each ventricle (*fig. 624*).

The Atrioventricular Valves. The right atrioventricular valve is *tricuspid*, and the left, *bicuspid*. The bicuspid valve was likened in the 16th century to a bishop's miter by Vesalius, the father of modern Anatomy, hence the left valve is referred to as the *mitral valve*.

The *chordae tendineae* are attached to the edges and ventricular surfaces of the cusps, thus avoiding any obstruction to the incoming blood. The chordae of each papillary muscle control the contiguous margins of two cusps (*fig. 617*). Hence, there are two papillary muscles on the left side and three, or groups of three, on the right.

Cusps. The bases of the cusps unite to form a short cuff which is attached to the fibrous atrioventricular orifice. The margins of the cusps are dentate where the chordae are attached. The edges and surfaces of the cusps must meet when the valve is closed, otherwise the valve will leak. Therefore, the two cusps of the mitral valve are parallel to each other; they are also parallel to the septum, and also to the septal cusp of the tricuspid valve. *All four structures* face forward and to the right, and backward and to the left (*figs. 615, 619*).

The cusps of the mitral valve are called *anterior* and *posterior*. The anterior cusp is interposed between the atrioventricular and the aortic orifice. Accordingly, the current of blood flows over both surfaces of this cusp, so the chordae are largely confined to its margin. The clinician prefers to think of this anterior cusp of the mitral valve as *the aortic cusp* because it lies near the aorta.

The cusps of the tricuspid valve are named *anterior*, *posterior*, and *septal*.

Structure. The papillary muscles, chordae tendineae, and cusps of the atrioventricular valves are developed from the (primitive) muscular spongework of the heart (*fig. 618*); and in fetal life the cusps are fleshy and vascular; but before birth they become fibrous and lose their blood vessels (unlike those of most domestic animals, e.g., dog, cat, pig, sheep). However, inflamed valves do become vascularized and may remain so for many years (e. g., after severe rheumatic fever—Gross; Harper). Before birth the muscle fibers and the vessels undergo regression. In adult life muscle fibers are found in the bases of all five cusps; they include smooth nonstriated fibers.

Surface Anatomy of the Four Cardiac Orifices guarded by functioning valves—pulmonary, aortic, mitral, and tricuspid:

These lie behind the sternum on an oblique line joining the 3rd left sternocostal joint to the 6th right: thus, the *orifice of the pulmonary trunk* is deep to the left 3rd sternocostal joint; the *aortic orifice*, being slightly lower, more medial, and more posterior, is behind the sternum at the level of the 3rd intercostal space; the *mitral orifice*, is still lower and more medial at the level of the 4th costal cartilage; and the *tricuspid orifice* is on the right of the median plane at the level of the 4th and 5th spaces.

Heart Sounds. While the above para-

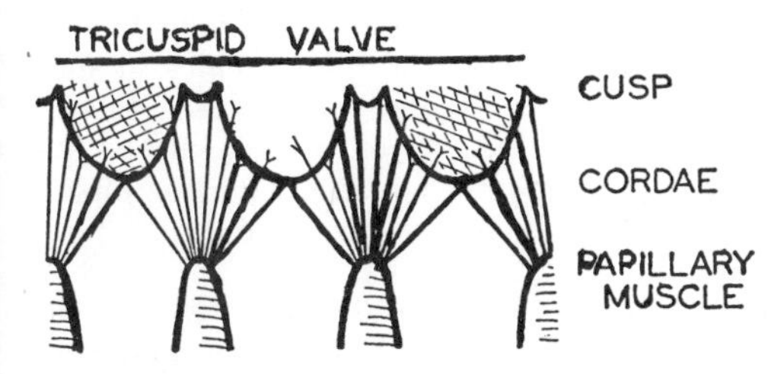

FIG. 617. Right atrioventricular valve spread out.

Fig. 618. The primitive muscular spongework of the ventricles.

graph describes anatomical locations, the heart sounds originating from the valves are heard on the chest wall in places somewhat removed from the direct surface projections. The reasons are complex and the details are given in clinical textbooks.

STRUCTURES OF WALLS OF HEART

Skeleton of the Heart (so-called) (*fig. 620*). The ventricles are emptied during *systole* by the blood being wrung from the cavities like water from a wet cloth. This requires a point of purchase or "skeleton", and so a fibrous ring surrounds each of the four orifices (*figs. 619, 620*). The aortic ring is the strongest and is like a cuff. The rings are joined to each other and to the pars membranacea septi. In some animals, e.g., the sheep, there is a central bone, the *os cordis*.

The Musculature of the Heart is called the *myocardium*. (Definitions on p. 31.)

Atrial Musculature. The atrial walls are translucent. The superficial muscle fibers run transversely; the deep fibers arch over the atrium from front to back and are attached to the skeleton by both ends; other fibers encircle the mouths of the great veins.

Ventricular Musculature (*figs. 621 to 623*). The ventricular *musculature* is composed of three layers—(1) superficial, (2)

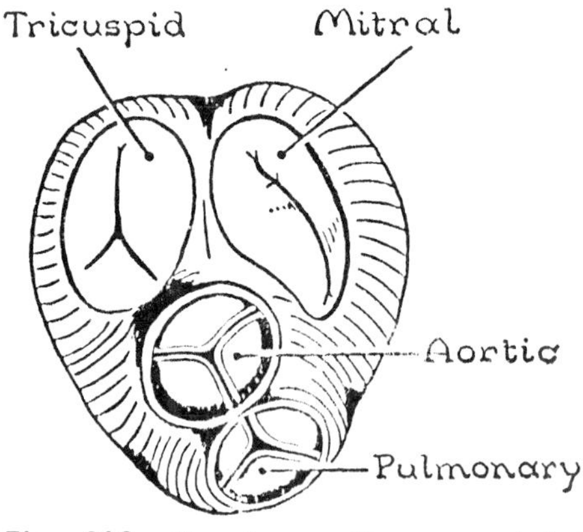

Fig. 619. The four orifices guarded by valves, showing the cusps, also the superficial muscle layer of the ventricle. (After Spalteholz.)

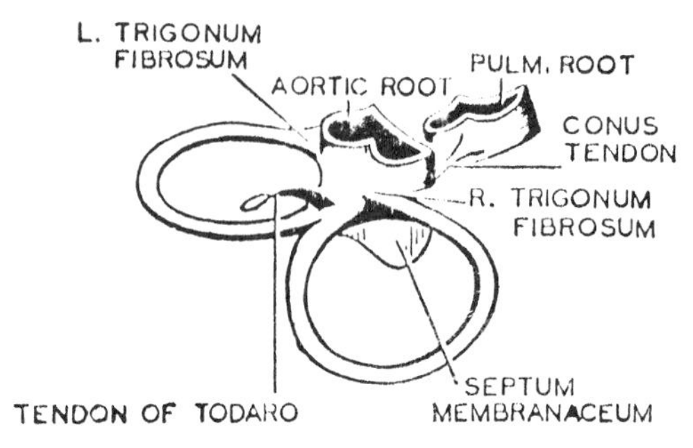

Fig. 620. The fibrous skeleton of the heart. (From Walmsley, after Ungar).

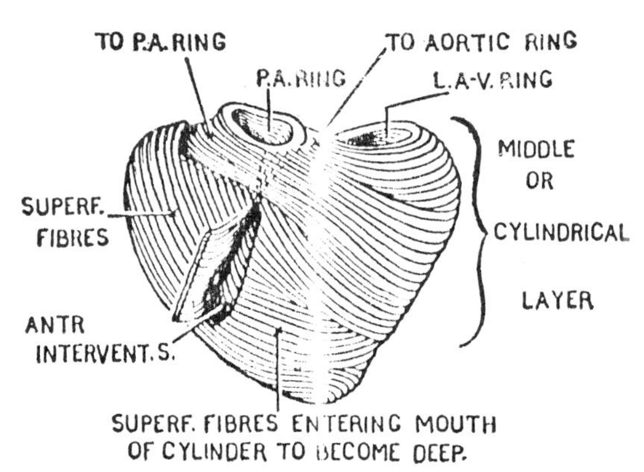

Fig. 621.

middle, and (3) deep.

All the fibers of the ventricles arise from the skeleton of the heart, and eventually they return to be inserted into the skeleton.

The following details of the arrangement of muscular fibers should be read to be appreciated, not to be memorized.

The Superficial Layer. If you twist your coat sleeve to the left, the spiral creases thereby produced will indicate the direction in which all superficial heart fibers run, whether at the front, sides, or back of the heart.

The Middle Layer. The *middle layer of the left ventricle* is a cylinder that surrounds the cavity of the left ventricle. It is the thickest and most basic layer of the whole heart. This cylindrical left middle layer is distinguished by possessing a free lower border. The fibers forming this lower border turn or roll in upon themselves.

The fibers of the middle layer arise from around the left atrioventricular ring and run from left to right across the front of the heart to be inserted around the pulmonary ring, conus

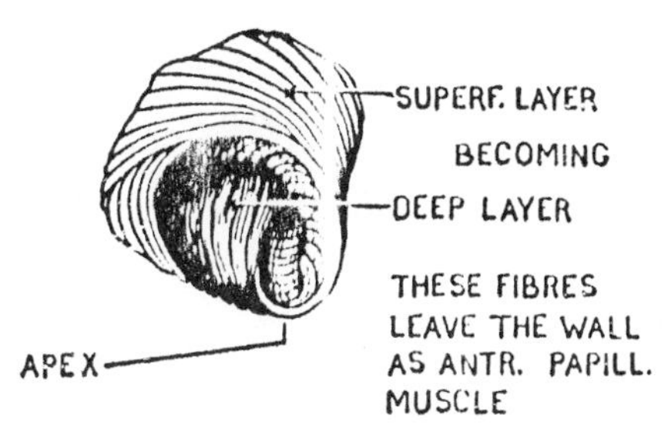

Fig. 622.

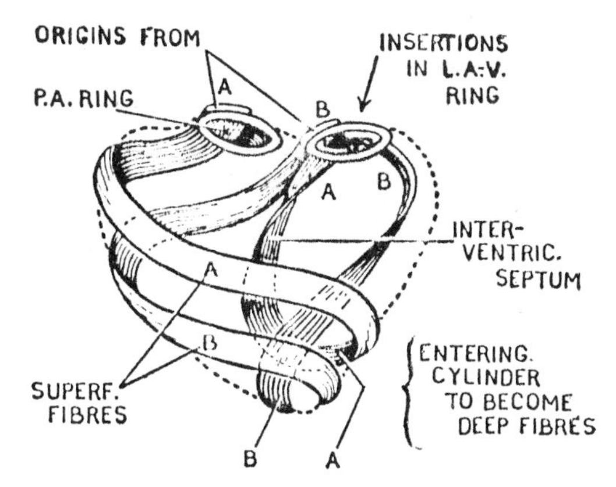

Fig. 623. Dissections of the ventricles of the heart of the sheep. (By B. Leibel.)

tendon, and septal side of the aortic ring.

The *middle layer of the right ventricle.* Its fibers arise from the left a-v. ring, run round the back of the left ventricle to the posterior interventricular sulcus where (1) some run vertically downward in the interventricular septum, while (2) others run round the ventricle to meet and interdigitate with the septal fibers at the anterior interventricular sulcus, and from there they continue over the anterior surface of the left ventricle.

The Deep Layer. The deep layer is a direct continuation of the superficial layer which, after running obliquely downward like a twisted sleeve on the superficial surface of the middle layers, turns around the lower border of the cylinder (or penetrates to the deep surface of the right ventricle's middle layer) and then ascends to gain attachment to the skeleton either directly, or indirectly through the papillary muscles, chordae tendineae and cusps of the valves.

As the superficial fibers are turning or twisting around the lower border of the cylinder, they skirt it for a third of a circle before proceeding upward as the fibers of the deep layer. As the mouth of the cylinder becomes more and more filled by these entering fibers, the orifice becomes progressively narrower and the third of a circle, which the most superficial fibers describe, becomes diminishingly smaller until ultimately the apex is represented by a fibrous pinpoint.

Owing to its attachment to the pulmonary ring, etc., the middle cylindrical layer on contracting tends to pull the left ventricle forward and to the right, so in systole the heart rotates anteriorly and to the right and strikes the chest wall. (This description is based on work done by B. Liebel).

Impulse Conducting System (*fig. 624*). This comprises—sinu-atrial node, atrioventricular node, and atrioventricular bundle and its two crura, right and left.

1. The *sinu-atrial node* initiates the heart beat. Composed of peculiar, longitudinally striated cells, it is 2 cm long by 2 mm wide, and is situated along the upper end of the sulcus terminalis. It is supplied by the right or left coronary artery (anastomoses are free) and by the right vagus nerve.

2. The *atrioventricular node* has the same structure and is situated in the interatrial septum beside the mouth of the coronary sinus. It also is usually supplied by the right coronary artery (and anas-

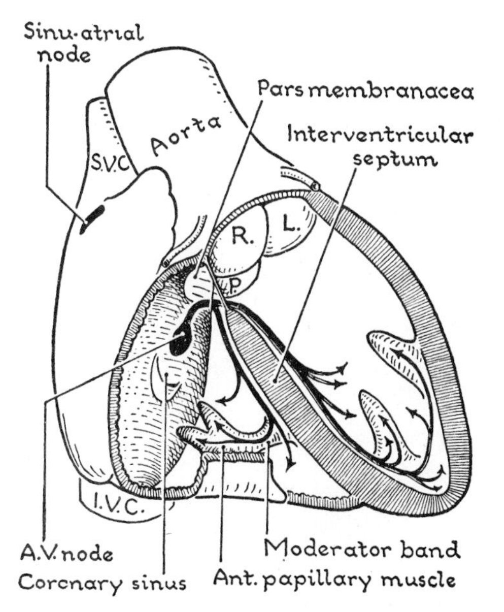

Fig. 624. The conducting system of the heart.

tomoses are free and ample) but by the *left* vagus nerve (James and Burch).

3. The *atrioventricular bundle* (of His) is a pale bundle of peculiar muscle fibers about 2 mm thick, enveloped in a loose sheath. This slender bundle is the sole muscular connection between the musculature of the atria and the musculature of the ventricles. It extends from the a-v. node, through the fibrous skeleton, to the interventricular septum. It skirts the hinder part of the membranous septum and, at the upper part of the muscular septum, it divides into a *right and a left crus.*

The crura descend in their sheaths, subendocardially, to the bases of the papillary muscles, that of the right side passing through the septomarginal (or moderator) band. When the anterior papillary muscle arises near the septum, the band need not free itself from the septal wall in order to reach the papillary muscle (Truex and Warsham).

Purkinje fibers, in connection with the fibers of the a-v. bundle, ramify both subendocardially and throughout the myocardium of the ventricles (Davis and Francis).

Nerve Supply to the Heart

(1) *Sympathetic*, via one to three *cervical cardiac* branches arising at variable levels from the cervical part of the sympathetic trunk; two or three *cervicothoracic branches* arising from the region of the cervicothoracic (stellate) ganglion; and two to four *thoracic branches* from the upper four thoracic levels of the sympathetic trunk (*fig. 634*). (2) *Vagus*, via a single *cervical cardiac branch* in its cervical course; one or two *cervicothoracic cardiac branches* from the main nerve at the inlet to the thorax (or its right recurrent laryngeal branch); and two to four *thoracic cardiac branches* from the thoracic part of the vagus nerve (and its left recurrent branch). Most of the cardiac nerves tend to fuse with each other early in their descent to the cardiac plexus (Mizeres).

Cardiac Plexus. See p. 441 for details.

Distribution. Branches of the *cardiac plexus* find a ready path in front of and behind the right pulmonary artery to the back of the atria, which they supply, the right nerves controlling the s-a. node; the left nerves, the a-v. node. Other branches descend in front of the pulmonary trunk to join the *right coronary plexus*. They also pass forward, mainly on the sides of the pulmonary trunk, and as *coronary plexuses* are distributed with the coronary arteries to the ventricles.

The vagal fibers are cardio-inhibitory; the sympathetic fibers are cardio-accelerator, vasodilator, and sensory.

33 SUPERIOR AND POSTERIOR MEDIASTINA

Prelude to Study of Region

Fate of Primitive Aortic Arches; Comparative Anatomy and Anomalies

CONTENTS OF SUPERIOR MEDIASTINUM

Boundaries; Contents.
Retrosternal Structures.
Thymus; Great Veins: Left and Right Brachiocephalic Veins and Tributaries.
Prevertebral Structures.
Esophagus; Trachea and Its Relationships; Extrapulmonary Bronchi; Tracheobronchial Lymph Nodes; Cardiac Plexus.
Intermediate Structures.
Aortic Arch and Its Three Branches; Anomalies of Arch; Vagus Nerve; Phrenic Nerve.

POSTERIOR MEDIASTINUM

Contents; Relations; Descending Aorta.
Thoracic Duct: course, communications, comparative anatomy, valves, variations.
Azygos and Hemiazygos Veins.
Esophagus—Course; Four Constrictions; Curvatures; Thoracic Relationships: right margin, left margin, posterior surface; Anterior Relations; Vessels; Nerves; Sphincters.

Prelude to Study of Region

Development of Great Arteries. To appreciate the relations of the superior and posterior mediastina in the adult, you should have some knowledge of the symmetrical arrangement in the embryo. Without this knowledge the anatomy of the region is unintelligible. The facts are simple and interesting and without them you will be forced to rote memorization.

Veins. The disappearance of the left s. v. cava was described on pages 429–430 (*fig. 611*).

Fate of the Aortic Arches (*figs. 625, 626*). Transitionally, in the early embryo, six primitive aortic arches (which are comparable with, though not identical with, the gill vessels of the fish) pass through the six pharyngeal arches on each side of the neck.

1. They connect the right and left primitive ventral aortae with the corresponding right and left primitive dorsal aortae, which fuse to form the descending aorta (*fig. 625*).

2. The vagus nerves descend on the side of the pharynx and esophagus (and of the trachea after it has taken form), the six primitive aortic arches alone intervening. The recurrent laryngeal nerve on each side runs medially below the VI primitive arch to supply the larynx (*fig. 626*).

3. Arches I, II, and V are transient, disappearing in human development.

4. The VI pair may be thought of as *right and left pulmonary arches*, for, after the spiral splitting of the truncus arteriosus into the ascending aorta and pulmonary trunk, this pair remains connected to the pulmonary trunk. From each VI or pulmonary arch branches sprout into the corresponding lung. Thereafter, the ventral portion of each VI arch becomes the stem of the corresponding right and left pulmonary artery of adult anatomy. The dorsal part of the VI left arch is the ductus arteriosus (*figs. 625, 635*).

5. The portion of the right primitive dorsal aorta caudal to the III right arch disappears and the dorsal end of the VI right arch, which is thereby rendered use-

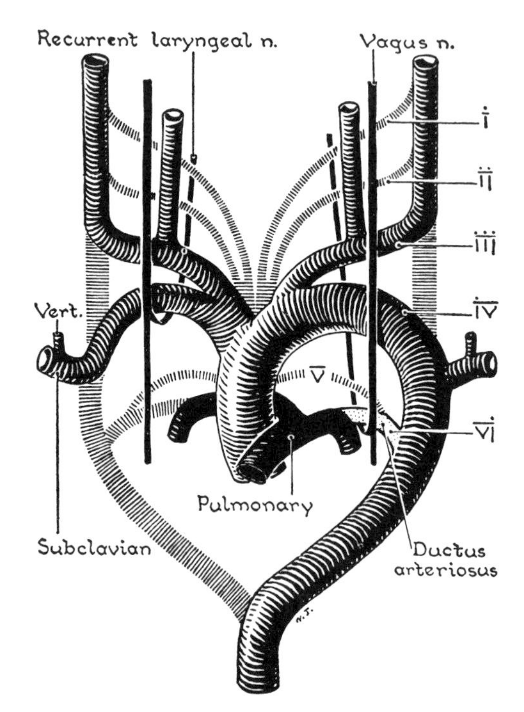

Fig. 625. The six paired primitive aortic arches.

less, disappears also. The left arteries, having now to do double duty, enlarge.

6. The *definitive aortic arch* is derived from the entire IV left primitive aortic arch and the adjacent parts of the primitive ventral and dorsal aortae.

7. The brachiocephalic trunk is derived from the right primitive ventral aorta. It is equivalent on the right side to part of the definitive aortic arch. The right subclavian artery includes the IV right aortic arch (*fig. 625*).

Comparative Anatomy and Anomalies. See figure 627 and p. 442.

Asymmetrical Results: (1) The left aortic arch enlarges and displaces the esophagus slightly and trachea markedly to the right. (2) When the heart descends and the neck elongates, the recurrent laryngeal nerves are dragged down by the lowest persisting aortic arches (*fig. 625*). Hence, the left recurrent laryngeal nerve recurs around the lig. arteriosum (VI arch) which is overshadowed by the enlarged definitive left aortic arch; but the right recurrent laryngeal nerve, in the absence of a right lig. arteriosum, recurs around the right subclavian artery (IV arch).

In cases where the right IV arch is absorbed and the arterial channel to the upper limb is maintained by the caudal part of the right dorsal aorta (*fig. 627*, anomaly), there being nothing to drag the right recurrent laryngeal nerve down, it does not recur but passes directly to the larynx.

CONTENTS OF SUPERIOR MEDIASTINUM

1. Retrosternal structures (*fig. 628*):
   - a. Thymus.
   - b. Great veins.
2. Prevertebral structures:
   - a. Trachea
   - b. Esophagus
   - c. Left recurrent nerve
   - d. Thoracic duct

   } Unit of four parallel structures
3. Intermediate structures:
   - a. Aortic arch and its three great branches.
   - b. Vagus nerves.
   - c. Phrenic nerves.

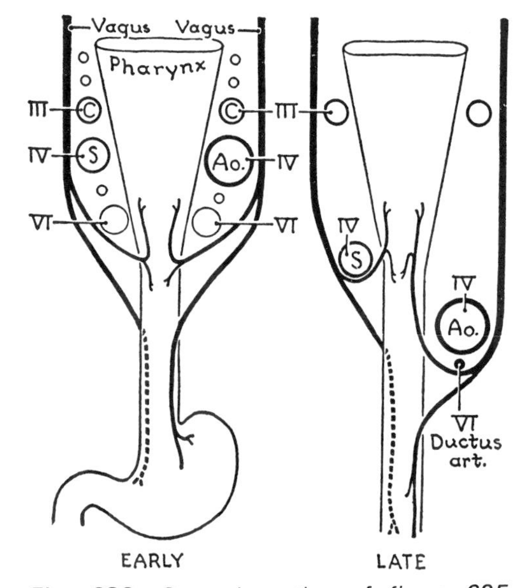

Fig. 626. Coronal section of *figure 625* showing: that only the arches separate the vagus nerves from the digestive tract (*early*). A later stage of explaining the asymmetrical courses of the recurrent laryngeal nerves (*late*).

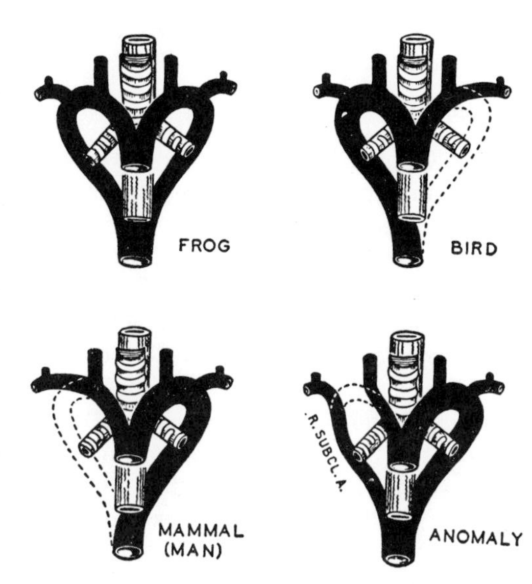

Fig. 627. Variations of the aortic arch.

Retrosternal Structures

The Thymus in the adult is an elongated, encapsulated, fatty, and lymphoid mass, lying in the loose tissue behind the manubrium sterni (*fig. 628*). On each side of it are the diverging anterior borders of the lungs and pleurae. Behind it are the left brachiocephalic vein and the aortic arch.

The thymus consists of two, loosely joined, asymmetrical lobes. At birth, its upper end (or ends) reaches nearly to the thyroid gland (*fig. 629*), while its lower end covers the upper part of the pericardium.

It is relatively largest and most extensive at birth; absolutely it is largest at puberty, after which it diminishes rapidly.

Great Veins (*fig. 630*). The **left brachiocephalic vein** is formed by the confluence of the left internal jugular and subclavian veins behind the sternal end of the clavicle; it passes behind the upper half of the manubrium to unite with the right brachiocephalic vein half way down the right margin of the manubrium to form the s.v. cava (*fig. 628*).

Below lies the aortic arch and behind are the three great branches of the arch.

At its origin it grooves the left lung and pleura. In youth and when engorged, its upper border rises above the jugular notch into the neck where it is in surgical danger.

The **right brachiocephalic vein** is formed similarly behind the sternal end of the right clavicle. It descends vertically. After being joined by the left brachiocephalic vein, it continues vertically, as the **superior vena cava,** to the 3rd right costal cartilage where it joins the right atrium.

This vertical venous channel projects beyond the right margin of the sternum. Along its right side runs the phrenic nerve (*fig. 630*). On its left side lie the ascending aorta, which is overlapping, and the brachiocephalic trunk (innominate artery). It is covered with pleura on three sides—in front, on the right, and behind. Before entering the pericardium and descending in front of the upper part of the root of the right lung, the s.v. cava is joined from behind by the arch of the azygos vein.

On each side, the phrenic and vagus nerves enter the thorax behind the brachiocephalic vein (*fig. 630*).

The Tributaries of the brachiocephalic veins fall into three groups:

1. Internal jugular and subclavian veins.
2. Thoracic or right lymph duct.
3. Veins returning blood delivered by the four branches of the subclavian artery, viz.,

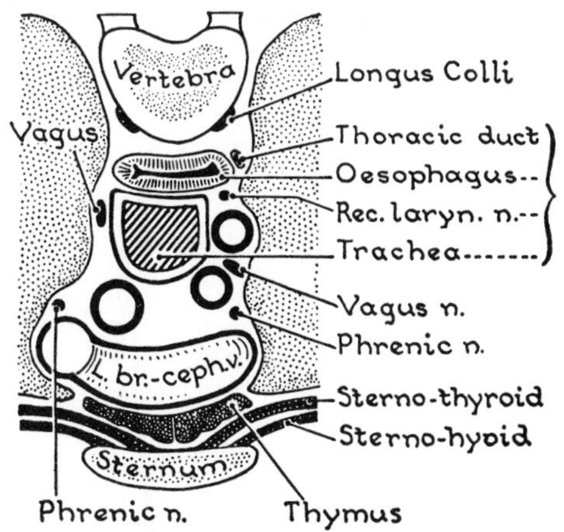

Fig. 628. Cross section of the superior mediastinum showing the arrangement of the contents.

vertebral, internal thoracic, inferior thyroid, and highest intercostal veins. The left brachiocephalic veins receives also the left superior intercostal vein (*fig. 726*). (These veins are discussed on p. 518).

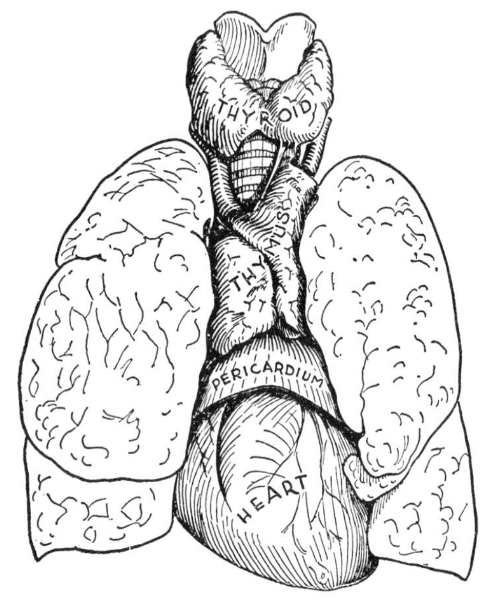

Fig. 629. The thymus gland of a child.

Prevertebral Structures

The esophagus, trachea, left recurrent nerve, and thoracic duct run through the superior mediastinum as a bundle or unit of four parallel structures (*fig. 631*). They have been described on page 407.

The Trachea (*fig. 632*) begins where the larynx ends on a level with the 6th cervical vertebra. Half its 10-cm length lies within the neck, half in the superior mediastinum.

About 20 U-shaped rings of hyaline cartilage keep its lumen patent; the ring at the bifurcation has a *carina* or keel that supports the "crotch" of the trachea.

It occupies the median plane except at its lower end where the aortic arch deflects it to the right.

Relationships of Trachea (*figs. 630, 632*). *The right side* is subpleural except where

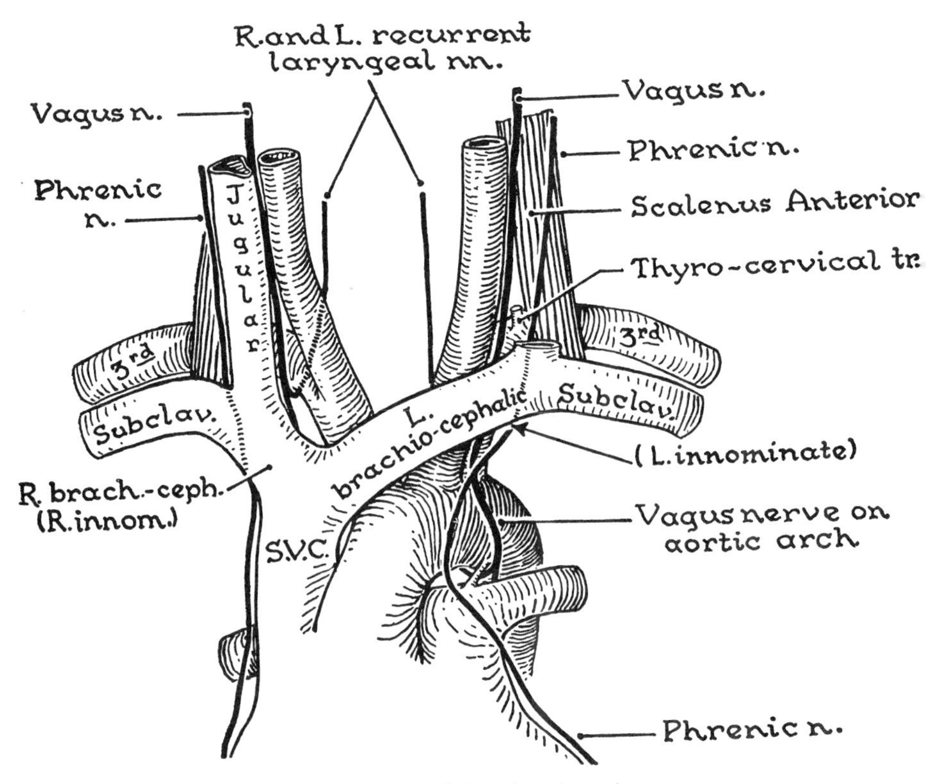

Fig. 630. The courses of the phrenic and vagus nerves.

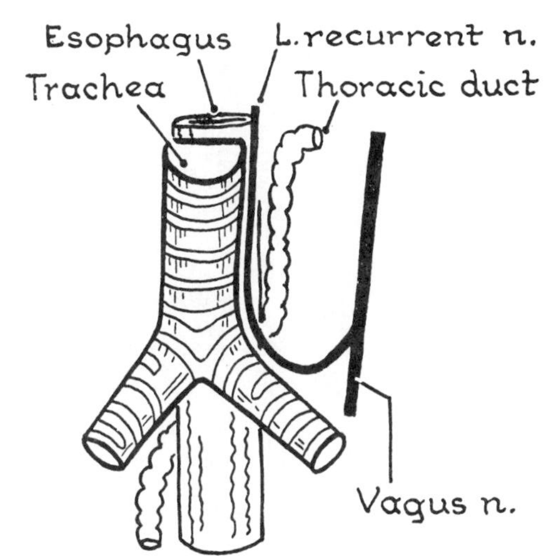

Fig. 631. A unit of four parallel structures runs through the superior mediastinum.

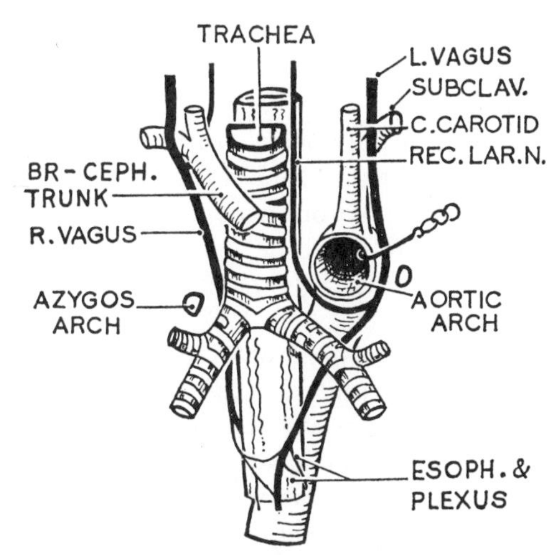

Fig. 632. Trachea and extrapulmonary bronchi, and their lateral relations.

the brachiocephalic trunk, right vagus nerve, and azygos arch intervene.

The left side is excluded from contact with pleura by the left subclavian and left common carotid arteries and by the arch of the aorta.

Further, tracheobronchial lymph nodes occupy the three angles at the bifurcation of the trachea (*fig. 633*), and the cardiac plexus of nerves lies in front of the bifurcation.

Extrapulmonary Bronchi. Right and Left (*fig. 633*). Of these the right bronchus is much the larger because it supplies the larger lung; it is the more vertical because the aortic arch deflects the trachea to the right. Hence, foreign bodies are more commonly aspirated into the right lung than into the left.

Tracheobronchial Lymph Nodes (*fig. 633*).

The Cardiac Plexus is a plexus of sympathetic and vagal fibers situated in front of the bifurcation of the trachea, above the bifurcation of the pulmonary trunk, and therefore below the arch of the aorta (*fig. 634*). Subdivision of the cardiac plexus is artificial (Mitchell; Mizeres) and should be abandoned; but extensions of it are the *right* and *left pulmonary plexus* and the *plexus on the thoracic aorta*.

The right and left pulmonary plexuses are formed by the cervical, cervicothoracic, and thoracic cardiac branches of both sides. At the hilus of each lung the pulmonary plexuses receive (both anteriorly and posteriorly) branches of the vagus nerve and sympathetic trunk.

Cardiac Nerves. On each side, the plexus receives slender cardiac nerves from the sympathetic trunk and the vagus nerve, as described on page 436. Most of the vagal branches are interconnected with sympathetic branches before joining the plexus. Some branches are mere filaments or groups of filaments.

The origin and course of the cervical cardiac nerves, and occasionally lower branches, are extremely variable on both sides, arising from any part of the vagus nerve and cervical sympathetic trunk, and crossing either anterior or posterior to the arch of the aorta, in contrast to the usual teaching that the left cervical cardiac nerves tend to be the only ones that remain on the left (anterior) side of the aortic arch (Mizeres).

Functions and Cell Stations. All the cardiac nerves have both efferent and afferent fibers, except the superior cervical sympathetic cardiac nerve which is believed to have only efferent fibers.

The postganglionic efferent fibers of the sympathetic cardiac nerves produce acceleration of the heart's action and dilatation of the coronary arteries. Their cell stations are in the three cervical and upper four (or five) thoracic sympathetic ganglia. Afferent pain fibers from the heart and aorta run mainly, or entirely, in the sympathetic cardiac nerves to the sympathetic

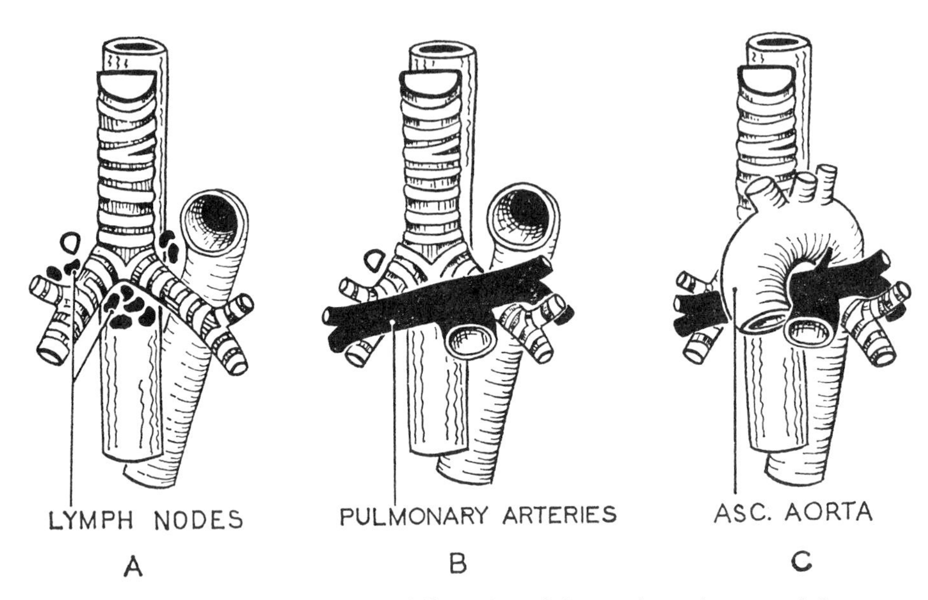

Fig. 633. Relations at the bifurcation of the trachea, shown serially.

ganglia, and through these ganglia and their rami communicantes to the upper four (or five) spinal ganglia where, like somatic afferent nerves, they have their cell stations.

The preganglionic efferent fibers of the vagal cardiac nerves produce slowing of the heart's action and contraction of the coronary arteries. They synapse with postganglionic fibers in the ganglia of the cardiac plexus and in the intrinsic cardiac ganglia, which are practically confined to the atria and interatrial septum and areas near the roots of the great vessels. Afferent vagal fibers from the heart, ascending aorta, and great veins are concerned in reflexes that depress the heart. Their cell stations are in the inferior vagal ganglion (G. A. G. Mitchell).

Intermediate Structures

Aortic Arch and Its Three Branches. The following paragraphs are in large measure a review.

Surfaces and Relations. The Left Anterior Aspect is touched by the right and covered by the left mediastinal pleura and lung. It is crossed by four nerves (*fig. 635*)—left phrenic, left vagus, and two cardiac nerves—and by the left superior intercostal vein.

The Right Posterior Aspect curves past the "unit" of four parallel structures (trachea, esophagus, recurrent nerve, and thoracic duct) and the nerves to the deep cardiac plexus that descend on the side of the trachea.

Below, the pulmonary trunk bifurcates into right and left branches. The lig. arteriosum joins the left pulmonary artery to the concavity of the arch beyond the origin of the left subclavian artery. On the left of the ligament is the left recurrent nerve; on the right is the superficial cardiac plexus.

The left recurrent nerve, therefore, arises on the left, passes below, and ascends on the right of the aortic arch.

Above, its three branches arise, and in front of their stems lies the left brachiocephalic vein.

Branches. The right and left coronary arteries being the first two branches of the aorta, the three great vessels arising from the arch are the 3rd, 4th, and 5th branches.

The brachiocephalic trunk (innominate a.) arises behind the center of the manubrium and ends behind the right sternoclavicular joint by dividing into right common carotid and right subclavian aa. The *left common carotid artery* arises close to the trunk, and the left subclavian artery arises behind the left carotid; both arteries ascend behind the sternoclavicular joint.

Anomalies of the Aortic Arch. 1. Rarely both the right and left arch persist, as in

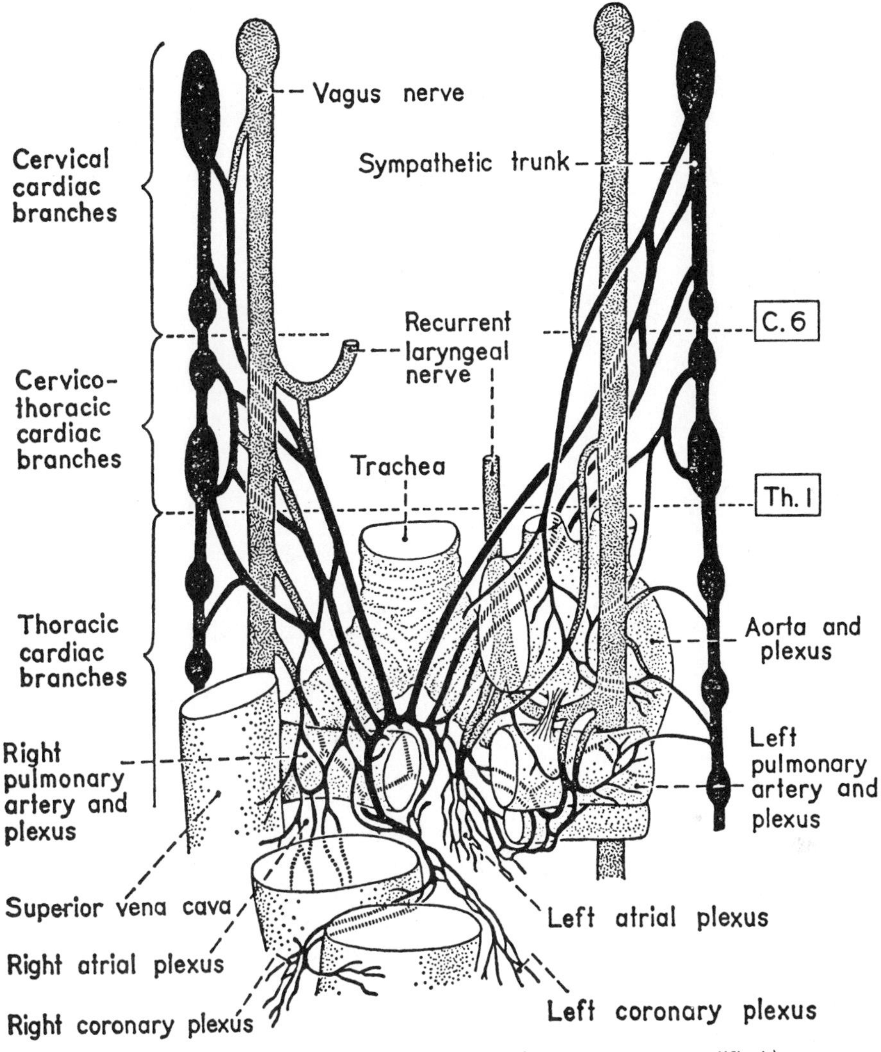

Fig. 634. Cardiac nerves and plexuses. (After Mizeres, modified.)

amphibia, forming an arterial ring through which the esophagus and trachea pass (*fig. 627*).

2. Rarely the right arch persists and the hinder part of the left arch disappears, as in birds, thus transposing the normal human scheme (*fig. 627*).

3. Sometimes (about 1 per cent) the anterior part of the right arch disappears and the posterior part persists to form the stem of the right subclavian artery. This stem becomes the 4th branch of the aortic arch and crosses behind the esophagus and trachea (*fig. 627*), which it may constrict —giving symptoms of obstruction.

4. The isthmus of the aorta (i.e., the segment of the aortic arch between the left subclavian artery and the ductus arteriosus) is narrow at birth, but it enlarges soon afterward as the ductus closes (*fig. 636*). Occasionally, it fails to enlarge—a condition called coarctation of the aorta, which requires surgical correction.

5. The ductus arteriosus may remain patent, generally in combination with cardiac anomalies.

6. Commonly the left common carotid artery arises from the stem of the brachiocephalic trunk, as in many primates.

7. Commonly the left vertebral artery

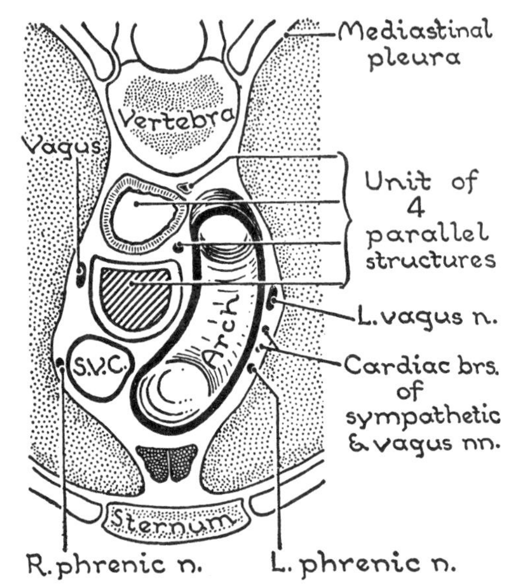

Fig. 635. Cross section of superior mediastinum showing the relations of the aortic arch.

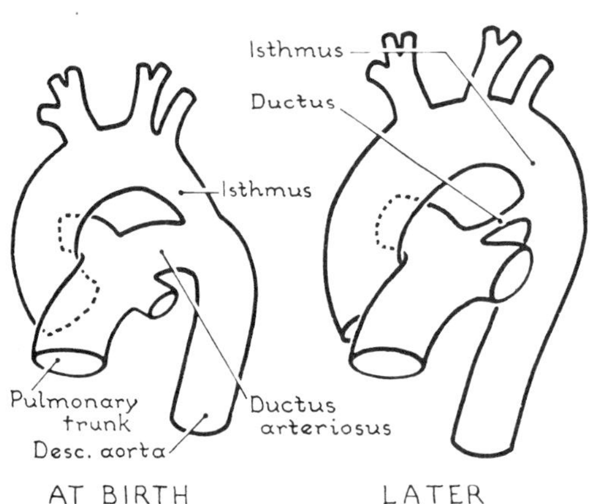

Fig. 636. Aortic isthmus and ductus arteriosus, at birth and a few months later. (After Patten.)

(usually a branch of the subclavian) arises from the aortic arch.

Each Vagus Nerve descends through the neck, applied to the posterolateral side of the carotid arterial stem. Continuing into the thorax, it passes behind the brachiocephalic vein (*figs. 630, 632*), and passes backward to the posterior pulmonary plexus.

Branches of the Vagi in the Superior Mediastinum. Here each vagus is responsible for recurrent, cardiac, tracheal, and esophageal branches. The right recurrent nerve gives off a cardiac branch as it hooks round the right subclavian artery; the other right branches, including one cardiac branch, spring directly from the right vagus. All the left branches spring from the left recurrent nerve.

The *right vagus* must cross the origin of the right subclavian artery (in front) and the brachiocephalic trunk (laterally) in order to reach the trachea, which conducts it subpleurally to the back of the root of the right lung. Above the root, the azygos vein arches forward, lateral to the right vagus. The *left vagus* continues to descend along the posterolateral side of the left carotid stem (and therefore in the angle between it and the left subclavian artery) to the aortic arch which it crosses far back in order to reach the back of the root of the left lung.

Each Phrenic Nerve enters the thorax between the subclavian artery and the beginning of the brachiocephalic vein (*fig. 630*) where it lies lateral to the thyrocervical trunk, which separates it from the vagus—and each passes a finger's breadth in front of the root of the lung. The *right phrenic* follows subpleurally along the side of the great vertical venous channel (*fig. 569*). The *left phrenic nerve*, in its strictly subpleural course, is the most anterior of the four nerves that cross the aortic arch (*fig. 635*).

POSTERIOR MEDIASTINUM

Contents (*fig. 637*).

A. Longitudinal tubular structures:
  1. Descending aorta.
  2. Thoracic duct.
  3. Azygos and hemiazygos veins.
  4. Esophagus (with vagus nerves).

B. Transverse tubular structures:
  1. Aortic intercostal arteries.
  2. Thoracic duct (from right to left).
  3. Certain posterior intercostal veins.
  4. Hemiazygos veins (terminal parts).

Postulate. The transversely running structures cling to the thoracic wall; they

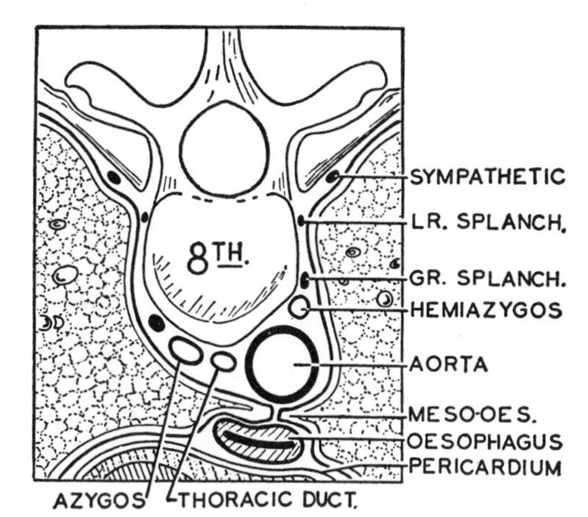

Fig. 637. The posterior mediastinum (in transverse section).

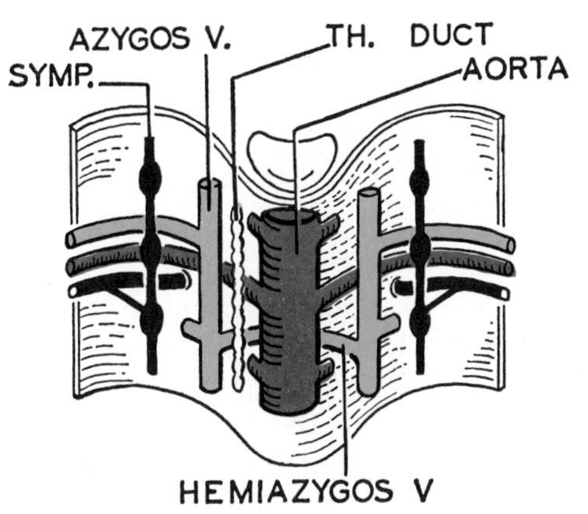

Fig. 638. Postulate: horizontal structures in the posterior mediastinum belong to the thoracic wall and pass external to longitudinal structures.

supply the wall and, like the ribs, are to be regarded as part of the wall (*fig. 638*).

Descending Aorta. Being the continuation of the aortic arch, the descending aorta runs the first part of its course on the left side of the bodies of the vertebrae (5, 6, and 7) and commonly it grooves or erodes them (*fig. 637*). Lower down, it gains the median plane and lies in front of the vertebrae (8 to 12).

Relations. Crossing it posteriorly are the terminal parts of the hemiazygos veins. Throughout its course, the thoracic duct and the azygos vein lie on its right posterolateral side, and accompany it through the aortic hiatus in the diaphragm where it becomes the abdominal aorta. The hemiazygos veins lie on its left posterolateral side.

On the left are the mediastinal pleura and lung. *On the right* are the esophagus in its upper part; the right mediastinal pleura and lung in its lower part.

In front are: (1) the root of the left lung; (2) the pericardium, which separates it from the oblique pericardial sinus and left atrium; (3) the esophagus, which was to its right opposite vertebrae 5, 6, and 7, is in front and passing to the left side opposite vertebrae 8, 9, and 10; (4) the diaphragm, at the level of the vertebrae 11 and 12.

Branches (*fig. 639*)—(1) *visceral:* one to

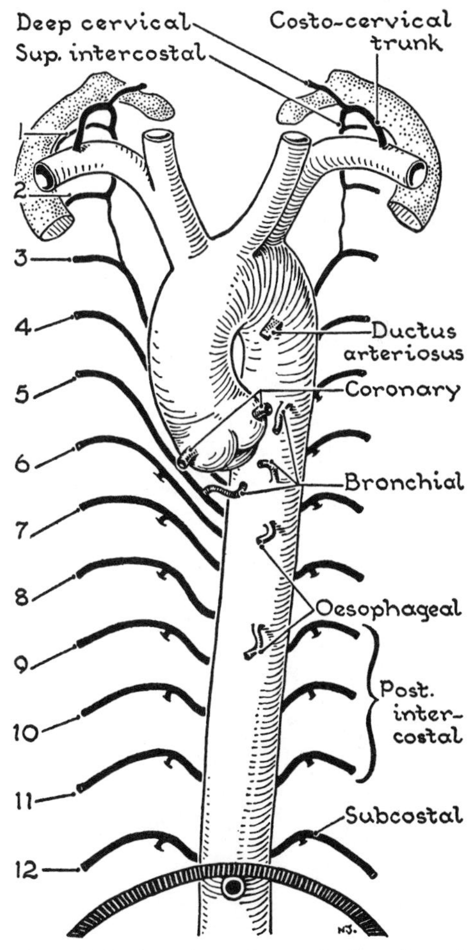

Fig. 639. The branches of the thoracic aorta.

three bronchial aa., one to three esophageal aa., and twigs to the pericardium and diaphragm; (2) *parietal:* lower nine pairs of posterior intercostal and one pair of subcostal arteries.

Thoracic Duct (*figs. 640 to 642*).

The thoracic duct begins as the effluent from the cisterna chyli, enters the thorax through the aortic hiatus in the diaphragm, ascends the whole thorax, and then at the level of vertebra C. 7, curves laterally in the neck (p. 518) to end in the left jugulo-subclavian angle (*fig. 640*).

Communications. If the thoracic duct be cut deliberately or by accident, chyle (the product of the digestion of fat) escapes from the cut end. Provided the duct is tied, evil results seldom follow, which indicates that accessory lymphaticovenous communications must exist (see p. 36).

Valves. The thoracic duct has few valves. There are usually two (or one) at the cephalic end; when only one, it is always inadequate (Rouviere).

Variations. Most of the variations possible from *figure 641* have been observed.

The *primitive* lymph ducts were phylogenetically paired (*figs. 641, 642*). Of the various

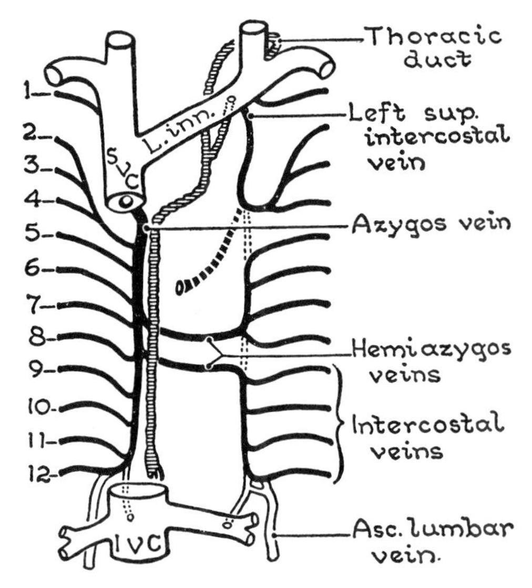

Fig. 640. The thoracic duct and the intercostal veins. (Continued from *figure 677*.)

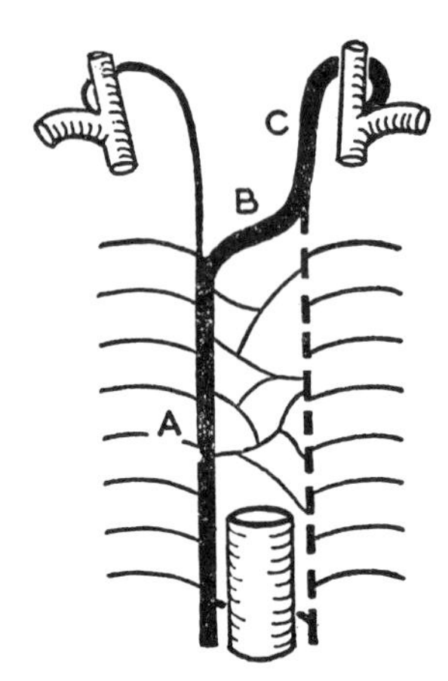

Fig. 641. The thoracic duct

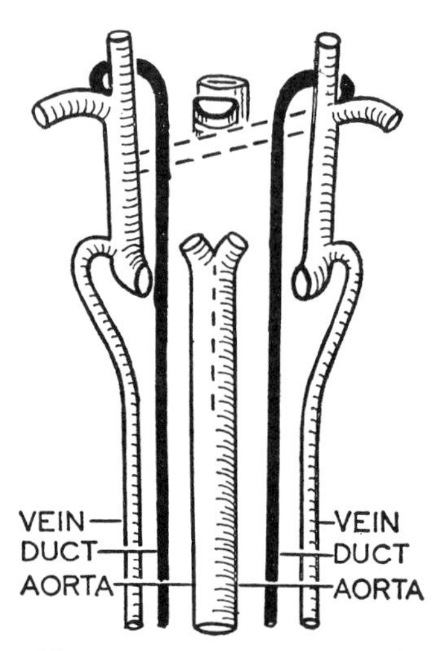

Fig. 642. Three symmetrically placed embryonic vessels.

prevertebral cross-communications, one lying on the plane between the posterior and superior *mediastina enlarged* (*fig. 641-B*).

Comparative Anatomy. In the New World monkeys the lymph vessels below the diaphragm empty into the veins near the kidneys, not the thoracic duct (Silvester). In rats lymph trunks communicate with numerous veins and in cats, with the azygos and intercostal veins.

Azygos and Hemiazygos Veins (*figs. 640, 642*). These veins developed from the posterior cardinal system of veins (*fig. 611*).

Each vein is covered with pleura and is applied to the vertebral column. The *azygos vein*, usually swings to, or beyond, the median plane and returns to arch over the root of the right lung to join the superior vena cava (Nathan).

The left vein breaks into three segments (*hemiazygos*, *accessory hemiazygos*, and *left superior intercostal veins*) after two or more cross-branches have united it to the azygos vein. Alternatively, it may persist without breaks; it may even retain connection with the coronary sinus, when it is called the left superior vena cava.

Tributaries: posterior intercostal veins, vertebral venous plexus (*fig. 676*), and mediastinal, esophageal, and bronchial veins.

Esophagus

The esophagus extends from the pharynx to the stomach, and has, therefore, cervical, thoracic, and abdominal portions. It pierces the diaphragm behind the 7th left costal cartilage at the level of the 10th thoracic vertebra, and joins the stomach just beyond (*fig. 643*).

It has *Four Constrictions*—at its origin in the neck, at the aortic arch in the superior mediastinum, at the tracheal bifurcation, and where it passes through the diaphragm.

Thoracic Relationships of the Esophagus. When considering these, have special regard to: (1) the mediastinal pleurae and lungs, (2) the heart and great arteries, (3) the respiratory tract, (4) the vertebral column, and (5) the thoracic duct. Ask yourself "Where could a sharp foreign body piercing its walls enter these structures?"

Its right margin is in contact throughout with right mediastinal pleura and lung, except where the arch of the azygos vein crosses it.

Its left margin is separated throughout from left mediastinal pleura and lung (except at two areas) by the great arteries (*fig. 643*).

Of the two areas in contact with pleura (1) one is in the superior mediastinum (*fig. 573*); (2) the other is below the level of the

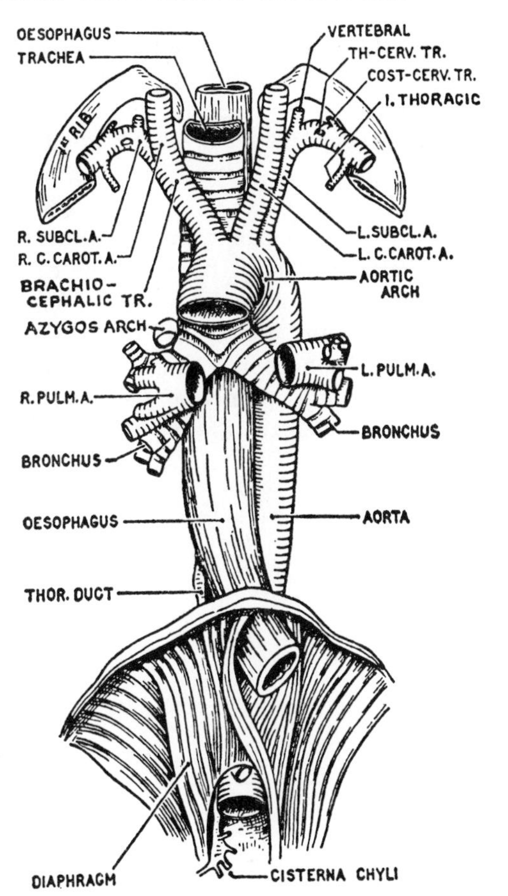

Fig. 643. The esophagus, the aorta, and the three branches of the aortic arch.

heart; there the esophagus is "suspended" from the front of the aorta by a meso-esophagus which allows it to curve forward (*fig. 637*).

Its posterior surface is in contact with vertebral bodies 1–4; then the thoracic duct, azygos vein, some right intercostal arteries, and cross-channels of the hemiazygos veins (*fig. 640*). At the level of vertebrae 8, 9, and 10, the descending aorta insinuates itself between the esophagus and the vertebrae.

Its anterior relations are as follows: in the superior mediastinum—the trachea and left recurrent nerve; in the posterior mediastinum, the bifurcation of the trachea, the bronchi, and nodes, all crossed ventrally by the right pulmonary artery;

and then the pericardium, oblique pericardial sinus, and left atrium; lastly, the diaphragm.

Vessels. *The Arteries of the Esophagus* are a rather casual series it picks up as it descends.

Nerves. The recurrent nerves and the sympathetic trunks in the neck; the right vagus and left recurrent nerve in the superior mediastinum. Below the bronchi the vagi, joined by branches from the sympathetic trunks and splanchnic nerves, form the esophageal plexus around the esophagus. From this plexus two nerves, the *anterior* and *posterior vagal trunks,* descend on the esophagus to the stomach.

Sphincters. There is a sphincter at both ends of the esophagus. The *gastric or cardiac sphincter* is a physiological one, supplied (1) by the vagus which conducts opening impulses (relaxes), and (2) by the sympathetic which conducts closing impulses (contracts).

At the *pharyngeal end* is the Cricopharyngeus (p. 554).

SECTION SEVEN

HEAD AND NECK

34 Front of Skull, Face, and Scalp 451

35 Posterior Triangle of Neck 463

36 Back 469

37 Interior of Cranium 479

38 Orbital Cavity and Contents 493

39 Anterior Triangle of Neck 505

40 Root of Neck 516

41 Side of Skull, Parotid, Temporal, and Infratemporal Regions 521

42 Cervical Vertebrae, Prevertebral Region, and Exterior of Base of Skull 533

43 Great Vessels and Nerves of Neck: Review and Summary 542

44 Pharynx and Palate 552

45 Mouth, Tongue, and Teeth 566

46 Nose and Related Areas 575

47 Larynx 584

48 Ear 590

49 Lymphatics of Head and Neck 599

50 Bones of Skull: Details for Reference 602

34 FRONT OF SKULL, FACE, AND SCALP

SKULL ON FRONT VIEW AND FACE

Warning; Orientation, Frankfort Plane

Muscles and Features of Face

Muscles of Rima Oris; of Lips; of Chin; of Cheek; Nerve Supply.
Lips, External Nose. Auricle.
Eyelids, Conjunctival Sac, Tear Apparatus—Inspection; Orbital Septum; Tarsi.
Muscles of Eyelids and Forehead.
Lacrimal Apparatus: Nasolacrimal Duct.

SENSORY NERVES OF FACE AND COMPANION ARTERIES

Ophthalmic Nerve (and Its Arteries); Maxillary Nerve (and Its Arteries); Mandibular Nerve (and Its Arteries).

MOTOR NERVE OF FACE (FACIAL NERVE)

BLOOD SUPPLY OF FACE

Facial Artery; Superficial Temporal Artery; Transverse Facial Artery.
Facial Vein.

SKULL FROM ABOVE AND SCALP

Cephalic Index; Sutures; Fonticuli.
Dangerous Area; Muscles and Aponeurosis; Nerves; Arteries.

SKULL ON FRONT VIEW

Warning: keep your fingers out of the orbital cavities or you will certainly break their medial walls which are papery in thinness.

Orientation: At a convention of anthropologists held in Frankfort (1882) it was agreed to examine skulls when so placed that the lower margins of the orbital apertures and the upper margins of the external acoustic (auditory) meatuses lie on a horizontal plane.

The Frankfort Plane. This most nearly approximates the Anatomical Position (*fig. 1*), in which the eyes look straight forward as, of course, they do in life.

The *zygomatic arches* lie at the widest parts of the face (*fig. 645*). Above them, the outline of the skull is rounded because it is formed by the *cranium* or brain case, and it bulges a few millimeters beyond the zygomatic arches. Below them, the skull is angular and is outlined by the *posterior border of the ramus*, the *angle*, and the *base* or lower border of the mandible.

At Birth (*fig. 644*) a median suture line bisects the skull vertically, separating the *parietal*, *frontal*, *nasal*, *maxillary*, and *mandibular* bones of opposite sides. During the 2nd year, the two halves of the human mandible fuse at the *symphysis menti* unlike most mammals.

The two halves of the frontal bone likewise fuse about the 2nd year, but in some skulls they remain separate, i.e., the interfrontal or *metopic suture* persists. The interparietal or *sagittal suture* is usually obliterated by the age of 35 years (Todd and Lyon).

The *bregma* is the point at which the sagittal and coronal (frontoparietal) sutures intersect. It is situated 2.5 cm in front of the *vertex* or highest point. The *nasion* is the point at the root of the nose where the frontonasal suture crosses the median plane.

Entrance to the Orbit. [Aditus Orbitae]. Each of three bones—frontal, maxillary, and zygomatic—forms approximately one-third of the orbital margin (*fig. 645*).

The fullness above the medial part of the supra-orbital margin of the frontal bone is the *superciliary arch*, well marked in the male. The elevation between the superciliary arches is the *glabella* because the overlying skin is bald or glabrous.

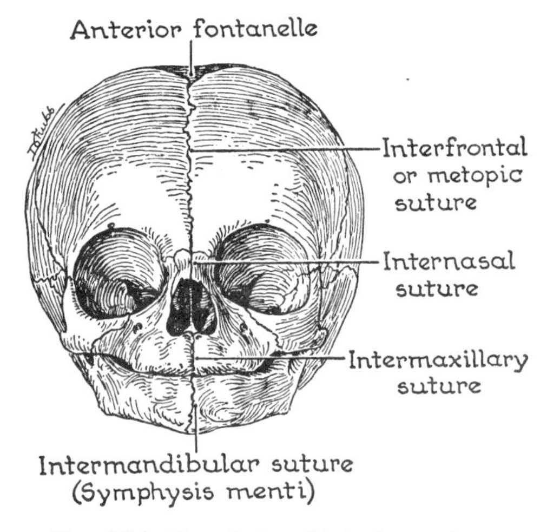

Fig. 644. The skull at birth, front view.

The *piriform aperture* (anterior nasal aperture) is formed by the nasal bones above; by the maxillae laterally and below. A median spine of bone, the *anterior nasal spine*, juts forward from the maxillae and helps to support the septal cartilage of the nose.

Lateral to the orbit, the *zygomatic arch* bends sharply upward and then, as the *temporal line*, it curves backward across the side of the cranium (*figs. 646, 647*).

Teeth. There are 32 *teeth* in all, 16 in each jaw. Of the eight upper and lower teeth on each side, two are *incisors* or cutting teeth, one is a *canine*, two are *premolar* or bicuspid teeth for their crowns have two cusps, and three are *molar* or

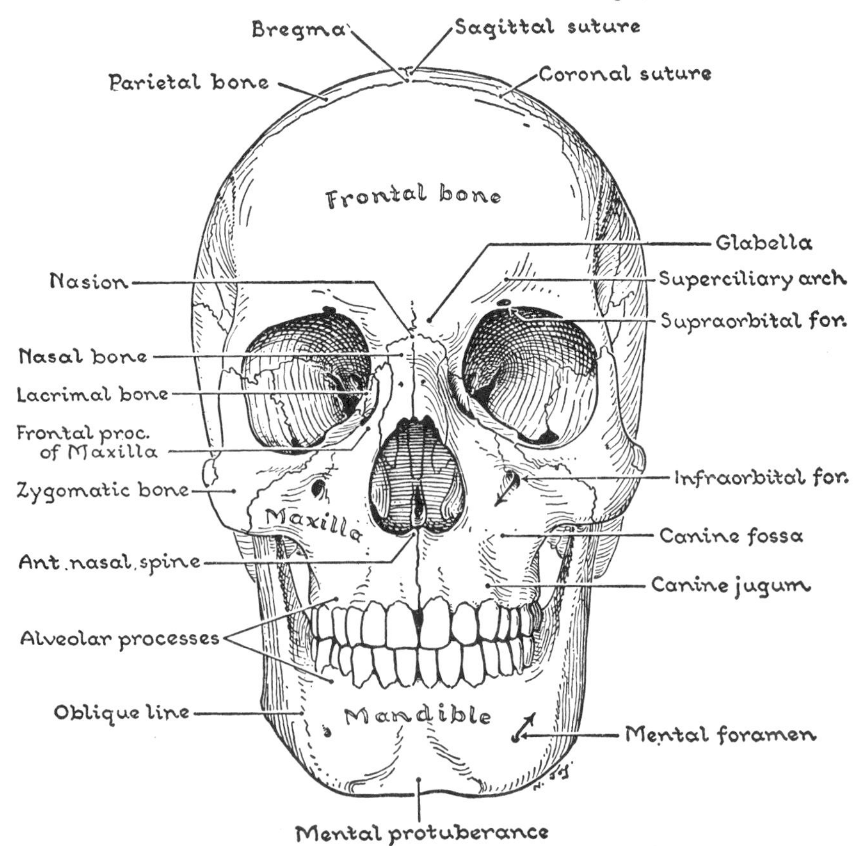

Fig. 645. The skull, on front view (Norma frontalis of Anthropologists).

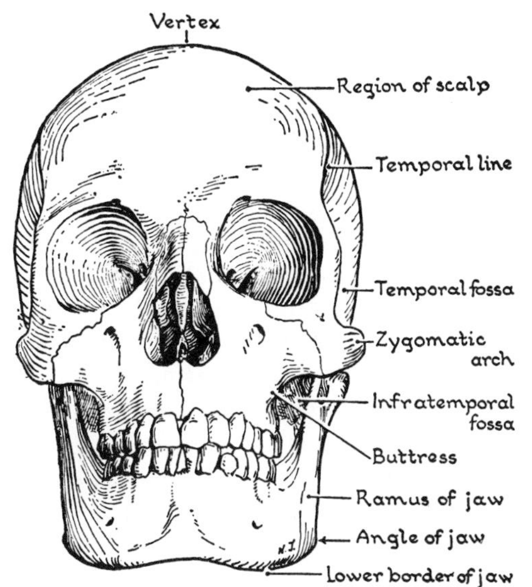

Fig. 646. The skull, on front view.

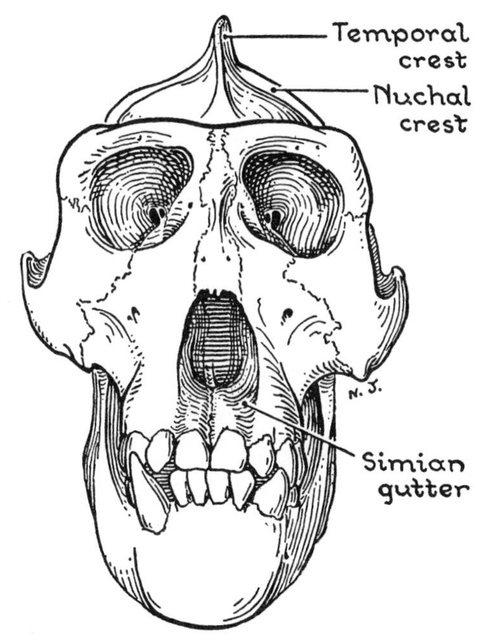

Fig. 647. Skull of gorilla (After Gregory.) The temporal lines meet to form the temporal crest because the Temporalis muscles are enormous.

millstone teeth. The roots of the teeth are embedded in little troughs or alveoli contained in the *alveolar processes* of the maxilla and mandible. (Also see p. 571.)

Man alone has a *chin* and an *anterior nasal spine* (*fig. 648*).

The skull has but one pair of movable joints, the *temporomandibular* or *jaw joints.* When the teeth of the lower jaw close on the upper teeth, a force between them equivalent to a 100-kg weight can be exerted. Therefore, a buttress or strengthening bar is provided in the form of a ridge that extends upward from the (1st or) 2nd molar tooth to the zygomatic bone. The force is then largely transmitted up the strong lateral orbital margin to be dispersed through the dome of the cranium. Stresses from the front teeth are mainly transmitted via the frontal process of the maxilla along the medial orbital margin.

At the point of the jaw, there is a slightly raised triangular area, the *mental protuberance.* From its lateral angle an *oblique line* runs upward and backward to become continuous with the *anterior border of the ramus* of the jaw.

Foramina. Three intra-osseous foramina —*supra-orbital, infra-orbital,* and *mental*—open onto the face on a vertical line that passes between the premolar teeth. They penetrate the frontal, maxillary, and mandibular bones, transmitting sensory branches of the 1st, 2nd, and 3rd divisions, respectively, of the trigeminal nerve and their companion vessels (*fig. 658*).

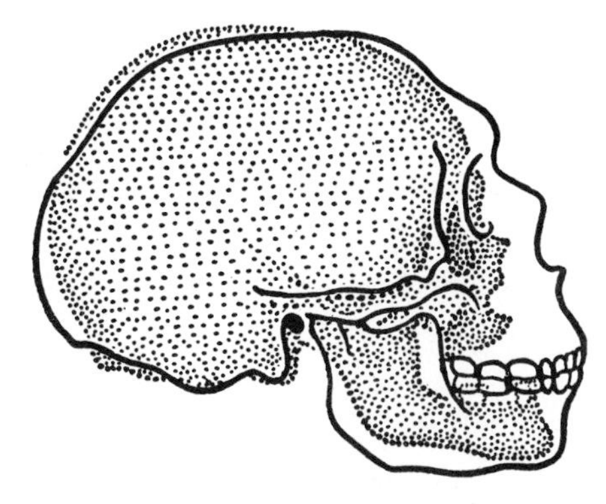

Fig. 648. Superimposed profiles of a modern white skull (stippled) and a prehistoric skull (clear). Note the recession of the face and the appearance of a chin in the modern skull. (After Boule.)

FACE

Muscles and Features of the Face

The facial muscles or muscles of expression are disposed around the orifices of the mouth, eye, nose, and ear, as sphincters and dilators (*fig. 649*). Developmentally, they took origin in one bar of tissue supplied by the facial nerve (hyoid or 2nd pharyngeal arch) and from it spread as a sheet over the face, dragging branches of the nerve after them. The *Platysma*, which spread downwards over the neck, and the *Epicranius*, which spread upward over the cranium, have the same source and nerve.

Muscles of Rima Oris (Aperture of Mouth). The *Orbicularis Oris* lies within the lips and encircles the oral aperture.

Converging on the angle of the mouth are five muscles—the *Levator Anguli Oris*, which arises below the infra-orbital foramen; the *Zygomaticus Major*, or smiling muscle, which arises from the bone of the same name; the *Risorius*, or grinning muscle, which arises from the fascia and is joined by the posterior fibers of the *Platysma* (*fig. 708.2*), and the *Depressor Anguli Oris* (Triangularis), which arises from an oblique line on the mandible.

Muscles of the Lips (L. Labium, sing., labia, pl.; labii = of the lip; labiorum = of the lips). Attached to the upper lip are three bands, which arise from the medial and lower borders of the orbital margin. They are: *Levator Labii Superioris Alaeque Nasi*, **Levator Labii Superioris,** and *Zygomaticus Minor*. (L., que = and; alae = of the wing; nasi = of the nose.)

Attached to the lower lip is the *Depressor Labii Inferioris* (Quadratus), which arises from the oblique line of the mandible. Between the right and left muscles, the space is occupied by the paired Mentales.

Muscles of the Chin (L. Mentum). Each *Mentalis* puckers the skin of the chin.

Muscles of the Cheek (L. Bucca). The *Buccinator* is a flat muscle whose inner surface is lined with the mucous membrane of the cheek. Above and below, it arises

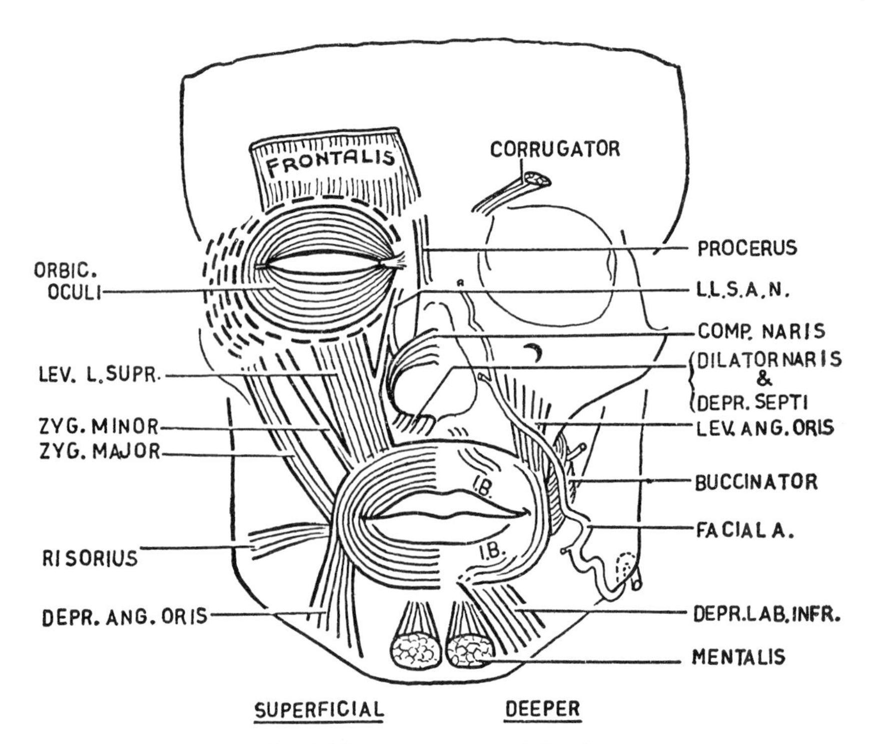

Fig. 649. The muscles of the face.

lateral to the molar teeth. Posteriorly, it is continuous with the Superior Constrictor of the pharynx, the line of union being marked by a faint fibrous suture (the *pterygomandibular raphe*) (*figs. 770, 771*). Anteriorly, it extends into the upper and lower lips to blend with the Orbicularis Oris.

Actions. The Buccinator aids in mastication by pressing the cheeks against the teeth, thereby preventing food from collecting in the vestibule of the mouth (*fig. 649.1*). It also acts variably in blowing and in sucking.

Nerve Supply: Facial nerve (nerve VII).

Structures Piercing: Buccal branches of nerve V^3 (sensory) and parotid duct.

The space between the Buccinator medially and the ramus of the jaw laterally is occupied by the *buccal pad* of fat.

Lips

If you run the tip of your tongue across the back of your lower or upper lip, you will feel the small nodular *labial glands* that here form an incomplete subepithelial tunic. Grasp the margin of either lip between your finger and thumb to feel the pulsations of the *superior* and *inferior labial arteries* between the labial muscles and the tunic of labial glands.

The lip margins are red partly because the skin is translucent and partly because the vascular papillae or thelia are unusually long. Historically, the term "epithelium" was first applied to the cells covering the thelia of the lip (E. A. Schafer).

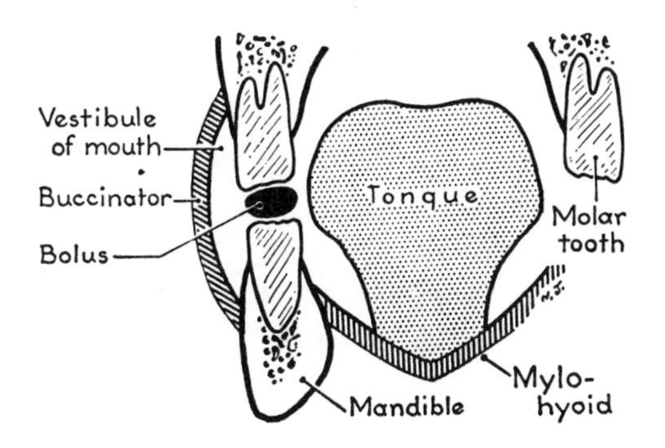

Fig. 649.1. The Buccinator and the tongue hold the food between the teeth.

External Nose (*figs. 650, 651*)

The framework of the external nose is made of the paired nasal bones, hyaline cartilage, and fibro-areolar tissue. The cartilages are: *septal*, *lateral*, and *alar*. The right and left *lateral cartilages* are not entities but are winglike expansions of the septal cartilage, which passes far back between the right and left nasal cavities. They are firmly united to the nasal bones above and maxillae behind, but are connected only loosely with the paired alar cartilages below.

Auricle

The framework of the *auricle* is made of a single piece of elastic cartilage except at its most dependent part, the lobule, which is fibro-areolar. The cartilage is continuous with the cartilage of the external acoustic meatus (meatus. L = a canal) (see p. 590). *Figures 652* and *653* provide the names of its elevations and depressions. *Darwin's tubercle* represents the primitive apex.

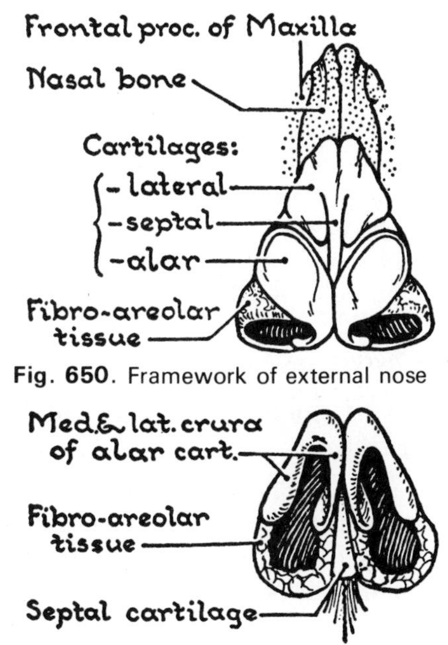

Fig. 650. Framework of external nose

Fig. 651. Framework of the external nose (from below).

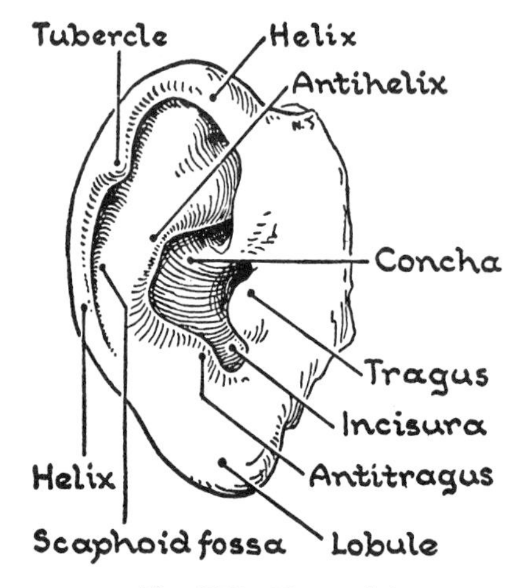

Fig. 652. The auricle.

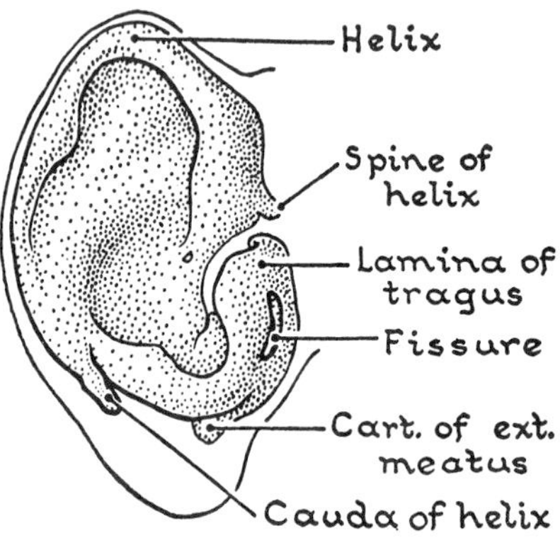

Fig. 653. The cartilage of the auricle.

Muscles. The auricle has several rudimentary intrinsic muscles and three extrinsic muscles—Auriculares Posterior, Superior, and Anterior—all supplied by the facial nerve.

Sensory Nerves—auriculotemporal (V^3), great auricular (C. 2, 3), and lesser occipital (C. 2, 3), also twigs *of the vagus* and *facial nerves.*

Eyelids, Conjunctival Sac, and Tear Apparatus

Definitions. The upper and low eyelids, or *palpebrae*, are united at the medial and lateral angles by the corresponding *palpebral commissures.*

The posterior five-sixths of the outermost coat of the eyeball is white, tough, and called the *sclera;* its anterior one-sixth is transparent and called the *cornea.* Through the cornea the varicolored *iris* is seen, and in the center of the iris is the *pupil.*

The potential space between the eyeball and the eyelids is the *conjunctival sac.* The membrane lining the sac is the *conjunctiva.* At the upper and lower limits of the sac, called the *fornicės*, the conjunctiva is reflected from eyeball to eyelid.

Inspection. Examine your eye in a mirror:

1. The margin of the lower lid crosses the lower limit of the cornea; the margin of the upper lid encroaches on the cornea (*fig. 654*).

2. The lateral five-sixths of the margins of the lids are flat and carry eyelashes or *cilia.* The medial one-sixth is devoid of hairs and rounded, containing the *canaliculus* that drains away the tears.

3. At the medial angle there is a triangular area, the *lacus lacrimalis*, bounded laterally by a free crescentic fold of conjunctiva, the *plica semilunaris.* In the lacus there is a reddish area, the *caruncle.*

Gently pull down the lower lid to note:

4. A *papilla* on which the *punctum*, or entrance to the inferior lacrimal canaliculus, can easily be seen.

Evert the upper lid over a match stick:

5. A sulcus lies near and parallel to the margin. In it foreign particles are commonly caught.

6. The hairs or cilia projecting from the lid margins are in two or three irregular rows.

7. Hairs imply the presence of sebaceous glands, and these open into each hair

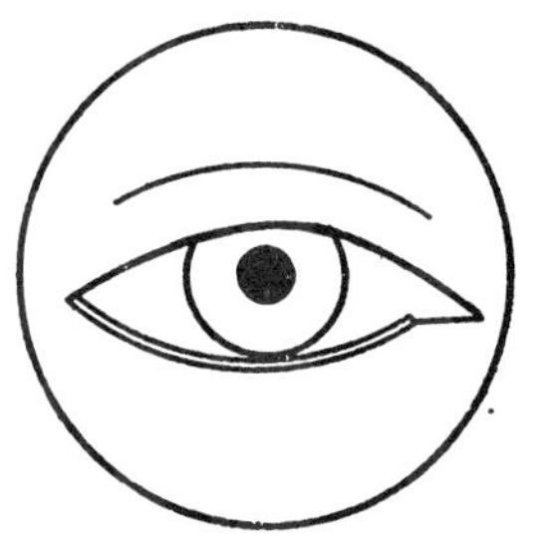

Fig. 654. The margins of the eyelids.

follicle. Sweat glands likewise open into or beside the hair follicles (*fig. 655*).

8. The *tarsal glands*, which waterproof the lids, are embedded in the tarsus, a tough fibrous plate; they are visible as yellow streaks through the conjunctiva.

An obstructed and inflamed hair follice or stye will project on the front of the lid; an obstructed tarsal gland on to the globe of the eye.

Orbital Septum and Tarsi (*fig. 656*). The eyelids develop as folds of skin which come together and adhere along their edges during the middle 3 months of intra-uterine life. When they become free again, the palpebral fissure is re-established. (In kittens, the lids remain adherent for some days after birth.)

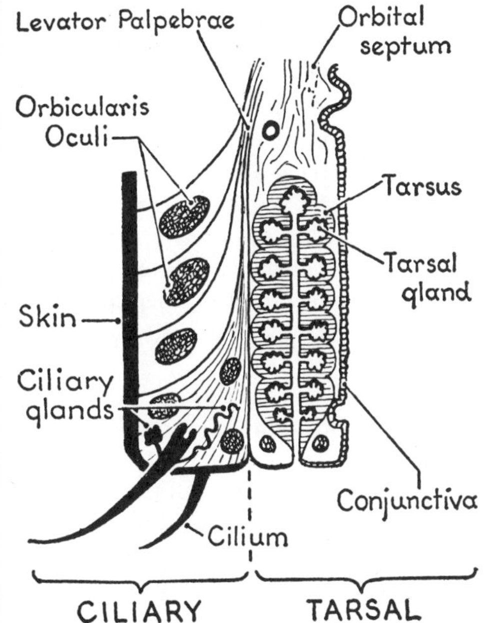

Fig. 655. Section through the upper eyelid. (After Whitnall.)

While the lids are closed, the orbital septum, which runs from the orbital margin into the eyelids, forms a complete diaphragm for the orbital cavity, for it is attached to the orbital margin all around; medially it passes behind the tear sac to gain attachment to the lacrimal bone, there creating a sharp ridge, the *posterior lacrimal crest* (*fig. 845*). Hence, operations on the sac are performed without invading the orbital cavity.

Condensation and thickening of the septum takes place in the lids, resulting in the formation of an upper and a lower *tarsus*. These plates are anchored to the orbital margin by the *medial* and *lateral palpebral ligaments*. The medial lig. is a strong band that crosses in front of the tear sac (*fig. 656*).

Muscles of Eyelids. The *sphincter* of the palpebral fissure is the *Orbicularis Oculi*. The fibers within the lids are the *palpebral portion*.

The *orbital portion* makes a complete circle from the medial palpebral ligament, having no lateral attachment. It is responsible for the "crow's foot" wrinkles.

[Some muscle fibers, the *Pars Lacrimalis* (Tensor Tarsi), are carried medially behind the tear sac (*fig. 657*).]

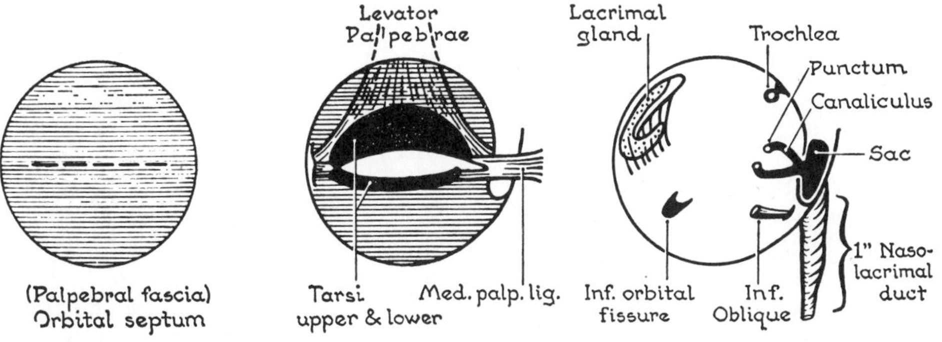

Fig. 656. (A) the orbital septum. (B) the tarsi, ligaments, and Levator Palpebrae. (C) features at the four corners of the orbital margin, and the tear apparatus (schematic).

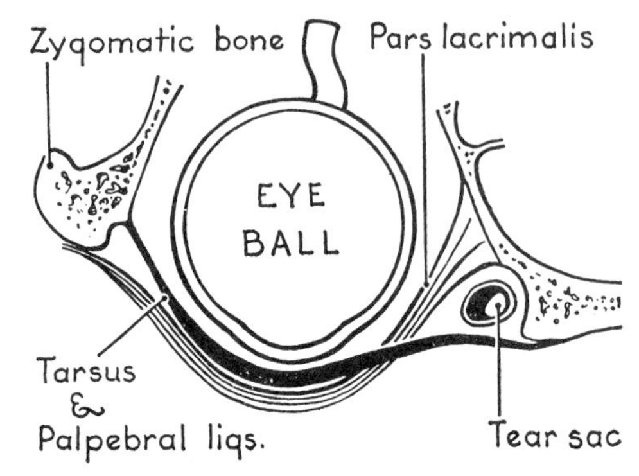

Fig. 657. Horizontal section of globe and upper lid and the Pars Lacrimalis. (After Whitnall.)

The "*dilator*" of the palpebral fissure, the *Levator Palpebrae Superioris* (p. 495) is not a facial muscle. Involuntary muscle fibers in both lids, the superior and inferior *tarsal muscles*, also widen the fissure (*fig. 698*).

Muscles of Forehead and Eyebrows. (*fig. 649*). The *Frontalis* (*fig. 661*) causes the transverse wrinkles on the forehead, associated with a surprised and supercilious look.

The *Corrugator Supercilii* causes the short vertical wrinkles in front of the glabella on frowning, etc. (*fig. 649*).

Lacrimal or Tear Apparatus (*figs. 656C, 703.1* on p. 500). The upper and lower *lacrimal canaliculi* are about 10 mm long and run near the free margin of the lid from *lacrimal punctum* to lacrimal sac. The *lacrimal sac* is the blind upper end of the *nasolacrimal duct* (*fig. 656C*), lying behind the medial palpebral lig. The *duct* runs downward in the bony *canal* to open into the inferior meatus of the nose (*fig. 799*).

SENSORY NERVES OF THE FACE AND COMPANION ARTERIES

The face develops from three rudiments, the *frontonasal* (unpaired), *maxillary*, and *mandibular processes*, each of which is supplied by one of the three divisions of the trigeminal, or 5th cranial, nerve (N. V). The divisions are named the ophthalmic, maxillary, and mandibular nerves; they will be indicated by the signs V^1, V^2, and V^3, hereafter.

The great auricular nerve (C. 2 and 3) encroaches on the face, sending branches across the parotid gland and the Masseter. The facial nerve mediates deep sensibilities. (see Nervus intermedius, page 597.)

The Ophthalmic Nerve (V^1). Of its five cutaneous branches, four are related to the upper lid; one to the nose (*fig. 658*).

The *supra-orbital branches* emerges through the supra-orbital notch or foramen. Its branches reach to the back of the skull. In contrast, the *lacrimal branch* is a mere twig in the lateral part of the lid.

The *supratrochlear* and *infratrochlear branches* are also small. They emerge above and below the trochlea for the Superior Oblique. The four preceding nerves supply the whole thickness of the upper lid. The cornea is supplied by important twigs that run to the back of the eyeball and will be discussed later (p. 499).

The *external nasal branch* emerges at the lower border of the nasal bone, which is easily palpated with the fingernail, and descends on the nasal cartilages to the tip of the nose.

Arteries. Branches of the ophthalmic art. from deep in the cranium accompany these five nerves (*fig. 703*).

The Maxillary Nerve (V^2). Of the 3 cutaneous branches of V^2, the **infra-orbital branch** emerges through the infra-orbital foramen. Its branches pass to the upper lip and to the mucous membrane of the cheek and upper gums, the skin and conjunctiva of the lower eyelid, and both the outside and the inside of the nostril.

Two twigs, the *zygomaticofacial* and *zygomaticotemporal* branches exit from the foramina with the same names. They supply the anterior part of the temporal region.

Arteries of the same names accompany all these nerves.

The Mandibular Nerve (V^3) (*fig. 658*) has several motor and 3 sensory branches. The motor nerves are discussed later (p. 529). The *mental branch* runs from the mental foramen to the skin and mucous mem-

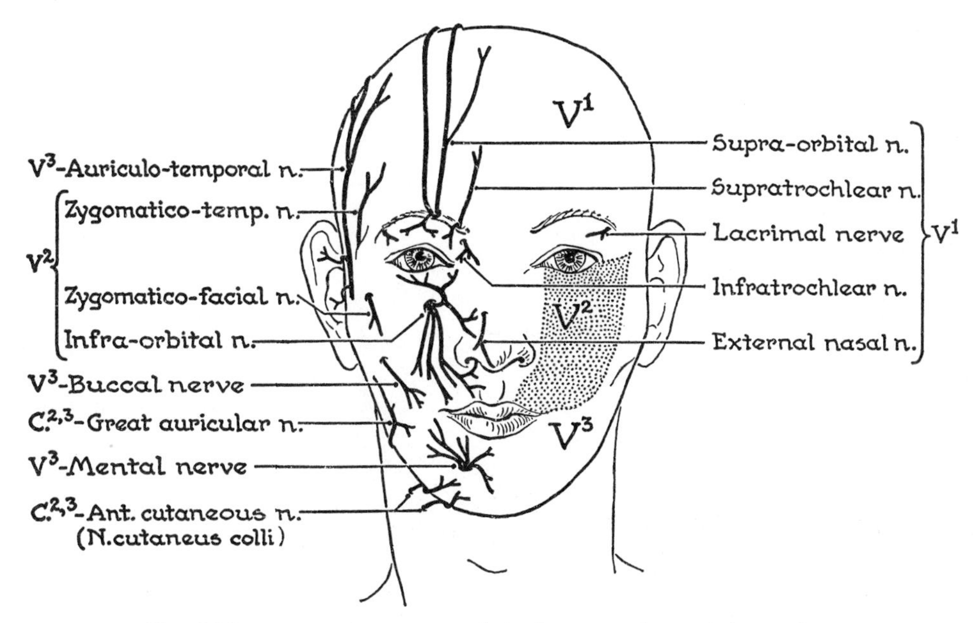

Fig. 658. The sensory nerves of the face and front of the scalp.

brane of the lower lip, the gums and the chin (*fig. 793*).

The *buccal branch* appears from the depths (lateral to Buccinator) and runs to the angle of the mouth. It supplies both the skin and the mucous membrane of the cheek, the latter by branches that pierce the Buccinator to supply the mucous membrane of the cheek.

The *auriculotemporal branch* crosses the zygoma just in front of the ear and ascends behind the superficial temporal artery. Its terminal distribution is to auricle and temporal region, but it also supplies the tympanic membrane, the external meatus, and the mandibular joint.

Arteries. Branches of the maxillary artery (mental and buccal) accompany the mental and buccal nerves.

MOTOR NERVE OF THE FACE

The facial, or 7th cranial, nerve (N. VII) supplies the muscles of the face, auricle, and scalp, the Platysma, and, to be seen later, the Digastric (post, belly), Stylohyoid, and Stapedius.

Its terminal branches appear at the margins of the parotid gland and fan out to supply the facial muscles (*fig. 659*). *Cervical branches* cross the angle of the jaw into the neck to supply the Platysma; one is vulnerable during minor surgery where it curves forward, crosses the jaw with the *facial artery* and helps to supply the muscles of the lower lip.

Loops and communications are common, but otherwise the arrangement is simple. Even the names (given below) have little practical significance.

The *temporal branches* cross the zygomatic arch and supply all the muscles above that level. The *zygomatic branches* pass forward above the parotid duct to supply the muscles of the infra-orbital region. The *buccal branch* supplies Buccinator and other muscles in the cheek. The *mandibular branch* supplies the muscles of the lower lip and chin.

The *posterior auricular branch* is the only branch to pass backward. If cut, it hardly would be missed.

The terminal branches of the facial nerve receive communications from the sensory branches of the trigeminal nerve on the face and from C. 2 and 3 about the neck and ear.

BLOOD SUPPLY TO THE FACE

This is very free and anastomoses are numerous. In addition to the many branches of the *ophthalmic* and *maxillary arteries* that accompany the various

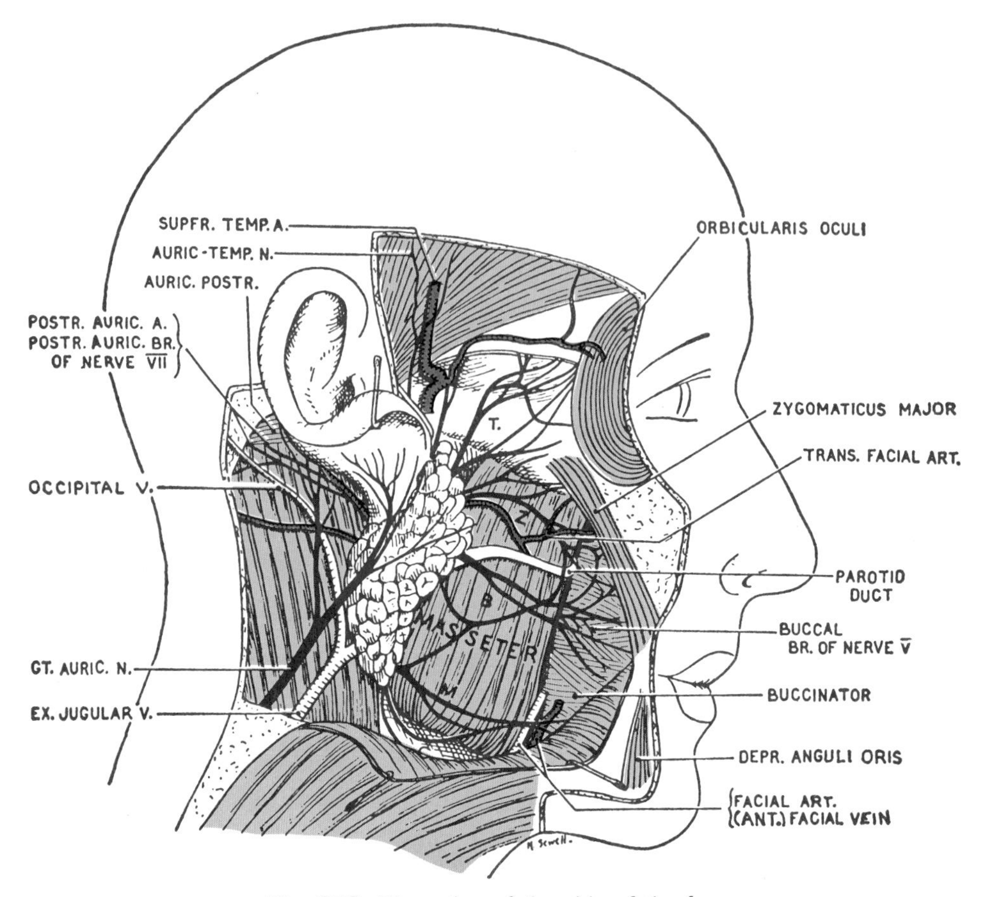

Fig. 659. Dissection of the side of the face.

branches of the 5th nerve onto the face, there are *facial* and *transverse facial arteries* (*fig. 659*).

Facial Artery (Ext. maxillary art.). The facial artery appears on the face at the base of the jaw, immediately in front of the Masseter. A cervical branch of the facial nerve enters the face with it; one or two lymph nodes lie beside it. In its sinuous course it passes near the angle of the mouth and it ends at the side of the nose by dividing into the angular and the lateral nasal arteries. Its vein is not a faithful companion (see below).

The facial artery crosses, in turn, the lower jaw, Buccinator, upper jaw, and Levator Anguli Oris (*fig. 649*). And, it is crossed superficially by all the other muscles it meets (viz., Platysma, Risorius, Zygomaticus Major and Minor, and Levator Labii Superioris).

Branches: An **inferior** and a **superior labial** art. encircle the mouth between the Orbicularis Oris and the layer of labial glands. A *septal branch* springs from the superior labial art.

A large *unnamed branch* runs on the jaw toward the mental foramen.

Small branches pass posteriorly.

The *lateral nasal art.* follows the upper border of the alar cartilage.

The *angular art.* ascends to the medial palpebral commissure and anastomoses with the dorsal nasal branch of the ophthalmic artery.

The *submental art.* arises in the submandibular region to enter the face near the chin.

The Superficial Temporal Artery is one of the two terminal branches of the external carotid artery (*fig. 712*). It arises deep to the parotid gland and, crossing the zygomatic, it ascends with the auriculotemporal nerve to divide into a **frontal** and a **parietal branch.** These are sinuous, and their pulsations may be visible. They enter the subcutaneous layer of the scalp (*fig. 662*).

BRANCHES: The *transverse facial art.* may be large (*fig. 659*).

Auricular branches anastomose on the ear.

The *middle temporal art.* pierces the temporal fascia and ascends in a groove on the temporal squama, supplying muscle and bone.

The Facial Vein (*fig. 664*) is the distant companion of the facial artery. It runs behind the artery in a straighter and more superficial course. It begins by the union of the supra-orbital and supratrochlear veins, and its tributaries correspond to the branches of the artery.

Its *Connections* are important (*fig. 664*): (1) through the *deep facial vein* with the *pterygoid plexus* of veins around the pterygoid muscles; these in turn communicate through the foramina at the base of the skull with the cavernous sinus; (2) through the *superior ophthalmic vein* with the *cavernous sinus* bringing the supra-orbital and supratrochlear veins into communication with the cavernous sinus; blood can flow in both directions because valves are lacking; (3) with the *frontal diploic vein*. This vein emerges from a pinpoint hole in the supra-orbital notch.

SKULL FROM ABOVE AND SCALP

Three **sutures** related to three sides of the parietal bone are visible: the *coronal* (frontoparietal), the *sagittal* (or interparietal), and the *lambdoid* (parieto-occipital) (*fig. 660*). The point of intersection of sagittal and coronal sutures is the *bregma*. The point of intersection of sagittal and lambdoid sutures is the *lambda* (= the Greek letter λ).

The bones of the roof of the skull develop in membrane. **Ossification** of the frontal and parietal bones begins during the 2nd fetal month at their points of greatest fullness, called the *frontal* and *parietal tubers* (eminences).

At birth, ossification has not reached any of the four corners of the parietal bones; so, at these sites, called **fonticuli** (*fontanelles*), the brain is covered with membrane. Of these, the *anterior* fonticulus lasts 1½ years; at birth it is 3–5 cm in diameter and is shaped like a flat kite (*fig. 660*) the long angle tapering forward. It provides the clue to an obstetrician's palpating fingers that is needed to identify the position of rotation of the child's head in the birth canal. It is obliterated before the end of the 2nd year—if not, there is something amiss. The bregma marks the site.

Cephalic Index. Viewed from above, a skull is roughly oval in outline, but it may have one of many shapes (*fig. 660*). When the maximum width of a skull is less than 75 per cent of its maximum length it is said to be dolichocephalic or long headed; when more than 80 per cent it is called brachycephalic or broad headed; when between 75 and 80 per cent it is mesaticephalic.

Scalp

The scalp is regarded as a unit composed of skin, dense subcutaneous tissue, and Epicranius. All three are firmly bound

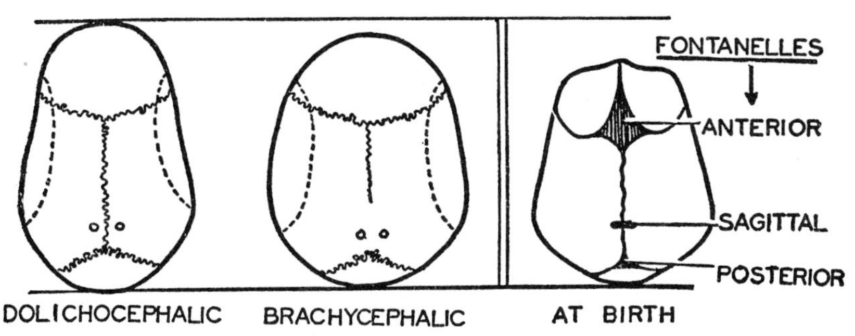

Fig. 660. Skulls viewed from above. Fonticuli or fontanelles.

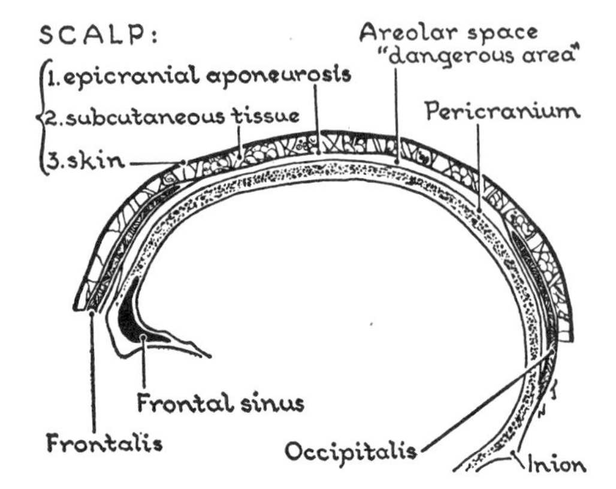

Fig. 661. Sagittal section of skull cap and overlying tissues. Note the fibrous bands, fat, vessels, and nerves in the scalp.

together and are separated from the periosteum of the skull by a very loose areolar space, "the dangerous area"—dangerous if infection gets in (*fig. 661*).

Phylogentically, the Epicranius was a continuous sheet of muscle, now aponeurotic except at its anterior and posterior ends. Its posterior muscular end, the *Occipitalis*, is attached close above the superior nuchal line. Its anterior end, the *Frontalis*, extends right across the forehead blending with its fellow in the median plane and interdigitating in front with the Orbicularis Oculi.

The Occipitalis has a fixed bony attachment, so its contraction draws the scalp backward. Contraction of the Frontalis causes horizontal wrinkles on the forehead as it lifts the eyebrows but not the lids. Just above the eyebrows, fascial connections with the bone close off the spread of fluid within the "dangerous area."

In the subcutaneous tissue, fascial fibers criss-cross in all directions, uniting skin to Epicranius. Between the fibers fat is imprisoned, and there the vessels and nerves run.

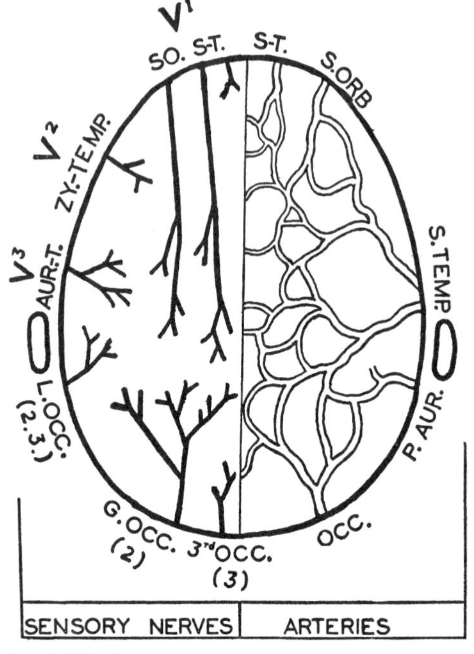

Fig. 662. Sensory nerves and arteries of scalp.

Nerves (*fig. 662*). The sequence of the cutaneous nerves in the scalp is orderly from front to back—V^1, V^2, V^3; ventral rami of cervical nerves 2 and 3; and dorsal rami of cervical nerves 2 and 3.

Arteries (*fig. 662*). The arteries anastomose freely. They are derived either indirectly from the *internal carotid* through its ophthalmic branch (viz., supratrochlear and supra-orbital) or directly from the *external carotid* (viz., superf. temporal, post. auricular and occipital).

35 POSTERIOR TRIANGLE OF NECK

Boundaries; Sternomastoid; Trapezius.
Investing Fascia; Accessory Nerve; Cervical Nerves 3 and 4; Omohyoid; External Jugular Vein; Floor of Triangle; Fascial Carpet; Lymph Nodes.

BLOOD VESSELS IN TRIANGLE

Subclavian Vein and Artery and Branches; Other Arteries in Triangle.

NERVES IN POSTERIOR TRIANGLE

Brachial Plexus.
Nerves to Four Muscles—Levator Scapulae; Rhomboids; Serratus Anterior; Diaphragm (Phrenic Nerve).
Cutaneous Nerves.

The posterior triangle of the neck (a "political" subdivision) is really the root of the upper limb, the clavicle being an artificial border (*fig. 59*). You cannot fully understand one without the other.

Boundaries

The middle third of the clavicle is the base of the triangle; the posterior border of the Sternomastoid is the anterior side; the anterior border of the Trapezius is the posterior side; the point where these two muscles meet on the superior nuchal line is the apex.

Sterno(cleido)mastoid and **Trapezius** split from one sheet of embryonic muscle. Above, these two muscles have a continuous attachment extending from the mastoid process (feel it behind your ear) to the inion in the midline (*fig. 663*). This attachment is aponeurotic, so it produces a ridge, the *superior nuchal* line.

Below, the two muscles have a discontinuous attachment to the clavicle. A part of the Sternomastoid crosses the sternoclavicular joint to the sternum.

Action. Flex your head and neck against resistance, and you bring into action both Sternomastoids. If now you rotate your bent head and neck, so that you look sideways up, the Sternomastoid of the opposite side will become still more prominent.

Trapezius (see p. 91 and *fig. 77*).

The Platysma is superficial to the lower part of the triangle (p. 506).

The Investing Deep Fascia, which encircles the neck, covers the posterior triangle, splits to envelop the Sternomastoid and Trapezius, and attaches to the clavicle. Embedded in it is the *accessory nerve*.

Contents

Accessory Nerve, External Branch (N. XI). The accessory nerve is the nerve to the Sternomastoid and Trapezius, and it spans the gap between them. It is quite superficial—embedded in the investing fascia. It divides the triangle into two nearly equal parts: in the part above the nerve you may explore *carefree* for there is no important structure to damage; but below you must be *careful*.

This part of the accessory nerve has a curious origin: arising from the upper five or six segments of the spinal cord (hence its old name, "spinal accessory n."), it enters the cranial cavity by the foramen magnum only to leave again by the foramen jugulare (*fig. 764*, p. 548).

In the neck it passes obliquely downward

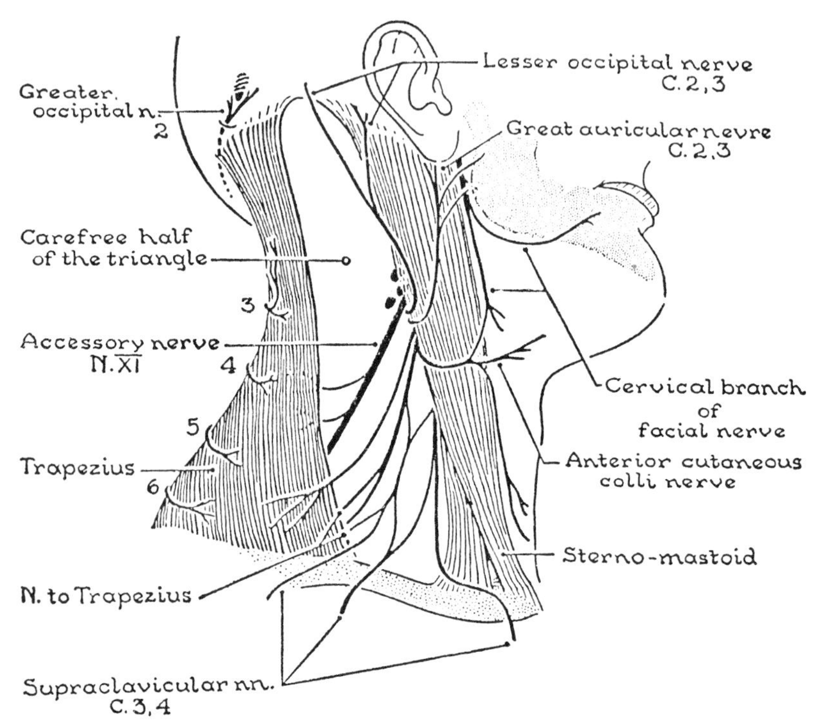

Fig. 663. The superficial nerves of the neck. Of these, the facial and accessory are motor. (*Nerves of neck: anterior cutaneous = transverse.*)

and backward from the transverse process of the atlas to the superior angle of the scapula (*fig. 77*). En route, it burrows through the Sternomastoid a short distance below the mastoid process. Here lymph nodes surround it.

It crosses the posterior triangle and disappears under cover of the Trapezius, just above the clavicle. Sensory (and possibly motor) twigs from C. 2, 3, and 4 join it.

Branches of Cervical Nerves 3 and 4 running below the accessory nerve are apt to be mistaken for it. Some also pass deep to the Trapezius and enter it (p. 91); others, *lateral supraclavicular nerves*, pass superficial to it (*fig. 663*).

(See *Supraclavicular Nerves*, p. 468.)

Omohyoid (*fig. 664*). The inferior belly of this straplike muscle passes from the anterior to the posterior triangle and gains attachment to the upper border of the scapula. It runs one or two fingers' breadth above the clavicle to which it is bound by an inverted sling of fascia.

External Jugular Vein (*fig. 664*). This large vein descends subcutaneously across the Sternomastoid, pierces the deep fascia (and the fascial sling of Omohyoid) to end in the subclavian vein.

Tributaries in Triangle. The *transversa colli*, *suprascapular*, and *anterior jugular veins* communicate with each other and end in the external jugular vein.

Floor of Triangle. The "floor" is formed by several muscles whose fibers run obliquely downward and backward (*figs. 664, 666*). *Levator Scapulae* occupies a middle position deep to the accessory nerve and parallel to it.

Above the Levator lies the *Splenius*. Below, are the three Scaleni (*fig. 749*) which arise from transverse processes and insert into ribs 1 and 2.

Levator Scapulae arises from the posterior tubercles of the transverse processes of C. 1—4 vertebrae, and it is inserted into the medial border of the scapula at the superior angle.

The *Scalenus Medius* arises from all the cervical transverse processes (Cave); and it is

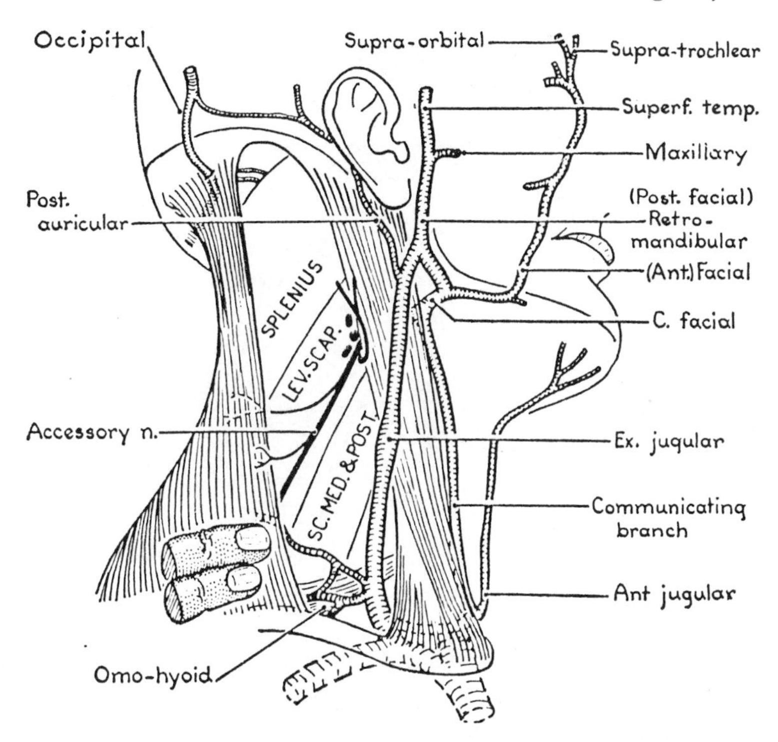

Fig. 664. The superficial veins of the face and neck.

inserted into the upper surface of the 1st rib between its neck and the groove for the subclavian artery. The *Scalenus Posterior* may be considered as the portion of the Medius that passes on to the 2nd rib.

The *Scalenus Anterior* arises from the anterior tubercles of all the transverse processes that have them (3, 4, 5, and 6). It runs downward to a tendinous insertion into the scalene tubercle, situated on the 1st rib between the grooves for the subclavian artery and vein.

A small part of the *Semispinalis Capitis* may appear at the apex of the triangle, above the Splenius (*fig. 666*).

Fascial Carpet. The muscular floor of the triangle is carpeted with a layer of fascia. This fascia is part of the fascia that envelops the vertebral column and the pre- and postvertebral muscles of the neck.

The carpet covers the subclavian vessels and the roots of the brachial plexus and it provides them with a covering, the *axillary sheath*, as they enter the axilla (*fig. 665*). It also covers and guards the motor nerves to four important muscles—Levator Scapulae, Rhomboids, Serratus Anterior, and Diaphragm.

Lymph Nodes. In the fat between the fascial roof and the fascial carpet, lie the *lateral group of inferior deep cervical* (supraclavicular) *lymph nodes*. They drain the back of the scalp and neck and they usually receive some efferent vessels from the upper deep cervical, axillary, and deltopectoral nodes. These nodes empty into the jugular lymph trunk. The axillary nodes empty into the subclavian lymph trunk, which follows the subclavian vein to the thoracic (or right lymph) duct.

BLOOD VESSELS IN POSTERIOR TRIANGLE

The Subclavian Vein, the continuation of the axillary, begins at the lateral border of the 1st rib and ends medial to the Scalenus Anterior. It joins the internal jugular vein to form the brachiocephalic vein. The vein is separated from its artery by the Scalenus Anterior; both vessels groove the 1st rib. The rib is obliquely set, so the vein lies antero-inferior to the artery, and behind the clavicle.

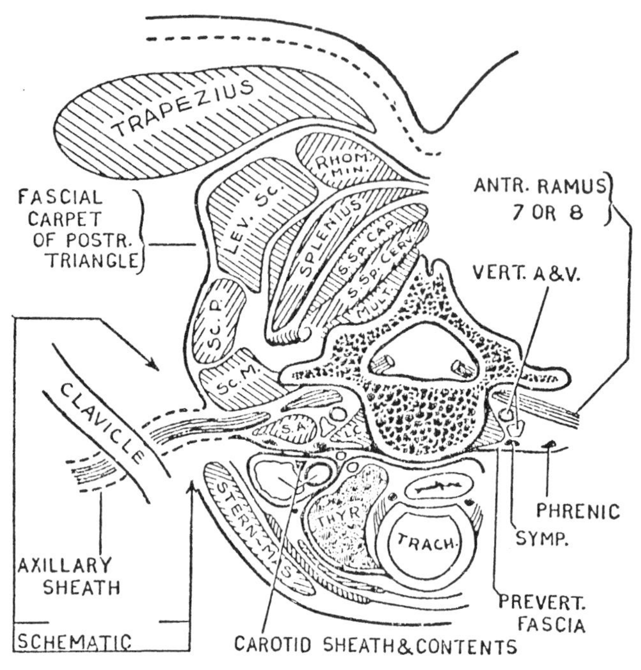

Fig. 665. Transverse section of the neck showing the pre- and postvertebral fasciae prolonged into the axilla as a tubular covering (axillary sheath) for the brachial plexus and subclavian vessels. The nerves shown in figure 666 remain deep to this fascia.

Tributaries. Ext. jugular vein and (variable) others (*figs. 664, 726, 759*).

The Subclavian Artery becomes the axillary artery. Its only branch in the posterior triangle is the dorsal scapular a. The branches from the subclavian a. usually come from its 2nd part (behind the Scalenus Anterior) and 1st part (medial to that muscle). Its relations:

Below—1st rib and the pleura.

Behind—Scalenus Medius and the lowest trunk of the brachial plexus.

In front—Platysma, cutaneous nerves, three layers of fascia, and a plexus of veins; the clavicle and the suprascapular artery.

Antero-inferiorly—subclavian vein.

Branch. Dorsal scapular art. usually.

Other Arteries in the Posterior Triangle. The *suprascapular* and *transversa colli arteries*, which spring from the 1st part of the subclavian art., via the thyrocervical trunk (*fig. 725*), pass laterally in front of the Scalenus Anterior and its fascia and the phrenic nerve. The following *details* should *not* be memorized except by professional anatomists.

The *Suprascapular Art.* runs a retroclavicular course to the (supra)scapular notch.

The *Transversa Colli* (*Cervicis*) *Art.*, lying at a higher level, crosses the floor of the triangle and accompanies the accessory nerve to the Trapezius. It commonly (30 per cent) gives off the dorsal scapular art.

The *Dorsal Scapular Art.* usually (67 per cent; Huelke) springs from the 2nd or 3rd part of the subclavian art. and follows the nerve to the Rhomboids [dorsal scapular n.] deep to the Levator Scapulae. Sometimes (20 per cent) it gives off the transversa colli art.

The *Occipital Art.* flits across the apex of the triangle.

NERVES IN POSTERIOR TRIANGLE

Terminology. In the thoracic, lumbar, and sacral regions the spinal nerves are named numerically after the vertebrae above them. In the cervical region, however, they are named after the vertebrae below them. The nerve between vertebrae C. 7 and Th. 1 is grouped with the cervical nerves, making it C. 8.

The Transverse Processes of the cervical

vertebrae have a downward tilt due to the pull of muscles, and a groove for the ventral ramus around which the process is molded (*fig. 668*). Each possesses a posterior tubercle. Anterior tubercles are present only on the 3rd, 4th, 5th, and 6th.

Brachial Plexus

This plexus (*fig. 70*) is formed by the ventral rami of nerves C. 5, 6, 7, 8 and Th. 1. The ventral ramus just beyond each end of the series, namely, C. 4 and Th. 2, commonly contributes branches.

Summary Review: The plexus is composed as follows (*fig. 70*): rami 5 and 6 unite to form an uppermost trunk; ramus C. 7 continues as a middle trunk; rami C. 8 and Th. 1 unite at the neck of the 1st rib to form a lowest trunk. The lowest trunk lies on the 1st rib behind the subclavian artery.

Each of the three trunks divides into an anterior and posterior division. Behind the clavicle the three posterior divisions unite to form the posterior cord of the plexus, whereas the upper and middle anterior divisions unite to form the lateral cord, and the lowest anterior division continues as the medial cord.

The roots of the plexus supply the Rhomboids, Serratus Anterior, and Diaphragm. They also supply the neighboring prevertebral muscles. They receive gray rami from the middle and inferior cervical and Th. 1 and 2 sympathetic ganglia (*fig. 44.4*).

If you follow along the lateral border of the plexus, you will be led away by the most lateral branch of the plexus, the *suprascapular nerve*, to the (supra)scapular notch of the scapula.

Nerves to Four Muscles—Levator Scapulae (C. 3, 4), Rhomboidei (C. 5), Serratus Anterior (C. 5, 6, 7), and Diaphragm (C. 3, 4, 5)—lie deep to the fascial carpet (*fig. 666*).

The Phrenic Nerve is not within the limits of the triangle, nor is the internal jugular vein, but they are so close that it would be

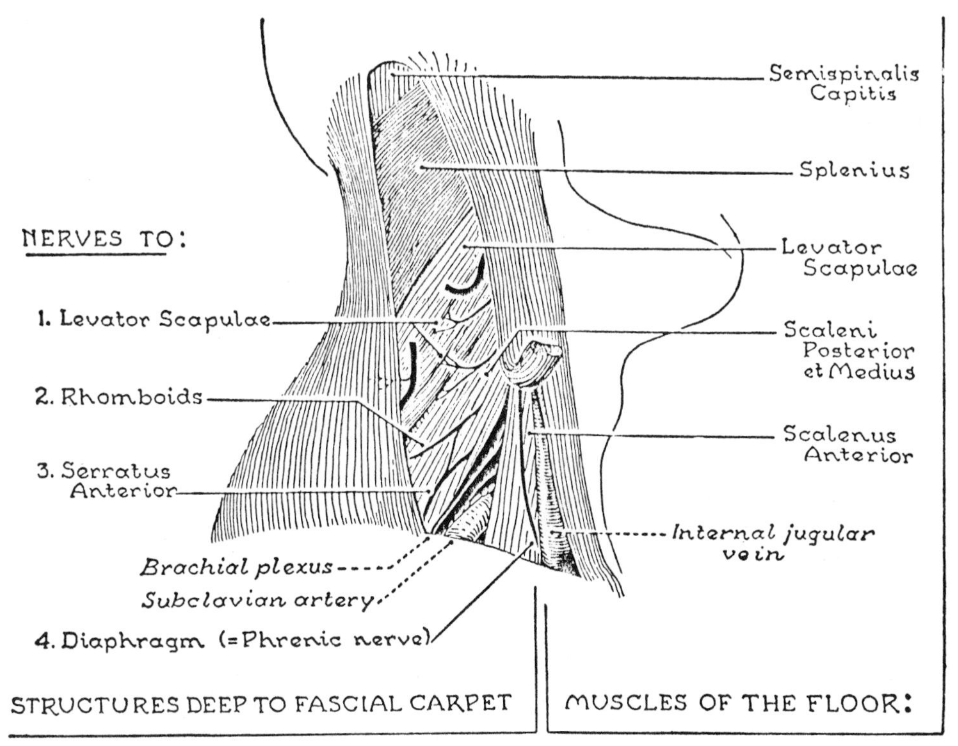

Fig. 666. Floor of the posterior triangle and nerves that have no occasion to pierce its fascial carpet.

an omission not to mention them.

The phrenic nerve arises from the ventral ramus of C. 4 and gets twigs from C. 3 and 5. It descends vertically on the Scalenus Anterior, reaching the medial border at its insertion on the 1st rib.

Nerve to Subclavius (C. 5 and 6), otherwise insignificant, may deliver to the phrenic nerve its root (a twig) from C. 5.

Cutaneous Nerves. Nerve C. 1 has no cutaneous branch and the ventral rami of C. 5—Th. 1 form the brachial plexus. Thus C. 2, 3, and 4 supply the cutaneous territory between the trigeminal nerve above and the 2nd thoracic nerve below (*fig. 98*). This they do by means of four nerves that radiate from about the middle of the posterior border of the Sternomastoid (*fig. 663*):

1. *The Lesser Occipital N.* (C. 2, 3) hooks round the accessory nerve and ascends near the Sternomastoid to supply the scalp.
2. *The Great Auricular N.* (C. 2, 3) runs vertically across the Sternomastoid to the lobe of the ear, with the external jugular vein. It has *mastoid, auricular*, and *facial branches*.
3. *The Anterior Cutaneous Colli N.* [N. transversus colli] (C. 2, 3) crosses the Sternomastoid and ext. jugular vein, and supplies the skin between jaw and sternum.
4. *The Supraclavicular Nerves* (C. 3, 4) are three sensory branches to the lower part of the posterior triangle and upper chest wall down to the level of the 2nd rib.

36 BACK

SKELETAL PARTS

Skull from Behind
Outline; Surface; Nuchal Area.
Vertebral Column from Behind.
Spinous Processes; Ligamentum Nuchae; Laminae; Articular Processes; Transverse Processes; Dorsal Sacral Foramina.

Ribs. Iliac Crest

MUSCLES OF BACK

Organization; Serrati Posteriores; Splenius; Thoracolumbar Fascia.
Deep or Intrinsic Muscles.
Erector Spinae; Transversospinalis; Interspinales; Intertransversarii.

SUBOCCIPITAL REGION

Bony Limits; Contents; Muscles.
Nerves and Arteries; Suboccipital Triangle.

Actions of Deep Muscles of Back

NERVES OF THE BACK

(Dorsal Rami of Spinal Nerves)

Sacral and Coccygeal Nerves.

ARTERIES OF BACK

Vertebral; Occipital; Profunda Cervicis.

VEINS OF BACK

(Vertebral Venous System)

Clinical Significance.

SKELETAL PARTS

Skull from Behind

The **Outline** is horseshoe-shaped and extends from one mastoid process over the vault to the other (*fig. 667*). On each side it crosses the mastoid and the parietal bone. Across the base of the skull the outline is nearly horizontal, crossing the occipital condyle and the foramen magnum.

The **Surface** is convex and includes parts of the parietal, occipital, and temporal bones. At the center is the *lambda*. From it, the sagittal (interparietal) suture runs up and over the vertex, and the lambdoid (parieto-occipital) suture down to the postero-inferior corner of the parietal bone on each side.

Midway between the lambda and the foramen magnum is the *inion* or *external occipital protuberance*. From it the *superior nuchal line* curves to the rough outer surface of the *mastoid process*. All three are for Trapezius and Sternomastoid. The surface below the superior nuchal line is

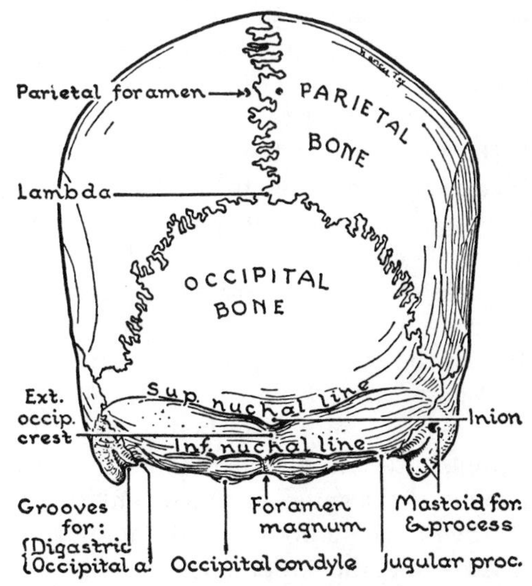

Fig. 667. The skull, from behind. (Norma occipitalis of Anthropologists)

the **nuchal area** (nucha = neck); it is subdivided by the inferior nuchal line.

The *external occipital crest* is a sharp median crest that gives attachment to the *ligamentum nuchae* and runs from the inion to the foramen magnum.

Foramina. On each side there are three *foramina* for emissary veins: (1) *parietal foramen*, (2) the *mastoid foramen*, and (3) the *condylar canal*, which opens onto the condylar fossa.

Vertebral Column from Behind

Spinous Processes. C. 1 (the atlas) has only a posterior tubercle; C. 2–C. 6 spines are bifid in white races; C. 7 is prominent but less so than Th. 1; Th. 5–8 are almost perpendicular, and are markedly overlapping; Th. 10 is often disproportionately small; the lumbar spines are oblong plates; the upper three or four sacral spines form an irregular *median* crest.

Supraspinous and interspinous ligaments unite the spines. In the neck the supraspinous ligament forms the **ligamentum nuchae,** which has an anterior border attached to the cervical spines; its superior border is attached to the inion and external occipital crest; its posterior is subcutaneous.

The Laminae are band-like and tend to overlap the laminae below like shingles. They are united to each other by *ligamenta flava*. Their inferolateral angles project down as the inferior articular processes.

Above and below the slender posterior arch of the atlas there is a wide interlaminar space closed by membranes (posterior atlanto-occipital and atlanto-axial). The lumbar interlaminar spaces also are wide. Flexion of the spine, of course, enlarges all of them. The upper sacral laminae are fused while the lower ones and the coccygeal laminae are absent (*fig. 669*).

The Articular Processes for the atlanto-occipital and atlanto-axial joints lie on a plane anterior to the subsequent articular processes which are not serially homologous with them (*fig. 675.1*). The subsequent cervical articular processes are segments of a column cut obliquely (*fig. 668*).

The Transverse Processes of C. 1 and C. 7 project far beyond those of C. 2–6 (*fig. 748*). Th. 1–12 diminish progressively in projection. L. 1–5 project farther than C. 1, L. 3 being the most projecting of all transverse processes.

The sacral transverse and articular processes fuse to form irregular crests.

The Dorsal Sacral Foramina (*fig. 669*) communicate with the sacral canal.

Ribs and Iliac Crest

The distance between the angles of the ribs and the tubercles diminishes from below upward until they meet on the 1st rib. The angles are marked by the attachment of the Iliocostalis.

A lumbar transverse process is morphologically a costal or rib element (*fig. 23.1*); a small tubercle, the *accessory process*, situated behind the root of a lumbar transverse process, is in series with the thoracic transverse processes.

On the back of the rim of each lumbar superior articular process there is a large tubercle, the *mamillary process*. It gives origin to the Multifidus.

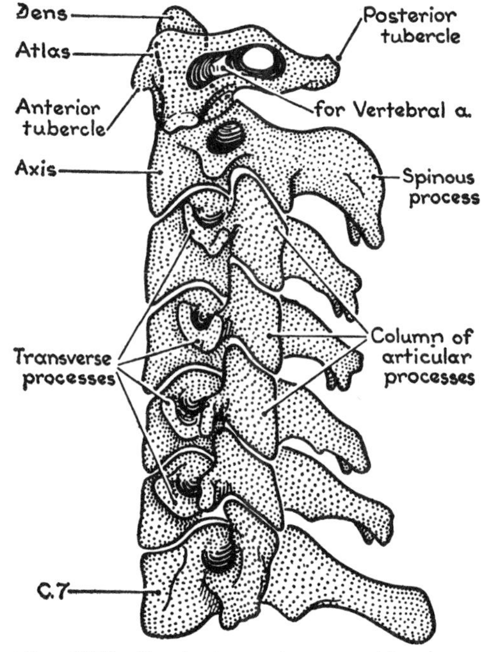

Fig. 668. Cervical vertebrae on side view.

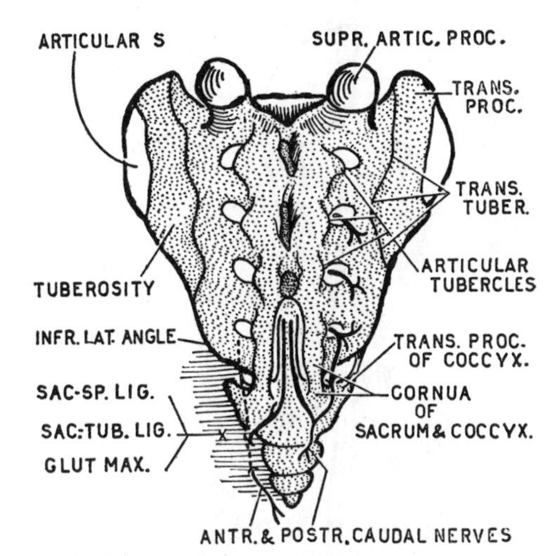

Fig. 669. Dorsal aspect of sacrum and coccyx.

Iliac Crest. The hinder 8 cm of the iliac crest gives part origin to the Erector Spinae.

MUSCLES OF THE BACK

The muscles of the back are arranged in superficial, intermediate, and deep groups. The *superficial group* acts upon the upper limb and has been described on pages 90 to 92. The *intermediate group* is respiratory in action; and both have migrated or spread backward across the deep group of muscles. The *deep group* is intrinsic to the back; it is composed of "native" muscles supplied by dorsal nerve rami (*figs. 670, 671*).

Superficial

1. Trapezius and Latissimus Dorsi.
2. Levator Scapulae and Rhomboidei.

Intermediate

3. Serrati Posteriores (superior et inferior) (*fig. 670*).

Deep

4. Splenius (cervicis et capitis).
5. *Longitudinal muscles*
   Erector Spinae:
Iliocostalis, Longissimus, Spinalis.
6. *Oblique muscles*
   Transversospinalis:
Semispinalis, Multifidus, Rotatores.
7. *Remaining deep muscles:*
Interspinales, Intertransversarii, Levatores Costarum, Suboccipital muscles.

It is **profitless to memorize** even for a short period the textbook description of the attachments of the deep muscles of the back because you can make *no use of the particular information*. The back is, however, a region of such great importance that you should not neglect to obtain a grasp of the scheme of things.

Intermediate Muscles and Splenius

The deep muscles are bridged by the *Serrati* in the thoracic region, by the *lumbar fascia* in the lumbosacral region, and by the *Splenius* in the neck.

The Serratus Posterior (*fig. 670*) is divided into the *Serratus Posterior Superior* and the *Serratus Posterior Inferior*. The latter blends with the thoracolumbar fascia and through it gains attachment to the lower thoracic and upper lumbar spines.

Both parts of this muscle elongate the thoracic cavity and may act as muscles of inspiration. They are migrants, still supplied by ventral nerve rami.

The Splenius (*figs. 666, 670*) is wrapped around the other deep muscles in the neck, as its name implies (splenius L. = a bandage). It arises from the lower half of the ligamentum nuchae and from the upper thoracic spines. It separates into two parts—Splenius Cervicis and Splenius Capitis.

The Splenius Cervicis joins the Levator Scapulae to share its attachments to transverse processes C. 1–4. *The Splenius Capitis* shares the attachments of Sternomastoid to the superior nuchal line and mastoid process.

Thoracolumbar Fascia

The thoracolumbar fascia has two parts: thoracic and lumbar.

The Lumbar Fascia is the dorsal aponeurosis of the Transversus Abdominis (*fig. 215*); it splits to form a superficial and a

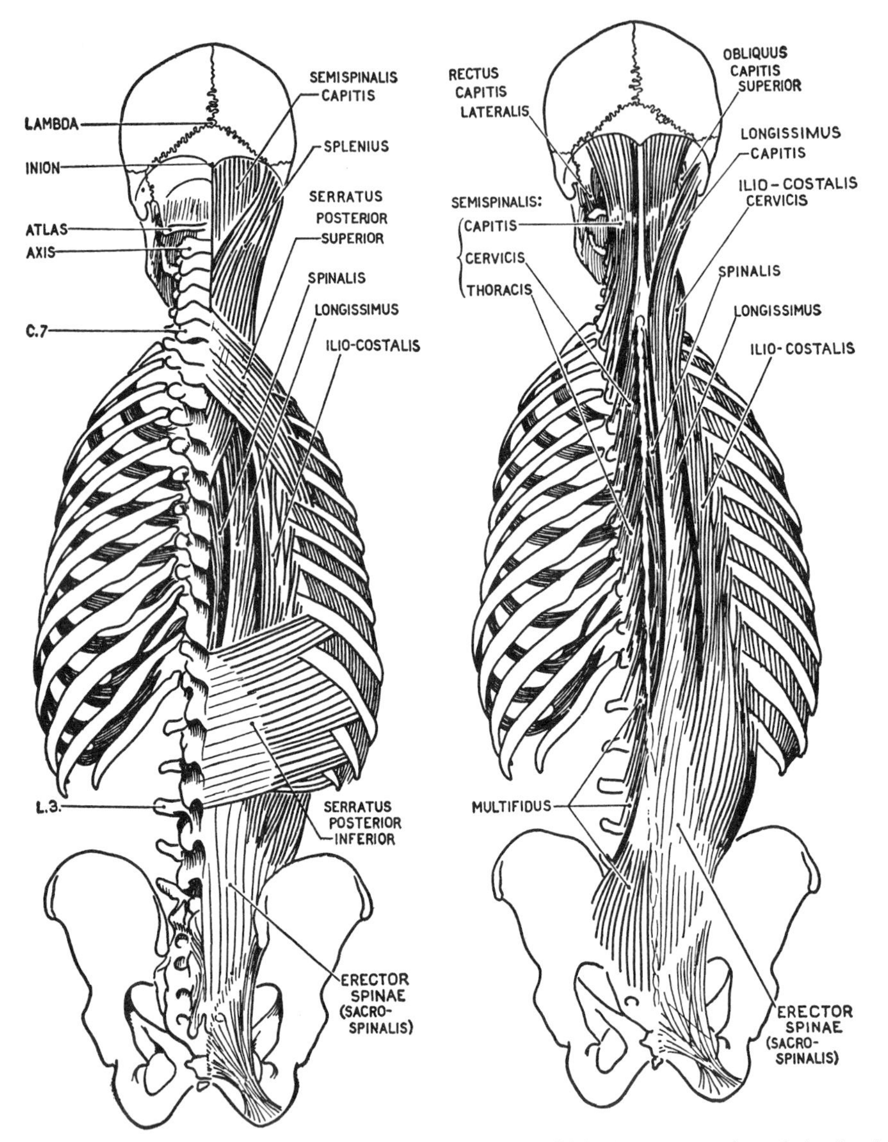

Fig. 670. Intermediate muscles of the back.

Fig. 671. Deep muscles of the back.

deep layer. *The superficial layer* attaches to the lumbar spines. *The deep layer* attaches to the tips of the lumbar transverse processes.

Traced upward, as the **thoracic part** of *the thoracolumbar fascia*, it becomes a delicate sheet that stretches from the vertebral spines to the angles of the ribs, spanning the deep dorsal muscles.

Deep or Intrinsic Muscles

Erector Spinae (Sacrospinalis) (*fig. 672*). This muscle extends from the pelvis to the skull. It has: (1) a dense aponeurotic origin which covers the fleshy origin of the Multifidus and arises as shown in *figure 672.1.*

A little below the last rib the Erector Spinae splits into 3 columns: Iliocostalis, Longissimus, and Spinalis (*fig. 672.1*).

Details of the crowded muscular attachments to cervical transverse and articular processes are shown in *figures 673* and *674.*

The Iliocostalis is inserted into the angles of the ribs and into the cervical transverse processes (C. 4–6) by a series of

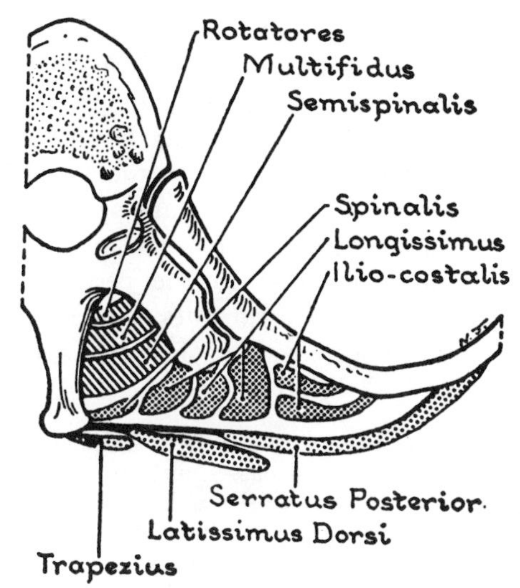

Fig. 672. Muscles of back on cross section.

relayed bundles that extend over about six segments—where one slip is inserted another slip arises on its medial side.

The Longissimus (thoracis, cervicis, and capitis) is inserted into lumbar accessory and transverse processes, and into thoracic transverse processes and nearby parts of the ribs (Th. 2–12). Bundles arising medial to these (Th. 1–4) are relayed to the cervical transverse processes (C. 2–6). Other bundles arising medial to these again (upper Th. transverse and lower C. articular processes) extend as a broad fleshy band to be attached to the mastoid process deep to the Splenius Capitis and Sternomastoid.

The Longissimus is the only column of the Erector Spinae to reach the skull.

The Spinalis is largely aponeurotic, flat, and narrow. It extends from the upper lumbar to the lower cervical spines.

Transversopinalis. This oblique group of muscles is concealed by the Erector Spinae. Its fibers pass obliquely upward and medially from transverse processes to spines. It is disposed in three layers: (1) *Semispinalis*, (2) *Multifidus*, and (3) *Rotatores*. The superficial layer (a) spans more segments than the intermediate layer (b), and therefore takes a more vertical course, and (c) arises nearer the tips of the transverse processes and is inserted nearer the tips of the spinous processes.

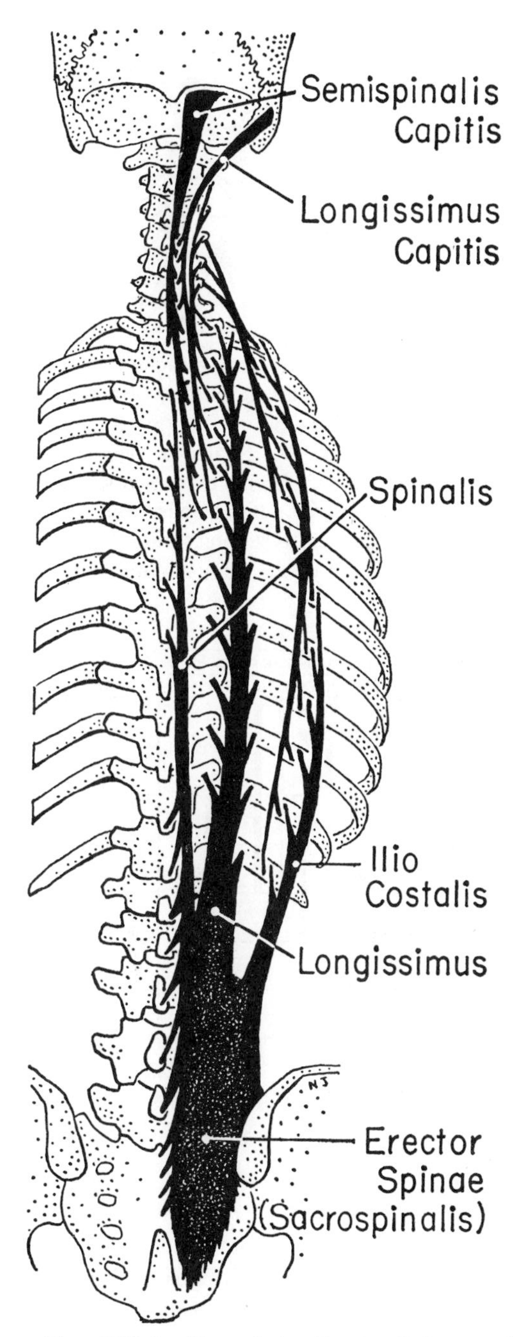

Fig. 672.1. Plan of the Erector Spinae and Semispinalis Capitis. (After Cunningham's Anatomy.)

The superficial layer spans about five segments; the intermediate about three;

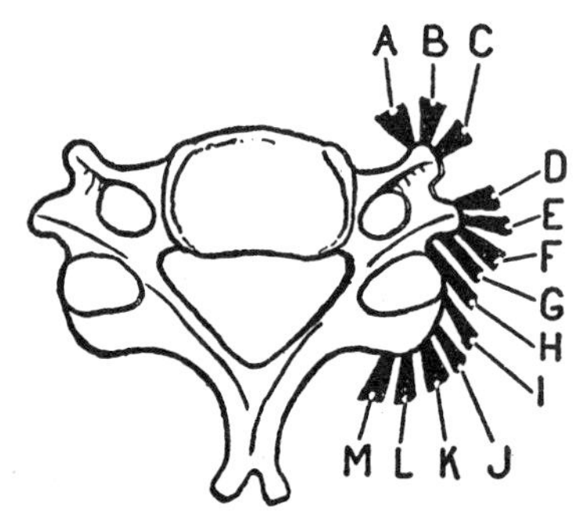

Fig. 673. The muscles attached to the transverse and articular processes of a cervical vertebra. *A*, Longus Colli. *B*, Longus Capitis. *C*, Scalenus Anterior. *D*, Scalenus Medius. *E*, Scalenus Posterior. *F*, Levator Scapulae. *G*, Splenius Cervicis. *H*, Iliocostalis Cervicis. *I*, Longissimus Cervicis. *J*, Longissimus Capitis. *K*, Semispinalis Capitis. *L*, Semispinalis Cervicis. *M*, Multifidus.

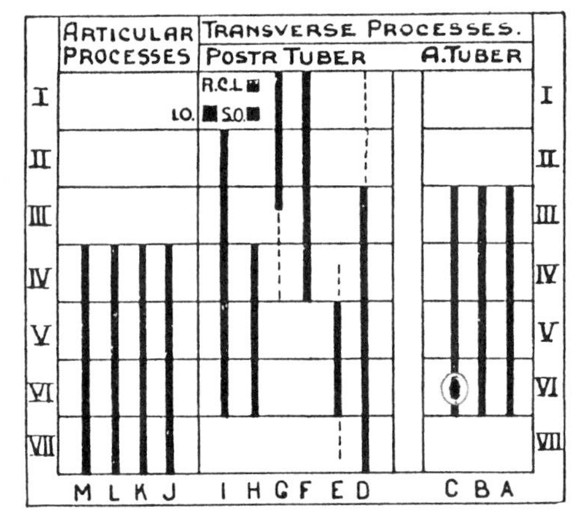

Fig. 674. Graphic representation of *figure 673.*

the deepest connects adjacent segments.

The Semispinalis (thoracis, cervicis, capitis) forms the superficial layer. *The Semispinalis Capitis* passes from the upper thoracic transverse processes and lower cervical articular processes (C. 4—Th. 5) to the occipital bone between the superior and inferior nuchal lines. The fibers of this massive muscle in the nape of the neck run nearly vertically. The medial border is free and is separated from its fellow by the lig. nuchae.

The Multifidus arises as a thick, fleshy mass from the dorsal aspect of the sacrum between its spinous and transverse crests, from adjacent ligaments, from the dense aponeurosis of the overlying Erector Spinae, and from all transverse processes up to C. 4 (actually from mamillary processes in the lumbar region, transverse in the thoracic, and articular in the cervical). It spans about three segments to be inserted into the lower border of every spinous process (C. 2—L. 5).

The Rotatores bridge one interspace. They are small slips that pass from the root of one transverse process to the root of the spinous process or lamina next above. They are best marked in the thoracic region.

OTHER DEEP MUSCLES. The *Interspinales* and *Intertransversarii* are well developed in the cervical and lumbar regions, but mostly absent from the thoracic region. The *Levatores Costarum*, though hidden by the Erector Spinae, are grouped with the thoracic muscles (p. 397).

Interspinales are well developed median paired muscles. They unite the bifid tubercles of adjacent cervical spinous processes, and adjacent borders of the oblong lumbar spinous processes.

Intertransversarii (Anteriores and Posteriores) unite adjacent anterior tubercles and adjacent posterior tubercles of the cervical transverse processes; the highest Posterior Intertransverse muscle being the Rectus Capitis Lateralis (p. 535). In the lumbar region they are well marked.

Suboccipital Muscles, see below.

Actions of Deep Muscles of Back. The deep dorsal muscles acting together *extend* the vertebral joints. They also prevent, or regulate, *flexion* of these parts. Much, however, of the apparent extensor and flexor movement of the vertebral column takes place actually at the hip joints. When the deep muscles of one side act, *lateral bending* and *rotation* occur, the oblique muscles being the chief rotators along with oblique muscles of the abdomen (see Functions etc., p. 181).

SUBOCCIPITAL REGION

The large dorsal muscles ascend as far as the spine of the axis; others span the atlas to reach the skull; deep to them is the suboccipital region (*fig. 675*). Thus it lies at the apex of the posterior triangle (*fig. 666*) and extends on both sides of it. The Semispinalis Capitis forms its main lid, aided by the Splenius.

Bony Features of the Region (*fig. 675*). *Above*, inferior nuchal line of occipital bone; *below*, axis; *laterally*, mastoid process, transverse process of atlas, and transverse process of axis; *medianly*, the spine of the axis and, above, the posterior tubercle of the atlas.

Contents of the Region:

- Four muscles: Two oblique; Two straight
- Two nerves: Greater occipital; Suboccipital
- Two arteries: Vertebral; Occipital

Vertebral Artery. The thin *posterior atlanto-occipital membrane* unites the posterior arch of the atlas to the margin of the foramen magnum. Here the *vertebral artery* winds medially, behind the superior articular process of the atlas to its groove on the arch with the dorsal ramus of C. 1. Its earlier course is given soon (p. 476).

Only the *course of the vertebral* artery should be memorized—it is important! The section below is given for appreciation and reference.

Muscles (*fig. 675*). The *Obliquus Capitis Inferior* passes from the spine of the axis obliquely upward and forward to the tip of the transverse process of the atlas. It bounds the region inferiorly.

The Obliquus Capitis Superior passes from the tip of the transverse process of the atlas obliquely upward and backward to be inserted between the two nuchal lines of the occipital bone lateral to the Semispinalis Capitis. It bounds the region laterally.

The Rectus Capitis Posterior Minor arises from the posterior tubercle of the atlas; the *Rectus Capitis Posterior Major* from the spine of the axis. These two are attached side by side to the occipital bone between the inferior nu-

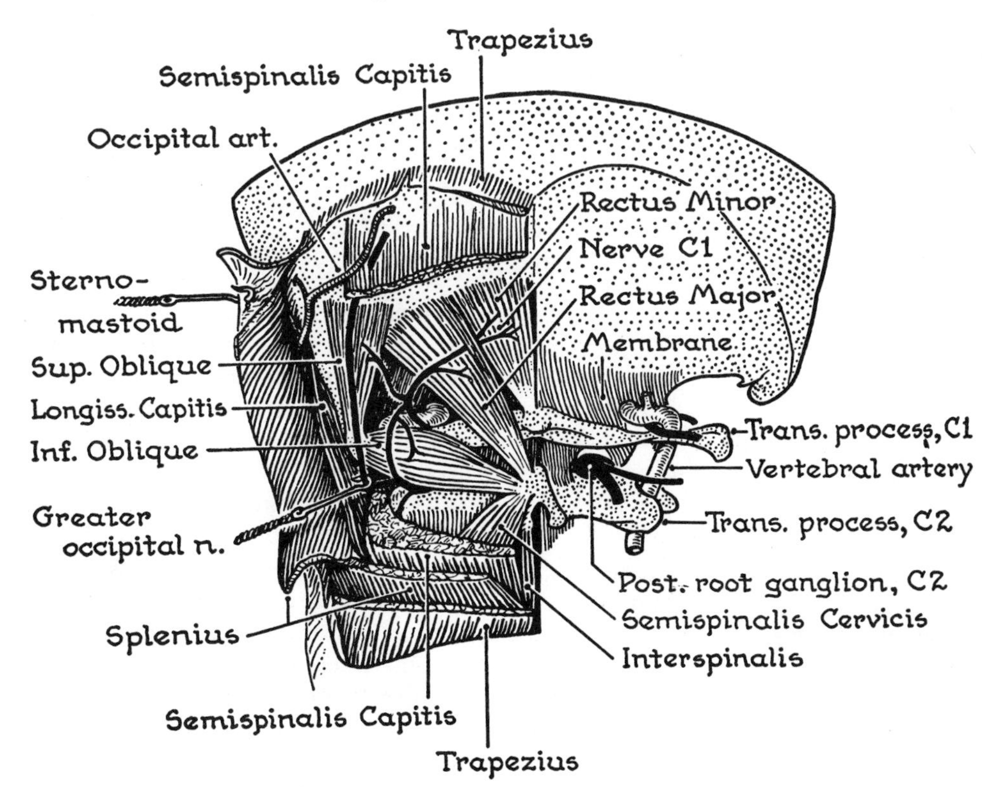

Fig. 675. The suboccipital region, containing the suboccipital triangle.

chal line and the foramen magnum.

Nerves and Other Vessels. The suboccipital nerve (i.e., dorsal ramus of C. 1) supplies the four muscles of the region and sends a branch to the muscle covering the region. It is never sensory because C. 1 nerve is unique in not having a spinal ganglion and dorsal root. It emerges through the **suboccipital triangle**—bounded by the Rectus Capitis Posterior Major and the two Oblique muscles.

Muscular twigs of the vertebral and occipital arteries pass through the triangle; and commonly a large vein from the vertebral plexus pierces the membrane and passes through the triangle to the *suboccipital plexus of veins*.

NERVES OF THE BACK

(Dorsal Rami of Spinal Nerves)

General Considerations. The dorsal rami of each of the 31 pairs of spinal nerves are similar in size, course, relations, and distribution, with the exception of the first two and last three.

A *typical dorsal ramus* takes origin just beyond a spinal ganglion (*fig. 72*), and passes backward on the side of a superior articular process. The dorsal sacral rami emerge from the dorsal sacral foramina.

The dorsal rami supply a serially segmented territory. They supply the skin and the "native" deep muscles of the back medial to the angles of the ribs. Hence, they are much smaller than the corresponding ventral rami (except for C. 1 and 2).

Certain cutaneous branches trespass beyond the angles of the ribs and a few wander afar, thus: C. 2 (greater occipital nerve) ascends to the top of the head: Th. 2 extends toward the acromion; L. 1, 2, and 3 descend to the buttock.

Each dorsal ramus divides into a medial and a lateral branch (except C. 1; S. 4, 5 and Co. 1), and one or other ends as a cutaneous nerve (except C. 1; C. 6, 7, 8; and L. 4, 5). Above the midthoracic region the medial branches become cutaneous; below it, the lateral branches.

Courses of Nerve Trunks C. 1 and 2. *Explanatory:* These are peculiar because the "true" articular processes of the atlas and upper ones of the axis disappear; new joints appear in front of the nerve trunks (*fig. 675.1*). So now the anterior rami wind forward around the two joints.

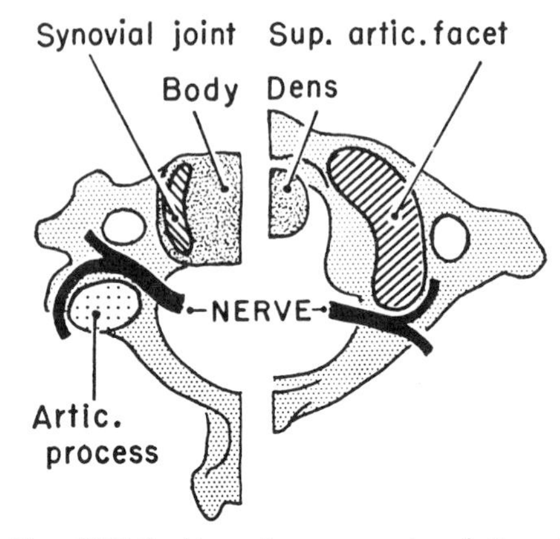

Fig. 675.1. Homologous parts of the atlas and vertebrae C. 3–6 are shown, for they explain the courses of nerves C. 1 and 2.

Sacral and Coccygeal Nerves. The nerve trunks of S. 5 and Co. 1 lie exposed at the lower end of the sacral canal (*fig. 669*). The dorsal rami of S. 4, 5 and Co. 1 unite to form a small descending cutaneous nerve that supplies the skin over the coccyx. This nerve is the homologue of the dorsal nerve of the quadruped's tail.

ARTERIES OF THE BACK

1. **The Vertebral Artery** is a branch of the subclavian (*fig. 725*). It has been seen in the suboccipital triangle (*fig. 675*), on the posterior arch of the atlas, where it enters the subarachnoid space. To get to the foramen transversarium of the atlas, it has threaded its way up the foramina transversaria of C. 6 to C. 1. With the internal carotid artery, it supplies the brain.

2. **Branches of the Vertebral, Intercostal, Lumbar, and Lateral Sacral Arteries** accompany the dorsal rami of the cervical, thoracic, lumbar, and sacral nerves.

3. **The Occipital Artery** (*fig. 675*) is a branch of the external carotid (*fig. 712*), which takes a direct course back to supply both superficial and deep structures of the occipital and suboccipital regions, anastomosing freely with its neighbors.

4. **The Profunda Cervicis Artery** is one of the two terminal branches of the deep-lying costocervical trunk in the root of the neck. It enters the back above the neck of

the 1st rib and anastomoses with branches of the occipital and vertebral aa. This may augment a collateral circulation to the brain when the carotid a. gets blocked.

VEINS OF THE BACK

These veins are large and they form liberal plexuses which tend to follow the arteries. In the neck they mainly descend around the vertebral and profunda cervicis arteries.

Veins of the Vertebral Column (Vertebral Venous System). *Because they are difficult to dissect and to demonstrate, these veins are not treated in classrooms with the great respect they deserve. They must not be ignored!* The vertebral canal contains a dense plexus of thin-walled, valveless veins which surrounds like a basketwork the spinal dura mater. Above, this plexus communicates through the foramen magnum with the occipital and basilar sinuses of the cranium. Anterior and posterior longitudinal channels (venous sinuses) can be discerned in this internal *vertebral venous plexus* (*fig. 676*).

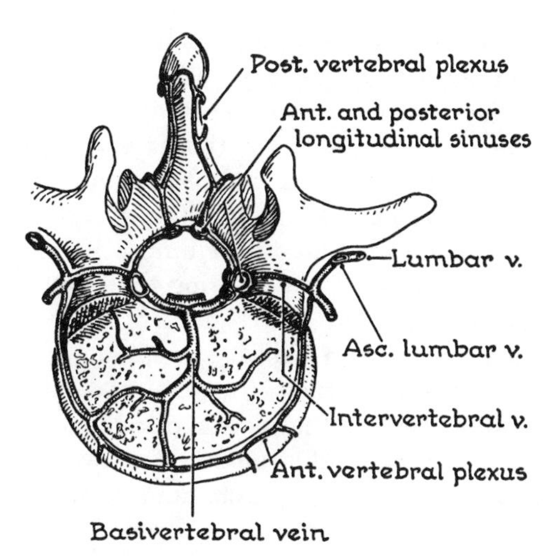

Fig. 676. The veins of the vertebral column.

At several spinal segments (but not all) the plexus receives a vein from the spinal cord and at each segment it receives a vein, the basivertebral vein, from the body of a vertebra; and in turn it is drained by intervertebral veins which pass through the intervertebral (and sacral) foramina to the vertebral, intercostal, lumbar, and lateral sacral veins.

Through the body of each vertebra come veins which form a meager *anterior vertebral plexus*, and through the lig. flava pass veins which form a well marked *posterior vertebral plexus*.

Other Longitudinal Channels. In the cervical region, these plexuses communicate freely with the occipital, vertebral, and profunda cervicis veins; and in the thoracic, lumbar, and pelvic regions, segment is linked to segment by the azygos (or hemiazygos), ascending lumbar, and lateral sacral veins.

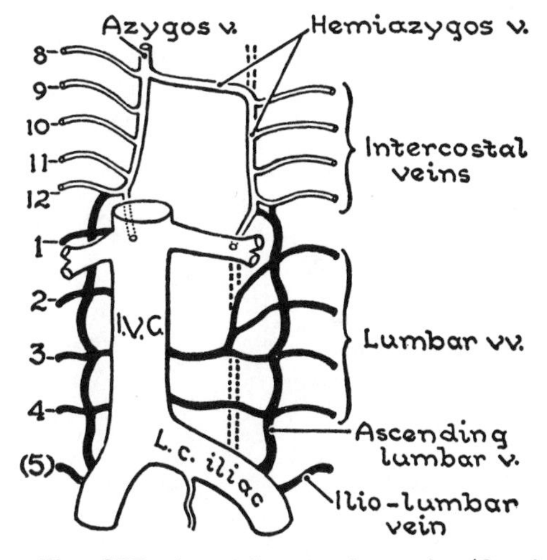

Fig. 677. Ascending lumbar vein. (Cont'd. *from fig.* **641**.)

The *ascending lumbar vein* (*fig. 677*) is an anastomotic vein that ascends in front of the lumbar transverse processes linking one lumbar vein to another, and connecting the common iliac vein with the azygos (or hemiazygos) vein.

Significance. Batson's ingenious experiments forced us to add to the recognized pulmonary, portal, and caval venous systems a fourth or **vertebral venous system.** This system may be considered a separate, although overlapping, system of veins. It comprises the veins of the brain, skull, neck, viscera, vertebral column (and their valveless connections in the limb girdles), and the veins of the body wall.

The Batson experiments indicate (1) that compression of the thorax and abdomen with the larynx and other sphincters closed, as occurs in straining, coughing, and lifting with the upper limbs, not only prevents blood from entering the thoraco-abdominal veins but squeezes it out of them into the vertebral system; (2) that the increase in the intraspinal and intracranial pressure that occurs during coughing, sneezing, and straining is active—not passive; (3) that tumors and abscesses having connection with this venous system may spread anywhere along this system without involving the portal, pulmonary, or caval systems; and (4) that the cranial and spinal parts of the system, as well as being pathways, are blood depots or storage lakes of blood; further, (5) they reveal the channels through which blood from the lower limbs and pelvis may in favorable circumstances return to the heart after the i. v. cava has been obstructed below the renal veins. In addition: (6) Eckenhoff shows that the C.S.F. pressure is regulated by the vertebral venous pressure.

37 INTERIOR OF CRANIUM

SKULL CAP OR CALVARIA

Structure; Diploic Veins (Blood Supply).

MENINGES

Dura Mater; Falx Cerebri; Falx Cerebelli; Tentorium; Diaphragma Sellae.
Meningeal Arteries and Veins; Nerves.
Cerebrospinal Fluid; Arachnoid Granulations.
Venous Sinuses—Emissary Veins.

CRANIAL NERVES

Arteries to Brain

Their Relationships to the Cranial Nerves: Internal Carotid; Vertebral.

FLOOR OR BASE OF SKULL

Boundaries; Lesser Wing of Sphenoid.
Anterior Cranial Fossa—Foramen Cecum; Orbital Plate of Frontal Bone.
Posterior Cranial Fossa—Parts; Petrous Bone; Cerebellar Surface.

MIDDLE CRANIAL FOSSA

Median Part: Features; Chiasmatic Groove; Sella Turcica; Hypophyseal Fossa; Foramen Lacerum.
Lateral Part: Greater Wing of Sphenoid; Crescent of Foramina.
Squama Temporalis.
Petrous Bone—cerebral surface.
Surface Anatomy of Skull—Pterion.

CONTENTS OF MIDDLE CRANIAL FOSSA

Trigeminal Nerve; Cavernous Sinus; Internal Carotid Artery; Abducent Nerve; Trochlear Nerve; Oculomotor Nerve.
Middle Meningeal Vessels.
Petrosal Nerves; Internal Carotid Nerve.

SKULL CAP OR CALVARIA

Structure. The bones of the roof of the skull consist of an *outer* and an *inner plate*, or *lamina*, of compact bone with a layer of spongy bone, called the *diploe*, sandwiched in between.

At birth, the bone consists of a single compact layer. Into this layer veins grow, branch, rebranch, and unite with neighboring veins; and marrow is laid down around them to form the diploe. There are four *diploic veins* on each side (*fig. 678*).

Hollow buds of mucous membrane sprout from the nasal cavity and mastoid antrum into the diploic layer of certain bones and, by replacing it, form *air sinuses*.

Diploe does not form in bone covered with thick, fleshy muscle: (1) the temporal squama and (2) the nuchal part of the occipital. In these parts the bone remains "infantile"—thin and translucent.

Diploic Veins (*fig. 678*). The frontal diploic vein emerges from a small foramen in the supra-orbital notch. The remainder drain to

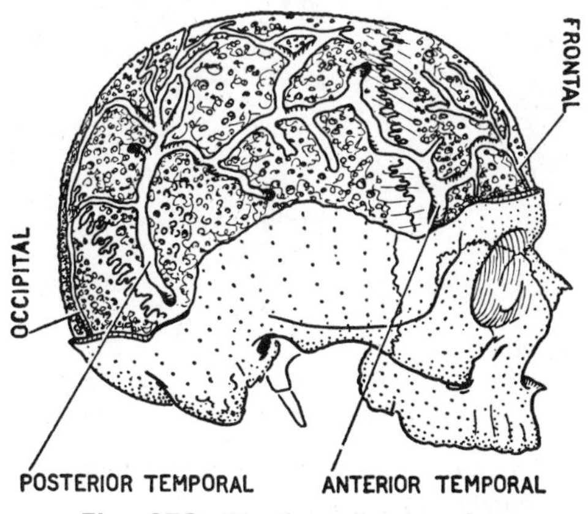

Fig. 678. The four diploic veins.

sinuses inside the cranial cavity as a rule. The anterior temporal diploic vein emerges from the greater wing of the sphenoid; the posterior temporal diploic vein, from the postero-inferior angle of the parietal bone. The occipital diploic vein emerges near the internal occipital protuberance.

MENINGES

Three meninges or membranes envelop the brain—*dura mater, arachnoid mater,* and *pia mater* (*fig. 679*). Names given these "mothers" of the brain indicate their qualities: the dura is tough, the arachnoid is like a spider's web, and the pia clings faithfully to the brain surface like a skin, following all its irregularities. Between the dura mater and arachnoid mater there is a potential space, the *subdural space.*

Between the arachnoid mater and pia mater there is an actual space, the *subarachnoid space,* filled with *cerebrospinal fluid.* The arachnoid mater is attached to the pia mater by loose scattered threads. The pressure of the fluid within the subarachnoid space makes the subdural an empty, potential space. The arteries of the brain travel in the subarachnoid space.

The Dura Mater consists of two closely adherent fibrous layers—an outer and an inner. The *outer layer* or *endocranium* is periosteum.

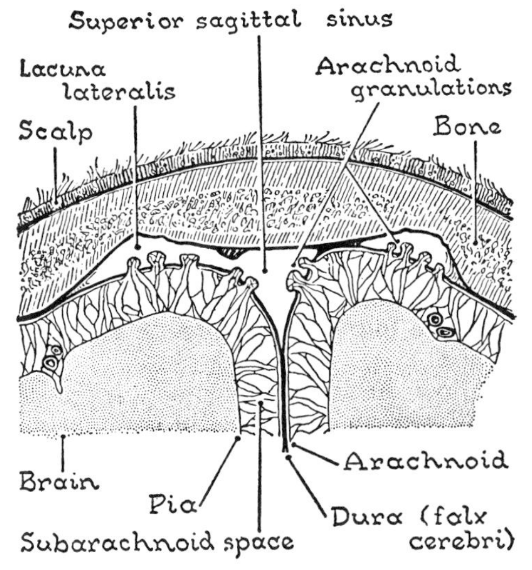

Fig. 679. The arachnoid granulations return the cerebrospinal fluid.

In the parietal region particularly two structures stand out from it in relief: (1) Cauliflower-like masses, *arachnoid granulations,* covered with a film of dura, grow widely (about 2 cm) from each side of the median plane (*fig. 679*). They are responsible for the marked pits on the parietal bone. (2) Branches of the *meningeal vessels* run in the outer layer of the dura and make grooves on the bones, notably the parietal bone.

The *inner layer of the dura* is smooth like a serous membrane. It is reduplicated to form **four inwardly projecting folds,** which partially subdivide the cranial cavity into compartments, and being taut they prevent shifting of the cranial cargo, the brain. Two of the folds, the *falx cerebri* and *falx cerebelli,* are sickle-shaped occupants of the median plane. The other two, the *tentorium cerebelli* and the *diaphragma sellae,* form roofs for the cerebellum and hypophysis cerebri, respectively. Between the layers of the dura lie certain canals and spaces, *venous sinuses,* lined with endothelium, filled with blood, and continuous with veins.

The Falx Cerebri is sickle-shaped, and it hangs between the two cerebral hemispheres (*fig. 680*). Its convex upper border is attached to the lips of the sagittal sulcus on the frontal, parietal, and occipital bones, and so contains the superior sagittal venous sinus. The anterior part of its lower border is free and contains the inferior sagittal sinus; the posterior part suspends the tentorium cerebelli and contains the straight sinus (*fig. 680*).

The Falx Cerebelli is a slight fold attached to the internal occipital crest. The occipital sinus lies in its attached posterior border.

The Tentorium Cerebelli, shaped like a bell tent with open flaps, forms a roof for the cerebellum and a floor for the hinder parts of the cerebrum. Its *medial border* is free, bounding a large oval opening, the

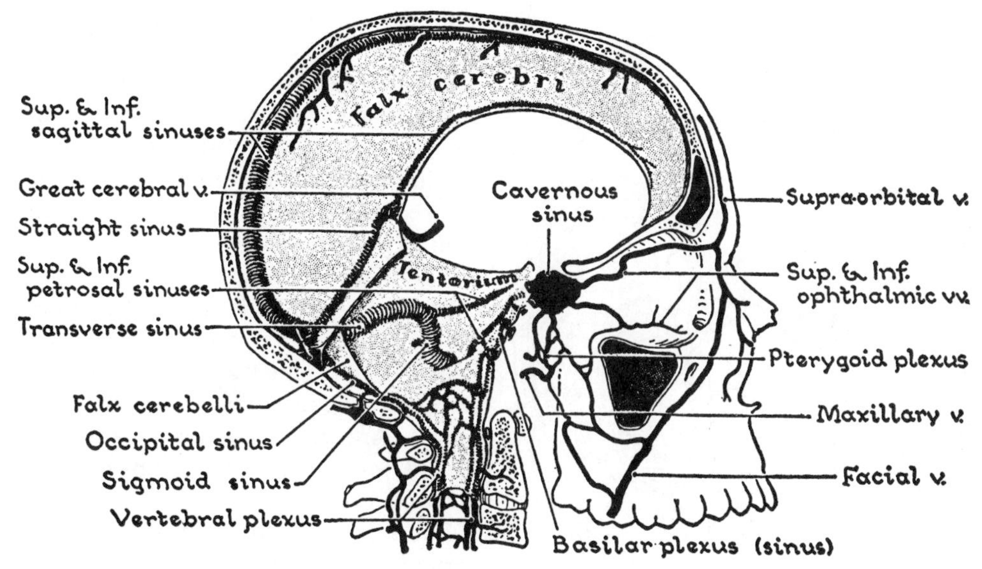

Fig. 680. The folds of the dura mater. Venous sinuses. Vertebral venous plexus.

incisura tentorii; and it lightly hugs the sides of the midbrain. Its **attached border** is occupied on each side by the superior petrosal and transverse sinuses.

The *attached border* begins on each side at the posterior clinoid process (which it drags laterally, downward, and backward) and attaches to the lips of the groove on the superior border of the petrous bone, postero-inferior angle of the parietal bone, transverse groove on the occipital bone, and to the internal occipital protuberance.

Its *medial border* is fixed in front to the anterior clinoid process. To reach it, the free border crosses above the anterior end of the peripheral border and here forms a *triangular field* at the side of the diaphragma sellae (*fig. 681*).

The Diaphragma Sellae forms a "tentorium" for the hypophysis cerebri and it has a large central aperture for the stalk of the hypophysis cerebri. (This gland lies in a depression, the *sella turcica* or Turkish saddle, hence the name diaphragma sellae.)

Meningeal Arteries being periosteal arteries, lie embedded in the outer layer of the dura mater. They supply dura mater, inner table of the skull, and diploe. The **middle meningeal a.,** assisted by a branch of the *anterior ethmoidal a.* in the anterior cranial fossa, caters for the territory above the level of the tentorium cerebelli.

Other arteries make small contributions: the *posterior ethmoidal a.* in the anterior fossa; the *accessory meningeal* and *internal carotid aa.* in the middle fossa; the *occipital a.* (via jugular and mastoid foramina), the *ascending pharyngeal a.* (via jugular foramen and hypoglossal canal), and the *vertebral a.* in the posterior fossa.

Meningeal Veins accompany the arteries and communicate with the venous sinuses and with the diploic veins.

Meningeal Nerves. The cranial dura is supplied by the trigeminal nerve (N. V^1, V^2, V^3) and cervical nerve(s) (1), 2, (3) in a pattern comparable to the overlying skin areas (*fig. 680.1; cf. fig. 662*). The brain itself is insensitive to pain.

The Cerebrospinal Fluid (CSF) is produced in and fills the ventricles of the brain. It escapes through three orifices in the 4th ventricle into the subarachnoid space. The fluid diffuses over and around the brain and spinal cord and, though the spinal cord ends at the 2nd lumbar vertebra, the subarachnoid space with its contained fluid extends to the 2nd sacral

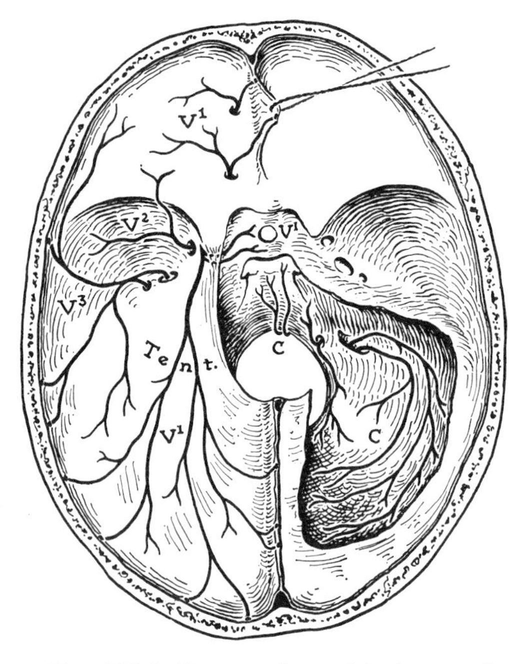

Fig. 680.1 Nerves of cranial dura mater (Kimmel).

vertebra (*fig. 42*, p. 40). The CSF returns to the blood stream via the arachnoid granulations and villi.

Arachnoid Granulations are aggregations of hollow, nipple-like processes of arachnoid tissue. They occur along the venous sinuses, and bulge into them through the inner layer of dura (*fig. 679*). Small in youth, large and cauliflower-like in the aged, they are best seen in the superior sagittal sinus.

Venous Sinuses of Dura Mater (*fig. 680*):

1. Unpaired median sagittal sinuses and their continuations: (1) superior sagittal and (right) transverse and sigmoid; (2) inferior sagittal, straight, and (left) transverse and sigmoid; and (3) occipital.

2. Sinuses associated with the paired cavernous sinuses—cavernous, sphenoparietal, superior petrosal, inferior petrosal; intercavernous and basilar.

The blood sinuses have no valves. Enclosed by dura, the superior sagittal, transverse, sigmoid, petrosal, and occipital sinuses groove the bones on which they lie. Their tributaries mostly come from the neighboring parts of the brain; the middle meningeal veins partly end in the superior sagittal sinus; three of the four diploic veins of each side commonly end in the sinuses. All sinuses drain ultimately into the internal jugular veins. *Emissary veins* also connect the sinuses with the extracranial veins (see below).

The superior sagittal sinus (unpaired) occupies the entire length of the attached margin of the falx cerebri, and it usually becomes the right transverse sinus. Six or so cerebral veins and large meningeal veins join it. Arachnoid granulations bulge into it and into its lateral expansions, called *lacunae laterales*.

The inferior sagittal sinus (unpaired) is very small. It occupies the free edge of the falx cerebri. It joins the *great cerebral vein* (of Galen) from the brain just behind its corpus callosum, to become the straight sinus.

The straight sinus (unpaired) lies in the line of union between the tentorium cerebelli and the falx cerebri; it usually becomes the left transverse sinus.

The transverse sinus occupies the attached margin of the tentorium and grooves the occipital bone. Leaving the tentorium it becomes the *sigmoid sinus* which takes a sigmoid course downward through the jugular foramen to become the internal jugular vein (*figs. 680, 684*). The transverse sinus is joined by the superior petrosal sinus, and several veins.

The cavernous sinus (paired) is trabeculated. It lies at the side of the hypophyseal fossa and of the hollow body of the sphenoid. It receives the sup. and inf. ophthalmic veins (they have no valves), the superficial middle cerebral vein, and the sphenoparietal sinus. It is drained by the sup. and inf. petrosal sinuses and emissary veins. (Contd. on p. 490.)

The *inferior petrosal sinus* (paired) drains the cavernous sinus. It runs down to and through the jugular foramen and ends in the internal jugular vein just below the skull.

Laterally, the cavernous sinus extends to the maxillary nerve and trigeminal ganglion; anteriorly, to the superior orbital fissure; and posteriorly, to the apex of the petrous bone (*fig. 680*).

The superior petrosal sinus (paired) occupies the groove on the superior border of the petrous bone in the attached margin of the tentorium. It drains the cavernous sinus into the transverse sinus.

The basilar sinus is a wide trabeculated space behind the dorsum sellae and basi-occipital. It unites the cavernous and the inferior petrosal sinuses of opposite sides and communicates below with the vertebral venous plexus (*fig. 680*).

The sphenoparietal sinus (paired) lies below the lesser wing of the sphenoid.

The intercavernous sinus connects the cavernous sinuses of opposite sides around and below the hypophysis cerebri.

The right lateral sinus is usually larger than the left, but this is variable (Browning).

The occipital sinus (unpaired) communicates with the vertebral venous plexus (*fig. 680*).

The confluens sinuum is the dilatation commonly found where the right and left transverse sinuses communicate and the occipital sinus begins.

Emissary Veins pass through the foramina of the skull and connect the sinuses inside with the veins outside. They are multiple and variable.

Foramina that transmit emissary veins include the following: f. cecum, parietal f., mastoid f., condylar canal, f. ovale, carotid f., f. magnum, and superior orbital fissure.

CRANIAL NERVES

On the floor of the anterior cranial fossa, about 0.5 cm from the median plane, lies a long (3 cm), fragile stalk with a slightly enlarged anterior end, the *olfactory tract* and *bulb* (*fig. 681*). The bulb is fixed to the floor by delicate *olfactory nerves* which pierce the cribriform plate to supply a small area on the roof and walls of the nasal cavity.

The *optic nerve* passes on each side to the optic canal from the *optic chiasma*. Lateral to it, the *internal carotid artery* emerges from the dura.

[On each side, the *posterior cerebral artery* winds around the midbrain, and the *great cerebral vein* emerges in the median plane from the brain and runs to the

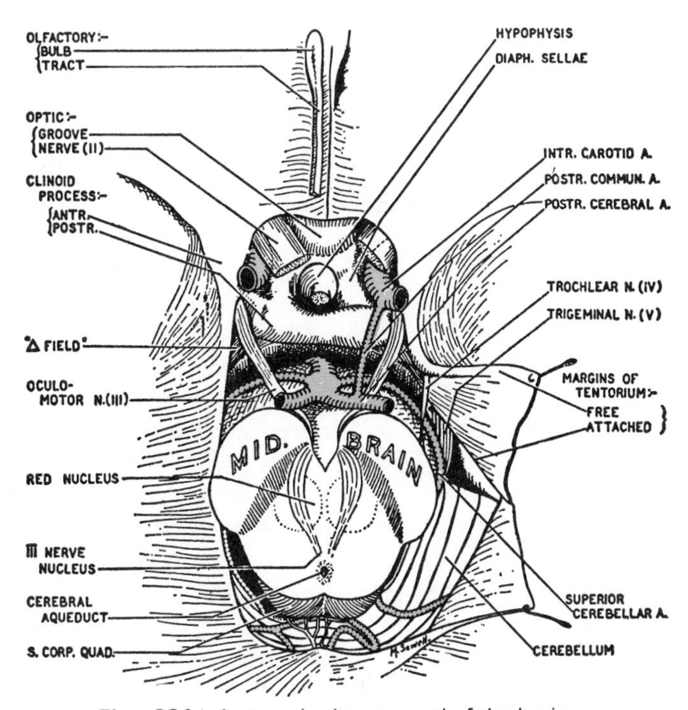

Fig. 681. A stage in the removal of the brain.

tentorium cerebelli.]

The *oculomotor nerve* (N. III) is large (*fig. 681*). It enters an arachnoid and dural cul-de-sac far forward in the triangular field between the attachments of the tentorium to the anterior and posterior clinoid processes. It may be traced backward and medially, past the posterior clinoid process and between the posterior cerebral and superior cerebellar arteries to the front of the midbrain just above the pons.

The *trochlear nerve* (N. IV), the most delicate of the nerves (*fig. 682*), pierces the dura in the same triangle as nerve III, but far back, and it also passes between the same two arteries. Followed backward around the midbrain, under shelter of the free edge of the tentorium—which requires to be raised to bring it into view—it leads to the back of the midbrain, where it arises.

The *trigeminal nerve* (N. V), the largest of the cranial nerves, arches over the most medial part of the upper border of the petrous bone below the attached margin of the tentorium. It curves backward and slightly downward to the side of the pons.

The *abducent nerve* (N. VI). It arises at the lower border of the pons in line with N. III, ascends clamped to the pons by the ant. inf. cerebellar artery, and pierces the dura overlying the inferior petrosal sinus (*fig. 682*).

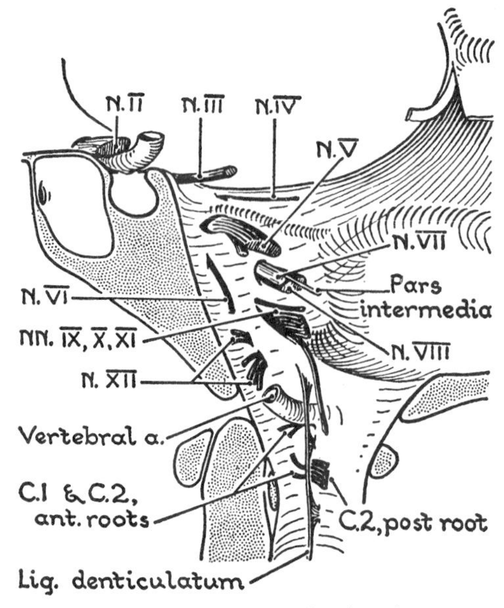

Fig. 682. Cranial nerves, piercing dura mater.

The *facial* (N. VII) and *vestibulocochlear* (N. VIII) *nerves* with the *nervus intermedius* (pars intermedia) between them arise at the lower border of the pons, abreast of N. VI and almost in line with N. V. They pass laterally and slightly upward into the internal acoustic meatus.

The *glossopharyngeal* (N. IX), *vagus* (N. X), and *accessory* (N. XI) *nerves* arise by a row of fila from the medulla and spinal cord just below and in line with the NN. VII and VIII. They converge laterally, pierce the dura below the internal acoustic meatus, and enter the jugular foramen.

The *hypoglossal nerve* (N. XII) arises from the medulla between the pyramid and olive in line with the origin of nerves III and VI above and with the ventral root of the 1st cervical nerve below. Its root fila converge laterally, pierce the dura through two apertures, and enter the hypoglossal canal.

[The vertebral artery and the spinal root of the accessary nerve enter on each side through the foramen magnum where the spinal cord becomes the medulla oblongata.]

ARTERIES TO THE BRAIN AND THEIR RELATIONSHIP TO THE CRANIAL NERVES

Two paired arteries—*internal carotid* and *vertebral*—alone supply the brain.

The Internal Carotid Artery (*fig. 683*) pierces the dura medial to the anterior clinoid process, which it grooves. It at once gives off the *ophthalmic a.* which runs below the optic nerve through the optic canal, and the *posterior communicating a.* which runs backward, medial to N. III to unite with the posterior cerebral a. It then ascends in the angle between the optic nerve and tract, and ends as the *anterior* and *middle cerebral aa.* Of these, the ant. cerebral a., after running medially above

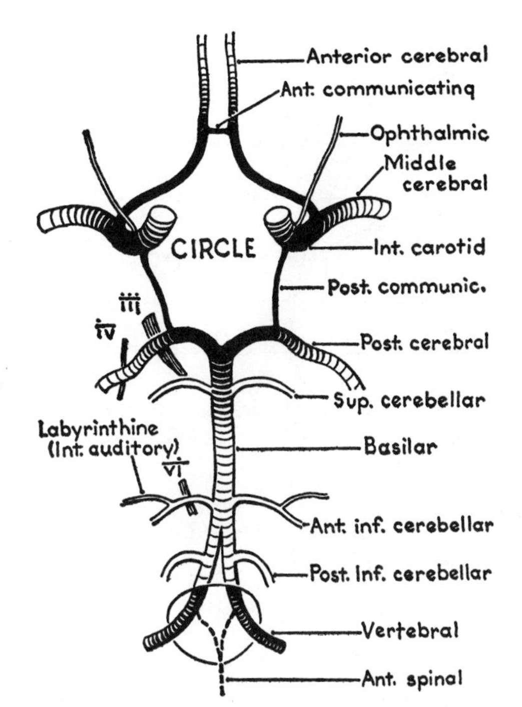

Fig. 683. Two vertebral and two internal carotid arteries supply the brain and form an arterial circle.

the optic nerve, is joined to its fellow by the *ant. communicating a.*, whereas the middle cerebral a. runs laterally in the stem of the lateral cerebral sulcus.

The Vertebral Artery (*fig. 683*) pierces the dura behind the occipital condyle and grooves the margin of the foramen magnum. It passes forward to the lower border of the pons where it unites with its fellow to form the basilar a.

The Posterior Inferior Cerebellar A is the largest branch of the vertebral a. On its course to the cerebellum it loops around or among nerves IX, X, XI.

The Basilar Artery ascends the bony slope in the midline from the lower border of the pons to the upper border where it ends in a T-shaped bifurcation, the *right* and *left posterior cerebral aa.* above nerves III and IV. Two other paired branches proceed horizontally from it: (1) the *anterior inferior cerebellar a.*, above, below, or between nerves VII and VIII, which carry its labyrinthine branch to the inner ear; and (2) the *superior cerebellar a.*

FLOOR OR BASE OF THE SKULL

Three Cranial Fossae

Boundaries (*fig. 684*). The interior of the base of the skull has three terraces, or downward steps, called fossae—an *anterior*, a *middle*, and a *posterior*.

The anterior cranial fossa is sharply marked off from the middle fossa by three free concave crests, a median and two lateral, which are separated from each other by two prominent backwardly projecting tubercles, the *anterior clinoid processes*. The median concave crest connects the anterior clinoid processes of opposite sides above the *optic canals* and across the body of the sphenoid bone. Each lateral crest is formed by the *lesser wing of the sphenoid*, and it fits into the lateral sulcus of the brain.

The middle cranial fossa is marked off from the posterior cranial fossa by a median, rectangular plate, the *dorsum sellae*, at whose free upper angles are the *posterior clinoid processes*. On each side, it is marked off by the *superior border of the petrous bone*.

Anterior Cranial Fossa (*fig. 685*). The upward extension of the nasal septum, called fancifully the *crista galli* or cock's comb, rises in the median plane and gives attachment to the falx cerebri.

For 3 mm on each side, the floor is depressed and forms the roof of the nasal cavities; it is perforated like a sieve by numerous olfactory nerves that stream through it; hence, it is called the *cribriform plate* of the ethmoid bone (cribrum, L. and ethmos, Gk., both = a sieve).

Laterally, the fossa is formed by the *orbital plate of the frontal bone*, the roof of the orbit.

Middle Cranial Fossa (pages 487–492).

Posterior Cranial Fossa (*figs. 684, 686*). This fossa lodges the hind brain, which comprises the cerebellum, medulla, and

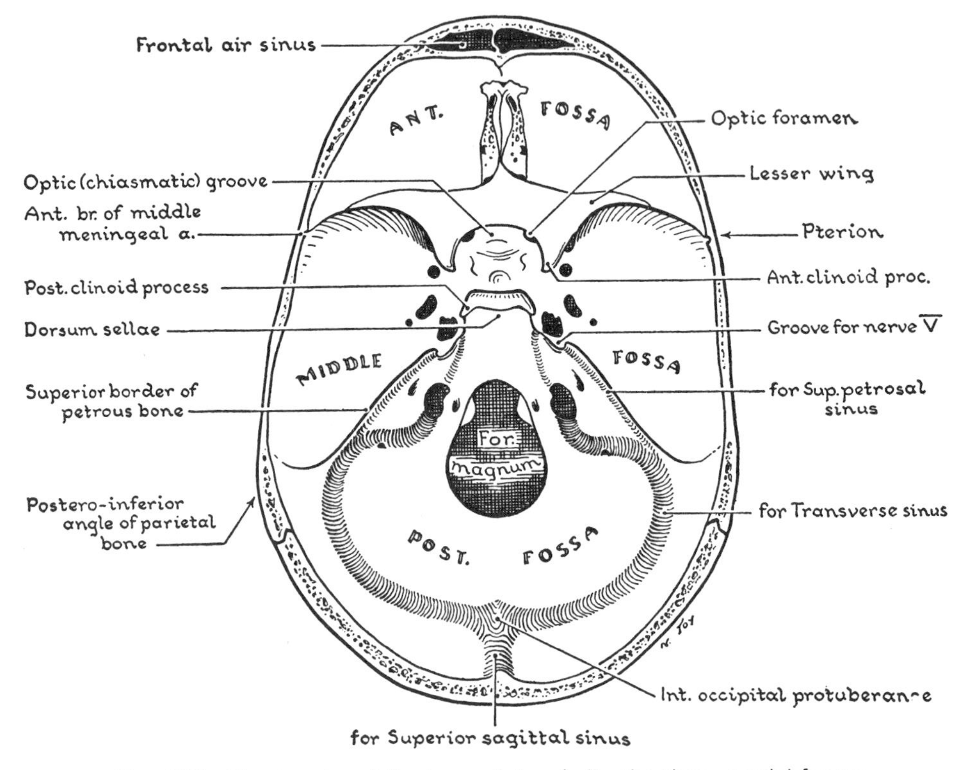

Fig. 684. The interior of the base of the skull—the three cranial fossae.

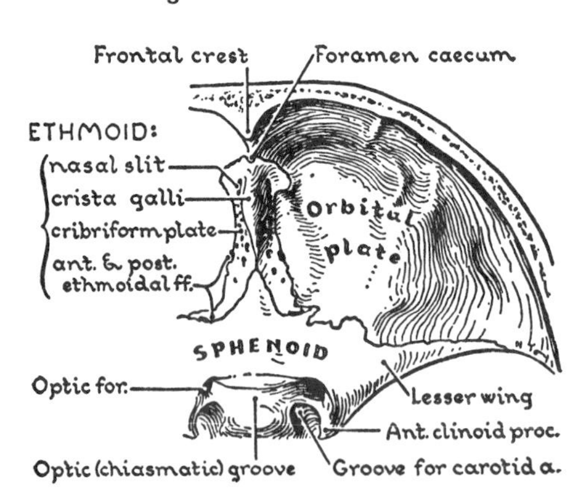

Fig. 685. The anterior cranial fossa. (*Optic foramen and optic groove = optic canal and chiasmatic sulcus.*)

pons. It is roofed in by the tentorium cerebelli. The *foramen magnum* is enormous, unpaired and oval. It lies at the lowest part of the posterior fossa.

Stepping upward and laterally in a coronal plane, are the essential features of the fossa: (1) the hypoglossal canal, which transmits nerve XII, (2) the *jugular foramen* for nerves XI, X, and IX; and (3) a finger's-breadth above this is the *internal acoustic meatus* for nerves VIII and VII.

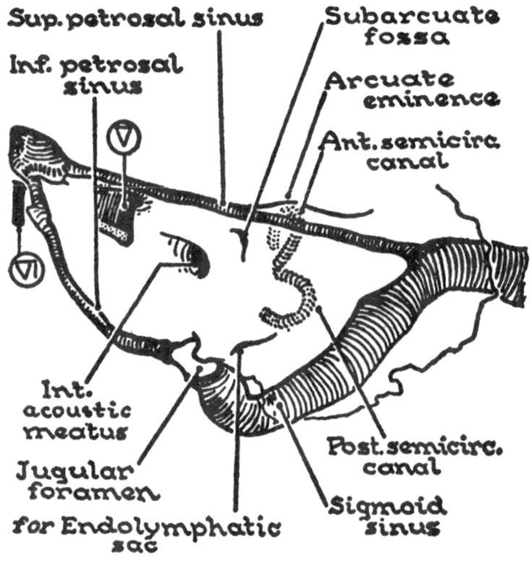

Fig. 686. The posterior surface of the petrous bone is bounded by venous sinuses.

Surface Anatomy. A line connecting both *external acoustic meatuses* is in the plane of the last six cranial nerves.

The *jugular foramen* (*fig. 686*) transmits the inferior petrosal sinus anteriorly. Descending sinuously to its posterior part is the groove for the *sigmoid sinus*. Nerves IX, X, and XI pass through its middle part. The *groove for the transverse sinus* becomes the groove for the sigmoid sinus.

Between the foramen magnum and the internal occipital protuberance the *internal occipital crest* gives attachment to the falx cerebelli. On each side, the bone is concave for a cerebellar hemisphere.

The inclined surface rising from the foramen magnum to the dorsum sellae is the **clivus.**

In youth the sphenoid and occipital are united just below the dorsum sellae by a synchondrosis which is completely ossified by the 21st year (McKern and Stewart).

The Posterior Surface of the Petrous Bone (*fig. 686*) is triangular. Its main feature is the internal acoustic meatus. *Details—*

Posterosuperior to the internal acoustic meatus is a small laterally directed pit (*fig. 834C*, p. 611) over which at birth the anterior (superior) semicircular canal arched, hence called the *subarcuate* fossa.

Postero-inferior to the acoustic meatus is a medially directed *vertical slit*, the opening of the aqueduct of the vestibule, which lodges the *saccus endolymphaticus* (*fig. 815*). A *pyramidal notch* lies on the lower border of the petrous bone, just above the anterior end of the jugular foramen. At the apex of the pyramid is the pinpoint opening of the cochlear canaliculus, which lodges the perilymphatic duct (aqueduct of cochlea).

MIDDLE CRANIAL FOSSA

The middle cranial fossa (*fig. 687*), shaped like a butterfly, has a median part and two lateral parts.

Median Part. The median part is likened

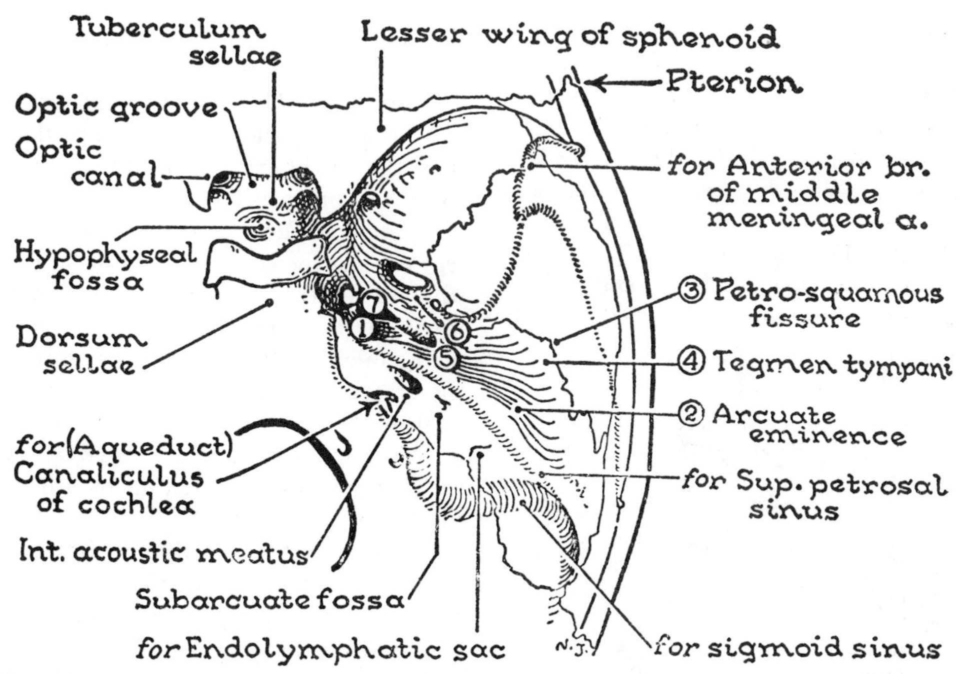

Fig. 687. Middle cranial fossa. Seven details of petrous bone: (*1*) depression for trigeminal ganglion, (*2*) elevation for anterior semicircular canal, (*3*) remains of petrosquamous fissure, (*4*) tegmen tympani, (*5*) hiatus for greater petrosal nerve, (*6*) hiatus for lesser petrosal nerve, (*7*) roof of carotid canal, commonly membranous.

to a bed with four clinoid processes, or bed posts (Kline, Gk. = a bed). It lies above the body of the sphenoid, which is inflated by the sphenoidal air sinuses.

Features. (1) The *optic canal* lies between the inflated body of the sphenoid and the two roots of the lesser wing of the sphenoid. The posterior root separates the canal from the groove for the internal carotid artery. The *chiasmatic* (*optic*) *groove* connects the optic canals, but does not lodge the optic chiasma.

(2) The *sella turcica* lies behind the chiasmatic groove. It has three parts: the *tuberculum sellae* or pommel of the saddle; the *hypophyseal fossa* for the hypophysis cerebri (pituitary gland); and (3) behind this rises the *dorsum sellae* or back of the saddle (*fig. 687.1*).

Foramen Lacerum. This ragged foramen in a dry skull is an artifact. It lies at the apex of the petrous temporal bone. The carotid artery turns up out of its canal here; in life, the f. lacerum has a floor of fibrocartilage below the artery.

Lateral Part. Each lateral part is limited in front by the lesser wing of the sphenoid, and behind by the superior border of the petrous bone. It includes the *greater wing* of the sphenoid and parts of the temporal bone.

Greater Wing of Sphenoid. The *feature of the cerebral surface of the greater wing is a* **crescent** *of foramina* (*fig. 688*): (1) The *superior orbital fissure* is comma-shaped and lies between the greater and lesser wings. It leads into the orbit. (2) The *foramen rotundum* opens forward into the deep-lying pterygopalatine fossa. (3) The *foramen ovale* and (4) the *foramen spinosum* open downward into the infratemporal fossa.

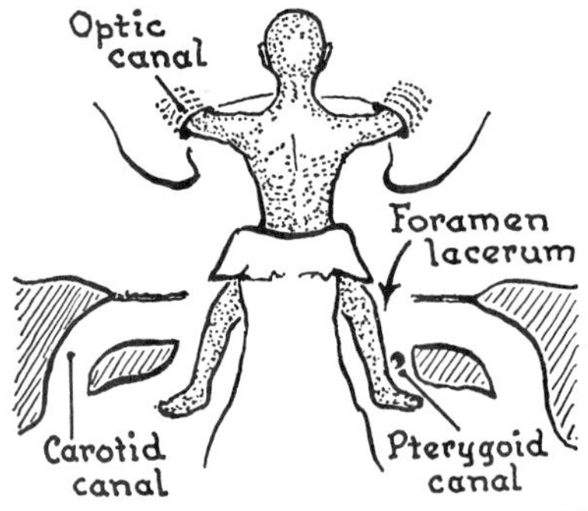

Fig. 687.1. The foramen lacerum. The rider sits in the saddle, his arms stretched forward through the optic canals, and his legs dangling through the lacerate foramina.

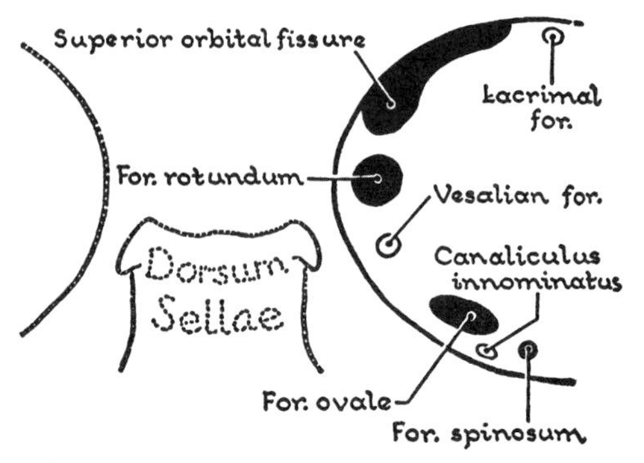

Fig. 688. The crescent of foramina within the greater wing. Of these foramina, 4 are constant and 3 (lacrimal; Vesalian, for an emissary v.; and innominatus, for a small nerve) are not.

Each of the first three foramina transmits a division of the trigeminal nerve, V^1, V^2, and V^3. The fourth transmits the middle meningeal artery; its name derives from the neighboring spine of the sphenoid (on the inferior surface).

Squamous Part of Temporal Bone. The feature of the cerebral surface of this squama is the **groove for the middle meningeal artery** (*fig. 687*) from the foramen spinosum. Its *anterior branch* regains the greater wing and at the antero-inferior angle of the parietal bone (*pterion*) where it may become a bony canal (*fig. 687*). Here the artery can be sheared off by a blow to the temple (*fig. 689*).

Petrous Part of Temporal Bone. The *tegmen tympani* is the roof of the bony auditory tube, tympanum (middle ear), and mastoid antrum (*fig. 817*). It is thin and can be broken with the point of the forceps.

Two foramina open 4 to 8 mm, respectively, behind the foramen spinosum, and

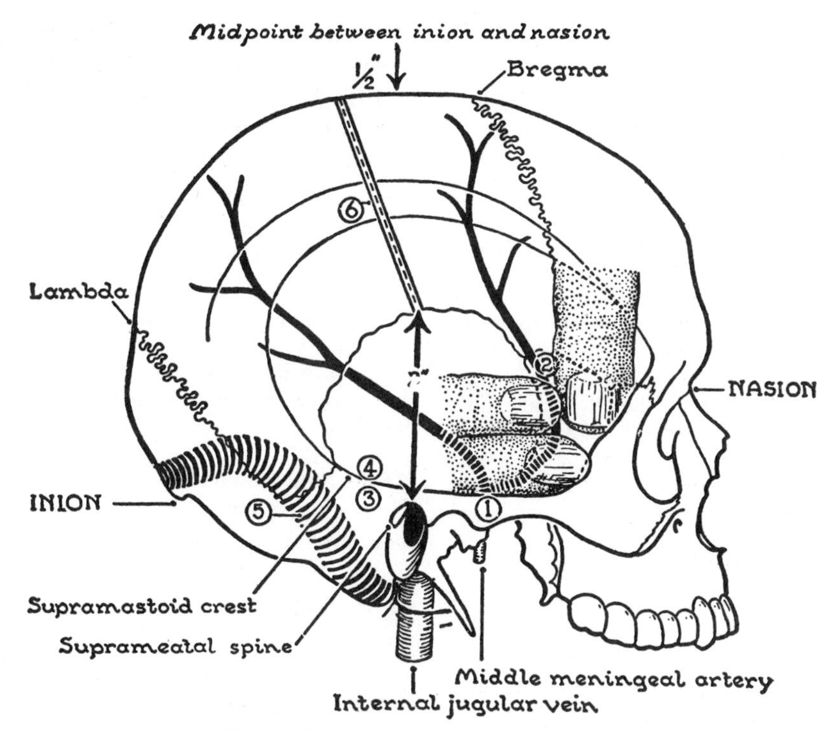

Fig. 689. Surface anatomy of the skull.

1. The stem of the *middle meningeal artery*, passing through the foramen spinosum, deep to the head of the mandible, which is readily located by palpation on opening and closing the mouth.

2. The *anterior branch* of the middle meningeal artery crossing the pterion. The **pterion** is the point where four bones meet (parietal, frontal, greater wing of sphenoid and temporal squama). To locate it on the skin, place the thumb behind the frontal process of the zygomatic bone and two fingers above the zygomatic arch, and mark the angle so formed (Stiles).

This great landmark overlies anterior branch of the middle meningeal art., and the lateral cerebral sulcus.

3. The *Suprameatal Triangle* lies below the supramastoid crest and behind the suprameatal spine; a hole drilled here enters the *mastoid antrum*.

4. A hole drilled above the supramastoid crest enters the *middle cranial fossa*.

5. The *"lateral sinus,"* i.e., transverse and sigmoid sinuses, passing from the inion to a point 2 cm or less behind the external acoustic meatus to become the internal jugular vein deep to the anterior border of the mastoid process.

6. The *central sulcus* of the cerebrum running from a point 1 cm (½") behind the mid inion-nasion point to a point 5 cm above the external acoustic meatus.

Application of Key Figure 688. The *crescent of foramina* on surface projection runs from the head of the mandible to the pterion.

from each a faint groove runs anteromedially, parallel to the superior border: (1) The lateral groove conducts the lesser petrosal nerve to the canaliculus innominatus (p. 608). (2) The medial groove conducts the greater petrosal nerve across the depression for the trigeminal ganglion to the foramen lacerum (*fig. 687*).

Surface Anatomy. *See figure 689.*

CONTENTS OF MIDDLE CRANIAL FOSSA

1. *Inside both layers of dura mater:*
Optic chiasma and nerve.
Hypophysis cerebri (pituitary gland).
Temporal lobe of cerebrum.

2. *Between two layers of dura mater:*
Cavernous sinus.

Nerve V.
Nerves III, IV, VI.
Int. carotid art. and sympathetic plexus.

3. *Embedded in outer layer of dura:*
Middle meningeal vessels.
(Superficial) Petrosal nerves.

Review. Nerves III and IV were seen to pierce the dura mater in the triangle between the free and attached borders of the tentorium. Nerve V was seen ascending from the pons, curving over and therefore rounding off the medial part of the superior border of the petrous bone, and passing through the elliptical mouth of an evagination of arachnoid and dura maters, called the *trigeminal cave*. Nerve VI was seen piercing the dura in the posterior cranial fossa over the inferior petrosal sinus.

Trigeminal Nerve (N. V). In the trigeminal cave, the loose parallel sensory fibers of nerve V form a swelling, called the **trigeminal ganglion** (or semilunar ganglion) (*fig. 690*), homologous with a spinal ganglion. It lies in a "fingertip" depression above the apex of the petrous bone.

The ganglion gives off three divisions:

The *ophthalmic nerve* (V^1) runs forward to divide into three branches, *nasociliary*, *frontal*, and *lacrimal*, which pass through the superior orbital fissure into the orbital cavity. The nasociliary nerve is very important for it is sensory to the eyeball, including the cornea.

The *maxillary nerve* (V^2) passes forward through the foramen rotundum into the pterygopalatine fossa (deep in the skull) and ultimately appears on the face as the infra-orbital nerve.

The *mandibular nerve* (V^3), joined by the motor root of N. V, drops down through the foramen ovale, as through a trap door, into the infratemporal fossa.

The motor root arises from the pons beside the sensory root, and crossing inferior to the ganglion, it joins the sensory part, V^3 to supply the muscles of mastication attached to the mandible.

Cavernous Sinus (described on p. 482). Through it passes the internal carotid artery surrounded with sympathetic fibers; and applied to the lateral side of the carotid artery are the nerves of the ocular muscles—III, IV, and VI (*fig. 690*). Nerve V^1 is a lateral relation of the anterior half of the sinus. These nerves pass through the superior orbital fissure (*fig. 691*).

The Internal Carotid Artery lies in the carotid canal (*fig. 692*), separated from the trigeminal ganglion by dura. It emerges from the apex of the petrous bone and bending upward enters the middle cranial fossa (*fig. 693*). Within the cavernous sinus, it makes a right angled turn, and

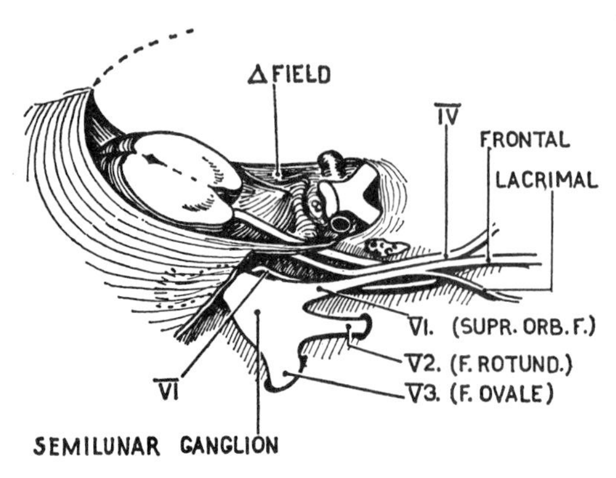

Fig. 690. The 3 divisions of the trigeminal nerve (N. V). (*Ganglion: semilunar = trigeminal.*)

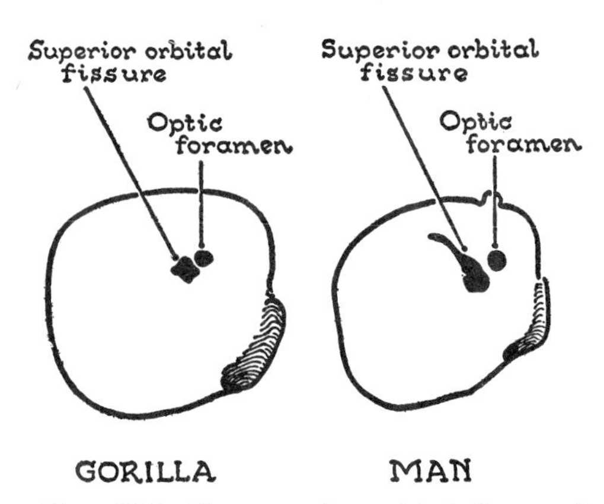

Fig. 691. The superior orbital fissure in the ape and in man. (*Optic: foramen = canal.*) In the ape this fissure is round—it has no lateral extension, as in man. In man it is the medial or primitive part of the fissure that the nerves *crowd through*.

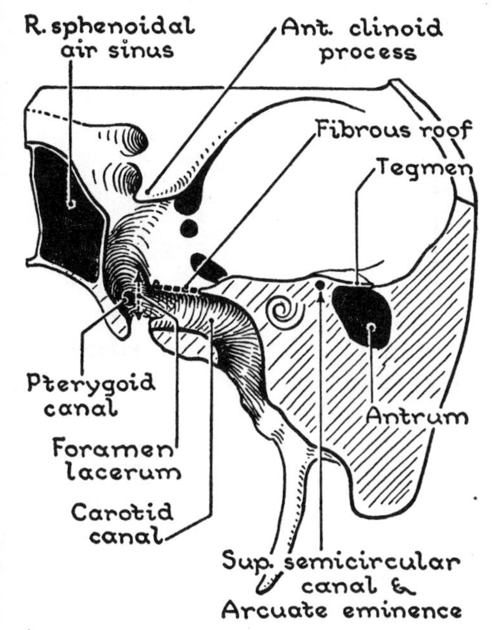

Fig. 692. The carotid canal, shaped like an inverted L, begins on the under surface of the petrous bone and ends at the apex by entering the foramen lacerum.

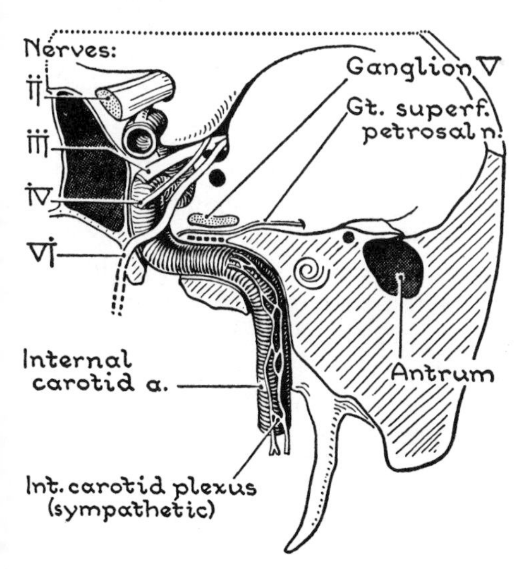

Fig. 693. The course and relations of the intrapetrous and intracranial parts of the internal carotid artery.

passes horizontally forward to the anterior clinoid process. Here it makes an acute angled turn upward, medially, and backward below the anterior clinoid process, pierces dura and arachnoid maters, and divides into the *anterior* and the *middle cerebral artery*. Its intimate relationship to the optic nerve and the three motor nerves to the ocular muscles are shown in *figure 693*.

It sends twigs to the trigeminal ganglion and hypophysis, gives off the ophthalmic and posterior communicating arteries, and a rich anastomotic network across the midline, joining the two internal carotid arteries as they pass through the cavernous sinuses (Parkinson).

The Abducent Nerve (N. VI) supplies the chief abductor of the eyeball. It arises at the lower border of the pons, ascends clamped to the pons by the ant. inf. cerebellar artery, passes through the inf. petrosal sinus, and bends sharply forward. In its horizontal course, it *curves tightly round* the lateral side of the int. carotid artery within the cavernous sinus, with nerves III, IV, and V^1.

It crowds through the superior orbital fissure between the two heads of the Lateral Rectus (*fig. 702*). Sympathetic fibers join it in the sinus.

If the Lateral Rectus is paralyzed, the oblique muscles can abduct the eye to, but not beyond, the normal resting position.

The Trochlear Nerve (N. IV) supplies the ocular muscle that plays in a trochlea (pulley). It is the only nerve to rise from the back of the midbrain, just below the inferior colliculus. It runs forward under the free edge of the tentorium cerebelli (*fig. 681*), pierces the dural triangle, enters the cavernous sinus, and runs forward lateral to the int. carotid art. with nerves III and VI (*fig. 693*).

Nerve IV passes through the superior orbital fissure, above the origin of the muscles and enters the upper border of the Superior Oblique. If paralyzed, double vision (diplopia) results on trying to look downward.

The Oculomotor Nerve (N. III) is motor to all the ocular muscles including Levator Palpebrae, but excluding the two muscles supplied by nerves IV and VI. It also carries parasympathetic fibers to smooth muscles of the eyeball.

It arises from the front of the midbrain and enters the triangular field of the dura mater (*figs. 681, 693*). Then, within the cavernous sinus, it crosses the int. carotid a. and crowds through the superior orbital fissure.

[It dips down medial to the trochlear and frontal nerves, in order to enter the orbit with the abducent and nasociliary nerves, between the two heads of the Lateral Rectus. This it does in two branches.]

Middle Meningeal Artery. This branch of the maxillary artery (*fig. 743*) appears through the foramen spinosum and occupies its grooves (with its veins) (see p. 488).

Just before entering the skull, the middle meningeal artery sends the *accessory meningeal artery* through the foramen ovale to the trigeminal ganglion and the dura. Just after entering the skull, it sends two large twigs, which accompany the two (superficial) *petrosal nerves*, into the petrous bone: one to supply the facial nerve and anastomose with the stylomastoid branch of the post. auricular art.; the other to the tympanum. The anterior branch of the artery commonly anastomoses through the superior orbital fissure with the lacrimal artery (*fig. 703*).

The **middle meningeal veins** pass through the foramen spinosum to the pterygoid plexus; they also join the superior sagittal sinus.

(Superficial) Petrosal Nerves. These two fine nerves, a greater and a lesser, appear in the middle cranial fossa through foramina in the petrous bone. Both (1) run forward and medially to the for. lacerum and for. ovale, respectively; (2) are embedded in the dura mater; (3) are secretory; and (4) are relayed in parasympathetic ganglia, situated beneath the base of the skull (*fig. 797*). Only specialists should be expected to learn the following details:

The greater petrosal nerve, a branch of the nervus intermedius (i.e., of the facial nerve), passes below the trigeminal ganglion, and descends through the foramen lacerum where it is joined by a sympathetic twig from the carotid plexus (the deep petrosal nerve). Thereupon, it passes through the pterygoid canal (as the nerve of the pterygoid canal) to the pterygopalatine ganglion; whence it is relayed to the lacrimal, nasal, and palatine glands (*fig. 693.1*).

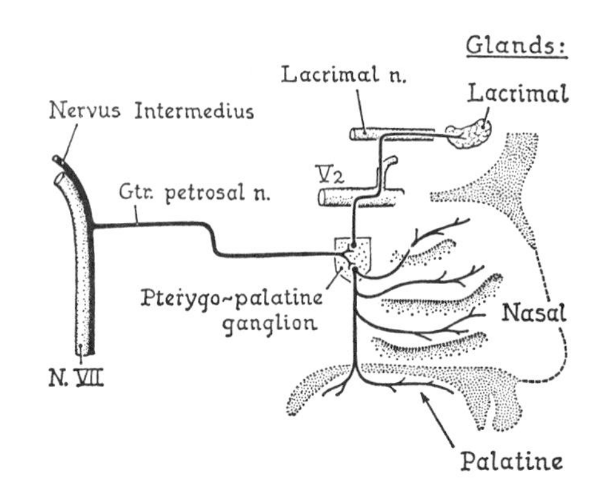

Fig. 693.1. Distribution of fibers in the greater petrosal nerve.

The lesser petrosal nerve is the continuation of the tympanic branch of the glossopharyngeal nerve (*fig. 761*). It appears in the middle cranial fossa lateral to the greater nerve, and leaves the skull through the foramen ovale to join the otic ganglion. Thence it is relayed to the auriculotemporal nerve, which conducts it to the facial nerve, and so it reaches the parotid gland (*fig. 797*).

Internal Carotid Nerve and Plexus (*fig. 693*). This branch of the superior cervical ganglion forms a plexus on the int. carotid artery, as it runs through the carotid canal and cavernous sinus, to be distributed with the branches of the artery (*fig. 766*).

In the cavernous sinus it sends twigs to nerves III, IV, V^1, and VI; other twigs pass through the superior orbital fissure to the eyeball.

Stimulation of these fibers causes the pupil to dilate and the involuntary muscles in the eyelids (sup. and inf. tarsal muscles, *fig. 698*) to contract, thereby widening the palpebral fissure.

While in the carotid canal, it sends two twigs, **caroticotympanic nerves,** to the tympanic plexus, and one twig, **deep petrosal nerve,** to unite with the greater petrosal nerve to form the N. of the pterygoid canal and so to the pterygopalatine ganglion.

38 ORBITAL CAVITY AND CONTENTS

Bony Cavity—Orbital Margin; Optic Canal; Fissures and Suture Lines; Walls and Beyond; Angles; Foramina.

CONTENTS OF ORBITAL CAVITY

Bulbus Oculi; Definitions. Optic Nerve.
Muscles of Eyeball—Four Recti; Two Obliqui; Cone of Muscles.
Muscles of Eyelids—Levator Palpebrae; Tarsal Muscles.
Vagina Bulbi (Fascia Bulbi).
Actions of the Six Muscles.
Nerves of the Orbit—Special Sense; Motor; Sensory; Autonomic (Ciliary Ganglion); Nerves.
Ophthalmic Artery and Branches; Ophthalmic Veins.
Lacrimal Gland. Other Contents.

THE EYEBALL

Composition of Eyeball or Bulb—Three Outer Concentric Coats; Structures Revealed by Dissection of Cow's Eyeball.
Development of Eyeball.

Bony Cavity

Each pear-shaped cavity is fundamentally pyramidal, having four walls, an apex, and an *orbital margin* at its base.

The parallel medial walls are separated by the nasal cavities. The lateral walls are at right angles to each other. The apex is at the optic canal (*fig. 694*).

Orbital Margin. Three bones—*frontal, maxillary*, and *zygomatic*—contribute nearly equal thirds to the orbital margin (*fig. 695*). The lower margin, traced medially, becomes the *anterior lacrimal crest;* the upper margin becomes the *posterior lacrimal crest*. The crests bound the fossa for the lacrimal sac in front and behind.

Optic Canal. This lies between the body of the sphenoid and the two roots of the lesser wing. It is 3 to 9 mm long.

Fissures and Suture Lines (*fig. 695*). A narrow bar of bone separates the optic canal from the upper limb of a V-shaped fissure. This is the *superior orbital fissure*. The lower limb is the *inferior orbital fissure*, which lies between the lateral wall and floor of the orbit (*fig. 695*).

The infra-orbital groove on the orbital plate of the maxilla runs forward from the inferior orbital fissure to open on the face as the infra-orbital foramen.

The anterior and posterior *ethmoidal foramina* lie in, or near, the fronto-ethmoidal suture (*fig. 695*).

Walls and Beyond (*fig. 696*). The medial wall is thin and papery; the lateral wall is the strong wall, transmitting forces from the molar teeth upwards.

Comparative Anatomy. In mammals lower than primates, there is no bony lateral wall; in many even the lateral orbital margin is missing, but the lines of force transmission are different as well.

A perforation made (1) in the *roof* leads to the anterior cranial fossa (if made anteriorly, it first traverses the frontal air sinus); (2) in the *floor*, to the maxillary air sinus; (3) in the *medial wall*, to the ethmoidal air cells; behind them, to the sphenoidal air sinus; and, *in front of them, to the atrium of the nasal cavity;* and (4) in the *lateral wall*, to the temporal fossa or, behind it, to the middle cranial fossa.

The *periorbita* (periosteum) is tough and easily detached, especially from the roof and medial wall—a matter of surgical importance.

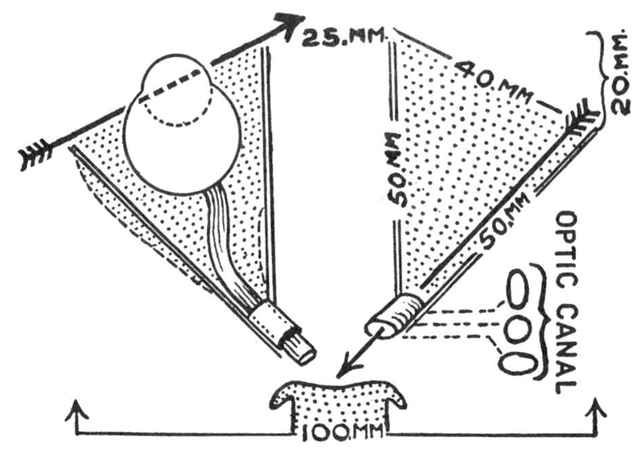

Fig. 694. The orbital cavities on horizontal section, and their dimensions.

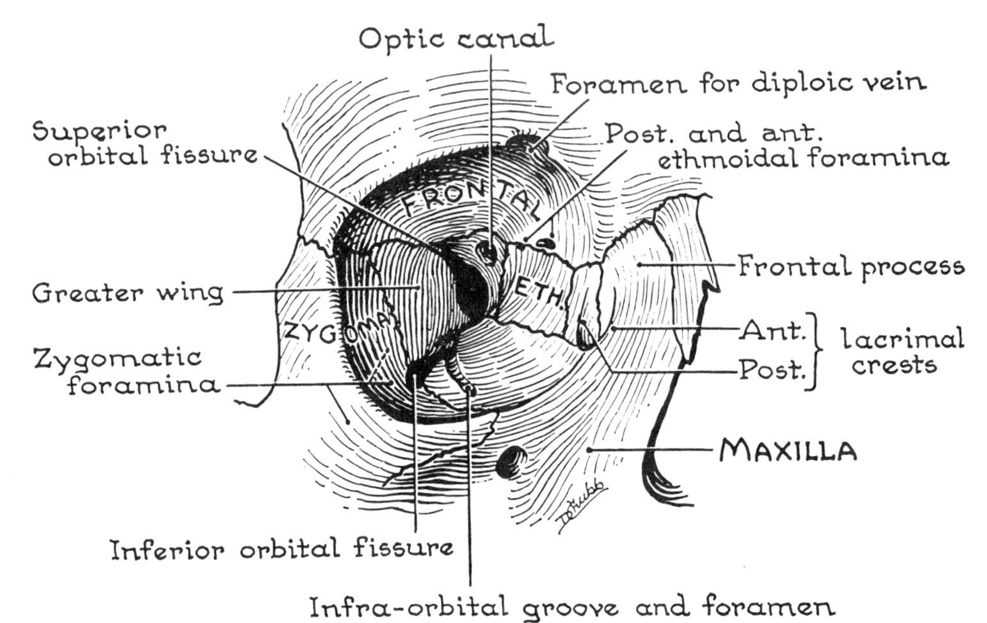

Fig. 695. The bony walls of the orbit (orbital cavity).

Angles. Behind each of the 4 angles of the orbital margin there is a feature of note (*fig. 656*). Thus:

Superolaterally—fossa for the lacrimal gland.

Superomedially—fovea (or spine) for the trochlea of the Superior Oblique (*fig. 702*).

Inferomedially—nasolacrimal canal and origin of the Inferior Oblique.

Inferolaterally—end of the inferior orbital fissure, 2 cm back.

Foramina. The *optic canal* is the royal entrance to the orbital cavity; through it pass the optic nerve (N. II) within its three meningeal tubes (*fig. 698*), and the ophthalmic artery which supplies this region (*fig. 703*).

The superior orbital fissure is the general entrance; through it pass motor nerves III, IV, and VI, sensory nerve V^1, sympathetic fibers, and ophthalmic veins.

The inferior orbital fissure is a "tradesmen's" entrance; through it pass branches from V^2 (infra-orbital and zygomatic nerves), the infra-orbital artery, and a communicating vein from the inf. ophthalmic vein to the pterygoid plexus.

Other openings serve as exits—nasolacrimal canal, infra-orbital groove, ant. ethmoidal, post. ethmoidal, zygomatic, and supra-orbital foramina.

CONTENTS OF ORBITAL CAVITY

Bulbus Oculi

The eyeball is 24.5 mm long. It occupies the anterior half of the cavity; muscles and fat largely fill the posterior half. It projects slightly beyond the orbital margin (*fig. 694*). The projecting supra-orbital margin and the bridge of the nose prevent a flat surface, such as a book, from striking the eye.

Definitions. The white fibrous posterior five-sixths of the eyeball is the *sclera;* the transparent anterior one-sixth is the *cornea.* These are structurally continuous at the *corneoscleral junction.* The center of the corneal curvature is the *anterior pole;* opposite it is the *posterior pole.*

A line joining the poles is the *anteroposterior, sagittal,* or *optic axis.* The *equator* encircles the bulb midway between two poles in the coronal plane. *The orbital axis,* or long axis of the orbital cavity, is the axis around which the Recti are arranged.

Vision is most acute where rays of light come to a focus on the retina at the posterior pole. This part of the retina is the yellow spot or *macula.* The optic nerve pierces the sclera 3 mm to the medial or nasal side of the posterior pole at the *optic disc* (papilla), which is the blind spot.

Optic Nerve

The optic nerve develops as part of the brain and, so is surrounded with meninges and bathed in cerebrospinal fluid. It takes a sinuous course in the orbit and, so, does not restrain movements of the eyeball. Its dural sheath extends to the sclera and blends with it.

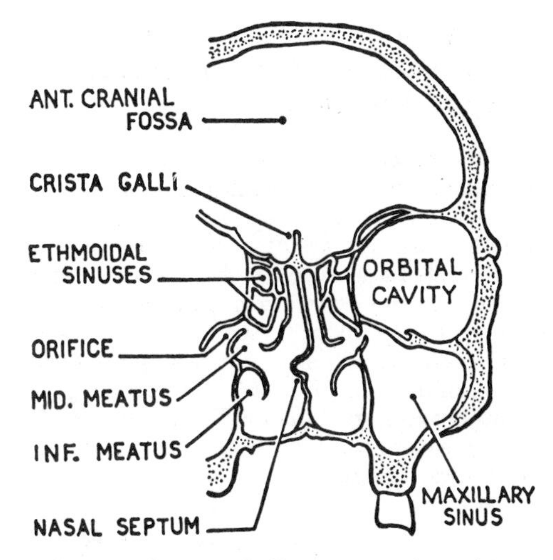

Fig. 696. The skull, on coronal section.

If the nerve is cut across close behind the sclera, a central black spot—the *central artery and vein of the retina*—can be seen.

Muscles of the Eyeball

Four are straight and 2 oblique:

The Recti (superior, inferior, medial, lateral) arise from the margin of a fibrous cuff, which encircles the optic canal and the medial end of the superior orbital fissure (*fig. 702*). They are inserted by band-like aponeuroses into the sclera, 6 to 8 mm behind the corneoscleral junction (*fig. 697*) and are there loosely covered with conjunctiva.

The Recti spread like the staves of a barrel: behind they are applied to the four walls of the orbital cavity; in front they hug the globe. Each Rectus has an areolar sheath, and adjacent sheaths are united by areolar tissue forming "*the muscle cone*" (*fig. 698*). A hammock of condensed areolar tissue, which is slung between the points of attachment of the palpebral ligaments to the orbital margin, supports the bulb, and extensions sent to the Medial and Lateral Recti act as "*check ligaments.*"

The Levator Palpebrae Superioris is delaminated from the upper part of the Superior Rectus, so they share the same nerve. When the Rectus raises the eye, the Levator raises the lid. It is inserted in three layers (*fig. 698*). *Details:*

(1) The anterior layer passes through the Orbicularis and is attached to the skin of the lid. Its edges extend to the medial and lateral palpebral ligs. and are attached with them. (2) The intermediate layer is a sheet of involuntary muscle, the *Superior Tarsal Muscle*, which is attached to the upper tarsus. (3) The posterior layer is fascial and passes to the superior fornix of the conjunctiva (*figs. 655, 656, 698*).

The involuntary *Inferior Tarsal Muscle* (*fig. 698*), spread forward from the Inferior Rectus to the lower tarsus.

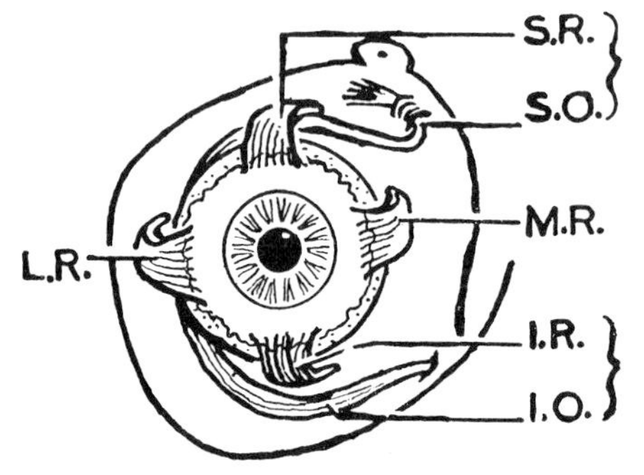

Fig. 697. Six muscles of eyeball (from front).

Vagina Bulbi (Fascia Bulbi) (*fig. 698*). The eyeball is invested in a bursal sheath, which extends from the optic nerve to the corneoscleral junction. This sheath of the bulb is necessarily pierced by the six tendons acting on the bulb, and it is reflected for a short distance along each of them; in the case of the Superior Oblique—backward beyond the trochlea.

The Obliques (Superior and Inferior) are directed backward and laterally from just behind the superomedial and inferomedial angles, respectively, of the orbital margin, to be inserted by fan-shaped tendons into the superolateral quadrant of the posterior half of the bulb.

The Inferior Oblique arises from the floor, lateral to the entrance to the nasolacrimal canal (*fig. 702*). The Superior Oblique arises from the common fibrous cuff. Its tendon passes through a fibrocartilaginous loop, the *trochlea*, at the superomedial angle, so the direction of its pull remains like that of the Inferior Oblique.

[The two Obliques cross below the corre-

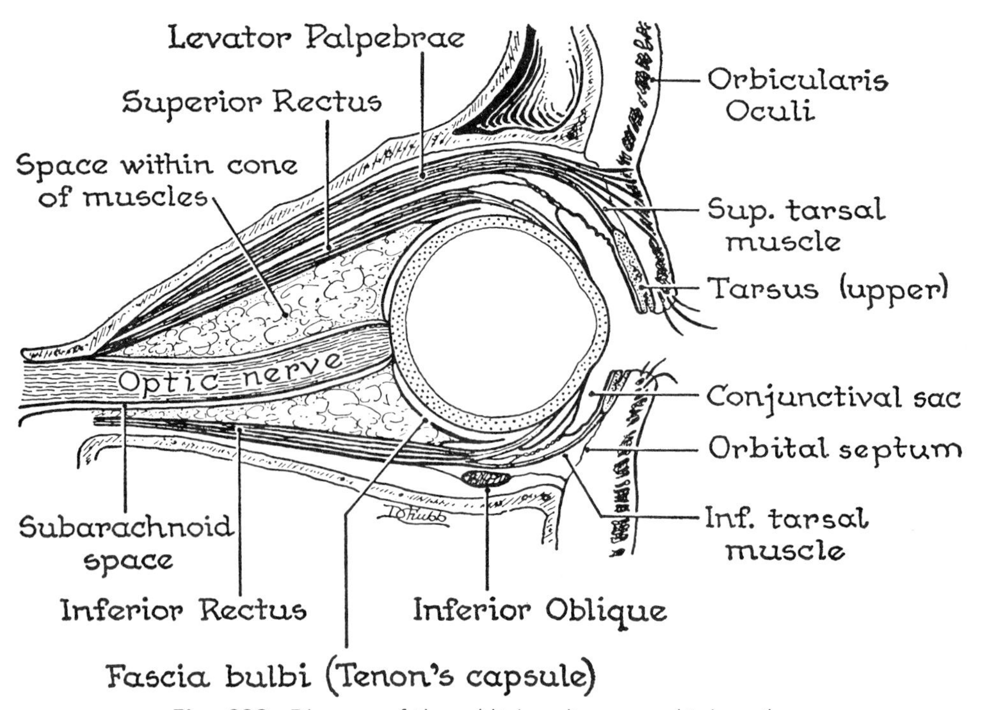

Fig. 698. Diagram of the orbital cavity, on sagittal section.

sponding Recti—S. Oblique below S. Rectus; I. Oblique below the I. Rectus.

The nerves enter the four Recti on their bulbar surfaces behind their midpoints, and the two Obliqui on their borders (upper and posterior; *see figure 702*).]

Actions of the Six Muscles (*fig. 699*). The actions should not be memorized. They can be worked out readily, provided it is appreciated: (1) that the eyeball has three axes on which to rotate (sagittal, horizontal, vertical); (2) that the four Recti are arranged around the orbital axis—not around the optic axis (p. 495)—hence, M., S., and I. Recti act as adductors; and (3) that the two Oblique muscles pass behind the vertical axis and are inserted behind the equator. Hence, they act as abductors.

The Medial and Lateral Recti act on one axis only; each of the other muscles act on multiple axes.

Optic Nerve and Nerves of the Orbit

1. Special Sense—II.
2. Motor—III, IV, and VI.
3. Sensory—V[1] (frontal, lacrimal, and nasociliary branches).
4. Autonomic—(1) sympathetic fibers from the carotid plexus, and (2) para-

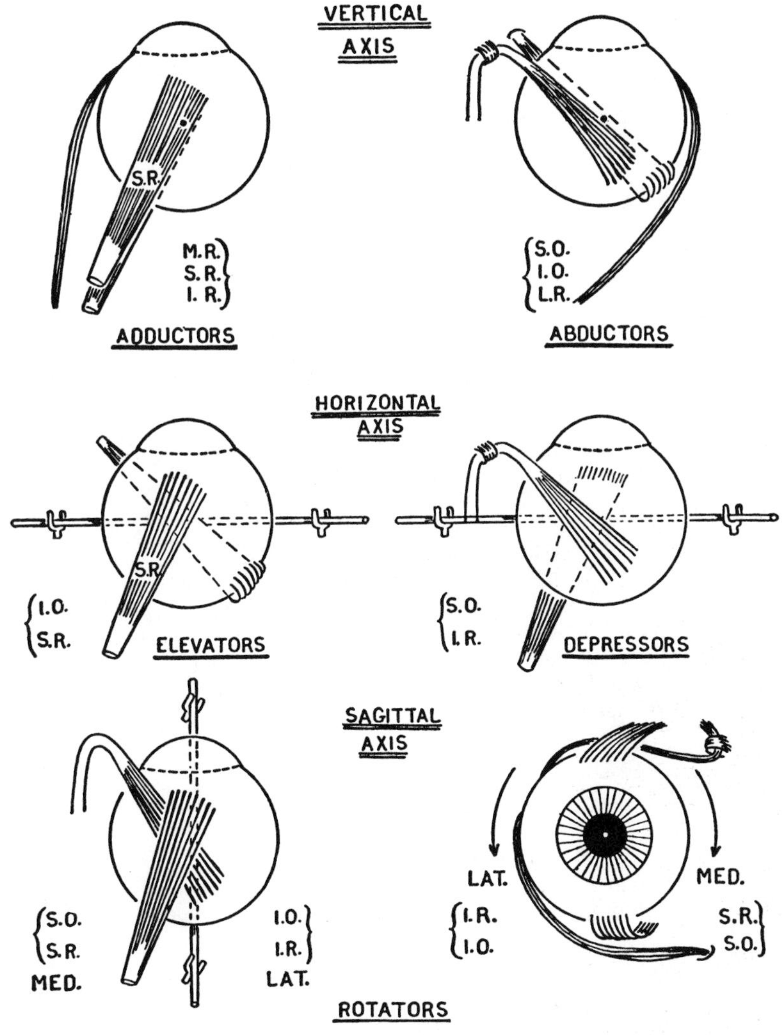

Fig. 699. The actions of the six muscles of the right eyeball represented graphically.

sympathetic fibers traveling with N. III.

All but the optic nerve crowd through the superior orbital fissure: three of these, the *trochlear*, *frontal*, and *lacrimal*, pass above the origin of the Lateral Rectus and lie between the roof of the orbit and the cone of muscles; the others pass as a bundle through the fibrous origin of the Lateral Rectus, and lie within the cone of muscles (*figs. 700–702*).

The Three Motor Nerves (*fig. 702*). *The Abducent Nerve* (VI) clings to the ocular surface of the Lateral Rectus and enters it behind its midpoint.

The Trochlear Nerve (IV) enters the upper border of the Superior Oblique far back.

The Oculomotor Nerve (III) supplies three Recti, Inferior Oblique, and Levator Palpebrae; and its conveys parasympathetic fibers to the ciliary ganglion, thence they are relayed to the ciliary muscle and Sphincter Pupillae.

Nerve III is employed when an object is examined close at hand, as in reading, because it causes convergence of the eyes (adductor muscles), accommodation of the lens or focusing (ciliary muscle), and contraction of the pupil (circular fibers of iris), thereby shutting out peripheral light.

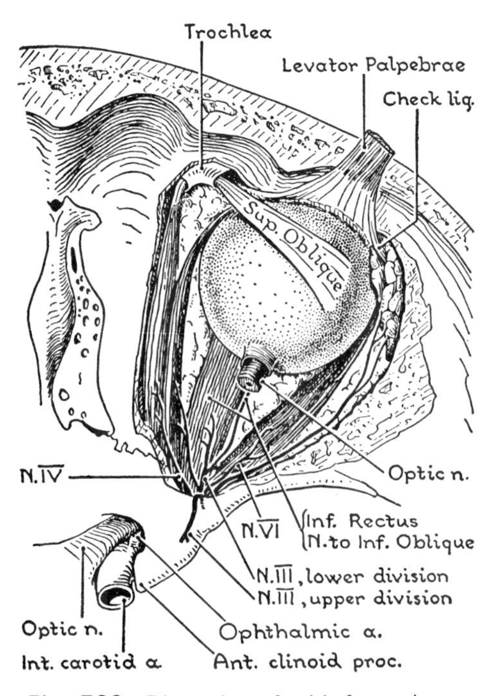

Fig. 700. Dissection of orbit from above.

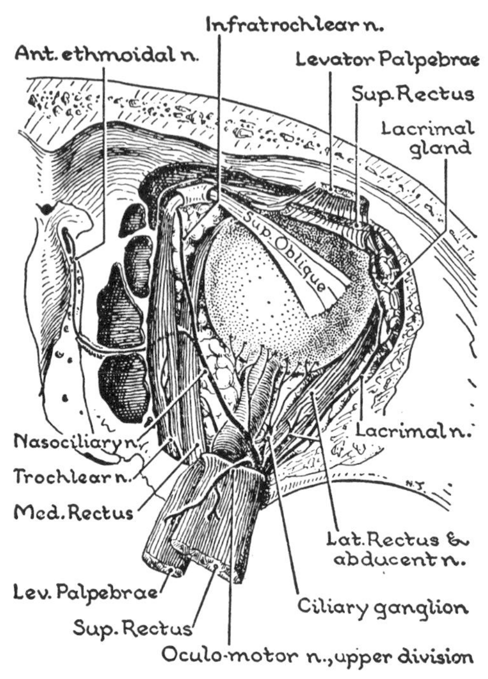

Fig. 701. Dissection of orbital contents.

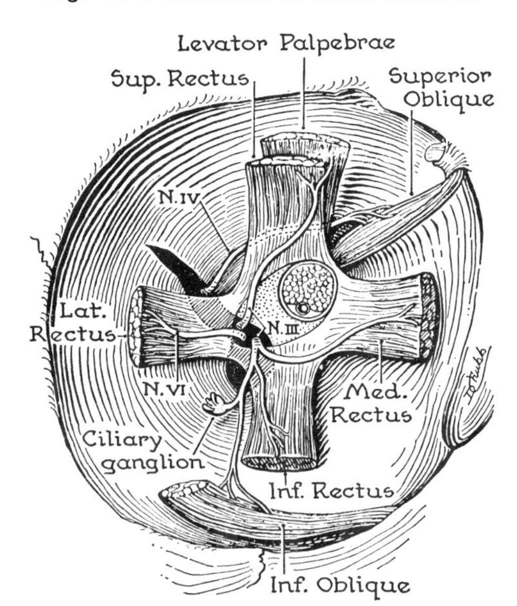

Fig. 702. Sketch of the distribution of cranial nerves III, IV, and VI.

Nerve III passes between the two heads of the Lateral Rectus as an upper and a lower division: of these, the upper supplies the Superior Rectus and Levator Palpebrae: the lower sends a branch below the optic nerve to the Medial Rectus, a branch to the Inferior Rectus, and a branch that runs along the lateral border of the Inferior Rectus to the Inferior Oblique. The nerve to the Inferior Oblique delivers the preganglionic fibers (motor root) to the ciliary ganglion.

Ophthalmic Nerve (V^1). This sensory nerve passes through the superior orbital fissure as three branches: frontal, lacrimal, and nasociliary. The *frontal nerve* passes above the Lateral Rectus, runs between the orbital plate of the frontal bone and the Levator Palpebrae, and divides into supra-orbital and supratrochlear branches. These turn round the supra-orbital margin to be distributed to scalp and eyelid (*fig. 658*).

The *lacrimal nerve* follows the upper border of the Lateral Rectus and ends in the upper lid. It accepts from the zygomatic nerve secretory fibers, relayed in the pterygopalatine ganglion, and delivers them to the lacrimal gland (*fig. 693.1*).

The **nasociliary nerve** (*fig. 701*) is of supreme importance on account, not of the nasal branch, but of the ciliary. The nasociliary nerve passes between the two heads of the Lateral Rectus, crosses above the optic nerve, and runs on the medial wall of the orbit between the Superior Oblique and Medial Rectus.

Two **long ciliary nerves** arise from the nasociliary nerve, as it crosses above the optic nerve; they accompany the short ciliary nerves to the eyeball. They are mixed—sensory and sympathetic.

Sensation from the tip of the nose and from the cornea is subserved by the respective divisions of the nasociliary nerve (*fig 702.1*). *Details*—

On the medial wall of the orbit, the nasociliary nerve sends (1) a twig through the posterior ethmoidal foramen to the sphenoidal and ethmoidal cells and (2) the *infratrochlear nerve* forward, below the trochlea, to the tear sac and the region all around it, and it continues as (3) the *anterior ethmoidal nerve*.

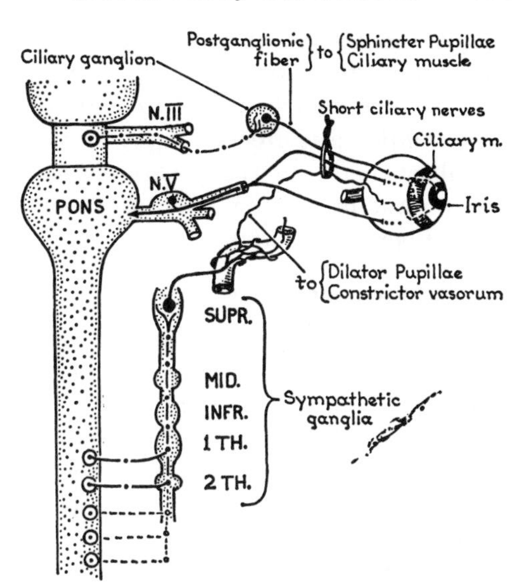

Fig. 702.1. The nerve supply to the eyeball.

The *anterior ethmoidal nerve* passes through the anterior ethmoidal foramen, and appears extradurally in the ant. cranial fossa. It then descends through a slit at the side of the crista galli into the nasal mucosa, and supplies ethmoidal cells and the front of the nasal cavity. Descending behind the nasal bone, it passes between nasal bone and cartilage to appear on the dorsum of the nose, where, as the *ext. nasal nerve*, it extends to the tip (*fig. 658*).

Autonomic and Ciliary Nerves (*fig. 701*). Far back between optic nerve and Lateral Rectus there is a small brown ganglion, the **ciliary ganglion,** the relay station for the parasympathetic fibers of nerve III. These fibers pass to the ganglion via the nerve to the Inferior Oblique and leave it via 12 or more *short ciliary nerves*. The short ciliary nerves pierce the scleral coat around the optic nerve and run forward between the scleral and choroidal coats. They are motor to the ciliary muscle and Sphincter Pupillae (circular fibers of the iris).

Short ciliary nerves, however, are not entirely parasympathetic. Thus, a branch from the sympathetic and another from the nasociliary nerve enter the ciliary ganglion and pass right through in the short ciliary nerves: the sympathetic fibers are vasoconstrictor; the nasociliary fibers are sensory.

The two *long ciliary nerves*, described with the nasociliary nerve (above), also carry sympathetic fibers. These join the nasociliary nerve in the cavernous sinus and are motor to the Dilator Pupillae (radial fibers of the iris).

Ophthalmic Artery (*fig. 703*)

This artery supplies the contents of the orbital cavity and sends branches beyond the cavity. It arises from the int. carotid artery, passes through the optic canal and, piercing the dural sheath of the nerve, finds itself free within the cone of muscles. It then crosses above the optic nerve.

Branches within the orbit are:
(1) The **central artery of the retina,** which, with its companion vein, enters the sheath of the nerve 1 cm behind the eyeball, runs in the center of the optic nerve through the sclera to the retina. Obstruction leads to instant and total blindness.
(2) **posterior ciliary arteries** (six or more).

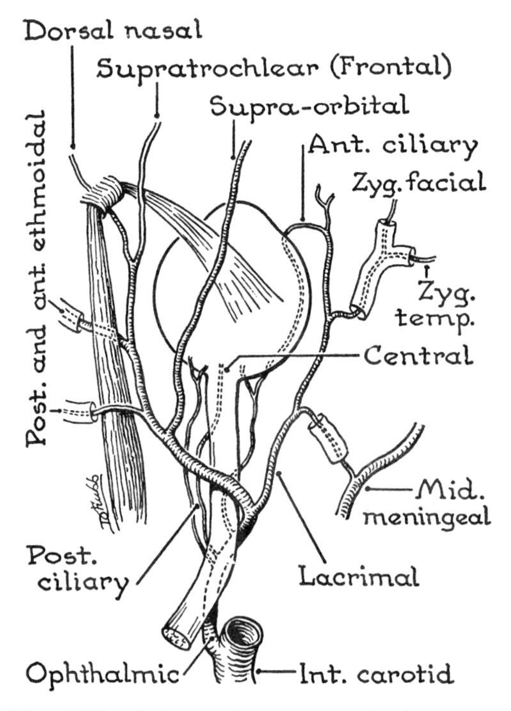

Fig. 703. Ophthalmic artery and its branches.

(3) **anterior ciliary arteries,** derived from the muscular branches to the Recti.
(4) Six branches of the ophthalmic a. that stream out of the orbit in company of correspondingly named nerves (*fig. 658*) and anastomose with branches of the ext. carotid artery.

Ophthalmic Veins

The superior ophthalmic vein anastomoses with the facial vein and, since it has no valves, the blood can flow in either direction (*fig. 680*). It crosses above the optic nerve, passes through the sup. orbital fissure and ends in the cavernous sinus.

The inferior ophthalmic vein on the floor of the orbit, communicates through the inf. orbital fissure with the pterygoid plexus, crosses below the optic nerve, and ends either in the superior vein or in the cavernous sinus.

Lymph Vessels

Lymph vessels have not been described definitely within the orbital cavity, except those from the lacrimal gland, which pass to the parotid nodes.

Lacrimal Gland (*fig. 703.1*)

This serous gland is placed behind the *superolateral* angle of the orbital margin, between the orbital plate of the frontal bone and the conjunctiva. It is indented by the lateral border of the Levator Palpebrae

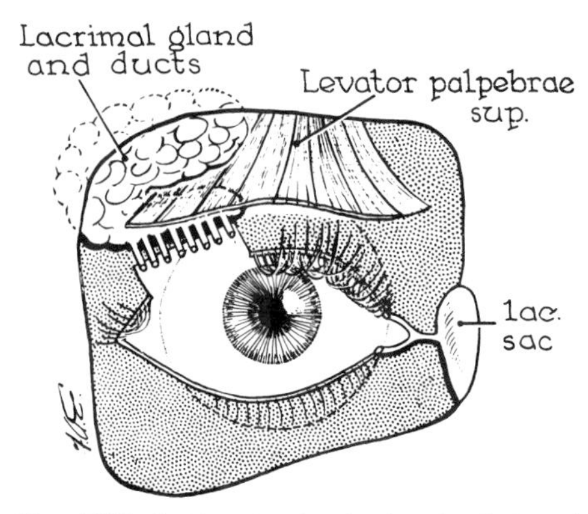

Fig. 703.1. The lacrimal gland of the right orbit (schematic).

and thereby made J shape. So, it has a deep (or orbital) lobe and a superficial (or palpebral) lobe. Less than a dozen ducts open near the superior fornix of the conjunctiva.

The **Lacrimal Nerve** (sensory) conveys secretory and sympathetic fibers received from the zygomatic nerve (*fig. 795*). *Details—*

The secretory fibers travel via intermediate, greater petrosal, and pterygoid canal nerves to the pterygopalatine ganglion; thence to be relayed by N. V^2, zygomatic nerve, and a communicating branch to the lacrimal nerve (*fig. 693.1*).

Other Contents of Orbit. Three branches of the maxillary nerve (V^2) enter the orbit through the inf. orbital fissure: (1) the *infra-orbital nerve*, lies embedded in the floor; (2) the *zygomatic nerve* clings to and perforates the lateral wall and sends secretory and sympathetic fibers to the lacrimal gland; and (3) the *orbital branch* of the pterygopalatine ganglion supplies the sphenoidal and ethmoidal sinuses and the *Orbitalis muscle* (involuntary) which bridges the inf. orbital fissure.

THE EYEBALL

Three Concentric Coats and their parts (*figs. 704, 705*):

1. Outer or fibrous coat:
   Sclera and corneal.
2. Middle or vascular coat:
   Choroid, ciliary body, and iris.

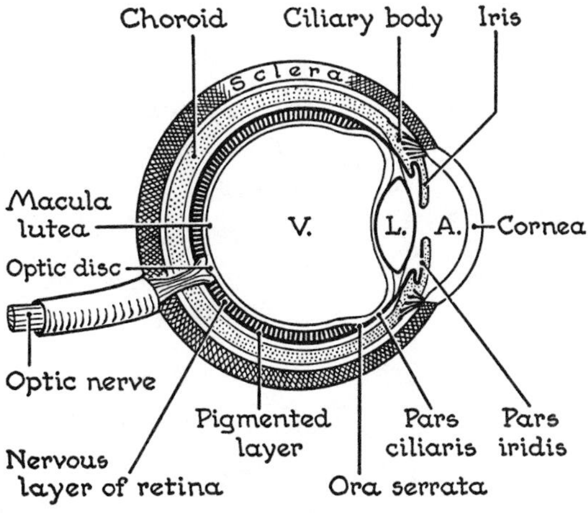

Fig. 704. Scheme of an eyeball, sagittal section.

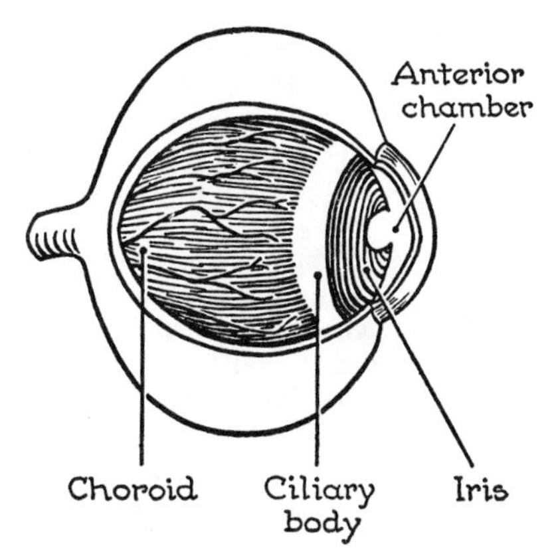

Fig. 705. Eyeball of cow, middle coat exposed.

3. Inner or retinal coat:
   Outer pigmented.
   Inner nervous.

Three Enclosed Refractive Media: aqueous humor—lens—vitreous body.

Dissection of Cow's Eyeball

A convenient way to acquire a general knowledge of the anatomy of the human eyeball is through the dissection of the eyeball of a cow, because it is large and obtainable without difficulty. It is necessary to remove the adnexa preparatory to hardening the ball in formalin or alcohol overnight.

The *conjunctiva* is only loosely attached to the sclera and is easily separated from it up to the corneal margin. The epithelium of the *cornea* is firmly adherent and cannot wrinkle.

Holding the bulb firmly in the left hand, shave off from the neighborhood of the equator thin slices of the sclera, which is grayish in color, until the choroid, which is told by its jet blackness, is exposed. Then holding the bulb loosely so as to relieve tension carefully insert the point of a probe through the hole just made and detach the choroid all round it. With scissors snip away the part of the sclera so freed. And so, with the alternate aid of probe and scissors, and working meridionally, between anterior and posterior poles, remove a third or more of the outer coat. This will expose an underlying elliptical area of the choroid. The point of the probe must be directed continuously against the sclera in order that the delicate middle and inner coats shall not be punctured. With a sweeping movement detach the middle coat from the corneoscleral junction where it is firmly attached.

Cut away the remaining part of the cornea and, after freeing structures with the probe, carefully enter the blade of the scissors through the pupil and cut a large square flap in the iris. Raise the flap and enlarge it backward through the ciliary body into the choroid, thereby exposing the jelly-like vitreous body.

Using probe and scissors, cut circularly through each of the three coats at the equator: first through the remaining part of the sclera, next through the choroid, and then through the retina. The bulb then can be separated into anterior and posterior halves; the vitreous body remains with the anterior half.

The space behind the cornea is the *anterior chamber*. It is continuous through the *pupil* with the *posterior chamber*. Both chambers contained *aqueous humor* in life. The lens touches the back of the *iris* at the pupillary margin.

In man the *cornea* forms one-sixth of the outer coat of the bulb; the *sclera* five-sixths. The cornea is more convex than the sclera, i.e., it is a segment of a smaller sphere. The cornea is about 1 mm thick; the sclera is thinner, especially at the equator. The two, however, are structurally continuous at the *corneoscleral junction*, where there are several features of note:

1. The edge of the cornea is overlapped by the sclera like a watch glass in its case.

2. Running circularly around the sclera there is a pinpoint canal, the *scleral sinus* (*fig. 706*).

3. The middle coat has a white zone, 6 mm broad, the *ciliary body*, which is continuous in front with the iris and behind with the choroid. The ciliary muscle arises from the sclera at the corneoscleral junction and is there firmly attached.

4. The acute angle between the cornea and the iris (actually between sclera and iris) is the *iridocorneal angle*.

The angle is crossed by interlacing strands, the *pectinate ligament*, that pass backward from the edge of the posterior elastic membrane of the cornea to the region of the iris and sclera, and the spaces enclosed are lined with the mesothelium of the anterior chamber. The anterior chamber communicates through the spaces with the scleral sinus which in turn communicates with the scleral veins.

The *ciliary processes*, about 70 short, black, finger-like projections from the ciliary body, occupy the peripheral part of the posterior chamber. They may reach to the periphery of the lens (*fig. 706*).

The *hyaloid membrane*, which encloses the *vitreous body* behind the lens, passes forward in folds, which the ciliary processes occupy, to near the margin of the lens where it divides into an anterior and a posterior layer. The posterior layer continues to encapsule the vitreous (*fig. 706*). The anterior and stronger layer, the *suspensory ligament of the lens*, blends with the front of the lens capsule. A triangular canal, the *zonular spaces*, encircles the margin of the lens between the two layers.

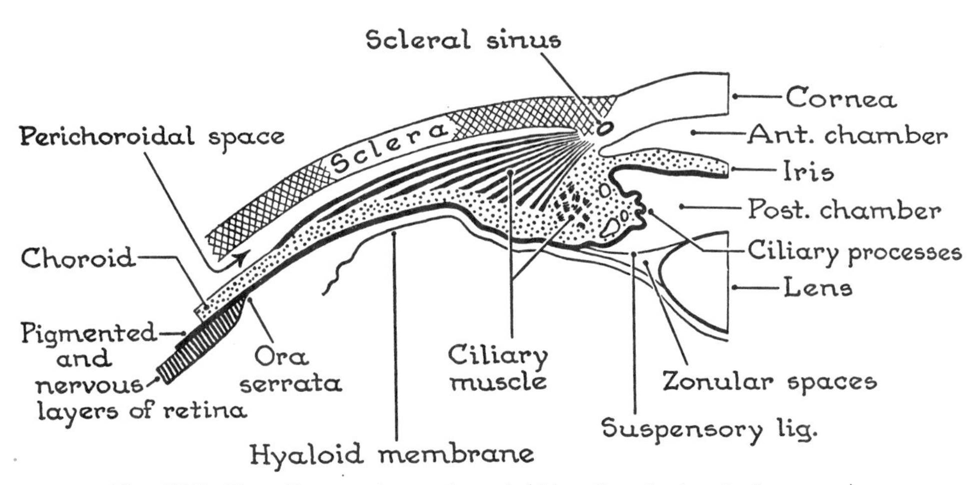

Fig. 706. The ciliary region, enlarged. (After Cunningham's Anatomy.)

The *ciliary muscle* (*fig. 706*) is triangular on cross-section. Its fibers radiate backward from the corneoscleral junction to the choroid which they pull forward when they contract, thereby causing relaxation of the suspensory ligament and allowing the lens to become more convex. The muscle also possesses an inner bundle of circular fibers.

The *Optic Nerve* in man pierces the sclera in bundles, 3 mm to the nasal side of the posterior pole of the bulb, and at this point the sclera is cribriform and weak. Next, the fibers pierce the choroid and the outer layer of the retina, which adheres to the choroid; and then after forming a circle, the *optic disc* (papilla), which is a blind spot (1.5 mm in diameter), they spread out as the inner layer of the retina. The *central artery* and *vein* each pass through the disc as two vessels, which bifurcate and pass to the four quarters of the inner layer of the retina (*fig. 706.1*).

After death the retina is gray and lusterless like an exposed photographic film. No longer supported by the vitreous it detaches itself from the choroid everywhere except at the disc, and it becomes wrinkled and perhaps broken.

The *choroid*, loosely adhering to the inner surface of the sclera, is easily detached and a *perichoroidal space* opened up.

The *retina* becomes thin along a wavy line, the *ora serrata*, a short distance be-

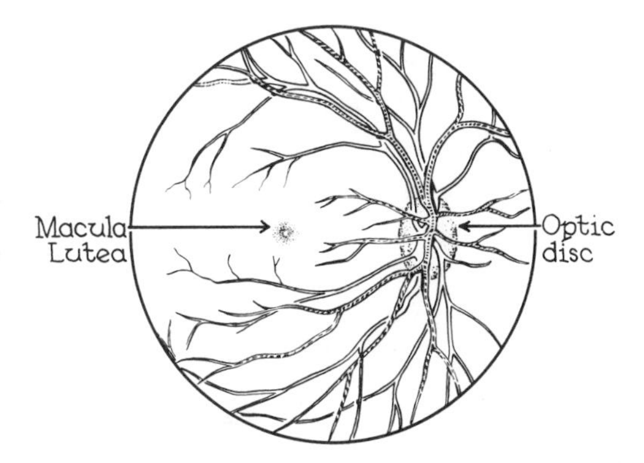

Fig. 706.1. The retina and its arteries and veins as seen in the right eye through an ophthalmoscope.

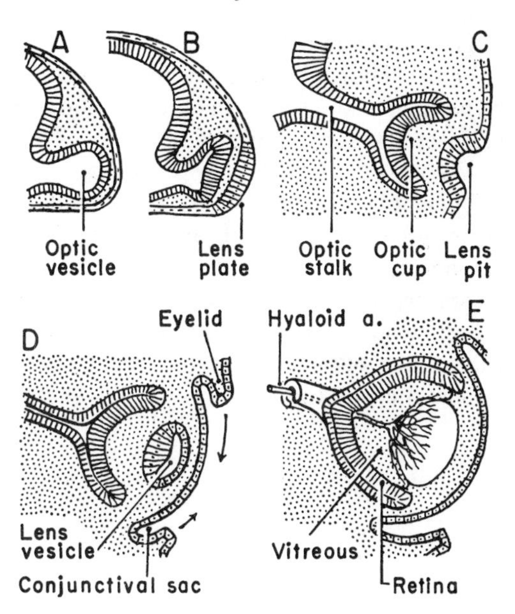

Fig. 706.2. Stages in the development of retina, lens, and conjunctival sac. (After Mann.)

hind the ciliary muscle. Behind this the retina is true optic retina, but in front (where rays of light cannot reach) the retina is represented by two layers of cubical cells, the outer of which is pigmented. These two layers are carried forward over the ciliary body and iris to the margin of the pupil (*fig. 706*).

The Refractive Media. The *cornea* does the chief focusing; the lens is for fine adjustment. The *lens*, derived from the same layer as skin, hardens or cornifies with advancing age. The hardening begins at the center, and accommodation becomes increasingly difficult. When hardened, a lens may be split into layers somewhat like an onion. The *vitreous body* is embryonic tissue comparable to that in the umbilical cord.

Arteries. Ciliary arteries, page 500.

The *central a.* supplies the inner layers of the retina; it is an end artery (page 500).

The function of the vascular choroid is to nourish by diffusion the outer layers of the retina; hence, where the retina is thin (i.e., in front of the ora serrata) the choroid is much less vascular.

The cornea is bloodless.

Veins. Four choroidal veins, *vorticose veins*, pierce the sclera behind the equator to enter the ophthalmic vein.

Nerves (pp. 498–500).

Development. The retina developed as an outgrowth from the brain and, so, is of ectodermal origin. In its early stage it resembled an inflated rubber balloon with a hollow stalk which was continuous with the cavity of the 3rd ventricle. On the appearance of the lens, which developed from the overlying skin surface and likewise is of ectodermal origin, the balloon was deflated and invaginated and made cup-shaped, hence the two layers of the retina (*fig. 706.2*).

The central artery originally passed through the hyaloid canal in the vitreous body and anastomosed in the capsule of the lens, which early was vascular. Before birth, this capsule atrophies and the artery, ceasing to supply it, confines itself to the retina and becomes an end artery.

39 ANTERIOR TRIANGLE OF NECK

ANTERIOR TRIANGLE AND MEDIAN LINE OF NECK

Subdivisions; Landmarks.

Superficial Structures: Platysma; Deep Fascia; Superficial Veins and Nerves.

Median Line of the Neck

Boundaries; Median Structures.

Infrahyoid Muscles; Cervical Visceral.

Pharynx and Constrictors; Thyroid Gland.

Carotid Sheath.

CAROTID TRIANGLE

Posterior Belly of Digastric in Key Position.

Stylohyoid; Nerve Supply; Relations.

Nerves in Carotid Triangle—XI; XII; Ansa Cervicalis; X; Ext. and Int. Laryngeal Branches.

Arteries in Carotid Triangle—Three Carotid Arteries; Six Collateral Branches of Ext. Carotid: Superior Thyroid; Lingual; Facial; Ascending Pharyngeal; Occipital and Posterior Auricular.

Veins in Carotid Triangle.

SUBMANDIBULAR TRIANGLE

Boundaries; Muscular Floor.

Contents—Submandibular Gland.

ANTERIOR TRIANGLE AND MEDIAN LINE OF THE NECK

Subdivisions (*fig. 707*). The anterior triangle of the neck is bounded by the median line from chin to manubrium, by the anterior border of the Sternomastoid, and by the lower border of the jaw together with the hinder part of the posterior belly of the Digastric. It is subdivided into three subsidiary triangles—*submandibular* (digastric), *carotid*, and *muscular*—by the anterior and posterior bellies of the Digastric and the superior belly of the Omohyoid.

The region of the neck bounded below by the body of the hyoid bone and on each side by the anterior belly of the Digastric is the *submental triangle*. (The *parotid region* above the posterior belly of the Digastric and behind the ramus of the jaw is beyond the confines of the anterior triangle.)

Landmarks (*fig. 708*). You can feel the tip of the transverse process of the atlas (*fig. 715*) between the angle of the jaw and the mastoid process by pressing upward. The other transverse processes cannot be identified individually.

If you run your fingers downward in the midline of the neck, you will palpate in succession the *body of the hyoid bone*, the *laryngeal prominence* of the thyroid cartilage (Adam's apple), the *arch of the cricoid cartilage*, and 5 cm of trachea—its upper half.

The hyoid bone lies above the level of the chin, at the angle between the floor of the mouth and the top of the neck. While palpating your hyoid bone perform the act of swallowing and note that the hyoid is pulled upward and forward. The thyroid and cricoid cartilages also are felt to rise.

Run your index fingers along the hyoid bone to the tips of its greater horns, and note that the bone can be moved from side to side like a shuttle. The right and left laminae of the thyroid cartilage and its superior horns are also readily felt provided the opposite side is steadied. (Cornu, L. = horn; plural = cornua.)

The hyoid bone lies at the level of the body of the 3rd cervical vertebra, the thyroid cartilage at the level of the 4th and 5th, the **cricoid cartilage** at the very important level of the **6th.**

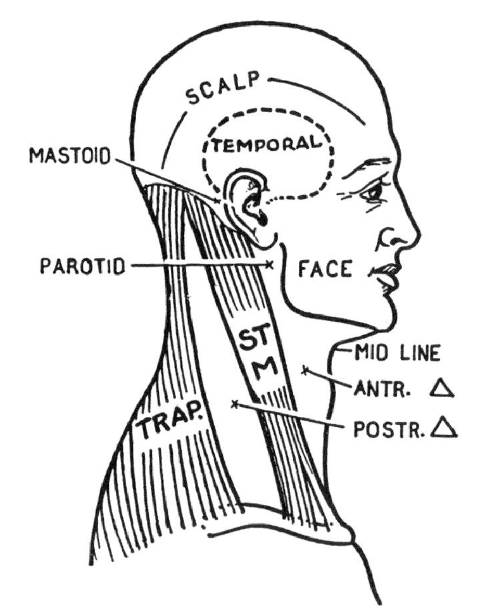

Fig. 707. Superficial regions of head and neck.

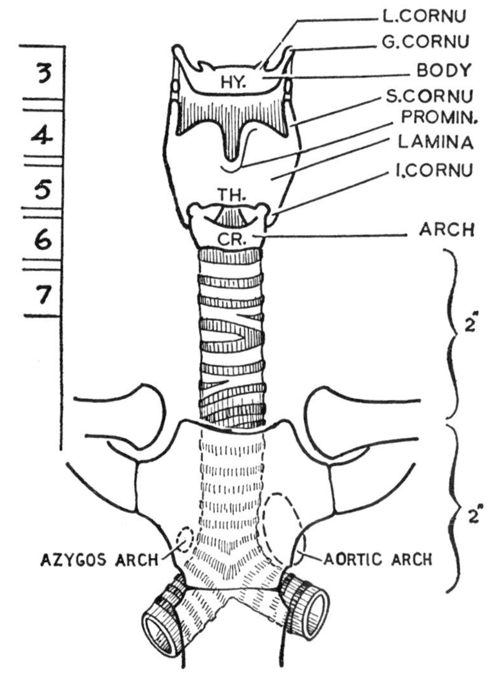

Fig. 708. Landmarks and vertebral levels.

Superficial Structures

Platysma (Gk. = a plate). The Platysma, a subcutaneous sheet of muscle, extends from the face to the level of the 2nd rib. It is continuous above with the facial muscles, but the most anterior fibers are attached to the lower border of the mandible and others decussate with the opposite Platysma for 2 to 3 cm behind the chin. Its anterior border slopes from this point to the sternoclavicular joint. Hence, the Platysma leaves the midline of the neck and the lowest part of the anterior triangle uncovered; whereas it covers the lowest part of the posterior triangle (*figs. 708.1, 708.2*). The *facial nerve* supplies it.

Action. When it contracts, it eases a tight collar. Its posterior fibers pull the angle of the mouth downward (see also p. 454).

Deep Fascia. The neck is enveloped in a collar of *investing deep fasia.* According to rule, it is attached to all the exposed bony parts and ligaments it encounters (*fig. 708.1*).

Superficial Veins (*fig. 664*). The right and left *anterior jugular veins*, unequal in size and asymmetrically placed, begin in the submental region, run near the median line, pierce the deep fascia about 2 cm or so above the manubrium to enter the suprasternal space. Here a cross-channel unites them. Each then turns laterally and runs along the upper border of the clavicle between Sternomastoid and the infrahyoid ("strap") muscles (*fig. 709*) to end in the external jugular vein.

Not infrequently a *communicating vein*,

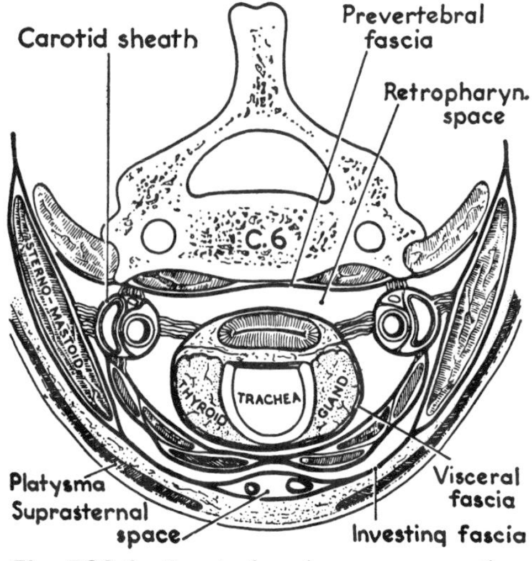

Fig. 708.1. Front of neck, on cross-section.

Fig. 708.2. The Platysma muscle.

lying along the anterior border of the Sternomastoid, connects the common facial and anterior jugular veins. It may equal in size the int. jugular vein and be mistaken for it.

Three Superficial Nerves appear in the anterior triangle (*fig. 663*): (1) the *great auricular nerve*, (2) the *transversus colli nerve* (ant. cutaneous nerve of neck), and (3) the *cervical branch of the facial nerve*.

Median Line of the Neck

Boundaries and Subdivisions (*fig. 709*). The median line is a broad strip bounded above by the slightly diverging *anterior bellies of the Digastrics*, below by the converging *Sternothyroids*, and between these by the nearly parallel *Sternohyoids*. The Mylohyoids form the floor of the *submental triangle* which is bounded by the two Digastrics.

Median Structures. The *thyrohyoid membrane* passes from the upper border of the thyroid cartilage to the body and greater cornua of the hyoid. The median *cricothyroid ligament* unites the adjacent borders of the cricoid and thyroid cartilages. The *isthmus of the thyroid gland* generally covers the 2nd, 3rd, (and 4th) tracheal rings. Only those planning to do tracheostomies should *memorize* the following fearsome *details:*

In front of the trachea in the lower neck, large veins are in surgical danger: (1) one or more cross communications between the anterior jugular veins within the suprasternal space, (2) the brachiocephalic trunk and (3) the left brachiocephalic, or innominate, vein may peep above the suprasternal notch. (4) The inferior thyroid veins descend to the brachiocephalic veins; and (5) an occasional thyroidea ima artery ascends from the brachiocephalic trunk to the thyroid gland.

The Infrahyoid Muscles

These 4 paired muscles (*fig. 709*) are depressors of the larynx and hyoid bone. Often referred to as "strap muscles," they belong to the same superficial ventral sheet of muscles as the *Rectus Abdominis*, and it is helpful to think of them as the "*Rectus Cervicus*." They are supplied by the ventral rami of cervical nerves 1, 2, and 3 via

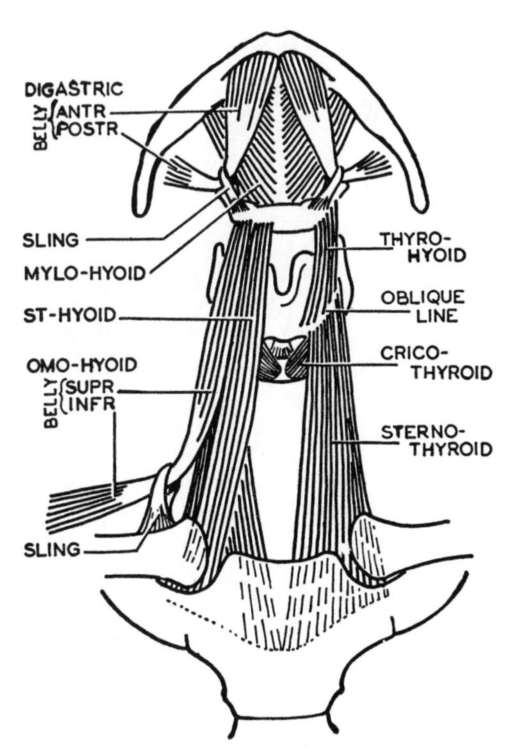

Fig. 709. Muscles bounding midline of neck.

the hypoglossal nerve and the ansa cervicalis (ansa hypoglossi) (*figs. 710, 765*).

The **Sternohyoid** and **Omohyoid** attach side by side to the hyoid body. The Sternohyoid runs down to the posterior aspect of the capsule of the sternoclavicular joint and adjacent bone. The Omohyoid leaves the Sternohyoid abruptly below the level of the cricoid cartilage, passes deep to the Sternomastoid, crosses the posterior triangle to the upper border of the scapula. The fascia covering and joining these muscles across the midline covers the larynx and trachea, and attaches to the carotid sheath laterally.

The **Thyrohyoid** extends upward to the greater horn and body of the hyoid (*fig. 709*).

The **Sternothyroid** converges on its fellow as it descends until their medial borders meet at the center of the posterior surface of the manubrium (*fig. 709*).

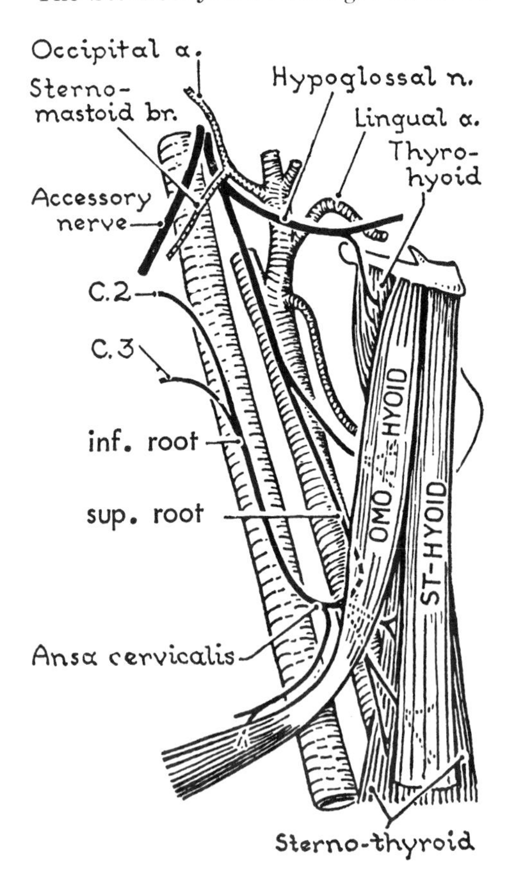

Fig. 710. Ansa cervicalis, its roots and its branches to Infrahyoid muscles.

Cervical Viscera vs. Abdominal. The Rectus Abdominis extends from pubis to chest wall. It is a segmental muscle, supplied by somatic segmental (intercostal) nerves. Similarly, the "*Rectus Cervicis*" extends from chest wall to mandible. It likewise is a segmental muscle, supplied by somatic segmental (C. 1, 2, and 3) nerves.

Through the neck run the pharynx and esophagus (with their off-shoot, the larynx and trachea). These **four visceral tubes** and the *thyroid gland* are the **"cervical viscera."**

PHARYNX

The pharynx descends on the vertebrae from the base of the skull to the level of the 6th cervical vertebra; here, at the cricoid cartilage, it is continuous with the esophagus. At this level the larynx becomes the trachea (*fig. 711*). The back of the trachea is applied throughout to the front of the esophagus.

There are three **Constrictors,** *Superior, Middle,* and *Inferior*. Each is fan-shaped, and each is fixed anteriorly by its narrower end or handle. The bases of the fans overlap, and those of opposite sides meet in the median plane behind. By "superimposing" *figure 712* on *figure 711*, one gains an appreciation of important lateral relations of the Constrictors. On the side walls of the pharynx there are spaces above and below their narrow handles of origin. Through these spaces pass vessels, nerves, muscles, and the auditory tube.

Origins of Constrictors (*fig. 711*):

The *Middle Constrictor* arises in the angle between the greater and lesser horns of the hyoid bone and attached stylohyoid ligament.

The *Inferior Constrictor* arises from the thyroid and cricoid cartilages.

The thyroid part arises from the oblique line on the thyroid cartilage and from a fibrous bridge over the Cricothyroid which extends from the lower border of the thyroid cartilage to

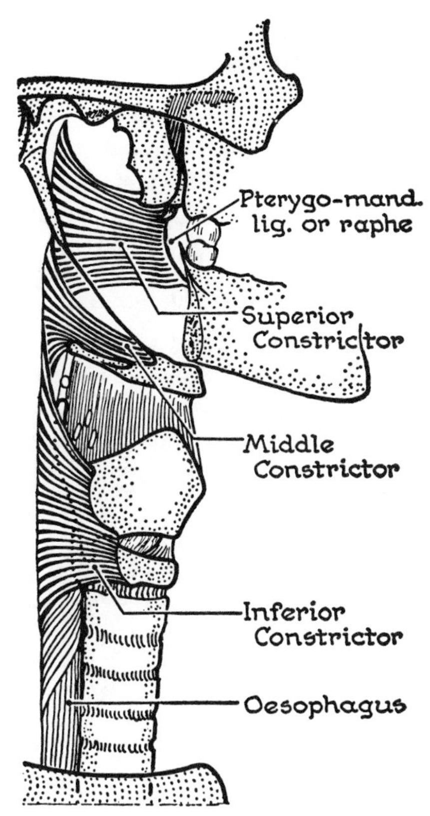

Fig. 711. The three Constrictors of the pharynx.

the lower cornu. It is concerned with swallowing.

The cricoid part, **Cricopharyngeus,** arises from the cricoid and runs downward. It keeps the esophagus closed but relaxes during the act of swallowing.

The *Superior Constrictor* is literally continuous with the Buccinator anteriorly.

It arises from the pterygomandibular raphe and from the bone at each end of the raphe, i.e., the lowest part of the posterior border of the medial pterygoid lamina, and the mandible behind the last molar tooth. Its upper border reaches the pharyngeal tubercle at the center of the under surface of the skull in front of the foramen magnum (*fig. 753*).

Nerve Supply of Constrictors: Cranial nerve XI via pharyngeal and external and recurrent laryngeal branches of the vagus (N. X).

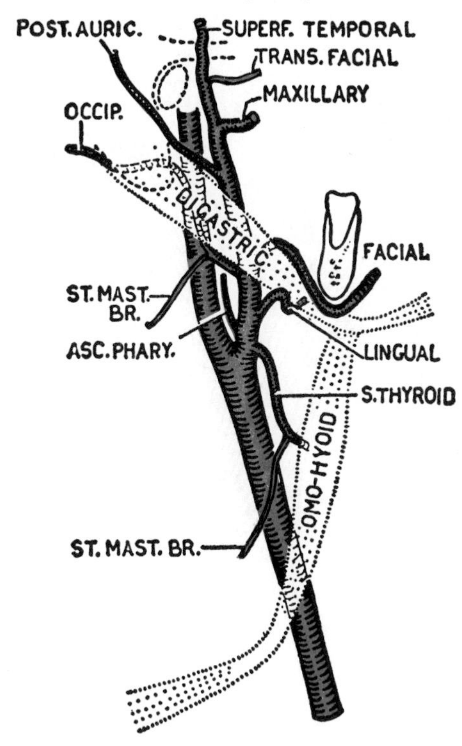

Fig. 712. The three carotid arteries and the branches of the external carotid.

Carotid Sheath. The common and internal carotid arteries, the internal jugular vein, and the vagus nerve extend from the cranial cavity to the thorax and in so doing traverse the neck in some condensed areolar tissue, called the *carotid sheath* (*fig. 714*). The arterial stem is medial, the vein lateral, and the vagus nerve posterior in the angle between the artery and the vein.

Behind the arterial stem, but outside the sheath, is a smaller nerve, the *sympathetic trunk:* it likewise traverses the neck. The sheath is applied to the side of the cervical viscera, and is partly under the cover of the Sternomastoid (*fig. 713*).

Thyroid Gland (*fig. 715*). This gland is wrapped around the front and sides of the four cervical "visceral tubes." It consists of a right and a left lobe, connected near their lower poles by an isthmus, which crosses the (1st), 2nd, 3rd, (and 4th) rings of the trachea. Each lobe lies on the side of the

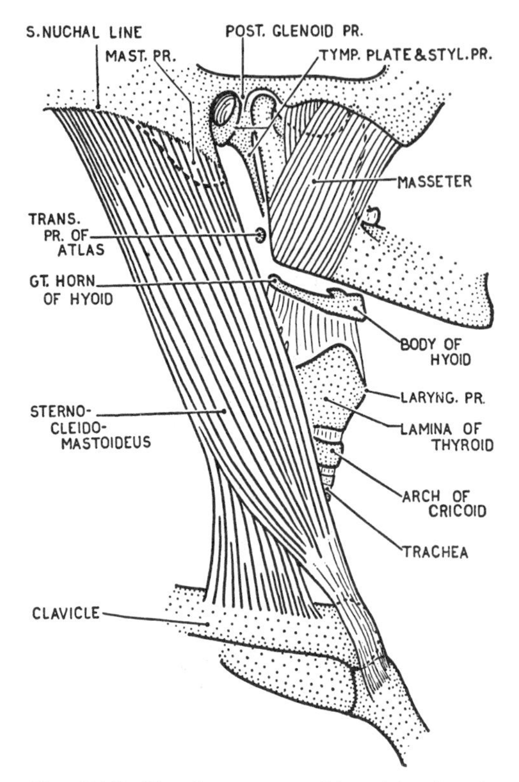

Fig. 713. The Sternomastoid and landmarks.

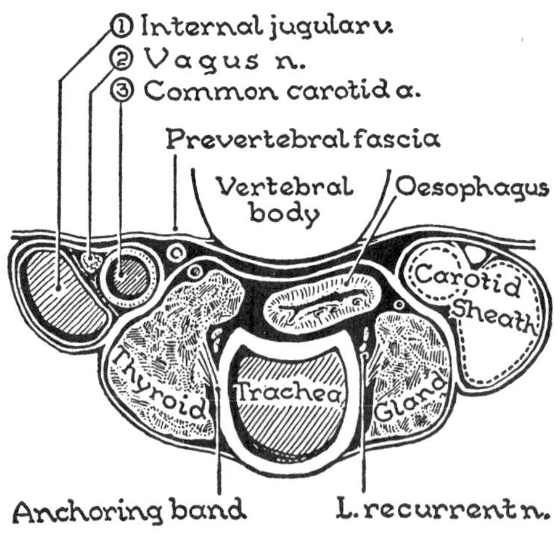

Fig. 714. The thyroid gland and the carotid sheath, on cross section.

trachea and esophagus, and extends upward on the side of the Inf. Constrictor until arrested by the attachment of the Sternothyroid to the oblique line of the thyroid cartilage (*fig. 713*). It abuts against the carotid sheath, and when enlarged it either overlaps the sheath or displaces it laterally (*fig. 714*). (See also pp. 519–520).

CAROTID TRIANGLE

The carotid triangle is bounded by the anterior border of the Sternomastoid, the superior belly of the Omohyoid, and the posterior belly of the Digastric.

Posterior Belly of the Digastric

This belly occupies a **key position** in the neck (*fig. 716*). It arises deep to the mastoid process and is united to the anterior belly by an intermediate tendon which is held down to the hyoid by a fascial sling.

The insignificant *Stylohyoid* is a slip of the posterior belly of the Digastric that has moved forward onto the styloid process. It splits to let the intermediate tendon pass through it, and it inserts into the greater horn of the hyoid bone (*fig. 721*).

Nerve Supply to Posterior Belly (and Stylohyoid)—the facial nerve.

Relations. The parotid and submandibular glands overflow the posterior belly; three relatively unimportant structures cross it and should be avoided in exposing the muscle—cutaneous veins, twigs of a

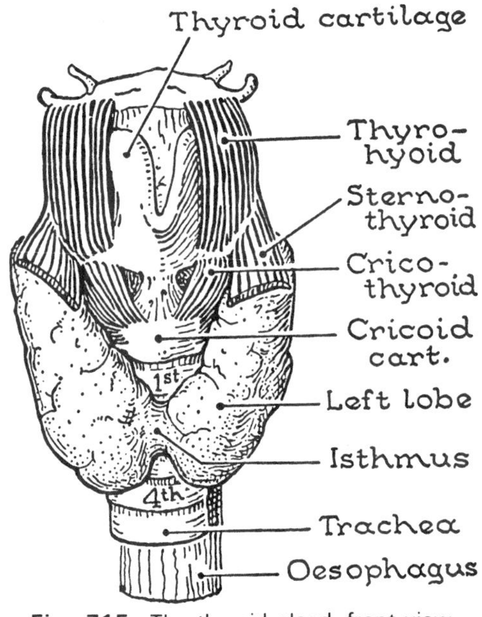

Fig. 715. The thyroid gland, front view.

cutaneous nerve (*great auricular nerve*) and the *cervical branch of the facial nerve* (*figs. 663, 664*).

In contrast, **passing deep** to the posterior belly of the Digastric and thereby placing it in a commanding or key position are three great vessels, the last three cranial nerves, and the sympathetic trunk. In fact, *all structures in the carotid triangle that reach to higher levels pass deep to the posterior belly of the Digastric* (*fig. 716*).

General Disposition of the Vessels and Nerves. The int. jugular vein, and the int. and ext. carotid arteries ascend side by side deep to the posterior belly. The vein is sheltered by the anterior border of the Sternomastoid and comes into view only when this is retracted. Nerves X, XI, and XII descend together from the skull and separate at, or just above, the lower border of the Digastric; nerve XI passes downward and backward; nerve XII curves forward superficial to the arteries.

Nerves in the Carotid Triangle

Accessory Nerve (N. XI), *external branch* (*fig. 764*), the motor nerve of the Sternomastoid and Trapezius, appears from under cover of the Digastric. It crosses the int. jugular vein at the transverse process of the atlas—which is palpable (*fig. 713*)—and disappears into the deep surface of the Sternomastoid, surrounded by lymph nodes.

Commonly (58 of 197 specimens) the nerve crosses behind the int. jugular vein and lies in contact with the transverse process of the atlas.

Hypoglossal Nerve (N. XII) (*fig. 716*). Because it is the motor nerve to the tongue, it swings forward and passes under cover of the posterior belly of the Digastric a second time. It enters the submandibular (digastric) triangle, where it disappears between the Mylohyoid and the Hyoglossus.

The hypoglossal nerve is most readily found just above the posterior end of the greater cornu of the hyoid bone. Keep to its upper

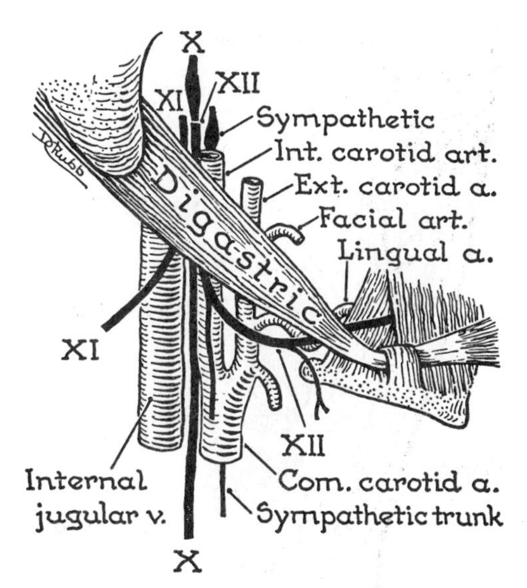

Fig. 716. Key: posterior belly of Digastric.

border lest you damage two branches that spring from its lower or convex border.

It curves forward superficial to every artery it meets—int. carotid, ext. carotid, and lingual arteries always, and commonly either the superior thyroid or the facial artery, but never deep to these branches or they would arrest its ascent.

Ansa Cervicalis (Ansa Hypoglossi) (*fig. 710*). The hypoglossal nerve gives off the superior root of the ansa (descendens hypoglossi), composed of fibers picked up from nerve C. 1. It joins the *inferior root of the ansa* (descendens cervicalis, C. 2 and 3) to form a loop, the *ansa cervicalis*, which supplies infrahyoid muscles (*fig. 710*). (Ansa, L. = pitcher-handle, i.e., loop.)

Vagus Nerve (N. X) (*figs. 716, 717*). It runs vertically within the carotid sheath behind and between the int. jugular vein and the great arteries.

Branches appearing in carotid triangle: (1) *cardiac branch* which fuses with a *cardiac branch* of the sympathetic trunk (Mizeres), (2) the two terminal branches of the superior laryngeal nerve, the *internal laryngeal nerve* (sensory) and the *external laryngeal nerve* (motor).

The *Internal Laryngeal Nerve* is easily found where it pierces the thyrohyoid

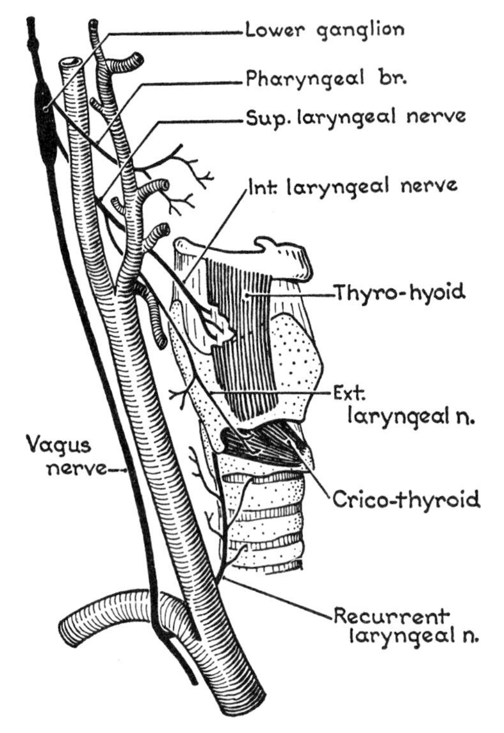

Fig. 717. Pharyngeal and laryngeal branches of vagus.

membrane. It is sensory to the larynx above the level of the vocal cords.

The *External Laryngeal Nerve*, loosely bound to the sup. thyroid artery, descends obliquely to reach the Cricothyroid, which is the tensor muscle of the vocal cord.

The upper pole of the thyroid gland (depending upon its size) pushes the sup. thyroid a. either nearly to or against the ext. laryngeal nerve. Hence, the nerve is liable to be damaged in goitre operations with resulting weakness of the voice.

Arteries in the Carotid Triangle

The arteries in this triangle are: (1) parts of the common, internal, and external carotid arteries, and (2) the stems of most of the six collateral branches of the external carotid artery (*fig. 718*).

Common Carotid Artery. This artery ascends through the carotid triangle to the level of the "Adam's apple" of the thyroid cartilage where it ends by dividing into two terminal branches of nearly equal size, the *internal* and *external carotid aa.*, the internal carotid to be distributed inside the skull, and the external outside.

These two arteries ascend side by side, the internal artery being posterolateral, the external artery, anteromedial. Both pass deep to the posterior belly of the Digastric—the external carotid to enter the parotid gland where it divides into its two terminal branches; the internal carotid to pass deep to the parotid gland and styloid process to the base of the cranium.

Branches. The common and internal carotid aa. give off no collateral branches in the neck; so, it falls to the external carotid, assisted by the inferior thyroid artery, to supply the "cervical viscera."

External Carotid Artery. This "outside" artery is applied to the Inferior and Middle Constrictors, where it has 5 branches; after giving off a sixth collateral branch, in the parotid gland it divides at the neck of the

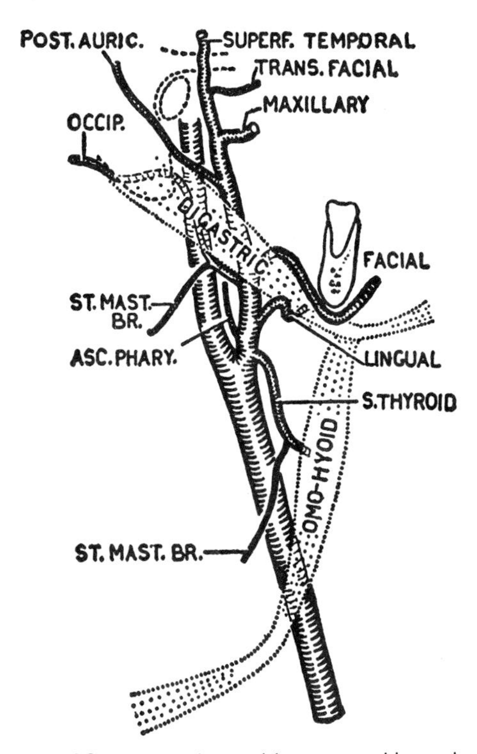

Fig. 718. External carotid artery and branches.

mandible into its terminal branches—the *maxillary* and *superficial temporal aa.*

Collateral Branches, six in all, radiate from the neighborhood of the posterior belly of the Digastric to reach the parts they supply.

Three of these—*superior thyroid, lingual*, and *facial*—arise from its anterior aspect close together below the Digastric (*fig. 718*). In this mobile part of the body all three take sinuous courses.

The **Superior Thyroid Artery** must pass deep to the three long infrahyoid muscles in order to reach the thyroid gland at its upper pole, where it divides into three glandular branches. The details are significant only to a small number of specialists:

One branch ramifies on the lateral surface of the thyroid gland; one runs along the medial border to the isthmus where it anastomoses with its fellow.

Other Branches are: (1) *superior laryngeal branch*, which pierces the thyrohyoid membrane below the internal laryngeal nerve and supplies the larynx; (2) a *cricothyoid branch*, which passes either superficial or deep to the Sternothyroid, crosses the Cricothyroid muscle and ligament, anastomoses with its fellow, and sends twigs through the ligament into the larynx; (3) a *sternomastoid branch*, which follows the Omohyoid across the carotid sheath to the Sternomastoid.

The **Lingual Artery** loops upward and forward and, passing deep to the posterior belly of Digastric, enters the submandibular triangle where it at once passes deep to the Hyoglossus. The hypoglossal nerve, looping downward and forward, crosses it superficially (*fig. 716*). (Con't on p. 570).

[In 20 per cent of 211 specimens the lingual and facial arteries spring from a common linguo-facial stem (G. F. Lewis).]

The **Facial Artery** (Ext. Maxillary a.) enters the face on the jaw just in front of the Masseter. It does not run a direct course to this point, but loops over the posterior belly of the Digastric, passes down deep to the ramus of the jaw, and then up. Hence, the "~" shaped course of the artery. Again, the following details are for specialists:

The Middle and Superior Constrictors lie medial to the 1st, or cervical, loop of the artery, the Sup. Constrictor separating it from the tonsil (*fig. 719*).

CERVICAL BRANCHES: ascending palatine, tonsillar, glandular, and submental.

The *ascending palatine and tonsillar branches* arise at the summit of the bend and ascend on the pharyngeal wall. The tonsillar branches pierce the S. Constrictor and descend with the Levator Palati to supply the soft palate and bed of the tonsil.

Glandular branches: to submandibular gland. The *submental artery* follows the end part of the mylohyoid nerve, sends twigs through the Mylohyoid to the floor of the mouth, and then turns round the lower border of the jaw to end in the chin.

(FACIAL BRANCHES: inf. labial, sup. labial, lat. nasal, and angular: see page 460).

The Ascending Pharyngeal Artery arises from the deep aspect of the ext. carotid artery near its origin, and ascends on the side wall of the pharynx as far as the base of the skull. It supplies the pharynx, soft palate, auditory tube, and meninges.

[The meningeal twigs traverse the hypo-

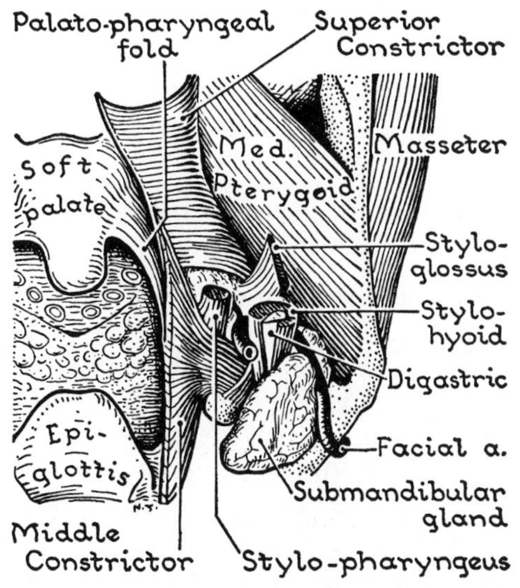

Fig. 719. Details of S-shaped course of facial artery, postero-inferior view.

glossal, jugular, and lacerate foramina, and a tympanic twig follows the tympanic branch of the glossopharyngeal nerve.]

The Occipital and Posterior Auricular Arteries follow the lower and upper borders of the Digastric backward.

The **Occipital Artery** ends in the scalp (*fig. 675*).

Branches. Sternomastoid, meningeal, descending, and terminal, see page 476.

The **Posterior Auricular Artery** (*fig. 662*).

Branches. Only two are important:
(1) the *stylomastoid branch*, which supplies the facial nerve in the facial canal; and (2) *auricular branches*.

Veins of the Anterior Triangle (*figs. 664, 759*)

The *superior thyroid*, *lingual*, and *facial veins* as well as a *middle thyroid vein* cross superficial to the carotid arteries and join the int. jugular vein. The facial vein unites with the anterior branch of the retromandibular vein (post. facial vein) to form the *common facial vein* (*fig. 664*). The *pharyngeal plexus* is drained by several veins into the int. jugular vein (*fig. 759*).

[The *occipital vein* partly ends in the internal jugular vein and partly in the posterior auricular vein but mostly in the vertebral plexus. The *posterior auricular vein* joins the posterior branch of the retromandibular vein to form the *external jugular vein* (*fig. 664*).]

SUBMANDIBULAR TRIANGLE

(Digastric Triangle)

Its boundaries are the two bellies of the Digastric and the lower border of the jaw (*fig. 720*). A broad band of fascia, stretching from the styloid process to the posterior border of the ramus of the mandible, and hence called the *stylomandibular ligament*, separates it from the parotid region behind.

The Floor of the Triangle is formed by parts of three flat muscles—*Mylohyoid*, *Hyoglossus*, and *Middle Constrictor* (*fig. 720*).

The Mylohoid arises from the mylohyoid line of the mandible (*fig. 735*). It is the floor of the mouth; so it runs inferomedially to be inserted into the body of the hyoid bone and a median raphe that extends from the hyoid bone to the mandible.

The Hyoglossus arises from the greater cornu of the hyoid (*fig. 721*), and runs upward into the side of the tongue.

The Middle Constrictor at its origin is deep to the Hyoglossus (*fig. 720*).

Contents of submandibular triangle:
(1) submandibular gland and lymph nodes,

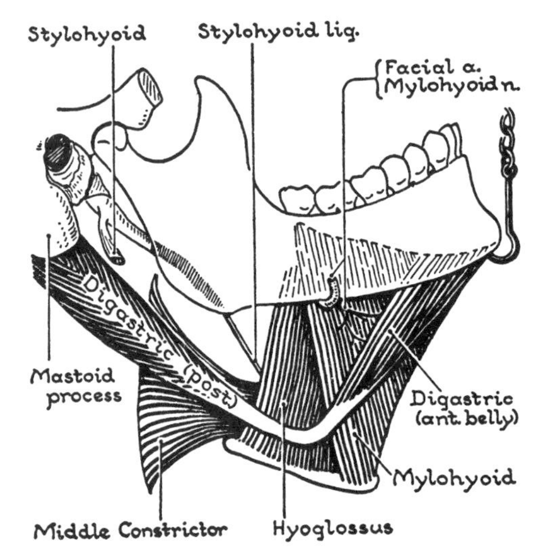

Fig. 720. Floor of submandibular triangle.

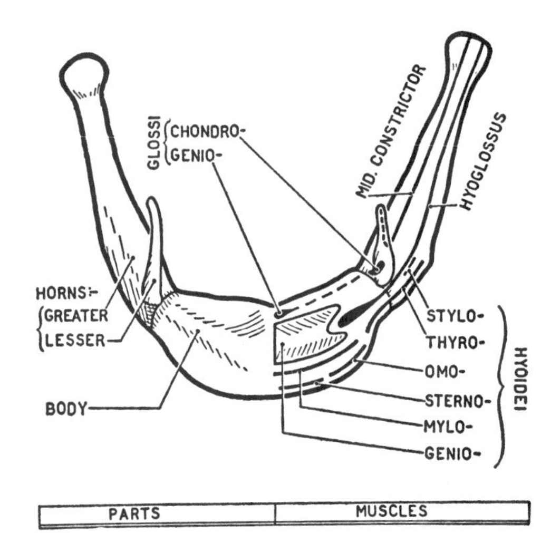

Fig. 721. Hyoid and details of its muscles.

(2) hypoglossal and (3) mylohyoid nerves, (4) lingual and (5) facial arteries, and (6) facial vein.

The **Submandibular Gland** (Submaxillary Gland) fills the submandibular triangle to overflowing; behind it abuts against the parotid gland, only the stylomandibular lig. intervening. A process insinuates around the free posterior border of the Mylohyoid and along the floor of the mouth as far as the sublingual gland (*fig. 721.1*).

Its 5 cm-long duct, seen in *figures 785* and *786*, runs forward between the Mylohyoid and the Hyoglossus, and finally opens under the tongue (p. 567).

Superficial to the gland are: skin, Platysma, cervical branch of nerve VII (*fig. 659*), deep fascia and the facial vein. The facial artery, however, is deep. The lingual artery and hypoglossal nerve are also deep as they run forward into the tongue.

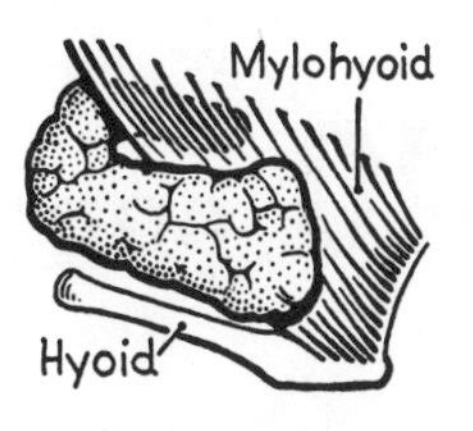

Fig. 721.1. Two U-shaped glands and the two responsible muscles, compared.

[The *mylohyoid nerve* (to Mylohyoid and ant. belly of Digastric), accompanied by the *submental branch* of the facial artery, runs deep to the gland to the two muscles from the interval between the anterior free border of the Medial Pterygoid and the ramus of the jaw.]

Secretory Nerve Supply: the chorda tympani relayed in the submandibular ganglion (*fig. 786*, and p. 567).

40 ROOT OF NECK

Great Vessels: Carotid; Subclavian; Internal Jugular; Brachiocephalic.
Surface Anatomy.
Triangle of Vertebral Artery—Carotid Tubercle; Contents; Relations.
Subclavian Artery—Course; Relations; Branches.
Brachiocephalic Vein and Tributaries.
Phrenic Nerve.
Thoracic Duct.
Thyroid Gland—Parts; Development; Relations; Blood and Nerve Supply; Anomalies.
Parathyroid Glands—Structure; Function; Development.
Trachea and Esophagus—Relations.

The right common carotid and subclavian arteries arise from the brachiocephalic trunk behind the right sternoclavicular joint. The left common carotid and subclavian arteries arise from the aortic arch, and enter the neck by passing behind the left sternoclavicular joint.

Subclavian Artery

Arching over the pleural cupola and 1st rib, it becomes the axillary artery at the lower border of this rib. It is divided into three parts—1st, 2nd, and 3rd—by the Scalenus Anterior which crosses in front of the 2nd part, separating it from its vein.

Surface Anatomy. A curved line running from the sternoclavicular joint and rising 2 to 3 cm above the clavicle and then crossing that bone near its middle marks the course of the artery (*fig. 722*).

Subclavian Vein

This vein (*fig. 725*) lies within the concavity of its companion artery. The int. jugular vein lies lateral to its companion, the common carotid artery, and conceals the 1st part of the subclavian artery and the stems of branches arising from it.

"Triangle of the Vertebral Artery" (*figs. 723, 725*)

Contents: *The vertebral artery* and its *vein* ascend from base to apex and there enters the foramen transversarium of vertebra C. 6.

The sympathetic trunk with its ganglia and branches is shown in *figure 724*, and described on pages 550–551.

Ascending in front of the Triangle are:

1. The carotid sheath and contents.
2. The phrenic nerve.

Arching in front of the Triangle are:

3. The inferior thyroid artery (*fig. 725*).
4. The thoracic duct, on the left.

The *phrenic nerve* is nearby (*fig. 725*).

Landmarks. The anterior tubercle of the transverse process of vertebra C. 6 at the apex of the triangle is a landmark of importance. The common carotid artery passes in front of the tubercle and may be compressed against it. Hence, it is called the **carotid tubercle** (*fig. 723*).

Subclavian Artery (cont'd)

Relations of 1st Part:

Below and *behind* are pleural cupola and 1st rib (*fig. 725*) and the lowest root (Th. 1) of the brachial plexus.

Antero-inferiorly are the large veins, but intervening are phrenic, vagus, and cardiac nerves. The *right recurrent laryngeal nerve* arises from the vagus while it is crossing the subclavian artery, and it recurs below and behind the artery. The *left vagus nerve*

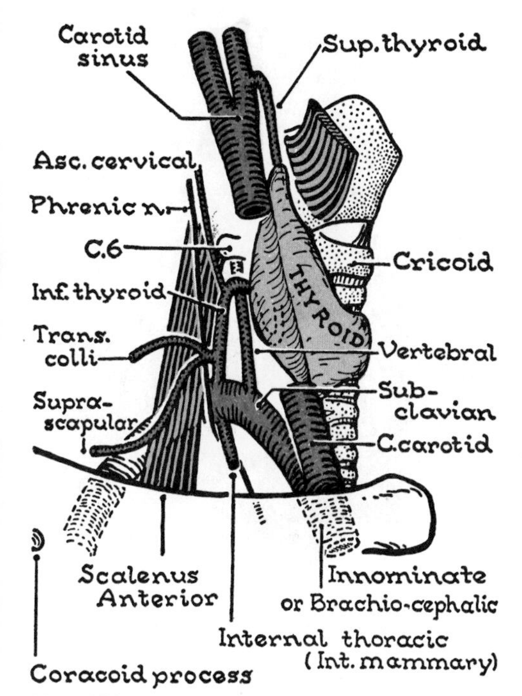

Fig. 722. The arteries at the root of the neck.

descends into the thorax. On the left side the *thoracic duct* is an immediate anterior relation.

Branches of Subclavian Artery (*fig. 725*). These mostly arise from the 1st part (*fig. 725*). They are:

1. Vertebral.
2. Thyrocervical trunk.
   Inferior thyroid.
   Transversa colli (cervicalis).
   Suprascapular.
3. Internal thoracic (Int. mammary).
4. Costocervical trunk.
   Deep cervical (p. 476).
   Highest intercostal.
   1st posterior intercostal.
   2nd posterior intercostal.
5. Dorsal scapular (p. 466).

The Vertebral Artery is described on pages 476, 516, and 536.

The Thyrocervical Trunk at once ends as three branches. Of these, the *Suprascapular* and *Transversa Colli Arteries*, run laterally across the Scalenus Anterior to the posterior triangle and "clamp down" the phrenic nerve, (*See* page 466).

The Inferior Thyroid Artery takes an S-shaped course across the sympathetic trunk to the lower pole of the thyroid gland; it also supplies twigs to its neighbors, including the *inferior laryngeal art.* and a large muscular branch, the ascending cervical artery.

The Internal Thoracic Artery descends on the pleura behind the subclavian vein, and crosses the phrenic nerve. (For its thoracic part, see p. 398.)

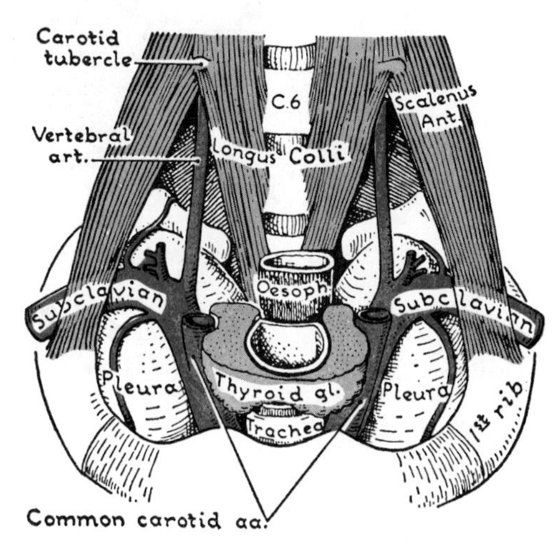

Fig. 723. "The triangle of the vertebral artery". The "base" is the 1st part of the subclavian artery.

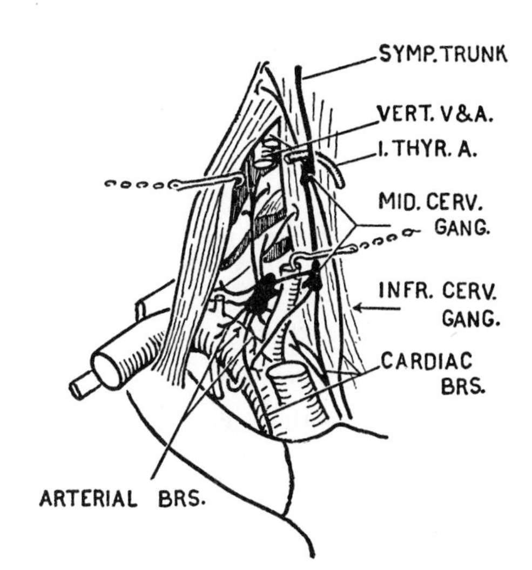

Fig. 724. The sympathetic trunk, ganglia, and branches, at the root of the neck.

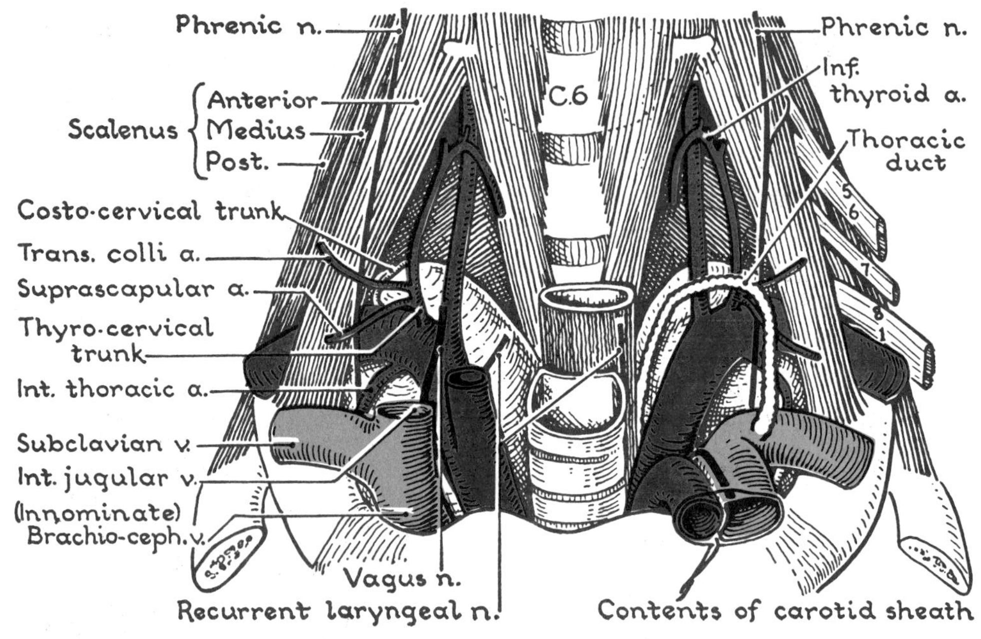

Fig. 725. The root of the neck.

Brachiocephalic Vein and Tributaries

The general pattern is shown in *figure 726*. The internal jugular and subclavian veins each possess one double cusped **valve** (*fig. 759*).

Phrenic Nerve

This branch of the cervical plexus (C. 4) receives contributions from C. 3 and C. 5, descends nearly vertically on the Scalenus Anterior, crosses the subclavian artery and enters the thorax.

Thoracic Course, see ..pm 403, 404 and 407.

Distribution, see p. 250 and *figure 326*.

Variations. The slender branch from C. 5 commonly travels via the nerve to the Subclavius.

Thoracic Duct (continued from p. 446)

The duct enters the neck on the left side of the esophagus and at once arches laterally on the pleural cupola to open into the angle between the left internal jugular and subclavian veins.

It goes behind the 3 structures within the

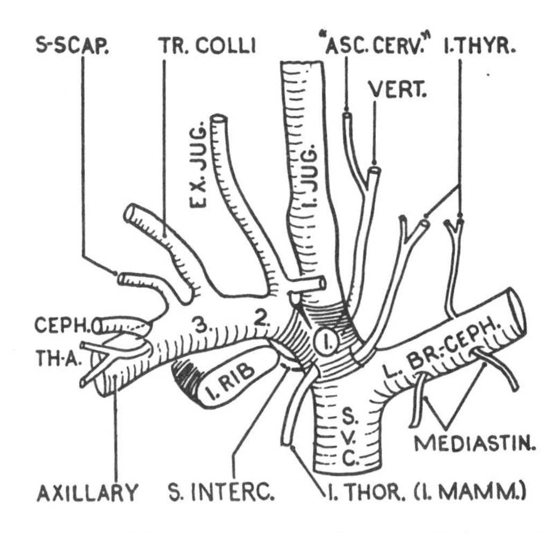

Fig. 726. The veins at the root of the neck. *1.* = right brachiocephalic vein.

carotid sheath, and in front of "the plane of the subclavian artery and its branches" (*fig. 726.1*).

On each side there are three lymph trunks, the *internal jugular*, *subclavian*, and *bronchomediastinal*, which open into the large veins or the thoracic duct. On the right side, they may unite to form a short stem, the *right lymph duct*.

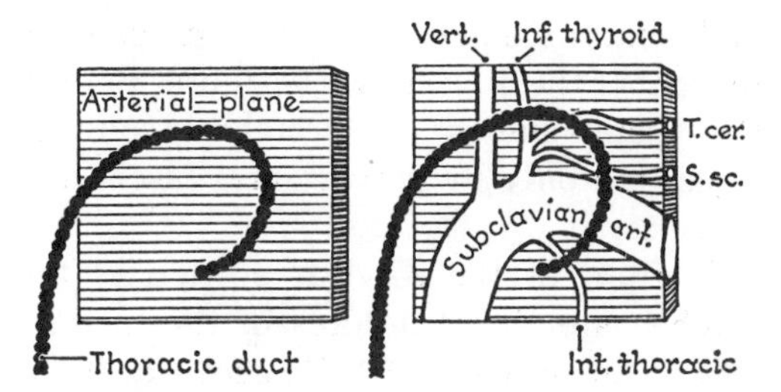

Fig. 726.1. Thoracic duct and "arterial plane."

Thyroid Gland (*fig. 715*)

The thyroid gland consists of right and left *lobes* and a narrow *isthmus*; from the isthmus a process, the *pyramidal lobe*, commonly ascends to the hyoid bone.

Development (*fig. 727*). The thyroid gland arose as a median outgrowth of the pharynx between the anterior and posterior rudiments of the tongue. The foramen cecum of the tongue marks its site of origin; the pyramidal lobe when present indicates its course.

Relations. The isthmus crosses the upper two or three tracheal rings; and each lobe expands downward on the side of the trachea, backward on the esophagus to abut against the carotid sheath, and upward onto the pharynx and larynx (*fig. 728*).

The *recurrent laryngeal nerve* ascends on the side of the trachea through branches of the interior thyroid art. to the back of the cricothyroid joint where it plunges deep.

The upward expansion of the gland is arrested by the attachment of the Sternothyroid to the oblique line of the thyroid cartilage. When enlarged, it overflows laterally in front of the carotid sheath, and downward retrosternally in front of the great vessels and pleura.

Vascular and Nervous Supply. Its freely anastomosing *Arteries and Veins* are the paired superior and inferior thyroid, and the occasional median unpaired thyroidea ima art. A middle thyroid vein runs unaccompanied from the lateral border of the gland.

Lymph Vessels pass from extensive lymph plexuses to the deep cervical, pretracheal, and paratracheal nodes, and even directly to the thoracic duct.

Nerves: Sympathetic fibers from the superior and middle cervical ganglia on the vessels and in laryngeal nerves.

Parathyroid Glands (*fig. 768*)

These 4 small (about 5 to 6 mm) but

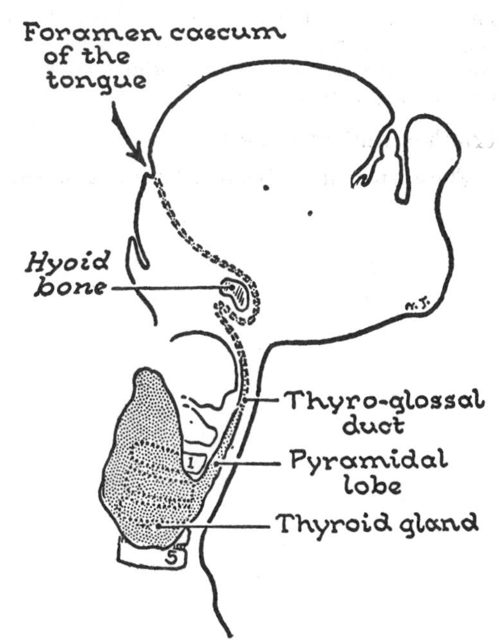

Fig. 727. The course of a developing thyroid gland. (From data by J. E. Frazer.)

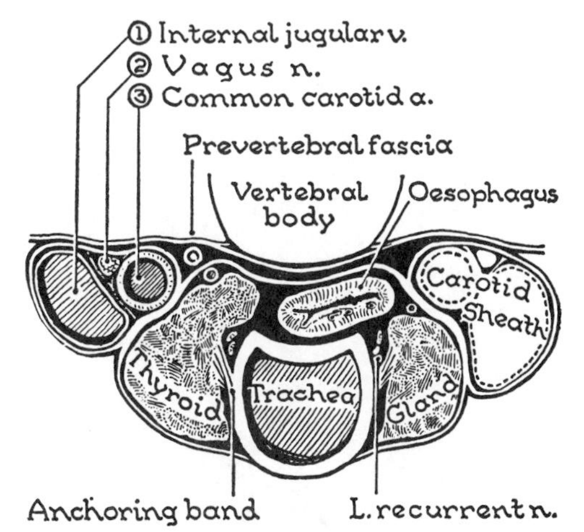

Fig. 728. Trachea and esophagus, on cross section. Note dense fibrous band attaching each lobe to trachea.

vital, glands are yellowish-brown like a bumble bee and lie along the posterior border of the thyroid gland between its capsule and sheath. There are two on each side, an upper and a lower. A branch of the inferior thyroid art. supplies both.

Development. The inferior gland, like the thymus, developed from the 3rd pharyngeal pouch; it follows the thymus to a lower level than the superior gland, which is developed from the 4th pouch—and sometimes ends up entirely below the thyroid.

Trachea and Esophagus

These two tubes begin where the larynx and pharynx end—at the cricoid cartilage in front of the 6th cervical vertebra. Behind the esophagus is the vertebral column, separated only by prevertebral fascia and muscles.

In front of the cervical portion of the trachea are the structures in the median line of the neck (p. 507). *On each side* is the common carotid artery. The recurrent laryngeal nerve ascends in the angle between the trachea and the esophagus and it supplies them. Briefly the thoracic duct on the left side is a lateral relation.

41 SIDE OF SKULL, PAROTID, TEMPORAL, AND INFRATEMPORAL REGIONS

LATERAL ASPECT OF THE SKULL

Cranium, on Lateral View; Contour Lines; Three Ovoid Areas. Stylomastoid Region.

THREE KEY BONES

Parietal Bone. Zygomatic Bone.
Mandible—Borders of Body; Ramus; Mandibular Canal; Surfaces of Body.

MASSETER AND PAROTID GLAND

Masseter—Attachments; Nerves; Vessels.
Parotid Mold and Gland—Parotid Duct; Radiating from Margin of Gland.
Passing through the Gland.
Facial Nerve; Nerve Supply of Parotid.
External Carotid Artery; Lymph Nodes.

TEMPORAL AND INFRATEMPORAL REGIONS

Temporalis.
Infratemporal Region. Bony Boundaries.
Contents: Pterygoid Muscles.
Mandibular Nerve—Inferior Alveolar (Dental) Nerve; Lingual Nerve; Chorda Tympani; Auriculotemporal Nerve.
Maxillary Artery: Branches; Vein.
Otic Ganglion.
Temporomandibular Joint (Jaw Joint).
Bony Parts; Ligaments; Palpation; Actions of Muscles; Relations; Nerve Supply; Accessory Ligaments or Bands.

LATERAL ASPECT OF THE SKULL

A straight line divides the skull into cranial and facial parts fairly well (*fig. 729*).

The Cranium on Lateral View presents three approximately "concentric" ovoid contour lines (*fig. 730*).

The Outermost Ovoid Contour Line is the outline of the cranium. From *nasion* to *foramen magnum* the frontal, parietal, and occipital bones contribute nearly equally to the outline.

From the inion the *superior nuchal line* curves to the *mastoid process*. To these are attached the Trapezius, Sternomastoid Splenius and Longissimus Capitis. A smooth, triangular part of the mastoid bone behind the external acoustic meatus covers the mastoid antrum (p. 595), and gives the surgeon access to deep structures.

The Innermost Ovoid Contour Line is the border of the squama of the temporal bone. Side by side below its middle are twin cavities, (1) *external acoustic meatus* and (2) *mandibular fossa* (*figs. 729* to *731*). From the squama projects the *zygoma* or

Fig. 729. Plan of the skull, on side view. The face is angular. The cranium is oval and is divided into three zones, placed above twin depressions (*fig. 731*).

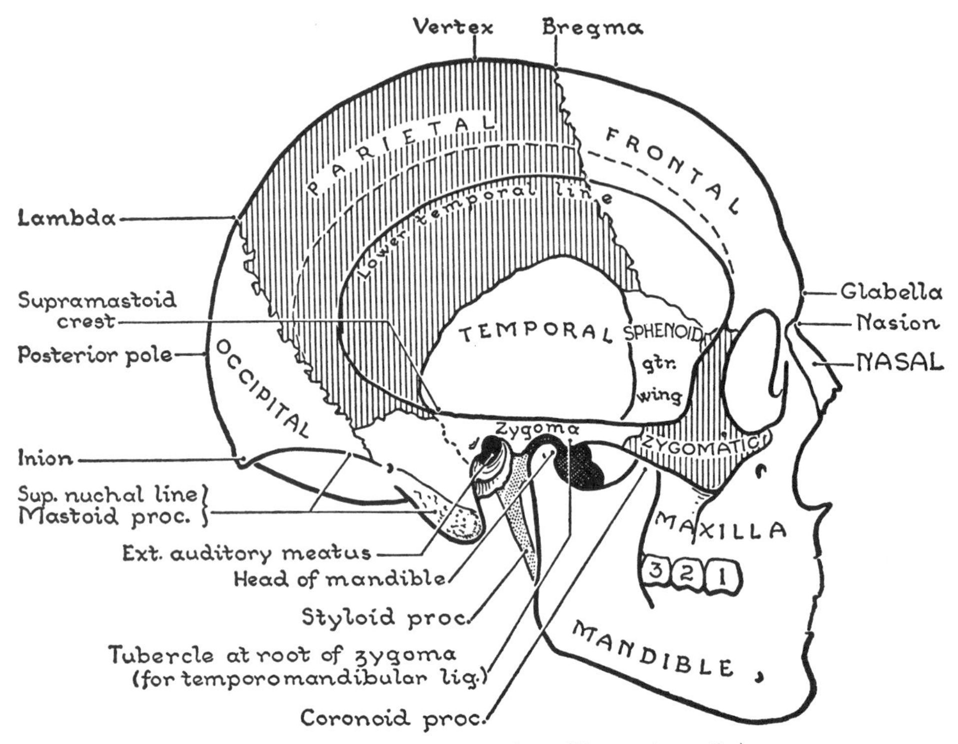

Fig. 730. The skull, on side view. (Norma lateralis.)

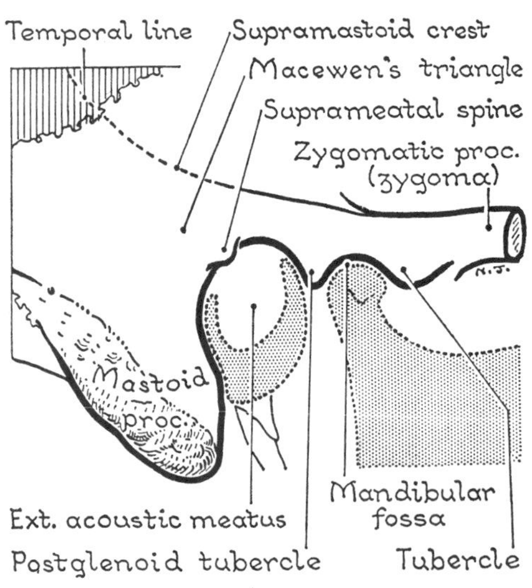

Fig. 731. Twin depressions (external acoustic meatus and mandibular fossa) and postglenoid tubercle between. The suprameatal spine is the surgeon's guide to the mastoid antrum deep to the triangle behind.

zygomatic process of the temporal bone. This process joins the zygomatic bone to form the zygomatic arch.

The *lower border of the zygomatic arch* begins at the 2nd molar tooth and ascends as a buttress to the zygomatic process of the maxilla, separating the facial aspect of the skull from the infratemporal fossa. It then curves backward along the lower border of the zygomatic arch (*fig. 730*).

The Intermediate Ovoid Contour Line surrounds the **temporal fossa** and gives attachment to the temporal fascia, which covers the contents, Temporalis muscle. The fossa is deepest where the Temporalis is thickest—antero-inferiorly. Here its bony anterior wall separates it from the orbital cavity.

Pterion. Four bones—temporal squama, greater wing of sphenoid, parietal, and frontal—take part in its medial wall and meet at a +-shaped or H-shaped suture called the pterion. The pterion is a landmark full of importance (*see fig. 689* on p. 489).

Stylomastoid Region. Anteromedial to the *mastoid process* is the elongated *styloid process*; between them is the *stylomastoid* *foramen*, i.e., the orifice of the facial canal. Out of it comes the facial nerve.

The U-shaped *tympanic bone* (*stippled* in *fig. 731*) forms two-thirds of the circumference of the *ext. acoustic meatus*. Its anterior part, the *tympanic plate*, separates meatus from *mandibular* fossa.

THREE KEY BONES

Due to their key positions and simple features the **parietal bone** in the cranium and the **zygomatic bone** in the face will get individual treatment; as will the **mandible.**

Parietal Bone

The parietal bone, a flat bone with two surfaces, four borders, and four angles, is molded on the brain. Its features are borders and angles (*figs. 730, 732*).

Borders. Sutures outline the bone (*fig. 730*).

Angles. Each of the four angles is in relation on its medial surface with an important vessel (*fig. 732*).

Surfaces. The external surface is convex. The point of greatest fullness is the *parietal tuber*, where ossification started and spread in the embryo. The internal surface is grooved for branches of the middle meningeal vessels.

Zygomatic Bone

The cheek bone is shaped like a conventional diamond (*figs. 730, 733*). The *lateral* (*facial*) *surface* is separated from the medial aspect by four angles and four borders. The medial aspect shares in the formation of the walls of the temporal and infratemporal fossae (*fig. 733 C*).

All four angles and the *maxillary border* (really a surface) are articular; the other borders are palpable.

Functions of the Zygomatic Bone. Its frontal process, which is strong, transmits to the frontal bone for dispersion the force of impacts delivered during mastication (*fig. 733*). Its temporal process is one of the buttresses of the face (*fig. 739*). It also gives origin to the Masseter and forms part of the orbital wall.

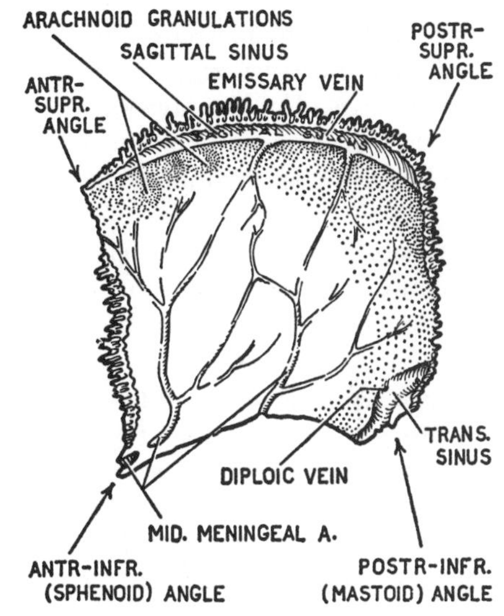

Fig. 732. The medial surface of the parietal bone.

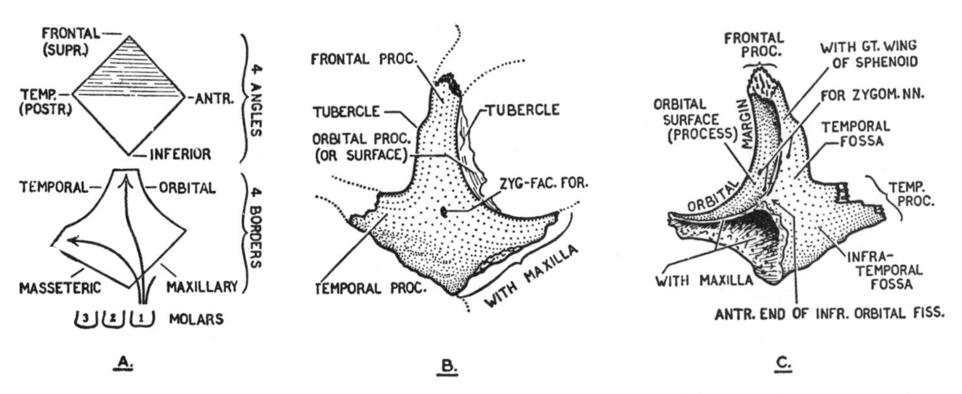

Fig. 733. Zygomatic bone. *A*, 4 angles, 4 borders, and lines of force. *B*, lateral surface. *C*, medial aspect.

The bone is traversed by the zygomatic nerve and artery and their zygomaticofacial and zygomaticotemporal branches (*fig. 703*).

Mandible or Lower Jaw

Each half of the horseshoe-shaped jaw (*figs. 734, 735*) is L-shaped. The horizontal parts of the two sides fuse at the *symphysis menti* in the median plane during the 2nd year to form the *body* of the jaw (although in most mammals they remain unfused); the vertical parts are the *rami*.

Borders of the Body. The *lower border* of the body is thick and rounded and is continuous behind with the lower border of each ramus. The *alveolar process*, or upper part of the body, carries eight teeth on each side. The roots of the teeth (except the 2nd and 3rd molars) cause rounded ridges on the thin front wall of the alveolar process, that of the canine being the most prominent. Medial to the canine ridge is the *incisive fossa*.

The Ramus is an oblong, vertical plate. It is surmounted by the *head* and the *coronoid process*, separated by the *mandibular notch*.

The head, supported by a neck, articulates with the mandibular fossa of the temporal bone, and is like a short segment of a pencil set horizontally. In front of the neck is a *fossa* for the insertion of the tendon of the Lateral Pterygoid.

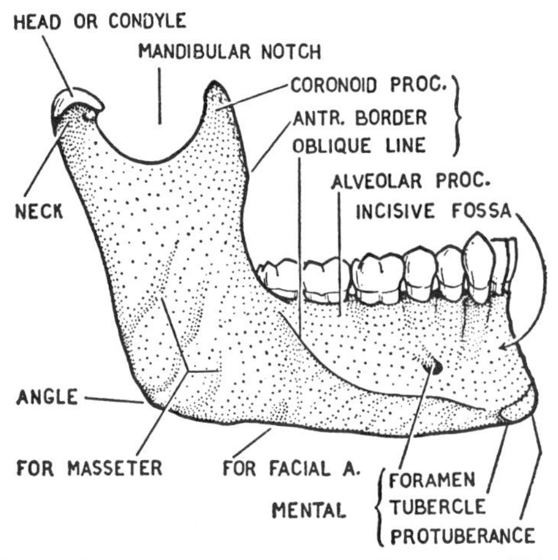

Fig. 734. The lateral aspect of the mandible.

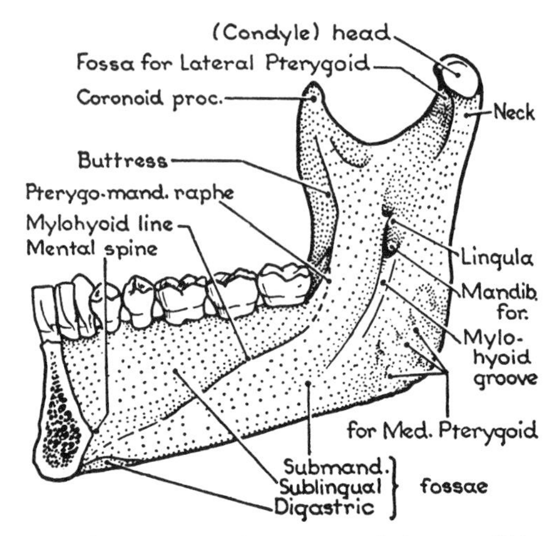

Fig. 735. The medial aspect of the mandible.

The coronoid process is a traction process for the Temporalis. Its tip is sheltered by the zygomatic arch (*fig. 730*).

The *angle* of the jaw is ridged for tendinous septa of the Masseter laterally and the Medial Pterygoid, medially.

Mandibular Canal. Through the bone runs the *mandibular canal*. Its entrance, the *mandibular foramen*, lies at the center of the medial surface of the ramus, above the level of the crowns of the molar teeth. Here the nerve that enters it—inferior alveolar—is accessible for injection by the dental surgeon.

Guarding the foramen in front is the *lingula*, which gives attachment to the sphenomandibular lig. A narrow groove, the *mylohyoid groove*, which lodges the mylohyoid nerve and artery, begins behind the lingula and runs obliquely downward and forward on the ramus.

The canal, conveying vessels and sensory nerves to the teeth, also opens on the outer surface of the body at the *mental foramen*, situated 3 cm from the symphysis, below the premolar or bicuspid teeth.

THE SURFACES OF THE BODY. *External.* The *mental protuberance* forms the chin. From it the *oblique line* runs upward and backward to become continuous with the sharp anterior border of the ramus and coronoid process. To this line are attached the Depressor Labii Inferioris and Depressor Anguli Oris.

Internal. Projecting backward in the midline is the *mental spine* (genial tubercles), for the origin of the Geniohyoid and Genioglossus. From below the spine, the *mylohyoid line* crosses the body diagonally to a strengthening *buttress* on the medial aspect of the ramus and coronoid process. To this line the Mylohyoid is attached, forming the floor of the mouth. Between the mylohyoid line and buttress the pterygomandibular raphe is attached (p. 455). and also the two muscles united by this raphe, viz., the Superior Constrictor behind and the Buccinator in front—the Buccinator extending forward on the aveolar process, lateral to the molar teeth.

Fossae are related to the mylohyoid line: (1) the *digastric fossa*, for the attachment of the anterior belly of the Digastric (*fig. 735*); (2) the *sublingual fossa*, for the sublingual gland is above the line, and (3) the *submandibular fossa* for the submandibular gland is below it.

MASSETER AND PAROTID GLAND

Masseter. This rhomboidal muscle arises from the lower border of the zygomatic arch, covers the lateral surface of the ramus and is inserted into it.

Nerves, p. 529. *Actions*, p. 532.

The Parotid Region (Gk. para = near; ous (otos) = the ear) is a mold lined with fascia and filled with **Parotid Gland.** The soft gland fills the irregularities of the mold and overflows its brim in front and below (*fig. 736*). Through it pass certain vessels and nerves. The parotid gland is invested in deep cervical fascia.

This is the "mumps gland," but the virus also infects other glands. It also is site of cancer that surgeons can remove, but they must "know their anatomy" or they will surely damage the facial nerve or other important transients.

The Parotid Mold or Bed (*figs. 715; 736*):

This comprises—*Behind:* mastoid process and anterior border of the Sternomastoid. *In front:* posterior border of the ramus of the jaw and of the two muscles (Masseter and Medial Pterygoid) inserted into the ramus. Here the gland overflows onto the Masseter. *Below:* Stylohyoid and the posterior belly of the Digastric, which the gland overflows.

The *bottom of the mold* is formed by: (1) the styloid process and the three muscles that arise

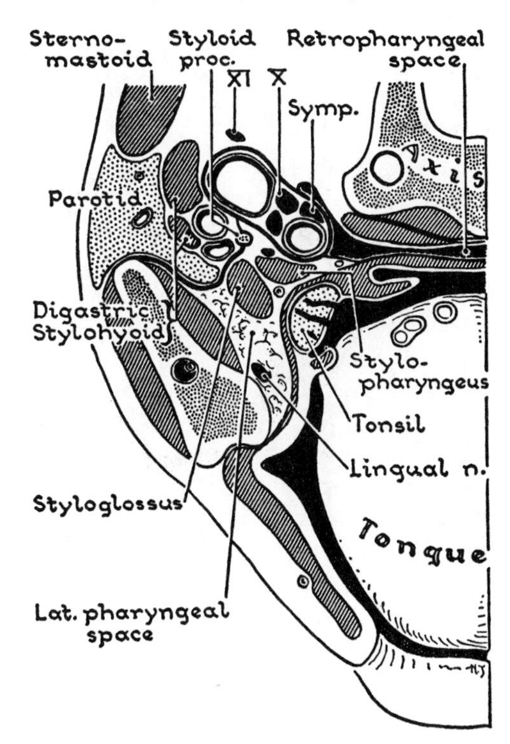

Fig. 736. Cross-section at level of parotid gland and tonsil.

from it. They intervene between the gland and the int. jugular vein, int. carotid artery, and the last four cranial nerves; (2) the fascial lining, which adheres to the styloid process and to the posterior border of the ramus of the mandible. Between these two bony parts the fascia is ballooned forward, like a sail before the wind, deep to the Medial Pterygoid. The lower part of this sheet, the *stylomandibular lig.*, is thickened and it separates the parotid from the submandibular gland.

Several tongue-like processes from the cast fill several crevices in the mold. Of these, (1) one passes forward between the upper part of the ramus and the Medial Pterygoid; (2) another passes upward between the external meatus and the capsule of the jaw joint; (3) another passes medially in front of the internal carotid artery to the Superior Constrictor; but (4) the largest projection is the facial process. It covers the hinder part of the Masseter (and the temporomandibular lig. and neck of the jaw) and is prolonged above the parotid duct as the *accessory parotid gland*.

Radiating from the margin of the gland are the following (*fig. 659*):

1. The superficial temporal artery and

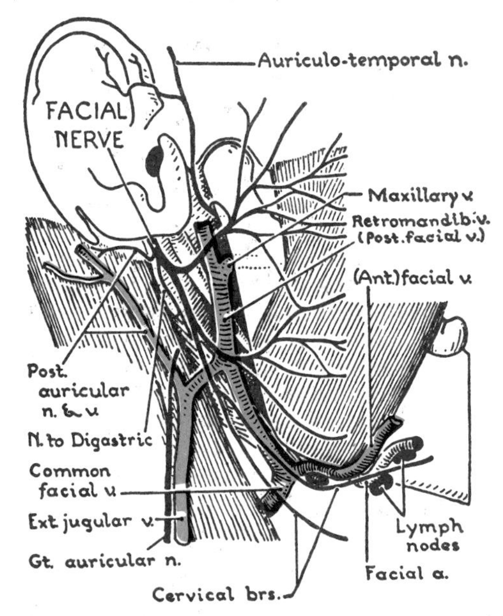

Fig. 737. Facial nerve and veins in parotid region.

vein, and the auriculotemporal nerve (*fig. 737*).

2. Branches of the facial nerve (temporal, zygomatic, buccal, mandibular, cervical and post. auricular), the transverse facial artery, and the parotid duct, crossing the Masseter.

3. The posterior auricular artery.

The Parotid Duct crosses the Masseter horizontally a finger's breadth below the zygomatic arch, turns round the anterior border of the Masseter, and then pierces the buccal pad of fat, Buccinator, and mucous membrane to open in the vestibule of the mouth. There its orifice, constricted as orifices usually are, can be seen at the level of the crown of the 2nd upper molar tooth.

Passing through the Gland (*fig. 737*):

1. Facial nerve and its branches—foremost in importance and in great danger.

2. Retromandibular (post. facial) vein—no great loss if tied off.

3. External carotid artery—can be tied off with impunity if necessary.

Facial Nerve (N. VII)

The facial nerve enters the parotid region by the stylomastoid foramen, and within the gland it divides into branches of a plexiform nature (*fig. 737*).

At birth the child has no mastoid process, so the stylomastoid foramen is subcutaneous; consequently, the facial nerve may be severed accidentally by a deep skin incision made behind the ear. As the process develops the foramen and the stem of the nerve become submerged.

The gland is divided into two almost equal parts, superficial and deep, by the plane of the facial nerve and its branches. The veins form a plexus on a plane immediately deep to that of the nerves.

This combined fasciovenous plane is found by following the superficial temporal vein downward, or else the ext. jugular vein upward, and, in so doing, splitting the gland (Patey and Ranger). The gland is likened to a creeping plant that weaves itself into the meshes of a supporting trellis of nerve and vein, as shown in *figure 737.1* (McKenzie).

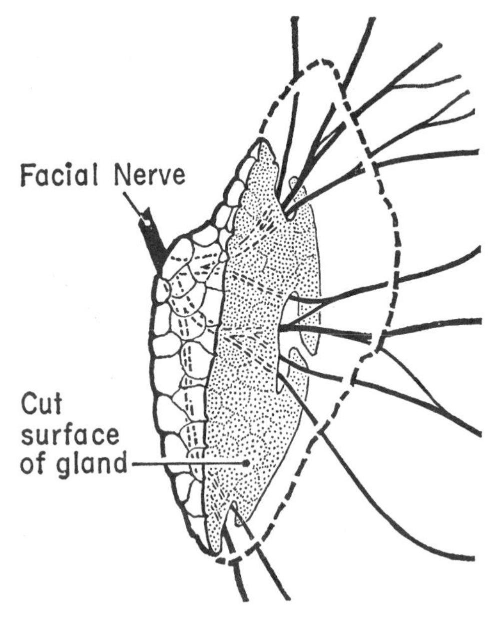

Fig. 737.1. To show the relation of the facial nerve to the parotid gland. (After McKenzie.)

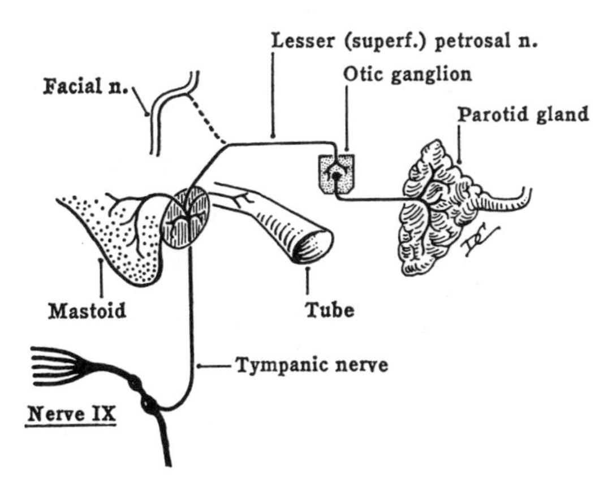

Fig. 737.2. Secretomotor nerve to parotid gland.

Local Branches: n. to Digastric (post. belly) and Stylohyoid.

Nerves Communicating with Facial N.: (1) auriculotemporal n., behind the neck of the jaw, transfers to N. VII secretory fibers received from the otic ganglion; and (2) great auricular n., superficial to the Masseter, sends sensory fibers to N. VII.

Nerve Supply of Parotid Gland. The glossopharyngeal nerve (via its tympanic br., lesser petrosal nerve, otic ganglion, auriculotemporal nerve, and facial nerve) provides the secretory fibers (*figs. 737.2, 797*). Sympathetic fibers reach the gland via the external carotid plexus.

The External Carotid Artery appears from under cover of the Digastric. Deeply grooving the deep surface of the parotid gland, it ascends under shelter of the posterior border of the ramus of the jaw to the neck where it divides into its two terminal branches: (1) the *maxillary artery* which, hugging the neck of the jaw, passes forward into the infratemporal region; and (2) the *superficial temporal artery* which, after giving off the transverse facial artery, crosses the root of the zygoma, where its pulsations can be felt (*fig. 712*).

Lymph Nodes are described in Chapter 49. They are important!

TEMPORAL AND INFRATEMPORAL REGIONS

Temporalis

The Temporal muscle arises by fleshy fibers from the temporal fossa and the covering of tough temporal fascia (*cf.* origin of Gluteus Medius, *fig. 424.1*).

The handle of this fan-shaped muscle is necessarily mostly tendinous; it is attached to the coronoid process.

Infratemporal Region

This region lies below the temporal fossa and deep to the mandibular ramus.

Bony Boundaries (*figs. 738, 740*):

The Lateral Wall is the ramus of the mandible. Near the center of its medial surface, just above the plane of the molar crowns is the *mandibular foramen* (*fig. 735*) overlapped by the lingula to which a slender ligament attaches (*fig. 738*).

The Anterior Wall is formed by the inflated body of the maxilla. It is separated from the facial surface of the maxilla by the *buttress* that descends from the zygomatic arch to the 2nd molar tooth. It is limited above by the *inferior orbital fissure;* and medially by the *pterygopalatine fossa* (*fig. 740*).

The Medial Wall is the *lateral plate* of the pterygoid process (*fig. 740*). This muscular lamina (for the origin of the two pterygoid muscles) is about 1 cm wide. The *pterygoid process* of the sphenoid bone is a

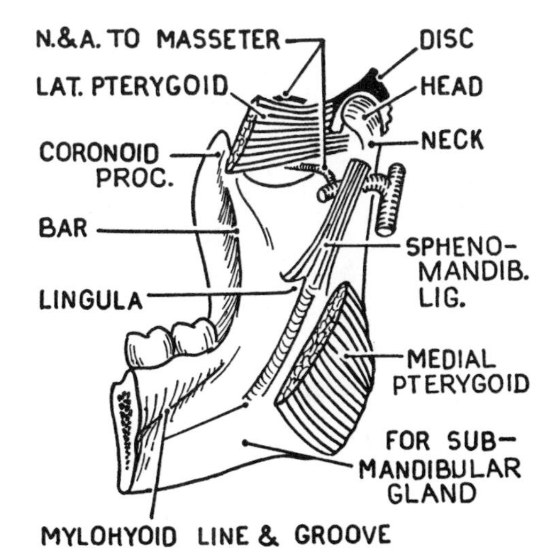

Fig. 738. The lateral wall of the infratemporal fossa: i.e., the ramus of the jaw.

flying buttress for the anterior wall (*fig. 739*).

The triangular cleft above the site of abutment is the *pterygopalatine fossa*. It is the hiding place of the *pterygopalatine ganglion* and of the 3rd part of the maxillary artery. The opening into the pterygopalatine fossa is the *pterygomaxillary fissure*, which joins the inferior orbital fissure at a right angle (*fig. 740*).

The **Roof** of the infratemporal fossa has a ragged edge, the *infratemporal crest*, separating it from the medial wall of the temporal fossa.

Two Important Foramina: (1) The *foramen ovale* at the posterior border of the lateral pterygoid plate—and the plate is the guide to it. (2) Just behind it, the *foramen spinosum* at the *spine of the sphenoid*, the medial limit of the mandibular fossa (*fig. 740*).

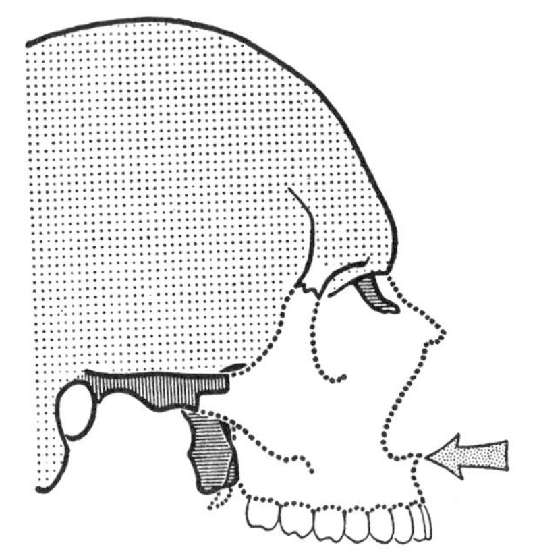

Fig. 739. The chief buttresses of the face—the pterygoid and zygomatic processes.

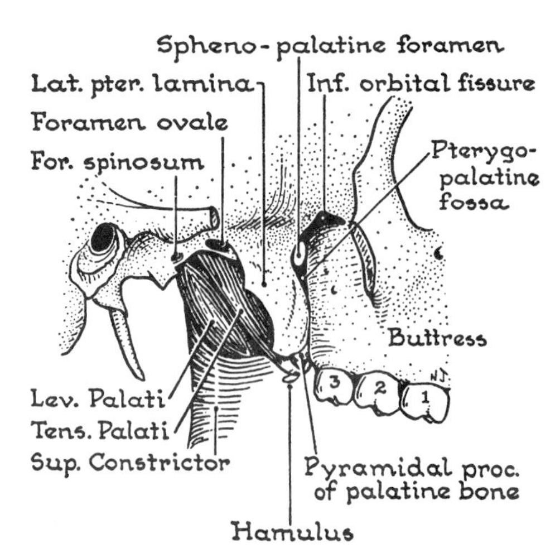

Fig. 740. Anterior wall, medial wall, and roof of the infratemporal fossa.

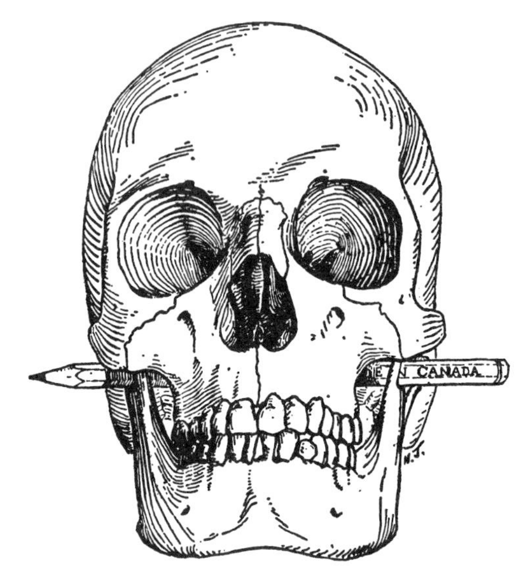

Fig. 741. A pencil can be passed through both mandibular notches, behind the pterygoid plates.

A Key Feature: See Figure 741.

Contents of Infratemporal Region:

1. Pterygoideus Medialis.
2. Pterygoideus Lateralis.
3. Mandibular nerve.
4. Maxillary artery.
5. Maxillary vein.
6. Otic ganglion.

The Medial Pterygoid is the most medial structure in the region (*fig. 742*), arising from the medial surface of the lateral pterygoid lamina. It crosses down and back to the lateral wall of the fossa to be attached to the medial surface of the mandibular angle (*fig. 738*).

The Lateral Pterygoid has a continuous origin from almost the entire roof and medial wall of the infratemporal fossa. The fibers converge as they pass backward and laterally to be inserted into the pit in front of the neck of the jaw and into the articular disc (*fig. 738*).

The *Maxillary Artery* enters the region

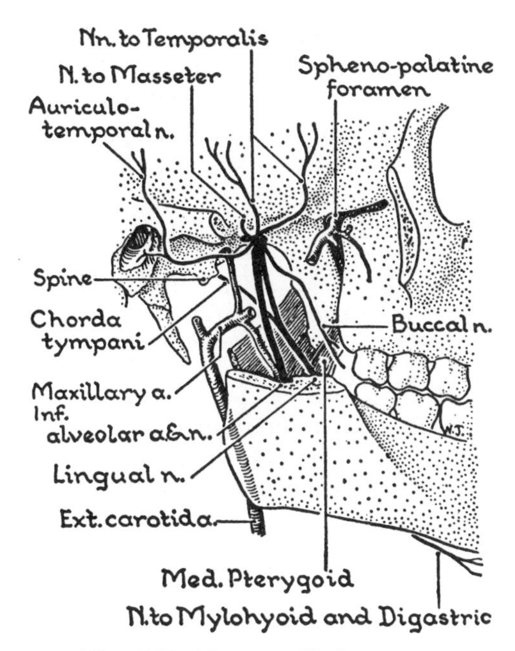

Fig. 742. The mandibular nerve.

in contact with the neck of the jaw (*fig. 742*). Its name indicates its distribution, but it also supplies this muscular area through which it passes.

The Mandibular Nerve (V^3) (*fig. 742*) drops through the foramen ovale—i.e., through the most medial part of the roof of the fossa. Developmentally, the mandibular nerve is the nerve of the mandibular arch, carrying both sensory fibers and motor, unlike V^1 and V^2 (entirely sensory).

Branches supply the muscles of mastication (but not the Buccinator) and are sensory to the full thickness of the mandibular region (*see* table 18).

1. Branches deep to Lateral Pterygoid: (1) motor to *two Pterygoids* and indirectly, via the otic ganglion, to *two Tensors* (Palati and Tympani) and (2) a *sensory twig* to the dura, via the foramen spinosum.

2. Branches that appear at the borders of the Lateral Pterygoid or perforate it: two branches to *Temporalis* and one to *Masseter*. The latter crosses the mandibular notch and sends a twig to the joint.

3. *The Buccal Branch* is the sensory nerve to the whole thickness of the cheek, including the mucous membrane of the cheek and lower gums (*fig. 793*), and the skin at the angle of the mouth (*fig. 658*).

The Inferior Alveolar (*Inferior Dental*) *and the Lingual Nerves* appear at the lower border of the Lateral Pterygoid and continue between the Medial Pterygoid and the ramus of the jaw (*fig. 742*).

The **Inferior Alveolar Nerve** enters the mandibular foramen to traverse the mandibular canal; one branch appears on the face as the mental nerve. En route, it sends twigs to the teeth (including the incisors) and to the gums.

[Before entering the canal it gives off the nerve to the *Mylohyoid* and anterior belly of the *Digastric*, which runs in a groove to the muscles.]

The **Lingual Nerve** supplies the anterior two-thirds of the tongue, the floor of the mouth, and the gums (p. 567), so naturally it lies in front of the inferior alveolar nerve. It enters the mouth between the Medial Pterygoid and the ramus, and, applied to the jaw, it runs forward submucously just below the 3rd molar tooth (*fig. 794*).

The *Chorda Tympani* ((*fig. 742*) joins the lingual nerve from behind, escaping through the floor of the middle ear (tympanic cavity) where it is seen as a cord (*fig. 820*). Its afferent fibers are taste fibers from the anterior two-thirds of the

TABLE 18

The Branches of the Mandibular Nerve

Muscular Branches	Sensory Branches	Other Branches
Temporalis and Masseter	Auriculotemporal	Taste
Medial and Lateral Pterygoids	Inferior alveolar	Secretory
Tensores Palati and Tympani	Lingual	Articular
Mylohyoid and Digastric (anterior)	Buccal	

tongue; the efferent fibers are secretory to the submandibular and sublingual salivary glands. (Continued on p. 567).

The **Auriculotemporal Nerve** (*fig. 742*) winds laterally behind the capsule of the joint, crosses the posterior root of the zygoma, and accompanies the superficial temporal artery. It carries sensory and secretory fibers.

Distribution:

1. Cutaneous fibers to the auricle and the temporal region (*fig. 737*), and to the external meatus and the outer surface of the ear drum.
2. *Articular twigs* to the jaw joint.
3. *Secretory fibers* to the parotid gland (*fig. 797*).

Maxillary Artery (*Int. maxillary art.*) (*fig. 743*). This crosses the Lateral Pterygoid, and disappears into the pterygopalatine fossa to break up into its end branches.

Distribution. The branches are distributed with the branches of both the mandibular (V^3) and maxillary (V^2) nerves.

Branches. In the Infratemporal Fossa: muscular branches; the **inferior alveolar artery**; the **middle meningeal a.,** which ascends through the foramen spinosum; and the *deep auricular* supplies the skin of the external acoustic meatus.

(*Terminal branches:* p. 578).

[The *accessory meningeal a.* ascends through the foramen ovale. The *anterior tympanic a.* passes through a fissure into the middle ear.]

Maxillary Vein (Internal). Veins corresponding to branches of the maxillary artery form a plexus, the **pterygoid plexus,** around the pterygoid muscles (*fig. 680*).

It has important connections (1) with the cavernous sinus via the foramen ovale, (2) with the facial vein via the deep facial vein, (3) with the pharyngeal plexus, and (4) the retromandibular vein (post. facial vein).

Otic Ganglion (*fig. 737.2*). This small ganglion is a parasympathetic relay station

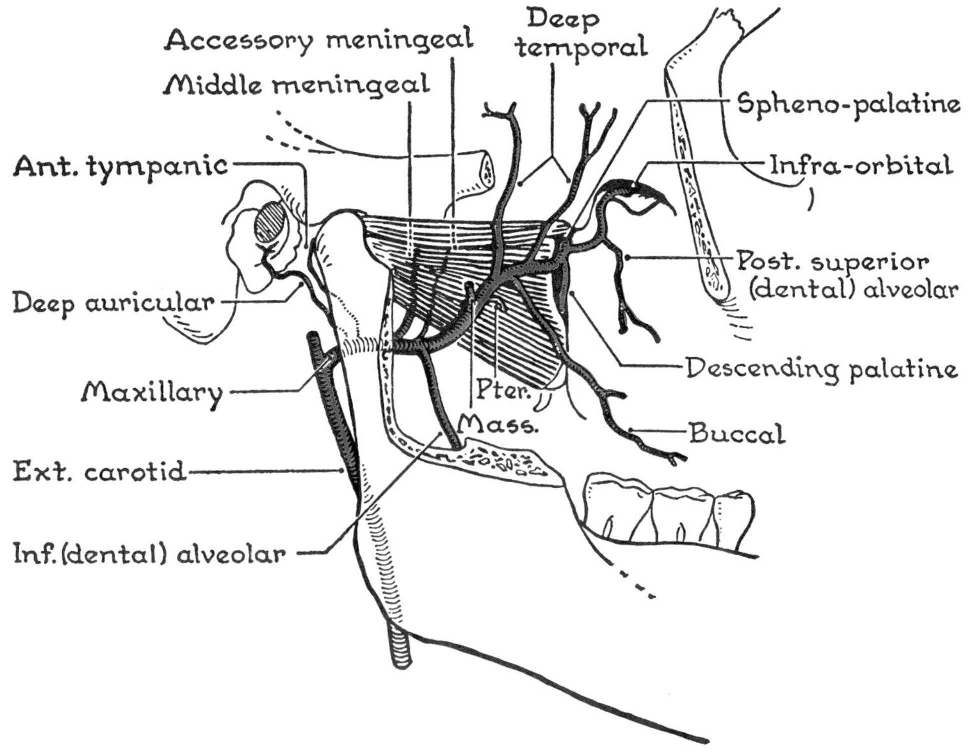

Fig. 743. The maxillary artery.

on the course of the *secretory fibers* of nerve IX (and nerve VII) to the parotid gland. It is situated below the foramen ovale, deep to the mandibular nerve.

The lesser petrosal nerve descends through the foramen ovale bringing preganglionic fibers to the ganglion; the postganglionic fibers travel with the auriculotemporal nerve as far as the neck of the jaw where they joint the facial nerve to be distributed with it (*figs. 737.2, 797*).

Motor fibers from the mandibular nerve (via the n. to the Medial Pterygoid) pass through the ganglion en route to the Tensor Palati and Tensor Tympani. *Sympathetic fibers* also pass through.

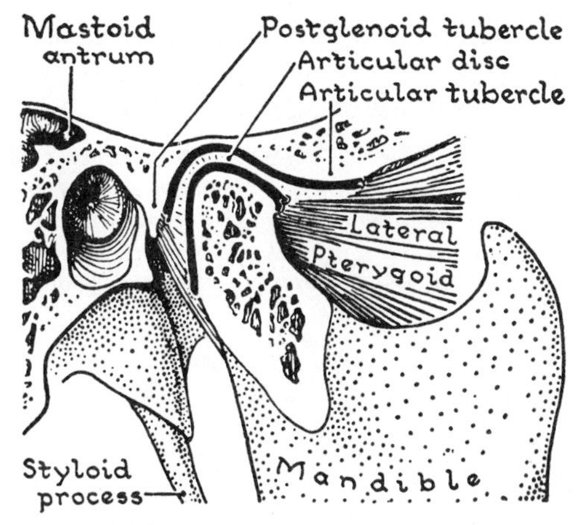

Fig. 744. Mandibular joint, on saggittal section.

Temporomandibular Joint (Jaw Joint).

Bony Parts. The bony parts of this joint are: (1) the head of the jaw; (2) the articular surface of the mandibular fossa of the temporal bone (*fig. 744*).

The Mandibular Fossa lodges the head of the jaw and a process of the parotid gland. It has an anterior articular part and a posterior nonarticular part. The spine of the sphenoid is directly medial. (Cont'd on p. 540).

Ligaments. The *strength* of this joint depends obviously on the bony conformation and on muscles. The *capsule* is necessarily relaxed. It is thickened laterally to form the *lateral* or *temporomandibular ligament*.

An articular disc caps the head of the jaw and projects forward under the articular tubercle, dividing the joint cavity into an upper and a lower compartment. This disc is firmly fixed to the medial and lateral ends of the condyle; the capsule blends

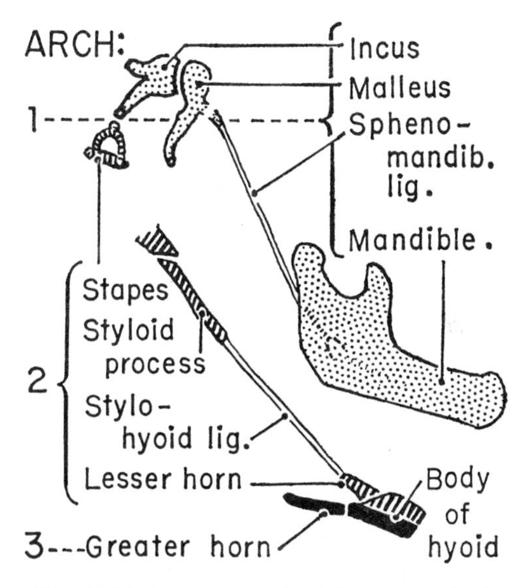

Fig. 744.1. Two vestigial ligaments (sphenomandibular and stylohyoid) derived from the cartilages of the 1st and 2nd pharyngeal arches.

TABLE 19
Movements of the Mandible

Depress (Open Mouth)	Elevate (Close Mouth)	Protract (Protrude Chin)	Retract (Withdraw Chin)	Side to Side (Grinding, Chewing)
L. Pterygoid Digastric (anterior) Mylohyoid Geniohyoid Infrahyoid muscles Gravity	Temporal Masseter M. Pterygoid	L. Pterygoid M. Pterygoid	Temporal (mid and post.) Masseter (deep part)	Temporal (mid. and post.) of same side Pterygoids of opposite side Masseter

with its circumference and the tendon of the Lateral Pterygoid is partly inserted into its anterior margin.

***Axiom:** In man, an articular disc implies two types of movement, one on each side of the disc. In the lower cavity simple hinge movements between the head and disc occur; in the upper cavity the disc and head together glide on the articular tubercle.*

Palpation—Surface Anatomy. The heads can be palpated either through the skin or by placing a finger in each cartilaginous external meatus, and their movements analyzed on opening and closing the mouth, on protruding and retracting the chin, and on performing grinding movements. You should do this.

Actions of Muscles (table 19).

The various movements of the jaw are produced by cooperative activity of several muscles bilaterally or unilaterally. (See Latif; Moyers; Carlsöö.)

Note: If the mouth is opened too widely (as in yawning) the head may move over the articular tubercle into the infratemporal fossa, undergoing dislocation.

Relations of the Joint. *Lateral*—subcutaneous; *anterior*—insertion of Lateral Pterygoid; *posteriorly*—parotid gland, auriculotemporal nerve, and superficial temporal vessels; *medial*—spine of the sphenoid (which is crossed on its medial side by the chorda tympani and on its lateral side by the auriculotemporal nerve) and just in front of the spine is the foramen spinosum for the middle meningeal vessels.

Nerve Supply. V^3 via (1) auriculotemporal n. and (2) nerve to Masseter.

Accessory Bands. (1) The *sphenomandibular ligament* is a vestige of the cartilage of the 1st pharyngeal arch. It extends from the spine of the sphenoid to the lingula of the mandible (*fig. 744.1*). (2) The *stylomandibular ligament* is a sheet of fascia condensed between the parotid and submandibular glands, and connecting the styloid process to the angle of the mandible. Do not confuse it with the stylohyoid lig., a remnant of the 2nd arch (*fig. 744.1*).

42 CERVICAL VERTEBRAE, PREVERTEBRAL REGION, AND EXTERIOR OF BASE OF SKULL

CERVICAL VERTEBRAE

Typical; Atypical: Atlas and Axis, and 7th; Anomalies.

DEEP CERVICAL STRUCTURES

Articulated Cervical Vertebrae.
Muscles; Prevertebral Fascia.
Vertebral Vessels; Ventral Rami of Cervical Nerves; Cervical Plexus.

CRANIOVERTEBRAL JOINTS

Transverse Ligament of Atlas; Alar Ligament.
General Ligaments and Homologues; Membrana Tectoria; Ligamentum Nuchae.

EXTERIOR OF THE BASE OF THE SKULL

Transverse Lines, Anterior and Posterior.

Anterior Area.
Bony Palate (Alveolar Process and Tuber Maxillae); Foramina (Greater and Lesser Palatine and Incisive).
Pyramidal Process; Incisive Bone (Premaxillae).
Choanae.
Pterygoid—Laminae; (Fossa); Hamulus and Canal.
Zygomatic Arch.

Intermediate Area.
Basi-occipital, (Pharyngeal Tubercle).
Oblique Line; Spine of Sphenoid.
Mandibular Fossa; Tympanic Plate; Articular Tubercle; Tympanosquamous Fissure; Postglenoid Tubercle.
Carotid Canal and Jugular Foramen.

Posterior Area.
Features; Occipital Bone.

CERVICAL VERTEBRAE

Man, giraffe, mouse and, indeed, all mammals (the manatee and certain sloths excepted) have seven cervical vertebrae. The 1st and 2nd are peculiar; the 3rd, 4th, 5th, and 6th are typical; the 7th is transitional. All have one distinguishing feature—their transverse processes are perforated.

Typical Cervical Vertebra (*fig. 745*). The *body* is slightly elongated from side to side. The upper surface resembles a shallow seat in having a raised ridge at the sides and in being rounded in front. The inferior surface is the counterpart.

The *pedicles* arise from the side of the body and project laterally as well as backward—so the *vertebral foramen* is triangular.

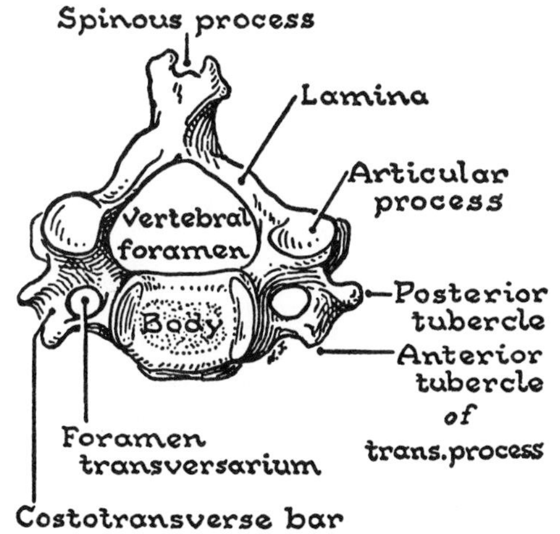

Fig. 745. Typical cervical vertebra, from above.

The *articular processes* are placed at the junction of the pedicles and laminae. They form a bony column cut obliquely into segments that permit flexion and extension of the neck, i.e., the lower rounded facet faces downward and forward and therefore can glide forward on the upper facet of the process of the vertebra below.

The *spinous process* is short, downturned, bifid, and V-shaped on cross-section.

The *transverse processes* have two roots, a perforation, and two tubercles. Its anterior root is a costal or rib element and, like a rib, it is attached to the side of the body and therefore is in front of the vertebral notch (*fig. 23.1*). The foramen transversarium transmits the vertebral artery. It is closed by the *costotransverse bar*, which forms a gutter on which lies the ventral ramus of a spinal nerve. The transverse processes are directed slightly downward and forward in the directions of the nerves.

[Muscles are attached to the posterior tubercles of all cervical transverse processes, but only the 3rd, 4th, 5th, and 6th have anterior tubercles (*figs. 673, 674*).]

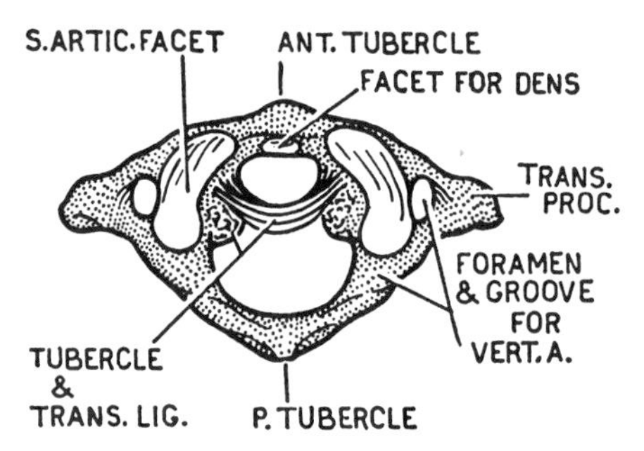

Fig. 746. The atlas and the transverse ligament. (Posterosuperior view.)

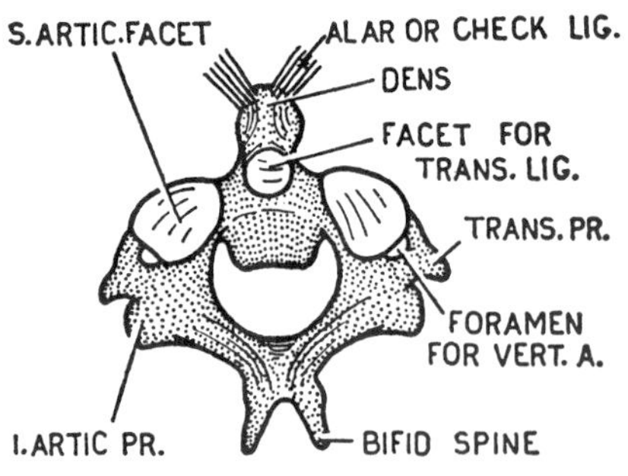

Fig. 747. Axis (posterosuperior view).

The 1st Cervical Vertebra supports the skull, so it is called the **atlas.** It loses its body and acquires an anterior arch. The lost body becomes the *dens* of the axis, so the atlas rotates around its own lost body (*figs. 746, 747*).

The atlas (*fig. 746*) is a ring divided into 5 equal areas—the anterior arch, right and left lateral masses, and a posterior arch composed of right and left halves. A transverse process projects from each lateral mass.

The *anterior arch* is anterior to the dens and has a facet behind for the dens. The *posterior arch* is grooved for the vertebral artery, just behind the lateral mass. It encircles a capacious vertebral foramen.

Each lateral mass has an upper and a lower weight-bearing facet. The upper facet fits the occipital condyle and so is oval and concave. A *tubercle* for the transverse lig. of the atlas, which retains the dens in position, projects medially from each mass (*fig. 746*).

The *transverse process* is a long lever that helps to rotate the atlas—only lumbar transverse processes are longer (*fig. 748*).

The 2nd Cervical Vertebra or Axis is typical below and atypical above. The *dens* is constricted at its root where the transverse ligament grips it, and it has a facet in front for the atlas (*fig. 747*).

The *superior articular facet*, being weight-bearing, is large and lies entirely in front of the plane of the anterior articular process. The bifid *spine* and *laminae* are massive.

The 7th Cervical Vertebra has a long nonbifid spine—nearly as prominent as that of Th. 1. The foramen transversarium transmits small veins (but not the vertebral a.).

Anomalies. (1) The occipital bone and atlas are occasionally fused. (2) The axis and vertebra C. 3 are commonly fused. (3)

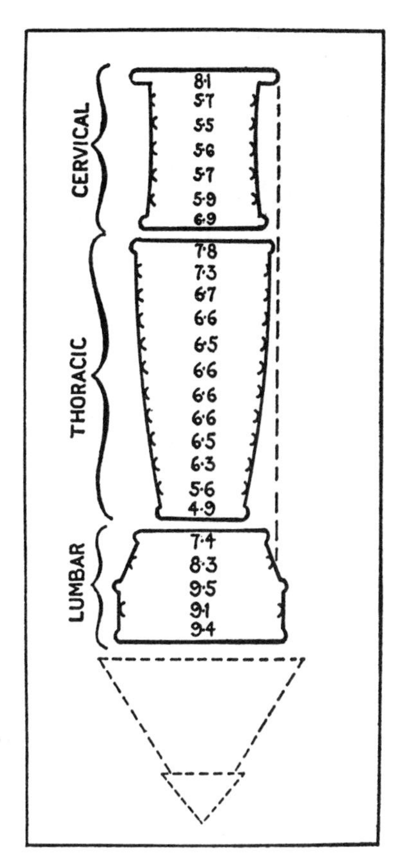

Fig. 748. Diagram showing the mean spread of the transverse processes (to scale).

Vertebra C. 7 may carry cervical ribs (*fig. 556*).

DEEP CERVICAL STRUCTURES

Articulated Cervical Vertebrae, when clothed with the muscles attached to the anterior tubercles of the transverse processes, present a relatively flat prevertebral surface.

Deep Anterior Cervical Muscles grouped according to their relations to the roots of the cervical and brachial plexuses (*fig. 749*):

Muscles Medial to the Plexuses:

1. Rectus Capitis Anterior.
2. Longus Colli (Cervicis).
3. Longus Capitis.
4. Scalenus Anterior.

Muscles Lateral to the Plexuses:

1. Rectus Capitis Lateralis.
2. Scalenus Medius and Posterior.
3. Levator Scapulae.

The following section recapitulates facts scattered elsewhere and previously learned; it also presents matter that requires appreciation but not memorization.

The *Longus Colli* extends from the body of the 3rd thoracic vertebra to the anterior tubercle of the atlas, and it is attached to the bodies of the vertebrae in between. It sends slips to the 3rd, 4th, 5th and 6th anterior tubercles and it receives slips from them.

The *Longus Capitis* arises from the 3rd, 4th, 5th, and 6th anterior tubercles and ascends to the basi-occipital to be attached behind the plane of the pharyngeal tubercle. It fills the hollow between the bodies and transverse processes.

The *Scalenus Anterior* arises from the 3rd, 4th, 5th, and 6th anterior tubercles and descends to the scalene tubercle of the 1st rib.

The *Rectus Capitis Anterior* covers the atlanto-occipital joint. It extends from the front of the lateral mass of the atlas to the basi-occipital.

The *Rectus Capitis Lateralis* extends from the transverse process of the atlas to the jugular process of the occipital bone.

The *Scalenus Medius* arises from all, or most of, the posterior tubercles and costotransverse bars of the transverse processes. It descends to the upper surface of the 1st rib between the groove for the subclavian artery and the tubercle on the neck. The *Scalenus Posterior* is its most posterior part combined to the 2nd rib behind the impression for the Serratus Anterior.

There is an angular gap between the transverse process of the atlas and the Scalenus Medius. This gap is occupied by the *Levator Scapulae* (*fig. 749*), which arises from the 1st, 2nd, 3rd, and 4th (posterior) tubercles.

Note that: (1) the jugular process is the "transverse process" of the occipital bone. (2) Well developed Ant. and Post. Intertransverse muscles unite the respective ant. and post. tubercles of the transverse processes. (3) The Rectus Capitis Anterior and Rectus Capitis Lateralis represent modified Anterior and Posterior Intertransverse muscles, and the ventral ramus of C. 1 appears between them. (4) Structures descending from the jugular foramen cross in front of the jugular process, Rectus

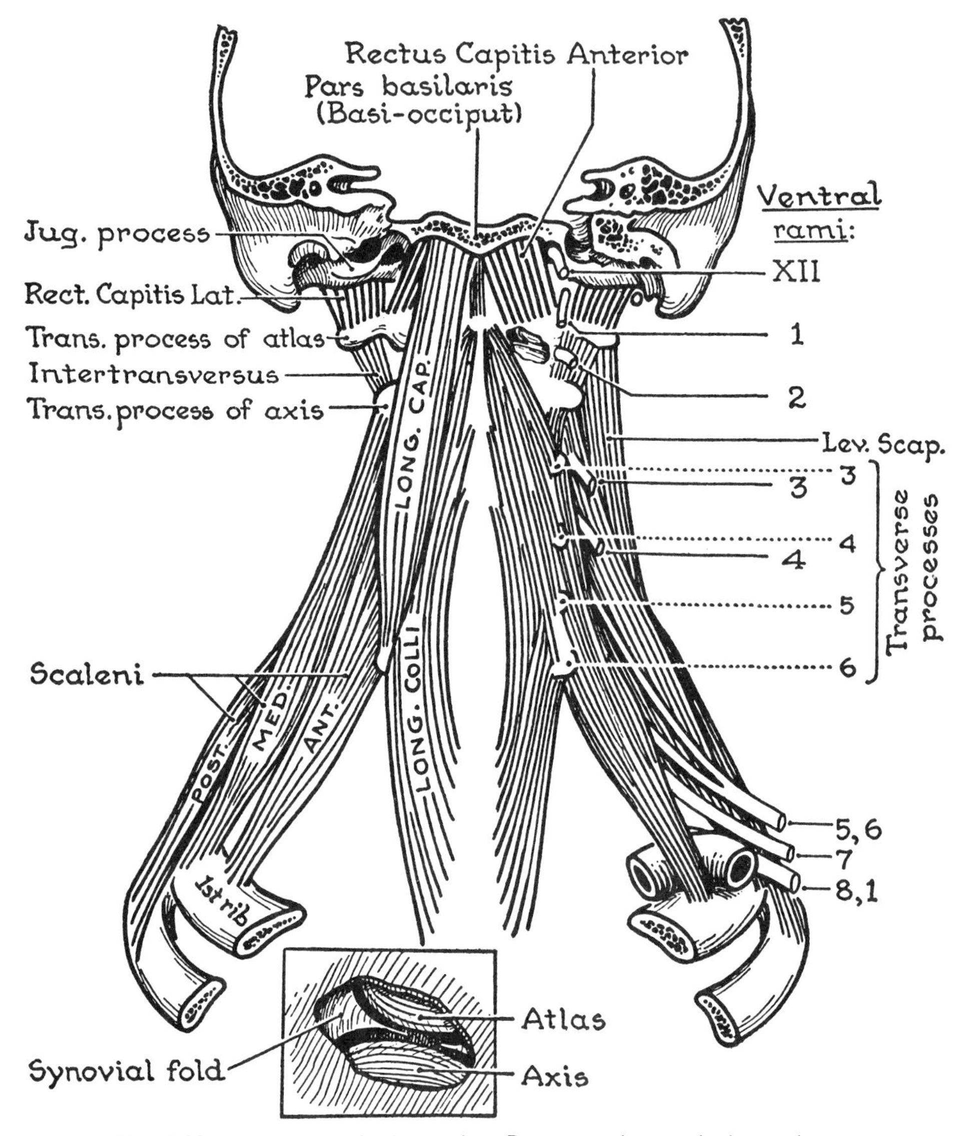

Fig. 749. The prevertebral muscles: Deep anterior cervical muscles.

Capitis Lateralis and the transverse process of the atlas—and lower down, Levator Scapulae, Scalenus Medius and Scalenus Anterior. (5) The Scaleni are modified Intercostal muscles; and (6) the brachial plexus represents enlarged and modified lateral branches of intercostal nerves piercing them.

The **Prevertebral Fascia** covers the prevertebral muscles and is continuous with the deep fascia that forms the carpet for the posterior triangle of the neck (p. 465). It is part of a strong sleeve that envelops the deep muscles of the neck. Above, it is attached to the base of the skull behind the jugular foramina. Below, it is lost in front of the thoracic vertebrae.

The roots of the cervical and brachial plexuses emerge from the vertebral column deep to the fascia, and of course the branches to the four muscles described above (p. 467 and *fig. 666*) remain deep to it. And, it must be pierced by all cutaneous branches (lesser occipital, great auricular, transversus colli, and supraclavicular), by branches of C. 1 and 2 to nerve **XII**, of C. 2 and 3 to ansa cervicalis, and of C. 2, 3, and 4 to nerve **XI.**

Vertebral Artery. *Its Origin and Course* are given in *figure 725* and page 516. (*See also fig. 750.*)

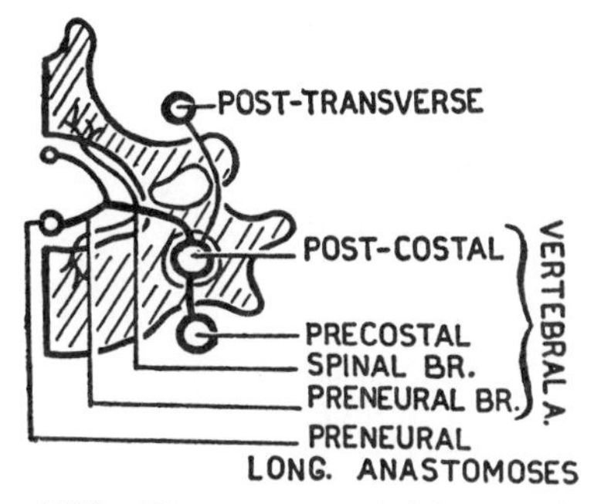

Fig. 750. Diagram to explain on developmental grounds the zig-zag course of the vertebral artery.

Its Function is to supply the cervical segment of the cord and part of the brain.

Its branches in the neck are spinal and muscular, including branches to the suboccipital muscles.

Variant. The artery sometimes passes in front of the 6th transverse process and enters the 5th, 4th, or 3rd foramen.

Ventral Rami of Cervical Nerves. The cervical plexus is formed by the ventral rami of C. 1, 2, 3, 4; the brachial plexus by the rami of C. 5, 6, 7, 8, and Th. 1. The courses of nerves C. 1 and 2 are peculiar, as explained by *figure 675.1*, and on p. 476. Interesting but unimportant details:

The ventral rami of C. 1 and 2 curve forward lateral to the superior articular processes of the atlas and axis, respectively; that of C. 1 then passes medial to the vertebral artery and appears between the adjacent borders of Rectus Capitis Anterior and Rectus Capitis Lateralis; whereas that of C. 2 passes lateral to the vertebral artery, appears between (Ant. and Post.) Intertransverse muscles, and divides into ascending and descending branches. This ascending branch of C. 2 joins most of C. 1 in front to the transverse process of the atlas. This united stem joins the hypoglossal nerve (*fig. 765*).

Distribution of the Cervical Plexus. The *phrenic nerve* is the chief nerve of the plexus. It is a mixed nerve from C. 3, 4 and 5.

Cutaneous Branches (see page 468).

Motor Branches supply the local muscles; branches of C. 2, 3, and 4 join nerve XI, branches of C. 2, 3, and 4 join the ansa cervicalis; and branches of C. 3 and 4 supply Levator Scapulae.

The ventral rami of the lower six cervical nerves run laterally behind the vertebral artery, between Ant. and Post. Intertransverse muscles, and appear between the Longus Capitis or Scalenus Anterior in front and the Scalenus Medius behind (*fig. 749*).

The rami of C. 8 and Th. 1 are related to the neck of the 1st rib; they unite at the medial border of the rib to form the lowest trunk of the brachial plexus.

CRANIOVERTEBRAL JOINTS

Joints between Skull, Atlas, and Axis

Five synovial joints are involved, two paired and one median. The paired joints are between the articular processes of the atlas and axis, and between the superior articular processes of the atlas and the occipital condyles; the median joint is between the dens and the anterior arch of the atlas.

At the *atlanto-occipital joints* we nod our heads, "Yes;" at the *atlanto-axial joints* we shake our heads, "No." At other cervical joints we flex and extend the neck and look up sideways.

The movements between the skull, atlas and axis have wide range, so the usual ligaments are loose up to a point where they get taut and check further movements. Near the centers of movement strong and peculiar ligaments are essential:

1. Transverse ligament of the atlas.
2. Alar ligaments.

Special Ligaments (*figs. 746, 747, 751*). *Transverse Lig. of the Atlas.* This strong band extends between the tubercles on the lateral masses of the atlas. It passes behind the root of the dens and embraces it. There is a synovial joint between the front of the dens and the anterior arch of the atlas and another between the back of the dens and the transverse lig. The head of the dens cannot easily be withdrawn from its ring.

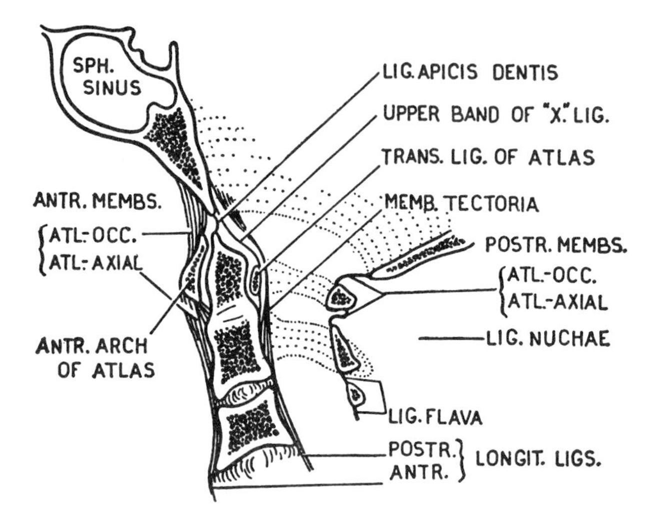

Fig. 751. Ligaments connecting the skull to the vertebral column (paramedian section).

[Weak upper and lower bands pass from the transverse lig. to the occipital above and to the body of the axis below, giving the whole the appearance of a cross; hence, called the *cruciform lig.*]

Alar Ligament (*fig. 747*). This short, very stout cord (one on each side) passes from the side of the apex of the dens laterally and slightly upward to the tubercle on the occipital condyle.

Together the two alar ligs. hold the skull tightly applied to the atlas and axis, while checking rotation of the skull and atlas on the axis.

[A pair of less important special ligaments—the *accessory atlanto-axial*—takes the same direction as the alar ligaments and joins the lateral mass of the atlas to the body of the axis. They likewise check rotation.]

General Ligaments and Derivatives (*fig. 751*). *The Anterior Longitudinal Ligament* of the bodies of the vertebrae becomes a cord at the axis, and as a cord it ascends to the anterior tubercle of atlas and on to the basi-occipital. Above the atlas its side parts are membranous, the *anterior atlanto-occipital membrane.*

As the **membrana tectoria** (*fig. 751*), the *Posterior Longitudinal Ligament* of the bodies passes from the posterior surface of the body of the axis to the inner surface of the occipital, spanning the atlas and the transverse lig. It is broad and strong.

[The intervertebral disc between the dens and the basi-occipital persists as a vestigial thread, the *lig. apicis dentis*, which transmitted the notochord.]

The supraspinous and interspinous ligs. form the *ligamentum nuchae* (p. 470).

The ligamenta flava become weak closing *fibrous membranes* between occipital bone, atlas, and axis (*see* suboccipital region, *fig. 675* on p. 475).

EXTERIOR OF BASE OF SKULL

The under surface has three areas: —anterior, intermediate, and posterior —separated by an "*anterior* and a *posterior transverse line*" (*fig. 752*).

These two imaginary lines cross most of the foramina at the base of the skull. Accordingly, they serve as reliable keys to the relationships of the nerves and vessels transmitted by these foramina; without them, the area may remain a jumble in your mind forever! They can be used in conjunction with *figures 755* and *688*.

The Anterior Transverse Line is found by passing a pencil through both mandibular notches and across the base of the skull (*figs. 741, 752*). On removing the mandible, the line is seen to cross: (1) foramen ovale

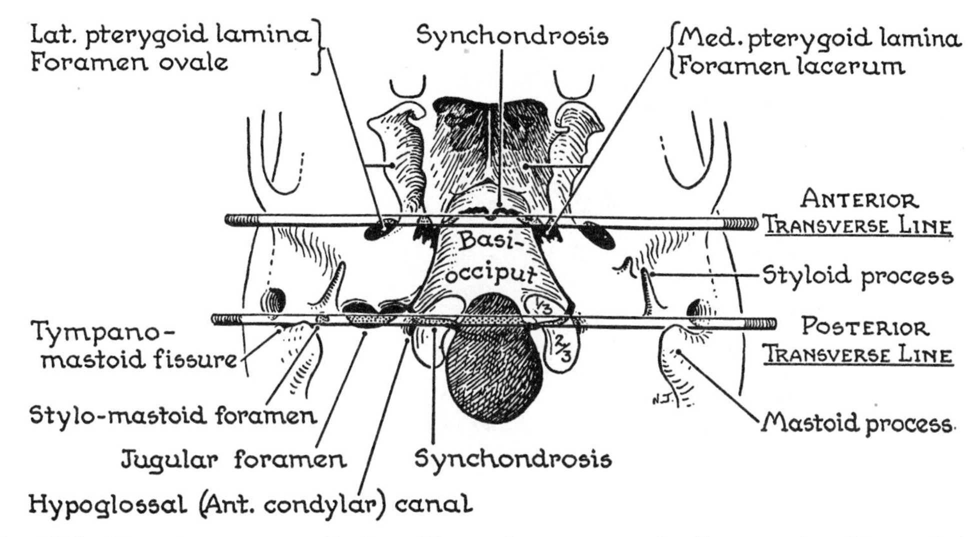

Fig. 752. "Anterior transverse line" and "posterior transverse line" on exterior of base of skull.

at (2) the root of the lateral pterygoid lamina, (3) foramen lacerum and (4) entrance to the pterygoid canal at (5) the root of the medial pterygoid lamina, and (6) the late-fusing synchondrosis between the basi-occipital part of the occipital bone and the sphenoid.

The Posterior Transverse Line unites the anterior margins of the right and left mastoid processes. It crosses: (a) stylomastoid foramen between (b) the styloid and mastoid processes, (c) posterior margin of jugular foramen, (d) hypoglossal canal, and (e) junction between anterior ⅓ and posterior ⅔ of occipital condyle. [Features *d* and *e* lie on an earlier fusing synchondrosis between the basilar and lateral (condylar) parts of the occipital bone.]

"Anterior Area" (*fig. 753*)

The **bony palate** is bounded by the alveolar process except behind where it has a free, sharp, crescentic border on each side separated by the *posterior nasal spine*. The *superior alveolar process* is U-shaped, carries 16 teeth, and has on each side a free posterior end, the *tuber maxillae*, palpable within your own mouth.

The foramen of the *greater palatine canal* lies medial to the 3rd molar tooth and from it a groove runs forward. The *incisive foramen* is situated just behind the incisor teeth.

[Two branches of the greater palatine canal, the *lesser palatine canals*, open just behind the greater palatine canal, descend through the pyramidal process.]

Choanae (Post. Nasal Apertures). These two oblong apertures are twice as deep as they are wide. Each is bounded on three sides by a sharp, free edge belonging to the *medial pterygoid lamina*, the *horizontal plate of the palatine bone*, and the *vomer*.

The medial pterygoid lamina is the hind part of the lateral wall of the nasal cavity. Its free border ends below in a hook, the *hamulus* (*fig. 754*).

The medial lamina ends above at the anterior border of the lower end of the foramen lacerum and the entrance to the *pterygoid canal* (*fig. 754*), which leads forward to where the pterygopalatine ganglion is situated in the pterygopalatine fossa.

Lower Border of the Zygomatic Arch. This arch is described on p. 522.

"Intermediate Area"

Medianly is the basi-occipital, which widens as it passes backward. Near its

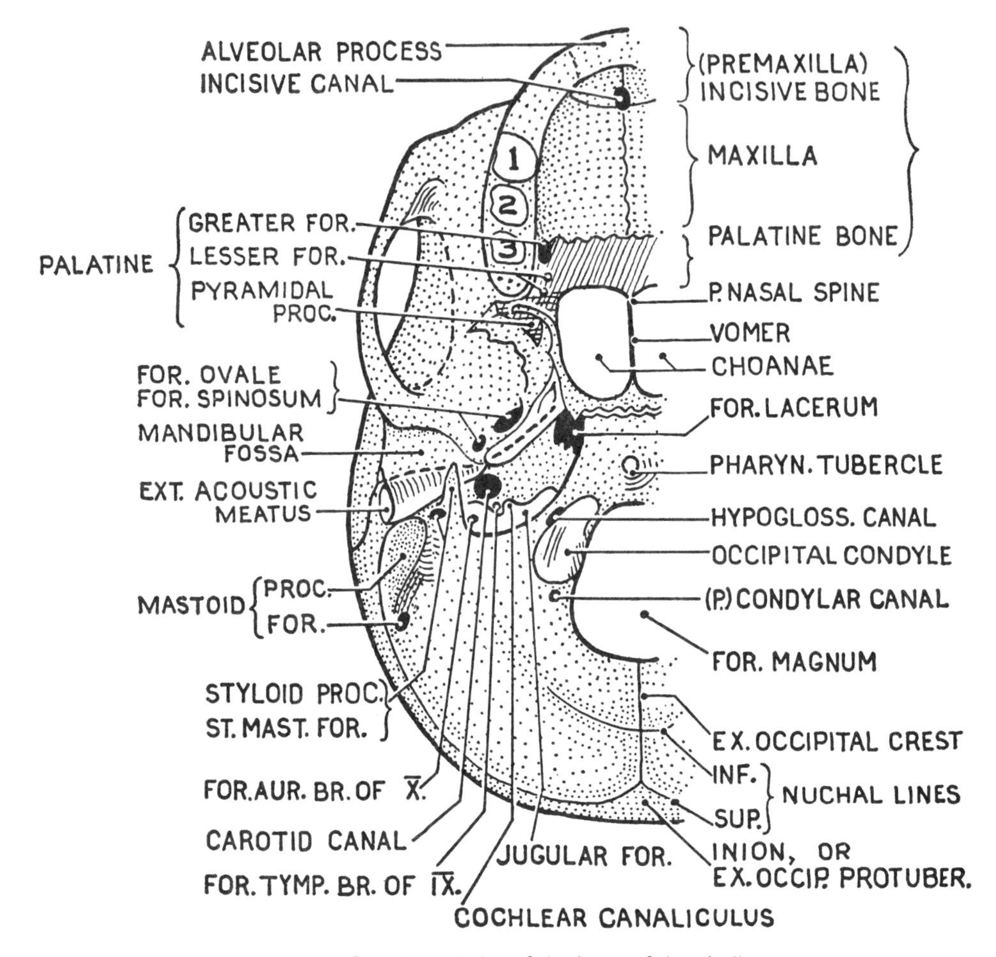

Fig. 753. The exterior of the base of the skull.

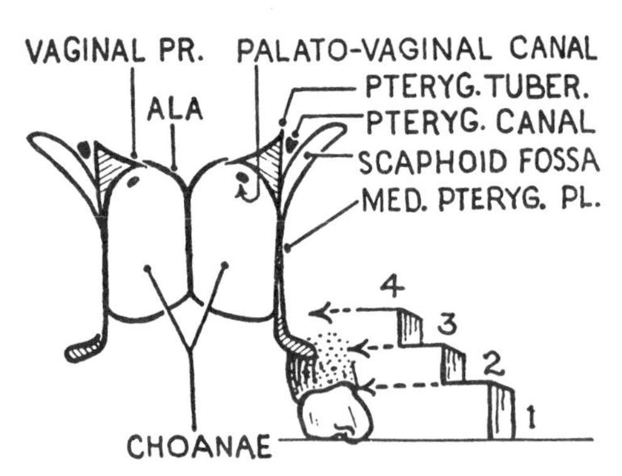

Fig. 754. (*1*) 3rd molar tooth, (*2*) tuber maxillae, (*3*) hamulus, and (*4*) bony palate in echelon.

center is the *pharyngeal tubercle* for the raphe of the constrictors of the pharynx (*fig. 753*).

"Oblique Line" at Base of Skull (*fig. 755*). This imaginary line extends from the root of the medial pterygoid plate on the anterior transverse line to the front of the ext. acoustic meatus. At its midpoint, the **spine of the sphenoid** stands out like a sentinel for: (1) anteriorly, the foramen spinosum for the middle meningeal vessels; (2) posteriorly, the opening of the carotid canal; (3) medially, the orifice of the bony auditory tube (a wire inserted here emerges through the ext. acoustic meatus); and, (4) laterally, the mandibular fossa.

Mandibular Fossa (*fig. 755*). The antero-inferior border of the fossa, the *articular tubercle* (eminence), is rounded, for it too is articular when the mouth is open. The dome is thin and translucent and it

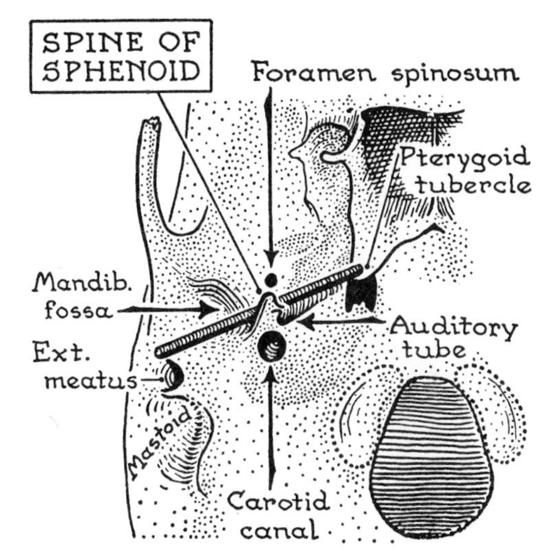

Fig. 755. Oblique line on exterior of base of skull.

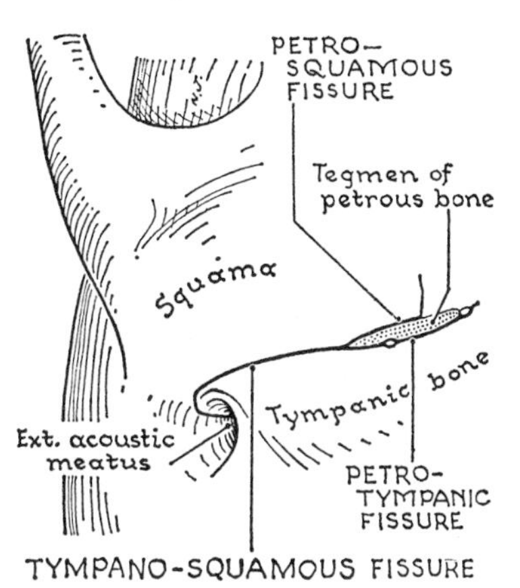

Fig. 756. Explanation of terms. The tympanosquamous fissure is bifurcated by the tegmen tympani reaching the surface here from the roof (tegmen) of the middle ear. This is a detail of interest to ear surgeons.

may be fractured by a blow on the chin. The forces of chewing are absorbed by the teeth, not this surface. *Details:*

The *tympanic plate* forms the square, posterior wall of a non-articular part of the fossa. At the upper limit of this plate, and lying on the "oblique line," is the tympanosquamous fissure. At the lateral end of the fissure, the squama of the temporal bone is prolonged downward behind the head of the mandible as the *postglenoid tubercle. See figure 756.*

Carotid Canal and Jugular Foramen. The entrance to the carotid canal lies immediately behind the bony auditory tube. It is separated from the jugular foramen (compartment for int. jugular vein) by a *bony wedge* (*fig. 753*), because the carotid curves forward and medially in the long axis of the petrous bone to open into the middle cranial fossa at the foramen lacerum (*fig. 692*) while the jugular foramen opens into the posterior fossa. Guarding the jugular foramen laterally is the styloid process, and separated from it medially by a wedge of bone is the hypoglossal canal.

***AXIOM**: Canals separated by wedges of bone transmit structures |vessels or nerves| that are either converging or diverging.*

"Posterior Area"

The posterior area was considered with the skull from behind, on page 469.

Occipital Bone. For details, see Chap. 50.

43 GREAT VESSELS AND NERVES OF NECK: REVIEW AND SUMMARY

General Dispositions.
Common and Internal Carotid Arteries—Relations; Branches; Carotid Sinus; Posterior and Lateral Relations.
Internal Jugular Vein—Relations; Bulbs; Tributaries.
Last Four Cranial Nerves—Glossopharyngeal; Sinus Nerve; Vagus, Accessory; Hypoglossal.
Sympathetic Trunk—Ganglia; Branches; Horner's Syndrome.

General Dispositions

The structures deep to the parotid region are:

1. Internal jugular vein.
2. Internal carotid artery.
3. Last four cranial nerves.
4. Sympathetic trunk.

These deep structures enter or leave the skull through one of *three openings:*

1. Jugular foramen—internal jugular vein and nerves IX, X, and XI.
2. Hypoglossal canal—nerve XII.
3. Carotid canal—internal carotid artery and nerve (sympathetic plexus).

The *jugular foramen* has three compartments: (1) a lateral for the internal jugular vein, (2) an intermediate for nerves IX, X, and XI, and (3) a medial for the inf. petrosal sinus.

The *hypoglossal canal* transmits the hypoglossal nerve (XII). It converges on the jugular foramen, below which nerves IX–XII gather into a compact bundle with the nerves between the artery and the vein (*fig. 768*).

The *carotid canal* is obviously in front of the jugular foramen since one opens into the middle, and the other into the posterior cranial fossa.

All *four nerves* descend together for a short distance between the int. jugular vein and the int. carotid artery (*fig. 757*). The X or *vagus nerve* continues vertically through the neck between the great artery and vein. The IX or *glossopharyngeal nerve*, in order to reach the pharynx and posterior third of the tongue, passes forward superficial to

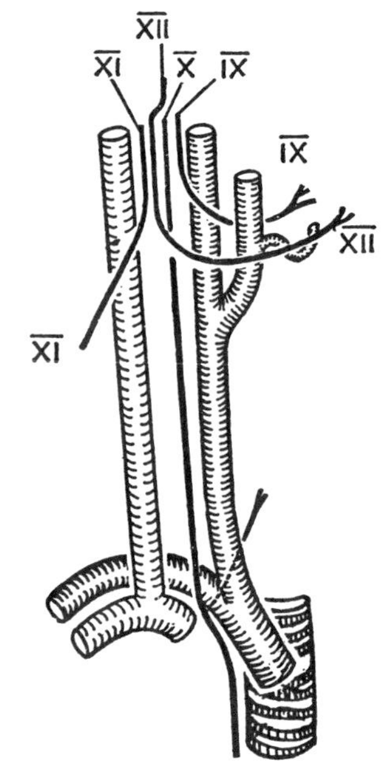

Fig. 757. Relations of nerves IX–XII to great vessels.

the int. carotid artery; the XI or *accessory nerve*, in order to reach the Sternomastoid, passes backward either superficial or deep to the int. jugular vein. The XII or *hypoglossal nerve*, in order to reach the tongue, passes forward superficial to the int. and ext. carotid arteries.

Common and Internal Carotid Arteries

This great arterial stem (*fig. 758*), the internal jugular vein, and the vagus nerve travel through the neck in the carotid sheath. The artery is medial, the vein lateral, and the nerve posterior in the angle between them. The sympathetic trunk runs behind the artery but outside the sheath.

Medial Relations. The arterial stem is applied to the side of the digestive and respiratory tubes. The width of the esophagus (2 cm) separates the right and left common carotids at the root of the neck; the width of the pharynx, more than twice that of the esophagus, separates the right and left internal carotids at the base of the skull.

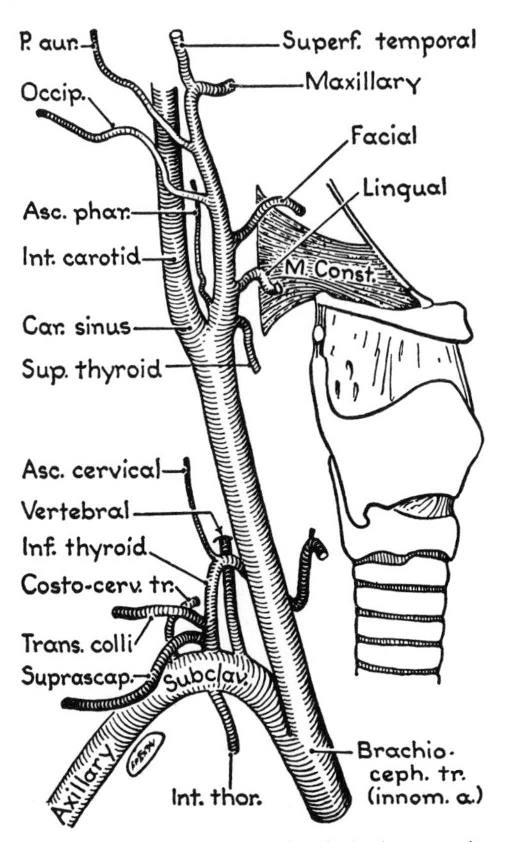

Fig. 758. The carotid and subclavian arteries.

The rounded posterior border of the *thyroid gland* usually insinuates itself medial to the common carotid artery and forces it laterally (*fig. 728*).

The *superior laryngeal nerve* and its two end branches, the *internal* and *external*, descend applied to the pharyngeal wall. They pass medial to the int. and ext. carotids (*fig. 763*).

Branches. The common carotid artery bifurcates into internal and external carotid arteries at about the level of the "Adam's apple" of the thyroid cartilage. Just above the bifurcation, the external carotid lies between the internal carotid and the pharyngeal wall.

In this description the pharyngeal fascia, pharyngeal plexus of veins and nerves, the ascending pharyngeal artery and the ascending palatine and the tonsillar branches of the facial artery are considered as constituents of the pharyngeal wall—which they are.

The internal carotid artery gives off no branches in the neck. At its origin there is a fusiform dilatation, the **carotid sinus.** The sinus is a blood pressure regulating mechanism which receives twigs from the glossopharyngeal nerve and also from the vagus and sympathetic. Some *interesting details*—

Passing between the external and internal carotids are the following:

1. The styloid process, Stylopharyngeus and Styloglossus—but not Stylohyoid which, being a derivative of the Digastric, passes superficial to both arteries.
2. The glossopharyngeal nerve and pharyngeal branches of the vagus—but not the superior laryngeal nerve—for reasons given in *figure 763*.
3. A portion of the parotid gland.

Posterior Relations—prevertebral muscles and fascia. The great arterial stem may be compressed against the prominent *carotid tubercle* of vertebra C. 6 (*fig. 723*).

Below the tubercle, the vertebral artery and vein lie behind the great arterial stem; and the inferior thyroid artery and the thoracic (or right lymph) duct arch between the carotid sheath and the vertebral vessels; on the right side, the right recurrent laryngeal nerve is an additional posterior relation (*figs. 725, 763*).

Above the tubercle, the prevertebral fascia and prevertebral muscles separate the artery from the transverse processes. At the base of the skull, the last four cranial nerves are behind the artery. The sympathetic trunk is posterior throughout, except below, where it passes behind the subclavian artery.

Lateral Relations. Laterally are the int. jugular vein and the vagus nerve, but the vein is posterior at the jugular foramen. Cardiac branches of the vagus and sympathetic accompany the artery.

Internal Jugular Vein

The continuation of the sigmoid sinus, it begins at the jugular foramen and ends after crossing the subclavian artery. Here it joins the subclavian vein to form the brachiocephalic vein behind the sternoclavicular joint (*fig. 709*).

Bulbs. The vein has a bulb at both ends. The *upper bulb* is an outpouching of the wall of the vein (*fig. 759*). The *lower bulb* is a dilatation of the vein below the bicuspid valve situated about 1 cm or so above the clavicle.

At the base of the skull the vein lies on the *posterior transverse line* (*fig. 752*), medial to the styloid process and the facial nerve; it is behind the carotid canal and therefore behind the int. carotid artery and the cranial prolongations of the sympathetic trunk; and it is posterolateral to the last four cranial nerves.

In the carotid sheath, it is separated by prevertebral fascia from the deep cervical muscles and the cervical plexus (*fig. 760*).

At the Root of the Neck (*fig. 725*), it crosses in front of: (1) the subclavian artery and its branches; (2) the phrenic and vagus nerves separated from each other by the thyrocervical trunk; (3) the thoracic (or right lymph) duct, which arches between the carotid sheath and the vertebral vessels, and (4) it makes contact with the pleural cupola.

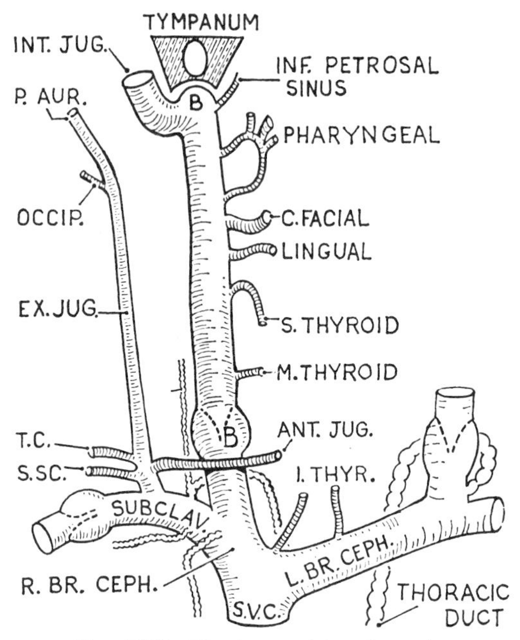

Fig. 759. The internal jugular vein.

The accessory nerve and the inf. root of the ansa cervicalis (descendens cervicalis) (C. 2, 3) cross either superficial or deep to the vein.

Tributaries (*fig. 759*). (1) The first tributary and the last, not being veins, are apt to be overlooked. They are the *inf. petrosal sinus* and the *thoracic* (or the *right lymph*) *duct*. (2) The *middle thyroid vein* joins the int. jugular vein at the root of the neck; it has no companion artery. (3) The four other tributaries accompany, more or less closely, four of the six collateral branches of the ext. carotid artery (*fig. 758*), and, accordingly, they are named: *superior thyroid*, *lingual*, and *common facial veins* (which join at and below the level of the hyoid bone), and the *pharyngeal veins* (which join at and above the level of the Digastric).

The terminations of the *occipital vein* (p. 477) and of the *posterior auricular*, *maxillary*, and the *superficial temporal veins* are shown in *figure 664*, on p. 465).

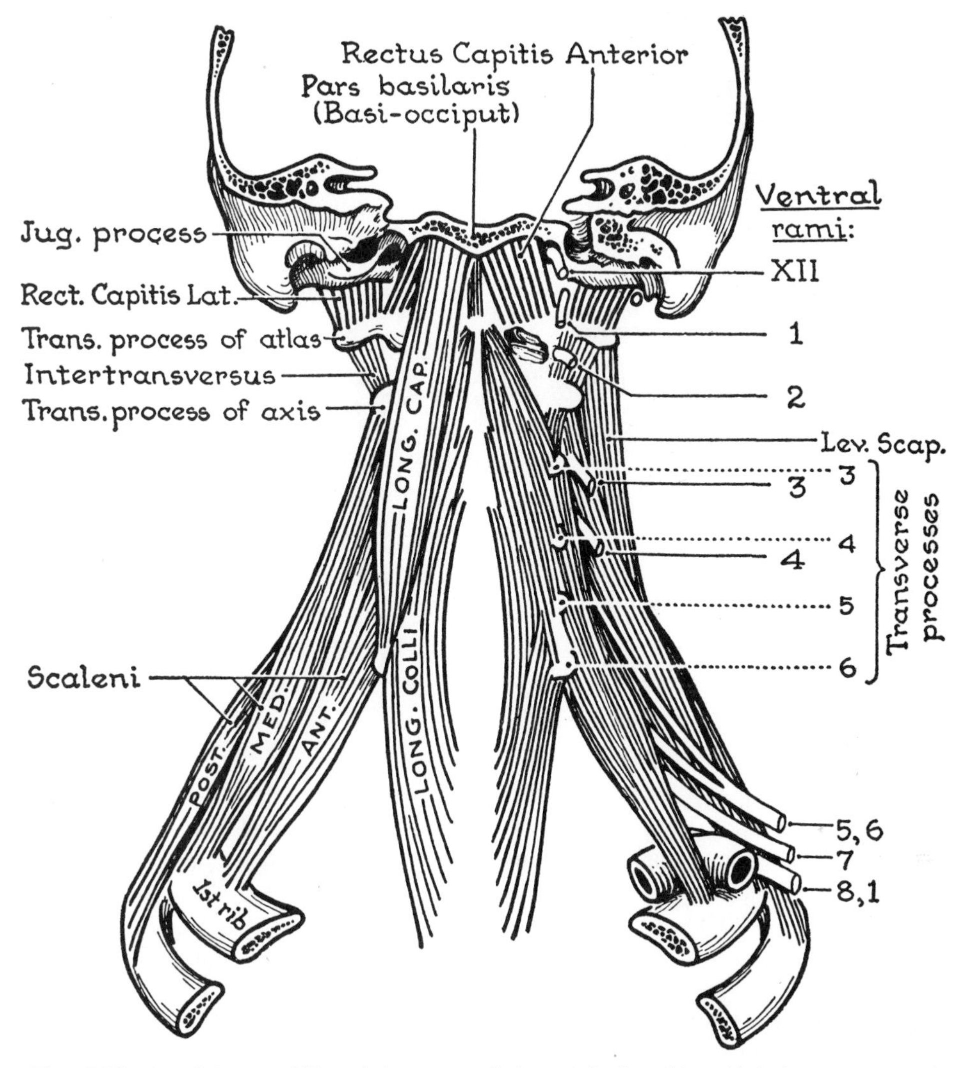

Fig. 760. Cranial nerve XII and the roots of the cervical and brachial plexuses emerging.

The Last Four Cranial Nerves (*Extracranial Courses*)

The general dispositions of these four nerves on leaving the skull are given on page 542.

Nerves IX and X, at the base of the skull, each possess two ganglia, an upper and a lower. These ganglia are the equivalent of spinal ganglia to spinal nerves and of the trigeminal ganglion to nerve V. Both nerves conduct efferent and afferent impulses to and from viscera.

Nerves XI and XII conduct efferent impulses only, their fibers supplying striated muscles, and both receive contributions from the cervical plexus.

Glossopharyngeal Nerve (N. IX). This nerve (*summarized in fig. 761*) leaves the skull through the jugular foramen with the vagus and accessory nerves, but in its own sheath of dura mater. It descends between the int. jugular and int. carotid vessels to the posterior border of the Stylopharyngeus which conducts it into the pharynx.

Winding superficially round Stylopharyngeus, it passes forward between the int. and ext.

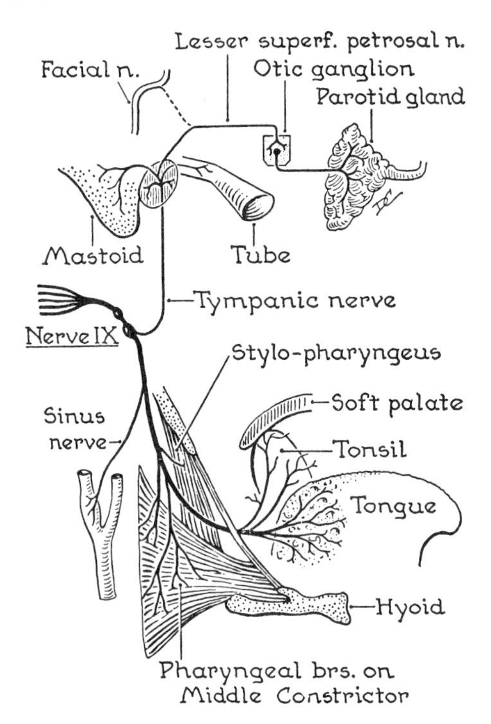

Fig. 761. Distribution of the glossopharyngeal (IX) nerve.

carotids (*fig. 763*). It then follows the upper border of the Middle Constrictor deep to the Hyoglossus and so enters the pharynx where it spreads out submucously over the posterior third of the tongue.

Distribution: Its one *muscular* branch supplies the Stylopharyngeus.

Fibers of *general sensation* pass to the pharyngeal plexus—which is formed on the Middle Constrictor by branches of the vagus, sympathetic, and glossopharyngeal—and through it supply most of the pharyngeal wall (*fig. 784*). One branch, the **sinus nerve,** consists of the afferent fibers from the carotid sinus and carotid body; when stimulated it brings about a reduction of the blood pressure.

The terminal fibers spread over the posterior third of the tongue, extending forward beyond the vallate papillae as fibers of general sensation and of *taste* (*fig. 778*); other branches are sensory to the tonsil, palatine arches, and soft palate. These fibers initiate the swallowing reflex.

Tympanic Nerve. This branch of nerve IX ascends into the tympanum through the tiny canal between the carotid canal and jugular foramen (*fig. 818*); it is sensory to the auditory tube, tympanum, medial surface of the ear drum, mastoid antrum, and mastoid air cells.

It is joined in the tympanum by sympathetic twigs from around the carotid artery. It leaves the tympanum and in the petrous bone is joined by a twig from the geniculate ganglion to form the *lesser petrosal nerve,* which passes to the otic ganglion, there to be relayed via the auriculotemporal nerve to the parotid gland as its *secretory nerve.*

Vagus Nerve (N. X). This vagrant or wandering nerve (*fig. 762*) leaves the skull

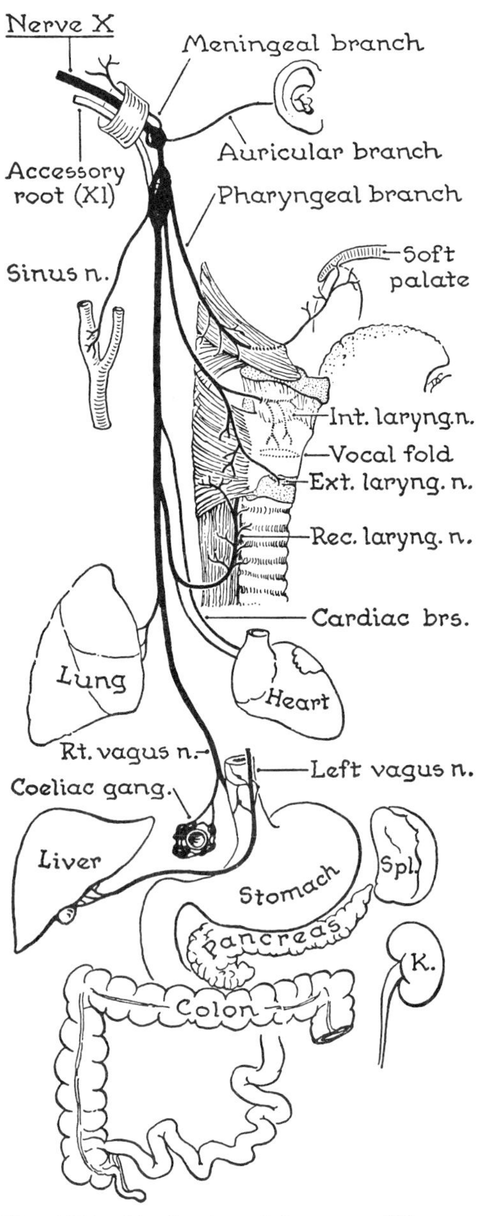

Fig. 762. Distribution of the vagus (X) nerve.

through the middle compartment of the jugular foramen in the same sheath of dura mater as nerve XI. It descends through the *neck* within the carotid sheath, occupying the posterior angle between the internal jugular vein and the great carotid stem as far as the sternoclavicular joint. There it crosses behind the brachiocephalic vein to enter the superior mediastinum.

In the Superior Mediastinum the courses are subpleural, but they differ on the two sides (*fig. 630*). *On the right side*, the vagus, having crossed the subclavian artery (IV primitive aortic arch), descends to the back of the root of the lung, lying first on the side of the brachiocephalic trunk (innominate art.), and then on the trachea. *On the left side*, the vagus continues its descent along the side of the carotid artery to the aortic arch (IV primitive aortic arch), and crosses the left side of the arch to reach the back of the root of the lung.

In the Posterior Mediastinum the courses are similar on the two sides. Each nerve forms a posterior pulmonary plexus, from which one or two stems emerge and pass to the esophagus, around which they form the esophageal plexus. From the plexus an anterior and a posterior vagal trunk emerge and descend on the respective surfaces of the esophagus, through the esophageal hiatus in the diaphragm, to the stomach.

In the Abdomen the two nerves lie close to each other at the lesser curvature of the stomach. The *anterior* (left) *gastric nerve* supplies the anterior surface of the stomach, and sends branches to the liver, pylorus, and first part of the duodenum. The *posterior gastric nerve* supplies the posterior surface of the stomach and sends one or two large branches alongside the stem of the left gastric artery to the celiac plexus, whence it travels with blood vessels to be distributed to the intestine as far as the left colic flexure and to the pancreas and other abdominal viscera.

Of the *two ganglia* of the vagus, the superior is small, the inferior is 2 or 3 cm long and lies just below the base of the skull. The cranial or *accessory root* of the accessory nerve joins the vagus at and beyond the ganglia and brings to it the motor fibers for the muscles of the pharynx, soft palate, and larynx.

Branches arise from the vagus thus:

In the jugular fossa:

Meningeal and auricular.

In the neck:

Pharyngeal, superior laryngeal, sinus, cardiac, and right recurrent laryngeal.

In the thorax:

Cardiac, left recurrent laryngeal, pulmonary, esophageal, and tracheal.

In the abdomen:

To most abdominal viscera.

Distribution. The vagus supplies: (1) the *striated muscles* of the pharynx (except Stylopharyngeus), soft palate (except Tensor Palati), and larynx; (2) the *heart* muscle, and the *smooth muscle* of the esophagus, stomach, and intestines down to the left colic flexure, and the gall bladder; and it contains (3) the *secretory* fibers for these organs. It also has (4) *afferent* fibers from these organs and from the ear, larynx and lower respiratory passages, and (5) a few *taste* fibers from the region of the epiglottis. Some readers may require the following *details*—

The *meningeal twigs* are derived from C. 1 and 2 and sup. cervical ganglion, and they are distributed to the dura of the posterior cranial fossa. The *auricular branch* crosses behind the int. jugular vein and enters a canal in the lateral wall of the jugular fossa, which conducts it past the facial nerve to the tympanomastoid fissure, through which it emerges. It assists the auriculotemporal n. to supply the outer surface of the tympanic membrane and the external acoustic meatus. It also sends twigs to the cranial surface of the auricle.

Explanatory of figure 763. In the neck, the chief duty of the vagus is to supply the alimentary and respiratory tubes. This it does via three branches: pharyngeal, superior laryngeal, and recurrent laryngeal. In the embryo, these three pass between the primitive ventral and dorsal cephalic aortic arches. Postnatally, the pharyngeal branch continues this course.

With the breaking down of the segment of the primitive dorsal aorta between the 3rd and 4th arches, the *superior laryngeal nerve* is enabled to rise to a higher level and to slip behind the int. and ext. carotids.

On the disappearance of the right 5th and 6th primitive aortic arches, the *right recurrent laryngeal nerve* rises to the 4th primitive arch (subclavian artery), recurs below it, and passes

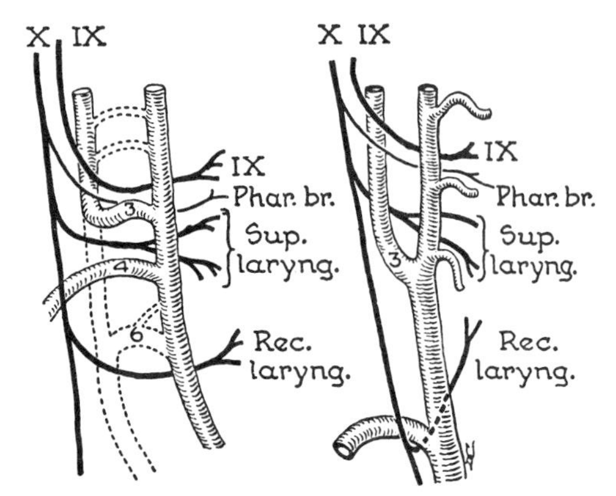

Fig. 763. Developmental explanation of relationship of glossopharyngeal, superior laryngeal, and recurrent laryngeal nerves to carotid arteries.

behind the common carotid artery. The *left recurrent laryngeal nerve* continues its original course round the primitive 6th arch (ductus arteriosus, then ligamentum arteriosum).

The *Pharyngeal Branch* pierces the Superior Constrictor and supplies all the muscles of the pharynx and soft palate, except Stylopharyngeus, Tensor Palati, and Inferior Constrictor.

The *Superior Laryngeal Nerve* passes medial to the int. and ext. carotids and, lying on the Middle Constrictor, divides into an internal and an external branch; the *internal branch* pierces the thyrohyoid membrane and is sensory to the larynx above the level of the vocal cords and to the region of the pharynx around the entrance to the larynx (p. 589); the *external branch*, after partly supplying the Inferior Constrictor, ends in the Cricothyroid.

The *Recurrent Laryngeal Nerves* are mixed nerves. The *right nerve* arises from the vagus where it crosses in front of the subclavian artery. It recurs below the subclavian and is there in contact with the pleural cupola; it then crosses behind the common carotid artery and ascends in the angle between the trachea and esophagus (*fig. 717*).

The *left nerve* arises from the vagus where it crosses the aortic arch. It recurs around the lig. arteriosum, passes below and medial to the aortic arch and may there be surrounded by tracheobronchial lymph nodes. It then ascends in the angle between the trachea and esophagus, as on the right side.

Both recurrent laryngeal nerves give off cardiac, esophageal, and tracheal branches and branches to the Inferior Constrictor. They supply all the muscles of the larynx (Cricothyroid excepted), and they are sensory to the larynx below the vocal cords.

A branch of the vagus passes to the *carotid sinus* and carotid body.

Cardiac branches are described on page 436; *pulmonary* branches on page 444.

Accessory Nerve (N. XI) (Spinal accessory nerve). The accessory nerve (*figs. 764, 768*) has a double origin—*spinal* and *cranial.*

The *spinal root* arises from the anterior gray column of the upper five or six segments of the spinal cord, ascends behind the lig. denticulatum, and enters the posterior cranial fossa through the foramen magnum. The *cranial root* arises from the medulla as several fila in line with those of the vagus. The two roots unite as they enter the middle compartment of the jugular foramen (within the same dural sheath as the vagus), and as they leave the foramen, they separate.

The *cranial part* at once joins the vagus,

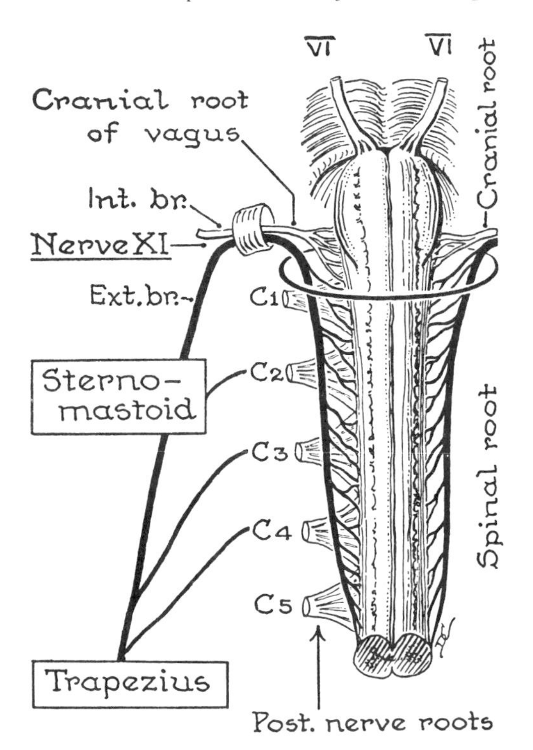

Fig. 764. Origin and distribution of the accessory (XI) nerve. (Ventral nerve roots have been cut away.)

bringing to it fibers for the muscles of the pharynx, soft palate, and larynx—it is accessory to the vagus.

The *spinal part* courses downward and backward superficial to the int. jugular vein (or deep to it) and the transverse process of the atlas. It makes a brief appearance in the carotid triangle between Digastric and Sternomastoid. Surrounded by lymph nodes and accompanied by the sternomastoid branch of the occipital artery, it enters the deep surface of the Sternomastoid 3 to 6 cm below the tip of the mastoid process.

It next appears at the posterior border of the Sternomastoid near its midpoint and here again it is surrounded by lymph nodes. Lying on the Levator Scapulae, it crosses the posterior triangle within the deep fascia, above and parallel to branches of C. 3 and 4 (*fig. 663*). It passes under cover of the anterior border of the Trapezius two or three fingers' breadths above the clavicle. Continuing deep to the Trapezius it crosses the superior angle of the scapula (*fig. 77*).

Distribution. The spinal part of the accessory nerve supplies the Sternomastoid and Trapezius. Within the Sternomastoid it is joined by a branch of C. 2, and deep to the Trapezius by branches of C. 3 and 4. Fine branches of the accessory n. may get cut where they cross the upper ("safe") ½ of the post. triangle.

Hypoglossal Nerve (N. XII) (*fig. 765*). This, the motor nerve of the tongue, emerges from its canal between the atlanto-occipital joint and the jugular foramen. It descends, making a half spiral turn behind the vagus and picking up a motor branch from C. 1 and a sensory branch from C. 2 (Pearson). It continues between the internal jugular vein and internal carotid artery to the lower border of the Digastric and Stylohyoid and so enters the carotid triangle. There it hooks around the occipital artery and, while doing so, gives off its *descendens branch, the superior root*

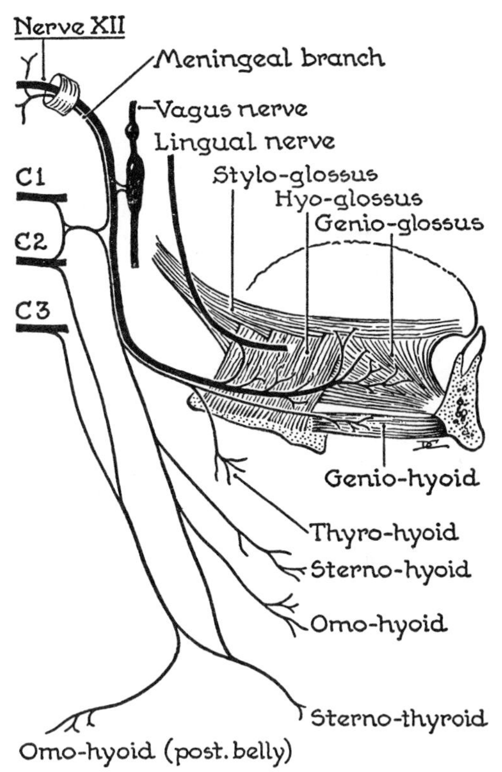

Fig. 765. Distribution of hypoglossal (XII) nerve.

of the ansa cervicalis.

It makes a gentle loop downward and forward superficial to the "arterial plane" (i.e., the carotid, lingual and facial arteries) and, passing deep once again to the Digastric and Stylohyoid, it enters the submandibular triangle. There, concealed by the submandibular gland, it continues forwards superficial to the Hyoglossus, which now separates it from the lingual artery, and passes deep to the free posterior border of the Mylohyoid. Arriving at the anterior border of the Hyoglossus it swings medially and its branches ascend in the substance of the tongue.

Distribution. The hypoglossal nerve supplies all three extrinsic muscles of the tongue (Stylo-, Hyo-, and Genioglossus) and all the intrinsic muscles. The contribution from C. 1 is distributed to Geniohyoid, Thyrohyoid and long infrahyoid muscles.

Its meningeal branches come from C. 1 and

2 and sup. cervical ganglion, and pass to the post. cranial fossa.

Sympathetic Trunk, *cervical part.*

This part of the trunk is an upward extension of the thoracic part (*fig. 766*). It ascends through the neck behind the common and internal carotid arteries, but outside the carotid sheath. It is medial to the vagus, which lies inside the carotid sheath. It is also thinner than the vagus, except at the sites of its ganglia (*fig. 768*). Entering the carotid sheath at the base of the skull, it becomes the *internal carotid nerve or plexus* which accompanies the int. carotid artery through the carotid canal (*fig. 693*).

Followed downwards into the root of the neck, the trunk usually splits to encircle, or throw loops (ansae) around, three arteries: (1) inferior thyroid, (2) vertebral, and (3) subclavian, the last loop being the *ansa subclavia*.

The Ganglia. On the trunk there are three (or four) ganglia: superior, middle, (vertebral), and inferior. *The superior ganglion* is fusiform and more than 3 cm long. It descends to the level of the greater horn of the hyoid bone.

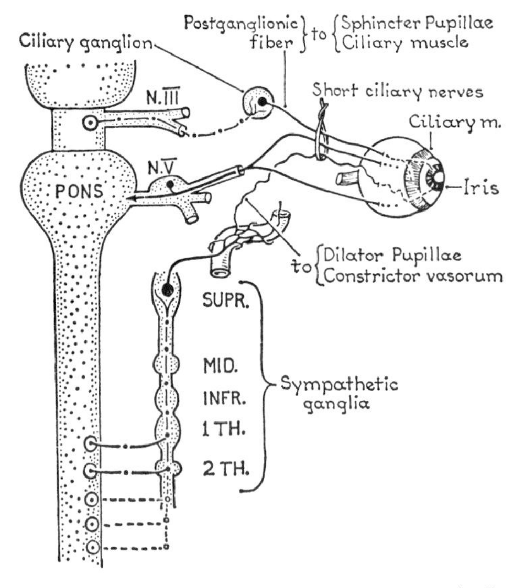

Fig. 766. The nerve supply to the eyeball.

The middle ganglion (inconstant), the size of a large pin's head, lies above the arch of the inf. thyroid artery. *The vertebral ganglion*, also inconstant and the size of a pin's head, lies in front of the vertebral artery, below the arch of the inf. thyroid artery. (The middle and vertebral ganglia are both present in about 50 per cent of specimens, but singly the vertebral ganglion is present much more often than the middle. (Becher and Grunt.)

The inferior ganglion lies behind the vertebral artery and in front of the 7th cervical transverse process. It commonly fuses in front of the neck of the 1st rib with the 1st thoracic sympathetic ganglion to form the *cervicothoracic* or *stellate ganglion*, which is large, nodular, and bristling with branches.

Branches (*fig. 766*) *Rami Communicantes.* The cervical part of the sympathetic trunk receives no white rami communicantes (preganglionic fibers) from the cervical segments of the spinal cord, but it delivers gray rami communicantes (postganglionic fibers) to each of the 8 cervical nerves.

It does so roughly thus: 1–4 spring from the superior ganglion; 5 and 6 from the middle ganglion; 6 or 7 from the vertebral ganglion; and 7 and 8 from the inferior ganglion.

Cardiac Branches, up to three in number, descend from the cervical ganglia or intervening parts of the trunk to the cardiac plexus (*fig. 634*).

Vascular Branches, carrying both efferent and afferent fibers, pass from each ganglion to blood vessels.

From the **superior cervical ganglion** postganglionic fibers pass to everything in its neighborhood—to the first *four cervical nerves* in gray rami, mainly to be distributed with branches of the cervical plexus (p. 537); to the last *four cranial nerves* or their branches; to the *pharyngeal plexus*, to the larynx via laryngeal branches of the vagus; to the *cardiac plexus*; to the *orbital cavity* (see below); and to the *carotid sinus*, and the meninges.

Vascular branches of the superior ganglion follow the int. and ext. carotids and their branches. *The details —*

With the *superior thyroid artery* to the thyroid gland; with the *facial artery* to the submandibular ganglion and on to the submandibular and sublingual glands, and to the cutaneous structures on the face (blood vessels, sweat glands, and arrectores); and with the *asc. pharyngeal* and *mid. meningeal arteries* to the dura (p. 492), and, via the otic ganglion, to the parotid gland.

The **internal carotid nerve** (plexus) (*figs. 692, 693*) accompanies the int. carotid art. It sends the deep petrosal n. to join the greater petrosal n. and travel with it through the pterygoid canal to the pterygopalatine ganglion to be distributed with branches of the maxillary nerve; it sends (caroticotympanic) twigs to the *tympanic plexus* in the middle ear; to the *dura*; to the *four nerves* in the cavernous sinus (III, IV, V, and VI); to the *trigeminal ganglion;* to the *hypophysis cerebri;* to the *terminal branches* of the int. carotid artery (mid. cerebral, ant. cerebral, and ophthalmic) and their branches; to the tarsal muscles; and by a branch that either joins the nasociliary nerve or runs independently through the superior orbital fissure to the ciliary ganglion and through it, via ciliary nerves, to the Dilator Pupillae and to the vessels of the eyeball. (The plexus around the ophthalmic artery may carry some of these fibers.)

From the **middle, vertebral,** and **inferior ganglia,** gray rami pass to the *brachial plexus* (*figs. 724, 44.4*); cardiac branches to the *heart*; and vascular branches to be distributed thus: from the middle ganglion with the *inferior thyroid artery*; from the inferior ganglion to the *subclavian artery* (*fig. 724*); and from the vertebral and inferior ganglia to the *vertebral* (sympathetic) *plexus*. As this plexus ascends through the foramina transversaria, it receives twigs from the middle and superior ganglia, and it delivers "accessory gray rami communicantes" to the cervical nerves.

Cutting the Cervical Sympathetic Trunk results in **Horner's syndrome**—drooping of the upper eyelid (ptosis) due to paralysis of the superior tarsal muscle (*fig. 698*), contraction of the pupil due to the unopposed action of the oculomotor nerve, and vasodilatation and absence of sweating on the affected side of the face.

44 PHARYNX AND PALATE

EXTERIOR OF PHARYNX

Structure of Pharynx—Buccopharyngeal Fascia; The Three Constrictors; Longitudinal Muscles; Pharyngobasilar Fascia; Mucous Membrane.
Structures Crossing Borders of Constrictors.

INTERIOR OF PHARYNX AND THE PALATE

Nasal, Oral, and Laryngeal Parts; Orifice of Tube; Pharyngeal Recess and Tonsil.
Palatine Arches; Fauces.
Entrance to Larynx; Valleculae; Piriform Recesses; Epiglottis.
Submucous Course of Laryngeal Nerves.
Palate.
Development; Exploration of Sphenopalatine Foramen.
Structure of Palate: Hard and Soft; Glands; Mucoperiosteum; Aponeurosis.
Palatoglossus; Palatopharyngeus; Stylopharyngeus.
Palatine Tonsil.
Removal of Tonsil; Lingual Tonsil, Tonsillar Bed; Relations; Vessels and Nerves.
Side Wall of Nasal Pharynx.
Upper Border of Superior Constrictor; Pharyngobasilar Fascia.
Auditory Tube—Structure; Function; Nerve Supply.

SOFT PALATE: DETAILS

Levator and Tensor Palati; Hamulus.
Structure of Soft Palate—Aponeurosis; Muscles; Vessels and Nerves; Sucking and Swallowing.
Mechanism of Swallowing.
Spaces: Retropharyngeal and Lateral Pharyngeal.

EXTERIOR OF PHARYNX

The posterior wall of the pharynx attaches to the base of the skull well in front of the foramen magnum, and has the great vessels and nerves lying posterolateral to it. And lying between these and the Medial Pterygoid are: the styloid process, the 3 muscles that arise from the process, and the Digastric (post. belly). (*See figs. 767, 768.*)

Pharynx, Definition and Structure

This fibromuscular tube extends from the base of the skull to the lower border of the cricoid cartilage where, at the level of vertebra C. 6, it becomes the esophagus. Above, its posterolateral angles reach almost to the carotid canals. At its junction with the esophagus it is the narrowest and least dilatable part of the alimentary canal, thus a foreign body that passes the cricoid is not likely to be arrested farther on.

The Pharyngeal Wall has four coats or tunics: (1) areolar, (2) muscular, (3) fibrous, and (4) mucous (described with the Interior).

The Areolar Coat is continuous with the areolar coat of the Buccinator and is called the *buccopharyngeal fascia*. It contains the pharyngeal plexus of veins and of nerves. The *venous plexus* drains the pharynx including the soft palate and tonsil; it communicates with the pterygoid plexus, and it ends in the internal jugular vein near the angle of the jaw. The *nervous plexus* is formed by pharyngeal branches of the vagus, glossopharyngeal, and sympathetic nerves, which are motor, sensory, and vasomotor, respectively.

The Muscular Coat comprises five paired voluntary muscles, namely—

1. Superior } Constrictors
2. Middle } Constrictors
3. Inferior } Constrictors

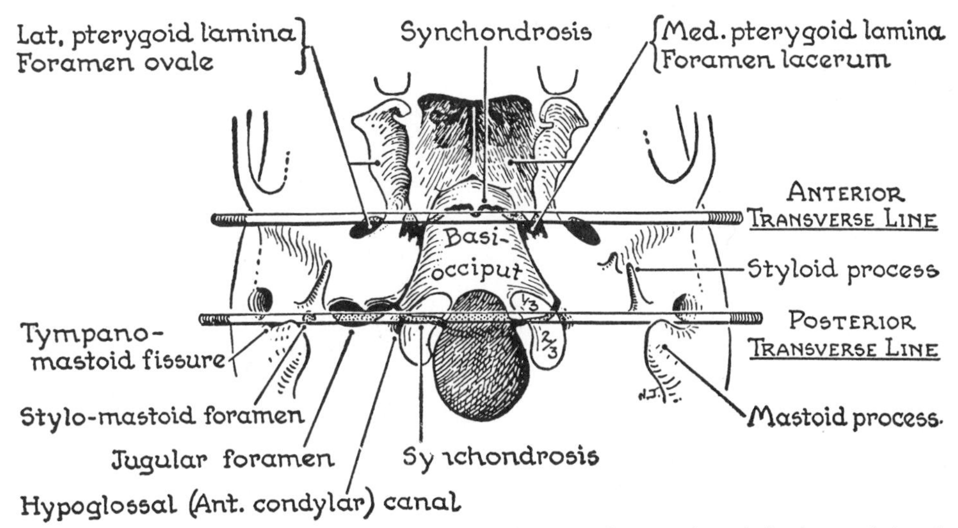

Fig. 767. The anterior and posterior transverse lines on the exterior of the base of the skull.

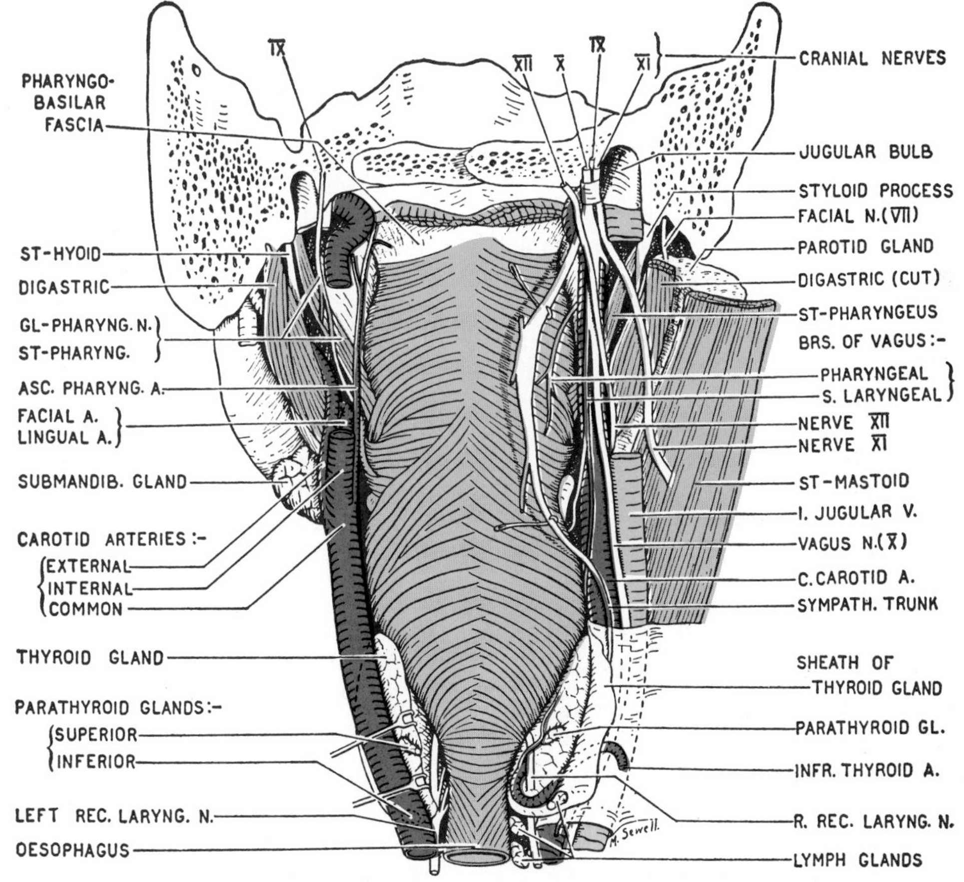

Fig. 768. The pharynx, last four cranial nerves, sympathetic trunk and great vessels—from behind. (The skull has been sectioned in "the posterior transverse line.")

which represent an outer "circular" coat,

4. Stylo-
5. Palato- } pharyngeus

which represent an inner "longitudinal" coat, are described on p. 558.

The **Three Constrictors** (*figs. 769, 770*). (Also see p. 508.) It is easiest to begin by placing the Middle Constrictor, thus:

The **Middle Constrictor** (*fig. 769*) arises from the angle between the greater and lesser cornua of the hyoid and from the lower end of the stylohyoid ligament.

The **Inferior Constrictor** has a continuous origin extending from the upper border of the thyroid cartilage to the lower border of the cricoid (*fig. 770*).

The cricopharyngeal part, the **Cricopharyngeus,** being normally in a state of contraction, guards the esophagus like a sphincter and prevents air from being sucked into it during inspiration (Negus; Raven).

The **Superior Constrictor** arises from the pterygomandibular raphe and from the bony point at each end of the raphe, i.e., the lower end of the medial pterygoid plate and the mandible behind the 3rd molar tooth.

The Fibrous Coat is the **pharyngobasilar fascia** and corresponds to a tunica submucosa. It is strong above, where it serves to anchor the pharynx to the posterior border of the medial pterygoid plate and base of the skull.

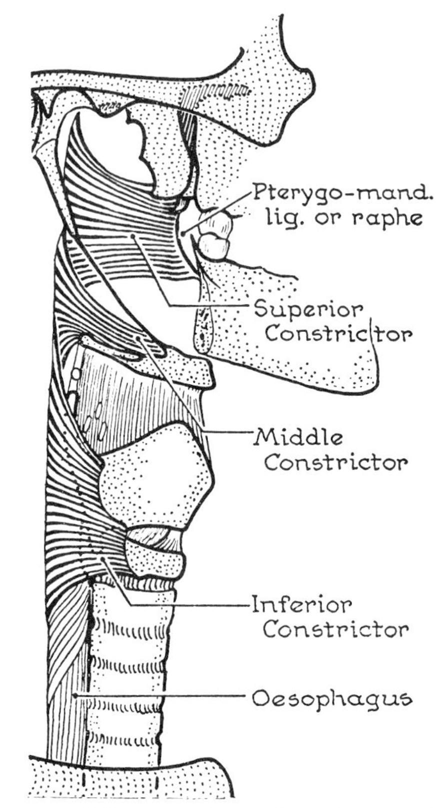

Fig. 770. The three Constrictors of the pharynx.

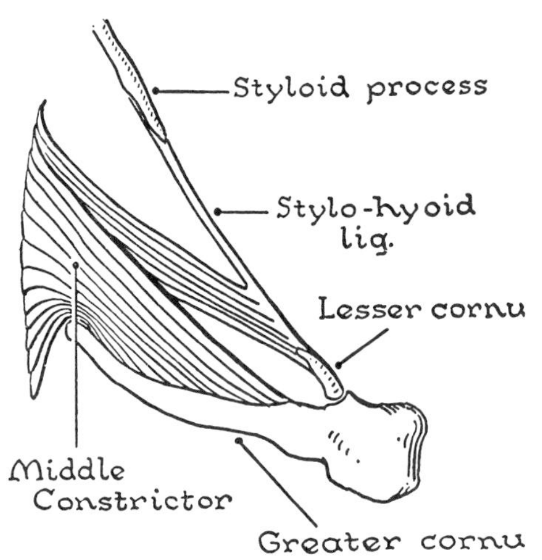

Fig. 769. Angular origin of Middle Constrictor.

Structures Crossing the Borders of the Constrictors (*fig. 771*):

1. *The recurrent laryngeal nerve* and its companion artery, the *inferior laryngeal a.*, pass through the gap between the esophagus and the Inf. Constrictors, closely applied to the back of the cricothyroid joint.
2. *The internal laryngeal nerve* and *superior laryngeal vessels* pierce the thyrohyoid membrane in the gap between the Inf. and Mid. Constrictors.
3. *The Stylopharyngeus*, accompanied by the *glossopharyngeal nerve*, passes through the gap between the Mid. and Sup. Constrictors, amalgamates with the Palatopharyngeus, and gains attachment to the greater horn of the hyoid and posterior border of the thyroid cartilage.
4. The *auditory tube*, the *Levator Pala-*

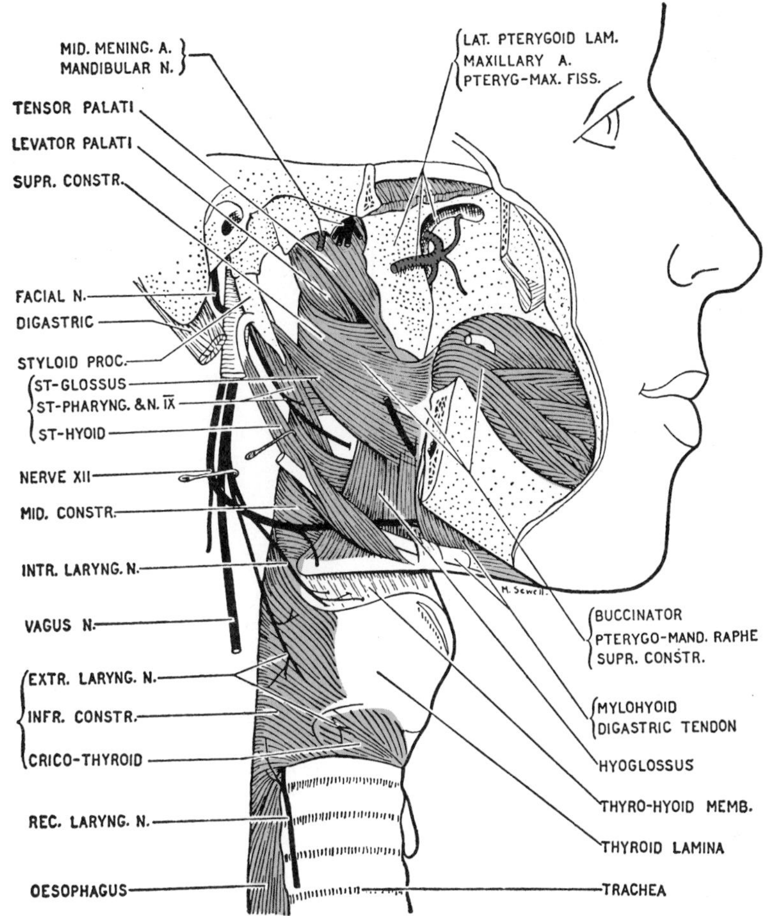

Fig. 771. The lateral aspect of the pharynx.

ti, and the *ascending palatine artery* pass through the gap between the Sup. Constrictor and the base of the skull.

INTERIOR OF PHARYNX AND THE PALATE

Opening into the pharynx anteriorly are orifices leading from the cavities of the nose, mouth, and larynx (*fig. 772*). Thus the pharynx is divided into three parts: the *nasal pharynx*, *oral pharynx*, and *laryngeal pharynx*. The soft palate, ending in the uvula, hangs down and separates the nasal pharynx above from the oral pharynx below.

The Nasal Pharynx (Nasopharynx) lies above the soft palate and behind the nasal cavities. It is, in fact, the backward extension of the nasal cavities and it cannot be shut off from them.

In front, are two oblong rigid orifices, the *choanae* (post. nasal apertures), described on page 539. On looking through these apertures the posterior ends of the middle and inferior conchae are seen.

On the side wall of the pharynx, 1 cm behind the inferior concha, is the orifice of the *auditory tube*. Its upper and posterior lips are prominent and cartilaginous. A fold of mucous membrane, the *salpingopharyngeal fold*, overlying a muscle of the same name, descends from the postero-inferior part of the orifice giving it the appearance of a hook. Behind the orifice of the tube there is a vertical cleft, the *pharyngeal recess*.

The *roof* is bone lined with mucous membrane, which, curving downward, becomes the posterior wall in front of the atlas and axis, prevertebral fascia and prevertebral muscles. On the roof there is some lymphoid tissue, the *(naso-)pharyngeal tonsil,* which when overgrown is known as "adenoids" (*fig. 798*). This tissue extends into the pharyngeal recess (behind the auditory tube) and, when hypertrophied, it may interfere with access of air to the middle ear with resulting deafness.

The Oral Pharynx is placed below the soft palate and behind the mouth and the posterior one-third of the tongue.

From the soft palate two folds of mucous membrane arch downward on each side. The anterior fold, the *palatoglossal arch*, overlies a muscle of the same name and descends to the junction of the anterior two-thirds and posterior one-third of the tongue. It lies at the dividing line between mouth and pharynx (oropharyngeal isthmus). The posterior fold, the *palatopharyngeal arch*, also overlying a muscle of the same name, arches downward to be lost on the side wall of the pharynx.

On each side, the two palatine arches and the triangular area between them, occupied by the tonsil, is called the *fauces*. The space between the right and left fauces is called the *isthmus of the fauces*, and the soft palate is its roof.

The Laryngeal Pharynx lies behind and around the freely projecting upper end of the larynx.

The *inlet of the larynx* is oval and obliquely placed. In front it is formed by the free, curved upper end of the *epiglottis*; behind, by the mucous membrane clothing the apices of the *arytenoid cartilages* and the *Arytenoideus* which unites these cartilages; on each side, by the *aryepiglottic fold* which extends from epiglottis to arytenoid. Slightly in front of the apex of the arytenoid cartilage, which is surmounted by the *corniculate cartilage*, is the rounded end of the *cuneiform cartilage*.

Three folds of mucous membrane leave the epiglottis: one, the *median glosso-epiglottic fold*, connects it in the median plane with the back of the tongue; one on each side, the *lateral glosso-epiglottic fold* (pharyngo-epiglottic fold), connects it with the pharyngeal wall. Between these three folds are two fingertip depressions, the *valleculae* (*fig. 772*).

On each side, behind the lateral fold there is a space, the *piriform recess*, which is bounded by the thyroid cartilage and thyrohyoid membrane laterally and by the free upper end of the larynx medially. On the posterior wall of the pharynx numerous *lymphoid follicles* are scattered. They may become enlarged.

SUBMUCOUS COURSE OF LARYNGEAL NERVES. These two nerves are readily exposed (*fig. 773*). The *internal laryngeal nerve*, having pierced the thyrohyoid membrane, runs transversely in a fold across the front of the piriform recess; it is sensory to mucous membrane. The *recurrent laryngeal nerve* runs vertically, applied to the back of the cricothyroid joint. It supplies all the muscles of the larynx except the Cricothyroid and is also sensory to the mucous membrane.

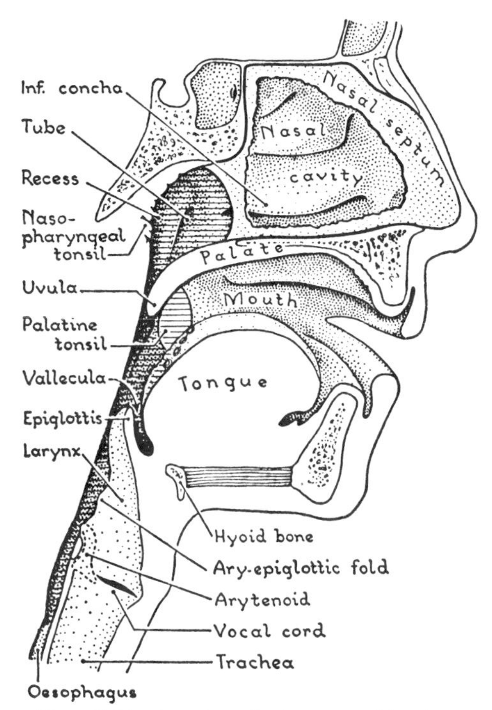

Fig. 772. Interior of pharynx, side view.

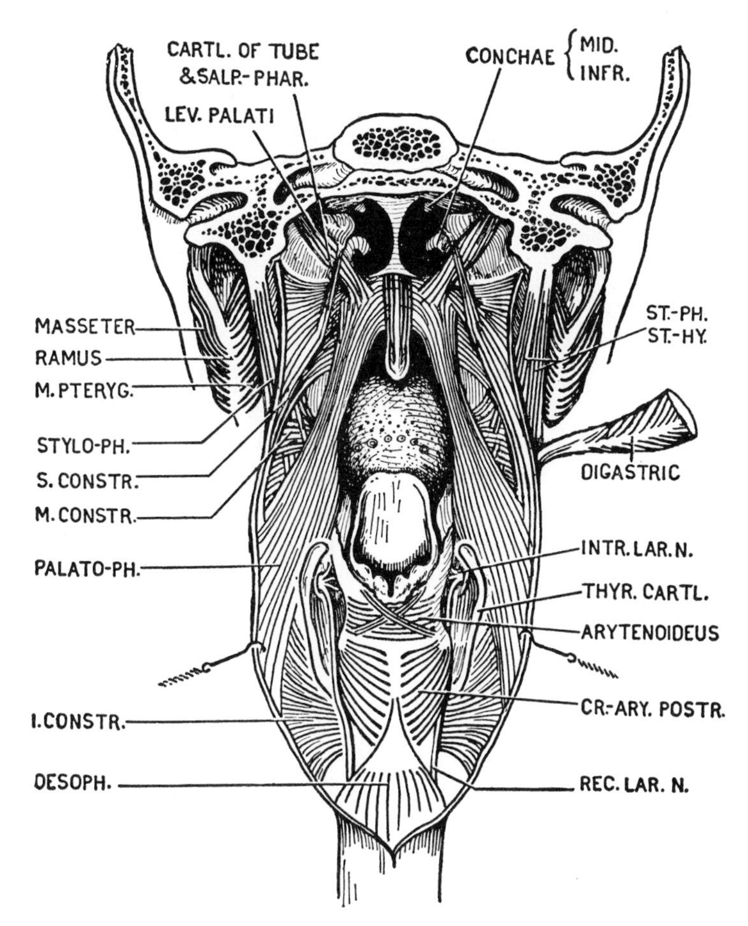

Fig. 773. The muscles of the pharynx, from behind.

Palate

Development. Before the palate appeared the nasal and oral cavities were one, the *stomodaeum*, and the greater part of its side wall developed in the *maxillary process* (*fig. 775*), the nerve supply of which is the maxillary nerve (V^2). Subsequently, as the palate takes form, this nerve becomes the nerve of the palate also.

The maxillary artery (3rd part) was seen to disappear into the pterygopalatine fossa (*fig. 743*). There it meets the maxillary nerve and breaks up into branches that accompany the branches of the maxillary nerve through various bony passages, including the sphenopalatine foramen and the greater and lesser palatine canals (*fig. 796*).

Sphenopalatine Foramen. A needle pushed through the mucous membrane and lateral wall of the nasal cavity just above the middle concha near its posterior end, flush with the roof of the nose, slips through the *sphenopalatine foramen*.

The sphenoid bounds it above and the palatine bone bounds it in front, below, and behind. The chief vessels and nerves of the nasal cavity pass through this foramen; so, in a sense it is for them the "porta" or door of the nasal cavity (*figs. 774, 799*). The clinician uses the above information for local anesthesia.

Structure of the Palate. The anterior two-thirds of the palate is called the *hard palate*; it is bone. The posterior third, the *soft palate*, is composed of muscles and *palatine aponeurosis*, which continues the plane of the bony palate and is the pliable basis of the soft palate.

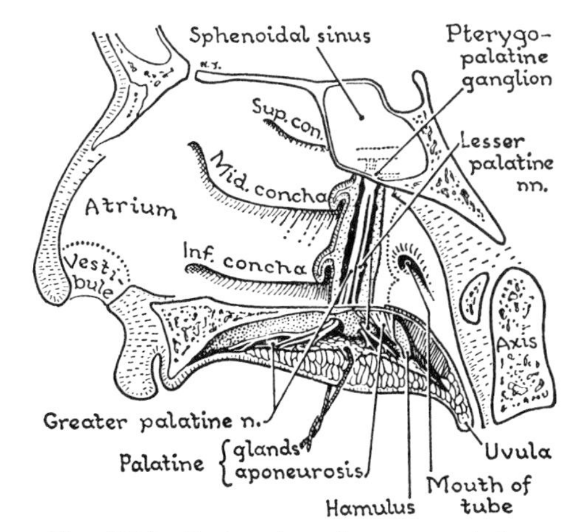

Fig. 774. Exploration of greater palatine canal. Dissection of under surface of palate.

Medial to the 3rd molar tooth is the greater palatine foramen, vertically below the sphenopalatine foramen (*fig. 774*). The greater palatine vessels and nerve emerge from this foramen and run forward in two deep grooves just medial to the alveolar process.

The greater palatine nerve drops down in a bony canal from the *pterygopalatine ganglion* (sphenopalatine g.) which hangs from the maxillary nerve in the deeply placed pterygopalatine fossa (*fig. 774*).

Piercing the bony palate behind the incisor teeth is the *incisive canal.*

Development of incisive canal. It lies between (1) the *primary palate* which developed from a (frontonasal) process which is part of the premaxilla or bone bearing the incisor teeth (*fig. 775*); and (2) the *secondary palate* which developed from the right and left maxillary processes. A process of the vomer in the nasal septum (*fig. 798*) descends into the incisive canal dividing it into right and left sides. Through each side a branch of the *greater palatine artery* ascends to anastomose on the nasal septum, and the *nasopalatine nerve* (*fig. 794*) descends to the premaxilla.

Muscles of Soft Palate

Levator Palati and Tensor Palati (*fig. 780*). These two muscles arise close together from the base of the skull, one on each side of the auditory tube. Both muscles descend to the soft palate. The Levator elevates and pulls the posterior part backward; the Tensor depresses and tenses the anterior part because it turns around the pterygoid hamulus and spreads out as the palatine aponeurosis—it also opens the tube. (Details of both muscles are given on p. 562.)

Palatoglossus, Palatopharyngeus, and Stylopharyngeus. Deep to the mucous membrane of the palatoglossal arch, the *Palatoglossus* is a small bundle. Its fibers reach (and even cross) the midline above and below, and so act as a sphincter that guards the entrance to the pharynx or isthmus of the fauces.

The *Palatopharyngeus* is a detached inner sheet of the Sup. Constrictor (*fig. 778*). From the palate it drops vertically to form an inner longitudinal sheet of muscle in the pharynx.

Explanatory. A palate is peculiar to mammals including man. Mammals require lips with which to grasp the nipple and suck. This also demands a hard and a soft palate to shut off the nasal cavities and nasopharynx from the mouth when sucking—otherwise more air than milk is sucked in, as happens in children born with cleft palates.

New muscles are not called into being but pre-existing ones are modified—*native* or *immigrant*. All the muscles of the soft palate, save one, are native, belonging to the same group as the Superior Constrictor and have the same nerve supply—the accessory nerve via the pharyngeal plexus. The *Tensor Palati* is the immigrant; its nerve comes from the mandibular nerve (V^3) via the otic ganglion.

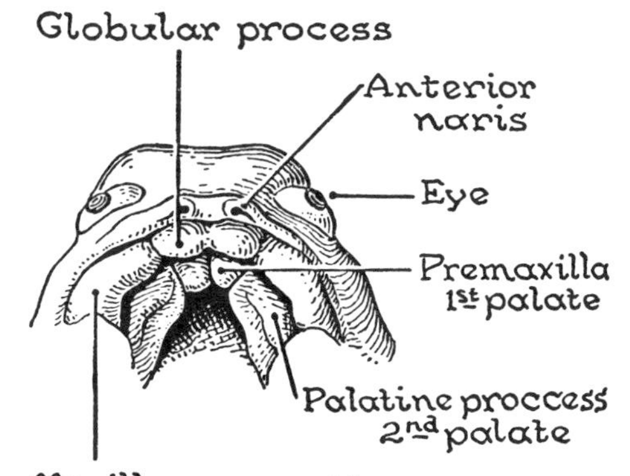

Fig. 775. The palate develops from three shelves—anterior, right, and left.

Details. Palatopharyngeus separates above into three distinct parts—tubal, palatine, and tonsillar. Its *tubal fibers*, the Salpingopharyngeus, form a slender bundle that ascends in the salpingopharyngeal fold to the lower edge of the cartilage of the auditory tube. Its *tonsillar fibers* spread out within the tonsillar bed.

The *Stylopharyngeus* arises from the styloid process—pharyngeal side, of course. It passes through the gap between the Sup. and Mid. Constrictors to blend with Palatopharyngeus (*fig. 771, 773, 778*). It is the only muscle supplied by the glossopharyngeal nerve, primarily a sensory nerve.

Palatine Tonsil

The tonsil (*figs. 776, 777*) is embedded in the side wall of the pharynx in the triangular fossa between the palatoglossal and palatopharyngeal arches and the posterior third of the tongue.

Removal of the Tonsil (*figs. 779, 780*) is

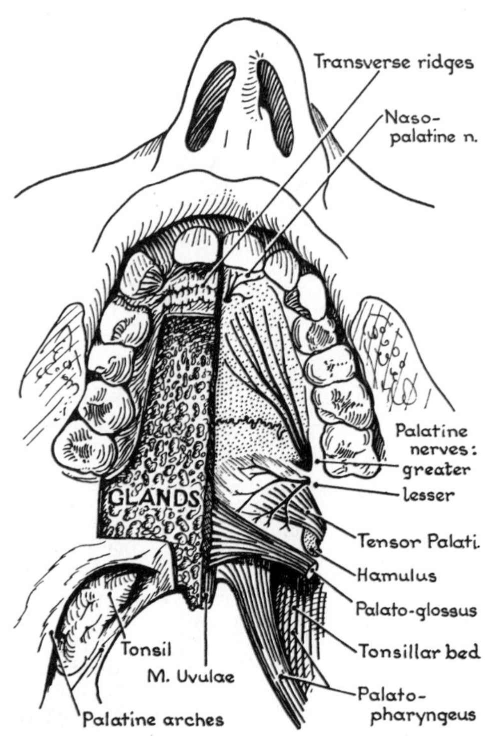

Fig. 776. Dissection of the palate and palatine arches. The upper pole of the tonsil is deeply buried.

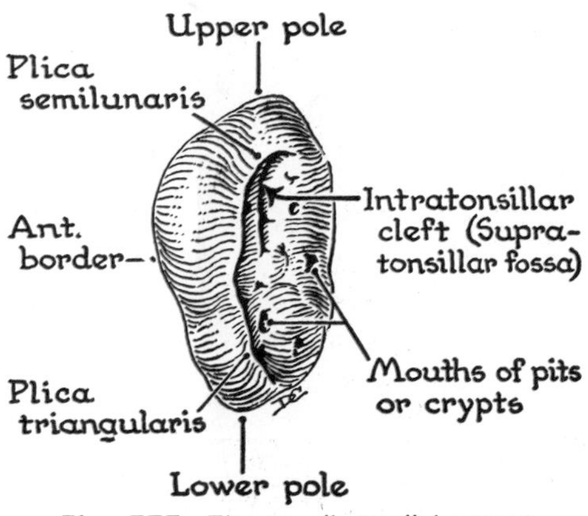

Fig. 777. The tonsil, medial aspect.

easily done by either anatomist or surgeon because the rounded lateral aspect of the gland has a *fibrous capsule*, separated from the pharyngobasilar fascia and the muscular wall of the pharynx by a bed of *loose areolar tissue*. Tonsillitis may obliterate the areolar space, making dissection difficult.

The lower pole is continuous with the **lingual tonsil,** the lymphoid follicles on the dorsum of the tongue. Also it is moored to the tongue by a fibrous band and by some muscle fibers. These and the vessels and nerves, which enter near the lower pole, must be severed during the removal of the tonsil.

White test-tube-like *crypts* extend from its free surface almost to the very capsule.

Tonsillar Bed. Lateral to the loose areolar bed or space there are four thin sheets—two areolar and two fleshy—which constitute the tonsillar bed proper.

Details. From within outward they are (1) the *pharyngobasilar fascia* which forms a complete filmy sheet, (2) the *Palatopharyngeus* and (3) the *Superior Constrictor*, both of which are deficient below (*fig. 778*), and (4) the *buccopharyngeal fascia* (p. 552).

A large vein, the *paratonsillar vein*, descending from the soft palate and receiving tributaries from the tonsil, pierces the lower part of the bed to join the pharyngeal plexus; it is inconspicuous, unless engorged.

Two structures passing to the tongue, (1) *Styloglossus* and (2) *glossopharyngeal nerve*, form immediate lateral relations of the lower third of the tonsillar bed (*fig. 778*).

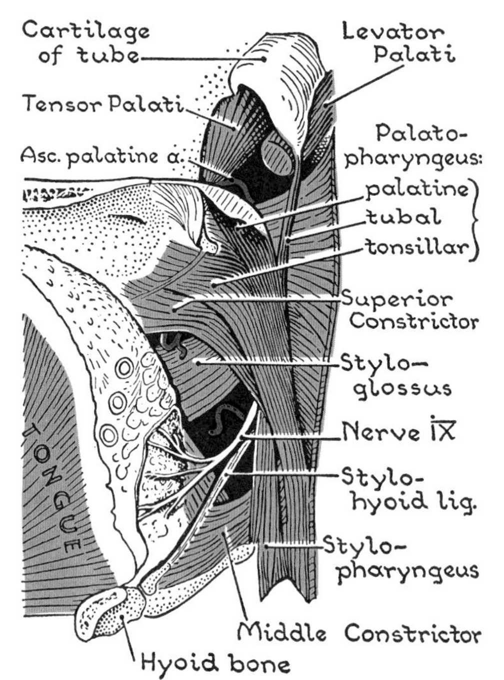

Fig. 778. A stage in the dissection of the side wall of the pharynx from within, showing particularly the relations of the tonsil. (By Dr. B. L. Guyatt.)

The Glossopharyngeal Nerve passes downward, medially, and forward in the tonsillar bed to spread out submucously over the vertical part of the tongue (i.e., posterior one-third).

More Distant Relations of Tonsil. The posterior belly of the Digastric and the submandibular gland, with the facial artery arching over them, are lateral relations of the lowest part of the bed. Farther laterally are the Medial Pterygoid and the angle of the jaw.

Vessels and Nerves of Tonsil Significant to Surgeons (Only): the *tonsillar branch of the facial artery* (*figs. 779, 780*); veins which pass through the tonsillar bed to the *pharyngeal plexus* of veins and to the common facial vein; *lymph vessels*, which pass through the bed to a *deep cervical gland* below the angle of the jaw (*fig. 824.1*); and nerve twigs from *nerve IX* and the *lesser palatine nerves*.

The *Tonsillar Bed* receives arterial twigs from the *tonsillar* and *ascending palatine branches* of the facial a., the *dorsales linguae aa.*, *ascending pharyngeal a.*, and *lesser palatine aa.* In the event of hemor-

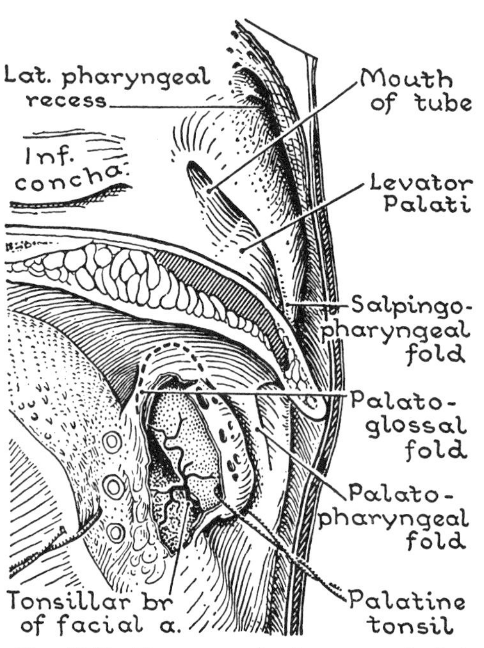

Fig. 779. First stage in the removal of the tonsil. The side wall of the pharynx.

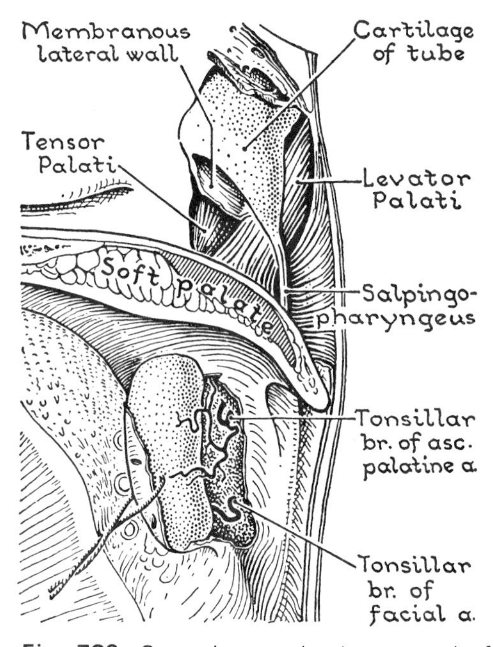

Fig. 780. Second stage in the removal of the tonsil. Dissection of nasopharynx. (Dissections by Dr. P. G. Ashmore.)

rhage, they are controlled by tying the external carotid a. at its origin.

Side Wall of Nasal Pharynx

The following have been observed:

1. Orifice of the auditory tube, just behind the inferior concha (*fig. 779*).
2. Pharyngeal recess, behind the orifice of the tube (*fig. 779*), and
3. (Naso-)pharyngeal tonsil, on the posterior wall of the pharynx and extending into the recess.

Now consider:

1. Upper border of Sup. Constrictor,
2. Pharyngobasilar fascia,
3. Auditory tube,
4. Levator and Tensor Palati,
5. Ascending palatine artery.

Upper Concave Border of the Superior Constrictor and Pharyngobasilar Fascia (*fig. 781*). The border runs from the hamulus and lowest limit of the medial pterygoid plate to the pharyngeal tubercle on the basi-occipital (*fig. 753*). The fascia closes the gap between the Sup. Constrictor and the base of the skull. The *tube*, the *Levator Palati*, and the *ascending palatine artery* pass through the gap (*fig. 781*).

The Auditory Tube (Pharyngotympanic Tube of Eustachius). This air duct is developmentally continuous with the tympanum (middle ear); so, its direction is backward, laterally, and slightly upward (*fig. 817*).

The medial ⅔ of the tube is *cartilaginous*; the lateral ⅓ is *bony*. The narrowest part, called the *isthmus*, is where bone and cartilage meet medial to the spine of the sphenoid (which, you should recall, is just medial to the jaw joint). Here the lumen is only 2 to 3 mm high and 1.0 to 1.5 mm wide (*fig. 781*).

The cartilaginous part has a membranous lateral wall or "floor." The cartilage, curved like an inverted J, forms only the upper and medial walls of the tube (*fig. 780*). It is firmly bound to the posterior border of the medial pterygoid plate.

Except at the funnel-shaped *mouth* or *pharyngeal orifice*, the membranous lateral wall is applied to the cartilaginous medial wall, so that the lumen is closed to form a vertical slit. The Levator Palati runs submucously below the tube, raising its floor (*figs. 780, 781*).

Function. When relaxed, the slit-like lumen of this air duct is closed, but during swallowing, yawning and sneezing (and by no other natural means) it is opened reflexly through the action of the Tensor Palati (Rich). As a result, the atmospheric pressure on each side of the eardrum is maintained in equilibrium.

While awake, one swallows once every minute; while asleep, once every 5 minutes (Graves and Edwards). Hence, while ascending

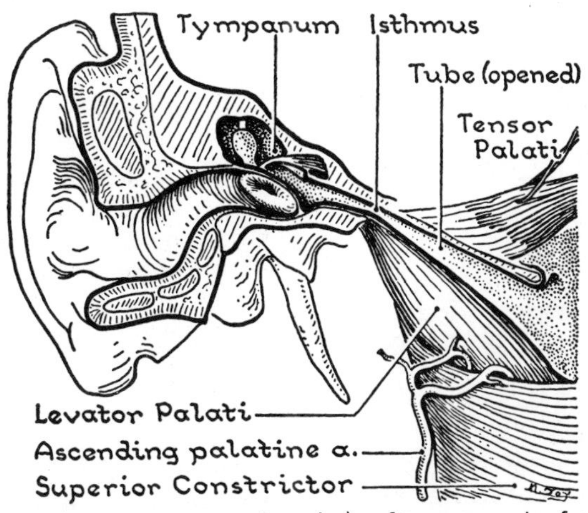

Fig. 781. Auditory tube (pharyngotympanic tube), after removal of membranous and bony lateral wall.

and descending in an aeroplane it may be wise to be awake.

Nerve Supply. Via its tympanic branch the glossopharyngeal nerve (IX) is sensory to the tube, and via its pharyngeal branch it is sensory to the mouth of the tube. Perhaps the pharyngeal branch of the maxillary nerve (V^2) helps to supply the mouth of the tube.

SOFT PALATE: DETAILS

The Levator Palati is as stout as a lead pencil. It arises from the under surface of the apex of the petrous bone in front of the carotid canal. It runs beneath the whole length of the membranous floor of the tube, accompanying it downward, forward, and medially and across the upper border of the Sup. Constrictor. Below the mouth of the tube it enters the upper surface of the soft palate and there spreads out to join its fellow and the palatine aponeurosis (*fig. 782*).

The Tensor Palati is thin and fanshaped. It arises from the *scaphoid fossa* at the root of the medial pterygoid plate and also from the whole length of the membranous lateral wall ("floor") of the tube. It descends lateral to the Sup. Constrictor and medial pterygoid plate to below the level of the hard palate. Then, after piercing the attachment of the Buccinator to the pterygomandibular raphe, it utilizes the hamulus as a pulley and takes a recurrent course to its insertion into the palatine aponeurosis.

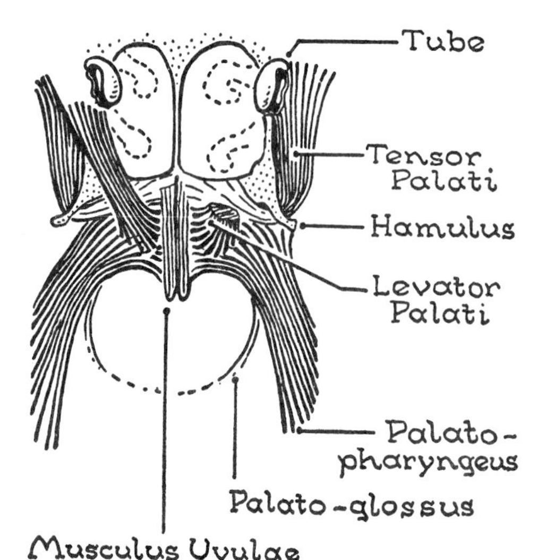

Fig. 782. Five muscles of soft palate on each side.

The Tensor must on principle be tendinous where it turns around the pully; and on principle there must be a bursa to facilitate the play of the tendon on the pulley.

Palpation. The hamulus, or hook-like lower end of the medial pterygoid plate, is situated about a centimeter behind the greater palatine foramen. If your gag reflex permits, you can palpate it by pressing upward immediately posteromedial to the maxillary tuberosity just in front of the palatoglossal arch.

It is important to note that the anterior one-third of the soft palate is *aponeurotic* and the posterior two-thirds *fleshy*; but it is not important to know the detailed arrangement of the fleshy fibers by layers. There is, then, a *bony palate*, an *aponeurotic palate*, and a *fleshy palate*. The aponeurosis is continuous in front with the sharp, posterior border of the hard palate. It is the aponeuroses of the two tensor muscles, but the other palatine muscles also gain partial attachment to it.

The Muscles, 5 in number, are paired:

Tensor Palati and Levator Palati
Musculus Uvulae
Palatoglossus and Palatopharyngeus (p. 558)

The two **Musculi Uvulae** (*fig. 782*) arise beside the posterior nasal spine and, like two closely applied fingers, they descend near the dorsum of the soft palate into the uvula, which they stiffen when they contract.

Structure of Soft Palate. As a general principle, free surfaces subjected to friction, pressure, or other rough treatment are lined with *stratified squamous epithelium*. The oral aspect of the soft palate and the part of its upper surface that strikes the posterior pharyngeal wall are lined with stratified squamous epithelium. The remainder of its upper aspect is lined with *ciliated epithelium*, like the nasal cavity and nasopharynx.

A thick carpet of mucous *glands* covers the under surface of the soft palate (*fig. 783*).

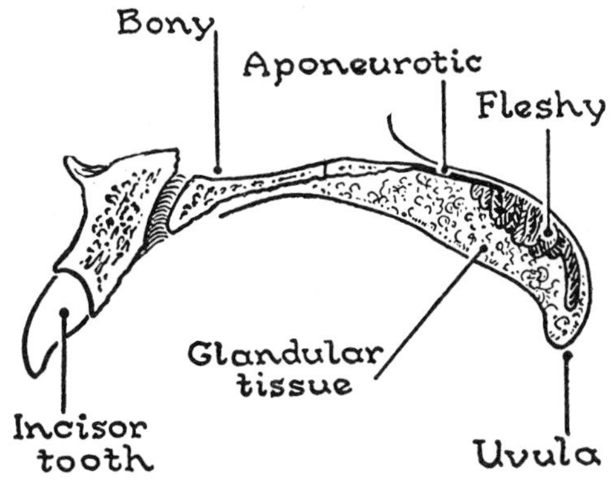

Fig. 783. The palate on sagittal section.

Details of Vessels and Nerves of Pharynx and Soft Palate. The following details are of importance to specialists.

Arteries. The pharynx is supplied mainly by the *ascending pharyngeal*, sup. and inf. thyroid, and pharyngeal aa.; the soft palate by the *ascending palatine branch of the facial a*; and the tonsil by the *tonsillar branch of the facial a*. These anastomose with each other and with the *lesser palatine and dorsales linguae aa.*

Veins go to the pharyngeal plexus, and thence to the internal jugular vein.

Lymph vessels pass to the upper deep cervical nodes; those from the nasopharynx pass to nodes between the pharynx and the prevertebral fascia; those from the tonsil to a node below the angle of the jaw (*fig. 824.1*).

Motor Nerves. Nerve XI (through the vagus) via the pharyngeal plexus supplies all the pharyngeal and palatine muscles, except:

Stylopharyngeus (nerve IX)
Tensor Palati (nerve V^3)

The external and recurrent laryngeal branches of the vagus also supply the Inferior Constrictor.

The Sensory Nerve of the pharynx, including the soft palate and tonsil, is the glossopharyngeal (IX). But the maxillary nerve (V^2), via its pharyngeal br., supplies the roof of the pharynx; and, via the lesser palatine nerves, helps to supply the soft palate and the adjacent part of the tonsil. The vagus (X) via the int. laryngeal nerve, supplies the region round the entrance to the larynx (*fig. 784*).

SUCKING AND SWALLOWING

Sucking. Place a finger in your mouth and, while sucking it, note that your lips grasp it, that a groove forms along the middle of the tongue, and that the tongue recedes from the palate, thereby creating a vacuum. Note also that, while sucking fluid through a straw, you can breathe in and out through the nose. This is possible, since the Palatoglossi shut the mouth off from the pharynx. A child with a cleft palate cannot suck effectively, because air, drawn in through the cleft, prevents the formation of a vacuum within the mouth.

Mechanism of Swallowing (Deglutition). Swallowing begins as a voluntary movement and continues as an involuntary one. Thus, the lips are closed and usually the Buccinators are pressed against the teeth. The lingual muscles pass the bolus of food backward on the dorsum of the tongue to the palatoglossal arch; there, at the entrance to the pharynx, it may rest, the voluntary stage of deglutition being completed.

If, however, some saliva, fluid or portion of food enters the pharynx, the involuntary stage is started reflexly by stimulation of the glossopharyngeal nerve. The jaws are held closely by the Masseters and Temporals, while the hyoid bone and the larynx rise, as by palpation you can determine on yourself (by the action of the Digastrics and Mylohyoids). The tongue, like a piston

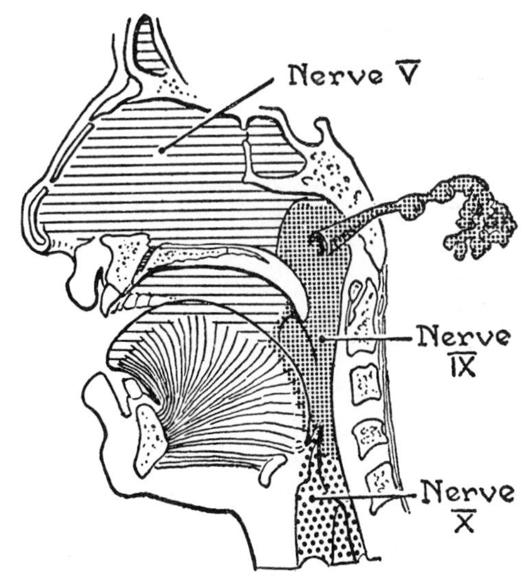

Fig. 784. The sensory distribution of the glossopharyngeal nerve. (After Edwards.)

thrust forcefully backward and upward against the soft palate, forces the bolus into the pharynx (by the action of the intrinsic muscles of the tongue, the Mylohyoids and the Styloglossi); thereupon a peristaltic wave propels it through the pharynx and into the esophagus. During this phase the larynx rises slightly more (by the action of the Stylopharyngei).

The erect epiglottis, inclined slightly backwards by the backward movement of the tongue, is now swept farther backward by the oncoming bolus which fills the valleculae. In fact, it is swept backward and downward till it covers the laryngeal aperture like a lid. Since the aryepiglottic folds can be seen to shorten, it is not unlikely that the Aryepiglottic muscles pull the epiglottis while the bolus pushes it.

The entrance to the larynx and the vestibule are tightly closed by the sphincteric muscles (*fig. 811*), which also tilt the apices of the arytenoid cartilages against the tubercle of the epiglottis and approximate the vocal cords. Although the epiglottis closes like a lid, the sphincteric mechanism is alone sufficient to prevent the entrance of food into the larynx (Saunders *et al.*).

The movements that take place during the act of swallowing are too rapid for the eye to perceive. For example, the epiglottis bends from the erect position to 60° below the horizontal and recovers in a fifteenth of a second. Hence, high speed cineradiography using 30 and 60 frames per second was employed by Saunders, Davis, and Miller, who in more senses than one threw new light on the mechanism of deglutition.

During the act of swallowing two apertures are closed and two are opened:

1. The entrance to the nasopharynx, called the *pharyngeal isthmus*, rendered narrower by the contracting Palatopharyngeus, is closed by the Levatores Palati which draw the soft palate upward and also backward against the wall of the narrowing pharynx. Lubricating mucus renders it air tight.

2. The Tensores Palati open the auditory tubes, and render taut the anterior half of the soft palate.

3. The entrance to the larynx, as stated above, is closed by the sphincteric muscles (p. 587).

4. The cricopharyngeus relaxes its guard over the esophagus to let the bolus pass.

5. The three pharyngeal constrictors contract vigorously only for 1/3 of a second each. Their contractions occur in series but overlap, resulting in a peristaltic wave that continues on to the esophageal muscles (Basmajian and Dutta).

Two Deep Cervical Spaces of Surgical Importance

Retropharyngeal Space (*fig. 784.1*). During the act of swallowing, the pharynx and esophagus must have freedom of movement. Accordingly, between the prevertebral fascia, which covers the prevertebral muscles, and the buccopharyngeal fascia, there is an areolar space called the *retropharyngeal space*. This potential space is closed above by the base of the skull, and on each side by the carotid sheath; caudally it opens into the superior mediastinum. Because of its important neighbors,

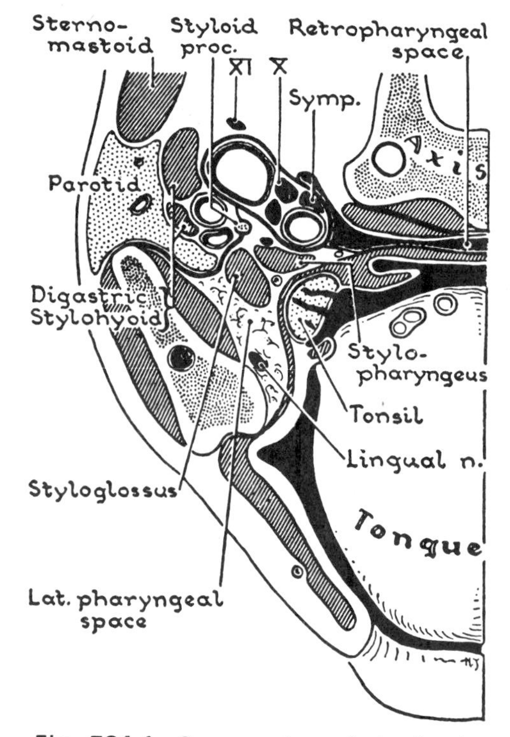

Fig. 784.1. Cross-section of the head, at the level of the parotid gland and the tonsil.

the space has great surgical interest (Cole and Baylin).

Lateral Pharyngeal Space (*fig. 784.1*). This is a space lined with areolar fascia, filled with fat and containing branches of the maxillary nerve and maxillary vessels.

Its practical importance is surgical: (1) infection may spread to the space from the tonsil; (2) the dentist when injecting the lingual and inferior alveolar nerves may carry infection to it, and (3) when infected, infection may spread by the veins to the internal jugular vein.

The space communicates with the parotid space and with the submandibular space, but not with the retropharyngeal space.

It extends from the base of the skull and the auditory tube above, where it is widest, to the level of the hyoid bone below, where its apex lies. The pharynx is situated medially; the parotid gland posterolaterally; the Medial Pterygoid and the ramus of the jaw anterolaterally; and the styloid process and its muscles largely separate the space from the carotid sheath, which is posterior.

45 MOUTH, TONGUE, AND TEETH

MOUTH

Vestibule and Cavity Proper; Inspection and Palpation.
Sublingual Region—Sublingual Gland; Lingual Nerve; Chorda Tympani Nerve; Submandibular Ganglion.
Hypoglossal Nerve.
Muscles—Mylohyoid; Genioglossus; Geniohyoid.

TONGUE

Structure; Parts; Mucous Membrane and Papillae; Anterior Lingual Glands.
Muscles—Extrinsic, Intrinsic, and Accessory.
Vessels and Nerves—Lingual Artery; Nerves and Lymphatics.
Development Explains Nerve Supply.

TEETH

Parts; Permanent; Deciduous; Eruption.
Descriptive Terms; Crowns; Roots, Occlusion.
Development; Growth and Calcification.
Nerve Supply to Teeth and Gums.

MOUTH

Parts

The cavity of the mouth has two parts—a *vestibule* and a *cavity proper*. These are separated from each other by the teeth, alveolar processes, and gums; and they communicate with each other on each side through a space between the last molar teeth and the ramus of the mandible.

The Vestibule is bounded externally by the lips and cheeks. It opens onto the skin surface of the *aperture of the mouth*. The upper and lower lips are attached to the gums in the median plane by folds of mucous membrane, the *frenula*. The constricted orifice of the parotid duct opens opposite the 2nd upper molar tooth.

The Lips and Cheeks have four layers: *cutaneous*, *muscular*, *glandular*, and *mucous*. Between the muscular and glandular layers lies an *arterial circle*, formed by the upper and lower labial branches of the facial artery. Its pulsations can be felt on grasping the lip between the finger and thumb. The facial artery ascends on the Buccinator a short distance from the angle of the mouth. The *glands* can be felt as velvety bumps with the tip of the tongue or fingers.

Palpate with the index finger in the vestibule (1) the *Masseter*, which is rendered prominent when the teeth are alternately clenched and relaxed; (2) the lower border of the *zygomatic arch*, and the *maxilla*; and (3) the anterior border of the *ramus of the jaw*, and trace it to the *coronoid process* and perhaps to the *Temporalis tendon*.

Nerve Supply of Lips and Cheek. The *motor* nerve is the facial. When it is paralyzed the lips cannot be puckered to whistle, and food collects in the vestibule.

Sensory: The skin and mucous surface of the upper lip, lower lip, and cheek near the angle of the mouth are supplied by the infra-orbital (V^2), mental (V^3), and buccal (V^3) nerves, respectively. *Lymph vessels* (p. 600).

Cavity Proper of the Mouth [Oral Cavity]. *Inspection*. The cavity is roofed in by the hard and soft palates. The *soft palate* ends medianly in the *uvula*. Two folds on each side arch downward from the soft palate: the anterior fold, the *palatoglossal arch*, ends at the side of the tongue and marks the entrance to the pharynx. On looking beyond it into the pharynx the posterior fold, the *palatopharyngeal arch*, is seen to pass from the margin of the uvula to the side wall of the pharynx.

Between these two palatine arches and the posterior one-third of the tongue a portion of the *palatine tonsil* is seen with the aid of a tongue depressor. The anterior two-thirds of the *tongue* rises from the floor and covers the structures on the sides and front of the floor, known as the sublingual region.

When the tip of the tongue is raised, a short median fold of mucous membrane, the *frenulum linguae*, is seen running from the tongue to the floor of the mouth. On each side, just in front of the frenulum there is a papilla on which opens the duct of the submandibular gland. Running posterolaterally from this orifice is a rounded ridge, the *plica sublingualis*, which overlies the upper border of the sublingual salivary gland.

Palpation. You have already palpated the tuber of the maxilla and the pterygoid hamulus (p. 562). With the index finger try to roll the lingual nerve against the jaw medial to the root of the lower 3rd molar tooth (*fig. 785*).

Sublingual Region (*Floor of the Mouth*)

The Sublingual Gland empties by a row of a dozen short ducts along its upper edge. It is enveloped in an areolar sheath and running diagonally across its medial aspect is the *submandibular duct*.

Laterally, it occupies the sublingual fossa of the jaw. In front, it reaches its companion of the other side; behind, the submandibular gland. Medially, the submandibular duct and the lingual nerve run forward between the gland and the Genioglossus (and Hyoglossus). Below, is the Mylohyoid.

The Lingual Nerve (*fig. 785*) is a general sensory branch of nerve V^3 with added fibers (taste and secretory) from the chorda tympani. It appears from behind the last molar tooth, clamped to the ramus of the jaw by the Medial Pterygoid. It passes forward to the side of the tongue spiralling around the submandibular duct. It spreads out within the anterior two-thirds of the tongue and the sublingual region, including the floor of the mouth and the gums (*fig. 786*).

The Chorda Tympani joined the lingual nerve in the infratemporal fossa. The *secretory fibers* are preganglionic parasympathetic fibers whose relay station is the submandibular ganglion (*fig. 786.1*).

The *taste fibers* accompany the lingual nerve to the anterior two-thirds of the tongue where they end in the taste buds; their cell station is the geniculate ganglion of the nervus intermedius (*fig. 786.1*).

The Submandibular (Submaxillary) Ganglion is a parasympathetic ganglion—a relay station on the course of the efferent fibers of the chorda tympani. It is suspended from the lingual nerve by two roots, and it lies (on the Hyoglossus) medial to the submandibular gland (*figs. 785, 786*).

Branches of the ganglion are distributed to the submandibular and sublingual glands (*fig. 785, 786*).

The Hypoglossal Nerve runs forward between the submandibular gland and the Hyoglossus well below the lingual nerve (*fig. 785, 786*). Fibers radiate to supply the extrinsic muscles of the tongue; and finally it plunges into the tongue to supply the intrinsic muscles.

Muscles. The paired **Mylohyoids** constitute the *Diaphragma Oris*. (See p. 514 and *figs. 720* to *721.1*.)

The **Genioglossi** (paired) arise from the mental spine (genial tubercle) and spread like a fan backward into the tongue, their medial surfaces in contact (Doran and Baggett).

The **Geniohyoids** (paired) also arise from the mental spine (*fig. 786*) and pass below the tongue to the body of the hyoid bone, medial borders in contact.

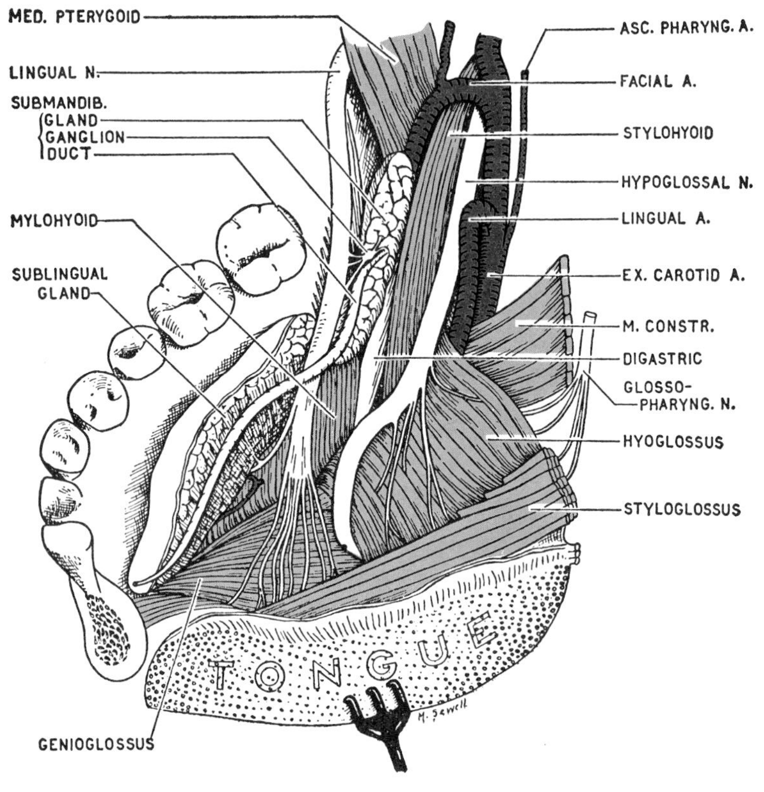

Fig. 785. Dissection of right side of floor of mouth.

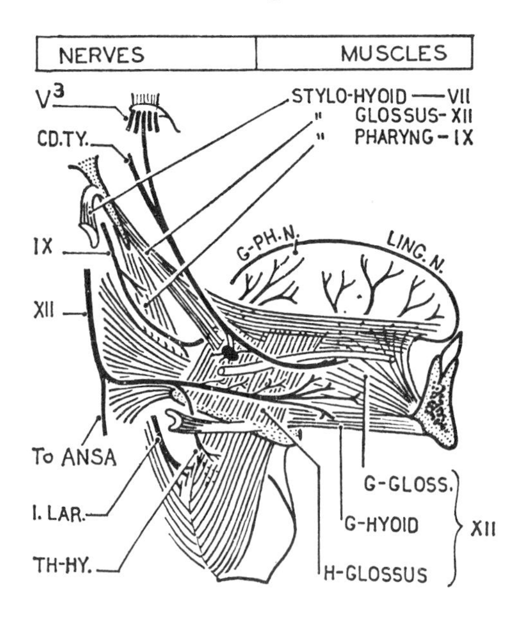

Fig. 786. Nerves and muscles of tongue.

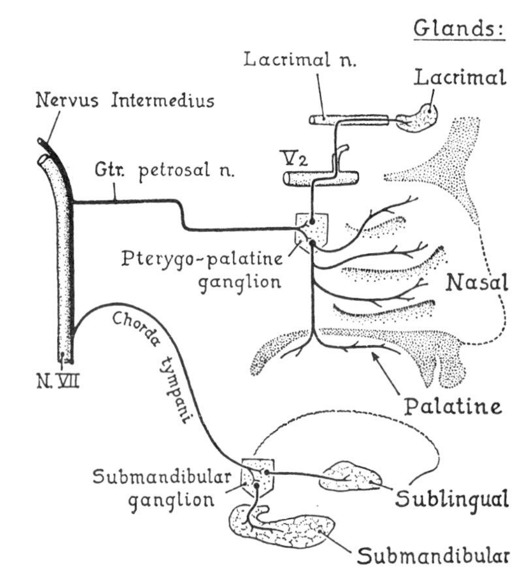

Fig. 786.1. Distribution of parasympathetic fibers of the nervus intermedius.

Nerve supply. Mylohyoid n. from V^3 (*fig. 742*). Genioglossus by N. XII. Geniohyoid by C. 1 via N. XII (*fig. 765*).

TONGUE

The tongue is a muscular organ concerned with mastication, deglutition, speech, and taste. It has two parts—an anterior two-thirds and a posterior one-third. These differ topographically, developmentally, structurally, functionally, in nerve supply, and in appearance. (Lingua, L = glossa, Gk = tongue.)

The anterior ⅔ rises from the floor of the mouth and hence is called the *oral part* (body); the posterior ⅓ forms part of the anterior wall of the pharynx and hence is called the *pharyngeal part* (root). The boundary between the oral and pharyngeal parts is marked on the dorsum of the tongue by a V-shaped line, the *sulcus terminalis*, which runs medially and backward on each side from where the palatoglossal arch joins the side of the tongue to a median pit, the *foramen cecum linguae*.

Mucous Membrane. The mucous membrane differs conspicuously on the two parts of the dorsum (*fig. 787*): on the *oral part* it is covered with papillae and is velvety; on the *pharyngeal part* it is studded with tubercles between which it is smooth and glistening.

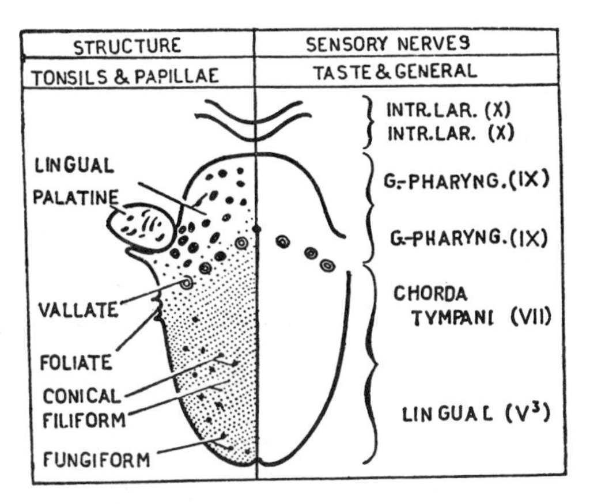

Fig. 787. The dorsum of the tongue showing structure (*left*) and sensory nerve supply (*right*).

On the *dorsum*, or upper surface, of both parts of the tongue, as well as the tip, lateral margins, and under surface of the oral part, a mucous membrane of stratified squamous epithelium rests on a fibrous stroma, the tunica propria. It is firmly adherent except on the posterior third of the dorsum where a submucous coat is present. The papillae consist of a core derived from tunica propria, covered with stratified squamous epithelium.

The Papillae are of four varieties: filiform, fungiform, vallate, and foliate.

Filiform Papillae are tapering and thread-like and are arranged in V-shaped rows that cover the dorsum of the oral part of the tongue. They contain touch corpuscles. Their epithelium is scaly and in some animals (e.g., cat, cow) it is cornified and is used as a rasp to grasp food.

Fungiform Papillae have globular heads and are red because the core is more vascular and the epithelium not scaly. Like daisies on a lawn, they lie scattered singly among the filiform papillae at the tip and margin of the tongue, but they do not rise above them.

Vallate Papillae are circular, about 2 mm in diameter, and are surrounded by a moat which is 2 to 3 mm deep. Twelve or less in number, they also are arranged in a V-shaped row just in front of the sulcus terminalis. Their flat tops hardly rise above the general surface.

Foliate Papillae, rudimentary in man, are three to four short, vertical folds at the hinder part of the sides of the tongue.

Taste Buds occur on most fungiform papillae, on the opposed sides of the foliate, and on both walls of the vallate. They also occur sparsely on the soft palate, epiglottis, and posterior wall of the pharynx.

On the Pharyngeal Third of the tongue there are no papillae. The numerous tubercles seen there are encapsuled *lymphoid nodules*. Each nodule surrounds a crypt, which receives the ducts of underlying mucous glands and opens conspicuously on the surface at the center of the nodule. The nodules are known collectively as the *lingual tonsil*. They are indefinitely separated from the lower pole of the palatine tonsil.

The mucous membrane on the under surface of the tongue is smooth, and on each side the lingual vein shines purple through it. Lateral to the vein is the fimbriated fold.

The Anterior Lingual Gland (paired) is a small cluster of mucous and serous glands, situated under the apical part of the tongue.

Muscles of Tongue. There are 3 extrinsic and 3 intrinsic muscles on each side—all supplied by the hypoglossal nerve. The former move the tongue bodily and alter its shape; the latter can only alter its shape.

The Extrinsic Muscles (*fig. 788*) are—Genio-, Hyo-, and Stylo-glossus. Their actions are obvious. The Genioglossi as protruders of the tongue are "*safety muscles*", and if put out of action—the result of paralysis, fracture of the jaw, or during anesthesia—the tongue falls back, threatening suffocation. If only one Genioglossus is paralyzed, the protruded tongue points *to the paralyzed side* (*Fig. 814*).

The Intrinsic Muscles—Longitudinal, Vertical, and Transverse, decussate in bundles with each other and extrinsic fibers. The tongue has a median *fibrous septum* which is relatively "bloodless." Strangely there are *areas of fat* among the posterior muscle fibers.

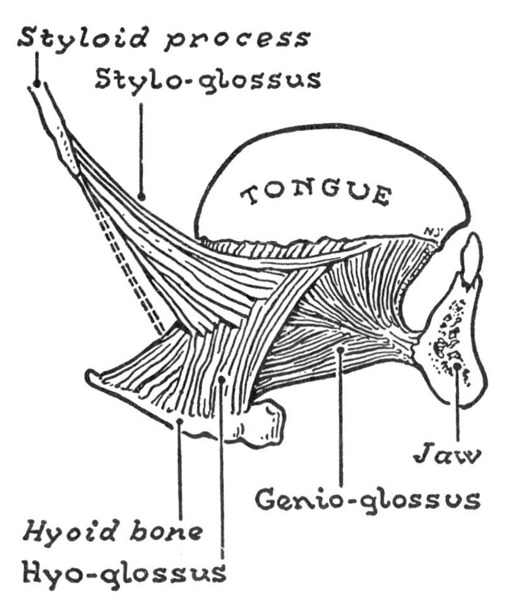

Fig. 788. The 3 extrinsic muscles of the tongue and their 3 bony origins.

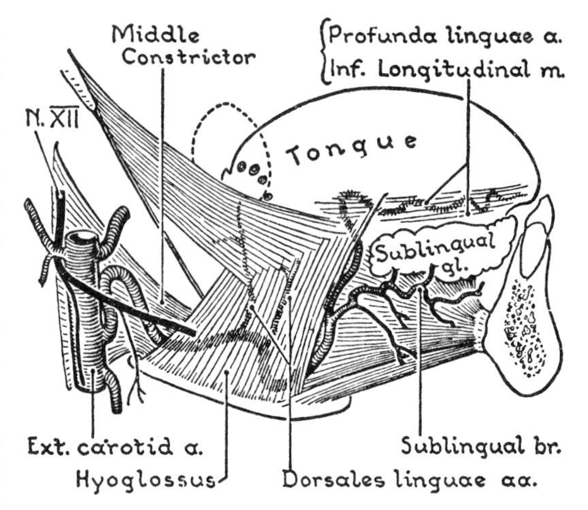

Fig. 789. The lingual artery.

The *Palatoglossus*, supplied by nerve XI via the vagus is primarily a palatine muscle; it helps to narrow the isthmus of the fauces during swallowing.

Vessels and Nerves of Tongue

The Lingual Artery alone supplies the tongue. It arises from the external carotid at the level of the hyoid (*fig. 789*), and runs forward above it—in a necessarily sinuous path—to the tip of the tongue applied to the Genioglossus.

Details. Its other relations are superficial ones. In the carotid triangle it arches upward and is crossed by nerve XII which arches downward. It passes deep to the posterior belly of the Digastric and Stylohyoid and enters the submandibular triangle, where nerve XII crosses it again. It runs deep to the Hyoglossus, which now separates it from nerve XII, and under cover of the anterior border of the Hyoglossus, it ascends on the Genioglossus.

In the last part of its course (between the Genioglossus and Longitudinalis Inferior) it is called the *profunda artery*. Its only anastomosis with its fellow is at the tip of the tongue; so, the tongue can be bisected almost bloodlessly.

BRANCHES: *Two dorsales linguae aa.* supply

the posterior third of the tongue and anastomose in the tonsil bed (p. 560); numerous *muscular twigs*; and a *sublingual a.*, which anastomoses with the submental branch of the facial artery.

The chief vein of the tongue (the profunda vein) is the conspicuous submucous vein. It follows nerve XII and is then joined by the venae comitantes that accompany the arteries to form the lingual vein, which ends in the internal jugular vein.

Lymphatics of the Tongue (p. 601).

Development Explains Nerve Supply. The anterior two-thirds of the tongue develops from two ingrowing shelves of the 1st or mandibular arch. This explains the nerve supply from the mandibular nerve (V^3). With this part is incorporated posteriorly a median eminence (*tuberculum impar*) which makes its appearance between the 2nd arches. This briefly explains the nerve supply from the chorda tympani branch of the facial nerve (VII).

The posterior one-third of the tongue is unpaired. It develops from a median bar, placed between the ends of the 3rd and 4th arches. This explains the nerve supply from the glossopharyngeal (IX) and superior laryngeal nerves (X).

TEETH

Parts (*fig. 790*). Each tooth has a root buried in the jaw, a *crown* projecting beyond the gum, and a *neck* encircled by the gum. At the apex of each root a pinpoint foramen, the *apical foramen*, leads through a widening *root canal* to the *tooth cavity*.

Each tooth is composed of *dentine* which is the exquisitely sensitive, yellowish basis of the tooth; *enamel*, which is the white insensitive covering of the crown, *cement* which is a bony covering for the root and neck of the tooth; and *pulp* which is a fibrous material containing the nerves and vessels that pass through the pinpoint, root canal. The pulp occupies the tooth cavity within the dentine.

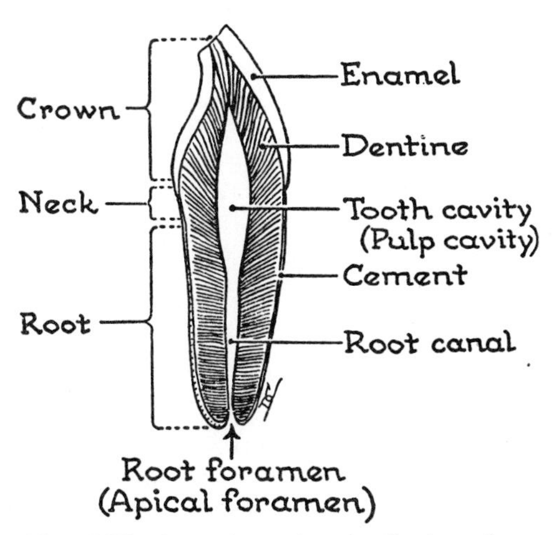

Fig. 790. A tooth, on longitudinal section.

Each tooth lies in a bony socket or *alveolus* which narrows toward its bottom; thus a large pressure surface is afforded the tooth and extraction of a tooth made possible. Between the tooth and the socket there is a vascular membrane, the *periodontal membrane*. This modified periosteum is continuous with the lamina propria of the gum and is attached both to the cement and to the alveolar wall.

There are *32 permanent teeth*, 16 in the upper dental arch and 16 in the lower. Of the eight on each side of each arch—two are incisors, one canine (cuspid), two premolars (bicuspids), and three molars. The formula therefore reads:

$$\frac{3.\ 2.\ 1.\ 2.}{3.\ 2.\ 1.\ 2.}\Bigg|\frac{2.\ 1.\ 2.\ 3.}{2.\ 1.\ 2.\ 3.}$$

All except the molars are preceded by deciduous or primary teeth.

There are *20 deciduous teeth* (primary, temporary, or milk teeth), two incisors, one canine, and two molars on each side of each arch. The formula therefore reads:

$$\frac{2.\ 1.\ 2.}{2.\ 1.\ 2.}\Bigg|\frac{2.\ 1.\ 2.}{2.\ 1.\ 2.}$$

Eruption. At birth, the jaws are rigid bony bars, suitable to grasping a nipple. Between the 6th and 8th months the deciduous, lower, medial incisors erupt through the gums, and eruption proceeds as follows (Schour and Massler), the process being completed by the 24th month:

Order and Time of Eruption of Deciduous Teeth

Teeth	Med. Incisor	Lat. Incisor		1st Molar	Canine	2nd Molar
Months	6–8	8–10	10–12	12–16	16–20	20–24

Then comes an *interval of 4 years*. At the 6th year the permanent teeth begin to erupt, and, because the 1st molars are the first permanent teeth to erupt, they are commonly called the *6th year molars*. The deciduous teeth are next replaced by permanent teeth in the following order: medial incisors, lateral incisors, 1st premolars, canines, and 2nd premolars. The 2nd molars erupt about the 12th year and the 3rd molars about the 18th year (15th–21st), but not uncommonly they fail to erupt.

Order and Time of Eruption of Permanent Teeth

Teeth	1st Molar	Med. Incisor	Lat. Incisor		1st Pre-molar	2nd Pre-molar Canine	2nd Molar
Years	6	7	8	9	10	11	12

Descriptive Terms. It is convenient to refer to the anterior surfaces of the front teeth and lateral surfaces of the side teeth as *labial* (or buccal) surfaces; and to refer to the opposite surfaces as *lingual* surfaces: to refer to the medial surfaces of the front teeth and anterior surfaces of the side teeth as *proximal surfaces*, and to the opposite surfaces as *distal* surfaces. The biting surfaces may be referred to as the *occlusal* surfaces. With the exception of the distal surfaces of the last molars, proximal and distal surfaces are *contact* surfaces.

Crowns. There is evidence that the crowns of the human teeth have evolved from a tritubercular or tricuspid tooth. And two labial tubercles and one lingual tubercle are detectable on each tooth.

In the *incisors*, the labial tubercles fuse to form a cutting edge which is joined to an indistinct lingual tubercle by two faint lines (the cingulum) that enclose a triangular space. Among the North American Indians these are pronounced and give the incisors a shovel-like appearance.

In the *canines*, the labial tubercles fuse to form a single large cone and a lingual tubercle is often well defined.

In the *premolars* or bicuspids, the labial tubercles fuse to form a medium sized cone and the lingual tubercle or cusp is pronounced.

All *molars* have as a basis two labial tubercles and a proximal lingual tubercle. The *upper molars* characteristically have an additional lingual tubercle placed distally, making four in all—the 1st molar always has four, the 2nd commonly, and the 3rd variably. The *lower molars* characteristically have five tubercles, two being labial, two lingual, and a fifth distal—these tend to be reduced on the 3rd lower molar.

Roots (*fig. 791*). The roots of the incisors, canines, and premolars are single. (The first upper premolar has commonly a bifid or even a double root.) The lower molars

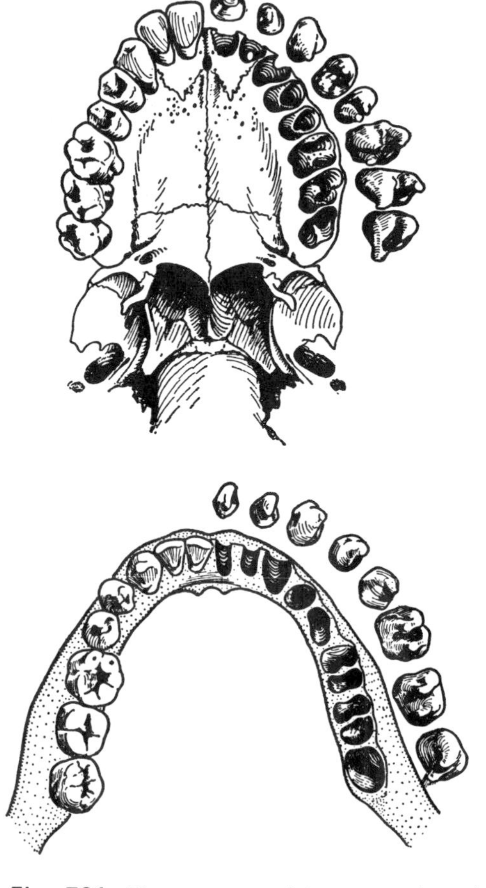

Fig. 791. The upper and lower teeth and their sockets. (The 2nd upper premolar or bicuspid, as well as the 1st, happens to have two roots.)

have two flattened roots, a proximal and a distal; the upper molars have three conical roots, two smaller labial and one larger lingual.

The roots of all teeth tend to be flattened proximodistally, and in all lower teeth the flattening is pronounced. In the upper teeth there is a compromise between being rounded and conical on the one hand and being flattened on the other. The upper medial incisor has the roundest root; the canines have the longest roots; and the roots of the molars are commonly recurved.

Occlusion. The teeth of the upper arch project labially beyond the teeth of the lower arch. As a result, the labial borders of the masticatory or occlusal surfaces of the lower premolars and molars tend to be worn off and rounded and the lingual borders are sharp. The reverse is true of the upper premolars and molars. The upper incisors in most races "overbite" the lower incisors and do not come into occlusion (*figs. 792, 793*).

The upper and lower dental arches end flush with each other posteriorly (*fig. 792*). The upper medial incisors are relatively large and the 3rd upper molars relatively small; so, when the arches are in occlusion most teeth bite on two teeth.

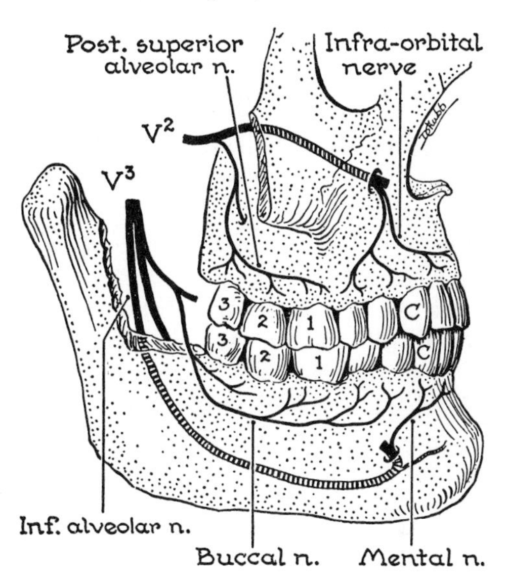

Fig. 793. Nerve supply to the outer aspect of the gums. Teeth in occlusion.

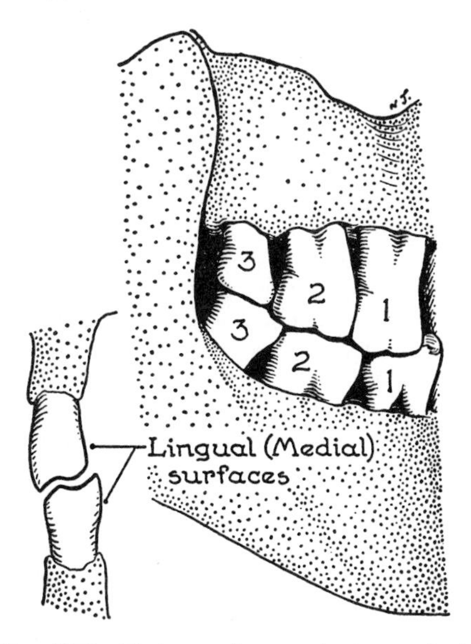

Fig. 792. Right molar teeth in occlusion. Teeth, when in occlusion, bite on two teeth, except the distal maxillary tooth and proximal mandibular tooth.

Development. Enamel is of ectodermal origin. It begins to develop during the 3rd fetal month from buds that sprout from an ingrowing plate (the primary dental lamina) of ectodermal cells. Each bud takes the form of a cap that covers the mesodermal papilla from which the remainder of the tooth is formed. At the same time buds, from which the enamel of the corresponding permanent teeth arises, sprout from the lingual surface of the dental plate, but they remain quiescent temporarily. The three permanent molars develop similarly from a backward extension of the plate.

Growth and Calcification of the Permanent Teeth begin in the 1st molars (6th year molars) about the time of birth; in the incisors and canines (the upper lateral incisors excepted) from 4th to 6th month; in upper lateral incisors from 10th to 11th month; and in premolars and 2nd molars early in the 2nd year.

Therefore, metabolic disturbances occurring in early infancy (from gastro-intestinal and other causes), while affecting the anterior teeth, will omit the upper lateral incisors. And, no amount of dietary regulation or calcium therapy can ever correct enamel defects once they have occurred (Schour and Massler).

Nerve Supply to the Teeth and Gums. The maxillary nerve (V^2) supplieds the teeth and gums of the upper jaw; the mandibular nerve (V^3) supplies those of the lower (*figs. 793, 794, 795*).

The teeth of the *upper jaw* and also their periodontal membranes are supplied by the post. and ant. superior alveolar nerves. The lingual part of the upper gums related to molars and premolars is supplied by the greater (anterior) palatine nerve; the part related to the canine and incisors by the nasopalatine nerve. The labial part of the upper gums related to the molars is supplied by branches of the post. superior alveolar nerve (which descends on the infratemporal surface of the maxilla); the part related to the premolars, canine, and incisors by the infra-orbital nerve. (See the infra-orbital nerve, p. 575).

The teeth of the *lower jaw* and their periodontal membranes are supplied by the inf. alveolar nerve. The entire lingual part of the lower gum is supplied by the lingual nerve. The labial part related to the molars and premolars is supplied by the buccal nerve; the part related to the canine and incisors by the mental nerve (*fig. 793*).

Variations in Distributions:

1. The pulp of the lower teeth may retain residual sensation after injection of the inf. alveolar nerve. There is evidence to show that this is mediated by fibers of the lingual and buccal nerves that pierce the alveolar walls to reach the pulp;
2. branches of the inferior alveolar nerve to the incisor teeth may decussate in the mandibular canal with those of the opposite side and supply the opposite incisors;

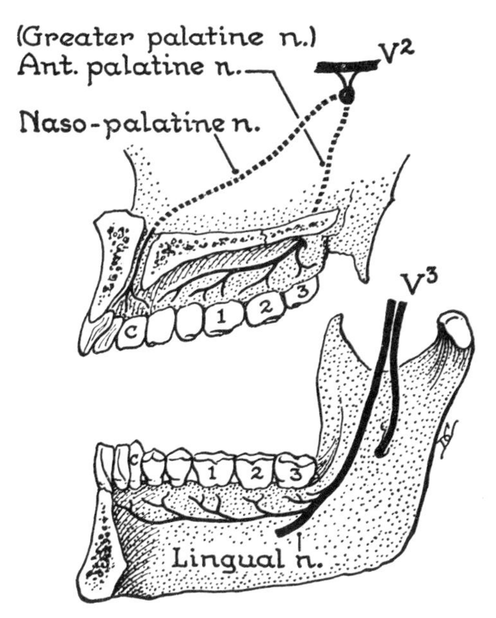

Fig. 794. Nerve supply to the inner aspect of the gums. Branches also reach the gums between the tooth sockets (*fig. 795*).

3. similarly, branches of the mental nerves of opposite sides decussate across the midline to supply the gum of the opposite side;
4. the anterior limits of the distribution to the gums of the lingual and buccal nerves varies, as is evidenced by the variable loss of sensation on injecting these nerves. Thus, the lingual nerve, though usually extending to the median plane, sometimes stops abreast of the canine tooth, and the lingual nerve of the opposite side crosses to supplement it. Similarly, the buccal nerve distribution may cease at the 2nd molar or extend to the canine; the posterior limits of the mental nerve vary inversely with this (Stewart and Wilson).
5. Apparently, the buccal nerve may help to supply the upper gum, and the post. sup. alveolar nerve the lower.

The Lymph Vessels of the pulp of the upper and lower teeth pass to the submandibular and upper deep cervical nodes.

46 NOSE AND RELATED AREAS

Pterygopalatine Fossa—Boundaries; Foraminae and Canals.
Maxillary Nerve—Infra-orbital Nerve; Pterygopalatine Ganglion; Roots and Nerve of Pterygoid Canal.
Maxillary Artery (3rd Part)—Branches.

NASAL CAVITIES

Boundaries; Nostril; Choana; Floor; Roof.
Septum—Structure; Deflected; At Birth; Vessels and Nerves.
Lateral Wall—Conchae; Meatuses; Vestibule; Apical Recess; Atrium; Nasolacrimal Canal and Duct.
Blood Supply; Nerve Supply.

PARANASAL AIR SINUSES

Sphenoidal; Maxillary, Ethmoidal and Frontal.
Development of Sinuses; Dimensions.
Nerve Supply.

The **pterygopalatine fossa** (*fig. 740*) is the elongated triangular space between the rounded posterior border of the maxilla and its buttress, the *pterygoid process.* Its contents are:

1. Maxillary nerve (in part).
2. Pterygopalatine ganglion.
3. Maxillary artery (3rd part) and veins.

Interesting but Unessential Details. The fragile vertical plate of the palatine bone forms its medial wall; the greater wing of the sphenoid forms its roof. If each of its walls were included in its name, it would be designated pterygo-(maxillospheno-)palatine fossa.

The *sphenopalatine foramen,* as might be inferred from its name, is situated at the junction of the roof and medial wall (*figs. 796, 799*). It is large and round, and is the main door for the vessels and nerves to the nasal cavity. A wire passed through the foramen follows the roof of the nasal cavity onto the nasal septum. Thin wires passed up the *greater* and the *lesser palatine canals* enter the fossa from below. Wires passed through the *foramen rotundum* and *pterygoid canal* enter it from behind. The pterygoid canal is demonstrated by passing a wire forward from a point immediately below the foramen lacerum (*figs. 692, 755*).

The fossa communicates with the orbit through the hinder part of the *inferior orbital fissure,* and with the infratemporal fossa through the *pterygomaxillary fissure.*

The Maxillary Nerve (*figs. 795, 796*), (V^2), or 2nd division of the trigeminal nerve, arises from the trigeminal ganglion, and is purely sensory to what developmentally is the maxillary process.

The wavy route taken by the nerve and its "continuation" is duplicated by a wire threaded through the foramen rotundum, across the pterygopalatine fossa to the inferior orbital fissure, and thence through the infra-orbital sulcus, canal, and foramen to the face. (On entering the orbit, it becomes the infra-orbital nerve.)

In the pterygopalatine fossa it is surrounded by branches of the maxillary artery and veins, and the pterygopalatine ganglion is suspended from it. Its other main branches supply the nasal cavity, nasopharynx, and palate.

The Infra-Orbital Nerve (*fig. 795*) ends on the face by dividing into many branches (*fig. 658*). These radiate to the skin and conjunctiva of the lower lid, the side and vestibule of the nose, the mobile part of the nasal septum, the skin and mucous sur-

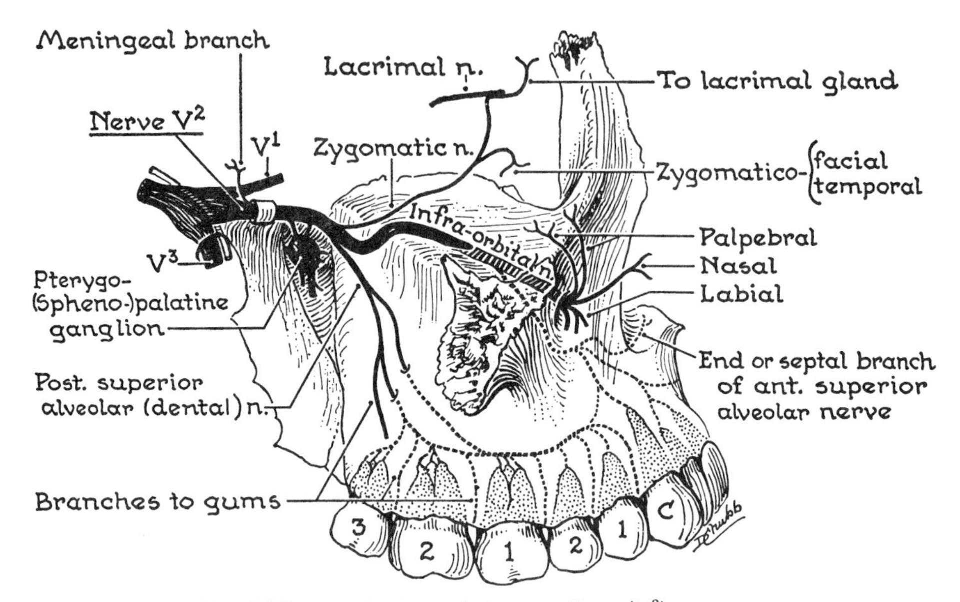

Fig. 795. Distribution of the maxillary (V^2) nerve.

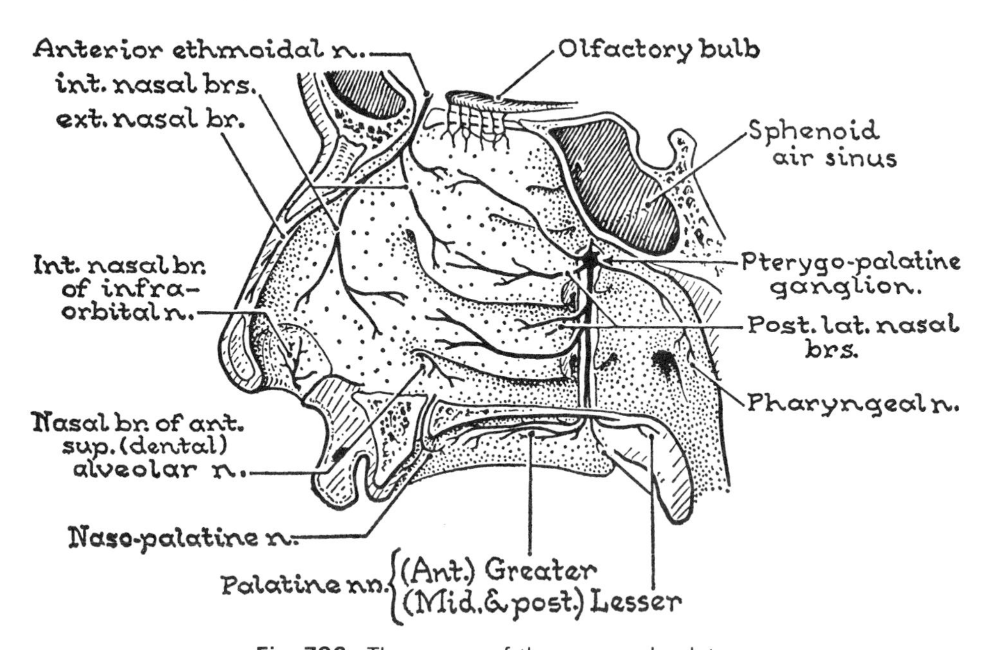

Fig. 796. The nerves of the nose and palate

faces of the upper lip and cheek, and to the gum.

In its course, it gives off two branches, the post. and ant. *superior alveolar* (dental) *nerves* (*fig. 795*). They supply teeth, periodontal membranes and gum (in part), and mucous membrane of the maxillary sinus.

The anterior superior alveolar nerve is apt to be damaged in opening into the maxillary sinus from the front; it sends a twig to the inferior meatus, floor, and septum of the nose (*figs. 795, 796*).

The Greater Palatine Nerve descends in its *canal* to the under surface of the hard palate medial to the 3rd molar tooth.

Thereafter, as two branches, it runs forward in the mucoperiosteum near the alveolar process and supplies the remainder of the hard palate and the adjacent portion of the gums. *Details:*

It gives off (1) *posterior lateral nasal nerves*, which pierce the vertical plate of the palatine bone and pass forward to the inferior concha and inferior meatus; and (2) *two lesser palatine nerves* traverse the lesser palatine canals, appear close behind the parent nerve and supply the mucous membrane of the soft palate and anterior tonsillar region.

Pterygopalatine Ganglion (parasympathetic). It lies in the pterygopalatine fossa, and is a relay station on the secretory pathway of the greater petrosal nerve. The fibers reach the ganglion in the nerve of the pterygoid canal (*see* below for details). Sensory and sympathetic fibers flit through the ganglion without interruption (*fig. 797*).

The **sensory fibers** from nerve V^2 to the ganglion are distributed to the mucosa of the nasal cavity, nasopharynx, and palate *via* the posterior lateral nasal nerves, nasopalatine nerve, pharyngeal n., and greater and lesser palatine nn. *Details* for specialists follow:

The **sympathetic fibers** (mainly vasoconstrictor) from the superior cervical ganglion run on the int. carotid artery to the foramen lacerum, where (as the deep petrosal nerve) they join the **secretory fibers** (greater petrosal nerve) to form the nerve of the pterygoid canal.

The Nerve of the Pterygoid Canal passes forward, through its canal, to the pterygopalatine ganglion and, after being relayed, is conveyed with the various branches of V^2 to the glands of the nose, nasopharynx, and palate; and some fibers travel with the zygomatic nerve, thence to the lacrimal nerve, and so the lacrimal gland (*fig. 797*).

Through the Sphenopalatine Foramen pass three nerves: (1) *Posterior lateral nasal nerves* run forward to the upper parts of the side wall of the nasal cavity and to ethmoidal cells. (2) the *nasopalatine nerve* (long sphenopalatine n.) crosses the roof to the nasal septum, descends in the mucoperiosteum in a groove on the vomer to the incisive foramen, through which it passes. It supplies the septum, front of the hard palate, and the gum (*fig. 776*). (3) *The pharyngeal nerve* runs backward to supply the roof of the nasopharynx and the sphenoidal sinus. It runs in a groove or canal.

Twigs of Maxillary Nerve: (1) *A meningeal*

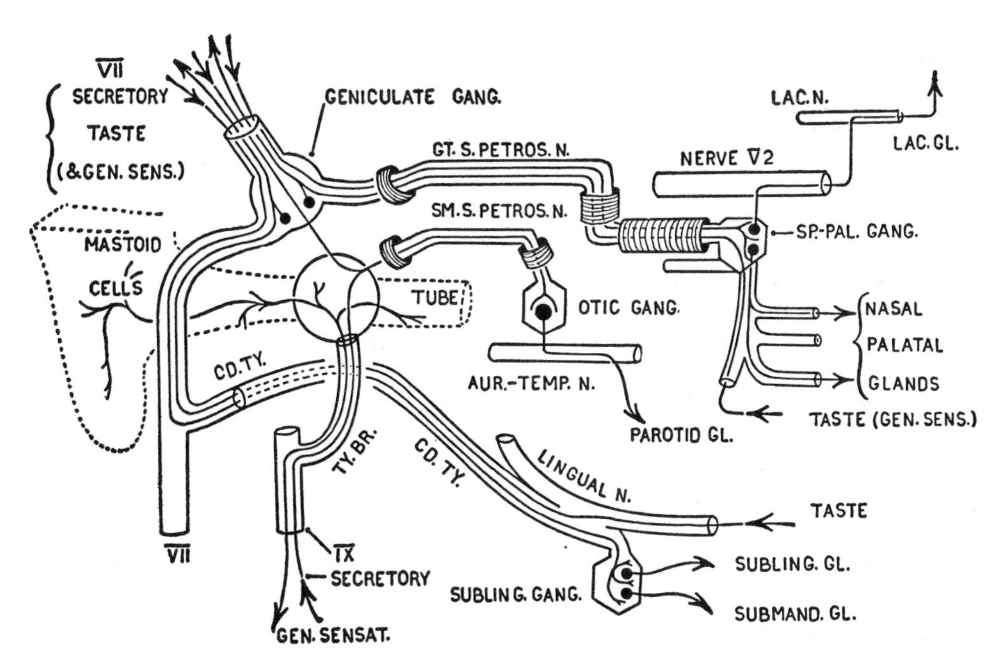

Fig. 797. Parasympathetic ganglia on the branches of nerves VII and IX: pterygo-(spheno-)palatine, otic, and submandibular (sublingual).

twig goes to the dura mater. (2) *An orbital twig* supplies the periorbita and conveys sympathetic fibers to the *Orbitalis muscle*, which bridges and extends beyond the margins of the inferior orbital fissure. (3) *The zygomatic nerve* passes through the inferior orbital fissure to the orbit, enters a V- or Y-shaped canal in the zygomatic bone, from which it emerges as two cutaneous branches, the *zygomaticofacial* and *zygomaticotemporal*—one on the cheek, the other in the temporal fossa (*fig. 658*). The secretory fibers to the lacrimal gland travel with the zygomatic nerve into the orbit, and there, leaving it, ascend on the lateral wall of the orbit to join the lacrimal nerve, which conveys them to the gland.

Maxillary Artery, 3rd part (continued from page 530 and *fig. 743*). This artery passes through the pterygomaxillary fissure into the pterygopalatine fossa and there gives off its branches.

Branches. The branches all escape through foramina (*fig. 743*) in company of the branches of the maxillary nerve and pterygopalatine ganglion, shown in *figures 795* and *796*, namely:

- Posterior superior alveolar
- Infra-orbital
  - anterior superior alveolar
- Artery of pterygoid canal
- Descending palatine
  - greater palatine
  - lesser palatine
- Sphenopalatine
  - pharyngeal
  - posterior lateral nasal
  - posterior septal

Corresponding *veins* form the pterygoid plexus, from which the maxillary vein emerges.

NASAL CAVITIES

The right and left nasal cavities are situated above the hard palate and are separated from each other by the nasal septum. Each cavity has an anterior and a posterior aperture, a floor, roof, median wall or septum, and a lateral wall.

Each Naris (nostril, or anterior nasal aperture) is kept patent by the U-shaped alar cartilage and is controlled by muscles (*figs. 650, 651*).

The Choana or posterior nasal aperture is oblong, rigid, bounded by bone, 2 cm high and 1 cm wide (*fig. 754*).

The Floor, formed by the horizontal hard palate, is concave, 1 cm (or more) wide, and 8 cm long from the tip of the nose to the posterior border of the septum (*fig. 798*).

The Roof has: (1) an *anterior part* whose slope corresponds to the slope of the bridge of the nose; (2) an *intermediate part*, formed by the cribriform plate, which is horizontal, 2 to 3 mm wide, thin, delicate, and perforated by the olfactory nerves and ethmoidal vessels; (3) a *posterior part* is formed by the anterior and the inferior surface of the body of the sphenoid. This part is less than 1 cm wide and confluent with the roof of the nasopharynx.

The Nasal Septum has three main parts (*fig. 798*):

1. *Perpendicular plate of ethmoid*, above.
2. *Vomer*, below and behind.
3. *Septal cartilage*, in front.

The inferior border of the septal cartilage extends from the anterior nasal spine to the tip of the nose (*fig. 650*).

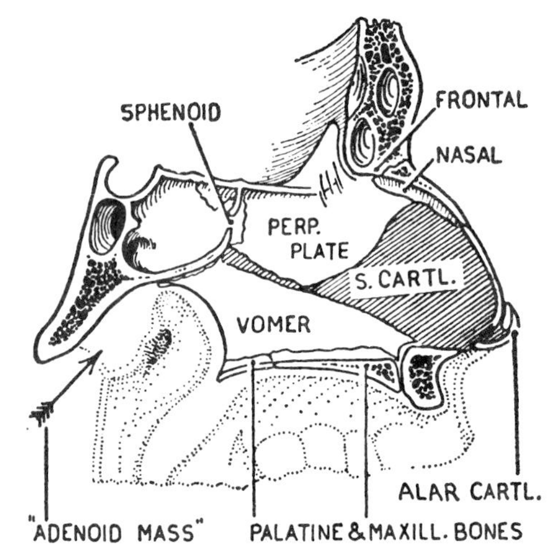

Fig. 798. Bones and cartilages of nasal septum.

Deflected Septum. If the septum is not straight, the deflection usually takes place at the junction of the cartilage with the vomer, and the middle concha on the concave side enlarges, as though to fill excess space.

Lateral Wall (*figs. 799, 800*). Two downwardly curved shelves, the *inferior* and *middle conchae*, project from the lateral wall of the cavity and conceal the *inferior and middle meatuses*. A third shelf, the *superior concha*, quite short and oblique, projects from the posterosuperior part of the lateral wall and conceals the *superior meatus*. The slit above and behind the superior concha is the *spheno-ethmoidal recess*.

The inferior concha almost reaches the auditory tube behind. The lower border of the middle concha almost reaches the lower surface of the body of the sphenoid.

The part of the nasal cavity between the conchae and the septum is the *common meatus*. The part anterior to the conchae is divided into two areas—a *vestibule* and an *atrium* (*fig. 800*).

The *vestibule*, or entrance chamber, is lined with skin and guarded by hairs.

The *atrium*, behind the nasal bones and lateral nasal cartilages, is lined with the mucoperiosteum.

Inferior Meatus. A probe passed from the orbital cavity down the *nasolacrimal duct* into the inferior meatus traverses the mucous membrane obliquely, like a ureter entering the bladder; hence a flap *valve*.

The (*bony*) *nasolacrimal canal* opens at the summit of attachment of the inferior concha.

A thin area—the maxillary process of the inferior concha (*fig. 846*) lies about 6 cm from

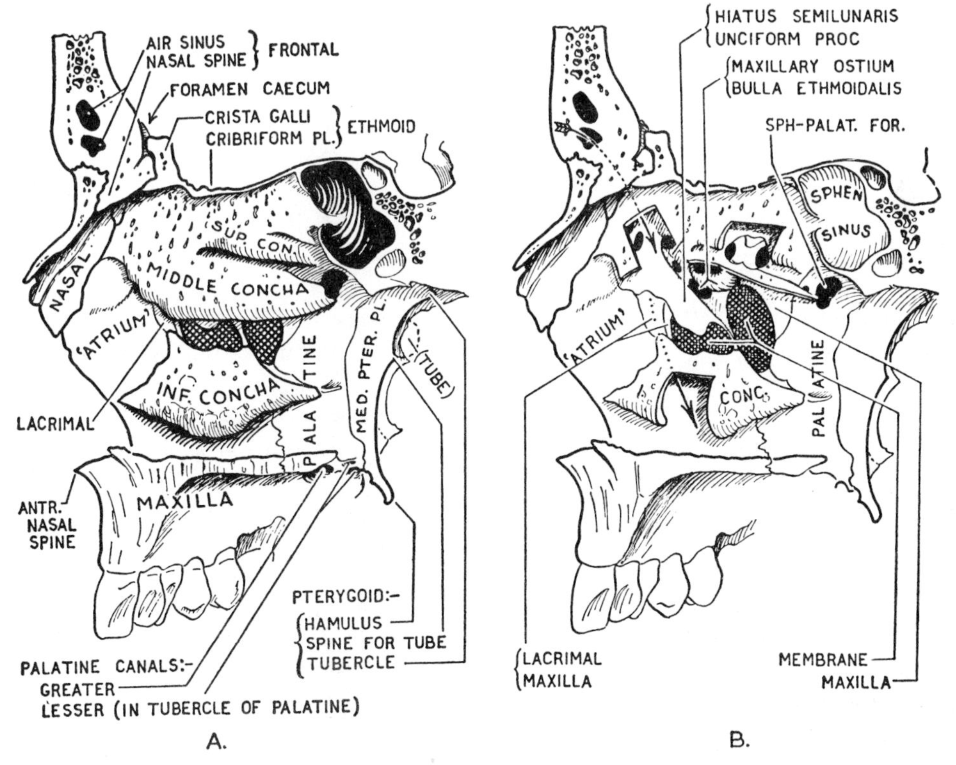

Fig. 799. *A*, bones of lateral wall of nasal cavity. *B*, orifices of air sinuses and nasolacrimal canal, revealed by cutting away parts of the conchae. (*Arrows* lead from frontal sinus and nasolacrimal canal.)

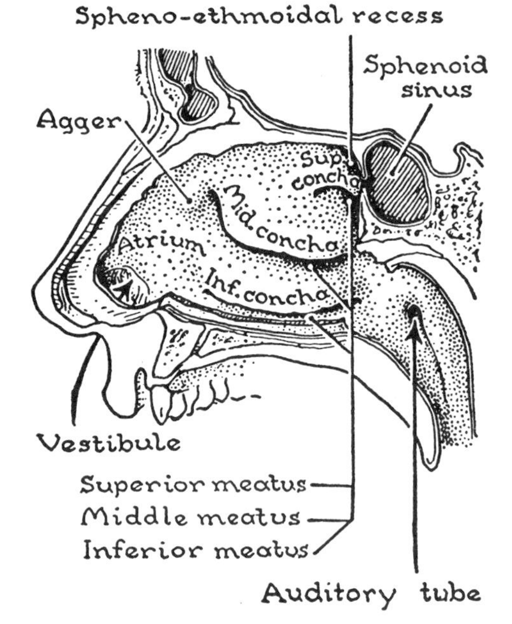

Fig. 800. The lateral wall of the nasal cavity.

the tip of the nose, or 3 cm from the anterior bony aperture. Through this thin bone the surgeon can push a hollow needle into the maxillary sinus.

Middle Meatus. A probe passed from *the lowest part of the frontal sinus* into the nose enters the middle meatus at the anterior end of a curved groove, the *hiatus semilunaris*. The hiatus has a sharp lower edge; above it there is an ovoid swelling, the *bulla ethmoidalis*. Behind its midpoint the hiatus opens through an *ostium* of variable size into the *maxillary sinus at its highest part* (*fig. 803*).

The orifices of several anterior ethmoidal cells are seen under cover of the middle concha.

Superior Meatus. The large orifices of one or more posterior ethmoidal cells open here.

Spheno-Ethmoidal Recess. The circular orifice of the sphenoidal air sinus opens here, high up (*fig. 800*).

Blood Supply to Nasal Cavity. The arterial network is very free and is derived from:

1. The *ophthalmic artery*, via its two ethmoidal branches, supplies the upper and front parts of the lateral wall and septum.

2. The *maxillary artery*, via its sphenopalatine branch, which traverses the sphenopalatine foramen, supplies the posterior parts of the lateral wall and septum.

3. The *facial artery*, via the superior labial art., supplies the antero-inferior part of the septum.

The *venous network* forms a distensible cavernous tissue, especially over the inferior and middle conchae, its function being to warm and humidify the inspired air.

Following the arteries, the veins drain: (1) upward via the ethmoidal veins into the superior ophthalmic vein; one vein, however, piercing the cribriform plate, joins the veins beneath the frontal lobe of the brain; (2) mainly backward through the sphenopalatine foramen to the pterygoid plexus; and (3) forward to the (anterior) facial vein.

The *lymphatics* (p. 601).

Nerve Supply of Nasal Cavity. The upper parts of the septal and lateral walls (over a total area of 2 cm^2) are supplied by 20 or more *olfactory nerves*, which pierce the cribriform plate to end in the olfactory bulb.

Perineural Spaces. Around the filaments of the olfactory nerve distributed in the nasal mucosa are perineural spaces. These have been shown by experiment to be prolongations of the cranial subarachnoid space and to have no connection with lymph vessels (Faber).

Ordinary sensation is carried mostly by V^2 with some help anteriorly from V^1. *Figure 796* provides a summary. *Applied Surgical Details:*

To cut off the sensory nerve supply: (1) inject with a curved hollow needle an anesthetic through the sphenopalatine foramen; (2) put a plug of gauze soaked in anesthetic in the angle between the nasal bone and the septum (3) and another under the anterior end of the inferior concha (*fig. 796*).

The 1st anesthetizes the branches of the pterygopalatine ganglion, viz., post. lateral nasal and post. septal (nasopalatine) nerves (p. 577). The 2nd catches the internal nasal branches of the anterior ethmoidal nerve. The

3rd catches the nasal branch of the anterior superior alveolar nerve which pierces the medial wall of the maxilla deep to the inferior concha.

The branches of the *infra-orbital nerve* to the vestibule and mobile part of the septum would remain intact (*fig. 658*).

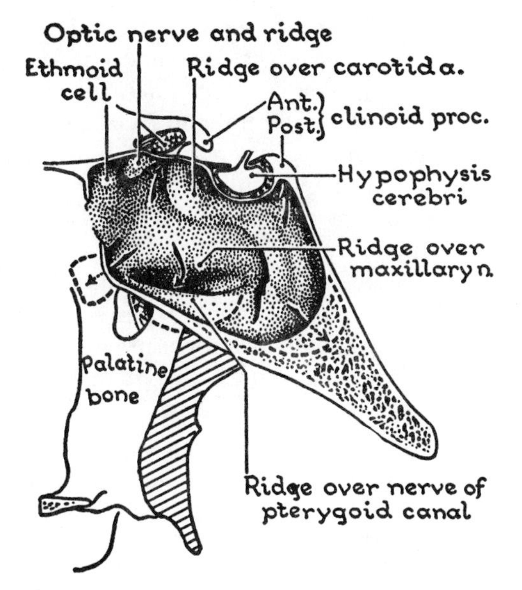

Fig. 801. A large sphenoidal sinus with many diverticula. Note its important relationships.

PARANASAL AIR SINUSES

The paranasal sinuses (sphenoidal, ethmoidal, frontal, and maxillary) are paired but asymmetrical (*figs. 801* to *805*).

Sphenoidal Sinuses. These are two in

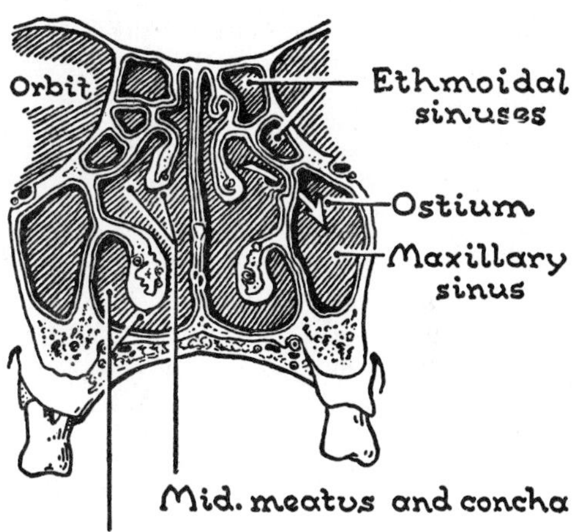

Fig. 802. The nasal cavities and adjacent air sinuses, on coronal section.

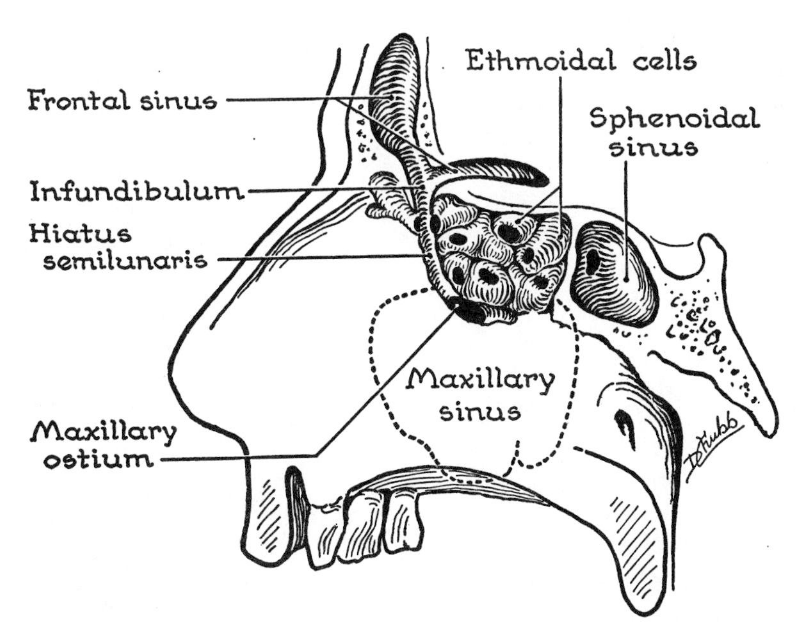

Fig. 803. Diagram of the paranasal sinuses.

Note that if there were fluid in the frontal sinus, it would drain down the infundibulum, along the hiatus, and into the maxillary sinus, for the orifice of the frontal sinus is at its floor and that of the maxillary sinus at its roof. The sphenoidal ostium is in the upper half of its anterior wall. The ethmoidal ostia are variable.

number—a right and a left (*fig. 803*). They are variable in extent, rarely symmetrical, and the partition between them is usually deflected. They occupy the body of the sphenoid, but may greatly expand (*fig. 801*).

The circular ostium of the sinus is near the upper part of the anterior wall.

Maxillary Sinus (*Antrum*). Between the inferior concha and the ethmoidal bulla the lateral wall of the nose is largely membranous (*fig. 799B*), and a needle pushed through it will enter the sinus.

The maxillary *ostium* (*fig. 803*) opens into the hiatus semilunaris, near the roof of the sinus. It is usually an oval or slit-like canal, 4.5 mm long. Hence, when the mucous membrane lining it is congested, the canal may be obstructed temporarily. It may, however, be very large.

The maxillary sinus (*fig. 802*) is a *three-sided, hollow* pyramid. The base contributes to the lateral wall of the nasal cavity (*fig. 799*). The apex stretches toward, or even into, the zygomatic bone. The three sides are translucent: one facing the infratemporal fossa, one the face, and one the orbital cavity. The *infra-orbital canal* creates a ridge on the orbital and facial walls.

The *ant. and post. superior alveolar vessels* and *nerves* occupy bony grooves, as they loop downward, and are covered with mucous membrane.

The floor or lower border of the sinus lies below the level of the floor of the nose. The roots or root of any *tooth* might project into the bony maxillary sinus, except those of the incisors.

Those of the molars commonly do; hence, infection may readily spread from a decaying tooth to the sinus. Again, during the extraction of such a tooth the mucous membrane over the projecting root may be torn with the result that the empty socket connects the sinus to the mouth.

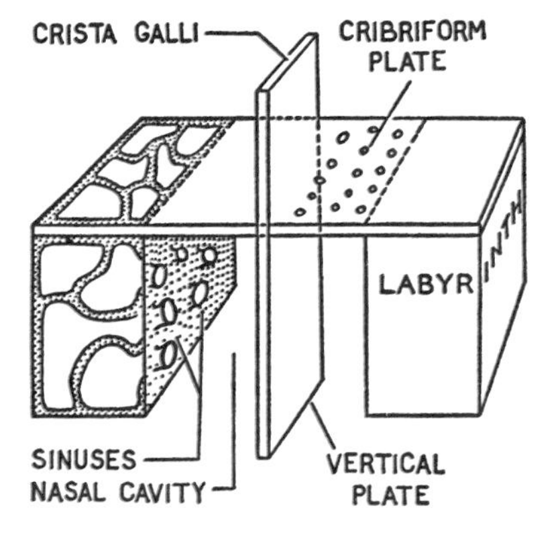

Fig. 804. Scheme of the ethmoid bone

The Ethmoidal Sinus or Cells (*figs. 803, 804*) may be likened to an oblong mass of 8 or 10 inflated balloons, but the shape of a particular balloon (or cell) depends upon the degree to which it and its neighbors are inflated. The ethmoidal cells are limited laterally by the orbital plate of the ethmoid, but the surrounding bones (lacrimal, frontal, sphenoid, palatine, and maxillary) help to close the cells.

(Ethmoid bone, details, p. 609).

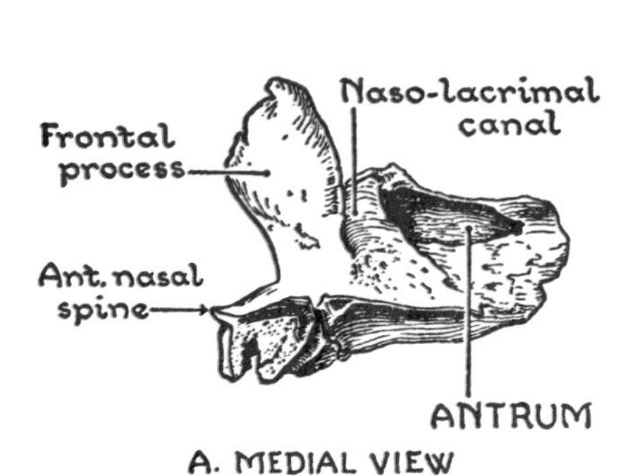

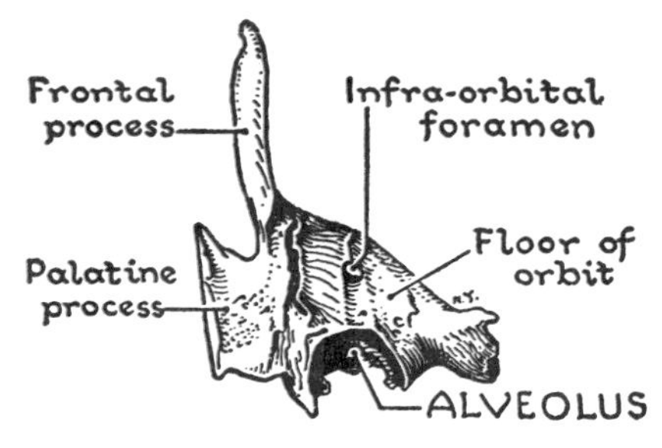

Fig. 805. The maxillary bone at birth, showing (*A*) the developing sinus (antrum), which (*B*) has not yet separated the alveolus from the floor of the orbit.

The Frontal Sinus is merely an anterior ethmoidal cell that has extended beyond the ethmoid into the frontal bone. Topographically, it is frontal; developmentally, it is ethmoidal. It drains downward (*fig. 803*).

Development of Sinuses. The upper jaw and face require to enlarge progressively as the teeth erupt. Enlargement is economically achieved by the development of the air sinuses. They appear prenatally as evaginations, being shallow depressions at birth. The *frontal sinus* begins to invade the bone in the first year but remains small until the 3rd year. The *sphenoidal sinus* at the 6th year is the size of a small pea and the *maxillary sinus* is 2 cm in all dimensions (*fig. 805*).

The *Mucoperiosteum* of the sinuses has a ciliated epithelium like that of the nasal cavities. Being less glandular, it is less vascular and thinner.

Nerve Supply. The sinuses are supplied by the ophthalmic (V^1) and the maxillary (V^2) nerves—the *frontal sinus* via the supra-orbital nerve; the anterior *ethmoidal sinuses* via the anterior ethmoidal nerve; the posterior ethmoidal and *sphenoidal sinuses* via the posterior ethmoidal nerve and the pharyngeal and other branches of the maxillary nerve traveling through the pterygopalatine ganglion; and the *maxillary sinus* via the superior alveolar nerves.

47 LARYNX

EXTERNAL STRUCTURES

Cricoid Cartilage; Lamina and Arch.
Thyroid Cartilage; Lamina; Cornua and Oblique Line.
Cricothyroid Joint; Ligaments and Muscles.

INTERNAL FRAMEWORK

Arytenoid Cartilage; Crico-arytenoid Joint.
Conus Elasticus (Cricothyroid, Median Cricothyroid, and Vocal Ligaments).
Epiglottic Cartilage; Thyro-epiglottic Lig.
Quadrangular Membrane and Vestibular Lig.
Corniculate and Cuneiform Cartilages.
Mechanics of Crico-arytenoid Joint.

INTRINSIC MUSCLES

Lateral Cricoarytenoid; Thyro-arytenoid; Vocalis; Thyro-epiglotticus; Ary-epiglotticus; Posterior Crico-arytenoideus; Arytenoideus.
Muscle Actions and Uses. Act of Swallowing.

INTERIOR OF LARYNX

Aditus; Vestibular and Vocal Folds; Rima Glottidis.
Parts: Vestibule; Middle Part (Ventricle and Saccule); Infraglottic Cavity.
Nerve supply. Blood Supply. Structure.

The larynx is the upper end of the lower respiratory passages, the pharynx and the nasal cavities being the upper respiratory passages. The *superior laryngeal aperture*, or *aditus laryngis*, rises freely into the pharynx and is separated on each side from the hinder part of the thyroid cartilage by a space, the *piriform recess*, and from the back of the tongue by a depression, the *vallecula*. The valleculae of the two sides are separated by a median fold of mucous membrane, the *median glosso-epiglottic fold*.

EXTERNAL STRUCTURES

The Cricoid Cartilage (Gk. = like a ring) has two parts: a *lamina* or signet plate placed behind, and an *arch* with horizontal lower border and oblique upper border. Its circular lumen barely admits a finger (*fig. 806*).

It lies in front of the vertebra C. 6.

The Thyroid Cartilage (Gk. = like a shield) consists of two quadrilateral plates, the *right and left laminae* (*fig. 807*), which converge to meet in the median plane in front at an angle. After puberty the angle varies with the sex, being like the subpubic angle—greater in the female than in the male (*fig. 809*). The upper border in the male projects forward, thereby creating a projection, the *laryngeal prominence* or Adam's apple. The *posterior border* is prolonged into an *upper* and a *lower horn* (*fig. 807*).

The upper border is attached to the hyoid by the *thyrohyoid membrane*, which is thickened to form the (*lateral*) *thyrohyoid ligament* between the adjacent horns of the thyroid cartilage and hyoid bone.

The lateral surface is crossed by an *oblique line*, which gives attachment to 3 muscles—Sternothyroid, Thyrohyoid, and Inferior Constrictor.

The Cricothyroid Joint and Muscle. A facet on the tip of the inferior horn of the thyroid cartilage articulates with a corresponding facet on the side of the cricoid.

The *median cricothyroid ligament* is all that its name implies; and the cricoid is attached to the first tracheal ring by the *cricotracheal ligament*.

The Cricothyroid (*fig. 808*) arises from the arch of the cricoid, and is inserted into the lower border of the thyroid cartilage and the inferior cornu. Its contraction tenses the vocal folds and its nerve is the ext. laryngeal nerve.

INTERNAL FRAMEWORK

The (paired) **Arytenoid Cartilage** (*fig. 807*) has 3 angles—a sharp anterior, the *vocal process;* a blunt lateral, the *muscular process;* and the *apex* (*fig. 809.1*).

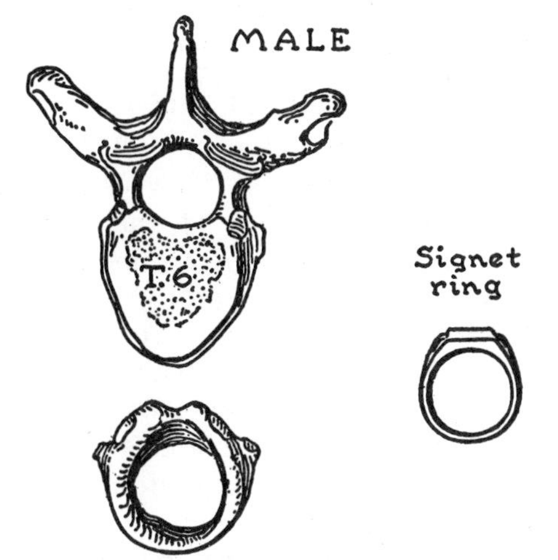

Fig. 806. Compare the lumina of a cricoid cartilage, a vertebral foramen, and a signet ring.

Vocal and Cricothyroid Ligaments. From the vocal process the *vocal ligament*—the fibrous core of the *vocal fold* (or *cord*)—passes forward to the angle between the thyroid laminae. It forms the upper border of a triangular membrane, the *cricothyroid ligament* or *conus elasticus*. This membrane is attached to the upper border of the arch of the cricoid; in front it blends with the median cricothyroid ligament.

The (unpaired) **Epiglottic Cartilage** (*fig. 812*) is a soft, curved, leaf-shaped cartilage. Its stalk is attached to the angle of the thyroid laminae above the vocal ligaments (*fig. 810*). Its rounded tip rises above the level of the hyoid bone. From each lateral border a fibro-elastic sheet, the *quadrangular membrane*, runs to the lateral border of the arytenoid cartilage. Its free upper edge is the *ary-epiglottic lig.*; the free lower border is the *vestibular ligament*, which is the basis of the vestibular fold (false vocal cord).

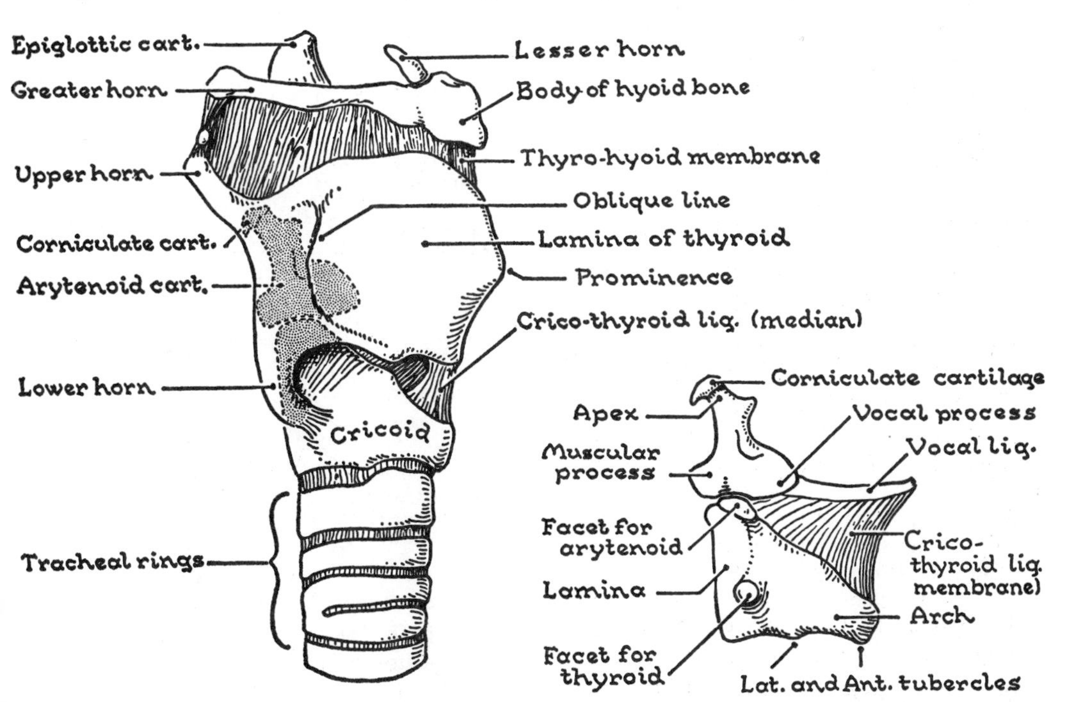

Fig. 807. The cartilages of the larynx, side view. Three ligaments—median cricothyroid, cricothyroid and vocal—constitute the conus elasticus.

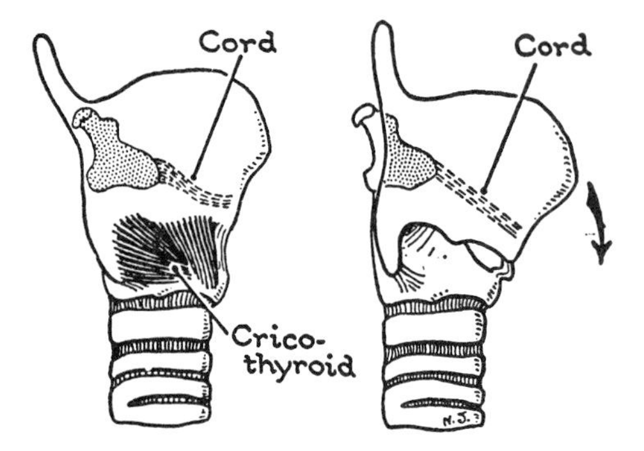

Fig. 808. The Cricothyroids render tense the vocal cords.

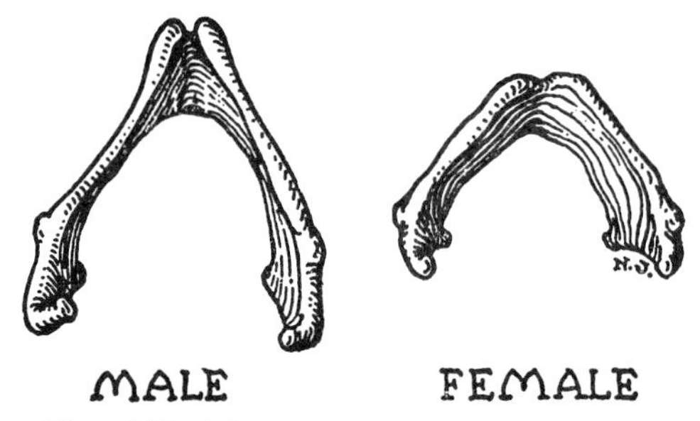

Fig. 809. The angle at which the laminae of the thyroid cartilage meet varies with the sex.

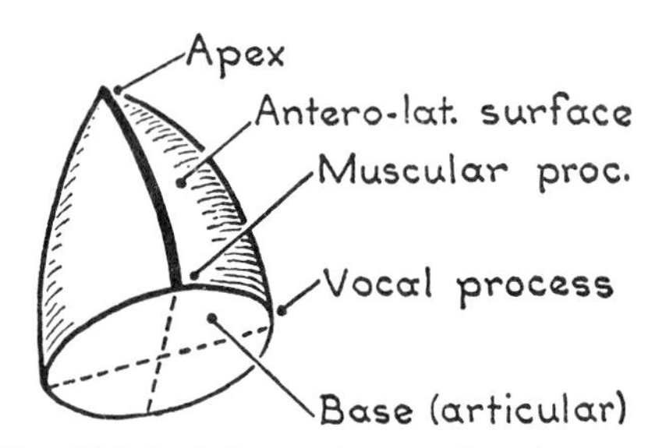

Fig. 809.1 Scheme for naming the parts of the arytenoid cartilage (see text).

This internal framework is lined with mucous membrane.

The apex of the arytenoid, the *corniculate cartilage*, is separate from the main cartilage. In the hinder part of the quadrangular membrane a detached portion of the side of the epiglottic cartilage was stranded during development—the *cuneiform cartilage* (*fig. 810*).

Mechanics of Crico-Arytenoid Joint. Articulating with the upper border of the cricoid is the (deeply) grooved base of the arytenoid. The movements permitted are rather complex: (1) gliding down the sloping surface on abduction, and up on adduction; (2) rocking (rotating forward and backward) so that, on abduction, the vocal process of the arytenoid (which dips downward and forward on adduction) moves upward, laterally, and backward; (3) a very slight pivotal movement, on a vertical axis, around a strong, steadying posterior crico-arytenoid lig. (von Leden and Moore).

INTRINSIC MUSCLES

The intrinsic muscles (*fig. 811, 812*) do not include Cricothyroid. They are applied to the sides and back of the internal framework like a sphincter, imperfectly subdivided into several parts.

Posteriorly, there are two parts (*figs. 811, 812*): (1) *The Arytenoideus* is a transverse muscle that covers the arytenoid cartilages posteriorly. It extends from one arytenoid to the same parts of the opposite arytenoid. Some *oblique fibers* continue as the *Aryepiglotticus*. (2) *The Posterior Crico-arytenoid* is *perhaps the most important of all.* Its action is to separate the vocal cords, thereby widening the rima glottidis, the

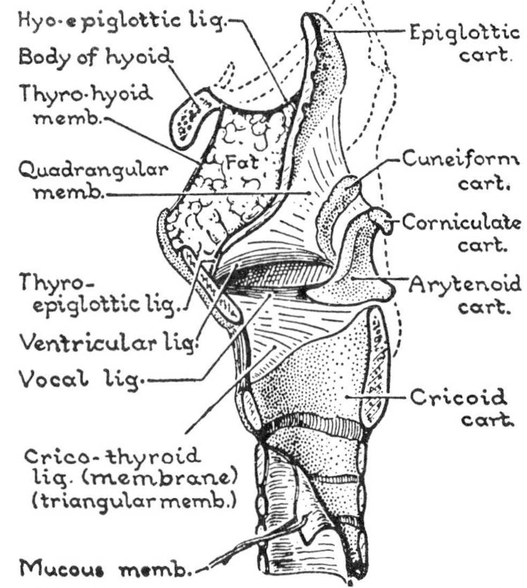

Fig. 810. The cartilaginous and membranous skeleton of the larynx, on median section. (Cricothyroid lig. or membrane = cricovocal membrane = ½ conus elasticus.)

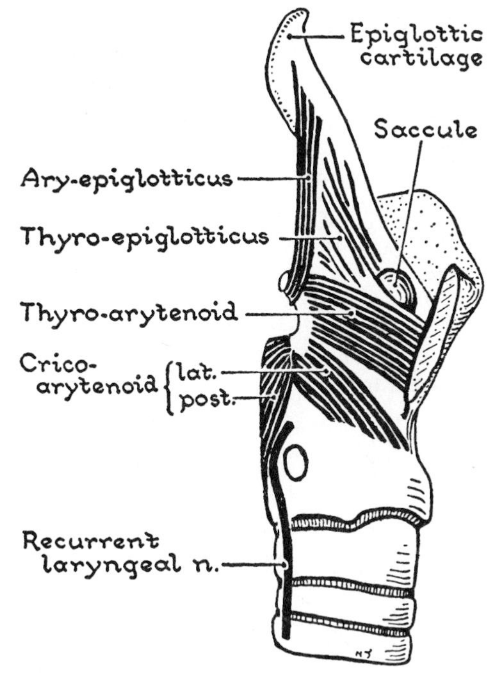

Fig. 811. Intrinsic muscles of larynx, side view.

space between. All other intrinsic muscles have a sphincteric action on the larynx. The Posterior Crico-arytenoids are, therefore, "*safety muscles.*" Bilateral paralysis results in closure of the rima glottidis with the attendant risk of suffocation. Each Posterior Crico-arytenoid arises from its own half of the posterior surface of the lamina of the cricoid to insert on the muscular process of the arytenoid.

Laterally, there is a sheet in five variable parts:

1. *The Lateral Crico-aryntenoid* inserted into the muscular process of the aryntenoid.
2. *The Thyro-arytenoid* inserted into the lateral border of the arytenoid.
3. *The Vocalis* is a slip of the Thyro-arytenoid applied to the vocal ligament. It is inserted chiefly into the vocal process, not the vocal ligament (von Leden).

4 and 5. *The Thyro-epiglotticus* and *Ary-epiglotticus* are wisps on the quadrangular membrane.

Muscle Actions and Uses (*fig. 812, 813*). The muscles of the larynx have three functions, (1) to open the rima glottidis, allowing passage of air; (2) to close the rima glottidis and vestibule, denying entry to food during a swallow; (3) to regulate the tension of the vocal cords for speaking.

The first two actions are automatic and are controlled by the medulla; the third is voluntary and is controlled by the cerebral cortex.

Muscles Abducting and Adducting the Cords, i.e., opening and closing the rima glottidis (*fig. 813*). The paired *Posterior Crico-arytenoids* are the only abductors, i.e., they widen the rima glottidis.

The *Lateral Crico-arytenoids* and the *Thyro-arytenoids* rotate the arytenoids medially and, so, adduct the cords. The *Arytenoideus* approximates the arytenoid cartilages.

Muscles Closing the Vestibule, as in the act of swallowing. The *Thyro-arytenoids* with the help of the *Ary-epiglottic* and *Thyro-epiglottic* muscles close the vestibule of the larynx, and tilt the arytenoid cartilages forward.

Muscles Affecting the Tension of the

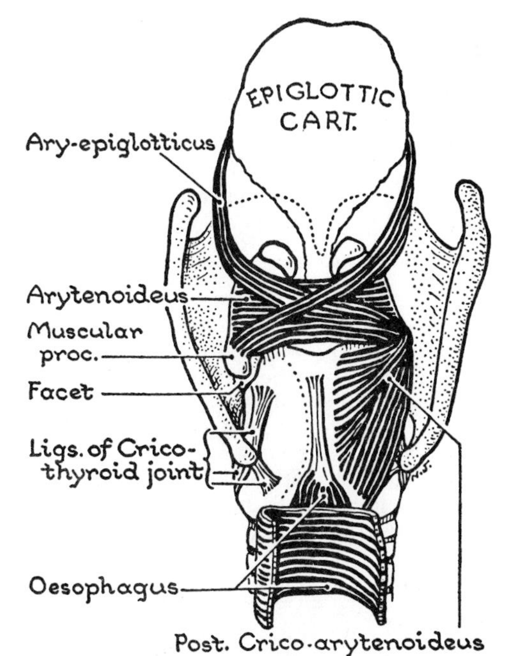

Fig. 812. Intrinsic muscles of larynx, post view.

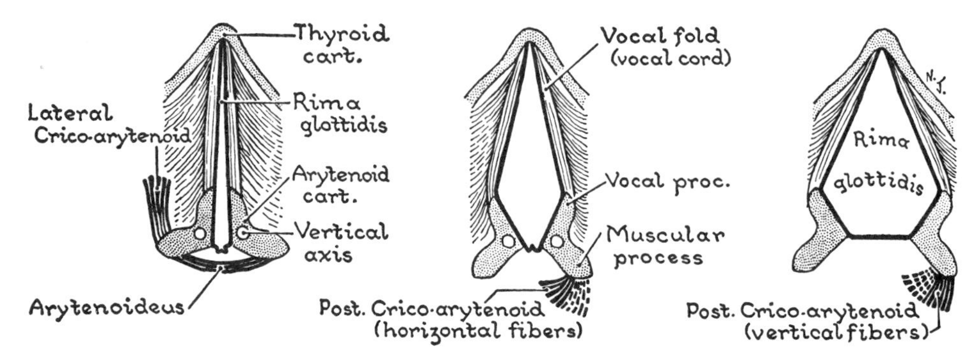

Fig. 813. Scheme of glottis, from above, to explain the actions of the Lateral and Posterior Crico-arytenoids. (The vertical axis is not a stationary one—see text.)

Cords. The cords being adducted, the *Cricothyroids* rotate the thyroid cartilage downward and forward, away from the arytenoid cartilages (*fig. 808*). At the same time the *Posterior Crico-arytenoids steady* the arytenoid cartilages on the cricoid cartilage. The Vocales muscles control the fine adjustment of tension.

During the Act of Swallowing the *Arye-piglottics* constrict the entrance to the larynx; the *Thyro-arytenoids* constrict the vestibule and tilt the arytenoid cartilages forward to touch the tubercle of the epiglottis; the *Arytenoideus* brings the arytenoid cartilages together and the Lateral *Crico-arytenoids* swing the vocal processes medially, thereby closing the rima glottidis.

If one Cricothyroid is paralyzed, the result (*fig. 814*) is reminescent of unilateral tongue paralysis.

INTERIOR OF THE LARYNX

The larynx extends from the tip of the epiglottis, which projects above the level of the hyoid bone, to the lower border of the cricoid cartilage.

The Aditus or Entrance to the larynx is oblique. It is bounded by the upper border of the epiglottis, the ary-epirlottic fold with the contained upper ends of the cuneiform and corniculate cartilages, and the mucous membrane covering the upper border of Arytenoideus.

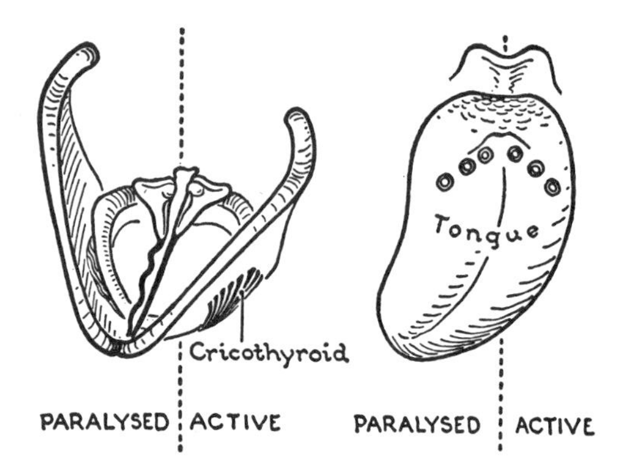

Fig. 814. Unilateral action of the Cricothyroid and of the Genioglossus projects the thyroid cartilage and the tongue to the contralateral side, i.e., inactive or paralyzed side.

The Vestibular (Ventricular) and Vocal Folds are also called the false and true vocal cords. From each side of the larynx these two anteroposterior folds project medially. They overlie the ligaments of the same names (*fig. 810*) and are about 5 mm apart. The vocal folds are visible from above because their attachments are medial to those of the vestibular folds.

Three Parts of Laryngeal Cavity (*fig. 814.1*). The vestibular (ventricular) and vocal folds divide the cavity of the larynx into three parts:

1. The upper part or *vestibule.*

2. On each side of the *middle part* the canoe-shaped depression between the vestibular and vocal folds is called the *ventricle.* From its anterior end a cul-de-sac of mucous membrane, the *saccule*, extends upward for about 1 cm (*fig. 814.1*). It lies

between the quadrangular membrane and the Thyro-arytenoid (*fig. 811*). Occasionally, it extends above the thyroid cartilage and through the thyrohyoid membrane. In the gorilla, the inflatable saccule is of enormous size, extending to the axilla and chest wall.

3. The inferior part or *infraglottic cavity*. The vocal folds are (together) the *glottis* and the slit between them, the *rima glottidis*.

Nerve Supply. The vagus nerve, via its superior and recurrent laryngeal nerves and aided by the sympathetic (vasomotor), supplies the larynx, providing motor, secretomotor, and sensory fibers. The superior nerve supplies the Cricothyroid muscle and sensation to the larynx above the vocal folds; the recurrent supplies all the internal muscles and sensation below the vocal folds.

The *recurrent laryngeal nerve* enters the pharynx in contact with the back of the cricothyroid joint.

It divides into: an *anterior branch* which ascends on the lateral sheet of muscles and supplies them; and a *posterior branch* which supplies the two posteriorly placed muscles and communicates with the internal laryngeal nerve.

The *superior laryngeal nerve* divides into the internal and the external laryngeal nerve (*fig. 762*). The *internal laryngeal nerve* perforates the thyrohyoid membrane as several branches, crosses the anterior wall of the piriform recess, reaches the lateral sheet of muscles, and is sensory to the larynx above the glottis, and to the region immediately around the entrance of the larynx. The *external laryngeal nerve* supplies the Cricothyroid.

Blood Supply. *Laryngeal branches* of both thyroid arteries.

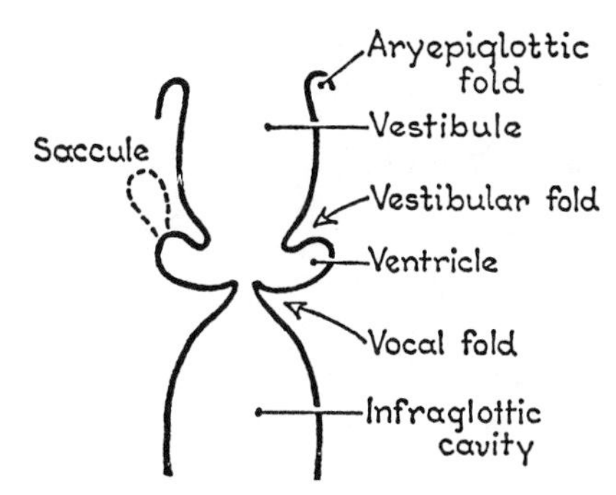

Fig. 814.1. The three parts of the laryngeal cavity, coronal section.

Lymph vessels (p. 601).

Structure. *The Mucous Membrane* has a *stratified ciliated epithelium*. Goblet cells, mucous glands, lymph follicles, and diffuse lymphoid tissue are present, as elsewhere in the respiratory passage. *Ciliated* epithelium gives place to *nonciliated* where vibration and impact occur, i.e., over the *vocal folds;* and to *stratified squamous epithelium* over the upper part of the back of the epiglottis due to the friction of passing food. The mucous membrane is *adherent* over the epiglottis and vocal folds; elsewhere and especially behind, where movements are most free, it is loose. *Mucous glands* are abundant in the pits of the epiglottic, cuneiform, and arytenoid cartilages, about the ventricles, and in the saccules.

The Vocal Folds comprise the vocal ligament, the Vocalis muscle, and a covering of thick nonciliated epithelium. They are nonvascular and therefore pale. They contain little submucous tissue and no glands and therefore they do not easily become swollen with risk of suffocation.

Types of Cartilage. The thyroid, cricoid, and arytenoid cartilages and the tracheal rings are *hyaline* in type. They tend to calcify before middle life and later to turn into bone, as does the hyoid bone at an earlier age. The thyroid cartilage in youth can be cut with a knife; in later life it requires to be sawn. The epiglottic and cuneiform cartilages are formed of *elastic cartilage* and, like the cartilages of the external ear and auditory tube, the corniculate cartilages and the vocal processes of the arytenoid cartilages, they neither calcify nor ossify.

48 EAR

EXTERNAL EAR

Auricle; External Meatus; Structure; Vessels and Nerves.

MIDDLE EAR OR TYMPANUM

Bony Surroundings in Detail; Tegmen Tympani.
Tympanic Cavity and Its Walls.
Tympanic Membrane.
Auditory Ossicles—Malleus; Incus and Stapes; Joints and Muscles: Tensor Tympani and Stapedius.
Amplification of Sound.
Vessels and Nerves.
Mastoid Antrum and Cells.

INTERNAL EAR

Bony Labyrinth—Cochlea; Vestibule; Semicircular Canals.
Membranous Labyrinth—Cochlear Duct; Saccule; Utricle; Semicircular Ducts.
Vessels and Nerves.

FACIAL NERVE (Intrapetrous Part)

Course; Geniculate Ganglion; Petrosal Nerves.
Nervus Intermedius; Chorda Tympani.

EXTERNAL EAR

The external ear consists of an auricle and an external acoustic meatus.

Auricle

This is described on p. 455.

Meatus

This canal is 24 mm long from the bottom of the concha (or 3 cm from the surface of the auricle) (*fig. 815*). It ends at the *tympanic membrane* or ear drum. It is oval on cross-section.

The Lateral Third of the meatus (8 mm) is chiefly *cartilaginous*, and it is attached to the thick outer edge of the *bony* meatus. The cartilage of the auricle and the cartilage of the meatus are one piece of elastic cartilage.

The meatus is sinuous, being convex upward and convex backward. So, pulling the auricle upward, backward, and laterally straightens the cartilaginous part, thereby making inspection of the drum, through a speculum, possible.

The Medial Two-Thirds of the canal (16 mm) is *bony*.

Structure. The cartilaginous part of the meatus is lined with skin in which there are hairs, sebaceous glands, and modified sweat glands which secrete cerumen or wax; hence, boils may occur here. The glands may extend for a short distance along the posterosuperior part of the bony meatus. Otherwise the bony meatus is lined with thin stratified squamous epithelium which is adherent to the periosteum and to the ear drum.

Vessels and Nerves of the Meatus. Arteries: posterior auricular, superficial temporal, and deep auricular branch of the maxillary. *Lymph vessels:* to mastoid, parotid, and superficial cervical nodes. *Sensory nerves:* auriculotemporal and auricular branch of the vagus (which emerges through the tympanomastoid fissure).

MIDDLE EAR OR TYMPANUM

The auditory tube and middle ear are derived from the 1st and 2nd pharyngeal pouches. A chain of three ossicles (malleus, incus, and stapes) develops from the 1st and 2nd pharyngeal arches. They pass from the tympanic membrane across the cavity to a membrane which closes an oval window, the *fenestra vestibuli*, leading to the internal ear (*fig. 816*). By this means vibrations in the air are amplified and conducted to the internal ear. The aditus to the mastoid antrum is a backward continuation of the middle ear (*figs. 816.1, 817*).

Tympanic Cavity and Its Walls

The cavity bears remote resemblance to a red blood cell—in being narrow and rounded, compressed at the center, and enlarged peripherally. It is 15 mm in vertical diameter; 2 mm across at the center, 4 mm at the floor, and 6 mm at the roof.

The Roof or tegmen tympani is thin and sloping (*fig. 817*), and the dura mater is adherent to it; and when broken through from above, the malleus and incus are seen rising into the epitympanic recess (*figs. 816, 818*).

The *Floor* also is thin and it rests upon the jugular bulb (*fig. 818*). As the internal jugular vein and the internal carotid ar-

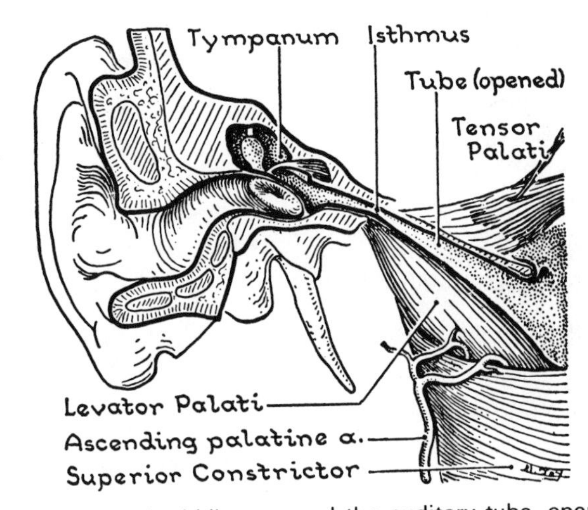

Fig. 815. The external and middle ears and the auditory tube, opened throughout.

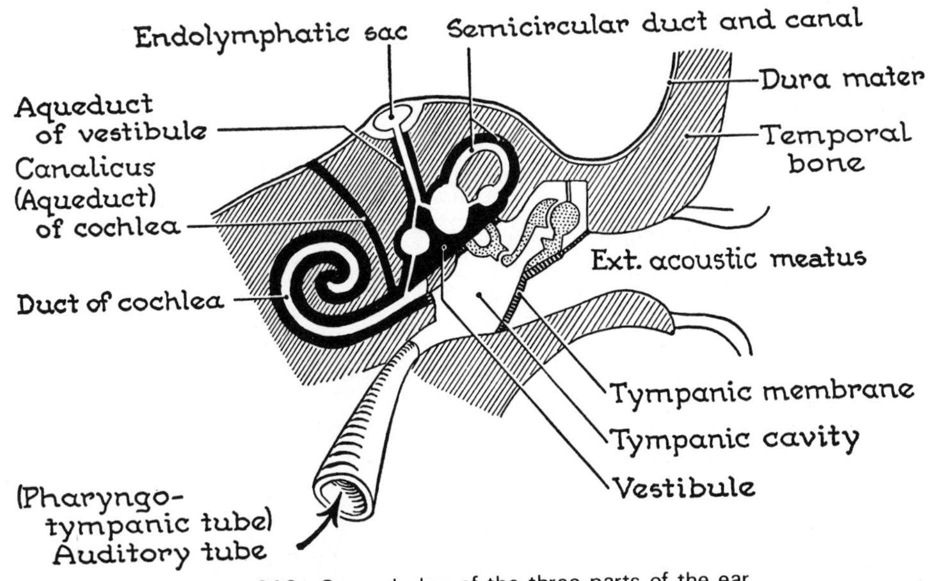

Fig. 816. General plan of the three parts of the ear.

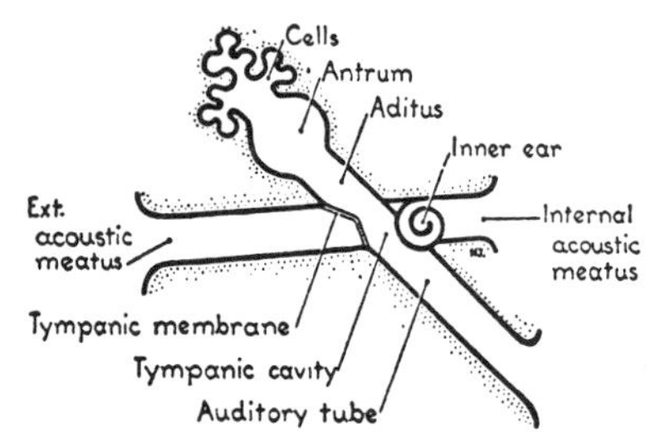

Fig. 816.1 The two meatuses, which have blind ends, and the line of the airway (tube, cavity, aditus, and antrum), which passes between them, viewed from above.

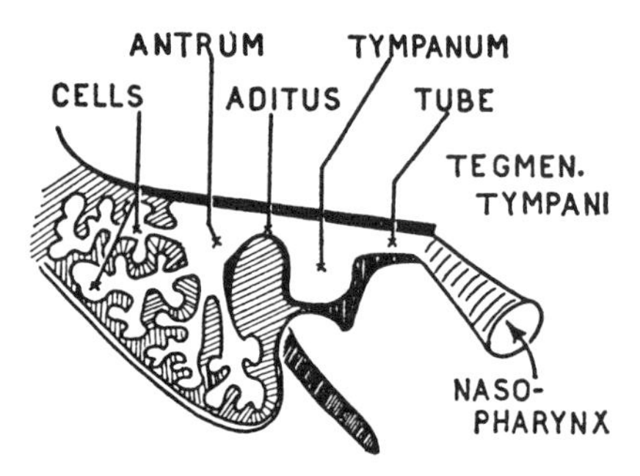

Fig. 817. Tegmen tympani and passages it covers.

tery, both lying within the carotid sheath, are followed upward to the base of the skull, they diverge below the middle ear (*fig. 818*).

Running through the wedge of bone between the diverging vein and artery is the minute canal for the tympanic branch of the glossopharyngeal nerve (*fig. 818*).

The Lateral Wall of the tympanic cavity is closed by the tympanic membrane (*fig. 819*), but the cavity rises a few millimeters above the membrane, this part being the *epitympanic recess* (*fig. 820*).

Anteriorly, the roof and floor converge to become the auditory tube. A thin bony shelf divides the tube and provides an upper compartment occupied by the *Tensor Tympani* (*fig. 820*). The free lateral edge of the shelf projects into the middle ear and curls upward to form a pulley (the processus cochleariformis) around which the tendon of the muscle turns to run laterally to its insertion. The ascending part of the *carotid canal* forms the anterior wall below the orifice of the tube (and it ascends medial to it), a delicate plate of bone intervening (*fig. 820*).

The Posterior Wall near the roof has a tunnel, the *aditus*, through which the epitympanic recess communicates with the mastoid antrum. The aditus is only a few millimeters long and wide. Below this, the *facial nerve* descends in the posterior wall; and jutting forward from this wall is a tiny elevation, the *pyramid* (*fig. 821*). At the

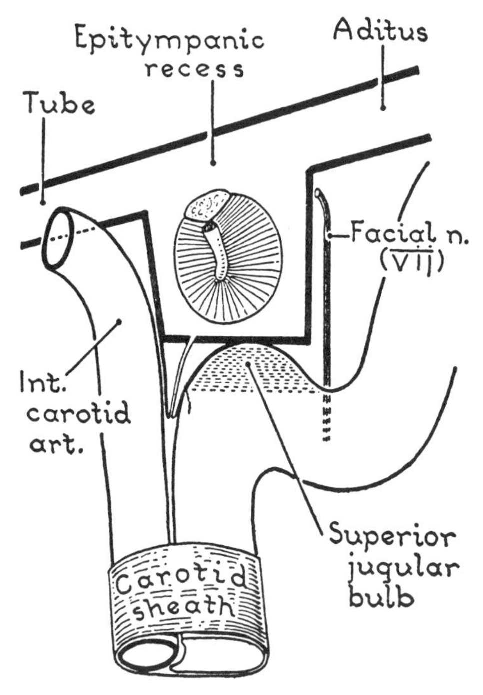

Fig. 818. Tympanic cavity and its neighbors.

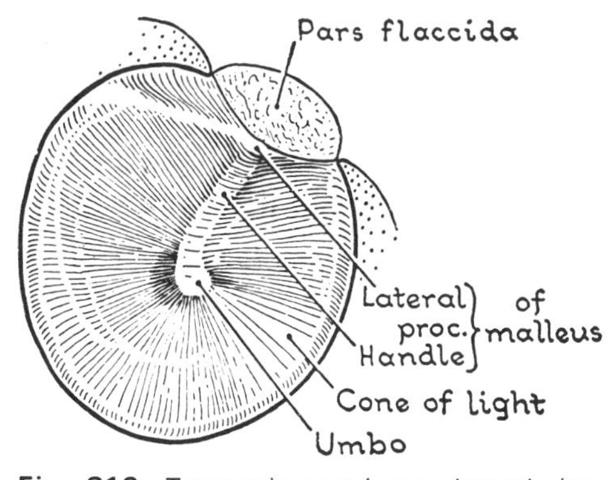

Fig. 819. Tympanic membrane, lateral view.

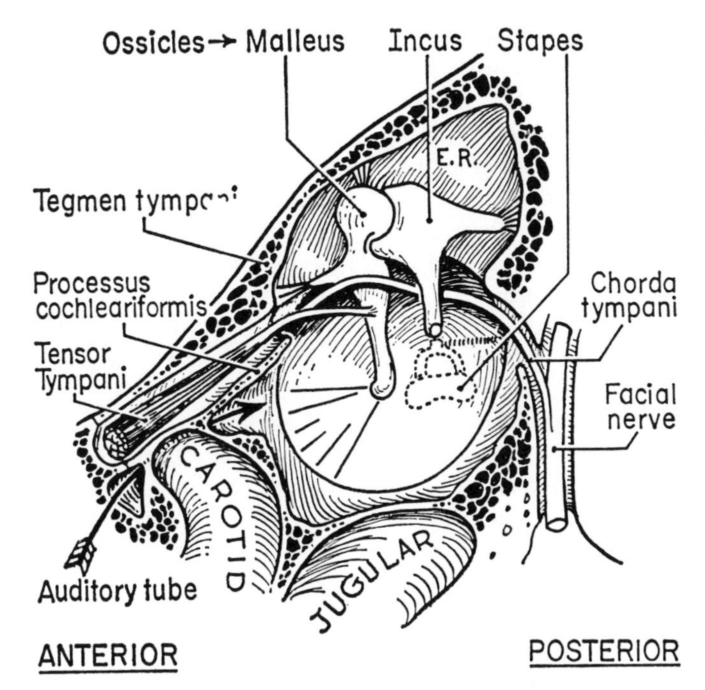

Fig. 820. Lateral wall of the tympanic cavity and the ossicles.

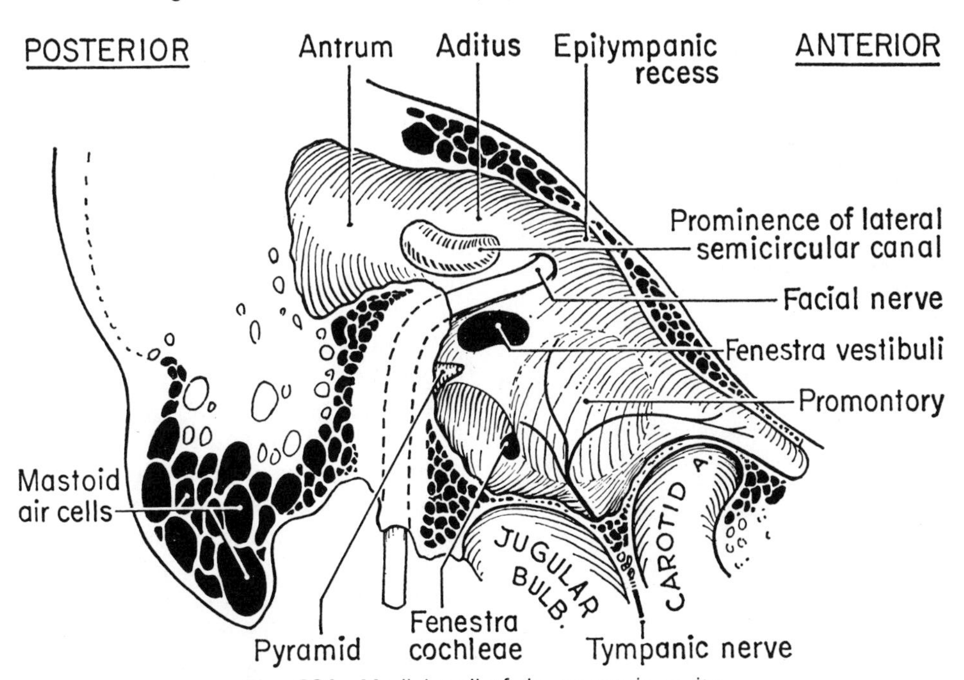

Fig. 821. Medial wall of the tympanic cavity.

apex of the pyramid there is a pinpoint orifice through which the tendon of the Stapedius passes forward to the neck of the stapes.

The Medial Wall (*fig. 821*) has at its center a swelling, the *promontory*, which overlies the first coil of the cochlea of the inner ear.

On an oblique line between the promontory and the roof of the aditus are (1) the fenestra vestibuli, (2) the canal for the facial nerve, and (3) the lateral semicircular canal. The *fenestra vestibuli* is an oval window opening into the vestibule of the inner ear. It is three mm in the horizontal axis and it is closed by the footplate of the stapes. The *canal for the facial nerve* curves backward on the medial wall (between the oval window and pyramid below and the lateral semicircular canal above) and descends within the posterior wall. The bony wall of the canal is thin and may be absent in its curved part. The *lateral semicircular canal* lies horizontally, bulges into the aditus and extends forward above the fenestra vestibuli.

The *fenestra cochleae* is a round window opening into the scala tympani of the cochlea, and closed by the *secondary tympanic membrane*. It lies postero-inferior to the promontory, but it is not seen immediately because it lies sheltered at the bottom of a depression that faces backward.

Bony Surroundings in Detail. On the *under surface of the skull* (*fig. 755*) a fissure runs from the pterygoid process to the anterior border of the external meatus. The projecting spine of the sphenoid divides the fissure into two halves: the anteromedial half lodges the cartilaginous part of the tube; the posterolateral half encloses the bony part of the tube and extends to the tympanum.

The upper part of the mastoid process is called the *postauditory process* (*fig. 834A*), is smooth and triangular; developmentally, it is a down-growth from the squama of the temporal bone and it closes the mastoid antrum laterally, just behind the suprameatal spine. The *suprameatal spine* lies just behind the posterosuperior part of the orifice of the external meatus.

Tympanic Membrane

This membrane or eardrum (*fig. 819*) is nearly circular, being 9 mm high and 8 mm wide. Set in the sulcus of the tympanic bone, it faces laterally, forward, and downward as though to catch sounds reflected from the ground as one advances. It is composed of circular and radial fibers and is lined with epidermis laterally and mucous membrane medially, and the handle of the malleus is incorporated in it. In fact, the radial fibers of the membrane radiate everywhere from the handle except over a triangular area, the *pars flaccida*, between the upper end of the handle (lateral process) and the roof of the meatus. At the lower end of the handle the membrane is indrawn: this point, called the *umbo*, lies antero-inferior to the center of the membrane.

The *chorda tympani* (*fig. 820*) crosses medial to the upper end of the handle of the malleus and is slung from the membrane by an anterior and posterior fold of mucous membrane.

Auditory Ossicles, Joints, and Muscles

Ossicles. There are three ossicles—malleus (hammer), incus (anvil), and stapes (stirrup) (*figs. 815, 820*). The **Malleus** is 8 mm long. It has a round head with a facet posteriorly, which ends below in a cog; a neck; a long handle embedded in the membrane and ending above in a lateral process, which is short and conical; and an anterior process which is very slender.

The **Incus** is shaped like a molar tooth. It has a body, two diverging processes, a short horizontal one and a long vertical one. The body articulates with the head of the malleus; the long process is parallel to the handle of the malleus and from its end a (lentiform) nodule projects medially to articulate with the stapes.

The **Stapes** has a head with a concave socket for the incus; a short neck; anterior and posterior limbs, which are attached to an oval foot plate. The footplate is attached to the margin of the fenestra vestibuli by an anular ligament—at least in youth; later an articular cavity appears in the ligament (Bolz and Lim).

Joints of the Ossicles. The joints between the malleus and incus, and the incus and stapes have synovial cavities. The head of the malleus and the body and short process of the incus lie within the epitympanic recess (*fig. 820*). A ligament suspends the head of the malleus from the roof of the cavity, another suspends the body of the incus. Ligaments attach the anterior process of the malleus and the short process of the incus to the front and back of the cavity; and around this axis the ossicles move.

When the vibrating tympanic membrane moves medially, the handle of the malleus moves too and with it the incus, forced by the cog. The stapes is driven into the perilymph within the inner ear, which, being incompressible, causes the secondary tympanic membrane closing the fenestra cochleae to bulge.

When the tympanic membrane is bulged laterally (e.g., when holding the nostrils and inflating the middle ear), the malleus moves also; the cog disengages, and the incus is released. This ensures against the stapes being torn from the window.

Sound Conduction. The effective vibratory area of the tympanic membrane is 55 mm^2 and the average size of the footplate of the stapes is 3.2 mm^2. Hence, the hydraulic ratio (membrane to footplate) is 17 to 1.

The lever ratio of malleus to incus being 1.3 to 1.0, the total transformer ratio is 22 to 1 (von Békésy).

The average size of the round window is 3.0 mm^2.

Muscles of the Ossicles. *The Tensor Tympani* is about 2 cm long. As its tendon leaves its canal it turns laterally round the processus cochleariformis to the upper end of the handle of the malleus; this it pulls medially thereby rendering the membrane tense. It is supplied by the mandibular nerve (V^3) (see otic ganglion, p. 530).

The Stapedius occupies the hollow pyramid and is but a few millimeters long. Its tendon, on leaving the foramen at the apex of the pyramid, passes forward to the neck of the stapes, which it "damps." It is supplied by the facial nerve. This muscle and the Tensor have various anomalies including absence and being ectopic (Hishino and Paparella).

Mucous Membrane lines the walls, bridges the hole in the stapes, and covers the ossicles, ligaments, and tendons, forming many folds and pockets. It has no mucous glands, although there are goblet cells in the auditory tube.

Details of Vessels and Nerves. The *Tympanic Membrane* has two surfaces, each supplied by different nerves and different arteries. Lateral surface—auricular br. of vagus and auriculo-temporal nerve; deep auricular br. of maxillary art. Medial surface—tympanic nerve (a branch of nerve IX; *see fig. 761*); and tympanic branches of the maxillary and stylomastoid arteries.

The Tympanum, Mastoid Antrum, and *Mastoid Air Cells* are supplied by the tympanic nerve (*fig. 821*). This ascends through the minute canal between the jugular foramen and the carotid canal and ramifies on the promontory, which it grooves and where it is joined by two caroticotympanic twigs of the internal carotid plexus. It supplies the *tube* also.

The arterial supply is (1) the tympanic br. of the maxillary art. which enters through the tympanosquamous fissure (*fig. 756*), (2) the tympanic br. of the ascending pharyngeal art. which follows the tympanic nerve, (3) caroticotympanic twigs of the internal carotid art., which pierce the wall of the carotid canal, (4) the stylomastoid br. of the posterior auricular art., and (5) petrosal brs. of the middle meningeal art.

Veins pass to the pterygoid plexus and inferior petrosal sinus.

Lymph vessels pass to the parotid and retropharyngeal nodes.

Mastoid Antrum and Cells

The antrum is slightly smaller than the tympanic cavity of which it is a backward extension through the aditus into the petromastoid bone. Its roof is thin and separates it from the middle cranial fossa. The *sigmoid sinus* is very close behind it.

The **Mastoid Cells** may be few and small or many and well inflated (*fig. 821*).

Development. The *lateral wall* is only 1 mm thick at birth and it increases by about 1 mm a year until it is about 15 mm thick. The antrum is large *at birth*, but there are no mastoid air cells and there is no mastoid process. Cells sprout from the antrum soon after birth and grow like a bunch of grapes into the enlarging bone.

Another collection of air cells, the *tubal* or *tympanic cells*, sprout from the medial wall of the tympanum near the orifice of the auditory tube and the carotid canal, reaching the apex of the petrous bone.

INTERNAL EAR

The internal ear is concerned with the reception of sound and with balancing. It lies within the petrous bone and has two parts, (1) the *bony labyrinth*, which contains (2) the *membranous labyrinth*.

Bony Labyrinth (*fig. 822*). This has three parts—*cochlea*, *vestibule*, and *semicircular canals*. The cochlea lies deep to the promontary, the vestibule to the fenestra vestibuli, and the lateral semicircular canal to the aditus. The whole apparatus is only 17 mm long.

The Cochlea resembles a snail's shell with two and a half coils. It has a central pillar, the *modiolus*, whose base lies at the bottom of the internal acoustic meatus; and from it an *osseous spiral lamina*, like the thread of a screw nail, projects half way across the canal of the cochlea and, with the *basilar membrane*, which stretches across the other half, it divides the canal into two: (1) the *scala vestibuli*, which opens into the vestibule, and (2) the *scala tympani*, which is separated from the tympanic cavity, at the fenestra cochleae, by the *secondary tympanic membrane*.

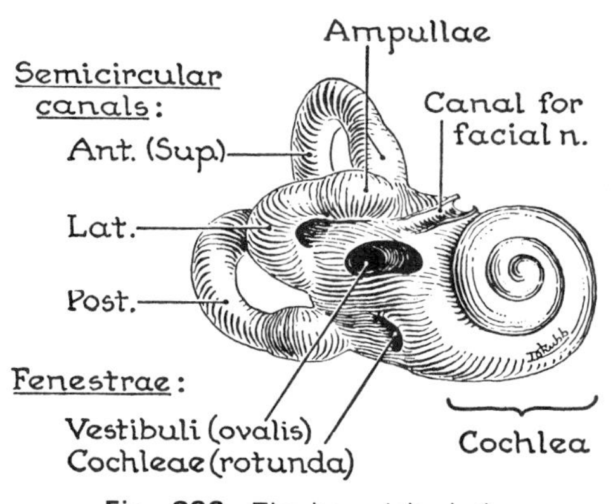

Fig. 822. The bony labyrinth.

The two scalae are continuous at the apex of the cochlea, called the *helicotrema*.

A minute duct, the *canaliculus* (aqueduct) *of the cochlea*, runs from the scala tympani through the petrous bone to the notch at the anterior margin of the jugular foramen straight below the internal acoustic meatus. The aqueduct brings the perilymph within the bony labyrinth and the cerebrospinal fluid within the subarachnoid space into communication—such is the general teaching.

Careful investigations indicate that the aqueduct of the cochlea is closed by a membrane and not open as depicted in *figure 815* (Wharton Young).

The Vestibule communicates *in front* with the cochlea; *behind* with the three semicircular canals; and *medially* with the posterior cranial fossa through the aqueduct of the vestibule (*fig. 815*). When the stapes is removed, the vestibule communicates through the fenestra vestibuli (oval window) with the tympanic cavity.

The Three Semicircular Canals—anterior (superior), posterior, and lateral—are set at right angles to each other and occupy three planes in space. The lateral canals of opposite sides are horizontal and lie in the same plane. The anterior canal of one side is parallel with the posterior canal of the other side, but they vary somewhat from specimen to specimen.

They are from 12 to 22 mm long, the lateral being the shortest. Each is less than 1 mm in diameter, except at one end where there is a swelling, the *ampulla*. They communicate at both ends with the vestibule. There are, however, but five openings, the anterior and posterior canals having a *crus commune*.

The *anterior* (superior) *canal* lies at a right angle to the posterior surface of the petrous bone and produces the arcuate eminence on the anterior (cerebral) surface (*fig. 687*). The *posterior canal* is immediately deep to the posterior (cerebellar) surface of the petrous bone and is parallel to it. It comes to within a centimeter away from the sigmoid sinus (*fig. 686*). The *lateral canal* is deep to the medial wall of the aditus and it runs above the canal for the facial nerve.

Membranous Labyrinth (*fig. 815*). This comprises (1) the cochlear duct, (2) the saccule, and utricle, and (3) the three semicircular ducts. It is a closed system containing endolymph. A stalk, the *ductus endolymphaticus*, passes from the saccule and utricle through a canal (aqueduct of the vestibule) in the petrous bone to the fissure lateral to the internal acoustic meatus. There the duct, which acts as a "safety" expansion sac, is placed extradurally. The *cochlear duct* (which is fully described in Neuroanatomy textbooks) lies on the vestibular side of basilar membrane. It is triangular on section and has a blind end. The *saccule* and *utricle* occupy the vestibule. The *semicircular ducts* only partially fill their canals.

The *Acoustic Nerve* has two branches. The cochlear branch is auditory; the vestibular branch (distributed to the utricle, saccule, and semicircular canals) is for equilibrium.

Vessels pass through the internal acoustic meatus with nerve VIII. The *labyrinthine artery* (int. auditory a.) is a branch of the ant. inf. cerebellar a. The *labyrinthine vein* passes to the inferior petrosal sinus.

FACIAL NERVE

Intrapetrous Part

In the internal acoustic meatus, the facial nerve lies above the acoustic nerve and is there joined by its sensory root, the *nervus intermedius*, which enters the meatus with it. At the lateral end ("bottom") of the meatus, the nerve enters the facial canal and travels through it to the stylomastoid foramen, whence it issues into the parotid region (p. 526; *fig. 737*).

Its Course takes it laterally, above the vestibule of the bony labyrinth (*fig. 823*), to the medial wall of the tympanic cavity. There it makes an abrupt bend, the *genu*.

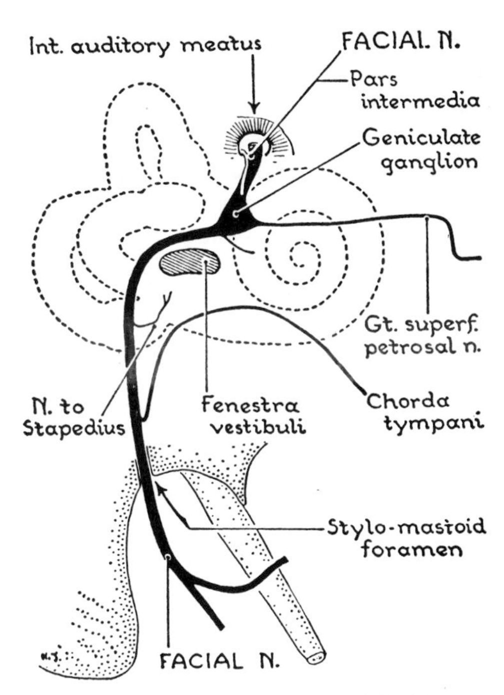

Fig. 823. Intrapetrous course of facial nerve.

It then arches backward and downward above the fenestra vestibuli on the medial wall and finally descends in the bony posterior wall of the cavity.

At the genu lies the *geniculate ganglion* (ganglion of the facial nerve) and there the *greater petrosal* and one *root of the lesser petrosal* leave it and *sympathetic twigs* (external petrosal nerve) join it. In its descending part it supplies the *Stapedius* and gives off the *chorda tympani*.

The Geniculate Ganglion is the cell-body station of the taste fibers in the greater petrosal nerve and chorda tympani. The secretory fibers of the nervus intermedius pass "nonstop" through the ganglion (*fig. 797*).

The Nervus Intermedius (Pars Intermedia) or so-called *sensory root* of the facial nerve, has both afferent and efferent fibers. The *afferent fibers* subserve taste and deep sensibility—their cell station is in the geniculate ganglion. The *efferent fibers* are secretory (autonomic)—their relay stations are in the pterygopalatine and submandibular ganglia (*fig. 797*).

The Chorda Tympani is distributed with the lingual nerve, so it leaves the facial nerve a few millimeters above the stylomastoid foramen, passes forward between the mucous and fibrous layers of the tympanic membrane, crosses medial to the handle of the malleus, and exits through a crack in the floor of the middle ear.

The chorda tympani is to lower jaw (mandibular nerve) territory what the greater petrosal nerve is to upper jaw (maxillary nerve) territory.

It emerges through the medial end of the petrotympanic fissure. It then crosses medial to the spine of the sphenoid and joins the lingual nerve some distance below on the surface of the Medial Pterygoid (*figs. 786, 786.1, 742, 820*).

49 LYMPHATICS OF HEAD AND NECK

LYMPH NODES

Main Chain of Nodes.
Horizontal Series.

LYMPHATICS

Lymph Vessels of Various Parts—Tongue; Tonsil; Teeth; Gums; Larynx; Ear; Nasal Cavity.

LYMPH NODES

The Main Chain of Lymph Nodes of the head and neck, called the *deep cervical nodes*, extends along the internal jugular vein from the base of the skull above to the clavicle below (*fig. 824*). Here it forms a *jugular lymph trunk*, which either opens independently into the angle between the internal jugular and subclavian veins or else joins the thoracic duct on the left side (right lymph duct on the right). Though the deep cervical nodes are largely covered by the obliquely set Sternomastoid, a few of them spread forward into the upper part of the anterior triangle and many spread backward beyond the posterior border of the Sternomastoid into the posterior triangle.

All these nodes lie superficial to the prevertebral fascia and the roots of the cervical and brachial plexuses. The inferior belly of the Omohyoid subdivides them into an upper and a lower group. A few nodes of the *upper group*, which extend medially behind the nasopharynx, are called the *retropharyngeal nodes*. Their afferents come from the nasopharynx and soft palate, middle ear and auditory tube.

Two nodes are specially to be noted (*fig. 824.1*): (1) the *jugulo-digastric node*, which lies below the posterior belly of the Digastric where the common facial vein enters the internal jugular; and (2) the *jugulo-omohyoid*, which lies above the inferior belly of the Omohyoid where it crosses the internal jugular. The accessory nerve is surrounded by nodes both where it enters the Sternomastoid and where it leaves it. The upper group of nodes drains into the lower group. The *lower group* (supraclavicular nodes) communicates with the nodes of the axilla and with the lymph vessels of the mamma.

All parts of the head and neck drain through the deep cervical chain. The chain has a few forward outposts in the neck, e.g., *infrahyoid* (on the thyrohyoid membrane) whose afferents follow the superior laryngeal artery and come from the larynx above the vocal cords; *prelaryngeal* (on the cricothyroid lig.) and *paratracheal* (in the groove between the trachea and esophagus) follow the inferior thyroid artery. These receive afferents from the larynx below the vocal cords and from the thyroid gland and adjacent parts.

A Horizontal Series of superficial nodes surrounds the junction of head and neck. These nodes are placed on the stem of named blood vessels and they receive afferents from corresponding territories. Thus:

One or two *occipital nodes* lie on the Trapezius where it is pierced by the occipital artery, 2 or 3 cm inferolateral to the inion. Their afferents are from the scalp; their efferents pass deep to the posterior

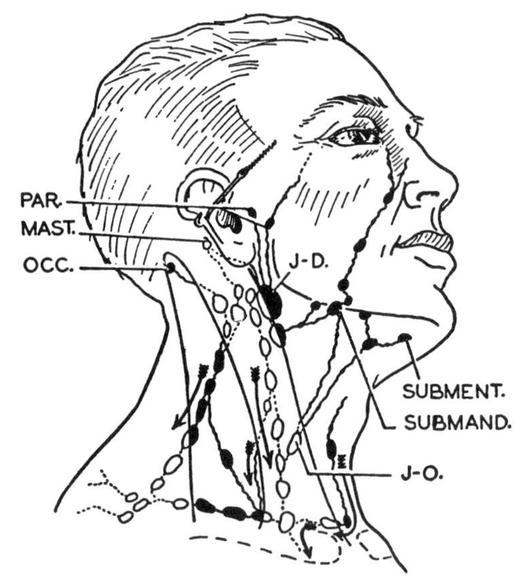

Fig. 824. Lymphatics of head and neck. (Rouvière.)

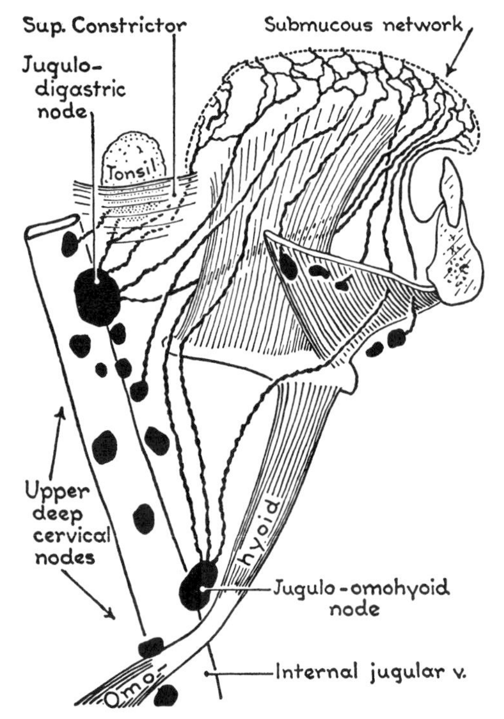

Fig. 824.1. Lymphatics of tongue and tonsil. (After Jamieson and Dobson.)

border of the Sternomastoid. They are palpable in German Measles.

One of two *retro-auricular nodes* (mastoid nodes) lie on the mastoid with the posterior auricular artery. Their afferents come from the scalp and auricle; efferents pass to the deep cervical nodes.

Several *superficial parotid nodes* (preauricular nodes) lie superficial to the parotid fascia near the superficial temporal and transverse facial arteries. Their afferents come from the scalp, auricle, eyelids, and cheek. Their efferents pass to the deep parotid and superficial cervical nodes.

[The *deep parotid nodes* may be conveniently described now. Embedded in the parotid salivary gland, they receive afferents from the superficial parotid nodes and from the external acoustic meatus, tympanum, deep parts of the cheek, soft palate, and posterior part of the nasal cavity. Their efferents pass to the deep cervical nodes.]

The superficial cervical nodes are small and are placed beside the external jugular vein on the upper part of the Sternomastoid. They are an offshoot of the superficial parotid nodes.

Half a dozen *submandibular nodes* lie on the surface of the submandibular salivary gland and also between it and the lower jaw, beside the facial artery. They have two extensions: (1) upward in the face along the course of the facial artery; the *facial nodes* are small and inconstant, except one or two at the lower border of the jaw: (2) forward along the submental artery; the *submental nodes* lie on the Mylohyoid below the symphysis menti. They receive afferents from the lower lip and chin and also from the tip of the tongue by vessels that pierce the Mylohyoid in company with anastomotic branches of the sublingual artery. The efferents pass to the submandibular nodes and also to the jugulo-omohyoid nodes.

The *submandibular nodes* receive afferents from their 2 extensions, and from the face, cheek, nose, upper lip, gums, and tongue. The efferents pass to the upper deep cervical nodes. To examine these nodes the subject should be told to drop his chin in order to slacken the cervical fascia.

One index finger should then be placed below the tongue, and the fingers of the other hand should be placed below the jaw and the structures between them palpated.

LYMPHATICS

The Lymph Vessels of the Tongue (*fig. 824.1*) spring from the extensive submucous plexus and all vessels drain ultimately into the deep cervical nodes alongside the internal jugular vein, between the levels of the Digastric and the Omohyoid, the uppermost node being the *jugulo-digastric node*; and the lowest the *jugulo-omohyoid node* (*fig. 824*).

The nearer the tip of the tongue the vessels arise, the lower is the recipient node; and the farther back, the higher the node.

Course. The vessels from the apex of the tongue pierce the Mylohyoid and are mostly intercepted by the submental nodes. The marginal or lateral vessels of the anterior two-thirds partly pierce the Mylohyoid to end in the submandibular nodes, and partly follow the blood vessels across both surfaces of the Hyoglossus to the deep cervical nodes. The medial vessels, however, descend in (or near) the septum, between the Genioglossi, and, after either piercing or passing below that muscle, follow the lingual artery to the deep cervical nodes. The vessels from the posterior one-third pass through the pharyngeal wall below the tonsil.

Crossing in part to Nodes of the Opposite Side are vessels near the median plane, and also vessels leaving the submental nodes (*fig. 824.2*).

The Lymph Vessels of the Tonsil pierce or run below the Sup. Constrictor mainly to the jugulo-digastric node.

The Lymph Vessels of the Upper Teeth pass through the infra-orbital foramen and run with the facial artery to the submandibular nodes. Those from the **Lower Teeth** run through the mandibular canal to the deep cervical nodes.

The vessels from the buccal surfaces of the **Upper and Lower Gums** run to the submandibular nodes; those from the lingual part of the lower gums end in the submandibular and deep cervical nodes;

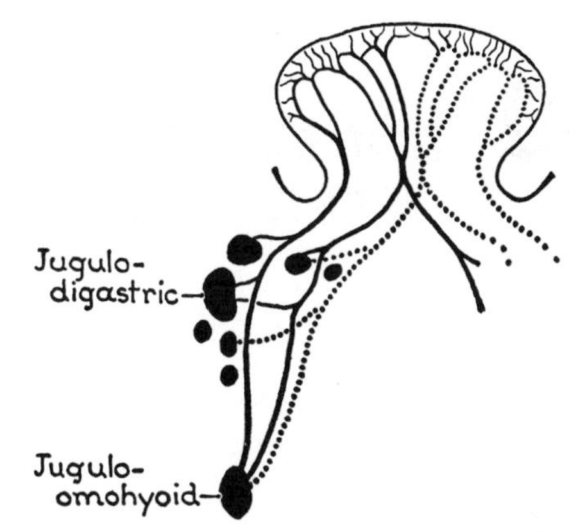

Fig. 824.2. Lymphatics of the tongue decussating. (After Jamieson and Dobson.)

those from the lingual part of the upper gums run dorsally with the palatine vessels to the deep cervical or retropharyngeal nodes.

The Lymph Vessels of the Larynx *above the vocal cords* follow the superior laryngeal artery through the thyrohyoid membrane to the upper deep cervical nodes after partial interception by the infrahyoid nodes; those *below the vocal cords* pierce the cricothyroid and cricotracheal ligaments and pass to the deep cervical nodes after partial interception by the prelaryngeal and paratracheal nodes.

Vessels of the upper and lower parts of the larynx anastomose submucously in the posterior wall of the larynx but not in the region of the cords, which act as a barrier, comparable to the one erected at the pyloric sphincter.

The Lymph Vessels of the Ear. Those of the auricle and external meatus pass to the retro-auricular, upper cervical, and parotid nodes; those of the tympanic membrane and lateral wall of the tympanum pass to the parotid nodes; those of the auditory tube and medial wall of the tympanum pass to the retropharyngeal and deep cervical nodes.

The Lymph Vessels of the Nasal Cavity from the anterior part run with those of the external nose to the submandibular nodes; those from the posterior pass to the retropharyngeal, deep parotid, and deep cervical nodes.

50 BONES OF SKULL: DETAILS FOR REFERENCE

FRONTAL BONE

Supra-orbital Margin and Outer Aspect of Squama; Under Surface; Inner or Cerebral Surface; Articulations; Ossification.

PARIETAL BONE

Ossification.

OCCIPITAL BONE

Squama; Condylar Parts; Basilar Part or Basi-Occipital; Articulations; Variations.

SPHENOIDAL BONE

Parts; Body and Lesser Wings; Greater Wing and Its Foramina; Pterygoid Process and Laminae; Ossification; Variations.

ETHMOID BONE

Shape; At Birth; Labyrinth and Cells; Conchae.

TEMPORAL BONE

Squamous Part; Tympanic Part [and Ring]; Petromastoid Part.

MAXILLA

Facial Surface; Frontal Process; Infratemporal (Posterior) Surface; Orbital (Superior) Surface; Nasal (Medial) Surface; Ossification.

ZYGOMATIC BONE

Ossification.

PALATINE BONE

Plates; Processes; Greater Palatine Canal; Ossification.

NASAL BONE

Surfaces and Markings.

LACRIMAL BONE

INFERIOR CONCHA

VOMER

MANDIBLE

Ossification.

HYOID BONE

Development and Ossification; Variations; Muscle Attachments; Fibrous Attachments.

SKULL AT BIRTH

It is, generally speaking, much more important to be familiar with the skull as a whole than with the individual bones that comprise it, because (except in the cases of the mandible and the ossicles of the ear) the bones are united to each other either by suture or synchondrosis and there is no movement between them. Muscle attachments, bony fossae, bony lines and ridges, blood sinuses, fasciae, and so on extend from bone to bone without respect to such joints, so the locations of the immovable joints outline the individual bones are of little account.

The bones of the skull are classified as:

1. *Bones of the cranial cavity:* frontal, parietal (paired), occipital, sphenoid, ethmoid, and temporal (paired).

2. *Bones of the face and nasal cavities:* maxilla, zygomatic, palatine, nasal, lacrimal, inferior concha, vomer (unpaired), and mandible (unpaired).

The following pages offer a convenient source for quick reference to **details** *that normally should not be memorized. They should be considered carefully only when special problems are confronted.*

FRONTAL BONE (DETAILS)

The frontal bone (os frontale) (*figs. 825,*

826) is shaped like a cockle shell and has two parts: a vertical part, the *squama*, in the forehead; and a horizontal part, the two *orbital parts* or *plates*, which forms the greater part of the roof of each orbit. Between the two orbital plates there is an oblong space, the *ethmoidal notch*.

Supra-Orbital Margin and Outer Aspect of Squama. Between the squama and each orbital plate is the *supra-orbital margin;* this is concave, forms a third of the margin of the orbit, and has either a notch, foramen, or canal, the *supra-orbital notch* (f. or c.) 3 cm from the median plane. The supra-orbital margin ends laterally in a stout projection, the *zygomatic process;* medially it ends at a point (*medial angular process*).

Between the right and left medial angular processes there is a broad semilunar surface, the *nasal margin* or *notch*, for articulation with the nasal bones and frontal processes of the maxillae. The point in the median sagittal plane between nasal and frontal bones, i.e., on the nasal notch, is the *nasion*.

The prominence 1 to 2 cm above the nasion is the *glabella*, so called because it is situated between the eyebrows and is bald or glabrous. Lateral to it on each side a fullness, the *superciliary arch*, extends to, or beyond, the supra-orbital notch.

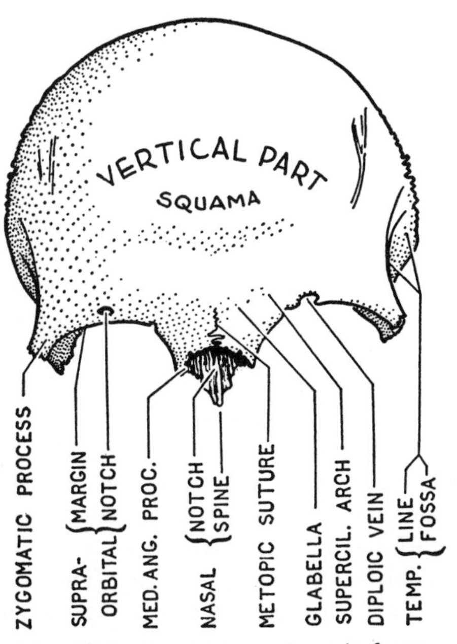

Fig. 825. Frontal bone–from in front.

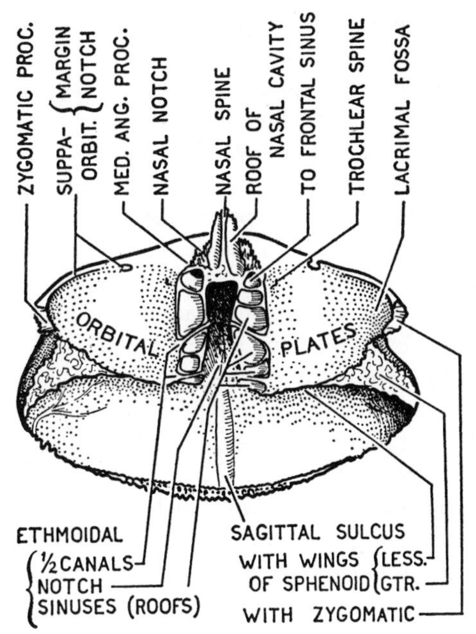

Fig. 826. Frontal bone–from below.

A sharp line, the *temporal line*, curves upward and backward from the zygomatic process and separates the temporal fossa below from the region of the scalp above. A fullness at the center of each half of the squama, the *tuber frontalis* (or *frontal eminence*), marks the site where ossification began. Vertical grooves for branches of the supra-orbital nerves are sometimes seen on the squama.

The Under Surface. Each orbital plate is very thin, and has laterally just behind the supra-orbital margin a *fossa* for the lacrimal gland, and medially a *spine* (or depression) for the trochlea of the Obliquus Oculi Superior.

The *ethmoidal notch* lodges the cribriform plate of the ethmoid. Skirting the notch on each side are broken cells which overlie the ethmoidal labyrinth and form the roofs of the *ethmoidal air cells* or *sinuses*. Two half canals, *anterior* and *posterior ethmoidal canals* (foramina), run in the walls of the cells from orbit to ethmoidal notch. The most anterior cell (sometimes the second most anterior) opens into the *frontal air sinus*.

Descending from the nasal notch is a broad triangular process, the *nasal spine*. In front it buttresses the nasal bones; behind these bones it articulates on each side with the frontal process of the maxilla; posteriorly it has a median crest for the vertical plate of the ethmoid, and on each side of this a longitudinal groove (4 mm wide) forms part of the roof of the nasal cavity. The latter facts are best appreciated on inverting a skull and looking at the narrow roofs of the nasal cavities just above the nasal bones.

The Inner or Cerebral Surface takes part in the anterior cranial fossa. A median ridge, the *frontal crest* extends upward from the ethmoidal notch to a broad shallow groove, the *sagittal sulcus*. The upper surface of each orbital plate is convex and is marked by ridges which occupy sulci of the brain.

Articulations. Just as you raise your hat from your head, so you may raise a frontal bone from off the other bones of the skull (nasal, maxillary, lacrimal, ethmoid, sphenoid, and zygomatic), because it rests on them; it is true that at the upper part of the coronal (frontoparietal) suture the frontal bone overlaps the parietal; however, at the lower part it is overlapped by the parietal.

Ossification is in membrane; it begins during the 7th fetal week at the frontal tubers. At birth the frontal bone is in two halves; these fuse about the 2nd year at the *frontal* or *metopic suture*. Remnants of this suture persist at the glabella.

PARIETAL BONE (DETAILS)

The parietal bone is described as a "key bone" on page 523.

Ossification is in membrane; it begins about the 7th fetal week at the *tuber parietale* or parietal eminence, that is at the point of fullness at the center of the bone. The parietal bone therefore corresponds closely to the frontal, but the two parietal bones do not fuse till the 3rd decade. Areas around the margins of the parietal bone may ossify separately giving rise in the sutures to small independent bones, commonly the size of a finger nail, called *sutural (Wormian) bones*.

OCCIPITAL BONE (DETAILS)

The occipital bone (unpaired) lies at the back and base of the spheroidal brain case. *At birth (fig. 827)* and until the 3rd or 4th year, it consists of four pieces, disposed around the *foramen magnum* thus: the squamous part or *squama* behind, a *lateral (condylar) part* on each side, and the *basilar part* or basi-occipital in front. These names are retained for parts of the adult bone.

Squama. Near the center of the outer surface of the squama is a bump, the *external occipital protuberance* or more briefly the *inion*. From the inion a line, the *superior nuchal line*, curves on each side to the lateral border, separating the area for the scalp above from the area for the muscles of the neck, the *nuchal area*, below. A median crest, the *external occipital crest*, runs from inion to foramen magnum; and from near the midpoint on this crest an *inferior nuchal line* curves laterally on each side.

Below the center of the inner surface is an elevation, the *internal occipital protuberance*. From it a cruciate arrangement of lines radiates (*fig. 828*). The upper lines bound the *sagittal sulcus;* two transverse lines on each side bound the *transverse sulcus;* and, a prominent median line, the *internal occipital crest* descends to the foramen magnum, occasionally splitting below to enclose a triangular depression, the *vermian fossa*.

The cruciate lines divide the inner surface of the squama into four fossae—two upper ones for the occipital lobes of the cerebrum, and two lower ones for the hemispheres of the cerebellum. The upper fossae are covered externally merely with

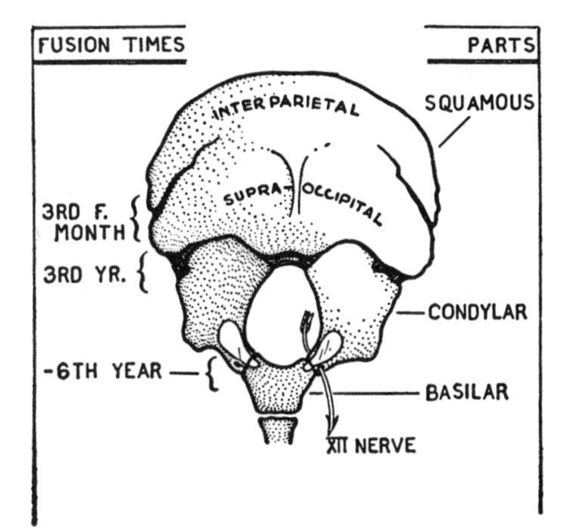

Fig. 827. Occipital bone at birth, exterior views. (*Condylar part = lateral part.*)

scalp, and the bone is thick; the lower fossae are protected externally by nuchal muscles, and the bone is thin and translucent.

Lateral (Condylar) Parts. On the under surface of each lateral part (and extending onto the basilar part) an oval articular eminence, the *occipital condyle*, skirts the anterior half of the foramen magnum. Behind each condyle there is a *condylar fossa* into which usually opens a *condylar canal* for an emissary vein. Lateral to the posterior two-thirds of each condyle projects a bar of bone, the *jugular process*, which is homologous with the transverse process of a vertebra.

The anterior one-third of each condyle extends forward onto the basilar part of the bone. The site of union between the basilar and condylar parts is marked by the *hypoglossal (anterior condylar) canal*, for the transmission of the hypoglossal nerve. The external orifice of this canal lies anterolateral to the condyle; the internal orifice lies within the margin of the foramen magnum above the middle of the condyle and is overhung by the *jugular tubercle*. The jugular process is grooved both above and in front by the *sigmoid sinus*, which here becomes the internal jugular vein. Posteriorly it is continuous with the squama.

Basilar Part or Basi-Occipital is a bar of bone that extends upward and forward from the foramen magnum to the sphenoid. It is thin and wide at the foramen magnum, but narrow and nearly square on cross-section where it joins the sphenoid. Its cerebral surface, concave from side to side, supports the pons and medulla, and along each side has a half of the groove for the *inferior petrosal sinus*, the petrous temporal having the other half. Each lateral margin is united by synchondrosis to the petrous temporal bone.

The under surface carries one-third of a condyle on each side, and in front of these are rough markings for the attachments of the Longus Capitis and Rectus Capitis Anterior. At the center is the *pharyngeal tubercle* for the attachment of the fibrous median raphe of the pharynx.

The basi-occipital and the sphenoid are united by cartilage which is usually completely ossified by the 19th year and never later than the 21st year (McKern and Stewart).

Articulations: With both parietals, both petrous temporals, the sphenoid and the atlas.

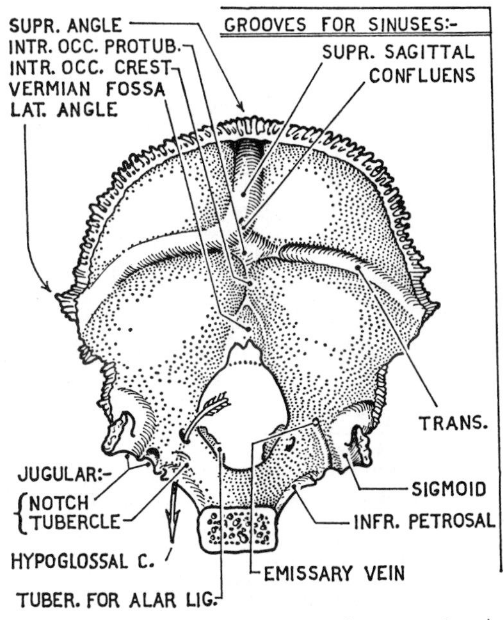

Fig. 828. Adult occipital bone (inner surface).

Variations. The occipital bone develops in cartilage, except the portion above the superior nuchal line, which develops in membrane; this superior part may fail to fuse with the rest of the bone thereby constituting an *interparietal bone.* The interparietal bone itself develops from several centers any of which may remain discrete, thereby simulating a large sutural bone. The *paramastoid process* is an occasional bar of bone that descends from the jugular process toward the transverse process of the atlas. The *3rd occipital condyle* is an occasional tubercle that projects from the anterior border of the foramen magnum to articulate with the dens of the axis. A *median cleft* may extend from the foramen magnum backward into the squama. It is due apparently to the nonappearance of an ossific center (*fig. 828*). *Fusion of the atlas* and *occipital bone* may occur.

SPHENOID BONE (DETAILS)

Viewed from the front the sphenoid resembles a bat or an owl with wings outstretched and legs dependent. It extends across the base of the skull, articulates with numerous bones, takes part in many fossae and possesses many foramina. It comprises a body, two lesser wings, two greater wings and two pterygoid processes (*fig. 829*).

At Birth it is in three pieces, the body and lesser wings forming one piece, the greater wing and pterygoid process on each side forming two others (*fig. 829.1*). It becomes one bone during the 1st year. The cubical body contains two air sinuses, the right and left sphenoidal air sinuses (*fig. 829*).

During the 1st year the anterior roots meet and fuse above the anterior part of the body, thereby forming a yoke, the *jugum sphenoidale* (*fig. 829.1*). The posterior edge of the jugum is the anterior edge of the *chiasmatic (optic) groove.*

Body and Lesser Wings—viewed from above. The attenuated lesser wing of each side has a free concave posterior border which ends medially in a blunt triangular spine, the *anterior clinoid process.* Medial to this it is attached to the anterior half of the body by two roots which bound the *optic canal* (foramen) above and below. After birth the upper root of each side extends like a sliding door across the upper surface of the anterior half of the body and,

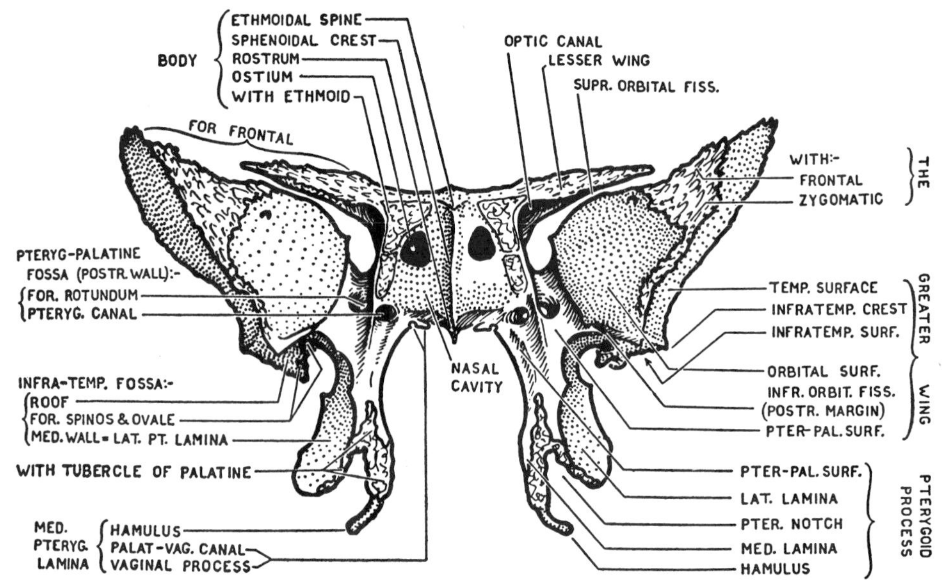

Fig. 829. Sphenoid bone–front view. (*Of palatine: tubercle = pyramidal proc.; Surface; pterygopalatine = maxillary.*)

joining, they form a yoke, the *jugum*, which conceals this part of the body. The upper surface of the lesser wings and jugum forms the hinder part of the anterior cranial fossa. This surface is narrow and flat, and is pointed at each lateral end; in front it articulates with the orbital plates of the frontal bone and between these with the ethmoid, which it also overlaps (*fig. 829*).

The posterior edge of the jugum is also the anterior edge of a groove, the *chiasmatic sulcus* (*optic groove*), that connects the optic canals of opposite side. The part of the body behind the groove is the *sella turcica* (Turkish saddle); it is subdivided into a pommel, a seat, and a back. The pommel of the saddle is a transversely set olive-like eminence, the *tuberculum sellae*, each end of which may form a spine, the *middle clinoid process*. Behind the tuberculum sellae is the excavated seat, the *hypophyseal* or *pituitary fossa*. Behind this rises a square plate of bone, the *dorsum sellae*, whose upper angles are tubercular, the *posterior clinoid processes*.

The side of the hollow body gives attachment antero-inferiorly to the greater wing; postero-inferiorly it articulates with the apex of the petrous temporal; and between these lies the foramen lacerum where the internal carotid artery enters the skull. Between the foramen and the anterior clinoid process the bone is faintly grooved by the artery. The side of the body extends forward beyond the optic canal, superior orbital fissure, and foramen rotundum.

The hinder part of the body is a square "epiphyseal" surface which fuses with the basi-occipital, usually by the 19th year.

The anterior surface of the body has a median crest, the *sphenoidal crest*, which forms part of the nasal septum. This crest begins above in a spine, the *ethmoidal spine*, and ends below in a beak, the *rostrum*, which is received between the alae of the vomer. A vertical triangular area at the side of the sphenoidal crest forms part of the roof of the nasal cavity. Near the midpoint of this area is the orifice

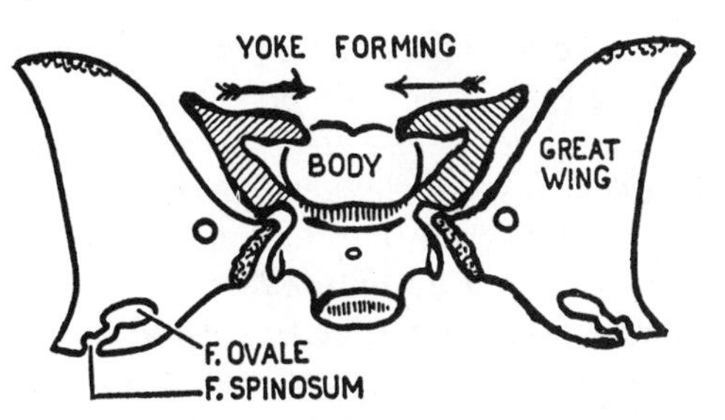

Fig. 829.1. The sphenoid bone at birth is in three parts. Note the lesser wings spreading above the body to a form a yoke (jugum).

of the *sphenoidal air sinus*. Lateral to this the body articulates with the ethmoidal labyrinth.

The mouths of five *bony passages* lie grouped at the side and under aspect of the front of the body. They are: (1) optic canal (foramen) between the roots of the lesser wing, (2) foramen rotundum at the root of the greater wing, (3) pterygoid canal at the root of the pterygoid process, (4) superior orbital fissure between the lesser and greater wings, and (5) palatinovaginal canal below the vaginal process. The anterior (antero-inferior) surface of the lesser wing forms the hinder part of the roof of the orbit.

Greater Wing. This projects from the side of the body and its inner or cerebral surface forms the anterior part of the lateral subdivision of the middle cranial fossa. It is separated from the lesser wing by a comma-shaped fissure, the *superior orbital fissure*, which opens from the middle cranial fossa to the orbit.

The *foramen rotundum* is situated below the medial end of the superior orbital fissure, and passes forward through the roof of the greater wing to the pterygopalatine fossa. This fossa lies below the level of the orbit and is seen from the side of the skull.

Behind its site of attachment to the body, the wing has a posterior border which ends posterolaterally in an angle. On the under surface of the angle there is a spine, the *spine of the sphenoid*. The posterior border grew around and engulfed the mandibular nerve thereby forming the *foramen ovale* and more laterally, at the root of the angular spine, it engulfed the middle meningeal artery thereby forming the *foramen spinosum*. These two foramina open downward into the infratemporal fossa.

The lesser (superficial) petrosal nerve may be similarly engulfed resulting in the formation of a third foramen, the *canaliculus innominatus*—a minute foramen between the foramen ovale and foramen spinosum. The superior orbital fissure and three foramina—rotundum, ovale, spinosum—lie on a crescent (*fig. 688*). The wing is grooved near its tip by the anterior branch of the middle meningeal artery.

The greater wing forms part not only of the middle cranial fossa, but also of the orbit, and of the temporal, infratemporal and pterygopalatine fossae. Between the temporal and infratemporal surfaces, which are set at a right angle to each other, is the sharp and often spinous *infratemporal crest*.

Pterygoid Process. A stout process, it descends obliquely from the junction of the body and greater wing (*fig. 830*). It consists of two plates, the *medial* and *lateral pterygoid laminae*. They are fused in front, but free behind and below. Between them is the *pterygoid fossa*. At the lower end of the posterior border of the medial pterygoid lamina is a delicate hook, the *hamulus*, at the upper end is a conical tubercle, *pterygoid tubercle*. This tubercle is the guide to the *pterygoid canal*, which lies just above and passes forward to the pterygopalatine fossa, where lies the ptergopalatine ganglion.

From the root of the medial lamina a plate, the *vaginal process*, runs medially toward the ala of the vomer. On the under surface of this process a groove or canal, the *palatinovaginal canal*, runs backward.

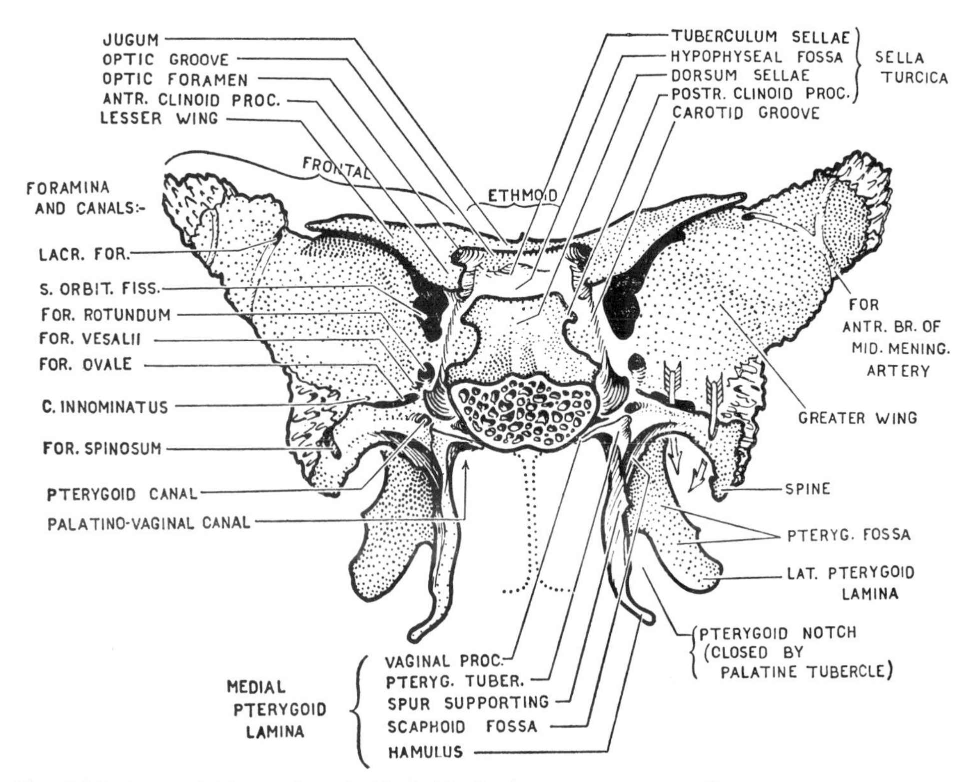

Fig. 830. Sphenoid bone–from behind. (*Optic, foramen = canal; Groove: optic = chiasmatic; Palatine: tubercle = pyramidal proc.*)

The superomedial part of the pterygoid fossa is carried backward toward the spine of the sphenoid as a sharply defined fusiform fossa, the *scaphoid fossa*.

The medial lamina forms part of the lateral wall of the nasal cavity. Its posterior border is free and sharp for the pharyngeal aponeurosis and has a spine for the support of the mouth of the auditory (pharyngotympanic) tube.

The lateral lamina forms the medial wall of the infratemporal fossa. Its posterior border is free and serrated; followed upward it leads to the foramen ovale. Its medial surface gives origin to the Medial Pterygoid, its lateral surface to the Lateral Pterygoid—so it is a muscular process.

Ossification (*fig. 829.1*). The body and lesser wings develop in cartilage; so does the root of the greater wing and its downgrowth, the lateral pterygoid lamina. The remainder of the greater wing develops in membrane. The medial pterygoid lamina also develops in membrane, and its line of fusion with the body and greater wing is usually obvious; it runs above the vaginal process and pterygoid tubercle and crosses through the pterygoid canal. Further, two paired fragments ossify independently in cartilage: one is a curved plate of bone, the *lingula*, which lies above the posterior orifice of the pterygoid canal and sweeps laterally in front of the carotid artery; the other is a triangular plate, the *sphenoidal concha*, which is applied to the anterior and inferior surfaces of the body of the sphenoid. About the 3rd year the mucous membrane of the nasal cavity bursts through the right and left sphenoidal conchae into the body of the sphenoid, thereby forming the right and left *sphenoidal air sinuses*.

Variations. The septum between the right and left sinuses is usually greatly deflected. The sinus is commonly over-inflated, so to speak, with the result that it partly surrounds the optic canal, the pterygoid canal, and the foramen rotundum so that they project as ridges within it. The walls of the ridges may be resorbed; the optic nerve, nerve of the pterygoid canal, the maxillary nerve and also the cavernous sinus and carotid artery are then brought close to the mucoperiosteum of the sinus.

ETHMOID BONE (DETAILS)

The ethmoid bone may be likened to a St. George's cross made of planks and having an oblong box suspended from each end of the cross-piece (*fig. 831*). The boxes are the *ethmoidal labyrinths;* the cross-piece is the *cribriform plate* (lamina cribrosa); the part of the upright above the cribriform plate is the *crista galli;* and the part below is the *vertical* or *perpendicular plate* (*fig. 832*).

The ethmoid is developed from the cartilaginous nasal capsule. *At birth* it is in

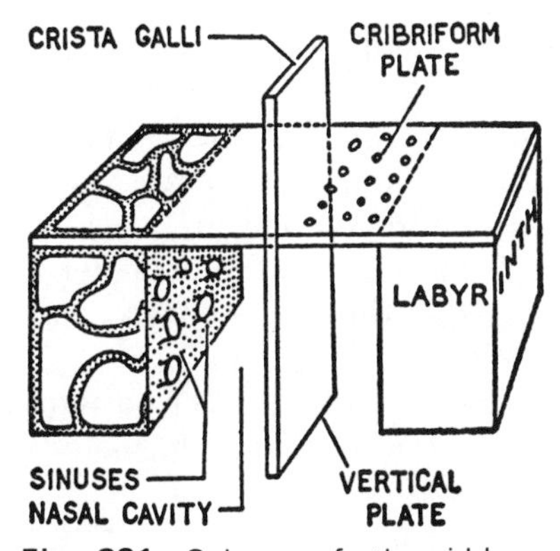

Fig. 831. Scheme of ethmoid bone.

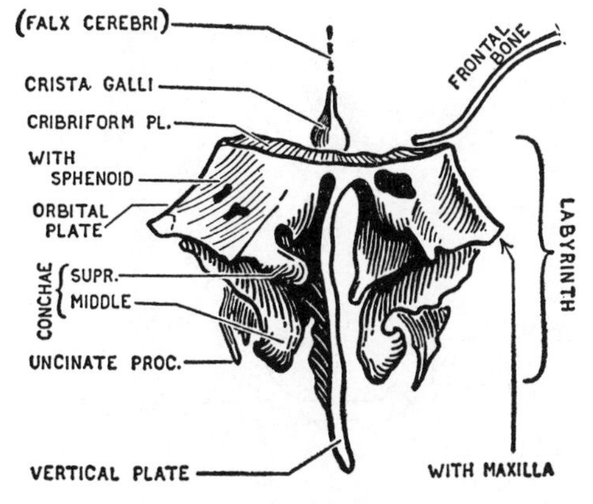

Fig. 832. Ethmoid bone—from behind.

three pieces—a median plate, and a right and a left labyrinth. The median plate, which forms part of the nasal septum, and the crista galli (cock's comb), which is its upward extension into the anterior cranial fossa, are cartilaginous; these begin to ossify during the 1st year. The labyrinths, however, are bony at birth and are joined to the median plate by a fibrous lamina cribrosa. Fusion is complete by the 5th or 6th year.

The *crista galli* is thick and triangular. The falx cerebri is attached to its posterior border and apex; the anterior border splits into two *alae* which, with the frontal bone, enclose the *foramen cecum* (*fig. 833*).

The *cribriform plate* is a fragile, sieve-like plate lying at each side of the crista galli and occupying the ethmoidal notch of the frontal bone. It forms part of the floor of the anterior cranial fossa and of the roof of the nasal cavities. Through the perforations pass the olfactory nerves in their arachnoid coverings, also the anterior ethmoidal nerve and nasal branches of the anterior and posterior ethmoidal arteries—the anterior ethmoidal nerve and artery passing through a special opening, the *nasal slit*.

The *vertical plate* forms the posterosuperior third of the nasal septum (*fig. 798*).

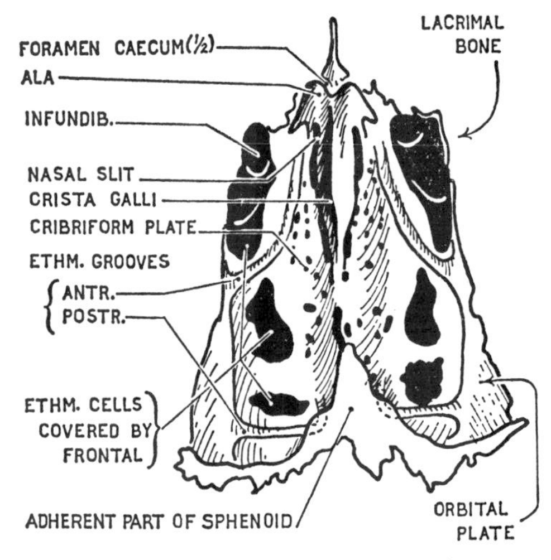

Fig. 833. Ethmoid bone–from above.

Labyrinth. Each box-like labyrinth is composed of a dozen or less air cells, the *ethmoidal cellules* or *sinuses*, which open medially into the nasal cavity; laterally, it has a smooth, oblong, fragile wall, the *orbital plate* (lamina papyracea); above, it is covered by the medial part of the orbital plate of the frontal bone and slightly by the sphenoid.

The cells in places break through the (bony) walls of the labyrinth; those that break through the roof proper adopt the orbital plate of the frontal as their new roof, and one cell (or more) constantly extends even into the frontal bone itself and becomes the *frontal air sinus*, its stalk being the *infundibulum;* others are limited by the surrounding lacrimal, maxillary, palatine and splenoid bones.

From the hinder part of the medial surface of the labyrinth a scroll, the *superior concha*, hangs downward; and the medial surface itself continues downward as the *middle concha*. A hook of bone, the *uncinate process*, curves backward from the anterior end of this surface to meet the corresponding process of the inferior concha, thereby forming the lower limit of the hiatus semilunaris. The oblong posterior surface of the labyrinth abuts against the anterior surface of the body of the sphenoid.

TEMPORAL BONE (DETAILS)

The temporal bone is a composite bone, situated at the base and side of the skull between the sphenoid in front and the occipital behind. **At birth** it is in 3 parts—the squamous, tympanic, and petrous (petromastoid)—which fuse during the 1st year (*fig. 834*). The inner ear lies within the petrous part. The tympanic cavity, which developed from the first and second pharyngeal clefts, is enclosed by the three parts of the bone; it communicates with the mastoid antrum behind, and with the nasopharynx in front via the auditory (pharyngotympanic) tube. Through the bone runs the facial nerve. The squamous and tympanic parts develop in membrane; the petrous in cartilage.

Squamous Part. The squama, resembling a pilgrim's shell, forms part of the lateral wall of the cranium (*fig. 835*). Its medial surface, described fully on page 488, is grooved by the middle meningeal artery. From the lower part of its lateral surface the finger-like *zygoma* (zygomatic process) curves forward (p. 522). On the under surface is a translucent, oval socket, the *mandibular* (articular) *fossa* for the head of the lower jaw (p. 531). An angular part, the *postauditory process*, described in the adult bone as part of the mastoid, projects downward for about 1 cm below the level of the middle cranial fossa, closing the mastoid antrum laterally.

Tympanic Part. At birth it is a *ring* open above. It is grooved for the tympanic membrane, and attached to it laterally is the cartilage of the external acoustic (auditory) meatus. During the early years of life the ring becomes oval and elongated to form the anterior wall, floor, and lower part of the posterior wall of the bony external acoustic meatus. It also extends downward into a plate, the *tympanic plate*, which forms the posterior wall of the mandibular fossa and, splitting below to form the *vaginal process*, partly ensheaths the styloid process.

The Petrous Part (*fig. 836*) is the most important and also the most difficult part of the bone to understand. Its surfaces have been dealt with (pp. 487, 488, 520–523). In brief: it is pyramidal; its base is lateral; its apex is medial lying at the foramen lacerum. It has three surfaces, an anterior and a posterior which form parts of the middle and posterior cranial fossae, respectively, and an inferior which forms part of the undersurface of the base of the skull.

The *carotid canal* begins on the under surface of the bone (*fig. 837*) and takes an inverted L-shaped course through it, opening into the foramen lacerum at the apex.

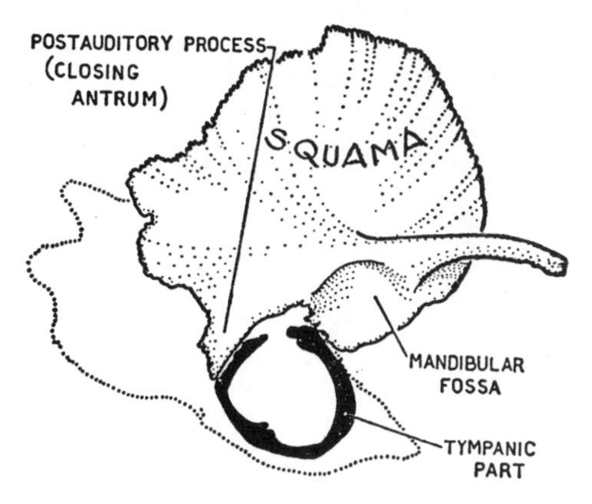

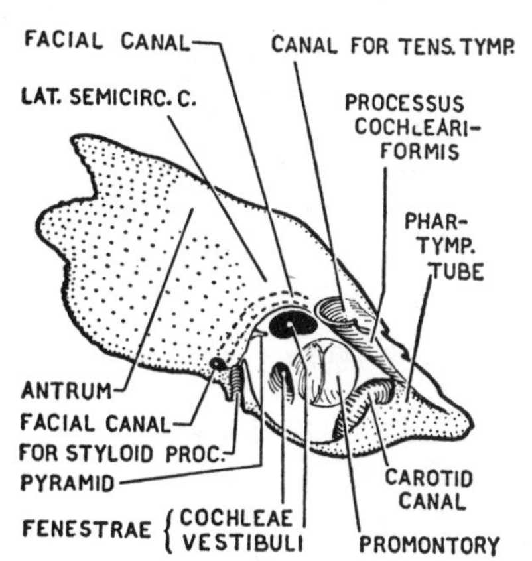

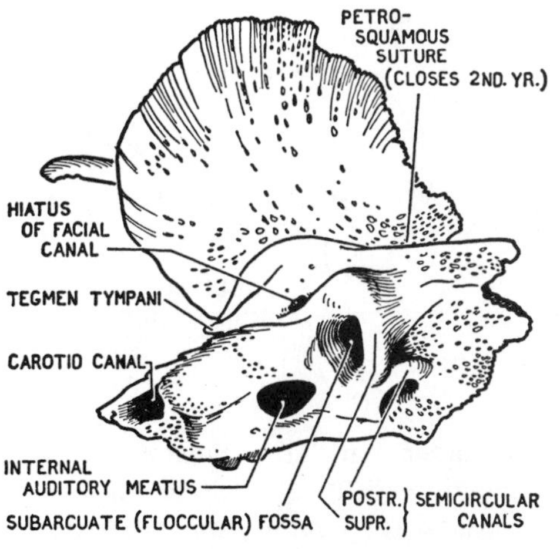

Fig. 834. The three parts of the temporal bone at birth. Upper, lateral aspect of squamous and tympanic parts. *Middle*, lateral aspect of petromastoid part. (The squamous and tympanic parts have been removed to show the medial wall of the tympanum and the mastoid antrum.) *Lower*, inner aspect.

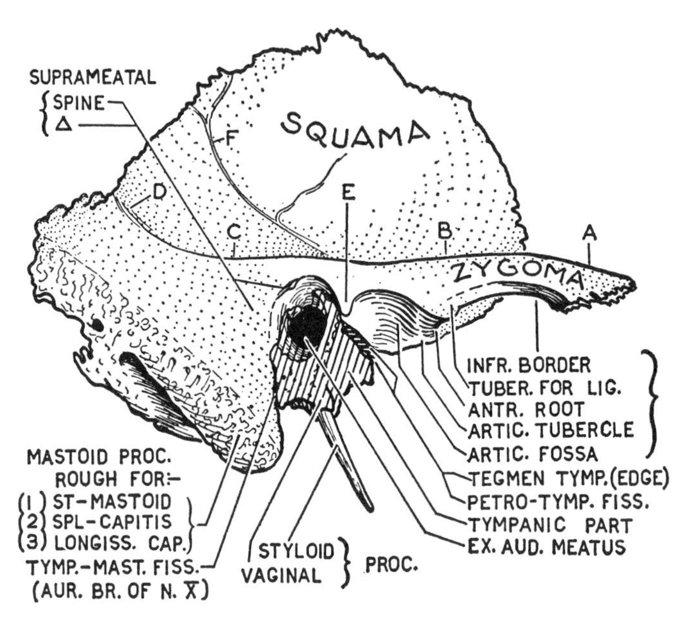

Fig. 835. Temporal Bone—lateral aspect.

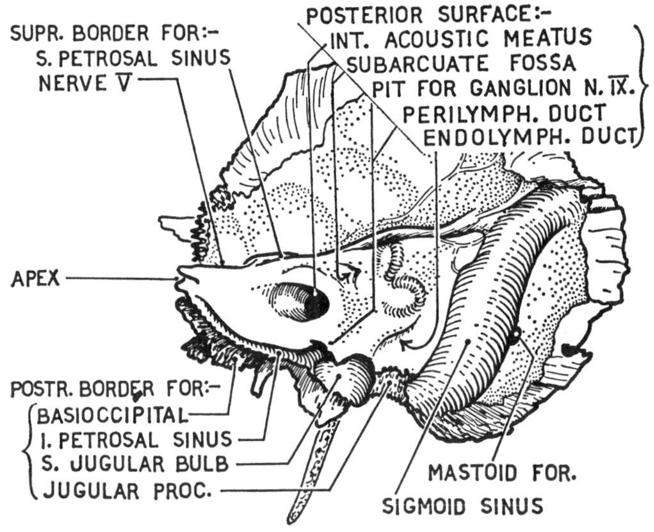

Fig. 836. Petromastoid part—cerebellar surface.

The hinder part, the *mastoid bone*, is grooved internally, at its junction with the petrous, for the *sigmoid sinus;* externally it is prolonged downward into a nipple, the *mastoid process*, but this is not present during the 1st year of life, so the *stylomastoid foramen*, situated where its name suggests, opens subcutaneously and there discharges the facial nerve.

For features of the anterior or cerebral surface see *figure 838* and also *figure 687*, for features of the posterior or cerebellar surface see *figure 686*, and for features of the inferior surface see *figure 753*.

MAXILLA (DETAILS)

The maxilla or upper jaw (paired) has a body and four processes. The body is a hollow pyramid with three surfaces, an apex, and a base.

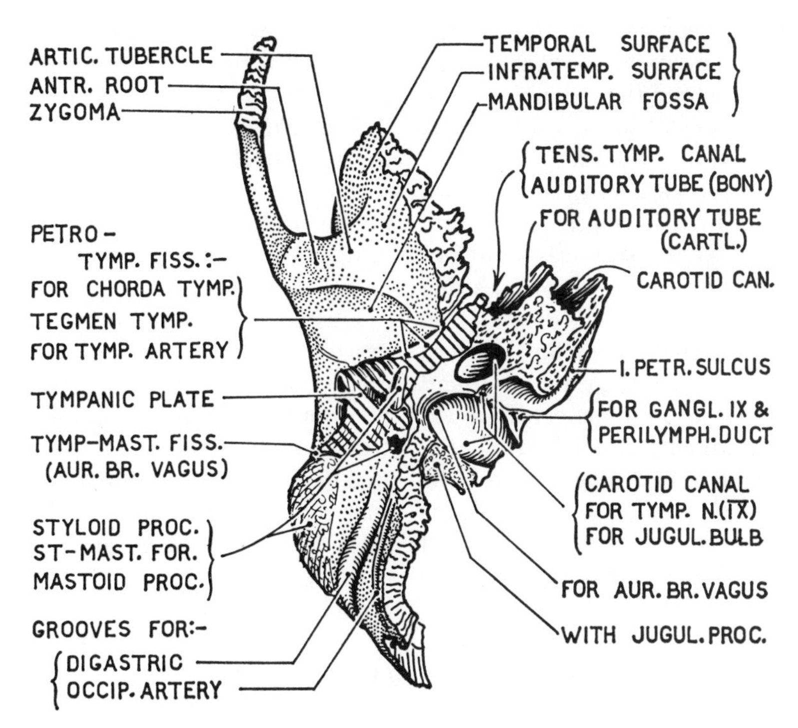

Fig. 837. Temporal bone–inferior aspect.

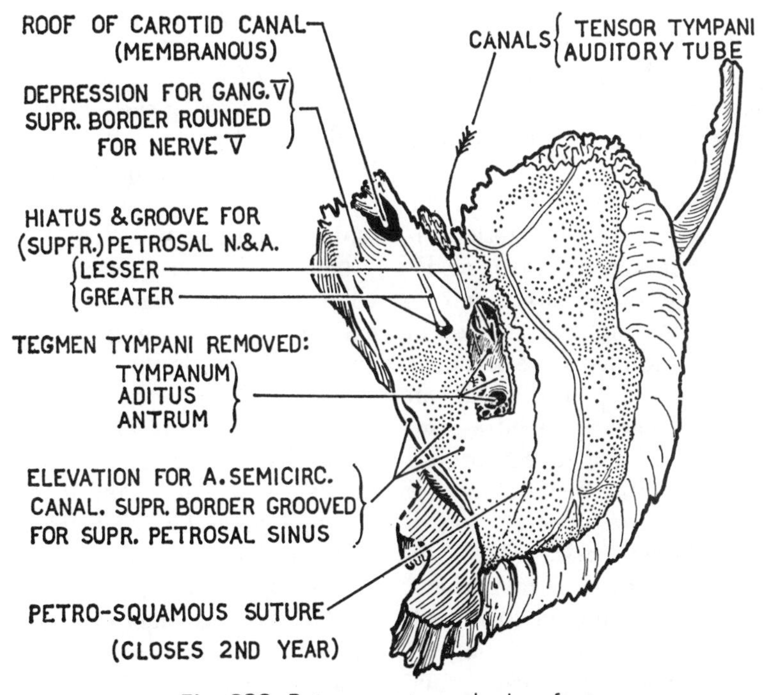

Fig. 838. Petrous part–cerebral surface.

The most conspicuous feature of the bone is the *alveolar process*. It carries eight teeth, and ends behind the 3rd molar tooth in a free rounded part, the *maxillary tuberosity (fig. 839)*. As in the lower jaw so in the upper, the roots of the teeth (2nd and 3rd molars excepted) cause ridges on the thin outer wall of the alveolar process, but not on the thick inner wall, that for the canine tooth being the largest.

From the 1st or 2nd molar tooth a rounded buttress ascends to the lower end of a large triangular prominence, the *zygomatic process*, which forms the truncated apex of the bone and is placed where the three surfaces (orbital, facial, and infratemporal) meet. It ends in a rough triangular area for articulation with the zygomatic bone. The buttress and the process separate the facial (anterior) surface from the infratemporal (posterior) surface.

Facial (Anterior) Surface. The facial surface is flat on a powerful skull, like that of the Eskimo, but concave in skulls of white races. The area medial to the ridge for the canine tooth, i.e., between the incisor teeth and the anterior nasal orifice, is the *incisive fossa*; the area lateral to this ridge is the *canine fossa*.

Opening on to the canine fossa is the *infra-orbital foramen*, i.e., the anterior orifice of the infra-orbital canal. It is placed 1 cm below the infra-orbital margin and its direction contrasts with that of the mental foramen in the mandible, since it opens inferomedially. The *infra-orbital canal* can be followed backward below the infra-orbital margin to the middle of the orbital surface where, ceasing to have a roof, it becomes the *infra-orbital groove*; this in turn may be followed backward to the upper border of the infratemporal surface.

The facial surface is limited superiorly by the infra-orbital margin and is continuous superomedially with the lateral surface of the frontal process.

Frontal Process. This is triangular. Its apex articulates with the nasal notch of the frontal bone; its anterior border supports the nasal bone; its posterior border articulates with the lacrimal bone.

The infra-orbital margin is continued, as the *anterior lacrimal crest*, on to the lateral surface of the frontal process, dividing it into a convex area, which is part of the bridge of the nose, and a concave area, the *lacrimal groove*.

The medial or nasal surface of the frontal process is partly crossed by an oblique crest, the *ethmoidal crest*, for the attachment of the middle concha. About two centimeters below this, on the body of the maxilla, there is a second oblique crest, the *conchal crest*, for the attachment of the inferior concha. The area between the crests is part of the *atrium* of the middle meatus of the nose. Above the upper crest a small area forms the anterior wall of an ethmoidal cell.

Infratemporal (Posterior) Surface. This smooth, convex surface is perforated near its center by one or more *alveolar* (or *posterior dental*) *foramina*. The part of this surface just above the tuberosity is buttressed by the pterygoid laminae (the pyramidal process of the palatine bone intervening as a buffer), and the part above this is the anterior wall of the *pterygopalatine fossa*, wherein resides the pterygopalatine (sphenopalatine) ganglion.

Orbital (Superior) Surface. This is smooth, triangular, and slightly concave. Its apex extends on to the zygomatic process; its anterior border is the infraorbital margin; its posterior border is the anterior margin of the inferior orbital fissure; its medial border is formed by the margin of the nasolacrimal notch and behind this by articular areas for the lacrimal, ethmoid, and palatine bones. The surface is crossed posteriorly by the *infra-orbital groove*. Just lateral to the nasolacrimal notch there is a *depression* for the origin of the Obliquus Oculi Inferior.

Nasal (Medial) Surface (*fig. 840*). This is the base of the hollow pyramidal body. Its anterior two-thirds is separated from the

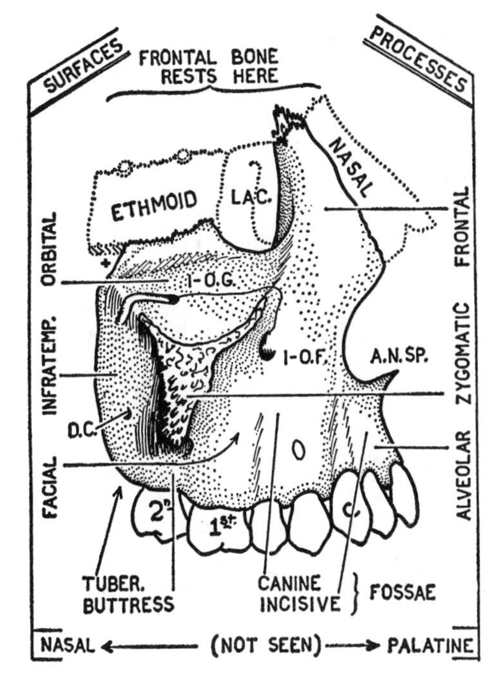

Fig. 839. Maxilla–lateral aspect.

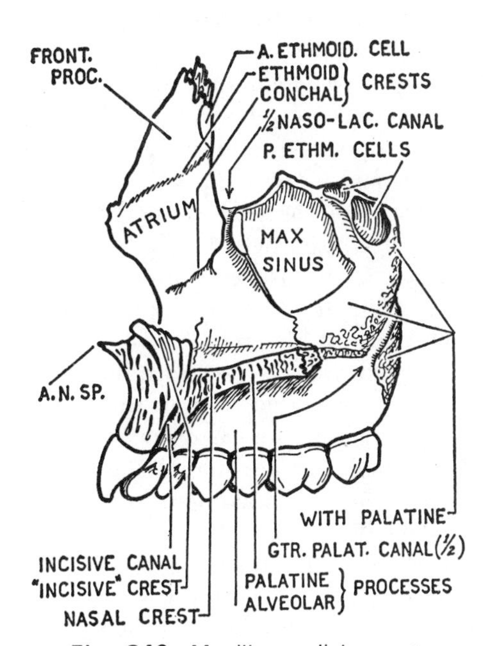

Fig. 840. Maxilla–medial aspect.

D.C. = dental canals; I-O.G. and I-O.F. = infra-orbital groove and canal; A.N.Sp. = Anterior nasal spine; + = space for orbital process of palatine bone.

alveolar process by a horizontal plate, the *palatine process*. The nasal surface presents, in the disarticulated bone, an opening large enough to admit the thumb. This, the *maxillary hiatus*, is the bony *orifice of the maxillary air sinus* or antrum. Between this and the frontal process is a groove (lacrimal groove) which forms one-half of the circumference of the *nasolacrimal canal*.

The part of the nasal surface behind the maxillary orifice is overlaid by the palatine bone; and when this is in position a perpendicular canal, the *greater palatine canal*, is formed. The maxillary half of this canal is continued forward as two grooves on the under surface of the palatine process of the maxilla.

The medial border of the palatine process is slightly raised and with its fellow forms the *nasal crest*, which articulates with the vomer and forms part of the nasal septum. Its most anterior part is markedly raised to form the "*incisive crest*" which ends anteriorly on the face as the *anterior nasal spine*.

The part of the bone carrying the incisor teeth is the *incisive bone* (or *premaxilla*), which in most mammals is an independent, paired bone. It extends backwards to the junction of the nasal and "incisive" crests where a canal, the *incisive canal*, passes from its nasal to its oral surface.

Ossification. The maxilla proper ossifies in membrane from a single center; the premaxilla ossifies from two centers or perhaps more. At birth the infra-orbital nerve lies free on the floor of the orbit just as the supraorbital nerve lies free on the roof. As the maxilla enlarges the nerve sinks into a groove whose lateral edge then folds over the nerve, thus forming the infra-orbital canal and foramen.

ZYGOMATIC BONE

The zygomatic bone is described on page 523. It is a "key bone" of the face.

Ossification is in membrane.

PALATINE BONE (DETAILS)

The palatine bone (paired) gives many students undue concern. It is a fragile L-shaped bone comprising an oblong vertical plate, a square horizontal plate, and three processes—the pyramidal, orbital, and sphenoidal (*figs. 841, 842*).

Plates. *The vertical or perpendicular plate* forms the portion of the lateral wall of the nasal cavity just in front of the medial pterygoid lamina. It is applied to the hinder part of the nasal surface of the maxilla, but projects backward behind this so as to form the medial wall of the pterygopalatine fossa (seen from the side of the skull) and projects forward closing the hinder part of the opening into the maxillary sinus.

The horizontal plate articulates with its fellow to form the posterior third of the bony palate, the site of union being raised to form a *nasal crest*, for articulation with the vomer and ending behind in the *posterior nasal spine*. The anterior border of this plate articulates with the palatine process of the maxilla; the posterior border is sharp and concave.

Processes (*fig. 843*). An inverted pyramid, the *pyramidal process* (*tubercle*) projects from behind the lower part of the vertical plate and interposes itself like a buffer between the posterior border of the maxilla (just above the tuberosity) and the medial and lateral pterygoid laminae of the sphenoid. The process has a gutter for each lamina, and between the gutters a triangular area forms the lowest part of the pterygoid fossa.

Surmounting the upper border of the vertical plate are the orbital and sphenoidal processes, separated from each other by the U-shaped sphenopalatine notch—much as the head and coronoid process of the

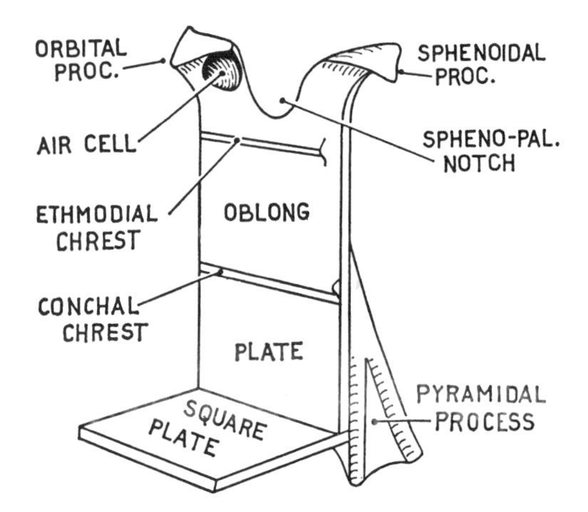

Fig. 842. Scheme of palatine bone.

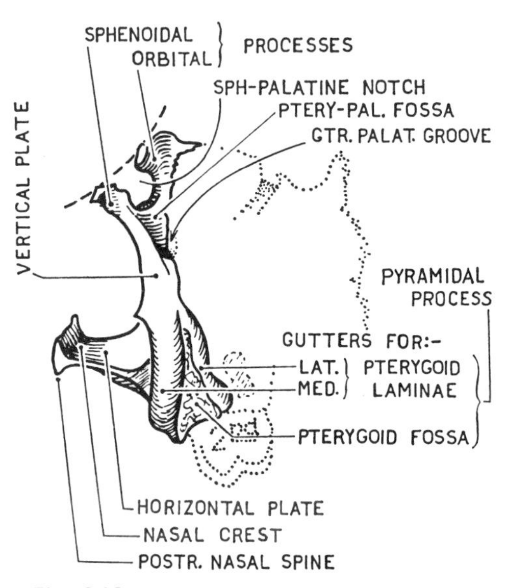

Fig. 843. Palatine bone—posterior aspect.

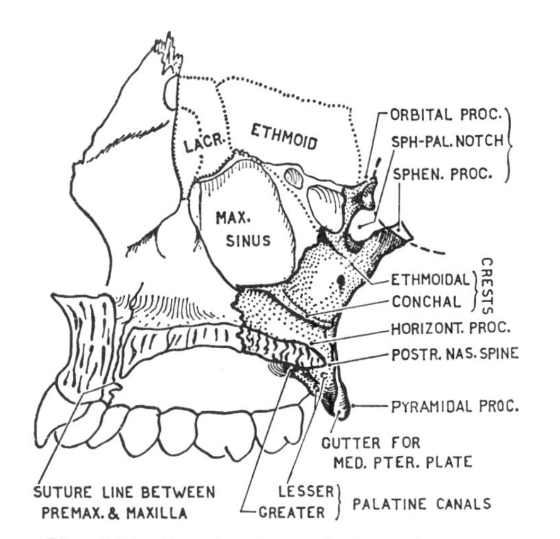

Fig. 841. Palatine bone (*stippled*) in articulation with maxilla—medial aspect.

mandible are separated by the mandibular notch.

The orbital process forms the hindermost 3 to 4 mm of the floor of the orbit. It is a hollow box, hollowed by an extension from the sphenoidal sinus or the maxillary sinus or a posterior ethmoidal sinus.

The sphenoidal process is a small plate applied to the upper surface of the body of the sphenoid and reaching to the ala of the vomer. It converts the groove below the vaginal process of the sphenoid into the *palatinovaginal canal* (pharyngeal canal). *The sphenopalatine notch* is converted by the sphenoid into the sphenopalatine foramen; it is the gateway to the nasal cavity.

Greater Palatine Canal. This canal drops between the body of the maxilla and the vertical plate of the palatine bone and opens between the alveolar process of the maxilla and the horizontal plate of the palatine bone as the greater palatine foramen, while two *lesser palatine canals* descend from the greater palatine canal through the pyramidal process and open on to its under surface (*fig. 841*).

Ossification is from a single center in the membrane in the lateral wall of the nasal cavity. The vertical plate is the primitive plate, the horizontal is secondary.

NASAL BONE (DETAILS)

The nasal bone (paired) is small and stout; it is triangular with truncated apex above. It has two surfaces (inner and outer), two borders, an apex, and a base.

The *apex* is blunt, thick, and serrated; it articulates with the nasal notch of the frontal bone (*fig. 844*). The *base* is broad, thin, and notched; attached to it is the lateral nasal cartilage. The *lateral* (posterolateral) *border* is thin; it articulates with the frontal process of the maxilla. The *medial border* is flat, triangular area which articulates with its fellow.

The outer or *facial surface* of the paired bones is saddle-shaped, being convex from side to side and concave from above downward. Near the center is a foramen for an emissary vein from the nasal mucosa.

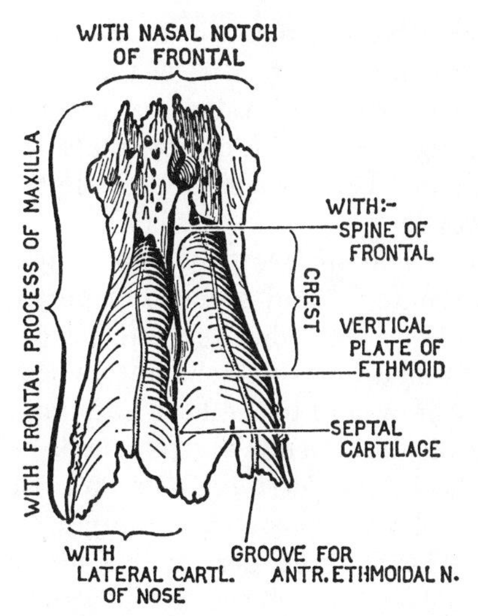

Fig. 844. Nasal bones articulated—posterior aspect.

The inner or *nasal surface* of the paired bones presents a median *crest* which articulates with the nasal spine of the frontal bone above, with the vertical plate of the ethmoid below, and perhaps with the septal cartilage still lower—depending upon how far the vertical (perpendicular) plate of the ethmoid has replaced the septal cartilage (*see fig. 798*). Lateral to the crest, the nasal surface of each bone is concave from side to side, grooved longitudinally for the anterior ethmoidal nerve, and covered with mucous membrane.

The paired bones form the upper part of the bridge of the nose. The brunt of a blow on the nose is transmitted from the nasal bones to the frontal processes of the maxillae, the nasal notch and spine of the frontal, the vertical plate of the ethmoid, and septal cartilage. The nasal bone develops in the membrane covering the cartilage of the nasal capsule.

LACRIMAL BONE (DETAILS)

The lacrimal bone (paired) resembles a fingernail but is much thinner. It has two surfaces (lateral and medial) and four borders which articulate thus—in front with the frontal process of the maxilla, behind with the orbital plate of the ethmoid, above with the orbital plate of the frontal, and below with the orbital plate of the maxilla.

The lateral surface (*fig. 845*) is divided into an anterior and a posterior part by a razor-like crest, the *posterior lacrimal crest*, which gives attachment to the orbital septum and to the Pars Lacrimalis of the Orbicularis Oculi. The crest ends below in a hook, the *lacrimal hamulus*, which may or may not reach the margin of the orbit. The posterior part of this surface is flat and forms part of the medial wall of the orbit. The anterior part, the *lacrimal groove*, together with the grooved surface on the frontal process of the maxilla, forms a half-tube in which lodges the lacrimal sac. This part is prolonged downward into a *descending process* which articulates with the lacrimal process of the inferior concha and with it forms the medial wall of the nasolacrimal canal.

The medial or nasal surface of the lacrimal bone is covered with mucous membrane. A needle perforating it from the lateral surface will enter the atrium of the middle meatus of the nose, unless the perforation is made posteriorly when it will enter an ethmoidal cell, or made above when it will enter either the infundibulum of the frontal sinus or an intervening ethmoidal cell.

The lacrimal bone ossifies in membrane.

INFERIOR CONCHA (DETAILS)

The inferior concha (turbinate bone) hangs downward like a scroll from the side wall of the nasal cavity. In the articulated skull it can be seen from the anterior and posterior apertures of the nose because it extends from a crest on the frontal process of the maxilla to a crest on the vertical (perpendicular) plate of the palatine bone. The lower border is thickened and gently curved, and the ends are pointed.

There are three fragile processes—one is large and downturned; two are small and upright (*fig. 846*). The *maxillary process* curves downward and laterally and forms part of the medial wall of the maxillary sinus; it is large, thin and triangular.

The *lacrimal process* ascends from near the anterior end of the maxillary process and, by joining the descending process of the lacrimal bone, completes the nasolacrimal canal medially.

The *ethmoidal process* ascends from near the posterior end of the maxillary process and, by joining the uncinate process of the ethmoid, completes the lower border of the hiatus semilunaris of the middle meatus of the nose.

The inferior concha ossifies in the cartilage of the nasal capsule.

VOMER (DETAILS)

The vomer or plowshare (unpaired) forms the entire postero-inferior third of

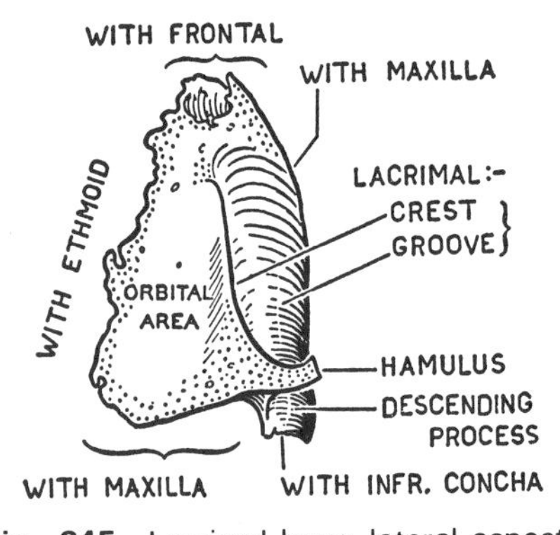

Fig. 845. Lacrimal bone–lateral aspect.

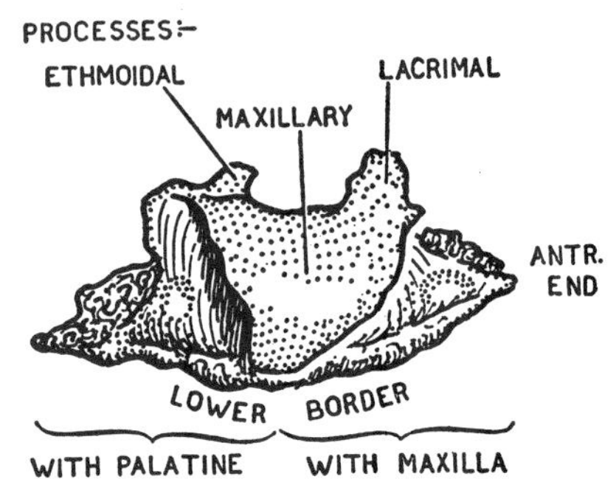

Fig. 846. Inferior concha–lateral aspect.

the nasal septum. When the skull is viewed from behind, the free posterior border is seen dividing above into right and left alae. Viewed from the front the oblique anterior border is seen to be grooved in its lower part to receivė the septal cartilage, and to be thin in its upper part for articulation with the vertical plate of the ethmoid. Each surface is grooved longitudinally by the nasopalatine (long sphenopalatine) nerve and companion vessels (*fig. 847*).

The vomer develops in the postero-inferior part of the membrane that lies on each side of the "primitive" septal cartilage. The intervening cartilage is absorbed thus allowing the membrane bones of the opposite sides to fuse; evidence of the bilateral origin of the vomer is still seen, however, in the groove for the septal cartilage and in the alae.

MANDIBLE (DETAILS)

The mandible is described on page 524.

Ossification. The lower jaw is the second bone in the body to start ossifying (6th fetal week), the clavicle being the 1st. Each half of the jaw ossifies from a single center which appears in the membrane overlying the anterior half of Meckel's cartilage, i.e., the cartilage of the first or mandibular arch. In front of the mental foramen, however, ossification involves a small part of Meckel's cartilage, and posteriorly the condyle and part of the coronoid process passed through a cartilaginous stage.

At birth each half of the jaw is a fragile trough in which the five milk teeth and 6th year molar (i.e., first permanent molar) lie buried (*fig. 848*). The mandibular canal —in part open above—runs along the bottom of the trough. The ramus meets the body at a very obtuse angle, the two being almost in line. The eruption of the teeth separates the upper and lower jaws; hence the angle decreases; conversely, it increases again if the jaws become edentulous. Growth takes place mainly through additions to the outer surface of the bone and to the posterior and inferior borders.

HYOID BONE (DETAILS)

The hyoid bone is shaped like the letter U, hence its name (Gk. (H)U-eidos = U-like). It comprises a quadrate middle part, the *body*, and two processes on each side, the *greater* and *lesser horns* (cornua) (*fig. 849*). Muscles ascend to the hyoid and muscles descend to the hyoid, but no muscle crosses it, so the entire length of the bone (from the tip of one greater horn to the tip of the other) is subcutaneous externally and submucous internally (*fig. 810*).

It is readily palpated at the angle where the upper part of the neck meets the floor of the mouth (pp. 505, 514). Theoretically, the simplest way to open the pharynx is

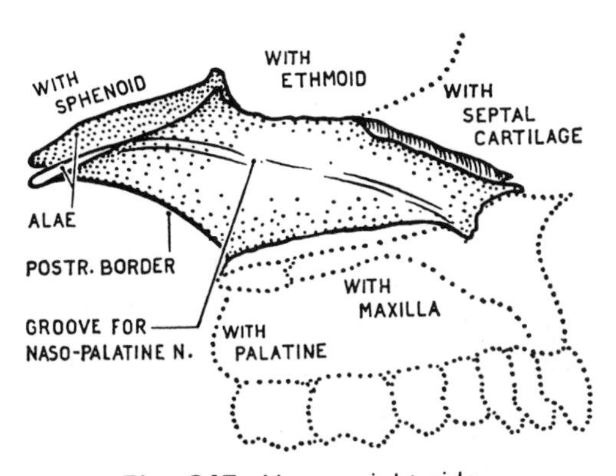

Fig. 847. Vomer–right side.

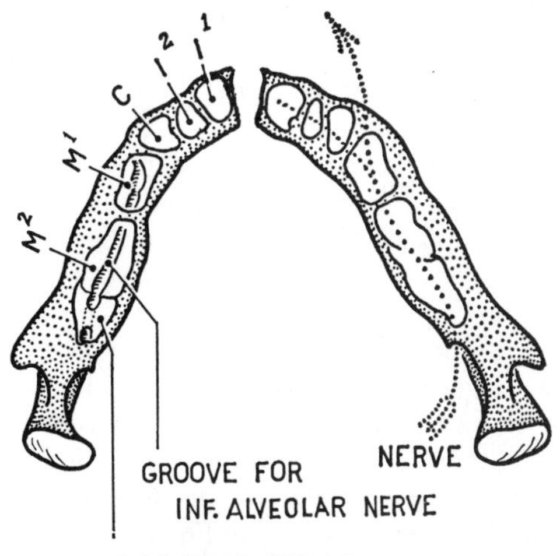

Fig. 848. Mandible at birth.

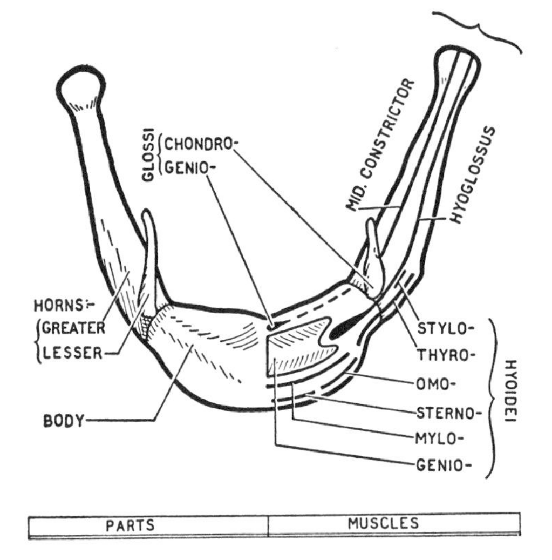

Fig. 849. Hyoid bone–anterosuperior aspect.

(1st) to cut transversely through the skin, (2nd) to saw through the hyoid, and (3rd) to cut transversely through the mucous membrane.

Attached to the entire length of the body and greater horns is the thyrohyoid membrane. Oddly, the attachment is along the upper border of the body—not the lower—and the posterior aspect is free, smooth and in contact with a bursa.

Development and Ossification. The hyoid bone is one of the structures developed from the six paired cartilages of the branchial (pharyngeal) arches (*fig. 744.1*).

Each 1st cartilage (Meckel's) is converted into the incus, malleus, sphenomandibular ligament, and one-half of the body of the mandible.

Each 2nd cartilage (Reichert's) is converted into the stapes, styloid process of the temporal bone, stylohyoid ligament, lesser horn of the hyoid, and upper part of the body of the hyoid.

The dorsal halves of the 3rd, 4th, and 6th cartilages are resorbed. The ventral halves of the 3rd become the greater horns and lower part of the body of the hyoid; the ventral halves of the 4th become the thyroid cartilage; the ventral halves of the 6th become the arytenoid cartilages and (?) the cricoid cartilage.

The body and the greater horns begin to ossify independently about the time of birth, and remain united by synchondroses until middle life, when synostosis occurs. The lesser horns articulate by synovial joints with the junction of the body and greater horns, and are partly cartilaginous at middle life; synostosis may occur.

Variations. The upper part of the stylohyoid ligament may ossify, that is, the styloid process may be unduly long and be a lateral relation of the tonsil bed. The lower part may ossify, that is, the lesser horn may be unduly long (as normally in many mammals).

Muscles Attachments (*fig. 849*).

Fibrous Attachments are: the thyrohyoid membrane, lateral thyrohyoid ligaments, hyo-epiglottic ligament, the deep fascia of the neck and the septum of the tongue.

THE SKULL AT BIRTH

The teeth of the newborn child are rudimentary and unerupted; the child can suck but cannot chew. Accordingly, the facial or masticatory portion of the skull is very small, being about one-seventh the size of the cranium or brain case (in the adult it is one-half the size); the ramus of the mandible is almost in line with the body (*fig. 850*); the mandibular (articular) fossa is very shallow (*fig. 834*); the air sinuses, which enlarge as the teeth erupt, are rudimentary; and the nasal cavities are small. The orbits are nearly circular.

There being no mastoid process, the

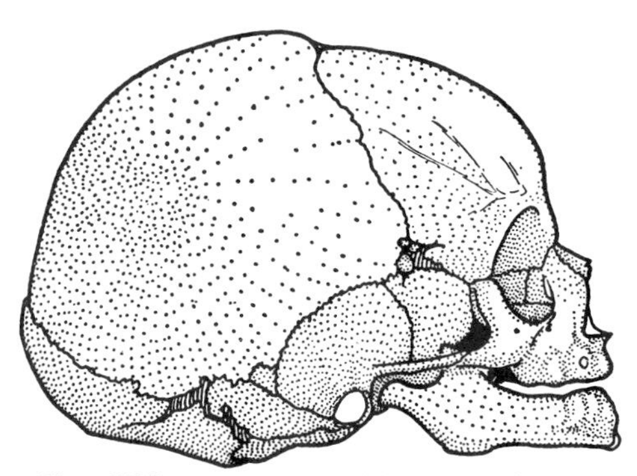

Fig. 850. The skull at birth–lateral aspect.

stylomastoid foramen, through which the facial nerve emerges, is exposed. The tympanic bone being but a ring, the eardrum is exposed. At the sites where ossification began in the frontal and parietal bones, there are eminences, the *tubers*, which are almost conical.

Ossification not having reached the four angles of the parietal bone, the skull is still membranous at these areas; that is, there is a fontanelle or *fonticulus* at each angle (*fig. 660*). There is also a median fontanelle (sagittal fontanelle) in the interparietal suture, and another (metopic fontanelle) in the frontal (or metopic) suture.

The bones of the skull are thin, easily bent, and, having no diploe, consist of a single plate. The occipital bone is in four pieces; (*fig. 827*); the sphenoid is in three (*fig. 829.1*); the temporal is in three (*fig. 834A*); the ethmoid is in three; and the frontal bone and the mandible are each in two halves.

REFERENCES

GENERAL CONSIDERATIONS

Barclay, A. E., Barcroft, J., Barron, D. H., Franklin, K. J., and Prichard, M. M. L.: Studies of the foetal circulation and of certain changes that take place after birth. Am. J. Anat., *69:*383, 1941.

Barnett, C. H., Davies, D. V., and MacConaill, M. A.: *Synovial Joints.* Longmans, Green and Company, London, 1960.

Basmajian, J. V.: *Primary Anatomy.* 6th edition. The Williams & Wilkins Company, Baltimore, 1970.

Basmajian, J. V.: *Muscles Alive: Their Functions Revealed by Electromyography.* 3rd edition. The Williams & Wilkins Company, Baltimore, 1974.

Book, M. H.: The secreting area of the glomerulus. J. Anat., *71:*91, 1936.

Brash, J. C.: Some problems in the growth and developmental mechanics of bone. Edinburgh M. J., *41:*305, 363, 1934.

Brash, J. C.: *Neuro-Vascular Hila of Limb Muscles.* E. & S. Livingstone Ltd., Edinburgh, 1955.

Bridgman, C.: Changes in intramuscular pressure during contraction (abstract). Anat. Rec., *148:*263, 1964.

Brookes, M., Elkin, A. C., Harrison, R. G., and Heald, C. B.: A new concept of capillary circulation in bone cortex: some clinical applications. Lancet, *1:*1078, 1961.

Charnley, J.: Articular cartilage. Brit. M. J., *2:*679, 1954.

Charnley, J.: How our joints are lubricated. Triangle, Sandoz J. M. Sc., *4:* no. 5, 1960.

Charnley, J.: Arthroplasty of the hip: a new operation. Lancet, *1:*1129, 1961.

Clark, E. R., and Clark, E. L.: Further observations on living lymphatic vessels in the transparent chamber in the rabbit's ear. Am. J. Anat., *52:*273, 1933.

Clark, W. E. Le Gros: *The Tissues of the Body,* Ed. 2. Clarendon Press, Oxford, 1945.

Coventry, M. B. *et al.:* The intervertebral disc, etc. J. Bone & Joint Surg., *27:*105, 1945.

Cronkite, A. E.: The tensile strength of human tendons. Anat. Rec., *64:*173, 1936.

Cox, H. T.: The cleavage lines of the skin. Brit. J. Surg., *29:*234, 1941.

Davies, D. V.: Observations on the volume, viscosity and nitrogen content of synovial fluid, etc. J. Anat., *78:*68, 1944.

Dawson, B. H., and Hoyte, D. A. N.: Observations on premature fusion of the sutures of the cranial vault (abstracts). J. Anat., *91:*590, 583, and 613, 1957.

Digby, K. H.: The measurement of diaphysial growth, etc. J. Anat., *50:*187, 1916.

Duchenne, G. B. A.: *Physiologie des mouvements, etc.,* (translated by Kaplan, see below.) Paris, 1867.

Edwards, E. A.: The orientation of venous valves in relation to body surfaces. Anat. Rec., *64:*369, 1936.

Ekholm, R.: Nutrition of articular cartilage. Acta anat., *24:*329, 1955.

Gardner, E.: Physiology of movable joints. Physiol. Rev., *30:*127, 1950.

Gardner, E.: The anatomy of the joints. Am. Acad. Orthop. Surgeons, Instruction course. *9:*149, 1952. (Edwards, Ann Arbor, Mich.)

Girgis, F. G., and Pritchard, J. J.: Effects of skull damage on the development of sutural patterns in the rat. J. Anat., *92:*39, 1958.

Haines, R. W.: On muscles of full and of short action. J. Anat., *69:*20, 1934.

Haines, R. W.: The laws of muscle and tendon growth. J. Anat., *66:*578, 1932.

Ham, A. W.: *Histology,* Ed. 5. J. B. Lippincott Company, Philadelphia, 1965.

Harris, H. A.: *Bone Growth in Health and Disease.* Oxford University Press, London, 1933.

Hughes, H.: The factors determining the direction of the canal for the nutrient artery in the long bones of mammals and birds. Acta anat., *15:*261, 1952.

Inman, V. T., Saunders, J. B. deC. M., and Abbott, L. C.: Observations on the function of the shoulder joint. J. Bone & Joint Surg., *26:*1, 1944.

Inman, V. T., and Saunders, J. B. deC. M.: Anatomicophysiological aspects of injuries to the intervertebral disc. J. Bone & Joint Surg., *29:*461, 1947.

Kaplan, E. B.: *Physiology of Motion.* J. B. Lippincott Company, Philadelphia, 1949. (This is a translation of *Physiologie des mouvements* by Duchenne, see above.)

Keegan, J. J., and Garrett, F. D.: The segmental distribution of the cutaneous nerves in the limbs of man. Anat. Rec., *102:*409, 1948.

Keith, A.: *Menders of the Maimed.* Froude, London, 1919.

Kimmel, D. L.: Innervation of spinal dura mater and dura mater of the posterior cranial fossa. Neurology, *11:*800, 1961.

Kimmel, D. L.: The nerves of the cranial dura mater and their significance in dural headache and referred pain. Chicago M. Sch. Quart., *22:*16, 1961.

Langworthy, O. R. *et al.: Physiology of Micturition.* The Williams & Wilkins Company, Baltimore, 1940.

Learmonth, J. E.: A contribution to the neurophysiology of the urinary bladder in man. Brain, *54:*147, 1931.

Le Double, A. F.: *Traité des variations du système*

musculaire de l'homme. Paris, 1897.
MacConaill, M. A.: The movements of bones and joints. The synovial fluid and its assistants. J. Bone & Joint Surg., *32B:*244, 1950.
MacConaill, M. A., and Basmajian, J. V.: *Muscles and Movements: A Basis for Human Kinesiology*. The Williams & Wilkins Company, Baltimore, 1969.
MacDonald, I. B. *et al.:* Anterior rhizotomy. The accurate identification of motor roots at the lower end of the spinal cord. J. Neurosurg., *3:*421, 1946.
McKern, T. W., and Stewart, T. D.: Skeletal age changes in young American males, analysed from the standpoint of age identification. Smithsonian Institution, 1957.
Mednick, L. W., and Washburn, S. L.: The role of the sutures in the growth of the braincase of the infant pig. Am. J. Phys. Anthropol., *14:*175, 1956.
Mitchell, G. A. G.: *Anatomy of the Autonomic Nervous System*. E. & S. Livingstone, Ltd., Edinburgh, 1953.
Mortensen, O. A., and Guest, R. L.: The absorption of thorium dioxide by the reticuloendothelial system in the dog. Anat. Rec., *70:*58, 1938.
Patten, B. M.: *Human Embryology*, Ed. 2. McGraw-Hill Book Company, Inc., New York, Toronto, London, 1953.
Petter, C. K.: Methods of measuring the pressure of the intervertebral disc. J. Bone & Joint Surg., *15:*365, 1933.
Phemister, D. B.: Bone growth and repair. Ann. Surg., *102:*261, 1935.
Pressman, J. J., and Simon, M. B.: Experimental evidence of direct communications between lymph nodes and veins. Surg. Gynec. & Obst., *113:*537, 1961.
Pressman, J. J., Simon, M. B., Hand, K., and Miller, J.: Passage of fluids, cells, and bacteria via direct communications between lymph nodes and veins. Surg. Gynec. & Obst., *115:*207, 1962.
Radin, E. L., and Paul, I. L.: A consolidated concept of joint lubrication. J. Bone & Joint Surg., *54A:* 607, 1972.
Rappaport, A. M. *et al.:* Subdivision of hexagonal liver lobules into a structural and functional unit. Anat. Rec., *119:*11, 1954.
Rappaport, A. M.: *The Liver*, Vol. 1. Academic Press, Inc., New York, 1963.
Rau, R. K.: Skull showing absence of coronal suture. J. Anat., *69:*109, 1934.
Roofe, P. G.: Innervation of annulus fibrosus, etc. J. Neurol. Neurosurg. & Psychiat., *44:*100, 1940.
Rusznyák, I., Földi, M., and Szabo, G.: *Lymphatics and Lymph Circulation*. Pergamon Press, London, 1960.
Smith, C. G.: Changes in length and position of the segments of the spinal cord, etc. Radiology, *66:*259, 1956.
Smorto, M. P., and Basmajian, J. V.: *Clinical Electroneurography*. The Williams & Wilkins Company, Baltimore, 1972.
Sunderland, S.: Blood supply of the nerves of the upper limb in man. Arch. Neurol. & Psychiat., *53:*91, 1945.
Todd, T. W., and Pyle, S. I.: A quantitative study of the vertebral column, etc. Am. J. Phys. Anthropol., *12:*321, 1928.
Trueta, J., and Cavadias, A. X.: A study of the blood supply of the long bones. Surg. Gynec. & Obst., *118:*485, 1964.
Walls, E. W.: The fibre constitution of the human gastrocnemius and soleus muscles. J. Anat., *87:*437, 1953.
Walmsley, T.: The articular mechanism of the diarthroses. J. Bone & Joint Surg., *10:*40, 1928.
Weinmann, J. P., and Sicher, H.: *Bone and Bones*, Ed. 2 C. V. Mosby Company. St. Louis, 1955.
White, J. C., and Smithwick, R. H.: *The Autonomic Nervous System*. The Macmillan Company, New York, 1946.
Woodburne, R. T.: The sphincter mechanism of the urinary bladder and the urethra. Anat. Rec., *141:*11, 1961.

UPPER LIMB

Basmajian, J. V., and Latif, A.: Integrated actions and functions of the chief flexors of the elbow. J. Bone & Joint Surg., *39A:*1106, 1957.
Beevor, C. E.: Croonian lecture on muscular movements, etc. Brit. M. J., *1:*1357, 1417, 1480, *2:*12, 1903.
Bing, R.: *Compendium of Regional Diagnosis in Lesions of the Brain and Spinal Cord*, Ed. 11, translated and edited by W. Haymaker. C. V. Mosby Company, St. Louis, 1940.
Braithwaite, F. *et al.:* The applied anatomy of the lumbrical and interosseous muscles of the hand. Guy's Hosp. Rep., *97:*185, 1948.
Bunnell, S.: *Surgery of the Hand*. J. B. Lippincott Company, Philadelphia, 1944.
Corbin, K. B., and Harrison, F.: The sensory innervation of the spinal accessory and tongue musculature in rhesus monkey. Brain, *62:*191, 1939.
Cummins, H., and Midlo, C.: *Finger Prints, Palms, and Soles*. Blakiston Company, division of Doubleday & Company, New York, 1943.
Flecker, H.: Time of appearance and fusion of ossification centers as observed by roentgenographic methods. Am. J. Roentgenol., *47:*97, 1942. Also with similar title in J. Anat., *67:*118, 1932.
Forrest, W. J.: Motor innervation of human thenar and hypothenar muscles in 25 hands: A study combining electromyography and percutaneous nerve stimulation. Can. J. Surg., *10:*196, 1967.
Gardner, E.: The innervation of the elbow joint. Anat. Rec., *102:*161, 1948.
Garn, S. M. *et al.:* A rational approach to the assessment of skeletal maturation. Ann. radiol., V-VI, 1964.
Greulich, W. W., and Pyle, S. I.: Radiographic atlas of skeletal development of the hand and wrist, Ed. 2. Stanford University Press, Stanford, Calif., 1959.
Haines, R. W.: The mechanism of rotation at the first carpo-metacarpal joint. J. Anat., *78:*44, 1944.
Haines, R. W.: The extensor apparatus of the finger. J. Anat., *85:*251, 1951.
Halls, A. A., and Travill, A.: Transmission of pressures across the elbow joint. Anat. Rec., *150:*243, 1964.

Harness, D., and Sekeles, E.: The double anastomotic innervation of thenar muscles. J. Anat., *109:*461, 1971.

Haymaker, W., and Woodhall, B.: *Peripheral Nerve Injuries: Principles of Diagnosis*, Ed. 2. W. B. Saunders Company, Philadelphia and London, 1953.

Inman, V. T., Saunders, J. B. deC. M., and Abbott, L. C.: Observations on the function of the shoulder joint. J. Bone & Joint Surg., *26:*1, 1944.

Johnson, G. F., Dorst, J. P., Kuhn, J. P., Roche, A. F., and Dávila, G. H.: Reliability of skeletal age assessments. Am. J. Roentgenol., *118:*320, 1973.

Jones, F. W.: *The Principles of Anatomy as Seen in the Hand*, Ed. 2. Baillière, Tindall & Cox, London, 1941.

Kanavel, A. B.: *Infections of the Hand*, Ed. 7. Lea and Febiger, Philadelphia, 1939.

Landsmeer, J. M. F.: The anatomy of the dorsal aponeurosis of the human finger, etc. Anat. Rec., *104:*31, 1949.

Rowntree, T.: Anomalous innervation of the hand muscles. J. Bone & Joint Surg., *31B:*505, 1949.

Salsbury, C. R.: The interosseous muscles of the hand. J. Anat., *71:*395, 1937.

Stopford, J. S. B.: *Sensation and the Sensory Pathway*. Longmans, Green & Co., Inc., London, 1930.

Sunderland, S.: The innervation of the first dorsal interosseous muscle of the hand. Anat. Rec., *95:*7, 1946.

Sunderland, S.: Voluntary movements and the deceptive action of muscles in peripheral nerve lesions. Australian & New Zealand J. Surg., *13:*160, 1944.

Wilder, H. H.: *The History of the Human Body*, Ed. 2. Henry Holt & Co., New York, 1923.

ABDOMEN

Alvarez, W. C.: *An Introduction to Gastro-enterology*. Heinemann, London, 1940.

Anson, B. J., and McVay, C. B.: Inguinal hernia. The anatomy of the region. Surg. Gynec. & Obstet., *66:*186, 1938.

Basmajian, J. V.: The marginal anastomoses of the arteries to the large intestine. Surg. Gynec. & Obst., *99:*614, 1954.

Basmajian, J. V.: The main arteries of the large intestine. Surg. Gynec. & Obst., *101:*585, 1955.

Benjamin, H. B., and Becker, A. B.: A vascular study of the small intestine. Surg. Gynec. & Obst., *108:*134, 1959.

Boyden, E. A.: The accessory gall-bladder. Am. J. Anat., *38:*202, 1926.

Boyden, E. A.: The anatomy of the choledochoduodenal junction in man. Surg. Gynec. & Obst., *104:*641, 1957.

Boyden, E. A.: *Gallbladder. McGraw-Hill Encylopedia of Science and Technology*. New York, 1960.

Cullen, T. S.: *Embryology, Anatomy and Diseases of the Umbilicus*. W. B. Saunders Company, Philadelphia, 1916.

Curtis, G. M., and Movitz, D.: The surgical significance of the accessory spleen. Ann. Surg., *123:*276, 1946.

Daseler, E. H., Anson, B. J., Hambley, W. C., and Reimann, A. F.: The cystic artery and constituents of the hepatic pedicle. A study of 500 specimens. Surg. Gynec. & Obst., *85:*45, 1947.

Dawson, W., and Langman, J.: An anatomical-radiological study on the pancreatic duct pattern in man. Anat. Rec., *139:*59, 1961.

Doyle, J. F.: The superficial inguinal arch: a reassessment of what has been called the inguinal ligament. J. Anat., *108:*297, 1971.

Drummond, H.: The arterial supply of the rectum and pelvic colon. Brit. J. Surg., *1:*677, 1914.

Edwards, E. A.: Functional anatomy of the portasystemic communications. A. M. A. Arch. Int. Med., *88:*137, 1951.

Farkas, L. G.: Basic morphological data of external genitals in 177 healthy Central European men. Am. J. Phys. Anthrop., *34:*325, 1971.

Falconer, C. W. A., and Griffiths, E.: The anatomy of the blood-vessels in the region of the pancreas. Brit. J. Surg., *37:*334, 1950.

Finlayson, J.: Herophilus and Erasistratus. Glasgow M. J., May, 1893.

Franklin, K. J.: *A Monograph on Veins*. Charles C Thomas, Springfield, Ill., 1937.

Frigerio, N. A., Stowe, R. R., and Howe, J. W.: Movement of sacroiliac joint. Clin. Orthop., *No. 100:*370, May, 1974.

Graves, F. T.: The anatomy of the intrarenal arteries and its application to segmental resection of the kidney. Brit. J. Surg., *42:*132, 1954.

Halbert, B., and Eaton, W. L.: Accessory spleens: a pilot study of 600 necropsies (abstract). Anat. Rec., *109:*371, 1951.

Hardy, K. J.: Involuntary sphincter tone in the maintenance of continence. Aust. N. Z. J. Surg., *42:*48, 1972.

Harrison, R. G.: The distribution of the vasal and cremasteric arteries to the testis, etc. J. Anat., *83:*267, 1949.

Healey, J. E., and Schroy, P. C.: Anatomy of the biliary ducts within the human liver. Arch. Surg., *66:*599, 1953.

Healey, J. E., Schroy, P. C., and Sorensen, R. J.: The intrahepatic distribution of the hepatic artery. J. Internat. Coll. Surgeons, *20:*133, 1953.

Hjortsjo, C.-H.: The topography of the intrahepatic duct systems (and of the portal vein). Acta anat., *11:*599, 1951.

Hjortsjo, C.-H.: The intrahepatic ramifications of the portal vein. Lunds Universitets Arsskrift., *52:*20, 1956.

Hyde, J. S., Swarts, C. L., Nicholas, E. E., Snead, C. R., and Strasser, N. F.: Superior mesenteric artery syndrome. Am. J. Dis. Child., *106:*25, 1963.

Jamieson, J. K., and Dobson, J. F.: The lymphatic system of the stomach, and of the caecum and appendix. Lancet, April 20 and 27, 1907.

Jamieson, J. K., and Dobson, J. F.: The lymphatics of the testicle. Lancet, Feb. 19, 1910.

Jamieson, J. K., and Dobson, J. F.: The lymphatics of the colon. Proc. Roy. Soc. Med., March, 1909.

Jay, G. D. III *et al.*: Meckel's diverticulum: survey of 103 cases. Arch. Surg., *61:*158, 1950.

Lofgren, F.: *Some Features in the Renal Morphogenesis and Anatomy with Practical Considerations.* Institute of Anatomy, University of Lund, Sweden, 1956.

Lofgren, F.: An attempt at homologizing different types of pyelus (renal pelvis). Urologia Internat., *5:*No. 1, 1956.

Lytle, W. J.: The internal inguinal ring. Brit. J. Surg., *32:*441, 1945.

Maisel, H.: The position of the human vermiform appendix. Anat. Rec., *136:*385, 1960.

Mann, C. V., Greenwood, R. K., and Ellis, F. H., Jr.: The esophagogastric junction. Surg. Gynec. & Obst., *118:*853, 1964.

Michels, N. A.: *Blood Supply and Anatomy of the Upper Abdominal Organs.* J. B. Lippincott Company, Philadelphia, 1955.

Michels, N. A., Siddharth, P., Kornblith, P., and Parke, W. W.: The variant blood supply to the small and large intestine: its import in regional resections. J. Internat. Coll. Surgeons, *39:*127, 1963.

Millbourn, E.: On the excretory ducts of the pancreas, etc. Acta anat., *9:*1, 1950.

Mitchell, G. A. G.: *Anatomy of the Autonomic Nervous System.* E. & S. Livingstone, Ltd., Edinburgh, 1953.

Moody, R. O., and Van Nuys, R. G.: Some results of a study of roentgenograms of the abdominal viscera. Am. J. Roentgenol., *20:*348, 1928.

Moody, R. O., Van Nuys, R. G., and Kidder, C. H.: The form and position of the empty stomach in healthy young adults. Anat. Rec., *43:*359, 1929.

Moody, R. O., and Van Nuys, R. G.: The position and mobility of the kidneys in healthy young men and women. Anat. Rec., *76:*111, 1940.

Oh, C., and Kark, A. E.: Anatomy of the external anal sphincter. Brit. J. Surg., *59:*717, 1972.

Patey, D. H.: Some observations on the functional anatomy of inguinal hernia, etc. Brit. J. Surg., *36:*264, 1949.

Pierson, J. M.: The arterial blood supply of the pancreas. Surg. Gynec. & Obst., *77:*426, 1943.

Reeves, T. A.: A study of the arteries supplying the stomach and duodenum and their relation to ulcer. Surg. Gynec. & Obst., *30:*374, 1920.

Rienhoff, W. F., and Pickrell, K. L.: Pancreatitis. An anatomic study of the pancreatic and extrahepatic biliary systems. Arch. Surg., *51:*205, 1945.

Roberts, W. H. B., and Taylor, W. H.: Inferior rectal nerve variations as it relates to pudendal block. Anat. Rec., *177:*461, 1973.

Roche, M. B., and Rowe, G. G.: The incidence of separate neural arch, etc. J. Bone & Joint Surg., *34A:*491, 1952.

Ross, J. A.: Vascular patterns of small and large intestine compared. Brit. J. Surg., *39:*330, 1952.

Rowe, G. G., and Roche, M. B.: The etiology of separate neural arch. J. Bone & Joint Surg., *35A:*102, 1953.

Shah, M. A., and Shah, M.: The arterial supply of the vermiform appendix. Anat. Rec., *95:*457, 1946.

Sheehan, D.: The afferent nerve supply of the mesentery, etc. J. Anat., *67:*233, 1933.

Solanke, T. F.: The blood supply of the vermiform appendix in Nigerians. J. Anat. *102:*353, 1968.

Solanke, T. F.: The position, length and content of the vermiform appendix in Nigerians. Brit. J. Surg., *57:*100, 1970.

Stephens, F. D.: Nervous pathways in anorectal control. Aust. N. Z. J. Surg., *42:*45, 1972.

Steward, J. A., and Rankin, F. W.: Blood supply of the large intestine: its surgical considerations. Arch. Surg., *26:*843, 1933.

Stewart, T. D.: The age incidence of neural arch defects in Alaskan natives. J. Bone & Joint Surg., *35A:*937, 1953.

Tobin, C. E., and Benjamin, J. A.: Anatomic and clinical re-evaluation of Camper's, Scarpa's, and Colles' fasciae. Surg. Gynec. & Obst., *88:*545, 1949.

Underhill, B. M. L.: Intestinal length in man. Brit. M. J., *2:*1243, 1955.

Varma, K. K.: The role of the voluntary anal sphincter in the maintenance of faecal continence in normal and abnormal states. Aust. N. Z. J. Surg., *42:*52, 1972.

Wakeley, C. P. G.: The position of the vermiform appendix, etc. J. Anat., *67:*277, 1933.

Wells, L. J.: Descent of the testis: anatomical and hormonal considerations. Surgery, *14:*436, 1943.

Wells, L. J.: Observations on the development of the diaphragm in the human embryo. Anat. Rec., *100:*778, 1948.

Wells, L. J.: Contributions to embryology, Carnegie Institute, *35:*107, 1954.

Wharton, G. K.: The blood supply of the pancreas, etc. Anat. Rec., *53:*55, 1932.

Wilkie, D. P. D.: The blood supply of the duodenum, etc. Surg. Gynec. & Obst., *13:*399, 1911.

Woodburne, R. T., and Olsen, L. L.: The arteries of the pancreas. Anat. Rec., *111:*255, 1951.

PERINEUM AND PELVIS

Blair, J. B., Holyoke, E., and Best, R. R.: A note on the lymphatics of the middle and lower rectum and anus. Anat. Rec., *108:*635, 1950.

Braithwaite, J. L.: Vesiculo-deferential artery. Brit. J. Urol., *24:*64, 1952.

Braus, H.: *Anatomie des menchen,* Ed. 2. Springer-Verlag, Berlin, 1929.

Caldwell, W. E., and Moloy, H. C.: Anatomical variations in the female pelvis, etc. Am. J. Obst. & Gynec., *26:*479, 1933.

Derry, D. E.: The innominate bone and the determination of sex. J. Anat., *43:*266, 1908.

Greulich, W. W., and Thomas, H.: A study of pelvic type and its relationship to body build in white women. J. A. M. A., *112:*485, 1939.

Greulich, W. W., and Thomas, H.: The dimensions of the pelvic inlet of 789 white females. Anat. Rec., *72:*45, 1938.

Keith, A.: *Human Embryology and Morphology,* Ed. 6. Edward Arnold & Company, London, 1948.

Langworthy, O. R. *et al.*: *Physiology of Micturition.* The Williams & Wilkins Company, Baltimore, 1940.

Leaf, C. H.: *The Lymphatics by Pouirier and Cuneo.*

(translation) Constable, London, 1903.
Learmonth, J. E.: A contribution to the neurophysiology of the urinary bladder in man. Brain, *54:*147, 1931.
Mitchell, G. A. G.: *Anatomy of the Autonomic Nervous System.* E. & S. Livingstone, Ltd., Edinburgh, 1953.
Moloy, H. C.: *Evaluation of the Pelvis in Obstetrics.* W. B. Saunders Company, Philadelphia, 1951.
Phenice, T. W.: A newly developed visual method of sexing the os pubis. Am. J. Phys. Anthrop., *30:*297, 1969.
Ricci, J. V. *et al.:* The female urethra: a histological study, etc. Am. J. Surg., N. S. *79:*499, 1950.
Rouvière, H.: Anatomie des lymphatiques de l'homme. Masson & Cie, Paris, 1932 (translated into English by Tobias, see below).
Sheehan, D.: *Annual Review of Physiology,* Vol. 3, 1941.
Stopford, J. S. B.: The autonomic nerve supply of the distal colon. Brit. M. J., *1:*572, 1934.
Thomas, H.: *Pelvimetry.* Paul B. Hoeber, Inc., 1956.
Tobias, M. J.: *Anatomy of the Human Lymphatic System.* Edwards, Ann Arbor, Mich., 1938 (translation of the work of Rouvière, see above).
Washburn, S. L.: Sex differences in the pubic bone. Am. J. Phys. Anthropol., N. S. *6:*199, 1948.
Hanna, R. E., and Washburn, S. L.: The determination of the sex of skeletons as illustrated by a study of the Eskimo pelvis. Human Biol., *25:*21, 1953.
Wilde, R. F.: The anal intermuscular septum. Brit. J. Surg., *36:*279, 1949.

LOWER LIMB

Basmajian, J. V.: The distribution of valves in the femoral, external iliac and common iliac veins, etc. Surg. Gynec. & Obst., *95:*357, 1952.
Basmajian, J. V., and Bentzon, J. W.: An electromyographic study of certain muscles of the leg and foot, etc. Surg. Gynec. & Obst., *98:*662, 1954.
Basmajian, J. V., and Lovejoy, J. F., Jr.: Functions of popliteus: a multifactorial electromyographic study. J. Bone & Joint Surg., *53A:*557, 1971.
Bing, R.: *Compendium of Regional Diagnosis in Lesions of the Brain and Spinal Cord,* Ed. 11, translated and edited by W. Haymaker. C. V. Mosby Company, St. Louis, 1940.
Doyle, J. F.: The superficial inguinal arch: a reassessment of what has been called the inguinal ligament. J. Anat., *108:*297, 1971.
Gardner, E.: The innervation of the hip joint. Anat. Rec., *101:*353, 1948.
Gardner, E.: The innervation of the knee joint. Anat. Rec., *101:*109, 1948.
Harty, M.: Anatomic features of the lateral aspect of the knee joint. Surg., Gynec. & Obst., *130:*11, 1970.
Haxton, H.: The functions of the patella and the effects of its excision. Surg. Gynec. & Obst., *80:*389, 1945.
Haymaker, W., and Woodhall, B.: *Peripheral Nerve Injuries: Principles of Diagnosis,* Ed. 2. W. B. Saunders Company, Philadelphia and London, 1953.
Hicks, J. H.: The mechanics of the foot: The joints. J. Anat., *87:*345, 1953. The plantar aponeurosis and the arch. J. Anat., *88:*25, 1954. The foot as a support. Acta anat., *25:*34, 1955.
Hughston, J. C., and Eilers, A. F.: The role of the posterior oblique ligament in repairs of acute medial (collateral) ligament tears of the knee. J. Bone & Joint Surg., *55A:*923, 1973.
Jack. E. A.: Naviculo-cuneiform fusion in the treatment of flat foot. J. Bone & Joint Surg., *35B:*75, 1953.
Jones, F. W.: *The Foot, Structure and Function.* Ballière, Tindall & Cox, London, 1949.
Jones, R. L.: The human foot...the role of its muscles and ligaments in the support of the arch. Am. J. Anat., *68:*1, 1941.
Keegan, J. J., and Garrett, F. D.: The segmental distribution of the cutaneous nerves in the limbs of man. Anat. Rec., *102:*409, 1948.
Lambert, E. H.: The accessory deep peroneal nerve: a common variation in innervation of extensor digitorum brevis. Neurology, *19:*1169, 1969.
Last, R. J.: The popliteus muscle and the lateral meniscus. J. Bone & Joint Surg., N. S. *32B:*93, 1950.
Marshall, J. L., Girgis, F. G., and Zelko, R. R.: The biceps femoris tendon and its functional significance. J. Bone & Joint Surg., *54A:*1444, 1972.
Mitchell, G. A. G.: *Anatomy of the Autonomic Nervous System.* E. & S. Livingstone, Ltd., Edinburgh, 1953.
Morton, D. J.: *The Human Foot.* Columbia University Press, New York, 1937.
Noyes, F. R., and Sonstegard, D. A.: Biomechanical function of the pes anserinus at the knee and the effect of its transplantation. J. Bone & Joint Surg., *55A:*1225, 1973.
O'Rahilly, R.: A survey of carpal and tarsal anomalies. J. Bone & Joint Surg., *35A:*626, 1953.
Robichon, J., and Romero, C.: The functional anatomy of the knee joint, with special reference to the medial collateral and anterior cruciate ligaments. Can. J. Surg., *11:*36, 1969.
Singer, C.: *The Evolution of Anatomy.* Kegan Paul, etc., 1925.
Trueta, J.: The normal vascular anatomy of the femoral head during growth. J. Bone & Joint Surg., *39B:*358, 1957.
Trueta, J., and Harrison, M. H. M.: The normal vascular anatomy of the femoral head in adult man. J. Bone & Joint Surg., *35B:*442, 1953.
Tucker, F. R.: Arterial supply to the femoral head and its clinical importance. J. Bone & Joint Surg., *31B:*82, 1949.
Walmsley, T.: The articular mechanism of the diarthroses. J. Bone & Joint Surg., *10:*40, 1928.
Weinert, C. R., Jr., McMaster, J. H., and Ferguson, R. J.: Dynamic function of the human fibula. Am. J. Anat., *138:*145, 1973.
Wolcott, W. E.: The evolution of the circulation in the developing femoral head and neck. Surg. Gynec. & Obst., *77:*61, 1943.

THORAX

Boyden, E. A.: *Segmental Anatomy of the Lungs.* McGraw-Hill Book Company, Inc., New York,

1955.
Bradley, W. F. *et al.*: Anatomic considerations of gastric neurectomy. J. A. M. A., *133:*459, 1947.
Brock, R. C.: *The Anatomy of the Bronchial Tree.* Oxford University Press, London, 1946.
Foster-Carter, A. F.: Broncho-pulmonary abnormalities. Brit. J. Tuberc., Oct., 1946.
Gradwohl, R. B. H.: *Clinical Laboratory Methods and Diagnosis*, Vol. II, Ed. 4. Mosby, St. Louis, 1948.
Grant, R. T.: Development of the cardiac coronary vessels in the rabbit. Heart, *13:*261, 1926.
Gross, L.: *The Blood Supply to the Heart.* Paul B. Hoeber, Inc., New York, 1921.
Harper, W. F.: The blood supply of human heart valves. Brit. M. J., *2:*305, 1941.
Hayek, H. von: *The Human Lung.* Translated by Krahl, V. E. Illustration based on figure 216, by courtesy of Hafner Publishing Company, Inc., New York, 1960.
Jackson, C. L., and Huber, J. F.: Correlated applied anatomy of the bronchial tree and lungs with a system of nomenclature. Dis. Chest, *9:*319, 1943.
James, T. N.: The arteries of the free ventricular walls in man. Anat. Rec.,*136:*371, 1960.
James, T. N.: Anatomy of the human sinus node. Anat. Rec., *141:*109, 1961.
James, T. N., and Burch, G. E.: The atrial coronary arteries in man. Circulation 2, *17:*90, 1958.
Jones, D. S., Beargie, R. J., and Pauly, J. E.: An electromyographic study of some muscles of costal respiration in man. Anat. Rec., *117:*17, 1953.
Krahl, V. E.: Translation of Hayek's *The Human Lung*, Hafner Publishing Company, Inc., New York, 1960.
Lachman, E.: The dynamic concept of thoracic topography, etc. Am. J. Roentgenol., *56:*419, 1946.
Lachman, E.: A comparison of the posterior boundaries of lungs and pleura, etc. Anat. Rec., *83:*521, 1942.
Macklin, C. C.: Bronchial length changes and other movements. Tubercle, Oct.-Nov., 1932.
Macklin, C. C.: The dynamic bronchial tree. Am. Rev. Tuberc., *25:*393, 1932.
Mainland, D., and Gordon, E. J.: The position of organs determined from thoracic radiographs, etc. Am. J. Anat., *68:*457, 1941.
Merklin, R. J.: Position and orientation of the heart valves. Anat. Rec., *125:*375, 1969.
Miller, W. S.: *The Lung*, Ed. 2. Charles C Thomas, Springfield Ill., 1921.
Mitchell, G. A. G.: *Anatomy of the Autonomic Nervous System.* E. & S. Livingstone, Ltd. Edinburgh, 1953.
Mizeres, N. J.: The cardiac plexus in man. Am. J. Anat., *112:*1963.
Morris, E. W. T.: Some features of the mitral valve. Thorax, *15:*70, 1960.
Nathan, H.: Anatomical observations on the course of the azygos vein. Thorax, *15:*229, 1960.
Nelson, H. P.: Postural drainage of the lungs. Brit. M. J., *2:*251, 1934.
Reed, A. F.: The origins of the splanchnic nerves. Anat. Rec., *109:*81, 1951.
Ross, J. K.: Review of the surgery of the thoracic duct. Thorax, *16:*207, 1961.
Rouvière, H.: *Anatomie des lymphatiques de l'homme.* Masson & Cie, Paris, 1932.
Silvester, C. F.: On the presence of permanent communications between the lymphatic and the venous system at the level of the renal veins in South American monkeys. Am. J. Anat., *12:*447, 1912.
Singer, R.: The coronary arteries of the Bantu heart. South African M. J., *33:*310, 1959.
Tobin, C. E.: The bronchial arteries and their connections with other vessels in the human lung. Surg. Gynec. & Obst., *95:*741, 1952.
Tobin, C. E.: Human pulmonic lymphatics. Anat. Rec., *127:*611, 1957.
Tobin, C. E., and Zariquiey, M. O.: Arteriovenous shunts in the human lung. Proc. Soc. Exper. Biol. & Med., *75:*827, 1950.
Thoracic Society: The nomenclature of bronchopulmonary anatomy. Thorax, *5:*222, 1950.
Trotter, M.: Synostosis between manubrium and body of the sternum in whites and negroes. Am. J. Phys. Anthropol., *18:*439, 1934.
Truex, R. C., and Warshaw, L. J.: The incidence and size of the moderator band, etc. Anat. Rec., *82:*361, 1942.
Walls, E. W.: Dissection of the atrio-ventricular node and bundle in the human heart. J. Anat., *79:*45, 1945.
Walmsley, T.: The Heart, in *Quain's Anatomy*, 1929.
White, J. C., and Smithwick, R. H.: *The Autonomic Nervous System*, Ed. 2. The Macmillan Company, New York, 1946.
Woodburne, R. T.: The costomediastinal border of the left pleura in the precordial area. Anat. Rec., *97:*197, 1947.
Zoll, P. M., Wessler, S., and Schlesinger, M. J.: Interarterial coronary anastomoses, etc. Circulation, *4:*797, 1951.

HEAD AND NECK

Basmajian, J. V., and Dutta, C. R.: Electromyography of the pharyngeal constrictors and levator palati in man. Anat. Rec., *139:*561, 1961.
Batson, O. V.: The function of the vertebral veins and their role in the spread of metastases. Ann. Surg., *112:*138, 1940.
Békésy, G. P. See von Békésy, G., *below.*
Bolz, E. A., and Lim, D. J.: Morphology of the stapediovestibular joint. Acta Otolaryngol., *73:* 10, 1972.
Browning, H.: The confluence of dural venous sinuses. Am. J. Anat., *93:*307, 1953.
Carlsöö, S.: Nervous co-ordination and mechanical function of mandibular elevators, Acta odont, Scandinav., *10:* suppl. 11, 1952.
Cave, A. J. E.: A note on the origin of the m. scalenus medius. J. Anat., *67:*480, 1933.
Cole, T. B., and Baylin, G.: Radiographic evaluation of the prevertebral space. Larynogoscope, *83:*721, 1973.
Doran, G. A., and Baggett, H.: The genioglossus muscle: a reassessment of its anatomy in some mammals, including man. Acta anat., *83:*403, 1972.

Eckenhoff, J. E.: The physiologic significance of the vertebral venous plexus. Surg. Gynec. & Obst., *131*:72, 1970.

Graves, G. O., and Edwards, L. F.: The eustachian tube. Arch. Otolaryngol., *39*:359, 1944.

Hoshino, T. and Paparella, M. M.: Middle ear muscle anomalies. Arch. Otolaryngol., *94*:235, 1971.

Jamieson, J. K., and Dobson, J. F.: The lymphatics of the tongue, etc. Brit. J. Surg., *8*:80, 1920.

Kimmel, D. L.: Innervation of spinal dura mater and dura mater of the posterior cranial fossa. Neurology, *11*:800, 1961.

Kimmel, D. L.: The nerves of the cranial dura mater and their significance in dural headache and referred pain. Chicago M. Sch. Quart., *22*:16, 1961.

Latif, A.: An electromyographic study of the temporalis muscle, etc. Am. J. Orthodontics, *43*:577, 1957.

Leden, H., and Moore, P.: See von Leden, H. and Moore, P., *below*.

Lewinsky, W., and Stewart, D.: An account of our present knowledge of the innervation of the teeth and their related tissues. Brit. Dent. J., Dec. 1, 1938.

Mann, Ida: *The Development of the Human Eye*. Cambridge University Press, 1928.

McKenzie, J.: The parotid gland in relation to the facial nerve. J. Anat., *82*:183, 1948.

Mitchell, G. A. G.: *Anatomy of the Autonomic Nervous System*. E. & S. Livingstone, Ltd., Edinburgh, 1953.

Moyers, R. E.: An electromyographic analysis of certain muscles involved in temporomandibular movement. Am. J. Orthodontics, *36*:481, 1950.

Negus, V. E.: *The Comparative Anatomy and Physiology of the Larynx*. William Heinemann, Ltd., London, 1949.

Parkinson, D.: Collateral circulation of cavernous carotid artery: anatomy. Can. J. Surg., *7*:251, 1964.

Pearson, A. A.: The hypoglossal nerve in human embryos. J. Comp. Neurol., *71*:21, 1939.

Pearson, A. A. *et al.*: Cutaneous branches of the dorsal (primary) rami of the cervical nerves. Am. J. Anat., *112*:169, 1963.

Powell, T. V., and Brodie, A. G.: Closure of spheno-occipital synchondrosis. Anat. Rec., *147*:15, 1963.

Pressman, J. J., and Simon, M. B.: Experimental evidence of direct communications between lymph nodes and veins. Surg. Gynec. & Obst., *113*:537, 1961.

Saunders, J. B. deC. M., Davis, C., and Miller, E. R.: The mechanism of degluition as revealed by cine-radiography. Ann. Otol. Rhin. & Larying., *60*:897, 1951.

Schour, I., and Massler, M.: The development of the human dentition. J. Am. Dent. A., *28*:1153, 1941.

Stiles, H. J.: In *Cunningham's Text-Book of Anatomy*. Oxford University Press, London, 1913.

Stewart, D., and Wilson, S. L.: Regional anaesthesia and innervation of the teeth. Lancet, *Oct. 20*:809, 1928.

Sunderland, S.: The meningeal relations of the human hypophysis cerebri. J. Anat., *79*:33, 1945.

von Békésy, G.: The ear. Scientific American, August, 1957.

von Békésy, G.: *Experiments in Hearing*, translated and edited by E. G. Wever. McGraw-Hill Book Company, Inc., New York, 1960.

von Leden, H. and Moore, P.: The mechanics of the cricoarytenoid joint. Arch. Otolaryngol., *73*:541, 1961.

von Leden, H., and Moore, P.: The mechanics of the Otolaryngol., *74*:660, 1961.

Watt, J. C., and McKillop, A. N.: Relation of arteries to roots of nerves in posterior cranial fossa. Arch. Surg., *30*:336, 1935.

Young, M. W.: The termination of the perilymphatic duct. Anat. Rec., *112*:404, 1952.

INDEX

A

Abdomen, 173
Acetabulum, 364
Acid, hyaluronic, 19
Acromion, 89, 96, 101
  ossification, 96
  parts covered by, 96
Adenoids, 556
Acinus, hepatic, 54
Aditus to antrum, 592
Age, skeletal, 131
Agger nasi, *fig. 800*
Agonist. See Prime mover
Ala of sacrum, 272
Allantois (*fig. 258*), 204, 257
Alveolus, 414
Ampulla—
  of bile duct, 229
  of ductus deferens, 282
  of rectum, 283
  of semicircular canals, 596
  of uterine tube, 293
  of Vater, 229
Anastomoses, 27, 34
  arteriovenous, 35
  cruciate, 334
  of elbow, see *fig. 128*
  genicular, 375, *fig. 438*
  of hand, 139
  of heart, 427-428
  of liver, 215
  portacaval (accessory portal system), 216
  posterior interosseous, 145
  of profunda brachii, 112
  scapular, 92
  of wrist, *fig. 155*
Angle—
  acromial, 90, 101
  carrying, 164
  of jaw, 524
  of ribs, 396
  iridocorneal, 502
  sternal, 77, **393**
Ansa cervicalis (hypoglossi), 511, 549
  hypoglossi. See Ansa cervicalis
  subclavia, 550; *fig. 579*
Antagonist, 27
Antrum—
  of Highmore. See Sinus, maxillary

Antrum—*Continued*
  mastoid, (tympanic), 592
  pyloric, 194
Anulus (Annulus) fibrosus, 18
Anus, 52, 193, 259
Aorta—
  abdominal, 238
    surface anatomy, 238
  arch of, 437, **442**
    definition, 406
    definitive, 438
    relations, 442
    surface anatomy, 422
    variations, 442
  ascending, 422, 425
  branches of, in abdomen, 238
    in thorax, 445
    relation to i.v. cava, 241
  coarctation of, 443
  descending thoracic, 406, 445
  development, note on, 443
Aperture—
  of larynx, superior, 584
  nasal
    anterior. See A. piriform
    posterior. See Choanae
  orbital, 451, 493
  piriform, (nasal, anterior), 452
Aponeurosis, 22
  bicipital (lacertus fibrosus), 111, 112
  epicranial or Galea, 462
  palatine, 557, 558
  palmar, 136
  plantar, 356, 385
Appendix, Appendices—
  of epididymis, 187
  epiploicae, 195
  testis, 187
  vermiform, 192, 195-196
Aqueduct, of cochlea. See Canaliculus of cochlea, 487, 596
  of vestibule, 597
Arc, reflex, 43
Arc of Riolan, 225
Arcades (arterial), 195, 223
Arches and Arcus—
  of aorta. See Aorta
  of atlas, 476, 534, 537
  axillary, 87
  carpal
    dorsal (posterior), 141

Arches and Arcus—*Continued*
    palmar (anterior), 141
  coraco-acromial, 158
  costal. See Ribs
  of foot, 352, 382
    support of, 384
  glossopalatine. See A., palatoglossal
  lumbocostal (arcuate ligaments), 234, 247, **249**
  neural, 13
  palatoglossal, 556, 567
  palatopharyngeal, 556, 567
  palmar
    deep, 141
    superficial, 139
  plantar, 359
  pubic, 299, 322
  superciliary, 452
  tendinous (arcuate line), 270
  venous, dorsal
    of foot, 309
    of hand, 107
  vertebral, 12
  zygomatic, 451, 452, 522
Area—
  nuchal, 470, 604
Arrectores pilorum, 69
ARTERIES—
  of abdominal wall, 177
  abnormal, 35
  acromiothoracic. See A., thoraco-acromial
  adrenal, 232
  alveolar
    inferior, 529, 530
    superior, 578
  angular, 460
  aorta. See Aorta
  appendicular, 224
  arcuate, 339
  auditory, internal. See A., labyrinthine
  auricular
    deep, 530
    posterior, 514, 526
  axillary, **82,** 83, 92
  of back, 476
  basilar, 485
  brachial, 112
    high division of, *fig. 111*
  brachiocephalic (innominate), 422

Arteries—*Continued*
to brain, 485
bronchial, 416, 446
of bulb, 264
calcanean
lateral, 348
medial, 348
carotid, common, 512, **543**
external, 512, 527
internal
in neck, 512, 543
in skull, **490,** 543
left, 442
in thorax, 406
carpal
dorsal (posterior), 141
palmar (anterior), 141
cecal (caecal), 224
celiac (See Trunk, celiac), **213,** 231, 239
central of retina, 500, 503
cerebellar
anterior inferior, 485
posterior inferior, 485
superior, 485
cerebral, anterior, 485
communicating
anterior, 485
middle, 485
posterior, 485
cervical
ascending, 517
deep (profunda), 476, 517
transverse, 466, 517
ciliary, 500
circumflex
femoral
lateral, 315, 333
medial, 315, 333
variations in, see *fig. 114*
humeral, 82, 95
iliac
deep, 182
superficial, 178, 310
scapular, 82
colic, 223–225
variations in marginal, 225
colli, transversa, see transversa colli
communicating, posterior, 485
of posterior tibial, 348
coronary, 426
variations in, 426
costal, lateral, 399
cremasteric, 178, 187
cricothyroid, 513
to crus penis. See A., deep, of penis
cystic, 215, 221
variations of, 215

Arteries—*Continued*
deep cervical, 476, 517
circumflex iliac, 178, 182, 184
of clitoris or penis, 264
deferent, (of ductus deferens), 187, 283
dental. See A. alveolar
development of, in lower limb, 334, 339
digital
of foot, 339, 359
of hand
dorsal, 141, *fig. 157*
palmar, 139, 140
dorsalis indicis. See A., dorsal digital
linguae, 570
pedis, 338
variations in, 339
penis, 263, 264
pollicis, indicis. See A., dorsal digital
end-arteries, 35
epigastric
inferior, 177, 184
superficial, 178, 310
superior, 177, 399
epiploic, 215
ethmoidal, 481, 580
facial (ext. maxillary), in face, 460
in neck, 512, 513
branches, 460, 513
transverse, 460, 461
femoral, 314, 315, 317
gastric
left, 214
right, 215
gastroduodenal, 215
gastro-epiploic, 213, 215
to G. I. Tract & Co., 238
genicular, 375, *fig. 438*
to glands, three paired, 239
gluteal
inferior, 285, 323, 326
superior, 286, 323, 326
"great arteries," note on development, 437
of hand, 139
hemorrhoidal. See A., rectal
hepatic, 215
aberrant, 214
accessory, 216
collateral anastomoses, 215
variations of, 215
hypogastric. See A., iliac, internal
obliterated. See A., umbilical
ileal, 223

Arteries—*Continued*
ileocolic, 224
iliac, common, 238
external, 238, 269
internal, 238, 262, 285
iliolumbar, 286
infra-orbital, 578
inguinal, superficial, 178, 310
innominate. See Trunk, brachio-cephalic
intercostal, 446, 476
anterior, 398
posterior, 398, 446
supreme (superior), 408, 517
interlobar, renal, 61
interosseous
anterior, 122
common, 122, 145
posterior, 122, **145**
recurrent, 122
intralobar, renal, 61
jejunal, 223
labial
of face, 455, 460
of vulva, 265
laryngeal, 589
superior, 513
lingual, 513, **570**
lumbar, 239, 476
malleolar, 339
mammary, internal. See A., thoracic, internal
marginal, of Drummond, 225, 226
maxillary
internal, 527, 528, **530,** 578, 580
third part of, 578
external. See A., facial
median, 122
median sacral, 239, 285
meningeal, 492
accessory, 530
middle, 488, 492, 530
mesenteric
inferior, **224,** 239
superior, **223,** 239
metacarpal
dorsal, 141, see *fig. 157*
palmar, 141
metatarsal, dorsal, 339
musculophrenic, 250, 398
nasal, lateral, 460
nutrient (medullary), 8
obturator, 286, 333, 366
accessory (abnormal), 286
occipital, 502, 466, **476,** 514
ophthalmic, 484, 500
ovarian, 231, 297
palatine

Arteries—*Continued*
descending, 578
greater, 558, 578
palmar arch, deep, 141
superficial, 123, 139
pancreatic, inferior, 229
pancreaticoduodenal, inferior, 213, 229
superior, 213, *fig. 300*
of pelvis, 285
of penis
deep, 264
dorsal, 264
perforating
of foot, 361
of hand, 141, see *fig. 157*
of thigh, 334
pericardiacophrenic, 250, 399, 404, 407
perineal, 264
transverse, 260, 264
peroneal, 348
perforating, 348
pharyngeal, 578
ascending, 513
phrenic, 239
plantar, 359
popliteal, 330
profunda brachii, 112, **114**
variations in. See *fig. 114*
cervicis, 476, 517
of dorsalis pedis, 339, 361
femoris, 315, **333**
of tongue, 570
pubic
of inferior epigastric, 178
of obturator, 286
pudendal
external, 178, 310
internal, 257, 259, 286
pulmonary, 414, 425, 429
radial, 122, 140
palmar digital, 140, 141
rectal (hemorrhoidal), 284
inferior, 259, 260, 264, 284, 285
middle, 284, 285
superior, 224, 284, 285
renal, 61, 231, 234–235
accessory (multiple), 237. See *fig. 311*
of retina, central, 500, 503
retro-duodenal, 215, 229
sacral
lateral, 286, 476
median (middle), 239, 285
saphenous, 317
of scalp, 462. See *fig. 662*
scrotal, posterior, 260, 264
septal, post. nasal, 578, 580

Arteries—*Continued*
sigmoid, 225
spermatic. See A., testicular
sphenopalatine, 578, 580
splenic, 213, **214**
sternomastoid, 514
structure, 33
stylomastoid, 514
subclavian, 516
branches of, 517
left, 406, 442
third part, 466
subcostal, 446
sublingual, 571
submental, 460
subscapular, 82
superficial
circumflex iliac, 178
epigastric, 178, 310
external pudendal, 178, 310
supraduodenal, 215, 229
suprarenal, 232
suprascapular, (*fig. 79*), 91, 466, 517
supratrochlear, see ulnar collateral, inferior, 112
tarsal, 339
temporal
middle, 461
superficial, 460, 461, 525
testicular, 186, **187, 231,** 238
thoracic, internal (int. mammary), **398,** 517
branches of, 398–399
lateral, 82, 86
lateral costal, 399
supreme, *fig. 68*
thoraco-acromial, 82
thyroid
inferior, 517, 519
superior, 513, 519
thyroidea ima, 507, 519
tibial, anterior, 338–339
recurrent branch of, 339
variations, 339
posterior, 348
tonsillar, 513
transversa colli, 92, **466,** 517
facial, 461
perineal, 264, *fig. 338*
tympanic, anterior 530, 595
ulnar, 122
carpal, 122
collateral
inferior, 112
superior, 112
recurrent, 122
superficial, 108
umbilical (obliterated hypogastric), 204, 270, 285, 286
uterine, 286, **297**
vaginal, 297
of vas deferens. See A., deferent
vertebral, 516, 536
development, see *fig. 750*
in neck, 476, 516, 517, **536**
in skull, 485
vesical
inferior, 283, 285, 286
superior, 286
vesiculodeferential, 286
vitelline, 207
Arteriolae rectae of kidney (vasa recta), 61
Arterioles, 32
Articulations, **16.** See also Joints
Atlas, 470, **534,** 537
transverse process, 475, 505
Atrium
of heart
left, 420, **431**
right, 420, 421, **430**
of nose, 579, 614
Auricle—
of ear, 455
nerves, 456
of heart. See Atrium
Autonomic Nervous System, 43
Axilla, 76–88
vessels and nerves of, 81
walls of, 79
Axis—
of eyeball, 495
of orbital cavity, 495
vertebra (epistropheus), 534
visual, 495
Axon, 40

B

Back—
arteries of, 476
muscles of, 471
deep, actions of, 474
nerves of, 476
Band—
iliotibial. See Tract, iliotibial
moderator. See B., septomarginal
septomarginal (Band, moderator), 432
Bar, costotransverse, 534
Barber's sign, 108
Basi-occipital bone (basi-occiput). 605
Beds (moulds)—
of bladder, 278

Beds —*Continued*
of parotid, 525
of stomach, 219. See *fig. 282*
of tonsil, 559
Bile passages. 212. See Passages, bile
Bladder—
gall. See Gall bladder
urinary, 62, **278**
bed (or mould), 278
development, 257
interior of, 279
lymphatics, 304
nerves, 62, 302
vessels, 279
Blood, 30
circulation of, 31
distribution, nervous regulation of, 34
vessels, 32
structure, 33
Blow, striking a, 128
Body, Malpighian (corpuscle) of kidney, 61
perineal, 258, 291
vitreous, of eye, 503
Body types, 201
Boils, 70
Bone, 3
blood vessels, 8, 9
borders of 116, 337
diaphysis, 7, 8
epiphysis, 7, 8
of foot, articulated, 350
functions, 4
historical, 9
incisive, 615
markings on, 6
marrow, 8
metaphysis, 8
nerves, 3, 9
ossification, 8
parts of a young, 7
properties, 3, 4
structure, 4
Bones. See under individual names.
accessory, 6, 385
classification, 5
flat, 6
interparietal, 604
irregular, 6
long (tubular), 5
markings on, 6
pneumatic, 6
sesamoid, 6
short (cubical), 5
supernumerary (accessory), 6, 385
sutural (Wormian), 604
Brachycephaly, 461
Bregma, 451, 461
Brim, pelvic, 272
Bronchial tree, 413
Bronchiole, 58, 414
Bronchus, 58, 413, 414
eparterial, 405
extrapulmonary, 441
movements of, 417
structure of, 414
Buds, taste, 569
Bulb—
of aorta, 425
olfactory, 483
of penis, 260
of vestibule, 265
Bulbus oculi. See Eyeball, 495
Bulla ethmoidalis, 580
Bundle—
atrioventricular, 436
Bursa, omental (lesser sac), 198, 210
mouth of epiploic foramen, 197, 212
Bursae, synovial (mucous), **24**
of Achilles. See B., of calcaneus
articular, 24
of biceps brachii, 110, 157
of biceps femoris, 375
communicating with knee joint, 374
of ext. carpi radialis br., 145
of gastrocnemius, 375
infrapatellar, superficial, 368, 374
about knee, 374, 375
between metacarpal heads, 168
of obturator internus, 326
olecranon, 162
patellar. See *fig. 517*; 374
pectineal. See Bursa, of psoas
of popliteus, 375
of psoas, 365
of quadriceps femoris, 374
of sartorius, 375
of semimebranosus, 375
subacromial, 96, 158
subcutaneous, 24
of subscapularis, 158
subtendinous, 24
of tensor palati, 562
tubular. See Sheaths, synovial
Buttresses of face. See *fig. 739*
of maxilla, 528

C

Caecum. See Cecum, 195
Calcaneus, 346, **352,** 353, 354, 356, 380, 382
Calvaria, 479
Calices of kidney, 235
Canaliculus, bile, 54
of cochlea (aqueduct), 487, 596
innominatus, 489, 608
lacrimal, 456, **458**
Canals—
adductor (subsartorial), 316
contents of, 317
walls of, 317
alimentary, 52
parts of, 191
anal, 52, **283**
pecten, 284
carotid, 541, 611
condylar, 470, 486, 605
for facial nerve, 594, 597
femoral, 312
gastro-intestinal, parts of, 191
Hunter's. See C., adductor, 316
hyaloid, 504
hypoglossal (ant. condylar), 486, 542, 605
incisive, 615
infra-orbital, 614
inguinal, 183
innominatus. See Canaliculus
intestinal, 192–197
mandibular, 524
nasolacrimal, 579, 615
of Nuck (processus vaginalis), 267
nutrient, 8
optic, (foramen, optic), 485, 488, 493, 495, 606
palatine
greater, 539, 558, 615, **617**
lesser, 539, 617
palatinovaginal, 617
of Petit. See Space, zonular
pharyngeal. See C., palatovaginal
pleuroperitoneal, 250
pterygoid, 539, 608
pudendal (of Alcock), 260
pyloric, 194
sacral, 272
semicircular, 596
subsartorial. See C., adductor
vertebral, 12, 16
Capillaries—
bile. See Canaliculi, bile
blood, 30, 32
lymph, 36
Capitate bone, 126, 130, 167
Capitulum of humerus, 109, 162, 165
Capsule—
fibrous, of joint, 19
of elbow, 162

Capsule—*Continued*
of knee, 369
of shoulder, 156
glomerular (Bowman's), 60
prostatic, 281
renal, fibrous and fatty, 233
synovial, of joint 19
of ankle, 378
of elbow, 162
of hip, 364
of knee, 369
Tenon's. See Sheath of eyeball
of tonsil, 559
"Carpet," fascial of neck, 465
Carpus, **125.** See also individual bones
accessory bones in, 386
homologies (comparisons), 385
Cartilage, 11
articular (hyaline), 8, 11, 19
arytenoid, 556, 585
branchial, first. See C., Meckel's
corniculate, 556, 585
cricoid, 584
cuneiform, 556, 585
elastic, 11
of epiglottis, 585
epiphyseal. See Epiphyses
fibrocartilage, 11
hyaline, 11
of larynx, 584-585
Meckel's, 619, 620
nasal, 455
Reichert's, 620
semilunar. See Menisci, 371
septal, 455, 578
thyroid, 584
types, 11, 589
Caruncle, lacrimal, 456
Cauda equina, 41
Cave, trigeminal (semilunar), 490
Cavernous tissue, 35
Cavity—
abdomino-pelvic, 190
articular (see C., synovial), 19
glenoid, 89, 99, 155
infraglottic, 589
mouth, 566
nasal, 55, 578
blood supply of, 580
lymph vessels of, 601
nerves of, 601
wall, lateral of, 579
oral, see Mouth
orbital, 493
pericardial, 420
peritoneal, 197; see also Peritoneum, 205
synovial, 19
Cavity—*Continued*
thoracic, 391
tympanic, 590
Cecum, (caecum), 192, **195**
Cells. See also Sinus
air
ethmoidal, 582
mastoid, 595
tubal, tympanic, 596
connector (intercalated), 42, 44
excitor, 44
of Kuppfer, 54
Centrum of vertebra, 13
Cervix uteri, 67, 294
Chambers of eyeball, 502
Cheek, 566
Chiasma, optic, 483
Chin, 454
Choanae, (Aperture, nasal posterior), 539, 555, 578
Chorda tympani, 567, 594
Chordae tendineae, 433
Chromaffin bodies, 51
Choroid, 501
Cilia of eyelids, 456
Circulation, 31
collateral. See Anastomoses
fetal, 37
Circumduction, 21
Cisterna (receptaculum) chyli, 251
Clavicle, 75, 77, **78, 98-99,** 152
functions of, 75, 98, 152, 153, 161
ossification, 99
Clitoris, 265
Clivus, 487
Cloaca, 257
Coarctation of the aorta, 443
Coccyx, 257, 271, **272**
Cochlea—
canaliculus (aqueduct) of, 487, 596
Cohesion of peritoneum, 208
Colliculus of prostatic urethra, 280
Colon, 192, 195-197
descending, nerve supply of, 302
to mobilize, 210
sacculations (haustra) of, 195
sigmoid (pelvic), 196
Column, anal (rectal), 284
renal (of Bertin), 234
vertebral, 11
from behind, 470
veins of, 477
Commissure, palpebral, 456
Communications, lymphaticovenous, 36
Conchae—
inferior, 618
nasal, 579, 610
sphenoidal, 609
Conduction, sound, 595
Condyle—
of femur, 313, 371
of jaw. See Head of jaw
occipital, 539
"Cone, muscle, ocular," 495
Confluens sinuum, 483
Conjunctiva, 456
Constrictors of pharynx, 508, 554
Conus arteriosus or infundibulum of right ventricle, 432
Conus elasticus, (see also Ligaments, cricothyroid), 585
Cor, see Heart
Coracoid, 77, 100, 101
epiphysis, 96, 102
Cord—
of brachial plexus, 83
oblique, 165
spermatic, 185, 186
arteries of, 187
coverings of, 185
spinal, or spinal medulla, 40
umbilical, 204
vocal. See Fold, vocal
Cornea, 456, 495, 499, 501, 504
sensation, 499
Corpus cavernosum clitoridis, 265
penis, 65, 260
spongiosum penis (urethrae), 65, 260
Corpuscle, renal (Malpighian), 61
Ruffini, 20
Costa, 394
Coughing, 184
Cranium, 451, 469
Creases, skin, 131
Crest—
conchal, 614
ethmoidal, 614
frontal, 604
iliac, 190, 244, 322
incisor (incisive), 615
infratemporal, 528, 608
intertrochanteric, 313
lacrimal
of lacrimal bone, 457, 618
of maxilla, 614
nasal, of maxilla, 615
occipital
external, 470, 604
internal, 487, 604
pubic, 322. See *fig. 418*
of rib neck, 396

Crest—*Continued*
sphenoidal, 607
supramastoid, *fig. 731*
of tubercle
greater (groove, bicipital, lateral lip), 80, 93
lesser (groove, bicipital, medial lip), 80, 93
urethral, 280
Crista galli, 485, 610
terminalis, 430
Crus (crura) of diaphragm, 249, 250
penis, *fig. 334*
Crypts of tonsil, 559
Cupola, pleural (cervical pleura), 400, 408
Curves of vertebral column, 14
Cusp, aortic, 425, 433
terminology of, 429
Cylinder, axis, 42

D

Dartos, 185
Dendrite, 42
Dens of axis, 534
Dermatomes, 103. See *fig. 43.3*
Dermis, 68
Development, notes on. See individual names
Diaphragm
pelvic, 273
functions, 274
nerve supply, 274
"The," 249
development, 250
nerve supply, 250
relations of, 250
structures, piercing, 250
urogenital, 257, 258, 262
Diaphragm oris, 567
sellae, 481
Diaphysis, 7
Digital formula, 131
Digits, movements of, 132
Diploë, 6, 479
Disc—
articular (meniscus), 20
acromioclavicular, 153
inferior radio-ulnar (triangular), 165, 166, 167
intervertebral (fibrocartilage), 13, 18
intra-articular. See D., articular
of jaw joint (temporomandibular), 531
of knee joint (semilunar cartilages), 371–372
sternoclavicular, 152

Disc—*Continued*
optic. See Papilla, optic, 495, 503
Diverticulum, allantoic, 257
ilei, (Meckel's), 205
Dolichocephaly, 461
Door. See Porta
Dorsum—
ilii, 320
ischii, 320
sellae, 485, 488, 607
Ducts, Ductules, and **Ductus**—
aberrans testis, 189
allantoic, 204
alveolar, 414
arteriosus, 408, 422
closure of, 39
bile, to display, 230. See also Passages bile
bronchomediastinal, lymph, 518
cochlear, 597
of Cuvier. See Vein, cardinal
cystic, 212, 222
deferens (Vas deferens), 64, **186.** See also *fig. 225*
in abdomen, 231, 282
in cord, 186
development, note on, 257
deferens in pelvis, 269, 285
vessels and nerves, 186
ejaculatory, 280, 283
endolymphatic, 597
of epididymis, 187
of epoöphoron (Gartner), 264, 297
hepatic, 212, 222
accessory, 222
variations, 221, 228
jugular, lymph, 518
lacrimal, 458
lymphatic, right, 518
mesonephric, 63, 257, 296
Müllerian. See D., paramesonephric
nasolacrimal, 458, 579
pancreatic, 229, 230
variations in, 227, 228
paramesonephric, 63, 296
parotid, 458
prostatic, 281
semicircular, 597
Stenson's. See D., parotid
subclavian, lymph, 518
sublingual, 567
submandibular, 567
tear, 458
of testis, 64, 187
efferent, 187
thoracic

Ducts, Ductules, and Ductus—*Continued*
in neck, 518
in thorax, 446
venosus, 39, 199
vitello-intestinal, 205
Wolffian. See D., mesonephric
Duodenum, 192, 194, **226**
blood supply of, 228
development, note on, 227
interior of, 229
position of, 194, 203
relations of, 227, 228
Dura mater, 40, 480

E

Ear—
auricle of, 455
external, 590
lymph vessels, 601
internal (or labyrinth), 596
middle (or tympanum), 591
vessels and nerves, 595
Ejaculation, 63
Elbow, structures around, 112
tennis, 145
Elevation of the upper limb, 161
Eminence—
articular. See Tubercle, articular, 540
frontal. See Tuber, frontal, 461, 603
iliopubic (iliopectineal), 271
intercondylar, 373
parietal. See Tuber, parietal, 461, 523
End-arteries, 35
Endocardium, 31
Endocranium, 480
Endometrium, 294
Epicardium, 31, 420
Epicondyle—
of femur, 314
of humerus, 109, 112
Epidermis, 68
Epididymis, 64, **187**
anomalies of, (rudimentary), 187
structure of, 187
vessels and nerves of, 187
Epiglottis, 556, 585
movements of, 563
Epiphyses, 7. See also Ossification
acromion, 96, 102. See *fig. 188*
appearance and fusion times. See table 8, p. 171 and table 17, p. 387.
atavistic, 8
calcaneus, see *fig. 457*

Epiphyses—*Continued*
cartilaginous, 8
clavicle, 99, 160, 171
coracoid, 96, 102, 160, 171
femur, 314, 364, 375, 387
fibula, 344, 379, 387
foot bones, 354, 387
hip bone (os coxae), 320, 361; and *fig. 416*
iliac crest and ant. inf. spine, 320
humerus, 159, 165
pressure, 7
radius, 165, 166
scapula, 102, 160
thumb bones. See Table 8, p. 171
tibia, 375, 379
traction, 7
ulna, 165, 166
vertebral, 12
Epistropheus. See Axis vertebra
Epoöphoron, 297
Equator of eyeball, 495
Erection, 262, 264
Esophagus (oesophagus)—
abdominal portion, 191, 250
cervical portion, 520
sphincters of, 448
thoracic portion of, 403, 406, **447**
vessels and nerves, 448
Ethmoid, 485, 582, **609**
Eversion of foot, 350, 355, 379
Expansion, extensor (dorsal), 146
Expiration. See Respiration
Eye—
accommodation of, 498
convergence of, 498
development, note on, 504
Eyeball, 501
blood vessels, 500
development, 504
dissection, 501
fascia (fascial sheath), 496
muscles, 495–497
nerves, 497–500
Eyelids, 456
muscles of, 457, 458, 495–496

F

Face—
blood supply, 459
bones, 451, 602
muscles of, 454–455
nerve supply of, 458
Falx—
cerebelli, 480
cerebri, 480
inguinalis, 180
Fascia—
of abdomen, 175, 182, 183
of arm, 109
axillary, 86
of back. See F., thoracolumbar
buccopharyngeal, 552
Buck's. See F., of penis
bulbi. See Sheath of eyeball
Camper's. See F., of abdomen
cervical. See F., of neck
clavipectoral, 79, 80, 86
Colles'. See F., of perineum
cremasteric, 183, 185
cribriform, 310
deep, 72. See also individual names.
Denonviliers, see prostatic, 264, 281
dorsal. See F., lumbar
of foot. See F., plantar
of forearm, 145
of glutei muscles, *fig. 424.1*
iliaca, 247
investing (enveloping). See F., deep
lata, 247, **309,** 323
of leg
back of, 348
front of, 341
lumbar, 471
of neck (investing), 463, 506
posterior triangle of, 463
obturator, 259, 270, 274
omohyoid, 464
palpebral. See Septum, orbital, 457
of pelvis, 270
of penis, 260
of perineum, 258
pharyngobasilar, 554, 561
plantar, 356
popliteal, 327, 329
of popliteus, 329
postvertebral, 465
prevertebral, 465, 536
prostatic, 264, 281
psoas, 247
rectovaginal (septum), 291
rectovesical, 271, *fig. 348*
renal, 233
Scarpa's. See F., of abdomen
sheath. See Sheath
Sibson's. See Membrane, suprapleural, 397
spaces. See Spaces
spermatic
external, 183, 185
internal, 183, 185
superficial. See F., spermatic, external
Fascia—*Continued*
superficial, 71
temporal, 527
of thigh, 309
thoracolumbar (F. lumbodorsal), 471
transversalis, 183
of urogenital diaphragm. See Membrane, perineal
vesical, 271
Fat. See also Tissue, adipose
buccal pad, 455
extraperitoneal (subperitoneal), *fig. 194*
fat-pads, synovial, 20, 163
infrapatellar pad, 374
perinephric, 233
Fauces, 556
Femur, 312
head of, blood supply, 367
orientation, 313
ossification, 364, 387
platymeria, 314
Fenestra—
cochleae, 594
vestibuli, 594
Fibers—
intercrural, 180
postganglionic, 44, 47
preganglionic, 44, 47
Purkinje, 436
Fibula, 342
functions of, 342
ossification, 344, 387
"Field, triangular," of dura, 481, 484
Fila, nerve, 41
Filum terminale, 41
Finger-print, 131
Fingers, movements of, 132, 169–171
Fissures. See also Sulci
"H-shaped" of liver, 219
of lung, 412
orbital
inferior, 493, **494,** 527
superior, 488, 493, **494,** 607
palpebral, 457
petrotympanic, *fig. 756*
pterygomaxillary, 528
squamotympanic, *fig. 756*
tympanosquamous, 541
Fixator, 28
Flexures of colon, 192, 193, 195
Fluid—
cerebrospinal, 40, 41, 482
synovial, 19
tissue, 35

Folds. See also **Plicae**
alar, 374
aryepiglottic, 556, 588
of Douglas. See Line, arcuate
glosso-epiglottic, 556, 584
infrapatellar, 373, 374
interureteric (torus), 279
labioscrotal (genital), 185, *fig. 224*
peritoneal, 205
salpingopharyngeal, 555
synovial, 20
vestibular (ventricular), 588
vocal (cord), 588
Follicles (nodules), lymphatic, aggregated and solitary, 226
Fonticuli (fontanelles), 461, 621
Foot—
arch of, 352, 384
support, 384
articulated, from above, 350
from below, 356
door of, 361
from the sides, 353
bearing points of, 356, 384
compared with hand, 385
muscles of the sole, 356–361
units of, 352
vessels and nerves, comment on, 363
weight distribution of, 356, 384
Foramen—
alveolar, 614
cecum of skull, 610
of tongue, 569
condylar, posterior. See Canal
"crescent of", in sphenoid, 488
dental. See F., alveolar
epiploic (mouth of lesser sac), 197, 212
ethmoidal, 485, 495, 603
incisive, 539
infra-orbital, 453, 614
intervertebral, 13
jugular, 486, 487, 539, 541
lacerum 488, 489, 539
magnum, 470, 486, 604
mandibular, 524, 527
mastoid, 470
mental, 453, 524
nutrient, 8
obturator, 273, 286, 322
optic. See Canal, optic
in orbit, 494
ovale
of heart, 431
closure of, 39
of skull, 488, 528, 531, 607
Foramen—*Continued*
palatine
greater, 617
lesser, 617
parietal, 470
in pelvis, 273
for petrosal nerves, 488
rotundum, 488, 606
sacral, 273
sciatic
greater, 273, 286, **323.** See Porta to gluteal region
lesser, 273, 323
of skull, norma frontalis, 453
sphenopalatine, 557, 577
structures traversing, 577
spinosum, 488, 530, 608
stylomastoid, 522, 539, 612
supra-orbital, 453, 495
transversarium, 516, **534**
vertebral, 12, 16, 533
of Vesalius, *fig. 688*
Force couples, 161
Foreskin (prepuce) of penis, 260
Formula, digital, 131
Fornices—
of conjunctiva, 456
of vagina, 291, **294**
Fossa. See also Pouch
acetabular, 364
articular of jaw. See F., mandibular
canine, 614, *fig. 645*
cecal, 206
coronoid, 161
cranial
anterior, 485
middle, 485, 487–492
posterior, 485–487
cubital, 112
digastric, 525
duodenal, 205
hypophyseal, 488, 607
iliac, 273
incisor (incisive), 524, 614
infraclavicular (deltopectoral), 79
infraspinous, 89, 101
infratemporal (region), 527
intercondylar, 313
intersigmoid, 206
ischiorectal, 259
malleolar, 344
mandibular (articular), 531, 540, 611
navicular, of urethra, 281
olecranon, 109, 161
ovalis, of heart, 431
paracolic, 206
pararectal, 268
Fossa—*Continued*
paravesical, 268
pelvic, 206
peritoneal, 206, 268
pituitary. See F., hypophyseal
piriform, 556, 584
popliteal, 327, 329
pterygoid, 608
pterygopalatine (pterygomaxillary), 528, 575, 614
rectovesical, 206, 268
retrocecal, 195, 206
retrocolic, 195, 206
retro-omental, 206
scaphoid, 609
subarcuate, 487
sublingual, 525
submandibular, 525
subscapular, 96, 101
supraspinous, 89, 101
temporal, 524
terminal. See F., navicular of urethra, 281
vermian, 604
Frenulum—
clitoridis, 265
of tongue, 567
Frontal bone, 602

G

Galea aponeurotica. See Aponeurosis, epicranial, 462
Gall-bladder, 197
anomalies of, 221, 222
comparative anatomy, 197
development, 221
functions of, 221
structure of, 222
surface anatomy of, 221
vessels and nerves of, 221
Ganglia—
aorticorenal, 410
celiac, **243,** 250
cervical
inferior, 550, 551
middle, 550, 551
superior, 550, 551
cervicothoracic (stellate), 409
ciliary, 499
coccygeal. See G., impar
facial. See G., geniculate
geniculate (facial n.), 597
impar (coccygeal), 302
otic, 530
paravertebral, 44
posterior root, See G., spinal
prevertebral, 44
pterygopalatine (spheno-), 558, 577
sacral, 302

Ganglia—*Continued*
semilunar. See G., trigeminal
sphenopalatine. See G., pterygopalatine
spinal (dorsal root), 42, 85, 476
stellate. See G., cervicothoracic
submandibular, 567
submaxillary. See G., submandibular
sympathetic,
cervical, 550, 551
visceral, 44
trigeminal, 490
vertebral, 550, 551
Genitals, internal, female, 292
Genu, of facial nerve, 597
Girdle
pectoral (shoulder), 76
movements of, 154
pelvic, 274–278, 320
Glabella, 452
Glands—
adrenal. See G., suprarenal
bulbo-urethral, 64, 262
ciliary, 456
of Cowper. See G., bulbourethral
endocrine, 49
exocrine, 49
"Haversian." See Fat-pads
labial, 455
lacrimal, 500
nerve to, *fig. 795*
lingual, anterior, 570
lymph. See Nodes, lymph
mammary, 86
lymph supply of, 87
of Meibomeus. See G., tarsal
"paired, three," 231
oral, 53
parathyroid, 50, 519
development of, 520
para-urethral, 267
parotid, 53, 525
accessory, 525
nerve supply, 527
pineal, 49
pituitary. See Hypophysis cerebri, 49, 481
prostate. See Prostate
salivary, 53
sebaceous, 69
sublingual, 53, 567
submandibular, 53, 515
suprarenal, 50, 231
vessels and nerves of, 232
sweat, 70
tarsal, 457
thymus. See Thymus, 472, 50
thyroid, 50, 519

Glands—*Continued*
development, 519
vessels and nerves of, 519
of urethra, 281
vestibular, greater (of Bartholin), 66, 266
Glans—
clitoridis, 265
penis, 260
Glomerulus, renal, 60
Glottis, definition, 589
Gomphosis, 17
Granulations, arachnoid, 480, **482**
Gravity, line of, 15
Grooves. See also **Sulcus**
bicipital. See G., intertubercular
chiasmatic, 488, 607
costal (subcostal), 394
infra-orbital, 614
intertubercular, 80, 93
lacrimal, of maxilla, 614
optic. See G., chiasmatic
for petrosal nerves, 489
for radial nerve (spiral), 94, 113
semimembranosus, *fig. 428.2*
spiral, 80, 93
subcostal. See G., costal
for transverse sinus, 487
Growth. See Development
Gubernaculum ovarii, 267, 295
testis, 184, 265
Gums, nerve supply of, 574
Gut, rotation of, 207, 227
Gutter. See Fossa

H

Habitus, bodily, 201
Hairs, 69
Hamate bone, 126, 127, 130, 134, 135, 167
Hamulus—
lacrimal, 618
pterygoid, 539, 562, 608
Hand, 125
Hand bones compared with foot, 385
pentadactyl hand, 385
Head—
anatomy, surface, *fig. 689* on p. 489
of femur, 314, 364
of fibula, 343, 371
of humerus, 93, 155
of jaw, 524, 531
of metacarpal, 129, 168
of metatarsal, 353
of radius, 116, **162**
of talus, 350, 380
of ulna, 116, **165**

Heart, 420
blood supply of, 425
circulation of, collateral, 428
myocardial, 427
development, 428
musculature, 427
nerve supply of, 436
rotation of, 429
skeleton of, 434
structure of, 31
surface anatomy, 420, 433
surface
diaphragmatic, 424
posterior, 424
sterno-costal, 420
Helicotrema, 596
Hemorrhoids (haemorrhoids). See **Piles**
Hernia—
diaphragmatic, 250
femoral, 314, *fig. 436*
inguinal, 184, *fig. 436*
obturator, 332, *fig. 436*
Hiatus. See also Orifice in Adductor Magnus, 317
aortic, 250
esophageal, 250
maxillary, 615
saphenous, 310
semilunaris, 580
Hip bone, 320
functions of, 274–277
Hormones, 49
Humerus, 75, 79, 92
head of, 93, 155
lower end of, 109, 161
lower half of body, 109
upper end of, 92
Humor, aqueous, 502
Hymen, 266
Hyoid bone, 505, 507, **619**
development of, 620
variations, 620
Hypophysis cerebri, 481, 483, 489

I

Ileum, 192, **194,** 223
ilium, 320, 321
auricular surface, 276
tuberosity, 276
Incisive bone (premaxilla), 615
Incisura, tentorrii, 481
Incus, 594
Index—
cephalic, 461
ischium-pubis, 301
Infundibulum—
of right ventricle, 432
of uterine tube, 66, 293

Inion (Protuberance, external occipital), 469, 604
Innominate bone. See Hip bone
Inspiration. See Respiration
Insulin, 51, 55, 227
Interparietal bone, 606
Intestines, 55
development, note on, 207
features of, 195
functions of, 55
large, parts of, 192, 195
lymphatics, 251
nerve, supply of, 243-244
rotation of. See Rotation of gut
small, parts of, 192
structure of, 54, 195, 229
Inversion of foot, 350, **379**
Iris, 490, 542
Ischium, 320, **321,** 322, 323, 327
Islets of Langerhans, 227
Isthmus—
of aortic arch, 433
of auditory tube, 561
of fauces (oropharyngeal), 556
of pharynx, 564
of thyroid gland, 509, 519
of uterine tube, 293

J

Jaw. See Mandible
Jejunum, 192, **194,** 223
JOINTS—
hinge (ginglymus), general features, 21, 161
lubrication of, 20
strength of, 157
synovial, 19
types (varieties, classification) of, 17
of Axial Skeleton—
of atlas, 537
of costosternal, 396
costotransverse, 396
costovertebral, 396
craniovertebral, 537
crico-arytenoid, 586
cricothyroid, 584
interchondral, 396
intervertebral, 13, 17
lumbosacral, 275
mandibular, or temporomandibular, 453, **531**
nerve supply, 531
manubriosternal, 393
synostosis of, 393
of ossicles of ear, 594
of pelvis, 275
of ribs, 395
sacrococcygeal, 275

Joints—*Continued*
sacro-iliac, 276
development, 278
facets, accessory articular, 278
variations, 278
sternoclavicular, 152
sternocostal, 395
symphysis pubis, 275
temporomandibular or mandibular, 453, 531
vertebral, 13, 17
xiphisternal (junction), 394
of Lower limb—
ankle, 376
nerve supply of, 376
calcaneocuboid, 381
of foot, 379
end-to-end, 383
side-to-side, 383
hip, 364
anterior approach, 318
movements, 367
nerve supply of, 367
relations, 367
synovial membrane of, 365
intermetatarsal, 383
interphalangeal, 383
intertarsal, 383
inversion and eversion, 379
knee, 367
nerve supply of, 376
relations, 375
synovial cavity of, 374
metatarsophalangeal, 383
patellar, 369
subtalar (talocalcanean), 351, 380
talocalcanean, 380
talocalcaneonavicular, 381
talonavicular, 381
tarsal, transverse, 350, 355, 380
tarsometatarsal, 383
tibiofemoral, 369
tibiofibular, 342
of Upper Limb—
acromioclavicular, 153
carpometacarpal, 128, 168
of thumb, 127, 128
elbow, 161
nerve supply of, 165
intercarpal, 128, 168
intermetacarpal, 128, 168
interphalangeal, 130, 169
metacarpophalangeal, 128, 130, 169
midcarpal, 126, 127
radiocarpal. See J., wrist
radio-ulnar

Joints—*Continued*
distal, 165
intermediate, 165
proximal, 162
shoulder, 155
nerve supply of, 161
sternoclavicular, 152
function, 153
transverse carpal (midcarpal), 120, 127
wrist, 166
nerve supply of, 168
Jugum sphenoidale, 607
Junction—
corneoscleral, 495, 502
duodenojejunal, 192
xiphisternal, 394

K

"Key positions," axilla (Pectoralis Minor), *fig. 59*
digastric, 514
elbow, *fig. 108*
gluteal region (Piriformis), 323
heart, sternocostal surface, 420
neck (Digastric), 514
suprascapular region, 91
wrist, *fig. 124*
Kidney, 60, 231, **233**
anomalies of, 235, 236
blood supply of, 61, 234, 235
function of, 60
lower poles of, 196, 233
migrations of, 231
nerves of, 62, 234
pelvis of, 235
relations of, 233, 234
segments, 235
structure of, 60, 234
surface anatomy, 233
unit (nephron), 60
vessels of, 61, 234

L

Labia—
majora, 265
minora, 265
Labra, 20
acetabular, 364
glenoidal, 155
Labyrinth—
bony, 596
ethmoidal, 609, 610
membranous, 597
Lacertus fibrosus. See Aponeurosis, bicipital
Lacrimal bone, 617
Lacteals, 251
Lacunae—
laterales, 482

Lacunar—*Continued*
urethrales, 281
Lacus lacrimalis, 456
Lambda, 461, **469**
Laminae. See also Plates
pterygoid, 528, 539, 608
spiral, 641
vertebral, 12, 15
Landmarks. See Surface anatomy
Larynx, 57, **584**
lymph vessels, 601
mucosa of, 589
muscle actions of, 587
nerves and blood vessels, 589
Lens, 501, 502
LIGAMENT(UM)—
alar, 538
of ankle
lateral, 378
medial (deltoid), 378
anular (annular) of radius, 162
apicis dentis, 538
arcuate
of diaphragm, median, 249
lateral and medial. See Arches, lumbocostal, 234, 250
of pubis (inferior pubic), 257
arteriosum, 408, 422
ary-epiglottic, 585
atlanto-axial, 537
of atlas, 537
bifurcate, 381
broad, of uterus, 67, 292, 293, 294, 297
calcaneocuboid, 381
calcaneofibular, **342,** 378, 380
calcaneonavicular, plantar, 354, 381
calcaneotibial, 378, 380
cardinal, 298
carpal, transverse. See Retinaculum, flexor, of wrist
cervical lateral, 298
check. See L., alar
collateral
of elbow, 162
of interphalangeal, 129
of knee, 370
of metacarpophalangeal joint, 129
of wrist, 168
conoid, 153
of Cooper. See L., pectineal of breast, 86
coraco-acromial, 158
coracoclavicular, 153
function of, 153
coracohumeral, 156

Ligament (um)—*Continued*
coronary
of knee, *fig. 515;* 369, 374
of liver, 199; (*fig. 251*)
costoclavicular, 153
costotransverse, 396
cricothyroid lig. or memb., 584, 585
median, 507, 584
cricotracheal, 584
cricovocal lig or memb. See Lig., cricothyroid and Conus Elasticus
cruciate
of atlas, 537
crural. See Retinaculum, extensor of ankle
of knee, 373
deltoid, of ankle, 378
denticulatum, 40
definition, 20, 191
dorsal carpal. See Retinaculum, extensor of wrist
of elbow, 162
falciform, 198, 205
fibular collateral, 341, 371, 372
flavum, **19,** 537
of foot, fore part, 383
gastrocolic, 203; *fig. 235.1*
gastrolienal (-splenic), 193, 198, 209
gastrophrenic, 193
glenohumeral, 156
of head of femur, 366
hepatorenal. See L., coronary, 199
iliofemoral, 365
iliolumbar, 246, **275,** 277
infrapubic. See L., arcuate of pubis
infundibulopelvic. See L., suspensory of ovary
inguinal (of Poupart), 178, 181, 183
intercarpal, 128
interclavicular, *fig. 173*
interosseous
talocalcanean, 380
tibiofibular, 337, 343
interspinous, 19, 538
intertransverse, 19
intra-articular. See also Disc, articular
of rib head, 395
ischiofemoral, 365
of knee
lateral (fibular collateral), 341, **371,** 372, 374
medial (tibial collateral), 336, **370,** 372, 374

Ligament(um)—*Continued*
oblique popliteal, 370, 372
lacinate. See Retinacula, flexor of ankle
lacunar, 179
lienorenal, 197, 209, 210
link, of digits, 170
longitudinal, 18, 538
metacarpal, deep transverse, 130; *figs. 149, 150, 170*
metatarsal, deep transverse, 130, 383
mucosum. See Fold, infrapatellar
of neck of rib. See costo-transverse, 396
nuchae, 470; 538
ovario-uterine. See L., of ovary
of ovary, 205, 292
palmar, (palmar plate), 129
palpebral, 457
patellae, 318, 369, 375
pectinate, of iris, 502
pectineal, 273
peritoneal, 191
phrenicocolic, 196, 205
plantar
calcaneocuboid (short plantar), 381
calcaneonavicular (spring), 354, 381
long, 382
short, 381
Poupart's. See L., inguinal
pterygomandibular. See Raphe
pubic, inferior. See Lig., arcuate, of pubis, 257
pubofemoral, 365
puboprostatic, 269, 271. See L., pubovesical
pubovesical, 269, 271
pulmonary (pulmonale), 400, 405
radiate
costovertebral, 396
sternocostal, 396
radiocarpal, 168
retinacular, of digits, 170
round. See L., teres
sacro-iliac, 276, 278
sacrospinous, 273, 277
sacrotuberous, 273, 277
sphenomandibular, 524, 532
spring, 354, 381, 382
stylohyoid, 554; *fig. 769*
stylomandibular, 514, 525, 532
suprascapular. See Lig., transverse, scapular, 91
supraspinous, 19, 538

Ligament(um)—*Continued*
suspensory
of lens, 502
of ovary (infundibulopelvic), 206 (*Table 10*), 292
of penis, 261
talocalcanean, 380
talofibular, 378
temporomandibular, of lateral lig., 531
teres
of femur. See L., for head of femur
of liver, 198, 205
of uterus (round), 265, 292, 298
thyro-arytenoid. See L., vocal
thyrohyoid, 584
tibial collateral, 336, **370,** 372, 374
tibiofibular, distal (inferior), 343
transverse, 378
transverse
of acetabulum, 364
of atlas, 537
carpal. See Retinaculum, flexor of wrist
crural. See Retinaculum, extensor
deep, of foot, 130, 383
deep, of palm, 130
humeral, 157
scapular (suprascapular), 91
trapezoid, 153
triangular
of liver, 199, 205
of perineum. See Membrane, perineal
of tubercle of rib. See Lig., costotransverse, lateral, 396
uterosacral, 292, 298
venosum, 199, 205
vestibular (ventricular), 585
vocal, 585
volar, accessory. See L., palmar, 129
"Y". See L., iliofemoral
Limb, upper, elevation of, 161
to measure, 90
Linea and **Line.** See also Ridge.
alba, 176
aspera, 313, 318, 327, 332
axial
of foot, 359
of hand, 132, 147
cleavage of skin (Langer's), 68
epiphyseal. See Epiphyses
of gravity, 15

Linea and Line—*Continued*
iliopectineal, 272
internervous (boundary), 117, **123,** 142, 318
intertrochanteric, 313
Langer's, 68
median, of neck, 505
milk, 87
mylohyoid, 525
nuchal
inferior, 470, 604
superior, 469, 521, 604
"oblique," of base of skull, 540
oblique
of jaw, 453, 524
of radius, 116, 117, 143
of thyroid cartilage, 508
(pectineal) pecten pubis, 178, 272, 312
of safety. See L., internervous and Nerves, sides of safety
soleal (oblique), 346; *fig. 455*
supracondylar, of humerus, 109
temporal, 452, 522
transverse
anterior, 538
posterior, 539
Lingula of lung, 412
of mandible, 524
of sphenoid, 609
Lips of mouth, 455, 566
Liver, 54, 197, 198–199, **219**
acinus of, 54
blood flow through, 54, 215, 216
development of 209, 217, 227, 251
ducts of. See Ducts, hepatic and bile
functions of, 55
lobes, 221
lobule of, 54
peritoneal attachments, 198, 199
porta, 197, 219
segments of, 221
structure of, 54, 222
surface anatomy of, 203, 221
surfaces, 197
posterior, 220
visceral or inferior, 197, 220
Lobe—
of liver
caudate, 221
quadrate, 221
of lung, 412
of azygos vein, 412
infracardiac, *see fig. 584*
lingular, 412
of thyroid, pyramidal, 519

Lobule
of liver, 54
of lung, 414
Lubrication boundary, 20
Lunate bone, 126, 127, 128
ossification, 130
Lung, 58, **411**
arteries, 414
bronchopulmonary segment, 413
fissures of, 412
lingula of, 412
lobes of, 412
lobule of, 414
lymphatics of, 416
movements of, 418
nerves of, 417
notch, cardiac, of, 402, **411**
root of, 403, 405, 406, 411
structure of, 414
surface anatomy, 411, 417
variations in, 412
vessels of, 414
Lymph, 36
Lymphatics, 36. See also Duct, Node, Trunk
of abdomen, 251
of abdominal wall, superficial, 178
of bladder, 304
follicles (nodules), lymph, solitary and aggregated, 226
of genitalia, 187, 303, 304
of intestine, 251
of liver, 253
of pelvis, 303
of rectum, 304
of stomach, 252
of axilla, 87
of head and neck, 599
of lower limb
inguinal, 311, 315
popliteal, 331
of lung or pulmonary, 416
of mammary gland, 87
parotid, 600
pelvic, 303
of posterior triangle of neck, 465, 599–600
of testis, 187
of tongue, 601
tracheo-bronchial, 441
of upper limb, axillary, 87
mammary gland, 87
supratrochlear, 108

M

Macula lutea, 495; *fig. 706.1*

Malleoli, tibial and fibular, 351, 354, 376, 377, 378
Malleus, 594
Mamma, 87
  accessory, 87
Man, Java, 372
Mandible, 451, **524,** 531, 691
Manubrium sterni, **392,** 393
Margin, orbital, 451, 493
Marrow of bones, 8
Mass, lateral, atlas, 534
  of sacrum. See Sacrum, pars lateralis, 272
Mastication, 52
Mastoid bone, 612
Mater
  arachnoid, 40
  dura, 40, 480
    nerve supply of, 481
    of spinal cord, 40
  pia, 40
Maxilla, 451, **612**
  zygomatic process of, 613
Meatus—
  acoustic (auditory)
    external, 521, **590,** 611
    internal, 486; *fig. 686*
    of nose, 579, 580
Media, refractive, 501, 503
Mediastinum of thorax, 400
  anterior, 400
  middle, 400
  posterior, 400, 437, **444**
  superior, 400, 437, **438**
Mediastinum testis, 186
Medulla, spinal, 40
Membranes—
  atlantooccipital
    anterior, 538
    posterior, 475
  basilar, 596
  costocoracoid, 79
  cricothyroid, memb. or lig., 584, 585
  cricovocal. See Memb., cricothyroid
  hyaloid, 502
  intercostal
    external, 397
    internal, 397
  interosseous
    of forearm, 76, 116, 165
    of leg, 343
  mucous, of larynx, 589
  obturator, 273
  perineal, 258, 260, 266
  periodontal (peridental), 571
  pleuroperitoneal, 250
  quadrangular, 585
  synovial, 19
Membranes—*Continued*
  tectorial, 538
  thyrohyoid, 507, 584
  tympanic, 590, 592, **594**
    secondary, 594, 595
Meninges, See also Mater, 40, **480**
  arteries, 481
  nerves of, 481
Menisci. See Disc, articular
  of knee (semilunar cartilages), 371–372
Mesentery, 191, 205
  of appendix. See Meso-appendix
  primitive, 195, 207, 209
  root of, 194
  "of umbilical vein" (falciform ligament), 209
  vessels of, 195
  windows in, 195
Meso-appendix, 196
Mesocardium, dorsal, 428, 429
Mesocolon, sigmoid (pelvic), 196, 206
  transverse, 196, 211
Mesoduodenum, 209, 210
Meso-esophagus, 403, 408, 447
Mesogastrium, 209, 210
Mesotendon, 137
Mesovarium, 292
Metacarpus, 128
  bases of, 128
Metaphysis, 8
Metatarsal bones, 352, 382, 385, 387
  bases of, 352
Modiolus, 596
Mons pubis (veneris), 265
Mold, parotid, 525
Mouth, 52, **566**
  floor of, 567
  of lesser sac. See Foramen, epiploic
Movements. See Respiration and Joints
Mucoperiosteum, 583
Multangular bones:
  greater. See Trapezium
  lesser. See Trapezoid
Muscle—
  action, 26
  antagonists, 27
  architecture (internal structure), 24
  attachments (origin, insertion), 22, 23
  blood vessels, 27
  contraction, 25
  couples, Force, 161
  fixators, 27
Muscle—*Continued*
  insertion, 23
  investigation of, 25
  mooring, 75
  motor unit, 26
  nerves, 27
  nomenclature, 28
  origin, 23
  parts, 22
  prime movers (agonists), 27
  sphincter, 22
  structure, 23, 24
  supernumerary (accessory), 29
  synergists, 27
  synovial sheath, 24
  types of, 21
  variations, 29
  voluntary (skeletal), 21, 22
MUSCLES—
  abdominal, functions of, 176. See also individual muscles.
  abductor, digiti minimi, of hand, 134
    of foot
      insertion, 357
      nerve supply, see *fig. 491*
      origin, 357
    hallucis
      insertion, 357
      nerve supply, 359
      origin, 357
    pollicis brevis
        attachments, 133, 134
        nerve supply, 134, 135
      longus
        insertion, 145
        nerve supply, 143. See also fig. *171*
        origin, see *fig. 161*
  adductor hallucis,
    attachments, 359
    nerve supply, 361
    pollicis, 134
      nerve supply, 134–135
    of thigh
      brevis
        insertion, 332
        nerve supply, 332
        origin, 332
      longus, 314, 332
        insertion, 332
        nerve supply, 332
        origin, 332
  adductor magnus, 327, 332
    insertion, 332
    nerve supply, 328, 332
    origin, 327, 332
  anconeus, 145
  arrectores pilorum, 69

Muscles—*Continued*
articularis genu, 318
aryepiglotticus, 587
arytenoideus, 556, **586,** 588
auriculares, 456
axillary arch, 87
of the back, 471
deep (intrinsic), 471
action, 474
nerve supply, 476
biceps brachii
actions, 111, 164. See also table 5, p. 164
attachments, 110, 111
nerve supply, 111
biceps femoris, 327
nerve supply, 327
of bile duct, 222
brachialis
action, 111, 164. See also table 5 on p. 164
attachments, 111
nerve supply, 111
relations, 164
brachioradialis, 144
action, 164. See also table 5, p. 164
attachments, 144
nerve supply, 145
buccinator, 454
bulbospongiosus, 258, 262, 266
caninus. See M., levator anguli oris
cervical, deep, 535
ciliary, 503
nerve supply, 499
coccygeus, 274
nerve supply, 286
"cone of ocular", 495
constrictors of pharynx, 508, **554**
nerve supply, 509. See *fig. 762*
upper border of superior constrictor, 561
coracobrachialis
attachments, 110
nerve supply, 111
corrugator supercilii, 458
cremaster, 181, 185
crico-arytenoideus
lateralis, 587, 588
posterior, 586, 587, 588
cricopharyngeus, 509, 554, 564
cricothyroid, 508, 584, 586, 587
nerve supply, 511, 589
dartos, 185
deltoid, 93, **94**
nerve supply, 96
structure, 94

Muscles—*Continued*
depressor anguli oris, 454
(quadratus) labii inf., 454
diaphragm. See Diaphragm
digastric, 505, 510, 514; *fig. 720*
anterior belly, 505, 507; *fig. 735*
nerve supply, 529
posterior belly, 505, **510,** 514
nerve supply, 510
dialator pupillae, 500
nerve supply, 47, **500,** 551
of ear. See M., auriculares
epicranius, 454, 461
erector spinae, (sacrospinalis), 471–473
nerve supply, 476
extensor of wrist and hand
common origin, 144
extensores
carpi
radiales
attachments, 144
nerve supply, 144
ulnaris
insertion, 146
nerve supply, 144
digiti minimi (V) of hand, 145
insertion, 146
nerve supply, 144. See *fig. 171*
origin, 145
digitorum (communis)
insertion, 146
nerve supply, 144. See *fig. 171*
origin, 145
brevis (of foot)
attachments, 341
nerve supply, 340
longus (of foot)
attachments, 340
nerve supply, 340
hallucis
brevis, 341
longus
attachments, 341
nerve supply, 340. See also *fig. 491*
indicis
origin. See *fig. 161*
insertion, 146
nerve supply, 144
pollicis
origin, see *fig. 161.*
insertion, 145
nerve supply, 144. See also *fig. 171*
of eyeball, 495

Muscles—*Continued*
actions of, 497; *fig. 699*
cone of, ocular, 495
of face, 454
flexor accessorius. See M. quadratus plantae
carpi radialis, 119, 120
insertion, 120–121
nerve supply, 123
origin, 119
ulnaris, 119
insertion, 120–121
nerve supply, 123
origin, 120
digiti quinti (minimi) (brevis), of foot
insertion, 357
nerve supply, see *fig. 491*
origin, 357
of hand, 134
digitorum brevis, 357, 358
nerve supply, 361
longus, 347, 357
insertion, 357
nerve supply, 347. See also *fig. 491*
origin, 347
profundus
insertion, 137
nerve supply, 123
origin, 118
superficialis (sublimis)
insertion, 137
nerve supply, 123
origin, 118, 119
hallucis brevis
insertion, 358
nerve supply, 361
origin, 358
longus
insertion, 347, 358
nerve supply, 347. See also *fig. 491*
origin, 347
pollicis brevis, 134
pollicus longus
insertion, 13
nerve supply, 123, 135
origin, 117
frontalis, 458
gastrocnemius
insertion. See "Tendon of Achilles", 346
nerve supply, 330
origin, 345
gemelli, 325
genioglossus, attachments, 567, 570
function, 570
nerve, 549

Muscles—*Continued*
geniohyoid, 567
nerve, 549
gluteus
maximus, 322
nerve supply, 322
medius and minimus, 324
insertion, 324
nerve supply, 324
origin, 417, 324
gracilis
insertion, 317, 332
nerve supply, 332
origin, 332
hamstring, 327
"hybrid" or composite, 327
hyoglossus, 511, 514, 570
nerve supply, 549
hypothenar, 134
nerve supply, 134
iliacus and iliopsoas, 246
nerve supply, 247, 319
iliococcygeus, 274
iliocostalis, 471, **472**
infrahyoid, 507
nerve supply, 511
infraspinatus, 97, 157
insertion, 97, 157
nerve supply, 100
origin, 97
intercostal, 397
action, 418
interossei
of foot, 359
of hand, 147, 149
actions of, 147, 169
nerve supply, 149
interspinales, 474
intertransverse (intertransversarii), 247, 474
ischiocavernosus, 262
ischiococcygeus, 274
of larynx, extrinsic. See M., cricothyroid
intrinsic, 586
nerve supply, 589
latissimus dorsi
action, 90
origin, 90
insertion, 80
nerve supply, 80; table 2 on p. 84
levator, anguli oris, (Caninus), 454
ani, 258, 259, 264, **270,** 284, 298
nerve supply, 288
costae, 418, 474
labii superioris, 454
labii super. alaeque nasi, 454

Muscles—*Continued*
palati, 554, **558, 562**
nerve supply, 563
palpebrae superioris, 458, 491, **495**
nerve supply, 491
scapulae
insertion, 92, 464
nerve supply, 92, 467
origin, 92, 464
longissimus, 473
longus capitis, 535
colli (cervicis), 535
lumbricales
of foot
attachments, 358
nerve supply, *fig. 491*
of hand
actions of, 147, 169
insertion, 137, 169
nerve supply, 137
origin, 137
masseter, 525
nerve supply, 529
mentalis, 454
"mooring upper limb," 75
multifidus, 474
mylohyoid, 514, 563
nerve supply, 515
"native", of back, 471
of (soft) palate, 558
obliquus
abdominis externus, 176, **178,** 180, 181, 393
function, 181, 419
nerve supply, 177
abdominis internus, 176, 177, 178, **180,** 181, 393
function, 181, 419
nerve supply, 177
capitis
attachments, 475
nerve supply, 476
of eyeball, 495
nerve supply, 497
obturator externus
insertion, 326
nerve supply, 332
obturator internus, 323, 325
nerve supply, 289
origin, 273
occipitalis, 462
omohyoid, 464, 510
nerve supply, 511
opponens pollicis, 134
nerve supply, 135
digiti V, 134
orbicularis oculi, 457
nerve supply. See Muscles of the face, 454

Muscles—*Continued*
orbicularis oris, 454, 458
orbitalis (of Muller), 501
nerve supply, 578
palatoglossus, 558, 570
palatopharyngeus, 554, 558, 570
palmaris
brevis, 135
longus, 121
insertion, 136
nerve supply, see median nerve, 123
origin, 119
palpebral. See M., tarsal
papillary, 433
pars lacrimalis. See M., orbicularis oculi
pectinati, 430
pectineus, 312, 316, 317, 319, 322, 332
action. See table 15 on p. 368
insertion, 332
nerve supply, 249, 316, 328
origin, 332
pectoralis major,
action, 79, 154, 418; see table 4 on p. 160
anomaly, 88
attachments, 79
nerve supply, see *figs. 171 and 172*
pectoralis minor, 78
action, 154, 419; see table 4 on p. 160
attachments, 78
nerve supply, see *figs. 171 and 172*
perineal
deep transverse. See Diaphragm, urogenital
superficial, 260
peroneus brevis
attachments, 342
nerve supply, 340
peroneus longus,
insertion, 342, 359
nerve supply, 340
origin, 342
peroneus tertius, 340
piriformis, 273, 289, **323,** 326
insertion, 324
nerve supply, 288
origin, 273
plantaris, 29, 345, **346**
nerve supply, 330
platysma, 78, 454, **506**
nerve supply, 454, 506
popliteus, 345, 346
nerve supply, 330

Muscles—*Continued*
prevertebral, 535
procerus, *fig. 649*
pronator
quadratus
attachments, 117
nerve supply, 119
teres,
attachments, 116, 119
nerve supply, see median nerve, 123
psoas major, 246
nerve supply, 248, 319
tendon of, 316
psoas minor, 246
pterygoid
lateral (external), 528
nerve supply, 529
medial (internal), 528
nerve supply, 529
pubococcygeus, 274
puborectalis (sling), 274; *fig. 274*
pyramidalis, 176
quadratus
femoris, 325
nerve supply, 289
labii inferioris. See M., depressor labii inf.
superioris, 454
lumborum, 246, 418
nerve supply, 247
plantae, or flexor accessorius, 357, 358, 363
insertion, 358
nerve supply. See Nerve, lateral plantar, 361
quadriceps femoris, 318
nerve supply, 319
rectus
abdominis, 176
nerve supply, 177
capitis
anterior, 535
lateralis, 535
posterior, 475
nerve supply, 476
"cervicis", 507
femoris, 318
nerve supply, 319
oculi, 495
nerve supply, 497
rhomboid
attachments, 92
nerve supply, 92, 467
risorius, 454
rotatores, 471, 473, **474**
sacrospinalis. See Erector Spinae, 472
"safety"

Muscles—*Continued*
of larynx, 587
of tongue, 570
salpingopharyngeus, 559
sartorius, 314
insertion, 316, 319
nerve supply, 316
origin, 316
scalenus
anterior, 465, 535
medius, 465, 535
posterior, 465, 535
semimembranosus, 327, 330
nerve supply, 328
semispinalis, 465, 471, 474, 475
semitendinosus, 327, 330
nerve supply, 328
serratus anterior
action, 80
attachments, 80–81
function of, 80
nerve supply, see table 2 on p. 84, 467
serratus posterior, 418, 471
"shut-off." See M., sphincter urethrae
soleus
insertion. See Tendon of Achilles, 346
nerve supply, 330
origin, 346
sphincter
ani externus, **259**, 260, 284
nerve supply, 259
ani internus, 284; see *fig. 370*
nerve supply. See N., pelvic splanchnic, 288
cloacae, 255
nerve supply, see *fig. 329*
of esophagus, 448
ileocecal (ileocolic), 225
of pancreatic duct, *see fig. 284.1*
pupillae, 499
nerve supply, 499; *fig. 702.1*
pyloric, 54, 193
urethrae, 258, 262
nerve supply, 63, 302, 264
vaginae, 265
vesicae, 288
nerve supply, 63
spinalis, 471, 472, 473
splenius
attachments, 464, 471, 473, 475
nerve supply, 476
spurt and shunt, 27
stapedius, 593, 595
nerve supply, 459, 597

Muscles—*Continued*
sternalis, 87
sternocleidomastoid. See sternomastoid (below)
sternohyoid, 508
nerve supply, 511
sternomastoid, 463
nerve supply, 463, 449
sternothyroid, 508
nerve supply, 511
"strap." See M., infrahyoid
styloglossus, 563, **570**
nerve supply, 549
stylohyoid, 510
nerve supply, 510, 549
stylopharyngeus, 554, 558, 559; *fig. 778*
nerve supply, 563
subclavius, *fig. 173*
nerve to, 468
origin, 397
subcostalis, 397
subscapularis
insertion, 80, 96
nerve supply, see table 2, p. 84 and *fig. 171*
origin, 100
supinator, 116
insertion, 116, 118, 143
nerve supply, 144
origin, 143
supraspinatus, 97, 157
nerve supply, 100
tarsal, 458
nerve supply, 47
temporalis, 527
nerve supply, 529
tensor
fasciae latae (femoris)
insertion, 323, 325
nerve supply, 324
origin, 324
palati, 558, 561, **562**
nerve supply, 563
tarsi. See M., orbicularis oculi
tympani, 592, **595**
nerve supply, table 19 on p. 531
teres major
actions, 97
insertion, 80
nerve supply, see table 2, p. 84 and *fig. 171*
origin, 97, 102
teres minor
actions, 97
insertion, 97, 157
nerve supply, 96
origin, 96, 97

Muscles—*Continued*
thenar, 133
nerve supply, 134
of thumb, "outcropping", 145
thyro-arytenoid, 587, 588
thyro-epiglotticus, 587
thyrohyoid, 508
nerve supply, 511, 549
tibialis anterior
insertion, 341, 355, 359
nerve supply, 339
origin, 340
tibialis posterior,
insertion, 355, 359
nerve supply, 347
origin, 346
of tongue, 570
transversospinalis, 471, **473**
transversus
abdominis, 176, 177, 178, **180,** 181, **246,** 397, 419
aponeurosis of, 246
function, 181, 419
nerve supply, 177
perinei profundus. See Diaphragm, urogenital
superficialis, 258, 262
thoracis, 397
trapezius, 91
function, 91
nerve supply, 91, 463
triangularis. See M., depressor anguli oris
triceps
brachii
attachments, 111
function, 111, 164
nerve supply, 112
surae, 346
trigonal, 281
urorectalis. See M., recto-urethral
vasti, 318
nerve supply, 319
vocalis, 587, 588
zygomaticus, 454
Myocardium, 31, 434
circulation in, 427

N

Nails, 69
Naris (aperture, nasal, anterior), 578
Nasal bones, 617
Nasion, 451
Nasopharynx, 555, 561
Neck—
of femur, 313
of humerus, anatomical, 93
surgical, 93
Neck—*Continued*
of jaw, 524
median line of, 505
of radius, 116
root of, 516. See also Triangles of the neck
Nephron, 60
Nerve—
blood supply of, 43
roots of, 41
"safe sides", 114, 318, 323
unit (neuron), 40, 42
NERVES—
of abdominal wall, 177, 184
abducent, 484, 491, 498
accessory, 91, **463,** 484, 511, **548**
acoustic (auditory), see vestibulocochlear, 484, 597
alveolar (dental),
inferior, 529
superior, 576, 581
anococcygeal. See N., caudal
ansa cervicalis (hypoglossi), 511
hypoglossi. See N., ansa cervicalis
subclavia, *fig. 579*
auditory. See N., vestibulocochlear
auricular, great, **468,** 507, 527
posterior, 459, 526
auriculotemporal, 459, 526
autonomic, 43
abdominal, 242
pelvic, 302
axillary (circumflex), 83, 95
of back, 90, **476**
buccal, 459, 529
cardiac, 436, 441
caroticotympanic, 492
carotid, internal, 492
caudal (anterior and posterior), 289, 476
cervical
anterior (ventral) rami, 537
branch of facial, 459, 507
ventral, 537
chorda tympani, 529, **567, 598**
ciliary, 499
circumflex. See N., axillary
coccygeal, 476
colli, transversus, 468, 507
cranial, within the skull, 483
last four, 545
relations to great arteries, 484
cutaneous—
of arm, 103
of axilla, 86
of back, 90
Nerves—*Continued*
of ear, 456
of forearm, 104
of hand, 106
variations in, 107
of leg, 348
of neck, anterior. See N. colli, transversus
perforating (of S. 2 and 3), 260, 289
of shoulder, 103
of sole, 361
of thigh (femoral) cutaneous, 248, **311**
posterior, 260, 289, 323, **326,** 329
of upper limb, 103
dental. See Nn., alveolar
descendens cervicalis, 511, 549
hypoglossi, 511, 549
digital of hand, 106, 136, 137
dorsal, of penis, 261, 264
scapular (n. to Rhomboids), 92, 467
of epididymis, 187
erigentes. See N., splanchnic, pelvic
ethmoidal, anterior, 499
of eyeball, 497
of face
motor, 459
sensory, 458
facial (VII), 459, 484, 506, 510, **526, 597**
buccal branch, 459
cervical branch, 459
intrapetrous portion, 597
sensory root. See N., intermedius
femoral, 248, 311, 316
frontal, 498
gastric. See Trunk, vagal
genitofemoral, 249, 312
glossopharyngeal (IX), 49, 484, 487, **545,** 554, 559, 563
in pharynx, 554
gluteal
inferior, 289, 323
superior, 289, 323, **324**
hemorrhoidal, inferior. See N., rectal inf.
hypogastric, 302
hypoglossal (XII) 484, 486, **511, 549,** 567
distribution of, 549
iliohypogastric, 177, 248
ilio-inguinal, 177, 248
infra-orbital, 458, **575**
infratrochlear, 458, 499
intercostal, 397

Nerves—*Continued*
intercostobrachial, 86, 103; *fig. 97.1*
intermedius, 49, 484, 567, 597
interosseous
anterior, 119
posterior, 114, 143
Kuntz's, *fig. 579*
lacrimal, 458, 499
laryngeal
external, 511, 512, 548, 589
internal, 511, 563, 589
recurrent, 548, 563, 589
development, 438
in neck, 516
in pharynx, 563, 589
in thorax
left, 408, 438, 444
right, 408, 444
superior, 511, 548, 589
lingual, 529, 567, 571
of lower limb, motor distribution, 362 (*fig. 491*)
lumbar, 247
first, 177
lumbo-inguinal. See N., genitofemoral
mandibular (V^3) 458, 490, **529**
branches of, 529. Table 18 on p. 529
maxillary (V^2), 457, 490, 574, **575**
median, 83
in arm, 113
in forearm, 119, 121, 123
in hand, 134, 135, 136
motor distribution and variations in, 135
meningeal, 481
mental, 459
musculocutaneous, of arm, 83, 112
of leg. See N., peroneal, superficial, 340
musculospiral. See N., radial
nasal
external, 458, 499
internal, see *fig. 796*
posterior lateral, 577, 580
nasociliary, 499, 500
nasopalatine, 577, 580
obturator, 248, 270, 332
accessory, 249, 333
in thigh, 312, 332
occipital
greater (great). (C. 2, post. ramus) 475; *fig. 662*
lesser (small), 468
oculomotor (III)
in cranium, 484, 491

Nerves—*Continued*
in orbit, 498
olfactory (I), 483
ophthalmic (V^1), 457, **490, 499**
optic (II), 483, 489, **495, 497**
palatine, 576
parasympathetic, 48
of parotid gland, 530
pectoral, table 2 on p. 84
of penis, 261, 264
perineal, 263
branch of fourth sacral, 260, 289
peripheral, 42
blood supply, 43
structure of, 42
to peritoneum, 211
peroneal
common (popliteal, lateral), 330, 339
deep (tibial, anterior), 339
recurrent (tibial, anterior) 339
superficial (musculocutan. of leg), 340
petrosal, deep, 492
(superficial) greater, 48, 489, **492**
(superficial) lesser, 48, 489, **492,** 546
pharyngeal, 577
phrenic
distribution of, 250, 404
in neck, 467, 518
in thorax, 403, 404, 406, 407, **444**
variations in, 407
plantar, 348, 361
plexuses, see Plexuses
pneumogastric. See N., vagus
popliteal
lateral. See N., peroneal, common
medial. See N., tibial, 330
presacral. See Plexus, hypogastric, superior
of pterygoid canal, 492, 577
pudendal (internal), 260–264, 289
radial, 83, 113
deep. See N., interosseous, post., 114
superficial branch, 114, 123
rami, spinal, dorsal (posterior), 42, 85, 476
communicantes, 45
ventral (anterior), 42, 85
rectal (hemorrhoidal),

Nerves—*Continued*
inferior, 263
recurrent. See N., laryngeal, recurrent
sacral, 476
fourth, perineal branch, 260, 289
saphenous, (long), 312, 316
scapular, dorsal, 92
sciatic, 288, **323, 328**
scrotal, posterior (cutaneous), 260, 264
segmental (spinal), 43, 84
sensory. See also N., cutaneous
of face, 458
of scalp, 462
"sides of safety", 114, 318, 323
sinus, 546
spermatic, external. See N., genitofemoral
spheno-palatine, 577
spinal, 84
accessory. See N., accessory
splanchnic, 45, 243, 250, **410**
greater, lesser, lowest, 250, **410**
lumbar, 45, 243
pelvic (n. erigens), 49, 244, 288, **302**
to find, 271
stimulation of, 62
thoracic, 45, **410**
stato-acoustic, see N. vestibulocochlear
subcostal, 249, 398
suboccipital, (C. 1, dorsal branch), 476
supraclavicular, 78, 103, 464
supra-orbital, 538, 458
suprascapular, 91, 464
supratrochlear, 458, 499
sural, 331, 340, *fig. 425*
communicating, 331, *fig. 425*
sympathetic system, 44
testicular, 187
thoracic, 177, 187. See also N., intercostal and subcostal
tibial, (med. popliteal and post. tibial), 330, **347**
anterior. See N., peroneal, deep, 339
recurrent, 339
posterior. See N., tibial
of tongue, 570
tonsillar, 560
transversus colli, 464, 507
trigeminal (V), 484, **490**

Nerves—*Continued*
trochlear (IV), 484, 491, 498
tympanic, 546
ulnar
in arm, 112
in axilla, 83; table 2 on p. 84
in forearm, 123
in hand, 134, 136
motor distribution and variations, 134, 135
ulnar, collateral, 113; *fig. 114*
of upper limb, motor distribution. See *figs. 171, 172*
vagus, 49, 484, 511, **546**
in abdomen. See Trunk, vagal
branches of, 436, 441, 444, 484, **547**
in cranium, 484
distributions, 547
in neck, 511, **547**
in thorax, 436, 441, 444
vestibulocochlear **(VIIIth)**, 484, 485, 597
zygomatic, 578
zygomaticofacial, 458, 578
zygomaticotemporal, 458, 578
Neurilemma (primitive sheath), 42
Neuron, 40
Node, atrioventricular, 435, 436
sinu-atrial, 435, 436
lymph, 36. See also Lymphatics.
abdominal, 251–254
axillary, 87
cervical, 465, 599–601
cubital, 108
iliac, 303
inguinal, 303, 311
jugulodigastric, 599, 601
jugulo-omohyoid, 599, 601
parotid, 600
pelvic, 303–305
popliteal, 311, 331
retropharyngeal, 600
submandibular (submaxillary), 600
supraclavicular. See N., lymph, cervical
supratrochlear, 108, *fig. 102*
of testis, 187
tracheobronchial, 441; *fig. 633*
Nodules, lymph. See Follicles, lymph
Nose. See also Cavity, nasal
aperture of, anterior. See Aperture, piriform posterior. See Choanae
Nose—*Continued*
external, structure of, 455
Notch, or Incisura—
acetabular, 364
cardiac, 411
ethmoidal, 603
fibular, of tibia, 337. See *fig. 443*
intercondylar, See Fossa, 313
jugular, (suprasternal), 79, 392
mandibular, 524
nasal, 603
pyramidal, of petrous bone, 487
radial, of ulna, 116, 162
scapular, (suprascapular), 91, 100, 466
sciatic
greater, 277, 321
structures traversing, 323
lesser, 277, 321
sphenopalatine, 616
spinoglenoid, 89, 101
supra-orbital. See Foramen
suprascapular. See Notch, scapular, 91, 100, 466
suprasternal. See Notch, jugular, 79, 392
trochlear, of ulna (semilunar), 162
ulnar, of radius, 165
vertebral (intervertebral), 12

O

Occipital bone, 469, 539, **604**
anomalies of, 606
ossification of, 604, 605
Oesophagus. See Esophagus
Olecranon, 162
Omentum. See *fig. 235.1*
development, notes on, 207
greater, 193; table 10 on p. 206
gastrocolic lig., 196
gastrolienal (-splenic) lig., table 10 on p. 206
gastrophrenic lig., table 10 on p. 206
lesser, 193, 199; table 10 on p. 206
Opening: See also Hiatus
into lesser sac. See Foramen, epiploic
saphenous (fossa ovalis), 310
Opposition, 135
Ora serrata, 503
Orbit, 493
blood supply of, 500
cavity of, 493
fissures and sutures, 493
margin of, 493
nerve supply, 497
Orifice, atrioventricular, 431, 433
bicuspid. See O., atrioventricular
cardiac (esophageal), 191–192, 193
of coronary sinus, 431
ileocecal (ileocolic), 192
mitral, 431, 433
tricuspid, 431, 433
urethral, 281
primitive, 267
Os (bone)
centrale, *fig. 540*
cordis, 434
coxae, See "Hip bone"
intermedium, 386
trigonum, 386
Vesalianum, 386
Os (mouth)
of uterus
external, 294
internal, 294
Ossicles of ear, 594
Ostium, 7
maxillary, 582
of uterine tube, 66
Ossification, 8. See also Epiphyses
of acromion, 96, 102
of clavicle, 99
of femur, 364, 375
of fibula, 344
of foot bones, 354
of frontal bone, 604
of hand bones, 130
of humerus, 165
of hyoid, 620
of mandible, 619
of maxilla, 615
of palatine bone, 617
of parietal bone, 604
of patella, 375
of ribs, 396
of sacrum, 272
of scapula, 102
of sphenoid, 606, 609
of sternum, 394
of temporal bone, 610
of tibia, 375
of vertebrae, 13
of wrist bones, 130
of zygomatic bone, 615
Ovary, 51, 66, 292
vessels and nerves, 231, 297–298

P

Pacchionian bodies. See Granulations, arachnoid
Pad, buccal (sucking), 455

Pad—*Continued*
fat-pad, synovial, 20
infrapatellar fat, 374
Palate, 557
bony, 539
development, note on, 557, 558
soft, 558
vessels and nerves, 563
structure, 557, 562
Palatine bone, 616
Palm, skin of, 131
Palpation. See Surface anatomy
Palpebrae. See Eyelids
Pancreas, 51, 55, **226,** 228
blood supply of, 223, 228
development, note on, 227-228
functions of, 51, 55
position of, 203
relations of, 227, 228
Papillae—
duodenal, 229
lacrimal, 456
optic, (disc), 495, 503
of tongue, 569
Paradidymis, 189
Parathyroid. See "Gland, parathyroid"
Parietal bone, 523, 604
Paroophoron, 297
Passages, bile, 221
structure, 222
Patella, 314, 319, **316,** 374
bipartite. See *fig. 518*
emarginate, 375
ossification of, 375
Pecten pubis or Line, pectineal, **273,** 312
of anal canal, 284
Pelvis, 268-305
arteries of, 285
bony, interior of, 271
brim, 272
diameters of, 299-301
false (greater), See P., major
female, 290
diameters of, 299-301
identification of, 301
in median section, 290
types of, 301
foramina in the walls, 273
joints of, 275
of kidney, 235
major (greater, false), 273
mechanism of, 274
minor (lesser, true), 273
muscles and fasciae, 273
nerves of (autonomic), 302
orientation of, 274
sex differences, 298
side wall of, 269
Pelvis—*Continued*
true (lesser), See P., minor
of ureter. See P., of kidney
veins of, 286
Penis, 64, 260
nerves to, 261
Pericardium, 400, **420,** 424
fibrous, 420
serous, 424
Perineum, 257
of female, 265
Perineurium, 42
Periorbita, 493
Periosteum, 7
Peritoneum, 191
absorption of. See Cohesion, 210
development, notes on, 207
folds of, 205
fossae of, 205
morphology of, 207
nerve supply of, 211
pelvic
female, 291-292
male, 268
Petrous bone, 487, 488. See also Temporal bone
Phalanges—
of foot, 350
of hand, 130
Pharynx, 53, 508, **552**
constrictors of. See Muscles
from behind, *fig. 768*
interior of, 555
laryngeal, 556
nasal, 555
oral, 556
structure of, 552
structures crossing borders of, 554
vessels and nerves of, 563
Piles (varices)
esophageal, 217
Pinna. See Auricle of ear
Pisiform bone, 126
Planes
Frankfort, 451
intertubercular, *fig. 191*
transpyloric, 190
transumbilical, 190
Plates. See also Laminae
cribrosa (cribiform), 485, 610
epiphyseal, 8, 17
orbital, of frontal bone, 485, 603
palatine, 616
palmar. See Ligaments, palmar
pterygoid. See Laminae
tympanic, 541, 611
Plates—*Continued*
vaginal, 296
vertical
of ethmoid, 609
of palatine bone, 616
Platysma. See Muscles
Pleura, 58, **400**
cupola (cervical), 400, 408
mediastinal
left, 406
right, 403
nerve supply of, 417
recesses (sinuses) of, 401-402
reflexions of, 401-402
PLEXUSES—
Nerve, 43
on aorta, 441
brachial, 83, 467
cardiac, 405, 410, 436, **441,** 442
carotid, internal, 492
celiac, 213, 219, 232, 243
cervical, 537
coccygeal, 289
coronary, 436
esophageal, 448
hypogastric, 62, 302
inferior (plexus, pelvic), 302
superior (n. presacral), 243, 302
intermesenteric, 243
lumbar, 247
pelvic. See P., hypogastric, inferior
pharyngeal, 563
presacral. See P., hypogastric, superior
pulmonary, 405, 410, 436, 441
renal, 234
sacral, 287, 288
solar. See P., celiac
Venous,
pampiniform, 187
pharyngeal, 514, 563
prostatic, 262, 286
pterygoid, 461, **530**
rectal, 286
suboccipital, 477
uterine, 286, 297
vaginal, 286, 297
vertebral
anterior and posterior, 477
vesical, 279, 286
Plicae. See also Folds
circulares, 194, 229
fimbriata, 570
longitudinalis duodeni, 230
semilunaris, 456
sublingualis, 567
transverse of rectum, 283

Point—
  central of perineum. See Body, perineal
  midinguinal, *fig. 231*
Porta—
  "of foot", 361
  "of gluteal region", 323
  of liver (hepatis), 197, 219
Position, anatomical, xii
Postaxial, definition of, 75, 103
Pouches. See also Fossae and Recesses of Colles. See P., perineal, superficial of Douglas. See P., recto-uterine
  hepato renal, 205
  of Morison. See P., hepato-renal
    perineal, deep, 258, 262
    superficial, 258, 260
      contents of, 260
  recto-uterine, 292
  vesico-uterine (uterovesical), 316, 292, 294
Preaxial, definition of, 75, 103
Premaxilla or incisive bone, 615
Prepuce—
  clitoridis, 265
  penis, 65, 260
Prime mover, 27
Prints, finger, 131
Processes—
  accessory of vertebrae, 470
  alveolar, 453, 524, 539, 614
  of calcaneus, 352
  caudate of liver, 199
  cervical, transverse, 470
  ciliary of eye, 502
  clinoid
    anterior, 481, 485, 606
    middle, 607
    posterior, 485, 607
  cochleariformis, 592, 595
  coracoid. See Coracoid, 77, 91, 101
  coronoid
  of mandible, 524
  of ulna, 162
  frontal, of maxilla, 614
  funicular, 186
  jugular, 535, 605
  mammillary, 470
  mastoid, 463, 469, 521, 522, 539, 612
  odontoid. See Dens of axis, 534
  palatine, 615
  paramastoid, 606
  postauditory, 611
  pterygoid, 527, 528, 538, **608**
  pyramidal
    of palatine bone, 616
Process—*Continued*
  styloid, of radius, 116, 145
    of temporal bone, 522, 539, 541, 611
    ulna, 116, 145
  transverse, morphology of, 15
  uncinate
    of ethmoid, 610
    of pancreas, 226
  vaginalis
    peritonei, 184, 186
    in female (of Nuck), 295
    of temporal bone, 608
  vermiform. See Appendix
  xiphoid, 393
  zygomatic
    of frontal, 603
    of maxilla, 614
    of temporal (zygoma), 611
Prominence, laryngeal, "Adam's apple," 505, 584
Promontory—
  of sacrum, 272
  of tympanic cavity, 593
Pronation and supination of foot, 379
  of forearm, 166
Prostate, 64, 279, **281**
  exposure of, 264
  nerves of, 302
  structure of, 282
  vessels of, 283
Protuberance—
  mental, 524; *fig. 645*
  occipital
    external. See Inion
    internal, 604
Pterion, *fig. 689; 522*
Pubis, 321
  arch, 322
  functions of, 275
  symphysis, 275
Pulmones, see Lungs
Punctum lacrimale, 456, 458
Pupil, 456
  contraction of, 498
  dilator pupillae, 500, 551
Pylorus, 54, 192, 194, 218
Pyramid of tympanic cavity, 592, 595

R

Radius, 116
  dorsal surface of, 145
  head of, 116, 162
  palpable parts of, 142
  tuberosity, 116
Rami. See Nerves, rami, spinal
Ramus—
  of ischium, 321
Ramus—*Continued*
  of mandible, 524
  of pubis, 321
Raphe—
  anococcygeal, 259
  pterygomandibular, 455
Receptaculum chyli. See Cisterna chyli
Recess. See also Fossa and Pouch
  epitympanic, 591
  peritoneal. See Fossae, 205
  pharyngeal, 555, 556
  piriform (fossa), 556
  pleural (sinus), 401
  spheno-ethmoidal, 579
Rectum, 197, **283**
  ampulla of, 283
  development, note on, 257
  nerves of, 302
  relations of, 285
  vessels of, 284-285
Reflexion, pleural, 401-402
Regions—
  anal, 257, **259**
  deltoid, see R., shoulder, 92
  fibular, 341
  gluteal, 322
  infratemporal, 521, 527
  mastoid. See R., stylomastoid
  parotid, 525
  pectoral, 75
  perineal, 257
  peroneal (lateral crural), 341
  scapular, 89
  shoulder, 92
  stylomastoid, 522
  sublingual, 567
  suboccipital, 474
  temporal, 521, 527
  of thigh, 318
  urogenital, 257, **260, 265**
Respiration, 59, 418
Rete testis, 187
Retina, 503
Retinacula, 72
  extensor
    of ankle, 341
    of wrist, 146
  flexor
    of ankle, 348
    of wrist, 127, **132**
  peroneal
    inferior, 342
    superior, 342
Ribs, 394, 470
  angles of, 396
  variations, 397
Ridge. See also Lines
  interureteric. See Fold
  pronator, 117

Ridge—*Continued*
supracondylar, of humerus, 109
Rima glottidis, 589
Ring—
femoral, 312
inguinal
deep (abdominal), 183
superficial (subcutaneous), 180, 183
Roots, nerve, 41
Rostrum, 607
Rotation—
of duodenum, 228
of gut, 207
of heart, 429
of limbs, 75
of stomach, 210

S

Sac—
alveolar, 56, 414
conjunctival, 456
endolymphatic, 487, 597
lacrimal, 458
pericardial. See Pericardium
of peritoneum
lesser. See Bursa, omental, 198, 205
mouth of. See Foramen, epiploic
Saccule—
of ear, 597
laryngeal, 588
Sacrum, **271,** 275
auricular surface, 272, 276
pars lateralis, 272
tuberosity, 276
Scala—
tympani, 596
vestibuli, 596
Scalp, 461
Scaphoid bone, 126
Scapula, 89, 99–102
dorsum of, 89
inferior angle of, 90, 99, 100
lateral border of, 90, 99, 100
medial border of, 90, 100
movements, 154–155
orientation of, 99
upper border of, 91
variations, 102
winged, 92
Sclera, 495, 501, 502
Scrotum, 65, **184**
Segments—
bronchopulmonary, 413
of liver, 221
Sella turcica, 488, 607
Semen, 64
Septa—
crural (peroneal). See *fig. 441*
"intercondylar", 372
intermuscular, 72
anal, 284
of arm, 109
of leg. See fig. 441
of thigh, 323; *fig. 437*
interventricular, 432
nasal, 578
deflected, 579
nerves and vessels, 580–581
orbital, 457
peroneal (fibular). See S., crural
primum, 431
secundum, 431
spiral, in truncus arteriosus, 429
of tongue, 570
transversum, 251
urorectal, 257, 268
vesicovaginal, 298
Sesamoid bone, 6, 310
in ball of great toe, 355, 356, 358, 384
in Gastrocnemius, lateral head, 346
patella, 319
Sex differences in pelvis, 298
Sheath—
axillary, 81, 465
carotid, 509, 516, 543
of eyeball, fascial, 496
femoral, 312
fibrous flexor digital, 136
myelin, 42
of rectus abdominis, 176
contents of, 176–177
synovial, 23, 137
of biceps brachii, 157
development, note on, 137
flexor (carpal) common, 138
in hand, 137–138
Sinuses—
air, 581
anal. See Valve, anal
aortic, 425
basilar, 483
carotid, 543, 546
cavernous, 482, 489, 490
nerves traversing, 490
coronary, 426, 428, 430, 431
of epididymis, 187
ethmoidal, 582, 609, 610
frontal, 583, 604
intercavernous, 483
lactiferous. See *fig. 73*
lateral venous, 482
Sinuses—*Continued*
longitudinal. See S., sagittal
maxillary, 582
occipital, 483
paranasal, 581
development of, note on, 583
nerve supply of, 583
pericardial
oblique, **424,** 429
transverse, 424, 429
petrosal, 482, 483
pleural. See Recesses, pleural
prostatic, 280
renal, 233
sagittal
inferior, 482
superior, 482
scleral, 502
sigmoid, 482
sphenoidal, 582, 607
sphenoparietal, 483
straight, 482
tarsal, 380
transverse, venous, of cranium, 482
urogenital, 296
venarum (venosus), 428, 430, 431
venous
in dura mater, 482
vertebral, longitudinal, 477
Sinusoids, 35
of liver, 54
Skeleton—
age, 125, 130–131
divisions of, 4
of heart, 434
Skin, 67
blood vessels, 70
creases, 131
at wrist, 121; *fig. 125*
lines of cleavage (Langer's), 68
nerves of, 70
Skull—
base of
exterior, 538
interior (floor), 485
at birth, 451, **620**
bones of, 602 (See also individual bones)
cap (calvaria), 479
facial portion of, 451, 522, 602
norma
frontalis, 451
lateralis, 521
occipitalis, 469
verticalis, skull from above, 461
Slit, nasal, *fig. 685*

"Snuff box", 142
Sound conduction, 595
Spaces—
of Burns. See S., suprasternal
intercostal, 397
palmar, 138
perichoroidal, 503
pharyngeal, lateral, 565
quadrangular, 80, 95
retro-esophageal, 408
retropharyngeal, 564, 565
retropubic of Retzius, 268
subarachnoid, 40, 480
subdural, 40, 480
subphrenic, 205
suprasternal, 506
triangular, 80, 95
zonular, 502
Spermatozoa, 64
Sphenoid bone, 606
wing
greater, 488, 607
lesser, 485, 488, 606
Sphincters, 22. See Muscles, sphincter
Spines—
ethmoidal (of sphenoid), 607
iliac
anterior inferior, 320, 321
anterior superior, 320; *fig. 398*
posterior inferior, 320
posterior superior, 320, 321
ischial, 321
lumbar, 4th, 322
mental (tubercle, genial), 525
nasal
anterior, 452, 615
of frontal bone, 604
posterior, 539, 616
sacral, second, 322
of scapula, 89, **101**
suprameatal, *fig. 731*
of sphenoid, 528, 540, 607
Spleen, 55, 199
accessory, 201
functions of, 201
"pedicle" of, 198
peritoneal attachments of, 197
structure of, 200
surface anatomy, 200
Spondylolisthesis, 246
Squama—
of frontal bone, 603
of occipital bone, 604
of temporal bone, 488, 611
Standing, weight distribution, 384–385
Stapes, 594
Stature, 14
Sternebrae, 393
Sternum, 392
anomaly of, 394
Stomach, 54, 192, 193, **217**
arteries of, 218
lymph plexuses of, 252
nerves of, 218
position of, 193
relations of, 219
rotation of, 210
structure of, 54, 217
Sucking, 563
Sulci. See also Grooves
bicipital (groove). See S., intertubercular, 80, 93, 157
chiasmatic, optic, 488, 607
of heart, 420, 422, **424,** 426
intertubercular (bicipital), 80, 93, 157
pre-auricular of sacrum, 301
sagittal, 604
terminalis
of heart, 430, 435
of tongue, 569
transverse, 604
Supination and pronation, of foot, 380
of forearm, 166
Surface anatomy, see under individual items
Surfaces (aspects), of bones, 116; *fig. 117*
of femur, popliteal, 313
of heart, diaphragmatic, 424
posterior, 424
sternocostal, 420
Sustentaculum tali, 352
Sutural bones, See Bones, sutural
Suture, 17
coronal (frontal), 461
lambdoid, 461
metopic, 451, 604
petrotympanic. See Fissure
sagittal, 451, 461, 523
Swallowing, 53, **563,** 588
Sympathetic nervous system, 44
Symphysis, 17
menti, 451, 524
pubis, 275
"vertebral," 17
Synchondrosis, 17, 393
neurocentral, 13
occipitosphenoidal, 487, 539, 605
Syndesmosis, 17, 19
Syndrome, Horner's, 551
Synergists, 27, 170
Synostosis, 17
of sacro-iliac joint, 278
of sternomanubrial joint, 393
Synovia. See Fluid, synovial, 19
Synovial sheaths. See "Sheaths, synovial"
Systems—
blood-vascular, 30
circulatory. See blood-vascular, 30
conducting of heart, 435
digestive, 52
genital, 63
locomotor, 3
lymphatic, 35
nervous, 39
autonomic, 43
parasympathetic, 48
portal, development, 217
respiratory, 55
epithelium of, 59
sympathetic, 44
urinary, 60
urogenital, 60
vertebral venous, 477
clinical significance of, 477

T

Talus, 345, **350,** 378, 380, 381
movement about talus, 379
Tarsal bones, 350. See also individual bones
accessory (supernumerary), 385
Tarsus, of eyelid, 457
Teeth, 452, **571**
eruption, 571
growth and calcification, 573
lymph vessels, 574, 601
nerves of, 574
occlusion of, 573
Tegmen tympani, 488, 591. See also "Petrous bone"
Tela subcutanea (superficial fascia), 71
Temporal bone, 610
Tendo calcaneus (Achilles), 346
Tendon, 22
of Achilles, 34
axioms regarding, 119; *fig. 158.1*
conjoint (falx inguinalis), 180
control of, 184
nerve, 180
Teniae coli, 195
Tentorium cerebelli, 480
Testis, 51, 63, 185, **186**
descent of, 184
development of, 184, 187, 296

Testis—*Continued*
ducts of, 64
ectopic, 185
migrations of, 185, 231
structure of, 63, 186–187
vessels and nerves of, 187
Thorax, bony, 391
Thymus, 50, 439
Tibia, 336, 337
posterior aspects of, 344
Tissue—
adipose (fat), 71
areolar, 71
extraperitoneal fatty, table 9 on p. 182
cavernous, 35
lymphoid, 36
vascularity of, 35
Toes, function of, 357
Tongue, 569
blood supply, 570
development of, note on, 571
lymph vessels, 601
muscles, 567, **570**
nerve supply of, 570
Tonsil—
lingual tonsil, 559, 570
palatine tonsil, 57, 556, **559,** 567
bed of, 559
lymphatics, 601
to remove, 559
structure, 559
vessels and nerves, 560
pharyngeal tonsil, 556
adenoids, 556
Torus, interureteric. See Folds, 279
Trabeculae carneae, 432
Trachea, 58, 404, 406, 413, 417, **440**
cervical part of, 520
Tracts—
"G. I. Tract & Co.," 207. See also Canal gastrointestinal, 191
examination of, 193
parts and general disposition of, 191
iliotibial (band), 309, 318, 323, 370
olfactory, 483
urinary, structure of, 281
"U. G. Tract & Co.," 207
Tragus, *fig. 652*
Trapezium (greater multangular) bone, 126
Trapezoid (lesser multangular) bone, 126
Tree, bronchial, 413
Triangles and **Trigones**—
anal, 257, **259**
of auscultation, 91, 397
of bladder (trigone of), 279
carotid, 505, **510**
deltopectoral. See Fossa, infraclavicular, 79
digastric. See Tr., submandibular
femoral (of Scarpa), 314
lumbar (of Petit), 91
lumbocostal, 250
of neck
anterior, 505
posterior, 463
of Petit. See T., lumbar
of Scarpa. See T., femoral
submandibular (digastric), 505, **514**
submental, 505, 507
suboccipital, 476
suprameatal, *fig. 689*
urogenital, 257, 260
female, 265
"of vertebral artery," 516
vertebrocostal, 250
Triquetrum, 126
Trochanter—
greater, 313
lesser, 313
third, 314
Trochlea—
definition of, 7
of femur (patellar), 314
of humerus, 162
peroneal, of calcaneus. See *fig. 466*
of superior oblique, 496
Truncus arteriosus, 428, 429
Trunk—
brachiocephalic (art., innominate), 442
celiac (art., celiac), **213,** 218, 243
costocervical, 517
gastro-intestinal (lymphatic), 252
lumbar (lymphatic), 251
lumbosacral (nerve), 248, 287
lymph, with thoracic duct, 518
pulmonary (artery, pulmonary), **425,** 429, 433
sympathetic, composition of, 47
in abdomen and pelvis, 242, 302
in neck, 516, **550**
paralysis of, 551
Trunk—*Continued*
in thorax, **408**
thyrocervical, 517
vagal (anterior and posterior), 218–219, 244, 448, 547
Tube—
auditory (pharyngotympanic), 56, 540, 555, 556, 558, 592
Eustachian. See T., auditory
uterine, 66, 293
development, 296
Tuber (tuberosity)—
calcaneal, 352
frontal (eminence), 461, 603, 621
ischii, 257, 321, 327
maxillary, 539, 614
parietal (eminence), 461, 523, 621
Tubercle, tuberosity—
adductor, 314; *fig. 398*
articular (eminence), 540
carotid (on C. 6), 516, 543
conoid, 99
Darwin's, 455
deltoid, 94
genial. See Spine, mental, 525
of humerus
greater (tuberosity), 93, 97, 98
lesser (tuberosity), 93, 97
of iliac crest, 320
infraglenoid, 100
jugular, 605
of palatine bone. See Process, pyramidal
peroneal. See Trochlea, peroneal
pharyngeal, 540, 605
postglenoid, 541
pterygoid, 608
pubic, *fig. 211*
quadrate, 313
radial (bicipital), 110, 116
dorsal (of Lister), 116, 145
sacral, 276
scalene, 397
sellae, 488, 607
supraglenoid, 100
of talus, 351
tibial, 318
of ulna, 162
Tubules, renal (uriniferous), 61
mesonephric, 296
seminiferous, 187
Tunica albuginea testis, 186
vaginalis testis, 186
Turbinate bone. See Concha

Tympanum. See Ear, middle
Types, body. See Habitus

U

Ulna, 116
  palpable parts, 142
  posterior surface, 116, 145
  upper part of, 162
Umbilicus, 204
  plane of (vertebral level), 190
Umbo, 638
"Units" of foot, *fig. 465*
  of kidney, 60
  of lung, 414
  motor, 26
Urachus, 204, 257; *fig. 330*
Ureter, 233, **235,** 269
  anomalies, 236
  arteries, 236
  development, 237
  female, 297
  nerve supply of, 62, 236
  in pelvis, 269, 279
Urethra—
  female, 267, 290
  glands of, 281
  male, 62, 279
  nerve supply of, 62
  structure of, 281
Uterus, 67, 290, 292, **294**
  development, 296
  nerve supply of, 298
  structure of, 67, 294
Utricle of ear, 597
  prostatic, 282
Uvula, of palate, 562
  of bladder, 279

V

Vagina, 67, 295
  bulbi, (fascia bulbi), 67, 496
  development, 296
  structure of, 295
  vestibule of, 266
Vallecula, epiglottic, 556
Valves—
  anal, 284
  of heart
    aortic, 425
      terminology of cusps, 429
    atrioventricular, 433
      structure of, 433
    blood supply, 433
    of coronary sinus, 431
    mitral, 433
    pulmonary, 425
      terminology of cusps, 429
    tricuspid, 433
    of vena cava, inferior, 431

Valves—*Continued*
  ileocecal (ileocolic), 225
  of veins, 32
    femoral, 316
    iliac, 316
    jugular, internal, 518
    saphenous, 316
    subclavian, 518
Varicocele, 187
Varix, esophageal, 216
Vas, Vasa—
  aberrans. See Ductus
  afferens and efferens, 61
  deferens. See Ductus deferens, and *fig. 225*
  recta
    intestinal, 195, 223
    renal (arteriolae rectae), 62
  vasorum, 34
VEINS and VENAE. See also Arteries
  adrenal. See V., suprarenal
  anastomotic. See Anastomoses
  auricular, posterior, 514
  axillary, 83
  azygos, 446
    arch of, 403, **405,** 446
    lobe of, (lung), 412
  of back, 477
  basilic, 107
  basivertebral, 477
  brachiocephalic (innominate), 421, 439
    tributaries, 439, 518
  cardiac, 426
  cardinal, 239, 240, 446
  cava
    inferior, 239, 421
      development, 239
      obstruction of, 478
      relations, 239–241
      tributaries, 240
    superior, 421, 439
      development of, 429
      persisting left, 430
    surface anatomy of, 421
  cephalic, 79, 108
  cerebral, great, 482
  cervical
    ascending, *fig. 726*
    superficial, of neck, 506
    transverse, see V., colli, transversa
  choroidal, 504
  colli, transversa, 464; *figs. 664, 726, 759*
  comitantes (comites), 32
  cubital, median, 108
  cutaneous, of upper limb, 107

Veins—*Continued*
  cystic, 221
  deferent (of ductus deferens), 187
  digital, 107
  diploic, 461
    frontal, 461, 479
  dorsal of penis
    deep, 261
    superficial, 261, 262
  emissary (emerging diploic), 479
  facial (anterior), 461, 514
    common, 507, 514
    deep, 461
    posterior. See V., retromandibular
  femoral, 312, 315, 317
  gastric, left, 217
  of "G. I. Tract & Co.," 207
  of glands, "three paired", 241
  hemi-azygos, 446
  hemorrhoidal. See V., rectal
  hepatic, 241
  iliolumbar, 241
  inguinal, superficial, 178, 310
  innominate. See V., brachiocephalic
  intercostal, 398. See *fig. 641*
    anterior, 447
    posterior, 398, 447
    superior
      left, 447
      right, 447
  jugular
    anterior, 464, 514
    external, 464, 514
    internal, 514, **544**
  lingual, 514, 571
  lumbar, 241
    ascending, 241, 477
  mammary, int. See V., thoracic, int.
  maxillary (internal), 514, 530
  median
    of forearm, 108
    sacral, 241
  meningeal, 481, 492
  mesenteric
    inferior, 216, 225
    superior, 216, 225
  minimae cordis (Thebesian), 427
  of neck. See v., cervical
    communicating (connecting), 514
    at root of, *fig. 726*
    superficial, 514
  oblique, of left atrium, 427

Veins—*Continued*
occipital, 477
ophthalmic, 461, **500**
ovarian, 231, 241, 297
pampiniform plexus of, 187
paratonsillar, 559
para-umbical, 217
of pelvis, 286
of penis, deep and superficial, 261
pharyngeal, 514
popliteal, 330
portal, 207, **216,** 217
portacaval anastomosis (collateral circulation), 216
profunda femoris, 315
pulmonary, 414, 425, 431
rectal, 285
renal, left, 241
retromandibular (facial, posterior), 514, 526
saphenous
great (long), 309, 337
small (short), 331, 349
splenic, 216; *fig. 282*
structure of veins, 34
subclavian, 465, 516
suprarenal (adrenal), 232, 241
suprascapular, 464, 514
temporal, superficial, 514
testicular (spermatic), **187,** 231, 241
left, 241
Thebesian. See Venae minimae, 427
thoracic, internal (mammary, int.), *fig. 726*
thyroid
inferior, 517, 519
middle, 514, 519, 544
superior, 514, 519, 544
of tongue. See V., lingual
transversa colli (trans. cervical), 464, 514; *figs. 664, 726, 759*
umbilical (obliterated), **38, 198,** 205, 216
of upper limb, superficial, 107
uterine, 297
vaginal, 297
valves of, 33
vertebral, 477
vitelline, 217
vorticose, 504
Ventricles
of heart, 424, 431-433
Vertebrae, 11
articulated, from behind, 470
articulations, 13, 15
atlas, 534
axis (epistropheus), 534
bone and disc, proportions of, 13
cervical, 533
anomalies, 534
seventh (prominens), 534
coccyx, 271, 272
functions, 12
lumbar, 244
variations in, 245
ossification of, 13
parts, 12
thoracic, 391
typical, 11
sacrum, 271, 272, 470
Vertex, of skull, 451; *fig. 730*
Vesalius, foramen of, fig. 688
illustration from, *fig. 55*
Vesicles, seminal, 64, 282
nerves of, 63
vessels of, 283
Vessels. See Arteries, Lymphatics, Veins and Blood vessels
Thebesian. See Veins
Vestibule, aortic, 432
of internal ear, 596
of larynx, 588
of mouth, 566
of nose, 579
of vagina, 266
Villi, intestinal, 225
Vincula, of tendons, 138
Viscera—
abdominal
changing positions of, 201
protection, 190
"cervical", 508
pelvic, female, 292
support of, 298
Vomer, 539, 578, **618**

W

Wall, abdominal—
anterior, 175
arteries of, 177
layers of, 181
posterior, 244
Weight, distribution of, 356, **384**
Wing of sphenoid. See Sphenoid
Wormian bones. See Bones, sutural
Wrist. See also Carpus
tendons at back of, 146

X

Xiphoid (process), 176

Z

Zona orbicularis, 365
Zygoma (processes, zygomatic of temporal bone), 522, 611
Zygomatic bone, 523
arch, 522
ossification, 615

Grant's DISSECTIONS

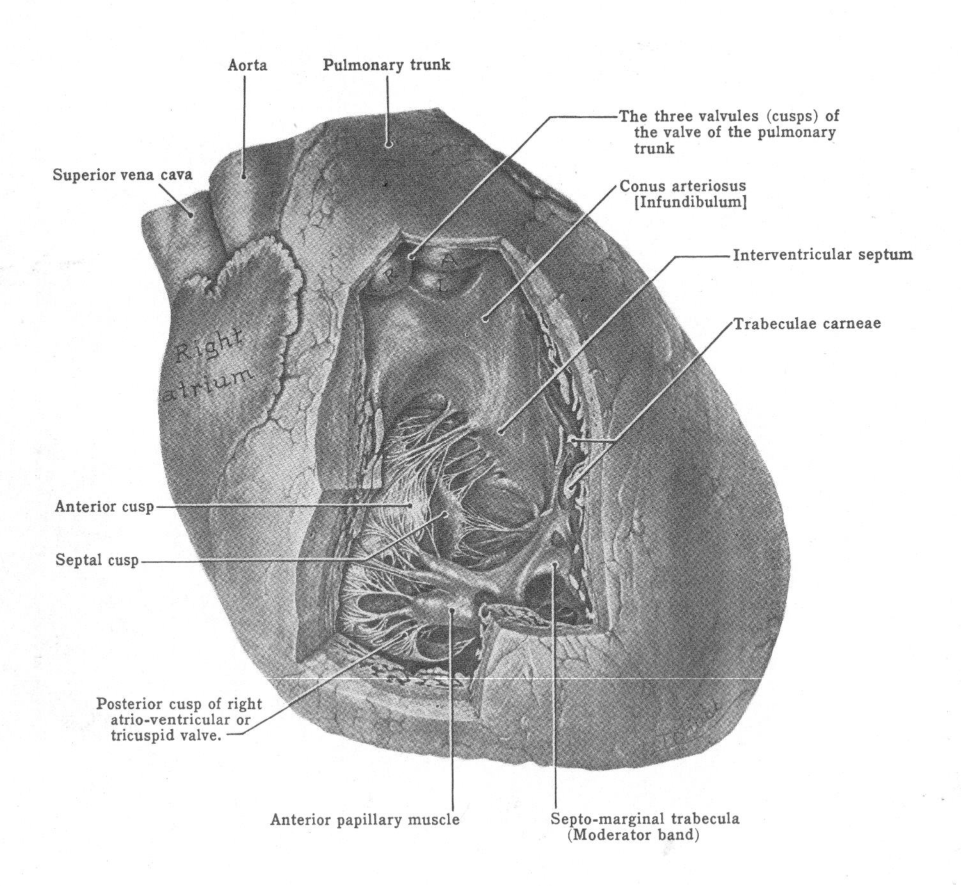